Boundary Elements XIII

THIRTEENTH INTERNATIONAL CONFERENCE
ON
BOUNDARY ELEMENT METHODS
BEM 13

SCIENTIFIC COMMITTEE

Boundary Elements XIII

Editors: C.A. Brebbia, Wessex Institute of Technology, U.K.
G.S. Gipson, Oklahoma State University, U.S.A.

Computational Mechanics Publications
Southampton Boston

Co-published with

Elsevier Applied Science
London New York

C.A. Brebbia
Wessex Institute of Technology
Ashurst Lodge
Ashurst
Southampton
SO4 2AA
UK

G.S. Gipson
School of Civil Engineering
314 Engineering South
Oklahoma State University
Stillwater
OK 74078
USA

Co-published by

Computational Mechanics Publications
Ashurst Lodge, Ashurst, Southampton, UK

Computational Mechanics Publications Ltd
Sole Distributor in the USA and Canada:

Computational Mechanics Inc.
25 Bridge Street, Billerica, MA 01821, USA

and

Elsevier Science Publishers Ltd
Crown House, Linton Road, Barking, Essex IG11 8JU, UK

Elsevier's Sole Distributor in the USA and Canada:

Elsevier Science Publishing Company Inc.
655 Avenue of the Americas, New York, NY 10010, USA

British Library Cataloguing-in-Publication Data

A Catalogue record for this book is available
from the British Library

ISBN 1-85166-696-6 Elsevier Applied Science, London, New York
ISBN 1-85312-145-2 Computational Mechanics Publications, Southampton
ISBN 1-56252-072-5 Computational Mechanics Publications, Boston, USA

Library of Congress Catalog Card Number 91 73255

PREFACE

Since its origin in 1978, the International Conference on Boundary Element Methods has provided the recognized and established forum for innovations in boundary element research. Practically all new ideas on boundary elements have been presented at these conferences and the resulting papers can be found in the published books. The conference brings together the most renowned scientists and engineers working on boundary element research throughout the world.

A unique feature of these meetings is that the participation of younger researchers is actively encouraged by the organizers in an effort to bring forward to the attention of the international community an ever expanding range of new ideas.

This book contains the edited version of the papers presented at the XIIIth BEM Conference held in Tulsa, Oklahoma in August of 1991. The meeting attracted a large number of participants and many excellent contributions which have been divided into nineteen different sections, i.e. Potential Problems; Diffusion and Convection Problems; Fluid Mechanics; Fluid Flow; Wave Propagation; Groundwater Flow; Heat Transfer; Electrical Problems; Geomechanics; Plates and Shells; Inelastic Problems; Damage Tolerance; Contact Mechanics; Industrial Applications; Design Sensitivity and Optimization; Inverse Problems; Special Techniques; Numerical Aspects and Computational Aspects.

The editors are grateful to the Members of the International Scientific Committee who have more actively supported the meeting, i.e. M.H. Aliabadi (Wessex Institute of Technology); A.D.H. Cheng (University of Delaware); R. Ciskowski (IBM); J.J. Connor (Massachusetts Institute of Technology); S. Crouch (University of Minnesota); J. Dominguez (University of Seville); S. Grilli (University of Liege, Belgium); W.S. Hall (Teesside Polytechnic, UK); N. Kamiya (Nagoya University, Japan); D. Karabalis (University of South Carolina); K. Onishi (Fukuoka University); F. Paris (University of Seville); H. Power (Central University of Caracas); T.J. Rudolphi (Iowa State University); J.J. Rencis (Worcester Polytechnic Institute); R.P. Shaw (State University of New York at Buffalo); P. Skerget (University of Maribor, Yugoslavia); G. Symm (National Physical Laboratory, UK); M. Tanaka (Shinshu University, Japan); M. Vable (Michigan Technological University); L.C. Wrobel (Wessex Institute of Technology).

The editors trust that this book will be useful to all scientists and engineers who wish to be aware of the state-of-the-art in boundary element research. They also hope that the book will help to encourage further research into a method which has gained such a wide audience since 1978.

C.A. Brebbia
G.S. Gipson
August, 1991

CONTENTS

SECTION 1: POTENTIAL PROBLEMS

SECTION 2: DIFFUSION AND CONVECTION PROBLEMS

SECTION 3: FLUID MECHANICS

SECTION 5: WAVE PROPAGATION

SECTION 6: GROUNDWATER FLOW

SECTION 7: HEAT TRANSFER

SECTION 8: ELECTRICAL PROBLEMS

SECTION 16: INVERSE PROBLEMS

SECTION 17: SPECIAL TECHNIQUES

SECTION 18: NUMERICAL ASPECTS

SECTION 19: COMPUTATIONAL ASPECTS

SECTION 1: POTENTIAL PROBLEMS

Treatment of Corners in the Complex Variable Boundary Element Method

R.T. Bailey, C.K. Hsieh

Department of Mechanical Engineering, University of Florida, Gainsville, Florida, U.S.A.

ABSTRACT

The treatment of corners in the complex variable boundary element method for simply-connected domains is considered. A methodology is developed for all possible combinations of Dirichlet, Neumann, and Robin boundary conditions at the corner. It is shown that no special augmentation of the resulting matrix equations is necessary when one follows the guidelines set forth in this paper. The methodology is successfully applied to four example problems with available analytical solutions, and comparisons are also made with published results from a specialized real variable boundary element method corner-handling formulation.

INTRODUCTION

The complex variable boundary element method (CVBEM) has been shown to be a powerful tool for obtaining solutions to the two-dimensional (2-d) Laplace and Poisson equations [1]. The CVBEM transforms Cauchy's integral formula into a boundary element method (BEM) which possesses the following two significant advantages over the traditional real variable boundary element methods (RVBEMs): 1) approximations are made only at the boundary of the domain, and 2) all integrations are carried out analytically without the need for numerical integration. These two factors combine to give the CVBEM excellent potential for both accuracy and efficiency.

A well-known problem which is common to many BEMs (and which has yet to be addressed in the CVBEM) is the proper treatment of the corners or edges in the boundaries of the domain. A corner (in 2-d) or edge (in 3-d) is defined as a point or locus of points where the tangent to the boundary possesses a sharp discontinuity. This discontinuity must be due solely to the problem geometry and not due to any discretization scheme associated with the BEM. At such a corner or edge, quantities associated with the direction normal to the boundary are double-valued, and the conventional RVBEMs can produce systems of equations which are underconstrained.

In 2-d potential problems, the three traditional boundary conditions are 1) the Dirichlet condition where ϕ is known and $\partial\phi/\partial n$ is unknown, 2) the Neumann condition where $\partial\phi/\partial n$ is known and ϕ is unknown, and 3) the Robin (or convective) condition where both ϕ and $\partial\phi/\partial n$ are unknown. Here, ϕ represents the potential function, while n represents the direction normal to the boundary. The application of these three conditions logically leads to nine possible boundary-condition combinations at any corner; see Table 1. Here, the subscripts U and D refer to the "upstream" and "downstream" sides of the corner, respectively, with the direction of travel along the boundary always placing the domain to the left. For the optimum situation where both the upstream and downstream conditions are available at the corner node, the unknowns and corresponding number of equations which can be applied in ordinary RVBEM formulations are given in Table 2. In all nine cases, one boundary element nodal equation is always available for application. Thus, with only one unknown, Cases 1 through 3 pose no problem. Cases 4 and 5 have two unknowns, but in both cases an extra equation is available from the Robin condition. Case 6 has three unknowns, but two additional equations are available from the Robin conditions. Cases 7 and 8 have two unknowns each, but once again, the Robin condition can be used as an extra equation in either case. Finally, Case 9 has two unknowns, but no additional equation is available for application. This renders Case 9 underconstrained, and it is here that investigators have been forced to seek methods for special treatment of corners in the RVBEMs.

A number of different approaches have been employed in the RVBEMs in order to handle the corner problem, including

- rounding the corners [2];
- setting the upstream and downstream gradients equal [3];
- using double nodes or "binodes" [4];
- using specialized corner elements [5];
- using an additional equation to properly constrain the corner [6-10];
- using discontinuous or partially discontinuous elements [11,12].

Each of these approaches has met with some success, but the latter two seem to provide the most consistently accurate results.

Table 1. The nine traditional boundary condition combinations at a corner node in a potential problem.

	Upstream Nodes		Downstream Nodes	
Case	Condition	Quantity Specified	Condition	Quantity Specified
1	Dirichlet	ϕ	Neumann	$(\partial\phi/\partial n)_D$
2	Neumann	$(\partial\phi/\partial n)_U$	Dirichlet	ϕ
3	Neumann	$(\partial\phi/\partial n)_U$	Neumann	$(\partial\phi/\partial n)_D$
4	Dirichlet	ϕ	Robin	none
5	Robin	none	Dirichlet	ϕ
6	Robin	none	Robin	none
7	Neumann	$(\partial\phi/\partial n)_U$	Robin	none
8	Robin	none	Neumann	$(\partial\phi/\partial n)_D$
9	Dirichlet	ϕ	Dirichlet	ϕ

Table 2. The unknowns and corresponding number of equations available in the standard RVBEMs for the nine possible corner node boundary condition combinations listed in Table 1.

Case	Unknowns	Number of Equations Available
1	$(\partial\phi/\partial n)_U$	1
2	$(\partial\phi/\partial n)_D$	1
3	ϕ	1
4	$(\partial\phi/\partial n)_U$ and $(\partial\phi/\partial n)_D$	2
5	$(\partial\phi/\partial n)_U$ and $(\partial\phi/\partial n)_D$	2
6	$(\partial\phi/\partial n)_U$, $(\partial\phi/\partial n)_D$, and ϕ	3
7	$(\partial\phi/\partial n)_D$, and ϕ	2
8	$(\partial\phi/\partial n)_U$, and ϕ	2
9	$(\partial\phi/\partial n)_U$ and $(\partial\phi/\partial n)_D$	1

By formulating 2-d potential problems with complex variables instead of real variables, the CVBEM avoids the problem of gradient discontinuity at boundary corners. Instead of solving for the double-valued gradients at a corner node, one now solves for a single-valued stream function. Still, Neumann and Robin conditions can lead to special circumstances at corners, and a methodology for the treatment of corners in the CVBEM is needed. This paper addresses this need.

DESCRIPTION OF THE CVBEM

The CVBEM makes use of the complex potential, $\omega(z) = \phi(z) + i\psi(z)$, which is analytic so that its real and imaginary components (ϕ and ψ) satisfy the two-dimensional Laplace equation. Cauchy's integral formula holds for ω as

$$\omega(z_0) = \frac{1}{2\pi i}\int_{\Gamma} \frac{\omega(\zeta)}{\zeta - z_0} d\zeta \tag{1}$$

This expression relates the value of ω at point z_0 located inside the simply-connected domain, Ω, to a contour integral (containing ω) along the boundary, Γ; see Figure 1.

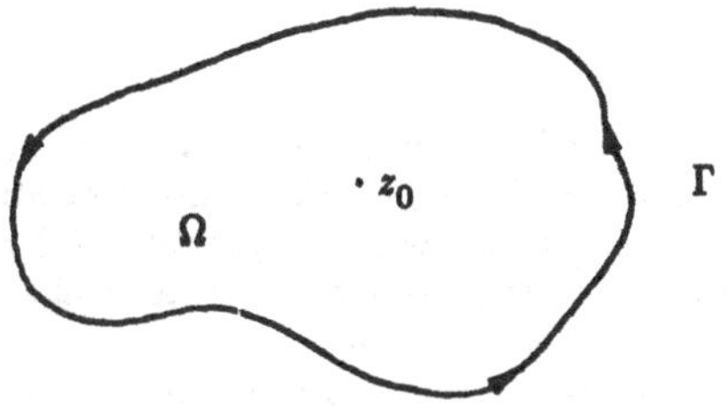

Figure 1. Domain Ω with boundary Γ and internal point z_0.

Equation (1) is used in the CVBEM with two approximations: 1) the boundary Γ is replaced by finite-length elements and 2) the function $\omega(z)$ is replaced by a global trial function $G(z)$ composed of piecewise continuous polynomial basis functions. By introducing these approximations into Equation (1) and performing the contour integration over each element analytically, the following expressions for the CVBEM approximation to $\omega(z_0)$, designated $\hat{\omega}(z_0)$, are obtained:

$$\hat{\omega}(z_0) = \frac{1}{2\pi i} \sum_{j=1}^{N} [A_j \omega_j + A_{j+1} \omega_{j+1}] \tag{2}$$

for linear elements (N elements and nodal points), and

$$\hat{\omega}(z_0) = \frac{1}{2\pi i} \sum_{\substack{k=1 \\ j=2k-1}}^{M} C_j\, \omega_j + C_{j+1}\, \omega_{j+1} + C_{j+2}\, \omega_{j+2} \tag{3}$$

for quadratic elements (M elements and N=2M nodal points). Here, ω_j is the value of ω at boundary nodal point j. The reader is directed to Ref. 14 for a complete description of the coefficient terms (A_j, C_j, etc.) and their derivation.

Equations (2) and (3), being expressed in complex variables, actually embody two equations each—one for the real part and one for the imaginary part. If the values of ϕ and ψ (and thus ω) are known at each boundary node, Equation (2) or Equation (3) can be used to calculate $\hat{\phi}$ and $\hat{\psi}$ at any interior point, z_0. In most potential problems, however, boundary conditions specify either ϕ , ψ, or neither of them explicitly. To solve for the unknown values of ϕ and ψ, it is necessary to derive extended versions of Equations (2) and (3) by moving z_0 to the position of z_j on the boundary. In this effort, one takes the limit as z_0 approaches z_j with the result

$$\hat{\omega}_j = \frac{1}{2\pi i} \left\{ B_j \omega_j + \sum_{\substack{i=1 \\ i,i+1 \neq j}}^{N} [A_i \omega_i + A_{i+1} \omega_{i+1}] \right\} \tag{4}$$

for linear elements and

$$\begin{aligned} \hat{\omega}_j = \frac{1}{2\pi i} \Big\{ & F_{j-2}\, \omega_{j-2} + F_{j-1}\, \omega_{j-1} + F_j\, \omega_j \\ & + F_{j+1}\, \omega_{j+1} + F_{j+2}\, \omega_{j+2} \\ & + \sum_{\substack{i=1 \\ i,i+1 \neq k \\ j=2k-1 \\ l=2i-1}}^{M} C_l\, \omega_l + C_{l+1}\, \omega_{l+1} + C_{l+2}\, \omega_{l+2} \Big\} \end{aligned} \tag{5}$$

for quadratic elements. The reader is again directed to Ref. 14 for a complete description of the coefficient terms (B_j, F_j, etc.) and their derivation. Equations (4) and (5) are known as nodal equations, and like Equations (2) and (3), they each embody two equations—one for $\hat{\phi}_j$ and one for $\hat{\psi}_j$. These real and imaginary parts are used to estimate the values of ϕ and ψ not specified by the boundary conditions. Before this estimation process is discussed, the implementation of the boundary conditions in the CVBEM is described.

Boundary Conditions
Four types of boundary conditions are considered in the CVBEM.

Dirichlet Condition The potential at node j is known and specified, i.e.,

$$\phi_j = \phi_{specified,\, j} \tag{6}$$

Notice that for such a condition ψ_j is unknown.

Neumann Condition The normal gradient of the potential at node j is known. The stream function there is then given by

$$\psi_j = \psi_{j-1} + \frac{1}{2}\left\{\left(\frac{\partial\phi}{\partial n}\right)_j + \left(\frac{\partial\phi}{\partial n}\right)_{j-1}\right\}|z_j - z_{j-1}| \tag{7}$$

Here, it is assumed that $\partial\phi/\partial n$ varies linearly from node $j-1$ to node j [13]. For this condition, both ϕ_j and ψ_j are unknown.

Robin Condition The normal gradient of the potential and the potential itself are related by $\partial\phi/\partial n = \mathrm{H}(\phi - \phi_\infty)$. The stream function there is then given by

$$\psi_j = \psi_{j-1} + \frac{1}{2}\left\{\left(\mathrm{H}(\phi_\infty - \phi)\right)_j + \left(\mathrm{H}(\phi_\infty - \phi)\right)_{j-1}\right\}|z_j - z_{j-1}| \tag{8}$$

Here, it is assumed that $\mathrm{H}\phi$ and $\mathrm{H}\phi_\infty$ vary linearly from node $j-1$ to node j [13]. As in the Neumann condition, both ϕ_j and ψ_j are unknown.

Stream Function Condition The stream function at node j is known and specified, i.e.,

$$\psi_j = \psi_{specified,\, j} \tag{9}$$

For this condition, ϕ_j is unknown.

Solution Methodology
As mentioned previously, the real and imaginary parts of Equations (4) or (5) are used to estimate the unknown values of ϕ and ψ at the boundary nodes. Three approaches are possible [1], and of these, the Class II (implicit) approach is preferred. (See Refs. 1 and 14 for a complete discussion of the implicit approach.)

Application of the implicit approach leads to a system of linear algebraic equations which can be represented in a matrix form as

$$\mathbf{C}\begin{Bmatrix}\hat{\phi}\\ \hat{\psi}\end{Bmatrix} = \mathbf{R} \tag{10}$$

Here, **C** is the square matrix of coefficients on the unknown values of $\hat{\phi}$ and $\hat{\psi}$ at the boundary nodes and **R** is a vector of known constants. A detailed description of how **C** and **R** are formed is given in Ref. 14. The system can be solved by any direct method such as Gaussian elimination. Once the unknown values of $\hat{\phi}$ and $\hat{\psi}$ are found at each boundary node, one can use these values together with the specified values as the boundary nodal values needed by Equations (2) and (3) for calculating $\hat{\omega}$ at any interior point.

Calculation of the Normal Gradient of the Potential
After ϕ and ψ have been evaluated at the boundary nodes and interior points, the values of ψ may be used to determine $\partial\phi/\partial n$ at the boundary nodes. Use is made of

the Cauchy-Riemann conditions,

$$\frac{\partial \phi}{\partial n} = \frac{\partial \psi}{\partial s} \tag{11}$$

where n is taken as the outward normal to Γ with s as the tangential coordinate along Γ, assumed positive in the direction of the contour integral in Equation (1). The partial differential $\partial\psi/\partial s$ (and thus $\partial\phi/\partial n$) can be approximated by using finite-difference formulae. One-sided formulae should be used at corner nodes and at any others where the normal gradients are different on either side of the node. Central-differencing can be used at all remaining nodes.

Additional Comments and the Psi Reference Node

Two more points are worthy of note. First, nodal points which are located at corners can be considered as a special case which is studied in great detail in the next section. Second, no matter what boundary conditions exist, one must specify a reference value of ψ at some nodal point along the boundary in order to serve as the constant of integration in Equation (1). Numerically speaking, the associated matrices become singular if no ψ value is provided. This anchor node will be referred to as the "psi reference node". Numbering it as node j, one has

$$\psi_j = \psi_{specified,j} \tag{12}$$

The implicit method can then be used to determine the unknown $\hat{\phi}_j$ as was discussed earlier.

If, however, the value of ϕ is also known at node j (i.e., a Dirichlet condition), one can specify both ϕ and ψ at this node as

$$\phi_j = \phi_{specified,j} \tag{13}$$

and

$$\psi_j = \psi_{specified,j} \tag{14}$$

The complex potential ω is thus fully specified, and no nodal equation needs to be applied. This node will be called a "completely specified node".

CORNER TREATMENT METHODOLOGY

It has been stated that the corners in a domain boundary can cause problems for the RVBEMs since $\partial\phi/\partial n$ is double-valued at a corner node. The CVBEM avoids this problem by solving for ψ instead of $\partial\phi/\partial n$. Since ψ is both continuous and single-valued at any corner of a simply-connected domain in which Laplace's equation holds, corner nodes can generally be handled more easily than in the RVBEMs. A different "corner" problem arises in the CVBEM, however, due to the way that Neumann and Robin boundary conditions are implemented.

When using the CVBEM to solve potential problems with corners in the boundary, the same nine traditional boundary-condition combinations in Table 1 are possible; the unknowns in the CVBEM for Cases 1 through 9 are listed in Table 3.

A nomenclature is now introduced whereby a particular node is identified by the boundary condition imposed there. Thus, four types of nodes are identified—Dirichlet nodes, Neumann nodes, Robin nodes, and stream function nodes. A fifth type of node—the completely specified node—was also discussed previously. Now, a

Table 3. The unknowns and corresponding number of equations available in the implicit CVBEM for the nine possible corner node boundary condition combinations listed in Table 1.

Case	Unknowns	Number of Eqns. Available
1	ψ	2
2	ψ	2
3	ϕ and ψ	3
4	ψ	2
5	ψ	2
6	ϕ and ψ	3
7	ϕ and ψ	3
8	ϕ and ψ	3
9	ψ	1

question arises as to whether in each of the nine different corner cases, the corner node can be treated as one of the five nodal types above. In order to address this question, the application of the boundary condition equations at arbitrary corner node m will be investigated.

The Neumann and Robin equations (Equations (7) and (8)) are "upstream" equations for calculating ψ_m, meaning that they require information from the previous upstream node, $m-1$. Due to this "upstream" nature, applying either Equation (7) or (8) at a corner node which is the first node of a Neumann or Robin stretch of boundary is undesirable since it forces the downstream Neumann or Robin condition back to the previous node $m-1$ where the condition is not imposed. (Here, the upstream and downstream boundary conditions are assumed to terminate and begin, respectively, at the corner node itself.) Thus, in Cases 1, 3, 4, 6, 7, and 8, although two conditions are available, the corner node should be treated using the upstream condition rather than the downstream Neumann or Robin condition.

In contrast to the Neumann and Robin equations, the equation for imposing a Dirichlet boundary condition (Equation (6)) involves direct specification of ϕ_m. Therefore, Case 9 is easily handled as a Dirichlet node.

This leaves Cases 2 and 5 which are grouped together as an upstream equation for ψ_m followed by a direct specification of ϕ_m. For these cases, ϕ_m can be specified directly using Equation (6), and the upstream Neumann or Robin equation (Equation (7) or (8)) can then be applied at node m. Clearly, this is not one of the five nodal types described above. This new type of node will be referred to as a Neumann-Robin/Dirichlet node.

To summarize, it has been shown that, of the nine corner cases, seven can be handled using previously defined nodal types. Only two—Cases 2 and 5—require the addition of a new type of node.

Rules for the Nine Corner Cases

A list is now presented which gives the proper treatment for each of the nine cases in the implicit CVBEM at arbitrary corner node m.

Cases 1, 4, and 9: Treat the corner node as a Dirichlet node; ψ_m is then the only unknown. Apply the implicit nodal equation for ψ_m.

Cases 2 and 5: Treat the corner node as a Neumann-Robin/Dirichlet node; ψ_m is then the only unknown. Apply the Neumann equation for ψ_m (Equation (7)).

Cases 3 and 7: Treat the corner node as an upstream Neumann node. Both ϕ_m and ψ_m are unknown. Apply the implicit nodal equation for ϕ_m and the upstream Neumann equation for ψ_m(Equation (7)).

Cases 6 and 8: Treat the corner node as an upstream Robin node. Both ϕ_m and ψ_m are unknown. Apply the implicit nodal equation for ϕ_m and the upstream Robin equation for ψ_m (Equation (8)).

An exception to the above rules involves the psi reference node.

Exception: The first nodal point of the longest stretch of boundary imposed with a Neumann or Robin boundary condition (traveling from the initial point along the boundary, keeping the interior of the domain on the left) should be treated as the psi reference node.

This important exception can supersede the rules listed for Cases 1, 3, 4, 6, 7, and 8. Consider a case where a Neumann or Robin condition exists along some part of the boundary, say from node j to node $j+k$. Equations (7) and (8) show that, in either case, the value of ψ at nodes $j+1$ through $j+k$ depend intimately upon the value of ψ_j. In fact, any error in ψ_j will be carried over as errors in ψ_{j+1} through ψ_{j+k}. The errors in ψ along the Neumann or Robin stretch of boundary will, in turn, cause errors in the estimated values of ϕ and ψ at all other nodes along the boundary due to the coupled nature of the equations in matrix equation (10). If the Neumann or Robin part of the boundary comprises a significant portion of the total boundary length, this can result in a large amount of error. Thus, it is imperative that the value of ψ at the first node of a Neumann or Robin stretch of boundary be as accurate as possible.

EXAMPLES AND RESULTS

In order to assess the accuracy of the CVBEM in solving problems with corners in the domain boundary, several examples were investigated. The ones shown in Figures 2 and 3 were taken from a paper by Walker and Fenner [8]. For Example 1 (Figure 2), the potential distribution over the domain was taken to be

$$\phi = \sin(x)\cosh(y) \tag{15}$$

while for Example 2 (Figure 3), the somewhat more complicated expression

$$\phi = \ln\{[(x-2)^2 + (y-2)^2]^{-0.5}\} + x^2 - y^2 \tag{16}$$

was assumed. In both examples, Dirichlet boundary conditions were imposed along all parts of the boundary (except at one node, where both ϕ and ψ were specified). Sixteen nodes were used in Example 1, while 32 nodes were used in Example 2.

Examples 3 and 4 involve square domains and they differ only in the boundary conditions imposed as shown in Figures 4 and 5. Here, the potential and stream function distributions over the domains were assumed to be

$$\phi = \exp(x)\cos(y) \tag{17}$$

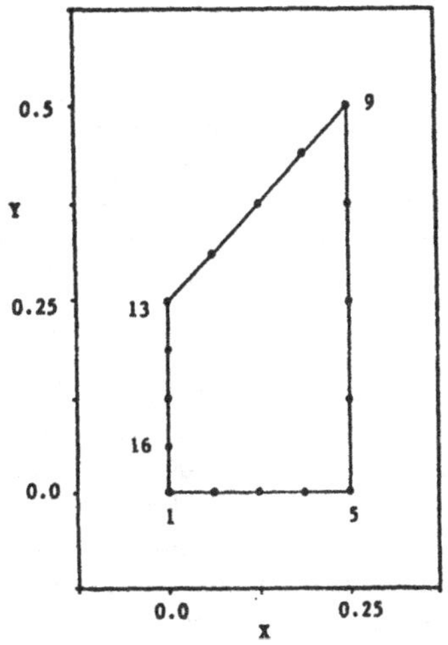

Figure 2. The geometry and nodal point distribution for Example 1.

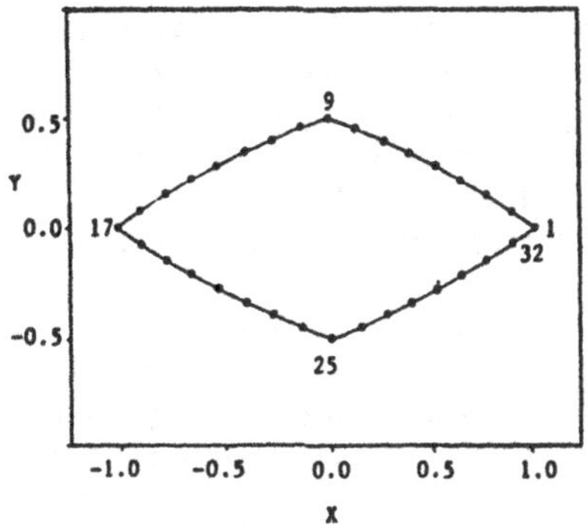

Figure 3. The geometry and nodal point distribution for Example 2.

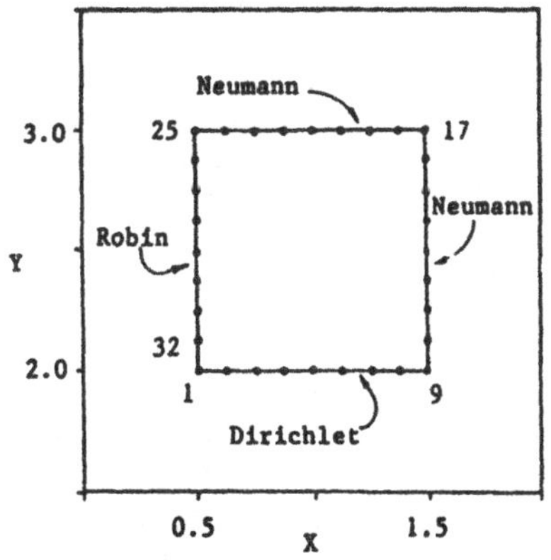

Figure 4. The geometry, nodal point distribution, and boundary conditions for Example 3.

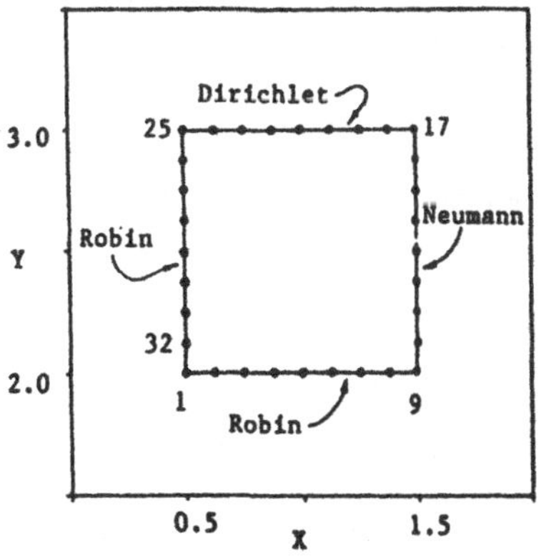

Figure 5. The geometry, nodal point distribution, and boundary conditions for Example 4.

and

$$\psi = \exp(x)\sin(y) \tag{18}$$

and they were used to generate the conditions imposed on the boundaries. These examples were chosen to test the CVBEM corner methodologies developed for the treatment of boundary-condition-combination Cases 1 through 8 in Table 1. (Case 9 has been covered by Examples 1 and 2.) The four corners in Example 3 represent Cases 1, 3, 5, and 7, while those in Example 4 cover Cases 2, 4, 6, and 8. Thirty-two boundary nodes were used for both examples.

The results for Example 1 are given in Figure 6. This figure allows a comparison between the errors in $\partial\phi/\partial n$ at the boundary nodes as calculated by the linear-element CVBEM and a quadratic-element RVBEM. For Examples 1 and 2, the RVBEM solutions were taken from the paper by Walker and Fenner [8], who successfully used an "extra equation approach" to handle the corners. As for the present work, the CVBEM gradient values were calculated from the stream function nodal values using 3-point finite-difference formulae. At nodes 2 through 4, $\partial\phi/\partial n$ is zero so that no percent errors are shown.

As plotted in Figure 6, the CVBEM errors are less than the RVBEM errors at 11 of the 13 boundary nodes shown, and the maximum error along the boundary is 67 percent lower in the CVBEM than in the RVBEM. This is encouraging since linear elements were used in the CVBEM, while quadratic elements were used in the RVBEM.

Figure 7 presents the results for Corner Example 2. This time, the CVBEM results were generated using quadratic elements, and the CVBEM gradient values were calculated using five-point rather than three-point finite-difference formulae. One can see that the percent errors in the CVBEM solution are less than the percent errors in the RVBEM solution at 31 of the 32 nodes, with the maximum error being 0.83 percent in the RVBEM and 0.32 percent in the CVBEM.

The results for Corner Examples 3 and 4 are presented in Figure 8. To save space, only the average and maximum percent errors in ϕ, ψ, and $\partial\phi/\partial n$ along the boundary are shown. The results for both examples are good, with errors in ϕ below 0.2 percent and errors in $\partial\phi/\partial n$ below 4.0 percent.

CONCLUSIONS

A complete methodology for handling corners in the CVBEM was presented. It was found that the traditional RVBEM problem of dealing with the double-valued normal gradient of the potential at a corner does not appear in the CVBEM due to the use of the stream function ψ. Instead, problems associated with the proper implementation of Neumann and Robin boundary conditions at a corner node were encountered and resolved in a series of rules covering the nine possible combinations of Dirichlet, Neumann, and Robin conditions at a corner. An important exception to the rules regarding the placement of the reference value of the stream function was noted.

The methodology presented for handling corners using the CVBEM was validated by application to four example problems. It was determined that the CVBEM was able to generate solutions with greater accuracy than a well-established RVBEM [8] when boundary elements of the same order were used.

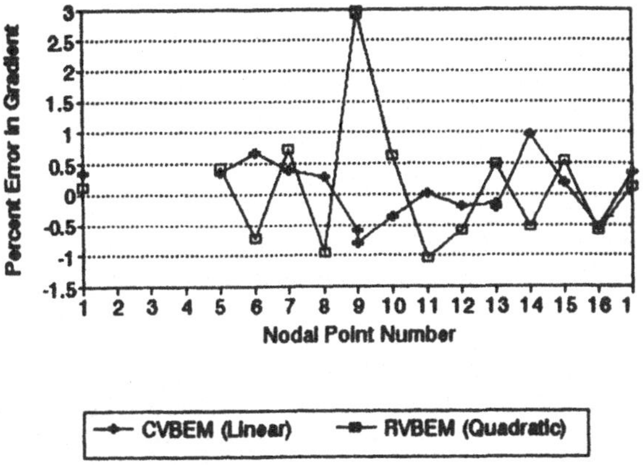

Figure 6. CVBEM and RVBEM results for Example 1.

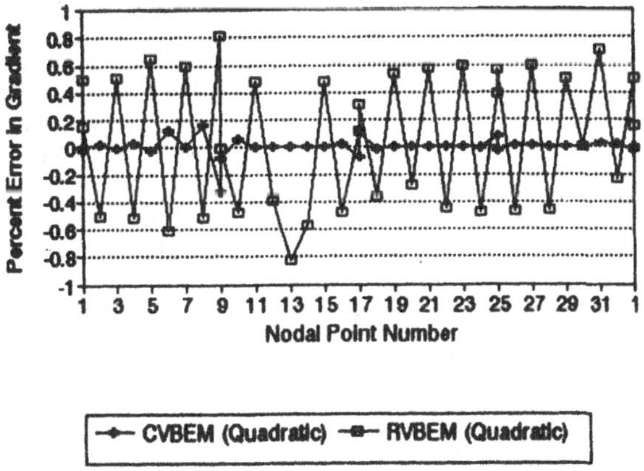

Figure 7. CVBEM and RVBEM results for Example 2.

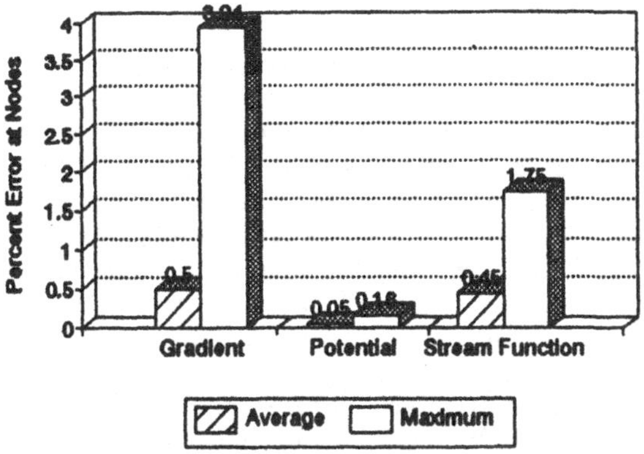

Figure 8. CVBEM results for Example 3.

REFERENCES

1. Hromadka II, T. V. and Lai, C. The Complex Variable Boundary Element Method in Engineering Analysis, Springer-Verlag, New York, 1987.
2. Symm, G. T. Integral Equation Methods in Potential Theory II, Proceedings of the Royal Society, Vol. 275(A), pp. 33-46, 1963.
3. Lachat, J. C. and Watson, J. O. Effective Numerical Treatment of Boundary Integral Equations: A Formulation for Three-Dimensional Elastostatics, International Journal for Numerical Methods in Engineering, Vol. 10, pp. 991-1005, 1976.
4. Brebbia, C. A. and Dominguez, J. Boundary Element Methods for Potential Problems, Applied Mathematical Modelling, Vol. 1(7), pp. 372-378, 1977.
5. Detournay, C. On a Cauchy Integral Element Method for Potential Flow with Corner Singularities, in Computational Engineering with Boundary Elements (Eds. Grilli, S., Brebbia, C. A., and Cheng, A. H-D.), Vol. 1, pp. 119-130, Proceedings of the Fifth International Conference on Boundary Element Technology, Newark, Delaware, 1990. Computational Mechanics Publications, Southampton and Boston, 1990.
6. Alarcon, E., Martin, A., and Paris, F. Boundary Elements in Potential and Elasticity Theory, Computers and Structures, Vol. 10, pp. 351-362. 1979.
7. Paris, F., Martin, A., and Alarcon, E. Potential Theory, Chapter 3, Progress in Boundary Element Methods, (Ed. Brebbia, C. A.), Vol. 1, pp. 45-83, John Wiley and Sons, New York, 1981.
8. Walker, S. P. and Fenner, R. T. Treatment of Corners in BIE Analysis of Potential Problems, International Journal for Numerical Methods in Engineering, Vol. 28, pp. 2569-2581, 1989.
9. Bruch, E. and Lejeune, A. Treatment of Corner Node Problems in Laplace Fields by the BEM, in Computational Engineering with Boundary Elements (Eds. Grilli, S., Brebbia, C. A., and Cheng, A. H-D.), Vol. 1, pp. 131-141, Proceedings of the Fifth International Conference on Boundary Element Technology, Newark, Delaware, 1990. Computational Mechanics Publications, Southampton and Boston, 1990.
10. Gray, L. J. Electroplating Corners, in Computational Engineering with Boundary Elements (Eds. Grilli, S., Brebbia, C. A., and Cheng, A. H-D.), Vol. 1, pp. 63-72, Proceedings of the Fifth International Conference on Boundary Element Technology, Newark, Delaware, 1990. Computational Mechanics Publications, Southampton and Boston, 1990.
11. Patterson, C. and Sheikh, M. A. Interelement Continuity in the Boundary Element Method, Chapter 6, Topics in Boundary Element Research, (Ed. Brebbia, C. A.), Vol. 1, pp. 123-141, Springer-Verlag, New York, 1984.
12. Danson, D. J., Brebbia, C. A., and Adey, R. A. BEASY. A Boundary Element Analysis System, Finite Element Systems Handbook, Springer-Verlag and Computational Mechanics Publications, Southampton, 1982.
13. Kassab, A. J. and Hsieh, C. K. Application of the Complex Variable Boundary Element Method to Solving Potential Problems in Doubly-Connected Domains, International Journal for Numerical Methods in Engineering, Vol. 29, pp. 161-179, 1990.
14. Bailey, R. T. Extensions and Refinements to the Complex Variable Boundary Element Method Including Its Application to Numerical Grid Generation, Ph.D. Dissertation, University of Florida, Gainesville, 1991.

On Using the Delta-Trigonometric Method to Solve the 2-D Neumann Potential Problem

R.S.-C. Cheng

Computational Mechanics Division, David Taylor Research Center, Bethesda, Maryland, U.S.A.

ABSTRACT

The Neumann problem for Laplace's equation is often solved by means of the single layer potential representation, leading to a Fredholm integral equation of the second kind. We propose to solve this integral equation using a Petrov-Galerkin method with trigonometric polynomials as test functions, and a span of delta distributions centered at the boundary points as trial functions. For the exterior boundary value problem, the approximate potential converges exponentially away from the boundary and algebraically up to the boundary. We show that these convergence results hold even when the discretization matrices are computed via trapezoidal integration and present numerical examples to confirm our theory. For the interior boundary value problem, we suggest that the approximate potential also converges exponentially away from the boundary and algebraically up to the boundary, and present numerical examples to confirm our conjecture.

INTRODUCTION

The two-dimensional Neumann potential problem is

$$\Delta u = 0 \ \text{ on } \mathbf{R}^2 \backslash \Gamma, \qquad \frac{du}{d\nu} = G \ \text{ on } \Gamma, \tag{1}$$

where Γ is a simple closed analytic curve, ν is the outward normal, G is an analytic function, and u is bounded at infinity. From Folland [5, theorem 3.41], the differential system (1) has a unique solution except for a constant if and only if the Neumann data G satisfy

$$\int_{\Gamma} G(y)\, d\Gamma_y = 0. \tag{2}$$

Moreover, the solution is the single-layer potential representation,

$$u(z) = \frac{1}{2\pi} \int_{\Gamma} \Phi(y) \log|z-y| \, d\Gamma_y + C \quad \text{for } z \in \mathbf{R}^2 \backslash \Gamma, \tag{3}$$

where C is a constant, and the density Φ solves

$$G(z) = \pm\frac{1}{2}\Phi(z)+\frac{1}{2\pi}\int_\Gamma \Phi(y)\frac{d}{d\nu_z}\log|z-y|\,d\Gamma_y \quad \forall z\in\Gamma. \tag{4}$$

In equation (4), the plus and minus signs correspond to the exterior and interior Neumann problems, respectively, and the $\Phi(z)/2$ term arises from the normal derivative jump discontinuity of the single-layer potential representation across the boundary Γ. For the exterior Neumann problem, equation (4) is uniquely solvable for all G (Fredholm alternative theorem and Folland [5, theorem 3.38]). For the interior Neumann problem, equation (4) has a nonunique solution if and only if equation (2) holds (Fredholm alternative theorem and Folland [5, theorem 3.36]).

In this paper, we seek integral methods to approximate the density in equation (4) such that the potential in equation (3) is approximated with exponential convergence based on the mesh discretization. These methods reduce the matrix system significantly, thus reducing the need for tedious matrix operations. Previous works were confined to solving the two-dimensional Dirichlet potential problem using the single-layer potential representation. D.N. Arnold [1] showed that the spline-trigonometric method obtains the potential on compact sets away from the boundary with exponential convergence. W. McLean [6] showed that the trigonometric-trigonometric method obtained the potential everywhere with exponential convergence. Neither Arnold nor McLean accounted for numerical integration. Cheng [3] and Cheng and Arnold [4] showed that the delta-trigonometric method using numerical integration obtains potential on compact sets away from the boundary with exponential convergence.

In this paper, we extend the previous work by investigating the use of the delta-trigonometric method to solve the two-dimensional Neumann potential problem. This method is a Petrov-Galerkin method applied on equation (4) in which the trial space is a span of a finite set of Dirac delta functions, and the test space is a span of trigonometric polynomials. For the exterior problem, we show that the semidiscrete and the fully discrete delta-trigonometric methods obtain approximate potentials which converge exponentially on compact sets away from the boundary and algebraically up to the boundary. Then, we present two numerical examples to confirm the theory. For the interior problem, we conjecture that the semidiscrete and fully discrete delta-trigonometric methods obtain approximate potentials which converge exponentially on compact sets away from the boundary and algebraically up to the boundary. Finally, we present two numerical examples to confirm our conjecture. Here, we numerically solve the singular integral equation (4) by finding the eigenvalues and eigenvectors of the discretization matrix.

In the next two sections, we review some standard notations and define the delta-trigonometric method. In the following section, we present convergence analyses on using the semidiscrete delta-trigonometric method to solve the exterior problem. Then, we show that the matrix condition number for the exterior problem is bounded proportionally to the number of mesh intervals. In a later section, we extend the convergence analyses for the fully discrete solution. Afterward, we present two numerical examples to confirm the theory. In the final section, we discuss the use of the delta-trigonometric method for the interior problem, and present two numerical examples to confirm our conjecture.

PRELIMINARIES

Define the space of trigonometric polynomials with complex coefficients as

$$T := \text{span}\,\{\, \exp(2\pi i\, k\, t) \;\big|\; k \in \mathbf{Z} \,\}. \tag{5}$$

Any function f in this space can be represented as

$$f(t) = \sum_{k \in \mathbf{Z}} \hat{f}(k) \exp(2\pi i\, k\, t), \tag{6}$$

where

$$\hat{f}(k) := \int_0^1 f(t) \exp(-2\pi i\, k\, t)\, dt \tag{7}$$

are arbitrary complex numbers, for all but finitely many zeros. For $f \in T$, $s \in \mathbf{R}$, and $\epsilon > 0$, we define the Fourier norm as

$$\|f\|_{s,\epsilon} := \sum_{k \in \mathbf{Z}} |\hat{f}(k)|^2\, \epsilon^{2|k|}\, \underline{k}^{2s}, \tag{8}$$

where

$$\underline{k} = \begin{cases} 1, & \text{if } k = 0, \\ 2\pi\, |k|, & \text{if } k \neq 0. \end{cases} \tag{9}$$

We denote by $X_{s,\epsilon}$ the completion of T with respect to this norm. As in [4], the L^2 innerproduct extends to a bounded binear form on $X_{s,\epsilon} \times X_{-s,\epsilon^{-1}}$ for all $s \in \mathbf{R}$, $\epsilon > 0$ in which $X_{-s,\epsilon^{-1}}$ is the dual space of $X_{s,\epsilon}$. For $\epsilon = 1$, $X_{s,\epsilon}$ is the usual periodic Sobolev space of order s, H^s, with norm $\|\cdot\|_s$. See Arnold [1,§3] for a more complete discussion of these spaces.

We denote the standard Euclidean vector and matrix norms by $\|\cdot\|$. The constants C and ϵ are generic and are not necessarily the same in each occurrence.

THE DELTA-TRIGONOMETRIC METHOD

Let $x : \mathbf{R} \to \Gamma$ be a 1-periodic analytic function which parameterizes Γ and has nonvanishing derivatives, and define

$$\phi(t) := \Phi(x(t)), \quad g(t) := G(x(t)). \tag{10}$$

Next, define three operators

$$B\phi(s) := \int_0^1 \phi(t)\, K(s,t)\, dt, \tag{11}$$

$$V\phi(s) := \pm \frac{1}{2}\, \phi(s), \tag{12}$$

$$A\phi(s) := B\phi(s) + V\phi(s), \tag{13}$$

where $K : \mathbf{R}^2 \to \mathbf{R}$ is a smooth kernel defined by

$$K(s,t) := \begin{cases} |x'(t)| \dfrac{[x(s)-x(t)]\cdot \nu(s)}{2\pi\,[x(s)-x(t)]^2}, & \text{if } s \neq t, \\ \dfrac{-x''(s)\cdot \nu(s)}{4\pi\,|x'(s)|}, & \text{if } s = t, \end{cases} \tag{14}$$

and $\nu : \mathbf{R} \to \Gamma$ is the 1-periodic outward normal. For the operator V, the plus and minus signs correspond for the exterior and interior problems, respectively. The operator B has a smooth bounded kernel and is compact; thus the operator A is a Fredholm operator of the second kind. For the exterior problem, the operator A is an isomorphism on $C(\Gamma)$ by Folland [5, theorem 3.36 and 3.38] and the Fredholm alternative theorem. For the interior problem, the operator A has a null space of dimension 1. Define

$$M := \{\phi \in C(\Gamma) \mid A\,\phi = 0\}, \tag{15}$$

and

$$N := \{g \in C(\Gamma) \mid \int_0^1 g(t)\,|x'(t)|\,dt = 0\}. \tag{16}$$

Then, for the interior problem, M and $N^{\perp}$ have dimension 1. Moreover, A is an isomorphism from $C(\Gamma)\backslash M$ to N, and the general solution is $\phi = \phi^1 + \phi^2$ such that ϕ^1 is orthogonal to M, ϕ^2 is any vector in M, and $A\,\phi^1 = g$. For the convergence analysis, we choose to restrict ourselves to the exterior problem so that A is an isomorphism on $C(\Gamma)$.

The single-layer potential representation (3) becomes

$$u(z) = \frac{1}{2\pi}\int_0^1 \phi(t)\,\log|z - x(t)|\;|x'(t)|\,dt \quad \text{for } z \in \mathbf{R}^2\backslash\Gamma, \tag{17}$$

and the boundary integral equation (4) becomes

$$A\phi(s) = g(s) \quad \text{for } s \in [0,1]. \tag{18}$$

Restrict n to be a positive odd number and define

$$\Lambda_n := \{\, k \in \mathbf{Z} \;\big|\; |k| \leq (n-1)/2 \,\}. \tag{19}$$

The trial space is chosen as

$$S_n := \text{span}\,\{\, \delta(t - j/n) \;\big|\; j = 0,\ldots,n-1 \,\}, \tag{20}$$

where $\delta(t - j/n)$ is the 1-periodic extension of the Dirac mass at j/n for $j = 1,2,..,n$. The test space is chosen as

$$T_n := \text{span}\,\{\, \exp(2\pi i k t) \;\big|\; k \in \Lambda_n \,\}, \tag{21}$$

the space of trigonometric polynomials with degree $\leq n$.

The semidiscrete delta-trigonometric method seeks $\phi_n \in S_n$ such that

$$\int_0^1 A\phi_n(s)\,\psi(s)\,ds = \int_0^1 g(s)\,\psi(s)\,ds \quad \forall\,\psi \in T_n, \tag{22}$$

and then computes the approximate potential as

$$u_n(z) = \frac{1}{2\pi}\int_0^1 \phi_n(t)\,\log|z - x(t)|\ |x'(t)|\,dt \quad \text{for } z \in \mathbf{R}^2\backslash\Gamma. \tag{23}$$

Now define the matrix equations with and without numerical quadratures for the delta-trigonometric method. We now use

$$\phi_n(t) = \sum_{j=0}^{n-1} \alpha_j\,\delta(t - j/n), \tag{24}$$

where α_j's are unknown coefficients, and

$$\psi_k(s) = \exp(2\pi i\,k\,s), \quad k \in \Lambda_n. \tag{25}$$

Define $n \times n$ matrices **A**, **B**, **V** and n-vectors **g**, $\boldsymbol{\alpha}$ as

$$\mathbf{V}_{kj} := \pm\frac{1}{2}\,\psi_k(j/n), \tag{26}$$

$$\mathbf{B}_{kj} := \int_0^1 K(s, j/n)\,\psi_k(s)\,ds, \tag{27}$$

$$\mathbf{A}_{kj} := \mathbf{V}_{kj} + \mathbf{B}_{kj}, \tag{28}$$

$$\mathbf{g}_k := \int_0^1 g(s)\,\psi_k(s)\,ds, \tag{29}$$

$$\boldsymbol{\alpha} := (\alpha_1, \ldots, \alpha_n)^T, \tag{30}$$

for $k \in \Lambda_n$, $j = 1,\ldots,n$. Then the matrix form of the semidiscrete delta-trigonometric method is

$$\mathbf{A}\,\boldsymbol{\alpha} = \mathbf{g}, \tag{31}$$

and the approximate potential is expressed as

$$u_n(z) = \frac{1}{2\pi}\sum_{j=0}^{n-1} \alpha_j\,\log|z - x(j/n)|\ |x'(j/n)| \quad \text{for } z \in \mathbf{R}^2\backslash\Gamma. \tag{32}$$

The fully discrete method is obtained by using the trapezoidal rule to evaluate **B** and **g** Define

$$\tilde{B}_{kj} := \frac{1}{n}\sum_{l=0}^{n-1} K(l/n, j/n)\, \psi_k(l/n), \tag{33}$$

$$\tilde{A}_{kj} := V_{kj} + \tilde{B}_{kj}, \tag{34}$$

$$\tilde{g}_k := \frac{1}{n}\sum_{l=0}^{n-1} g(l/n)\, \psi_k(l/n), \tag{35}$$

$$\tilde{\alpha} := (\tilde{\alpha}_1, \ldots, \tilde{\alpha}_n)^T, \tag{36}$$

for $k \in \Lambda_n$, $j = 1,\ldots,n$. The delta-trigonometric method with numerical quadrature seeks coefficients $\tilde{\alpha}_j$'s such that

$$\tilde{A}\tilde{\alpha} = \tilde{g}. \tag{37}$$

Then the approximate potential is

$$\tilde{u}_n(z) = \frac{1}{2\pi}\sum_{j=0}^{n-1} \tilde{\alpha}_j \log|z - x(j/n)|\, |x'(j/n)| \quad \text{for } z \in \mathbf{R}^2\backslash\Gamma. \tag{38}$$

For the exterior problem, the matrix equations (31) and (37) are nonsingular and can be solved directly. For the interior problem, these matrix equations are singular and solvable, and have exactly one zero eigenvalue. The density vectors α and $\tilde{\alpha}$ are found as a linear combination of eigenvectors, and are unique except for the eigenvector corresponding to the zero eigenvalue.

CONVERGENCE FOR THE SEMIDISCRETE DELTA-TRIGONOMETRIC METHOD

In this section, we show convergence for the approximate potentials to the exterior Neumann problem obtained by the delta-trigonometric method by revising the convergence analyses used for the analogous Dirichlet potential problem given by Cheng [3] and Cheng and Arnold [4]. For all convergence analysis, we assume that the operators V and A correspond to the exterior problem only.

Theorems 1 and 3 state the inf-sup condition for the operators V and A, respectively. Theorem 4 states the existence and quasioptimality of the approximate potential for the exterior problem. Theorem 5 states the optimal convergence for the approximate density, while Theorem 6 states that the approximate potential converges exponentially on any compact set in the exterior. Theorem 7 states that the approximate potential converges algebraically in the entire exterior region using a weighted L^2 norm.

THEOREM 1. *Let $s \le s_0 < -1/2$. Then there exists a constant C depending only on s_0 such that*

$$\inf_{0 \ne \rho \in S_n} \sup_{0 \ne \sigma \in T_n} \frac{(V\rho, \sigma)}{\|\rho\|_{s,\epsilon}\, \|\sigma\|_{-s,\epsilon^{-1}}} \ge C \tag{39}$$

for all $\epsilon \in (0,1]$ and $n \in \mathbf{Z}^+$.

PROOF. From Cheng [3, theorem 3.1.1] or Cheng and Arnold [4, theorem 3.1], there exists constant C_1 depending only on s_0 such that

$$\|\rho\|^2_{s,\epsilon} \le C_1 \sum_{k\in\Lambda_n} |\hat{\rho}(k)|^2\, \epsilon^{2|k|}\, \underline{k}^{2s} \quad \forall\, \rho \in S_n. \tag{40}$$

Choose

$$\sigma(t) = \sum_{k\in\Lambda_n} \hat{\rho}(k)\, \epsilon^{2|k|}\, \underline{k}^{2s} \exp(2\pi i\, k\, t). \tag{41}$$

Then

$$\|\sigma\|^2_{-s,\epsilon^{-1}} = \sum_{k\in\Lambda_n} |\hat{\rho}(k)|^2\, \epsilon^{2|k|}\, \underline{k}^{2s}, \tag{42}$$

and

$$(V\rho, \sigma) = \frac{1}{2} \sum_{k\in\Lambda_n} |\overline{\hat{\rho}(k)}|\, \epsilon^{2|k|}\, \underline{k}^{2s} \int_0^1 \rho(t) \exp(-2\pi i\, k\, t)\, dt \tag{43}$$

$$= \frac{1}{2} \sum_{k\in\Lambda_n} |\hat{\rho}(k)|^2\, \epsilon^{2|k|}\, \underline{k}^{2s} \tag{44}$$

$$\ge \frac{1}{2\sqrt{C_1}}\, \|\rho\|_{s,\epsilon}\, \|\sigma\|_{-s,\epsilon^{-1}}. \tag{45}$$

□

The next lemma states that the Fourier coefficients of the analytic kernel K have exponential decays. This lemma is similiar to Cheng [3, lemma 3.1.3] and Cheng and Arnold [4, lemma 3.2], and its proof is omitted.

LEMMA 2. *Let $S_\delta := \{z \in \mathbf{C} \mid |\mathrm{Im}(z)| < \delta\}$. Then the kernel K defined in (14) is a real 1-periodic analytic function in each variable and extends analytically to $S_\delta \times S_\delta$ for some $\delta > 0$. Moreover, there exists constants C and $\epsilon_K \in (0,1)$ such that*

$$\left|\hat{K}(p,q)\right| \le C\, \epsilon_K^{|p|+|q|} \quad \forall\, p,\ q \subset \mathbf{Z}. \tag{46}$$

The next theorem provides the stability result for the operator A.

THEOREM 3. *Let $s \le s_0 < -1/2$, $\epsilon \in (\epsilon_K, 1]$ (ϵ_K being determined in Lemma 2). Then for sufficiently large n, there exists a constant C depending only on s_0 and Γ such that*

$$\inf_{0\ne\rho\in S_n} \sup_{0\ne\sigma\in T_n} \frac{(A\rho, \sigma)}{\|\rho\|_{s,\epsilon}\, \|\sigma\|_{-s,\epsilon^{-1}}} \ge C. \tag{47}$$

PROOF. From similiar arguments as in Arnold [1, theorem 4.6 and corollary 4.7], B is a compact operator from $X_{s,\epsilon}$ to $X_{s,\epsilon}$ for all $s \in \mathbf{R}$, $\epsilon \in (\epsilon_K, 1]$. From similiar arguments as in Arnold [1, corollary 4.8], A is an isomorphism on $X_{s,\epsilon}$. By Theorem 1, there exists $\beta > 0$ such that for all n in $\mathbf{Z}^+$ and ρ in S_n, there exists $\sigma \in T_n$ satisfying

$$(A\rho, \sigma) \ge \beta\, \|\rho\|_{s,\epsilon}\, \|\sigma\|_{-s,\epsilon^{-1}} - (B\rho, \sigma). \tag{48}$$

The inf-sup condition for the operator A is obtained by using a compactness argument (e.g. [2]). □

The next theorem gives the existence and quasioptimality of the approximate solution for the exterior problem. Its proof is omitted and can be invoked from the standard theory of Galerkin methods.

THEOREM 4. *There exists a constant N, depending only on Γ, such that for all $n \geq N$ and g in $\bigcup\{X_{s,\epsilon} \mid s \in \mathbb{R}, \epsilon > 0\}$, the delta-trigonometric method (22) obtains unique solutions, $\phi_n \in S_n$, for the exterior problem. Moreover, if $s < -1/2$, $\epsilon \in (\epsilon_K, 1]$ (ϵ_K being determined in Lemma 2), $g \in X_{s,\epsilon}$, and $n \geq N$, then there exists a constant C, depending only on ε, s, and Γ such that*

$$\|\phi - \phi_n\|_{s,\epsilon} \leq C \inf_{\rho \in S_n} \|\phi - \rho\|_{s,\epsilon}. \tag{49}$$

The approximate solution is quasioptimal, and its convergence is established by bounding the error of any approximation from S_n of the exact solution. As in [1], [3], and [4], we choose $P_n\,\phi$ in S_n such that

$$\widehat{P_n\,\phi}(k) = \hat{\phi}(k) \quad \forall\, k \in \Lambda_n. \tag{50}$$

From Cheng [3, theorem 3.1.6] or Cheng and Arnold [4], if $s < -1/2$, $t \in [s, 0]$, and $\epsilon \leq 1$, then

$$\|\phi - P_n\,\phi\|_{s,\epsilon} \leq C\, \epsilon^{n/2}\, (\pi n)^{s-t}\, \|\phi\|_t \quad \forall\, \phi \in H^t. \tag{51}$$

Combining this with Theorem 4, we obtain:

THEOREM 5. *Let $s < -1/2$, $t \in [s, 0]$, $n \geq N$, and $\phi \in H^t$. Then for $\epsilon \in (\epsilon_K, 1]$ (ϵ_K being determined in Lemma 2), there exists a constant C depending only on ε, s, and Γ, such that*

$$\|\phi - \phi_n\|_{s,\epsilon} \leq C\, \epsilon^{n/2}\, (\pi n)^{s-t}\, \|\phi\|_t. \tag{52}$$

For our convergence analyses, we compute the approximate exterior potential in equation (3) with the constant C as zero. Since the single-layer potential was also used in the two-dimensional Dirichlet potential problem, the next theorem is similiar to Cheng [3, theorem 3.1.7] and Cheng and Arnold [4, theorem 3.6], and will not be proved.

THEOREM 6. *Let $n \geq N$, $\phi \in H^t$, and Ω_K be a compact set in the exterior region. Then, for any multiindex β, there exist constants C and $\epsilon \in (0,1)$ depending only on t, N, Ω_K, and Γ, such that*

$$\|\partial^\beta (u - u_n)\|_{L^\infty(\Omega_K)} \leq C\, \epsilon^n\, \|\phi\|_t. \tag{53}$$

Although the convergence on compact sets away from the boundary is exponential, the approximate potential converges on the entire exterior region at an algebraic rate. Convergence of order 3/2 holds in L^2 on bounded sets. Let Ω and Ω_e denote the interior and exterior regions, respectively. Define the weighted norm as

$$|||v|||^2 = \int_{\Omega_e} \frac{|v(z)|^2}{1+|z|^4}\, dz \tag{54}$$

to cover the case of convergence near infinity.

THEOREM 7. *Let $-3/2 \le t \le 0$, $n \ge N$, and $\phi \in H^t$. Then there exists a constant C depending only on Γ such that*

$$|||u - u_n||| \le C\, n^{-t-3/2}\, \|\phi\|_t \tag{55}$$

PROOF. Without loss of generality, assume that the origin is in Ω. By Theorem 5, it suffices to prove

$$\int_{\Omega_e} |(u-u_n)(z)|^2 \frac{dz}{|z|^4} \le C\, \|\phi - \phi_n\|_{-3/2}. \tag{56}$$

Let $v = u - u_n$ and $q = dv/d\nu\,|_\Gamma$. Then v solves the exterior Neumann problem

$$\Delta v = 0 \ \text{ on } \Omega_e, \quad \frac{dv}{d\nu} = q \ \text{ on } \Gamma. \tag{57}$$

We use the Kelvin transform, $\kappa(z) = z/|z|^2$ to map Γ analytically to $\bar{\Gamma}$. Let $\bar{\Omega}$ denote the bounded component of $\bar{\Gamma}$. Also, set $\bar{v} = v \circ \kappa$ and $\bar{q} = d\bar{v}/d\nu\,|_{\bar{\Gamma}}$. Then $\bar{v}$ is harmonic on $\Gamma \backslash \{0\}$, and the singularity at the origin is removable (since v is bounded). Thus,

$$\|\bar{v}\|_{H^s(\bar{\Omega})} \le \|\bar{q}\|_{H^{s-3/2}(\bar{\Gamma})} \tag{58}$$

In particular,

$$\|\bar{v}\|_{L^2(\bar{\Omega})} \le \|\bar{q}\|_{H^{-3/2}(\bar{\Gamma})} \le \|q\|_{H^{-3/2}(\Gamma)}. \tag{59}$$

We also know that $q = A\,(\phi - \phi_n)$ implies that

$$\|q\|_{H^{-3/2}(\Gamma)} \le C\, \|\phi - \phi_n\|_{-3/2}. \tag{60}$$

Therefore,

$$\|\bar{v}\|_{L^2(\bar{\Omega})} \le C\, \|\phi - \phi_n\|_{-3/2}. \tag{61}$$

We conclude the proof by noting that

$$\|\bar{v}\|_{L^2(\bar{\Omega})} = \int_{\Omega_e} |(u-u_n)(z)|^2 \frac{dz}{|z|^4}. \tag{62}$$

□

CONDITION NUMBER

We now show that the condition numbers of the discretization matrices (corresponding to the exterior problem) are bounded linearly proportional to the number of subintervals. Lemma 8 states the relationship between $\|\phi_n\|$ and $\|\alpha\|$ defined in equation (24) and (30), respectively. Then Theorem 9 states the bounds for the matrix condition number.

LEMMA 8. *There exists a constant C such that*

$$\|\phi_n\|_{-1} \leq C\sqrt{n}\ \|\alpha\| \tag{63}$$

and

$$\|\alpha\| \leq C\sqrt{n}\ \|\phi_n\|_{-1}. \tag{64}$$

PROOF. See Cheng [3, lemma 3.2.1] or Cheng and Arnold [4, lemma 3.8]. □

THEOREM 9. *There exists a constant C depending only on Γ such that*

$$\|\mathbf{A}\| \leq C\sqrt{n}, \quad \|\mathbf{A}^{-1}\| \leq C\sqrt{n}. \tag{65}$$

Moreover, the condition number $\kappa(\mathbf{A})$ *satisfies*

$$\kappa(\mathbf{A}) \leq C\, n. \tag{66}$$

PROOF. The operator A (for the exterior problem) is an isomorphism on L^2 and, therefore, $\|A\|_{L^2}$ and $\|A^{-1}\|_{L^2}$ are bounded. The rest of the proof follows by standard argument as in Cheng [3, theorem 3.2.2]. □

CONVERGENCE FOR THE FULLY DISCRETE DELTA-TRIGONOMETRIC METHOD

In this section, we modify the convergence analyses in §4 for the fully discrete delta-trigonometric method. The theorems in this section are similiar to that in Cheng [3] and Cheng and Arnold [4], and their proofs are referenced. Again, we assume that the operator A refers to the exterior problem only.

THEOREM 10. *Let f be an analytic 1-periodic function and define*

$$f_k := \int_0^1 f(t)\exp(2\pi i k t)\,dt \tag{67}$$

and

$$\tilde{f}_k := \frac{1}{n}\sum_{l=0}^{n-1} f(l/n)\exp(2\pi i k l/n) \quad \forall k \in \mathbb{Z}. \tag{68}$$

Then there exist constants C and $\epsilon \in (0,1]$ *depending only on f such that*

$$|f_k - \tilde{f}_k| \leq C\,\epsilon^n \quad \forall k \in \Lambda_n. \tag{69}$$

PROOF. See Cheng [3, theorem 3.3.2] or Cheng and Arnold [4, theorem 4.2]. □

THEOREM 11. *There exist constants C and* $\epsilon \in (0,1]$ *depending only on g and Γ such that*

$$\|\mathbf{g} - \tilde{\mathbf{g}}\| \leq C\,\epsilon^n, \quad \|\mathbf{B} - \tilde{\mathbf{B}}\| \leq C\,\epsilon^n. \tag{70}$$

PROOF. See Cheng [3, theorem 3.3.3] or Cheng and Arnold [4, theorem 4.3]. □

THEOREM 12. *Let Ω_K be any compact set in the exterior region. Then, for any multiindex β, there exist constants C and $\epsilon \in (0,1)$ depending only on g, Γ, and Ω_K, such that*

$$\|\partial^\beta(u_n - \tilde{u}_n)\|_{L^\infty(\Omega_K)} \leq C\,\epsilon^n. \tag{71}$$

PROOF. See Cheng and Arnold [4, theorem 4.4]. □

THEOREM 13. *Let $-3/2 \leq t \leq 0$, $n \leq N$, and $\phi \in H^t$. Then there exists a constant C depending only on g and Γ such that*

$$|||u - \tilde{u}_n||| \leq C\,n^{-t-3/2}. \tag{72}$$

PROOF. See Cheng and Arnold [4, theorem 4.5]. □

NUMERICAL IMPLEMENTATION OF THE EXTERIOR PROBLEM

In this section, we present two numerical examples to confirm the theoretical results. For numerical purposes, we use real test functions,

$$\tilde{\psi}_k = \begin{cases} \sin(k\,\pi s), & \text{for } k = 2,4,..,n-1, \\ \cos((k-1)\,\pi s), & \text{for } k = 1,3,\ldots,n. \end{cases} \tag{73}$$

The fully discrete delta-trigonometric method using trapezoidal rule seeks coefficients $\tilde{\alpha}_j$'s such that

$$\pm\frac{1}{2}\sum_{j=0}^{n-1} \tilde{\alpha}_j\,\tilde{\psi}_k(j/n) + \frac{1}{n}\sum_{l=0}^{n-1}\sum_{j=0}^{n-1} \tilde{\alpha}_j\,K(l/n, j/n)\,\tilde{\psi}_k(l/n) \tag{74}$$

$$= \frac{1}{n}\sum_{l=0}^{n-1} g(l/n)\,\tilde{\psi}_k(l/n) \quad \text{for } k = 0,..,n-1. \tag{75}$$

The approximate potential is

$$u(z) = \frac{1}{2\pi}\sum_{j=0}^{n-1} \alpha_j \log|z - x(j/n)|\;|x'(j/n)| \quad \text{for } z \in \mathbf{R}^2. \tag{76}$$

Fast Fourier transform is used to compute right hand side vector $\tilde{\mathbf{g}}$ and for each column of the matrix $\tilde{\mathbf{B}}$.

EXAMPLE 1. *Circle with analytic data.* The circle was chosen as a test problem to verify the numerical implementation of the delta-trigonometric method because the analytic solution is easily derived using separation of variables.

Boundary: $x_1^2 + x_2^2 = a^2$

Data: $g = x_1 / a$

Exact exterior solution: $u = -a^2 x_1 / (x_1^2 + x_2^2)$

Figure 1 shows the absolute error of u on the line $x_1 = 2x_2$ for $a = 2$. As expected, the absolute error decays exponentially with respect to the mesh discretization until machine error is reached except near the boundary ($x_1 = 1.414$). Table 1 shows the CPU times used on the APOLLO DN 3000. As n becomes large, we expect the total time to be proportional to n^3.

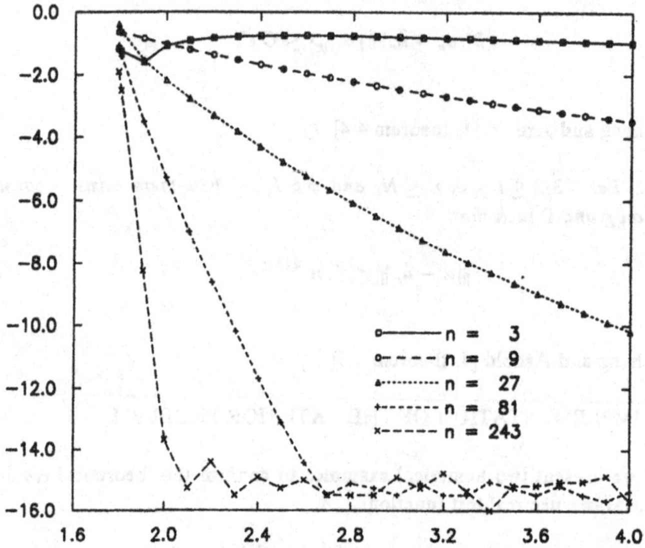

Figure 1. Log(error) versus x_1 on the line $x_1 = 2x_2$ for example 1.

Table 1. CPU times for example 1.

n	3	9	27	81	243
CPU	0.779	0.834	2.831	25.003	334.506

EXAMPLE 2. ***Ellipse exterior with analytic data.*** This example is a slight modification of the first example. The analytic data were derived given the exact solution.

Boundary: $\dfrac{x_1^2}{a^2} + \dfrac{x_2^2}{b^2} = 1$

Data:

$$g(s) = \frac{b\cos(2\pi s)}{\sqrt{b^2\cos^2(2\pi s) + a^2\sin^2(2\pi s)}} \cdot \left[\frac{a^2 + (a^2 - b^2)\sin^2(2\pi s)}{[a^2\cos^2(2\pi s) + b^2\sin^2(2\pi s)]^2}\right]$$

Exact exterior solution: $u = -4x_1 / (x_1^2 + x_2^2)$

Figure 2 shows the absolute error of u on the line $x_1 = 2x_2$ for $a = 2$ and $b = 1$. Again, the absolute error decays exponentially with respect to the mesh discretization until machine error is reached except near the boundary. Table 2 shows the CPU times which is similiar to the previous example.

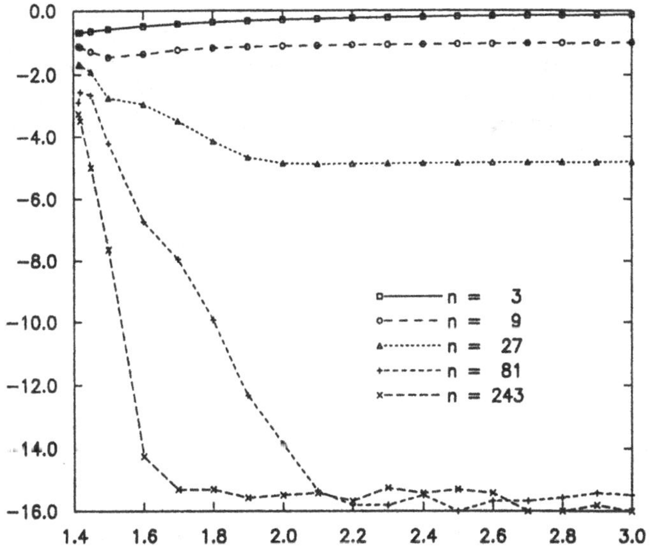

Figure 2. Log(error) versus x_1 on the line $x_1 = 2x_2$ for example 2.

Table 2. CPU times for example 2.

n	3	9	27	81	243
CPU	0.664	0.730	2.730	24.972	332.598

NUMERICAL IMPLEMENTATION OF THE INTERIOR PROBLEM

In this section, we consider the numerical implementation of the interior Neumann problem. We conjecture that the semidiscrete and fully-discrete delta-trigonometric methods obtains the potential with exponential convergence on interior compact sets.

To approximate numerically the density vector, we first symmetrize the matrix equation $\tilde{A}\tilde{\alpha} = \tilde{g}$ by multiplying by $\tilde{A}^T$, and then solve for the eigenvalues, $\lambda_1, \ldots, \lambda_n$, and eigenvectors, $e^1, \ldots, e^n$ of $\tilde{A}^T\tilde{A}$. The matrix $\tilde{A}^T\tilde{A}$ has exactly one zero eigenvalue, say $\lambda_1 = 0$. The approximate density vector is

$$\tilde{\alpha} = \sum_{k=2}^{n} \frac{(\tilde{A}^T g, e^k)\, e^k}{\lambda_k}. \tag{77}$$

We now present two numerical examples to confirm our conjecture.

EXAMPLE 3. ***Circle with analytic data.*** **The circle was chosen as a test problem to verify the numerical implementation of the delta-trigonometric method because the analytic solution is easily derived using separation of variables.**

Boundary: $x_1^2 + x_2^2 = a^2$

Data: $g = x_1 / a$

Exact interior solution: $u = x_1$

Figure 3 shows the absolute error of u on the line $x_1 = 2x_2$ for $a = 2$. As expected, the absolute error decays exponentially with respect to the mesh discretization until machine error is reached except near the boundary ($x_1 = 1.414$). Table 3 shows that the CPU times are longer and approach $O(n^3)$ more quickly than for the exterior problem because the eigensystem must be solved.

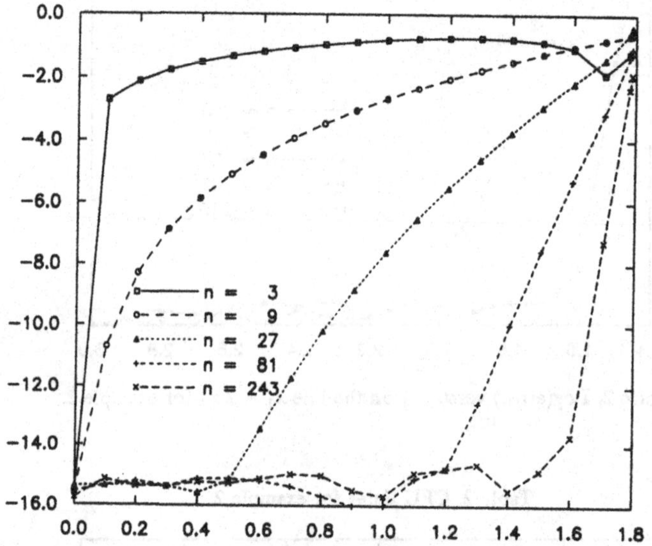

Figure 3. Log(error) versus x_1 on the line $x_1 = 2x_2$ for example 3.

Table 3. CPU times for example 3.

n	3	9	27	81	243
CPU	0.734	0.870	5.158	88.929	2058.364

EXAMPLE 4. ***Ellipse interior with analytic data.*** **This example is a slight modification of the previous example. Given the exact solution, the analytic data were derived.**

Boundary: $\dfrac{x_1^2}{a^2} + \dfrac{x_2^2}{b^2} = 1$

Data:

$$g(s) = \frac{b\cos(2\pi s)}{\sqrt{b^2\cos^2(2\pi s) + a^2\sin^2(2\pi s)}}.$$

Exact interior solution: $u = x_1$

Figure 4 shows the absolute error of u on the line $x_1 = 2x_2$ for $a = 2$ and $b = 1$. Again, the absolute error decays exponentially with respect to the mesh discretization until machine error is reached except near the boundary. Table 4 shows the CPU times used. The computational time differed slightly from the previous example and is due to the number of iterations needed to solve the eigensystem.

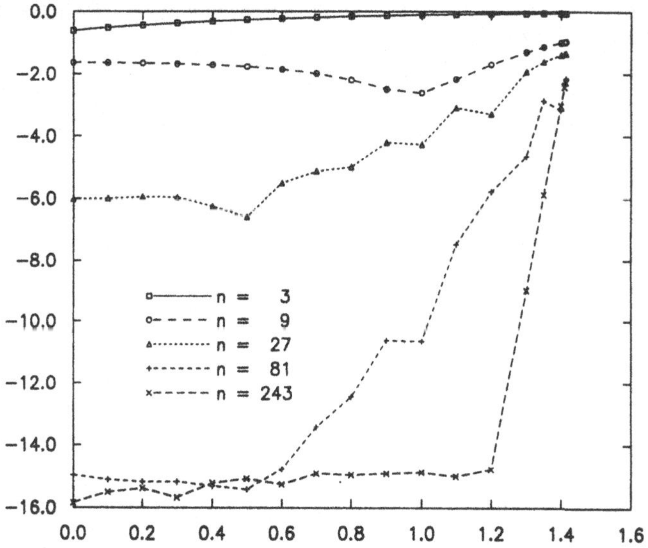

Figure 4. Log(error) versus x_1 on the line $x_1 = 2x_2$ for example 4.

Table 4. CPU times for example 4.

n	3	9	27	81	243
CPU	0.715	0.845	5.445	73.874	1732.693

ACKNOWLEDGEMENT

This work was supported in Fiscal Year 1991 by the David Taylor Research Center (DTRC) Independent Research Program, sponsored by the Office of Naval Research and administered by the DTRC Research Director. This work was performed under United States government sponsorship, it is publicly available information with no copyrights associated with it.

REFERENCES

1. Arnod, D.N. A Spline-Trigonometric Galerkin Method and an Exponentially Convergent Boundary Integral Method, Math. Comp, Vol. 41, pp. 383-397, 1983.

2. Aziz, A. and Kellogg, B. Finite Element Analysis of a Scattering Problem, Math. Comp. Vol. 37, pp. 261-272, 1981.

3. Cheng, R.S.-C. Delta-Trigonometric and Spline-Trigonometric Methods using the Single-Layer Potential Representation, Dissertation, University of Maryland - College Park, 1987.

4. Cheng, R.S.-C. and Arnold, D.N. The Delta-Trigonometric Method using the Single-Layer Potential Representation, J. Integral Eqns. Applic., Vol. 1, n. 4, 1988.

5. Folland, G.B. Introduction to Partial Differential Equations, Princeton Lecture Note Series, Princeton University Press, 1976.

6. McLean, W. A Spectral Galerkin Method for a Boundary Integral Equation, Math. Comp., Vol. 47, pp. 597-607, 1986.

Non-Linear Material Problems with BEM: Dual Reciprocity v. The Kirchhoff Transformation

P.W. Partridge

Departamento de Engenharia Civil, Universidade de Brasília, 70910 Brasília, Brazil

ABSTRACT

In this paper a Dual Reciprocity, (DRM), solution to non-linear material problems will be presented as an alternative to the Kirchhoff transformation in BEM analysis. The computational formulation and results for both approaches will be presented for cases with and without a boundary condition of the third kind. In the former case a solution procedure involving a DRM approximation to second derivatives of space is also presented.

INTRODUCTION

In practical problems of heat transfer, non-linear materials for which the conductivity is a function of temperature are quite common. Boundary Element analysis of such problems is convenient as in many cases the interior solution is of no interest.

The Kirchhoff transformation was introduced for this type of problem by Bialecki and Nowak, [1]. This transformation (a) eliminates the domain integrals which would otherwise arise due to the non-linear material, (b) linearizes the problem if only Neumann and Dirichlet type boundary conditions are present. However in the case of a convective or third kind boundary condition the resulting boundary equation is non-linear requiring an iterative solution procedure, [2].

The Dual Reciprocity Method is a powerful technique for taking domain integrals to the boundary in BEM analysis. It was introduced by Nardini and Brebbia in 1982, [3] and has been recently generalized to a wide range of engineering problems, [4]. The method is straightforward to apply and uses simple fundamental solutions, being capable of transforming any remaining terms to the boundary.

Applied to a non-linear material equation, DRM will always lead to an iterative solution, however a boundary condition of the third kind can be taken into account in a simple way.

The Dual Reciprocity method and the Kirchhoff transformation will be applied first to a problem with boundary conditions of the first and second kind only and then to a case involving one of the third kind also. It will be shown that due to the complexity of the governing equation in this case alternative DRM formulations are possible, and one will be presented in which second derivatives of space are approximated using the method.

COMPUTATIONAL FORMULATIONS

The steady-state heat transfer equation, neglecting heat sources, can be written

$$\frac{\partial}{\partial x}\left\{k(u)\frac{\partial u}{\partial x}\right\}+\frac{\partial}{\partial y}\left\{k(u)\frac{\partial u}{\partial y}\right\}=0 \tag{1}$$

The conductivity, k, is considered to be a known function of temperature, u.
Three types of boundary condition will be considered:-

$$u=\overline{u} \quad \text{on} \quad \Gamma_1$$

$$q=k\frac{\partial u}{\partial n}=\overline{q} \quad \text{on} \quad \Gamma_2$$

$$q=h(u_f-u) \quad \text{on} \quad \Gamma_3 \tag{2}$$

where q represents heat flux, h is a heat transfer coefficient and u_f is the temperature of a surrounding medium. n is the outward normal at the boundary. The total boundary, $\Gamma=\Gamma_1+\Gamma_2+\Gamma_3$. Equations (2) are also known as boundary conditions of the first, second and third kind respectively. A fourth type of boundary condition, the radiation condition, will not be considered here.

The Kirchhoff Transformation

The Kirchhoff transformation, [1], when applied to equation (1), introduces a new variable ϕ such that

$$\frac{\partial \phi}{\partial x}=k\frac{\partial u}{\partial x}$$

$$\frac{\partial \phi}{\partial y}=k\frac{\partial u}{\partial y} \tag{3}$$

which reduces (1) to the Laplace equation:-

$$\frac{\partial^2 \phi}{\partial x^2}+\frac{\partial^2 \phi}{\partial y^2}=0 \tag{4}$$

Boundary conditions of the first and second kind become:-

$$\phi=\overline{\phi}$$

$$\frac{\partial \phi}{\partial n}=\overline{q} \tag{5}$$

The transformation of $\overline{u}$ into $\overline{\phi}$ on the boundary before solution and the transformation of results for ϕ back into u after solution depend on the nature of the function $k=k(u)$, [1,2]. The former is known as a Kirchhoff transformation and the latter an inverse Kirchhoff transformation.

$$\phi=K(u)$$

$$u=K^{-1}(\phi) \tag{6}$$

An example will be given later.
Applying standard boundary element procedures [2,5] to equations (4) and (5) produces the boundary integral equation for the point i

$$c_i\phi_i + \int_\Gamma \phi q^* d\Gamma = \int_\Gamma q u^* d\Gamma \tag{7}$$

where u^* is the fundamental solution to the Laplace equation, $u^* = \frac{1}{2\pi}\ln(\frac{1}{r})$, r being the usual boundary element distance function. $q^* = \frac{\partial u^*}{\partial n}$ and c_i depends on the geometry of the boundary at i. In what follows constant boundary elements will be used, these have been shown to be accurate for this type of problem, [2], and avoid special treatment of the boundary conditions at corners. In this case $c_i = 1/2$ and equation (7) becomes:-

$$\frac{1}{2}\phi_i + \sum_{k=1}^{N}\left(\int q^* d\Gamma\right)\phi_k = \sum_{k=1}^{N}\left(\int u^* d\Gamma\right) q_k \tag{8}$$

the summation being done over elements. Evaluating the integrals and writing (8) for each node i produces the usual matrix equation

$$\mathbf{H}\phi = \mathbf{G}\mathbf{q} \tag{9}$$

Equation (9) may be reordered and solved for unknown values of ϕ and q on the boundary. No iterations are necessary and a standard boundary element program for the Laplace equation [4,5] may be used.

Incorporation of Boundary Conditions of the Third Kind The incorporation of boundary conditions of the third kind or convective conditions into the analysis requires a modification of equation (7), [2]. The final term is split into a part referring to the Γ_{1+2} boundary and a part referring to the Γ_3 or convective boundary. The substitution of the third equation (2) into this term produces

$$c_i\phi_i + \int_\Gamma \phi q^* d\Gamma = \int_{\Gamma_{1+2}} q u^* d\Gamma + \int_{\Gamma_3} h u_f u^* d\Gamma + \int_{\Gamma_3} hK^{-1}[\phi] u^* d\Gamma \tag{10}$$

This equation may be discretized in the usual way, however the final term will have to be iterated upon. A separate code is needed to solve (10), due to the addition of the new boundary condition.

The Dual Reciprocity Method

This technique is described in detail in [4], here application to equation (1) only will be outlined.

In the case of equations which contain the Laplace operator, this is isolated on the left hand side and all other terms are transferred to the right hand side to form an equation of the type

$$\nabla^2 u = b(x, y, u) \tag{11}$$

The form of b for equation (1) depends on the nature of the function $k = k(u)$. Initially the case

$$k = k_0(1 - \beta u) \tag{12}$$

will be considered. A further example will be given later. In (12) k_0 and β are material constants.

Substituting (12) into (1) and simplifying, one finds that there are several different possible expressions b in (11), for example

$$\nabla^2 u = -\frac{\beta}{(1+\beta u)}\left(\frac{\partial u}{\partial x}\frac{\partial u}{\partial x} + \frac{\partial u}{\partial y}\frac{\partial u}{\partial y}\right) = b$$

$$\nabla^2 u = -\frac{1}{\beta u}\nabla^2 u - \frac{1}{u}\left(\frac{\partial u}{\partial x}\frac{\partial u}{\partial x} + \frac{\partial u}{\partial y}\frac{\partial u}{\partial y}\right) = b \tag{13}$$

DRM can be used take either right hand side to the boundary, and it will be shown that the results are very similar in both cases. The former equation (13) needs the approximation only of first derivatives of space, described in [4,6], the latter needs the approximation of second derivatives which will be considered after the fundamentals of the method have been described.

In order to take a given right hand side b to the boundary, the following approximation is proposed:-

$$b_i = \sum_{j=1}^{N+L} \alpha_j f_{ij}. \tag{14}$$

where b_i is the value of the function b at node i. The f_{ij} are approximating functions and the α_j unknown coefficients. The approximation is done at $(N+L)$ nodes, (N boundary nodes and L internal nodes, see figure 1). These nodes are called the DRM collocation points. The use of internal nodes is optional in most cases, however they have been shown to improve the accuracy of the solution, [4]. Internal nodes are usually defined at places where the solution is required on the domain.

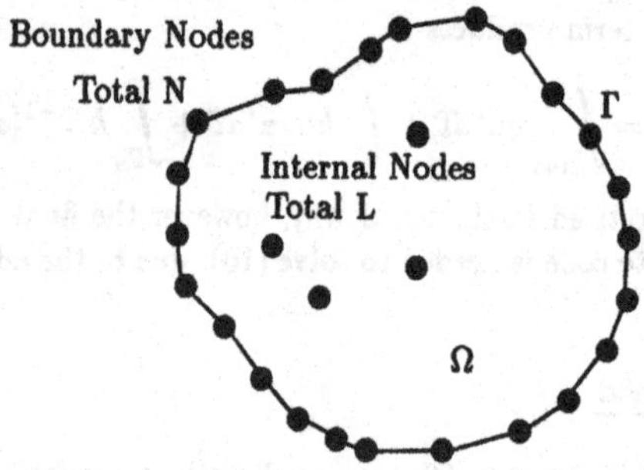

Figure 1: Boundary and Internal Nodes

The functions f are defined by

$$\nabla^2 \hat{u} = f \tag{15}$$

where $\hat{u}$ is a particular solution. The functions f and $\hat{u}$ will be considered shortly. Combining (11), (14) and (15) one obtains

$$\nabla^2 u = -\sum_{j=1}^{N+L} (\nabla^2 \hat{u}_j)\alpha_j. \tag{16}$$

Applying the boundary element technique in its usual form, equation (16) is multiplied by the fundamental solution u^* and integrated, *i.e.*

$$\int_\Omega (\nabla^2 u) u^* d\Omega = -\sum_{j=1}^{N+L} \alpha_j \int_\Omega (\nabla^2 \hat{u}_j) u^* d\Omega. \tag{17}$$

Integrating by parts produces

$$c_iu_i - \int_\Gamma u_i^* \left(\frac{\partial u}{\partial n}\right) d\Gamma + \int_\Gamma q_i^* u d\Gamma = \sum_{j=1}^{N+L} \left\{\alpha_j \left(\int_\Gamma \hat{u}_j q_i^* d\Gamma - \int_\Gamma \hat{q}_j u_i^* d\Gamma + c_i \hat{u}_{ij}\right)\right\}. \quad (18)$$

In equation (18) $\hat{q} = \frac{\partial \hat{u}}{\partial n}$. Note also that $\frac{\partial u}{\partial n}$ is written as $q = k\frac{\partial u}{\partial n}$ for this problem. After discretization this equation becomes

$$c_iu_i + \sum_{k=1}^{N} H_{ik}u_k - \sum_{k=1}^{N} G_{ik}\left(\frac{\partial u}{\partial n}\right)_k = \sum_{j=1}^{N+L}\left\{\alpha_j\left(\sum_{k=1}^{N} H_{ik}\hat{u}_{kj} - \sum_{k=1}^{N} G_{ik}\hat{q}_{kj} + c_i\hat{u}_{ij}\right)\right\} \quad (19)$$

where H_{ik} and G_{ik} are the usual resultants of integration over the boundary elements. In (19), i are the source nodes, k the elements and j the DRM collocation points. Note that $\hat{q}$ is not defined for $i > N$ while $\hat{u}$ is defined for all points i.
Equation (19) is written for each of the $(N + L)$ nodes i. The terms in $c_i = 1/2$ for constant elements, are incorporated onto the diagonal of $\mathbf{H}$.
The resulting matrix equation has N values of u and N values of $(\frac{\partial u}{\partial n})$ on the boundary and L values of u at interior nodes *i.e.*

$$\mathbf{Hu} - \mathbf{G}\left(\frac{\partial \mathbf{u}}{\partial n}\right) = (\mathbf{H}\hat{\mathbf{U}} - \mathbf{G}\hat{\mathbf{Q}})\boldsymbol{\alpha} \quad (20)$$

The vector $\boldsymbol{\alpha}$ in (20) can be calculated writing equation (15) at all $(N + L)$ nodes

$$\mathbf{b} = \mathbf{F}\boldsymbol{\alpha}$$

$$\boldsymbol{\alpha} = \mathbf{F}^{-1}\mathbf{b} \quad (21)$$

thus (20) becomes

$$\mathbf{Hu} - \mathbf{G}\left(\frac{\partial \mathbf{u}}{\partial n}\right) = (\mathbf{H}\hat{\mathbf{U}} - \mathbf{G}\hat{\mathbf{Q}})\mathbf{F}^{-1}\mathbf{b} \quad (22)$$

The matrices $\hat{\mathbf{U}}$, $\hat{\mathbf{Q}}$ and $\mathbf{F}^{-1}$ are all known if f is defined.
The usual approximating function is [3,4]

$$f = 1 + r \quad (23)$$

such that, from (15)

$$\hat{u} = \frac{r^2}{4} + \frac{r^3}{9} \quad (24)$$

where r is the BEM distance function which appears in the fundamental solution.

The b Vector using First Derivatives The DRM approximation to the vector $\mathbf{b}$ will now be derived for the case, equation (13), in which only first derivatives of space are present, *i.e.*

$$\nabla^2 u = -\frac{\beta}{(1+\beta u)}\left(\frac{\partial u}{\partial x}\frac{\partial u}{\partial x} + \frac{\partial u}{\partial y}\frac{\partial u}{\partial y}\right) = b \quad (25)$$

The DRM approximation to first derivatives, [6], starts from

$$\mathbf{u} = \mathbf{F}\boldsymbol{\gamma} \quad (26)$$

where **F** is as in equation (21) and γ are a set of unknown coefficients, similar to α. Deriving equation (26) one obtains

$$\frac{\partial \mathbf{u}}{\partial x} = \frac{\partial \mathbf{F}}{\partial x}\gamma \tag{27}$$

but

$$\gamma = \mathbf{F}^{-1}\mathbf{u} \tag{28}$$

such that

$$\frac{\partial \mathbf{u}}{\partial x} = \frac{\partial \mathbf{F}}{\partial x}\mathbf{F}^{-1}\mathbf{u} \tag{29}$$

A similar expression holds for the derivative with respect to y. The $(N + L)$ values of $\frac{\partial u}{\partial x}$ may thus be evaluated using the values of u from the previous iteration. Defining a diagonal matrix $\mathbf{T}_x$ such that

$$T_x(i,i) = \frac{\beta}{(1+\beta u_i)}\left\{\frac{\partial u}{\partial x}\right\}_i \tag{30}$$

with a similar definition for $\mathbf{T}_y$, then

$$\mathbf{b} = -\left(\mathbf{T}_x\frac{\partial \mathbf{F}}{\partial x} + \mathbf{T}_y\frac{\partial \mathbf{F}}{\partial y}\right)\mathbf{F}^{-1}\mathbf{u} \tag{31}$$

In (31), $\mathbf{T}_x$ and $\mathbf{T}_y$ must be evaluated at each iteration.
Defining

$$\mathbf{S} = -(\mathbf{H}\hat{\mathbf{U}} - \mathbf{G}\hat{\mathbf{Q}})\left(\mathbf{T}_x\frac{\partial \mathbf{F}}{\partial x} + \mathbf{T}_y\frac{\partial \mathbf{F}}{\partial y}\right)\mathbf{F}^{-1} \tag{32}$$

Then (25) becomes

$$(\mathbf{H} - \mathbf{S})\mathbf{u} = \mathbf{G}\left(\frac{\partial \mathbf{u}}{\partial n}\right) \tag{33}$$

Equation (33) is reordered according to the known boundary values in the usual way and solved iteratively.

The b Vector using Second Derivatives The DRM approximation to the vector **b** will now be derived for the case, equation (13), which involves second derivatives:-

$$\nabla^2 u = -\frac{1}{\beta u}\nabla^2 u - \frac{1}{u}\left(\frac{\partial u}{\partial x}\frac{\partial u}{\partial x} + \frac{\partial u}{\partial y}\frac{\partial u}{\partial y}\right) \tag{34}$$

In should be noted that (34) is only valid for $\beta \neq 0$.
A DRM approximation to second derivatives is obtained differentiating (27):-

$$\frac{\partial^2 \mathbf{u}}{\partial x^2} = \frac{\partial^2 \mathbf{F}}{\partial x^2}\gamma \tag{35}$$

again

$$\gamma = \mathbf{F}^{-1}\mathbf{u} \tag{36}$$

such that

$$\frac{\partial^2 \mathbf{u}}{\partial x^2} = \frac{\partial^2 \mathbf{F}}{\partial x^2}\mathbf{F}^{-1}\mathbf{u} \tag{37}$$

A similar expression may be written for the second derivative with respect to y.
It should be noted that in this case $f = 1 + r$ proves to be an inadequate approximation function, as the second space derivatives of r are singular when $r = 0$. To overcome this the expansion

$$f = 1 + r^2 + r^3 \tag{38}$$

was adopted.
One can now write

$$\mathbf{b} = -\left\{\mathbf{U}\left(\frac{\partial^2\mathbf{F}}{\partial x^2} + \frac{\partial^2\mathbf{F}}{\partial y^2}\right)\mathbf{F}^{-1} + \left(\mathbf{U}_x\frac{\partial\mathbf{F}}{\partial x} + \mathbf{U}_y\frac{\partial\mathbf{F}}{\partial y}\right)\mathbf{F}^{-1}\right\}\mathbf{u} \tag{39}$$

Where $\mathbf{U}$ is a diagonal matrix containing nodal values of $1/(\beta u)$, $\mathbf{U}_x$ is a diagonal matrix containing nodal values of $\frac{1}{u}\frac{\partial u}{\partial x}$ and where a similar definition holds for $\mathbf{U}_y$. All of these are obtained using the values of u at nodes from the previous iteration.
With an appropriate definition of the $\mathbf{S}$ matrix, the final equation is the same as (33).

Inclusion of Convective Boundary Conditions Equation (33) is reordered before solving, given that at each boundary node either u or $\frac{\partial u}{\partial n}$ is known, ($\frac{\partial u}{\partial n} = \frac{q}{k}$). In the case of a boundary condition of the third kind

$$q = h(u_f - u) \tag{40}$$

such that

$$\frac{\partial u}{\partial n} = \frac{h}{k}(u_f - u) \tag{41}$$

Equation (33) thus becomes

$$\mathbf{Hu} - \mathbf{Ge} + \mathbf{GEu} = \mathbf{Su} \tag{42}$$

In (42) $\mathbf{e}$ is a vector containing nodal values of hu_f/k and $\mathbf{E}$ is a diagonal matrix containing nodal values of h/k. In the above, $k = (1 + \beta u)$, is evaluated using values of u from the previous iteration. (42) is only used on Γ_3, (33) is used on other boundaries.
For the first iteration, terms in β are ignored, and the Laplace solution is obtained. After solution, values of q are obtained on Γ_3 from (40), and on Γ_1 $q = k\frac{\partial u}{\partial n}$, in this case the values of u are known, $u = \overline{u}$.
As the application of DRM is similar in different cases, the above may be implemented using as a starting point a standard program, for example Program 3 in reference [4]. The difference in the code if third order boundary conditions are included is minimal.
It may be noted that if $\overline{q} \neq 0$ this term would also require iterations. In the examples considered below, $\overline{q} = 0$.

RESULTS

Problem Including Boundary Conditions of the First and Second Kind only.

The geometry of the problem considered is shown in fig. 2, and consists of a square plate of unit side. Nine equal constant boundary elements were used to discretize each side. The boundary conditions employed were $u = 300$ at $x = 0$, $u = 400$ at $x = 1$ and $q = 0$ at $y = 0$ and $y = 1$.

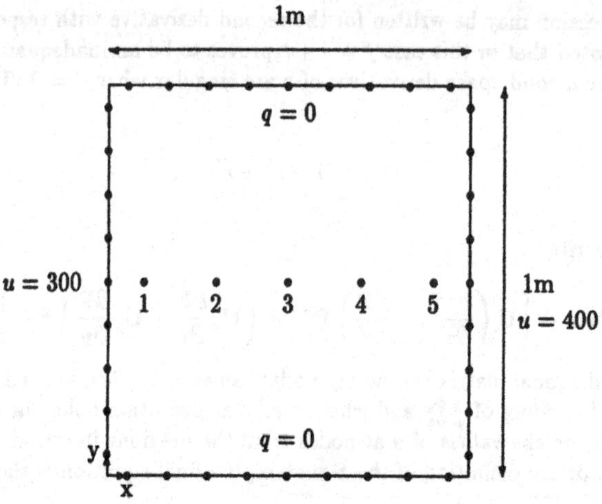

Figure 2: Square Plate Problem 1

Equation (1) was solved with

$$k = k_0(1 + \beta\left\{\frac{u - u_0}{u_0}\right\}) \tag{43}$$

$k_0 = 1$, $\beta = 3$ and $u_0 = 300$.
The Kirchhoff transformation and inverse transformation were taken from [2] for this situation,

$$\phi = \frac{k_0 u_0}{2\beta}\left\{1 + \beta\left(\frac{u - u_0}{u_0}\right)\right\}^2 - \frac{k_0 u_0}{2\beta} \tag{44}$$

such that for $\overline{u} = 300$ $\overline{\phi} = 0$ and for $\overline{u} = 400$ $\overline{\phi} = 150$.

$$u = u_0 + \frac{u_0}{\beta}\left\{1 + \frac{2\beta\phi}{k_0 u_0}\right\}^{\frac{1}{2}} - \frac{u_0}{\beta} \tag{45}$$

In the case of the DRM solution

$$b = -\frac{\beta}{u_0 + \beta(u - u_0)}\left(\frac{\partial u}{\partial x}\frac{\partial u}{\partial x} + \frac{\partial u}{\partial y}\frac{\partial u}{\partial y}\right) \tag{46}$$

Results are given in table 1 at the 5 nodes numbered in figure 2. The DRM solution converged in 4 iterations.
For this type of non-linearity, the difference between the results obtained and the Laplace solution *i.e.* $\beta = 0$, is small.
The DRM solution is seen to be in reasonable agreement with that obtained using the Kirchhoff transformation.

Results for a Problem with a Boundary Condition of the Third Kind

The problem shown in fig. 3 was published in ref [1] in which the Kirchhoff transformation solution was originally presented. The geometry and boundary discretization is as for the previous example, however in this case the boundary conditions are $q = 0$ at $x = 0$

node	x	y	DRM $\beta = 3$	Kirchhoff Transform	Laplace ($\beta = 0$)
1	.1	.5	314.15	314.00	310
2	.3	.5	338.34	337.82	330
3	.5	.5	358.49	358.11	350
4	.7	.5	376.27	376.08	370
5	.9	.5	392.43	392.36	390

Table 1: Results for u for Square Plate Problem 1

and $y = 0$, $u = 300$ at $x = 1$, and the convective or third kind boundary condition at $y = 1$.

$$q = h(u_f - u) \tag{47}$$

h was taken to be 10 and $u_f = 500$.

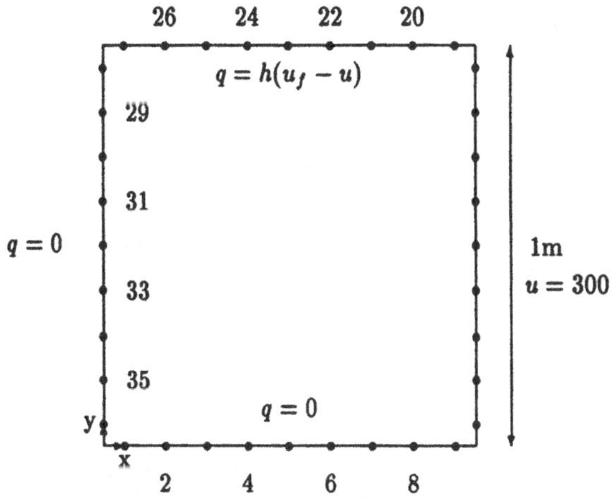

Figure 3: Square Plate Problem 2

For this problem, $k = k_0(1 + \beta u)$, $k_0 = 1$ and $\beta = 0.3$.
For the Kirchhoff transform solution, [7], no internal nodes were used. The DRM solutions used 25 evenly spaced internal nodes, and were done for each of the equations (25) and (34). That using (25) will be called DRM1 and that using (34) will be called DRM2. Results for u for the 12 nodes shown in figure 3 are given in table 2.
The DRM results converged in 5 iterations, a tolerance of 0.00001 was used. It can be seen that the agreement between both DRM results and the Kirchhoff transformation solution is excellent, particularly in the case of DRM2, which uses second derivatives.

CONCLUSIONS

The results presented show that the DRM method is able to obtain similar values to the Kirchhoff transformation method for non-linear material problems.

node	DRM1 $\beta = 0.3$	DRM2 $\beta = 0.3$	Kirchhoff Transform	Laplace $(\beta = 0)$
2	307.13	306.66	306.79	385.12
4	306.04	305.70	305.79	372.98
6	304.26	304.03	304.08	351.83
8	301.94	301.86	301.84	323.48
20	307.05	306.95	306.93	442.50
22	311.84	311.61	311.61	469.57
24	314.64	314.34	314.33	478.68
26	316.11	315.78	315.71	482.30
29	313.53	313.15	313.14	455.84
31	310.55	310.17	310.22	424.88
33	308.60	308.19	308.28	402.82
35	307.59	307.07	307.25	390.57

Table 2: Results for u for Square Plate Problem 2

This application of DRM is interesting in that the right hand side of the equation involves the multiple of a series of terms, the results show that this can be done without loss of accuracy. A surprising result is that the agreement is better for DRM2 which uses second derivatives.

The implementation of boundary conditions of the third kind in the DRM program is straightforward, and the general structure of the program very similar to that used for other potential problems.

It may also be noticed that if DRM is used for the non-linear material terms, the same approach can be applied to any other domain integrals in the equation, *i.e.* a time derivative, sources, etc. The Kirchhoff transformation technique applies to non-linear material terms only.

ACKNOWLEDGEMENT

The author would like to thank Dr. Luiz Wrobel for making available the correct Kirchhoff transformation results to the second problem. The results for this published in [1] and [2] are not correct.

REFERENCES

1. Bialecki, R and Nowak, A. J.
Boundary value problems for non-linear material and non-linear boundary conditions, Applied Mathematical Modelling, Vol. 5, pp 417-421, 1981.

2. Brebbia, C. A., Telles, J. and Wrobel, L. C.
Boundary Element Techniques: Theory and Applications In Engineering. Springer-Verlag, Berlin and N. York, 1984.

3. Nardini, D. and Brebbia, C. A.
A New Approach for Free Vibration Analysis using Boundary Elements, Boundary Element Methods in Engineering, (Ed C. A. Brebbia), pp 312-326, Springer-Verlag, Berlin and N. York, 1982.

4. Partridge, P. W., Brebbia, C. A. and Wrobel, L. C.
The Dual Reciprocity Boundary Element Method, Computational Mechanics Publications, Southampton and Boston, 1991.

5. Brebbia, C. A.
The Boundary Element Method for Engineers. Pentech Press, London, 1978.

6. Partridge, P. W. and Brebbia, C. A.
Computer Implementation of the BEM Dual Reciprocity Method for General field Problems. Communications in Applied Numerical Methods, Vol. 6, pp 83-92, 1990.

7. Wrobel, L. C., Private Communication, 1990.

Boundary Integral Equations for a Class of Nonlinear Problems by the Kirchhoff Transformation

R.P. Shaw

S.U.N.Y. at Buffalo, U.S.A.

ABSTRACT:

The boundary integral equation method (BIEM) appears to depend on the concept of linear superposition in its formulation, yet will treat nonlinear boundary conditions without significant difficulty, i.e. without the need for volume integrals in most cases. It is therefore capable of dealing with inherently nonlinear problems. There are also a number of problems governed by nonlinear differential equations which have been treated by BIEM using iterative solutions based on a linearized form. However, the use of the Kirchhoff transformation on a class of these nonlinear governing equations moves the nonlinearity from the governing equation to the boundary conditions, leaving a linear governing equation with nonlinear boundary conditions which is directly amenable to standard BIE methods.

KEYWORDS: Boundary Integral Equation, Boundary Element Methods, Nonlinear Problems, Kirchhoff Transformation

INTRODUCTION:

The boundary integral equation method is well established in the solution of linear problems for which fundamental Green's functions are known, e.g. Brebbia and Dominguez[1]. Since one of the interpretations of this method lies in the concept of the superposition of sources and/or doublets, BIEM is frequently taken to be restricted to linear problems and may be thought to be applicable to nonlinear governing differential equations only through some approximate linearization procedure. Yet it is equally well accepted that the BIE approach can deal with nonlinear boundary conditions for linear governing equations, e.g. Banerjee[2], who examined the effect of nonlinear boundary conditions and of linearization and iteration of nonlinear potential governing equations. In fact, there are a number of recent papers, e.g. by Tosaka et al[3,4], which have exemplified this BIEM treatment of problems governed by nonlinear equations as approximated by linearized equations with the nonlinear effect moved to a forcing function term and a solution found by iteration. While this approach may be appropriate for modest nonlinearity, it requires volume integrals which are the antithesis of the BIEM approach; in addition, the convergence of the iteration process may be questionable for severe nonlinearities.

Thus nonlinearity of the governing equations per se would appear to form a significant drawback to the general acceptence and efficient use of the BIEM and its numerical counterpart, the boundary element method (BEM), as engineering computational tools. It should be noted then by any researcher in BIE/BEM that there are a number of nonlinear equations which may be linearized exactly by appropriate transformations. These problems then lend themselves to standard BIE/BEM procedures, including the reduction of dimensionality, since the nonlinearity would have been moved from the governing equations to the boundary conditions where the BIE/BEM methods can deal with it. While this class of problems may appear somewhat limited, it does form a subset of physical problems of engineering interest and therefore appears worthwhile for study. There will be no attempt here to discuss problems of nonlinear solid mechanics, such as elasto-plasticity, or of nonlinear fluid mechanics, such as viscous compressible flow; emphasis in this study will be placed on potential and diffusion type problems which typify Kirchhoff transformation applications, although this is not the only transformation available for such reductions. The Kirchhoff transformation has been well known in conductive heat transfer for quite some time, e.g. it is mentioned in a limited context in such a classical textbook as Schneider[5].

NONLINEAR POTENTIAL PROBLEMS:

It may be surprising that this very simple category has experienced only limited and fairly recent attention from the BIEM community. Indeed some review papers on this very topic, e.g. Banerjee[2], have completely overlooked the Kirchhoff transformation for nonlinear potential problems, although this transformation has been available for quite some time and leads to extremely simple results. It may very well be that nonlinearity in the governing equations has represented a 'blind spot' for BIE researchers because of the 'superposition' interpretation mentioned above, although nonlinear boundary conditions have been considered since almost the beginning of BIE, e.g. Shaw[6] who formulated, without numerical examples, a nonlinear water wave BIE. Fortunately, over the last ten years this transformation has been gradually introduced to nonlinear potential BEM applications, e.g. by Khader and Hanna[7] and by Bialecki and Nowak[8] through journal papers in 1981, followed by Skerget and Brebbia[9] in a BEM review and by Ingham and Kelmanson[10] at a BEM conference in 1983. A description of the Kirchhoff transformation is readily available in the classic text on nonlinear differential equations by Ames[11], but the gist of the method will be described below.

Consider a nonlinear potential problem, e.g. steady heat conduction with a temperature dependent conductivity, K(U), and a heat source, $Q(\vec{r})$, as given by

$$[1] \quad \nabla \circ [\, K(U)\, \nabla U \,] = Q(\vec{r})$$

The Kirchhoff transformation uses a new dependent variable, W, defined through

$$[2] \quad \nabla W = K(U)\, \nabla U$$

Please note the (undoubtedly typographical) sign error in eq. 6.13 and the following equations, in the review article by Beskos[12] who discusses this transformation. Then, as an indefinite integral,

$$[3] \quad W = \int K(U)\, dU + a_0$$

where a_0 is arbitrary, usually zero. Eq. [1] then is the linear Poisson equation

$$[4] \quad \nabla^2 W = Q(\vec{r})$$

The boundary conditions on U may be Dirichlet, Neumann or Robin or in fact any linear or nonlinear combination of these, each defined on a segment of the boundary surface. When transformed to equivalent boundary conditions on W, these may become nonlinear even if the original boundary conditions on U were linear, but this is an aspect of nonlinearity which is readily treated by BIE/BEM. The Dirichlet boundary condition, $U = f_1(\vec{r})$ on S_1, simply requires an evaluation of W(U) from eq. [3] and the substitution of f_1 for U to determine W on S_1. The Neumann boundary condition, $\partial U/\partial n = f_2(\vec{r})$ on S_2, would lead to a nonlinear condition on W. However, this boundary condition is usually obtained from a flux condition rather than one on a normal derivative and would then be given as $K(U)\, \partial U/\partial n = f_2(\vec{r})$ on S_2 in which case the corresponding condition on W would take the simple form, $\partial W/\partial n = f_2(\vec{r})$ on S_2. Other forms of boundary conditions would also be readily transformed from U to W. Of even greater interest is the possibility of using the Kirchhoff transformation to eliminate the need for a volume integral in a BIE formulation even with a non-zero distributed source. Consider eq. [3] as a definite integral,

$$[5] \quad W = \int_{U_0}^{U} K(V)\, dV$$

such that eq. [4] becomes

$$[6] \quad \nabla^2 W = Q(\vec{r}) - \nabla \circ \{ K(U_0)\, \nabla U_0 \}$$

If U_0 is chosen, regardless of the boundary conditions, to satisfy

$$[7] \quad \nabla \circ \{ K(U_0) \nabla U_0 \} = Q(\vec{r})$$

then W satisfies an unforced Laplace equation with boundary conditions modified to account for the contribution of U_0 at the boundaries. U_0 may often be found analytically as a solution to the Kirchhoff transformation on U_0 such that

$$[8] \quad \nabla W_0 = K(U_0) \nabla U_0$$

Then W_0 satisfies a Poisson equation, eq. [7], but with any boundary conditions which may be convenient for such a simple analytical solution.

The details of the solution of eq. [4] or eq. [6], typically by BEM, are well known, even with nonlinear boundary conditions, and are not described further here. Since this deals with a fairly general form of the nonlinear potential problem as it occurs in engineering applications, it might be expected that this transformation would be widely used in all computational methods. However, volume methods, such as finite differences and finite elements, can readily deal with nonlinear calculations within the volume of the problem, but face more difficulty in incorporating nonlinear boundary conditions into their procedures. Thus this transformation, although still convenient, would not have the same impact in these methods as it would in BIEM where it allows the reduction of dimensionality to be maintained. A simple example is given in the appendix.

NONLINEAR DIFFUSION PROBLEMS:

The diffusion problem with material properties dependent on the unknown variable is another example of a nonlinear problem which may, in some special cases, be transformed into a linear form and solved by standard BIEM. This problem was considered by Skerget and Brebbia[13], but an (unwritten) approximation was made in their final formulation that should be clarified. Consider a diffusion equation with temperature dependent conductivity, K, density, ρ, and specific heat, c, as

$$[9] \quad \nabla \circ [\, K(U)\, \nabla U \,] - c(U)\, \rho(U)\, \partial U/\partial t = Q(\vec{r},t)$$

Here, the Kirchhoff transformation leads to

$$[10] \quad \nabla^2 W - \{c(U)\, \rho(U)/K(U)\}\, \partial W/\partial t = Q(\vec{r},t)$$

This form will be linear if and only if the ratio , $c(U)\, \rho(U)/K(U)$, is a constant, α^2, in which case the BIEM formulation would proceed as usual, or at most a function of the independent variables, $\vec{r}$ and t, in which case a Green's function for a heterogeneous material would be required. For $C(U) = c(U)\, \rho(U)$ proportional to the same power of U as K(U) is equivalent to a case given by Ames[11] in sec. 1.2; an exponential dependence such as exp(A + BU) for both C(U) and K(U) would follow the same pattern. {Other forms may be derived, e.g. involving transformation of the independent variables as described in sec. 3.5 of Rogers and Shadwick[14], but these are beyond the scope of this paper — e.g. involving the transformation of a nonlinear diffusion equation with fixed boundaries to a linear diffusion equation with moving boundaries.) The approximation made by Skerget and Brebbia was that this ratio was constant with constant ρ and c but with an exponentially temperature dependent K(U). This approximation has been continued in some later work, e.g. Pasquetti and Caruso[15] who use the Kirchhoff transformation by name but without definition and obtain a boundary integral equation for variable K and constant ρ and c. Their numerical solutions do not agree completely with a standard example which they explain by the statement "the BEM formulation itself is approximative in the nonlinear case" without discussion as to why it is an approximation. The asymptotic steady state solution will of course not involve ρ and c and thus not be affected by this approximation. Once eq. [10] is obtained, the usual transformation of boundary and initial conditions must be carried out but this presents no real difficulty for most problems, even though these conditions may be nonlinear in

W, e.g. Onishi and Kuroki[16]. An interesting result may again be obtained by a slight modification of the usual definition of W, i.e. as

$$[11] \quad W = \int_{U_0}^{U} K(V)\, dV$$

but now some limitations on U_0 must be introduced. If U_0 is equal to the initial state of $U(\vec{r},t)$ at time equal to zero, the new variable, W, is zero when U equals U_0, i.e. begins from a zero initial state at time zero. However, the differential equation is now modified since this definition of W requires that

$$[12] \quad \nabla W(\vec{r},t) = K(U)\,\nabla U(\vec{r},t) - K(U_0)\,\nabla U_0(\vec{r}, t)$$

and thus eq. [10] becomes

$$[13] \quad \nabla^2 W - [c(U)\rho(U)/K(U)]\,\partial W/\partial t = Q(\vec{r},t) - \nabla\circ\,[K(U_0)\nabla U_0(\vec{r}, t)] + [c(U)\rho(U)]\,\partial U_0/\partial t$$

This again requires a constant ratio $c(U)\,\rho(U)\,/K(U)$ to have a linear equation. However, since U_0 will not have the same form as U in general, the right hand side of this equation would be nonlinear unless U_0 is independent of time. Then it will vanish if and only if Q is also independent of time and satisfies

$$[14] \qquad Q(\vec{r}) = \nabla\circ\,[\, K(U_0)\,\nabla\, U_0(\vec{r})\,]$$

Otherwise a volume integral will be required in the BIE formulation for W. Again, a simple example is given in the appendix.

NONLINEAR "HELMHOLTZ" EQUATIONS:

The phrase "nonlinear Helmholtz equation" as used here a misnomer, i.e. this is no longer the time harmonic counterpart to a wave equation since the nonlinearity prevents the factor exp(i ω t) from being common to all terms (unless the equation is dimensionally homogeneous, i.e. all terms involve the same power of the dependent variable). Nevertheless it represents a form of some interest in its own right, i. e. a Laplacian plus a U dependent term equals a forcing function. For example, this could represent a potential problem with a source that is in part temperature dependent, $Q(U) = -Q_1(\vec{r}, U) + Q_0(\vec{r})$.

[15] $\nabla \circ [\, K(U)\, \nabla U \,] + Q_1(\vec{r}, U) = Q_0(\vec{r})$

The Kirchhoff transformation again leads to a linear equation

[16] $\nabla^2 W + \alpha(\vec{r})\, W = Q_0(\vec{r})$

but if and only if

[17] $Q_1(\vec{r}, U) = \alpha(\vec{r})\, W = \alpha(\vec{r}) \left[\int K(U)\, dU + a_0 \right]$

where a_0 is a constant. Integration from U_0 to U could again be introduced, but there appears to be little gain. While this puts a severe restriction on Q_1, it is possible to solve some problems in this fashion, e.g. if K(U) were expressed as a single power term in U, i.e. $b_n U^n$, then

[18] $Q_1 = [\alpha\, b_n/(n+1)]\, U^{n+1} + \alpha\, a_0 = \alpha\, [\, c_n U^{n+1} + a_0 \,]$

where both b_n and c_n are arbitrary constants. While α could be position dependent, it is probably simplest to consider it as a constant.

CONCLUSION:

The results presented here are for a limited class of problems governed by nonlinear partial differential equations, but a class which does have some practical application in a number of physical situations including heat conduction, fluid flow, flow in porous media, electromagnetic fields, etc. While the results given here are based on the Kirchhoff transformation, there are other transformations on both dependent and independent variables which may also make these nonlinear problems amenable to BIE/BEM. As long as the nonlinearity can be moved from the governing equations to the boundary conditions, the reduction of dimensionality is still accomplished and BIE/BEM methods have their unique place as a boundary solution technique, even though these problems remain inherently nonlinear.

REFERENCES:

1. Brebbia, C. A. and Dominguez, J., BOUNDARY ELEMENTS: AN INTRODUCTORY COURSE, CMP/McGraw Hill Book Co., N.Y., N.Y., 1989,

2. Banerjee, P. K., 'Nonlinear Problems of Potential Theory', Chap. 2, DEVELOPMENTS IN BOUNDARY ELEMENT METHODS - 1, edited by P. K. Banerjee and R. Butterfield, Appl. Sci. Publ., London, U.K., (1979),

3. Tosaka, N., and Kakuda, K., 'The Generalized Boundary Element Method for Nonlinear Problems', BOUNDARY ELEMENTS X, VOL. 1, edited by C. A. Brebbia, pp. 3-18, CMP/Springer Verlag, N.Y., N.Y., (1988),

4. Tosaka, N., 'Application of the Generalized Boundary Element Method to Nonlinear Problems', ADVANCES IN BOUNDARY ELEMENTS, VOL. 1, edited by C. A. Brebbia and J. J. Connor, pp. 3-12, CMP/Springer Verlag, N.Y., N.Y., (1989),

5. Schneider, P. J., CONDUCTION HEAT TRANSFER, Addison-Wesley Pub., Reading, Mass., (1955),

6. Shaw, R. P., 'Boundary Integral Equation Methods Applied to Water Waves', BOUNDARY INTEGRAL EQUATION METHOD - AMD Vol. 11, edited by T. A. Cruse and F. J. Rizzo, A.S.M.E., N.Y., N.Y., (1975),

7. Khader, M. S. and Hanna, M. C., 'An Iterative Boundary Numerical Solution for General Steady Heat Conduction Problems', J. Heat Transfer, Vol. 103, pp. 26-31, (1981),

8. Bialecki, R. and Nowak, A. J., 'Boundary Value Problems in Heat Conduction with Nonlinear Material and Nonlinear Boundary Conditions', Appl. Math. Mod., Vol. 5, pp. 417-421, (1981),

9. Skerget, P. and Brebbia, C. A., 'Non-Linear Potential Problems', Chap. 1, PROGRESS IN BOUNDARY ELEMENT METHODS, VOL. 2, edited by C. A. Brebbia, Pentech Press, London, (1983),

10. Ingham, D. B. and Kelmanson, M. A., 'Solution of Nonlinear Elliptic Equations with Boundary Singularities by an Integral Equation Method', in BOUNDARY ELEMENTS, edited by C. A. Brebbia, T. Futagami and M. Tanaka, pp. 61-71, Springer Verlag, N.Y., N.Y., (1983),

11. Ames, W. F., NONLINEAR PARTIAL DIFFERENTIAL EQUATIONS IN ENGINEERING, Academic Press, N.Y., N.Y., (1965),

12. Beskos, D. E., 'Potential Theory', Chap. 2 in BOUNDARY ELEMENT METHODS IN MECHANICS, edited by D. E. Beskos, North-Holland Publ. Co., N.Y., N.Y., (1987),

13. Skerget, P. and Brebbia, C. A., 'Time Dependent Non-Linear Potential Problems', Chap. 3, TOPICS IN BOUNDARY ELEMENT RESEARCH, VOL. 2, edited by C. A. Brebbia, Springer Verlag, N.Y., N.Y., (1985),

14. Rogers, C. and Shadwick, W. F., BACKLUND TRANSFORMATIONS AND THEIR APPLICATIONS, Academic Press, N.Y., N.Y., (1982),

15. Pasquetti, R. and Caruso, A., 'A New Software for the Modelization of Transient and Nonlinear Thermal Diffusion', BOUNDARY ELEMENTS X, VOL. 2, edited by C. A. Brebbia, pp. 29-43, CMP/Springer Verlag, N.Y., N.Y., (1988),

16. Onishi, K. and Kuroki, T., 'On Non-Linear Heat Transfer Problems", Chap. 6, DEVELOPMENTS IN BEM, VOL. 4, edited by P. K. Banerjee and J. O. Watson, Elsevier Applied Science Publ., London , (1986),

17. Shaw, R. P., 'The Boundary Element Method as a Weighted Mean Value', Math and Computer Modelling, (in press), (1991).

APPENDIX – EXAMPLES:

The nonlinear potential problem may be illustrated by the simple unit square problem with insulated sides at x = 0 and x = 1, a temperature of 0 at y = 0 and of 100 at y = 1. The nonlinearity will be in the conductivity, $K(U) = K_0 U^n$. Then

[A-1] $\nabla \circ [\, K(U) \nabla U \,] = \nabla \circ [\, K_0 U^n \nabla U \,] = 0; \quad 0 < x < 1, 0 < y < 1$

$K(U)\, \partial U / \partial x = 0; \; x = 0$ and $x = 1, \; 0 < y < 1$

$U = 0; \; y = 0, \; 0 < x < 1$

$U = 100; \; y = 1, \; 0 < x < 1$

which, under the Kirchhoff transformation

[A-2] $W = \int K(U)\, dU = K_0\, U^{n+1}/(n+1)$

leads to a linear equation

[A-3] $\nabla \circ [\, \nabla W \,] = \nabla^2 W = 0; \quad 0 < x < 1, 0 < y < 1$

$\partial W / \partial x = 0; \; x = 0$ and $x = 1, \; 0 < y < 1$

$W = 0 \,; \; y = 0, \; 0 < x < 1$

$W = K_0\, 100^{n+1}/(n+1) = W_1 \,; \; y = 1, \; 0 < x < 1$

This problem has been solved repeatedly as an example for BEM calculations; one of the latest is by Shaw[17] who used it to illustrate the concept of mean values rather than shape functions in these calculations. Once the solution for W is available, the solution for U follows from the inverse Kirchhoff transformation,

[A-4] $U = [\, (n+1)\, W / K_0 \,]^{1/(n+1)} = [\, (n+1)\, W_1\, y / K_0 \,]^{1/(n+1)}$

The same problem with a distributed heat source, e.g. $Q(\vec{r}) = 50$, is conveniently solved using the definite integral form of the Kirchhoff transformation. Here,

$$[A\text{-}5]\quad W = \int_{U_0}^{U} K(V)\, dV = K_0\, U^{n+1}/(n+1) - K_0\, U_0^{\,n+1}/(n+1)$$

where U_0 defines a Kirchhoff transformed function, W_0, which satisfies

$$[A\text{-}6]\quad \nabla^2 W_0 = 50$$

or, as a convenient solution,

$$[A\text{-}7]\quad W_0 = K_0\, U_0^{\,n+1}/(n+1) = 50\,(y^2/2)$$

Then W has the solution given in [A-4] with a modified W_1 as

$$[A\text{-}8]\quad \nabla \circ [\nabla W] = \nabla^2 W = 0;\ 0 < x < 1,\ 0 < y < 1$$

$$\partial W/\partial x = 0;\ x = 0 \text{ and } x = 1,\ 0 < y < 1$$

$$W = 0\ ;\ y = 0,\ 0 < x < 1$$

$$W = K_0\, 100^{n+1}/(n+1) - 25 = W_1\ ;\ y = 1,\ 0 < x < 1$$

The nonlinear diffusion equation may also be discussed in terms of this basic problem with a distributed source, $Q(\vec{r}, t)$, here taken equal to zero, insulated sides at $x = 0$ and $x = 1$ and temperatures of 0^0 on $y = 0$ and 100^0 on $y = 1$ for time greater than zero and an initial condition of $U(x, y, 0) = 50^0$. The same conductivity will be used as above, i.e. $K(U) = K_0\, U^n$ with $\rho(U)\, c(U) = C_0\, U^n$. Then the nonlinear governing equations are

$$[A\text{-}9]\quad \nabla \circ [K(U)\, \nabla U] - \rho(U)\, c(U)\, \partial U/\partial t = 0\ ;\ 0 < x < 1,\ 0 < y < 1$$

$K(U)\,\partial U/\partial x = 0;\ x = 0 \text{ and } x = 1,\ 0 < y < 1,\ t > 0$

$U = 0;\ y = 0,\ 0 < x < 1,\ t > 0$

$U = 100;\ y = 1,\ 0 < x < 1,\ t > 0$

$U = 50;\ t = 0,\ 0 < x < 1,\ 0 < y < 1$

Under the Kirchhoff transformation, the solution will be in terms of

$$\text{[A-10]}\quad W = \int_{U_0}^{U} K(U)\,dU = K_0\,U^{n+1}/(n+1) - K_0\,U_0^{\,n+1}/(n+1)$$

where the field U_0, with a Kirchhoff transformation to W_0, satisfies

$$\text{[A-11]}\quad \nabla\circ\{K(U_0)\,\nabla U_0\} = \nabla^2 W_0 = 0$$

In this simple case, the initial condition on U requires U_0 to be 50 which does satisfy the differential equation, [A-11]; this will not always be the case. Then

$$\text{[A-12]}\quad W_0 = K_0\,(50)^{n+1}/(n+1)$$

which alters the given boundary and initial conditions on W to

$$\text{[A-13]}\quad \nabla^2 W(\vec{r},t) - (C_0/K_0)\,\partial W(\vec{r},t)/\partial t = 0$$

$\partial W/\partial x = 0;\ x = 0 \text{ and } x = 1,\ 0 < y < 1,$

$W = -K_0\,[50]^{(n+1)}/(n+1);\ y = 0,\ 0 < x < 1,$

$W = K_0\,[100]^{(n+1)}/(n+1) - K_0\,[50]^{(n+1)}/(n+1);\ y = 1,\ 0 < x < 1,$

$W = 0;\ t = 0,\ 0 < x < 1,\ 0 < y < 1$

which is a relatively simple problem to solve by BEM and requires no volume integrals. If a distributed source were also present, the auxiliary solution W_0 would have to take care of both the initial condition and the distributed source to avoid a volume integral which is generally unlikely unless the initial condition itself arose from the steady state induced by the given time independent source, $Q(\vec{r})$, and the subsequent time dependence of U arose through the boundary conditions alone.

Green's Functions for Helmholtz and Laplace Equations in Heterogeneous Media

R.P. Shaw, N. Makris

Department of Civil Engineering, S.U.N.Y. at Buffalo, U.S.A.

ABSTRACT:

Fundamental Green's functions are developed for a class of heterogeneous materials for Helmholtz and Laplace equations. Although limited to specific heterogeneities, the resulting Green's functions are particularly simple and may be used directly in standard boundary integral equation methods.

KEYWORDS: Boundary Integral Equation, Green's Function, Heterogeneous Media

INTRODUCTION:

The Boundary Integral Equation Method (BIEM) and the corresponding Boundary Element Method (BEM) are based on knowledge of the fundamental solution or free space Green's function for the problem at hand. When the problem considered involves a heterogeneous medium, such fundamental solutions are rare. This paper addresses the derivation of such fundamental Green's functions for a class of specific heterogeneous behavior for Helmholtz and Laplace equations.

PROBLEM FORMULATION:

Consider the following heterogeneous Helmholtz equation,

[1] $\nabla \circ \{ K(\vec{r}) \nabla\Phi(\vec{r}) \} + N(\vec{r})\, \Phi(\vec{r}) = Q(\vec{r})$

where K and N are some position dependent material parameters. For example, in time harmonic elastodynamic SH waves, K would be a shear modulus, $\mu(\vec{r})$, and N would be $\rho\ \omega^2$ where $\rho(\vec{r})$ is the solid density and ω is the frequency, as given by Ewing, Jardetsky and Press[1], while in time harmonic underwater acoustics, K would be $1/\rho$ and N would be ω^2/κ where ρ is the fluid density and κ is the bulk modulus, as given by Brekhovskikh[2]. The 'Handbook of Oceanographic Tables', i.e. SP-68 of the U. S. Naval Oceanographic Office (1966), indicates monotonically increasing densities with depth but a sound speed which decreases and then increases in many locations, indicating that both K and N will vary with position. The forcing term, $Q(\vec{r})$, will eventually be taken as a point source, $-\ \delta(\vec{r} - \vec{r}_0)$, to yield $\Phi = \Phi^*$, the fundamental Green's function for this heterogeneous problem. The problem is assumed to have been nondimensionalized, e.g. $K = K(\text{dimensional})/K_{ref}$, $N = N(\text{dimensional})/N_{ref}$ and $(x,y,z) = (x,y,z)(\text{dimensional})/L_{ref}$ such that $N_{ref}\ L_{ref}^2/\ K_{ref} = 1$ to arrive at this form. In this form, the shape of the nondimensional curve will change with frequency, i.e. new coefficients will be required in the curve fit for each new frequency. However, these new coefficients are related to those for a reference frequency in a simple manner and need not actually be recalculated by curve fitting. A simple change of variable, $\Psi(\vec{r}) = K^m\ \Phi(\vec{r})$ turns eq. [1] into

[2] $K^{-m+1}\ \nabla^2\ \Psi(\vec{r}) + \{ -2\ m + 1 \}\ K^{-m}\ \nabla K \circ \nabla\Psi(\vec{r}) +$

$\{ m^2\ K^{-m-1}(\nabla K \circ \nabla K) - m\ K^{-m}\ \nabla^2 K + K^{-m} N \}\ \Psi(\vec{r}) = Q(\vec{r})$

By choosing $m = 1/2$, the term involving $\nabla\Psi$ will drop out. By further requiring

[3] $(\nabla K \circ \nabla K)/(4\ K^2) - \nabla^2 K/(2\ K) + N/K = \beta^2$

where β is a constant (which might be imaginary), eq. [2] reduces to a standard Helmholtz equation,

$$[4] \qquad \nabla^2 \Psi(\vec{r}) + \beta^2 \Psi(\vec{r}) = K(\vec{r})^{-1/2} Q(\vec{r})$$

If $Q(\vec{r})$ is taken to be $-\delta(\vec{r} - \vec{r}_0)$, the right hand side of eq. [4] may be replaced by $-K(\vec{r}_0)^{-1/2}\,\delta(\vec{r} - \vec{r}_0)$ since the delta function only operates at $\vec{r} = \vec{r}_0$. Then the equation on $\Phi(\vec{r})$ becomes

$$[5] \quad \nabla^2 \{ K^{1/2}(\vec{r})\, K^{1/2}(\vec{r}_0)\, \Phi^*(\vec{r}, \vec{r}_0) \} + \beta^2 \{ K^{1/2}(\vec{r})\, K^{1/2}(\vec{r}_0)\, \Phi^*(\vec{r}, \vec{r}_0) \} = -\delta(\vec{r} - \vec{r}_0)$$

which identifies $\{ K^{1/2}(\vec{r})\, K^{1/2}(\vec{r}_0)\, \Phi^*(\vec{r}, \vec{r}_0) \}$ with the standard homogeneous medium Helmholtz fundamental Green's function, $G^*(\vec{r}, \vec{r}_0)$, which is well known, e.g. in three dimensions, $G^* = \exp(i\,\beta\,|\vec{r} - \vec{r}_0|)/(4\,\pi\,|\vec{r} - \vec{r}_0|)$ for real β.

BOUNDARY INTEGRAL EQUATION FORMULATION:

Once Φ^* is known, the standard Green's theorem approach leads to

$$[6] \int_S [\hat{n}(\vec{r}_0)\{\Phi^*(\vec{r},\vec{r}_0) \circ K(\vec{r}_0) \nabla \Phi(\vec{r}_0) - \Phi(\vec{r}_0) \circ K(\vec{r}_0) \nabla \Phi^*(\vec{r},\vec{r}_0)\}] dS(\vec{r}_0) = c\Phi(\vec{r})$$

where c is the usual (0, 1/2, 1) for $\vec{r}$ (outside of V, on S at a smooth location, inside of V). An actual forcing term Q in the original equation would appear as the usual volume integral term. The only real change then is the inclusion of the factor K in the integral.

SPECIFIC MATERIAL VARIATION:

Eq. [3] forces a relationship between K and N in order to obtain this simple form for Φ^*; clearly this is a limited approach to such problems. This restriction, if it is not too severe, may be justified by the simple end result and the fact that the actual variations of material properties are generally only known at a few locations and fit by some convenient functional form anyway. The final judgement lies in how many degrees of freedom exist in this relationship. For example, if K were known to be constant, then N must also be constant and the homogeneous medium problem is recovered with no real gain. However, frequently the physical processes which cause N to vary with position also apply to K. Eq. [3] may be simplified by a transformation, $W = K^{1/2}$, leading to

$$[7] \quad \nabla^2 W - N/W + \beta^2 W = 0$$

as the required relationship. If W is prescribed, then this would force a specific form for N with one degree of freedom, β^2. If N were specified, this would lead to a nonlinear equation on W which is no real gain over the original problem. If N were zero, the original equation would have been a heterogeneous Laplace equation and the requirement on W would be that it satisfied a Helmholtz equation. If furthermore β^2 were zero, this requirement is that $W = K^{1/2}$ be a potential function, i.e. satisfy a homogeneous Laplace equation, as noted by Cheng[3], Clements[4], etc. The case of $\beta^2 \neq 0$ is not new; it provides a Helmholtz type Green's function for a Laplace type governing equation which appears odd but does fit this development. However, if it is possible to assume (within the limits of the data available) a relationship between N and W, one obvious choice would be

$$[8] \quad N = A(\vec{r})\, W + B_0 W^2$$

where A may be a function of position but B_0 for convenience is taken as a constant. This leads to

$$[9] \quad \nabla^2 W + (\beta^2 - B_0)\, W = A(\vec{r})$$

which is a linear Helmholtz equation on W with constant coefficients. There will be two degrees of freedom in the complementary solution to this equation plus however many degrees of freedom are chosen for $A(\vec{r})$.

STRATIFIED MEDIA — GENERAL CASE:

This approach may be illustrated by the special case of stratified media where K and N depend only on one spatial coordinate, z. Then eq. [9] is an ordinary linear differential equation with constant coefficients whose solution is well known even for general A(z). In fact, in this case, B_0 could even be some specific function of position leading to a complementary solution for W as one of the special functions of mathematical physics.

CONSTANT A(z) CASE:

Consider the special case of a constant $A(z) = A_0$. The solutions for W and N are then

$$[10]\quad W(z)=C_1\sin(qz) + C_2\cos(qz) + A_0/(\beta^2 - B_0);\ q^2 = \beta^2 - B_0 > 0$$

$$= D_1\exp(\lambda z) + D_2\exp(-\lambda z) + A_0/(\beta^2 - B_0)\ ;\ \lambda^2 = B_0 - \beta^2 > 0$$

$$[11]\quad N(z) = A_0\, W(z) + B_0\, W(z)^2$$

Consider a semi-infinite region with $z > 0$ as an example. Boundary conditions on the plane $z = 0$ may well be satisfied by reflection of this point source solution with some image for $z < 0$. The exponential form of solution seems most appropriate here, although the sinusoidal variation for an infinite domain gives rise to possible Floquet solutions which are of interest. If $\lambda^2 = B_0 - \beta^2 > 0$, then D_1 must vanish to have finite K as z approaches infinity. The remaining constants of the problem may be related to the four limiting values of K (or W) and N, i.e.

$$[12]\quad \begin{aligned} N_0 &= N(z = 0) \\ W_0 &= W(z = 0) \\ N_1 &= N(z \to \infty) \\ W_1 &= W(z \to \infty) \end{aligned}$$

$$[13]\quad \begin{aligned} D_2 &= W_0 - W_1 \\ A_0 &= (\, N_1\, W_0^2 - N_0\, W_1^2\,)\ /\ (\, W_1\, W_0\, [W_0 - W_1]\,) \\ B_0 &= (\, N_0\, W_1 - N_1\, W_0\,)\ /\ (\, W_1\, W_0\, [W_0 - W_1]\,) \end{aligned}$$

$\beta^2 = N_1/W_1^2$ or $\lambda^2 = (N_0 W_1^2 - N_1 W_0^2) / (W_1^2 W_0 [W_0 - W_1])$

leading to

[14a] $W(z) = (W_0 - W_1) \exp(- \lambda z) + W_1$

[14b] $N(z) = \{ (N_1/W_1) [1 - \exp(-\lambda z)] + (N_0/W_0) \exp(-\lambda z) \}$ $\{W_1[1 - \exp(-\lambda z)] + W_0 \exp(-\lambda z)\}$

Since this form requires $\lambda^2 > 0$, if $W_0 < W_1$, then $N_0/N_1 < (W_0/W_1)^2 = K_0/K_1 < 1$ and if $W_0 > W_1$, then $N_0/N_1 > (W_0/W_1)^2 = K_0/K_1 > 1$. These correspond to the local value of wave speed, $c = (K/N)^{1/2}$ being larger at the origin than at infinity or smaller at the origin than at infinity respectively. Note that the asymptotic value of c is $c_1 = 1/\beta$. These simple forms for W, or K, and N are exponentials beginning at specified values at $z = 0$ and asymptotically approaching specified values as z approaches infinity. The variation between these values however is forced to match the forms of eqs. [14]. However, the local speed c will not have any local extremum which makes this case of lesser interest than others and will not be pursued further here.

VARIABLE A(z) CASE:
Here A(z) may be taken as a sum of convenient functions. Since the physical parameters would be expected to approach constant values, or possibly some constant slope, as z approaches infinity, it would make sense to consider decaying exponentials plus a simple linear term, e.g.

$$[15] \quad A(z) = \alpha_0 + \beta_0 z + \sum_{j=1}^{M} \alpha_j \exp(-\gamma_j z)$$

with a corresponding solution for W, with $\lambda^2 = B_0 - \beta^2$, as

$$[16] \quad W(z) = -(\alpha_0 + \beta_0 z)/\lambda^2 + \sum_{j=1}^{M} \alpha_j \exp(-\gamma_j z)/(\gamma_j^2 - \lambda^2) + a_1 \exp(\lambda z) + a_2 \exp(-\lambda z)$$

where the a_1, a_2 terms represent the complementary solution. Again for a problem with $z > 0$, the a_1 term will be suppressed. Consider this case with a single exponential term in A, i.e. $M = 1$. Furthermore, set $\beta_0 = 0$; this term allows for a slowly increasing form of the material parameters as z approaches infinity, which will be assumed NOT to be the case here. Then

[17] $W(z)=K(z)^{1/2}=-\alpha_0/\lambda^2 + \alpha_1\exp(-\gamma_1 z)/(\gamma_1^2-\lambda^2) + a_2\exp(-\lambda z)$

[18] $N(z) = A(z)\, W(z) + (\lambda^2 + \beta^2)\, W(z)^2$

and the local speed, c, which is $(K/N)^{1/2}$, is

[19] $c(z) = [\, A(z)/W(z) + (\beta^2 + \lambda^2)\,]^{-1/2}$

The limiting values are now

[20] $W(0) = W_0 = a_2 - \alpha_0/\lambda^2 + \alpha_1/(\gamma_1^2 - \lambda^2)$

$W(\infty) = W_1 = -\alpha_0/\lambda^2$

$N(0) = N_0 = (\alpha_0 + \alpha_1)\, W_0 + (\beta^2 + \lambda^2)\, W_0^2$

$N(\infty) = N_1 = \alpha_0^2\, \beta^2/\, \lambda^4$

$c(0) = c_0 = [\beta^2 + \{a_2\lambda^2 + \alpha_1\gamma_1^2/(\gamma_1^2-\lambda^2)\}/\{a_2 - \alpha_0/\lambda^2 + \alpha_1/(\gamma_1^2-\lambda^2)\}]^{-1/2}$

$c(\infty) = c_1 = 1/\beta$

As an illustration, it is of interest to consider the special case when $c_0 = c_1$, i.e. the local speed returns asymptotically to its original value, although clearly other choices could equally well be used. This requires

[21] $a_2 = -\alpha_1\, \gamma_1^2/[\lambda^2\, (\gamma_1^2 - \lambda^2)\,]$

in which case the original values for W and N simplify to

$$[22] \quad W_0 = -(\alpha_0 + \alpha_1)/\lambda^2 \ ; \quad N_0 = (\alpha_0 + \alpha_1)^2 \beta^2/\lambda^4$$

which indicates that $\alpha_0 < 0$ to have $W_1 > 0$ and $(\alpha_0 + \alpha_1) < 0$ to have $W_0 > 0$, as required for real K. This leads to an expression for c(z) as

$$[23] \quad c(z)/c_0 = [\, 1 + (\gamma_1^2 \lambda^2/\beta^2) [\, \exp(-\gamma_1 z) - \exp(-\lambda z) \,] \,/$$

$$[\, -(\alpha_0/\alpha_1)(\gamma_1^2 - \lambda^2) + (\, \lambda^2 \exp(-\gamma_1 z) - \gamma_1^2 \exp(-\lambda z) \,] \,]^{-1/2}$$

In this case, c(z) will have an extremum at the z defined by

$$[24] \quad 0 = (-\alpha_0/\alpha_1)[-\gamma_1 \exp(-\gamma_1 z) + \lambda \exp(-\lambda z)] + (\gamma_1 - \lambda)\exp(-[\gamma_1 + \lambda]\, z)$$

A low velocity layer, with a decrease of about 15%, is illustrated in Fig. (1) for $\lambda = 2.0$, $\alpha_0 = -1.0$, $\beta_0 = 0.0$, $\beta = 1.0$, $\alpha_1 = -1.0$ and $\gamma_1 = 1.9$ while an analogous case of a high velocity layer, with an increase of about 15%, is shown in Fig. (2) with $\lambda = 2.0$, $\alpha_0 = -2.0$, $\beta_0 = 0.0$, $\beta = 1.0$, $\alpha_1 = 1.0$ and $\gamma_1 = 1.0$. These cases are hypothetical using simple numerical values and are given to indicate the potential uses of this approach; later studies will concentrate on more practical examples, including the use of several terms in the exponential series to fit observed data.

STRATIFIED MEDIA – LAPLACE EQUATION CASE:

The case of $N = 0$ everywhere, (a Laplace equation), leads to W of the form

$$[25] \quad W(z) = C_1 \sin(q\, z) + C_2 \cos(q\, z) \ ; \ q^2 = \beta^2 > 0$$

$$= D_1 \exp(\lambda\, z) + D_2 \exp(-\lambda\, z) \ ; \ \lambda^2 = -\beta^2 > 0$$

If $\beta^2 = 0$ as well as $N = 0$, this reduces to well known fundamental

Green's function solution to a heterogeneous Laplace equation, i.e.

[26] $W(z) = K(z)^{1/2} = C_1 + C_2 z$

leads to a heterogeneous Green's function

[27] $\Phi^*(\vec{r}, \vec{r}_0) = 1 / [\, 4 \pi \, (C_1 + C_2 z)\, (C_1 + C_2 z_0)\, R \,]$

If $\beta^2 = -\lambda^2 < 0$ and $N = 0$, then the standard Green's function solution to eq. [5] would be, with $R = |\vec{r} - \vec{r}_0|$,

[28] $G^*(\vec{r}, \vec{r}_0) = \exp(-\lambda R) / (4 \pi R)$

and the three dimensional heterogeneous (stratified) Green's function would be

[29] $\Phi^*(\vec{r}, \vec{r}_0) = (D_1 \exp(\lambda z) + D_2 \exp(-\lambda z))^{-1/2}$

$$(D_1 \exp(\lambda z_0) + D_2 \exp(-\lambda z_0))^{-1/2} \; \exp(-\lambda R) / (4 \pi R)$$

where again for a half space, $z > 0$, with bounded K the coefficient D_1 would be zero. This case in a way is more restricted than the Helmholtz case since it has only one degree of freedom, $K(\vec{r})$, rather than the two degrees of the Helmholtz case, $K(\vec{r})$ and $N(\vec{r})$.

CONCLUSION:

While the Green's functions developed here apply only to the case of specific forms of the material parameters, K and N, there is actually quite a bit of freedom in the fitting of observed data to these forms, especially in the heterogeneous Helmholtz case. Future work will deal not only with actual stratified material behavior but also with multi-dimensional variations, e.g. depth and range dependence in underwater acoustics. The hope here is that an entirely new class of problems may be opened up to boundary integral equation/element methods without the tedious volume integrals previously used for such problems.

REFERENCES:

1: Ewing, W. M., Jardetsky, W. and F. Press, ELASTIC WAVES IN LAYERED MEDIA, sec. 7-3, McGraw-Hill Book Co., N. Y., N. Y., (1957)

2: Brekhovskikh, L. M., WAVES IN LAYERED MEDIA, sec. 13.2, Academic Press, N. Y., N. Y., (1960)

3: Cheng, A. H-D, "Darcy's Flow with Variable Permeability", Water Resources Research, Vol. 20, pp. 980-984, 1984

4: Clements, D. L., "Green's Functions for the Boundary Element Method", pp. 14-20, BEM IX, Vol. 1, edited by C. A. Brebbia, W. L, Wendlund and G. Kuhn, CMP/Springer Verlag Publishers, Southampton, 1987,

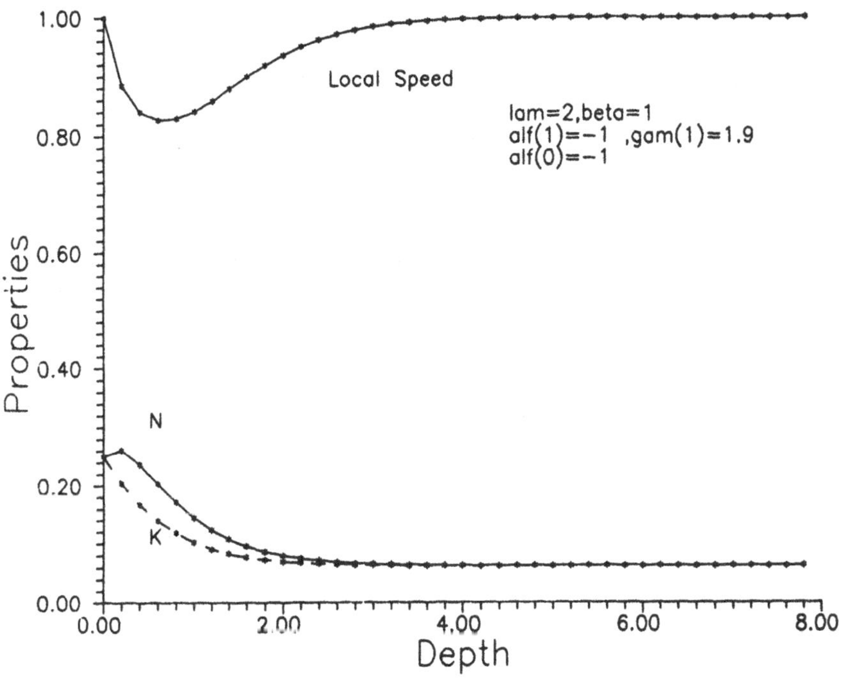

Fig. 1: Local sound speed (c) and material properties K and N for a low velocity layer (about 15% decrease)

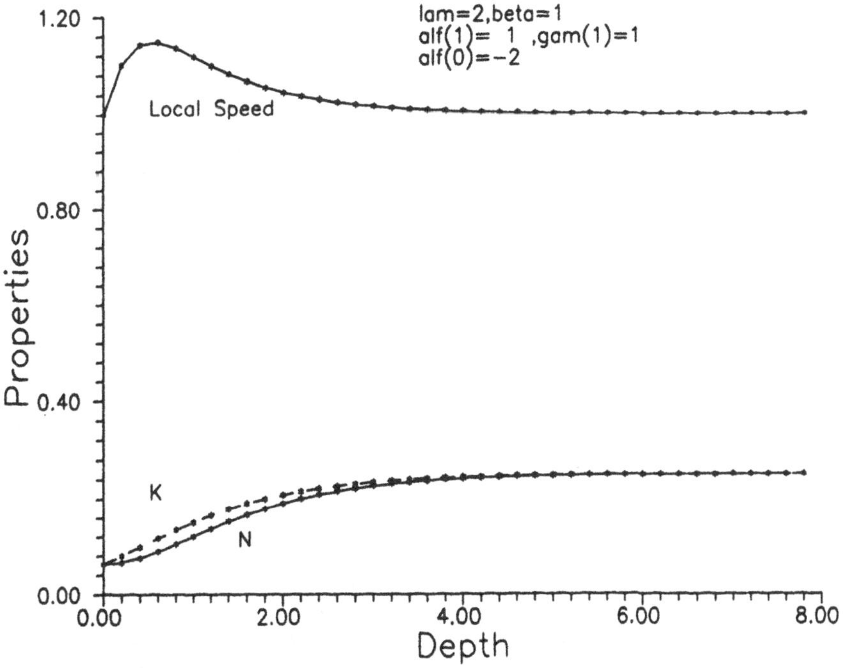

Fig. 2: Local sound speed (c) and material properties K and N for a high velocity layer (about 15% increase)

SECTION 2: DIFFUSION AND CONVECTION PROBLEMS

A Study of Flow Structures in a Cavity due to Double-Diffusive Natural Convection by Boundary Element Method and Sub-Domain Technique

I. Žagar, M. Hriberšek, P. Škerget, A. Alujevič
Faculty of Engineering, University of Maribor, Slovenia, Yugoslavia

ABSTRACT

This paper reports numerical simulation results of natural convection caused by combined heat and mass transfer in a square cavity with insulated top and bottom walls. General time dependent results show the development of velocity, temperature and molar fraction fields. Boundary-domain integral method is used and the sub-domain technique combined with block-solver is employed to reduce computer requirements.

INTRODUCTION

Many transport phenomena of natural convection motion that occur in nature or in engineering are driven or modified due to the simultaneous existence of temperature and concentration gradients in fluids. A practical example of such a complex phenomenon is a storage tank of liquified natural gas. Fluid properties of this gas ($Pr = 2.2$; $Sc = 130$; $Gr_t = 2500$; $Gr_m = -1000$) are taken into account in the numerical example of a closed cavity subjected to various kinds of boundary conditions.

In the past, the problem was studied by finite-difference technique for a steady-state case [2]. In the present work, the problem is studied by boundary-domain integral method as time dependent problem. The velocity-vorticity formulation for the mass and momentum transfers is used. The buoyancy effect is included through the heat and species concentration transfers and Boussinesq approximation.

Subdomain technique in connection with block-solver, a computer program for solving block-structured matrices, was developed to reduce CPU time and memory requirements of the method. Comparison between the consumption for variable numbers of subdomains has been made.

GOVERNING EQUATIONS

The partial differential equations set, governing the motion of viscous incompressible fluid is known as nonlinear Navier-Stokes equations expressing the basic conservation balances of mass, momentum, energy and species concentration. Introducing vorticity ω and stream function ψ of the solenoidal velocity field, the computation of the flow is divided into kinematics given by the Poisson's elliptic equation, writen in plane $x - y$ by

$$\frac{\partial^2 \psi}{\partial x^2} + \frac{\partial^2 \psi}{\partial y^2} + \omega = 0 \tag{1}$$

and into the kinetics described by the vorticity equation

$$\frac{\partial \omega}{\partial t} + v_x \frac{\partial \omega}{\partial x} + v_y \frac{\partial \omega}{\partial y} = \nu \left(\frac{\partial^2 \omega}{\partial x^2} + \frac{\partial^2 \omega}{\partial y^2}\right) + g_y \frac{\partial F}{\partial x} - g_x \frac{\partial F}{\partial y} \tag{2}$$

The buoyancy effect is included by energy and molar fraction equations

$$\frac{\partial T}{\partial t} + v_x \frac{\partial T}{\partial x} + v_y \frac{\partial T}{\partial y} = a \left(\frac{\partial^2 T}{\partial x^2} + \frac{\partial^2 T}{\partial y^2}\right) \tag{3}$$

$$\frac{\partial C}{\partial t} + v_x \frac{\partial C}{\partial x} + v_y \frac{\partial C}{\partial y} = D \left(\frac{\partial^2 C}{\partial x^2} + \frac{\partial^2 C}{\partial y^2}\right) \tag{4}$$

and Boussinesq approximation, given for linear normalised difference of density

$$\frac{\rho - \rho_o}{\rho_o} = F = -\beta_t \,(T - T_o) - \beta_m \,(C - C_o) \tag{5}$$

BOUNDARY-DOMAIN INTEGRAL EQUATIONS

The boundary domain integral statement for the flow kinematics can be derived from the vector elliptic equation for vector potential ψ_i [3] applying Green's theorem for the vector functions and the elliptic fundamental solution u^*, resulting in the following statement

$$c(\xi) v_x(\xi) + \int_\Gamma v_x \frac{\partial u^*}{\partial n} \, d\Gamma = \int_\Gamma v_y \frac{\partial u^*}{\partial t} \, d\Gamma - \int_\Omega \omega \frac{\partial u^*}{\partial y} d\Omega \tag{6}$$

$$c(\xi) v_y(\xi) + \int_\Gamma v_y \frac{\partial u^*}{\partial n} \, d\Gamma = -\int_\Gamma v_x \frac{\partial u^*}{\partial t} \, d\Gamma + \int_\Omega \omega \frac{\partial u^*}{\partial x} d\Omega \tag{7}$$

Describing the laminar transport of the vorticity, temperature and molar fraction in the integral statement, one has to take into account that the vorticity, temperature and species concentration obey nonhomogeneous parabolic equations [4],[5]. So the following boundary-domain integral formulations can be derived for a plane

$$c(\xi)\omega(\xi,t_F)+\nu\int_\Gamma\int_{t_{F-1}}^{t_F}\omega\frac{\partial u^*}{\partial n}\,dt\,d\Gamma=\nu\int_\Gamma\int_{t_{F-1}}^{t_F}\frac{\partial\omega}{\partial n}u^*\,dt\,d\Gamma$$

$$-\int_\Gamma\int_{t_{F-1}}^{t_F}(\omega v_n+g_t\,F)u^*\,dtd\Gamma$$

$$+\int_\Omega\int_{t_{F-1}}^{t_F}(\omega\,v_x+g_y\,F,\,\omega\,v_y-g_x\,F)\,(\frac{\partial u^*}{\partial x},\frac{\partial u^*}{\partial y})\,dtd\Omega+\int_\Omega\omega_{F-1}\,u^*{}_{F-1}\,d\Omega \quad (8)$$

$$c(\xi)T(\xi,t_F)+a\int_\Gamma\int_{t_{F-1}}^{t_F}T\frac{\partial u^*}{\partial n}\,dt\,d\Gamma=a\int_\Gamma\int_{t_{F-1}}^{t_F}\frac{\partial T}{\partial n}u^*\,dt\,d\Gamma$$

$$-\int_\Gamma\int_{t_{F-1}}^{t_F}T\,v_nu^*\,dt\,d\Gamma$$

$$-\int_\Omega\int_{t_{F-1}}^{t_F}(T\,v_x,\,T\,v_y)\,(\frac{\partial u^*}{\partial x},\frac{\partial u^*}{\partial y})\,dt\,d\Omega+\int_\Omega T_{F-1}\,u^*{}_{F-1}\,d\Omega \quad (9)$$

$$c(\xi)C(\xi,t_F)+D\int_\Gamma\int_{t_{F-1}}^{t_F}C\frac{\partial u^*}{\partial n}\,dt\,d\Gamma=D\int_\Gamma\int_{t_{F-1}}^{t_F}\frac{\partial C}{\partial n}u^*\,dt\,d\Gamma$$

$$-\int_\Gamma\int_{t_{F-1}}^{t_F}C\,v_nu^*\,dt\,d\Gamma$$

$$-\int_\Omega\int_{t_{F-1}}^{t_F}(C\,v_x,\,C\,v_y)\,(\frac{\partial u^*}{\partial x},\frac{\partial u^*}{\partial y})\,dt\,d\Omega+\int_\Omega C_{F-1}\,u^*{}_{F-1}\,d\Omega \quad (10)$$

SUBDOMAIN TECHNIQUE

One of the main problems arising in boundary-domain integral method is also a fully populated system matrix in case of just one subdomain. This presents a restriction while using some fast and efficient "solvers" like frontal and Cholesky method.

By the use of the subdomain technique a block structure of the matrices (Figure 1) may be obtained, which enables the application of the so called "block-solver". The fully populated matrix is thus replaced by a system matrix with the block-structure. Although the number of nodes increases with the growing number of subdomains, good effects can be noticed. Population of system matrix decreases with the number of subdomains and transforms itself into the block-structure, which enables the use of the block-solver. The new system of equations can thus be faster triangulized and quickly solved. One of the advantages is also a much smaller working memory, needed for block-solver to operate. If this is also associated with the use of out-of-core memory (sequential and direct-access files) a considerable saving in computer memory requirements can be achieved.

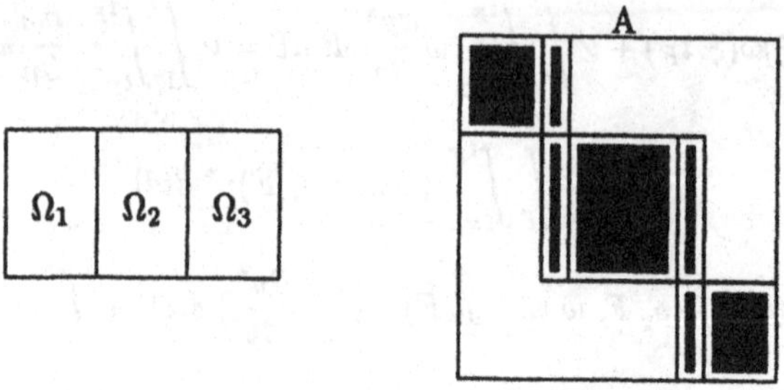

Fig. 1: Division into three subdomains and system matrix structure

The existent block-solver, used in the presented calculations, solves systems of equations with the use of Gauss-elimination method and out-of-core memory (a similar approach can be found in [1]). In the table below (1) a quick overview is given how CPU time decreases with rising number of subdomains. A percentage decrease in population of matrix is also presented. The table confirms pre-given information on the use of block-solver although it also shows that the decrease of time and memory vanishes at about six subdomains for the computed case. This means that the further increase in number of subdomains wouldn't result in the decrease of CPU time, because of the growth of nodes, caused by interface boundaries. It can be said, that for this example six subdomains present the best choice of dividing the original domain.

Number of subdomains	Total CPU time (%)	Pop. of system matrix (%)
1	100	100
2	57	58
3	43	37
4	38	34
5	33	29
6	31	25
7	31	23

Table 1: Comparison of CPU time and population for various numbers of subdomains

DOUBLE-DIFFUSION IN CLOSED CAVITY

Geometry and mesh discretization of buoyancy driven laminar flow in a closed square cavity are shown on Fig. 2. A mesh of 40 elements (80 nodes) and 100 internal cells has been used.

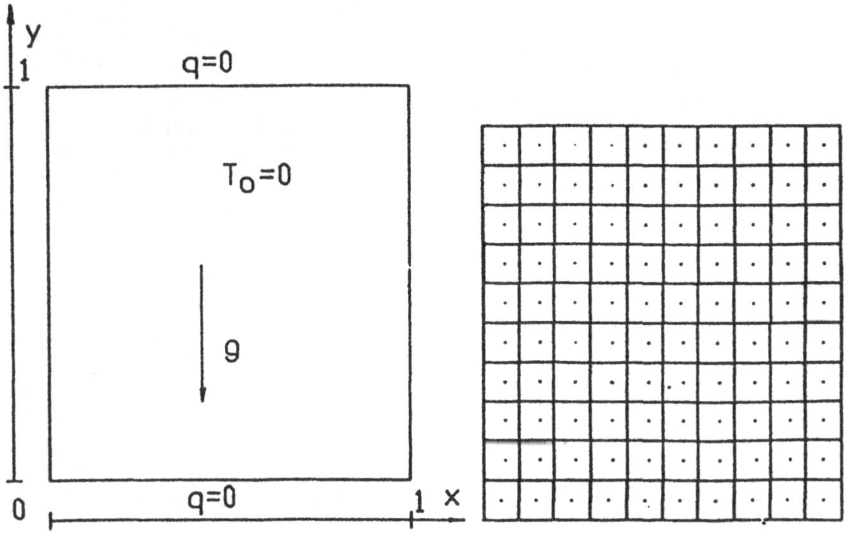

Fig. 2: Problem geometry and boundary conditions

Horizontal walls are insulated, on the vertical walls temperature and molar fraction are prescribed. Linear variations of temperature and molar fraction are imposed

$$X = 0\&1 : \quad T = 1 - Y \quad C = 1 - Y$$

The molar fraction boundary condition can be interpreted as diffusion of the component through the wall. Fluid was considered for which body forces due to heat and species transfer are opposed, $Ra_t = 5.5\,10^3$ ($Gr_t = 2500$, $Pr = 2.2$) and $Ra_m = -1.3\,10^5$ ($Gr_m = -1000$, $Sc = 130$) [2].

Figures 3,4 and 5 show development of velocity, temperature and molar fraction field. Time dependent development of all three fields shows the characteristic nature of transport phenomena which occur in the cavity due to double-diffusion. Starting velocity distribution is mostly a reflexion of the heat transfer in fluid flow. Namely, steady-state temperature field is almost completely formed after a few time steps, and later slightly changes due to the effect of molar fraction field development. New rotating cells which are created in the corners and then spread in the whole region (Fig. 3) are also affected by species concentration transfer. The numerical results agree with those obtained from calculations by other authors [2].

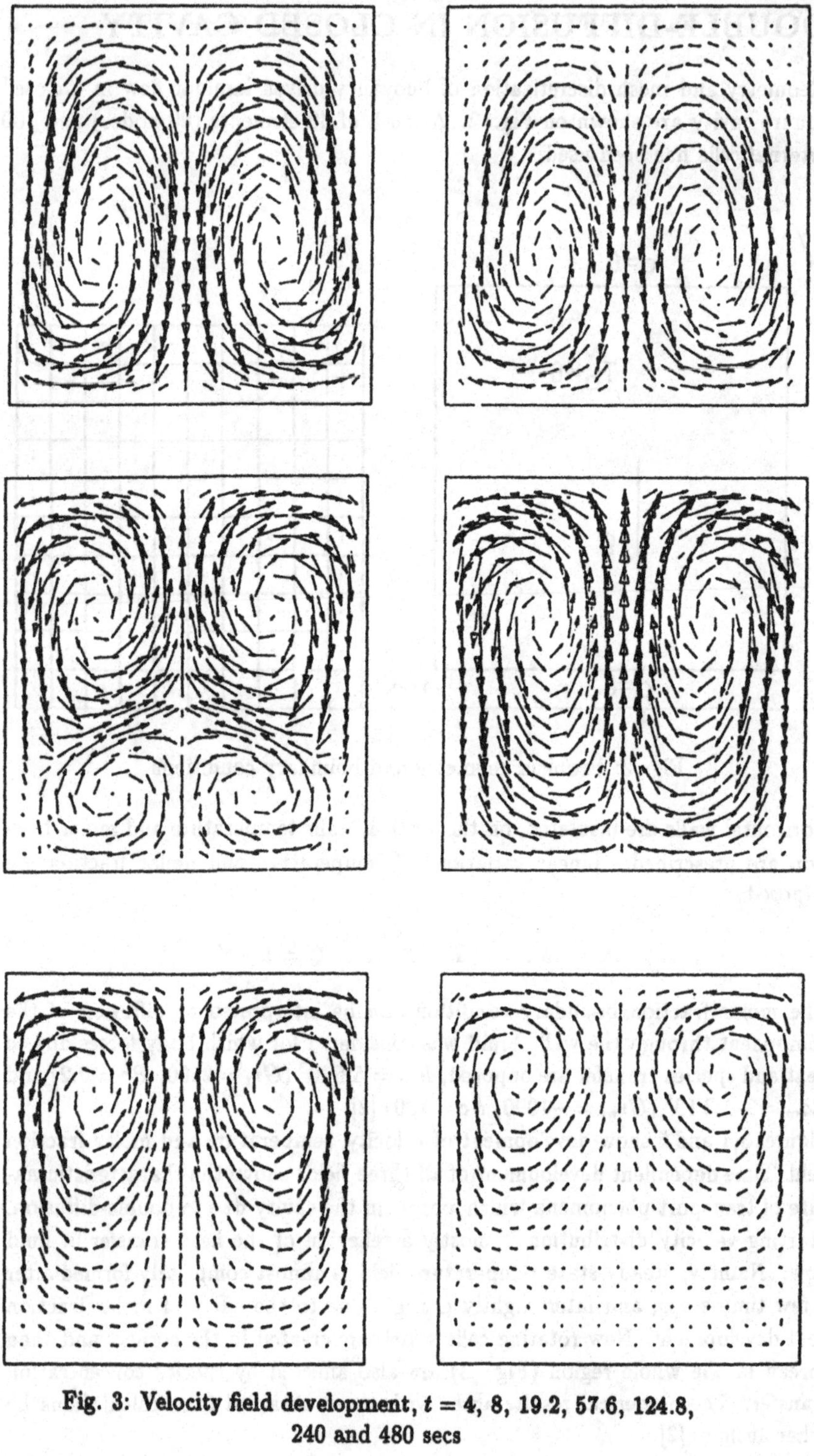

Fig. 3: Velocity field development, t = 4, 8, 19.2, 57.6, 124.8, 240 and 480 secs

Fig. 4: Temperature field development, $t = 4$, 8, 19.2, 57.6, 124.8, 240 and 480 secs

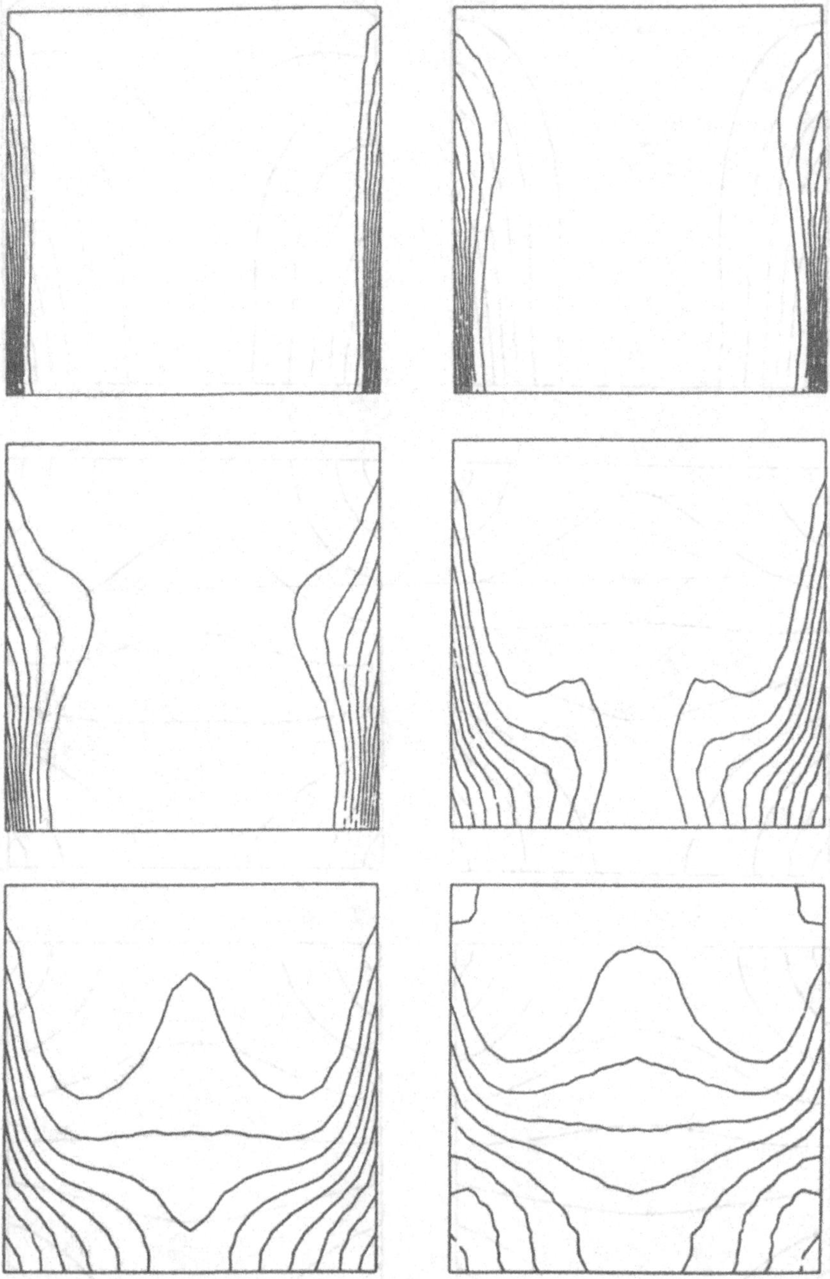

Fig. 5: Molar fraction field development, t = 4, 8, 19.2, 57.6, 124.8, 240 and 480 secs

REFERENCES

[1] Bozek D.G., Ciarelli D.M., Hodous M.F., Ciarelli K.J., Kline K.A., Katnik R.B.: *Vector Processing Applied to Boundary Element Algorithms on the CDC CYBER-205.* EdF-Bulletin de la Direction des Etudes et des Recherches - Serie C - No. 1 - 1983.

[2] Seveleder,V., Petit,J.P.:Flow Structures Induced by Opposing in Double-Diffusive Natural Convection in a Cavity. Numerical Heat Transfer, Part A, Vol.15, pp.431-444, 1989.

[3] Skerget,P., Alujevic,A., Zagar,I., Brebbia,C.A., Kuhn,G.: *Time Dependent Three Dimensional Laminar Isochoric Viscous Fluid Flow by BEM*, 10th Int. Conf. on BEM, Southampton, Springer-Verlag, Berlin, 1988.

[4] Skerget,P., Alujevic,A., Brebbia,C.A., Kuhn,G.: *Natural and Forced Convection Simulation Using the Velocity-Vorticity Approach*, Topics in Boundary Element Research (Ed. by Brebbia C.A.), Vol.5, Ch.4, Springer-Verlag , Berlin, 1989.

[5] Zagar,I., Skerget,P., Alujevic, A.: *Boundary-Domain Integral Method for the Space Time Dependent Viscous Incompressible Flow.* Symposium of the International Association for Boundary Element Method. IABEM-90, Roma, Italy, 1990.

The Laplace Transform Boundary Element (LTBE) Method for the Solution of Diffusion-Type Equations

G.J. Moridis, D.L. Reddell

Groundwater Research Program, Agricultural Engineering Department, Texas A&M University, College Station, TX 77843-2117, U.S.A.

ABSTRACT

A new numerical method, the Laplace Transform Boundary Element (LTBE) method, was developed for the solution of diffusion-type PDE's by eliminating the time dependency of the problem using a Laplace transform formulation. In essence, LTBE yields a solution numerical in space and semi-analytical in time. It consists of 4 steps: 1) a Laplace transform is performed on the PDE, 2) the transformed PDE is approximated using the Boundary Element Method (BEM), 3) the resulting system of simultaneous equations is solved and the transformed vector of unknowns is determined in the Laplace space, and 4) the solution vector obtained in step 3 is inverted numerically to yield the solution in time. The solution in the Laplace space renders the time interpolation and time marching schemes employed in the standard BEM irrelevant because time is no longer considered. The method was tested against results obtained from two- and three-dimensional test cases of heat conduction and groundwater flow using a standard BEM simulator and analytical models. For a single time-step, LTBE requires an execution time 6 times longer than the analogous BEM requirement without an increase in storage. This disadvantage is outweighed by the fact that LTBE allows an unlimited time-step size with a more accurate solution than the conventional BEM treatment. Execution times are reduced by orders of magnitude because calculations in the LTBE scheme are necessary only at the desired observation times, while the standard BEM requires calculations at all the intermediate times of the discretized time domain. Roundoff and material balance errors are stable, non-increasing, and much lower than in the standard BEM simulations.

INTRODUCTION

The basic concept of any numerical solution of a Partial Differential Equation (PDE) is the replacement of the original continuous PDE by a set of algebraic equations that are valid in the discretized solution domain Ω and/or the discretized boundary surface Γ of Ω. These equations are easier to solve, and provide a solution arithmetically 'close' to the true solution of the original problem which may be impossible to find. In diffusion-type equations, we seek an approximate solution to the problem governed by the PDE

$$\nabla^2 U(x_\kappa, t) = \frac{1}{k}\frac{\partial U(x_\kappa, t)}{\partial t} \qquad x_\kappa \in \Omega \tag{1}$$

with the boundary conditions

$$U(x_\kappa, t) = \overline{U}(x_\kappa, t), \qquad x_\kappa \in \Gamma_1 , \tag{2}$$

$$V(x_\kappa, t) = \frac{\partial U(x_\kappa, t}{\partial n(x_\kappa)} = \overline{V}(x_\kappa, t), \qquad x_\kappa \in \Gamma_2 , \tag{3}$$

and initial conditions

$$U(x_\kappa, t) = U_0(x_\kappa, t_0) . \tag{4}$$

Ω is the solution domain (a volume in three-dimensions and an area in two dimensions) of the dependent variable U, and x_κ ($\kappa \equiv 1, 2, 3$) are the space coordinates. Γ_1 and Γ_2 represent complementary segments of the exterior surface Γ (an area in three-dimensions and a line in two dimensions), and n is the unit outward vector normal to Γ. U_0 indicates known initial conditions, and $\overline{U}$ and $\overline{V}$ indicate known values of U and V on the boundaries. The term k is assumed to be constant in time and space, and its interpretation depends on the physical problem under study. Parameter invariability does not limit the generality of equation (1). Inhomogeneous and anisotropic conditions are easily included by using subregions (e.g. Brebbia [1]) and stretched coordinates (e.g.Liggett and Liu [2]).

The conceptual basis of the Boundary Element Method (BEM) is the minimization of the error E introduced by replacing $U(x_\kappa, t)$ by an approximate solution. Using Green's second identity, solutions are obtained on the boundary, from which solutions at any interior point can be calculated. The formulation of BEM starts with the development of a weighted residual equation, the boundary integral equation, which minimizes E in (1). Instead of solving for an analytical, continuous, smooth function $U(x_\kappa, t)$, numerical approximations ($\tilde{U}$) of the solution are obtained at a finite set of predetermined N points $(x_\kappa, t)_1, (x_\kappa, t)_2, \ldots, (x_\kappa, t)_N$ on the boundary Γ of the domain Ω. A PDE problem with a continuous smooth solution surface is thus reduced to a set of algebraic equations through the following general procedure described by Brebbia, Telles, and Wrobel [3]: a) the boundary Γ is discretized into a series of elements where U and V vary according to interpolation functions, b) the discretized equation is applied to a number of particular nodes within each elemnt, c) using numerical (usually Gaussian) quadrature schemes, the integrals over each boundary element are computed, d) a system of simultaneous linear equations is obtained by imposing the presecribed boundary conditions, and the solution of this set of equations supplies the remaining boundary information, and e) U and V can be determined at any internal point using the results from (d).

Two traditional methods are used to address the problem of time dependency in equation (1). The first uses a coupled boundary element-finite difference approach in which a finite difference approximation of $\frac{\partial U}{\partial t}$ is employed. Using weighted residuals, the boundary integral equation for this formulation is

$$C(\xi)U(\xi, t + \Delta t) + k \int_\Gamma U(x_\kappa, t + \Delta t) V^*(\xi, x_\kappa, \Delta t) d\Gamma(x_\kappa) =$$

$$k \int_\Gamma V(x_\kappa, t + \Delta t) U^*(\xi, x_\kappa, \Delta t) d\Gamma(x_\kappa) + \frac{1}{\Delta t} \int_\Omega U(x_\kappa, t) U^*(\xi, x_\kappa, \Delta t) d\Omega(x_\kappa) , \tag{5}$$

where $C(\xi)$ is a function of the angle of the boundary at point ξ, and Δt is the time step size. U^* represents an appropriate fundamental solution, and is interpreted as a weighting function, while $V^*(\xi, x_\kappa, t) = \partial U^*(\xi, x_\kappa, t)/\partial n(x_\kappa)$. Since the initial conditions U_0 at $t = t_0$ are known, solutions at time $t = t_0 + \Delta t$ are obtained by solving (5), and they serve as the pseudo-initial conditions for the next time step. This formulation, thoroughly investigated by Curan, Cross, and Lewis [4], needs small time steps to produce accurate results, and requires large computing times.

The second formulation accounts for time dependence by directly integrating over time a weighted residuals equation of the PDE in (1). The resulting boundary integral equation is given by Brebbia, Telles, and Wrobel [3] as

$$C(\xi)U(\xi, t_F) + k\int_{t_0}^{t^F}\int_\Gamma U(x_\kappa, t)\; V^*(\xi, x_\kappa, t_F, t)\; d\Gamma(x_\kappa)dt =$$

$$k\int_{t_0}^{t^F}\int_\Gamma V(x_\kappa, t)\; U^*(\xi, x_\kappa, t_F, t)\; d\Gamma(x_\kappa)dt + \int_\Omega U_0(x_\kappa, t_0)\; U^*(\xi, x_\kappa, t_F, t_0)\; d\Omega(x_\kappa)\,, \tag{6}$$

where $V^*(\xi, x_\kappa, t_F, t) = \partial U^*(\xi, x_\kappa, t_F, t)/\partial n(x_\kappa)$, and t_F is the current observation time. The variation of U and V over time requires the use of time interpolation functions, which may be constant, linear, quadratic, or even higher order. Moreover, since the time dependence of U and V is not known a priori, a time-marching procedure has to be introduced. The two available time-marching schemes (e.g. Wrobel and Brebbia [5]) were discussed in detail by Brebbia [1]. The first scheme treats the solution at the time t_{F-1} as a pseudo-initial condition for the solution at $t_F = t_{F-1} + \Delta t$, while in the second scheme all time integrations always start from the initial time $t = t_0$.

Discretization of equations (5) and (6) results in a system of linear equations of the type $\mathbf{A}\vec{Y} = \vec{R}$, where $\mathbf{A}$ is s fully populated $N \times N$ matrix, $\vec{Y}$ is the vector of unknown U's and V's, and $\vec{R}$ is the known right-hand side of the discretized equations. The time dependence, the complexity of determination of the elements of the matrix $\mathbf{A}$ (compounded by the elaborate time interpolation and time-marching schemes), and the fact that $\mathbf{A}$ is a fully populated matrix (unable to benefit from the wealth of fast matrix solvers available for the solution of banded matrices) results in a laborious and computer-time intensive numerical method which has limited the adoption BEM for the simulation of transient diffusion-type problems.

Laplace transforms have been used in the BEM to eliminate time dependency and simplify the solution of (1). The first numerical solution of the transient heat equation using the BEM was reported by Rizzo and Shippy [6], and involved a solution of the integral equation of heat conduction transformed via a Laplace transform. The method of Schapery [7] was used to numerically invert the transformed solution from the Laplace space into the original time space. This inversion method had several serious weaknesses. It was essentially a curve-fitting process, and as such presupposed knowledge of the expected solution. The Laplace space transform parameter s was arbitrarily chosen, and a poor choice resulted in unstable solutions or insufficient definition of the curve and reduced accuracy (e.g. Liggett and Liu [8]). The curve fitting scheme required the determination of N coefficients obtained from the computer-time consuming solution of a system of simultaneous equations. Moreover, this inversion method was impractical and time-consuming if the boundaries had a complex time history (e.g. Lachat and Combescure [9]). Because of these limitations, the Laplace transform-based BEM has found limited applicability in the solution of the diffusion equation until now.

THE LTBE NUMERICAL METHOD

A new numerical simulation technique, the Laplace Transform Boundary Element (LTBE) numerical method, is described in this paper. It belongs to a family of Laplace transform-based numerical methods introduced by Moridis and Reddell [10,11,12], and is not hampered by the limitations of the Schapery [7] inversion scheme. The method eliminates problems with stability and accuracy posed by the traditional treatment of the time derivative, and allows an unlimited Δt size without any loss of accuracy or stability because time is no longer considered. In the following sections the mathematical basis of LTBE is presented and the four steps involved in its application are described.

Step 1: The Laplace Transform of the PDE

Because of the properties of the Laplace transform, equation (1) becomes in the Laplace transform space

$$\nabla^2\Psi(x_\kappa, s) = \frac{1}{k}\left[s\Psi(x_\kappa, s) - U_0(x_\kappa, t_0)\right] \tag{7}$$

with the boundary conditions

$$\Psi(x_\kappa, s) = \overline{\Psi}(x_\kappa, s), \qquad x_\kappa \in \Gamma_1 , \tag{8}$$

$$\Theta(x_\kappa, s) = \frac{\partial\Psi(x_\kappa, s}{\partial n(x_\kappa)} = \overline{\Theta}(x_\kappa, s), \qquad x_\kappa \in \Gamma_2 , \tag{9}$$

where s is the Laplace space transform parameter,

$$\Psi(x_\kappa, s) = \mathcal{L}\{U(x_\kappa, t)\}, \qquad \Theta(x_\kappa, s) = \mathcal{L}\{V(x_\kappa, t)\} , \tag{10}$$

and $\mathcal{L}\{\}$ denotes the Laplace transform of the quantity in brackets. If sources or sinks of strength Q are included, then the term $\tilde{Q}(x_\kappa, s) = \mathcal{L}\{Q(x_\kappa, t)\}$ is added to the right-hand side of equation (7).

Step 2: The BEM in the Laplace Space

Ψ^* being an appropriate fundamental solution, the minimization of the weighted residuals using the BEM at all interior points ξ yields the statement

$$\begin{aligned}\Psi(\xi, s) =& k\int_\Gamma \Theta(x_\kappa, s)\, \Psi^*(\xi, x_\kappa, s)\, d\Gamma(x_\kappa) - k\int_\Gamma \Psi(x_\kappa, s)\, \Theta^*(\xi, x_\kappa, s)\, d\Gamma(x_\kappa) + \\ & \int_\Omega U_0(x_\kappa, t_0)\, \Psi^*(\xi, x_\kappa, s)\, d\Omega(x_\kappa) - \int_\Omega \tilde{Q}(x_\kappa, s)\, \Psi^*(\xi, x_\kappa, s)\, d\Omega(x_\kappa) ,\end{aligned} \tag{11}$$

where $\Theta^*(\xi, x_\kappa, s) = \partial\Psi^*(\xi, x_\kappa, s)/\partial n(x_\kappa)$. The fundamental solutions Ψ^* for the two- and the three-dimensional problem are given by Brebbia, Telles, and Wrobel [3] and Greenberg [13], and are respectively

$$\Psi^* = \frac{1}{2\pi k}\cdot K_0\left\{r\sqrt{s/k}\right\} \quad \text{and} \quad \Psi^* = \frac{(ks)^{1/4}}{r^{1/2}\cdot(2\pi k)^{3/2}}\cdot K_{1/2}\left\{r\sqrt{s/k}\right\} , \tag{12}$$

where K_m is the modified Bessel function of the second kind and of order m. The Ψ^* of the three-dimensional problem has a singularity as $r \to 0$, with a limiting form of $\Psi^* = 1/(4\pi kr)$ (e.g. Abramowitz and Stegun [14]). The Ψ^* of the two-dimensional problem is singular as $r\sqrt{s/k} \to 0$ [14], with a limiting form

$$lim\,\Psi^* = \frac{1}{2\pi k}\ln\frac{1}{r} - \frac{1}{4\pi k}\ln\frac{s}{k} \quad \text{for} \quad r\sqrt{s/k} \to 0 . \tag{13}$$

For the point ξ approaching Γ, the equation to be solved is

$$C(\xi)\ \Psi(\xi,s) = k\int_\Gamma \Theta(x_\kappa,s)\ \Psi^*(\xi,x_\kappa,s)\ d\Gamma(x_\kappa) - k\int_\Gamma \Psi(x_\kappa,s)\ \Theta^*(\xi,x_\kappa,s)\ d\Gamma(x_\kappa) + \int_\Omega U_0(x_\kappa,t_0)\ \Psi^*(\xi,x_\kappa,s)\ d\Omega(x_\kappa) - \int_\Omega \tilde{Q}(x_\kappa,s)\ \Psi^*(\xi,x_\kappa,s)\ d\Omega(x_\kappa)\ , \tag{14}$$

in which Θ^* is discontinuous as $\xi \to \Gamma$. Discretization of equation (14) on the boundary Γ yields the relationship between the node i at which the fundamental solution is applied and all the elements of the boundary, and has the form

$$C_i\ \Psi_i + \overbrace{\sum_{j=1}^{N}\int_j \Psi\Theta^* d\Gamma}^{\Delta_{1i}} - \overbrace{\sum_{j=1}^{N}\int_j \Psi^*\Theta d\Gamma}^{\Delta_{2i}} - \underbrace{\sum_{N_e}\left[\sum_{m=1}^{M} w_m\,(U_0\Psi^*)_m\right] A_e}_{B_i} + \underbrace{\sum_{N_e}\left[\sum_{m=1}^{M} w_m \left(\tilde{Q}\Psi^*\right)_m\right] A_e}_{T_i} = 0\ , \tag{15}$$

where Γ_j is the length of the element j, w_m are the integration weights, and N_e is the number of cells (of area A_e) into which Ω is subdivided. The quantities $(U_0\Psi^*)_m$ and $(\tilde{Q}\Psi^*)_m$ must be evaluated at the m integration points. The quantity B_i represents a domain integral resulting from the imposition of specific initial conditions. The quantity T_i represents the domain integral of source and sink singularities usually appearing in the interior of Ω, and allows an easy accounting for the effects of sources and sinks if a high definition of U and V in their vicinity is not needed. B_i and T_i do not introduce any more unknowns since both U_0 and Q are prescribed. If U_0 satisfies Laplace's equation or is harmonic in Ω, the domain integral of U_0 can be transformed into equivalent boundary integrals [10], and integration over the N_e cells of Ω to obtain B_i is avoided.

If constant boundary elements are employed, $C_i \equiv \frac{1}{2}$, and Ψ and Θ are constant within each element. Then Δ_{1i} and Δ_{2i} become

$$\Delta_{1i} = \sum_{j=1}^{N} \hat{H}_{i,j}\Psi_j, \quad \Delta_{2i} = \sum_{j=1}^{N} G_{i,j}\Theta_j, \text{ where } \hat{H} = \int_\Gamma \Psi^* d\Gamma \text{ and } G = \int_\Gamma \Theta^* d\Gamma\ . \tag{16}$$

Substituting in equation (15), we obtain

$$\sum_{j=1}^{N} H_{i,j}\ \Psi_j - \sum_{j=1}^{N} G_{i,j}\ \Theta_j - B_i + T_i = 0\ , \tag{17}$$

where $H_{ij} = \hat{H}_{i,j}$ if $i \neq j$ and $H_{ij} = \hat{H}_{i,j} + \frac{1}{2}$ if $i = j$. Caution must be used in the calculation of H_{ij} and $G_{i,j}$ in the vicinity of singularities (usually as $r \to 0$), in which case higher order integration schemes need to be used, e.g. Brebbia, Telles, and Wrobel [3], Ligget and Liu [2]. Collecting and rearranging terms, we arrive at the matrix equation

$$\mathbf{H}\vec{\Psi} - \mathbf{G}\vec{\Theta} = \vec{R}, \qquad \vec{R} = \vec{B} - \vec{T}\ . \tag{18}$$

The total number of unknown Ψ's and Θ's in equation (18) is N. Separating and reordering knowns and unknowns yields the final BEM system of simultaneous linear equations described by the matrix equation

$$\mathbf{A}\vec{Y} = \vec{R} \,, \tag{19}$$

in which $\vec{Y}$ is the vector of unknown Ψ's and Θ's, $\vec{R}$ is the vector of the known right-hand side, and $\mathbf{A}$ is the fully populated $N \times N$ coefficient matrix. The use of constant elements in our analysis does not affect the generality of the LTBE method. If linear or higher order elements are used, the resulting final equations are invariably of the type of equations (17), (18) and (19), and all subsequent procedures apply.

Step 3: The Solution in the Laplace Space

The computation of $\mathbf{A}$ and $\vec{R}$ necessitates arithmetic values of the s parameter of the Laplace space. For a desired observation time t they are provided by the first part of Stehfest's [15] algorithm as

$$s_\nu = \frac{\ln 2}{t} \cdot \nu \,, \quad \nu = 1, \ldots, N_S \,, \tag{20}$$

where N_S is the number of summation terms in the algorithm and N_S is an even number. Optimum values for N_S are discussed in a following section. The solution of (20) returns a set of N_S vectors of the unknown $\vec{Y}$'s as

$$\vec{Y}_\nu = \vec{Y}(s_\nu) = [\mathbf{A}(s_\nu)]^{-1} \vec{R}(s_\nu) \,, \quad \nu = 1, \ldots, N_S \,. \tag{21}$$

To obtain a solution at a time t, all vectors $\vec{Y}_\nu$, $\nu = 1, \ldots, N_S$ are needed, i.e. the system of simultaneous equations has to be solved N_S times. Once $\vec{Y}_\nu$ is known, the value of $[\Psi(s_\nu)]_i$ at any interior point i can be determined from the discretized integral relationship between i and the boundary values of $\Psi(s_\nu)$ and $\Theta(s_\nu)$ as

$$[\Psi(s_\nu)]_i = \sum_{j=1}^{N} G_{i,j} \; [\Theta(s_\nu)]_j - \sum_{j=1}^{N} \hat{H}_{i,j} \; [\Psi(s_\nu)]_j + B_i - T_i \,. \tag{22}$$

Step 4: The Numerical Inversion of the Laplace Solution

The unknown U's at any boundary or interior point i and at any time t are obtained by using the Stehfest [15, 16] algorithm to numerically invert the Laplace space solutions $\Psi_i(s_\nu)$. The procedure is described by the following equations:

$$[U(t)]_i = \frac{\ln 2}{t} \sum_{\nu=1}^{N_S} W_\nu \cdot [\Psi(s_\nu)]_i \,, \tag{23}$$

and

$$W_\nu = (-1)^{\frac{N_S}{2}+\nu} \sum_{\kappa=\frac{1}{2}(\nu+1)}^{\min\{\nu,\frac{N_S}{2}\}} \frac{\kappa^{\frac{N_S}{2}} (2\kappa)!}{(\frac{N_S}{2}-\kappa)!\kappa!(\kappa-1)!(\nu-\kappa)!(2\kappa-\nu)!} \,. \tag{24}$$

Although the accuracy of the method is theoretically expected to improve with increasing N_S, Stehfest [15] showed that with increasing N_S the number of correct significant figures increases linearly at first and then, due to roundoff errors, decreases linearly. Testing his algorithm against 50 equations with known inverse Laplace transforms, he determined that the optimum N_S was 10 for single precision variables (8 significant figures) and 18 for double precision variables (16 significant figures). However, our experience indicated no significant differences in the performance of LTBE for N_S between 6 and 20.

The solution in the Laplace space eliminates stability and accuracy problems caused by the treatment of the time derivative in standard BEM simulators, thus allowing an unlimited time-step size. The truncation error of the method is limited to the truncation error caused by the space discretization because the domain is not discretized in time, and provides a solution inherently more accurate than the standard BEM method for the same grid system. The ability to use an unlimited time-step size bounds the accumulation of roundoff error by an upper limit defined as the roundoff error accumulated after the N_S solutions required by the method. Thus, LTBE offers a stable, non-increasing roundoff error irrespective of the time of observation t_{obs} because calculations are performed at one time only by letting $\Delta t = t_{obs}$. Calculations in the standard BEM method have to be performed at all the intermediate times of the discretized time domain, continuously accumulating roundoff error in the process.

VERIFICATION AND TEST CASES

The LTBE numerical method was tested in three diffusion-type problems of heat conduction and groundwater flow which represented increasing levels of complexity. Analytical solutions existed for the first two test problems, and the LTBE solutions were verified through comparison to the analytical solutions. No analytical solution existed for the third test case. In all three cases the results obtained from the implementation of LTBE were tested against results obtained from a standard BEM simulator for the same space discretization. A direct matrix solver was used to solve the system of simultaneous equations arising in the LTBE and BEM methods. Linear space interpolation functions were used in all cases. The domain integral B_i in equation (15) was computed from its equivalent boundary integrals in the first two cases because the initial conditions satisfied Laplace's equation. In the BEM simulator the first time-marching scheme with a constant time interpolation was employed. Double precision variables with 20 significant figures were used in all simulations.

Verification & Test Case 1

Test case 1 involved the flow of heat in the finite rectangle $-a < x < a$, $-b < y < b$ with a unit initial temperature and boundaries kept at a zero temperature. The temperature distribution in the rectangle was provided by the analytical solution of Carslaw and Jaeger [17] as

$$U(x,y) = \frac{16}{\pi^2} \sum_{n=0}^{\infty} \sum_{m=0}^{\infty} L_{n,m} \cos\frac{(2n+1)\pi x}{2a} \cos\frac{(2m+1)\pi y}{2b} \exp\left(-D_{n,m} \cdot t\right) , \qquad (25)$$

where

$$L_{n,m} = \frac{(-1)^{n+m}}{(2n+1)(2m+1)} \quad \text{and} \quad D_{n,m} = \frac{k\pi^2}{4}\left[\frac{(2n+1)^2}{a^2} + \frac{(2n+1)^2}{a^2}\right] . \qquad (26)$$

The quantity k in equation (26) corresponds to the term k in equation (1), and is the thermal diffusivity of the substance with units of $[L^2][T^{-1}]$. Using equation (25) the temperature distribution was calculated along the x axis at $y = 0.025\ m$ in a glass square of size $a \times b = 0.2\ m \times 0.2\ m$. An extremely low thermal diffusivity of $k = 5.8 \times 10^{-7}\ m^2/sec$, resulting in a steep solution surface, was used to create adverse numerical conditions. For the LTBE and the BEM simulators the boundary was divided in 40 equally-sized linear elements, resulting in a total of $N = 40$ nodes. A single observation at $t_{obs} = 9000$ *sec* was made. A $N_S = 6$ was used in the LTBE simulator, while 1, 100, 1000, 5000, and 9000 equally-sized time steps were used in the BEM simulator to cover t_{obs}.

The temperature distribution results for both the analytical and the numerical solutions (LTBE and BEM) are presented in Figure 1. The analytical solution and the LTBE solution practically coincided. The BEM solution tended towards the LTBE and the analytical solutions with an increasing number of time-steps corresponding to smaller Δt's, and the three solutions coincided when 9000 time steps were used. In essence this means that to obtain an accurate solution, the system of linear equations of equation (19) had to be solved 9000 times when using BEM, as opposed to 6 times for the LTBE solution. This corresponds to an acceleration by a factor of 1500. This spectacular result is due to the extremely low value of k. In other problems such as the flow of groundwater, the high speed of propagation of the pressure wave results in smooth solution surfaces, and more modest accelarations are expected.

The effect of N_S on the accuracy of the LTBE scheme is shown in Figure 2. The difference between the analytical and the LTBE solutions was negligible for an N_S ranging between 6 and 20. The minimum difference was observed for $N_S = 6$, and increased for a $N_S > 6$. The solutions for an N_S between 8 and 20 were virtually identical and exhibited arithmetic differences beyond the 12th decimal place. The implications of the results in Figure 2 are that a) the accuracy of LTBE for this two-dimensional problem is practically insensitive to the value of N_S, and b) the number of summation terms N_S for an accurate solution may be as small as 6, far smaller than the $N_S = 18$ which Stehfest [15] suggested for double precision variables. This drastically reduces the execution time and makes the LTBE method even more efficient than expected.

<u>Verification & Test Case 2</u>
Test case 2 involved the flow of heat in the rectangular parallelepiped $-a < x < a$, $-b < y < b$, $-c < z < c$ with a unit initial temperature and boundaries kept at a zero temperature. The analytical solution of Carslaw and Jaeger [17] provides the temperature distribution in the parallelpiped as

$$U(x,y,z) = \frac{64}{\pi^3}\sum_{n=0}^{\infty}\sum_{m=0}^{\infty}\sum_{\ell=0}^{\infty}\frac{(-1)^{n+m+\ell}}{(2n+1)(2m+1)(2\ell+1)}\cdot\exp\left(-D_{n,m,\ell}\cdot t\right)\times \times\cos\frac{(2n+1)\pi x}{2a}\cos\frac{(2m+1)\pi y}{2b}\cos\frac{(2\ell+1)\pi z}{2c}, \tag{27}$$

where

$$D_{n,m,\ell} = \frac{k\pi^2}{4}\left[\frac{(2n+1)^2}{a^2}+\frac{(2m+1)^2}{b^2}+\frac{(2\ell+1)^2}{c^2}\right]. \tag{28}$$

Using equation (27) the temperature distribution was calculated along the x axis at $y = 0.025\ m$ and $z = 0.08\ m$ in a glass cube of size $2a \times 2b \times 2c = 0.4\ m \times 0.4\ m \times 0.4\ m$ and with a $k = 5.8 \times 10^{-7}\ m^2/sec$. For the LTBE and the BEM simulators the boundary was divided in 250 equally-sized triangular elements, resulting in a total of $N = 152$ nodes. A single observation was made at $t_{obs} = 9000$ *sec*. A $N_S = 6$ was used in the LTBE simulator, while 1, 100, 1000, 5000, and 9000 equally-sized time steps were used in the BEM simulator to cover t_{obs}.

The temperature distribution results for both the analytical and the numerical solutions (LTBE and BEM) are presented in Figures 3. The analytical solution and the

LTBE solution practically coincided. The BEM solution tended towards the LTBE and the analytical solutions with an increasing number of time-steps corresponding to smaller Δt's. When 9000 time steps were used in BEM, the three solutions coincided. As in case 1, we observed a dramatic reduction in the number of times equation (19) had to be solved for an accurate solution.

Test Case 3
Test case 3 was an extremely anisotropic and inhomogeneous two-dimensional cartesian (x, y) system of flow to two wells. The geometry, boundaries, and properties of the aquifer, as well as the location and pumping rates of the two wells, are shown in Figure 4. Initial and boundary pressures were equal to 6.0×10^5 Pa, and the specific storativity $S_S = 2.04 \times 10^{-10}$ Pa^{-1}. The aquifer had a uniform thickness $\Delta z = 50$ m. No analytical solution is possible for this problem. The formulation of the equations followed the procedure described by Liggett and Liu [2]. The domain was subdivided into 7 subregions, and stretched coordinates were used to describe the anisotropy of the problem.

The boundary of the solution domain was discretized into 186 unequally-sized linear elements, with $N = 174$. The interior domain was subdivided into 724 triangular elements for the computation of B_i and T_i. A single observation was made at $t_{obs} = 2$ $days$. The LTBE method was applied with $N_S = 6$, and 1, 2, 6, 12, 24, and 48 equally-sized time steps were used in the BEM simulation. The LTBE method was evaluated by comparing the distribution of the drawdown and of the percent drawdown difference between the LTBE and the BEM solutions along the x axis at $y = 2175$ m and passing through the well at $(x, y) = (2400$ $m, 2175$ $m)$. The distribution of drawdown and percent drawdown difference are shown in Figures 5(a) and 5(b) respectively. The capability of LTBE and the inability of BEM – caused by the averaging effect of the treatment of the time derivetive for larger Δt's – to accurately describe the effects of the presence of wells and zones of significantly different permeability are reflected in the difference between the two solutions. The location of wells and permeability zones can be identified by the existence of peaks and sharp variations in drawdown, variations which decrease in magnitude with a decreasing Δt size in the BEM solution. The power of the LTBE method is demonstrated in Figure 5(b). LTBE produced an accurate solution, a fact indicated by the realization that the BEM solution for a decreasing Δt size tended to the LTBE solution.

Energy and Mass Balance Error Considerations
A very important measure of the validity and accuracy of the LTBE numerical method was provided by the determination of the energy and mass balance error E_L, which was 3.862×10^{-6}%, 6.772×10^{-5}%, and 6.433×10^{-4}% respectively of the original energy (test cases 1 and 2) and fluid mass (test case 3). In all cases a $N_S = 6$ was used. The extremely small magnitude of these errors further testified to the power of the method. The relative mass balance error E_{RM} of the standard BEM method vs. the LTBE method was computed as

$$E_{RM} = \frac{E_{BEM}}{E_L} \quad . \tag{29}$$

E_{BEM} is the mass balance errors (as a perecentage of the inital fluid mass) when the BEM is implemented, and E_L is the mass balance error for LTBE discussed in the previous paragraph. As Figure 6 depicts, for the same domain discretization the LTBE scheme was in all cases consistently more accurate than the BEM scheme. This was expected because of a) the inability of the traditional BEM to handle large time steps, b) a smaller truncation error, and c) a smaller roundoff error due to fewer operations.

The effect of the number of summation terms N_S on accuracy is demonstrated in Figure 7. The relative mass balance error E_{R,N_s} was computed as

$$E_{R,N_s} = \frac{E_{L,N_s}}{E_L} , \tag{30}$$

where E_{L,N_s} is the error calculated with a variable N_S. It can be seen that the accuracy of the LTBE method, indicated by decreasing values of E_{R,N_s}, was either virtually constant or improved as N_S increased between 6 and 20. An improvement was observed in the most inhomogeneous and anisotropic case, i.e. case 3. The greatest accuracy was observed for N_S values between 16 and 20. However, the improvement over the range between $N_S = 6$ and $N_S = 20$ was marginal, and in no case exceeded 8%. This indicates that a) for $6 \leq N_S \leq 20$, the accuracy of LTBE is practically insensitive to the value of N_S, and b) the number of summation terms N_S for an accurate solution may be far fewer than the $N_S = 18$ which Stehfest [17] suggested. Therefore, a value of $N_S = 6$ or 8 is suggested for LTBE simulations, and drastically reduces the execution time requirements. The accuracy of the LTBE method may deteriorate for a $N_S > 20$ due to limits in computer accuracy when calculating the weighing factors W_ν in equation (24).

SUMMARY AND DISCUSSION

A new numerical method, the Laplace Transform Boundary Element (LTBE) method, was developed to eliminate the adverse effects of the treatment of the time derivative in the numerical approximation of the parabolic Partial Differential Equation (PDE) of transient, diffusion-type problems. The LTBE method is insensitive to the time-step size and consists of four steps: 1) A Laplace transform is performed on the PDE, 2) the transformed PDE is approximated using a BEM, 3) the resulting system of simultaneous equations is solved and the transformed vector of the unknowns is determined in the Laplace space, and 4) a numerical inversion is performed on the traansformed vector of unknowns obtained in step 3 using the Stehfest algorithm. The solution in the Laplace space renders the time-marching and time-interpolation schemes used in the BEM irrelevant because time is no longer a consideration.

Three test cases were investigated to evaluate the LTBE method. Due to its formulation, LTBE requires a solution of the resulting system of simultaneous equations N_S times, one for each of the N_S different approximations of the Laplace space variable s. Combination of the resulting N_S solutions using Stehfest's algorithm returns the actual solution in time. Compared to the standard BEM method, LTBE does not increase the storage requirement because a) the values of the unknowns at the previous time step are not needed, and b) the N_S sets of unknowns can be stored and summed in a single array. In the BEM this array is that of the unknowns at the end of the previous time-step.

It was determined that a $N_S = 6$ is sufficient to provide an extremely accurate solution. This means that the solution at any time is obtained by solving the system of simultaneous equations in LTBE 6 times and algebraically combining the solutions. Although the accuracy increases with increasing N_S for $6 \leq N_S \leq 20$, the improvement for $N_S > 6$ is marginal and insufficient to justify the additional execution time. With a smaller time-step size and more time-steps, the BEM solution tends to the LTBE solution. LTBE provides a solution consistently more accurate than the BEM solution for the same space discretization because the truncation error is reduced to the truncation error of the space discretization, and the roundoff error was reduced due to the limited number of operations needed. An unlimited time-step size with a stable, non-increasing error is thus possible. The disadvantage of having to solve the system of simultaneous equations

6 times for a single time-step is outweighed by a) an unlimited time-step size without any loss of accuracy, b) a superior accuracy, and c) a stable, non-increasing roundoff error. Therefore, calculations in a LTBE scheme are necessary only at the desired observation times, thus allowing 'snapshots' in time. On the other hand, in a standard BEM method calculations are needed at all the intermediate times of the discretized time domain.

REFERENCES

1. Brebbia, C.A. (Ed.). Topics in Boundary Element Research, Vol. 1, Basic Principles and Applications, Springer-Verlag, Berlin and New York, 1984.
2. Liggett, J.A. and Liu, P.L.-F. The Boundary Integral Equation Method for Porous Media Flow, George Allen Unwin, 1983.
3. Brebbia, C.A., Telles, J.C.F. and Wrobel, L.C. Boundary Element Techniques, Spinger-Verlag, Berlin and New York, 1984.
4. Curran, D.A.S., Cross, M. and Lewis, B.A. Solution of Parabolic Differential Equations by the Boundary Element Method Using Discretization in Time., Appl. Math. Modelling, Vol. 4, pp. 398-400, 1980.
5. Wrobel., L.C. and Brebbia, C.A. Time Dependent Potential Problems. Chapter 6, Progress in Boundary Elements Methods, (Ed. Brebbia, C.A.), Vol. 1, Pentech Press, London, 1981.
6. Rizzo, F.J. and Shippy, D.H. A Method of Solution for Certain Problems of Transient Heat Conduction, AIAA J., Vol. 8, 11, pp. 2004-2009, 1970.
7. Schapery, R. A. Approximate Methods of Transform Inversion for Visco-elastic Stress Analysis, Proc. Fourth U.S. National Congress on Applied Mechanics, Vol. 2, 1962.
8. Liggett, J.A. and Liu, P.L.-F. Unsteady Flow in Confined Aquifers: A Comparison of Two Boundary Integral Methods, Water Resources Research, Vol. 15, pp. 861-866, 1979.
9. Lachat, J.C. and Combescure, A. Laplace Transform and Boundary Integral Equation: Application to Transient Heat Conduction Problems, Innovative Numerical Analysis in Applied Engineering Science (Eds. Cruse, T.A. et al.), CETIM,Versailles, 1977.
10. Moridis, G.J. and Reddell, D.L. The Laplace Transform Finite Difference (LTFD) Method for Simulation of Flow Through Porous Media, Water Resources Research (in press), 1991.
11. Moridis, G.J. and Reddell, D.L. The Laplace Transform Finite Difference (LTFD) Numerical Method for the Simulation of Solute Transport in Groundwater, paper No. H42A-7, 1990 AGU Fall Meeting, San Fransisco, Dec. 3-7, EOS Trans. of the AGU, 71(43), 1990.
12. Moridis, G.J. and Reddell, D.L. The Laplace Transform Finite Element (LTBE) Numerical Method for the Solution of the Groundwater Equation, paper H22C-4, AGU 91 Spring Meeting, Baltimore, May 28-31, 1991, EOS Trans. of the AGU, 72(17), 1991.
13. Greenberg, M.D. Applications of Green's Functions in Science and Engineering, Prentice Hall, Englewood Cliffs, 1971.
14. Abramowitz, M. and Stegun, I.A. (Eds.). Handbook of Mathematical Functions, Dover, New York, 1965.
15. Stehfest, H. Numerical Inversion of Laplace Transforms, J. ACM, Vol. 13, pp. 47-49, 1970.
16. Stehfest, H. Numerical Inversion of Laplace Transforms, J. ACM, Vol. 13, pp. 56, 1970.
17. Carslaw, H.C. and Jaeger, J.C. Conduction of Heat in Solids, Oxford University Press, London, 1959.

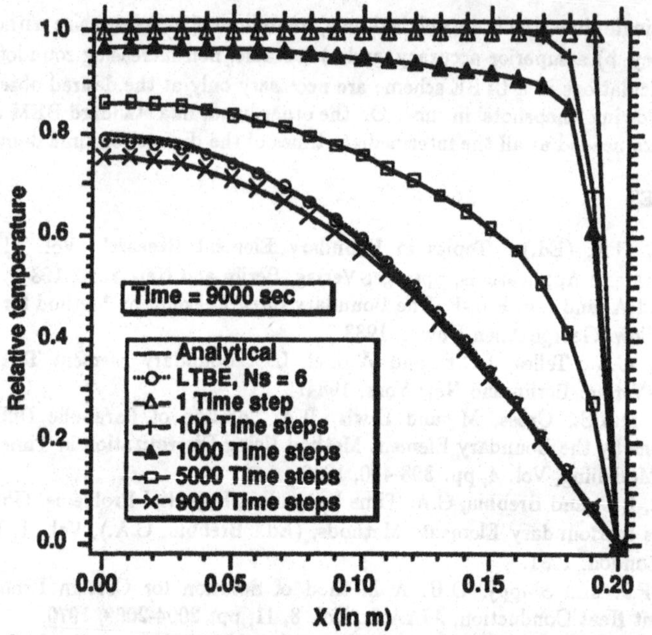

Fig. 1. Comparison of the LTBE solution for Test Case 1 at t = 9000 *secs* to 1) the analytical solution, and 2) the BEM solutions for various numbers of time-steps.

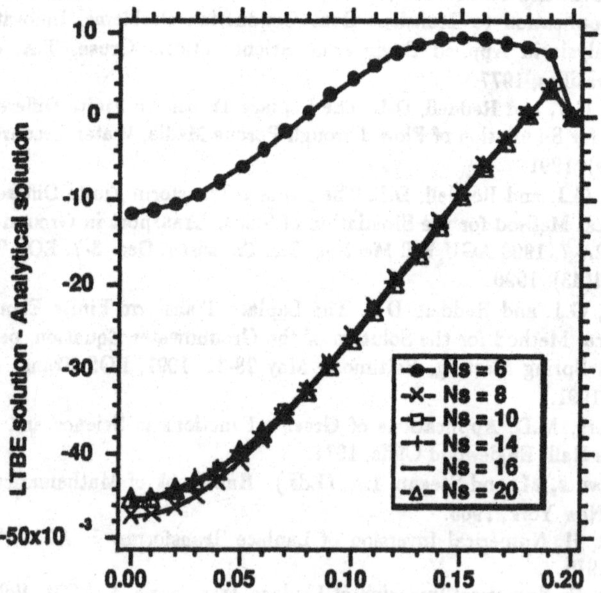

Fig. 2. Effect of N_S on the accuracy of the LTBE method in Test Case 1.

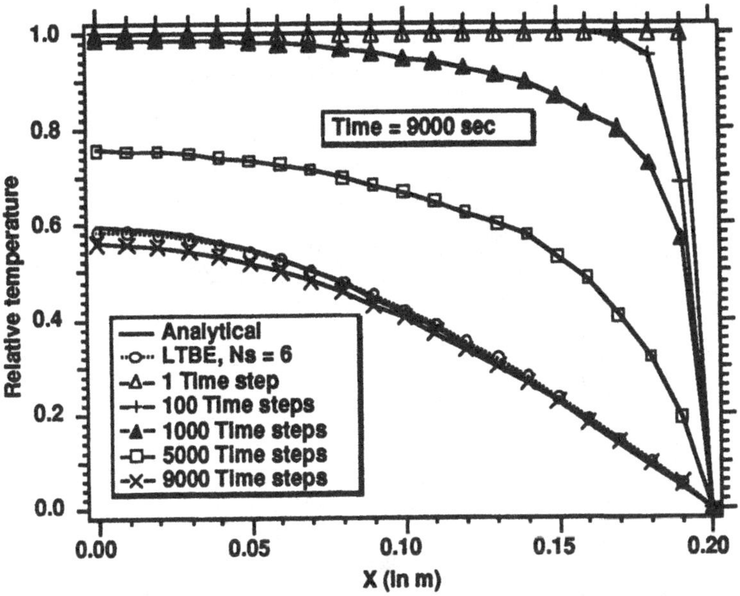

Fig. 3. Comparison of the LTBE solution for Test Case 2 at t = 9000 *secs* to 1) the analytical solution, and 2) the BEM solutions for various numbers of time-steps.

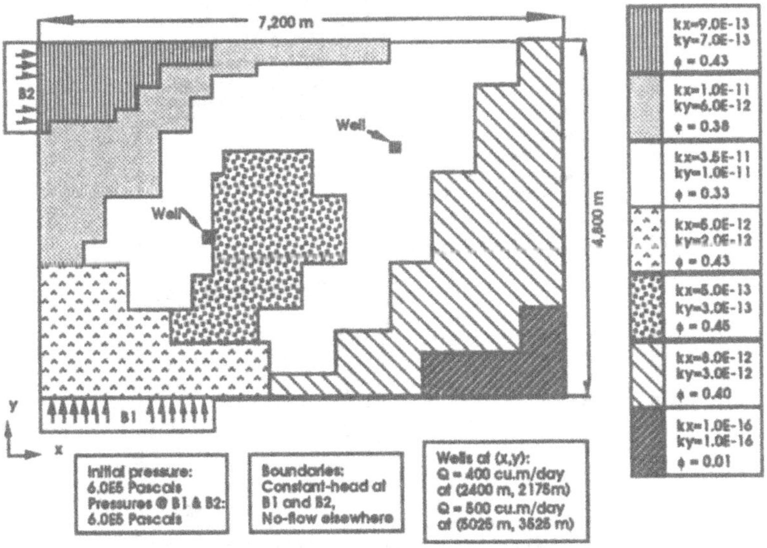

Fig. 4. Geometry, boundaries, and properties of the aquifer in Test Case 3. k_x and k_y are in m^2.

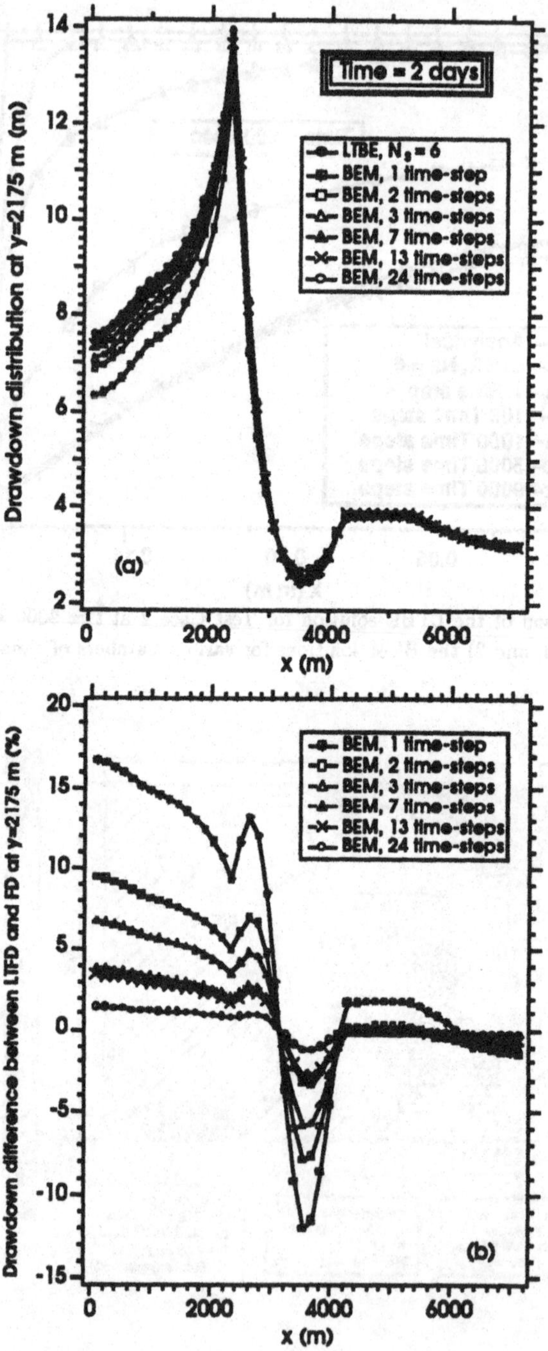

Fig. 5. Comparison of the LTBE solution for Test Case 3 at $t = 2$ *days* to the BEM solutions for various numbers of time-steps: (*a*) drawdown distribution, and (*b*) % drawdown difference distribution along x at $y = 2175$ *m*.

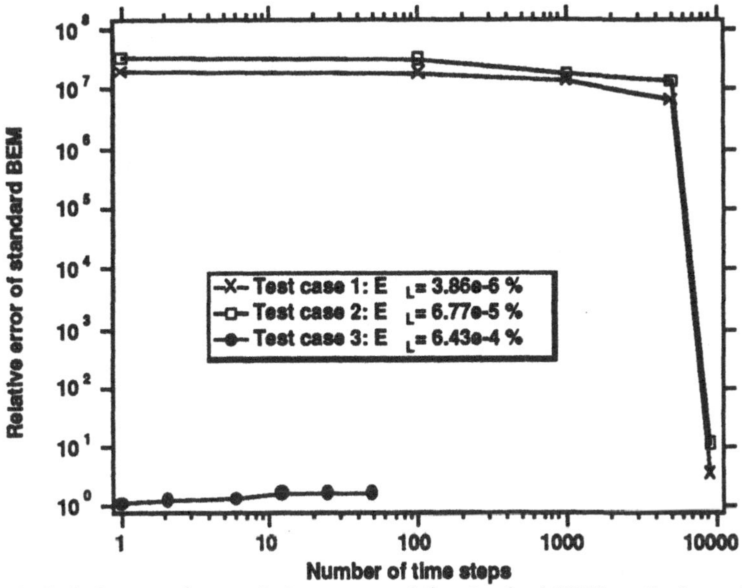

Fig. 6. Relative mass/energy balance error of the standard BEM method as a fraction of the error of the LTBE method.

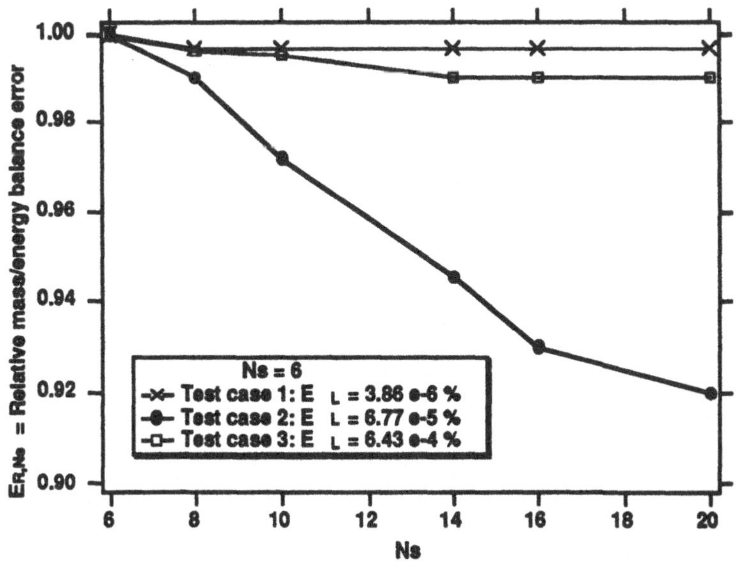

Fig. 7. Relative mass/energy balance error of the LTBE method for a varying N_S as a fraction of the error of the LTBE method for a $N_S = 6$.

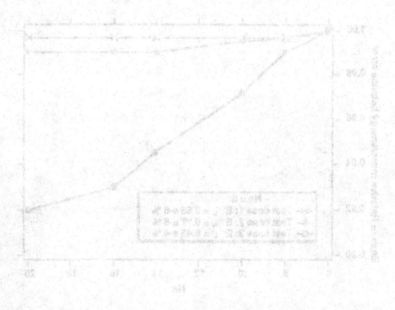

Solution of Variable Coefficients PDEs by Means of BEM and Perturbation Technique

R. Rangogni

ENEL-DSR-CRIS, Via Ornato 90/14,
20159 Milan, Italy

ABSTRACT

Numerical solution of variable coefficients PDEs by the BEM is quite difficult as the Green function is unknown and not general. In the paper a procedure is presented that permits the solution of some types of non-constant coefficients PDEs using the standard Green function used for the Laplace equation. This is achieved coupling BEM and perturbation technique. After stating the procedure two examples of application are numerically solved showing the efficiency and accuracy of the proposed technique.

INTRODUCTION

The Boundary Element Method (BEM) is, nowadays, a well established and powerful numerical tool to solve Partial Differential Equations (PDEs) of applied sciences (e.g. [1,2]). Amongst the most distinctive features of BEM we recall that it only has unknowns on the boundary of the computational domain; it shows high accuracy whether for the primary function or for its derivatives. But, unfortunately, the application range of BEM is generally limited to PDEs with constant coefficients as for non-constant PDEs the associated fundamental solution (Green function) is, in general, unknown, and also if it would be known it would change from one equation to another one. However some attempts have been made to overcome this drawbacks (see i.e. [3,4,5]). Here the procedure presented in [4], for the particular case of the generalized Laplace equation, is extended to more general second order elliptic PDEs with non-constant coefficients.

The procedure consists in coupling BEM and perturbation technique. This allows to transform the original equation into a sequence of Poisson's equations that are solved using standard BEM. Each of these problems contains only boundary unknowns leaving the advantages of BEM. It is not claimed that any problem can be solved in this way and further researches are required to unsure under which conditions convergence is achieved. However most of the real problems of engineering interest behave so well that convergence is attained.

The final aim of this work is to apply this technique to non-linear equations such as those arising in unsaturated groundwater flow problems and also to parabolic equations.

In the sequel, after stating the equation and outlining the procedure two examples of application are presented. The first problem is a theoretical one with known analytical solution while the second one is extracted from [5] and it concerns a lubrication problem. In both cases convergence is attained quickly and results are quite accurate.

MATHEMATICAL STATEMENT

We consider general second order elliptic PDEs that can be written in the following form:

$$\nabla^2 \phi + f(x,y)\, \phi_x + g(x,y)\, \phi_y = h(x,y) \qquad \text{on } D \quad (1)$$

where ϕ is the unknown function, ∇^2 is the laplacian operator

($\nabla^2 = \phi_{xx} + \phi_{yy}$), ϕ_u denotes the partial derivative of ϕ with respect to the indipendent variable u (u = x or u = y) and f,g,h are known functions. Associated to equation (1) are boundary conditions of Dirichlet ($\phi = \bar{\phi}$ on S_1) or Neumann type ($\partial\phi/\partial n = H$ on S_2, $S_1 \cup S_2 = S$ boundary of domain D) or other in order to define a well posed problem. All functions appearing in (1) are well behaved in the definition domain D of interest.

Equation (1) could be solved, directly, by BEM if the Fundamental Solution (FS) (Green function) would be known but also in this case one should change the computer code every time whenever functions f,g,h are changed. So here a different approach is adopted and instead of solving directly equation (1) we write the same equation in the form:

$$\nabla^2\phi + \epsilon\,[f\,\phi_x + g\,\phi_y] = h \qquad 0 \leq \epsilon \leq 1 \tag{2}$$

It is clear that for $\epsilon = 1$ we have again equation (1) while for $\epsilon = 0$ a Poisson equation is obtained.

Then we search for a solution ϕ in a series of powers of ϵ in the following form:

$$\phi = \phi_0 + \epsilon\,\phi_1 + \epsilon^2\,\phi_2 + \dots\dots\dots \tag{3}$$

Substituting function (3) and its partial derivatives into equation (2) and ordering according to the powers of ϵ gives:

$$\nabla^2\phi_0 + \epsilon\,[\nabla^2\phi_1 + f\phi_{0x} + g\phi_{0y}] + \epsilon^2\,[\nabla^2\phi_2 + f\phi_{1x}$$

$$+ g\phi_{1y}] + \dots\dots\dots = h \tag{4}$$

As ϵ is arbitrarily chosen in the interval [0,1] equation (4) is always satisfied if the following sequence of problems are identically satisfied (boundary conditions are supposed satisfied by the first term ϕ_0):

$$P_0 \quad \begin{cases} \nabla^2 \phi_0 = h(x,y) & \text{on } D \\ \phi_0 = \bar{\phi} \quad \text{on } S_1 & S_1 \cup S_2 = S \\ \dfrac{\partial \phi_0}{\partial n} = H & \text{on } S_2 \end{cases} \tag{5}$$

$$P_1 \quad \begin{cases} \nabla^2 \phi_1 = -f\phi_{0x} - g\phi_{0y} & \text{on } D \\ \phi_1 = 0 & \text{on } S_1 \\ \dfrac{\partial \phi_1}{\partial n} = 0 & \text{on } S_2 \end{cases} \tag{6}$$

....

It can be seen that each problem P_i ($i \geq 0$) is a Poisson problem so the numerical treatment is the same for each of them (only P_0 can be a Laplace problem when $h \equiv 0$). So it is possible to write any problem P_i in the form:

$$\nabla^2 \theta = F(x,y) \tag{7}$$

together with its boundary conditions. The important thing to note is that FS of each problem is the well known Green function for the Laplace equation

$$G = \log(r) \qquad r = |P - Q|, \qquad P,Q \in D \tag{8}$$

Using the second Green identity, it is possible to transform equation (7) in the following integral form:

$$\alpha\, \theta(P) = \int_S \left(G \frac{\partial \theta}{\partial n} - \theta \frac{\partial G}{\partial n}\right) ds + \int_D GF\, dD \tag{9}$$

where n denotes the inward normal vector on S and α is the angle in P described by the vector r when Q runs all over the boundary S.

Each problem P_i ($i > 0$) can be solved, in this way, when function ϕ_{i-1} is known.

The integral equation (9) is discretized, as usual, in BEM technique [1,2], subdividing boundary S into subelements S_k ($k=1,2....,N$) and approximating functions θ and $\partial\theta/\partial n$ on each element S_k by polynomials. The domain integral is computed by dividing domain D into subdomains D_k ($k=1,2,.....,M$) and approximating function F by suitable shape functions. For example, in the numerical problems that will be solved below all functions are chosen to be of parabolic shape. Without entering into details, the discretized form of equation (9) is:

$$\alpha\,\theta_i = \sum_{k=1}^{N} (\alpha_{ik} q_k - \beta_{ik}\theta_k) + \delta_i \,, \quad q_k = \frac{\partial\theta}{\partial n} \qquad (10)$$

It is worth noting that the domain integral δ_i does not contain the unknown function θ, so the number of unknowns is limited by the N boundary points.

When a sufficient number of problems P_j are solved the solution ϕ is obtained by writing expression (3) with $\epsilon = 1$ and, in practice, it turns out to be enough to compute only a few terms of the series. One can chose to stop the sequence when the contribution of the last term added is less than a small quantity fixed by the user.

NUMERICAL EXAMPLES.

In this section two problems are solved numerically using the procedure presented in the previous paragraph in order to show its performance.

The first one admits of an analytical solution while the second is extracted from [5].

Problem 1. It is required to solve the following problem:

$$\begin{cases} \nabla^2 \phi + \dfrac{2xy^2}{1+x^2y^2}\phi_x + \dfrac{2x^2y}{1+x^2y^2}\phi_y = \dfrac{1+2x^2y^2}{1+x^2y^2}6(x+y) \ , \text{ on } D \\ \phi = x^3 + y^3 \qquad \text{on } S \end{cases} \qquad (11)$$

where $D\equiv[(x,y) : 0 \le x \le 2, 0 \le y \le 2]$. The analytical solution of problem (11) is $\bar{\phi} = x^3 + y^3$. To check the numerical results we define two norms:

$$e_\infty = \text{Max} \mid \phi_i - \bar{\phi}_i \mid \quad i=1,2,\ldots,N+M$$

$$e_2 = [\sum_{i=1}^{N+M} (\phi_i - \bar{\phi}_i)^2 / (N+M)]^{1/2} \qquad (12)$$

Boundary S of domain D is discretized by 20 second order elements (40 boundary points) while domain D is subdivided by 25 internal equal square elements (56 internal points). After computing 5 terms of series (3) the following errors are obtained:

$e_\infty = 1.48 \ 10^{-2}$
$e_2 = 3.78 \ 10^{-3}$

It is worth noting that solution function $\bar{\phi}$ ranges from 0 to 16 so if we refer to an average value of 8 it means that the results are affected by an average error of:

e_∞ (%) = 1.4 % , e_2 (%) = 0.05 %.

It is also interesting to see how this errors change adding new terms of the series; this is shown in fig.1.

From fig. 1 it is quite evident how fast the numerical solution approaches the exact one. It is also to be noted that in this case the exact solution is a cubic function while all the approximations are of parabolic shape but nevertheless the convergence is fast. Besides the internal discretization is not the better one as it is expected, looking at the boundary imposed values, a greater variation near the point P=(2,2) so a better discretization could have been done.

Problem 2. This example concerns a real problem of lubrication and it is found in [5] where it is solved in an efficient way transforming the original equation into another more suitable for a BEM treatment and then imposing to a function (describing the film shape) to satisfy a partial differential equation in order to reduce the transformed equation to a constant coefficient one. Although from a practical point of view the results are acceptable and all the work is very good, from a mathematical point of view it suffers from the disadvantage of imposing particular functions to obtain a solution and also one has to make some transformation after getting the numerical solution. Instead the technique here presented solves directly the starting equation without choosing particular film shape.

The steady state incompressible lubrication problem for sliding bearings is governed by the Reynolds equation (see e.eg. [6]):

$$\frac{\partial}{\partial x}\left(h^3 \frac{\partial p}{\partial x}\right) + \frac{\partial}{\partial y}\left(h^3 \frac{\partial p}{\partial y}\right) = 6\mu U \frac{\partial h}{\partial x} \qquad (15)$$

where $p = p(x,y)$ is the lubrication pressure, $h = h(x,y) > 0$ is the film thickness, μ is the viscosity (considered constant throughout the bearing) and U is the velocity (assumed parallel to the x axis) of the driving surface. The boundary conditions are $p = 0$ all over the boundary of the domain D shown in fig. 2 together with the boundary and internal discretization adopted.

Equation (15) is certainly of type (1) so it is possible to apply the technique previously presented. Only 3 terms of the series (3) was

required to obtain accurate results. We compare our numerical results versus the ones extracted in [5] along the two lines y = 0.375 and y = 0.5625; the comparison is shown in fig. 3. In fig. 4 are shown the lines of equal pressure obtained using three terms of the series.
As it can be seen the results are quite accurate.

CONCLUSION

The procedure presented in the paper seems to be very promising in allowing to solve PDEs with variable coefficients. In all the examples solved up to now the convergence is achieved quite fast and the results are accurate. Author's aim is to go on and apply this technique also to non-linear elliptic and parabolic PDEs with variable coefficients.

REFERENCES

1. Brebbia, C.A., Telles, J.C.F. and Wrobel, L.C. Boundary Element Techniques, Springer-Verlag, Berlin and New York, 1984.
2. Beskos, D.E.(ed.) Boundary Element Methods in Mechanics, North-Holland, 1987.
3. Clements, D.L. The BEM for linear elliptic equations with variable coefficients, (Ed. Brebbia, C.A.), pp. 91-96, Proceedings of the X Int. Conf. on BEM, Southampton, UK, 1988, Springer-Verlag, Berlin and New York, 1988.
4. Rangogni, R. Numerical solution of the generalized Laplace equation by coupling BEM and the perturbation method, Appl. Math. Modell., Vol.1, pp. 266-270, 1986.
5. Bassani, R. and Guiggiani, M. Lubrication problems solved by the BEM, Proc. Japan Int. Tribology Conf., Nagoya, Japan, October 1990.
6. Cameron, A. The principles of lubrication, John Wiley & Sons, 1986.

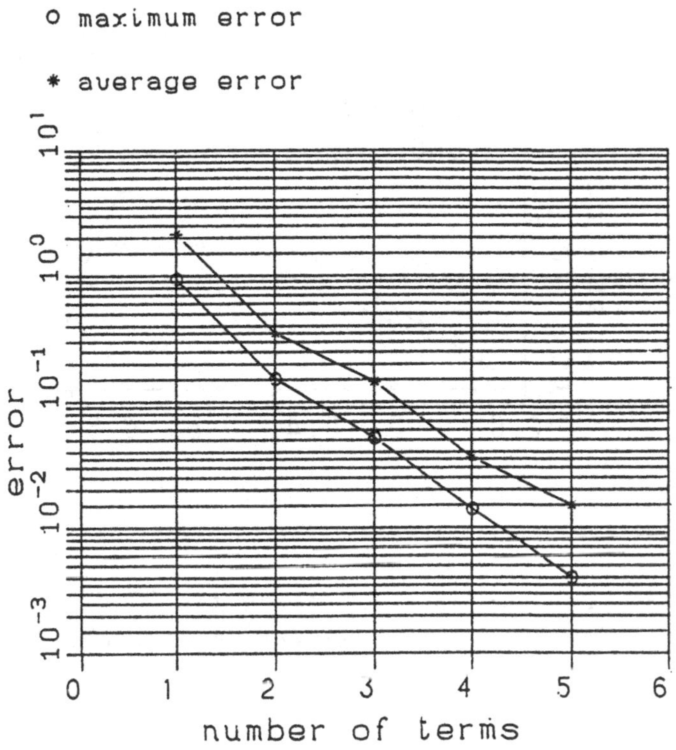

Fig. 1. Variation of errors versus the number of series terms.

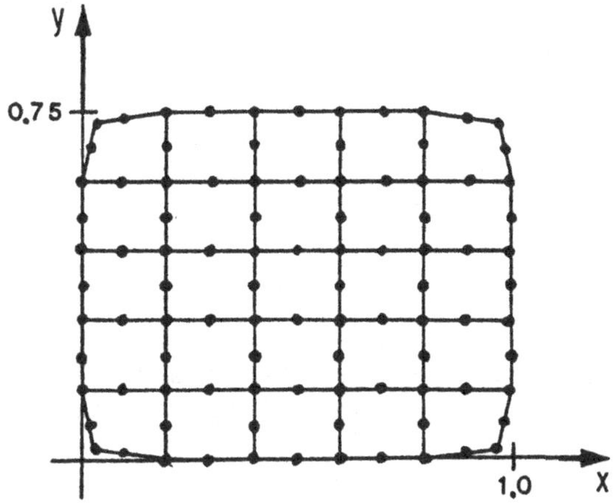

Fig. 2. Discretization of domain D for the second example.

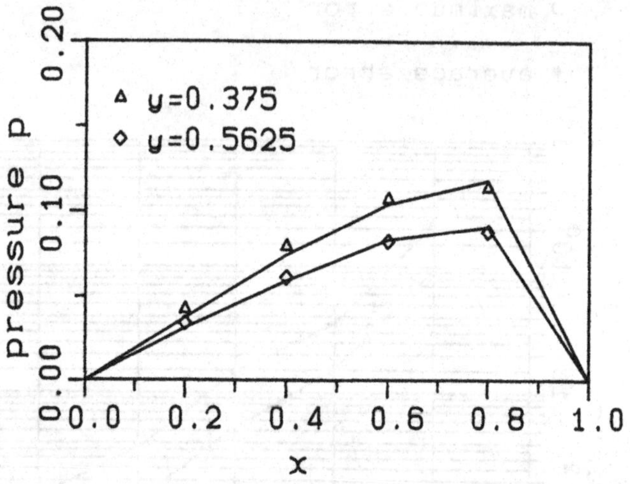

Fig. 3. Pressure profile along the lines y = 0.375 and y = 0.5625;——values from [5].

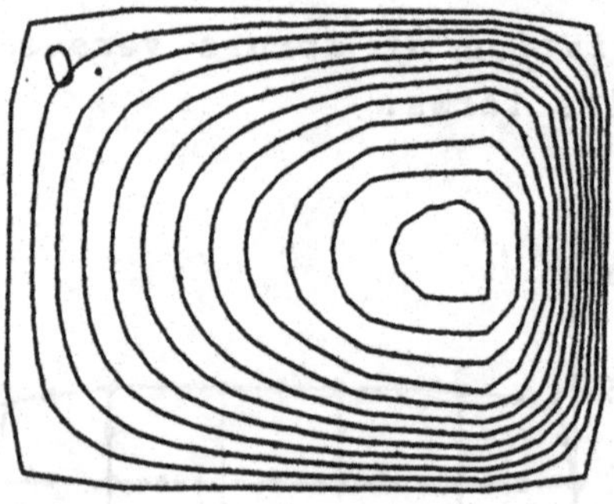

Fig. 4. Pressure profile as obtained by BEM using 3 terms of the series; line interval = 0.1.

SECTION 3: FLUID MECHANICS

Explicit Forms of the Fundamental Solution Tensor and Singular Integrals for the 2D Primitive-Variables Navier-Stokes Formulation

K. Darkovich (*), K. Kakuda (**), N. Tosaka (**)

(*) Institute for Environmental Chemistry, National Research Council Canada, Ottawa, Ontario, K1A OR6, Canada

(**) Department of Mathematical Engineering, College of Industrial Technology, Nihon University, 1-2 Izumi-cho 1-Chome, Narashino-shi, Chiba 275, Japan

ABSTRACT

This paper details the development of the explicit forms of the fundamental solutions for the two-dimensional incompressible unsteady time-difference primitive-variable Navier-Stokes boundary element formulation based on the method of Tosaka. Also, explicit forms of singular integrals are derived and presented. In this regard, series expansions of the fundamental solutions, which contain modified Bessel functions, are performed to enable integration of the singular cases. An algorithm for incorporating the singular solutions in to partitions of the affected boundary elements and domain cells is presented. Practical comments are offered in order to allow this method to be more readily used.

INTRODUCTION

Using Hörmander's weight-function procedure for fundamental solutions[1], Tosaka developed a primitive-variable global matrix form for incompressible unsteady Navier-Stokes flow[2]. In the course of current work on the simulation of a particle-bubble interaction in the context of mineral flotation [3,4], the formulations of Tosaka have been adopted for the two-dimensional case. Further, the discrete time-step version was necessary, as the problem at hand involves a two-body interaction producing geometrical changes at each time-level.

The objective of this paper is to present concrete forms of the fundamental solution tensor for the incompressible two-dimensional Navier-Stokes primitive-variable (velocity and pressure) discrete time-step Boundary Element formulation. That is, to make available in the literature, directly usable formulae, as well as to detail the development of solutions for singular-case integrals as they apply to this problem. The presentation of these results should facilitate adoption of this method for use by members of the technical community.

FORMULATION

Fundamental Equations

Given below are the non-dimensional equations which define the flow field under general conditions. Here, Re is the Reynolds number, u_i is the velocity vector, p is the pressure, t is the time and τ_{ij} is the stress tensor.

Equations of momentum:

$$\mathrm{Re}\left(\frac{u_i - u_i(t-\Delta t)}{\Delta t} + u_j u_{i,j}\right) = \tau_{ij,j} + f_i \text{ in } \Omega \tag{1}$$

Incompressibility condition:

$$u_{i,i} = 0 \qquad \text{in } \Omega \tag{2}$$

Constitutive equations:

$$\tau_{ij} = -\mathrm{Re}p\delta_{ij} + u_{i,j} + u_{j,i} \qquad \text{in } \Omega \tag{3}$$

Boundary Conditions:

$$\begin{aligned} u_i &= \hat{u}_i && \text{on } \Gamma_u \\ \tau_i \equiv \tau_{ij}n_j &= \hat{\tau}_i && \text{on } \Gamma_\tau \\ u_i(t=0) &= u_i^0 && \text{in } \Omega \end{aligned} \tag{4}$$

In equation 4, the vector n_j is the unit normal in the j-direction at the given boundary location.

Integral Equation

Let us define, $\lambda \equiv \frac{\mathrm{Re}}{\Delta t}$. Further, $D_i \equiv \partial/\partial x_i$ and Δ is the Laplacian operator equal to D_iD_i (not to be confused with the time step, Δt).

Composing the Navier-Stokes equations from equations 1, 2 and 3, they can be summarized in matrix form. Hence,

$$[\hat{L}_{\alpha\beta}]\,\{U_\beta\} = \{B_\alpha\} - \lambda\{\tilde{U}_\alpha(t-\Delta t)\} \tag{5}$$

where the subscripts $\alpha, \beta = 1,2,3$ and the unknown $U_3 \equiv p$. In particular the elements $\hat{L}_{\alpha\beta}$ can be expressed as:

$$\begin{aligned} \hat{L}_{ij} &= -\frac{\mathrm{Re}}{\Delta t}\delta_{ij} + \Delta\delta_{ij} + D_iD_j \qquad i,j = 1,2 \\ \hat{L}_{i3} &= -\mathrm{Re}D_i \\ \hat{L}_{3j} &= D_j \\ \hat{L}_{33} &= 0 \end{aligned}$$

Thus the matrix $[\hat{L}_{\alpha\beta}]$ may be assembled. In a complete form, the components of equation 5 can be written for the two-dimensional case.

$$[\hat{L}_{\alpha\beta}] = \begin{bmatrix} -\frac{\mathrm{Re}}{\Delta t} + \Delta + D_1^2 & D_1D_2 & -\mathrm{Re}D_1 \\ D_2D_1 & -\frac{\mathrm{Re}}{\Delta t} + \Delta + D_2^2 & -\mathrm{Re}D_2 \\ D_1 & D_2 & 0 \end{bmatrix}$$

$$\{U_\beta\} = [\ u_1\ u_2\ p\ ]^T$$

$$\{B_\alpha\} = \begin{Bmatrix} \mathrm{Re}u_ju_{1,j} - f_1 \\ \mathrm{Re}u_ju_{2,j} - f_2 \\ 0 \end{Bmatrix}$$

$$\{\tilde{U}_\alpha(t-\Delta t)\} \;=\; [\; u_1(t-\Delta t) \;\; u_2(t-\Delta t) \;\; 0 \;]^T$$

Now we can rearrange equation 5 to the form,

$$[\hat{L}_{\alpha\beta}]\,\{U_\beta\} - \{B_\alpha\} + \lambda\{\tilde{U}_\alpha(t-\Delta t)\} = 0$$

Let us now introduce a weighting function, $W_{\alpha\gamma}$. Similarly, the subscript $\gamma = 1, 2, 3$. We can write,

$$\int_\Omega (\hat{L}_{\alpha\beta}U_\beta - B_\alpha + \lambda\tilde{U}_\alpha(t-\Delta t))W_{\alpha\gamma}\, d\Omega = 0 \tag{6}$$

After some manipulation, a boundary integral equation was derived[2]. The coordinate y_i refers to the source point, and x_i denotes the field points. This equation is:

$$\begin{aligned} c(y)U_\gamma(y) = &\int_\Gamma u_i(x)\Sigma_{i\gamma}(x,y)\, d\Gamma(x) - \int_\Gamma \tau_i(x)W_{i\gamma}(x,y)\, d\Gamma(x) \\ &+ \int_\Omega B_\alpha(x)W_{\alpha\gamma}(x,y)\, d\Omega(x) - \int_\Omega \lambda\tilde{U}_\alpha(t-\Delta t)(x)W_{\alpha\gamma}(x,y)\, d\Omega(x) \end{aligned} \tag{7}$$

Here, the tensor $\Sigma_{i\gamma} = (-W_{3\gamma}\delta_{ij} + W_{i\gamma,j} + W_{j\gamma,i})n_j$. From the method of weighted residuals, the fundamental solution tensor was given as:

$$\begin{aligned} W_{ij} &= (\Delta\delta_{ij} - D_iD_j)\phi \\ W_{i3} &= \tfrac{1}{\mathrm{Re}}D_i(-\lambda+\Delta)\phi \\ W_{3j} &= -D_j(-\lambda+\Delta)\phi \\ W_{33} &= \tfrac{1}{\mathrm{Re}}(-\lambda+2\Delta)(-\lambda+\Delta)\phi \end{aligned} \tag{8}$$

Here ϕ is a scalar function. From the differential operator derived for the global matrix system[5], the fundamental solution is:

$$\phi = \frac{-1}{2\pi\lambda}\left(\ln r + K_0(\sqrt{\lambda}r)\right) \tag{9}$$

where K_0 is the modified Bessel function of the third kind of order zero. In equation 9, the variable r represents the distance between a source point and a field point.

Discretization Scheme

To allow a pseudo-linear system of equations to be formed for the unknowns (velocity u_i and traction τ_i) at each node, these variables are computed using interpolation functions. It should be noted that the formulations presented here, are done with a *linear* interpolation scheme. For example, on the boundary, the tractions would be expressed as:

$$\tau_i(\mathbf{x}) = \varphi_n\tau_i(\mathbf{x}_n) \tag{10}$$

Similarly, in the domain, the velocities would be expressed as:

$$u_i(\mathbf{x}) = \psi_q u_i(\mathbf{x}_q) \tag{11}$$

$\mathbf{x}_q$ and $\mathbf{x}_n$ refer to coordinates at the qth or nth node in the respective domain element or boundary element. In equations 10 and 11, the interpolation functions φ_n and ψ_q are given as:

$$\begin{aligned} \varphi_1 &= \frac{1}{2}(1-\xi) \\ \varphi_2 &= \frac{1}{2}(1+\xi) \end{aligned} \tag{12}$$

The local parameter ξ is determined as follows. (see Figure 1)

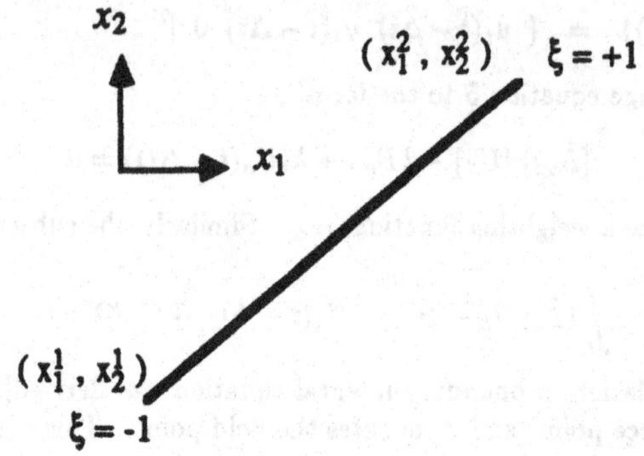

Figure 1: Schematic of Boundary Element

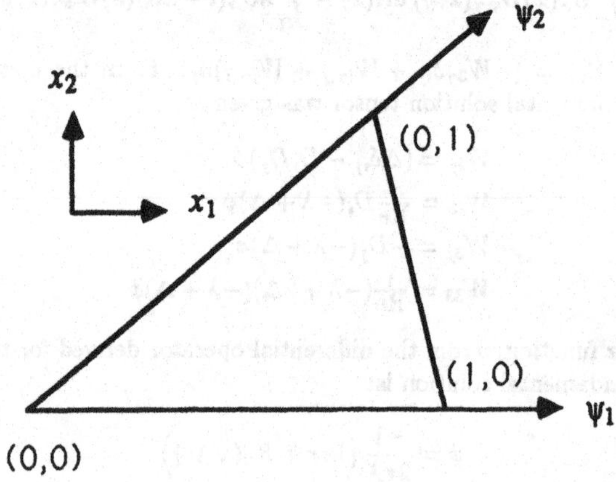

Figure 2: Schematic of Domain Element

$$\xi = \left(\frac{2\sqrt{(x_1 - x_1^1)^2 + (x_2 - x_2^1)^2}}{\sqrt{(x_1^2 - x_1^1)^2 + (x_2^2 - x_2^1)^2}} \right) - 1 \tag{13}$$

For domain elements[6]:

$$\psi_i = \frac{1}{2A}\left(2A_i^0 + b_i x_1 + a_i x_2\right) \tag{14}$$

Here (see Figure 2),

$$\begin{aligned} a_i &= x_1^K - x_1^J \\ b_i &= x_2^J - x_2^K \\ 2A_i^0 &= x_1^J x_2^K - x_1^K x_2^J \\ A &= \frac{1}{2}(b_1 a_2 - b_2 a_1) \end{aligned}$$

Above, when $i = 1, 2, 3$ (referring to shape function index), then $J = 2, 3, 1$ and $K = 3, 1, 2$. Above, x_i^J is the value of x_i at the Jth end point or corner of the respective domain element or boundary element. Equation 7 can be rewritten in a discretized form when the boundary and domain are divided up into separate elements. In addition, since

the unknowns, (velocity u_i and traction τ_i) are expressed as linear combinations of shape functions involving only nodal values, we obtain the following algebraic form:

$$\Lambda_{\gamma l}u_l + G_{\gamma in}\tau_{in} = H_{\gamma in}u_{in} - C_{\gamma iq}u(t-\Delta t)_{iq} + Z_{\gamma iqpa}u_{iq}u_{pa} \tag{15}$$

In the above equation, the coefficients $\Lambda_{\gamma k}$, are given by:

$$\Lambda_{\gamma l} = \begin{cases} (\frac{\theta}{2\pi})\delta_{\gamma l}(\mathbf{y}) & \text{for } y \in \Gamma \\ \delta_{\gamma l}(\mathbf{y}) & \text{for } y \in \Omega \end{cases} \tag{16}$$

In equation 16, the factor $(\frac{\theta}{2\pi})$ is the angle (in radians) formed between two boundary elements, when the node is at their junction. If the surface is smooth, this factor reduces to a value of $\frac{1}{2}$.

The other coefficients (for integration over the Jth element) from equation 15 are given as follows:

$$G_{\gamma in} = \int_{\Gamma_J} W_{i\gamma}(\mathbf{x},\mathbf{y})\varphi_n(\mathbf{x})\, d\Gamma(\mathbf{x}) \tag{17}$$

$$H_{\gamma in} = \int_{\Gamma_J} \Sigma_{i\gamma}(\mathbf{x},\mathbf{y})\varphi_n(\mathbf{x})\, d\Gamma(\mathbf{x}) \tag{18}$$

$$C_{\gamma iq} = \int_{\Omega_J} \mathrm{Re} W_{i\gamma}(\mathbf{x},\mathbf{y})\psi_q(\mathbf{x})\, d\Omega(\mathbf{x}) \tag{19}$$

$$Z_{\gamma iqpa} = \int_{\Omega_J} \mathrm{Re}\psi_{a,p}(\mathbf{x}) W_{i\gamma}(\mathbf{x},\mathbf{y})\psi_q(\mathbf{x})\, d\Omega(\mathbf{x}) \tag{20}$$

EXPLICIT FORMS OF THE FUNDAMENTAL SOLUTION TENSOR

To numerically implement equation 15, several of the fundamental tensors must be used. Using the values of 1 and 2 for the subscript l is sufficient to produce a velocity field solution. To achieve this, the tensors $W_{i\gamma}$ and $\Sigma_{i\gamma}$ are required. Recall that $\Sigma_{i\gamma} = (-W_{3\gamma}\delta_{ij} + W_{i\gamma,j} + W_{j\gamma,i})n_j$. Hence, with $\gamma = j$, the tensors W_{3j} and $W_{ik,j}$ $(k = 1,2)$ are also needed.

Explicit form of the Fundamental Solution Tensor, W_{ij}

Recall that,

$$W_{ij} = \delta_{ij}\Delta\phi - \phi_{,ij} \qquad \text{and,} \quad \Delta\phi = \phi_{,kk}$$

thus,

$$W_{ij} = \delta_{ij}\phi_{,kk} - \phi_{,ij}$$

Recall also,

$$\phi = \frac{-1}{2\pi}[\ln r + K_0(\sqrt{\lambda}r)]$$

By definition, $r^2 = (y_i - x_i)(y_i - x_i)$. Let $s_i = (y_i - x_i)$, so $r^2 = s_i s_i$. It will be convenient to have an explicit expression for the derivative of r, and it can be easily shown that $dr/dx_i = -s_i/r$.

Beginning with equation 9, we can write an expression for $\phi_{,ij}$.

$$\phi_{,ij} = \frac{-1}{2\pi\lambda}[(\ln r)_{,i} + K_0(\sqrt{\lambda}r)_{,i}]_{,j}$$

It will be convenient to combine $\sqrt{\lambda}r$ into one variable, z. From the discussion given in Lebedev[7], the derivatives of the Bessel functions conform to the following recurrence relations:

$$\frac{d}{dz}[z^\nu K_\nu(z)] = -z^\nu K_{\nu-1}(z)$$

and,

$$\frac{d}{dz}[z^{-\nu}K_\nu(z)] = -z^{-\nu}K_{\nu+1}(z) \tag{21}$$

Thus for example,

$$\frac{d(K_0(\sqrt{\lambda}r))}{dx_i} = \frac{d(z^{-0}K_0(z))}{dx_i} = \frac{d(z^{-0}K_0(z))}{dz}\frac{dz}{dr}\frac{dr}{dx_i}$$

Now,

$$\frac{d(K_0(\sqrt{\lambda}r))}{dx_i} = -z^{-0}K_1(z)(\sqrt{\lambda})(-s_i/r) = \lambda z^{-1}K_1(z)s_i$$

Returning to $\phi_{,ij}$, we have,

$$\begin{aligned}
\phi_{,ij} &= \frac{-1}{2\pi\lambda}(\frac{-s_i}{r^2} + \lambda z^{-1}K_1(z)s_i)_{,j} \\
&= \frac{-1}{2\pi\lambda}(\frac{\delta_{ij}}{r^2} + (-s_i)\frac{-2}{r^3}\frac{(-s_j)}{r} + \lambda s_i z^{-1}K_2(z)\sqrt{\lambda}\frac{s_j}{r} + \lambda z^{-1}K_1(z)(-\delta_{ij})) \\
&= \frac{-1}{2\pi\lambda}(\frac{\delta_{ij}}{r^2} - \frac{2}{r^4}s_is_j + \lambda^2 z^{-2}K_2(z)s_is_j - \lambda z^{-1}K_1(z)\delta_{ij})
\end{aligned}$$

Arranging terms gives,

$$\phi_{,ij} = \frac{-1}{2\pi\lambda}(\delta_{ij}(\frac{1}{r^2} - \frac{\sqrt{\lambda}}{r}K_1(\sqrt{\lambda}r)) + s_is_j(\frac{\lambda}{r^2}K_2(\sqrt{\lambda}r) - \frac{2}{r^4})) \tag{22}$$

Now, recall $W_{ij} = \delta_{ij}\phi_{,kk} - \phi_{,ij}$. Thus,

$$\begin{aligned}
W_{ij} &= \frac{-1}{2\pi\lambda}[\delta_{ij}(\frac{2}{r^2} - 2\frac{\sqrt{\lambda}}{r}K_1(\sqrt{\lambda}r) + \lambda K_2(\sqrt{\lambda}r) - \frac{2}{r^2}) \\
&\quad -\delta_{ij}(\frac{1}{r^2} - \frac{\sqrt{\lambda}}{r}K_1(\sqrt{\lambda}r)) - s_is_j(\frac{\lambda}{r^2}K_2(\sqrt{\lambda}r) - \frac{2}{r^4})]
\end{aligned}$$

Simplifying gives,

$$W_{ij} = \frac{-1}{2\pi\lambda}[\delta_{ij}(\lambda K_2(\sqrt{\lambda}r) - \frac{\sqrt{\lambda}}{r}K_1(\sqrt{\lambda}r) - \frac{1}{r^2}) - s_is_j(\frac{\lambda}{r^2}K_2(\sqrt{\lambda}r) - \frac{2}{r^4})] \tag{23}$$

<u>Calculation of $W_{ik,j}$</u>

We can begin by writing $W_{ik,j}$ in a general form.

$$\begin{aligned}
W_{ik,j} &= \frac{-1}{2\pi\lambda}[\delta_{ik}(\lambda K_2(z) - \lambda z^{-1}K_1(z) - \frac{1}{r^2})_{,j} \\
&\quad -(s_is_k)_{,j}(\lambda^2 z^{-2}K_2(z) - \frac{2}{r^4}) - s_is_k(\lambda^2 z^{-2}K_2(z) - \frac{2}{r^4})_{,j}] \\
&= \frac{-1}{2\pi\lambda}[\delta_{ik}((\lambda K_2(z))_{,j} - \lambda^2 z^{-2}K_2(z)s_j - \frac{2s_j}{r^4}) \\
&\quad +(\delta_{ij}s_k + \delta_{kj}s_i)(\lambda^2 z^{-2}K_2(z) - \frac{2}{r^4}) - s_is_k(\lambda^3 z^{-3}K_3(z)s_j - \frac{8s_j}{r^6})] \\
&= \frac{-1}{2\pi\lambda}[\delta_{ik}(z^2(\lambda z^{-2}K_2(z))_{,j} + z_j^2(\lambda z^{-2}K_2(z)) - \lambda^2 z^{-2}K_2(z)s_j - \frac{2s_j}{r^4}) \\
&\quad + (\delta_{ij}s_k + \delta_{kj}s_i)(\lambda^2 z^{-2}K_2(z) - \frac{2}{r^4}) - s_is_k(\lambda^3 z^{-3}K_3(z)s_j - \frac{8s_j}{r^6})]
\end{aligned}$$

Collecting like terms and rearranging gives,

$$W_{ik,j} = \frac{-1}{2\pi\lambda}[\delta_{ik}(\lambda^2 z^{-1}K_3(z)s_j - 3\lambda^2 z^{-2}K_2(z)s_j - \frac{2s_j}{r^4}) \\ +(\delta_{ij}s_k + \delta_{kj}s_i)(\lambda^2 z^{-2}K_2(z) - \frac{2}{r^4}) - s_i s_k(\lambda^3 z^{-3}K_3(z)s_j - \frac{8s_j}{r^6})] \quad (24)$$

Calculation of W_{3j}

In equation 8, W_{3j} was given in general form.

$$W_{3j} = -D_j(-\lambda + \Delta)\phi$$

As given in Appendix A, the scalar $(-\lambda + \Delta)\phi$ can be expressed as,

$$(-\lambda + \Delta)\phi = \frac{1}{2\pi}\ln r$$

Hence,

$$W_{3j} = -D_j\left(\frac{1}{2\pi}\ln r\right) = \frac{1}{2\pi}\left(\frac{s_j}{r^2}\right) \quad (25)$$

SINGULAR INTEGRALS

Solution of the system of equations 15 requires integration over all boundary elements and domain elements from each node. When the node in question is contained in the same element (ie; source and field point coincide, $r = 0$), singularities arise.

The integrals in equations 17 to 20 must be evaluated analytically to avoid substantial error in the flow field solution. Note that with linear shape functions, the term $\psi_{a,p}$ in equation 20 will be constant over a domain element. So in total, we must make singular formulations for three cases, namely equations 17, 18 and 19.

Singular Case for $\int_\Omega W_{ij}\psi_\alpha \, d\Omega$

Let us define a modified domain element, illustrated in Figure 3. Using the method of Brebbia *et al.*[6], we transform the system to polar coordinates, (r, φ). In the triangular element,

$$r(\varphi) = \frac{-2A}{b_\Xi \cos\varphi + a_\Xi \sin\varphi}$$

Above, Ξ is the triangular domain element corner index at the singular point. A, b_Ξ and a_Ξ are as defined in 14. Further, the element of area, $d\Omega = r\,dr\,d\varphi$. Using the subscript α as the shape function index, and transforming the linear triangular shape functions to (r, φ) coordinates, we produce,

$$\tilde{\psi}_\alpha = \psi_\alpha^\Xi + \frac{r}{2A}(b_\alpha \cos\varphi + a_\alpha \sin\varphi) \quad (26)$$

where,

$$\psi_\alpha^\Xi = \begin{cases} 0 & \text{when } \alpha \neq \Xi \\ 1 & \text{when } \alpha = \Xi \end{cases}$$

In order to successfully evaluate this singular integral, a limiting case must be considered. Along the radius $r(\varphi)$ away from the singular point, a maximum value of $r(\varphi) = R(\varphi)$ must be selected since the fundamental solution must be transformed to an infinite series format, which involves a convergence criterion on this variable. To exclude the singularity, the lower limit of integration is taken as, ε, an infinitesimal number. Recall the following transforms:

$$\begin{array}{ll} z = \sqrt{\lambda}r & \text{when } r = \varepsilon \Rightarrow z = \sqrt{\lambda}\varepsilon = a_1 \\ dz = \sqrt{\lambda}dr & \text{when } r = R(\varphi) \Rightarrow z = \sqrt{\lambda}R(\varphi) = a_2 \\ r = \frac{1}{\sqrt{\lambda}}z & r\,dr\,d\varphi = \frac{z}{\lambda}\,dz\,d\varphi \end{array}$$

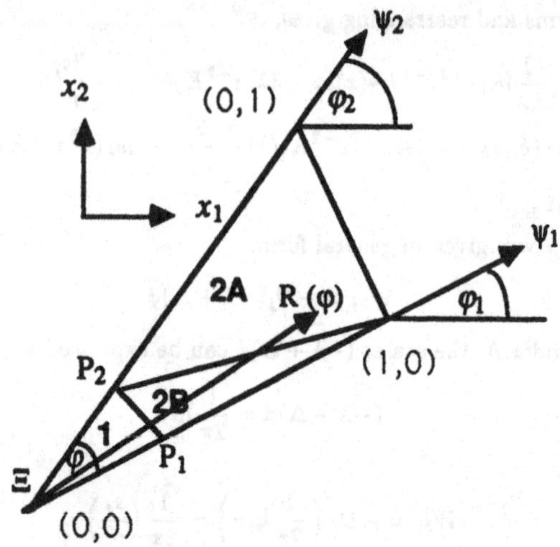

Figure 3: Schematic of Domain Element for singular case

In view of these transforms, the shape function becomes,

$$\tilde{\psi}_\alpha = \psi_\alpha^\Xi + \frac{z}{2A\sqrt{\lambda}}(b_\alpha \cos\varphi + a_\alpha \sin\varphi)$$

So, now we begin the evaluation of singular integral.

$$I_\Omega = \int_\Omega W_{ij}\psi_\alpha \, d\Omega = \lim_{\epsilon\to 0}\int_{\varphi_1}^{\varphi_2}\int_\epsilon^{R(\varphi)} W_{ij}\tilde{\psi}_\alpha r \, dr \, d\varphi$$

Inserting the explicit form of the fundamental solution gives,

$$\begin{aligned} I_\Omega &= \lim_{\epsilon\to 0}\int_{\varphi_1}^{\varphi_2}\int_\epsilon^{R(\varphi)}\left[\frac{-\delta_{ij}}{2\pi\lambda}\left(K_2(z) - z^{-1}K_1(z) - \frac{1}{z^2}\right)\right. \\ &\quad \left. + \frac{s_i s_j}{2\pi\lambda r^2}\left(K_2(z) - \frac{2}{z^2}\right)\right]\tilde{\psi}_\alpha z \, dz \, d\varphi \end{aligned}$$

Note that for the singular case, the ratio s_i/r will be constant in the domain element, equal to $\cos\varphi$ for $i = 1$ and $\sin\varphi$ for $i = 2$. Now, inserting the transformed shape function into the integral gives,

$$\begin{aligned} I_\Omega &= \frac{-1}{2\pi}\lim_{\epsilon\to 0}\int_{\varphi_1}^{\varphi_2}\left[\delta_{ij}\left\{\int_{a_1}^{a_2} zK_2(z)\tilde{\psi}_\alpha \, dz - \int_{a_1}^{a_2} K_1(z)\tilde{\psi}_\alpha \, dz - \int_{a_1}^{a_2}\frac{1}{z}\tilde{\psi}_\alpha \, dz\right\}\right. \\ &\quad \left. - \frac{s_i s_j}{r^2}\left\{\int_{a_1}^{a_2} zK_2(z)\tilde{\psi}_\alpha \, dz - 2\int_{a_1}^{a_2}\frac{1}{z}\tilde{\psi}_\alpha \, dz\right\}\right] d\varphi \end{aligned}$$

Let us express,

$$\tilde{\psi}_\alpha = \psi_\alpha^\Xi + \Theta_\alpha(\varphi)z$$

where,

$$\Theta_\alpha(\varphi)z = \frac{1}{2A\sqrt{\gamma}}(b_\alpha \cos\varphi + a_\alpha \sin\varphi)$$

We now insert $\psi_\alpha^\Xi + \Theta_\alpha(\varphi)z$ for $\tilde{\psi}_\alpha$ into the expression for I_Ω.

$$\begin{aligned}
I_\Omega = & \frac{-1}{2\pi}\lim_{\varepsilon\to 0}\int_{\varphi_1}^{\varphi_2}\Bigg[\delta_{ij}\left\{\psi_\alpha^\Xi\int_{a_1}^{a_2} zK_2(z)\,dz + \Theta_\alpha(\varphi)\int_{a_1}^{a_2} z^2K_2(z)\,dz\right.\\
& \left. - \psi_\alpha^\Xi\int_{a_1}^{a_2} K_1(z)\,dz - \Theta_\alpha(\varphi)\int_{a_1}^{a_2} zK_1(z)\,dz - \psi_\alpha^\Xi\int_{a_1}^{a_2}\frac{1}{z}\,dz - \Theta_\alpha(\varphi)\int_{a_1}^{a_2} dz\right\}\\
& -\frac{s_is_j}{r^2}\left\{\psi_\alpha^\Xi\int_{a_1}^{a_2} zK_2(z)\,dz + \Theta_\alpha(\varphi)\int_{a_1}^{a_2} z^2K_2(z)\,dz\right.\\
& \left. -2\psi_\alpha^\Xi\int_{a_1}^{a_2}\frac{1}{z}\,dz - 2\Theta_\alpha(\varphi)\int_{a_1}^{a_2} dz\right\}\Bigg]\,d\varphi
\end{aligned}$$

At this point we see that we have integrals of the form,

$$\lim_{\varepsilon\to 0}\int_{a_1}^{a_2} z^m K_n(z)\,dz$$

which must be evaluated. This procedure and a table of results for all such integrals required in this paper are detailed in Appendix B. Using these results for the integrations over z, we have,

$$\begin{aligned}
I_\Omega = & \frac{-1}{2\pi}\int_{\varphi_1}^{\varphi_2}\lim_{\varepsilon\to 0}\Bigg[\delta_{ij}\left\{\psi_\alpha^\Xi S(zK_2(z)) + 2\psi_\alpha^\Xi\int_{a_1}^{a_2}\frac{1}{z}\,dz\right.\\
& +\Theta_\alpha(\varphi)S(z^2K_2(z)) - \psi_\alpha^\Xi S(K_1(z))\\
& \left. -\psi_\alpha^\Xi\int_{a_1}^{a_2}\frac{1}{z}\,dz - \Theta_\alpha(\varphi)S(zK_1(z)) - \psi_\alpha^\Xi\int_{a_1}^{a_2}\frac{1}{z}\,dz - \Theta_\alpha(\varphi)a_2\right\}\\
& -\frac{s_is_j}{r^2}\left\{\psi_\alpha^\Xi S(zK_2(z)) + 2\psi_\alpha^\Xi\int_{a_1}^{a_2}\frac{1}{z}\,dz + \Theta_\alpha(\varphi)S(z^2K_2(z)) - 2\psi_\alpha^\Xi\int_{a_1}^{a_2}\frac{1}{z}\,dz\right.\\
& \left. -2\Theta_\alpha(\varphi)a_2\right\}\Bigg]\,d\varphi
\end{aligned}$$

It can be seen from the above equation that all the integrals of the form,

$$\int_{a_1}^{a_2}\frac{1}{z^m}\,dz$$

sum to zero.

Now, let $\varphi = \frac{\xi}{2}(\varphi_1 - \varphi_2) + \frac{1}{2}(\varphi_1 + \varphi_2)$. Thus, when $\varphi = \varphi_1$, $\xi = -1$, and when $\varphi = \varphi_2$, $\xi = +1$. Further, $d\varphi = ((\varphi_1 - \varphi_2)/2)\,d\xi$. Since I_Ω is a complicated function of φ, a transform to variable limits of -1 and +1 is employed to facilitate numerical integration over this variable by Gaussian quadrature. The singularity in I_Ω arises in the variable r, and the integration over r was done analytically in the interior integral of the double integral over the domain. Hence, the net integral procedure for I_Ω is considered semi-analytical.

Finally, we can write,

$$\begin{aligned}
I_\Omega = & \frac{-1}{2\pi}\int_{-1}^{+1}\left[\delta_{ij}\left\{\left[S(K_1(z)) - S(zK_2(z))\right]\psi_\alpha^\Xi\right.\right.\\
& \left. + \left[S(z^2K_2(z)) - S(zK_1(z)) - a_2\right]\Theta_\alpha(\xi)\right\}\\
& \left. -\frac{s_is_j}{r^2}\left\{\psi_\alpha^\Xi S(zK_2(z)) + \left[S(z^2K_2(z)) - 2a_2\right]\Theta_\alpha(\xi)\right\}\right]\frac{\varphi_2-\varphi_1}{2}\,d\xi \qquad (27)
\end{aligned}$$

Thus, equation 27 can be implemented numerically and solved by Gaussian quadrature in the interval -1 to +1 in ξ.

Singular Case for $\int_\Gamma \Sigma_{ij}\varphi_\alpha\, d\Gamma$

Again, let us define a modified boundary element, shown in Figure 4. In this element, a number of transforms are effected. Let $r = (R/2)(1+\xi)$. Thus, $\xi = (2r/R) - 1$. Here, R is a maximum value of r, selected for the convergence criteria of the series forms of the integrals of the modified Bessel functions. The singularity is avoided by choosing the lower limit of the integration to be $\xi = -1+\varepsilon$. Recall the boundary element shape functions,

$$\varphi_\alpha = \frac{1}{2}(1+(-1)^\alpha\xi) = \frac{1}{2}(1+C_\alpha\xi)$$

Here we set $C_\alpha = (-1)^\alpha$ and define $\tilde{C}_\alpha = 1 - C_\alpha$. Thus,

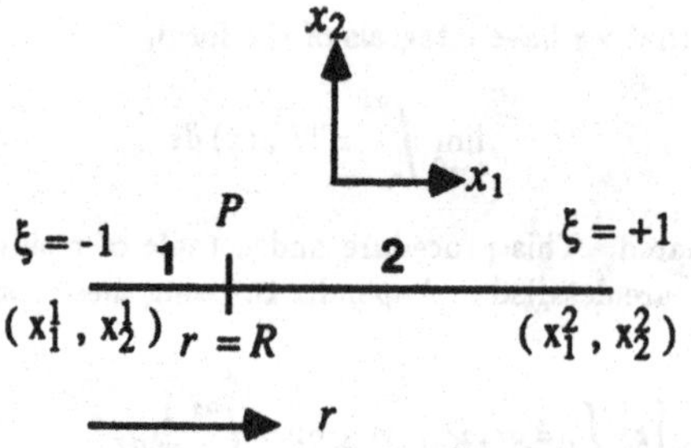

Figure 4: Schematic of Boundary Element for singular case

$$\varphi_\alpha = \frac{1}{2}\left(1 + C_\alpha\left(\frac{2r}{R} - 1\right)\right)$$
$$= \frac{1}{2}\left(\frac{2C_\alpha r}{R} + \tilde{C}_\alpha\right)$$

We make use of the following transforms:

$$\begin{aligned} z &= \sqrt{\lambda}r \quad \text{when } \xi = +1 \Rightarrow z = \sqrt{\lambda}R = a_2 \\ dz &= \sqrt{\lambda}dr \quad \text{when } \xi = -1+\varepsilon \Rightarrow z = \tfrac{\sqrt{\lambda}R\varepsilon}{2} = a_1 \end{aligned}$$

In view of these transforms, the shape function becomes,

$$\varphi_\alpha = \frac{1}{2}\left(\frac{2C_\alpha z}{\sqrt{\lambda}R} + \tilde{C}_\alpha\right)$$

Thus our singular integral can be stated as,

$$I_{\Gamma\Sigma} = \lim_{\varepsilon\to 0}\int_{a_1}^{a_2} \Sigma_{ij}\varphi_\alpha\, d\Gamma$$

At this point, before inserting the explicit form of the fundamental solution Σ_{ij} into the above equation, the following result is observed. For the singular case, (refer to Figure 4) the terms s_i are the components of $\vec{r}$. $\vec{r}$ itself, simply extends along the boundary element in question. By definition, $\vec{n}$ is the normal to the boundary element and hence, its components must form a zero dot-product with s_i. Thus, the terms $s_i n_i = 0$ and can be excluded from the calculations. Further, for the singular case, $s_1/r = \sin\theta$, and $s_2/r = \cos\theta$, which are constants. In this case θ is the angle formed between the linear boundary element and the x_1-axis.

So we proceed,

$$
\begin{aligned}
I_{\Gamma\Sigma} = \frac{-1}{2\pi\lambda} \lim_{\varepsilon\to 0} \int_{a_1}^{a_2} & \left[\frac{s_j n_i \lambda^{\frac{3}{2}}}{r} \left(\frac{1}{z}\right) \right. \\
& + \left(\frac{s_i n_j}{r}\right) \left(\lambda^{\frac{3}{2}} K_3(z) - 3\lambda^{\frac{3}{2}} z^{-1} K_2(z) - 2\lambda^{\frac{3}{2}} \left(\frac{1}{z^3}\right)\right) \\
& \left. + \left(\frac{2s_j n_i + s_i n_j}{r}\right) \left(\lambda^{\frac{3}{2}} z^{-1} K_2(z) - 2\lambda^{\frac{3}{2}} \left(\frac{1}{z^3}\right)\right)\right] \\
& \frac{1}{2} \left(\frac{2C_\alpha z}{\sqrt{\lambda} R} + \tilde{C}_\alpha \right) \frac{1}{\sqrt{\lambda}}\, dz
\end{aligned}
$$

$$
\begin{aligned}
= \frac{-1}{4\pi\lambda^{\frac{3}{2}}} \lim_{\varepsilon\to 0} \int_{a_1}^{a_2} & \left[\frac{2s_j n_i \lambda C_\alpha}{rR} + \frac{s_j n_i \lambda^{\frac{3}{2}} \tilde{C}_\alpha}{r} \left(\frac{1}{z}\right) \right. \\
& + \left(\frac{s_i n_j}{r}\right) \left(\frac{2\lambda C_\alpha}{R} z K_3(z) + \lambda^{\frac{3}{2}} \tilde{C}_\alpha K_3(z) \right. \\
& - \frac{6\lambda C_\alpha}{R} K_2(z) - 3\lambda^{\frac{3}{2}} \tilde{C}_\alpha z^{-1} K_2(z) \\
& \left. - \frac{4\lambda C_\alpha}{R} \left(\frac{1}{z^2}\right) - 2\lambda^{\frac{3}{2}} \tilde{C}_\alpha \left(\frac{1}{z^3}\right)\right) \\
& + \left(\frac{2s_j n_i + s_i n_j}{r}\right) \left(\frac{2\lambda C_\alpha}{R} K_2(z) + \lambda^{\frac{3}{2}} \tilde{C}_\alpha z^{-1} K_2(z) \right. \\
& \left.\left. - \frac{4\lambda C_\alpha}{R} \left(\frac{1}{z^2}\right) - 2\lambda^{\frac{3}{2}} \tilde{C}_\alpha \left(\frac{1}{z^3}\right)\right)\right] dz
\end{aligned}
$$

Now we can insert the results from Table 1 of the form:

$$
\lim_{\varepsilon\to 0} \int_{a_1}^{a_2} z^m K_n(z)\, dz \quad \text{expressed as:} \quad \mathcal{S}(z^m K_n(z))
$$

Upon the above substitution, all the integrals of the form,

$$
\int_{a_1}^{a_2} \frac{1}{z^m}\, dz
$$

will sum to zero as in the previous singular case. These integrals and their coefficients are not written for the sake of brevity. Finally, for $I_{\Gamma\Sigma}$ we have,

$$
\begin{aligned}
I_{\Gamma\Sigma} = \frac{-1}{4\pi\lambda^{\frac{3}{2}}} & \left[\frac{2s_j n_i \lambda C_\alpha a_2}{rR} \right. \\
& + \mathcal{S}(z K_3(z)) \left(\frac{s_i n_j}{r} \frac{2\lambda C_\alpha}{R} \right) + \mathcal{S}(K_3(z)) \left(\frac{s_i n_j \lambda^{\frac{3}{2}} \tilde{C}_\alpha}{r} \right) \\
& + \mathcal{S}(K_2(z)) \left\{ \frac{4\lambda C_\alpha}{R} \left(\frac{s_j n_i - s_i n_j}{r} \right) \right\} \\
& \left. + \mathcal{S}(z^{-1} K_2(z)) \left\{ 2\lambda^{\frac{3}{2}} \tilde{C}_\alpha \left(\frac{s_j n_i - s_i n_j}{r} \right) \right\} \right]
\end{aligned} \tag{28}
$$

Equation 28 can be implemented to give analytical results for integrations of $I_{\Gamma\Sigma}$ over singular boundary elements.

Singular Case for $\int_\Gamma W_{ij} \varphi_n \, d\Gamma$

Proceeding in an entirely parallel fashion as for $I_{\Gamma\Sigma}$, we can define the singular integral,

$$I_{\Gamma W} = \lim_{\epsilon \to 0} \int_{a_1}^{a_2} W_{ij}\varphi_\alpha \, d\Gamma$$

Employing the same transforms and operations as with $\int_\Gamma \Sigma_{ij}\varphi_\alpha \, d\Gamma$ we find that,

$$\begin{aligned} I_{\Gamma W} = \frac{-1}{2\pi\lambda} \Bigg[& \frac{C_\alpha}{R}\left(\delta_{ij} - \frac{s_i s_j}{r^2}\right) S(zK_2(z)) \\ & + \frac{\tilde{C}_\alpha}{2}\left(\delta_{ij} - \frac{s_i s_j}{r^2}\right) S(K_2(z)) - \frac{C_\alpha}{R}\delta_{ij} S(K_1(z)) \\ & - \frac{\tilde{C}_\alpha}{2}\delta_{ij} S(z^{-1}K_1(z)) \Bigg] \end{aligned} \tag{29}$$

Equation 29 can be implemented to give analytical results for integrations of $I_{\Gamma W}$ over singular boundary elements.

NOTES ON ELEMENT PARTITIONING

Domain Elements

In Figure 3, it can be seen that the element where the singularity occurs is sub-divided into 3 regions labelled 1, 2A and 2B.

The semi-analytical integration method cannot be used over the entire domain element since beyond a limiting value of $R(\varphi)$, the series expansions for the integrals of the modified Bessel functions will not converge properly.

For example, consider the tensor $C_{\gamma iq}$ as given in equation 19. Evaluation of this tensor produces coefficients for variables positioned at the domain element corners. Using the unmodified integration schemes would require the introduction of new nodes at points P_1 and P_2 in Figure 3. $C_{\gamma iq}$ is a function of geometry only. If the integration over the three regions uses the standard local triangular Gauss points, but values for the shape functions transformed to their values in the overall domain, then the tensor $C_{\gamma iq}$ will be simply the sum of the integrals in regions 1, 2A and 2B.

Integrations over regions 2A and 2B can be done by standard Gaussian quadrature as these domains do not contain singularities.

In the context of the entire domain, at P_1, $\psi_1 = N_1$ and at P_2, $\psi_2 = N_2$. Thus, when evaluating the singular integral in region 1, the shape functions are transformed with the following relations. Here $\bar{\psi}_i$ is the transformed shape function in the singular sub-domain, and ψ_i is the shape function for the whole domain.

$$\begin{aligned} \bar{\psi}_1 &= N_1\psi_1 \\ \bar{\psi}_2 &= N_2\psi_2 \\ \bar{\psi}_3 &= 1 - \bar{\psi}_1 - \bar{\psi}_2 \end{aligned} \tag{30}$$

In regions 2A and 2B, standard local Gauss points are used, but the shape functions are determined by equation 14 applied to the whole domain. This can work as ψ_i is used only as a multiplier in Gaussian quadrature, and must not be manipulated as a function of r as in the singular sub-domain.

Boundary Elements

Similarly as with domain elements, the boundary element must be partitioned into two regions to ensure series convergence. In Figure 4 at point P, $\varphi_1 = N_1$. To evaluate the singular integrals in region 1, the shape functions are transformed by:

$$\begin{aligned} \bar{\varphi}_1 &= N_1\varphi_1 \\ \bar{\varphi}_2 &= 1 - \bar{\varphi}_1 \end{aligned} \tag{31}$$

Here $\bar{\varphi}_i$ is the transformed shape function in the singular sub-element, and φ_i is the shape function for the whole boundary element.

In region 2, the quadrature uses local Gauss points but the shape functions from equation 12 are applied to the whole element in an analogous fashion as with the domain elements.

CONCLUSIONS

This paper details the development of the explicit forms of the fundamental solutions for the two-dimensional incompressible unsteady time-difference primitive-variable Navier-Stokes boundary element formulation. Also, explicit forms of singular intgrals are derived and presented. Comments on the interpolation functions and numerical implementation should allow the reader to more easily adapt this method for practical use.

ACKNOWLEDGEMENTS

The authors would like to express their appreciation to Drs. C.E. Capes and F.D.F. Talbot for their continued assistance in this project and providing the contract through the University of Ottawa which sponsored this work.

References

[1] Hörmander, L. Linear partial differential operators. 2nd revised printing, Berlin: Springer, 1964.

[2] Tosaka, N. Integral equation formulations with the primitive variables for incompressible viscous fluid flow problems. *Comp. Mech.* 1989, **4**, 89-103.

[3] Darcovich, K.; Talbot, F.D.F.; Capes, C.E. Hydrodynamic Interactions in Mineral Flotation: Moving Boundary Aspects. *ISCFD-Nagoya*, Nagoya, Japan, August 28-31, 1989; 1228-1233.

[4] Darcovich, K. Letter to the Editor on the "Integral equation formulations with primitive variables for viscous fluid flow", *Comp. Mech.*, 1990, **6**, 456.

[5] Tosaka, N.; Onishi, K. Boundary Integral equation formulations for unsteady incompressible viscous fluid flow by time-differencing. *Eng. Analysis* 1986, **3**, 2, 101-104.

[6] Brebbia, C.A.; Telles, J.C.F.; Wrobel, L.C. Boundary Element Techniques. Theory and Applications in Engineering. Berlin: Springer-Verlag, 1984.

[7] Lebedev, N.N. Special Functions and Their Applications. Dover: New York, 1972.

[8] Suh, I.-G. Application of the Boundary Element Method to Coupled Problems in Thermoelasticity, PhD. Thesis, Grad. School of Industrial Technology, Nihon U., 1990.

APPENDIX A

The two-dimensional differential operator equivalent to the Dirac function was given by Tosaka for the fundamental solution scalar ϕ[5]. It is:

$$(-\lambda + \Delta)\Delta\phi = \delta(x - y)$$

Now, let $\Delta\phi = \Psi$, so that $(-\lambda + \Delta)\Psi = \delta(x - y)$. $(-\lambda + \Delta)$ is the modified Helmholtz operator so that,

$$\Psi = -\frac{1}{2\pi}K_0(\sqrt{\lambda}r)$$

Let,

$$(-\lambda+\Delta)\phi = \Upsilon$$

So by the original equation,

$$\Delta\Upsilon = (-\lambda+\Delta)\Delta\phi = \delta(x-y)$$

Hence the differential operator on Υ is simply the Laplacian, and the corresponding Green's function is,

$$\Upsilon = (-\lambda+\Delta)\phi = \frac{1}{2\pi}\ln r$$

APPENDIX B

B.1 Sample Calculation

We begin by making a Taylor series expansion of the modified Bessel function[8]. In general,

$$K_n(z) = (-1)^{n+1}\sum_{l=0}^{\infty}\frac{\left(\frac{z}{2}\right)^{2l+n}}{l!\Gamma(n+l+1)}\left(\gamma+\ln\left(\frac{z}{2}\right)\right)$$
$$+\frac{(-1)^n}{2}\sum_{s=0}^{\infty}\frac{\left(\frac{z}{2}\right)^{n+2s}}{s!(n+s)!}\left[\sum_{m=1}^{s}\frac{1}{m}+\sum_{m=1}^{s+n}\frac{1}{m}\right]$$
$$+\frac{1}{2}\sum_{q=0}^{n-1}(-1)^q\frac{(n-q-1)!}{q!}\left(\frac{z}{2}\right)^{2q-n}$$

Above, $\gamma = 0.5772156649...$, which is Euler's constant.

So, for example,

$$K_2(z) = -\sum_{l=0}^{\infty}\frac{\left(\frac{z}{2}\right)^{2l+2}}{l!\Gamma(l+3)}\left(\gamma+\ln\left(\frac{z}{2}\right)\right)$$
$$+\frac{1}{2}\sum_{s=0}^{\infty}\frac{\left(\frac{z}{2}\right)^{2s+2}}{s!(2+s)!}\left[\sum_{m=1}^{s}\frac{1}{m}+\sum_{m=1}^{s+2}\frac{1}{m}\right]$$
$$+\frac{1}{2}\left(\left(\frac{z}{2}\right)^{-2}-1\right)$$

Now consider $\lim_{\epsilon\to 0}\int_{a_1}^{a_2}K_2(z)\,dz$. First, let $z/2 = x$, so $dz = 2\,dx$. Also, when $z = a_1 \Longrightarrow x = a_1/2$ and when $z = a_2 \Longrightarrow x = a_2/2$.

$$\lim_{\epsilon\to 0}\int_{a_1}^{a_2}K_2(z)\,dz = 2\lim_{\epsilon\to 0}\int_{a_1/2}^{a_2/2}K_2(2x)\,dx$$

Inserting the infinite series form of $K_2(z)$ into the above equation gives,

$$\lim_{\epsilon\to 0}\int_{a_1}^{a_2}K_2(z)\,dz = 2\lim_{\epsilon\to 0}\int_{a_1/2}^{a_2/2}\left\{-\sum_{l=0}^{\infty}\frac{x^{2l+2}}{l!\Gamma(l+3)}(\gamma+\ln x)\right.$$
$$\left.+\frac{1}{2}\sum_{s=0}^{\infty}\frac{x^{2s+2}}{s!(2+s)!}\left[\sum_{m=1}^{s}\frac{1}{m}+\sum_{m=1}^{s+2}\frac{1}{m}\right]-\frac{1}{2}\right\}dx$$
$$+2\lim_{\epsilon\to 0}\int_{a_1}^{a_2}\frac{1}{z^2}\,dz$$

$$= 2\lim_{\epsilon\to 0}\int_{a_1/2}^{a_2/2}\left\{-\sum_{l=0}^{\infty}\frac{x^{2l+2}}{l!\Gamma(l+3)}(\gamma+\ln x)\right.$$

$$\left.+\frac{1}{2}\sum_{s=0}^{\infty}\frac{x^{2s+2}}{s!(2+s)!}\left[\sum_{m=1}^{s}\frac{1}{m}+\sum_{m=1}^{s+2}\frac{1}{m}\right]-\frac{1}{2}\right\}dx$$

$$+2\lim_{\epsilon\to 0}\int_{a_1}^{a_2}\frac{1}{z^2}\,dz$$

$$= 2\lim_{\epsilon\to 0}\left\{-\sum_{l=0}^{\infty}\frac{1}{l!\Gamma(l+3)}\left\{\frac{\gamma}{2l+3}x^{2l+3}\Big|_{a_1/2}^{a_2/2}\right.\right.$$

$$\left.+\frac{1}{2l+3}(\ln x)x^{2l+3}\Big|_{a_1/2}^{a_2/2}-\frac{1}{(2l+3)^2}x^{2l+3}\Big|_{a_1/2}^{a_2/2}\right\}$$

$$+\frac{1}{2}\sum_{s=0}^{\infty}\frac{1}{s!(2+s)!}\frac{1}{2s+3}x^{2s+2}\Big|_{a_1/2}^{a_2/2}\left[\sum_{m=1}^{s}\frac{1}{m}+\sum_{m=1}^{s+2}\frac{1}{m}\right]$$

$$\left.-\frac{1}{2}x\Big|_{a_1/2}^{a_2/2}\right\}$$

$$+2\lim_{\epsilon\to 0}\int_{a_1}^{a_2}\frac{1}{z^2}\,dz$$

Recall that when $x = (a_1/2) \Longrightarrow z = (\sqrt{\lambda}R\varepsilon/2)$, and when $x = (a_2/2) \Longrightarrow z = (\sqrt{\lambda}R/2)$. Thus,

$$\lim_{\epsilon\to 0}\frac{a_1}{2} = 0 \quad \text{and,} \qquad \lim_{\epsilon\to 0}\frac{a_2}{2} = \frac{\sqrt{\lambda}R}{2}$$

Applying and evaluating the limits gives,

$$\lim_{\epsilon\to 0}\int_{a_1}^{a_2} K_2(z)\,dz =$$

$$2\left\{-\sum_{l=0}^{\infty}\frac{1}{l!\Gamma(l+3)}\left\{\frac{\gamma}{2l+3}\left(\frac{a_2}{2}\right)^{2l+3}\right.\right.$$

$$\left.+\frac{1}{2l+3}\ln\left(\frac{a_2}{2}\right)\left(\frac{a_2}{2}\right)^{2l+3}-\frac{1}{(2l+3)^2}\left(\frac{a_2}{2}\right)^{2l+3}\right\}$$

$$+\frac{1}{2}\sum_{s=0}^{\infty}\frac{1}{s!(2+s)!}\frac{1}{2s+3}\left(\frac{a_2}{2}\right)^{2s+2}\left[\sum_{m=1}^{s}\frac{1}{m}+\sum_{m=1}^{s+2}\frac{1}{m}\right]$$

$$\left.-\frac{1}{2}\left(\frac{a_2}{2}\right)\right\}$$

$$+2\lim_{\epsilon\to 0}\int_{a_1}^{a_2}\frac{1}{z^2}\,dz$$

Using the relation for integers that $\Gamma(n+1) = n!$, we can finally write,

$$\lim_{\epsilon\to 0}\int_{a_1}^{a_2} K_2(z)\,dz =$$

$$-2\left\{\sum_{l=0}^{\infty}\frac{1}{l!(l+2)!}\frac{1}{2l+3}\left(\frac{a_2}{2}\right)^{2l+3}\right.$$

$$\left.\left[\gamma+\ln\left(\frac{a_2}{2}\right)-\frac{1}{2l+3}-\frac{1}{2}\left[\sum_{m=1}^{l}\frac{1}{m}+\sum_{m=1}^{l+2}\frac{1}{m}\right]\right]\right\}$$

$$-\frac{a_2}{2}+2\lim_{\epsilon\to 0}\int_{a_1}^{a_2}\frac{1}{z^2}\,dz. \tag{32}$$

B.2 General Series Description

In general the singular integrals of the modified Bessel functions can be expressed as,

$$\lim_{\epsilon\to 0}\int_{a_1}^{a_2} z^m K_n(z)\,dz = \mathcal{S}(z^m K_n(z)) + F(z)$$

They have a series solution of the form,

$$\mathcal{S}(z^m K_n(z)) = C_1\left\{\sum_{l=0}^{\infty}\frac{1}{l!(l+C_2)!}\frac{1}{2l+C_3}\left(\frac{a_2}{2}\right)^{2l+C_3}\right.$$
$$\left.\left[\gamma+\ln\left(\frac{a_2}{2}\right)-\frac{1}{2l+C_3}-\frac{1}{2}\left[\sum_{m=1}^{l}\frac{1}{m}+\sum_{m=1}^{l+C_2}\frac{1}{m}\right]\right]\right\}$$
$$+C_4 \tag{33}$$

Table 1 summarizes the particular coefficients for all nine series required in this paper. Computational testing with the set of series detailed in Table 1 has shown that using $a_2 = 1.5$ gives satisfactory convergence for all the series, and produces suitable element partitioning.

integral	variable	C_1	C_2	C_3	C_4	$F(z)$
$\lim_{\epsilon\to 0}\int_{a_1}^{a_2} K_2(z)\,dz$	$\mathcal{S}(K_2(z))$	-2	2	3	$\frac{-a_2}{2}$	$2\lim_{\epsilon\to 0}\int_{a_1}^{a_2} \frac{1}{z^2}\,dz$
$\lim_{\epsilon\to 0}\int_{a_1}^{a_2} zK_2(z)\,dz$	$\mathcal{S}(zK_2(z))$	-4	2	4	$\frac{-a_2^2}{4}$	$2\lim_{\epsilon\to 0}\int_{a_1}^{a_2} \frac{1}{z}\,dz$
$\lim_{\epsilon\to 0}\int_{a_1}^{a_2} z^{-1}K_1(z)\,dz$	$\mathcal{S}(z^{-1}K_1(z))$	1	1	1	0	$\lim_{\epsilon\to 0}\int_{a_1}^{a_2} \frac{1}{z^2}\,dz$
$\lim_{\epsilon\to 0}\int_{a_1}^{a_2} K_1(z)\,dz$	$\mathcal{S}(K_1(z))$	2	1	2	0	$\lim_{\epsilon\to 0}\int_{a_1}^{a_2} \frac{1}{z}\,dz$
$\lim_{\epsilon\to 0}\int_{a_1}^{a_2} z^2K_2(z)\,dz$	$\mathcal{S}(z^2K_2(z))$	-8	2	5	$4(\frac{a_2}{2}-\frac{a_2^3}{24})$	0
$\lim_{\epsilon\to 0}\int_{a_1}^{a_2} zK_1(z)\,dz$	$\mathcal{S}(zK_1(z))$	4	1	3	a_2	0
$\lim_{\epsilon\to 0}\int_{a_1}^{a_2} K_3(z)\,dz$	$\mathcal{S}(K_3(z))$	2	3	4	$\frac{a_2^2}{48}$	$8\lim_{\epsilon\to 0}\int_{a_1}^{a_2} \frac{1}{z^3}\,dz - \lim_{\epsilon\to 0}\int_{a_1}^{a_2} \frac{1}{z}\,dz$
$\lim_{\epsilon\to 0}\int_{a_1}^{a_2} zK_3(z)\,dz$	$\mathcal{S}(zK_3(z))$	2	3	5	$-a_2+\frac{a_2^3}{72}$	$8\lim_{\epsilon\to 0}\int_{a_1}^{a_2} \frac{1}{z^2}\,dz$
$\lim_{\epsilon\to 0}\int_{a_1}^{a_2} z^{-1}K_2(z)\,dz$	$\mathcal{S}(z^{-1}K_2(z))$	-1	2	2	0	$2\lim_{\epsilon\to 0}\int_{a_1}^{a_2} \frac{1}{z^3}\,dz - \frac{1}{2}\lim_{\epsilon\to 0}\int_{a_1}^{a_2} \frac{1}{z}\,dz$

Table 1: Summary of series expansions of modified Bessel function singular integrals.

Lagrangian Approach of BEM for Incompressible Unsteady Viscous Flow

X. Jin, D.K. Brown

Department of Mechanical Engineering, University of Glasgow, Glasgow, G12 8QQ, U.K.

ABSTRACT

A Lagrangian approach of boundary element method is presented for the numerical simulation of unsteady viscous flow. The time-dependent fundamental solutions are used for the transformation of the Navier-Stokes equations into the boundary integral equations in primitive variables. The method is based on the Lagrangian description of the fluid motion which localizes the variables to a finite number of fluid particles; the material acceleration is then the time derivative considering the change of the location of a fluid particle. Since the boundary element method has the advantage of the unlimited analytical resolution within the fluid body, a velocity correction approach can be obtained for a more accurate approximation. While the advantage is based on the accuracy of the boundary elements, an approximation function for the singularity of $E_1(s)-S_1(s)$ and a replacing integral for the other singularity on the collocated element integral are given.

1.INTRODUCTION

Many developments in numerical simulation of fluid dynamics by boundary element methods have been made over the last decade. As an alternative approach, the methods have gradually received the same attention as that finite difference and finite element methods in solving the general problems in engineering. This is because the boundary element methods have some natural advantages like (1) the reduction of problem by one dimension; (2) the unlimited analytical resolution within the domain whose accuracy depends on the accuracy of discretization of the boundary elements. Although they have been very successful in linear equations like inviscid fluid flow problems, they need some special treatments in dealing with the severe nonlinearities because the integral equations are no longer " boundary-only", or we can say that "the fundamental solutions" are no longer those of the original equations.

Most developments of the boundary element method in fluid dynamics use the Eulerian description which unavoidably has the nonlinear convection terms, they are characterized by finding the values of the fluid variables at spatial points.

For problems with free surfaces or those concerned with the history of the fluid particles, however, the Lagrangian description has an advantage. The numerical techniques in Lagrangian form have been developed for fluid dynamics both in the finite difference method and the finite element method for many years. Since the grid system follows the fluid motion, their main advantages also include linearization of the governing equations. The Lagrangian techniques have a major disadvantage of mesh tangling at severe distortions in these numerical methods, and it has been thought to block their developments. Some efforts have been made to overcome the mesh distortion problems, such as various methods of remeshing, finite volume method and other mixed Lagrangian-Eulerian methods like PIC[1,2], MAC[3] and ALE[4]. The combinations of both the Lagrangian and Eulerian viewpoints become one of the main directions in the numerical simulations in fluid dynamics.

The disadvantage of pure Lagrangian techniques is the grid system in which most of the unknowns are defined and solved explicitly or implicitly. Since the unknowns to be solved in boundary element methods are those on the boundaries, the unknowns within the domain can be derived from the boundary values, though in nonlinear problems they have to be iterated out. The boundary element methods in Lagrangian description can give better solutions to the mesh distortion problem, and avoid the nonlinear convection terms. In this paper, we wish to present this Lagrangian approach of the boundary element method for incompressible unsteady viscous flow.

2.GOVERNING EQUATIONS AND FUNDAMENTAL SOLUTIONS

The governing equations for incompressible viscous unsteady fluid are given by the continuity equation

$$u_{i,i} = 0 \tag{1}$$

and the Navier-Stokes equation

$$Re\frac{Du_i}{Dt} = T_{ij,j} + f_i \tag{2}$$

in which the stress tensor is given by

$$T_{ij} = -Re\,P\delta_{ij} + u_{i,j} + u_{j,i} \tag{3}$$

where D/Dt is the substantial (or Lagrangian time) derivative, and u_i, ρ, $p=\rho P$ and Re are the velocity, density, pressure and Reynolds number respectively. The initial and boundary conditions about the fluid medium are given as follows

$$\left.\begin{array}{ll} u_i = u_i^o(\vec{x}), & T_{ij} = T_{ij}^o(\vec{x}) \\ u_i = \bar{u}_i(\vec{x}, t), & \tau_i = n_j\bar{T}_{ij}(\vec{x}, t) \end{array}\right\} \tag{4}$$

Equations (1)-(4) provide a complete system for solving the unsteady viscous flow problems. If we reduce the substantial derivative in equation (2) to a partial derivative respect to time, the equation system then becomes the unsteady state form of the creeping motion equations. The fundamental solutions of these equations can then be easily derived by Hormander's method, however, the significant work was first done by Oseen[5]. The solutions of velocity and stress component in the i-direction under a unit pulse force in the k-direction acting on the point x' at time t' can be given as follows:

$$u^*_{ik} = \delta_{ik}\Delta\Phi - \Phi_{,ik} \tag{5}$$

$$\sigma^*_{ik} = (\delta_{ik}D_j + \delta_{jk}D_i + \delta_{ij}D_k)\Delta\Phi - 2\Phi_{,ikj} - \mathrm{Re}\,\delta_{ij}\frac{\partial}{\partial t}\Phi_{,k} \tag{6}$$

where Φ is the fundamental solution of equation

$$\left(\Delta - \mathrm{Re}\frac{\partial}{\partial t}\right)\Delta\Phi = \delta(x - x')\delta(t - t') \tag{7}$$

From equations (5) and (6) we can see that it is enough that the derivative of Φ with respect to r is given. Its two dimensional form is provided as follows:

$$\frac{d\Phi}{dr} = \frac{H(t, t')}{2\,\mathrm{Re}\,\pi r}\left[1 - \exp\left(\frac{-\mathrm{Re}\,r^2}{4(t - t')}\right)\right] \tag{8}$$

in which H(t,t') is Heaviside function and r is the distance between x and x'. Velocity and stress components of the fundamental solutions are provided in detail in the Appendix.

3.INTEGRAL FORMULATION IN LAGRANGIAN DESCRIPTION

Having obtained the fundamental solutions of the unsteady creeping motion equations, we wish to establish a group of integral equations for replacing the original Navier-Stokes equations. Both groups of equations can be rewritten as following:

$$\left.\begin{aligned} &\mathrm{Re}\frac{\partial u^*_{ik}}{\partial t'} = -\sigma^*_{ikj,j} && 0 \le t' < t \\ &\mathrm{Re}\frac{\partial u_i}{\partial t'} + \mathrm{Re}(u_j - w_j)u_{i,j} = \sigma_{ij,j};\ x' \in \Omega(t');\ 0 \le t' \end{aligned}\right\} \tag{9}$$

where w_j are the velocity components of a reference system, and the left hand side of the second groups are the expansion of the substantial derivative. From equation (9) we can have

$$\begin{aligned} \mathrm{Re}\frac{\partial(u^*_{ik}u_i)}{\partial t'} &= \mathrm{Re}\left(u^*_{ik}\frac{\partial u_i}{\partial t'} + u_i\frac{\partial u^*_{ik}}{\partial t'}\right) = \\ &= u^*_{ik}[\sigma_{ij,j} - \mathrm{Re}(u_j - w_j)u_{i,j}] - u_i\sigma^*_{ikj,j} \end{aligned} \tag{10}$$

Integrating both sides of this equality over the time-spatial region $\Omega x(t_o,t_1)$, where Ω is the space which the medium concerned occupies at time t, and t_1=t-ε, ε is a small positive value, $t_o \le t' < t$, we obtain

$$\mathrm{Re}\int_{t_o}^{t_1}\int_{\Omega}\frac{\partial(u^*_{ik}u_i)}{\partial t'}\,d\Omega dt' = \int_{t_o}^{t_1}\int_{\Omega}\{u^*_{ik}[\sigma_{ij,j} - \mathrm{Re}(u_j - w_j)u_{i,j}] - u_i\sigma^*_{kj,j}\} \quad (11)$$

We assume that the reference system deforms with time, the left hand side of this equation can be expanded:

$$\int_{t_o}^{t_1}\int_{\Omega}\frac{\partial(u^*_{ik}u_i)}{\partial t'}\,d\Omega dt' = \int_{t_o}^{t_1}\left[\frac{\partial}{\partial t}\int_{\Omega}u^*_{ik}u_i d\Omega - \int_{\Gamma}(u_n - w_n)u^*_{ik}u_i d\Gamma\right]dt' \quad (12)$$

where w_n is the boundary moving velocity. The first integral on the right hand side of this expression can be further integrated

$$\int_{t_o}^{t_1}\frac{\partial}{\partial t}\int_{\Omega}u^*_{ik}u_i d\Omega\, dt' = \int_{\Omega(t-\varepsilon)}u^*_{ik}u_i d\Omega - \int_{\Omega_o}u^*_{ik}(x,t,x',t_o)u_i(x',t_o)d\Omega \quad (13)$$

in which the domain Ω_o is where the reference space at time t_o.

By using the velocity fundamental solution (5), we expand the first of this expression as

$$\int_{\Omega(t-\varepsilon)}u^*_{ik}u_i d\Omega = \int_{\Omega(t-\varepsilon)}\delta_{ik}\Delta\Phi u_i d\Omega - \int_{\Gamma(t-\varepsilon)}u_i n_i \Phi_{,k} d\Gamma \quad (14)$$

where equation (1) has been used. By a transformation, the first integral can be integrated as

$$\lim_{\varepsilon\to 0}\int_{\Omega(t-\varepsilon)}\delta_{ik}\frac{\exp\left(\frac{-\mathrm{Re}r^2}{4\varepsilon}\right)}{4\pi\varepsilon}u_i d\Omega = \frac{Cu_k(x',t)}{\mathrm{Re}} \quad (15)$$

where constant C depends on the location of x', it is one for an internal point, half for the point lying on the smooth boundary and zero for any point outside the domain. By using equation (8), the second integral can be rewritten as

$$\lim_{\varepsilon\to 0}\int_{\Gamma(t-\varepsilon)}u_i n_i\frac{x_k}{2\,\mathrm{Re}\,\pi r^2}\left[1 - \exp\left(\frac{-\mathrm{Re}\,r^2}{4\varepsilon}\right)\right]d\Gamma = \int_{\Gamma(t)}\frac{u_i n_i x_k}{2\,\mathrm{Re}\,\pi r^2}d\Gamma \quad (16)$$

The first and the third terms in the right hand side of equation (11) can produce boundary integrals after integrations by parts. The integral equations can now be rewritten as:

$$Cu_k(x;t) = \int_{t_o}^{t}\left[\int_{\Gamma}(u^*_{ik}\sigma_i - u_i\sigma^*_{ik})d\Gamma - Re\int_{\Omega}u^*_{ik}(u_j - w_j)u_{i,j}d\Omega\right]dt'$$

$$+\int_{\Gamma(t)}\frac{u_i n_i x_k}{2\pi r^2}d\Gamma + Re\left[\int_{t_o}^{t}\int_{\Gamma}(u_n - w_n)u^*_{ik}u_i d\Gamma dt' + \int_{\Omega_o}u^*_{ik}u_i d\Omega\right] \quad (17)$$

It can be seen that the second domain integral and the fourth boundary integral can be eliminated if the reference system is fixed with the deforming medium, that is, $w_j=u_j$. Under such a Lagrangian description, the domain Ω is now the medium which is time-dependent and the boundary moving velocity is then u_n, the following integration representations can be obtained:

$$Cu_k = \int_{t_o}^{t}\int_{\Gamma}(u^*_{ik}\sigma_i - u_i\sigma^*_{ik})d\Gamma dt' + \int_{\Gamma(t)}\frac{u_i n_i x_k}{2\pi r^2}d\Gamma + Re\int_{\Omega_o}u^*_{ik}u_i d\Omega \quad (18)$$

in which Ω_o is now where the medium occupies at time t_o.

This group of equations (18) provides a required relationship in Lagrangian description between the velocity-stress field inside the medium and the corresponding velocity and traction boundary values. Based on the equations the motion of a finite fluid medium or multimedia interaction can be solved, as well as infinite fluid motion. The major difference between its form in Eulerian description is the disappearance of the time-spatial integral about the convection terms.

4.NUMERICAL IMPLEMENTATION

Discretizations

Numerical implementation of equations (18) require discretization both in time and in the medium. Consider first the time integrals that appear on the right hand side of expression (18). If the time period from zero to T is divided into N small increments of interval Δt, and the increments are so small that the velocity and stress components can be assumed to be constants within each time increment, then the time integrals can be integrated analytically, consequently expression (18) can be rewritten as:

$$Cu_k(x;t) = \int_{\Gamma}\left[U^*_{ik}\sigma_i - u_i\left(\Sigma^*_{ik} - \frac{n_i x_k}{2\pi r^2}\right)\right]d\Gamma + Re\int_{\Omega_o}u^*_{ik}u_i d\Omega \quad (19)$$

where

$$U^*_{ik}(x - x') = \int_{t_o}^{t}u^*_{ik}(x,t,x';t')dt' \quad (20a)$$

$$\Sigma^*_{k}(x-x') = \int_{t_0}^{t} \sigma^*_{ik}(x,t,x',t')dt' \tag{20 b}$$

The detail of the expressions are provided in the Appendix.

After the time discretization, the medium can be discretized into small elements, both the body and the boundary. Being different from Eulerian description, these elements are moving, deforming in each time step as they are defined in the reference system which is fixed with the medium. Similarly, they are defined in terms of nodal coordinates and shape functions, and the velocity and traction components are also written as products of the nodal values and shape functions. The integrations over each element can be done by standard Gaussian formulae except the element which x and x' collocate each other.

Singularities

For linear elements, the integrations in such collocation elements contain singularities. Careful investigation shows that $E_1(s)$-$S_1(s)$ and $S_1(s)/r$ have singularities while others are all finite values when r approaches zero.

From the expanding of the exponential integral in series

$$E_1(s) = -c - \ln s + \sum_{n=1}^{\infty}(-1)^{n-1}\frac{s^n}{nn!} \tag{21}$$

it can be seen that when s approaches zero $E_1(s)$-$S_1(s)$ is a logarithmic type singularity. This gives rise to the idea to use a logarithmic function to approximate it. The function for linear boundary elements can be written as follow

$$F_1(r) = [E_1(Dr^2) - S_1(Dr^2)]\left(1 - \frac{r}{L}\right) \tag{22}$$

from which there may be a zero point r_o within the element. Since logarithmic function is zero at one, we have a reason to use the following three-point approximation scheme:

$$F_1(r) \approx \ln\left(\frac{r_o}{r}\right)(1 + a_1 r + a_2 r^2 + a_3 r^3) \tag{23}$$

where a_i (i=1,2,3) are the constants to be determined by the following conditions

$$\left.\begin{aligned} &F_1(r_1) = \ln\left(\frac{r_o}{r_1}\right)(1 + a_1 r_1 + a_2 r_1^2 + a_3 r_1^3);\ \text{where}\ \left.\frac{dF_1}{dr}\right|_{r=r_1} = 0 \\ &\ln\left(\frac{r_o}{L}\right)(1 + a_1 L + a_2 L^2) = 0 \\ &F_1(r_2) = \ln\left(\frac{r_o}{r_2}\right)(1 + a_1 r_2 + a_2 r_2^2 + a_3 r_2^3);\ \text{where}\ r_1 < r_2 < L \end{aligned}\right\} \tag{24}$$

With proper choice of r_2, the approximation can cover a range of $0.8 < ReL^2/(4dt) < 10$, which is large enough for the numerical calculation. By this approximation, the integral singularity can be easily worked out.

The second integral singularity, $S_1(s)/r$, can be changed by reconsidering equation (14) in following form:

$$\int_{\Gamma(t)} u_i n_i \frac{x_k}{2\,\mathrm{Re}\,\pi r^2}\left[1-\exp\left(\frac{-\mathrm{Re}\,r^2}{4\theta(t-t_o)}\right)\right]d\Gamma; \quad 0<\theta<1$$

in which case the function for the boundary integral is then replaced by $[S_1(s)-\exp(-s/\theta)]/r$ which is nonsingularity. For problems with condition $u_i n_i=0$ on all of the boundary, the singularity does not exist.

Time-marching and Iteration

The discretized equations are based on the Lagrangian coordinates. The coordinates, when viewed as functions of particles and time, are then expressed as the displacement functions $x_i=x_i(x_k{}^o,t)$, where $x_k{}^o$ are the initial coordinates at time t_o. Therefore the location of each particle concerned after time increment Δt is given by

$$x_i(P_k, t^{n+1}) = x_i^n + \frac{\Delta t}{2}\left(u_i^{n+1} + u_i^n\right) \tag{25}$$

The velocity at each time step can be rewritten as a function of the velocity and the location of each particle, that is,

$$u_i = f_i(x_k^n, u_k^n) \tag{26}$$

However, it can be seen that the location is not where the particle goes at time t^{n+1}. To find out the location and the velocity, iterations must be carried out. At the initial step of iteration, the velocity and the location are computed as

$$\left.\begin{aligned} u_i^{L(o)} &= f_i(x_k^n, u_k^n) \\ x_i^{L(o)} &= x_i^n + \Delta t u_i^n \end{aligned}\right\} \tag{27}$$

Since the boundary element methods have the advantage of the unlimited analytical resolution within the domain, we can calculate the velocity of any position inside the medium, thus we have

$$\left.\begin{aligned} u_i^{L(m)} &= f_i\left(x_k^{L(m-1)}, u_k^{L(m-1)}\right) \\ x_i^{L(m)} &= x_i^n + \frac{\Delta t}{2}\left(u_i^{L(m)} + u_i^{L(o)}\right) \end{aligned}\right\} \tag{28}$$

Equations (28) specify that the velocity and the location can be computed using the latest location and velocity of the fluid particles. The iteration is repeated until the computed velocity satisfies the convergence criterion

$$\left|u_i^{L(m)} - u_i^{L(m-1)}\right| \le \varepsilon \tag{29}$$

where ε is a previously defined small value. If the process converges then both the velocity and location values obtained at the mth iteration are assigned to (n+1)th time step.

5.CONCLUSION

The boundary integral equations in Lagrangian description for incompressible unsteady viscous flow are presented. The formulations are based on time-dependent fundamental solutions of unsteady creeping motion equations which were first given by Oseen. The equations can be implemented by boundary element method. In order to solve the two singularities existing on the collocation elements, an approximation and a replacement are used.

REFERENCES

1. Harlow, F. H., The Particle-In-Cell Computing Method for Fluid Dynamics, Methods in Computational Physics, Vol. 3, Alder, Fernbach, and Rotenberg, Eds., Academic Press, New York and London, p319.(1964)
2. Harlow, F. H. PIC and Its Progeny, LA-UR-87-1862, (1987)
3. Welch,J.E.;Harlow,F.H.;Shannon,J. P.; Daly, B. J., The MAC Method—A Computing Technique for Solving Viscous Incompressible Transient Fluid-Flow Problems Involving Free Surfaces, LA-3425, Nov. 1965
4. Amsden, A. A.; Hirt, C. W., YAQUI: An Arbitrary Lagragian-Eulerian Computer Program for Fluid Flow at All Speed, LA-5100, Mar. 1973
5. Oseen,C. W., Neuere Methoden und Ergebnisse in der Hydrodynamik,Akad. Verlagsgesellschaft, Leipzig, (1927)

APPENDIX

$$u^*_{ik} = \delta_{ik}(\Delta\Phi - S_o(r)) - \frac{x_i x_k}{r^2}(\Delta\Phi - 2S_o(r)) \qquad (A.1)$$

$$\begin{aligned}\sigma^*_{ikj} = &-\left(\delta_{ik}x_j + \delta_{ij}x_k + \delta_{jk}x_i\right)\left[\frac{\mathrm{Re}\,\Delta\Phi}{2(t-t')} + \frac{2}{r^2}(\Delta\Phi - 2S_o(r))\right] \\ &+ \frac{4x_i x_k x_j}{r^4}\left[2(\Delta\Phi - 2S_o(r)) + \frac{\mathrm{Re}\,r^2\Delta\Phi}{2(t-t')}\right] \\ &- \delta_{ij}\,\mathrm{Re}\,x_k\left\{\frac{\delta(t-t')}{2\pi\,\mathrm{Re}\,r^2}\left[1 - \exp\left(\frac{-\mathrm{Re}\,r^2}{4(t-t')}\right)\right] + \frac{\Delta\Phi}{2(t-t')}\right\}\end{aligned} \qquad (A.2)$$

where

$$\Delta\Phi = \frac{H(t,t')}{4\pi(t-t')}\exp\left(\frac{-\mathrm{Re}\,r^2}{4(t-t')}\right) \qquad (A.3)$$

$$S_o(r) = \frac{H(t,t')}{2\,\mathrm{Re}\,\pi r^2}\left[1 - \exp\left(\frac{-\mathrm{Re}\,r^2}{4(t-t')}\right)\right] \qquad (A.4)$$

The integrations over time range from t_o to t are given as follows

$$U^*_{ik} = \int_{t_o}^{t} u^*_{ik}\,dt = \frac{1}{8\pi}\left\{ \delta_{ik}[E_1(s) - S_1(s)] + \frac{2x_i x_k}{r^2} S_1(s) \right\} \tag{A. 5}$$

$$\Sigma^*_{ik} = \int_{t_o}^{t} \sigma^*_{ikj} n_j\,dt = \frac{1}{2\pi r^2}\Big\{ \left(\delta_{ik} x_j n_j + x_i n_k\right)[S_1(s) - e^{-s}]$$

$$- \frac{2x_i x_j n_j x_k}{r^2}[2S_1(s) - e^{-s}] + x_k n_i [S_1(s) - 1] \Big\} \tag{A. 6}$$

where

$$S_1(s) = \frac{1}{s}(1 - e^{-s}); \; E_1(s) = \int_s^{\infty} \frac{e^{-u}}{u} du \; ; \; s = \frac{Re\, r^2}{4(t - t_o)} \tag{A. 7}$$

Regular Perturbations for the Exterior Three-Dimensional Slow Viscous Flow Problem

H. Power (*), G. Miranda (**), V. Villamizar (**)

(*) Instituto de Mecánica de Fluídos, Universidad Central de Venezuela, Facultad de Ingeniería, Venezuela

(**) Departamento de Matemática, Universidad Central de Venezuela, Facultad de Ciencias, Venezuela

ABSTRACT:

A numerical method for solving the inhomogeneous Oseen's problems resulting from Finn's regular perturbation expansion, established for the three-dimensional steady flow of a viscous incompressible fluid past an arbitrary obstacle, at small Reynolds number, is developed here. This method is based on the numerical solution of a system of linear Fredholm's integral equations of the first kind derived from the representation formula of Green's type for the exterior Oseen's flow fields.

INTRODUCTION:

We consider a uniform flow of an incompressible viscous fluid past a body of arbitrary shape, in the limit of arbitrarily small Reynolds number.

The steady Navier-Stokes equation, in dimensionless form, is:

$$\frac{\partial^2 u_i}{\partial x_j \partial x_j} - \frac{\partial p}{\partial x_i} - R u_j \frac{\partial u_i}{\partial x_j} = 0 \tag{1-a}$$

and the continuity equation is:

$$\frac{\partial u_i}{\partial x_i} = 0 \tag{1-b}$$

where $x = (x_1, x_2, x_3) \in \Omega_e$ and Ω_e is the three-dimensional unbounded domain exterior to the body whose boundary is an arbitrary Lyapunov surface S, by Ω_i we will designate the complement of $\bar{\Omega}_e$. Equations (1-a,b) are non-dimensionalized with respect to the characteristic variables $u_c = V$ (magnitude of the prescribed uniform velocity at infinity), $l_c = a$ (characteristic body length), and $p_c = \mu V/a$, with μ as the fluid viscosity. The Reynolds number is defined as $R = Va/\nu$ with $\nu = \mu/\rho$ and ρ is the fluid density.

The fluid velocity $\vec{u}$ has to satisfy the nonslip boundary condition at the body surface:

$$u_i(x) = 0 \qquad \text{for all } x \in S \tag{2}$$

as well as the following asymptotic value at infinity

$$\lim_{|x|\to\infty} \vec{u} \to \vec{e}_1 \tag{3}$$

where we have assumed that the uniform velocity at infinity is in the x_1 coordinate direction.

A general approach to this problem has been developed by the method of matched asymptotic expansions due to Kaplun and Lagerstrom (1957) and Proudman and Pearson (1957). In this method, the space around the body is divided into two separate but overlapping regions, and an appropriate perturbation scheme relevant to each region is considered. In the inner region, since the inertial term is small (i. e. $R \ll 1$), it is assumed an inner expansion of the form:

$$(\vec{u},p) = (\vec{u},p)^0 + R(\vec{u},p)^1 + O(R), \qquad \text{as } R \to 0 \tag{4}$$

where the leading-order term satisfies the Stokes' problem:

$$\frac{\partial^2 u_i^0}{\partial x_j \partial x_j} - \frac{\partial p^0}{x_i} = 0 \tag{5-a}$$

$$\frac{\partial u_i^0}{x_i} = 0 \tag{5-b}$$

$$u_i^0 = 0 \quad \text{on S}; \qquad u_i^0 \to \delta_{ij},\ p^0 \to 0, \qquad \text{as } |\, x \,| \to \infty \tag{5-c}$$

and the first order term $(\vec{u},p)^1$ is solution of the following non-homogeneous Stokes' problem:

$$\frac{\partial^2 u_i^1}{\partial x_j \partial x_j} - \frac{\partial p^1}{\partial x_i} - u_j^0 \frac{\partial u_j^0}{\partial x_j} = 0 \tag{6-a}$$

$$\frac{\partial u_i^1}{\partial x_i} = 0 \tag{6-b}$$

$$u_i^1 = 0 \quad \text{on S}; \qquad u_i^1 \to 0,\ p^1 \to 0 \qquad \text{as } |\, x \,| \to \infty \tag{6-c}$$

The boundary value problem (5) is usually referred as the Stokes' problem, and it possesses a unique solution. Problem (6) is singular, in the sense that there exists no solution to it (The Whitehead Paradox). However, it is possible to obtain a solution if the condition at infinity is relaxed. The question of how to modify this condition at infinity, so that there exists a corresponding solution, has been solved formally using an outer expansion.

In the outer region $|\, x \,| \geq 0(R^{-1})$, where the inertia terms must be considered, expansion (4) is not valid, as is known. In this region, let the solution $(\vec{v},\hat{p})$

be expressed in the form of an asymptotic outer expansion (where $(\vec{v},\hat{p})$ is the flow field at the outer region).

$$(\vec{v},\hat{p}) = (\vec{v},\hat{p})^0 + R(\vec{v},\hat{p})^1 + O(R) \qquad \text{as } R \to 0 \tag{7}$$

The obvious choice for the first term in (7) is the uniform flow at infinity:

$$(\vec{v},\hat{p})^0 = (e_1, 0) \tag{8}$$

Then the governing equation for the next term $(\vec{v},\hat{p})^1$ will be Oseen's equations:

$$\frac{\partial^2 v_i^1}{\partial \hat{x}_j \partial \hat{x}_j} - \frac{\partial \hat{p}^1}{\partial \hat{x}_i} - R\frac{\partial v_i^1}{\partial \hat{x}_1} = 0 \tag{9-a}$$

$$\frac{\partial v_i^1}{\partial \hat{x}_i} = 0 \tag{9-b}$$

The inner expansion (4) satisfies the nonslip condition at the body surface, and the outer expansion (7) satisties the uniform stream condition at infinity, and the construction of the solution can be completed by the matching procedure applied in the overlapping domain of the two regions. This matching procedure yields further asymptotic conditions for each expansion and it enables to find its successive terms alternatively.

The validity of the above formal matched asymptotic expansion was established by Fisher et al. (1985) using an existing theory for a Fredholm's first kind integral equation arising when the solution of the problem corresponding to each term in the expansions is represented as a single-layer potential.

Youngren and Acrivos (1975) solved numerically the Stokes' problem (5) by means of a first kind Fredholm's integral equation derived from an integral representation formulae of the Green's type for the Stokes' equation. Lee and Leal (1986) studied the low Reynolds number flow past cylindrical bodies of arbitrary cross-sectional shape based on the method of matched asymptotic expansions, and there they use the two-dimensional version of Youngren and Acrivos' integral equation method to solve, numerically, the corresponding Stokes' problems associated with the expansion.

On the other hand, Finn (1968) was able to establish an existence proof for the solution of the Navier-Stokes problem (1)-(3) by considering the solution of (1) as a regular perturbation of the solution of Oseen's problem. More precisely, Finn shows that exists a unique physically reasonable solution of (1)-(3) for sufficiently small R and that this solution admits a regular expansion:

$$u_i(x,R) = \delta_{i1} + \Sigma_{k=1}^{n} Q_i^k(x,R) + 0(R^n) \tag{10-a}$$

$$p(x,R) = \sum_{k=1}^{n} P^k(x,R) + 0(R^n) \tag{10-b}$$

uniformly in Ω_e, where the $(\vec{Q}^k, P^k)$ are solutions of the corresponding Oseen's problems, non-homogeneous for $k > 1$. In particular, $\vec{Q}^1$ and $\vec{Q}^2$ are solutions of the following boundary value problems:

$$\left(\frac{\partial^2}{\partial x_j \partial x_j} - R\frac{\partial}{\partial x_1}\right)Q_i^1 - \frac{\partial P^1}{\partial x_i} = 0 \qquad \text{in } \Omega_e \qquad (11\text{-a})$$

$$\frac{\partial Q_i^1}{x_i} = 0 \qquad \text{in } \Omega_e \qquad (11\text{-b})$$

$$Q_i^1 = -\delta_{1i} \qquad \text{on S;} \qquad Q_i^1 \to 0,\ P^1 \to 0 \qquad \text{as } |x| \to \infty \qquad (11\text{-c})$$

$$\left(\frac{\partial^2}{\partial x_j \partial x_j} - R\frac{\partial}{\partial x_1}\right)Q_i^2 - \frac{\partial P^2}{\partial x_i} = RQ_j^1\frac{\partial}{\partial x_j}Q_i^1 \qquad \text{in } \Omega_e \qquad (12\text{-a})$$

$$\frac{\partial Q_i^2}{x_i} = 0 \qquad \text{in } \Omega_e \qquad (12\text{-b})$$

$$Q_i^2 = 0 \qquad \text{on S;} \qquad Q_i^2 \to 0,\ P^2 \to 0 \qquad \text{as } |x| \to \infty \qquad (12\text{-c})$$

With this recognized importance of solving the Oseen's equations subject to appropriate boundary conditions, we turn our attention to Finn's regular perturbation method instead of Kaplum's singular perturbation method.

The formulation of boundary value problems for Oseen's flows as integral equations goes back to Oseen's original work (1927) (for a good literature survey see Olmstead and Gautesen (1976)), but most of the actual solutions determined by this method, correspond to two-dimensional problems for simple geometries, where approximate analytic solutions via Fourier transform are obtained.

In this work, we present a numerical first kind Fredholm's integral equations method for solving Oseen' boundary value problem associated with Finn's regular perturbation method, developing for Oseen's boundary value problem the analog of Youngren and Acrivos' integral equation method for the Stokes' boundary value problem.

OSEEN'S SOLUTION

In this section, we will find a numerical integral equation solution of the general non-homogeneous Oseen's problem:

$$\mu\frac{\partial^2 u_i}{\partial x_j \partial x_j} - \rho U\frac{\partial u_i}{\partial x_1} - \frac{\partial p}{\partial x_i} = g_i(x) \qquad \text{in } \Omega_e \qquad (13\text{-a})$$

$$\frac{\partial u_i}{\partial x_i} = 0 \qquad \text{in } \Omega_e \qquad (13\text{-b})$$

$$u_i = -U\delta_{1i} \qquad \text{on S;} \qquad u_i \to 0,\ p \to 0 \qquad \text{as } |x| \to \infty \qquad (13\text{-c})$$

The reduction of the solution of (13) to the solution of a system of integral equations given originally by Oseen himself, and a complete account of the properties of the surface potentials and volume potential generated from Oseen's integral representation is given by Miranda and Power (1983). The fundamental singular solution of (13-a,b) (called Oseenlet) is:

$$u_i = V_i^k(x,y) = \frac{1}{4\pi\mu} e^{k(x_1-y_1+r)}\delta_{ik} + \frac{\partial\phi^k}{\partial x_i}(x,y) \tag{14-a}$$

$$p = q^k(x,y) = \frac{x_k - y_k}{4\pi r^3} \tag{14-b}$$

with

$$r = | x - y | \tag{14-c}$$

$$\phi^k(x,y) = -\frac{1}{4\pi\rho U}\left(1 - e^{k(x_1-y_1+r)}\right)\frac{\partial}{\partial x_k}\left(\ln(x_1 - y_1 + r)\right) \tag{14-d}$$

$$k = \frac{-\rho U}{2\mu} \tag{14-e}$$

here δ_{ik} is the Kronecker's delta and (V_i^k, q^k) is the fundamental solution of (13-a,b), it corresponds to the non-homogeneous term in (13-a) equal to $-\delta(x-y)\,\vec{e}^k$.

From Oseen's integral representation formulae, we can write the general solution of (13) as:

$$\begin{aligned} u_k(x) = &\int_S V_i^k(x,y)\sigma_{ij}(\vec{u}(y))\; n_j(y)\; dS_y - \int_S u_i(y)\Sigma_{ij}\left(\vec{V}^k(x,y)\right)\; n_j(y)\; dS_y \\ &- \rho U \int_S u_i(y)\; n_1(y) V_i^k(x,y)\; dS_y \\ &- \int_{\Omega_e} g_i(y) V_i^k(x,y)\; dy \qquad \text{for every } x \in \Omega_e \end{aligned} \tag{15}$$

if $\vec{g}(x)$ decays sufficiently rapid at infinity, condition that is satisfied by the non-homogeneous terms of equation (12), here:

$$\sigma_{ij}(\vec{u}) = -p\delta_{ij} + \mu\left(\frac{\partial u_i}{\partial x_j} + \frac{\partial u_j}{\partial x_i}\right)$$

$$\Sigma_{ij}(\vec{V}^k) = q^k\delta_{ij} + \mu\left(\frac{\partial V_i^k}{\partial x_j} + \frac{\partial V_j^k}{\partial x_i}\right)$$

and $\vec{n}$ is the exterior (with respect to Ω_i) normal to S at y. Equation (15) is the sum of two single-layer potentials, with $\sigma_{ij}(\vec{u}(y))\; n_j(y)$ and $\rho U\; u_i(y)\; n_1(y)$ as a density, plus a double-layer potential, with $u_i(y)$ as density plus a volume potential with density $g_i(y)$.

Substituting the boundary value of $u_i(x)$ on S into equation (15) and because a double-layer potential with constant density is equal to zero at a point $x \in \Omega_e$, we obtain

$$\begin{aligned} u_k(x) = &\int_S V_i^k(x,y)\sigma_{ij}(\vec{u}(y))\; n_j(y)\; dS_y - \rho U^2 \int_S \delta_{1i}\, n_1(y) V_i^k(x,y)\; dS_y \\ &- \int_{\Omega_e} g_i(y)\, V_i^k(x,y)\; dy \qquad \text{for every } x \in \Omega_e \end{aligned} \tag{16}$$

The above equation can be reduced to:

$$u_k(x) = \int_S V_i^k(x,y)\sigma_{ij}(\vec{u}(y))\, n_j(y)\, dS_y - \int_{\Omega_e} g_i(y) V_i^k(x,y)\, dy \qquad \text{for every } x \in \Omega_e \tag{17}$$

This reduction is a consequence of the following *LEMMA:*
Whenever a point $x \in \Omega_e$

$$\int_S \delta_{1i}\, n_1(y)\, V_i^k(x,y)\, dS_y = 0$$

for any shape of the surface S.

PROOF:

Let us consider a hypothetically Oseen's flow, solution of the homogeneous Oseen's equation, interior to the surface S, with constant velocity $u_i = U\delta_{1i}$ and zero pressure. Using Oseen's integral representation formulae for the bounded domain Ω_i, we have

$$U\delta_{1k} = U\delta_{1i} \int_S \Sigma_{ij}(\vec{V}^k(x,y))\, n_j(y)\, dS_y - \rho U^2 \int_S \delta_{1i}\, n_1(y) V_i^k(x,y)\, dS_y \qquad \text{for every } x \in \Omega_i$$

since $\sigma_{ij}(\vec{u}) = 0$ for every $x \in \bar{\Omega}_i$.

Using the fact that a double-layer potential with constant density is equal to its density at any point $x \in \Omega_i$, we can rewrite the above equation as:

$$U\delta_{1k} = U\delta_{1k} - \rho U^2 \int_S \delta_{1i}\, n_1(y) V_i^k(x,y) dS_y \qquad \text{for every } x \in \Omega_i$$

Therefore

$$\int_S \delta_{1i}\, n_1(y)\, V_i^k(x,y)\, dS_y = 0 \qquad \text{everywhere}$$

where it have been used the continuity property of a single-layer potential across the density carrying surface S, and its vanishing value at infinity. *q.e.d.*

Applying the Dirichlet boundary condition at S to the velocity field given by equation (17), we get the following system of Fredholm's integral equations of the first kind for the unknown local stress forces, $\sigma_{ij}(\vec{u}(y))\, n_j(y)$:

$$D_k(\xi) = U\delta_{1k} + \int_{\Omega_e} g_i(y)\, V_i^k(\xi,y)\, dy = \int_S V_i^k(\xi,y)\, \sigma_{ij}(\vec{u}(y))\, n_j(y)\, dS_y \qquad \text{for every } \xi \in S \tag{18}$$

The uniqueness of solutions of the system of integral equations (18) is given by Miranda and Power (1983). The numerical solution of this system becomes more complicated than the one corresponding to the Stokes' system, because the Oseen's kernel is more complex than the Stokes' one. However, it is known that the Stokeslet and the Oseenlet are related. After some manipulations, the Oseenlet can be written in terms of the Stokeslet as follows:

$$V_i^k(x,y) = \frac{1}{8\pi\mu}\left(\frac{\delta_{ik}}{r} + \frac{(x_i - y_i)(x_k - y_k)}{r^3}\right) + \tag{19}$$

$$+\frac{k}{8\pi\mu}\frac{\partial r}{\partial x_i}\frac{\partial r}{\partial x_k}\left(F'(z) - \frac{F(z)}{kr}\right) + \delta_{ik}\left(\frac{k}{8\pi\mu}(1-2z)\frac{F(z)}{kr} + \frac{k}{4\pi\mu}(\frac{\partial r}{\partial x_1} + 1)\right)$$
$$+\frac{k}{8\pi\mu}F'(z)\left(\delta_{1i}\frac{\partial r}{\partial x_k} + \delta_{1k}\frac{\partial r}{\partial x_i} + \delta_{1i}\delta_{1k}\right)$$

where

$$F(z) = \frac{1-e^z}{z} + 1, \qquad F'(z) = \frac{dF(z)}{dz} \qquad \text{and} \qquad z = k(x_1 - y_1 + r)$$

and

$$\frac{1}{8\pi\mu}\left(\frac{\delta_{ik}}{r} + \frac{(x_i - y_i)(x_k - y_k)}{r^3}\right)$$

is the fundamental singular solution of the Stokes' equation (commonly called Stokeslet).

Equation (18) can be solved numerically using a method similar to the one proposed by Youngren and Acrivos, to solve the corresponding first kind Fredholm's integral equation for the Stokes' problem. This method transforms the integral equations into a linear system of algebraic equations. This is accomplished by dividing S into N elements Δ_j, for $j = 1, 2, 3, \dots N$ all of them small relative to S and over which the components of the unknown local stress forces may be considered approximately constant and equal to their values at the center of the element. Then, we can write (18) approximately as:

$$D_k(\xi)^m = \frac{1}{2\mu}\sum_{j=1}^{N} f_i(\xi^j)\left(A_{ik}^j + B_{ik}^j\right) \qquad \text{for every } \xi^m \in S \tag{20}$$

where ξ^m is the location of the center of the element Δ_m; $f_i = \sigma_{ij}(\vec{u})\, n_j$, and

$$A_{ik}^j = \frac{1}{4\pi}\int_{\Delta_j}\left(\frac{\delta_{ik}}{r^m} + \frac{(\xi_i^m - y_i)(\xi_k^m - y_k)}{(r^m)^3}\right) dS_y \tag{21}$$

and the Oseen's correction B_{ik}^j is given by:

$$\begin{aligned} B_{ik}^j = \frac{k}{4\pi}\int_{\Delta_j} & \left(\frac{\partial r^m}{\partial \xi_i}\frac{\partial r^m}{\partial \xi_k}(F'(z^m) - \frac{F(z^m)}{kr^m})\right. \\ & + \delta_{ik}((1-2z^m)\frac{F(z^m)}{kr^m} + 2(\frac{\partial r^m}{\partial \xi_1} + 1)) \\ & \left. + F'(z^m)(\delta_{1i}\frac{\partial r^m}{\partial \xi_k} + \delta_{1k}\frac{\partial r^m}{\partial \xi_i} + \delta_{1i}\delta_{1k})\right) dS_y \end{aligned} \tag{22}$$

Since $F(z)/Kr$ and $F'(z)$ can be expanded in a power series for small z as:

$$\frac{F(z)}{kr} = -\left(1 + \frac{\partial r}{\partial x_1}\right)\left(\frac{1}{2!} + \frac{z}{3!} + \frac{z^2}{4!} + \cdots\right) \tag{23-a}$$

and

$$F'(z) = -\left(\frac{1}{2!} + \frac{2z}{3!} + \frac{3z^2}{4!} + \frac{4z^3}{5!} + \cdots\right) \tag{23-b}$$

it follows that the integrals in (22) are proper when $j = m$, so they do not present problems. On the other hand, the surface integrals in (21) present no problem, unless $j = m$, in that case, they are improper, because the integration is carried out over the element that contains ξ^m. Then, the integration is divided into two regions: in the first, the integration is carried out over a small neighborhood of ξ^m, where the surface is assumed locally flat and is approximated by the tangent plane at ξ^m. Then by transforming to a local polar coordinate system lying in this tangent plane, it is possible to integrate analytically the expression (21) in this region. The final expression is given by Youngren and Acrivos as:

$$A^m_{11,\varepsilon} = \frac{\varepsilon}{2\pi}\left(\ln\frac{\sqrt{2}+1}{\sqrt{2}-1}\right)\left(2 + \frac{1}{1+(S')^2}\right) \tag{24-a}$$

$$A^m_{12,\varepsilon} = \frac{-0.2805\, n_2\, S'\,\varepsilon}{(1+(S')^2)(n_2^2+n_3^2)^{1/2}} \tag{24-b}$$

$$A^m_{13,\varepsilon} = \frac{-0.2805\, n_3\, S'\,\varepsilon}{(1+(S')^2)(n_2^2+n_3^2)^{1/2}} \tag{24-b}$$

$$A^m_{22,\varepsilon} = 0.2805\varepsilon\left[2 + \frac{(S')^2(n_2^2+n_3^2)+n_3^2}{(n_2^2+n_3^2)(1+(S')^2)}\right] \tag{24-d}$$

$$A^m_{23,\varepsilon} = \frac{-0.2805\,\varepsilon\, n_2\, n_3}{(n_2^2+n_3^2)(1+(S')^2)} \tag{24-e}$$

$$A^m_{33,\varepsilon} = 0.2805\varepsilon\left[2 + \frac{(S')^2(n_2^2+n_3^2)+n_2^2}{(n_2^2+n_3^2)(1+(S')^2)}\right] \tag{24-f}$$

where $S(x_1, \theta)$ is the cylindrical radius describing the actual surface of the body, $S' = \partial S/\partial x_1$, 2ε is the length along a line of constant x_1 of a planar square S_ε (taken as the region of integration), with center at ξ^m, n_2 and n_3 are the components of the inward normal to the body at the point ξ^m in the x_2 and x_3 direction respectively.

The second region $\Delta_m - S_\varepsilon$ is the remainder of Δ_m, where the numerical calculation presents no problem. After finding the local surface-stress force, by solving the linear system of equation (20), the total force F_i is given by:

$$F_i = -\sum_{j=1}^{N} f_i(\xi)\int_{\Delta_j} dS_y \tag{25}$$

and the force coefficient can be defined as:

$$C_i = F_i/\pi\rho(Ua)^2 \tag{26}$$

To test the numerical method developed in the present work, the homogeneous Oseen's problem of a uniform flow past a sphere of unit radius centered at the origin is worked out. In figure 1 we show a comparison of the drag coefficient for different Reynolds numbers found by the above numerical solution with the analytical one, $C_D = (24/R)(1 + (3/16)R)$.

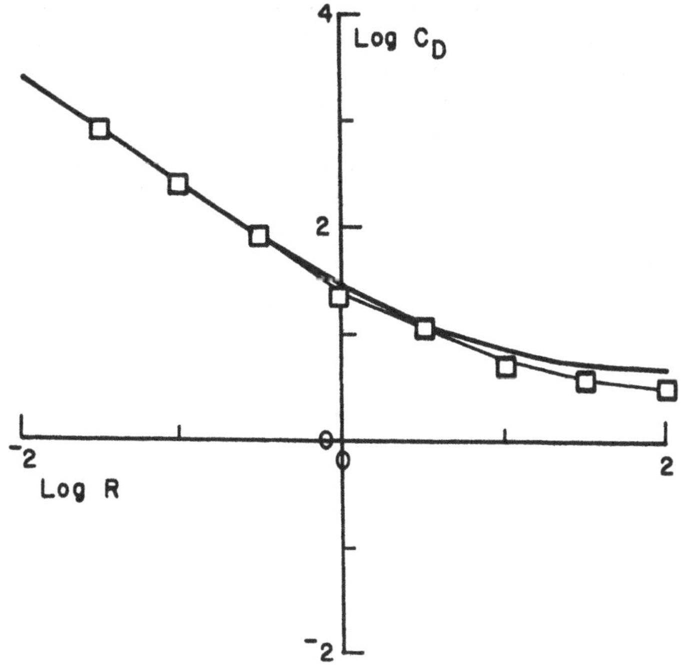

Figure 1. Comparison of the drag coefficient
—— Analytical solution and □ –□ Numerical solution $N = 16$

REFERENCES

Finn, R. (1965) "On the exterior stationary problem for the Navier-Stokes equations, and associated perturbation problems" Arch. Rational Mech. Anal. 19

Fisher, T.M.; G.C. Hsia and W.L. Wendland (1985) "Singular perturbations for the exterior three-dimensional slow viscous flow problem" J. Math. Appl. Vol. 110 No. 2

Kaplum, S. and P.A. Langerstrom (1957) "Asymptotic expansions of Navier-Stokes solutions for small Reynolds number" J. Math. Mech. 6

Lee, S.H. and Leal, L.G. (1986) "Low Reynolds number flow past cylindrical bodies of arbitrary cross-sectional shape" J. Fluid Mech. 164.

Miranda, G. and Power, H. (1983) "Integral equation solution of Oseen's flow with a free surface" Lecture Notes in Mathematics, 1005 Springer Verlag

Olmstead W.E. and A.K. Gautesen (1976) "Integral representations and the Oseen flow problem" Mechanics today, Vol. 3, edited by S. Nemat-Nasser, Pergamon

Oseen, C.W. (1927) "Neuere Methoden und Ergebnisse in der Hydrodynamik" Akademische Verlagsgesellschaft, Leipzig

Proudman, I. and J.R.A. Pearson (1957) "Expansions at small Reynolds numbers for the flow past a shere and a circular cylinder" J. Fluid Mech. 2

Youngren G.K. and A. Acrivos (1975) "Stokes flow past a particle of arbitrary shape: a numerical method of solutions" J. Fluid Mech. 69

A Boundary Element Study of the Motion of Rigid Particles in Internal Stokes Flow

J. Li, M.S. Ingber

Department of Mechanical Engineering, University of New Mexico, Albuquerque, New Mexico 87131, U.S.A.

ABSTRACT

The direct boundary element method is used to study the motion of a rigid particle or particles in Stokes flow. The method couples the quasi-static Stokes equations for the fluid with the equilibrium equations for the particles. We consider the problem of Jeffery's orbit in Couette flow. The results from our numerical calculation match those predicted by Jeffery's theory well. It is shown that both the shape of the particle and the proximity of the bounding walls can have a large influence on the orientation state of the particle, and thus on the period. We demonstrate that the interaction between particles changes their behavior in Couette flow. We also consider particle motions through contractions and expansions. A comparison is made between the trajectories of elliptical particles and rectangular particles.

INTRODUCTION

Particle motion in Stokes flow is of great interest in many engineering applications including pipeline transport, petroleum recovery, materials processing, blood flow, the study of the rheological properties, etc. Both analytical methods such as using special coordinates [1], the method of reflections [2], the unit cell technique [2], boundary collocation [3], and multipole expansion [4] and numerical methods such as finite difference [5], finite element [6], boundary element [7-9] have been developed in the past to study the above problems. The boundary element method (BEM) has proven to be an efficient technique for solving Stokes flow problems. The unique advantage of the BEM is that it requires only the discretization of the problem boundary while still being able to determine flow quantities within the fluid. The method is particularly attractive for problems involving infinite domains and for performing dynamic simulations [7].

We first briefly outline the numerical method used in the present investigation. The current study is based on the boundary element method. However, unlike traditional BEM's, the current method does not require knowledge of either the stresses or velocities on the surface of the particles. The kinematic equations relating the linear and angular velocities at the centroids of the particles to the surface velocities, the equilibrium equations, and the discretized boundary element equations are combined to generate a system of linear equations. This method has proven to be stable and efficient [7].

The prediction of particle orientation states in dilute suspensions is desirable in composite material manufacturing. Jeffery's theory is often used to determine the fiber orientation state. Some important assumptions are made in deriving the Jeffery's equation, resulting in deviations from actual physical flows. A careful study of Jeffery's orbit in Couette flow for both elliptical and rectangular particles is carried out. Significant differences between the case of an elliptical particle and the case of a rectangular particle are observed. Wall effects also influence the particle orientation states significantly.

BOUNDARY ELEMENT FORMULATION

Consider Stokes flow in domain Ω with boundary Γ. The governing differential equations in terms of the dimensionless fluid velocity u_i and the pressure p are given by:

$$\frac{\partial^2 u_i(\mathbf{x})}{\partial x_j \partial x_j} = \frac{\partial p(\mathbf{x})}{\partial x_i} \quad \mathbf{x} \in \Omega \tag{1}$$

$$\frac{\partial u_i(\mathbf{x})}{\partial x_i} = 0 \quad \mathbf{x} \in \Gamma + \Omega \tag{2}$$

These equations can be reformulated in terms of the following integral equations:

$$c_{ij}(\mathbf{x})u_j(\mathbf{x}) + \int_\Gamma q^*_{kji}(\mathbf{x},\mathbf{y})u_k(\mathbf{y})n_j(\mathbf{y})d\Gamma = -\int_\Gamma u^*_{ik}(\mathbf{x},\mathbf{y})f_k(\mathbf{y})d\Gamma \tag{3}$$

and

$$c^p(\mathbf{x})p(\mathbf{x}) = \int_\Gamma p^*_{kj}(\mathbf{x},\mathbf{y})u_k(\mathbf{y})n_j(\mathbf{y})d\Gamma - \int_\Gamma p^*_k(\mathbf{x},\mathbf{y})f_k(\mathbf{y})d\Gamma \tag{4}$$

We limit our attention here to the 2D problem for which

$$q^*_{kji} = \frac{(x_k - y_k)(x_j - y_j)(x_i - y_i)}{\pi|\mathbf{x}-\mathbf{y}|^4} \tag{5}$$

$$u^*_{ik} = -\left[\delta_{ik}\ln|\mathbf{x}-\mathbf{y}| - \frac{(x_i - y_i)(x_k - y_k)}{|\mathbf{x}-\mathbf{y}|^2}\right] \tag{6}$$

$$p_{kj}^* = -\frac{1}{\pi}\left[\frac{\delta_{kj}}{|\mathbf{x}-\mathbf{y}|^2} - \frac{2(x_k - y_k)(x_j - y_j)}{|\mathbf{x}-\mathbf{y}|^4}\right] \tag{7}$$

$$p_k^* = -\frac{x_k - y_k}{2\pi|\mathbf{x}-\mathbf{y}|^2} \tag{8}$$

where δ_{kj} is the Dirac delta function, the f_k's are the components of the traction along the surface Γ, and the n_j's are the components of the unit outward-normal vector to the boundary Γ. The coefficient tensor c_{ij} in equation (3) and the coefficient c^p in equation (4) can be determined from the geometry or by direct integration.

The boundary Γ can be discretized into NE elements within which the velocity, traction components, and the pressure are approximated as follows:

$$u_i(\mathbf{x})|_{\Gamma_e} \cong \sum_{j=1}^{n} u_{ij}^e N_j(\mathbf{x}) \tag{9}$$

$$f_i(\mathbf{x})|_{\Gamma_e} \cong \sum_{j=1}^{n} f_{ij}^e N_j(\mathbf{x}) \tag{10}$$

$$p(\mathbf{x})|_{\Gamma_e} \cong \sum_{j=1}^{n} p_j^e N_j(\mathbf{x}) \tag{11}$$

where u_{ij}^e and f_{ij}^e are the values of the ith components of velocity and traction, respectively, at the jth node within the eth element and p_j^e is the value of pressure at the jth node within the eth element. The N_j are the shape functions.

Thus, equation (3) for the field points at the M nodes within the NE elements becomes a $2M$ by $2M$ system of linear algebraic equations represented as

$$[H_{ij}]\{u_j\} = [G_{ij}]\{f_j\} \tag{12}$$

where u_j and f_j represent the components of the velocity and traction, respectively, at the collocation nodes.

Note, only the velocities on the fixed portion of the boundary are specified. Hence, on the particle boundaries, two additional relations are necessary to close the problem, namely, the kinematic equations and the equilibrium equations. The kinematic equations relate the velocities on the surface of the N particles to the three components of linear and angular velocity at the centroid of the particles. These equations can be written symbolically as

$$\{u_j\} = [K_{il}]\{U_l\} \tag{13}$$

where U_l represents the linear and angular velocities at the centroids. The equilibrium equations require that the resultant forces and moments on the particles generated by the surface tractions and body forces be zero. These equations can be written symbolically as

$$[M_{ri}]\{f_i\} = \{b_r\} \tag{14}$$

Equations (12), (13), and (14) are combined into the final set of equations to yield solutions for the particle velocities and surface tractions as follows:

$$\begin{bmatrix} [G_{ij}] & [H_{ij}][K_{il}] \\ [M_{ri}] & 0 \end{bmatrix} \begin{Bmatrix} \{f_j\} \\ \{U_l\} \end{Bmatrix} = \begin{Bmatrix} 0 \\ \{b_r\} \end{Bmatrix} \tag{15}$$

In order to perform dynamic simulations of particle motions consistent with the neglect of the fluid inertia, the particle inertia is also neglected. That is, we have assumed above that the particles are force and moment free. Upon solution of equation (15), the particles are repositioned in space using a fifth-order variable-time-step Runga-Kutta method. If desired, pressures within the fluid can be obtained by evaluating equation (4).

PARTICLE ORIENTATION IN COUETTE FLOW

In this section, we study the Jeffery's orbit of a particle in Couette flow. The geometry of the problem is shown in Fig. 1. The flow is generated by the motion of the top plane channel wall to the left and the bottom one to the right with equal velocity U. We choose $\mu = 1$ so that the dimensionless governing equation is the same as the dimensional one. Thus, $u = -H/2$ at $z = H/2$ and $u = H/2$ at $z = -H/2$. The velocity profile at the inlet and outlet of the channel is given by $u = z$.

The particle orientation state in dilute suspensions can be approximated by Jeffery's orientation equation [6]. In Jeffery's theory, one makes several assumptions. The important ones include that the particle may be represented as an ellipse, the velocity field is only locally perturbed by the motion of the particle, and there is no interaction between the particles or between the particles and the walls.

In the case of a particle in Couette flow, Jeffery's theory gives the following equation for the angular position of a particle:

$$\phi = -\tan^{-1}\left[\frac{1}{r_p}\tan\left(\frac{2\pi t}{T}\right)\right] \tag{16}$$

where T is the period of rotation and r_p is the particle aspect ratio. The period T is given by

$$T = \frac{2\pi}{|du/dy|}\left(r_p + \frac{1}{r_p}\right) \tag{17}$$

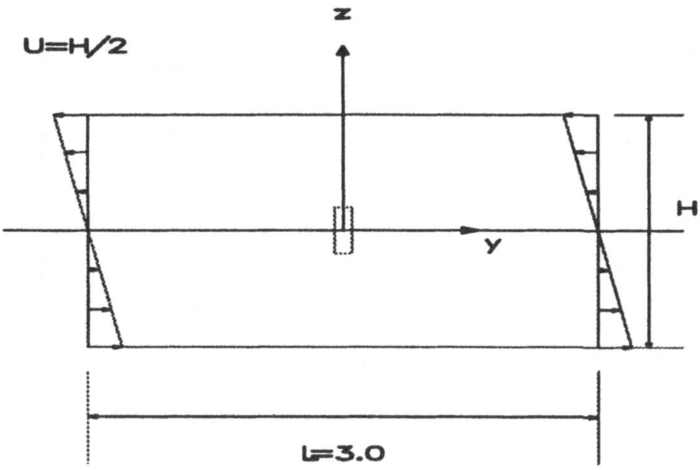

Figure 1: Problem geometry for a particle in Couette flow.

We first verify Jeffery's theory under the special conditions. Then, the effect of particle shape on the period of particle rotation is investigated. Finally, we consider the interaction between the particle and the walls, and between two particles.

In Fig. 2, we plot the angular orientation ϕ of an elliptical particle with aspect ratio equal to 10 as a function of normalized time for a quarter period ($T/4$) derived from both the boundary element calculations and Jeffery's theory. The dimensionless channel width (with respect to the major axis of the elliptical particle) is 15 so that the wall effect is negligible. The agreement between theory and the BEM calculation is evident.

The boundary element grid used in this example contained 24 quadratic boundary elements for the particle and 80 quadratic boundary elements for the channel. This grid was chosen after performing appropriate convergence tests. This level of discretization is typical of all the example problems considered in this paper.

We also compare the period T calculated using the BEM with that predicted by equation (17) for different aspect ratio ellipses in Fig. 3. It is seen that the two sets of results match each other quite well. In particular, the period T for an elliptical particle with aspect ratio equal to 10 calculated using the BEM is 61.66 as compared to the theoretical value 63.46 predicted from Jeffery's equation yielding an error of under 3%.

The orientations of an elliptical particle and a rectangular particle, respectively, are shown in Figs. 4 and 5 as a function of normalized time for values of the aspect ratio equal to 1, 10, and 50. It is seen that the orientation of both the elliptical and the rectangular particles exhibit simi-

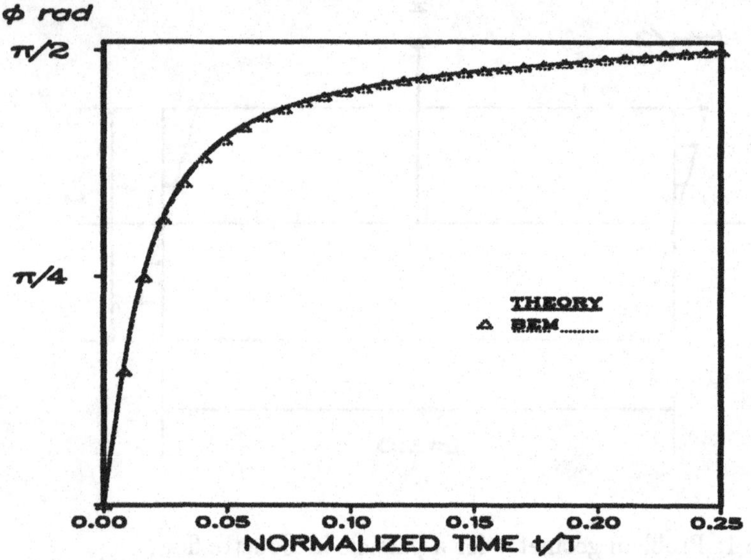

Figure 2: Comparison of the orientation states of a free elliptical particle in Couette flow as calculated by the BEM and predicted by Jeffery's theory.

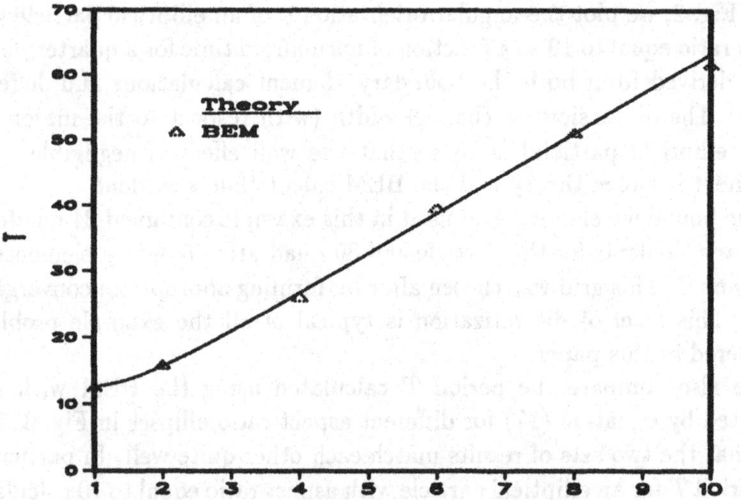

Figure 3: Comparison of the period of a free elliptical particle in Couette flow as calculated by the BEM and predicted by Jeffery's theory.

lar behaviors. We observe that for larger aspect ratios, the particle remains relatively oriented along the streamline for a large percentage of the period and flips over within a relatively short time.

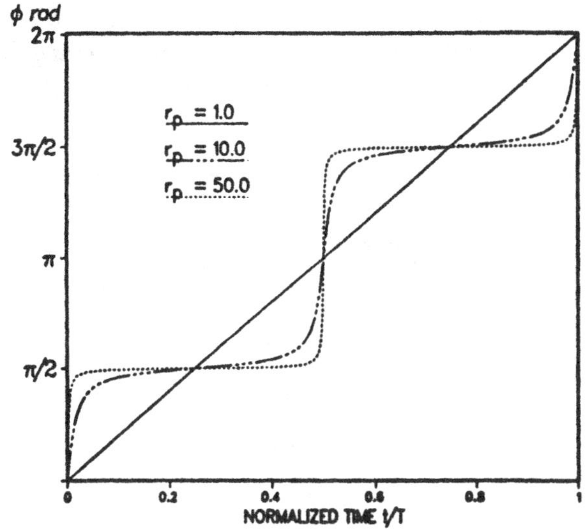

Figure 4: Angular position of a free elliptical particle in Couette flow as a function of normalized time.

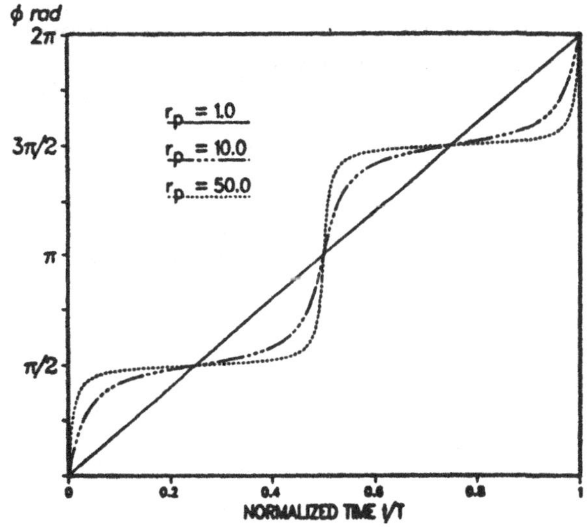

Figure 5: Angular position of a free rectangular particle in Couette flow as a function of normalized time.

The difference in the angular orientation states between the elliptical particle and the rectangular particle is shown in Fig. 6 for aspect ratio equal to 10. It is seen that the elliptical particle remains oriented along the streamline for a longer time and takes a shorter time to flip over. For the dimensionless channel width larger than 5, the period of an elliptical particle is found to be approximately twice that of a rectangular particle with the same aspect ratio ($r_p = 10$) as seen in Fig. 7.

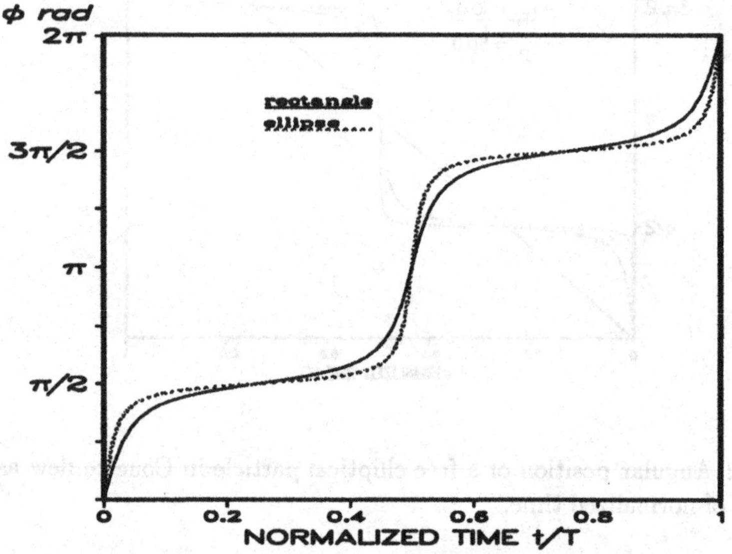

Figure 6: Comparison of the orientation state between a free elliptical particle and a free rectangular particle in Couette flow.

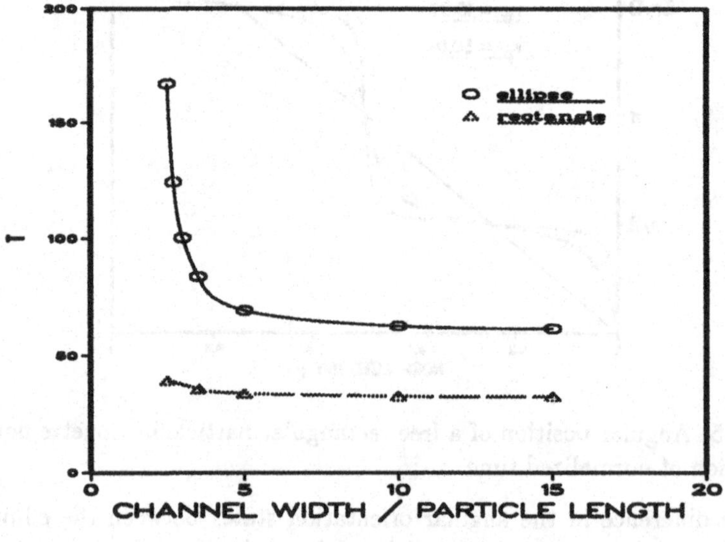

Figure 7: Period of a free elliptical particle and a free rectangular particle of aspect ratio 10 in Couette flow as a function of the ratio of the channel width to the particle length (cp).

The effect of the wall on the period experienced by a particle in Couette flow is shown in Fig. 7. The aspect ratios of the elliptical and rectangular particles are the same ($r_p = 10$). We notice that as the walls move closer to the particle and the ratio of the channel width to the length of particle major axis decreases, the period becomes larger. We will designate this ratio as "cp". The wall influence becomes larger as cp is reduced. A dramatic increase in the period is observed for the elliptical particle as compared to a mild increase in the period for the rectangular particle for cp less than 5.

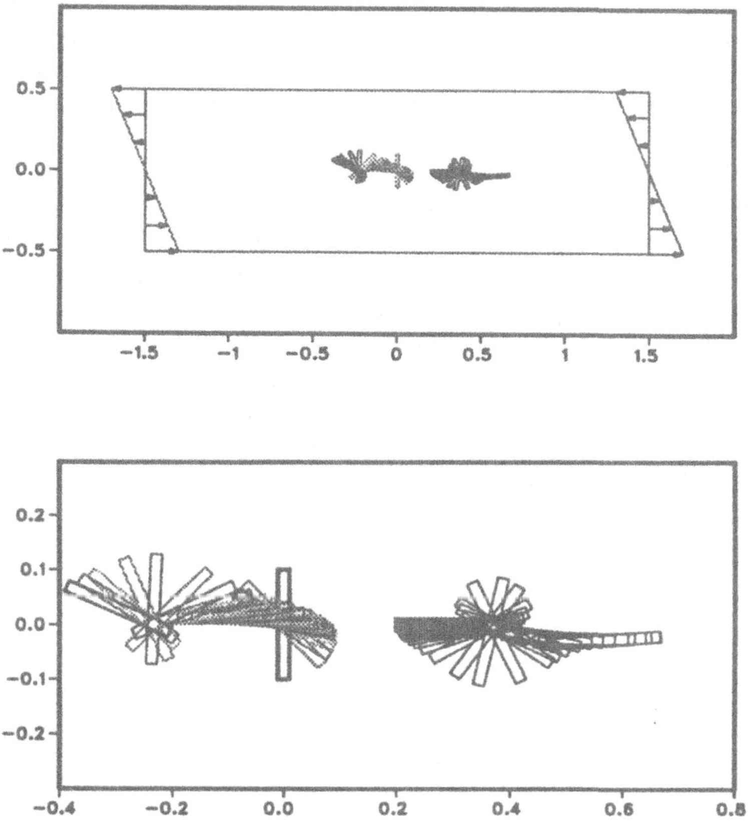

Figure 8: Trajectories of two free rectangular particles in Couette flow.

The effect of particle interactions in Couette flow is demonstrated in Fig. 8. Because of the presence of the other particle, both particles not only spin but also translate in the y and z directions. In a half period, particle 1 translates from (0.0,0.0) to (−0.23,0.028) and particle 2 translates from

(0.3,0.0) to (0.52,−0.025). The values of the half period ($T/2$) is 17.8 for particle 1 and 18.2 for particle 2 as compared to 17.0 for the case when a single particle is present in the Couette flow. The asymmetry of the flow field due to the particle interactions is evident.

PARTICLE MOTION THROUGH A CONTRACTION AND EXPANSION

In this section, we investigate the behavior of particle motions through a contraction or an expansion. The geometry of the contraction and expansion is shown in Fig. 9. and Fig. 10. Fully developed Poiseuille flow is specified at both the entry and exit sections. The maximum velocity ($Umax$) at the entry of the contraction or the exit of the expansion is 1. The aspect ratios for all the particles considered here are 2. Y_o, Z_o and ϕ_o in the figures specify the initial particle position, where $\phi_o = 0$ represents the angular orientation. The reference $\phi_0 = 0^0$ represents a particle whose minor axis is aligned with the z-axis.

The trajectories of a rectangular particle and an elliptical particle traveling through a contraction is compared in Fig. 9. Both particles have the same length major axis and aspect ratio. Their initial positions in the contraction are also the same. At $t = 8.59s$, the location and orientation of the rectangular particle is given by $y = -4.705$, $z = 0.3863$, and $\phi = -180.65^0$ whereas the location and orientation of the elliptical particle is given by $y = -5.176$, $z = 0.4404$, and $\phi = -165.87^0$. It is seen that the elliptical particle travels faster but rotates slower than the rectangular particle.

We compare the trajectories of three elliptical particles and three rectangular particles traveling through an expansion for a time period from

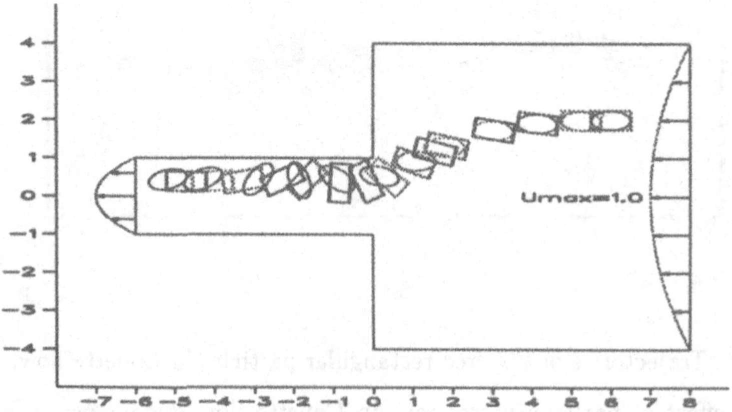

Figure 9: Comparison of the trajectories of a rectangular particle and an elliptical particle traveling through a contraction $Y_o = 6, Z_o = 2.0, \phi_o = 90^0$.

$t = 0.0s$ to $t = 7.07s$ in Fig. 10. Again, the particles have the same length major axis and the same aspect ratio. The two sets of particles start at the same positions and with the same orientations. It appears that the closer to the centerline of the channel the initial position of either shape of the particle, the faster it travels and spins. We notice that the final angular position of the particles depends on the particle shape. For example, for elliptical particle 1, $\phi^1_{final} = 9^0$ whereas for rectangular particle 1, $\phi^1_{final} = -26.87^0$.

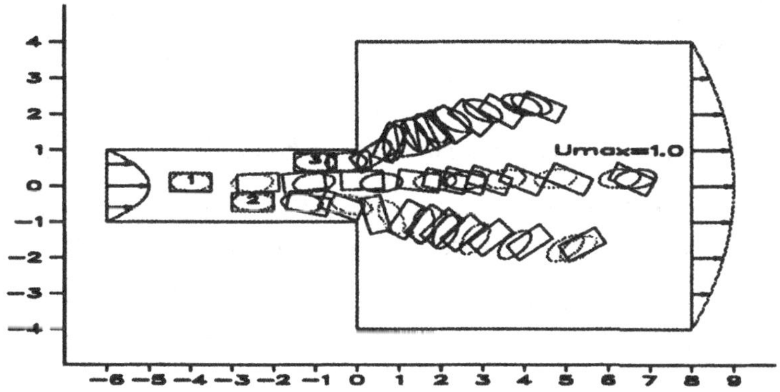

Figure 10: Comparison of the trajectories of three rectangular particles and three elliptical particles traveling through an expansion $Y^1_o = -4.0, Z^1_o = 0.1, Y^2_o = -2.5, Z^2_o = -0.45, Y^3_o = -1.0, Z^3_o = 0.65$.

CONCLUSION

A direct boundary element method has been applied to the study of Jeffery's orbits of a particle or particles in Couette flow and flows through a contraction or expansion. It has been shown that there are significant differences between the period of an elliptical particle and a similar aspect ratio rectangular particle. Further, the wall effects can have a large influence on the period of the Jeffery orbit. Comparisons between the trajectories of a rectangular particle and an elliptical particle traveling through a contraction, and the trajectories of three rectangular particles and three elliptical particles traveling through an expansion demonstrate that the approximation of rectangular particles with elliptical particles is sometimes not suitable.

REFERENCES

1. Stimson, M. & Jeffery, G. B., "The motion of two spheres in a viscous fluid," *Proc. Roy. Soc. A*, **111**: 110, 1926.

2. Happel, J. & Brenner, H., Low Renolds Number Hydrodynamics, Martinus Nijhoff Publishers, Dordrecht, 1983.

3. Ganatos, P., Pfeffer, R. & Weinbaum, S., "A numerical solution technique for three-dimensional Stokes flow, with applications to the motion of strongly interacting spheres in a plane," *J. Fluid Mech.*, **84**(1): 79, 1978.

4. Durlofsky, L., Brady, J.F. & Bossis, G., "Dynamic simulation of hydrodynamically interacting particles," *J.Fluid Mech.*, **180**: 21, 1987.

5. Dvinsky, A. S. & Popel A. S., "Motion of a rigid particle between parallel plates in Stokes flow," *Computers & Fluids*, **15**(4): 391, 1987.

6. Givler, R. C., Crochet, M. J. & Pipes, R. B., "Numerical prediction of fiber orientation in dilute suspensions," *Journal of Composite Materials*, **17**: 330, 1983.

7. Ingber, M. S., "Numerical simulation of the hydrodynamic interaction between a sedimenting particle and a neutrally buoyant particle," *Int. J. Num. Meth. Fluids*, **9**: 263, 1989.

8. Youngren, G. K. & Acrivos, A., "Stokes flow past a particle of arbitrary shape: A numerical method of solution," *J. Fluid Mech.*, **69**(2): 377, 1975.

9. Tran-Cong, T. & Phan-Thien, N., "Stokes problems of multiparticle systems: A numerical method for arbitrary flows," *Phys. Fluids*, A, **1**(3): 453, 1989.

Boundary Element Methods for Nonstationary Stokes Problem in Three Dimensions and Their Convergence Analysis

H. Wang

Department of Mathematics, University of Wyoming, Laramie, Wyoming 82071, U.S.A.

ABSTRACT

In this paper we develop boundary element methods for nonstationary Stokes problem in three dimensions. The method is based upon variational formulations for the integral equations of the first kind. Numerical schemes are developed, optimal-order error estimates and superconvergence estimates are obtained.

INTRODUCTION

During the last 20 years, boundary element methods (BEM) have been extensively used in practical computations of engineering, science and technology problems. BEM are popular with practising engineers because of the simplicity of the data required to run a problem and the versatility of boundary elements. BEM reduce the dimensions involved in the problem, leading to considerable economy in the numerical work. They tend to produce more accurate results than finite element methods and constitute a very convienient manner of treating adequately unbounded regions by numerical means. Generally, the dimension of the problem is reduced by one, but even part of the region is treated by BEM, the size of the discretized domain is reduced. About the applications and theoretical analysis of BEM, there have been a lot of work. For example, see [1-7].

Different boundary element methods have been successfully applied to incompressible viscous flow problems in the last few years, for example, see [8,9] among others. The core of these methods is the scheme for treating the linear Stokes problems, since the nonlinear problems would be reduced to linear ones by means of iterative or time-stepping procedures.

For the Stokes problem, two formulations are available for numerical procedures, i.e. the "stream function" form and the "velocity–pressure" form. Each of them can be transformed into boundary integral equations and so can be solved by BEM. Furthermore, incontrast to domain–type methods such as finite difference, finite element or mixed methods, the incompressibility constraint "$\nabla \cdot \mathbf{u} = 0$" is automatically satisfied when boundary integral formulations are used. Thus, BEM is well suited in solving the Stokes problem especially for the exterior problem.

In this paper, we present a boundary element method to solve the nonstationary Stokes problems in three dimensions. Our approach is based on the "velocity–pressure" formulation. The solution of the problem is expressed in terms of simple layer potential which is suitable to the interior as well as the exterior boundary value problems. The method is based upon variational formulations for the integral equations of the first kind. Also, we derive the corresponding discrete algebraic equation system, discuss the calculations of the singular integrals arising from the derived algebraic equations. Numerical experiments will be completely finished and be presented by the time of conference. For the numerical schemes developed in this paper, we obtain the optimal-order error estimates in energy norm and the superconvergence results in L^∞ norm. Moreover, for any order derivatives of the numerical solutions, we have the same order supreconvergence outside a neighbourhood of the boundary.

INTEGRAL REPRESENTATION OF THE SOLUTION AND VARIATIONAL PRINCIPLE

Consider the Dirichlet problem of the nonstationary Stokes equations in three dimensions:

$$\begin{cases} \dfrac{\partial \mathbf{u}}{\partial t} - \nu\Delta\mathbf{u} + \nabla p = \mathbf{f}(\mathbf{x},t), & (\mathbf{x},t) \in (\Omega \cup \Omega') \times (0,T], \\ \nabla \cdot \mathbf{u}(\mathbf{x},t) = 0, & (\mathbf{x},t) \in (\Omega \cup \Omega') \times (0,T], \\ \mathbf{u}(\mathbf{x},t) = \mathbf{w}(\mathbf{x},t), & (\mathbf{x},t) \in S \times [0,T], \\ \mathbf{u}(\mathbf{x},0) = \mathbf{u}_0(\mathbf{x}), & \mathbf{x} \in \Omega \cup \Omega', \\ \lim\limits_{|\mathbf{x}|\to\infty} \mathbf{u}(\mathbf{x},t) = 0, & t \in (0,T], \end{cases} \tag{2.1}$$

where Ω is a simply-connected domain with smooth boundary $S = \partial\Omega, \Omega' = R^3 - \bar{\Omega}$. The unknown variables are fluid rate $\mathbf{u} = (u_1, u_2, u_3)$ and pressure p of viscous incompressible fluid flows which fill Ω or Ω'. ν is the kinematic viscosity. $\mathbf{w}(\mathbf{x},t)$ and $\mathbf{u}_0(\mathbf{x})$ are given functions. Since a special solution of first two equations in (2.1) with general $\mathbf{f}$ can be obtained analytically [10], so in this paper, we only need to consider the case $\mathbf{f} \equiv 0$.

Introduce the following Sobolev spaces [10,11]:

$$W = \Big\{\mathbf{v}(\mathbf{x});\ \mathbf{v} \in \left(L^2(R^3)\right)^3,\ \nabla \cdot \mathbf{v} = 0,\ \ \mathbf{v} \cdot \mathbf{n}|_S = 0,$$

$$\text{where } \mathbf{n} = (n_1, n_2, n_3) \text{ is the outward normal to } S.\Big\},$$

$$W_0^1 = \left\{ \mathbf{v}(\mathbf{x});\ \mathbf{v}/(1+|\mathbf{x}|^2)^{\frac{1}{2}} \in \left(L^2(R^3)\right)^3, \nabla \mathbf{v} \in \left(L^2(R^3)\right)^9,\ \nabla\cdot\mathbf{v} = 0. \right\},$$

$$V(0,T) = \left\{ \mathbf{v}(\mathbf{x},t);\ \mathbf{v} \in L^2(0,T;W_0^1), \frac{\partial \mathbf{v}}{\partial t} \in L^2(0,T;(W_0^1)') \right\},$$

$$H^{p,q} = H^q(0,T;L^2(S)) \cap L^2(0,T;H^p(S)), \text{ for } p,q \geq 0,$$

$$H^{p,q} = (H^{-p,-q})', \text{ for } p,q \leq 0, \quad M^{p,q} = (H^{p,q})^3, \quad M_E^{p,q} = M^{p,q}/E,$$

$$\bar{M}^{p,q} = \left\{ \mathbf{w}(\mathbf{x},t);\ \mathbf{w}(\mathbf{x},t) \in M^{p,q}, \int_S \mathbf{w}\cdot\mathbf{n}\, ds(\mathbf{x}) = 0. \right\},$$

where E represents the equivalence relation:

$$\mathbf{g}(\mathbf{x},t) \sim \mathbf{g}'(\mathbf{x},t) \qquad \text{iff} \quad \mathbf{g}(\mathbf{x},t) - \mathbf{g}'(\mathbf{x},t) = c\,\mathbf{n}(\mathbf{x}), \text{ for } t \in [0,T],\ c \in R.$$

The fundamental solution of the Stokes problem is [10]:

$$\begin{cases} U_{ij}(\mathbf{x},t) = \mathcal{E}(\mathbf{x},t)\delta_{ij} + \dfrac{\partial^2}{\partial x_i \partial x_j} \dfrac{1}{2\pi^{\frac{3}{2}}|\mathbf{x}|} \displaystyle\int_0^{\frac{|\mathbf{x}|}{2\sqrt{\nu t}}} \exp(-\tau^2)d\tau, \\ P_i(\mathbf{x},t) = -\dfrac{\partial}{\partial x_i}\dfrac{1}{4\pi|\mathbf{x}|}\delta(t), \qquad i,j = 1,2,3, \end{cases} \tag{2.2}$$

where $\mathcal{E}(\mathbf{x},t) = (4\pi\nu t)^{-\frac{3}{2}} \exp(-\frac{|\mathbf{x}|^2}{4\nu t})$, δ_{ij} is the standard Kronecker notation, and $\delta(t)$ is the standard Dirac Delta function with respect to t.

Let $\mathbf{U}_i = (U_{1i}, U_{2i}, U_{3i})$, $\mathbf{P} = (P_1, P_2, P_3)$, $(\mathbf{v}, q) = (\mathbf{U}_i(\mathbf{x}-\mathbf{y}, t-\tau), 0)$, and $(\mathbf{u}, p)$ be the solution of problem (2.1), applying the Green formula to (2.1) over a domain G, we get

$$\int_0^{t-\eta}\int_G \left[\mathbf{u}\left(\frac{\partial \mathbf{v}}{\partial \tau} + \nu\Delta\mathbf{v} - \nabla q\right) - \mathbf{v}\left(\frac{\partial \mathbf{u}}{\partial \tau} + \nu\Delta\mathbf{u} - \nabla p\right)\right] d\mathbf{y}d\tau$$

$$= \int_0^{t-\eta}\left\{ \sum_{i,j=1}^{3} \int_{\partial G} [T_{ij}(\mathbf{v},q)n_j u_i - T_{ij}(\mathbf{u},p)n_j v_i]\, ds(\mathbf{y}) \right\} d\tau + \int_G \mathbf{u}\cdot\mathbf{v}d\mathbf{y}\, \Big|_{\tau=0}^{\tau=t-\eta}, \tag{2.3}$$

where $0 < \eta < t$, $T_{ij}(\mathbf{u},p) = -\delta_{ij}\ p + 2\nu e_{ij}(\mathbf{u})$ is the stress tensor, and $e_{ij}(\mathbf{u}) = \frac{1}{2}\left(\frac{\partial u_i}{\partial x_j} + \frac{\partial u_j}{\partial x_i}\right)$ is the deformation rate tensor.

For $\forall\ \mathbf{x} \in R^3$, we take a ball $B_\lambda = B(\mathbf{x}, \lambda)$, with radius λ and center $\mathbf{x}$, and another ball $B_\Lambda = B(0, \Lambda)$, with radius Λ and center origin such that $B_\Lambda \supset (\Omega \cup B_\lambda)$. Applying (2.3) over Ω and $\Omega' \cap B_\Lambda$, first we let $\lambda \to 0$ and $\Lambda \to \infty$, and then we let $\eta \to 0$, we obtain the integral representation of $(\mathbf{u}, p)$ as follows:

$$\begin{cases} u_i(\mathbf{x},t) = \int_0^t \int_S \mathbf{U}_i(\mathbf{x}-\mathbf{y},t-\tau)\cdot\mathbf{g}(\mathbf{y},\tau)ds(\mathbf{y})d\tau + \int_{R^3} u_{0i}(\mathbf{y})\mathcal{E}(\mathbf{x}-\mathbf{y},t)d\mathbf{y}, \\ p(\mathbf{x},t) = \int_0^t \int_S \mathbf{P}(\mathbf{x}-\mathbf{y},t-\tau)\cdot\mathbf{g}(\mathbf{y},\tau)ds(\mathbf{y})d\tau, \qquad i=1,2,3, \end{cases} \tag{2.4}$$

where $\mathbf{g} = (g_1, g_2, g_3)$, $g_i = \left(\sum_{j=1}^{3} \mathcal{T}_{ij}(\mathbf{u},p)n_j\right)\Big|_{\text{interior}} - \left(\sum_{j=1}^{3} \mathcal{T}_{ij}(\mathbf{u},p)n_j\right)\Big|_{\text{exterior}}$, represents the jump of $\sum_{j=1}^{3} \mathcal{T}_{ij}(\mathbf{u},p)n_j$ across S.

For $\mathbf{w}(\mathbf{x},t) = (w_1, w_2, w_3) \in \bar{M}^{\frac{1}{2},\frac{1}{4}}$, the integral equation:

$$w_i(\mathbf{x},t) = \int_0^t \int_S \mathbf{U}_i(\mathbf{x}-\mathbf{y},t-\tau)\cdot\mathbf{g}(\mathbf{y},\tau)ds(\mathbf{y})d\tau + \int_{R^3} u_{0i}(\mathbf{y})\mathcal{E}(\mathbf{x}-\mathbf{y},t)d\mathbf{y}, \tag{2.5}$$

($i = 1,2,3$) define a continuous mapping from $\mathbf{w}(\mathbf{x},t)$ to $\mathbf{g}(\mathbf{x},t)$. Since the pressure p can only be uniquely determined up to a constant in Ω, from the expressions of $\mathcal{T}_{ij}(\mathbf{u},p)$ and g_i, $\mathbf{g} = (g_1, g_2, g_3) \in M^{-\frac{1}{2},-\frac{1}{4}}$ can only be determined up to a vector $c\,\mathbf{n}$. Thus, (2.5) define a continuous mapping from $\bar{M}^{\frac{1}{2},\frac{1}{4}}$ to $M_E^{-\frac{1}{2},-\frac{1}{4}}$. This leads to the following variational problem: For $\mathbf{w}(\mathbf{x},t)$ and $u_0(\mathbf{x})$ given in (2.1), to find out $\mathbf{g} \in M_E^{-\frac{1}{2},-\frac{1}{4}}$, such that

$$b(\mathbf{g},\mathbf{g}') = F(\mathbf{g}'), \qquad \forall \mathbf{g}' \in M_E^{-\frac{1}{2},-\frac{1}{4}}, \tag{2.6}$$

where

$$b(\mathbf{g},\mathbf{g}') = \sum_{i,j=1}^{3} \int_0^T \int_0^t \int_S \int_S U_{ij}(\mathbf{x}-\mathbf{y},t-\tau) g_i(\mathbf{y},\tau) g_j'(\mathbf{x},t) ds(\mathbf{y}) ds(\mathbf{x}) d\tau dt,$$

$$F(\mathbf{g}') = \int_0^T \left(\mathbf{w}\cdot\mathbf{g}'\right)_S dt - \int_0^T \int_S \int_{R^3} \mathbf{u}_0(\mathbf{y})\cdot\mathbf{g}'(\mathbf{x},t)\mathcal{E}(\mathbf{x}-\mathbf{y},t)d\mathbf{y}ds(\mathbf{x})dt,$$

$$(\mathbf{u}(\mathbf{x}),\mathbf{v}(\mathbf{x}))_S = \int_S \mathbf{u}(\mathbf{x})\cdot\mathbf{v}(\mathbf{x})\, ds(\mathbf{x}).$$

THEOREM 1. The $b(\cdot,\cdot)$ defined in (2.6) is bounded and coercive on $M_E^{-\frac{1}{2},-\frac{1}{4}} \times M_E^{-\frac{1}{2},-\frac{1}{4}}$, i.e. there exist two positive constants C_1, C_2, such that for all $\mathbf{g}, \mathbf{g}' \in M_E^{-\frac{1}{2},-\frac{1}{4}}$,

$$|b(\mathbf{g},\mathbf{g}')| \le C_2 \|\mathbf{g}\|_{M_E^{-\frac{1}{2},-\frac{1}{4}}} \|\mathbf{g}'\|_{M_E^{-\frac{1}{2},-\frac{1}{4}}}, \tag{2.7}$$

$$b(\mathbf{g},\mathbf{g}) \ge C_1 \|\mathbf{g}\|^2_{M_E^{-\frac{1}{2},-\frac{1}{4}}}. \tag{2.8}$$

Proof. For a given $\mathbf{g} \in M_E^{-\frac{1}{2},-\frac{1}{4}}$, we find out $\mathbf{u} \in V(0,T)$, such that for all $\mathbf{v} \in Z$,

$$\int_0^T \left[-\left(\mathbf{u},\frac{\partial \mathbf{v}}{\partial t}\right) + \nu(\nabla\mathbf{u},\nabla\mathbf{v})\right] dt = \int_0^T (\mathbf{g},\mathbf{v})_S\, dt, \tag{2.9}$$

where $(\mathbf{u},\mathbf{v}) = \int_{R^3} \mathbf{u}\cdot\mathbf{v}dx$, $Z = \{\mathbf{v}(\mathbf{x},t) \in C^1(0,T;W_0^1); \mathbf{v}(T) = 0\}$.

By Lions theorem [11], (2.9) has a unique solution $\mathbf{u}$ to satisfy

$$\begin{cases} \left(\frac{\partial \mathbf{u}}{\partial t}, \mathbf{v}\right) + \nu(\nabla\mathbf{u},\nabla\mathbf{v}) = (\mathbf{g},\mathbf{v})_S, & \forall\, \mathbf{v}\in W_0^1,\ t\in[0,T], \\ \mathbf{u}(\mathbf{x},0) = 0. \end{cases} \tag{2.10}$$

Let $\mathbf{w}(\mathbf{x},t) = \mathbf{u}(\mathbf{x},t)\,|_S$, from [10–12], we have

$$\|\mathbf{w}\|_{M^{\frac{1}{2},\frac{1}{4}}} \le C\,\|\mathbf{g}\|_{M_E^{-\frac{1}{2},-\frac{1}{4}}}, \tag{2.11}$$

$$\|\mathbf{u}\|_{V(0,T)} \le C_3\,\|\mathbf{w}\|_{M^{\frac{1}{2},\frac{1}{4}}} \le C_4\,\|\mathbf{u}\|_{V(0,T)}. \tag{2.12}$$

By the equivalence between $\|\mathbf{u}\|_{W_0^1}$ and $\|\nabla\mathbf{u}\|_{(L^2(R^3))^9}$ on W_0^1, (2.9) and (2.12),

$$\|\mathbf{g}\|_{M_E^{-\frac{1}{2},-\frac{1}{4}}} = \sup_{\mathbf{v}\in M^{\frac{1}{2},\frac{1}{4}}} \frac{\left|\int_0^T \left[-\left(\mathbf{u},\frac{\partial\mathbf{v}}{\partial t}\right) + \nu(\nabla\mathbf{u},\nabla\mathbf{v})\right]dt\right|}{\|\mathbf{v}\|_{M^{\frac{1}{2},\frac{1}{4}}}} \le C\|\nabla\mathbf{u}\|_{L^2(0,T;(L^2(R^3))^9)}.$$

For all $\mathbf{g},\mathbf{g}' \in M^{-\frac{1}{2},-\frac{1}{4}}$, let $\mathbf{u},\mathbf{u}'$ be the solutions of (2.10) corresponding to $\mathbf{g},\mathbf{g}'$ respectively, by (2.5), (2.11), and the fact $\mathbf{u}(x,0) = 0$, we have

$$|b(\mathbf{g},\mathbf{g}')| = \left|\int_0^T (\mathbf{w},\mathbf{g}')_S dt\right| \le \|\mathbf{w}\|_{M^{\frac{1}{2},\frac{1}{4}}}\|\mathbf{g}'\|_{M_E^{-\frac{1}{2},-\frac{1}{4}}} \le C\|\mathbf{g}\|_{M_E^{-\frac{1}{2},-\frac{1}{4}}}\|\mathbf{g}'\|_{M_E^{-\frac{1}{2},-\frac{1}{4}}}.$$

$$b(\mathbf{g},\mathbf{g}) = \int_0^T (\mathbf{w},\mathbf{g})_S dt = \frac{1}{2}\|\mathbf{u}(T)\|^2_{(L^2(R^3))^3} + \nu\int_0^T \|\nabla\mathbf{u}\|^2_{(L^2(R^3))^9}dt \ge C_1\|\mathbf{g}\|^2_{M_E^{-\frac{1}{2},-\frac{1}{4}}}.$$

By Theorem 1 and Lax theorem, we conclude that (2.6) has a unique solution.

BOUNDARY ELEMENT APPROXIMATION

With the variational form developed, we now consider the BEM approximation. We assume the boundary surface S can be represented as $S = \cup_{i=1}^N S_i$, where $S_i = \mathcal{F}_i(G_i)$ and G_i is a bounded polygon domain in the $\mathbf{z}$ plane, $\mathcal{F}_i$ is a smooth bijection which maps G_i onto S_i. We partition each G_i into a set of triangles $\{G_{ij}\}_{j=1}^{N_i}$, and denote the diameter of G_{ij} by h_{ij}. We assume the partition is quasi-uniform, i.e., $h/\min_{i,j} h_{ij} \le C < \infty$, where $h = \max_{ij} h_{ij}$. We construct the finite-dimensional space $\mathbf{V}^h$ on S as follows:

CASE 1:

$$\mathbf{V}^h = \{\mathbf{g}^h(\mathbf{x});\ \mathbf{g}^h(\mathbf{x}) \in (L^2(S))^3,\ \mathbf{g}^h(\mathbf{x})\,|_{\mathcal{F}_i(G_{ij})} \in (P_0)^3,$$
$$j = 1,\ldots,N_i;\ i = 1,\ldots,N.\},$$

CASE 2:

$$\mathbf{V}^h = \{\mathbf{g}^h(\mathbf{x});\ \mathbf{g}^h(\mathbf{x}) \in (C(S))^3,\ \mathbf{g}^h(\mathcal{F}_i(\mathbf{z}))\,|_{G_{ij}} \in (P_\mu)^3,$$
$$j = 1,\ldots,N_i;\ i = 1,\ldots,N.\},$$

where P_μ is a polynomial space with degree less than or equal to μ.

Let K be a positive interger, $\kappa = T/K$, $t_i = i\kappa, i = 0,1,\ldots,K$. Define space-time finite-element spaces $\mathbf{V}^{h\kappa}$, $\bar{\mathbf{V}}^{h\kappa}$ as follows:

$$\mathbf{V}^{h\kappa} = \left\{\mathbf{g}^{h\kappa}(\mathbf{x},t);\ \ [0,T] \to \mathbf{V}^h,\ \ \mathbf{g}^{h\kappa}\,|_{[t_{i-1},t_i]} = \sum_{j=0}^{[\mu/2]} a_{ij}(\mathbf{x})t^j, a_{ij}(\mathbf{x}) \in \mathbf{V}^h.\right\},$$

$$\bar{\mathbf{V}}^{h\kappa} = \mathbf{V}^{h\kappa} \oplus \{\mathbf{n}\}.$$

Having defined the finite-element spaces: we can formulate the approximate variational problem as follows: To find out $\mathbf{g}^{h\kappa} \in \bar{\mathbf{V}}^{h\kappa}/E$, such that

$$b(\mathbf{g}^{h\kappa},\chi) = F(\chi), \quad \forall\chi \in \bar{\mathbf{V}}^{h\kappa}/E. \tag{3.1}$$

By Theorem 1, (3.1) has a unique solution $\mathbf{g}^{h\kappa} \in \bar{\mathbf{V}}^{h\kappa}/E$. Now we discuss the determination of $\mathbf{g}^{h\kappa} \in \mathbf{V}^{h\kappa}$.

(I) If $\mathbf{n} \in \mathbf{V}^{h\kappa}$, then $\bar{\mathbf{V}}^{h\kappa} = \mathbf{V}^{h\kappa}$. To determine $\mathbf{g}^{h\kappa} \in \mathbf{V}^{h\kappa}$ uniquely, we impose the condition:

$$\mathbf{g}^{h\kappa}(\mathbf{x},t)\cdot\mathbf{n}(\mathbf{x}) = 0, \quad \mathbf{x} \in S,\ t \in [t_{i-1},t_i],\ i = 1,\ldots,K. \tag{3.2}$$

Then, (3.1) can be reduced to follows: To find out $\mathbf{g}^{h\kappa} \in \mathbf{V}^{h\kappa}$, such that for all $\chi \in \mathbf{V}^{h\kappa}$, (3.1) and (3.2) hold. We can solve it by least square method.

(II) If $\mathbf{n}\bar{\in}\mathbf{V}^{h\kappa}$, the solution $\mathbf{g}^{h\kappa}$ of (3.1) can be expressed as:

$$\mathbf{g}^{h\kappa} = \tilde{\mathbf{g}}^{h\kappa} + c\,\mathbf{n}, \qquad c \in R, \tag{3.3}$$

where $\tilde{\mathbf{g}}^{h\kappa} \in \mathbf{V}^{h\kappa}$ is uniquely determined by $\mathbf{g}^{h\kappa} \in \bar{\mathbf{V}}^{h\kappa}/E$. (3.1) becomes: To find out $\tilde{\mathbf{g}}^{h\kappa} \in \mathbf{V}^{h\kappa}$, such that

$$b(\tilde{\mathbf{g}}^{h\kappa},\chi) = F(\chi), \quad \forall\chi \in \mathbf{V}^{h\kappa}. \tag{3.4}$$

Thus, $\mathbf{g}^{h\kappa} \in \bar{\mathbf{V}}^{h\kappa}/E$ is the unique solution of (3.1) iff $\tilde{\mathbf{g}}^{h\kappa} \in \mathbf{V}^{h\kappa}$ determined by (3.3) is the unique solution of (3.4). From the uniqueness of $\mathbf{g}^{h\kappa} \in \bar{\mathbf{V}}^{h\kappa}/E$ of (3.1), we deduce that the coefficient matrix of (3.4) is positive definite and obviously symmetric. We can directly solve (3.4) by standard method.

From the above discussion, we know that problem (3.1) only needs to be solved in $\mathbf{V}^{h\kappa}$ by either the least square method (when $\mathbf{n} \in \mathbf{V}^{h\kappa}$) or by standard methods for symmetric and positive definite matrix (when $\mathbf{n}\bar{\in}\mathbf{V}^{h\kappa}$).

Now we consider the computations of problem (3.1). For conciseness, we rewrite $S_{ij} = \mathcal{F}_i(G_{ij})$ as S_i ($i = 1, 2, \ldots, \mathcal{N}; \mathcal{N} = \sum_{i=1}^{N} N_i$). Also in the remaining part of this section, we denote $\mathbf{g}^{h\kappa}$ by $\mathbf{g}$. For simplicity, we only discuss the case $\mu = 0$ in detail. First, we select r_{lj} as follows:

$$r_{lj}|_{S_n \times [t_{m-1}, t_m]} = \delta_{jn}\delta_{lm}, \quad j, n = 1, \ldots \mathcal{N}; \ l, m = 1, \ldots K.$$

Then we choose 3 $\mathcal{N}$ K basis functions as follows:

$$\chi = (r_{lj}, 0, 0), \quad \chi = (0, r_{lj}, 0), \quad \chi = (0, 0, r_{lj}), \qquad j = 1, \ldots, \mathcal{N}; \ l = 1, \ldots, K. \tag{3.5}$$

The approximate solution $\mathbf{g}$ of (3.1) can be expressed as: $g_i = \sum_{n=1}^{\mathcal{N}} \sum_{m=1}^{K} (c_{mn})_i \, r_{mn}$ ($i = 1, 2, 3$). Thus, (3.1) can be rewritten as 3 $\mathcal{N}$ K algebraic equation system:

$$\begin{aligned}
&\sum_{i=1}^{3} \sum_{n=1}^{\mathcal{N}} (c_{kn})_i \int_{t_{k-1}}^{t_k} \int_{t_{k-1}}^{t} \int_{S_l} \int_{S_n} U_{ij}(\mathbf{x} - \mathbf{y}, t - \tau) ds(\mathbf{y}) ds(\mathbf{x}) d\tau dt \\
&= \int_{t_{k-1}}^{t_k} \int_{S_l} w_j(\mathbf{x}, t) ds(\mathbf{x}) dt - \int_{t_{k-1}}^{t_k} \int_{S_l} \int_{R^3} u_{0j}(\mathbf{y}) E(\mathbf{x} - \mathbf{y}, t) d\mathbf{y} ds(\mathbf{x}) dt \\
&- \sum_{m=1}^{k-1} \sum_{i=1}^{3} \sum_{n=1}^{\mathcal{N}} \left\{ (c_{mn})_i \int_{t_{k-1}}^{t_k} \int_{t_{m-1}}^{t_m} \int_{S_l} \int_{S_n} U_{ij}(\mathbf{x} - \mathbf{y}, t - \tau) ds(\mathbf{y}) ds(\mathbf{x}) d\tau dt \right\}, \\
&\quad l = 1, \ldots, N; \ j = 1, 2, 3; \ k = 1, \ldots, K.
\end{aligned} \tag{3.6}$$

(3.6) implies that for both cases with $\mathbf{n} \in \mathbf{V}^{h\kappa}$ and $\mathbf{n} \bar{\in} \mathbf{V}^{h\kappa}$, we can solve it level by level, i.e., for $k = 1, \ldots, K$, we solve (3.6) successively, by using either least square method or standard method. Moreover, we only need to restore information of current level and the last term on the right-hand side of (3.6) from last time level.

Now, we discuss the calculations of the integrals appearing in (3.6). We only consider the integrals on the left-hand side of (3.6). Others can be derived similarly. When S_n and S_l are neither coincident nor adjacent, we can directly calculate them by numerical quadrature formulas [14]. Now, we consider the case when S_n and S_l are either coincident or adjacent.

Let $\bar{A}_1, \bar{A}_2, \bar{A}_3$ be the vertices of G_m, where $S_m = \mathcal{F}(G_m)$. $A_j = \mathcal{F}(\bar{A}_j)$ ($j = 1, 2, 3$). Set up a local coordinate system $(\mathbf{e}_1, \mathbf{e}_2, \mathbf{e}_3)$ with origin A_1 such that $\mathbf{e}_1$ coincides with A_1A_2, $\mathbf{e}_2$ is in the plane determined by A_1, A_2 and A_3, and orthogonal to $\mathbf{e}_1$, $\mathbf{e}_3 = \mathbf{e}_1 \times \mathbf{e}_2$. In this system, S_n (or S_l) can be expressed as:

$$\beta_1 = A_1\mathcal{F}(\xi) \cdot \mathbf{e}_1, \quad \beta_2 = A_1\mathcal{F}(\xi) \cdot \mathbf{e}_2, \quad \beta_3 = A_1\mathcal{F}(\xi) \cdot \mathbf{e}_3 = \varphi(\beta_1, \beta_2),$$

where ξ varies in G_n (or G_l), (β_1, β_2) varies in the plane determined by A_1, A_2, and A_3.

From (2.2) and (3.6), we see

$$\sum_{i=1}^{3}(c_{kn})_i \int_{t_{k-1}}^{t_k}\int_{t_{k-1}}^{t}\int_{S_l}\int_{S_n} U_{ij}(\mathbf{x}-\mathbf{y},t-\tau)ds(\mathbf{y})ds(\mathbf{x})d\tau dt$$

$$= (c_{kn})_j \int_{t_{k-1}}^{t_k}\int_{t_{k-1}}^{t}\int_{S_l}\int_{S_n} \mathcal{E}(\mathbf{x}-\mathbf{y},t-\tau)ds(\mathbf{y})ds(\mathbf{x})d\tau dt$$

$$+\sum_{i=1}^{3}\frac{n_i(c_{kn})_i}{2\pi^{\frac{3}{2}}}\int_{t_{k-1}}^{t_k}\int_{t_{k-1}}^{t}\int_{S_l}\int_{\partial S_n}\left\{\frac{x_j-y_j}{2\sqrt{\nu(t-\tau)}|\mathbf{x}-\mathbf{y}|^2}\exp\left(-\frac{|\mathbf{x}-\mathbf{y}|^2}{4\nu(t-\tau)}\right)\right. \tag{3.7}$$

$$\left.-\frac{(x_j-y_j)}{|\mathbf{x}-\mathbf{y}|^3}\int_0^{\frac{|\mathbf{x}-\mathbf{y}|}{2\sqrt{\nu(t-\tau)}}}\exp(-q^2)dq\right\}ds(\mathbf{y})ds(\mathbf{x})d\tau dt.$$

As for the internal singular integrals, we have

$$\int_{S_l}\left\{\frac{(x_j-y_j)}{2\sqrt{\nu(t-\tau)}|\mathbf{x}-\mathbf{y}|^2}\exp\left(-\frac{|\mathbf{x}-\mathbf{y}|^2}{4\nu(t-\tau)}\right)-\frac{(x_j-y_j)}{|\mathbf{x}-\mathbf{y}|^3}\int_0^{\frac{|\mathbf{x}-\mathbf{y}|}{2\sqrt{\nu(t-\tau)}}}\exp(-q^2)dq\right\}ds(\mathbf{x})$$

$$= \int_{\bar{\bar{S}}_l} J(G_l)J(\bar{S}_l)\left\{\frac{\omega_j}{|\omega|^2}\exp\left[-\frac{|\omega|^2}{4\nu(t-\tau)}\right]-\frac{\omega_j}{|\omega|^3}\int_0^{\frac{|\omega|^2}{2\sqrt{\nu(t-\tau)}}}\exp(-q^2)dq\right\}ds(\omega), \tag{3.8}$$

where $J(G_l)$ is the Jacobian determinant from G_l to S_l, $J(\bar{S}_l)$ is the Jacobian determinant from $\bar{S}_l$ to G_l, where $\bar{S}_l$ is the triangle on the plane determined by A_1, A_2, and A_3, $\bar{\bar{S}}_l$ is the image of $\bar{S}_l$ under the transformation $\omega=\beta-\zeta$; $\omega=(\omega_1,\omega_2,\omega_3)$, $\beta=(\beta_1,\beta_2,\beta_3)$, $\zeta=(\zeta_1,\zeta_2,\zeta_3)$, $\beta_3=\varphi(\beta_1,\beta_2)$, $\zeta_3=\varphi(\zeta_1,\zeta_2)$, $\omega_3=\varphi(\zeta_1+\omega_1,\zeta_2+\omega_2)-\varphi(\zeta_1,\zeta_2)$.

Let the boundary $\partial\bar{\bar{S}}_l$ of $\bar{\bar{S}}_l$ can be expressed as:

$$\rho=\rho(\theta),\quad 0\le\theta\le 2\pi,\quad \rho^2=\omega_1^2+\omega_2^2 \tag{3.9}$$

then, (3.8) reduces to

$$\int_0^{2\pi}\left\{\int_0^{\rho(\theta)} J(G_l)J(\bar{S}_l)\left\{\frac{\rho\,\omega_j}{\bar{\varphi}(\zeta,\rho,\theta)}\exp\left[-\frac{\bar{\varphi}(\zeta,\rho,\theta)}{4\nu(t-\tau)}\right]\right.\right.$$

$$\left.\left.-\frac{\omega_j}{\bar{\varphi}(\zeta,\rho,\theta)^{\frac{3}{2}}}\int_0^{\varphi(\zeta,\rho,\theta)}2\sqrt{\nu(t-\tau)}\exp(-q^2)dq\right\}d\rho\right\}d\theta,$$

where $\bar{\varphi}(\zeta,\rho,\theta)=[\varphi(\zeta_1+\rho\cos\theta,\zeta_2+\rho\sin\theta)-\varphi(\zeta_1,\zeta_2)]+\rho^2$.

This time, since the internal integral is no longer singular, we can use the numerical quadrature formula to calculate it. The other term of (3.7) can be calculated similarly. In practice, we can compute all the integrals over the standard reference element. For the cases $\mu>0$, we can derive the equations similarly.

ERROR ESTIMATES

LEMMA [5,9,13]: For all $\mathbf{g}^h \in \mathbf{V}^h$, the following inverse estimates hold:

$$\|\mathbf{g}^h\|_{(H^r(S))^3} \leq Ch^{q-r} \|\mathbf{g}^h\|_{(H^q(S))^3}, \tag{4.1}$$

CASE 1: $-1 \leq q \leq r \leq 0$,

CASE 2: $-1 \leq q \leq 0, \quad q \leq r \leq 1$.

Furthermore, let R^h be the projection operator from $(L^2(S))^3$ onto $\mathbf{V}^h$, then

$$\|\mathbf{g} - R^h\mathbf{g}\|_{(H^q(S))^3} \leq Ch^{r-q} \|\mathbf{g}\|_{(H^r(S))^3}, \tag{4.2}$$

CASE 1: $-1 \leq q \leq 0 \leq r \leq 1$,

CASE 2: $-(\mu+1) \leq q \leq 1,\ q \leq r,\ 0 \leq r \leq \mu+1$.

THEOREM 2. Let $R^{h\kappa}$ be the projection operator from $(L^2(S \times [0,T]))^3$ onto $\bar{\mathbf{V}}^{h\kappa}$, then for all $\mathbf{g} \in M^{\mu+1,\frac{(\mu+1)}{2}}$, we have

$$\|\mathbf{g} - R^{h\kappa}\mathbf{g}\|_{M^{-r,-\frac{r}{2}}} \leq C\left(h^{\mu+1+r} + \kappa^{\frac{(\mu+1+r)}{2}}\right) \|\mathbf{g}\|_{M^{\mu+1,\frac{(\mu+1)}{2}}}, \quad 0 \leq r \leq \mu+1. \tag{4.3}$$

Proof:

$$\|R^h\mathbf{g} - R^{h\kappa}\mathbf{g}\|_{M^{0,0}} \leq \|R^h\mathbf{g} - IR^h\mathbf{g}\|_{M^{0,0}} \leq C\kappa^{\frac{q}{2}} \|\mathbf{g}\|_{M^{0,\frac{q}{2}}}, \quad 0 \leq q \leq \mu+1,$$

where I is the interpolation operator with respect to t and is defined on $[0,T]$. By (4.2) and triangular inequality, we get (4.3) for $r = 0$. For $0 < r \leq \mu+1$,

$$\|\mathbf{g} - R^{h\kappa}\mathbf{g}\|_{M^{-r,-\frac{r}{2}}} = \sup_{\mathbf{w}\in M^{r,\frac{r}{2}}} \frac{|\int_0^T (\mathbf{g} - R^{h\kappa}\mathbf{g}, \mathbf{w} - R^{h\kappa}\mathbf{w})_S dt|}{\|\mathbf{w}\|_{M^{r,\frac{r}{2}}}}$$

$$\leq C(h^r + \kappa^{\frac{r}{2}}) \|\mathbf{g} - R^{h\kappa}\mathbf{g}\|_{M^{0,0}} \leq C(h^{\mu+1+r} + \kappa^{\frac{(\mu+1+r)}{2}}) \|\mathbf{g}\|_{M^{\mu+1,\frac{(\mu+1)}{2}}}.$$

THEOREM 3. Let $\mathbf{g}$ and $\mathbf{g}^{h\kappa}$ be the solutions of problem (2.6) and (3.1) respectively, then we have optimal-order energy norm error estimates for $\mathbf{g} - \mathbf{g}^{h\kappa}$ as follows:

$$\|\mathbf{g} - \mathbf{g}^{h\kappa}\|_{M_g^{-r,-\frac{r}{2}}} \leq C\left(h^{\mu+1+r} + \kappa^{\frac{(\mu+1+r)}{2}}\right) \|\mathbf{g}^{(\mu+1)}\|_{M^{\mu+1,\frac{(\mu+1)}{2}}}, \quad \frac{1}{2} \leq r \leq \mu+1. \tag{4.4}$$

Furthermore, let $(\mathbf{u}, p)$ be given by (2.4), $(\mathbf{u}^{h\kappa}, p^{h\kappa})$ be defined below:

$$\begin{cases} u_i^{h\kappa}(\mathbf{x},t) = \int_0^t \int_S \mathbf{U}_i(\mathbf{x}-\mathbf{y}, t-\tau)\cdot \mathbf{g}^{h\kappa}(\mathbf{y},\tau) ds(\mathbf{y}) d\tau + \int_{R^3} u_{0i}(\mathbf{y})\, \mathcal{E}(\mathbf{x}-\mathbf{y},t) d\mathbf{y}, \\ p^{h\kappa}(\mathbf{x},t) = \int_0^t \int_S \mathbf{P}(\mathbf{x}-\mathbf{y}, t-\tau)\cdot \mathbf{g}^{h\kappa}(\mathbf{y},\tau) ds(\mathbf{y}) d\tau, \end{cases} \tag{4.5}$$

then we have optimal-order energy norm error estimates for $(\mathbf{u} - \mathbf{u}^{h\kappa}, p - p^{h\kappa})$ as follows:

$$\|\mathbf{u}-\mathbf{u}^{h\kappa}\|_{V(0,T)}+\|p-p^{h\kappa}\|_{L^2(\Omega\times[0,T])/R}+\|p-p^{h\kappa}\|_{L^2(\Omega'\times[0,T])}$$
$$\leq C\left(h^{\mu+\frac{3}{2}}+\kappa^{\frac{\mu}{2}+\frac{3}{4}}\right)\|\mathbf{g}^{(\mu+1)}\|_{M^{\mu+1,\frac{(\mu+1)}{2}}}, \tag{4.6}$$

where $\mathbf{g}^{(s)}$ satisfies

$$(\mathbf{g}^{(s)},\mathbf{n})_{M^{s+1,\frac{(s+1)}{2}}}=0. \tag{4.7}$$

Proof. From (2.6) and (3.1), $b(\mathbf{g}-\mathbf{g}^{h\kappa},\chi)=0$, for all $\chi\in\bar{\mathbf{V}}^{h\kappa}/E$.

$$C_1\|\mathbf{g}-\mathbf{g}^{h\kappa}\|^2_{M_E^{-\frac{1}{2},-\frac{1}{4}}}\leq b(\mathbf{g}-\mathbf{g}^{h\kappa},\mathbf{g}-\mathbf{g}^{h\kappa})=b\left(\mathbf{g}-\mathbf{g}^{hk},\mathbf{g}^{(\mu+1)}-R^{h\kappa}\mathbf{g}^{(\mu+1)}\right)$$
$$\leq C_2\|\mathbf{g}-\mathbf{g}^{h\kappa}\|_{M_E^{-\frac{1}{2},-\frac{1}{4}}}\|\mathbf{g}^{(\mu+1)}-R^{h\kappa}\mathbf{g}^{(\mu+1)}\|_{M^{-\frac{1}{2},-\frac{1}{4}}},$$

$$\|\mathbf{g}-\mathbf{g}^{h\kappa}\|_{M_E^{-\frac{1}{2},-\frac{1}{4}}}\leq C\left(h^{\mu+\frac{3}{2}}+\kappa^{\frac{\mu}{2}+\frac{3}{4}}\right)\|\mathbf{g}^{(\mu+1)}\|_{M^{\mu+1,\frac{(\mu+1)}{2}}}.$$

Thus, (4.4) is true for $r=\dfrac{1}{2}$. As for $r=\mu+1$, $\forall\mathbf{w}\in\bar{M}^{\mu+1,\frac{(\mu+1)}{2}}$, from [10–12], there exists a unique solution $\mathbf{v}\in M_E^{\mu,\frac{\mu}{2}}$ which satisfies:

$$w_i(\mathbf{x},t)=\int_0^t\int_S\mathbf{U}_i(\mathbf{x}-\mathbf{y},t-\tau)\cdot\mathbf{v}(\mathbf{y},\tau)ds(\mathbf{y})d\tau,\qquad i=1,2,3, \tag{4.8}$$

and the following estimate holds:

$$\|\mathbf{v}\|_{M_E^{\mu,\frac{\mu}{2}}}\leq C\|\mathbf{w}\|_{M^{\mu+1,\frac{(\mu+1)}{2}}}.$$

Thus, we have

$$\|\mathbf{g}-\mathbf{g}^{h\kappa}\|_{M_E^{-(\mu+1),-\frac{(\mu+1)}{2}}}=\sup_{\mathbf{w}\in\bar{M}^{\mu+1,\frac{(\mu+1)}{2}}}\frac{|b(\mathbf{g}-\mathbf{g}^{h\kappa},\mathbf{v})|}{\|\mathbf{w}\|_{M^{\mu+1,\frac{(\mu+1)}{2}}}}$$
$$\leq C\sup_{\mathbf{v}\in M_E^{\mu,\frac{\mu}{2}}}\frac{|b(\mathbf{g}-\mathbf{g}^{h\kappa},\mathbf{v})|}{\|\mathbf{v}\|_{M_E^{\mu,\frac{\mu}{2}}}}$$
$$\leq C\sup_{\mathbf{v}\in M_E^{\mu,\frac{\mu}{2}}}\frac{|b(\mathbf{g}-\mathbf{g}^{h\kappa},\mathbf{v}^{(\mu)}-R^{h\kappa}\mathbf{v}^{(\mu)})|}{\|\mathbf{v}^{(\mu)}\|_{M^{\mu,\frac{\mu}{2}}}}$$
$$\leq C\left(h^{2\mu+2}+\kappa^{\mu+1}\right)\|\mathbf{g}^{(\mu+1)}\|_{M^{\mu+1,\frac{(\mu+1)}{2}}},$$

where $\mathbf{v}^{(\mu)}$ satifies (4.7). Here we have used the fact $\|\mathbf{v}^{(\mu)}\|_{M_E^{\mu,\frac{\mu}{2}}}=\|\mathbf{v}^{(\mu)}\|_{M^{\mu,\frac{\mu}{2}}}$.

Thus, we have proved the case $r=\mu+1$. For the case $0<r<\mu+1$, the estimate can be obtained by use of the interpolation properties of the space $M^{-r,-\frac{r}{2}}$. (4.6) is a direct conclusion of prior estimates for Stokes problem and (4.4) with $r=\dfrac{1}{2}$.

The following theorem gives the L^∞ superconvergence estimates for $(\mathbf{u}-\mathbf{u}^{h\kappa}, p-p^{h\kappa})$ and its derivatives with respect to $\mathbf{x}$ and t outside the neighborhood of $S \times [0,T]$.

THEOREM 5. Let $(\mathbf{u},p),(\mathbf{u}^{h\kappa},p^{h\kappa})$ be defined by (2.4), (2.5) and (4.5) respectively, then for all $(\mathbf{x},t) \in R^3 \times [0,T]$ with $d = d(\mathbf{x},S) \geq d_1 > 0$, we have the maximum norm estimates:

$$\left|D_t^j D^\alpha(\mathbf{u}-\mathbf{u}^{h\kappa})(\mathbf{x},t)\right| \leq C(d,\alpha,j)(h^{2\mu+2}+\kappa^{\mu+1})\left\|\mathbf{g}^{(\mu+1)}\right\|_{M^{\mu+1,\frac{(\mu+1)}{2}}}, \tag{4.10}$$

$$\left|D_t^j D^\alpha(p-p^{h\kappa})(\mathbf{x},t)\right| \leq C(d,\alpha,j)(h^{2\mu+2}+\kappa^{\mu+1})\left\|\mathbf{g}^{(\mu+1)}\right\|_{M^{\mu+1,\frac{(\mu+1)}{2}}}, \tag{4.11}$$

where $\alpha = (\alpha_1,\alpha_2,\alpha_3)$, α_i– nonnegative integers; $d(\mathbf{x},S)$ is the distance between $\mathbf{x}$ and S, $C(d,\alpha,j)$ is a positive number dependent on $d(\mathbf{x},S)$, j and α.

Proof: First we have

$$\begin{aligned}\left|D^\alpha(u_i-u_i^{h\kappa})(\mathbf{x},t)\right| &= \left|\int_0^t\int_S D^\alpha\mathbf{U}_i(\mathbf{x}-\mathbf{y},t-\tau)\cdot(\mathbf{g}-\mathbf{g}^{h\kappa})(\mathbf{y},\tau)ds(\mathbf{y})d\tau\right| \\ &\leq \left\|D^\alpha\mathbf{U}_i(\mathbf{x}-\mathbf{y},t-\tau)\right\|_{M^{\mu+1,\frac{(\mu+1)}{2}}}\left\|\mathbf{g}-\mathbf{g}^{h\kappa}\right\|_{M_B^{-(\mu+1),\frac{(\mu+1)}{2}}}.\end{aligned}$$

Noticing that $\|D^\alpha\mathbf{U}_i(\mathbf{x}-\mathbf{y},t-\tau)\|_{M^{\mu+1,\frac{(\mu+1)}{2}}} \leq C(d,\alpha)$ and (4.4) with $r = \mu+1$, we directly get the (4.10) with $j = 0$. Other estimates can be derived similarly.

KEY WORDS: Boundary Elements, Stokes Problem, Convergence Analysis

ACKNOWLEDGEMENTS

This research was supported in part by the Office of Naval Research Contract No. 0014-88-K-0370, by National Science Foundation Grant No. DMS - 8922865, and by funding from the Institute of Scientific Computations at the University of Wyoming through NSF Grant No. RII-8610680.

REFERENCES

1. Brebbia, C.A. Telles, J.C.F. and Wrobel L.C. Boundary Element Techniques, Springer–Verlag, Berlin and New York, 1984.

2. Brebbia, C.A. (Ed.). Topics in Boundary Element Research, Springer–Verlag, Berlin and New York, 1985.

3. Du, Q. (Ed.). Boundary Elements, Pergamon Press, London, 1986.

4. Le Roux, M.N. Méthode d'éléments finis pour la résolution numérique d'un problème extérieur en dimension deux, *R.A.I.R.O., Série Analyse Numerique, Vol. 11*, pp. 27–60, 1977.

5. Nedelec, J.C. and Planchard, J. Une méthode variationnelle d'éléments finis pour la résolution numérique d'un problème extérieur dans R^3, *R.A.I.R.O., Vol. 7*, pp. 105–129, 1973.

6. Hsiao, G.C. and Wendland, W. A finite element method for some integral equations of the first kind, *J. Math. Appl., Vol. 58*, pp. 449-481, 1977.

7. Atkinson, K.E. and Chandler, G. BIE methods for solving Laplace's equation with nonlinear boundary conditions: The smooth boundary case, *Math. Comp.*, (to appear).

8. Zhu, J. A boundary integral equation method for the stationary Stokes problem in 3D, Boundary Elements, (Eds. Brebbia, C.A. *et al.*), pp. 283–292, Springer-Verlag, Berlin and New York, 1983.

9. Wang, H. The convergence of BEM for the stationary Stokes problem in three dimensions, Boundary Elements, (Ed. Du, Q.), pp. 143–150, Pergamon Press, London, 1986.

10. Ladyzhenskaya, O.A. The Mathematical Theory of Viscous Incompressible Flow, Gorden and Breach, New York, 1969.

11. Lions, J.L. and Magenes, E. Non-Homogeneous Boundary Value Problems and Applications, Springer-Verlag, Berlin and New York, 1972.

12. Temam, R. Navier-Stokes Equations, North-Holland, Amsterdam, 1977.

13. Ciarlet, P.G. The Finite Element Method for Elliptic Problems, North-Holland, Amsterdam, 1978.

14. Stroud, A.H. Approximate Calculation of Multiple Integrals, Prentice Hall, Englewood Cliffs, New Jersey, 1971.

The Motion of a Semi-Infinite Bubble between Parallel Plates

Y. Yuan, D.B. Ingham

Department of Applied Mathematical Studies, The University of Leeds, Leeds LS2 9JT, U.K.

ABSTRACT

In this paper we investigate the motion of a bubble moving between two parallel plates. The boundary element method with an iterative technique is presented and the numerical solutions are obtained for various values of the capillary number, Ca.

INTRODUCTION

Over the decades there has been much interest in the motion of bubbles in tubes. For example, when air is blown into one end of a horizontal tube which contains a viscous fluid it forms a bubble which travels down the tube and forces some of the liquid out at the far end of the tube. As the bubble moves along the tube a layer of fluid of thickness h_f forms between the bubble and the walls of the tube. The ratio, $h_\infty = h_f/h_0$, where h_0 is the radius of the tube, is directly related to the mean speed of the bubble. Fairbrother and Stubbs [1] presented the first experimental measurements of the relationship between h_∞ and the capillary number, $Ca = \mu U_0/T$, where μ is the viscosity of the fluid, U_0 is the velocity of the bubble and T is the surface tension. Their results for very low values of Ca indicated a square root relation between h_∞ and Ca and this work was extended by Taylor [2] to cover a larger range of values of the capillary number. Bretherton [3] provided the first theoretical work on the problem using a two dimensional model and derived a relationship between h_∞ and Ca, namely,

$$h_\infty = 0.643\ (3\ Ca)^{2/3} \qquad (1)$$

Recently some more precise measurements and analyses have been carried out by Schwartz *et al.* [4] and Saffman [5]. Numerical solutions of these problems have been obtained by Reinelt and Saffman [6], Shen and Udell [7], Homsy [8], and Saffman [9] by using finite difference or finite element methods and Homsy [8] and Saffman [9] give excellent reviews of the problem.

The Boundary Element Method (BEM) is especially useful for dealing with these problems and Ritchie [10] and Lu and Chang [11] have obtained numerical solutions by using the constant BEM. However, there is a subtle difference between ψ' and $\partial\psi/\partial n$ because two coordinate systems are being used, namely one fixed at a special point on the free surface and the other one is moving on the free surface. Kuiken [12] was first to observe this and derived the correct normal stress condition which has to be applied on the free surface. In this paper we develop the BEM in order to calculate the free surface problem of two-dimensional creeping flow [13] for the motion of a bubble between two parallel plates.

PROBLEM FORMULATION AND BOUNDARY CONDITIONS

A semi-infinite bubble is moving between two parallel plates a distance, $2h_0$, apart at a fixed speed, U_0, with a constant pressure, P_0, being maintained in the bubble, which has negligible viscosity, see Fig. 1. For convenience we change the co-ordinates such that the bubble is considered to be motionless and the plates to be moving past the bubble tip at a constant speed, $-U_0$. All variables in the problem are non-dimensioned using the half-distance between the plates, h_0, the speed of the bubble, U_0, and the viscosity of the fluid, μ, as follows

$$x = X/h_0 \qquad y = Y/h_0 \qquad u = U/U_0 \qquad p = P_0 h_0/\mu U_0$$

where U is the fluid velocity.

For steady two-dimensional flow of an incompressible Newtonian fluid the Navier-Stokes equation and continuity equation reduce to

$$\nabla p = \nabla^2 u \tag{2a}$$

$$\nabla \cdot u = 0 \tag{2b}$$

when the Reynolds number, $Re = U_0 h_0 /\upsilon$, is assumed to be very small. On introducing the stream function ψ, the x- and y-components of velocity are then given by

$$u = \psi_y, \qquad v = -\psi_x \tag{3}$$

respectively. From equations (2) and (3) it may be shown, see Batchelor [14], that ψ satisfies the biharmonic equation

$$\nabla^4 \psi = 0 \tag{4}$$

On introducing the vorticity, ω, equation (4) may be written in the form

$$\begin{cases} \nabla^2 \psi = \omega & (5a) \\ \nabla^2 \omega = 0 & (5b) \end{cases}$$

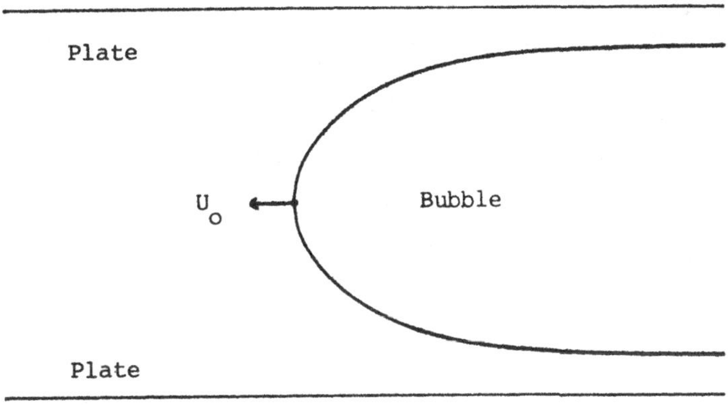

Fig. 1. The bubble moving between two parallel plates.

In order to solve the biharmonic equation (5) in the domain Ω we use the BEM. For any $p = (x,y) \in \bar{\Omega}$ and $q = (x_0,y_0) \in \partial\Omega$, let

$$F_1(p,q) = \ln|p - q| \tag{6a}$$

$$F_2(p,q) = |p - q|^2 (\ln|p - q| - 1) \tag{6b}$$

where $|p - q| = \{ (x - x_0)^2 + (y - y_0)^2 \}^{1/2}$. Applying Green's second identity we have

$$\eta(p)\psi(x) = \int_{\partial\Omega} \psi(q)\, F_1'(p,q)\, ds(q) - \int_{\partial\Omega} \psi'(q)\, F_1(p,q)\, ds(q)$$

$$+ \frac{1}{4}\int_{\partial\Omega} \omega(q)\, F_2'(p,q)\, ds(q) - \frac{1}{4}\int_{\partial\Omega} \omega'(q)\, F_2(p,q)\, ds(q) \tag{7}$$

$$\eta(p)\omega(x) = \int_{\partial\Omega} \omega(q)\, F_1'(p,q)\, ds(q) - \int_{\partial\Omega} \omega'(q)\, F_1(p,q)\, ds(q) \tag{8}$$

where

(i) ds(q) denotes differential increment of $\partial\Omega$ at q.

(ii) The prime (′) denotes differentiation with respect to the outward normal to Ω at q.

$$\text{(iii) } \eta(p) = \begin{cases} 2\pi & \text{when } x\epsilon\Omega \\ \theta & \text{when } x\epsilon\partial\Omega,\ \theta \text{ is the angle between the} \\ & \text{tangents to } \partial\Omega \text{ on either side of } x. \end{cases}$$

In order to obtain the numerical solution of the biharmonic equation from the integral equations (7) and (8) the simplest method to use is the classical BEM. However, for more accuracy we use the linear BEM in the present work. This is achieved by first subdividing the boundary $\partial\Omega$ into N segments $\partial\Omega_j$, where $j = 1, 2, \ldots, N$. On each segment $\partial\Omega_j$, ψ, ψ', ω and ω' are approximated by linear functions such that

$$\psi = (1-\xi)\, \psi(q_{j-1}) + \xi\, \psi(q_j)$$

$$\psi' = (1-\xi)\, \psi'(q_{j-1}) + \xi\, \psi'(q_j)$$

$$\omega = (1-\xi)\, \omega(q_{j-1}) + \xi\, \omega(q_j)$$

$$\omega' = (1-\xi)\, \omega'(q_{j-1}) + \xi\, \omega'(q_j)$$

where q_{j-1} and q_j are the end points of the interval $\partial\Omega_j$, ξ is a linear function which increases from zero at q_{j-1} to unity at q_j. With these approximations the integral equations (7) and (8) become

$$\eta(p)\psi(p) = \sum_{j=1}^{N} \{\psi_{j-1}\int_{\partial\Omega_j} (1-\xi)\, F_1'(p,q)ds(q) + \psi_j \int_{\partial\Omega_j} \xi\, F_1'(p,q)ds(q\ )\}$$

$$- \sum_{j=1}^{N} \{\psi'_{j-1} \int_{\partial\Omega_j} (1-\xi)\, F_1(p,q)ds(q) + \psi'_j \int_{\partial\Omega_j} \xi\, F_1(p,q)d(q)\} +$$

$$\frac{1}{4} \sum_{j=1}^{N} \{\omega_{j-1}\int_{\partial\Omega_j} (1-\xi)\, F_2'(p,q)ds(q) + \omega_j \int_{\partial\Omega_j} \xi\, F_2'(p,q)ds(q)\} -$$

$$\frac{1}{4} \sum_{j=1}^{N} \{\omega'_{j-1} \int_{\partial\Omega_j} (1-\xi)\, F_2(p,q)ds(q) + \omega'_j \int_{\partial\Omega_j} \xi\, F_2(p,q)d(q)\} \qquad (9)$$

$$\eta(p)\omega(p) = \sum_{j=1}^{N} \{\omega_{j-1}\int_{\partial\Omega_j} (1-\xi)\, F_1'(p,q)ds(q) + \omega_j\int_{\partial\Omega_j} \xi\, F_1'(p,q)ds(q)\}$$

$$\sum_{j=1}^{N} \{\omega'_{j-1} \int_{\partial\Omega_j} (1-\xi)\, F_1(p,q)ds(q) + \omega'_j \int_{\partial\Omega_j} \xi\, F_2(p,q)d(q)\} \qquad (10)$$

The integral equations (9) and (10) result in a 2N × 2N linear system equations with 4N variables. Hence, at every grid point two boundary conditions must be specified. However, since the location of the free surface Γ, which is assumed to be a function of x, is unknown, we shall adopt an iterative determination of the free surface. At each iteration two conditions on the free surface must be imposed and the remaining one, the shear stress condition, will be used to determine the free surface.

To derive the boundary conditions it is necessary to consider a closed contour problem domain, and therefore the infinity boundary conditions in x are applied at $x = -M_1$ and $x = M_2$ where M_1 and M_2 are two constants which have to be specified. Due to the symmetry we need only consider the solution in half of the original channel, see Fig. 2.

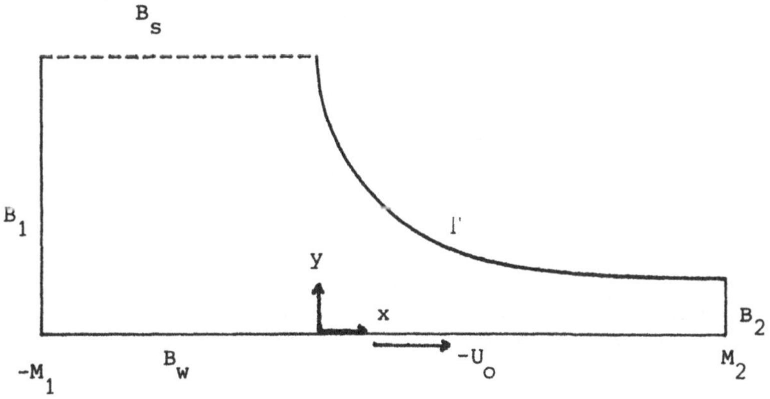

Fig. 2. The geometry of the problem.

The fluid flow far upstream of the origin is constrained to flow between two parallel plates, thus as $x \longrightarrow -\infty$ the flow is a combination of a uniform flow and Poiseuille flow which means that the fluid velocity profile across the channel is parabolic. This results in the stream function being a cubic function of y. Far downstream of the origin the fluid flow becomes plug flow, i.e. the fluid moves as a body with a uniform speed, U_0. The no-slip condition is specified on the wall, y = 0, and on y = 1 the symmetry boundary conditions, $\psi = 0$ and $\omega = 0$ are employed.

We now consider on the free surface the local right-handed coordinate system defined by the vectors t and n, where t denotes the tangent vector and n is the outward normal to the free surface vector. Then the boundary conditions on free surface may be expressed

(i) $\psi = q$ (11a)

(ii) $\psi'' - \psi_{tt} = 0$ (11b)

$$\text{(iii)} \quad -k/Ca = -p + \sigma'' \qquad (11c)$$

where q (= Q/h_0U_0, where Q is the fluid flow rate) is a non-dimensional fluid flow rate, (′) denotes the normal derivative, k is the local non-dimensional curvature of the free surface, p is the non-dimensional pressure and σ is the non-dimensional stress tensor, see Ingham and Kelmanson [15].

We may eliminate from the problem the pressure p. The t-component of equation (2a) may be expressed, in terms of the local coordinate system (t,n), as follows,

$$\partial p / \partial t = \nabla^2 u_t$$

or, in terms of the vorticity,

$$\partial p / \partial t = \nabla^2 (\partial \psi / \partial n) = \partial \omega / \partial n \qquad (12)$$

Differentiating equation (11c) with respect to t and substituting for $\partial p/\partial t$ from equation (12) into the boundary condition (11c) we obtain

$$2 \frac{\partial^3 \psi}{\partial n\, \partial t^2} + \frac{\partial \omega}{\partial n} + \frac{1}{Ca} \frac{dk}{dt} = 0 \qquad (13)$$

Instead of the t-derivatives appearing in the boundary conditions (11b) and (13) it is more convenient to have derivatives with respect to s, the arc length along the free surface. Coyne and Elrod [16] have shown that

$$\left.\frac{\partial}{\partial t}\right|_\Gamma = \frac{d}{ds}$$

$$\left.\frac{\partial^2}{\partial t^2}\right|_\Gamma = \frac{d}{ds^2} - k \frac{\partial}{\partial n}$$

which on substitution into the boundary condition (11b) gives

$$\omega = -2\, k\, \psi' \qquad (14)$$

If A is a fixed point on the free surface we observe that the derivative $\partial\psi/\partial n$ is taken with respect to a fixed local Cartesian coordinate system at A. On the other hand ψ' denotes the outward normal derivative at each separate point of Γ. Let B be an arbitrary neighborhood point of A on Γ, then we have

$$\partial\psi/\partial n\,|_B = \cos\theta\, \psi'\,|_B + \sin\theta\, \partial\psi/\partial s\,|_B = \cos\theta\, \psi'\,|_B$$

where θ is the angle between the tangent to Γ at B and the tangent to Γ at A. It is clear that as $B \longrightarrow A$ we have $\theta \longrightarrow 0$ and $\partial\psi/\partial n|_B \longrightarrow \psi'|_A$. We can therefore now write

$$\frac{\partial^2}{\partial t^2}\frac{\partial\psi}{\partial n}\Big|_{\Gamma,A} = \frac{d^2}{ds^2}\frac{\partial\psi}{\partial n}\Big|_{\Gamma,A} - k\,\frac{\partial^2\psi}{\partial n^2}\Big|_{\Gamma,A}$$

$$= \frac{d^2}{ds^2}\,(\ \cos\theta\ \psi'\ )\Big|_{\Gamma,A} - \frac{k}{2}\,\omega\,\Big|_{\Gamma,A}$$

$$= \frac{d^2}{ds^2}\,\psi'\Big|_{\Gamma,A}$$

where we have used $d\theta/ds = k$ and the boundary conditions (11b) and (14). The third boundary condition, (11c), may now be expressed in the form

$$-\frac{1}{Ca}\frac{dk}{ds} = \omega' + 2\,\frac{d^2}{ds^2}\,\psi' \qquad \text{on } \Gamma \tag{15}$$

The full boundary conditions may now be written in the form

$$\psi = q, \qquad \omega = 0 \qquad \text{on } y = 1,\ -M_1 \le x \le 0 \tag{16a}$$

$$\psi = 0, \qquad \psi_y = -1 \qquad \text{on } y = 0,\ -M_1 \le x \le M_2 \tag{16b}$$

$$\left.\begin{aligned} \psi &= \tfrac{1}{2}\,(1 - q)(y^3 - 3y^2) + y \\ \omega &= 3\,(1 - q)\,(y - 1) \end{aligned}\right\}\ \text{on}\ \begin{aligned} x &= -M \\ 0 \le\ &y \le 1 \end{aligned} \tag{16c}$$

$$\psi = y \qquad \omega = 0 \qquad \text{on } x = M\ ,\ 0 \le y \le q \tag{16d}$$

$$\left.\begin{aligned} &\psi = q \qquad \omega = -\,2k\,\psi' \\ &-k_s/Ca = \omega' + 2\,\psi'_{ss} \end{aligned}\right\}\ \text{on } \Gamma \tag{16e}$$

We note that in order to satisfy continuity then the fluid flow rate upstream and downstream gives $0 \le y \le q$.

NUMERICAL METHOD

For any given capillary number, Ca, the problem reduces to finding the downstream height, h_∞, and the location of the free surface which satisfies all three boundary conditions (16e). Once the location of the free surface is known then the stream function and vorticity may then be calculated everywhere within the fluid. Due to the symmetry the angle between the bubble and the axis of symmetry must be $-\pi/2$. Assuming that β is the angle between the tangent to the free surface and the x axis then we have

$$d\,\sin\beta\ /dx = k \tag{17a}$$

$$dy/dx = \tan\,\beta \tag{17b}$$

We parameterize the free surface as a collection of $N_1 + 1$ points connected by N_1 straight line segments. We then solve the problem using two of the three boundary conditions (16e) and the third boundary condition is used to determine how the free surface should be moved in order to satisfy all three boundary conditions. In this paper we choose the shear stress condition (14) as the iterative condition. The iteration steps required to find the free surface are detailed as follows:

Step 1. Specify a non-dimensional capillary number, Ca.

Step 2. Specify an initial free surface, $y = \Gamma(x)$, and the non-dimensional fluid flow rate, q.

Step 3. Solve the boundary value problem using the BEM. On the free surface the two boundary conditions (11a) and (15) are satisfied. Then we obtain all the values of ψ', ω and ω' on the boundary.

Step 4. Substitute the boundary values which have been obtained in step 3 into the boundary condition (14) on the free surface, then at each mesh point we calculate the residual

$$R_{i,m} = \omega + 2\,k\,\omega'$$

where m denotes the mth iteration, $i = 1, 2, \ldots, N_1$ and the total residual is given by

$$R_m = \sum_{i=1}^{N1} |\,R_{i,m}\,|$$

Step 5. Consider the equation

$$k = (\,\alpha\,R_{i,m} - \omega\,)/\,2\,\psi' \tag{18}$$

where k is the local curvature on free surface and $0 < \alpha < 1$ is the relaxation factor which depends on the total residual R_m. From the formula (18) we obtain the new curvature at all of the mesh points on free surface and then solve the equations (17a) and (17b) subject to the initial conditions $\beta_0 = -\pi/2$ and $y_0 = 1$. Thus we have obtained the next approximation to the free surface.

Step 6. The iteration is considered to have converged when both the residuals on the free surface and the change in position of the free surface between successive iterations are sufficiently small. That is

$$R < \varepsilon_1 \tag{19}$$

and

$$\sum_{i=1}^{N1} |\,y_{i,m} - y_{i,m-1}\,| < \varepsilon_2 \tag{20}$$

where $\varepsilon_1 > 0$ and $\varepsilon_2 > 0$ are two small preassigned constants. If both of the expressions (19) and (20) are satisfied then the process is complete and the free surface location has been determined and the downstream height $h_\infty = y_{N1}$, otherwise set $q = h_{N1}$ and return to step 3.

We note at step 5 that since near the tip of the bubble the angle β is very close to $-\pi/2$ then the value of tan (β) becomes very sensitive to small changes in β. Therefore we use an unequal mesh grid near this point on the free surface.

NUMERICAL SOLUTION

The iterative scheme requires some control parameters to be assigned values, namely, ε_1 and ε_2, and α. Typical values for these parameters are as follows

$$\varepsilon_1 = 0.01, \qquad \varepsilon_2 = 0.0001$$

$$\alpha = \begin{cases} 0.1 & R_m \geq 1 \\ 0.2 & 1 > R_m \geq 0.5 \\ 0.3 & 0.5 > R_m \geq 0.1 \\ 0.5 & 0.1 > R_m \end{cases}$$

In obtaining these values of the parameters we have considered many factors, e.g. the need to obtain accurate solutions, the rate of convergence of the iterative process and the CPU time. We have taken the fundamental shape of the free surface to be the same as that used by Ritchie [11] as the initial approximation, i.e.

$$y = 1 - (1 - q)(\tanh(x))^{1/2}$$

It is found that $M_2 = -M_1 = 5$ are sufficiently large so that the results do not significantly change by choosing a larger value. Table 1 shows the number of grid points on each boundary of the domain

Γ	B_1	B_2	B_w	B_s	Total
20	5	3	30	15	73

Table 1. The number of grid points on boundaries

where on the boundaries B_1, B_2, B_w and B_s equal mesh sizes are used, but on the free surface Γ we use an unequal mesh size which satisfies

$$x_i - x_{i-1} = d \qquad i = 1,\ldots, 20.$$

where $d = \dfrac{9}{20 \times 19}$ and $x_0 = 0$.

Fig.3 shows the bubble profiles for various values of the capillary number, Ca. A comparison of the values of h_∞ obtained using the BEM and Bretherton's theoretical predictions (1), which are valid if Ca << 1, is presented in Table 2. It is observed that the agreement between the two sets of results at very small values of the capillary number, Ca, is excellent. A simple calculation shows that if $h_\infty < 1/3$ (which corresponds in the present results to Ca < 0.1347) that flow separation exists in the upstream Poiseuille flow region and an additional stagnation point exists on the bubble profile other than at the tip of the bubble. Fig. 4 shows the speed of the fluid on the free surface. The stagnation point on the bubble moves towards the front as Ca increases until Ca = 0.1347.

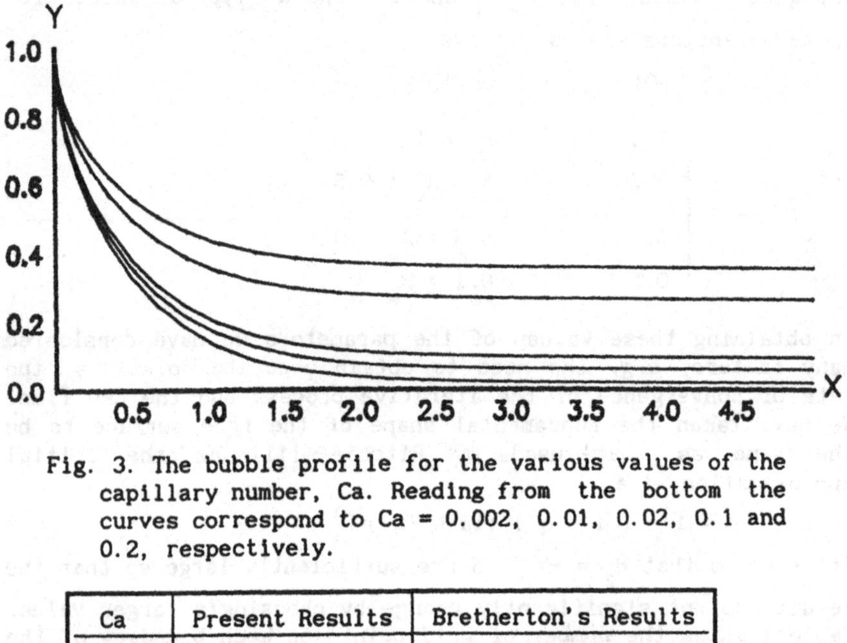

Fig. 3. The bubble profile for the various values of the capillary number, Ca. Reading from the bottom the curves correspond to Ca = 0.002, 0.01, 0.02, 0.1 and 0.2, respectively.

Ca	Present Results	Bretherton's Results
0.002	0.02114	0.02123
0.01	0.06193	0.06208
0.02	0.09785	0.09836
0.1	0.26768	0.28156
0.2	0.36473	0.45742

Table 2. The downstream height, h_∞, of the bubble for various values of the capillary number, Ca.

In conclusion, in this paper we have presented a very efficient iterative technique for solving the very slow viscous motion of a semi-infinite bubble. It is hoped that this technique may be used with confidence in other very slow viscous flow problems in which free surfaces occur, e.g. for axisymmetric bubbles, bubbles of finite size, the sintering problem, etc.

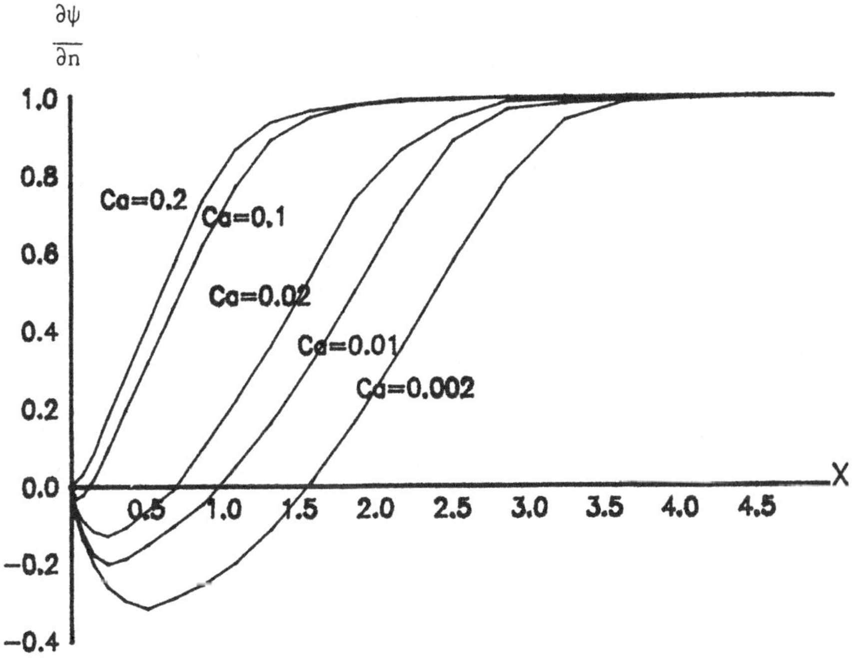

Fig. 4. The speed of the fluid on the surface of the bubble.

ACKNOWLEDGEMENT

The authors would like to thank the Sino-British Friendship Fellowship Scheme for the support of Y. Yuan.

REFERENCES

1. Fairbrother, F. and Stubbs, A.E. Studies in Electro-endosmosis, Part VI. The "Bubble-tube" Method of Measurement, J. Chem. Soc., Vol.1, pp. 527-529, 1935.

2. Taylor, G.I. Deposition of a Viscous Fluid on the Wall of a Tube, J. Fluid Mech., Vol.10, pp 161-165, 1961.

3. Bretherton, F.P. The Motion of Long Bubbles in Tubes, J. Fluid Mech. Vol.10, pp. 166-189, 1961.

4. Schwartz, L.W., Princen, H.M. and Kiss, A.D. On the Motion of Bubbles in Capillary Tubes, J. Fluid Mech., Vol.172, pp. 259-275, 1986.

5. Saffman, P.G. Prediction of Bubble Velocity in a Hele-Shaw Cell, Thin Film and Contact Angle Effects, Phys. Fluids A, Vol.1, pp. 219-223, 1989.

6. Reinelt, D.A. and Saffman, P.G. The Penetration of a Finger into a Viscous Fluid in a Channel and Tube, SIAM J. Sci. Statist. Comp., Vol.6, No.3, pp.542-561, 1985.

7. Shen, E.I. and Udell, K.S. A Finite Element Study of Low Reynolds Number Two Phase Flow in Cylindrical Tubes, J. Appl. Mech., Vol.52, pp. 253-256, 1985.

8. Homsy, G.M. Viscous Fingering in Porous Media, Annu. Rev. Fluid Mech., Vol.19 pp. 271-311, 1987.

9. Saffman, P.G. Viscous Fingering in Hele-Shaw Cells, J. Fluid Mech., Vol.173, pp. 73-94, 1986.

10. Ritchie, J.A. The Boundary Element Method in Lubrication Analysis, Ph.D. Thesis, Leeds University, 1989.

11. Lu, W.Q. and Chang, H.C. An Extension of the Biharmonic Boundary Integral Method to Free Surface Flow in Channels, J. Comp. Phys., Vol.77, pp.340-360, 1988.

12. Kuiken, H.K. Deforming Surface and Viscous Sintering, Proc. Conf. on the Math. and the Comp. of Deforming Surface, Oxford Univ. Press, 1990. (to appear)

13. Yuan, Y. and Ingham, D.B. The Numerical Solution of Viscous Flows with a Free Surface, Proc. of the 1st conf. on Comp. Mod. of Free and Moving Boundary Problems, Southampton, 1991.

14. Batchelor, G.K. An Introduction to Fluid Dynamics, Cambridge University Press, 1967.

15. Ingham, D.B. and Kelmanson, M.A. Boundary Integral Equation, Analyses of Singular, and Biharmonic Problems, Lecture Notes in Eng., Vol. 7, Springer-Verlag, Berlin Heidelberg and New York, 1984.

16. Coyne, J.C. and Elrod, H.G. Conditions for the Rupture of a Lubricating Film, Part 1: Theoretical Model, Trans. A.S.M.E. J. of Lub. Tech., Vol. 92, pp 451-456, 1970.

Moving Boundary Free Flow with Viscosity by BEM*

Z.-X. Feng, B.-Z. Wang
Wuhan University, China

ABSTRACT

Various moving boundary flows (with undetermined boundaries a priori) are of great interest in recent decades. Both steady & unsteady cases for potential flows with one or two free surfaces have been solved using BEM by author as well as the interaction between the flow and structures. In this paper,the flow is considered to be viscous or rotational so that the governing equation becomes biharmonic one. It is shown that the biharmonic equation could be formulated as the combination of a Laplace equation and a Poisson one. In such a way, the BEM discretized formulation will be similar to that of potential flow except the complicated boundary conditions which become difficult even to fit in an iterative scheme. Nevertheless, the numerical implementation of its BEM equation in this paper could still be compiled on a PC-computer with lower cost.

I. INTRODUCTION

Since 1980's, BEM has been widely extended to many engineering fields in which FEM has been used. In particular, BEM model shows more potential in analyzing problems with infinite domain, time dependence and various nonlinearities [1]. It is well known that the coupled BE/FE algorithm for fluid-structure interaction problem has a special advantage numerically [2].In the recent years, much more attention has been paid to the nonlinear flow with moving boundary free surfaces. This kind of project is quite important for fluid-induced strongly nonlinear structure vibration with finite amplitude. The author of this paper had done some work in last decade on the ship-wave interaction with small amplitude [3], the steady external flow with one or two moving boundaries by BEM [4], time-dependent (unsteady) moving boundary flow with finite free surface elevation distortion [5] etc. Owing to these, some numerical experience for the use of BE/FE coupled algorithm has been accumulated [6].

In this paper, the 2-D incompressible viscous flow with a free moving boundary is analyzed. The mathematical model could be used for the numerical simulation of industrial extrusion, for the fluid-structure interaction with viscous effect. However, The viscosity leads to much stronger nonlinearities geometrically and

*This project has been supported by The Chinese National Natural Science foundation.

dynamically. The treatment of singular integral including fundamental solution becomes more complicated as well.
In this paper, the fundamental solution of biharmonic operator is introduced together with that of Laplace one to construct a boundary integral equation and its corresponding BEM formulation. The special boundary conditions in relation with the viscosity are treated properly. The one-order BEM element incorporated with the double node scheme for corners are used to take into account the discontinuity of the normal derivatives. The least square method is used to make the potential function derivative on the moving free surface fit the engineering power-wise distribution. All these are shown of good efficiency.

II. THE GOVERNING EQUATIONS AND BOUNDARY CONDITIONS

1. Computational Model

In Fig.1 there is an example problem of the viscous creep flow in a pipe. It is assumed that the flow is 2-D,steady and incompressible viscous. The section BB' (or S,at the right side in Fig.1) is the inlet on which the Poiseuille velocity profile is taken. Section CC' is the outlet of the flow. From the exit of the pipe (AA' in Fig.1) the free surface AC (S_f)will be generated. The centre line of pipe is A'B' and A'C'(symbolizedas S_3).According toengineering experience, ifdis the diameter of the pipe, then the computational region could be taken as A'B'$\geq$ 3d length along the pipe and also A'C'$\geq$ 3d . For the boundary element distribution, only the upper half part is subdivided because of its symmetry. The result of calculation should be the flow field ($\phi, \phi_{,n}, \omega, \omega_{,n}$on the boundary counter ACC'A'B'BA and the geometric shape of free surface AC.

2.The governing Equations

With viscosity, the flow is no longer potential so that the stream function ψ instead of potential function ϕ is prefer to be used as the independent variable. Then, the continuity equation should be satisfied:

$$v_{i,i} = 0 \tag{1}$$

The relationship between the velocity components v_i and stream function ψ is:

$$v_1 = \psi_2, \quad v_2 = -\psi_1 \tag{2}$$

It can be shown that for a viscous creep flow,(viscous force $\gg$ inertia force), the equation of motion leads to:

$$\mu v_{i,jj} = p_i \tag{3}$$

It is easy to prove (see the appendix of this paper) that one can have following relations:

$$\nabla^2\psi = \psi,jj = v_{1,2} - v_{2,1} = \omega$$

$$\nabla^2\omega = \psi_{,iijj} = \nabla^4\psi = 0$$

in which ∇^2 means Laplace operator, ∇^4 means biharmonic operator. Hence,the basic differential equations for ψ and ω could be taken as:

$$\begin{cases} \nabla^2\psi = \omega \\ \nabla^2\omega = 0 \end{cases} \tag{4}$$

or

$$\begin{cases} \nabla^4\psi = 0 \\ \nabla^2\omega = 0 \end{cases} \tag{4a}$$

From these, the BIE form could be constructed.
For the biharmonic operator , the corresponding Green identity is [8]:

$$\int_V (\nabla^4\psi) \quad GdV = -\int_V \nabla(\nabla^2\psi)\nabla GdV + \int_{\partial V} \frac{\partial}{\partial n}(\nabla^2\psi)Gds$$

$$= \int_V (\nabla^2\psi)(\nabla^2 G)dV + \int_{\partial V} [\frac{\partial}{\partial n}(\nabla^2\psi)G - \frac{\partial G}{\partial n}(\nabla^2\psi)]dS \tag{5}$$

In the index form it becomes:

$$\int_V \psi_{,iijj} GdV = \int_V (\psi_{,ii} G_{,jj})dV +$$

$$\int_S (\psi_{,iij} G)n_j dS - \int_S (\psi_{,ii} G_{,j})n_j dS \tag{6}$$

In equation (5) and (6), function G is the fundamental solution of biharmonic operator and its particular expression is known as:

$$G = G_2 = r^2(1 - lnr) \tag{7}$$

It satisfies:

$$\nabla^4 G_2 + \delta(\bar{p} - \bar{q}) = 0 \tag{7a}$$

in which $\delta(\bar{p} - \bar{q})$ means Dirac δ function, i.e.

$$\delta(\bar{p} - \bar{q}) = \begin{cases} 0, & \bar{p} - \bar{q} \neq 0 \\ 1, & \bar{p} - \bar{q} = 0 \end{cases} \tag{7b}$$

Owing to equation (4a) $\nabla^4\psi = 0$,the left side of equation (5) reduced to zero. Moreover, one can introduce the fundamental solution of Laplace operator:

$$G_1 = ln\frac{1}{r} \tag{8}$$

It can be proved that there is a relationship between G_1 and G_2 :

$$\nabla^2 G_2 = 4G_1 \tag{9}$$

Therefore, the first term in the right side of (5) can be reduced as:

$$\int_V (\nabla^2\psi)(\nabla^2 G_2)dV = 4\int_V G_1\nabla^2\psi dV \tag{10}$$

On the other hand, the Green identity for Laplace operator is well known:

$$\int_V (G_1\nabla^2\psi - \psi\nabla^2 G_1)dV = \int_{\partial V} (G_1\partial\psi/\partial n - \psi\partial G_1/\partial n)dS \qquad (11)$$

Hence equation (5) can be reduced as:

$$-\alpha\psi + \int_{\partial V} [G_1\frac{\partial\psi}{\partial n} - \psi\frac{\partial G_1}{\partial n}]dS + \frac{1}{4}\int_{\partial V} [G_2\frac{\partial\omega}{\partial n} - \omega\frac{\partial G_2}{\partial n}]dS = 0 \qquad (12)$$

Of course,in equation (12) ψ,ω are all unknown so that this equation should be solved simultaneously with the BIE of $\nabla^2\omega = 0$ which according to conventional BEM leads to:

$$\alpha\,\omega = \int_{\partial V} [G_1\psi_{,n} - G_{1,n}\psi]ds + \frac{1}{4}\int_{\partial V} [G_2\omega_{,n} - G_{2,n}\omega]ds \qquad (13)$$

Then the BE discretization of (12) and (13) leads to:

$$\begin{bmatrix} HH & -GG \\ -H & G \end{bmatrix} \begin{Bmatrix} \omega \\ \omega_{,n} \end{Bmatrix} = \begin{bmatrix} -H & G \\ 0 & 0 \end{bmatrix} \begin{Bmatrix} \psi \\ \psi_{,n} \end{Bmatrix} \qquad (14)$$

where the elements for matrices [HH],[H],[GG],[G] are calculated as follows:

$$[HH] = \frac{1}{4}\int_{\partial\Omega} G_{2,n} N dS$$

$$[GG] = \frac{1}{4}\int_{\partial\Omega} G_2 N dS$$

$$[H] = \alpha\delta_{ij} + \int_{\partial\Omega} G_{1,n} N dS$$

$$[G] = \int_{\partial\Omega} G_1 N dS \qquad (13a)$$

One can express (14) briefly as

$$[GK]\{RX\} = [HK]\{R\} \qquad (14a)$$

If all the unknown variables in $\{RX\}$ and $\{R\}$ are collected in $\{RX\}$ and known values collected in R, the coefficients in [GK] and [HK] are interchanged correspondently, one can obtain:

$$[GK]\{RX\} = \{R_0\} \qquad (14b)$$

3.The Boundary conditions

The stream function ψ and vorticity ω are taken as the basic independent variables. The corresponding boundary conditions for viscous flow could be formulated as (See Fig. 1):

i) The upstream positionS_1:According to Poiseuille viscous creep flow in a pipe, the analyzed velocity profile is known and in terms of ψ it is:

$$\psi = \frac{1}{2}(3y - y^3) \tag{15a}$$

$$\psi_{,n} \equiv \psi_{,x} = 0 \tag{15b}$$

$$\psi_{,y} = \frac{3}{2}(1 - y^2) \tag{15c}$$

ii) The down-stream section S_2:

$$\psi_{,n} \equiv \psi_{,x} = 0 \tag{16a}$$

$$\psi_{,y} = \frac{1}{\eta_\infty} \tag{16b}$$

$$\psi = \frac{y}{\eta_\infty} \tag{16c}$$

where η_∞ is the elevation of free surface on S_2.
iii) On the centre line:

$$\psi = 0, \qquad \omega = 0 \tag{17}$$

iv) Upside wall S_3 :

$$\psi = 1, \qquad \psi_{,n} = 0 \tag{18}$$

v) Free surface S_f: [9]

$$\psi = 1 \tag{19a}$$

$$\omega = -2\kappa\psi_{,n} \tag{19b}$$

where κ means the curvature of the free surface and its determination will be described in the next section.

4.The Calculation of curvature κ

From the non-shear effect condition of free surface, one has in local coordinates (n,τ):

$$\frac{\partial v_\tau}{\partial n} + \frac{\partial v_n}{\partial \tau} = 0$$

i.e.

$$\psi_{,nn} - \psi_{,\tau\tau} = 0,$$

$$\psi_{,nn} = \psi_{\tau\tau} \tag{20}$$

But we have (See the appendix):

$$\omega = \psi_{,jj} = \psi_{,nn} + \psi_{,\tau\tau} = 2\psi_{,taur} \tag{20a}$$

Hence, if the arc length along the surface is S, one has [10],[11] :

$$\frac{\partial}{\partial \tau}\,|_{S_f} = \frac{d}{dS} = \cos\beta\frac{\partial}{\partial x}\,|_{S_f} \tag{21a}$$

$$\frac{\partial^2}{\partial \tau^2}\,|_{S_f} = \frac{d^2}{ds^2} - \kappa\frac{\partial}{\partial n}\,|_{S_f} \tag{21b}$$

in which $\cos\beta$ is the slope of free surface.
On the other hand, supposing the free surface curve is $\eta = \eta(x)$, then

$$\kappa = \frac{-\eta''}{(1+\eta'^2)^{3/2}} = \frac{-\eta''}{(1+\tan^2\beta)^{3/2}} = -\eta''\cos^3\beta \tag{21c}$$

This is the formula we want for the calculation of κ and the key is to obtain the value of η'' at each node on free surface.
To treat this problem, it is supposed that:

$$\eta = 1 + \alpha\, \tanh[x - \varepsilon(x)] \tag{22a}$$

in which α is a given parameter, ϵ is an function of x from experimental rule as:

$$\varepsilon(x) = \varepsilon^{\infty} + (\varepsilon^{0} - \varepsilon^{\infty})\exp(-\gamma x) \tag{22b}$$

Therefore, one has:

$$\eta'(x) = \alpha\, \mathrm{sech}^2[x - \varepsilon(x)]\,[x\varepsilon' + \varepsilon] = \alpha\, \mathrm{sech}[x - \varepsilon(x)][\varepsilon - \gamma x(\varepsilon - \varepsilon^{\infty})] \tag{22c}$$

$$\eta''(x) = \alpha \mathrm{sech}^2(x\varepsilon)\{x\gamma^2(\varepsilon - \varepsilon^{\infty}) - 2\gamma(\varepsilon - \varepsilon^{\infty}) - 2\tanh[\varepsilon - \gamma x(\varepsilon - \varepsilon^{\infty})]^2\} \tag{22d}$$

5.The Treatment of singular Integrals

The diagonal element of [H] could be treated just as that in the conventional BEM [1]:

$$H_{ii} = - - \Sigma_{i \neq j} H_{ij} \tag{23a}$$

The diagonal term of [G] and [GG] can be obtained analytically and the principal results are:

$$G_{ii} = \frac{1}{l}\int_0^l Sr^2(1 - \ln r)\, ds = \frac{1}{l}[D_2 I_1 - \frac{r^4}{4}(\ln r - \frac{5}{4})]|_j^{j+1} \tag{24a}$$

$$GG_{ii} = \frac{1}{l}\int_0^l r^2(1 - \ln r) ds = \frac{1}{l}[\frac{r^4}{4}(\ln r - \frac{5}{4}) - (l + D_2)I_1]|_j^{j+1} \tag{24b}$$

in which S is the length along the 2-D boundary element curve.
In the case of $i = j$:

$$\frac{1}{l}\int_0^l r^3(1 - \ln r) dr = \frac{l^3}{4}(\frac{5}{4} - \ln l) \tag{25a}$$

$$\frac{1}{l}\int_0^l r^2(l - r)(1 - \ln r) dr = \frac{l^3}{12}(\frac{19}{12} - \ln l) \tag{25b}$$

For the integration of$[G_1]$ in cases of $i \notin$ element j, one still has [4]

$$\frac{1}{l}\int_0^l r\ln\frac{1}{r} dr = \frac{1}{2}l(\frac{3}{2} - \ln l) \tag{26a}$$

$$\frac{1}{l}\int_0^l (l-r)ln\,\frac{1}{r}dr = \frac{1}{2}l(\frac{1}{2} - ln\,l) \tag{26b}$$

In the equation (24), the integral I_1 is:

$$I_1 = \int r^2(1 - ln\,r)ds =$$

$$\frac{1}{3}\{(D_2 + l)^3(\frac{4}{3} - ln\,r)\}|_j^{j+1} + D_1^2\{\sqrt{r^2 - D_1^2}ln\,r\}|_j^{j+1} + \frac{5}{3}l - \frac{2}{3}D_1\alpha \tag{27}$$

in which the definitions of D_1, D_2 can be found in Ref. [4] [5], l is the length of element, α is the angle extended by element with x.

III. NUMERICAL ITERATION PROCEDURE

The test problem in this paper is to solve the flow field (ψ, ω) and the geometric shape of the free surface S_f of a viscous flow as in Fig.1. The boundary element equation and boundary conditions are somewhat different from that in Ref.[13],[14], the iterative scheme used in this paper is much easier.

As mentioned above, when the initial geometry of whole boundary including free surface is given, there is no difficulty to construct the BEM formulation for ψ(satisfying the biharmonic equation) and for ω(satisfying Laplace equation) as (12). The special problems to be treated are:

i)On S_1, S_2 and S_4,the values of $\psi, \psi_{,n}$ are known and $\omega, \omega_{,n}$ are unknown, while on S_3, S_f,the value ψ is known, $\psi_{,n}$is unknown so that the coefficients of $\psi_{,n}$ andω should be exchanged.

ii) In the boundary condition (19b), the curvature of free surface is also unknown a priori and is dependent on the geometry factor $\eta'(x)$ as equation (21c). At the same time,κ and η should be solved in an iterative way. Therefore, the coefficients of $\psi_{,n}$ and ω in equation (14) for all nodes on S_f should be exchanged also.

In this paper, the iterative procedure for the calculation of η, β and κ is as follows.

i) The initial free surface curve is given according to the experience:

$$\eta^{(i)} = 1 + \alpha\ tanh[-x\epsilon(x)] \tag{28}$$

in which

$$\epsilon(x) = \epsilon^{\infty} + (\epsilon^0 - \epsilon^{\infty})e^{-\gamma x} \tag{28a}$$

ii)The curvature of free surface ,from equations (21c) and (22d):

$$\kappa^{(i)} = -\eta' cos^3\beta$$

iii)The BEM discretization of the initial boundary leads to the BEM equation as (14a) with fixed boundary geometry. The introduction of boundary conditions (15)-(19) leads to algebraic equation (14b).

iv)Solving (14b) to obtain the field information of $\psi, \psi_{,n}, \omega, \omega_{,n}$. From the value of ω on the free surface, the new value of $\kappa^{(i+1)}$ could be calculated by (18b):

$$\kappa^{(i+1)} = -\frac{\omega^{(i)}}{2\psi_{,n}{}^{(i)}} \tag{29}$$

v) From (21c), a new value of β can be obtained:

$$cos^{(i+1)}\beta = \sqrt[3]{\frac{-\kappa^{(i+1)}}{\eta^{*(i)}}} \tag{30}$$

Then new $\eta(x)$ can be calculated :

$$\eta^{(i+1)} = 1 + [x(I+1) - x(I)]\, tan\beta \tag{31}$$

If $max \mid \eta^{(i+1)} - \eta^{(i)} \mid \leq \epsilon$, the convergence is reached, otherwise the elevation of free surface should be corrected and next iteration will be taken until convergence.
The numerical test shows that the derivative $\psi_{,n}$ on free surface should be justified by fitting according to:

$$\psi_{,n}(x) = ax^b \tag{32}$$

In this paper, the fitting of this curve equation is realized by the least square scheme until

$$max|1 - \frac{\overline{\psi_{,n}}}{\psi_{,n}}| \leq 10^{-4} \tag{33}$$

to obtain the suitable value of a,b,then the corresponding $\psi_{,n}$ on free surface are calculated at last.

REFERENCES

1.Banerjee P.K.,Butterfield R., BOUNDARY ELEMENT METHOD IN ENGINEERING SCIENCES,McGraw-Hill Book COmpany (UK),1981.
2.Hsiao G.C.,"The coupling of BEM and FEM - A brief review ", BOUNDARY ELEMENTS X,Edited by C.A.Brebbia,Springer- Verlag,1988.
3.Feng Z.X., " 3-D response of ship-wave interaction by coupling BE/FE ", BOUNDARY ELEMENTS,Pergamon Press,1986.
4.Feng Z.X., " An alternative iteration algorithm for moving boundary free flow using BE ", 10-th International Conference on BEM, Vol.2,Springer-Verlag, 1988.

5. Feng Z.X., Li Z.X., Ye B.Q. and Shen C.W., ' Unsteady Moving Boundary Flow by BEM and its Interaction with Structure',BOUNDARY ELEMENT METHODS,principles and applications,Proceedings of 3-rd Japan-China Symposium on BEM,Pergamon Press,1990.
6. FengZ.X., " Mathematical Models for moving boundary problems ", THEORY AND APPLICATIONS OF BEM, Proceedings of 2- nd China-Japan Symposium on BEM, Tsinghua University Press,1988.
7. Feng Z.X., CONTINUUM MECHANICS AND ITS COMPUTER SIMULATION, textbook of Wuhan University, 1987.
8. MATHEMATICAL HANDBOOOK, Higher Educational Publishers,(in Chinese), 1979.
9.Brebbia C.A.,Wrobel L.C., " Viscous flow problems by the boundary element

method ", COMPUTATIONAL METHODS FOR FLUID FLOW, Edited by Taylor and Smith,1985.
10.Kelmanson M.A., " Boundary integral equation solution of viscous flows with free surfaces ", J. of Engineering Mathematics,Vol. 17,No.4,1983.
11.Kelmanson M.A., " An integral equation method for solution of singular slow flow problems ", Vol. 18,No. 2,1988.
12.Bush M.B., " 3-D viscous flows with a free surface flow out of a long square die ", J. of non-Newtonian fluid mechanics, Vol. 18, 1985.
13.Ingber M.S.and Mitra A.K., " The evaluation of the normal derivative along the boundary in the DBEM ", Applied Mathematical Modeling,Vol.13,No.1,1989.
14.Lu W.Q.,Chang H.C., " An Extension of the Biharmonic Boundary Integral Method to Free Surface Flow in Channels ", J. of Computational Physics, Vol. 77, No. 2 1988.

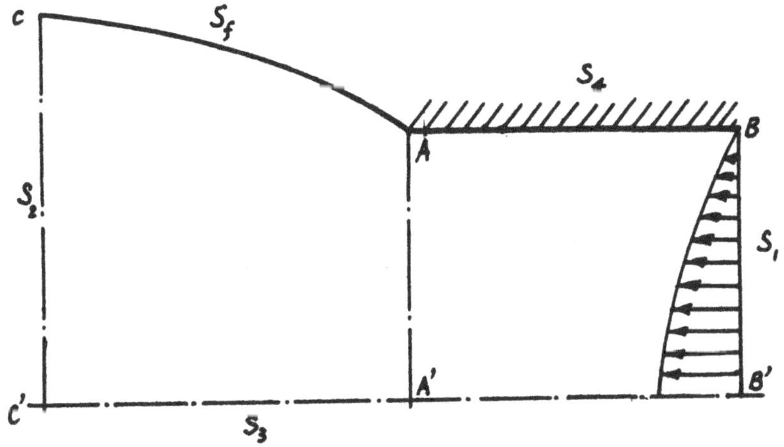

Fig. 1

method", COMPUTATIONAL METHODS FOR FLUID FLOW, Edited by Taylor and Smith, 1985.

10. Kelmanson M.A., "Boundary integral equation solution of viscous flows with free surfaces", J. of Engineering Mathematics, Vol. 17, No. 4, 1983.

11. Kelmanson M.A., "An integral equation method for solution of singular slow flow problems", Vol. 15, No. 2, 1983.

12. Bush M.B., "A 2-D viscous flows with a free surface flow out of a long square duct", J. of [illegible] Mechanics, Vol. [illegible], 1985.

13. [illegible] and [illegible] A.K., "The evaluation of the normal derivatives along the boundary in the DBEM", Applied Mathematical Modelling, Vol. 10, No. 1, 1986.

14. [illegible] Chang [illegible], "An Extension of the Boundary Integral Method to Free Surface Flow in Channels", J. of Computational Physics, Vol. 77, No. 1, 1988.

Fig. 1

SECTION 4: FLUID FLOW

Panel Methods and Their Applications to Sailboard Technology

C.A.M. Broers, M.M.A. Pourzanjani,
D.J. Buckingham, T.W. Chiu
School of Engineering, University of Exeter, Exeter, EX4 4QF, U.K.

ABSTRACT

Windsurfers appeared in the leisure market during the past two decades and have now been established as one of the most popular water sports. Their design has been dominated by the qualitative feedback from the users of such craft. They have been able to achieve a speed record of about 40 knots at the perfect sailing conditions. It is believed that a more scientific approach in design, using CFD, will result in a craft capable of breaking this speed record.

Studying the flow over planing Windsurfer hulls and quantifying it in terms of fluid dynamics is the ultimate goal of this project. The main tool for the study of the flow is a panel method program. A Second-order Doublet Panel Method has been developed for this purpose. Practical uses for the program are wide ranging but involve tailoring to the specific application. Incorporating the relevant boundary conditions the program is capable of modelling inviscid, potential flow over any body. By studying the flow in this manner insight into the performance of planing Windsurfer hulls can be made. The second-order Doublet method is used in order to model the lifting effects of planing hulls. Second-order Doublet sheets can be used to model the wake and free surfaces, therefore utilizing a single set of formulae for all the calculations. All of the formulae are in terms of a global coordinate system, so no local systems for the panels are necessary resulting in lower running times. Pressure coefficients are calculated by the program, thus lift and pressure drag are calculated. All of the programs are written in C for a dedicated computer which is a 486 PC with an I-860 board with 8 MB of RAM on board.

INTRODUCTION

The history of sailboards goes back to late 50's. The early versions are reported to have comprised a simple wooden board (some of these were perfect rectangles in shape, see for example ref(AYRS[3])) with an unstayed triangular sail. The first sailboard is reported to have been built in the back yard of an English sailing enthusiast and was capable of doing around 5 knots in favorable conditions, without it's huge market potential being realized at that time. Sailboards appeared in the leisure market in the United States as a serious water sport in early 70's, and have managed in this short period to become the most popular water sport on an international scale. The unit sales grow very rapidly in the early years as one would expect from a product of this potential, and as the sport has matured these have leveled at some 4,000,000 unit sales per year.

Windsurfers and sailboards refer to the same craft and are interchangable. However, since windsurfer also refers to the person driving the craft, throughout this paper we adopt the notion of referring to the driver as the windsurfer and the craft as the sailboard. Sailboards differ in some ways from other more traditional sailing craft. They consist of:

a) a board capable of planing at high speeds,
b) an unstayed mast (no rigging involved) attached to the hull by a movable joint and usually using a wishbone type boom and an inclined sail,
c) without any rudder, with the windsurfer controlling the speed and direction of movement through changing the exposed sail area and his own attitude on the craft.

The speed performance of a boat may be considered high when the ratio of $V/\sqrt{L}$ approaches or exceeds unity (Marchaj[8]). This traditional view on speed performance is mainly based on the experimental results compiled over the years on monohull sailing boats and suggests that the maximum speed achieved by a boat is related to the length of the vessel and all being the same one would expect a larger vessel to go faster. The following table compiled from (Marchaj[8], Wood-Rogers[11] and Farrar[4]) is a good indication of this school of thought.

Table 1. Speed performance recorded for some sailing craft

Boat	L (ft)	Speed (knots)
Crossbow II	74	36
'C' class	25	27.57
Tornado	20	20.6
Dart 18	18	18
Dart 15	15	15
Hawk	12	12

It is obvious that there are many factors other than the length of the boat which contribute to the speed performance of a sailing craft. These include such things as the weight to power ratio (mass of the craft, number of crew, sail area etc.), detailed design of the boat (such things as the stern profile, keel size and shape, winglets on the keel etc.), and crew fitness, training, and familiarity with the craft.

With the introduction of sailboards and some other 'fun' type boats the notion of "Larger = Faster"has to be reconsidered. In less than 15 years sailboards measuring less than 3 meters in length, have managed to improve their speeds records from around 15 knots to what today is the world speed sailing record at 42.91 knots *(Pascal Maka at St Maries De La Mer Feb 1990)*, an unofficial speed has also been recorded at 44.6 knots by the same windsurfer in April 1991. It may be that one can class the new entrants to the sailing world as 'unconventional' sailing craft for which the ratio of '$V/\sqrt{L}$'cannot be compared with the more traditional sailing boats.

This improvement in speed performance of sailborads has come about partly as a result of windsurfers getting better at driving these craft and more importantly as a result of improvement in the design of sailboards and their associated rigs and other accessories. This optimization of both the hull forms (including fin type and size) and their sails has been mainly on the qualitative feedback from the windsurfing experts to the manufacturers (usually their sponsors). In recent years some quantitative work on the design of sailboards and their rigs have been carried (e.g. see ref Jimenez[7] and Smith[10]). Most of the qualitative research seems to be taking place in the area of sail design and optimization, which is the reason for this project to concentrate mainly on the study of the behaviour and design of sailboards.

Work in this area has been going on for some years at the School of Engineering, University of Exeter, in the United Kingdom. The previous work included investigation in the design of fins for sailboards, study into the design of high efficiency sails for sailboards, and design of a multihull sailing craft based on sailboard technology.

The authors believe that the current speed record can be stretched even further with a highly optimized craft. This optimization can be achieved using advanced CFD methods now possible with what is now conventional technology in computing terms.

The main goals and objectives of this project as an academic piece of work are as follows:

1- Contribution to knowledge and understanding of the dynamic behaviour of planing surfaces.
2- To develop design guidelines for the design of sailboards and their associated accessories (e.g. mast, sail, fins etc.).
3- To design a craft which can achieve a higher speed than the current world record.

The major tasks in the early part of the project have been to appreciate the planing phenomena and develop the appropriate CFD code to cope with estimation of forces on planing hulls at finite yaw angle.

It was realized from the start that the computing requirements of this project will be quite demanding both in terms of computing power and the RAM size to handle large arrays involved in the CFD code. Two computers have been made available for the purpose of this project, one is a standard SUN SPARC station with 32 MB of RAM, and the other is a 486 based processor with a dedicated i860 board with 8 MB of RAM. This is believed to be sufficient for the conduct of this project, although the mainframe at the Exeter University can be used if necessary.

In the following sections we describe the rationale behind our approach and methodology and also report on the work carried out to date in development of the CFD code.

C.F.D. ANALYSIS

In order to try and better understand the Fluid Dynamics of Sailboards, we are using Panel Method programs to model the hull, fins and sails. This is not as far reaching as it may sound as the modelling of all three of the bodies use the same program with different fluid properties and boundary conditions. Criterion for the boundary conditions will come from both references and experimental data.

Second-Order Doublet Panel Method can model lifting bodies and inviscid potential flow. Second order doublet sheets can be used to model wakes and free surfaces, therefore utilising a single set of formulae for all the calculations. All of the formulae are in terms of a global coordinate system, so no local coordinate systems (see for example Hess and Smith[5]) are necessary, resulting in lower running times. Pressure coefficients are calculated by the program, thus lift and pressure drag are calculated. All of the programs are in C–language.

Three Dimensional Applications

The initial theory of the Panel Method is fairly straight foward, on the other hand the manipulation of these formulae into a usable form is more complicated. Two-dimensional calculations use line sources and or vortices, while in three–dimensions surface distributions of source, doublet and vortex can be used. The figure below shows a sphere with triangular panels:

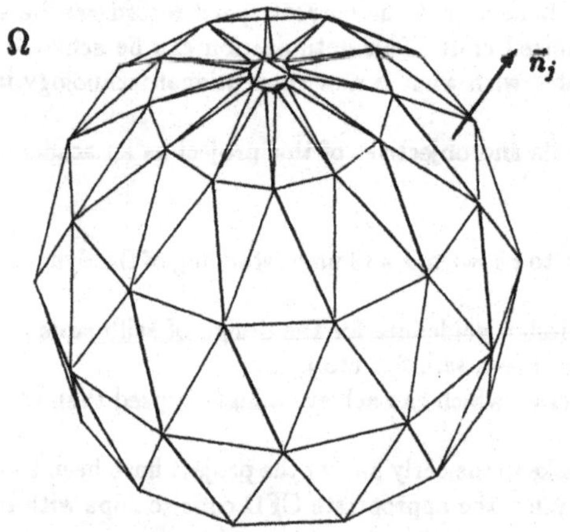

Figure 1. Diagram of Surface

Panel Method aims at solving the Laplace equation (continuity) by transforming it into a line integral equation using Green's Theorem. Green's Theorem can be used to express potentials in a 3–D mathematical domain (Ω):

$$\int\int\int_{\Omega}(U\nabla^2 W - W\nabla^2 U)\delta\Omega = \int\int_{S}(W\vec{n}.\nabla U - U\vec{n}.\nabla W)\delta S \qquad (1)$$

Volume integral Surface integral

$\vec{n}$ = normal pointing into Ω

U and W are two arbitrary scalar fields which are second order and continuous throughout Ω. For the idealised three–dimensional problem U can be replaced by the unknown velocity potential ϕ (where $\nabla^2\phi = 0$ throughout Ω) and W is replaced by $\frac{1}{r}$ (where r is the length of the vector r(P,Q)). Replacing the scalar field constants gives:

$$4\pi\phi_P = -\int\int_S \left[\left(\frac{1}{r}\right)\vec{n}.\nabla_Q\phi - \phi\vec{n}.\nabla_Q\left(\frac{1}{r}\right)\right]\delta S_Q \qquad (2)$$

This implies that the velocity potential at a point P (ϕ_p) in the flow field can be expressed as a function of the value of the velocity potential on the boundary surface

(S).The properties of this integral equation are thoroughly covered by B. Hunt[6]. By taking the gradient at the point P the velocity potential can be found.

$$\nabla_P \phi_P = V_P = \frac{1}{4\pi}\int\int_S \left[\vec{n}.\nabla_Q \phi \nabla_P \left(\frac{1}{r}\right) - \phi \nabla_P (\vec{n}.\nabla_Q \left(\frac{1}{r}\right))\right] \delta S_Q \qquad (3)$$

Replacing the velocity potential ϕ with a surface distribution of sources with strength $\sigma \equiv \vec{n}.\nabla\phi$ for the first integral and a doublicity $\mu \equiv -\phi$ for the second, the equation for 3–D potential becomes:

$$\phi_P = -\frac{1}{4\pi}\left(\int\int_S \left(\frac{\sigma}{r}\right)\delta s_Q + \int\int_S \mu\vec{n}.\nabla_Q \left(\frac{1}{r}\right)\delta s_Q\right) \qquad (4)$$

This expression contains the surface integral for the potential of both a distribution of sources and doublets. The surface of an obstacle is represented by a number of small surfaces (i.e the panels) on each of which a certain order of source and/or doublet distribution is assumed. Cantaloube and Rehbach[1] (1986) transformed these integrals mathematically into line integrals. For a plane panel with source and or doublet distributoin such that $\nabla\sigma$ and $\nabla\mu$ are constant:

Source

$$\begin{aligned}\phi_P^\sigma = &-\frac{1}{4\pi}\left(\vec{n}.\int_C \sigma\frac{\vec{r}}{r}\times d\vec{l}_Q + \vec{n}\times\int_c \frac{\sigma}{r}d\vec{l}_Q\right.\\ &+ (\vec{n}.\vec{r})(\vec{n}.\vec{e})\vec{n}.\left(\nabla\sigma\times\int_C ln(r+\vec{e}.\vec{r})d\vec{l}_Q\right)\\ &\left.-\vec{n}.\left(\nabla\sigma\times\int_C r d\vec{l}_Q\right)\right) \qquad (5)\end{aligned}$$

$$\begin{aligned}\vec{V}_P^\sigma = &-\frac{1}{4\pi}\left(\vec{n}\int_C \sigma\vec{A}.d\vec{l}_Q + \vec{n}\times\int_C \frac{\sigma}{r}d\vec{l}_Q\right.\\ &+ (\vec{e}.\vec{n})\vec{n}.\left((\vec{n}\times\nabla\sigma).\int_C ln(r+\vec{e}.\vec{r})d\vec{l}_Q\right)\\ &\left.+\nabla\sigma\left(\vec{n}.\int_C \frac{\vec{r}\times d\vec{l}_Q}{r} + (\vec{n}.\vec{r})\int_C \vec{A}.d\vec{l}_Q\right)\right) \qquad (6)\end{aligned}$$

Doublet

$$\begin{aligned}\phi_P^\mu = &-\frac{1}{4\pi}\left(-\int_C \mu\vec{A}.d\vec{l}_Q\right.\\ &\left.+(\vec{e}.\vec{n})(\vec{n}\times\nabla\mu).\int_c ln(r+\vec{e}.\vec{r})d\vec{l}_Q\right) \qquad (7)\end{aligned}$$

$$\begin{aligned}\vec{V}_P^\mu = &-\frac{1}{4\pi}\left(\int_C \mu\nabla_Q\left(\frac{1}{r}\right)\times d\vec{l}_Q\right.\\ &\left.+\nabla\mu\int_C \vec{A}.d\vec{l}_Q - (\vec{n}\times\nabla\mu)\times\left(\vec{n}\times\int_C \frac{d\vec{l}_Q}{r}\right)\right) \qquad (8)\end{aligned}$$

where C represents the boundary contour of the panel in question. These contour integrals can be evaluated analytically for a polygonal panel (Chiu[2] (1990)) so that the potential or velocity induced at a point P by the panel can be expressed in terms of the σ and μ distribution and the coordinates of the corners of the polygon.By imposing boundary conditions at a set of chosen 'control points', usually the centriods of the panels, a set of linear equations will be obtained with the source and doublet distributions as unknowns. Consider a general case where the velocities at the control points are prescribed, and $(\vec{U}_{\infty j})$ is the freestream velocity at the j-th control point:

$$\begin{pmatrix} \vec{n}_1.\vec{Q}_{11} & \vec{n}_1.\vec{Q}_{12} & \cdots & \vec{n}_1.\vec{Q}_{1n} \\ \vec{n}_2.\vec{Q}_{21} & \vec{n}_2.\vec{Q}_{22} & \cdots & \vec{n}_2.\vec{Q}_{2n} \\ \vdots & \vdots & \ddots & \vdots \\ \vec{n}_m.\vec{Q}_{m1} & \vec{n}_m.\vec{Q}_{m2} & \cdots & \vec{n}_m.\vec{Q}_{mn} \end{pmatrix} \begin{pmatrix} \sigma_1 \\ \sigma_2 \\ \vdots \\ \sigma_n \end{pmatrix} \tag{9}$$

$$+ \begin{pmatrix} \vec{n}_1.\vec{R}_{11} & \vec{n}_1.\vec{R}_{12} & \cdots & \vec{n}_1.\vec{R}_{1n} \\ \vec{n}_2.\vec{R}_{21} & \vec{n}_2.\vec{R}_{22} & \cdots & \vec{n}_2.\vec{R}_{2n} \\ \vdots & \vdots & \ddots & \vdots \\ \vec{n}_m.\vec{R}_{m1} & \vec{n}_m.\vec{R}_{m2} & \cdots & \vec{n}_m.\vec{R}_{mn} \end{pmatrix} \begin{pmatrix} \mu_1 \\ \mu_2 \\ \vdots \\ \mu_n \end{pmatrix} + \begin{pmatrix} \vec{n}_1.\vec{U}_{\infty 1} \\ \vec{n}_2.\vec{U}_{\infty 2} \\ \vdots \\ \vec{n}_m.\vec{U}_{\infty m} \end{pmatrix} = \begin{pmatrix} u_1 \\ u_2 \\ \vdots \\ u_m \end{pmatrix}$$

where the $m \times n$ matrices contain the influence coefficients and the $\vec{n}_1 \ldots \vec{n}_m$ are the unit normals for the panels. The prescribed normal velocities at each panel, $u_1 \ldots u_m$, are in the right hand side vector. Influence coefficients are found from the solutions to the integrals in equations (5) and (7), which allow the source and doublet strengths to be found. With this type of example the flow is modelled with both a source and doublet distribution. The source distribution is used to set the boundary conditions for the body and the doublet to model the wake and set the uniqueness of the solution. Neither of the flow solutions could represent a lifting flow on their own, source distribution can not impose circulation and the doublet sheet allows flow to permiate through its surface. However a *Second-order Doublet Panel Method* can be used on its own, the condition of zero normal velocity at the surface of the body can be set and circulation imposed. Hence, this is the method used.

<u>Second Order Doublet Only Formulation</u>

The second order doublet Panel Method allows lifting asymmetric flows to be modelled. The formation assumes that on each panel the singularity distribution varies linearly and there are no discontinuities in magnitude across the connecting edge of panels. In formulating the method, it is assumed that the *magnitude* of doublicity is constant at a nodal point (the corner of a panel, which can simultaneously be the corner of a few other panels) regardless of which panel is considered.

The panels are triangular, the j-th panel is shown in figure 2, and the doublicity at the corners are $\mu_{j,1}$, $\mu_{j,2}$, $\mu_{j,3}$ respectively. The equations from (7) can be calculated analytically. The position vectors of the points $\vec{Q}_1$ and $\vec{Q}_2$ and the observation point P are $\vec{X}_1$, $\vec{X}_2$ and $\vec{X}_p$ respectively.

Let:

$$a = ||\vec{X}_1 - \vec{X}_p||^2$$
$$c = ||\vec{X}_2 - \vec{X}_1||^2$$
$$b = (\vec{X}_1 - \vec{X}_p) \times (\vec{X}_2 - \vec{X}_1)$$
$$d = \vec{e}_j.\vec{r}$$

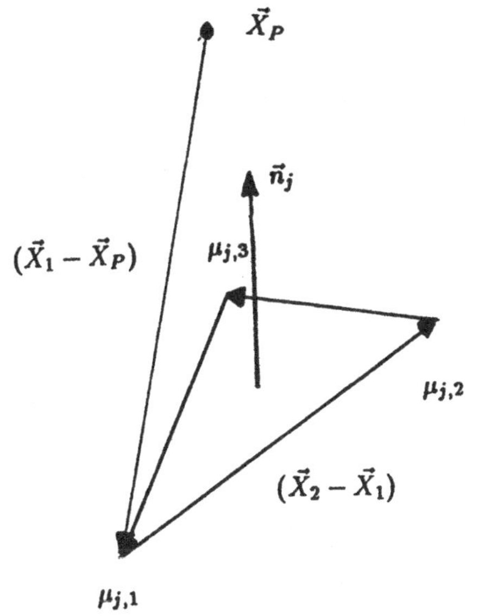

Figure 2. Diagram of Triangular panel

$$\begin{aligned}\int_{C_j} \mu\vec{A}.d\vec{l}_Q = \mu_{j,1}&\Big[(\vec{e}_j.\vec{G}_1)\big((1+\frac{b}{c})_1 M_{pj,1} - N_{pj,1}\big) + (\vec{e}_j.\vec{G}_3)\Big(N_{pj,3} - (\frac{b}{c})_3 M_{pj,3}\Big)\Big]\\ + \mu_{j,2}&\Big[(\vec{e}_j.\vec{G}_2)\big((1+\frac{b}{c})_2 M_{pj,2} - N_{pj,2}\big) + (\vec{e}_j.\vec{G}_1)\Big(N_{pj,1} - (\frac{b}{c})_1 M_{pj,1}\Big)\Big]\\ + \mu_{j,3}&\Big[(\vec{e}_j.\vec{G}_3)\big((1+\frac{b}{c})_3 M_{pj,3} - N_{pj,3}\big) + (\vec{e}_j.\vec{G}_2)\Big(N_{pj,2} - (\frac{b}{c})_2 M_{pj,2}\Big)\Big]\end{aligned} \tag{10}$$

where

$$M_{pj,1} = \frac{1}{\sqrt{ac-b^2-cd^2}}. \left(\tan^{-1}\frac{b^2-ac-cd\sqrt{a+2b+c}}{(b+c)\sqrt{ac-b^2-cd^2}} - \tan^{-1}\frac{b^2-ac-cd\sqrt{a}}{b\sqrt{ac-b^2-cd^2}}\right) \tag{11}$$

$$N_{pj,1} = \frac{1}{c}\log\left(\frac{\sqrt{a+2b+c}+d}{\sqrt{a}+d}\right), \text{ ect.} \tag{12}$$

$$\begin{aligned}\vec{G}_1 &= (\vec{X}_1 - \vec{X}_P) \times (\vec{X}_2 - \vec{X}_1)\\ \vec{G}_2 &= (\vec{X}_2 - \vec{X}_P) \times (\vec{X}_3 - \vec{X}_2)\\ \vec{G}_3 &= (\vec{X}_3 - \vec{X}_P) \times (\vec{X}_1 - \vec{X}_3)\end{aligned} \tag{13}$$

and the subsripts 1, 2 or 3 denote that the quantity is evaluated for the first, second or third side of the triangular panel. The second integral of equation (7):

$$(\vec{e}_j.\vec{n}_j)(\vec{n}_j \times \nabla\mu_j).\int_{C_j} \log(r + \vec{e}_j.\vec{r})d\vec{l}_Q$$
$$= \mu_{j,1}\left[-\frac{\vec{e}_j.\vec{n}_j}{||\vec{l}_{12} \times \vec{l}_{31}||}\Big((\vec{l}_{23}.\vec{l}_{12})I_{pj,1} + (\vec{l}_{23}.\vec{l}_{23})i_{pj,2} + (\vec{l}_{23}.\vec{l}_{31})I_{pj,3}\Big)\right]$$
$$+ \mu_{j,2}\left[-\frac{\vec{e}_j.\vec{n}_j}{||\vec{l}_{12} \times \vec{l}_{31}||}\Big((\vec{l}_{31}.\vec{l}_{12})I_{pj,1} + (\vec{l}_{31}.\vec{l}_{23})i_{pj,2} + (\vec{l}_{31}.\vec{l}_{31})I_{pj,3}\Big)\right]$$
$$+ \mu_{j,3}\left[-\frac{\vec{e}_j.\vec{n}_j}{||\vec{l}_{12} \times \vec{l}_{31}||}\Big((\vec{l}_{12}.\vec{l}_{12})I_{pj,1} + (\vec{l}_{12}.\vec{l}_{23})i_{pj,2} + (\vec{l}_{12}.\vec{l}_{31})I_{pj,3}\Big)\right] \quad (14)$$

where

$$I_{pj,1} = \frac{1}{c}(H_{pj,1} + (ac - b^2 - cd^2)Mpj,1 + F_{pj,1} - c) \quad (15)$$

$$H_{pj,1} = b\log\left(\frac{\sqrt{a+2b+c}+d}{\sqrt{a}+d}\right) + c\log(\sqrt{a+2b+c}+d) \quad (16)$$

$$F_{pj,1} = d\sqrt{c}\left(\sinh^{-1}\frac{b+c}{\sqrt{ac-b^2}} - \sinh^{-1}\frac{b+c}{\sqrt{ac-b^2}}\right) \quad (17)$$

Therefore, using equations (7), (10) and (14), the induced velocity potentials at any point P can be expressed in terms of the doublicity values at the n nodal points. If the Internal Dirichlet boundary conditions are imposed at the m control points (centroids of the panels), a system of linear equations can be obtained:

$$\begin{pmatrix} & & \\ & A_{kj} & \\ & & \end{pmatrix} \cdot \begin{pmatrix} \frac{\mu_j}{4\pi} \end{pmatrix} = -\begin{pmatrix} \Phi_{\infty k} \end{pmatrix} \quad (18)$$

where $[A_{kj}]$ is a $m \times n$ influence matrix, $[\mu_j/4\pi]$ is the uknown vector containing the doublicity at the n nodal points and $[\Phi_{\infty k}]$ contains the freestream velocity potential at the m control points. For three-dimensional objects represented by triangular panels, normally $m > n$. The system is overdetermined.

An attempt was made to select n panels on which boundary conditions were imposed so that the system became n equations and n unknowns. However, this gave unstable, and sometimes indeterminate solutions. A stable solution can be obtained using a least square approximation to solve the original system of m equations and n unknowns. The more stable method, from Numerical Recipes in C, of ***Singular Value Decomposition*** has been used to calculate the equivalent inverse of the influence coefficient matrix. The rectangular influence matrix A is decomposed into three matrices, U, W, and V:

$$\begin{pmatrix} \\ A \\ \\ \end{pmatrix} = \begin{pmatrix} \\ U \\ \\ \end{pmatrix} . \begin{pmatrix} w_1 & & & \\ & w_2 & & \\ & & \cdots & \\ & & \cdots & \\ & & & w_n \end{pmatrix} . \begin{pmatrix} \\ V^T \\ \\ \end{pmatrix} \tag{19}$$

Decomposition of the matix A allows the least square solution to the matrix $[\mu_j/4\pi]$ to be found using the following equation:

$$\begin{pmatrix} \frac{\mu_j}{4\pi} \end{pmatrix} = \begin{pmatrix} V \end{pmatrix} . \begin{pmatrix} diag(1/w_j) \end{pmatrix} . \begin{pmatrix} U^T \end{pmatrix} . - \begin{pmatrix} \\ \Phi_{\infty k} \\ \\ \end{pmatrix} \tag{20}$$

This method is found to be very stable for the application and to give an accurate solution. For a closed surface the first integral in equation (7) disappears. Therefore the velocity equation is:

$$\begin{aligned} \vec{U}_p &= \vec{U}_p^{\mu} + \vec{U}_{\infty p} \\ &= \frac{1}{4\pi}\sum_{j=1}^{m}\left[\nabla\mu_j \int_{C_j} \vec{A}.d\vec{l}_Q + (\vec{n}_j \times \nabla\mu_j) \times \left(\vec{n}_j \times \int_{C_j} \frac{d\vec{l}_Q}{r}\right)\right] + \vec{U}_{\infty p} \end{aligned} \tag{21}$$

The divergence of the doublicity $\nabla\mu_j$ and the term $\vec{n}_j \times \nabla\mu_j$ can be expressed in terms of $\mu_{j,1}$, $\mu_{j,2}$ and $\mu_{j,3}$ with the following geometrical form:

$$\nabla\mu_j = \frac{1}{||\vec{l}_{31} \times \vec{l}_{12}||^2} \begin{pmatrix} x_1 & x_2 & x_3 \\ y_1 & y_2 & y_3 \\ z_1 & z_2 & z_3 \end{pmatrix}$$

$$\begin{pmatrix} l_{23}^2 & -(l_{31}^2 + \vec{l}_{12}.\vec{l}_{31}) & -(l_{12}^2 + \vec{l}_{12}.\vec{l}_{31}) \\ -(l_{31}^2 + \vec{l}_{12}.\vec{l}_{31}) & l_{31}^2 & \vec{l}_{12}.\vec{l}_{31} \\ -(l_{12}^2 + \vec{l}_{12}.\vec{l}_{31}) & \vec{l}_{12}.\vec{l}_{31} & l_{12}^2 \end{pmatrix} \begin{pmatrix} \mu_{j,1} \\ \mu_{j,2} \\ \mu_{j,3} \end{pmatrix} \tag{22}$$

The term $\vec{n}_j \times \nabla\mu_j$:

$$\vec{n}_j \times \nabla\mu_j = \frac{1}{||\vec{l}_{31}.\vec{l}_{12}||} \begin{pmatrix} x_1 & x_2 & x_3 \\ y_1 & y_2 & y_3 \\ z_1 & z_2 & z_3 \end{pmatrix} \begin{pmatrix} 0 & -1 & 1 \\ 1 & 0 & -1 \\ -1 & 1 & 0 \end{pmatrix} \begin{pmatrix} \mu_{j,1} \\ \mu_{j,2} \\ \mu_{j,3} \end{pmatrix} \tag{23}$$

$$= \frac{1}{||\vec{l}_{31}.\vec{l}_{12}||}(\vec{l}_{23}\mu_{j,1} + \vec{l}_{31}\mu_{j,2} + \vec{l}_{12}\mu_{j,3})$$

Applying (22) and (23) to (21) the induced velocities for a closed surface can be calculated.

$$\vec{U}_p = \frac{1}{4\pi}\sum_{j=1}^{m}\left[\nabla\mu_j\Big(\sum_{i=1}^{h_j} J_{pj,i}\Big) + (\vec{n}_j \times \nabla\mu_j)\times\Big(\vec{n}_j \times \Big(\sum_{i=1}^{h_j} \vec{K}_{kpj,i}\Big)\Big)\right] + \vec{U}_{\infty p} \qquad (24)$$

This formulation of the second order doublet panel method has been tested on the potential flow around a sphere. The 'panelled'sphere (164 panels) is depicted in figure 1, and the results of the panel programs in figure 3. The results for the second order doublet distribution are far more accurate than the single order distribution, despite less panels being used, and they are consistant with the theoretical results.

Presently, the Second Order Doublet method has not been applied to lifting asymmetric bodies in a flow. For aerofoils the uniqueness of the boundary conditions will be set with the Kutta Condition and the wake with trailing vortices. Ultimately, the integrals will be applied to free surfaces, this involves including the first integral in equation (8).

CONCLUSION

Using C.F.D. to investigate the design parameters of planing hulls can allow the number of practical experiments to be reduced and lead to insight into the Fluid Dynamics of sailboards. *Second-Order Doublet Panel Method* is a very effective tool for the study of planing hulls. It is very flexible both in terms of its ability to model wakes and free surfaces and because it can be applied to many different flow calculations. As with other Panel Method programs the results can be linked with boundary layer programs to allow further application.

ACKNOWLEDGMENTS

This work is being conducted under a Science and Engineering Research Council grant through the Marine Technology Directorate Limited.

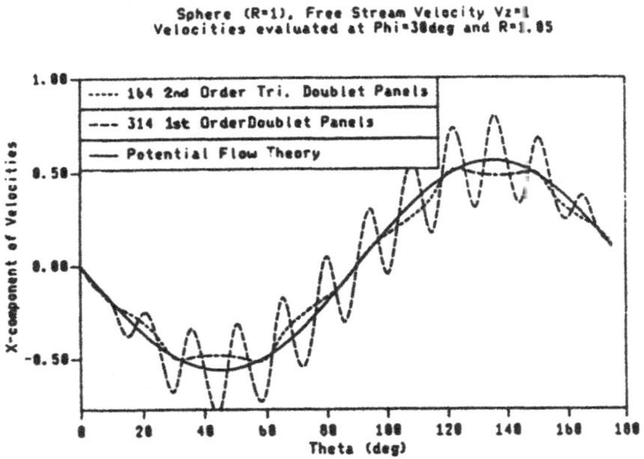

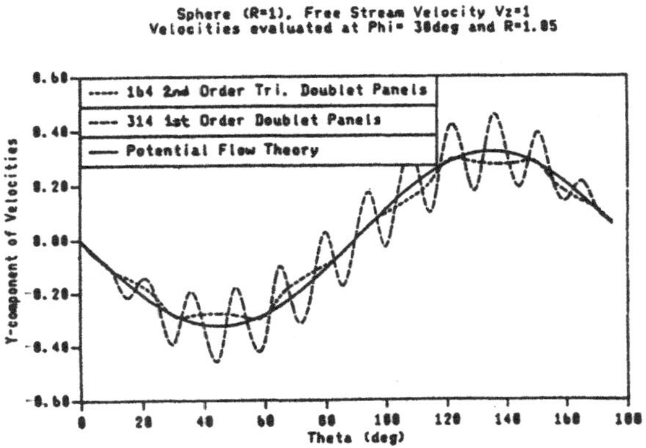

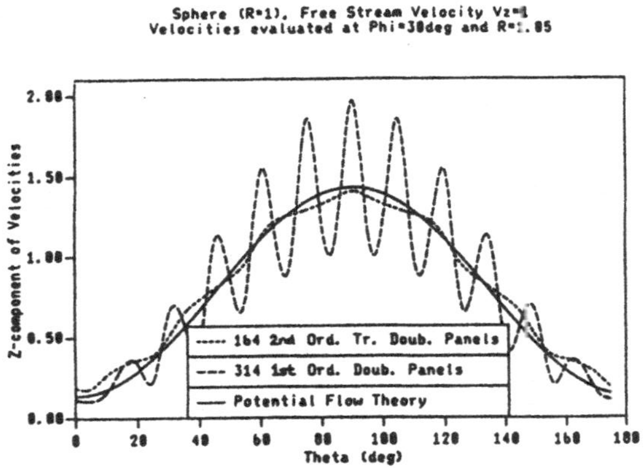

Coordinate System (R, θ, ϕ)

$x = R \sin\theta \cos\phi$

$y = R \sin\theta \sin\phi$

$z = R \cos\theta$

Figure 3. Comparison of the results of the calculation of the velocity flow field around a sphere of unit radius in a uniform flow of unit magnitude in the positive z-direction using *First* and *Second Order Doublet* formulation. Velocity shown in the graphs are evaluated at $\phi = 30°$ (i.e above the surface of the sphere) $R = 1.05$ and various θ. The Second Order formulation is more accurate.

REFERENCES

1- Cantaloube, B. and Rehbach, C.,"Calculation of the Integrals of the Singularities Method", Rech. Aerosp., 1986.

2- Chiu, T.W.,"Aerodynamic loads on a Railway Train in a cross–wind at large yaw angles", PhD thesis, Cambridge University Engineering Department, 1990.

3- Darby, S. N.,"Sailboarding: Exciting new water sport", Practical Hydrofoils, AYRS publication No 58, 1966.

4- Farrar, A.,"Fast Craft - Sail and Power", Ship & Boat International, RINA publication, U.K., December 1990.

5- Hess, J. L., and Smith, A. M. O.,"Calculation of Potential flow about arbitrary bodies", Progress in Aeronautical Sciences, Vol 8, 1966.

6- Hunt, B.,"The Panel Method for Subsonic Aerodynamic Flows", Von Kármán lecture series in Computational Fluid Dynamics, 1978–4.

7- Jimenez, C. R., "On the Hydrodynamic Performance of Sailboards", SNAME Publication, Northern California Section, United States.

8- Marchaj, C. A.,"Aero-Hydrodynamics of Sailing", 2nd Edition, published by Adlard Coles, 1990.

9- Press, W. H., etal,"Numerical Recipes in C, The Art of Scientific Computing", Cambridge University Press, 1988.

10- Smith, R. W., "An inviscid analysis of the flow about windsurfing sails", Proceedings of the 17th AIAA symposium on aerohydronautics of sailing, Stanford, California, USA, 1987.

11- Wood Rogers, A. R.,"The design and development of a planing multihull sailboat", MPhil thesis, University of Exeter, 1989.

Application of the Integral Equation Method to Flows Around a Wing with Circular-Arc Section

H. Hu

Department of Mathematics, Hampton University, Hampton, Virginia 23668, U.S.A.

ABSTRACT

An integral equation (or called field-panel, field-boundary element) scheme for solving the full-potential equation for transonic flows has been developed. The full-potential equation has been written in the form of the Poisson's equation. Compressibility has been treated as non homogeneity. The integral equation solution in terms of velocity field is obtained by the Green's theorem. The solution consists of surface (boundary elements) integral term(s) of vorticity/source distribution(s), wake surface (boundary elements) integral term(s) of free-vortex sheet(s) and a volume (field-elements) integral term of compressibility over a small limited domain around the source of disturbance. Solution procedure is an iterative procedure for non-linear flows. To consist with the mixed-nature of transonic flows, the Murman-Cole type-difference scheme is used to compute the derivatives of the density for non-linear flows. The present scheme is applied to flows around a rectangular wing with circular-arc section.

Key Words: integral equation method, full-potential equation, subsonic and transonic wing flows.

NOTATION

a	speed of sound
C_p	surface pressure coefficient
$\vec{d}$	distance vector pointed from sender to receiver
ds	infinitesimal surface area
$\vec{e}_d$	unit vector of $\vec{d}$
g	wing surface
G	compressibility
l	wing surface panel length
M_∞	free-stream Mach number
$\vec{n}$	surface normal unit vector
q	surface source distribution
$\vec{V}$	field velocity vector
$\vec{V}_\infty$	free-stream velocity vector

w	wake surface
α	angle of attack
$\vec{\gamma}$	surface vorticity distribution
κ	gas specific heat ratio
ρ	density
Φ	velocity potential

INTRODUCTION

The finite-difference (FD) and finite-volume (FV) methods for solving transonic flows have been well developed during the past twenty years. Although the Navier-Stokes equation formulation for the transonic flow computations has been understood as the best model and the FD and FV methods are successful in dealing with transonic flows, the computation of the unsteady Navier-Stokes equations over complex three-dimensional configurations is very expensive, particularly for time-accurated unsteady flow computations. There are also major technical difficulties in FDM and FVM for generating suitable grids for complex three-dimensional aerodynamic configurations.

The experience has shown that rather accurate solutions can been obtained for many transonic flows using the inviscid modeling of the full-potential equation. For transonic flows without strong shocks and massive separations, the full-potential equation is an adequate approximation to the Navier-Stokes equations. The integral equation method (IEM) for the potential equation is an alternative to the FDM and FVM. Moreover, the IEM has several advantages over the FDM and FVM. The IEM involves evaluation of integrals, which is more accurate and simpler than the FDM and FVM, and hence a coarse grid (field-elements) can been used in IEM. The IEM automatically satisfies the far-field boundary conditions and therefore only a small limited computational domain is needed. The IEM does not suffer from the artificial viscosity effects as compared to FDM and FVM for shock capturing in transonic flow computations. The generation of the three-dimensional grid for complex configuration is not difficult in the IEM, since the mapping from physical plane to computational plane is not required.

Integral equation methods for transonic flows have been developed by several investigators[1−18] during the past few years for steady airfoil, wing and aircraft configurations and unsteady airfoils and wings. In the present paper, a method for computing general 3-D flows is presented along with a simple numerical example, as an initial stage of a research program to develop a general integral equation method for steady and unsteady subsonic and transonic flows around complex configurations.

FORMULATION

Governing Equations

The non-dimensional steady full-potential equation is given by:

$$\nabla^2\Phi = G \tag{1}$$

with

$$G_1 = -\frac{\nabla\rho}{\rho}\cdot\vec{V} \tag{2}$$

and

$$\rho = [1 + \frac{\kappa - 1}{2}(1 - |\vec{V}|^2)]^{\frac{1}{\kappa-1}} \tag{3}$$

where the characteristic parameters, ρ_∞, a_∞ and l have been used; a is the speed of the sound, ρ the density, and l the wing surface panel length; and Φ is the velocity potential ($\vec{V} = \nabla\Phi$), G the compressibility, and κ the gas specific heat ratio.

Boundary Conditions

The boundary conditions are surface no-penetration condition, Kutta condition, infinity condition, and wake kinematic and dynamic conditions. They are described as follows:

$$\vec{V}\cdot\vec{n}_g = 0 \qquad \text{on} \quad g(\vec{r}) = 0 \tag{4}$$

$$\Delta C_p|_{sp} = 0 \tag{5}$$

$$\nabla\Phi \to 0 \quad \text{away from} \quad g(\vec{r}) = 0 \quad \text{and} \quad w(\vec{r}) = 0 \tag{6}$$

$$\vec{V}\cdot\vec{n}_w = 0 \qquad \text{on} \quad w(\vec{r}) = 0 \tag{7}$$

and

$$\Delta C_p = 0 \qquad \text{on} \quad w(\vec{r}) = 0 \tag{8}$$

where $\vec{n}_g$ is the unit normal vector of the wing (or a general body) surface, $g(\vec{r}) = 0$; C_p is the surface pressure coefficient; the subscripts sp refers to the edges of separation; and $w(\vec{r}) = 0$ is wake surface(s). It should be noticed that the infinity condition, Eq. (6), is automatically satisfied by the integral equation solution.

IE Solution

By using the Green's theorem, the integral equation solution of Eq. (1) in terms of the velocity field is given by

$$\begin{aligned}\vec{V}(x,y,z) &= \vec{V}_\infty \\ &+ \frac{1}{4\pi}\int\int_g \frac{q_g(\xi,\eta,\zeta)}{d^2}\vec{e}_d ds(\xi,\eta,\zeta) \\ &+ \frac{1}{4\pi}\int\int_g \frac{\vec{\gamma}_g(\xi,\eta,\zeta)\times\vec{d}}{d^3} ds(\xi,\eta,\zeta) \\ &+ \frac{1}{4\pi}\sum_{nw=1}^{NW}\int\int_w \frac{\vec{\gamma}_w(\xi,\eta,\zeta,t)\times\vec{d}}{d^3} ds(\xi,\eta,\zeta) \\ &+ \frac{1}{4\pi}\int\int\int_V \frac{G(\xi,\eta,\zeta)}{d^2}\vec{e}_d d\xi d\eta d\zeta\end{aligned} \tag{9}$$

where $\vec{V}_\infty$ is the free-stream velocity; q is the surface source distribution; $\vec{\gamma}$ is the surface vorticity distribution; NW is the total number of edges of separation; ds is the infinitesimal surface area; the vector $\vec{d}$ is given by $\vec{d} = (x-\xi)\vec{i} + (y-\eta)\vec{j} + (z-\zeta)\vec{k}$; and $\vec{e}_d$ is defined by $\vec{e}_d = \vec{d}/|\vec{d}|$.

COMPUTATIONAL SCHEME

Discretisation

The wing and its wake are represented by triangular vortex panels. A uniform rectangular parallelopiped type of volume elements are used throughout the flow field. A linear $\vec{\gamma}$-distribution is used over small surface panel, while a constant G-distribution is used over small field volume-element. The discretized integral equation solution becomes

$$\begin{aligned}\vec{V}(x,y,z) &= \vec{V}_\infty \\ &+ \frac{1}{4\pi}\sum_{i=1}^{LG}\sum_{k=1}^{NG}\int\int_{g_{i,k}} \frac{q_{g_{i,k}}(\xi,\eta,\zeta)}{d^2}\vec{e}_d ds(\xi,\eta,\zeta) \\ &+ \frac{1}{4\pi}\sum_{i=1}^{LG}\sum_{k=1}^{NG}\int\int_{g_{i,k}} \frac{\vec{\gamma}_{g_{i,k}}(\xi,\eta,\zeta,t)\times\vec{d}}{d^3} ds(\xi,\eta,\zeta) \\ &+ \frac{1}{4\pi}\sum_{nw=1}^{NW}\sum_{i=1}^{LW_{nw}}\sum_{k=1}^{MW_{nw}}\int\int_{w_{i,k}} \frac{\vec{\gamma}_{w_{i,k}}(\xi,\eta,\zeta,t)\times\vec{d}}{d^3} ds(\xi,\eta,\zeta,t) \\ &+ \frac{1}{4\pi}\sum_{i=1}^{LV}\sum_{j=1}^{MV}\sum_{k=1}^{NV} G_{i,j,k}\int\int\int_{V_{i,j,k}} \frac{1}{d^2}\vec{e}_d d\xi d\eta d\zeta\end{aligned} \tag{10}$$

with

$$\begin{aligned}\vec{\gamma}\times\vec{d} &= [\gamma_y(z-\zeta)-\gamma_z(y-\eta)]\vec{i} \\ &+ [\gamma_z(x-\xi)-\gamma_x(z-\zeta)]\vec{j} \\ &+ [\gamma_x(y-\eta)-\gamma_y(x-\xi)]\vec{k}\end{aligned} \tag{11}$$

where the linear distributed $(\gamma_x, \gamma_y, \gamma_z)$ are three components of $\vec{\gamma}$; the indices, i, j and k refer to the surface panels and field elements; $LG \times NG$ is the total number of wing (or a general body) surface panels; $LW \times MW$ is the total number of wake surface panels; and $LV \times MV \times NV$ is the total number of field elements.

Iterative Scheme

Due to the nature of the nonlinearity of transonic flows, the solutions are obtained through a iterative procedure, where the compressibility, G, and the wake shape and its strength are updated within each iteration. The solution procedure follows the successful forms of Ref. 10 and Ref. 18, hence only a brief description is given:

Equations (10) along with the boundary conditions, Eqs. (4), (5), (7) and (8), are solved iteratively until the solution converges. Here, two loops are used. The inner loop is used to calculate and check the convergence of the non-linear term, G. Here, G is set to be zero first to perform the incompressible flow calculations. The outer loop is used to update and check the convergence of the wake shape and wing surface pressure distribution. For transonic flows, the Murman-Cole type difference scheme is used. In subsonic flow region a central-differencing is used, while in supersonic flow region, a backward-differencing is used.

NUMERICAL EXAMPLES

The presesnt scheme has been applied to a rectangular wing of aspect ratio of 4 with a 6%-thick circular-arc section. The half-span of the wing (including upper and lower surfaces) is divided into 20 × 12 quadrilateral panels. Each quadrilateral panel consists of 2 triangular panels. The one-half of the computational domain is divided into 23 × 9 × 15 field volume elements in x, y and z directions, respectively.

To simplify the problem in this initial stage of the research program, a symmetric flow with zero angle of attack is considered, and no separation has been considered. A transonic flow case (at $M_\infty = 0.908$) has been tested and the results (Figures 1-2) shew that the shock predicted is not as strong as one expected (see Ref. 19). This is possibly due to the large grid size ($\Delta x = 0.1$ chord length) used. A successive grid refinement technique is being applied in the region around the shock to sharpen the shock. The further results will be presented in the near furture.

Figures 3-4 shows the present computational results for the incompressible flow around the same wing at zero angle of attack. The results look quite correct.

CONCLUDING REMARKS

An integral equation scheme based on the full-potential equation formulation for transonic flows has been developed. The scheme is capable of handling flows around general three dimensional configurations, although only a simple case is tested in the present paper. It is necessary to emphasize that the result presented here is just an preliminarly result in this initial stage of a research program of developing a general steady and unsteady subsonic and transonic IEM for flows around complex 3-D configurations. Presently, the research has been focused on transonic flow computations, and in the near furture effects of unsteadiness and wake(s) (separation(s)) will be added.

ACKNOWLEDGEMENTS

This work has been supported by NASA Langley Research Center under the Grant No.: NAG-1-1170. Dr. Carson Yates, Jr. is the technical monitor. The valuable discussions with Dr. Carson Yates, Jr. at NASA-LaRC and Dr. L-C Chu at Lockheed are gratefully acknowledged.

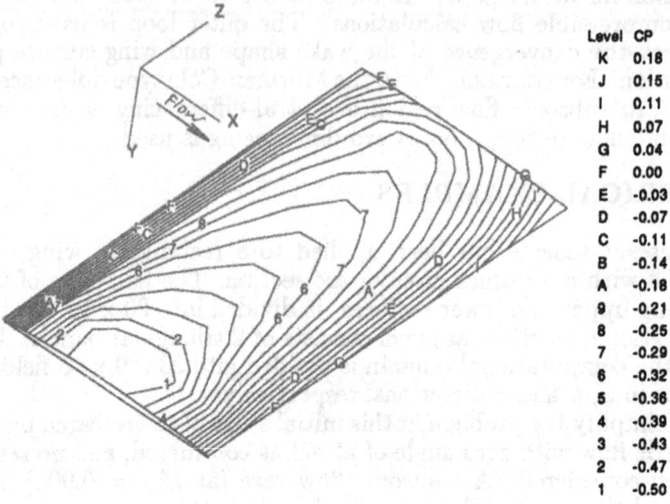

Figure 1. C_p **contour, rectangular wing,** $AR = 4$**,**
6%-thick circular-arc section, $\alpha = 0$**,** $M_\infty = 0.908$**.**

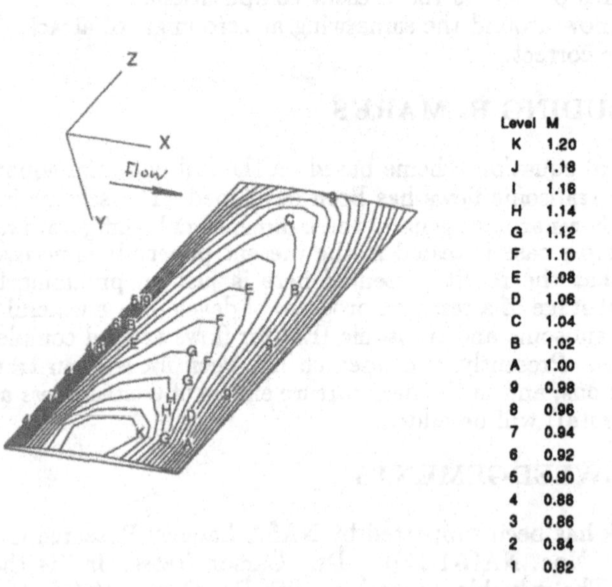

Figure 2. Mach contour, same case as Fig. 1.

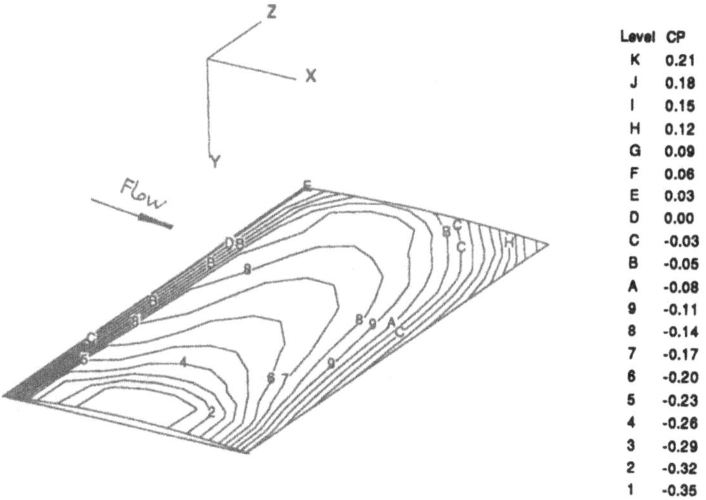

Figure 3. C_p contour, same case as Fig.1, but $M_\infty = 0$.

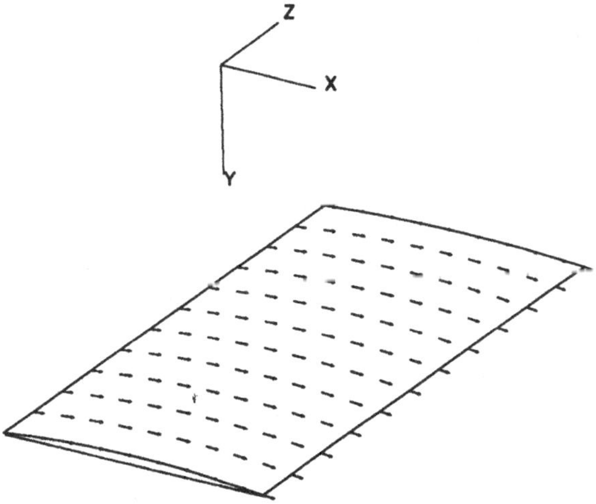

Figure 4. Velocity vector field, same case as Fig. 3.

REFERENCES

1. Piers, W. J. and Sloof, J. W. Calculation of transonic flow by means of a shock-capturing field panel method, AIAA Paper 79-1459, 1979.

2. Hounjet, M. H. L. Transonic panel method to determine loads on oscillating airfoils with shocks, *AIAA Journal*, 1981, Vol. 19, No. 5, pp.559-566.

3. Tseng, K. and Morino, L. Nonlinear Green's function methods for unsteady transonic flows, *Transonic Aerodynamics*, (Ed. Nixon, D), 1982, pp.565-603.

4. Oskam, B. Transonic panel method for the full potential equation applied to multicomponent airfoils, *AIAA Journal*, 1985, Vol. 23, No. 9, pp.1327-1334.

5. Erickson, L. L. and Strande, S. M. A theoretical basis for extending surface-paneling methods to transonic flow, *AIAA Journal*, 1985, Vol.23, No. 12, pp.1860-1867.

6. Sinclair, P. M. An exact integral (field panel) method for the calculation of two-dimensional transonic potential flow around complex configurations, *Aeronautical Journal*, June/July 1986, Vol. 90, No. 896, pp.227-236.

7. Kandil, O. A. and Yates, E. C., Jr. Computation of transonic vortex flows past delta wings-integral equation approach, *AIAA Journal*, 1986, Vol. 24, No. 11, pp.1729-1736.

8. Madson, M. D. Transonic analysis of the F-16A with under-wing fuel tanks: an application of the TranAir pull-potential code, AIAA Paper 87-1198, 1987.

9. Kandil, O. A. and Hu, H. Transonic airfoil computation using the integral equation with and without embedded Euler domains, *Boundary Elements IX*, Vol. 3, (Eds.: Brebbia, C. A., Wendland, W. L. and Kuhn, G.), Computational Mechanics Publications, Springer-Verlag, 1987, pp. 553-566.

10. Kandil, O. A. and Hu, H. Full-potential integral solution for transonic flows with and without embedded Euler domains, *AIAA Journal*, 1988, Vol. 26, No. 9, pp. 1074-1086.

11. Chu, L-C. Integral equation solution of the full potential equation for three-dimensional, steady, transonic wing flows, Ph.D. dissertation, Dept. of Mechanical Engineering and Mechanics, Old Dominion University, Norfolk, Virginia, March 1988.

12. Sinclair, P. M. A three-dimensional field-integral method for the calculation of transonic flow on complex configurations - theory and preliminary results, *Aeronautical Journal*, June/July 1988, Vol. 92, No. 916, pp. 235-241.

13. Kandil, O. A. and Hu, H. Unsteady transonic airfoil computation using the integral solution of full-potential equation, *Boundary Elements X*, Vol. 2, (Ed.: Brebbia, C. A.), Computational Mechanics Publications, Springer-Verlag, 1988, pp. 352-371.

14. Kandil, O. A. and Hu, H. Integral solution of unsteady full- potential equation for a transonic pitching airfoil, *Journal of Aircraft*, 1990, Vol. 27, No. 2, pp. 123-130.

15. Ogana, W. Transonic integro-differential and integral equations with artificial viscosity, *Engineering Analysis with Boundary Elements*, 1989, Vol. 6, No. 3, pp. 129-135.

16. Hu. H. Field-boundary element solution for a transonic pitching wing, *Topics in Engineering, Vol. 7: Advances in Boundary Elements Methods in Japan and USA*, (Eds.: Tanaka, M., Brebbia, C. A. and Shaw, R.), Computational Mechanics Publication, 1990, pp. 295-305.

17. Hu, H. Field-boundary element computation for transonic flows, *Computation Engineering with Boundary Elements*, Vol. 1, (Eds.: Grilli, S., Brebbia, C. A. and Cheng, A. H-D.), Computational Mechanics Publication, 1990, pp. 319-330.

18. Hu, H. and Chu, L-C. Unsteady Three-Dimensional Transonic Flow Computations Using Field Element Method, *Boundary Element Methods in Engineering*, (Eds.: Annigeri, B. S. and Tseng, K.), Springer-Verlag, 1990, pp.140-146.

19. Tseng, K. Application of Green's Function Method for 2- and 3- Dimensional Steady Transonic Flows, AIAA Paper 84-0425, 1984.

A Variational Symmetric Boundary Element Formulation for Fluid-Structure Interaction Problems

F. Erchiqui, A. Gakwaya

Department of Mechanical Engineering, Laval University, Quebec City, G1K 7P4, Canada

1. INTRODUCTION

The problem of interaction between a structure (solid body) and its surrounding fluid medium is of paramount importance in several technical domains as in the aerospace and naval industries or in transportation of gases and liquids in pipeline or in closed reservoirs. Recent studies [8,9,10] of vibratory behavior of structure in contact with a fluid medium, show the necessity of fully taking account of the interaction of the two media. In fact, experiments [23] have shown that the resonance characteristics of the structure are considerably modified by the presence of the fluid thus resulting in the fact that some deformation modes become critical and may cause fatal structural damage. Also, it has been observed that sound propagation in the fluid is altered by the presence of structure which engenders more noise problems [8,23].

This paper presents a recent development in analysis of vibrating structure in presence of viscous compressible fluids using a mixed symmetric variational formulation. Field equations and transmission conditions for the coupled dynamical system are briefly introduced. Then boundary integral equations valid for internal as well as for external flow are presented for viscous compressible fluids in small motions about a stationary reference state. The structural energy functional is coupled with a novel boundary element variational formulation of the viscous fluid that avoids the explicit evaluation of the finite parts of hypersingular integrals usually encountered if the conventional collocation BEM is used. The mixed variational functional is discretized using isoparametric boundary finite element procedure and results in a complex symmetrical algebraic system that is efficiently handled by existing numerical linear algebra routine for eigenvalue analysis. The efficiency of the formulation are assessed by applying the developed computer code to some examples related to acoustic radiation and scattering problems and to problems of fluid-structure interaction in

dynamic aeroelasticity. Due to limited space, details on computer implementation and numerical results will be presented in forthcoming papers.

2. FUNDAMENTAL FIELD EQUATIONS AND ASSUMPTIONS

We consider a simply connected bounded elastic body $\Omega_s \subset R^3$ with smooth boundary Γ_s immersed in a viscous compressible fluid medium occupying the exterior domain Ω_f and study the problem of determining the pressure and velocity field in Ω_f and displacement in Ω_s due to a given excitation in the fluid Ω_f. The fluid motion is governed by the usual time dependent Navier-Stokes equations [1,2] together with a state equation, $p = \rho c_f^2$. If, we consider small perturbations about an established stationnary flow (assumed initially at rest), then the governing equations reduce to the usual linearized time dependent equations [8,20]:

$$\frac{\partial \rho}{\partial t} + \rho_f^0 \frac{\partial v_i}{\partial x_i} = \frac{1}{c_f^2}\frac{\partial p}{\partial t} + \rho_f^0 \frac{\partial v_i}{\partial x_i} = 0 \qquad in\ \Omega_f \qquad (1)$$

$$\rho_f^0 \frac{\partial v_i}{\partial t} = -\frac{\partial p}{\partial x_i} + \mu_f\left[\frac{\partial^2 v_i}{\partial x_j^2} + \frac{1}{3}\frac{\partial}{\partial x_i}\left(\frac{\partial v_j}{\partial x_j}\right)\right] \qquad in\ \Omega_f \qquad (2)$$

where p, ρ, v are pressure, density and velocity; c is the speed of sound, and ρ_f^0, μ_f are time independent equilibrium density and viscosity.

The elastic solid body displacement is governed by the equations of elastodynamics [3,4]:

$$\mu_s \Delta \vec{u}_s + (\lambda_s + \mu_s)\nabla(\nabla.\vec{u}_s) = \rho_s^0 \frac{\partial^2 \vec{u}_s}{\partial t^2} \qquad in\ \Omega_s \qquad (3)$$

where λ_s, μ_s and ρ_s are the *Lamé* coefficients and density respectively. On the boundary Γ of Ω_s we have the usual traction operator defined by :

$$T_s(\partial x, n)\vec{u}_s = 2\mu_s \frac{\partial \vec{u}_s}{\partial n} + \lambda_s \vec{n}\nabla.\vec{u}_s + \mu_s \vec{n} \wedge \nabla \wedge \vec{u}_s \qquad (4)$$

with $T_f(\partial x, n)\vec{u}_f$ similarily defined for the fluid but with λ_f, μ_f replacing λ_s, μ_s.

For a specified incident velocity field V^{inc}, the fluid-structure interface condition becomes :

$$\vec{V}^{inc} + \vec{v}_f - \dot{\vec{v}}_s = 0 \qquad in\ \Gamma \qquad (5.a)$$

$$\vec{t}^{inc} + \vec{t}_f - \vec{t}_s = 0 \qquad in\ \Gamma \qquad (5.b)$$

and the initial condition are $\vec{v}_f = \vec{v}_s = \vec{0}$ at $t = 0$.

To obtain equations in a form suitable for coupling the fluid motion with the displacement of the elastic body, we consider the

4.3 COUPLED FLUID-STRUCTURE WEAK FORMULATION

Introducing eqn (27), expressing the surface stresses energy, into the structural variational eqn (24), we obtain:

$$a(\vec{u},\vec{v}) - \rho_s\omega^2(\vec{u},\vec{v}) = +\frac{1}{2} < \vec{t}_f^+,\vec{v} > + < \vec{t}^{inc}, v > + < \vec{v}(x), K_4\vec{u}^-(y) >$$
$$- < \vec{v}(x), K_3\vec{t}_f^+(y) > - < \vec{v}(x), K_4\vec{u}^{inc}(y) > \qquad (28)$$

Then the coupled fluid-structure problem is defined by adding equation (26) and (28) and can be stated as follows: For given $(\vec{u}^{inc}, \vec{t}^{inc}) \in (H^{1/2}(\Gamma))^3 \times (H^{-1/2}(\Gamma))^3$, find $\vec{u}^-$ and $\vec{t}_f$ satisfying the variational equation:

$$A((\vec{u},\vec{t}_f),(\vec{v},\vec{\tau})) = L(\vec{v},\vec{\tau}) \qquad (29)$$

where $A((.,.),(.,.))$ is a non symmetrical sesquilinear form defined by :

$$A((\vec{u},\vec{t}_f),(\vec{v},\vec{\tau})) = a(\vec{u},\vec{v}) - \rho_s\omega^2(\vec{u},\vec{v}) - < \vec{v}^-(x), K_4\vec{u}^-(y) >$$
$$+ < \vec{v}^-(x), K_3\vec{t}_f(y) > - < \vec{\tau}(x), K_2\vec{u}^-(y) >$$
$$+ < \vec{\tau}(x), K_1\vec{t}_f(y) > +\frac{1}{2} < \vec{\tau}(x), \vec{u}^-(x) >$$
$$-\frac{1}{2} < \vec{v}^-(x), \vec{t}_f(x) > \qquad (30)$$

and where the operators K_1, K_2, K_3 and K_4 are defined by:

$$K_1\{t\}(x)\} = \int_\Gamma [G(x-y,\omega)]\{t(y)\}\, d\Gamma_y$$

$$K_2\{u(x)\} = p.v \int_\Gamma [H^y(x-y,\omega)]^T\{u(y)\} d\Gamma_y$$

$$K_3\{t(x)\} = p.v \int_\Gamma [H^x(x-y,\omega)]\{t(y)\} d\Gamma_y$$

$$K_4\{u(x)\} = f.p \int_\Gamma [S^x(x-y,\omega)]^T\{u(y)\} d\Gamma_y$$

$L(.,.)$ is a linear form given by:

$$L(\vec{v},\vec{\tau}) = - < \vec{v}^-(x), K_4\vec{u}^{inc}(y) > + < \vec{v}^-(x), \vec{t}^{inc}(x) >$$
$$- < \vec{\tau}(x), K_2\vec{u}^{inc}(y) > +\frac{1}{2} < \vec{\tau}(x), \vec{u}^{inc}(x) > \qquad (31)$$

where the various integral operators are defined in the corresponding boundary integral equations (26-28). The following discussion helps to restore symmetry.

If, in the sesquilinear form A defined on $(H^1(\Omega_s))^3 \times (H^{1/2}(\Gamma))^3)$, $\vec{\tau}$ is time harmonic problem with state variable of the form [20,23]: $X_i = \Re\{X(r)\exp(-i\omega t)\}$. The above linearized equations take then the following form [3,12,13,20], for the fluid:

$$\overline{\mu}_f \Delta \vec{u}_f + (\overline{\lambda}_f + \overline{\mu}_f)\nabla(\nabla.\vec{u}_f) + \rho_f^0 \omega^2 \vec{u}_f = 0 \qquad in\ \Omega_f \qquad (6)$$

$$p = -\rho_f^0 c_f^2 \nabla.\vec{u}_f \qquad in\ \Omega_f \qquad (7)$$

where:

$$\overline{\mu}_f = -i\omega\mu_f \quad and \quad \overline{\lambda}_f = \rho_f^0 c_f^2 + \frac{2}{3} i\omega\mu_f \qquad (8)$$

and, for the solid :

$$\mu_s \nabla \vec{u}_s + (\lambda_s + \mu_s)\nabla(\nabla.\vec{u}_s) + \rho_s^0 \omega^2 \vec{U}_s = 0 \qquad in\ \Omega_s \qquad (9)$$

For a specified incident velocity field $\vec{V}^{inc}$ satisfying eqn.(6), conditions (5) must be satisfied on the fluid-solid interface and, as $x = r \longrightarrow \infty$, we must have the *Sommerfeld* radiation conditions which, upon decomposition into shear and dilatational waves, becomes

$$\lim_{r\to\infty} r\left(\frac{\partial \vec{u}_f^s}{\partial r} - ik_{s_f}\vec{u}_f^s\right) = 0\left(\frac{1}{r}\right) \qquad (10.a)$$

$$\lim_{r\to\infty} r\left(\frac{\partial \vec{u}_f^p}{\partial r} - ik_{p_f}\vec{u}_f^p\right) = 0\left(\frac{1}{r}\right) \qquad (10.b)$$

where:

$$\vec{u}_f^p = -\frac{1}{k_{p_f}}\nabla(\nabla.\vec{u}_f) \quad , \quad \vec{u}_f^s = \vec{u}_f - \vec{u}_f^p \qquad (11)$$

$$k_{p_f}^2 = \frac{\omega^2}{c_f^2}\frac{1}{1+\frac{4}{3}\frac{\overline{\mu}_f}{\rho_f^0 c_f^2}} \quad , \quad k_{s_f}^2 = \frac{\rho_f^0 \omega^2}{\overline{\mu}_f} \qquad (12)$$

The corresponding wave number k_{p_s} and k_{s_s} for the elastic solid are given by [3,8,20]:

$$k_{s_s}^2 = \frac{\rho_s \omega^2}{\mu_s} \quad , \quad k_{p_s}^2 = \frac{\rho_s \omega^2}{\lambda_s + 2\mu_s} \qquad (13)$$

Now the classical fluid-solid interaction problem can be stated as follows. For a given incident velocity field V^{inc} satisfying equation(6) almost everywhere in R^3, find the displacement $\vec{u}_f$ in Ω_f and displacement $\vec{u}_s$ in Ω_s such that equations (6) and (9) are satisfied and, the interface conditions, eqn.(5), together with the radiation condition (10) are also fullfilled [20].

3. BOUNDARY INTEGRAL EQUATIONS FOR THE FLUID

To present a unified integral equation formulation valid for interior and exterior fluid-structure interaction, we use an indirect approach

based on the theory of distribution. The problem is formulated in the whole space R^3, divided into interior domain Ω^- and exterior domain Ω^+, by a discontinuity surface Γ representing, for example, a thin shell structure (fig.1).

An integral equation representation of the harmonic elastic oscillator associated with the viscous compressible fluid (eqn.6) is thus sought in the infinite space R^3. Following [6,7,10,18] and fig.1, we introduce the jumps $[|\vec{u}|]$ and $[|\vec{t}|]$ on the discontinuity surface Γ. Then, upon using the Dirac delta distribution, the elastic oscillator equation, becomes in the sense of distributions:

$$[A(\partial_x,\omega)]\{u_x\} = \{A(\partial_x,\omega)\}\{u_x\}+[T(\partial_x,n_x]^T([|\{u_x\}|]\delta_\Gamma)+[|[T(\partial_x,n_x)]\{u_x\}|]\delta_\Gamma \qquad (14)$$

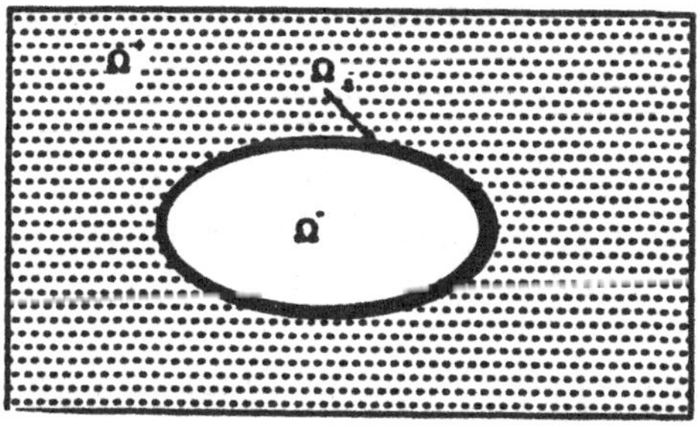

Fig 1 : Partition of R^3 into interior Ω^- and exterior Ω^+ by discontinuity surface Γ

for any point $x \in \Omega^+ \cup \Omega^-$ and for $\vec{u} \in C^2(R^3/\Gamma)$, where the traction operator has been used and $\{A(\partial x,\omega)\}$ represents the classical operator. An explicit integral representation, can be obtained by performing a convolution product of eqn(14) with the fundamental solution of the elastodynamic operator defined by [5,17]:

$$G(|x-y|,\omega) = \frac{1}{4\pi\overline{\mu}_f}g_1(|x-y|)\delta_{ij} - \frac{1}{4\pi\rho_f\omega^2}\frac{\partial}{\partial x_i}\frac{\partial}{\partial x_j}(g_1(|x-y|)-g_2(|x-y|)) \quad (15)$$

$$g_1(|x-y|) = \frac{\exp(ik_p|x-y|)}{|x-y|} \quad . \qquad g_2(x-y) = \frac{\exp(ik_s|x-y|)}{|x-y|} \quad (16)$$

$$H^y(|x-y|,\omega) = T(\partial y,n(y))G(|x-y|,\omega) \quad (17)$$

where δ_{ij} is the *Kronecker symbol*. Then, the resulting boundary integral equation for the displacement can be written as [20] ($\forall x,y \in \Gamma$):

$$\{u^\pm(x)\} = p.v\left\{\int_\Gamma [H^y(|x-y|,\omega)]^T[|\{u_y\}|]d\Gamma_y - \int_\Gamma [G(|x-y|,\omega)][|\{t_y\}|]\,d\Gamma_y\right\} \pm\frac{1}{2}|[\{u_x\}|] \quad (18)$$

Applying the traction operator $T(\partial x, n)$ to (18) written at interior points, and taking the limit as $x \in \Omega$ approaches $x_0 \in \Gamma$ we obtain ($\forall x, y \in \Gamma$):

$$\{t^{\pm}(x)\} = f.p \int_{\Gamma} [S^x(|x-y|,\omega)]\,[|\{u_y\}|]d\Gamma_y - p.v \int_{\Gamma} [H^x(|x-y|,\omega)]\,[|\{t_y\}|]\,d\Gamma_y \pm \frac{1}{2}|[\{t_x\}|] \qquad (19)$$

where $S^x(|x-y|,\omega) = T(\partial x, n_x) H^y(|x-y|,\omega)$. **The first term on the right hand side of eqn(19) must be taken in the** ***Hadamard finite*** **part sense, since taking the derivative inside the integral yields formally a divergent integral [3,21,22]. Eqn 19 is defined for any regular point $x \in \Gamma$, i.e. for points at which the tangent plane is continuous. Otherwise $[|\vec{t}|]$ is not uniquely defined.**

At the fluid-structure interface, the boundary conditions [5] can be written as:

$$[|\vec{u}|] = \vec{u}^{+} - \vec{u}^{-} = \vec{u}_f - \vec{u}_s = -\vec{u}^{inc} \qquad (20.a)$$

$$[|\vec{t}|] = \vec{t}^{+} - \vec{t}^{-} = \vec{t}_f - \vec{t}_s = -\vec{t}^{inc} \qquad (20.b)$$

so that eqn(18) and (19) reduce to the following boundary integral equations for external flows ($\forall x, y \in \Gamma$):

$$\frac{1}{2}(\{u_x^-\} - \{u^{inc}\}) = p.v \int_{\Gamma} [H^y(|x-y|,\omega)]^T \{u_y^+\} d\Gamma_y - \int_{\Gamma} [G(|x-y|,\omega)]\,\{t_y^+\}\,d\Gamma_y \qquad (21.a)$$

$$\{t_x^+\} = f.p \int_{\Gamma} [S^x(|x-y|,\omega)]\,\{u_y^+\} d\Gamma_y - p.v \int_{\Gamma} [H^x(|x-y|,\omega)]\,\{t_y^+\}\,d\Gamma_y + \frac{1}{2}\{t_x^+\} \qquad (22.a)$$

after applying the transmission condition (5) to introduce the fluid displacement $\vec{u}^+$ and the traction $\vec{t}_f^+$. Eqns (21) and (22) involve only physical variables of direct interest and are in contrast with the approach used by [9,10,14,18] where the evaluation of the surface layer densities was a premise to the determination of the physical variables. An interior coupling problem can be similarily defined and is given by:

$$\frac{1}{2}(\{u_x^+\} + \{u^{inc}\}) = -p.v \int_{\Gamma} [H^y(|x-y|,\omega)]^T \{u_y^-\} d\Gamma_y + \int_{\Gamma} [G(|x-y|,\omega)]\,\{t_y^-\}\,d\Gamma_y \qquad (21.b)$$

$$\{t_x^-\} = -f.p \int_{\Gamma} [S^x(|x-y|,\omega)]\,\{u_y^-\} d\Gamma_y + p.v \int_{\Gamma} [H^x(|x-y|,\omega)]\,\{t_y^-\}\,d\Gamma_y - \frac{1}{2}\{t_x^-\} \qquad (22.b)$$

4. VARIATIONAL FORMULATION OF THE EXTERIOR FLUID-STRUCTURE INTERACTION PROBLEM.

We present here a variational formulation that extends studies reported in [9,10,14] to a more general fluid flow by incorporating the viscous effects. The analysis reported in [12] is also extended by studying the regularisation and symmetry property of the coupled functional.

4.1 WEAK FORMULATION FOR THE STRUCTURE

For a solid structure, with negligeable body forces, the dynamic equilibrium equation with condition $\sigma_{ij}n_j = t_j$ on the interior boundary, and $t_j^- = t_j^+ + t_j^{inc}$, $u_j^- = u_j^+ + u_j^{inc}$, at the solid-fluid interface, the classical weak formulation can be written as [3,12,20]:

$$-a(\vec{u},\vec{v}) + \int_{\Gamma} \vec{v}.T(\partial x, n)\vec{u}_s d\Gamma + \rho_s\omega^2 \int_{\Omega_s} \vec{v}.\vec{u}_s d\Omega = 0 \qquad (23)$$

where $\vec{v}$ belong to the space V of test functions defined by: $V = \{\vec{v} \in (H^1(\Omega_s))^3, \vec{v} = 0 \; on\, \Gamma\}$. After introducing the boundary conditions (23) becomes:

$$a(\vec{u},\vec{v}) - \rho_s\omega^2(\vec{u},\vec{v}) =< \vec{t}_s, \vec{v} >_\Gamma =< \vec{t}^-, \vec{v} >_\Gamma =< \vec{t}^+, \vec{v} >_\Gamma + < \vec{t}^{inc}, v >_\Gamma \quad (24)$$

where $a(\vec{u},\vec{v})$ is a bilinear form given by:

$$a(\vec{u},\vec{v}) = \lambda_s \int_{\Omega_s} \nabla.\vec{v}\nabla.\vec{u}_s d\Omega_s + \frac{\mu_s}{2}\int_{\Omega_s} [(\nabla\vec{v} + \nabla^T\vec{v}) : (\nabla\vec{u}_s + \nabla^T\vec{u}_s)]\, d\Omega_s \quad (25)$$

and $(.,.)$ is the usual inner product in $L_2(\Omega) = H^0(\Omega)$, $< .,. >$ denotes a duality pairing of the functions defined on Γ with $\vec{t}^+ = \vec{t}_f$ and $\vec{t}^{inc} = T_f(\partial x, n)\vec{u}^{inc}$.

4.2 WEAK INTEGRAL EQUATION FORMULATION FOR THE FLUID (exterior flow)

Let us introduce the weighting functions $(\vec{v},\vec{r})$ belonging to a space V defined as $:V = \{(\vec{v},\vec{r}) \in (H^{1/2}(\Omega_f))^3 \times (H^{-\frac{1}{2}}(\Gamma^+))^3$, and apply a duality pairing on Γ to the basic integral equations (18) and (19). We then obtain the following weak form for external fluid:

$$< \vec{r}(x), \frac{1}{2}(-\vec{u}^{inc}(x) + \vec{u}^-(x)) >=< \vec{r}(x), \int_{\Gamma+} H^y(|x-y|,\omega)(-\{u_y^{inc}\} + \{u_y^-\})\Gamma_y >$$
$$- < \vec{r}(x), \int_{\Gamma+} G(|x-y|,\omega)\{t_{fy}\}\Gamma_y > \qquad (26)$$

$$< \vec{v}(x), \frac{1}{2}\vec{t}_f(x) >=< \vec{v}(x), \int_{\Gamma+} S^x(|x-y|,\omega)(\{-u_y^{inc}\} + \{u_y^-\})\Gamma_y >$$
$$- < \vec{v}(x), \int_{\Gamma+} H^x(|x-y|,\omega)\{t_{fy}\}\Gamma_y > \qquad (27)$$

A similar expression can be obtained for an interior fluid from eqns (21-22.b). If $v(x)$ is considered as a virtual displacement, eqn (27) is interpreted as the virtual work of the surface tractions.

replaced by $-\vec{\tau}$ in the fluid formulation and $A(-\vec{\tau})$ is then denoted by A', then A' is symmetric (see[20] for proof). Eqn (29) becomes:

$$A'((\vec{u},\vec{t}_f),(\vec{v},\vec{\tau})) = L'(\vec{v},\vec{\tau}) \tag{32}$$

where $A'((.,.),(.,.))$ and $L'(.,.)$ are formally obtained by replacing $\vec{\tau}$ by $-\vec{\tau}$ in eqn (30)-(31). Existence and uniqueness of a solution to the coupled system is guaranteed by the ellipticity of the sesquilinear form A' and the satisfaction of the *Garding's inequality (Lax-Milgram theorem)* [22].

If the elliptic sesquilinear form $A'((\vec{u},\vec{t}_f),(\vec{v},\vec{\tau}))$ defined $\forall(\vec{u},\vec{t}_f) \in H^1(\Omega_s) \times H^{-1/2}(\Gamma)$ is also symmetrical, then the solution to the variationnal problem :

$$A'((\vec{u},\vec{t}_f),(\vec{v},\vec{\tau})) = L'(\vec{v},\vec{\tau}) \tag{33}$$

minimize in $H^1(\Omega_s) \times H^{-\frac{1}{2}}(\Gamma)$ the following quadratic functional :.

$$J(\vec{u},\vec{t}_f) = \frac{1}{2}A'((\vec{u},\vec{t}_f),(\vec{u},\vec{t}_f)) - L'(\vec{u},\vec{t}_f) \tag{34}$$

A discretisation of the above functional by standard finite/boundary element technique will yield a stationnarity condition represented by a complex symmetric algebraic system [20].

4.4 REGULARISATION OF THE VARIATIONAL FUNCTIONAL

A direct computation based on eqn (30) and (31) will involve the numerical evaluation of finite part integrals. However, following ref [20], it is easy to see that, if we have a *Lipshitz* or *Lyapunov* surface (i.e. Γ is of class C^1), then the application of first and second *Stokes* theorem as well as the *divergence* theorem to continuous functions defined on Γ, can lead to fully regularized hypersingular integrals in the final variational functional, eqn(33). The following proposition is true:

If a function $f(=\{u\}\ or\ \{t\})$ is of class C^1 on Γ then the boundary integrals in the final variational equation are regular or weakly singular, and the final regularized coupled variational functional becomes:

$$\delta\frac{1}{2}\{A'((u,t_f),(u,t_f))\} = \delta\{L'(u,t_f)\} \tag{35}$$

with:

$$\frac{1}{2}A'((\vec{u},\vec{t}),(\vec{u},\vec{t})) = \frac{1}{2}\{\int_{\Omega_s}(\sigma_{kj}(\vec{u})\varepsilon_{kj}(\vec{u}) - \rho_s^0\omega^2(\vec{u}.\vec{u}))\,d\Omega_s$$

$$+\int_\Gamma\int_\Gamma u_k^-(x)\{\,B_{kj}^1(x,y) + B_{kj}^2(x,y) + B_{kj}^3(x,y) + B_{kj}^4(x,y)\,\}u_j(y)d\Gamma_y d\Gamma_y$$

$$+\int_\Gamma\int_\Gamma \frac{\partial u_k^-(x)}{\partial s_i(x)}B^5(x,y)\frac{\partial u_j^-(y)}{\partial s_i(y)}d\Gamma_y d\Gamma_y$$

$$+\int_\Gamma\int_\Gamma \frac{\partial u_k^-(x)}{\partial x_k} B_{kj}^6(x,y)\frac{\partial u_j^-(y)}{\partial y_j} d\Gamma_x d\Gamma_y$$

$$+\int_\Gamma\int_\Gamma t_{f,k}(x)\{C_{kj}^1(x,y)\,u_j^-(y)+C^2(x,y)\,M_{ji}(\partial y,n(y))u_i^-(y)\}\,d\Gamma_y d\Gamma_x$$

$$+\int_\Gamma\int_\Gamma t_{f,k}(x)\,C_{kj}^3(x,y)\frac{\partial u_j^-(y)}{\partial y_j}+\frac{\partial t_{f,k}(x)}{\partial x_k}C_{kj}^4(x,y)\,u_j^-(y)\,d\Gamma_y d\Gamma_x$$

$$+\int_\Gamma\int_\Gamma t_{f,k}(x)\,D_{kj}^1(x,y)\,t_{f,j}(y)+\frac{\partial t_{f,k}(x)}{\partial x_k}D^2(x,y)\frac{\partial t_{f,j}(y)}{\partial y_j}\,d\Gamma_x d\Gamma_y$$

$$-2\int_\Gamma t_{f,k}(x)u_j^-(x)\,d\Gamma\} \qquad (35.a)$$

where B^i $(i=\{1,..,6\})$, D^j $(j=\{1,2\})$, and C^k $(k=\{1,..,4\})$ represents various regularized kernels that are given in [20], and $L'(\vec{u},\vec{t}_f)$ is defined by:

$$L'(\vec{u},\vec{t}_f)=\int_\Gamma\int_\Gamma u_k^-(x)\{S_{kj}^1(x,y)+S_{kj}^2(x,y)+S_{kj}^3(x,y)+S_{kj}^4(x,y)\}\,u_j^{inc}(y)d\Gamma_y d\Gamma_y$$

$$+\int_\Gamma\int_\Gamma \frac{\partial u_k^-(x)}{\partial s_i(x)} S^5(x,y)\frac{\partial u_k^{inc}(y)}{\partial s_i(y)}\,d\Gamma_y d\Gamma_y$$

$$+\int_\Gamma\int_\Gamma \frac{\partial u_k^-(x)}{\partial x_k} S^6(x,y)\frac{\partial u_j^{inc}(y)}{\partial y_j} d\Gamma_x d\Gamma_y$$

$$+\int_\Gamma\int_\Gamma t_{f,k}(x)\{S_{kj}^7(x,y)\,u_j^{inc}(y)+S^8(x,y)\,M_{ji}(\partial y,n(y))u_i^{inc}(y)\}\,d\Gamma_y d\Gamma_x$$

$$+\int_\Gamma\int_\Gamma t_{f,k}(x)\,S_{kj}^9(x,y)\frac{\partial u_j^{inc}(y)}{\partial y_j}+\frac{\partial t_{f,k}(x)}{\partial x_k}S_{kj}^{10}(x,y)\,u_j^{inc}(y)\,d\Gamma_y d\Gamma_x$$

$$+\int_\Gamma u_k^-(x)t_{f,j}^{inc}(x)\,d\Gamma_x-\frac{1}{2}\int_\Gamma t_{f,k}(x)u_j^{inc}(x)\,d\Gamma_x \qquad (35.b)$$

where

$$M_{kj}(\partial x,n(x))=n_j(x)\frac{\partial}{\partial x_k}-n_k(x)\frac{\partial}{\partial x_j},\qquad \frac{\partial}{\partial s_k(x)}=\sum_{i,j=1}^{3}\epsilon_{ijk}n_i(x)\frac{\partial}{\partial x_j}$$

where ϵ_{ijk} is the *Levi – Civita* symbol and S^i $(i=\{1,2,..,10\})$ are given in [20].

5. FINITE ELEMENT DISCRETISATION

The structural bilinear form $a(\vec{u},\vec{v})$ yields a stiffness matrix $[K]$ and a mass matrix $[M]$ that are obtained by classical FE process [15]. However, the discretisation of the variational integral equation is not familiar and require special attention. These boundary terms yield autoinfluence and coupling matrices associated with the boundary integrals.

5.1 FINITE ELEMENT DISCRETISATION

Let Γ be a surface in R^3 endowed with a local frame with coordinates (η_1, η_2, η_3) such that (η_1, η_2) define the local tangent plane and η_3 is in the outward normal direction $\vec{e} = \vec{n}$. The surface is assumed to be parametrized by:

$$x_i = x_i(\eta_1, \eta_2) \qquad i = \{1,2,3\} \tag{36}$$

The regularized variational formulation, involves first order derivatives of the field variables $\vec{u}$ and $\vec{t}$, so that use of finite elements of class C^1 for their approximation is required. The structural domain and the fluid boundary are divided into isoparametric element and tractions and displacements are approximated by [15,17,18]:

$$\{u(x(\eta_1, \eta_2))\} = [N(\eta_1, \eta_2)]\,\{u^n\} \tag{37.a}$$

$$\{t(x(\eta_1, \eta_2))\} = [N(\eta_1, \eta_2)]\,\{t^n\} \tag{37.b}$$

where $[N]$ is the matrix of interpolation functions and $\{u^n\}$ or $\{t^n\}$ are vectors of element nodal variables. The above discretisation process is applied to the regularized variational integral functional. The structural contribution is given by:

$$J_s(\vec{u}, \vec{v}) = \frac{1}{2}\int_{\Omega_s} [\sigma_{ij}(\vec{u})\varepsilon_{ij}(\vec{v}) - \rho_s\omega^2(\vec{u}.\vec{v})]\, d\omega_s \tag{38.a}$$

which after standard finite element discretisation becomes:

$$J_s(\vec{u}, \vec{v}) = \frac{1}{2} < u^n > ([K] - \omega^2[M])\{u^n\} \tag{38.b}$$

The fluid contribution is obtained by evaluating the boundary integrals involved in equations for $A'(\vec{u}, \vec{t}_f)$ and $L'(\vec{u}, \vec{t}_f)$, and comprise (omitting the details) the displacement and traction autoinfluence matrices B and D and the coupling matrix C:

$$J_f(\vec{u}, \vec{t}) = \frac{1}{2}\omega^2 < u^n > [B]\{u^n\} + \frac{1}{2} < t^n > [D]\{t^n\} + < t^n > [C]\,\{u^n\} \tag{38.c}$$

and the source term from field $\vec{u}^{inc}$ and $\vec{t}^{inc}$:

$$L_f(\vec{u}, \vec{v}) = < u^n > \{F_u\} - < t^n > \{F_t\} \tag{38.d}$$

Combining (38.b,c,d), yield equation (38.e) for the global discrete quadratic functional of the coupled fluid-structure dynamical system:

$$A'(\vec{u}, \vec{t}) = \frac{1}{2} < u^n > ([K] - \omega^2([M] - [B]))\{u^n\} + \frac{1}{2} < t^n > [D]\{t^n\} + < t^n > [C]\,\{u^n\}$$
$$- < u^n > \{F_u\} - < t^n > \{F_t\} \tag{38.e}$$

where $[K]$ and $[M]$ are the structural stiffness and mass matrices respectively, $[B]$ and $[D]$ are autoinfluence matrices for the displacement and traction distribution respectively, $[C]$ is the coupling matrix between $\vec{u}$ and $\vec{t}_f$ on Γ, and vectors $\vec{F}_u$ and $\vec{F}_t$ represent the source terms from the incident field $\vec{u}^{inc}$ and $\vec{t}^{inc}$ respectively.

Applying the stationarity conditions to the above discretized functional then leads to the following system matrix equation:

$$\begin{bmatrix} [K] - \omega^2([M] - [B(k_1,k_2)]) & [C(k_1,k_2)]^T \\ [C(k_1,k_2)] & [D(k_1,k_2)] \end{bmatrix} \begin{Bmatrix} u^n \\ t^n \end{Bmatrix} = \begin{Bmatrix} F_u \\ F_t \end{Bmatrix} \qquad (39)$$

In the above equation, matrices $[B(k_1,k_2)]$ and $[D(k_1,k_2)]$ are complex and symmetric; its solution yields the unknowns $\vec{u}$ and $\vec{t}$ on the surface. Once $\vec{u}$ and $\vec{t}$ are determined, it is then possible to compute the pressure from eqn (7).

5.2 COMPUTER IMPLEMENTATION

The computer implementation of the above formulation involves the coupling of a classical FEM code for the evaluation of the structure matrices and a BEM code implementing a symmetric *Galerkin formulation* [20] that exploit the properties of various integral operators in eqn.(30)-(31). The computer process involves the following steps:

- A computer aided design system for geometric modeling and F.E. mesh generation.
- A finite element code that computes and stores the structural matrices for various element configurations and types.
- A boundary element code implementing the symmetric *Galerkin formulation* for the fluid/elastodynamic equations to evaluate and store the fluid matrices.
- A coupling module to deal with various combinations of structural and fluid degrees of freedom.
- A generalized eigenvalue/eigenvector module (available in public library LINPACK/EISPACK, or IMSL).
- A complex solver for forced response.
- A post processor for result visualisation and plotting.

As a pre and post processor CAD/CAE system, we use the PATRAN software. A modular computer program has been written for the F.E. and BEM parts and include over 150 subroutines. When the system's number of d.o.f. is large, a substructuring process can be applied to eliminate the traction variables. One then obtain the following matrix system for the structural displacement:

$$[K] - \omega^2([M] + [Z])\{u\} = \{F_u\} + \rho_f^0\omega^2[D(k_1,k_2)]^{-1}[C(k_1,k_2)]^T\{F_t\} \qquad (40)$$

where $\omega^2 Z$, are the acoustic impedance of the surface Γ defined by:

$$\omega^2[Z] = \rho_f^0\{\,[B(k_1,k_2)] + [C(k_1,k_2)]\,[D(k_1,k_2)]^{-1}[C(k_1,k_2)]^T\,\} \qquad (41)$$

The pressure can then be computed from the surface displacements using eqn(7).

Preliminary results obtained by applying the above approach to aeroelastic problems and acoustic radiation and scatterring in viscous compressible fluid, are encouraging and will be presented in forthcoming paper [25]. They allow one to determine the limitation of the inviscid compressible fluid model usually employed in acoustic and fluid-structure interaction problems.

6.CONCLUSION

We have presented an alternative to the usual practice of coupling the structural weak form with an integral equation formulation for the fluid using the collocation method. The formulation involves symmetric energy functionals for both the fluid and the solid and exhibits desirable features such as symmetry and positive definiteness.

7. BIBLIOGRAPHY

[1] Ryhming I. "Dynamique des fluides", Presses Polytechniques Romandes, Lausane 1985.

[2] Landau L. "Mécanique des fluides"- Edition Mir, Moscou 1971.

[3] Kupradze V.D, et al "Three-dimensional problems of the mathematical theory of elasticity and thermoelasticty", North-Holland, 1979

[4] Parton V., Perline P. "Méthodes de la théorie mathématique de l'élasticité", Tome 1, pp.221-230, Editions Mir, Moscou 1984

[5] Fritz J. "Plane wave spherical means applied to partial differential equation".International Publ.Inc., New York, 1955

[6] Vladimirov S.G."Equations of mathematical physics", Dekker, New-York 1971

[7] Vladimirov S.G. "Distribution en physique mathématique", Edition Mir, Moscou 1979

[8] Lesueur C. "Rayonnement acoustique des structures"- Edition Eyrolles, Paris 1988

[9] Ben Mariem J., Hamdi M.A. "A new boundary finite element method for fluid-solid interaction problems", Int.J. For Num.Met.In Engng, 1984, Vol.24, pp 1251-1267

[10] Hamdi A.H. "Formulation variationnelle par équations intégrales pour le calcul de champs acoustiques linéaires proches et lointains", Thèse de doctorat d'État, Université de Compiègne, France, 1981

[11] Sayhi M.N, Ousset Y., Verchery G. "Solution of radiation problems by collocation of integral formulation in terms af single and double potentiels", J.of.Sound and Vibration, 74, 2, pp 188:204, 1981

[12] Hsiao G.C., Kleinman R.E., Schuetz L.S. "On variational formulation of boundary value problems for fluid-solid interaction ", Elastic wave propagation M.F.McCarthy, M.A. Hayes, (editors) Elsevier Science Publisher B.V (North-Holand), pp.545:550, 1989

[13] Hsiao G.C. "The coupling of B.E.M and F.E.M", The mathematical and computational aspect, pp 431:445, 1988

[14] Jean P. " Une méthode variationnelle par équation inégrales pour la résolution numérique de problèmes interieurs et exterieurs de couplage elasto-acoustique", Thèse de Doctorat Université

de Compiègne, France, 1985
[15] Dhatt D., Touzot G. "Une présentation de la méthode des éléments finis", les Presses de l'Université Laval,Maloine S.A.éditeur, 1981
[16] Dohner J.L, Shoureshi R., Bernhard R."Transient analysis of three-dimensional wave propagation using the boundary element method"-International Journal for Numerical Methods in Engineering, Vol 24, pp621-634 (1987).
[17] Brebbia C.A., Telles J.C.F., Wrobel L.C. "Boundary element techniques"-Springer-verlag New York 1984.
[18] Gakwaya A. "Développement et application de la méthode des éléments finis de frontière à la mécanique du solide", Thèse de Ph.d, Université Laval 1983
[19] Morse P.M., Feshbach H. "Methods of theoretical physics", Part I and II, McGraw-Hill Book Compagny 1953.
[20] Erchiqui F., Gakwaya A. "Modélisation et simulation des interaction fluide-structure dans les systèmes aérospatiaux", Rapport de recherche CAO no 05/05/90 Dept de génie mécanique, Université Laval, 1990
[21] Schwartz L."Méthodes mathématiques pour les sciences physiques", Herman, Paris 1973
[22] Wendland W.L. "On asymptotic error estimates for combined BEM and FEM" in finite element and Boundary element techniques from mathematical and engineering point of view, CISM Course no.301 Springer verlag 1988
[23] Junger M.C., Feit D. "Sound, structures and their interaction " MIT Press, 1986
[24] Kobayashi S."Recent Progress in BEM for Elastodynamics" in BEM X (C.A. Brebbia (ed)) vol.4 pp333-347, 1988.
[25] Erchiqui F, Gakwaya A. "A 3D variational BEM formulation for fluid-structure interaction in viscous compressible flow: computational aspects and results" in preparation, 1991

Calculation of Two-Dimensional Potential Cascade Flow Using an Indirect Boundary Element Method

B. Massé, L. Marcouiller

IREQ Institut de recherche d'Hydro-Québec,
1800 montée Ste-Julie, Varennes, Québec,
J3X 1S1, Canada

INTRODUCTION

Turbomachinery blade design relies on computer codes to model the flow on the blade–to–blade surface. Finite–element codes are often employed but require an experienced user to manipulate the input data, boundary conditions and control parameters. Finite–element solutions can be quite accurate but the computer time required for a final solution is often prohibitive. A simple, stable and robust computer code has therefore been developed to provide a fast solution technique for blade–to–blade flows in order to give more versatility to the analyst. This code offers a means of screening several configurations with different input conditions without running complex codes for each solution.

The numerical method for calculating potential flow through a linear cascade of arbitrary blade shapes uses the indirect formulation of the boundary element method [1]. An integral equation for solving the Neumann problem is obtained by means of the Hess and Smith [6] approach. Circulation about the cascade and the Kutta–Joukowski condition are considered using a vortex distribution with a prescribed variation over the blade contour. The blade geometry is approximated using second–order elements. The algebraic system of equations is derived from a Fredholm equation of the second kind using first–order singularity distribution on the blade elements as the unknown. Fluid velocities on the boundaries are obtained by solving this linear system of equations. The flow properties in the fluid domain are calculated using the solution obtained on the boundaries.

This paper describes a numerical method for inviscid incompressible flow through a plane cascade. The fluid properties on the boundary and in the domain are obtained for an arbitrary blade profile and various cascade inlet and outlet conditions. Solutions were generated for different cascade configurations and good agreement was obtained when they were compared to other methods and experimental data.

CASCADE FLOW

Cascade flow can be very complex, especially when secondary and unsteady flows are present. In steady flow, a preliminary analysis can be done very rapidly using the potential flow assumption. In this case, the Laplace equation applies:

$$\nabla^2\phi = 0 \tag{1}$$

Several numerical techniques such as finite elements can be used for non viscous cascade flow but they require the fluid domain to be meshed. Boundary element techniques are attractive in this case and have proven their efficiency for potential problems. The elements are required on the boundaries only, with no need to map the fluid domain, so that the computer time needed to solve the system of equations is far shorter. Direct and indirect boundary element methods are available but, because of the periodicity of the cascade flow, an indirect method was selected for this study.

With boundary conditions of the Neumann type and singularity distributions, a Fredholm equation of the second kind is obtained:

$$\left(\frac{\delta\phi}{\delta n}\right)_p = 2\pi\sigma(p) + \oint_C \frac{\delta}{\delta n}\left(\frac{1}{r(p,q)}\right)\sigma(q)dS = \vec{U}\cdot\vec{n}_p \tag{2}$$

In this equation, $\sigma(q)$ is the intensity of the sources and $r(p,q)$ the distance between the source located at point q on the boundary to a point p on the same boundary where the influence of that source is calculated. The main stream velocity is $\vec{U}$ and C represents the contour of the cascade profiles.

In order to solve equation (2) for cascade flow, circulation about the cascade has to be considered. As seen in figure 1, the infinite cascade row is an infinite number of lifting blades acting at a great distance as a row of equally spaced vortices. The main stream flow is deflected by the cascade circulation which produces an upwash flow velocity upstream, V_{y+}, and a downwash velocity downstream, V_{y-}. These velocities are related to the cascade circulation [3] by equation (3):

$$V_{y+} = V_{y-} = -\frac{\Gamma}{2t} \tag{3}$$

The lift related to the circulation is defined as normal to the mean flow velocity through the cascade, $\vec{U}$, and equation (2) is solved using the average velocity through the blade row $\vec{U}$. If the inlet and outlet velocities $\vec{V}_i$ and

$\vec{V}_o$ and angles a_i and a_o are defined, the upwash and downwash velocities can be expressed as:

$$\begin{aligned} V_{y+} &= V_i \sin a_i - V \sin a \\ V_{y-} &= V_o \sin a_o - V \sin a \end{aligned} \tag{4}$$

The average angle of the flow through the cascade can be expressed as:

$$a = \tan^{-1}\left(\frac{\tan a_i + \tan a_o}{2}\right) \tag{5}$$

and the mean velocity as:

$$U = \frac{V_i \sin a_i + V_o \sin a_o}{2 \sin a} \tag{6}$$

Inlet and outlet velocities and angles are required to calculate the average velocity and angle and to solve the cascade flow problem. It is possible to

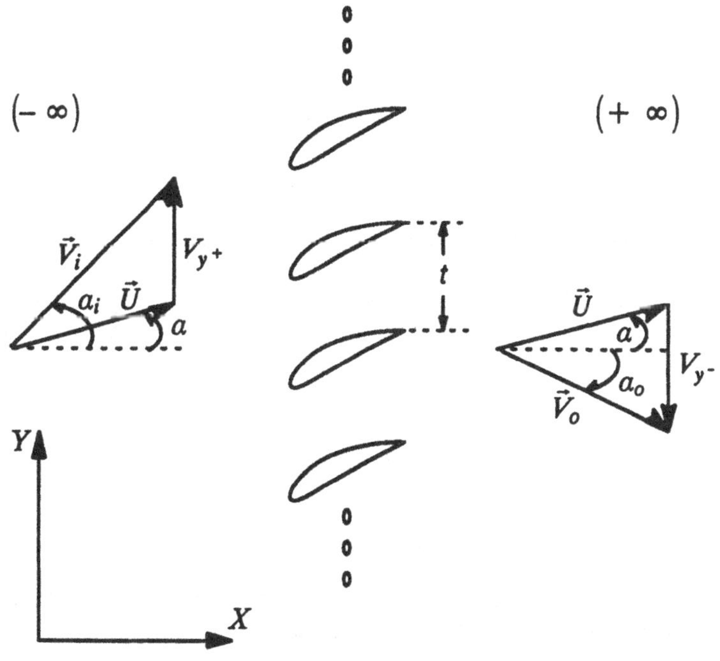

Figure 1: Velocities diagram for cascade flow.

define the required cascade flow conditions in terms of either the lift coefficient and the inlet angle or the flow deflection through the cascade but in both cases, the information must be sufficient to allow the average velocity and angle to be computed.

BASE ELEMENT DESCRIPTION

The geometry of the element used in the present model has been described by Hess [7] and consists of a second-order approximation on which a first-order source distribution is used. Figure 2 is a plot of the element with the

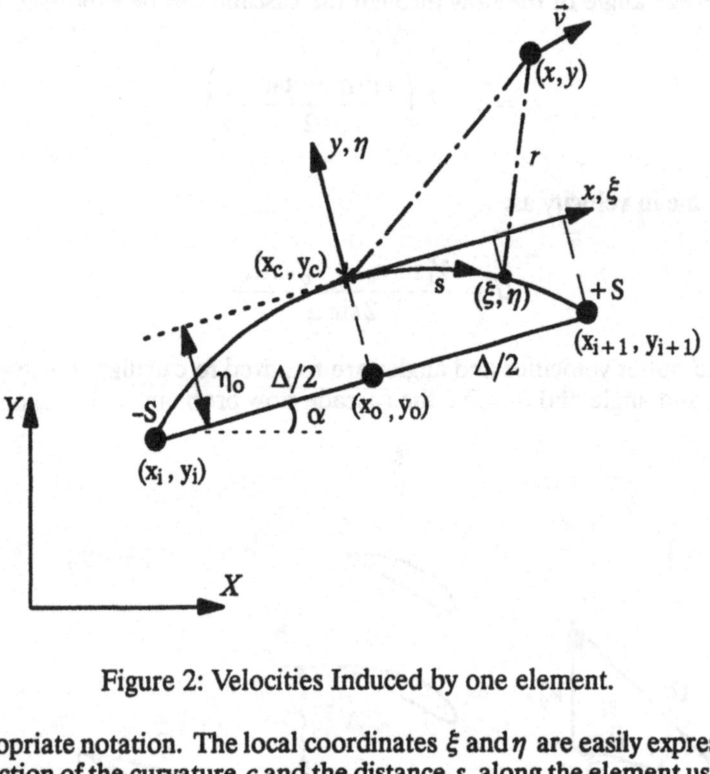

Figure 2: Velocities Induced by one element.

appropriate notation. The local coordinates ξ and η are easily expressed as a function of the curvature c and the distance s along the element using the expressions:

$$\xi = s - \frac{\frac{1}{6}c^2s^3}{1+\frac{1}{2}s^2c^2}$$
$$\eta = \frac{c}{2}\xi^2 \tag{7}$$

The central node in the global coordinate system is calculated as a function of the chord central point and element angle as:

$$x_c = x_o - \eta_o \sin a$$
$$y_c = y_o - \eta_o \cos a \tag{8}$$

Two curvatures are calculated using the two pairs of three points defined by the tip coordinates of the element considered and the two adjacent elements. The curvature of the element is defined as the arithmetic mean of the two curvatures if they have the same sign. Otherwise, the curvature is set to zero, the element being considered flat.

Integration over the element of the source distribution goes from $-S$ to $+S$, the half-length S being defined as:

$$S = \frac{1}{c}\sin^{-1}\left(\frac{c\Delta}{2}\right) \tag{9}$$

The source distribution is assumed linear and the intensity at the central node is σ^o. The source intensity at a location s on the element is:

$$\sigma(s) = \sigma^o + \sigma' s \tag{10}$$

where σ' is the first derivative. Using central finite differences, except for the first and last elements, where forward and backward finite differences are used, the distribution is easily expressed in terms of coefficients linked to the geometry in the form:

$$\begin{aligned}
\sigma_i(s) &= \sigma_i^o\ (1 + E_i\, s) + D_i\, s\ \sigma_{i-1}^o + F_i\, s\ \sigma_{i+1}^o \quad \text{for } i \neq 1 \text{ and } i \neq N \\
\sigma_i(s) &= \sigma_i^o\ (1 + E_i\, s) + D_i\, s\ \sigma_{i+1}^o + F_i\, s\ \sigma_{i+2}^o \quad \text{for } i = 1 \\
\sigma_i(s) &= \sigma_i^o\ (1 + E_i\, s) + D_i\, s\ \sigma_{i-1}^o + F_i\, s\ \sigma_{i-2}^o \quad \text{for } i = N
\end{aligned} \tag{11}$$

According to the potential theory, integration over the element of the source distribution gives the induced velocities at a field point (x,y) as:

$$\vec{v} = 2\int_{-\frac{\Delta}{2}}^{\frac{\Delta}{2}} \left[\left(\frac{x-\xi}{r^2}\right)\vec{i} + \left(\frac{y-\eta}{r^2}\right)\vec{j}\right] \sigma(s)\, \frac{ds}{d\xi}\, d\xi \tag{12}$$

where ξ and η represent a point on the element where the integration takes place.

Hess [7] has shown that the preceding integral has an analytical solution which avoids numerical integration on the elements. This gives the method both accuracy and performance.

EXTENSION TO THE CASCADE ELEMENT

In the case of cascade flow, a single blade can be considered as the base profile; the blades above and below it are equally spaced and of the same number, this number being infinity when a infinite cascade is used. In the same manner, an element on the base blade has corresponding elements on the other blades of the cascade in similar relative positions. Figure 3 shows three corresponding elements: element **0** is the base element with elements **1** and **2** in the same relative position on each side at a distance *t* from the base.

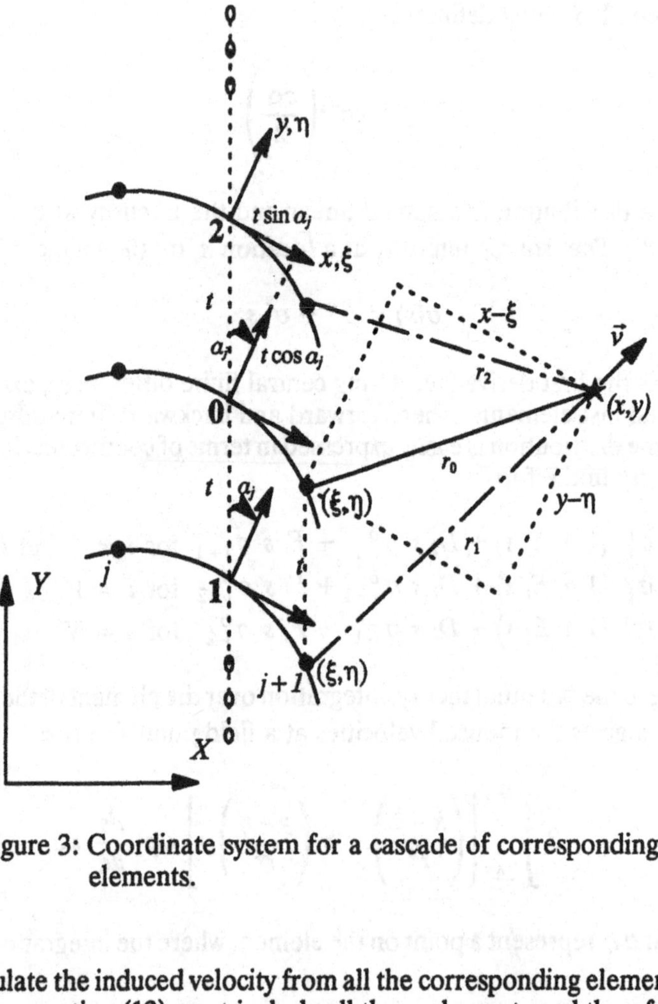

Figure 3: Coordinate system for a cascade of corresponding elements.

To calculate the induced velocity from all the corresponding elements of the cascade, equation (12) must include all these elements and therefore takes the form:

$$\vec{v} = 2\int_{-\frac{\Delta}{2}}^{\frac{\Delta}{2}} \left\{ \begin{aligned} &\left[\left(\frac{x-\xi}{r_0^2}\right) + \sum_{k=1}^{\infty}\left[\left(\frac{x-\xi_1}{r_1^2}\right) + \left(\frac{x-\xi_2}{r_2^2}\right)\right]\right]\vec{i} \; + \\ &\left[\left(\frac{y-\eta}{r_0^2}\right) + \sum_{k=1}^{\infty}\left[\left(\frac{y-\eta_1}{r_1^2}\right) + \left(\frac{y-\eta_2}{r_2^2}\right)\right]\right]\vec{j} \end{aligned} \right\} \sigma(s)\,\frac{ds}{d\xi}\,d\xi \tag{13}$$

where:

$$
\begin{aligned}
r_0^2 &= (x-\xi)^2 + (y-\eta)^2 \\
r_1^2 &= (x-\xi_1)^2 + (y-\eta_1)^2 \\
r_2^2 &= (x-\xi_2)^2 + (y-\eta_2)^2
\end{aligned}
\tag{14}
$$

$$
\begin{aligned}
\xi_1 &= \eta - k\ t \sin a_i \\
\eta_1 &= \eta - k\ t \cos a_i
\end{aligned}
\tag{15}
$$

$$
\begin{aligned}
\xi_2 &= \eta + k\ t \sin a_i \\
\eta_2 &= \eta + k\ t \cos a_i
\end{aligned}
\tag{16}
$$

The terms related to the base element in equation (13) are evaluated analytically. Integrals with summations can be expressed for an infinite cascade in terms of hyperbolic functions [3]. For the present model, numerical integration was preferred in order to be able to calculate a finite cascade of several albeit not an infinite number of blades. (Numerical integration is fast on modern computers and 20 blades on each side is a good approximation of the infinite cascade.)

NUMERICAL SOLUTION

To solve equation (2) and obtain the solution for the source intensity σ^o for each central node, the body contour is divided into N boundary elements of parabolic shape. Equation (2) is then written as:

$$\sum_{j=1}^{N} A_{ij}\sigma_j^o = \vec{U}\cdot\vec{n}_i \tag{17}$$

where A_{ij} is the normal component of the induced velocities at the central nodes i produced by a source distribution on element j and all the corresponding elements of the cascade from equation (13):

$$A_{ij} = \vec{v}_{ij}\cdot\vec{n}_i \tag{18}$$

The induced velocity $\vec{v}_{ij}$ must take into account the derivatives from the adjacent elements (see equations (11) and (12)). Equation (17) is a system of N linear equations to be solved for N σ_j^o. Without blade circulation, the tangential velocities are expressed as:

$$V_{t_i} = \sum_{j=1}^{N} B_{ij}\sigma_j^o + \vec{U}\cdot\vec{t}_i \tag{19}$$

where:

$$B_{ij} = \vec{v}_{ij} \cdot \vec{t}_i \tag{20}$$

To generate a circulation around the blade profile, a vortex distribution is superimposed on the source distribution and the Kutta–Joukowski criterion is applied at the blade trailing edge (first and last central nodes). A sinusoidal distribution on the blade contour is used to reduce the numerical error produced by vortex proximity near a sharp trailing edge. This distribution is introduced using a weighting factor w_k associated with the vortex intensity at each element k. A unique vortex distribution is used around the contour and the velocities on both sides of the blade at the trailing edge are assumed to be of the same magnitude. The vortex reference intensity being γ, the system of equations is now expressed as:

$$\sum_{j=1}^{N} A_{ij}\sigma_j^o + \sum_{k=1}^{N} -B_{ik}w_k \, \gamma = \vec{U} \cdot \vec{n}_i \tag{21}$$

The Kutta–Joukowski condition completes the equation system to be solved:

$$\sum_{j=1}^{N} (B_{1j} + B_{Nj})\sigma_j^o + \sum_{k=1}^{N} (A_{1k} + A_{Nk})w_k \, \gamma = -\left(\vec{U} \cdot \vec{n}_1 + \vec{U} \cdot \vec{n}_N\right) \tag{22}$$

The tangential velocities are then expressed as:

$$V_{t_i} = \sum_{j=1}^{N} B_{ij}\sigma_j^o + \sum_{k=1}^{N} A_{ik}w_k \, \gamma + \vec{U} \cdot \vec{t}_i \tag{23}$$

Velocities in the fluid domain are calculated similarly to tangential velocities, the preceding equations being valid for any point in the flow field.

RESULTS

A computer program CANIA has been developed to solve the inviscid incompressible potential flow for a cascade based on the above method. Examples correlating the numerical results with known theories and with measurements are given below.

Several calculations were done using a single body to test the accuracy of the method. Calculations on circular cylinders show very good agreement with as few as 24 elements. A solution on a cascade of ellipses with 20% thickness using 48 elements on the base profile and 40 adjacent ellipses to simulate an infinite cascade is compared with a solution from Martensen [2] in figure 4. For this case, the stagger angle of the ellipses is 30 degrees clockwise and the flow inlet and outlet velocity angles are 30 degrees counter-

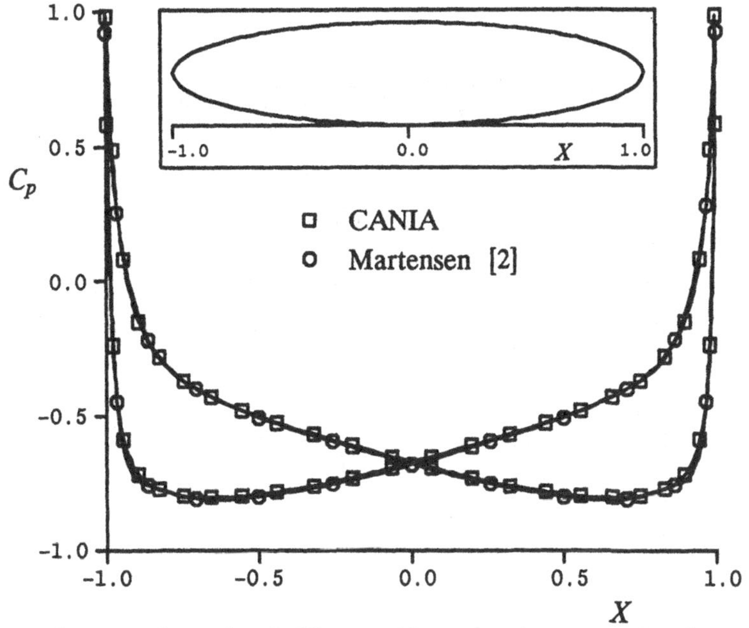

Figure 4: Cascade of ellipses with a zero degree angle of attack.

clockwise, so that the effective angle of attack is zero. The agreement between the two methods is excellent.

Figure 5 shows a similar calculation on an infinite cascade of ellipses but with a flow inlet velocity angle of 45 degrees and a flow outlet velocity of 30 degrees. Agreement with Martensen's method is still good.

Figures 6 and 7 compare results obtained by Martensen [2] and with CANIA and experimental measurements. Agreement is generally good, although the Martensen model shows some disagreement near the trailing edge, since sharp trailing edges are excluded from this model. The Kutta–Joukowski condition used in the present case gives good results.

Another cascade, from Gostelow [8] using a thinner blade, was calculated as an infinite cascade with a stagger of 37.5 degrees and an inlet flow angle of 53.5 degrees. Gostelow's outlet flow angle is 30.02 degrees. Figure 8 shows good agreement between the two methods. No experimental data are available for this case.

CONCLUSION

A numerical method for calculating potential flow through a linear cascade of arbitrary shapes using an indirect boundary element method has been derived. The approach combines high-order elements, linear singularity distribution, cascade circulation, Kutta–Joukowski criterion and calculations on the boundary as well as in the flow field. The inlet and outlet condi-

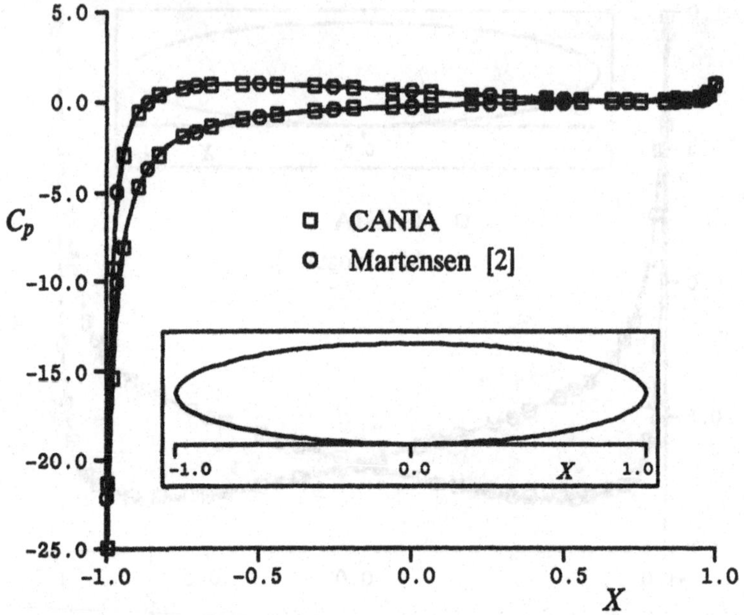

Figure 5: Cascade of ellipses subjected to a 45 degree inlet velocity.

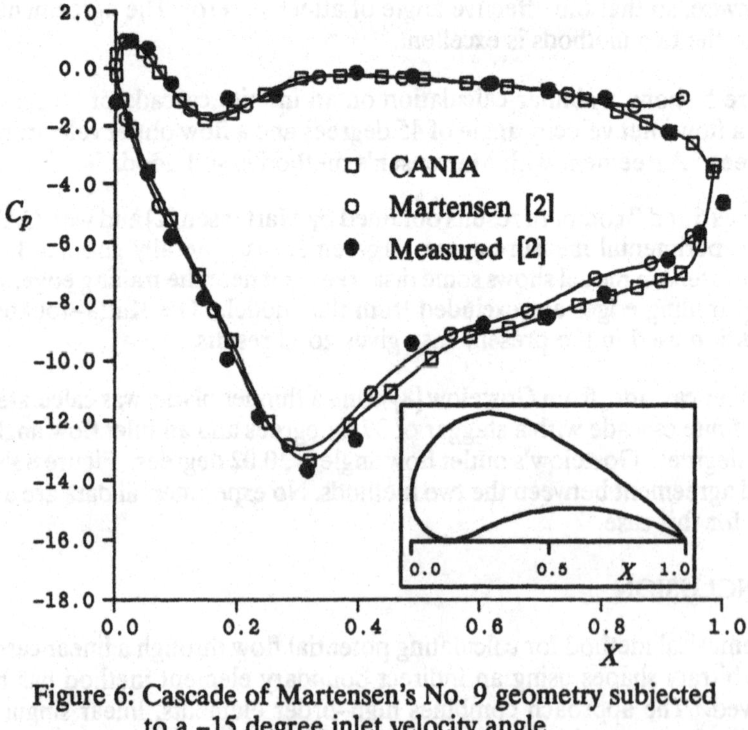

Figure 6: Cascade of Martensen's No. 9 geometry subjected to a –15 degree inlet velocity angle.

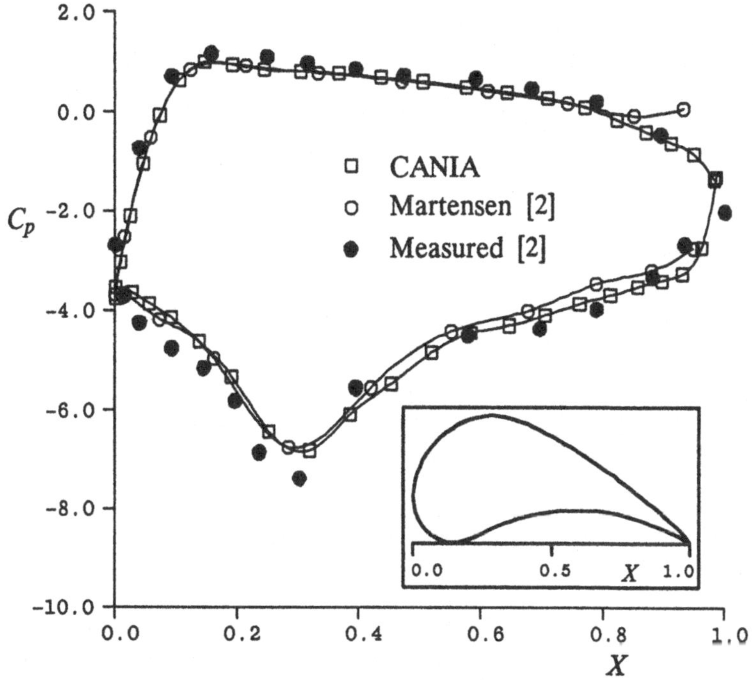

Figure 7: Cascade of Martensen's No. 9 geometry subjected to a 45 degree inlet velocity angle.

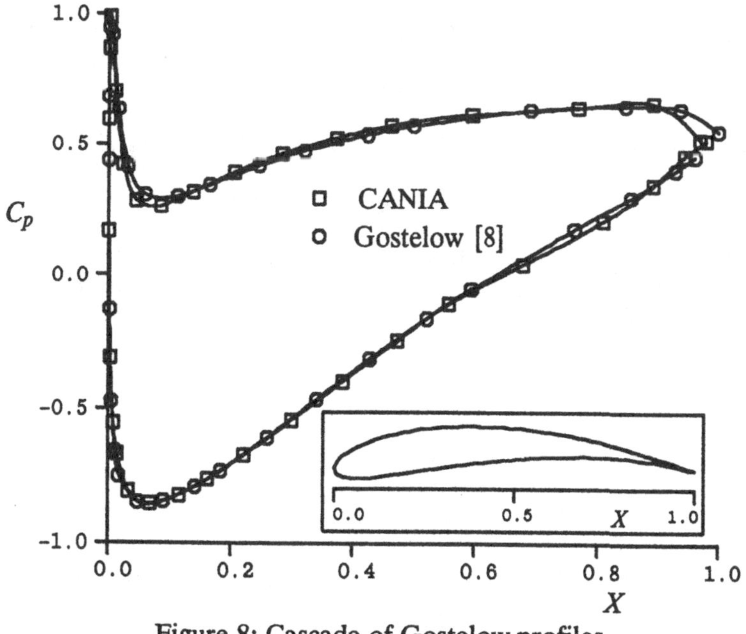

Figure 8: Cascade of Gostelow profiles.

tions must be specified in order to solve for cascade flow. Numerical simulations for infinite cascades of ellipses show good agreement with Martensen's method at both zero and non-zero angles of attack. Comparisons with measured values of pressure coefficients on the body contour of a cascade blade show the good accuracy of CANIA. The method agrees favorably with existing flow simulation methods.

REFERENCES

1. Brebbia, C.A., Telles, J.C.F. and Wrobel, L.C., Boundary Element Techniques, Springer–Verlag, Berlin, 1984.

2. Martensen, E., The Calculation of Pressure Distribution on a Cascade of Thick Airfoils by Means of Fredholm Integral Equations of the Second Kind, NASA TT F–702, 1971.

3. Giesing, J.P., Extension of the Douglas Neumann Program to Problems of Lifting Infinite Cascades, LB–31653, Douglas Aircraft Co. Inc., Long Beach, Calif., AD–605207, 1964.

4. Hess, J.L., and Stockman, N.O., An Efficient User–Oriented Method for Calculating Compressible Flow about Three–Dimensional Inlets, AIAA Paper No. 79–0081, Jan. 1979.

5. McFarland, E.R., Solution of Plane Cascade Flow Using Improved Surface Singularity Methods, ASME Journal of Engineering for Power, Vol. 104, No. 3, July 1982, pp. 668–674.

6. Hess, J.L., and Smith, A.M.O., Calculation of Potential Flow about Arbitrary Bodies, Progress in Aeronautical Sciences, Vol. 8, Kuchemann, K., ed. Pergamon Press, Oxford, 1967, pp. 1–138.

7. Hess, J.L., Higher Order Numerical Solution of the Integral Equation for the Two–Dimensional Neumann Problem, Computer Methods in Applied Mechanics and Engineering, Vol. 2, Feb. 1973.

8. Gostelow, J.P., Potential Flow through Cascades – A Comparison between Exact and Approximate Solutions, Aeronautical Research Council, London, C.P. No. 807, 1965.

A Combined Two- and Three-Dimensional Boundary Element Method for the Flow over a Train in a Cross-Wind at Large Angles of Attack

T.W. Chiu

School of Engineering, University of Exeter, Exeter EX4 4QF, U.K.

Abstract

An indirect formulation of the Boundary Element Method is developed to predict the aerodynamic loads on an idealized railway train model in a cross-wind at large yaw angles up to 90°. This is a combined two- and three-dimensional procedure in which a 2-D calculation simplifies dramatically the computation required for the 3-D calculation. The results compare positively with experiments.

1 Introduction

Modern developments in Material Sciences allow the use of lighter and stronger materials for train constructions. Lighter trains require less power to accelerate, but the reduction in weight could largely increase the likelihood of trains being blown over in a strong cross-wind.

Experimental studies and numerical modelling of the flow over a train in a cross-wind have, until recently, been focused on the cases of small angles of attack (i.e. yaw angles, α, see fig. 1) up to 40°, in which use was often made of the slender body analogy. The present paper reports a numerical modeling of the flow at large angles of attack above 60° using the Panel Method (an indirect formulation of the BEM).

Experiments on an idealized train model in a cross-wind at large yaw angles (60°-90°) suggested that the pressure distribution was essentially two-dimensional at locations away from the nose. This implied that the pressure distribution around most of the model at large yaw angles could be predicted by simply scaling the prediction for 90° yaw.

A separated flow model based on a two-dimensional second order vortex panel method using complex variables has thus been developed for 90° yaw. The computed pressure distribution around the train model and other two-dimensional bluff bodies agreed very well with experiments.

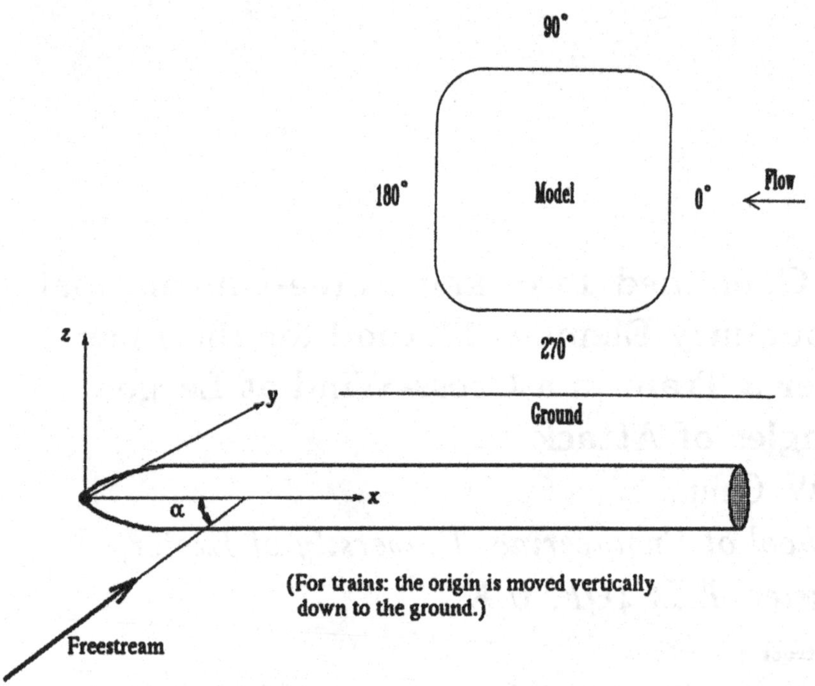

Fig 1 Definition of Yaw angle, α, and Angular Location round the cross-section.

The method was then extended to three-dimensions to calculate the pressure distribution close to the nose of the train. The three-dimensional scheme was basically a first order source panel method in which the wake was modeled by vortex panels, whose lengths had been determined in the two-dimensional calculation. The computed pressure distribution and overturning moment coefficients compared satisfactorily with experiments.

2 Two-Dimensional Model

Since the pressure distribution around the train model away from the nose at large yaw angles are essentially two-dimensional, a 2-D second order vortex panel method formulated using complex variables has been developed to predict the local force and moment coefficients at a yaw angle of 90°. The results could then be scaled by a factor of $\sin^2\alpha$ to give the corresponding coefficients at a yaw angle of α (Chiu[1]).

In the 2-D method (Chiu[2]), the surface of the cross-section is represented by a number of straight line panels on which is assumed a linearly varying distribution of vorticity (fig. 2). Two stationary vortex panels of finite lengths orientated in the freestream direction were introduced to model the wake. Everything has its image counterpart reflected about the 'ground'. The lengths of the wake panels were determined iteratively in the numerical scheme by the prescription of the base pressure coefficient, and their strengths were determined by a Kutta Condition (for details, see Chiu[2]). The separation positions are also determined in the scheme with the aid of a boundary layer code based on the lag-entrainment method (Green et al[3]).

This numerical scheme predicts satisfactorily the pressure distribution at a yaw angle of 90° in which the flow is essentially two-dimensional except very close to the nose (fig. 3).

The side force (C_S) and overturning moment (C_M) coefficients can then be obtained by integrating the pressure distribution. The C_S, C_M and C_p's are then scaled by a factor of $\sin^2 \alpha$ to give the corresponding coefficients at a yaw angle of α. That is,

$$C_M(\alpha) = C_M(90^\circ) \sin^2 \alpha, \text{ etc.} \quad [1]$$

Fig. 4 shows a comparison between the C_M's obtained from experiments at various yaw angles and the C_M's scaled from the computed result at 90° yaw. The prediction is very satisfactory.

Regarding the local C_S, C_M and C_p's close to the nose, it is believed that the 3-D effects of the nose span at most up to $4.5D$ from the nose, D being the model width. Within this region, the pressure distribution will only be predicted effectively by a 3-D scheme. In the 3-D scheme described in the next section, the separation position and the correct wake lengths determined in the 2-D scheme are made use of. This makes the 3-D calculation computationally simple and economical.

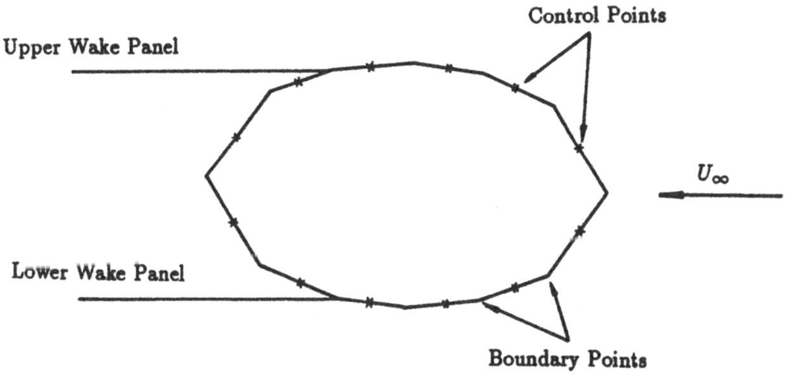

Fig 2 Schematic diagram of the 2-D model.

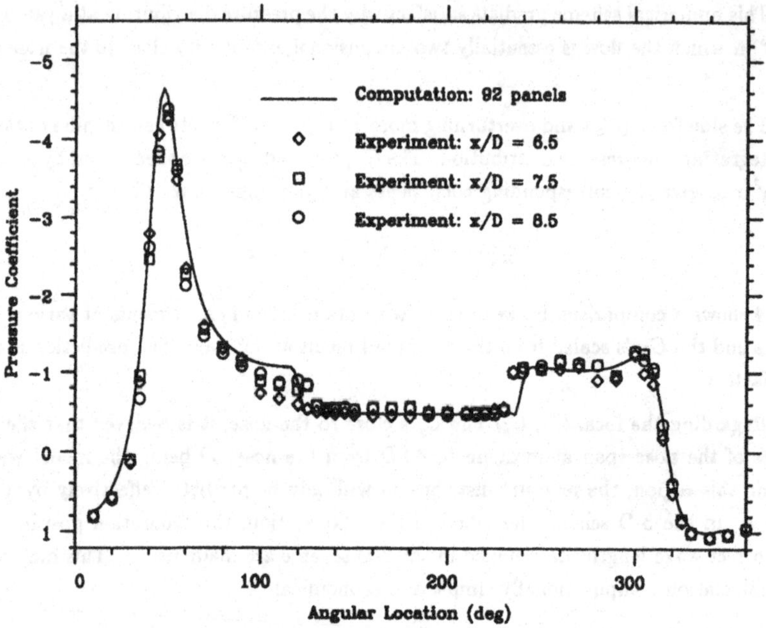

Fig 3 Computed result: $\alpha = 90°$, $Re = 3 \times 10^5$.

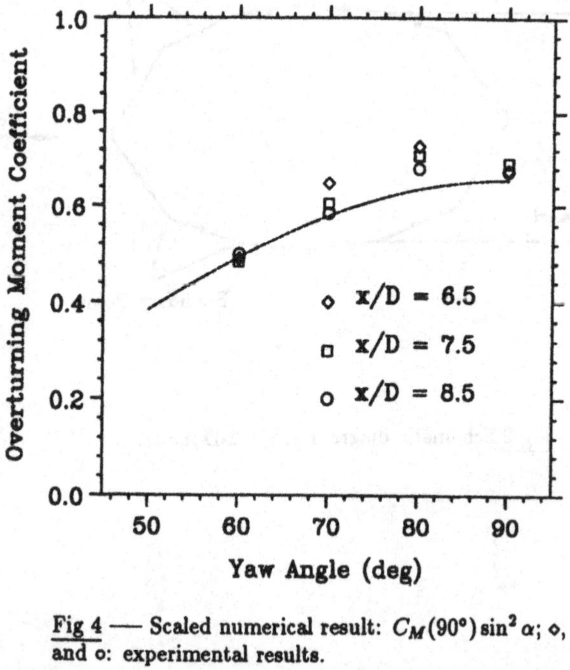

Fig 4 — Scaled numerical result: $C_M(90°)\sin^2\alpha$; ◇, □ and ○: experimental results.

3 The Three-Dimensional Model

3.1 Introduction

The 2-D model described in the previous section is now extended to the 3-D case in an attempt to predict the aerodynamic loads on the nose of the train. In the 2-D scheme, the use of vorticity distribution on the surface of the model allows the continuity of the surface vorticity to extend to the wake panels. In three dimensions, however, the direction as well as the magnitude of the vorticity distribution would appear as unknowns in the formulation. So the use of vortex panels in 3-D formulations is limited to cases where the direction of the vorticity distribution is known or prescribed. The present 3-D scheme is based on a first order source panel method, and vortex panels are added to model the wake.

The surface of the train model is represented by plane quadrilateral panels except at the two ends where triangular panels are used (fig. 5). Since, away from the nose, the aerodynamic loads experienced by the train model do not vary axially, the panels are made longer in this region and the tail is approximated by a *conical* shape. There are in total 540 panels.

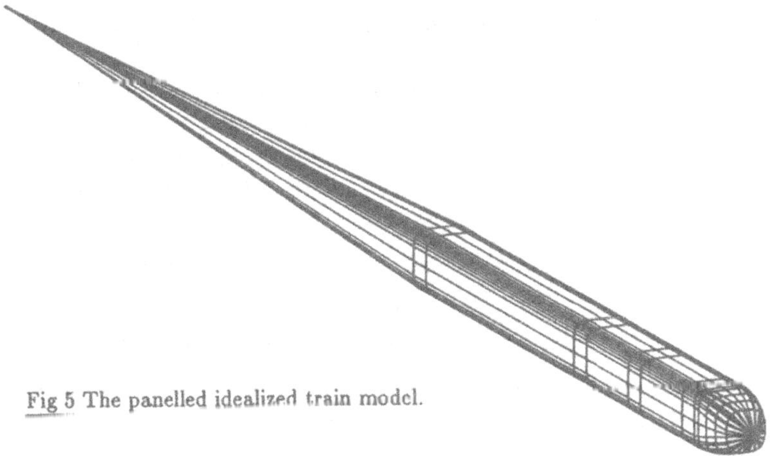

Fig 5 The panelled idealized train model.

3.2 Separated Shear Layers and Kutta Conditions

Similar to the 2-D method, the separated shear layer is approximated by stationary vortex panels with constant vorticities emerging from the separation lines parallel to the freestream direction. The correct lengths of the wake panels that can produce the same *time-averaged* blockage effect as the actual vortex shedding wake on the leeside as *felt* by the model has been determined using the 2-D calculation. These correct lengths are used in the 3-D calculation here and the wake panels are extended to the nose section without altering their lengths in the crossflow direction. Use is also made of the separation positions far from the nose determined by the 2-D calculation. Close to the nose, the observed separation positions are used.

The vortex panels terminated at the position corresponding to the positions of the groundside and roofside open separations. In the present problem, each separation line has 12 lengthwise segments according to the geometry of the panelled model.

The wake panels are oriented in the freestream direction as shown in fig. 6. The wake lengths are measured in the crossflow direction. The direction of the vorticity on a wake panel is taken as the same as the orientation of the segment of the separation line from which the wake panel emerges.

The vorticities of the wake panels are determined by a Kutta condition such that

$$\begin{aligned} | \vec{\gamma}_s | &= | U_o | \\ &= \| \mathbf{U}_t \times \mathbf{v}_s \| \end{aligned} \quad [2]$$

where $\mathbf{U}_t$ is the tangential velocity at the outer edge of the boundary layer at the separation line, and the unit vector $\mathbf{v}_s$ is the local direction of the separation line (see fig. 7). $\| \mathbf{U}_t \times \mathbf{v}_s \|$ thus represents the magnitude of the component of $\mathbf{U}_t$ which is perpendicular to the separation line. Hence,

$$\vec{\gamma}_s = \| \mathbf{U}_t \times \mathbf{v}_s \| \mathbf{v}_s \quad [3]$$

$\mathbf{U}_t$ is evaluated at the control point of the panel just upstream of the separation.

A line-vortex made up of 12 segments is put at the centre of the model, each segment corresponds to a segment of the separation lines. The circulation of each segment is determined by Kelvin's circulation constraint so that the total circulation of that segment of the centre line-vortex plus those of the wake panels emerging from the corresponding segment of the two separation lines is zero. Since the lift experienced by the model is mainly due to its circulation in the lengthwise direction, the Kelvin's constraint is only imposed on the lengthwise circulation component. Let

$\Gamma_i =$ the circulation of the i-th segment of the centre line-vortex,

$\vec{\gamma}_{ui} \cdot \mathbf{i} =$ the lengthwise (x-) component of the vorticity of the upper (roofside) wake panel emerging from the i-th segment of the separation line,

$\vec{\gamma}_{li} \cdot \mathbf{i} =$ the lengthwise (x-) component of the vorticity of the lower (groundside) wake panel emerging from the i-th segment of the seperation line,

$l_{uw}, l_{lw} =$ the length of the upper and lower wake panel measured in the crossflow direction.

Then,

$$\| \Gamma_i \| + (\vec{\gamma}_{ui} \cdot \mathbf{i}) l_{uw} + (\vec{\gamma}_{li} \cdot \mathbf{i}) l_{lw} = 0 \quad [4]$$

The base pressure is determined by taking into account the stagnation pressure drop across a shear layer (Chiu[2]). The correction to the pressure coefficient within the recirculation region is

$$\Delta C_p = (\gamma_i^2 / U_\infty^2) \quad [5]$$

where $| \gamma_i |$ is taken as $| \gamma_{ui} + \gamma_{li} | /2$. $| \gamma_{ui} |$ and $| \gamma_{li} |$ are found to be almost identical.

To produce the ground effects, everything has its image counterpart reflected below the ground, which is taken as the plane $y = 0$ in the present calculation.

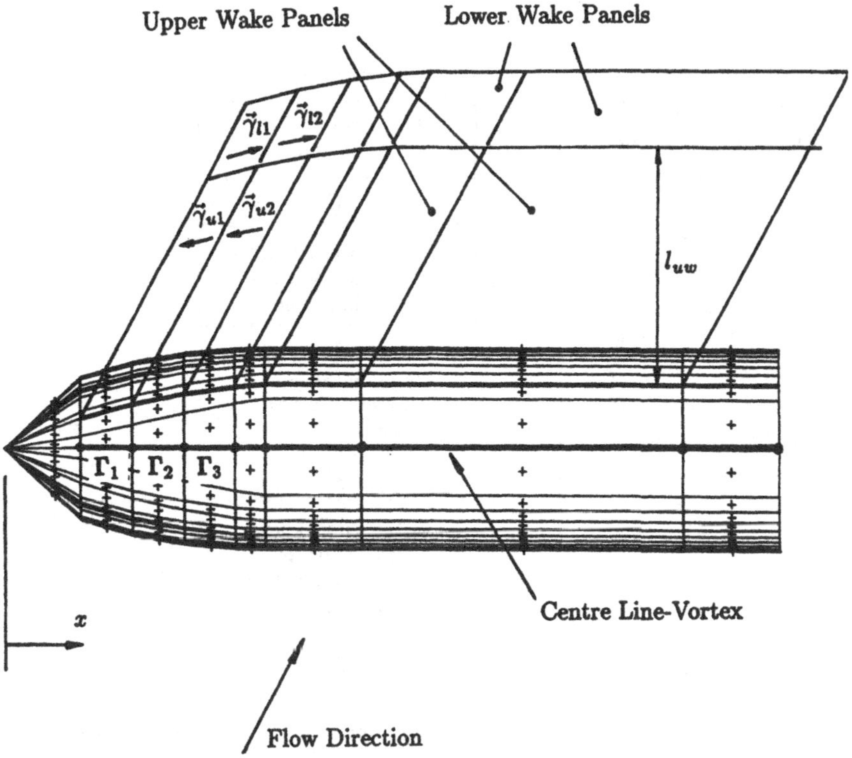

Fig 6 Modelling of the separated shear layer in the 3-D panel method: only the fore part is shown.

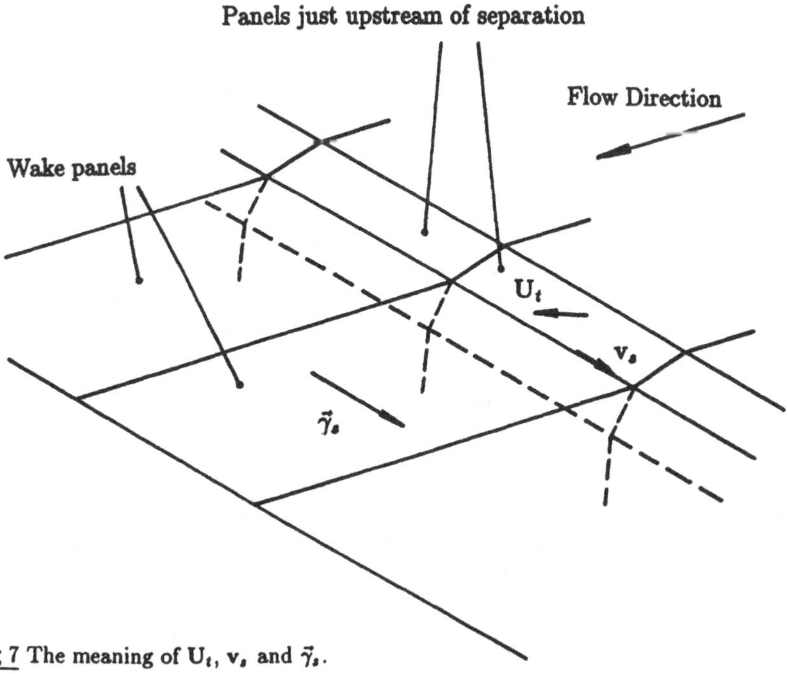

Fig 7 The meaning of U_t, v_s and $\vec{\gamma}_s$.

3.3 Iteration Procedure

It can be seen that, in the present model, the vorticity of the wake panels and the circulation of the centre line-vortex are determined from the velocities at the control points of the panels just upstream of the separation lines. However the correct velocity field around the model can be determined only if these vorticities and the circulation are known. So an iterative procedure as shown in fig. 8 is necessary. Normally four to six iterations would be sufficient.

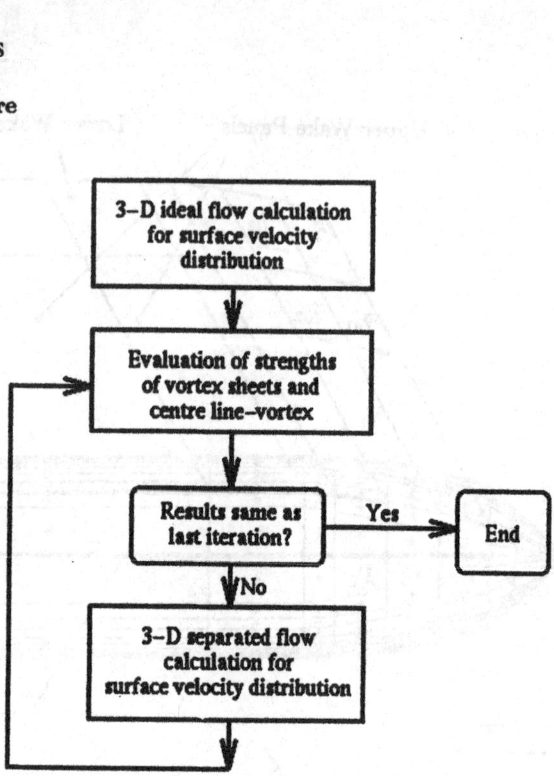

Fig 8 Iteration procedure in the 3-D calculation.

3.4 Results

Calculations have been done for yaw angles from 60° to 90°. Fig. 9 and fig. 10 compare the experimental pressure distribution profiles at yaw angles of 80° and 60° with the numerical ones obtained from the 3-D separated flow model, x being the longitudinal distance from the nose and D the width of the model. Better agreements are found at large distances from the nose. At higher yaw angles ($\geq 80°$) remarkable agreements occur at $x/D \geq 3.5$, while at 60°, these occur at $x/D \geq 5.5$. Closer to the nose, the method still predicts the roofside pressure distribution quite satisfactorily.

The pressure distribution profiles at various yaw angles are integrated to give the lengthwise distribution of local side force, lift and overturning moment coefficients: C_S, C_L and C_M respectively (fig. 11 a-d). At $\alpha \geq 80°$, where vortex shedding predominates over most of the length of the train model, the trends of the lengthwise distribution of the force and moment coefficients are predicted well, while the C_S's are generally underestimated and the C_L's overestimated. The C_M's, however, were predicted quite accurately.

At $\alpha \leq 70°$, where the steady nose vortices dominate the flow close to the nose, the method fails to predict the force and moment distribution in this region. This is mainly due to the fact that these nose vortices, where vorticities are concentrated along lines rather than planes (or panels), are not modelled. Without any local concentration of vortifcities, the computed force and moment distribution profiles varies in the lengthwise direction much less dramatically than the experimental ones. At $x/D \geq 5.5$, where vortex shedding predominates, the C_S's are slightly underestimated and the C_L's overestimated. As before, the C_M's are predicted quite satisfactorily.

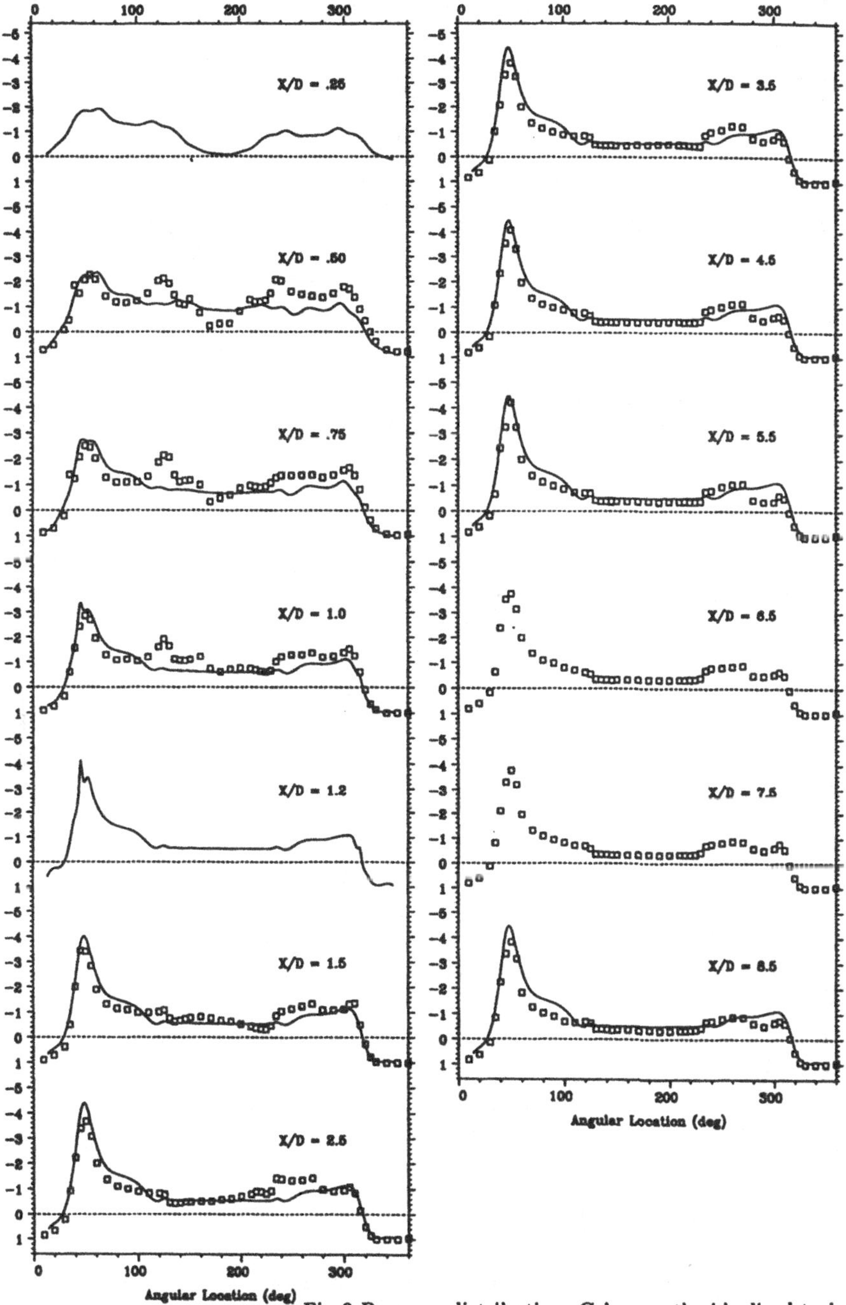

Fig 9 Pressure distribution, C_p's over the idealised train model, $\alpha = 80°$, $Re = 3 \times 10^5$: — 3-D computation; □ experiment.

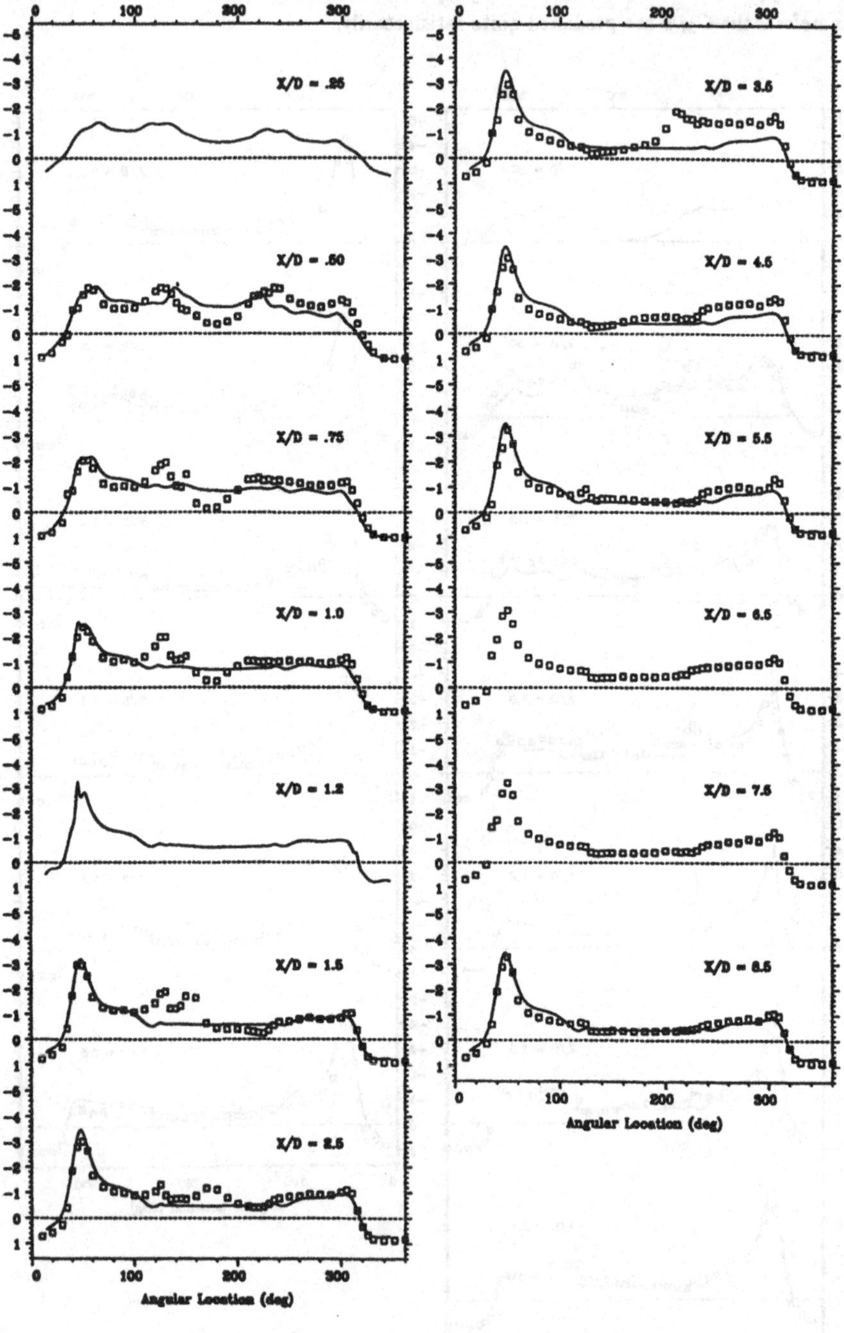

Fig 10 Pressure distribution, C_p's over the idealised train model, $\alpha = 60°$, $Re = 3 \times 10^5$: —— 3-D computation; □ experiment.

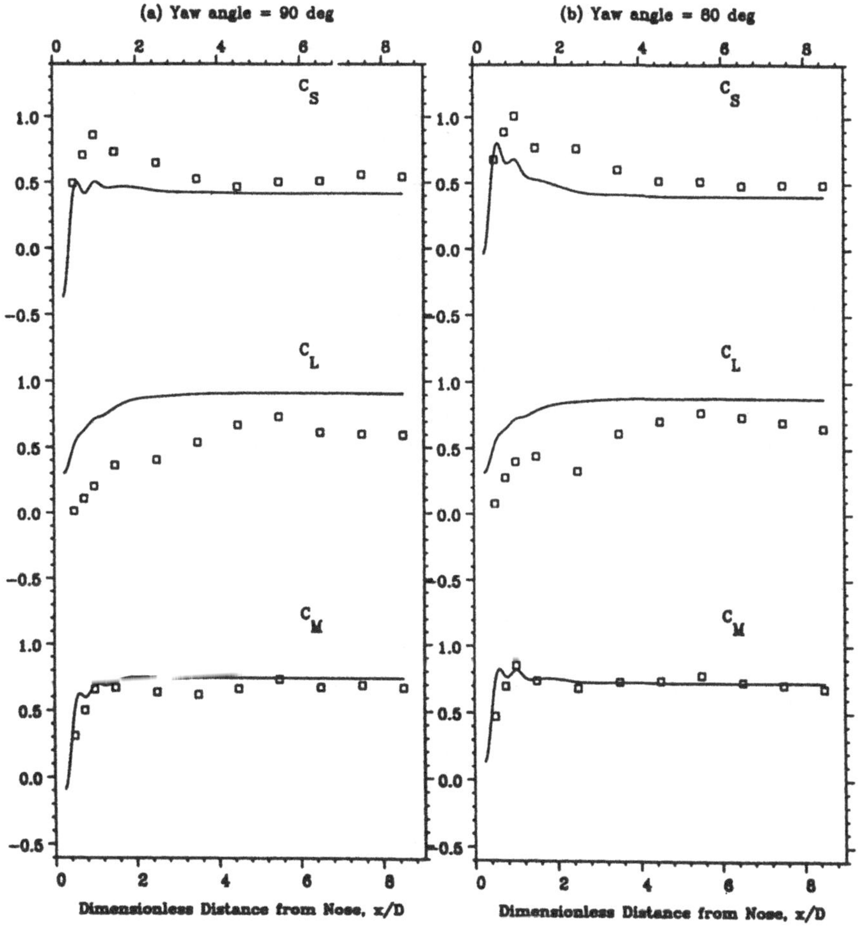

Fig 11 Lengthwise distribution of local side force (C_S), lift (C_L) and overturning moment (C_M) coefficients, $Re = 3 \times 10^5$: —— 3-D computation; □ experiment.

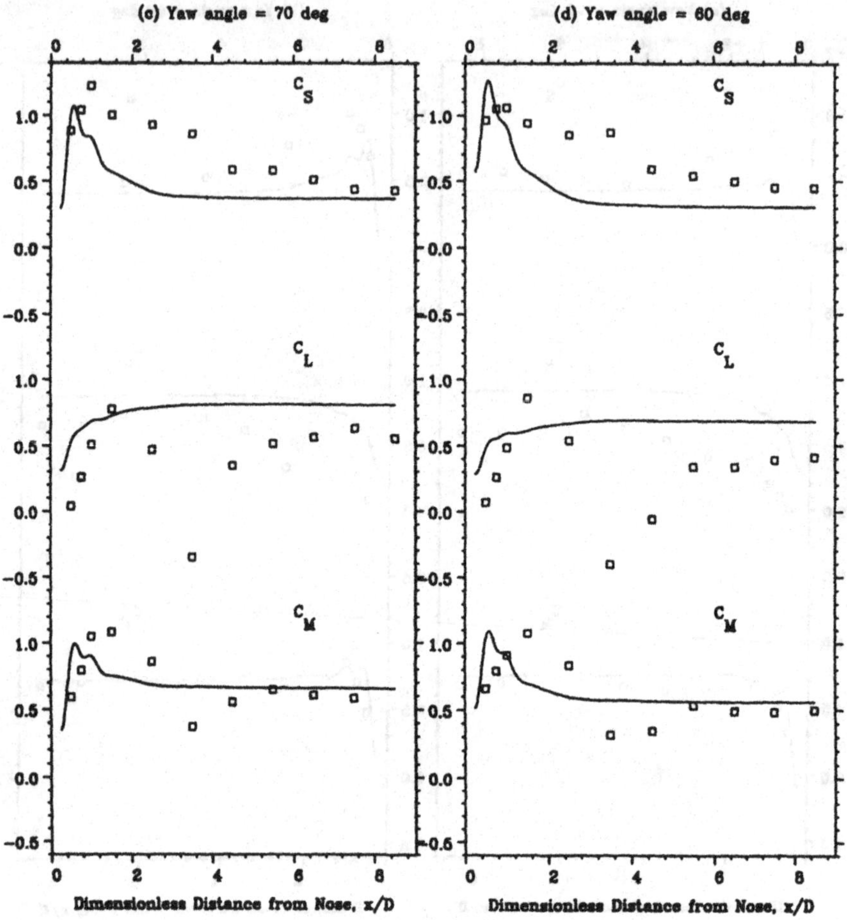

Fig 11 Lengthwise distribution of local side force (C_S), lift (C_L) and overturning moment (C_M) coefficients, $Re = 3 \times 10^5$: —— 3-D computation; □ experiment.

4 Conclusions

Comparisons with experiments have shown that the present 3-D scheme predicts the lengthwise overturning moment distribution quite satisfactorily when the flow is mainly of a vortex shedding type, such as at $\alpha \geq 80°$. At $\alpha \leq 70°$, where a vortex shedding régime occurs at a large distance from the nose, the overturning moment distribution in this region is also predicted satisfactorily.

A great advantage of the present method is that there are no control points on the wake panels. The wake vorticities are determined by a Kutta condition rather than a Neumann or Dirichlet boundary condition. This reduces the number of variables and hence helps to keep the computing time down. Furthermore, some previous researchers (Dvorak et al[4] and Ahmed et al[5]) who put control points on the wake panels in investigating the flow over 2- and 3-D bluff bodies found that this would retard the flow drastically just upstream of separation and make any boundary layer calculation inaccurate.

In the present 3-D calculations, the correct wake lengths and the positions of the separation lines far from the nose are determined from the 2-D calculation. (This simplifies dramatically the 3-D calculation). Close to the nose, the experimentally observed separation positions are used. The separation lines could have been determined numerically by a 3-D boundary layer code together with an iterative scheme as described in Chiu[2]. The computing time could, nevertheless, be increased considerably.

This method, however, fails to predict the pressure force and moment distribution where a steady line-vortex type wake structure predominates. Future development of the present method should incorporate a more effective modelling of the steady line-vortices emerging from the model surface.

Acknowledgements

I owe a debt of gratitude to Dr Len Squire of Cambridge University Engineering Department, who has given me much encouragement and valuable advice throughout the work. Thanks are due to British Rail, Barclays Bank plc and the Committee of Vice Chancellors and Principals for their financial support of the project.

References

1. Chiu, T.W., "Aerodynamic loads on a railway train in a cross-wind at large yaw angles." *PhD Thesis, Cambridge University Engineering Department, 1990.*

2. Chiu, T.W., "A two-dimensional second order vortex panel method for the flow over a train in a cross-wind and other two-dimensional bluff bodies." *Proceeding of the International Conference on Numerical Methods in Engineering: Theory and Application, Swansea, Jan. 1990, Elsevier Applied Science, London and New York.*

3. Green, J.E., Weeks, D.J. and Brooman, J.W.F., "Prediction of turbulent boundary layers and wakes in compressible flow by a lag-entrainment method." *TR72231, 1973, Royal Aircraft Establishment, now Royal Aerospace Establishment, Ministry of Defence, Farnborough, Hants.*

4. Dvorak, F.A., Maskew, B. and Woodward, F.A., "Investigation of three-dimensional flow separation on fuselage configurations." *Rep. TR-77-4, 1977, US Army Air Mobility Research & Development Laboratory, Fort Eustis, Va. 23604, USA.*

5. Ahmed, S.R. and Hucho, W.M., "The calculation of the flow field past a van with the aid of a panel method." *770390, 1977, Society of Automotive Engineers, 400 Commonwealth Drive, Warrendale, PA 15086, USA.*

SECTION 5: WAVE PROPAGATION

Numerical Study of Surface Waves in Channels

A.K. Mitra (*), M.S. Ingber (**)
(*) Department of Aerospace Engineering and Engineering Mechanics, Iowa State University, Ames, Iowa 50011, U.S.A.
(**) Department of Mechanical Engineering, University of New Mexico, Albuquerque, New Mexico 87131, U.S.A.

INTRODUCTION

This research concerns the application of the boundary element method (BEM) in the study of small-amplitude surface waves in channels. For the purposes of this paper, a channel will be considered a two-dimensional problem in which the length of the channel is very large compared to its depth and width. Further, the depth and width are assumed to be constant along the lengthwise direction.

Interest in surface waves comes from the importance of calculating the response of a contained liquid. Such calculations are employed in the design of aircraft fuel tanks, rocket fuel tanks, and railroad tank cars. Sloshing of liquids in a partially filled tank can be critical to the stability and structural integrity of the vehicle carrying the tank. Large forces and moments may be produced by the oscillating liquid, causing the failure of structural components or deviation from the planned path of the vehicle. Baffles are introduced in the tank to modify the frequencies of the excited modes and to enhance the viscous damping. A study of viscous damping is beyond the scope of this work. However, the shift of the frequencies brought about by the introduction of baffles is also of interest to the designer. The frequency of the oscillating liquid should not coincide with either the elastic body bending frequency or the dynamic control frequency of the vehicle. In this paper, the effects of two types of baffles on frequencies and mode shapes are presented.

Analytic results in the area of sloshing are sparse and are limited to a few simple geometries. Miles[1] employed an inverse method to determine the frequency associated with the dominant mode for a class of hyperbolic channels; Ghanimati and Naghdi [2], using the theory of a directed fluid sheet, determined the first even and odd modes for channels with parabolic and triangular cross sections.

Numerical investigations of this class of problems have been conducted by using the finite difference method[3] and the finite element method[4]. The disadvantage of these methods is that, in general, investigators are mainly interested in the natural frequencies and mode shapes and not in the additional information, concerning the interior of the domain, that these methods produce.

The BEM is well suited to problems of this type where the boundary information is of primary importance. Further, the BEM is able to resolve as many of the modes as desired.

FORMULATION

The derivation of the governing equations and boundary conditions assuming irrotational flow and assuming small amplitude waves can be found in several treatises (see, e.g., Ref. [5], p. 86). The velocity potential, ϕ, satisfies Laplace's equation.

$$\nabla^2 \phi = 0 \tag{1}$$

The no-penetration condition is imposed on the solid boundary as

$$\frac{\partial \phi}{\partial n} = 0 \tag{2}$$

where n is the unit outward normal to the solid surface. The condition on the free surface is derived by combining the kinematic condition and the linearized Bernoulli equation. The disturbance is assumed to contain the exponential factor $exp(-i\sigma t)$, where σ is the circular frequency and t is time. After such assumptions, the condition at the free surface becomes

$$\frac{\sigma^2 h}{g}\phi = \frac{\partial \phi}{\partial n} \tag{3}$$

where g is the gravitational constant. The characteristic length, h, is chosen as the maximum depth of the fluid as shown in Fig.1.

By using standard procedures[6], the differential equation (1) is converted to the following integral equation.

$$c(P)\phi(P) = \int_\Gamma [G'(P,Q)\phi(Q) - G(P,Q)\phi'(Q)]d\Gamma(Q) \tag{4}$$

where

$$c(P) = \int_\Gamma G'(P,Q)d\Gamma(Q)$$

In Eq.(4), $G(P,Q)$ is the fundamental solution, and the $'$ denotes derivative along the outward normal to Γ at point Q. The boundary Γ consists of two parts as

$$\Gamma = W + F$$

where W is the wetted perimeter of the channel (the solid boundary) and F is the free surface of the fluid.

By inserting the boundary conditions from Eqns.(2,3) in Eq.(4) one obtains

$$\begin{aligned} c(P)\phi(P) &= \int_W G'(P,Q)\phi(Q)d\Gamma(Q) + \int_F G'(P,Q)\phi(Q)d\Gamma(Q) \\ &\quad -\frac{\sigma^2 h}{g}\int_F G(P,Q)\phi(Q)d\Gamma(Q) \end{aligned} \tag{5}$$

The integral equation (5) is discretized and collocated by following standard procedures[6] to obtain

$$\begin{bmatrix} A & B \\ C & D \end{bmatrix} \begin{Bmatrix} \phi_W \\ \phi_F \end{Bmatrix} = \frac{\sigma^2 h}{g} \begin{bmatrix} K \\ L \end{bmatrix} \{\phi_F\} \tag{6}$$

where A, B, C, D, K, and L are block matrices; the vectors ϕ_W and ϕ_F are the nodal values of ϕ on the wetted perimeter and the free surface, respectively. By combining the top and bottom halves of Eq.(6) one can eliminate ϕ_W, and arrive at the eigenvalue problem

$$[M]\{\phi_F\} = \frac{\sigma^2 h}{g}[N]\{\phi_F\} \tag{7}$$

The eigenvalues are computed by using the IMSL subroutine called EIGZF.

By controlling the number of nodes on the free surface, one can control the number of eigenvalues that are determined. Further, the eigenvectors provide the nodal values of ϕ on F associated with the eigenvalues. These eigenvectors when inserted in Eq.(3) give the nodal values of velocity at the free surface.

SURFACE WAVES IN CHANNELS

Two channel geometries are examined in this section with the aim of establishing the reliability and accuracy of the BEM. The cross sectional geometries considered are shown in Fig.1. The numerical results are compared with results previously established in the literature.

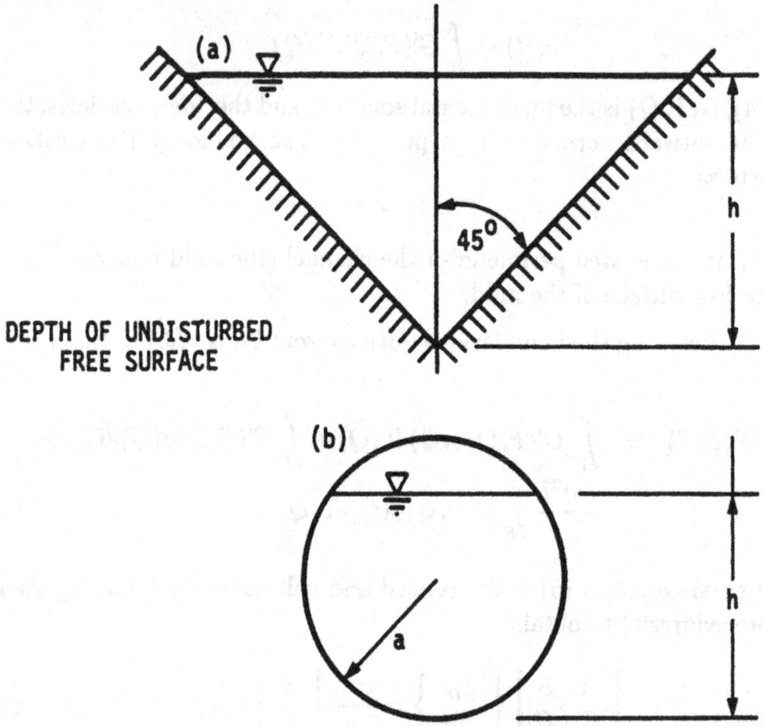

Figure 1(a,b): Triangular and circular channels.

A complete set of frequencies and mode shapes for a triangular cross section with an included angle of 90° (Fig. 1(a))were determined by Kirchhoff[7]. These results provide a benchmark test for the accuracy of the BEM. Three discretizations are considered in order to evaluate the convergence of the numerical method. The analytic results are compared to the boundary element results for the first five modes in Table I.

Budiansky[8], using an integral equation method not entirely different from the BEM, determined the first three antisymmetric modes for partially filled cylindrical containers (Fig. 1(b)). These results are compared to the boundary element results in Table 2 for various depth-to-diameter ratios. The agreement is seen to be very good.

Table 1: Eigenvalues for Kirchhoff's problem.
Mesh A: 10 elements on F, 12 elements on W
Mesh B: 20 elements on F, 24 elements on W
Mesh C: 40 elements on F, 48 elements on W

Mode Number	$\sigma^2 h/g$			
	Mesh A	Mesh B	Mesh C	Exact
1	0.975	0.991	0.996	1.000
2	2.243	2.288	2.309	2.324
3	3.797	3.855	3.899	3.924
4	5.376	5.394	5.450	5.498
5	6.999	6.945	7.003	7.067

Table 2: First three antisymmetric frequencies, f, in a circular channel at various depth-to-diameter ratios.
$(\sigma^2 a/g)_I = f_I$: Using 28 boundary elements.
$(\sigma^2 a/g)_B = f_B$: Budiansky's solution[8].

h/2a	First Mode		Second Mode		Third Mode	
	f_I	f_B	f_I	f_B	f_I	f_B
0.1	1.08	1.05	5.43	5.38	10.82	10.85
0.3	1.19	1.17	4.73	4.74	8.22	8.33
0.5	1.38	1.36	4.66	4.70	7.81	7.96
0.7	1.74	1.74	5.25	5.34	8.68	8.89
0.9	2.98	3.04	8.16	8.42	13.33	13.84

CHANNELS WITH BAFFLES

In this section, the effects of two kinds of baffles on mode shapes and frequencies are considered. A rectangular baffle piercing the free surface and a two-dimensional modification of a ring baffle are shown schematically in Fig. 2. In all calculations, the depth of the liquid and width of the tank are chosen to be unity.

First, the observations for the baffle shown in Fig. 2(a) are presented. The first five mode shapes for a baffle with w = 0.3, h = 0.2 placed symmetrically at $l = 0.35$ are shown in Fig. 3. The first mode (Fig. 3(a)) shows that a rise (positive velocity) in liquid level on one side of the baffle is accompanied by a fall (negative velocity) on the other side. The same magnitude of velocity on either side is the result of the conservation of mass. This mode is called the U-tube mode because of its similarity to the oscillation of a liquid in a U-tube (see Ref. [9], p.237). The four subsequent even and odd modes are shown in Fig. 3(b)-3(e).

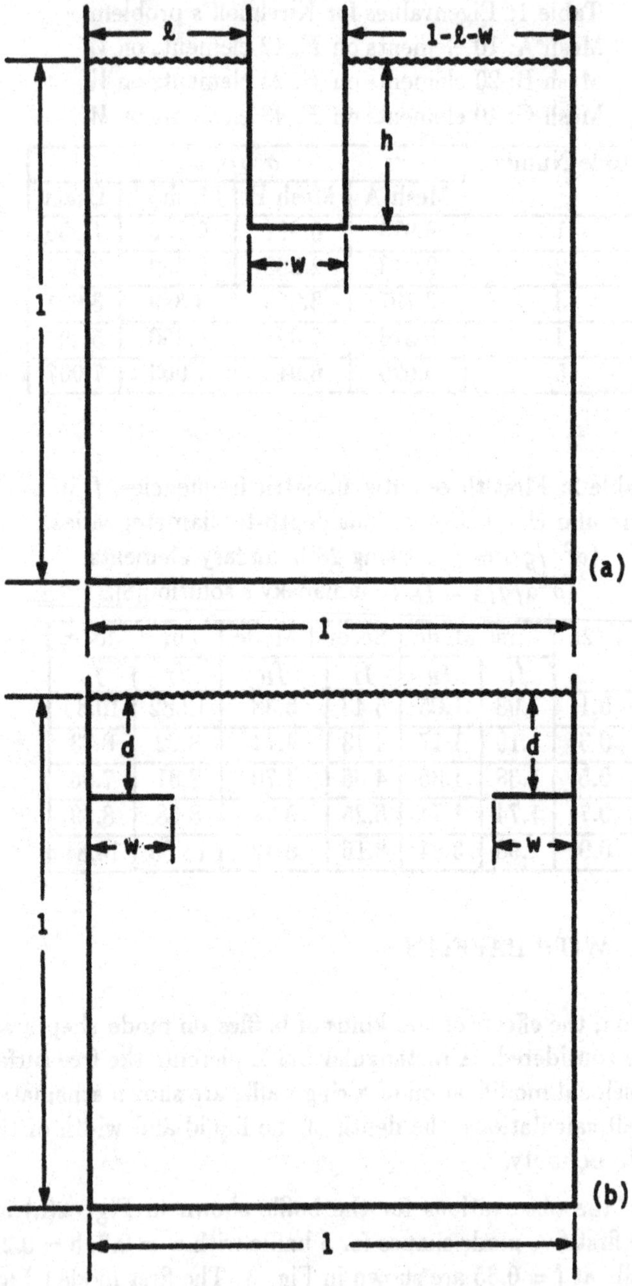

Figure 2(a,b) : Schematic showing the geometry of the baffles.

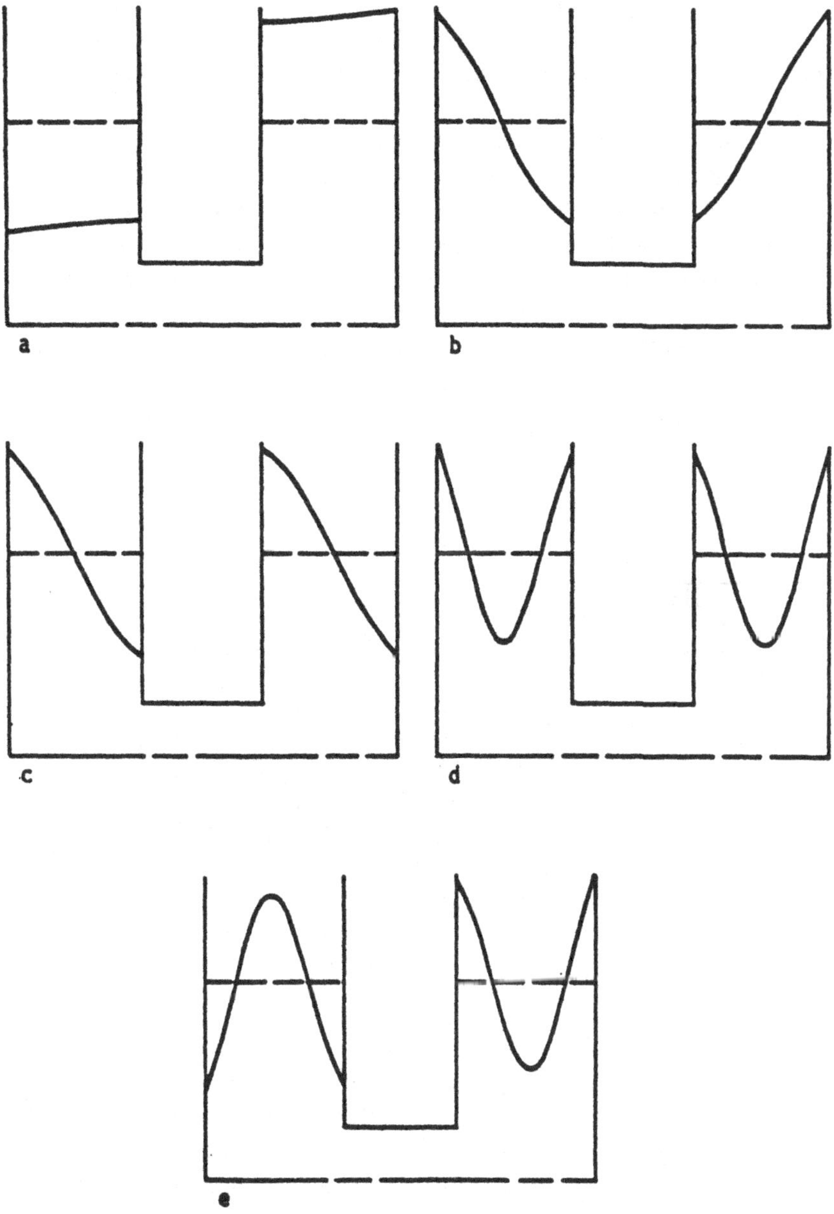

Figure 3(a,b,c,d,e) : The first five mode shapes for a baffle of w = 0.3, h = 0.2, and located symmetrically at l = 0.35.

To better understand the nature of these higher modes, it is important to study the effect of changing the height of the baffle on the natural frequencies. The first five frequencies for a baffle of constant width (w = 0.3) are given in Table 3 for several values of h. The frequency of the U-tube mode decreases as h is increased and should disappear in the limit h = 1.0 when the channel is completely partitioned. Since, both the parts of the channel, separated by the baffle, participate in this mode, this mode will be called the global mode. However, the frequencies for the second and higher modes do not have a strong dependence on h. The frequencies stay close to the eigenvalues for a 1 X 0.35 channel (which is very close to the size of each of the compartments) given in the last row of Table 3. This implies that the second and higher modes are compartmentalized; that is, the liquid columns oscillate independently in each part of the channel.

Table 3 : First five eigenvalues for baffle of width w = 0.3, symmetrically placed at $l = 0.35$.

Baffle	Eigenvalue				
Height (h)	1	2	3	4	5
0.05	2.737	8.160	8.682	15.86	15.90
0.1	2.427	7.974	7.166	15.70	15.70
0.2	1.938	7.869	7.901	15.66	15.68
0.4	1.357	7.848	7.852	15.66	15.68
0.6	1.016	7.848	7.850	15.66	15.67
0.8	0.7365	7.848	7.849	15.66	15.67
0.9	0.5500	7.848	7.849	15.66	15.67
(1 × 0.35) Channel	-	8.976	-	17.952	-

Next, the effect of varying the width of a symmetrically placed baffle is explored. The first five frequencies for a baffle of constant height (h = 0.2) are given in Table 4 for several values of w. The proper length scale associated with the U-tube mode is the height of the baffle. As a result, the first frequency is independent of w and remains approximately constant. The frequencies for the second and higher modes, which are compartmentalized, increase as the width of the free surface is reduced (w increased). This can be concluded from the analytic results of Lamb[10] for a rectangular channel. The analytic results are shown in parenthesis in Table 4.

Table 4 : First five eigenvalues for a baffle of height h = 0.2, placed symmetrically.

Baffle Width (w)	Eigenvalue				
	1	2	3	4	5
0.1 (1 × 0.45)	2.039	6.308 (6.981)	6.426	12.53 (13.963)	12.54
0.2 (1 × 0.4)	1.967	7.006 (7.854)	7.065	13.92 (15.71)	13.93
0.3 (1 × 0.35)	1.938	7.869 (8.976)	7.901	15.66 (17.95)	15.68
0.4 (1 × 0.3)	1.940	8.978 (10.47)	8.997	17.93 (20.94)	17.96
0.5 (1 × 0.25)	1.970	10.46 (12.57)	10.47	21.02 (25.13)	21.05
0.6 (1 × 0.2)	2.028	12.56 (15.71)	12.57	25.49 (31.42)	25.54

The same type of results as discussed above can be determined for an asymmetrically placed baffle. The first ten modes for a baffle with w = 0.4, h = 0.2 asymmetrically placed at l = 0.1 are shown in Fig. 4. For the U-tube mode in Fig. 4(a), the ratio of velocities of the free surface on either side of the baffle is 1:5, the same as the ratios of areas, thus satisfying the conservation of mass. The three subsequent modes in Figs. 4(b) - 4(d) show oscillations in the wider side with an unexcited narrow side. The first odd mode for the narrow side shown in Fig. 4(e) corresponds to a much higher frequency. This can be concluded for the analytic results (and also from the eigenvalues shown in Table 4) for a rectangular channel, recognizing the compartmentalized nature of the mode. The same pattern is repeated for the higher modes. The eigenvalues associated with these ten modes are shown in Table 5.

Table 5 : The eigenvalues for the first ten modes for a baffle of w = 0.4, h = 0.2, and asymmetrically placed at l = 0.1.

Eigenvalues									
1	2	3	4	5	6	7	8	9	10
2.19	5.78	11.4	17.1	21.4	23.0	28.9	34.9	41.1	47.0

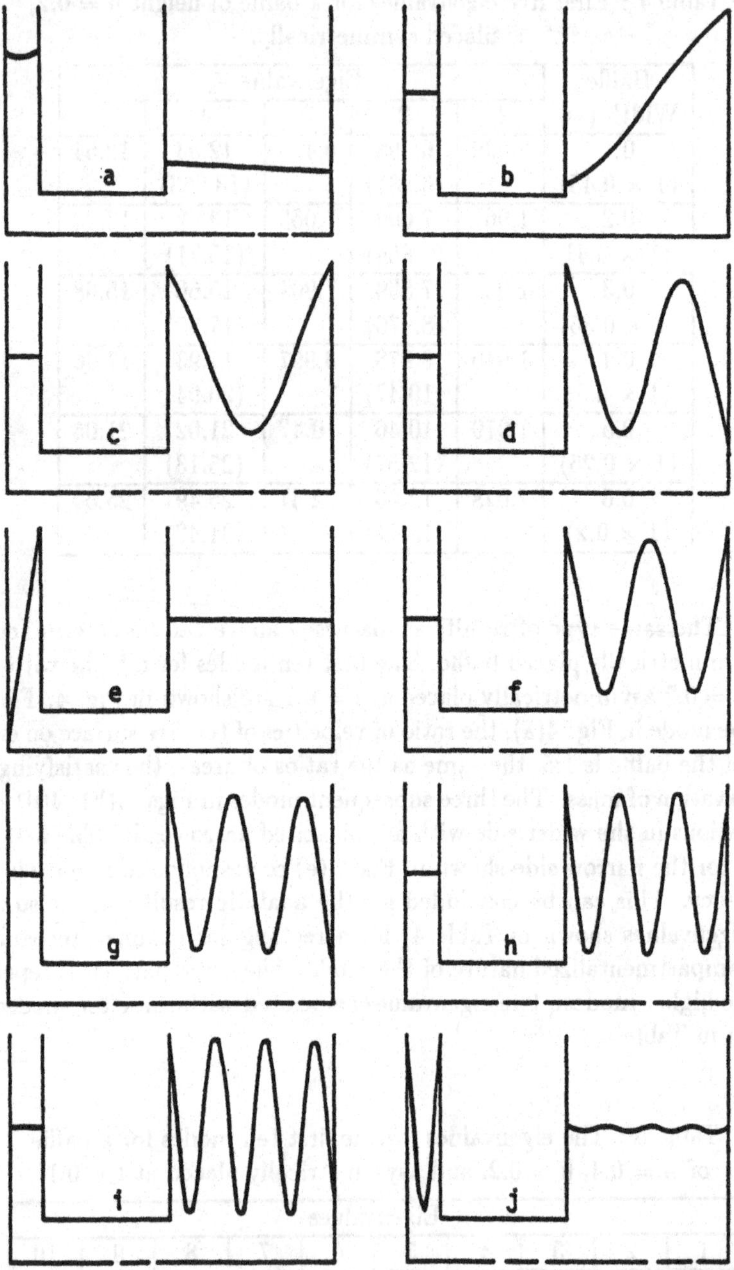

Figure 4 : The first ten mode shapes for a baffle of w = 0.4, h = 0.2, and asymmetrically placed at $l = 0.1$.

The effect of a ring baffle is considered next. The variation of the frequency for the first mode with the location of the baffle (controlled by d of Fig. 2(b)) for a constant width (w = 0.0785) is shown in Fig. 5. The frequency of the first mode is maximum when d = 0.0, i.e., when the baffle is placed at the free surface. Thus, the ring baffle could be an effective means of increasing the lowest frequency. A similar variation of frequency has been observed experimentally by Silverman and Abramson[10].

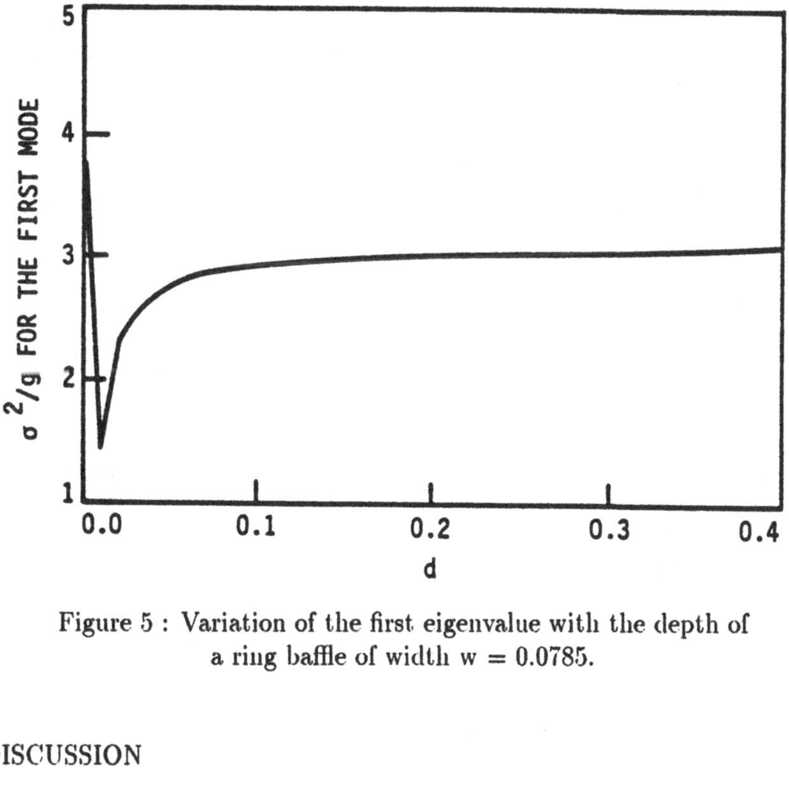

Figure 5 : Variation of the first eigenvalue with the depth of a ring baffle of width w = 0.0785.

DISCUSSION

There are many problems in which the motion of contained horizontal fluid surfaces are of interest. Except for a few simple geometries, analytic results are unavailable. The BEM has been shown to be a valuable tool in the analysis of these problems. The BEM can be applied to complicated geometries and can resolve as many modes as required. The method possesses some advantages over other discrete numerical schemes; one principal advantage is that attention is naturally focused on the free surface. In the course of this investigation, new results were obtained concerning channels with baffles.

REFERENCES

1. Miles, J., Journal of Fluid Mechanics, Vol.152, p. 379, 1985.
2. Ghanimati, G.R. and Naghdi, P.M., Journal of Fluid Mechanics, Vol. 164, p. 359, 1986.
3. Ehrlich, L.W., Riley, J.D., Strang, W.G. and Troesch, B.A., SIAM Jornal, Vol. 9, p. 149, 1961.
4. Yang, W.H. and Yih, C.S., Journal of Fluid Mechanics, Vol. 74, p. 183, 1976.
5. Yih, C.S., Stratified Flows, Academic Press, London, 1980.
6. Brebbia, C.A., Telles, J.C.F and Wrobel, L.C., Boundary Element Techniques, Springer Verlag, Berlin, 1984.
7. Kirchhoff, G., Mber. Akad. Wiss Berlin, 395, 1879.
8. Budiansky, B., Journal of Aerospace Sciences, Vol. 28, p. 161, 1960.
9. White, F.M., Viscous Fluid Flow, McGraw-Hill, New York, 1974.
10. Lamb, H., Hydrodynamics, Cambridge University Press, 1932.
11. Silverman, S. and Abramson, H.N., The Dynamic Behavior of Liquids in Moving Containers, NASA SP-106, p. 105, 1966.

Water Wave Modeling Using a Boundary Element Method with Overhauser Spline Elements

J.C. Ortiz, S.L. Douglass

Department of Civil Engineering, University of South Alabama, Mobile, AL 36688, U.S.A.

ABSTRACT

A BEM system for modeling nonlinear, periodic water waves in the physical space has been developed. The system uses Overhauser elements to eliminate discontinuities of the slope on the water surface. It is shown that the system developed accurate model breaking plungers without numerical instabilities. A shallow water wave is presented showing its behavior as it transforms in time from a sine wave to a breaking wave, ending with its jet meeting the trough.

INTRODUCTION

Wave breaking on a beach is one of the most easily observed and dramatic fluid phenomenon. However, very little is actually known about the hydrodynamics of wave breaking. In particular, a satisfactory mathematical description and solution of the breaking wave has not been found.

As society's coastal development and recreation desires increase in the face of a rising sea level, engineering analysis of surf zone dynamics related to beach erosion and pollutant dispersion issues are becoming more critical. Water waves are the primary forcing function in surf zone dynamics. Coastal engineering design must not only consider the forces of breaking waves for structural design but also the effects of the engineering alternatives on the incident waves and the wave-driven nearshore currents and sediment transport. Therefore, more complete methods of modelling the breaking wave phenomenom are being sought.

WAVE THEORY

The motion of an inviscid, incompressible fluid in irrotational flow is described using Laplace's equation:

$$\nabla^2\phi = 0 \tag{1}$$

where ϕ is a velocity potential. The fluid particle velocities u and v (in the x and y direction, respectively) are described in terms of the velocity potential ϕ as:

$$u = \frac{\partial\phi}{\partial x} \tag{2}$$

$$v = \frac{\partial\phi}{\partial y} \tag{3}$$

Equations 1, 2 and 3 are used with the boundary conditions described below to model the fluid motion through time as the water wave breaks.

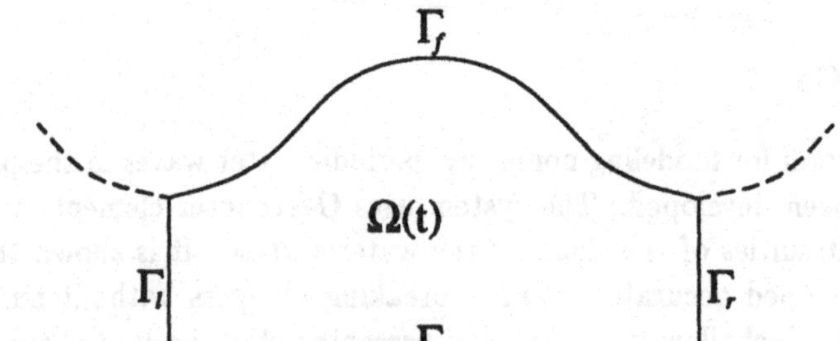

Figure 1: Periodic Water Wave Computational Domain $\Omega(t)$ for BEM analysis. Γ_b is the bottom surface, Γ_l and Γ_r are the left and right surfaces, and Γ_f is the free surface

A periodic water wave is shown in Figure 1. The boundary condition at the bottom surface (Γ_b) is one of no flow normal to the surface:

$$\left.\frac{\partial\phi}{\partial n}\right|_{\Gamma_b} = 0 \tag{4}$$

and the boundary conditions at the sides (Γ_l and Γ_r) are those of spacial periodicity:

$$\phi_{\Gamma_l} = \phi_{\Gamma_r} \quad \text{and} \quad \left.\frac{\partial \phi}{\partial n}\right|_{\Gamma_l} = \left.\frac{\partial \phi}{\partial n}\right|_{\Gamma_r} \tag{5}$$

where **n** is the outward pointing normal to the surface Γ which bounds the domain $\Omega(t)$.

The boundary conditions for the free surface, Γ_f in the figure, present unique challenges. The interface between the air and water is usually called the free surface regardless of whether or not it is forced by wind or uneven pressure. The determination of this surface and its deformation in time is our primary goal since fluid particle velocity and acceleration fields will follow directly. The free surface boundary conditions are twofold: a kinematic and a dynamic condition. The kinematic free surface boundary condition is usually taken to imply that a particle at the surface will remain at the surface [1]. Considering the total or material derivative with respect to time, $\frac{D}{Dt}$, of the horizontal and vertical particle locations, x and y, yields a very simple form of the kinematic free surface boundary conditions [2].

$$\frac{Dx}{Dt} = u \tag{6}$$

$$\frac{Dy}{Dt} = v \tag{7}$$

The dynamic condition is the satisfaction of Bernoulli's equation, with which the total derivative of the velocity potential following a surface particle becomes

$$\frac{D\phi}{Dt} = \frac{1}{2}(u^2 + v^2) - gy \tag{8}$$

where g is the acceleration of gravity and the pressure on the free surface is taken as zero.

The ensuing time-stepping procedure for the description of the water wave motion is as follows. Using an initial wave surface and velocity potential, and the boundary conditions set on the Γ_b, Γ_l and Γ_r surfaces (Equations 4 and 5), Equation 1 is solved for the unknown values of u and v for the fluid particle on the free surface. These values are then used in conjunction with the kinematic and dynamic boundary conditions on the free surface Γ_f (Equations 6, 7 and 8) to find the wave surface and velocity potential at the next time step.

PREVIOUS WORK

Most of the numerical solutions of breaking waves have used a boundary integral technique for velocity potential. Other approaches that depend on solving the interior flow problem (e.g. Harlow & Welch [3], Nichols, et.al [4]) have been less successful at producing realistic looking solutions. Longuet-Higgins & Cokelet [2] pioneered the use of boundary integral techniques on overturning waves. They developed the mixed Lagrangian-Eulerian technique of following specific surface particles as they move through time according to the free surface boundary conditions. The subsequent location of these surface particles defines the free surface. A boundary integral method calculates the normal and tangential gradients required for time-stepping the free surface boundary conditions.

Longuet-Higgins & Cokelet used a conformal mapping transformation with the complex form of the velocity potential. Vinje & Brevig [5,6] and Douglass [7] used a boundary integral method based on the Cauchy Integral for the complex velocity potential without mapping. They used linear elements to describe the surface location and potential for integral calculations. McIver & Peregrine [8] found that results from Vinje & Brevig's technique were in agreement with results from Longuet-Higgins & Cokelet's technique. Dold & Peregrine [9], New, et.al. [10], and Seo & Dalrymple [11] all used some form of conformal mapping with complex variables. A series of papers by Grilli & Svendsen [12–14] have presented applications of a Boundary Element Method (BEM) to the non-periodic breaking wave problem in the physical plane in terms of the velocity potential and its gradient using "quasi-spline" elements.

BEM SOLUTION

Time Stepping Scheme

The system developed uses an Adams-Bashford-Moulton (ABM) predictor-corrector scheme with automatic time step subdivision. It is initialized with a Runge-Kutta scheme. The initial time steps are chosen such that the Courant number C_0 is less than 0.5:

$$C_0 = \frac{\Delta x'}{\Delta t'} < 0.5$$

where $\Delta x'$ is the distance between the two closest nodes divided by the wave length and $\Delta t'$ is equal to the time step divided by the wave period. Subsequent time steps are subdivided automatically according to the following criteria: if any step forward in time by a node is larger than the distance to its nearest neighboring node, the time step is divided by two. This method insures that the Courant number C_0 does not increase significantly as the distance between nodes near the crest decreases.

The method used has proven to be very stable. No numerical instabilities were observed during the breaking process. Indeed, the calculations do not break down until the wave jet makes contact with the wave trough.

BEM Solution of Laplace's Equation

The boundary element method utilizes Green's third formula to obtain a solution to Laplace's equation over the domain Ω as a function of boundary integrals[15]:

$$C\,\phi(\mathbf{r}_\ell) = \int_\Gamma \phi(\mathbf{r})\frac{\partial W}{\partial n}W\,d\Gamma - \int_\Gamma \frac{\partial \phi(\mathbf{r})}{\partial n}W\,d\Gamma \tag{9}$$

where C is a constant, $\mathbf{r}_\ell$ is the location of the source point, $\mathbf{r}$ is a dummy point over which the integrations are performed, and W is the Green's function corresponding to Laplace's equation in two dimensions:

$$W = \frac{1}{2\pi}\log|\mathbf{r} - \mathbf{r}_\ell|$$

By discretizing the boundary and evaluating Equation 9 at every node, a system of simultaneous equations of the form

$$[H]\{\phi\} = [G]\left\{\frac{\partial \phi}{\partial n}\right\} \tag{10}$$

is obtained. The matrices $[H]$ and $[G]$ depend only on the boundary definitions. Vectors $\{\phi\}$ and $\{\frac{\partial \phi}{\partial n}\}$ represent the values for the potential and its normal derivative at the nodes.

With the boundary conditions defined in Equations 4 and 5, and an initial free surface potential, Equation 10 is rearranged into a standard system of linear equations which is then solved for the unknown $\{\phi\}$ and $\{\frac{\partial \phi}{\partial n}\}$. The important values for the time-stepping scheme are the normal derivatives of the potential on the free surface of the wave which are used in Equations 6, 7 and 8 to march the surface in time.

Boundary Discretization

The success of the time-stepping scheme depends on an accurate representation of the normal derivatives of the potential on the wave surface. Previous solutions to the wave breaking problem have used discretization of the boundary with elements which do not guarantee continuity of the derivative of the velocity potential. These discontinuities affect the accuracy of the normal derivatives at the wave surface and may lead to instabilities in the time-stepping scheme.

A more accurate method of representing the boundary and the normal derivatives at the free surface can be obtained using spline elements which guarantee derivative continuity throughout the boundary. This improved scheme may increase the efficiency of obtaining a solution by requiring fewer elements and allowing for larger time steps. In the BEM system used in this work, the wave surface is discretized with Overhauser spline elements [16–19], which present several advantages for the solution of breaking waves over non-splining, higher order elements and other splining elements:

- The Overhauser elements enforces C^1 continuity between each element, thus eliminating any errors due to derivative discontinuities present in BEM systems not using splining elements.

- The Overhauser element is a very stable element, practically eliminating the problems associated with using higher-order elements [20]. It is well known that highly irregular curves can be obtained when higher order polynomials are used to fit data points having non-equal spacing between the interpolating points. As a wave breaks, the distances between fluid particles change, coming together at the crest and moving farther apart at the trough. In numerical schemes, this means that the nodal point spacing also changes, becoming closer at the crest and further apart at the trough. This irregular node spacing normally leads to numerical instabilities when using higher-order elements [14]. It is observed that such instabilities do not occur when using Overhauser elements, even when the spacing between the nodes becomes highly irregular.

- The Overhauser element is very simple to implement [18]. B-Splines (a non-interpolating spline which has the advantage of being able to represent common geometric shapes exactly) have also been used in BEM analysis with success [21], but they require control points in addition to the boundary nodes. These control points, which help define the elements, would have to be updated and recalculated at each time step even though they are not part of the solution (the free surface itself). The additional computation time and storage required are not necessary for water wave problems.

All integrations over the elements (except for the singular integrals, on which a logarithmically weighted Gaussian quadrature method is used) are performed using standard 4-point Gaussian quadrature. The accuracy of the integrations seem to be adequate for the current type of problems being solved. This is probably due to the lack of near-singular integrals arising in the problems. One would expect difficulties at the tip of the jet formed when the wave breaks because nodes are coming closer together as the surface closes in on itself, but the spacing between the nodes along the surface also decreases. This makes the elements at the tip of the jet smaller and increases the accuracy of the integrations.

RESULTS

Figure 2 shows the results of a shallow water breaking wave similar to that shown in the work of Vinje & Brevig [5]. The initially sinusoidal wave with length $L = 2\pi$, amplitude $a = 0.13L/2$, and depth $d = 2a$ is given an initial potential on the free surface calculated from linear theory:

$$\phi|_{t=0} = \frac{aT}{2\pi}\sin(x/L)$$

where the period T, also calculated from linear theory, is given by:

$$T = \frac{2\pi}{\sqrt{\tanh d}}$$

The number of nodal points used on the surface is 60, initially equally spaced in the x direction, and the initial time step is $T/100$. It is seen that the wave front becomes vertical at approximately $t = 0.3T$, which is the same result seen in the work of Vinje & Brevig. Fluid then "jets" out from the wave front and hits the forward face of the wave at approximately $t = 0.5T$, after which the calculations break down, as expected. These results show that the BEM system developed accurately describes the development of a plunging breaker.

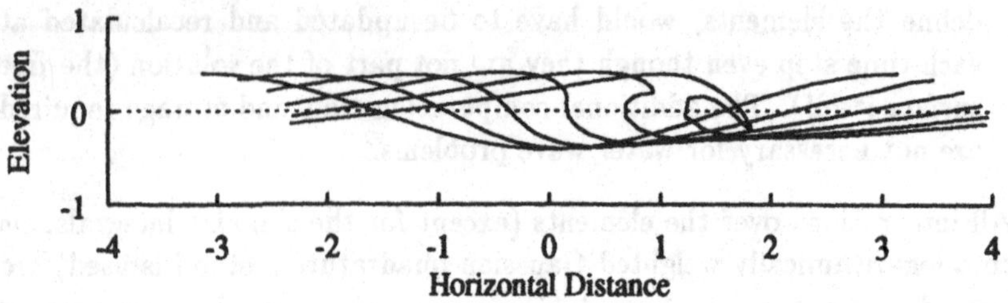

Figure 2: Linear Theory Wave with $d/L = 0.13$. Wave profiles are given at every $\frac{T}{100}$ seconds, beginning at $t = 0$.

CONCLUSIONS

The BEM system created to simulate water waves has been shown to provide an accurate model of plunging breakers. The results shown compare well with other models in the literature [5]. The system has proven to be very robust, showing no signs of numerical instabilities as seen in previous work [2], thus no "smoothing" of the wave as it moves in time was needed. Also, moving of the nodes along the surface as the wave breaks for better spacing of the nodes was not required for a solution — one of the advantages of using the Overhauser elements.

A more detailed presentation of this work is planned for the near future. Further work is also planned: improving the efficiency of the code by extending the Taylor expansion used in the calculation of the ϕ in the time-stepping procedure (i.e. $\phi(t+\Delta t) = \phi(t) + \frac{\partial\phi}{\partial t}\Delta t + \frac{\partial^2\phi}{\partial t^2}\frac{\Delta t^2}{2} + ...$); changing of the bottom surface to approximate beach slopes; modifying of the boundary conditions to simulate a wave tank; etc.

ACKNOWLEDGEMENTS

This work was supported by Cray Research Inc. and a computer time allocation grant from the Alabama Supercomputer Network Authority.

References

[1] B. Kinsman. *Wind Waves: Their Generation and Propagation on the Ocean Surface.* Prentice-Hall, Englewood Cliffs, New Jersey, 1965.

[2] Longuet-Higgins, M.S. and Cokelet, E.D. The Deformation of Steep Surface Waves on Water, I. A Numerical Method of Computation. *Proceedings of the Royal Society of London*, 350A:1–26, 1976.

[3] Harlow, F.H. and Welch, J.E. Numerical Calculation of Time-Dependent Viscous Incompressible Flow of Fluid with Free Surface. *The Physics of Fluids*, 8(12), 1965.

[4] Nichols, B.D., Hotchkiss, C.W., and Hurt, R.S. SOLA-VOF: A Solution Algorithm for Transient Fluid Flow with Multiple Free Boundaries. Technical Report LA-8355, Los Alomos Scientific Laboratory, 1980.

[5] Vinje, T. and Brevig, P. Numerical Simulation of Breaking Waves. In *Proceedings of the Third International Conference on Finite Elements in Water Resources*, pages 5.196–5.210. University of Mississippi, 1980.

[6] Vinje, T. and Brevig, P. Numerical Calculations of Forces from Breaking Waves. In *International Symposium on Hydrodynamics in Ocean Engineering*, pages 547–565. Norwegian Institute of Technology, 1981.

[7] S.L. Douglass. *The Influence of Wind on Nearshore Breaking Waves*. PhD thesis, Drexel University, Philadelphia, PA, 1989.

[8] McIver, P. and Peregrine, D.H. Comparison of Numerical and Analytical Results for Waves that are Starting to Break. In *International Symposium on Hydrodynamics in Ocean Engineering*, pages 203–215. Norwegian Institute of Technology, 1981.

[9] Dold, J.W. and Peregrine, D.H. Steep Unsteady Water Waves: An Efficient Computational Scheme. In B. Edge, editor, *Proceedings of the 19th Coastal Engineering Conference*, pages 955–967. American Society of Civil Engineers, New York, 1984.

[10] New, A.L., McIver, P., and Peregrine, D.H. Computation of Overturning Waves. *Journal of Fluid Mechanics*, 150:233–251, 1985.

[11] Seo, S.N. and Dalrymple, R.A. An Efficient Model for Periodic Overturning Waves. Submitted to: *Engineering Analysis with Boundary Elements*, 1990.

[12] Grilli, S. and Svendsen, I.A. The Modelling of Nonlinear Water Wave Interaction with Maritime Structures. In Brebbia, C.A. and Conner,

J.J., editors, *Boundary Elements XII, Vol. 2, Proceedings of the 11th International Conference on Boundary Element Methods*, pages 253–268. Computational Mechanics Publications, Boston, 1989.

[13] Grilli, S. and Svendsen, I.A. The Modelling of Highly Nonlinear Waves: Some Improvements to the Numerical Wave Tank. In Brebbia, C.A. and Conner, J.J., editors, *Boundary Elements XII, Vol. 2, Proceedings of the 11th International Conference on Boundary Element Methods*, pages 269–281. Computational Mechanics Publications, Boston, 1989.

[14] Grilli, S.T. and Svendsen,I.A. Corner Problems and Global Accuracy in the Boundary Element Solution of Nonlinear Wave Flows. *Engineering Analysis with Boundary Elements*, 7(4):178–195, 1990.

[15] C.A. Brebbia. *The Boundary Element Method for Engineers.* Pentech Press, London, 1978.

[16] Juan C. Ortiz. An Improved Boundary Element System for the Solution of Poisson's Equation. Master's thesis, Louisiana State University, 1986.

[17] H.G. Walters. Techniques for Boundary Element Analysis in Elastostatics Influenced by Geometric Modeling. Master's thesis, Louisiana State University, 1986.

[18] Walters, H.G., Ortiz, Juan C., Gipson G.S., and Brewer, J.A. Overhauser Splines as Improved Boundary Element Types. In T. A. Cruse, editor, *Advanced Boundary Element Methods*, pages 461–464. Springer-Verlag, New York, 1987.

[19] W.S. Hall and T.S. Hibbs. A Continuous Boundary Element for 3-D Elastostatics. In T. A. Cruse, editor, *Advanced Boundary Element Methods*, pages 135–144. Springer-Verlag, New York, 1987.

[20] J.A. Brewer and Anderson D.C. Visual Interaction With Overhauser Curves and Surfaces. *Computer Graphics II*, 2:132–137, 1977.

[21] Cabral, J.J.S.P., Wrobel, L.C., and Brebbia, C.A. A BEM Formulation Using B-Splines: I- Uniform Blending Functions. *Engineering Analysis with Boundary Elements*, 7(3):136–144, 1990.

SECTION 6: GROUNDWATER FLOW

Boundary Element Formulation for Time Harmonic Poroelastic Problems

J. Dominguez

Escuela Tecnica Superior de Ingenieros Industriales, Universidad de Sevilla, Av. Reina Mercedes s/n, 41012 Sevilla, Spain

ABSTRACT

A Boundary Element approach for dynamic poroelastic problems in the frequency domain is presented. The technique is based on an integral equation formulation in terms of solid displacements and fluid stress. A computer code for two-dimensional problems is implemented and validated by solving two basic problems. The obtained numerical results are compared with the corresponding exact solutions.

INTRODUCTION

In a recent paper, Dominguez [1] presented the basic formulation of the frequency domain integral representation for dynamic poroelasticity in terms of solid displacements and fluid stress. The present paper, based on that formulation, is focused on the Boundary Element solution of two-dimensional problems. The formulation emerges from a reciprocal relation established in terms of solid displacements, boundary tractions on the solid, fluid stresses, fluid displacements normal to the boundary, and body forces in both phases. The fundamental solution corresponds to unit point loads in the solid phase and also to a field of body forces in the solid and in the fluid, chosen in such a way that it gives rise to a simple integral representation of the fluid stress.

Very recently Chen et al.[2] have presented a Boundary Element solution in the frequency domain for dynamic poroelasticity. Their formulation depends on variables different to those of the present paper which is based on the previous formulation of the author. Consequently, the fundamental solution expressions and the BE formulation and implementation are different.

BASIC EQUATIONS

Following Biot [3], the constitutive and equilibrium equations for time harmonic linear dynamic poroelasticity are expressed as follows.

Equilibrium equations:

$$\mu\, \Delta u_i + \left(\lambda + \mu + \frac{Q^2}{R}\right) e_{,i} + Q\, \varepsilon_{,i} + X_i =$$

$$- \omega^2 \left(\rho_{11}\, u_i + \rho_{12}\, U_i\right) + i\, \omega\, b\, (u_i - U_i) \tag{1}$$

$$(Q\, e + R\, \varepsilon)_{,i} + X'_i = - \omega^2 \left(\rho_{12}\, u_i + \rho_{22}\, U_i\right) - i\, \omega\, b\, (u_i - U_i) \tag{2}$$

where u_i and U_i are the solid and fluid displacements, respectively, b is a dissipation constant, ρ_{11}, ρ_{12} and ρ_{22} are mass densities, and finally, X_i and X'_i are body forces acting in the solid and the fluid, respectively.

Constitutive equations:

$$\tau_{ij} = \left(\lambda + \frac{Q^2}{R}\right) \delta_{ij}\, e + 2\, \mu\, e_{ij} + Q\, \delta_{ij}\, \varepsilon \tag{3}$$

$$\tau = Q\, e + R\, \varepsilon \tag{4}$$

where δ_{ij} is the Kronecker delta function, $e_{ij} = 0.5\, (u_{i,j} + u_{j,i})$ are the solid strains, $\varepsilon = U_{i,i}$ and $e = u_{i,i}$ are the fluid and solid dilation, respectively, λ, μ, Q and R are elastic constants and τ is the stress in the fluid related to the fluid pressure by

$$\tau = - \beta\, p \tag{5}$$

Pressure-displacements relations

The six displacements in equations (1) are not independent and can be expressed in terms of only four variables, namely the solid displacements u_i and the fluid stress τ.

Using equations (2) and (4), the displacements in the fluid are written as

$$U_1 = \frac{\tau_{,1} + X'_1 + (i\,\omega\,b + \omega^2 \rho_{12})\,u_1}{i\,\omega\,b - \omega^2 \rho_{22}} \tag{6}$$

By substituting equations (4) and (6) in equation (1) the following is obtained:

$$\mu\,\Delta u_1 + (\lambda + \mu)\,e_{,1} + \tau_{,1}\left(\frac{Q}{R} + \frac{i\,\omega\,b + \rho_{12}\,\omega^2}{i\,\omega\,b - \omega^2\,\rho_{22}}\right) +$$

$$u_1\,\omega^2\left(\frac{\omega^2\,(-\rho_{11}\,\rho_{22} + \rho_{12}^2) + i\,\omega\,b\,(\rho_{11} + \rho_{22} + 2\,\rho_{12})}{i\,\omega\,b - \omega^2\,\rho_{22}}\right) +$$

$$X_1 + \frac{i\,\omega\,b + \rho_{12}\,\omega^2}{i\,\omega\,b - \omega^2\,\rho_{22}}\;X'_1 = 0 \tag{7}$$

Equation (7) only depends on the three components of the solid displacement and on the fluid stress. This equation can be written for i = 1,2 and 3.

The forth equation, which completes the system, is obtained by taking divergence of equation (2) and substituting equation (4).

$$\Delta\tau + \frac{\tau}{R}\,(-\,i\,\omega\,b + \omega^2\,\rho_{22}) +$$

$$e\left[\,i\,\omega\,b\left(1 + \frac{Q}{R}\right) + \omega^2\left(\rho_{12} - \rho_{22}\,\frac{Q}{R}\right)\right] + X'_{1,1} = 0 \tag{8}$$

BOUNDARY INTEGRAL FORMULATION

The boundary integral formulation for dynamic poroelasticity in the frequency domain can be obtained as usual from the corresponding reciprocal relation and fundamental solution.

Starting from the equilibrium equations for time harmonic behaviour , weighting the first equation with displacement type functions u_i^* and the second with U_i^*, adding the two equations, integrating over the body Ω, and using integration by parts twice the reciprocal relation is obtained

$$\int_\Gamma (t_i u_i^* + \tau U_n^*) d\Gamma + \int_\Omega (X_i u_i^* + X'_i U_i^*) d\Omega =$$

$$\int_\Gamma (t_i^* u_i + \tau^* U_n) d\Gamma + \int_\Omega (X_i^* u_i + X'^*_i U_i) d\Omega \tag{9}$$

where Γ is the boundary of the body Ω, $t_i = \tau_{ij} n_j$ and $U_n = U_i n_i$; n being the unit normal to the boundary. The variables denoted by superscript * are those associated with the weighting displacement fields.

The integral representation of the solid displacement u_i is obtained by substitution of the fundamental solution derived from the following body forces:

$$X_i^* = \delta(\underset{\sim}{x})\, \delta_{ij}$$

$$X'^*_i = 0 \tag{10}$$

in which $\delta(x)$ is the Dirac delta function and x indicates the point of application of the delta function.

The reciprocal relation gives:

$$c_{ij} u_i + \int_\Gamma t_{ij}^* u_i \, d\Gamma + \int_\Gamma \tau_j^* U_n \, d\Gamma =$$

$$\int_\Gamma u_{ij}^* t_i \, d\Gamma + \int_\Gamma \tau U_{nj}^* \, d\Gamma + \int_\Omega (X_i u_{ij}^* + X'_i U_{ij}^*) d\Omega \tag{11}$$

To obtain the integral representation of the fluid stress, the fundamental solution corresponding to the following body forces is used.

$$X'^{*}_{i} = \left(\frac{1}{2\pi} \ln r \right)_{,i} \qquad \text{for} \quad \text{2-D}$$

$$X'^{*}_{i} = \left(\frac{-1}{4\pi r} \right)_{,i} \qquad \text{for} \quad \text{3-D}$$

$$X^{*}_{i} = \frac{i\,\omega\,b + \omega^{2}\,\rho_{12}}{-\,i\,\omega\,b + \omega^{2}\,\rho_{22}}\, X'^{*}_{i} \tag{12}$$

where r is the distance to the point x at which the fluid stress is going to be represented. Notice that in this case forces in all directions exist simultaneously in the two phases. Using these body forces and equation (6), the last integral in equation (9) may be transformed by integration by parts to give the following integral representation.

$$\frac{c\,\tau}{-\,i\,\omega\,b + \omega^{2}\rho_{22}} + \int_{\Gamma} \left(t^{*}_{ij}\, u_{i} + \tau^{*}_{j}\, U_{n} \right) d\Gamma =$$

$$\int_{\Gamma} \left[t_{i}\, u^{*}_{ij} + \tau \left(U^{*}_{nj} + \frac{X'^{*}_{i}\, n_{i}}{-\,i\,\omega\,b + \omega^{2}\rho_{22}} \right) \right] d\Gamma \; +$$

$$\int_{\Omega} \left[X_{i}\, u^{*}_{ij} + X'_{i} \left(U^{*}_{ij} + \frac{X'^{*}_{i}}{-\,i\,\omega\,b + \omega^{2}\rho_{22}} \right) \right] d\Omega \tag{13}$$

where the j subindex corresponds in this case to the fundamental solution due to the second type of body forces, the integrals are in the sense of Cauchy's principal value, $c = 1$ for points inside Ω and $c = 0$ for points outside Ω. The value of c for points on the boundary Γ is computed from the Cauchy's principal value of the integrals. It is equal to 0.5 for points on Γ where the boundary is smooth.

It is worth noticing that the body forces used to obtain the fluid stress integral representation correspond to a point source of pressure, as can be seen from equation (8), and body forces X_{i}^{*} applied to the solid phase in such a way that the direct action of the fluid body forces transmitted to the solid by inertia and dissipation effects is counterpoised.

Equations (11) and (13) are the boundary integral equations for time harmonic poroelasticity. The volume integrals included in those equations involve only body forces and fundamental solution values, which are given.

It is possible to establish an analogy between poroelasticity and thermoelasticity for time harmonic regime.

Using this analogy the fundamental solution needed for the above integral equation formulation can be obtained. The close form expression of this solution can be seen in reference 4.

BOUNDARY ELEMENTS

The integral representations in equations (11) and (13) can be written for two-dimensional domains under zero body force conditions as

$$c_{\alpha\beta}\, u_\alpha + \int_\Gamma t^*_{\alpha\beta}\, u_\alpha \, d\Gamma + \int_\Gamma \tau^*_\beta\, U_n \, d\Gamma = \int_\Gamma u^*_{\alpha\beta}\, t_\alpha \, d\Gamma + \int_\Gamma \tau\, U^*_{n\beta} \, d\Gamma \qquad (14)$$

$$\int_\Gamma t^*_{\alpha 3}\, u_\alpha \, d\Gamma + \int_\Gamma \tau^*_3\, U_n \, d\Gamma = \int_\Gamma u^*_{\alpha 3}\, t_\alpha \, d\Gamma$$

$$+ \int_\Gamma \tau\, (U^*_{n3} - J\, X'^*_\alpha\, n_\alpha) \, d\Gamma + J\, c_{33}\, \tau \qquad (15)$$

Using vector notation, the integral representation for a point "i" can be written as:

$$\underset{\sim}{C}^i_u\, \underset{\sim}{u}^i + \int_\Gamma \underset{\sim}{p}^*\, \underset{\sim}{u} \, d\Gamma = \int_\Gamma \underset{\sim}{u}^*\, \underset{\sim}{p} \, d\Gamma + \underset{\sim}{C}^i_p\, \underset{\sim}{p}^i \qquad (16)$$

where **u** and **p** are the field variables vectors for displacements and stresses, respectively;

$$\underset{\sim}{u} = \begin{Bmatrix} u_1 \\ u_2 \\ U_n \end{Bmatrix} \qquad \text{and} \qquad \underset{\sim}{p} = \begin{Bmatrix} t_1 \\ t_2 \\ \tau \end{Bmatrix} \qquad (17)$$

and $\mathbf{p}^*$ and $\mathbf{u}^*$ are the fundamental solution tensors.

Consider now that the boundary of the domain under study is discretized using straight line constant elements, i.e. elements where the values of $\mathbf{u}$ and $\mathbf{p}$ are constant over the length and equal to the nodal value. Equation (16) can be written as

$$\underset{\sim}{C}_u^i \, \underset{\sim}{u}^i + \sum_{j=1}^{N} \left\{ \int_{\Gamma_j} \underset{\sim}{p}^* \, d\Gamma \right\} \underset{\sim}{u}^j = \sum_{j=1}^{N} \left\{ \int_{\Gamma_j} \underset{\sim}{u}^* \, d\Gamma \right\} \underset{\sim}{p}^j + \underset{\sim}{C}_p^i \, \underset{\sim}{p}^i \tag{18}$$

where each integral extends over a straight segment Γ_j corresponding to the element "j". Using the traditional BE notation, equation (18) can be written for all the boundary nodes as

$$\underset{\sim}{H} \, \underset{\sim}{u} = \underset{\sim}{G} \, \underset{\sim}{p} \tag{19}$$

where $\underset{\sim}{H}$ and $\underset{\sim}{G}$ are 3N x 3N matrices.

NUMERICAL EXAMPLES

In order to validate the proposed boundary element approach two basic geomechanical problems with known analytical solution are investigated. The problems are those studied by Chen et al. in reference 2. The first one corresponds to a square domain of saturated porous material excited by a uniform harmonic normal stress at the top. The surface is drained. Each side of the square is discretized into 6 equal constant elements. The boundary conditions are:

$t_1 = u_2 = U_n = 0$	at	$x_2 = 0$
$t_1 = 0$; $t_2 = -P_0$; $\tau = 0$	at	$x_2 = L$
$u_1 = t_2 = U_n = 0$	at	$x_1 = 0$
$u_1 = t_2 = U_n = 0$	at	$x_1 = L$

The material properties are those corresponding to the Berea Sandstone [2] which given in terms of Biot's parameters are: $\mu = 6x10^9$ N/m^2, $\lambda = 4x10^9$ N/m^2, R = $0.444x10^9$ N/m^2, Q = 1.399 N/m^2, b = $0.19x10^9$ Ns/m^4, β = 0.19, ρ_{11} = 2.418 Kg/m^3, ρ_{22} = 340 Kg/m^3 and ρ_{12} = - 150 Kg/m^3. The exact solution of this basic problem was obtained by Chen et al. with the aid of computer algebra.

Figure 1 shows the displacement at the top versus frequency.

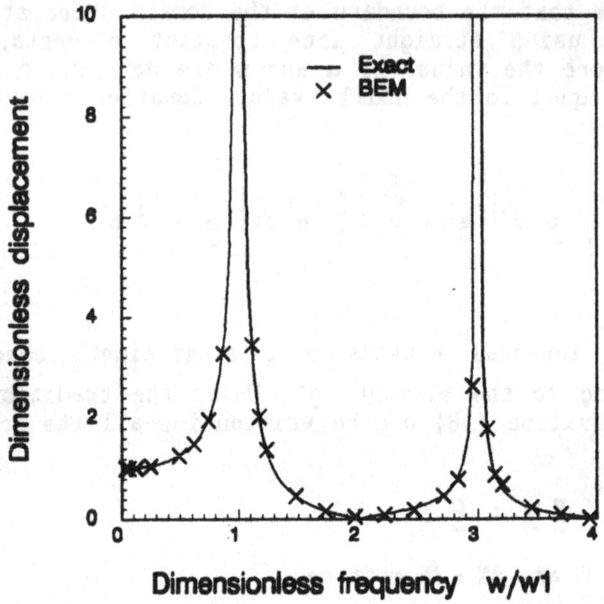

Figure 1.- Displacement at the top versus frequency for normal stress excitation.

The displacement is normalized by the static displacement of an elastic region with the undrained material properties (u_2E_u/P_0L). The frequency is normalized by the first natural frequency of the same elastic region

$$\omega_1 = \frac{\pi}{2L}\sqrt{\frac{E_u}{\rho}}$$

where

$$E_u = \frac{2\mu\,(1-\nu_u)}{(1-2\nu_u)} \quad ; \quad \rho = \rho_{11} + \rho_{22} + 2\rho_{12}$$

and ν_u is the undrained Poisson's ratio which is related to Biot's parameters Q and R. For the present material ν_u = 0.33. The above normalizations are done using the undrained properties because due to the low permeability of the rock the behaviour of the square region is basically elastic with solid and fluid working in a combined form. Only at very low frequencies a drained behaviour can be expected. The BE results and the analytical solution are compared in Figure 2 for a range of frequency ω/ω_1 going from o to 4. The figure shows a remarkable accuracy of the BE solution for the top displacement.

Figure 2 shows a comparison of the pore pressure profile computed for four different frequencies. Again the agreement between the analytical solution and the BE results is very good. The pore pressure at internal points has been obtained using the discretized version of equation (14) which gives τ in terms of the already computed boundary values and the integrals of the fundamental solution kernels over the boundary elements. These non-singular integrals are of the same type and are computed in the same form as those giving rise to the terms $\mathbf{H}^{ij}$ and $\mathbf{G}^{ij}$ of the system of equations when i=j. A four point Gauss quadrature has been used for the integrals over the boundary elements.

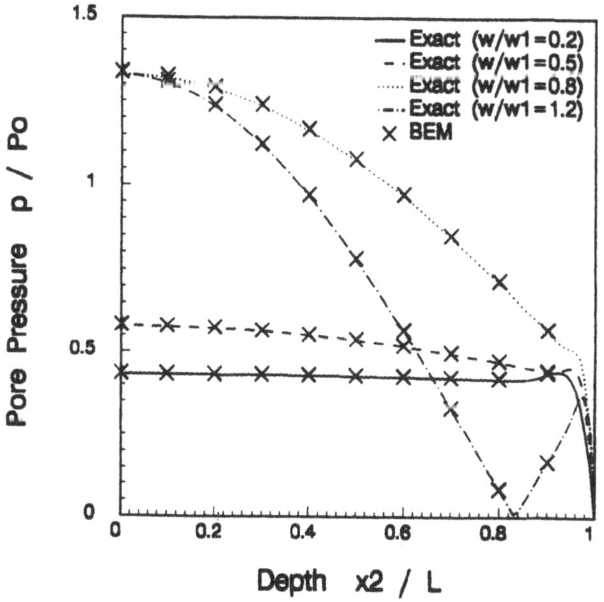

Figure 2.- Absolute value of the pore pressure along the vertical axis for normal stress excitation.

The second problem studied in the present paper deals with the same geometry and material properties as the first. The base of the square is now subject to a uniform vertical displacement harmonic in time with amplitude U_0. The boundary conditions for the BE system of equations are:

$t_1 = 0 \; ; \; u_2 = U_0 \; ; \; U_n = -U_0$ at $x_2 = 0$

$t_1 = t_2 = \tau = 0$ at $x_2 = L$

$u_1 = t_2 = U_n = 0$ at $x_1 = 0$

$u_1 = t_2 = U_n = 0$ at $x_1 = L$

The BE results are compared with the exact analytical solution obtained by Chen et al[2].

In Figure 3 the normalized vertical displacement of the solid u_2/U_0 at the top of the square is plotted versus the dimensionless frequency of excitation ω/ω_1. The resonance peaks and the displacement amplitudes computed by the BEM show a good agreement with the exact analytical solution. The pressure profiles for four different frequencies are shown in Figure 4. The normalized pore pressure pL/E_uU_0 is shown versus depth. The pore pressure values at internal points, obtained from the discretized version of the internal points integral representation given by equation (14), are very close to the exact solution.

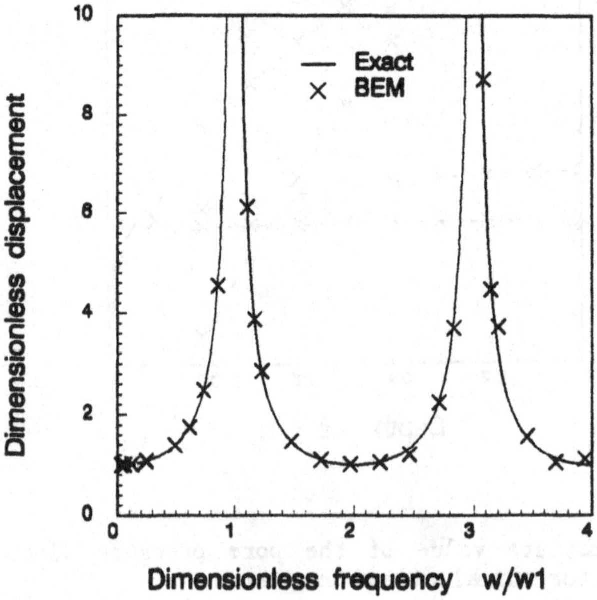

Figure 3.- Displacement at the top versus frequency for vertical displacement excitation of the base.

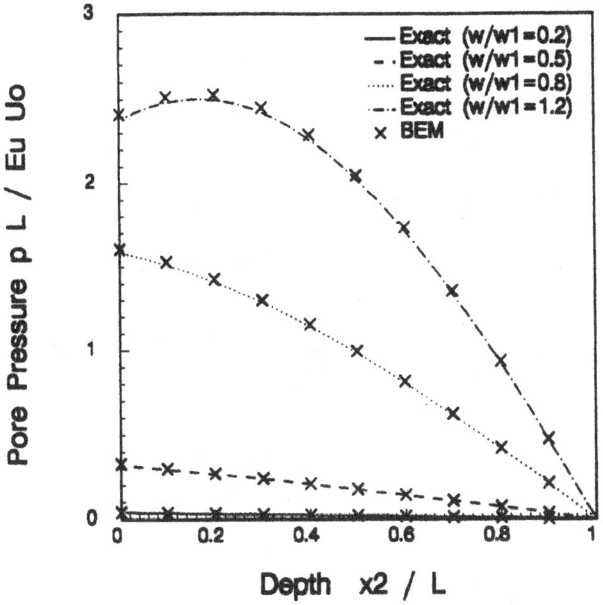

Figure 4.- Absolute value of the pore pressure along the vertical axis for vertical displacement excitation of the base

CONCLUSION

A Boundary Element approach for plane dynamic poroelastic problems has been presented in this paper. The Boundary Element formulation is based on an integral equation representation written in terms of solid displacements and fluid stress.

The formulation has been implemented in a code and validated by the analysis of two numerical examples. The accuracy of the obtained results reveals the great potential of the approach. It can be expected that, similarly to other BE fields of application, more complicated geometries and boundary conditions will require more refined meshes or higher order elements than those used in the present paper.

ACKNOWLEDGEMENT

This work is partially supported by the local goverment of Andalusia under a research grant. The financial support is gratefully acknowledged.

REFERENCES

1.- Dominguez,J.,"An Integral Formulation for Dynamic Poroelasticity", J.Appl.Mech., ASME, Vol.58, (1991).

2.- Cheng, A.H.D., Badmus, T. and Beskos, D.E., "Integral Equation for Dynamic Poroelasticity in Frequency Domain with Boundary Element Solution", J.Eng.Mech., ASCE, (In press),(1991).

3.- Biot, M. A., "Theory of Propagation of Elastic Waves in a Fluid-Saturated Porous Solid," J. Acoust. Soc. Am., Vol.28, pp.168-178, (1956).

4.- Dominguez, J., "Boundary Element Approach for Dynamic Poroeslastic Problems", Int. J. Num. Meth. Eng. (In press).

Saltwater Intrusion in Leaky Aquifers Modelled with B-Spline Elements

J.J.S.P. Cabral *, L.C. Wrobel

Computational Mechanics Institute, Wessex Institute of Technology, Ashurst Lodge, Ashurst, Southampton SO4 2AA, U.K.

* On leave from Water Resources and Environmental Engineering Group, Federal University of Pernambuco, Recife, PE, Brazil

ABSTRACT

Continuity of the derivatives of the main variable is an important feature to obtain an accurate representation of moving boundaries with iterative or time-marching schemes. A recently developed BEM formulation using cubic B-splines provides up to C^2 continuity between adjacent elements. This formulation is applied in this work to saltwater intrusion problems in leaky aquifers.

INTRODUCTION

Moving boundary problems generally demand an accurate computation of both direction and modulus of the velocity at points located on the moving line or surface. When using a discretization procedure with Lagrangian interpolation (even of higher order) oscillations may develop due to the appearance, during the motion, of points of geometric tangent discontinuity between adjacent elements at which the velocity is not unique.

A recently developed BEM formulation based on uniform and non-uniform cubic B-splines [1],[2], capable of preserving continuity of the tangent and curvature (C^2 continuity) between elements, is applied in the present work to study the motion of the interface between fresh and salt water in the context of saltwater intrusion problems in leaky aquifers.

Boundary elements have been applied to saltwater intrusion studies by Awater [3] using a formulation in the complex plane, with application to a lake reclamation problem. Liu *et al.* [4] developed a formulation using real variables and linear elements for the case of confined aquifers while Kemblowski [5], [6] applied a similar formulation to saltwater upconing problems. Taigbenu *et al.* [7] presented a different approach involving a modified potential and areal modelling.

Cabral and Wrobel [8] expanded the formulation of Liu *et al.* [4] by including drains and trenches for simulating water exploitation in confined coastal aquifers while Cabral and Cirilo [9] have used a similar approach to model a semiconfined aquifer. De Lange [10] applied a simplified BEM model to analyse the behaviour of the saltwater/freshwater interface in a three-dimensional problem.

Cubic B-splines are used in this work both for geometric representation and as interpolation functions. The numerical results obtained with the present formulation are compared with analytical and other numerical solutions showing accurate results with coarse discretizations.

PHYSICAL CONSIDERATIONS

In a coastal aquifer, the flow may be assumed everywhere to be orthogonal to the shore line, so that a two-dimensional vertical model may be used to analyse the saltwater intrusion problem.

In many practical cases the thickness of the transition zone between fresh water and salt water is relatively small compared to the aquifer dimensions. It is thus a common approximation to neglect the transition zone and assume a sharp interface separating the two regions. Bear [11] has given some examples of real aquifers in which experimental measurements have shown the salt concentration to vary sharply at a determined location, clearly establishing a region of low salt concentration (fresh water) and a region of high salt concentration (salt water).

Mualem and Bear [12] have shown that, for the flow under a semipervious layer, two different situations frequently occur in practice: the first, sketched in figure 1.a, is the case when the interface actually intersects the semipervious layer; the other case, shown in figure 1.b, is when the exceeding fresh water may leave the aquifer through an outflow face located after the end of the semipervious layer. Analytical and approximate solutions for the shape of the interface in a semiconfined aquifer, under the Dupuit assumption, have been found by Sikkema and van Dam [13], van Dam and Sikkema [14] for 13 different situations, according to the physical parameters of the problem.

A further approximation can be introduced into the mathematical model, by assuming a constant potential value in the saltwater region and analysing only the freshwater region (the Ghyben-Herzberg approximation, see [11]). This approach, although simpler, may lead to unsatisfactory results [10], [15].

In view of the above, the present paper adopts a sharp interface model in a two-dimensional vertical plane including the two regions by employing a standard BEM sub-regions technique [16].

MATHEMATICAL FORMULATION

Combination of Darcy's law and the mass balance equation for a two-dimensional flow through a homogeneous isotropic porous medium gives, for a vertical model,

$$K\left(\frac{\partial^2 u}{\partial x^2}+\frac{\partial^2 u}{\partial y^2}\right)=S\frac{\partial u}{\partial t} \tag{1}$$

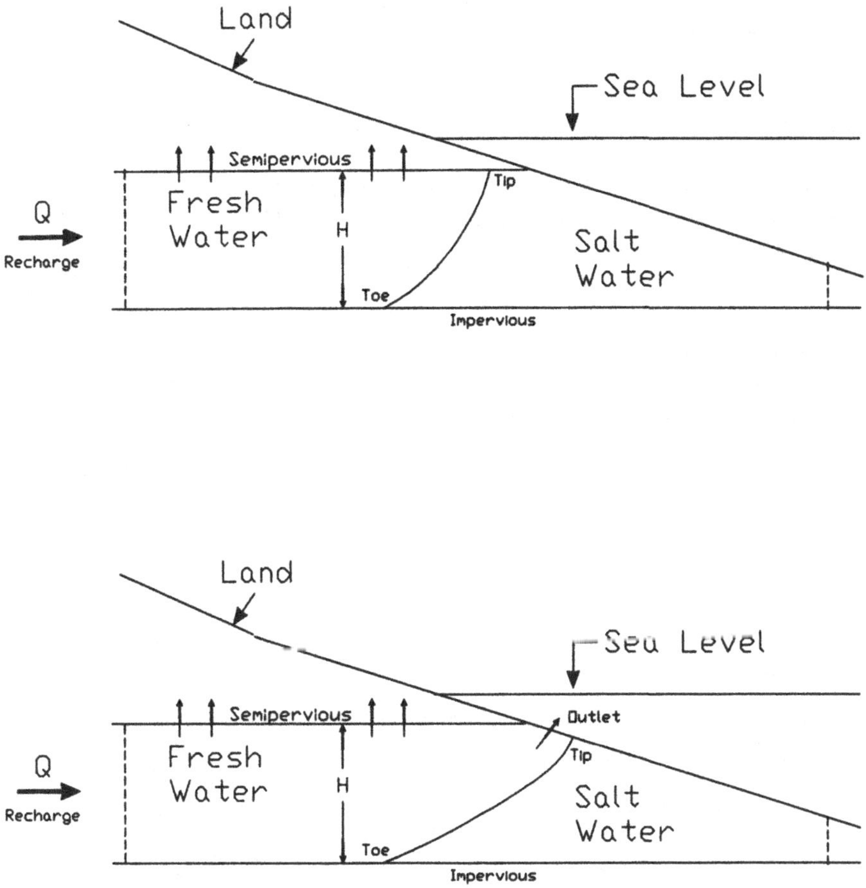

Figure 1: Saltwater intrusion in semiconfined aquifer: a) interface intersecting semipervious layer; b) outflow face

where K is the hydraulic conductivity, S the specific storage and the potential u is the piezometric head

$$u = y + \frac{p}{\gamma}$$

in which p is pressure, γ is specific weight and y is elevation.

For practical purposes, transient flows can be treated as a sequence of successive steady states in a step-by-step procedure. So, the problem can be modelled using Laplace's equation

$$\frac{\partial^2 u}{\partial x^2} + \frac{\partial^2 u}{\partial y^2} = 0 \tag{2}$$

In the present paper, this equation is applied to each region independently, and compatibility and equilibrium conditions employed to relate the potential and normal flux on the interface between regions [16].

The following integral equation, equivalent to (2), can be obtained after some transformations [16]

$$c^i u^i + \int_\Gamma u q^* d\Gamma = \int_\Gamma u^* q d\Gamma \tag{3}$$

where $c^i = \theta/2\pi$, θ is the internal angle at the point i on the boundary Γ, u^* is the fundamental solution of Laplace's equation, $q = \partial u/\partial n$ and $q^* = \partial u^*/\partial n$.

The values of the potential and its normal derivative using cubic B-splines as interpolation functions are given at any point by

$$u(\omega) = E_0(\omega)a_{i-1} + E_1(\omega)a_i + E_2(\omega)a_{i+1} + E_3(\omega)a_{i+2} \tag{4}$$

$$q(\omega) = E_0(\omega)b_{i-1} + E_1(\omega)b_i + E_2(\omega)b_{i+1} + E_3(\omega)b_{i+2} \tag{5}$$

This expression applies to the segment between nodes i and $i+1$; a_i are the coefficients (control points) of the potential representation and b_i the coefficients of the normal derivative; $E_k(\omega)$ are the blending functions for cubic B-splines given in [1]; and ω is a local parameter which varies between 0 and 1 within each segment. To represent corners, multiple knots are used and non-uniform blending functions must be applied as reported in [2].

The present formulation employs isoparametric boundary elements for which the coordinates x and y of each point are also expressed by B-spline functions

$$x = E_0 X_{i-1} + E_1 X_i + E_2 X_{i+1} + E_3 X_{i+2} \tag{6}$$

$$y = E_0 Y_{i-1} + E_1 Y_i + E_2 Y_{i+1} + E_3 Y_{i+2} \tag{7}$$

where X and Y are the coordinates of the control points.

Application of the boundary integral equation (3) to all nodal points, incorporating approximations (4) to (7) within each boundary element, generates a system of algebraic equations which, in the present case, will have the form

$$\mathbf{Ha} = \mathbf{Gb} \tag{8}$$

Evaluation of the coefficients of matrices **H** and **G** is discussed in detail in [1] and [2].

Boundary conditions
The boundary conditions for each region are of the following types:

- Specified potential:

$$u = \overline{u}$$

- Specified normal derivative:

$$q = \overline{q}$$

- Mixed condition (semipervious layer):

$$q = \frac{u - \overline{u}'}{c}$$

where $\overline{u}'$ is the specified potential above the semipervious layer and c its hydraulic resistance, *i.e.*

$$c = \frac{d}{K'}$$

in which d and K' are the thickness and the hydraulic conductivity of the semipervious layer, respectively.

Each of the above boundary conditions are related to the potential or flux, but generally the coefficients a_i and b_i are not known. So one must modify the system and substitute the coefficients by their relation with the potential and flux.

Writing expression (4) for all collocation points and collecting the result in matrix form gives

$$\begin{Bmatrix} u_1 \\ u_2 \\ \cdot \\ \cdot \\ u_n \end{Bmatrix} = \begin{bmatrix} E_{11} & E_{12} & \dots & E_{1n} \\ E_{21} & \dots & \dots & E_{2n} \\ \cdot & \cdot & \cdot & \cdot \\ \cdot & \cdot & \cdot & \cdot \\ E_{n1} & E_{n2} & \dots & E_{nn} \end{bmatrix} \begin{Bmatrix} a_1 \\ a_2 \\ \cdot \\ \cdot \\ a_n \end{Bmatrix} \tag{9}$$

Each row of matrix **E** of nodal values of blending functions has three non-zero terms (four blending functions but one of them is null at each node), and all the other terms are zero.

Inverting equation (9) and substituting for vector **a** in (8), adopting a similar procedure for vector **b**, produces

$$\mathbf{H}\mathbf{E}^{-1}\mathbf{u} = \mathbf{G}\mathbf{E}^{-1}\mathbf{q} \tag{10}$$

The next step in the formulation is to combine an equation of the above type for each region through the application of interface conditions. For each node along the interface between the freshwater and saltwater zones the following conditions are used:

- Equilibrium of pressure on both sides of the interface, leading to

$$u_s = \frac{u_f}{\chi} + \frac{\chi - 1}{\chi} y \tag{11}$$

where $\chi = \rho_s/\rho_f$, ρ is the density and the subscripts f and s are related to the fresh and salt waters.

- Compatibility of fluxes, which gives [17]

$$q_s = -\frac{\alpha}{\chi} q_f \tag{12}$$

where $\alpha = \mu_s/\mu_f$ and μ is the dynamic viscosity.

TRANSIENT PROCEDURE

After solving the BEM system of equations, the normal derivatives at the interface are used to compute the interface motion [17]. Assuming $x = \lambda(y, t)$, the following equation can be written:

$$\frac{\partial \lambda}{\partial t} = -\frac{1}{\sin\beta}\frac{K}{\epsilon} q_f \tag{13}$$

in which β is the angle between the interface and the horizontal axis, and ϵ is the effective porosity.

Applying a finite difference approximation to represent the interface motion equation (13) becomes

$$\lambda^{t+\Delta t} = \lambda^t - \frac{\Delta t\, K}{\epsilon \sin\beta^t}\left[\theta q_f^{t+\Delta t} + (1-\theta) q_f^t\right] \tag{14}$$

The above equation involves a linearization since the sine of β is calculated at time t although the equation is written for time $t + \Delta t$. The use of small time steps is generally sufficient to produce accurate results without iteration.

COMPUTATIONAL IMPLEMENTATION

Quasi-singularity

It is a common feature of groundwater analysis that the horizontal dimension of aquifers is generally much larger than the vertical one. Numerical problems may occur if, for example, a source point on the top surface of the aquifer is very close to a boundary element on the bottom surface or interface.

Furthermore, on dealing with moving interface problems, quasi-singularities may develop after some time steps have ellapsed due to a source point becoming very close to a neighbouring element or due to a very large length ratio between adjacent elements.

To avoid this problem, Telles' algorithm [18] for quasi-singular integrals was applied. In this algorithm, according to the smallest relative distance from the element to the source point, a cubic transformation is used and a higher concentration of integration points obtained in the region of the element nearest to the source point, improving the accuracy of standard Gaussian integration schemes.

Interpolated elements

When implementing Dirichlet boundary conditions (specified potential) for the outflow face in saltwater intrusion analyses, there are two unknown fluxes to be calculated at the interface tip.

The singularity of matrix **G** can be avoided using the so-called *interpolated element* [19] in which the source point is assumed somewhere inside the element rather than at the extreme points as usual. This procedure differs from the non-conforming elements [16] because in the interpolated element algorithm, the unknowns remain at the extreme points and only the collocation point is moved. So, one avoids the extrapolation of functions that could introduce inaccuracies.

Several tests [19] have shown that placing the source point at a quarter of the element length generally gives optimum results for the interpolated element implementation. In this work, the source point was assumed at a position corresponding to a B-spline parameter ω equal to 0.4 if the singularity is at the beginning of the element and at $\omega = 0.6$ if it is at the end of the element; this gives a position near a quarter element size.

For computing the coefficients of matrices **H** and **G** with the singularity of the collocation point inside the element, Telles' algorithm [18] was used with more integration points than in the case of singularity at the extreme. Accurate results were obtained with 24 integration points for the case of singularity inside the element and 10 integration points for the cases of non-singularity and singularity at the extreme point of the element.

Discharge computation

The discharge rate through each boundary element can be obtained by integrating the flux along the element length

$$Q_{elem} = \int_{\omega=0}^{\omega=1} -K\frac{\partial u}{\partial n} d\Gamma \tag{15}$$

Applying expressions (5) to (7) into equation (15), one obtains

$$Q_{elem} = -K \int_{\omega=0}^{\omega=1} [E_0(\omega)b_{i-1} + E_1(\omega)b_i + E_2(\omega)b_{i+1} + E_3(\omega)b_{i+2}] \mid G \mid d\omega \tag{16}$$

which can be evaluated using standard Gauss quadrature.

Time stepping

Cubic B-splines may provide superior spatial representation but present the same limitations of Lagrangian elements regarding temporal representation.

In the present implementation, small time steps are adopted at the beginning of the process; as time progresses, this value is increased based on the geometrical and physical characteristics of the problem and on mass continuity concepts, *i.e.* the equilibrium between the given rate of inflow and the calculated rate of outflow.

Tip motion

For the case in which the interface intersects the semipervious layer (figure 1.a), the normal derivative at the interface tip is computed similarly to the other interface nodes, and its motion follows the same procedure.

For the other case, when there is an outflow face (figure 1.b), the interface tip presents a singular behaviour. Bear [11] and Liu *et al.* [4] have shown that the interface should intersect the outflow face at a right angle at all times. From the BEM point of view, the potential is known at the intersection point but there are two unknown fluxes; thus, an interpolated element was used to overcome the problem. The normal derivative at this point is always null and, for calculating the location of the interface tip, an extrapolation was performed assuming a parabolic shape for the interface (following Bear [11]).

Remeshing

Due to the interface motion, the size of the freshwater and saltwater regions varies during the analysis. Thus, a node redistribution scheme ("remeshing") is needed after each time step. An efficient remeshing algorithm was developed which preserves the total number of elements and nodes of the initial discretization (although the number of elements and nodes in each region may change) and avoids that the elements located along the top and bottom boundaries become excessively large or excessively small, in comparison to their neighbours.

The basic ideas of the algorithm are as follows:

- The initial position of all nodes along the top and bottom boundaries is stored in memory;
- The interface toe displacement is calculated after each time step. The algorithm then finds the node in the *initial discretization* which is nearest to the displaced toe position, and moves this node;
- An analogous procedure is used for the interface tip;
- By always referring the interface toe and tip displacements to the initial discretization, elements and nodes are free to migrate from one region to the other. The algorithm keeps track of this movement, and assigns the boundary conditions at each node accordingly.

APPLICATION

Bear [11] considers the problem of saltwater intrusion in coastal aquifers as a management problem in which, by dealing with the rate of natural and artificial recharges and that of pumping, the extent of seawater intrusion can be controlled.

The present formulation has been applied to a leaky aquifer on a coastal plain where the phreatic groundwater table (p) above the semiconfined aquifer is at the mean sea level (h_s) (see van Dam [20], case d).

Van Dam presented an analytical solution to the one-dimensional steady flow under the Dupuit assumption which is as follows:

$$u = \chi \left[h_s + \frac{3E}{2} + \frac{(x - x_0)^2}{6Kc} \right]$$

$$Q = -\frac{\chi (x - x_0)^3}{18Kc^2} - \frac{\chi (x - x_0) E}{2c}$$

where Q is the groundwater flow rate per unit width (m^2/s) and E is the equilibrium depth. For the case under study, $-E = h_s = p$ and Q can be related to the normal derivative q in the BEM formulation by $Q = q \times K \times$ aquifer thickness.

The results presented in what follows refer to the case depicted in figure 1.a. The initial upstream recharge was assumed to be $Q = 6.78 \times 10^{-5}$ m^2/s ($q = 0.0339$); later, the recharge rate is increased to $Q = 1.7 \times 10^{-4}$ m^2/s ($q = 0.085$). Other physical characteristics of the problem are $\rho_f = 1000$ kg/m^3, $\rho_s = 1029$ kg/m^3, aquifer thickness = 200 m, $K = 10^{-5}$ m/s, $\epsilon = 1.0, \mu = 1.2 \times 10^{-3}$ kg/m.s and $c = 10^8$ s.

For the numerical analysis, an arbitrary position of the interface was initially assumed and a study undertaken to determine the initial steady-state configuration. Then, another analysis was performed, with the second recharge rate, to compute the transient behaviour of the saltwater tongue.

Figure 2 depicts the interface location at different times, obtained with the B-spline formulation and with linear boundary elements [9]. The number of nodes along the interface was 5 in the B-spline analysis and 7 with linear elements. It is interesting to notice that, from $t = 0$ to $t = 425$ days, a small change in the interface shape can be seen; at later times, the shape remains practically the same and only the interface position changes.

In figure 3, a comparison is made between the initial and final steady-state positions (for the two values of Q), for the linear and B-spline boundary elements and the analytical solutions of van Dam [20]. The agreement is very reasonable. It has been observed, however, that discrepancies between the numerical and analytical solutions appear as the value of Q increases, and become significant for large values of Q.

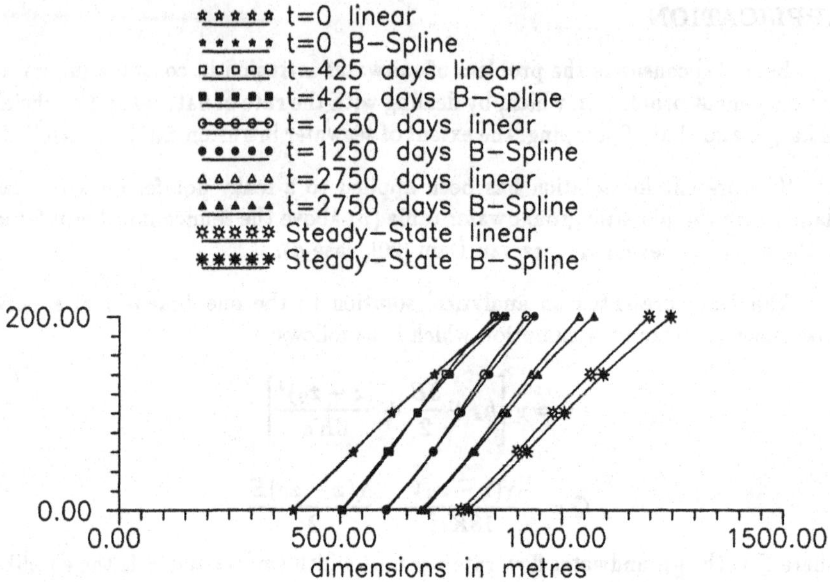

Figure 2: Transient behaviour of interface (different horizontal and vertical scales)

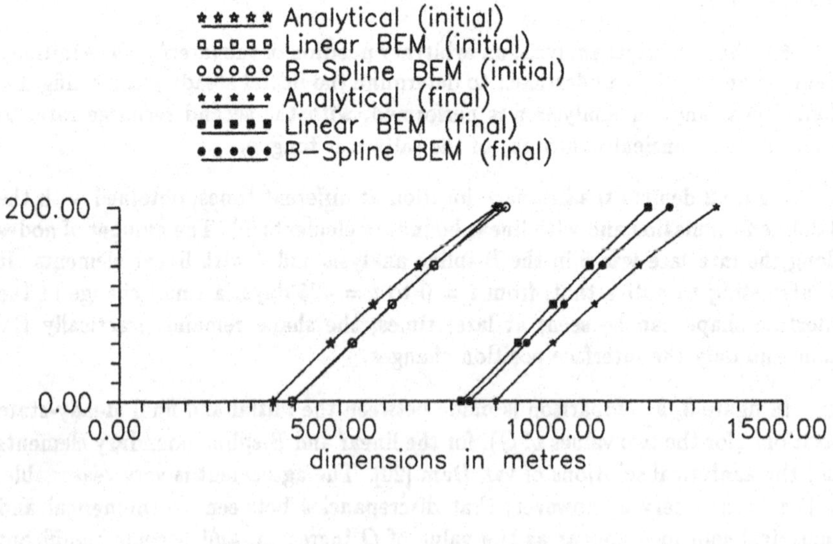

Figure 3: Steady positions of interface (different horizontal and vertical scales)

CONCLUSIONS

For transient flow in porous media with a moving interface, the normal derivative at points along the moving surface can be accurately computed using elements that provide derivative continuity. The results obtained suggest that, for the same level of accuracy, one can generally use less degrees of freedom with B-spline elements than other kinds of elements.

The model outlined in this paper can be used as a practical management tool to locate the saltwater wedge position by controlling the rate of upstream recharge.

ACKNOWLEDGMENT

The first author would like to acknowledge the financial support of CAPES/Ministry of Education, Brazil.

References

[1] Cabral,J.J.S.P., Wrobel,L.C. and Brebbia,C.A., A BEM formulation using B-splines: I- uniform blending functions, *Engineering Analysis*, Vol. 7, pp. 136-144, 1990.

[2] Cabral,J.J.S.P., Wrobel,L.C. and Brebbia,C.A., A BEM formulation using B-splines: II- multiple knots and non-uniform blending functions, *Engineering Analysis*, Vol. 8, pp. 51-55, 1991.

[3] Awater,R., The transient behaviour of a fresh/salt water interface, *New Developments in Boundary Element Methods*, Computational Mechanics Publications, Southampton and Springer-Verlag, Berlin, 1980.

[4] Liu,P.L-F., Cheng,A.H-D., Liggett,J.A. and Lee,J.H., Boundary integral equation solutions to moving interface between two fluids in porous media, *Water Resources Research*, Vol. 17, pp. 1445-1452, 1981.

[5] Kemblowski,M., Saltwater upconing under a river - a boundary element solution, *BEM VI*, Computational Mechanics Publications, Southampton and Springer-Verlag, Berlin, 1984.

[6] Kemblowski,M., Saltwater-freshwater transient upconing - an implicit boundary element solution, *Journal of Hydrology*, Vol. 78, pp. 35-47, 1985.

[7] Taigbenu,A.E., Liggett,J.A. and Cheng,A.H-D., Boundary integral solution to seawater intrusion into coastal aquifers, *Water Resources Research*, Vol. 20, pp. 1150-1158, 1984.

[8] Cabral,J.J.S.P. and Wrobel,L.C., A numerical analysis of saltwater intrusion in coastal aquifers, *Brazilian Engineering Journal, Water Resources Division*, Vol. 3, pp. 29-52, 1985 (in Portuguese).

[9] Cabral,J.J.S.P. and Cirilo,J.A., Saltwater-freshwater interface motion in leaky aquifers, *BETECH 87*, Computational Mechanics Publications, Southampton, 1987.

[10] de Lange,W.J., Application of the boundary integral element method to analyse the behaviour of a freshwater-saltwater interface calculating three-dimensional groundwater flow, *9th Salt Water Intrusion Meeting*, Delft, 1986.

[11] Bear,J., *Hydraulics of Groundwater*, McGraw-Hill, New York, 1979.

[12] Mualem,Y. and Bear,J., The shape of the interface in steady flow in a stratified aquifer, *Water Resources Research*, Vol. 10, pp. 1207-1215, 1974.

[13] Sikkema,P.C. and van Dam,J.C., Analytical formulae for the shape of the interface in a semiconfined aquifer, *Journal of Hydrology*, Vol. 56, pp. 201-220, 1982.

[14] van Dam,J.C. and Sikkema,P.C., Approximate solution of the problem of the shape of the interface in a semi-confined aquifer, *Journal of Hydrology*, Vol. 56, pp. 221-237, 1982.

[15] Kemblowski,M., The impact of the Dupuit-Forchheimer approximation on salt-water intrusion simulation, *Ground Water*, Vol. 25, pp. 331-336, 1987.

[16] Brebbia,C.A., Telles,J.C.F. and Wrobel,L.C., *Boundary Element Techniques*, Springer-Verlag, Berlin, 1984.

[17] Liggett,J.A. and Liu,P.L-F., *The Boundary Integral Equation Method for Porous Media Flow*, Allen and Unwin, London, 1983.

[18] Telles,J.C.F., A self-adaptive coordinate transformation for efficient numerical evaluation of general boundary element integrals, *International Journal for Numerical Methods in Engineering*, Vol. 24, pp. 959-973, 1987.

[19] Marques,E., Coupling of the finite element method and the boundary element method: an application to potential problems, *M.Sc. Thesis*, COPPE/UFRJ, Rio de Janeiro, 1986 (in Portuguese).

[20] van Dam,J.C., The shape and position of the salt water wedge in coastal aquifers, *Hamburg Symposium in Relation of Groundwater Quantity and Quality*, IAHS, Publication No. 146, 1983.

2-D Steady State and Transient Analysis of Salt Water Intrusion into Coastal Aquifers by a Direct Time Domain BEM

D.C. Rizos, D.L. Karabalis

Department of Civil Engineering, University of South Carolina, Columbia, SC 29208, U.S.A.

ABSTRACT

A direct time domain Boundary Element Method (BEM) is applied to the groundwater flow problem in a 2-D unconfined coastal aquifer. A time marching numerical scheme and quasi-steady state approximations are utilized in the proposed solution. Numerical examples and a comparison study is presented.

FORMULATION

Coastal aquifers are distinguished by the existence of a water table and an interface layer between the fresh and the salt water. The solution of the problem is for the steady and the transient location of the aforementioned moving boundaries due to tidal fluctuations. As it is shown in Figure 1, the solution domain consists of a vertical plane section through the fresh water zone of the aquifer. The dispersion zone of the fresh and the salt water along the interface layer has been neglected. It has also been assumed that there is no flow in the salt water region throughout time. The governing equation of the problem is the Laplace's equation, i.e.,

$$\nabla^2 \Phi(x,y) = 0 \tag{1}$$

where Φ is the piezometric head. The boundary conditions of the problem are dictated by the nature of the physical boundaries, as shown in Figure 1, and in the case where a water table exists, the dynamic and kinematic boundary conditions define the location of its free surface [1]. In the proposed model the interface layer is considered to be a zero flux boundary, due to the assumption of immiscibility of the two fluids and the quasi-steady state approach [2,3].

The Boundary Integral Equation (BIE) associated to Eq. (1) relates the values of the piezometric head Φ at any interior or boundary point, to the head Φ and its normal

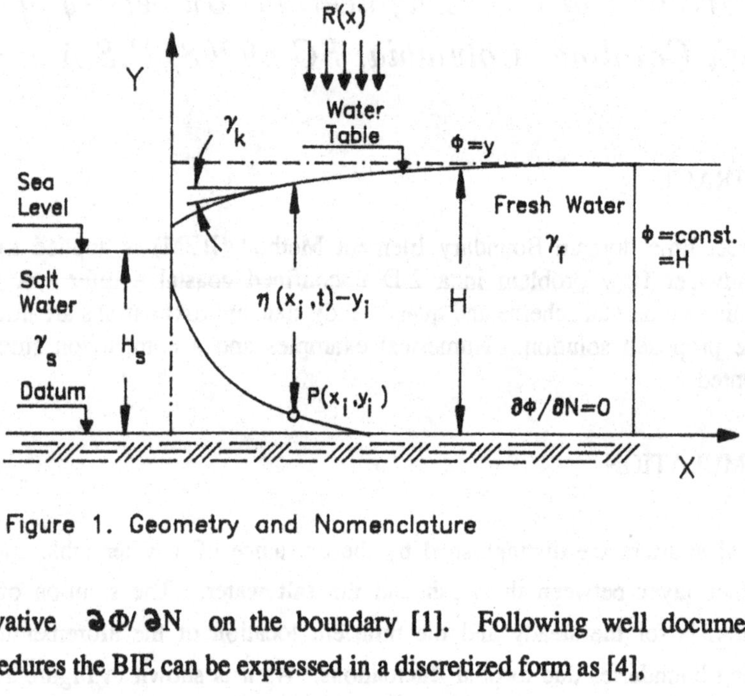

Figure 1. Geometry and Nomenclature

derivative $\partial\Phi/\partial N$ on the boundary [1]. Following well documented procedures the BIE can be expressed in a discretized form as [4],

$$-\delta_{km}\beta_k\Phi_k \sum_{i=1}^{M} [G_m]_k^i \{\Phi_m\}^i = \sum_{i=1}^{M} [H_m]_k^i \left\{\left(\frac{\partial\Phi}{\partial N}\right)_m\right\}^i \qquad (2)$$

$$k=1,\dots,N \qquad m=1,\dots,n$$

where N is the total number of boundary points, M is the total number of boundary elements and δ_{km} is the Kronecker delta, $\{\Phi_m\}^i$ and $\{(\partial\Phi/\partial N)_m\}^i$ are the vectors of the potential and its normal derivative defined on the n nodal points of the source element S_i, and $[G_m]_k^i$ and

$[H_m]_k^i$ are the vectors of the influence coefficients associated with the boundary point P_k and are defined by

$$[G_m]_k^i = \int_{(S_i)} F_m \frac{\partial \ln r(P_k,Q)}{\partial N} dS$$

$$[H_m]_k^i = \int_{(S_i)} F_m \ln r(P_k,Q)\, dS \tag{3}$$

with $r(P_k,Q)$ denoting the distance between the receiver point P_k and an arbitrary source point Q on the element S_i, and F_m being the shape functions associated with the nodal points of element S_i. In the proposed solution quadratic isoparametric elements are used for the spatial discretization. Eq. (2) can be cast in a matrix form as

$$[L]\{\Phi\}=[R]\{\partial\Phi/\partial N\} \tag{4}$$

where $\{\Phi\}$ and $\{\ \Phi/\ N\}$ are the boundary conditions and [L] and [R] are the associated coefficient matrices.

FREE SURFACE SOLUTION

In unconfined aquifers, the location of the water table is not known a priori, and, therefore, both the potential and its normal derivative are unknown. However, the combination of the dynamic with the kinematic boundary condition on the free surface can be expressed in a finite difference form as [1],

$$\Phi_k^{t+\Delta t} = \Phi_k^t - \frac{K\,\Delta t}{n_e \cos\gamma_k^t}\left[\Theta\left(\frac{\partial\Phi}{\partial N}\right)_k^{t+\Delta t} + (1-\Theta)\left(\frac{\partial\Phi}{\partial N}\right)_k^t\right] + \frac{\Delta t}{n_e}\left[\Theta\, R(x_k,t+\Delta t) + (1-\Theta)\, R(x_k,t)\right] \tag{5}$$

$$\eta_k^{t+\Delta t} = \Phi_k^{t+\Delta t}$$

where Δt is the time increment, K is the soil permeability, n_e is the effective porosity, γ_k is the angle formed by the tangent to the free surface and the horizontal, Θ is a weighting factor that positions the derivative between time levels t and $t+\Delta t$ and varies from 0.5 to 1.0 [1], the subscript k indicates a nodal point on the free boundary, and R represents the vertical recharge rate if surface accretion is to be taken into consideration, and η is the elevation of a point on the free surface as shown in Figure 1. The time marching scheme expressed by Eq. (5) relates the potential of the free surface at time $t+\Delta t$ to the potential at time t and its normal derivative at times t and $t+\Delta t$. Thus, assuming that the solution is known at time t, Eq. (5) can be used as the missing boundary condition in conjunction with Eq. (4), the solution of which yields the normal derivative of the potential on the free surface at time $t+\Delta t$. The new location of the water table can then be calculated from Eq. (5).

INTERFACE SOLUTION

In the proposed solution a successive steady state approach for the interface solution has been adopted. Thus, at each time step it is assumed that the interface has reached its steady state location which satisfies the hydrostatic condition of pressure equilibrium on both sides of it. The water pressure in the fresh (P_f) and salt (P_s) water region at a point (x_i,y_i) on the interface is expressed as

$$P_f(x_i) = (\eta(x_i)-y_i)\,\gamma_f$$

$$P_s(x_i) = (h_s-y_i)\,\gamma_s \tag{6}$$

where h_s is the sea level as shown in Figure 1, and γ_f and γ_s are the specific weights of the fresh and salt water, respectively. From the pressure equilibrium the location of the interface can be obtained as,

$$y_i = \frac{r\,h_s - \eta(x_i)}{r - 1} \qquad (7)$$

where $r=\gamma_s/\gamma_f$. The horizontal location of the interface toe is at $y_i=0.0$ and corresponds to the x coordinate of the free surface point for which $\eta(x_{toe}) = r\,h_s$.

NUMERICAL EXAMPLES

The proposed method has been applied to the problem of groundwater flow in two coastal aquifer models. In the first application, the unconfined coastal aquifer of Figure 2 is studied for the steady location of the interface layer. Figure 2 shows a good agreement of the obtained results with experimental results, analytical results based on the Dupuit's assumption, and numerical results based on a BEM formulation, all of which are presented by Fukuhara and Fukui [5].

In the second application the coastal aquifer of Figure 3 is studied. The problem is stated in a dimensionless form using as a characteristic length scale the height of the model H [6]. The dimensionless variables are $L_o^*=10.0$, $H_o^*=1.0$, $H_s^*=0.7$ and $\Delta t^*=0.2$. The system is initially left to reach its steady state. Thereafter, the sea elevation is allowed to oscillate in a sinusoidal fashion with amplitude $A^*=0.05$ and period $T^*=100\Delta t^*$. Figure 3 shows the transient location of the moving boundaries for high, mean and low tide positions.

CONCLUSIONS

A time marching BEM scheme is proposed for the location of the water table and the salt-fresh water interface of a 2-D unconfined coastal aquifer. The proposed model is based on successive steady state approximations and equilibrium of pressures on either side of the interface. The obtained results are in close agreement with experimental data, as well as other numerical and analytical studies.

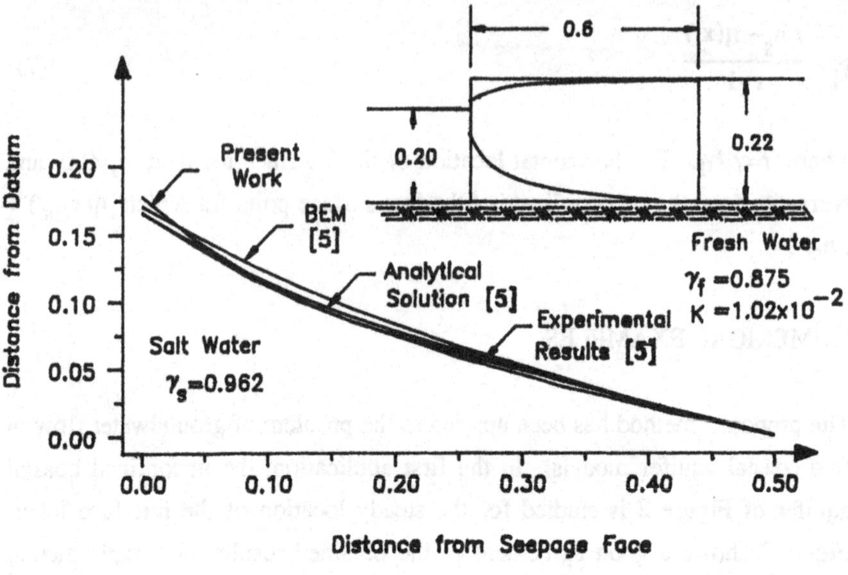

Figure 2. Location of Fresh–Salt Water Interface Under Steady State Conditions

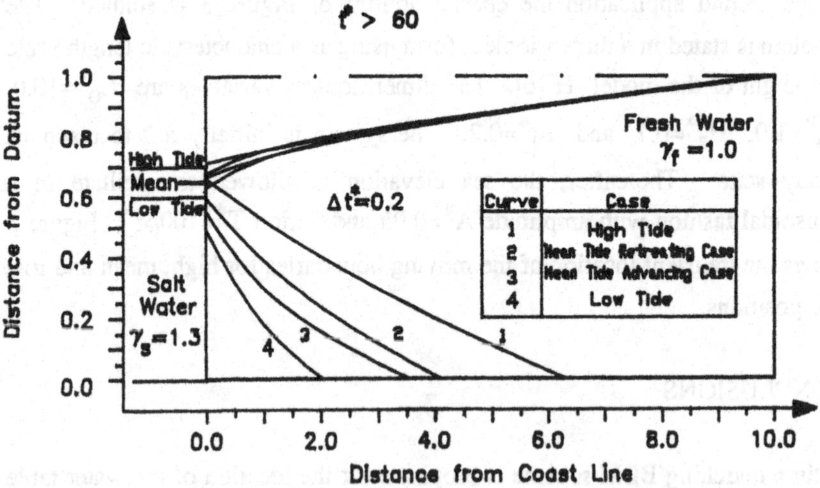

Figure 3. Coastal Aquifer During A Typical Tidal Cycle

REFERENCES

1. Liggett, J.A., and Liu, P.L-F., *The Boundary Integral Equation Method for Porous Media Flow*, George Allen & Unwin, London, 1983.
2. Glover, R.E., "The Patern of Fresh Water Flow in A Coastal Aquifer," *Journal of Geophysical Research*, **64**, No. 4, 457-459, April 1959.
3. McWhorter, D.B., and Sunada, D.K., *Ground Water Hydrology and Hydraulics*, Water Resources Publications, Fort Collins, Colorado, 1981.
4. Rizos, D.C., and Karabalis, D.L., "Transient Groundwater Flow in Unconfined Coastal Aquifers," *Water Resources Research* (in preparation).
5. Fukuhara, T., and Fukui, T., "Analysis of Fresh-Salt Water Interface in Coastal Aquifers," pp. 869-876, in : Tanaka, M., and Brebbia, C.A., Eds., *BoundaryElements VIII*, Computational Mechanics Publications, Southampton, 1986.
6. Rizos, D.C., "Direct Time Domain BEM Solutions of 2-D Groundwater Problems Governed by Laplace's Equation," M.S. Thesis, University of South Carolina, Columbia, SC, 1989.

SECTION 7: HEAT TRANSFER

Boundary Element Solution of Combined-Mode Heat Transfer in Radiatively Participating Media

A.J. Kassab, C.J. Saltiel

University of Florida Center for Advanced Studies in Engineering, 3950 RCA Blvd, Suite 5003, West Palm Beach Florida, 33410, U.S.A.

ABSTRACT

One of the major difficulties in the numerical solution of multi-mode heat transfer problems in radiatively participating media involves grid compatibility between conductive and radiative computational domains. Since the grid density of the conductive modes must often be much greater than that of the radiative grid, a two-grid approach is often employed, requiring some degree of interpolation between grids. Implementation of the boundary element method, as opposed to finite-difference or finite-element schemes, overcomes the incompatibility problem. An exchange factor method is used to provide the radiation flux vector in the energy equation, and the boundary element formulation for the energy equation in the presence of radiation involves both a surface and a domain integral. However, in the iterative solution of the problem, the domain integral is computed using the radiative grid break-up. Details of the formulation are presented. A benchmark problem is used to validate the approach.

INTRODUCTION

Resolution of the combined effects of multimode heat transfer in radiatively participating media is an important consideration in the engineering analysis of high temperature processes. The study of these interactions is quite difficult since solution of a nonlinear integro-differential equation, resulting from a First Law analysis, is required. There has been an extensive effort devoted to the solution of the combined-mode problem, with particular emphasis placed on the solution of combined conduction-radiation in planar [1], rectangular[2-8], and cylindrical [9] systems. These efforts relied upon iterative solutions based on finite difference methods (FDM) or finite element methods (FEM) for the solution of the energy equation. The radiative heat source was solved in a variety of ways, including the spherical harmonics [2], zonal [3], finite element [4], exponential integral function [5], discrete exchange factor [6,7], and discrete ordinate [8] methods. The resulting numerical schemes lead to the iterative use of two different computational grids, one tailored to the radiative flux vector evaluation and the other tailored to the FDM or FEM solution of the energy equation. The temperature field must then be interpolated between these two grids. This can lead to substantial inaccuracies and degradation of the solution in certain circumstances when one grid is very coarse and the other very fine. It is herein proposed, that the boundary element method be applied to the solution of this problem. The boundary element formulation for the energy equation involves both a weighted contour integral of the temperature and heat flux along the domain boundary and a weighted integral of the radiative flux vector over the domain. The integrand of the domain integral is the roduct of the Green's free space solution of the homogeneous form of the governing equation and the divergence of the radiative flux vector. The accurate evaluation of this domain integral is limited by the accuracy of the numerical approximation for the radiative flux vector. As this evaluation is computationally intensive, in practice a relatively coarse grid is used to discretize the domain and the radiative flux vector is often approximated as piecewise constant over regions of that grid [10]. This naturally leads to evaluating the domain integral in the boundary element formulation on discretized grid already generated for the radiative flux computation. There is no grid incompatibility between the two computational domains, and actually the advantage of the existing radiative computational grid is exploited. Detailed consideration is given to the formulation of the problem and the numerical implementation of the boundary element solution. The approach is validated by comparing the numerical solution to the benchmark problem originally presented by Ratzel and Howell [2]. In addition to addressing the radiative heat source term, we also included a non-radiative heat source, which may be present in chemically reactive systems and in certain manufacturing applications.

PROBLEM FORMULATION

The energy equation governing the temperature field in a two-dimensional domain in which conductive and radiative heat transfer modes are both present can be expressed as [6]

$$4N_s \left[\frac{\partial^2\Theta(\tau_x,\tau_y)}{\partial\tau_x^2} + \frac{\partial^2\Theta(\tau_x,\tau_y)}{\partial\tau_y^2} \right] + G(\tau_x,\tau_y)+Q_r(\Theta,\tau_x,\tau_y)=0 \tag{1}$$

The Stark number, $N_s= k_oK_a/4\sigma T_o^3$, is the radiation-conduction parameter, k_o is a reference thermal conductivity at a reference temperature T_o, K_a is the absorption coefficient for the medium, and σ is the Stefan-Boltsmann constant. The dependent variable $\Theta(\tau_x,\tau_y)=T(\tau_x,\tau_y)/T_o$ is the dimensionless temperature and the independent variables $\tau_x=K_tx$ and $\tau_y=K_ty$ are the optical distances of the media in x and y directions. The extinction coefficient K_t is equal to the sum of the absorption and scattering coefficients of the media. The dimensionless non-radiative volumetric generation term is $G(\tau_x,\tau_y)$. The dimensionless radiative volumetric generation term $Q_r(\Theta,\tau_x,\tau_y)$ is related to the divergence of the radiative heat flux vector as

$$Q_r(\Theta,\tau_x,\tau_y)=- \nabla \cdot \vec{q}^{\,r}(\Theta,\tau_x,\tau_y) \tag{2}$$

In the exchange factor formulation, $Q_r(\Theta,\tau_x,\tau_y)$ is expressed as an integral function of the dimensionless temperature to the fourth power [10]

$$Q_r(\Theta, \vec{r_i})=4(1-\omega_o)\Theta^4(\vec{r_i}) - \int_\Gamma \epsilon\, \Theta^4(\vec{r_j})\, DSV(\vec{r_j}, \vec{r_i})\, d\Gamma$$
$$- \int_\Omega 4(1-\omega_o)\, \Theta^4(\vec{r_j}) DVV(\vec{r_j}, \vec{r_i})\, d\Omega \tag{3}$$

where, $\vec{r_i}$ is the position vector in an arbitrary domain Ω bounded by a surface Γ,

ω_o is the scattering albedo, ϵ is the surface emissivity, and $DSV(\vec{r_j}, \vec{r_i})$ and $DVV(\vec{r_j}, \vec{r_i})$ are surface to volume and volume to volume total exchange factors between location $\vec{r_i}$ and location $\vec{r_j}$, respectively. The total exchange factor expressions account for direct radiative exchange between surfaces and volumes, and they account for the effects of multiple scattering within the media (isotropic and anisotropic scattering) and wall reflections. Clearly, equation (3) is the source of the nonlinearity in the integro-differential energy equation. Except for some specialized cases, this equation is not amenable to closed-form solution. Therefore iterative schemes must be resorted to.

NUMERICAL DISCRETIZATION AND BOUNDARY ELEMENT FORMULATION

A numerical solution to (1) is effected in two steps. First, a discretized form of (3) is needed for the numerical evaluation of the radiative source term. There are several methods available for this purpose. The direct exchange factor (DEF) method is a computationally efficient and flexible method. In the DEF method, the domain Ω is first discretized into N_Ω domain elements $\Omega^{(e)}$ and the boundary Γ is discretized into N_Γ boundary elements Γ_j. Radiation nodes are then placed in each domain and boundary element, see Figure 1. This constitutes the radiative computational domain. The integrals in (3) are then approximated and the radiative source term is modeled as constant over each element, and can be evaluated as

$$Q_r(\Theta_P, \vec{r_i}) = 4(1-\omega_o)\Theta_P^4 - \sum_{i=1}^{N_\Gamma} w_{\Gamma_i} \epsilon\, \Theta^4(\vec{r_P})\ DSV(\vec{r_i}, \vec{r_P})$$

$$- \sum_{i=1}^{N_\Omega} w_{\Omega_i} 4(1-\omega_o)\, \Theta^4(\vec{r_P})\, DVV(\vec{r_i}, \vec{r_P}) \qquad (4)$$

where w_{Γ_i} and w_{Ω_i} are surface and volume nodal weight factors proportional to the area of the element or the length of the boundary element surrounding the radiation nodes. Explicit expressions for the total exchange factors are given in[10]. The temperature Θ_P which is used in (4) is evaluated at the position $\vec{r_P}$ located at a radiation node. This then completes the discretization for the radiative source term.

A direct boundary element formulation for the solution of (1) can now be derived. The homogeneous form of (1) is Laplace's equation, and its Green's free space solution is $\Theta^*=(-1/2\pi)\ln r$, where $r=|\vec{r_k}-\vec{r_j}|$, $\vec{r_k}$ is the position vector for the source point, and $\vec{r_j}$ is the position vector for field point [11]. Using Θ^* and its normal derivative $q^*=\partial\Theta^*/\partial n$, a weighted residual statement can be used to derive the following boundary integral equation[11,12]

$$C_k\Theta_k + \oint_\Gamma \Theta\, q^* d\Gamma = \oint_\Gamma \Theta^* \frac{\partial\Theta}{\partial n} d\Gamma + \frac{1}{4N_s}\oint_\Omega [\, G(\tau_x,\tau_y)+Q_r(\Theta,\tau_x,\tau_y)]\Theta^*\, d\Omega \quad (5)$$

If the integral equation (5) is applied at a boundary point then $C_i=\phi/2\pi$, where ϕ is the internal angle by which the local tangent turns at the boundary point. Otherwise, if the integral equation (5) is applied at an internal point then $C_k=1$. The contour integrals in (5) are evaluated in their Cauchy principal value in the case that (5) is applied at the boundary. The boundary integral equation is next discretized. Taking advantage of the grid generated for the radiative computations, constant boundary elements are used for the contour integrals and parametric elements for the domain integrals [11] to yield the following equation

$$C_k\Theta_k + \sum_{j=1}^{N} H_{k,j}\Theta_j = \sum_{j=1}^{N} G_{k,j}q_j + B_k \quad (6)$$

where

$$B_k = \sum_{e=1}^{N_\Omega} \oint_{\Omega^{(e)}} [\; G(\tau_x,\tau_y)+Q_r(\Theta,\tau_x,\tau_y)]\Theta^*\, d\Omega \quad (7)$$

The integrals appearing in the influence coefficients $H_{k,j}$ and $G_{k,j}$ are evaluated for all elements (except for the ones corresponding to the node in question) by an adaptive quadrature based on the Gauss-Kronrod (G_{11},K_{21}) pair. It is noted that the domain integral in (7) is evaluated over the same domain elements already defined for the radiation computation in (4) with Θ_P evaluated at the element centroid. The domain integral was computed using Gauss-Legendre quadrature. Additionally, since the radiation computation models the temperature as constant over the radiation boundary elements, we used a constant element boundary model for the nodal equation (6). As such, $H_{ii}=0$ and G_{ii} was evaluated analytically [11]. Assembly of the nodal equations then yield a set of coupled nonlinear equations of the form

$$[H]\{\Theta\}=[G]\{q\}+\{B(\Theta_P)\} \tag{8}$$

Introducing the boundary conditions into (8), the known Θ's are moved to the right hand side and the unknown q's are moved to the left hand side. The resulting set of equations is nonlinear due to the dependence of the B vector on the fourth power of Θ. This set of equations must therefore be solved iteratively.

ITERATIVE SOLUTION PROCEDURE

A modified version of the flexible incremental loading procedure of Bialecki and Nowack[13] is used for the iterative solution of (8). In this method, the nodal equations are rewritten as

$$[H]\{\Theta\}=[G]\{q\}+\lambda_k\{B(\Theta_P)\} \qquad k=0,1...K \tag{9}$$

where λ_k is a parameter which is taken to vary from $\lambda_0=0$ to $\lambda_K=1$ in a series of steps controlled so as to incrementally increase the effect of the nonlinearity. By setting $\lambda=0$, the linear problem is first solved. This solution is then used to generate the centroidal values of Θ_P^0 to evaluate the right hand side vector $\{B(\Theta_P^0)\}$. Along with the solution to the linear problem, this vector then serves as an initial guess for the next problem for which λ_1 has been increased to a value greater than zero and less than or equal to one. That is, the set of equations

$$[H]\{\Theta^{n+1}\}=[G]\{q^{n+1}\}+\lambda_1\{B(\Theta_P^n)\} \tag{10}$$

are solved to iterative convergence. Here, the superscript "n" denotes the n-th iterative step. The relative convergence criteria used is defined in the L_∞ norm as

$$\frac{\left\|\{\Theta^{n+1}-\Theta^n\}\right\|_\infty+\left\|\{q^{n+1}-q^n\}\right\|_\infty}{\left\|\{\Theta^{n+1}\}\right\|_\infty+\left\|\{q^{n+1}\}\right\|_\infty}\le\delta \tag{11}$$

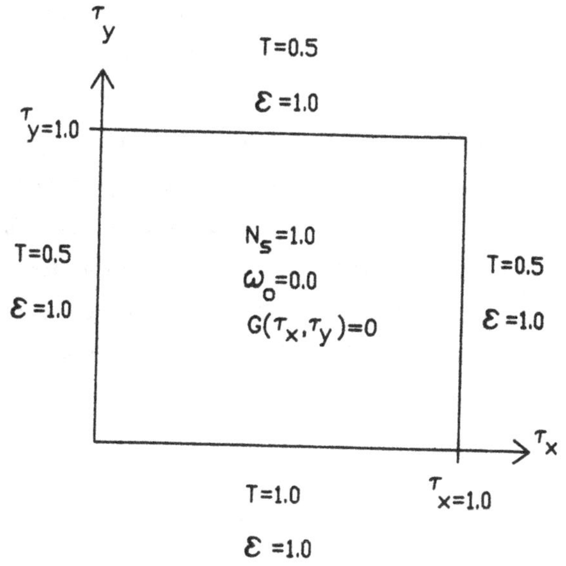

Figure 1. Radiation computational grid.

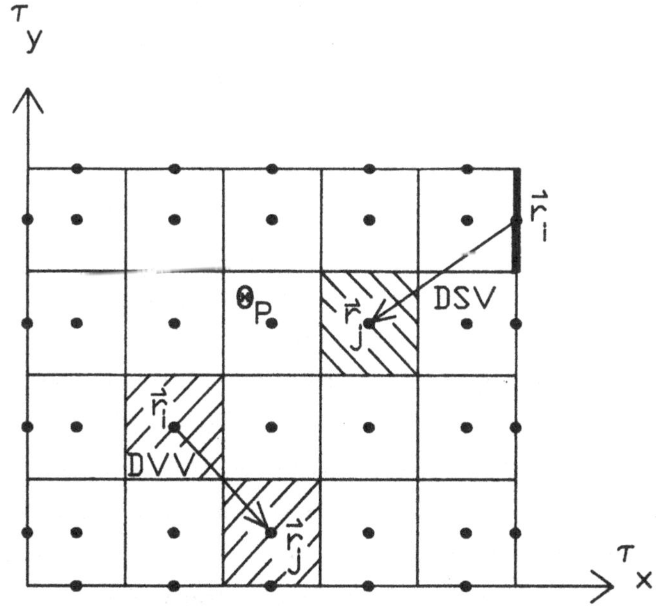

Figure 2. Benchmark problem of Ratzel and Howell[2].

A value of $\delta=10^{-3}$ was used in our computations. It is noted that LU decomposition was used to originally solve the linear problem (λ_o), and that the LU decomposition of the coefficient matrix was stored. As such, the iterative solution only involves a forward and a backward substitution at each iterative step. Once the problem for λ_1 is solved, its solution is in turn used as an initial guess for the next problem with a new $\lambda_k > \lambda_1$. This problem is then solved to convergence. The procedure is then repeated for a sequence of increasing λ_k. The number of steps K taken to reach a value of $\lambda_K=1$, i.e. the solution of (8), depends on the degree of the nonlinearity involved [13]. This method has the distinct advantage of being self-starting. The solution procedure outlined above was implemented in a computer code and results are discussed in the next section.

NUMERICAL RESULTS FOR BENCHMARK PROBLEM

The first results presented concern the solution of the benchmark problem depicted in Figure 2. Here, following Ratzel and Howell [2], the problem of conductive-radiative transfer in a square domain with black walls ($\epsilon=1$) is solved. The scattering albedo is taken as zero ($\omega_o=0$). Additionally, the thermal boundary conditions are taken to be constant dimensionless temperature $\Theta=0.5$ on all walls except for the wall of $\tau_y=0$ where $\Theta=1.0$. The source term is taken as $G(\tau_x,\tau_y)=0$ and the Stark number is set as $N_s=1$. The boundary was discretized into 40 equal size constant boundary elements, and the interior domain was discretized into 100 equally sized radiation elements. The iterative procedure converged using the sequence $\lambda_o=0$ and $\lambda_1=1$. The computed temperatures agree very well with those reported in the literature [2-8]. The isotherms generated for this problem are plotted in Figure 3. Comparative numerical studies in the radiative heat transfer literature customarily report the centerline temperatures obtained for the benchmark problem. The centerline temperatures ($\tau_x=0.5$) are compared with those values obtained by the finite element method (FEM) in Table 1. In addition to addressing the radiative heat source term, we also varied the values for the non-radiative heat source. The centerline temperatures are plotted in Figure 4 for the conditions imposed in Figure 3 above with the nonradiative-heat source term taking on values of $G(\tau_x,\tau_y)=0$, 1, 5, and 10. The presence of the non-radiative heat source term elevates the centerline temperature. For values of $G(\tau_x,\tau_y)$ of 5 and 10 the centerline temperature is

Table 1. Comparison of BEM and FEM results.

τ_y	BEM	FEM
0.1	0.909	0.899
0.2	0.823	0.820
0.3	0.746	0.736
0.4	0.684	0.676
0.5	0.634	0.630
0.6	0.596	0.593
0.7	0.565	0.564
0.8	0.541	0.540
0.9	0.519	0.519

In addition to addressing the radiative heat source term, we also varied the values for the non-radiative heat source. The centerline temperatures are plotted in Figure 4 for the conditions imposed in Figure 3 above with the nonradiative-heat source term taking on values of $G(\tau_x,\tau_y)$=0, 1, 5, and 10. The presence of the non-radiative heat source term elevates the centerline temperature. For values of $G(\tau_x,\tau_y)$ of 5 and 10 the centerline temperature is elevated above the highest temperature imposed at the boundary with a maximum exhibited around τ_y=0.5. To the authors' knowledge, these cases have not been studied in the literature.

CLOSING REMARKS

The solution of a conductive-radiative heat transfer problem by a boundary element method has been presented. An iterative solution procedure has been proposed. The results for the benchmark problem considered compared well with the data reported in the literature for finite difference and finite element solutions of the same problem. The boundary element approach however capitalizes on the presence of the radiation computational grid by using it to compute domain integrals. Additionally, identical discretization was used for boundary element and radiative computations performed in this study. The level of effort required in extracting a solution from the energy equation is minimized, thereby putting the onus on the radiation model. The proposed boundary element approach to solving conductive-radiative heat transfer is thus a positive step forward in the ongoing effort to analyze this difficult problem. Higher order approximations should produce more accurate results while reducing the number of iterations required for convergence. Different discretizations are now being studied and the results will be reported elsewhere.

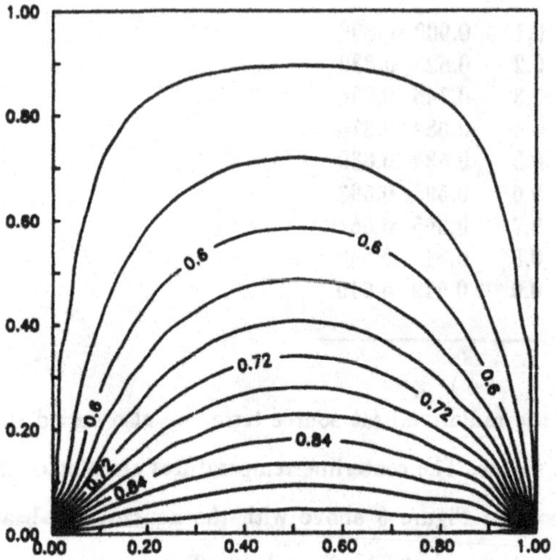

Figure 3. Isotherm plot for benchmark problem.

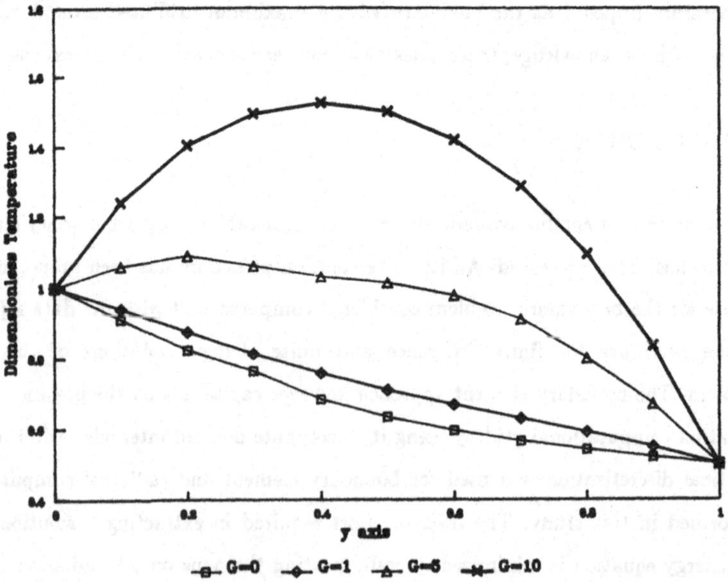

Figure 4. Plot of centerline temperatures for the benchmark problem of Ratzel and Howell[2]with G=0,1,5, and 10.

REFERENCES

1. Siegel, R., and Howell, H.R., Thermal Radiation Heat Transfer, 2nd Edition, Hemisphere Publishing, New York, 1981.
2. Ratzel, A.C., and Howell, J.R., Two-Dimensional Energy Transfer in Radiatively articipating Media with Conduction by the P-N Approximation, Vol. 2, pp. 535-540, Proceedings of the Seventh International Heat Transfer Conference, 1982.
3. Larsen, M.E., The Exchange Factor Method: An Alternative Zonal Formulation for Analysis of Radiating Enclosures Containing Participating Media, Ph.D. Dissertation, University of Texas at Austin, 1983.
4. Razzaque, M.M., Howell, J.R., and Klein, D.E., Coupled Radiative and Conductive Heat Transfer in a Two-Dimensional Rectangular Enclosure with Gray Participating Media Using Finite Elements, ASME Journal of Heat Transfer, Vol. 106, pp. 613-619, 1984.
5. Yuen, W.W., and Takara, E.E., Analysis of Combined Conductive-Radiative Heat Transfer in a Two-Dimensional Rectangular Enclosure with a Gray Medium, ASME Journal of Heat Transfer, Vol. 110, pp. 468-474, 1988.
6. Saltiel, C., and Naraghi, M.H.N., Combined-Mode Heat Transfer in Radiatively Participating Media Using the Discrete Exchange Factor Method with Finite Elements, Heat Transfer 1990, (Ed. Hestroni, G.), Vol. 6, pp. 391-396, Hemisphere, New York, 1990.
7. Naraghi, M.H.N., and Saltiel, C., Combined-Mode Heat Transfer in Radiatively Participating Media: Computational Considerations, Radiation Heat Transfer: Fundamentals and Applications, (Ed. Smith, T.F., Modest, M.F., Smith, A.M., and Thynell, S.T.), ASME Publication HTD-VOL. 137, p. 133-140, 1990.
8. Baek, S.W., and Kim, T.Y., The Conductive and Radiative Heat Transfer in Rectangular Enclosure Using the Discrete Ordinates Method, Heat Transfer 1990, (Ed. Hestroni, G.), Vol. 6, pp. 433-438, Hemisphere, New York, 1990.
9. Fernandes, R., and Francis, J., Combined Conductive and Radiative Heat Transfer in an Absorbing, Emitting, and Scattering Cylindrical Media, ASME Journal of Heat Transfer, Vol. 104, pp. 594-601, 1982.

10. Saltiel, C., and Naragi, M.H.N., Analysis of Radiative Heat Transfer in Participating Media Using Arbitrary Nodal Distribution, Numerical Heat Transfer-PartB, Fundamentals, Vol. 17, pp. 227-243, 1990.
11. Brebbia, C.A., Telles, J.C.F., and Wrobel, L.C. Boundary Element Techniques, Springer-Verlag, New York, 1984.
12. Brebbia, C.A. and Walker,S. Boundary Element Techniques in Engineering, Newnes-Butterworth, Boston, 1980.
13. Bialecki, R.B. and Nowak, A.J. Boundary value Problems in Heat Conduction With Nonlinear Material And Nonlinear Boundary Conditions, Applied Mathematical Modelling, Vol 5, pp.417-421, 1981.

Computational Aspects of the Boundary Element Method Applied to Frictional Heating

B. Vick, L.P. Golan II

Department of Mechanical Engineering, Virginia Polytechnic Institute and State University
Blacksburg, VA 24061-0238, U.S.A.

ABSTRACT

A variation of the boundary element method is developed to solve for the distribution of frictional heat and ensuing temperature rise caused by frictional heating between sliding solids. The problem under consideration is 3-dimensional and transient and involves the thermal coupling of two different materials at the true contact areas. By using a moving Green's function which naturally incorporates the combined effects of convection due to sliding and diffusion, the original differential formulation is reduced to a system of integral equations for the unknown boundary data. The integral equations are discretized using constant elements to numerically obtain the unknown boundary information.

This paper describes the basic methodology as well as some of the crucial numerical ingredients required to obtain accurate solutions. It is shown that the boundary element method is an accurate and viable method if the necessary Green's function integrals which are used to build the solutions are resolved accurately. In particular, the singular integrals are critical. Results demonstrating the effects of various numerical and physical parameters are presented.

INTRODUCTION

Tribology is the study of the friction, wear, and lubrication mechanisms. Tribological systems such as bearings, gears, cams and brakes are prevalent in many aspects of everyday life. Typical occurrences that characterize tribological phenomena are wear of the materials involved, transfer of material, the formation and evolution of contact area and the generation of energy at real areas of contact. The generation of energy and subsequent dissipation, mainly in the form of heat, can result in high surface and subsurface temperatures. Consequences of high temperatures include, but are not limited to, surface melting, deterioration and increased wear, oxidation, softening, and the breakdown of solid or liquid films.

The theoretical prediction of the surface temperature rise due to frictional heating was initiated by the pioneering works of Blok [1] and Jaeger [2]. These simple analyses however are limited to unidirectional sliding with a single contact area between semiinfinite regions. Also, the determination of the division of frictional heat in these works is approximate and does not capture the spatial and time variation. In the work of Lai [3], the variation of the division of frictional heat was determined accurately and multiple contacts were included. References [1-4] all use analytical methods, which tend to give high accuracy and good insight but are difficult or impossible to apply to more complicated situations. The finite difference and finite element methods have been used with varying degrees of success to handle more complicated problems. The finite difference method is quite flexible but requires an excessively fine grid and considerable computer time to account for steep temperature gradients near the contact area. Finite elements seem more efficient than finite differences and have been used by Kennedy [4] to study fairly general situations including unidirectional or oscillating velocity, multiple contacts and finite regions. However, elements are required over the entire domain. In addition, numerical oscillations or instabilities occur at high Peclet numbers, resulting in significant inaccuracies.

In view of the state of affairs of previous theoretical work, it appears that the boundary element method (BEM) is well suited for the thermal analysis of the frictional heating of solids. The method is numerically accurate yet flexible enough to include the important physical processes including 3-dimensional transient heat transfer with multiple, evolving contact areas. A variation of the boundary element method has been developed utilizing a moving Green's function as a basis. The purpose of this paper is to describe the basic methodology and to present some of the crucial numerical ingredients required to obtain accurate results.

GOVERNING EQUATIONS

The relative motion of one body upon another results in the conversion of mechanical energy into thermal energy. This process, referred to as frictional heating, occurs only at regions of intimate contact. Figure 1 shows the schematic of the theoretical model used in this analysis. Areas of intimate contact, designated by Γ_C, are shared by both regions while areas that are not in contact are designated as Γ_i, $i = 1,2$. Each of the two bodies presented in the model is assumed to have constant, yet unique thermal properties and hardness. The coordinate system is attached to the contact area, thus region 1 is stationary while region 2 moves in the x-direction with a time dependent velocity $v(t)$ relative to the contact area and coordinate system.

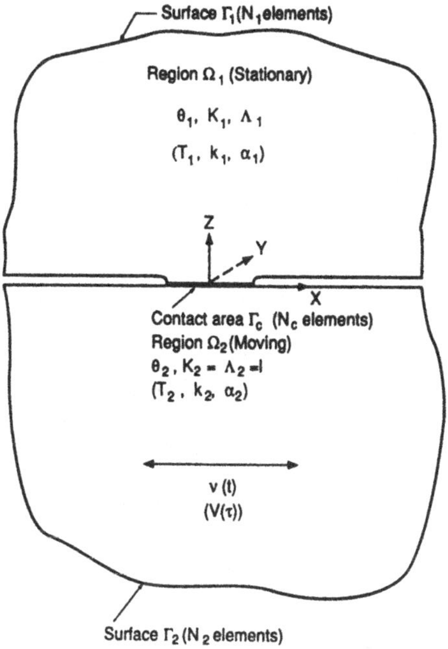

Fig.1.The theoretical model

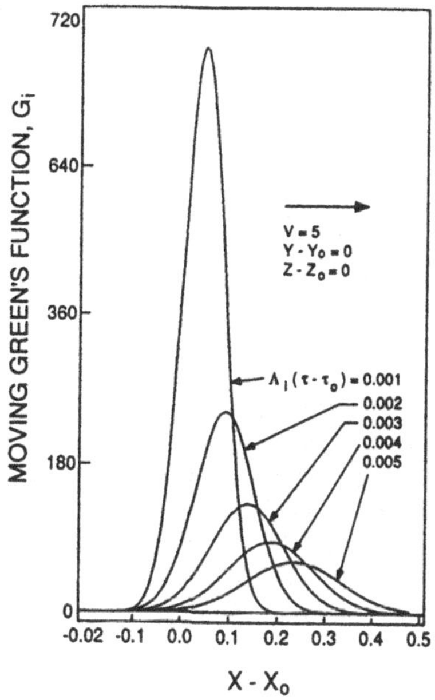

Fig.2.The Moving Green's Function

The amount of frictional heating that occurs in general is known and is given by

$$q_C'' = \mu P|v(t)| \tag{1}$$

where μ is the coefficient of friction and P is contact pressure. However, distribution of this frictional energy is generally not known and is dependent upon the relative motion, the thermal properties and geometry of the bodies involved. Other physical factors such as the thermal interaction of contact areas also affect the temperatures generated in the sliding contact.

The governing energy equations, boundary conditions and initial conditions are nondimensionalized using four reference quantities since the problem involves four independent dimensions (length, time, temperature rise, and energy). The reference quantities chosen are the thermal conductivity k_2 and thermal diffusivity $\alpha 2$ of the moving region, a characteristic velocity, V_r and a characteristic heat flux q_r''. The resulting nondimensional energy equations become,

$$\frac{\partial \theta_i}{\partial \tau} + V_i(\tau)\frac{\partial \theta_i}{\partial X} = \Lambda_i \nabla^2 \theta_i(\bar{r},\tau) \quad , \quad i = 1,2 \, . \tag{2}$$

This equation represents a balance between energy storage, convection in the sliding direction, and conduction in three dimensions. Note that V_i is defined as zero for i = 1 (stationary region) and becomes the dimensionless sliding velocity, $V(\tau) = v(t)/v_r$, for i = 2 (moving region). Various dimensionless quantities are defined as follows:

$$\theta_i = (T_i - T_o)k_2 v_r / q''_r \alpha_2 \quad , \quad Q'' = q''/q_r$$

$$X = x \cdot v_r/\alpha_2 \quad , \quad Y = y \cdot v_r/\alpha_2 \quad , \quad Z = z \cdot v_r/\alpha_2 \quad , \quad \tau = t \cdot v_r^2/\alpha_2$$

$$\Lambda_i = \alpha_i/\alpha_2 \quad , \quad K_i = k_i/k_2$$

The dimensionless temperature, θ_i represents the rise above the initial temperature, T_o. The boundary conditions at the non-contacting surfaces are expressed as,

$$-K_i \; \partial\theta_i/\partial\eta = Bi(\theta_i - \theta_\infty) \quad , \quad \bar{r} \text{ on } \Gamma_i \tag{3}$$

where $\partial/\partial\eta$ is the gradient with respect to the outward normal, θ_∞ is the environment temperature, and Bi is the Biot number or dimensionless heat transfer coefficient. From this generalized boundary condition special cases such as semiinfinite regions, insulated surfaces, or heat loss by convection can be extracted. The initial conditions for the dimensionless temperature rise are,

$$\theta_i = 0 \text{ at } \tau = 0, \; i = 1 \text{ or } 2 \; . \tag{4}$$

The coupling of the two regions at the contact area is governed by continuity of temperature and conservation of energy,

$$\theta_1 = \theta_2 = \theta_c \tag{5}$$

$$Q''_c = Q''_{c1} + Q''_{c2} \; , \; \bar{r} \text{ on } \Gamma_c \; . \tag{6}$$

Here $Q''_c = q''_c / q''_r$ is the total rate of frictional heating and Q''_{c1} and Q''_{c2} represent the division of the frictional heat which are expressed by Fourier's law as,

$$Q''_{ci} = (-1)^i K_i (\partial\theta_i/\partial Z)_{Z=0} \tag{7}$$

where normal gradients on the contact area are in the ±Z-direction. Generally Q''_c is known while Q''_{c1} and Q''_{c2} must be determined as part of the solution.

Equations (2)-(7) govern the temperature rise due to frictional heating. In the following section the solution using a variation of the boundary element method is described.

BOUNDARY ELEMENT SOLUTION

Temperature in Terms of a Moving Green's Function

The proposed solution method uses a moving Green's function as the weighting function which allows for an accurate and efficient incorporation of the convective effect due to the sliding velocity. Usually the stationary or pure diffusion Green's function has been used for boundary element solutions of thermal problems (see Wrobel and Brebbia [5] and Muzzio and Solaini [6]). The boundary element solution method reduces the original dimensionality of the problem by one, to include only unknown boundary data. The original problem is reduced to a set of integral equations for the unknown boundary data which must then be resolved numerically. Once the unknown boundary data has been determined, internal values of temperature can be calculated.

An advantage of the boundary element method, as compared to finite elements or finite differences, is the reduction in computer storage necessary to describe the solution domain. Other advantages of the boundary element method as applied to this particular problem are the incorporation of the convective effect in the weighting function, the ease with which an infinite domain is handled, and the ability to employ relatively large time and spatial elements.

The moving Green's function used in this analysis is governed by the energy equation (2) with the addition of a concentrated, pulsed heat source released at time τ_0 and location $\bar{r}_0$,

$$\frac{\partial G_i}{\partial \tau} + V_i(\tau)\,\frac{\partial G_i}{\partial X} = \Lambda_i \nabla^2 G_i + \delta(\bar{r} - \bar{r}_0)\,\delta(\tau - \tau_0) \quad (8a)$$

$$G_i = G_i(\bar{r},\tau/\bar{r}_0,\tau_0) = 0 \ , \ \tau < \tau_0 \quad (8b)$$

where $\delta(\bar{r} - \bar{r}_0)$ is the 3-dimensional delta function. The solution of Eqs. (8a,b) can be obtained using full space Fourier transforms and inversions to obtain,

$$G_i = \mathrm{EXP}\left[-\frac{(X-X_0-X_c)^2+(Y-Y_0)^2+(Z-Z_0)^2}{4\Lambda_i(\tau-\tau_0)}\right]/[4\pi\Lambda_i(\tau-\tau_0)]^{3/2} \quad (9)$$

where the convective effect is in the term

$$X_c(\tau,\tau_0) = \int_{\tau'=\tau_0}^{\tau} V_i(\tau')d\tau' \ .$$

The nature of the moving Green's function is displayed in Figure 2. As will be seen shortly, the solution for boundary data is dependent upon evaluations of the Green's function, its derivative, and their various spatial and time integrals. Of particular importance is the gradient with respect to Z_0,

$$\frac{\partial G_i}{\partial Z_0} = \frac{(Z - Z_0)}{2\Lambda_i(\tau - \tau_0)}\,G_i \quad (10)$$

The general temperature distribution is now constructed with the GF defined by Eq. (9) as a basis function. The governing energy equation (2) is written in the cause variables, X_0, Y_0, Z_0 and τ_0, multiplied by G_i and integrated over region Ω_i and over time to get,

$$\int_{\tau_0=0}^{\tau} \iiint_{\Omega_i} \left[\Lambda_i \nabla_0^2 \theta_i - \frac{\partial \theta_i}{\partial \tau_0} - V_i\,\frac{\partial \theta_i}{\partial X_0}\right] G_i(\bar{r},\tau|\bar{r}_0,\tau_0)\,d\Omega_0 d\tau_0 = 0. \quad (11)$$

After a lengthy but straightforward series of manipulations involving integration by parts and application of Green's theorem, the general solution for the dimensionless temperature rise in terms of the moving Green's function is obtained in the form,

$$\lambda\theta_i(\bar{r},\tau)/\Lambda_i = \int_{\tau_o=0}^{\tau} \iint_{\Gamma_c+\Gamma_i} \left[G_i \frac{\partial\theta_i}{\partial\eta} - \theta_i \frac{\partial G_i}{\partial\eta} \right] d\Gamma_o d\tau_o \qquad (12)$$

$$\text{where } \lambda = \begin{cases} \tfrac{1}{2}\ , & \bar{r} \text{ on smooth sections of boundaries } \Gamma_i \text{ or } \Gamma_c \\ 1\ , & \bar{r} \text{ inside region } \Omega_i \end{cases}$$

and $\partial/\partial\eta$ represents differentiation in the outward normal direction. The integration over the surface of each region has been broken into a contacting portion and a noncontacting portion since each portion needs to be handled individually. The unknowns are the distribution of frictional heat and the surface temperatures. The gradient of temperature on the noncontacting regions is not an additional unknown since it can be related to temperature through boundary condition (3).

Integral Equations for Unknown Boundary Data

The general temperature solution (12) is evaluated on the surfaces of region Ω_i to obtain,

$$\tfrac{1}{2}\,\theta_s(\bar{r},\tau)/\Lambda_i = \int_{\tau_o=0}^{\tau} \iint_{\Gamma_c+\Gamma_i} \left[G_i \frac{Q_s''}{K_i} - \theta_s \frac{\partial G_i}{\partial\eta} \right] d\Gamma_o d\tau_o,\ \bar{r} \text{ on } \Gamma_c,\Gamma_1,\Gamma_2 \qquad (13)$$

The temperatures and heat flux on the appropriate boundary surfaces are designated as follows,

$$\theta_s = \begin{cases} \theta_{si},\ \bar{r} \text{ on } \Gamma_i,\ i = 1 \text{ or } 2 \\ \theta_c\ ,\ \bar{r} \text{ on } \Gamma_c \end{cases} \qquad (14)$$

$$Q_s'' = \begin{cases} K_i \dfrac{\partial\theta_i}{\partial\eta} = B_i\,(\theta_\infty - \theta_{si})\ , & \bar{r} \text{ or } \Gamma_i\ ,\ i = 1 \text{ or } 2 \\ Q_{ci}'' = Q_c'' - Q_{c,3-i}'' & ,\ \bar{r} \text{ on } \Gamma_c \end{cases}$$

Note that on noncontacting surfaces Γ_i, Q_s'' and θ_s are related by the convective boundary condition (3) and only $\theta_s = \theta_{si}$ is considered as unknown. Also the surface temperature on the contact region Γ_c is shared by both regions as given by boundary condition (5).

Next the general temperature solution (12) is evaluated for both $i = 1$ and $i = 2$ on the contact area Γ_c and the resulting expressions are equated to one another as required by the continuity of temperature condition (5). Q_{c2}'' is replaced by $Q''_c - Q''_{c1}$ and the result is arranged into the following form,

$$\int_{\tau_o=0}^{\tau} \iint_{\Gamma_c} \left(\frac{\Lambda_1}{K_1} G_1 + G_2\right) Q''_{c1} \; d\Gamma_o d\tau_o \tag{15}$$

$$- \sum_{i=1}^{2} (-1)^i \Lambda_i \int_{\tau_o=0}^{\tau} \iint_{\Gamma_i} \left[G_i \, Bi(\theta_\infty - \theta_{si})/K_i - \theta_{si} \frac{\partial G_i}{\partial \eta}\right] d\Gamma_o d\tau_o$$

$$= \int_{\tau_o=0}^{\tau} \iint_{\Gamma_c} [G_2 Q''_c] \, d\Gamma_o d\tau_o \text{ , valid for } \bar{r} \text{ on } \Gamma_c.$$

The solution now hinges on finding the unknown surface temperatures and the distribution of frictional heat. We have just derived a set of simultaneous singular integral equations for this purpose, namely Eq. (13) for $\bar{r}$ on Γc, $\Gamma 1$ and $\Gamma 2$ and Eq. (15). These singular integral equations are of the Volterra type in the time variable and of the Fredholm type in the spatial variables since the original differential formulation was parabolic in time and elliptic in space. The equations are 2nd kind integral equations for the unknown surface temperatures but are 1st kind integral equations for the distribution of heat since Q''_{c1} is found under the integrals only. The numerical solution of the 1st kind singular Volterra integral equations can be tricky (see Linz [7]). The numerical discretization is addressed in the following section.

Numerical Integration

In order to numerically integrate Eqs. (13) and (15), the boundary domains are broken into boundary elements. As indicated in Figure 1, the noncontacting surfaces Γi, i = 1 or 2, are divided into N_i elements and the contact area Γc is broken up into N_c elements. Since the regions are thermally coupled through contact area Γc, both regions share the same elements over this area. The time variable is also discretized into elements with p indicating the time level.

The numerical integrations are performed using constant boundary elements over both time and space. In the subsequent derivations the following notation is used,

$$\theta^p_{s,n} \cong \theta_s \text{, over element n at time } \tau_p \tag{16}$$

$$Q^p_{c1,n} \cong Q''_{c1} \text{, over element n in time period } \tau_{p-1}<\tau<\tau_p$$

Note that temperatures are designated at various time levels since the required integral equations for temperature are 2nd kind Volterra equations in time. On the other hand, the distribution of heat and all heat fluxes are approximated over time intervals since the integral equations for heat fluxes are 1st kind Volterra equations in time.

Integral Eq. (13) for the surface temperatures is discretized at $\tau=\tau_{pp}$ to obtain,

$$\tfrac{1}{2}\,\theta_s(\bar{r},\tau_{pp})/\Lambda_i \tag{17}$$

$$= \sum_{p=1}^{pp}\sum_{n=1}^{N_c+N_i}\left[G^p_{i,n}(\bar{r},\tau_{pp})\,\frac{Q^p_{s,n}}{K_i} - \theta^p_{s,n}DG^p_{i,n}\right],\quad \bar{r}\text{ on }\Gamma_c,\Gamma_1,\Gamma_2$$

Definitions (16) have been used and the GF integrals are defined in the Appendix by Eqs. (A.1a,b). This equation can be integrated or averaged over a particular element, Γ_{nn}, with area A_{nn} to obtain,

$$\tfrac{1}{2}\,\theta^{pp}_{s,nn}\cdot A_{nn}/\Lambda_i \tag{18}$$

$$= \sum_{p=1}^{pp}\sum_{n=1}^{N_c+N_i}\left[\,G^p_{i,n,nn}(\tau_{pp})\,\frac{Q^p_{s,n}}{K_i} - \theta^p_{s,n}\,DG^p_{i,n,nn}\right].$$

The additional spatial integrals of the GF over the effect spatial variables are defined in the Appendix by equations (A.2a,b). Equation (18) can be written for each boundary element, thus providing $N_1 + N_2 + N_c$ independent equations.

Performing this discretization and averaging process on coupling Eq. (15) produces,

$$\sum_{p=1}^{pp}\sum_{n=1}^{N_c}\left(\frac{\Lambda_1}{K_1}\,G^p_{1,n,nn} + G^p_{2,n,nn}\right)Q^p_{c1,n} \tag{19}$$

$$- \sum_{i=1}^{2}(-1)^i\Lambda_i\sum_{p=1}^{pp}\sum_{n=1}^{N_i}\left[\,G^p_{i,n,nn}\,Bi(\theta_\infty-\theta^p_{si,n})/K_i - \theta^p_{si,n}DG^p_{i,n,nn}\right]$$

$$= \sum_{p=1}^{pp}\sum_{n=1}^{N_c} G^p_{2,n,nn}\,Q^p_{c,n}\quad,\quad \text{valid for } nn = 1,2,\ldots,N_c\,.$$

Since the numerical solution proceeds forward in the time variable one step at a time, only current values at τ_{pp} are considered as unknown. The unknown values are thus,

- $\theta^{pp}_{s1,n}$ for $n = 1,2,\ldots,N_1$ and $\theta^{pp}_{s2,n}$ for $n = 1,2,\ldots,N_2$
- $\theta^{pp}_{c,n}$ and $Q^{pp}_{c1,n}$ for $n = 1,2,\ldots,N_c$

The total number of unknowns at each time step is $N_1 + N_2 + 2N_c$. Equations (18) and (19) provide the appropriate number of simultaneous relations for these unknowns.

The system of equations (18) and (19) can be rearranged and expressed in matrix form as,

$$\underset{\sim}{G}\cdot\underset{\sim}{Q} = \underset{\sim}{F} \tag{20}$$

where $\tilde{Q}$ is the matrix of current unknown boundary data,

$$\tilde{Q} = [\ldots Q^{pp}_{c1,n} \ldots\ldots \theta^{pp}_{c,n} \ldots\ldots \theta^{pp}_{s1,n} \ldots\ldots \theta^{pp}_{s2,n} \ldots]^T$$

G is the square coefficient matrix involving singular integrals of the GF at the current time step and F is a column matrix involving past information. Matrix Eq. (20) can be assembled and solved for any particular case.

The special case of semiinfinite regions with insulated noncontacting boundaries is of particular importance. For this special case, integral Eq. (15) or discretized Eq. (19) for the distribution of frictional heat become uncoupled from the temperature solutions. Eq. (19) reduces to,

$$\sum_{p=1}^{pp} \sum_{n=1}^{N_c} \left(\frac{\Lambda_1}{K_1} G^p_{1,n,nn} + G^p_{2,n,nn}\right) Q^p_{c1,n}$$

$$= \sum_{p=1}^{pp} \sum_{n=1}^{N_c} G^p_{2,n,nn} Q^p_{c,n} \quad , \quad \text{valid for } nn = 1,2,\ldots,N_c. \qquad (21)$$

A key ingredient in obtaining accurate solutions is an accurate evaluation of the GF integrals, defined in the Appendix by Eq. (A.2a,b). As demonstrated in the Appendix, the spatial integrals required using constant elements can be evaluated analytically for rectangular boundary elements. Since solutions are sensitive to these integrals, the use of these exact analytical integrations is attractive. However, the required time integrals cannot be evaluated analytically and a numerical integration must be employed. This numerical integration is one of the most important ingredients in the overall numerical method. Of particular importance is the evaluation of the singular integrals. These matters will be demonstrated and discussed in the Results section.

RESULTS and DISCUSSION

Results showing some of the important numerical characteristics of the boundary element solution are presented. The case of two semiinfinite regions with insulated noncontacting surfaces and uniform frictional heat flux is considered. Also a constant sliding velocity is considered so that

$$V_i = \begin{cases} 0, & i=1 \\ 1, & i=2 \end{cases} \quad , \quad X_c = \begin{cases} 0 \;, & i=1 \\ \tau-\tau_0, & i=2. \end{cases}$$

The true contact area between sliding solids is complex and generally unknown [8,9]. However, optical and indirect observations suggest that the actual areas of contact are much smaller than the macroscopic dimensions of the solids. Thus, the semiinfinite assumption should be reasonable physical assumption. In addition, the semiinfinite case retains the fundamental behavior and difficulties associated with the general solution since integral Eq. (15) for the distribution of frictional heat, which reduces to discretized form (21), is still a first kind singular integral equation requiring the evaluation of the Green's function integrals. As a result, the numerical behavior of the general solution can be displayed by solving Eq. (21) for the distribution of heat and using these values to evaluate either local temperatures with Eq. (17) or average surface temperatures using Eqs (18).

For the following calculations, the contact area is taken as a square patch of size 2A by 2A, where A is the dimensionless half length of the contact area. The quantity A represents the sliding direction Peclet number or ratio of convection to conduction effects in the sliding direction. The true contact area between sliding solids is generally a collection of separate patches. The numerical representation of this true contact area with a collection of rectangular patches, each with any specified size and location, is fairly straightforward using the boundary element method developed in this paper. It was found that the true contact area is a major factor in the determination of the surface temperature rise. Some further results concerning the issue of a contact area consisting of multiple, interacting and evolving patches can be found in references [8,9].

In the following numerical computations, the square contact area is divided into rectangular elements with X and Y dimensions of ΔX and ΔY, respectively. Time is discretized using a constant time step, $\Delta\tau$. Boundary elements of variable size as well as variable time steps could also be implemented.

The evaluation of the GF integrals is now displayed. The integral of the GF given by Eq. (A.4) in the Appendix is particularly significant. Also the integral of the gradient normal to the contact area (surface $Z_0 = 0$) is important. The spatial integrations can be evaluated analytically for rectangular boundary elements. However, the time integrals over τ_0 cannot be evaluated analytically and must be computed numerically. Some typical results using several numerical integration procedures are shown in Figure 3. In order to evaluate these time integrals, the change of variable $\xi = \sqrt{4\Lambda_i(\tau-\tau_0)}$ was used to remove the $\sqrt{\tau-\tau_0}$ singularity. Numerical integrations for the singular time interval (τ_0 integrated from $\tau-\Delta\tau$ to τ) at $Z - Z_0 = 0.3$ using a 1 strip rule (trapezoid rule), 3 strip, 5 strip, and an adaptive quadrature are shown in Figure 3. The adaptive quadrature, as described in reference [10], tests and divides the region of integration as needed to achieve a desired accuracy and uses an 8 strip rule rule in each subinterval. It is clear that significant inaccuracy in the GF integrals can occur for a large range of the time integration (large $\Delta\tau$) without an adaptive quadrature. The reason for large errors is that the integrands exhibit humps and thus for any fixed integration rule and range of integration, the quadrature points fall at various locations on the humps. For a low order scheme combined with a large time integration range, $\Delta\tau$, the contribution of the hump in the integrand can be completely missed. Figure 3 shows that the use of an adaptive rule allows accurate GF integral evaluations even for extremely coarse time steps, $\Delta\tau$.

The important consequence of the evaluation of the GF integrals is the impact on temperature calculations. Inaccurate resolution of the GF integrals impacts internal levels of temperature as shown in Fig. 4. These internal levels of temperature are calculated after all unknown boundary fluxes and temperatures have been determined. Shown in Fig. 4 is the average temperature at a depth of $Z = 0.3$ directly below the contact area in the stationary region-1 as a function of time for each of the integration rules. The average temperature is calculated for an area the exact size of the frictionally heated contact area with $A = 10$. Figure 4 clearly shows that inaccurate GF integral evaluations can result in significant errors in temperature predictions. As a consequence, it is concluded that the boundary element method requires an adaptive integration for numerical accuracy.

The influence of the boundary element size, ΔX and ΔY, on the fraction of heat received by the moving region and the ensuing mean and maximum surface temperatures in the contact area is demonstrated in Fig. 5. Results are plotted for 1, 2, 3 and 4 divisions of the contact area in X and Y, corresponding to 1, 4, 9, and 16 total boundary elements, respectively. The adaptive integration is used for the GF integrals. The numerical results converge and show no noticeable difference between 9 and 16 elements for the division of heat and maximum temperature rise. On the other hand, the mean surface temperature is quite accurate using only one element.

The influence of time step is shown in Figure 6 for $\Delta\tau$ = 4, 8, 16, 32, 64, and 128. Although small inaccuracies occur in the first few steps using a course time step, all the time steps converge to the correct steady state. This is an inherent characteristic of the boundary element method. If any of the integral equations for distribution of heat or temperature are evaluated for a single time step with a sufficiently large value of $\Delta\tau$, the correct steady state is recovered.

In conclusion, the BEM based on a moving GF has been shown to be an accurate and efficient method to solve for the thermal response due to frictional heating between sliding solids. The overall numerical accuracy is sensitive to the evaluations of the required GF integrals used to build the solution. Once these GF integrals are evaluated accurately with an adaptive quadrative, excellent numerical accuracy for distribution of frictional heat and temperature rise can be obtained using large boundary elements in both space and time. In particular, the average steady state temperature of the contact area can be accurately obtained using only one spatial boundary element and a single large time step.

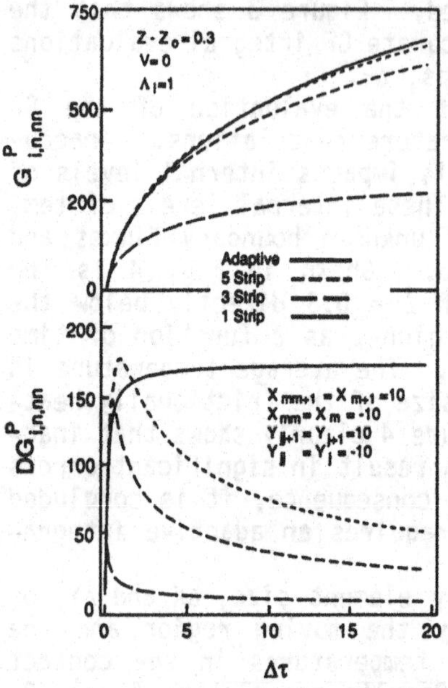

Fig.3.Effect of quadrature rule on the time integration of the GF

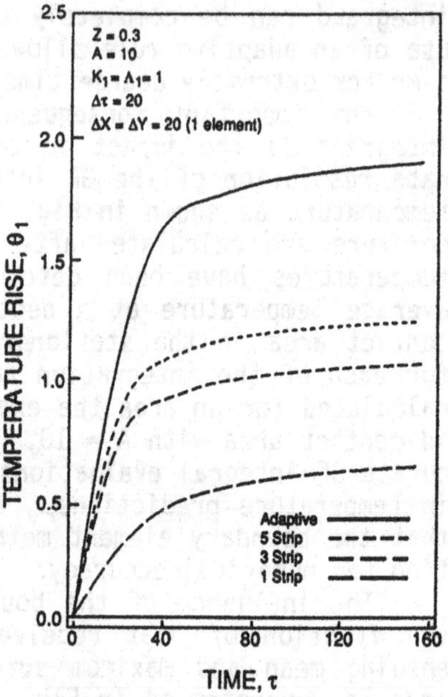

Fig.4.Effect of quadrature rule on the subsurface temperature rise

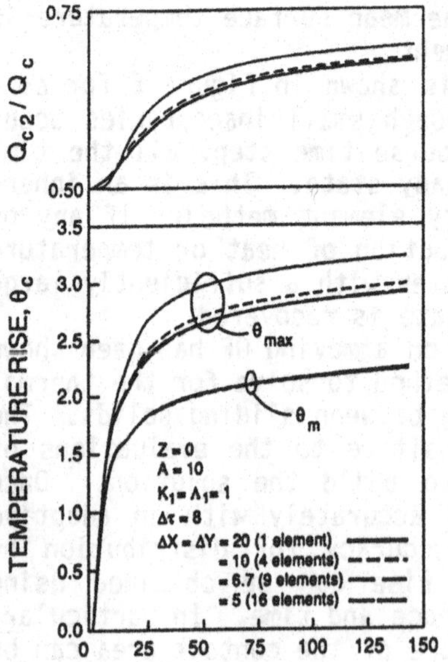

Fig.5.Effect of contact area mesh on distribution of heat and temperature rise of the contact area.

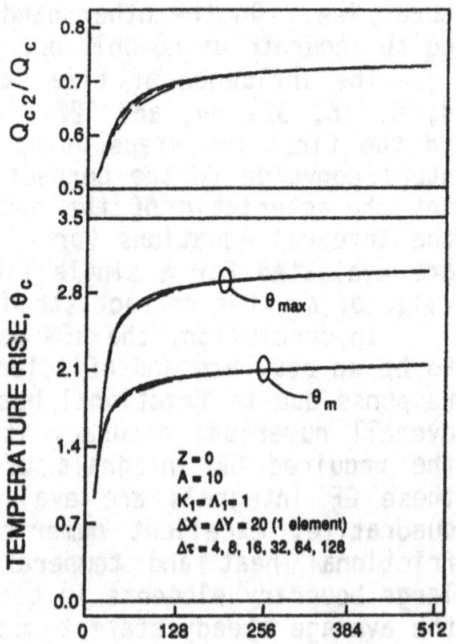

Fig.6.Effect of the time step on the distribution of heat and temperature rise of the contact area.

REFERENCES

1. Blok, H., "Theoretical Study of Temperature Rise at Surfaces of Actual Contact Under Oiliness Lubricating Conditions," Proc. Inst. of Mech. Engineers General Discussion of Lubrication, Institute of Mechanical Engineers, London, Vol. 2, 222-235, 1937.
2. Jaeger, J. C., "Moving Sources of Heat and the Temperature at Sliding Contacts, J. Royal Society of NS Wales, Vol. 76, 203-224, 1942.
3. Lai, W. T., "Temperature Analysis in Lubricated Simple Sliding Rough Contacts," ASLE Transactions, Vol. 25, 303-312, 1984.
4. Kennedy, F. E., "Surface Temperatures in Sliding Systems--A Finite Element Analysis," J. Lub. Tech., Vol. 103, 90-96, 1981.
5. Wrobel, L. C., C. A. Brebbia, "Boundary Elements in Thermal Problems," Numerical Methods in Heat Transfer, John Wiley, pp. 91-113, 1981.
6. Muzzio, A., and G. Solaini, "Boundary Integral Eq. Analysis of Three-Dimensional Transient Heat Conduction in Composite Media," Numerical Heat Transfer, Vol. 11, 239-247, 1987.
7. Linz, P., Analytical and Numerical Methods for Volterra Eq.s, SIAM, Philadelphia, 1985.
8. Golan, L. P., "Thermal Analysis of Sliding Contact Systems Using the Boundary Element Method," MS Thesis, Virginia Polytechnic Institute and State University, 1991.
9. Furey, M. J., B. Vick, S. J. Foo, and B. L. Weick, "A Theoretical and Experimental Study of Surface Temperatures Generated During Fretting," Proceedings of the Japan International Tribology Conference, Nagoya, 1990.
10. Forsythe, G. E., M. A. Malcolm, and C. B. Moler, "Computer Methods for Mathematical Computations," pp. 92-105, Prentice Hall, 1977.

APPENDIX: GREEN'S FUNCTION INTEGRALS

The GF integrals required in the numerical discretization are defined as follows,

$$G^p_{i,n}(\bar{r},\tau) = \int_{\tau_o=\tau_{p-1}}^{\tau_p} \iint_{\Gamma_n} G_i(\bar{r},\tau/\bar{r}_o,\tau_o)d\Gamma_o d\tau_o \quad (A.1a)$$

$$DG^p_{i,n}(\bar{r},\tau) = \int_{\tau_o=\tau_{p-1}}^{\tau_p} \iint_{\Gamma_n} \frac{\partial G}{\partial \eta}\, d\Gamma_o d\tau_o \quad (A.1b)$$

where Γ_n can be any boundary element on surface Γ_c, Γ_1 or Γ_2. The integrations are over the cause variables, $\bar{r}_o$ and τ_o. An addition spatial integration over the effect variable $\bar{r}$ required for the averaging process in equations (18) and (19) or Eq. (21) gives,

$$G^p_{i,n,nn}(\tau) = \iint_{\Gamma_{nn}} G^p_{i,n}(\bar{r},\tau)d\Gamma \quad (A.2a)$$

$$DG^p_{i,n,nn}(\tau) = \iint_{\Gamma_{nn}} DG^p_i(\bar{r},\tau)\, d\Gamma \tag{A.2b}$$

The subscripts n and nn represent integration over boundary elements while superscript p represents integration over time.

A vital step in the boundary integral equation method is an accurate evaluation of these Green's function integrals. Consider a rectangular boundary element located on a constant Z_o plane in the area defined by

$$\Gamma_n = \begin{cases} X_m < X_o < X_{m+1} = X_m + \Delta X \\ Y_j < Y_o < Y_{j+1} = Y_j + \Delta Y \;. \end{cases}$$

The spatial integration can be evaluated analytically for this rectangular patch and Eq. (A.1a) becomes

$$G^p_{i,n}(\bar{r},\tau) = \int_{\tau_o=\tau_{p-1}}^{\tau_p} \left\{ \frac{1}{4\pi^{\frac{1}{2}}\xi} \exp\left[-\frac{(Z-Z_o)^2}{\xi^2}\right] \right. \tag{A.3}$$

$$\left[\mathrm{erf}\left(\frac{X-X_{m+1}-X_c}{\xi}\right)-\mathrm{erf}\left(\frac{X-X_m-X_c}{\xi}\right)\right]\left[\mathrm{erf}\left(\frac{Y-Y_{j+1}}{\xi}\right)-\mathrm{erf}\left(\frac{Y-Y_j}{\xi}\right)\right]\left.\vphantom{\frac{1}{1}}\right\} d\tau_o$$

where $\xi = [4\Lambda_i(\tau-\tau_o)]^{1/2}$.

Equation (A.2a) includes the additional spatial integration required for calculating averages over elements. Consider a boundary element located on a constant Z plane in the rectangular patch defined by,

$$\Gamma_{nn} = \begin{cases} X_{mm} < X < X_{mm+1} \\ Y_{jj} < Y < Y_{jj+1} \end{cases}$$

The additional spatial integration can be performed analytically to obtain,

$$G^p_{i,n,nn}(\tau) = \int_{\tau_o=\tau_{p-1}}^{\tau_p} \left\{ \frac{1}{4\pi^{\frac{1}{2}}\xi} \exp\left[-\frac{(Z-Z_o)^2}{\xi^2}\right] \right. \tag{A.4}$$

$$[F(X_{mm+1}-X_{m+1}-X_c)-F(X_{mm}-X_{m+1}-X_c)-F(X_{mm+1}-X_m-X_c)+F(X_{mm}-X_m-X_c)]$$

$$[F(Y_{jj+1}-Y_{j+1})-F(Y_{jj}-Y_{j+1})-F(Y_{jj+1}-Y_j)+F(Y_{jj}-Y_j)]\left.\vphantom{\frac{1}{1}}\right\} d\tau_o$$

where $F(u) = u\,\mathrm{erf}(u/\xi) + \xi/\pi^{\frac{1}{2}} \exp(-u^2/\xi^2)$.

The integration of the gradients of the GF can be obtained immediately from (A.3) and (A.4) by observing the identity given by Eq. (10).

Thermal Simulations of Semiconductor Chips Using BEM

M. Driscart, G. De Mey
University of Ghent, Laboratory of Electronics,
St Pietersnieuwstraat 41, B-9000 Ghent, Belgium

ABSTRACT

This paper presents thermal simulations of a semiconductor chip in a ceramic chip-carrier. This structure is modelled in a mixed environment, using a combination of conductive volumes, conductive layers and contact resistances. The boundary element method is used because this technique is more suited than any other numerical technique to implement the solution in such a mixed environment.

INTRODUCTION

Since the development of the integrated circuit, circuit density and power dissipation has been increasing continually. As a result, thermal management of electronic components and circuits has become more important over the years, causing a growing interest in thermal simulations of electronic circuits (Mahalingam[1]) (Nakayama and Bergles[2]). Almost any numerical technique has been used one time or another, including the boundary element method (Lee and Palisoc[3]).

In this paper thermal simulations are presented of a semiconductor chip, mounted in a ceramic chip-carrier. Special attention will be paid to the advantages of flexible modelling, involving the combination of conducting volumes, conducting layers and contact resistances. The boundary element method is more suited than any other numerical technique to implement the solution in such a mixed environment.

The first part of the paper deals with the physical model and the simulated structure. In the second part the calculated temperature distribution is presented. The influence of several physical parameters is discussed in the third and last part.

THE PHYSICAL MODEL AND THE SIMULATED STRUCTURE

The simulated chip-carrier

The simulated structure consists of a semiconductor chip, mounted in a ceramic chip-carrier. The chip-carrier itself is mounted on a ceramic substrate and the whole circuit is cooled through convection. The structure is shown in figure 1. The chip-carrier is centrally located on the substrate and the substrate mea-

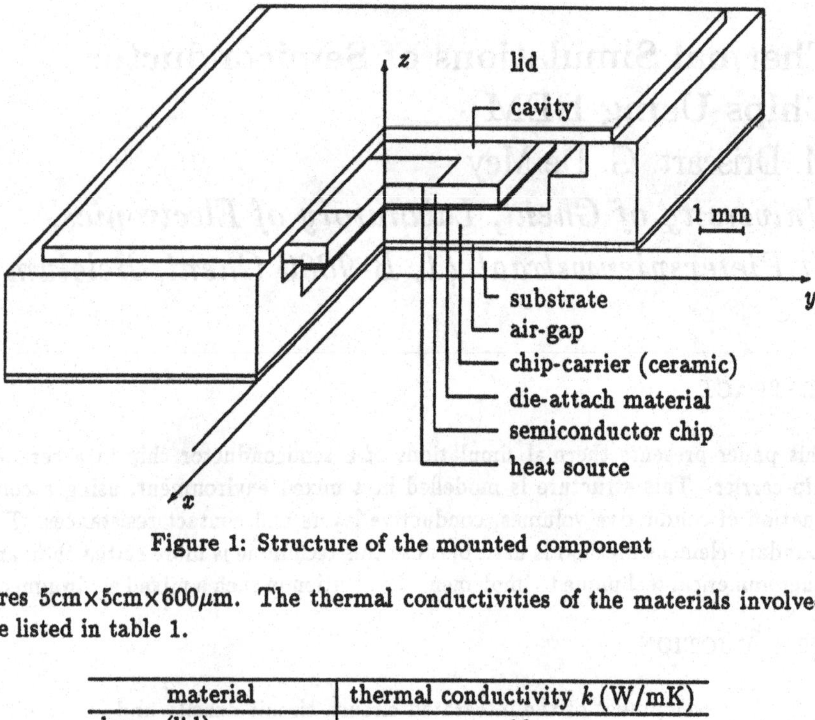

Figure 1: Structure of the mounted component

sures 5cm×5cm×600μm. The thermal conductivities of the materials involved are listed in table 1.

material	thermal conductivity k (W/mK)
kovar (lid)	16.
dry nitrogen (cavity)	.0286
Al_2O_3 (substrate)	25.
Al_2O_3 (chip-carrier)	16.7
AlN (substrate)	170.
semiconductor (Si)	127.
die-attach material	2.28
air	.0253

Table 1: Thermal conductivities

Modelling this structure some important simplifications can be made. First of all, if the perpendicular temperature gradient in the substrate is negligible, the substrate can be modelled as a 2D conductive layer. The perpendicular temperature gradient is negligible if the dissipated heat is transferred to the substrate over a sufficiently large area (Rottiers and De Mey[4]). This is the case since the entire bottom area of the chip-carrier is used to transfer the heat to the substrate. For the same reasons the metal lid of the chip-carrier can also be modelled as a

2D conductive layer.
Compared to the other materials involved the semiconductor material has a high thermal conductivity. Therefore it might be considered to model the semiconductor chip as a 2D conductive layer also. In the second part of the paper it will be shown under what circumstances this model yields good results.
The cavity of the chip-carrier is about 1 mm high. It is well known that no mass flow exists in such small cavities, therefore the cavity can be modelled as a conductive volume. Finally the body of the chip-carrier is also modelled as a conductive volume.
The models are represented in figures 2, 3 and 4. Figure 3 refers to the structure with a 3D model for the semiconductor chip. The structure with a 2D modelled chip is shown in figure 4. Because of symmetry, only a quarter of the structure is simulated.

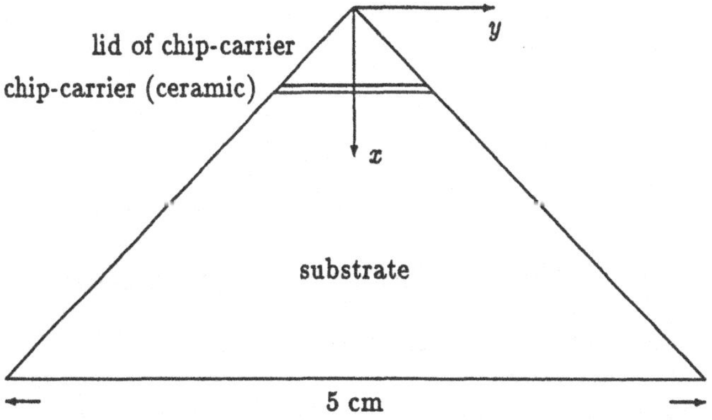

Figure 2: Top view of the simulated structure

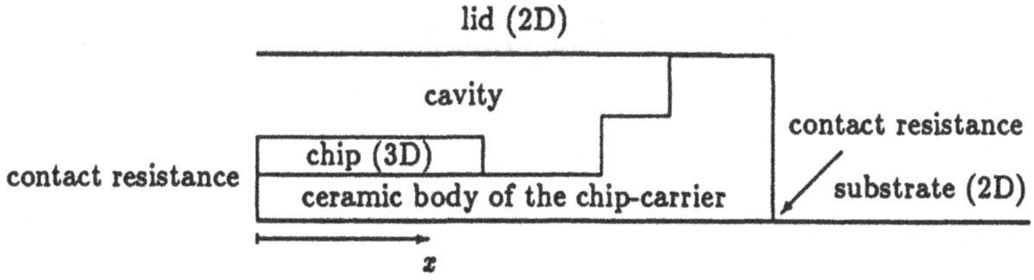

Figure 3: Cross-section of the structure with a 3D semiconductor chip

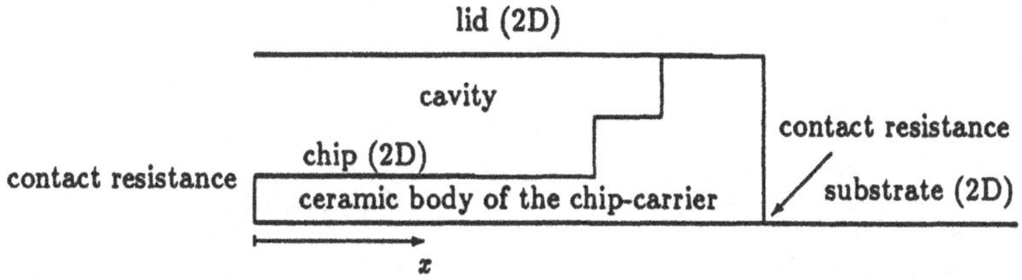

Figure 4: Cross-section of the structure with a 2D semiconductor chip

The mesh generation

In this section some relevant parts of the mesh generation are presented. Figures 5 and 6 show the mesh generation for the substrate. The substrate is modelled as a 2D conductive layer, therefore the sides of the surface need to be divided in elements. This mesh is shown in figure 5. The upper part of the

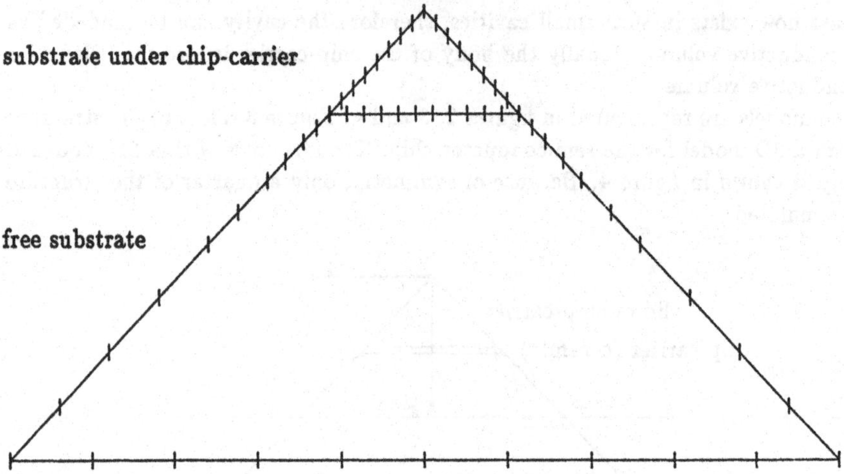

Figure 5: Mesh generation of the sides of the substrate

substrate is also a surface of the chip-carrier. This part also needs a 2D mesh generation, the chip-carrier being a 3D volume. The mesh is shown in figure 6. Similar meshes are used for the metal lid and the chip, when modelled as a 2D

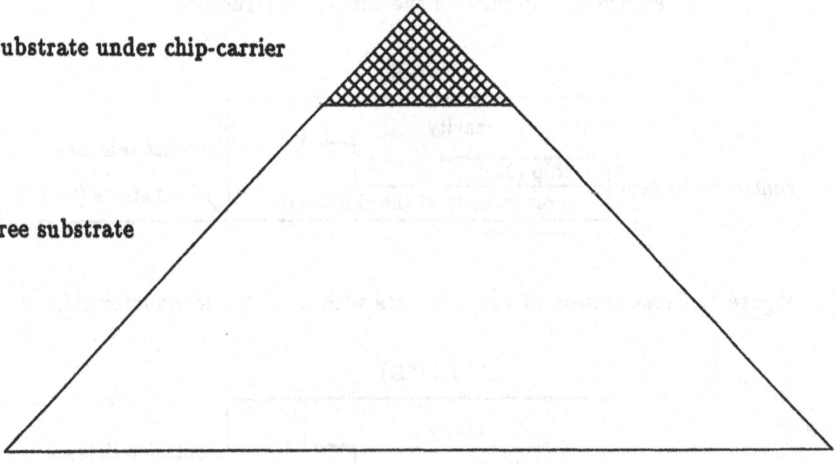

Figure 6: Mesh generation of the surface of the substrate

conductive layer.

Figure 7 shows the mesh generation of the sides of the simulated structure.

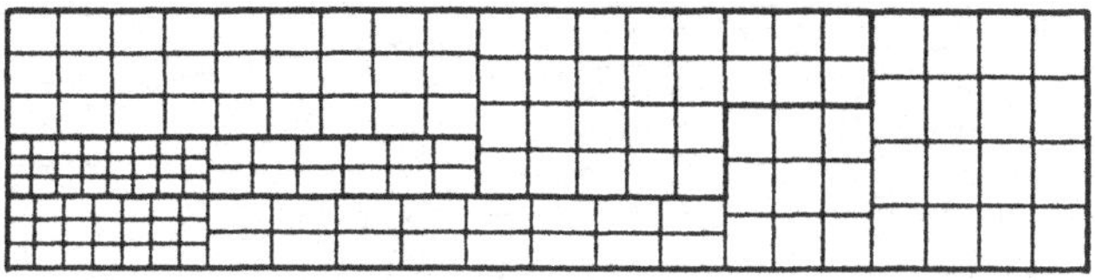

Figure 7: Mesh generation of the sides of the simulated structure

Advantages of flexible modelling

It is obvious that the combination of 3D conductive volumes and 2D conductive layers can reduce both the modelling and the computing effort significantly (Driscart and De Mey[5,6]). These simplifications are no luxury, since the simulation of the complex structures occurring in microelectronics, is a difficult and cumbersome task. Due to this complexity the simulated structures are often strong simplifications of reality. This can impede the straightforward interpretation of the results of the simulations.

The structures, encountered in microelectronics, often contain several layers which are modelled best as 2D conductive layers or as contact resistances. Examples are die-attach materials, substrates, lids, air-gaps, ... Because of these layers microelectronic components can be simulated best using a combination of conductive volumes, conductive layers and contact resistances. However, modelling in such a mixed environment gives rise to a wide variety of boundary conditions. The ability of the boundary element method to deal in a simple and straightforward manner with this variety of boundary conditions explains why the boundary element method is more suited than any other numerical method to implement the solution in such a mixed environment. For more details about the occurring differential equations and boundary conditions one is referred to the literature (Driscart and De Mey[5]).

TEMPERATURE DISTRIBUTION IN THE CHIP-CARRIER

This section deals with the temperature distribution in the chip-carrier. In this section a perfect contact is assumed between the semiconductor chip and the chip-carrier and also between the chip-carrier and the substrate. The substrate material is alumina and the heat transfer coefficient equals 13 W/m^2K.

Figure 8 shows the temperature distribution in the chip-carrier for a power dissipation of 1 W in a 1×1 mm^2 centrally located heat source. The indicated temperature is the rise above ambient temperature. As can be observed from this figure the temperature distribution in the semiconductor chip is clearly 3D, i.e. the temperature difference between the top surface and the backplane of the chip is not negligible under the given circumstances. The 3D character of the temperature distribution is again demonstrated in figure 9. This figure also shows the temperature profile obtained from a simulation with a 2D chip model. Since the temperature distribution in the semiconductor chip shows 3D character, this cannot possibly be considered a realistic model.

Figure 10 shows the temperature profiles for a power source of 3×3 mm^2. It can be observed from this figure that under these circumstances the 2D conductive

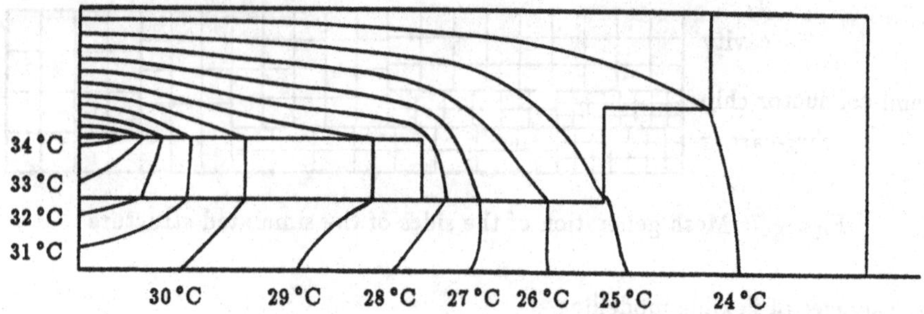

Figure 8: Temperature distribution in the chip-carrier

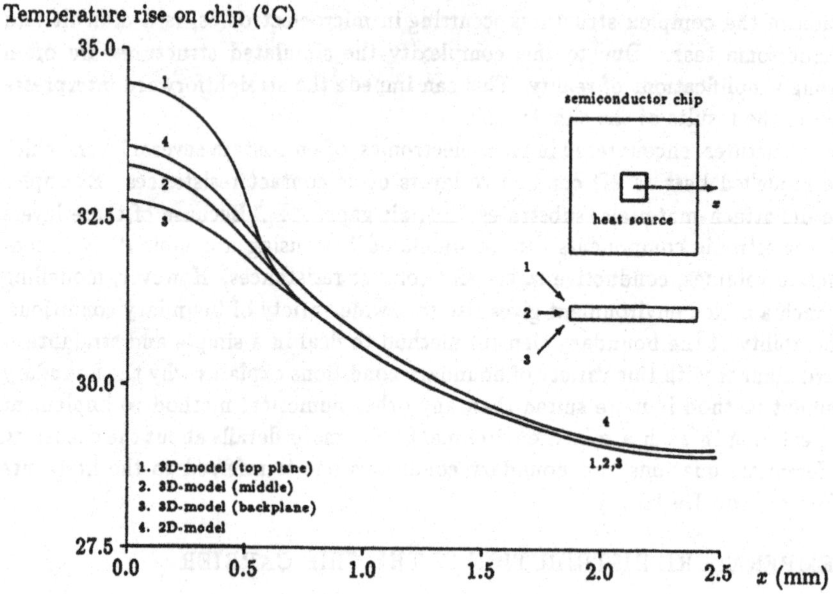

Figure 9: Temperature distribution for a 1×1 mm^2 heat source

layer becomes an adequate model for the chip.

INFLUENCE OF SOME PHYSICAL PARAMETERS ON THE TEMPERATURE DISTRIBUTION

Influence of the heat transfer coefficient

This section deals with the influence of the heat transfer coefficient on the temperature distribution in the circuit. Figure 11 shows the temperature profiles in the chip for a heat transfer coefficient of 13 W/m^{2}K (natural convection) and for a heat transfer coefficient of 75 W/m^{2}K (forced convection). The heat source measures 1×1 mm^2. First of all it can be observed that the temperatures are significantly lower in the case of forced convection cooling. However, it is interesting to notice that the shape of the temperature profiles is almost identical. This is not true for the temperature distribution in the substrate, as can be observed in

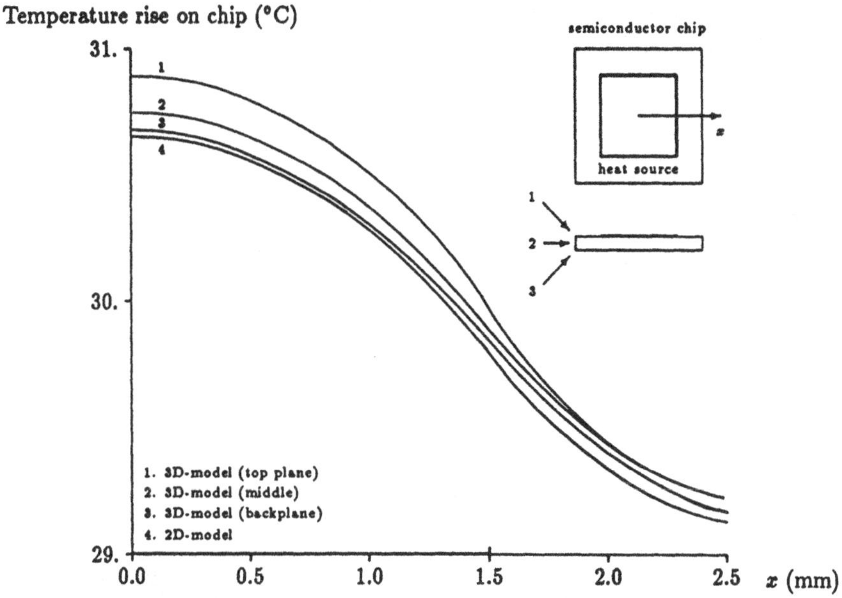

Figure 10: Temperature distribution for a 3×3 mm^2 heat source

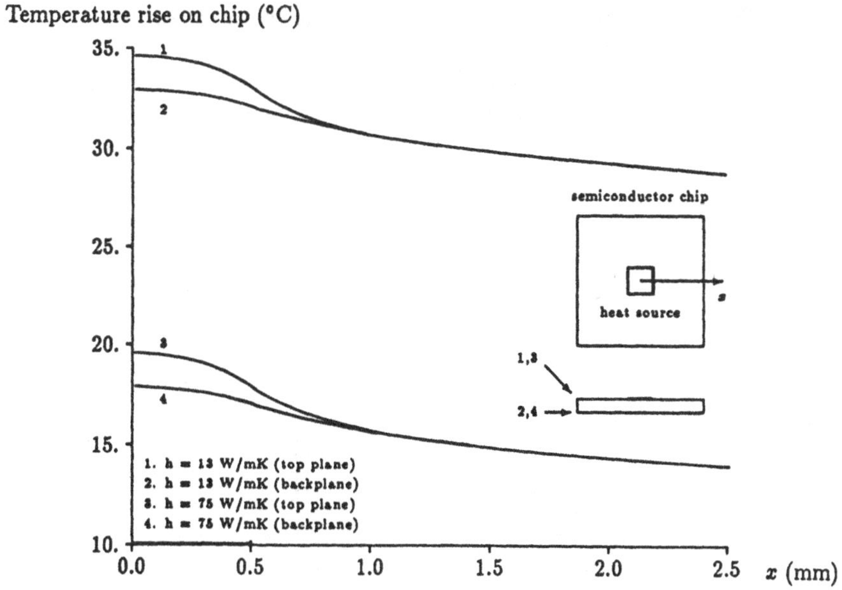

Figure 11: Influence of the heat transfer coefficient

figure 12.

Influence of the substrate material
In this section attention is paid to the influence of the substrate material on the temperature distribution in the circuit. Two substrates are considered, an Al_2O_3-

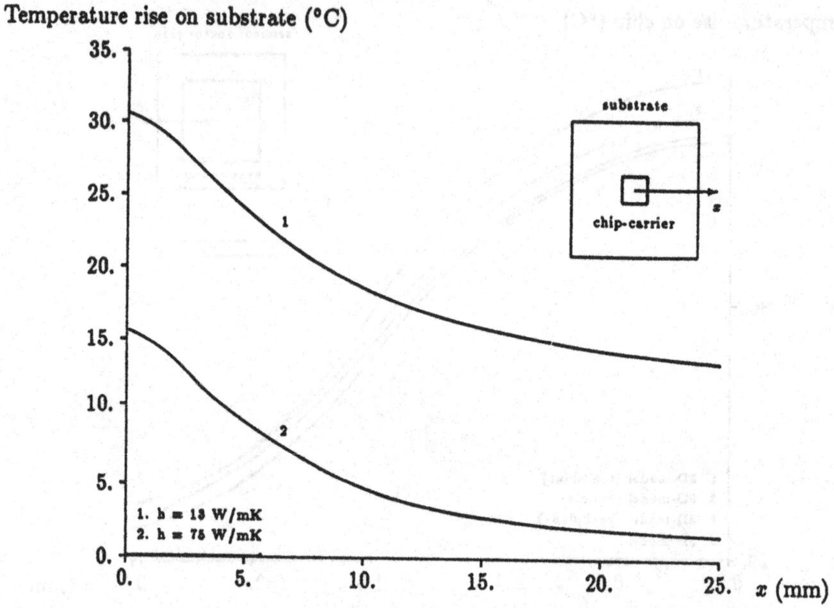

Figure 12: Influence of the heat transfer coefficient

substrate and an AlN-substrate. The heat transfer coefficient is 13 W/m^2K and the heat source measures 1×1 mm^2.
The results are quite similar to the influence of the heat transfer coefficient. Figure 13 shows the temperature profiles in the semiconductor chip. As might be

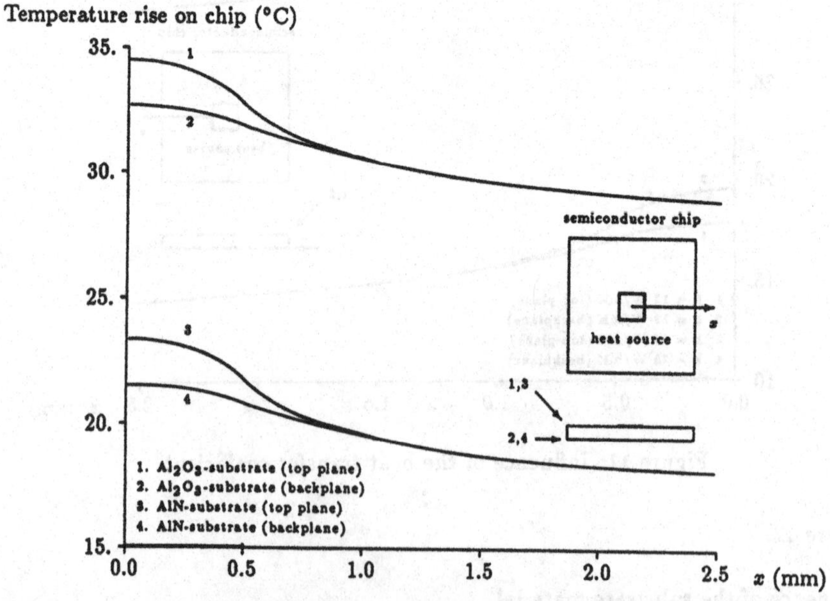

Figure 13: Influence of the substrate material

expected the temperatures are significantly lower in the case of an AlN-substrate. The shape of the temperature profiles is hardly influenced by the thermal conductivity of the substrate material, although the influence cannot be neglected. Figure 14 shows the temperature profiles in the substrate. This figure demon-

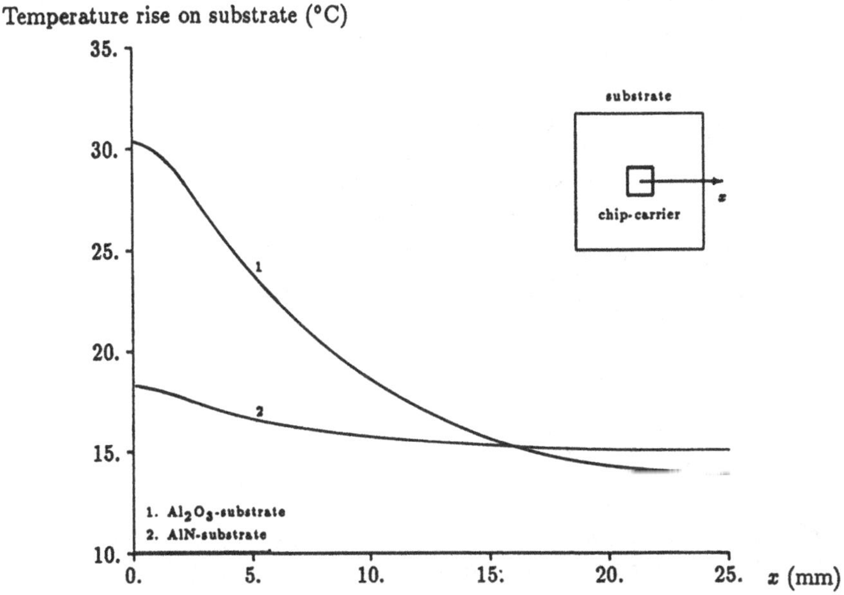

Figure 14: Influence of the substrate material

strates the strong influence of the thermal conductivity of the substrate material on the temperature distribution in the substrate. The heat spreading capability of the AlN-substrate is clearly demonstrated.

Influence of the die-attach material

This section deals with the influence of the die-attach material on the temperature distribution in the circuit. In the previous simulations a perfect contact was assumed between the semiconductor chip and the chip-carrier. This is a good model for an eutectic die bond. In this section a die-bond layer of 50 μm thickness is considered. The heat transfer coefficient is again 13 W/m^2K and the heat source measures 1×1 mm^2. The substrate material is Al_2O_3.

Figure 15 shows the temperature profiles in the semiconductor chip and the die-pad of the chip-carrier. It can be observed from this figure that the temperatures in the semiconductor chip are slightly higher in the presence of the die-bond, but the shape of the temperature profiles is hardly influenced. Furthermore, simulations have shown that the influence of the die-attach material on the temperature distribution in the substrate is almost non-exsitent, as could be expected.

Influence of the contact resistance between the package and the substrate

This last section deals with the influence of the contact between package and substrate on the temperature distribution. In this section a contact resistance of

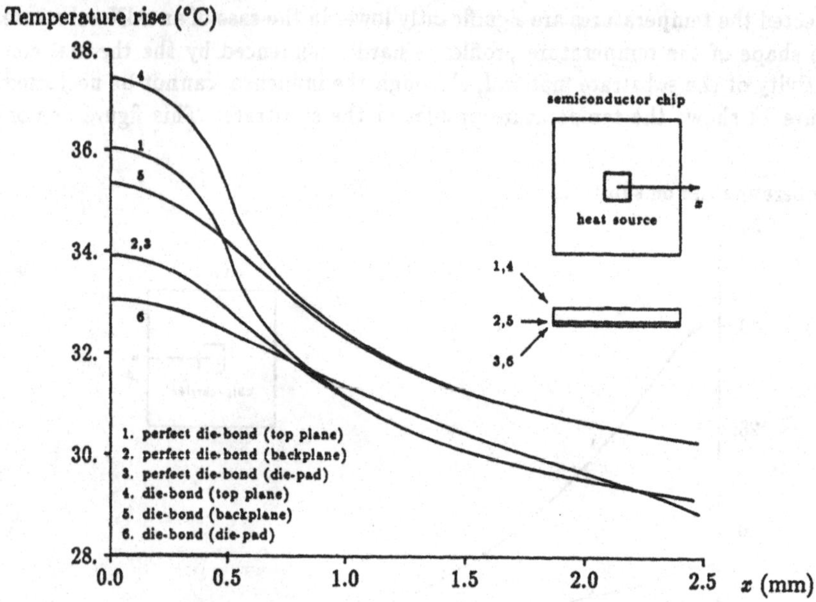

Figure 15: Influence of the die-attach material

2.2 10^5 Km^2/W is assumed between the package and the substrate. This contact resistance is equivalent to a uniform air-gap of 10 μm. The substrate material is again Al_2O_3, the heat transfer coefficient equals 13 W/m^2K and the heat source measures 1×1 mm^2.

The temperatures in the semiconductor chip are about 4 °C higher in the presence of this contact resistance, but the shape of the temperature profiles is indistinguishable. Figure 16 shows the temperature profiles in the substrate and the bottom of the chip-carrier. This figure shows that the influence of the contact resistance on the temperature distribution in the ceramic substrate is limited to a small region under the center of the chip-carrier.

CONCLUSION

In this paper the thermal simulations of a semiconductor chip in a ceramic chip-carrier were presented. This circuit was simulated in a mixed environment using a combination of conductive volumes, conductive layers and contact resistances. The boundary element method was used because it is more suited than any other numerical technique to implement the soluation in such a mixed environment.

REFERENCES

1. Mahalingam, M. Thermal management in semiconductor device packaging, Proc. IEEE, Vol. 73, pp. 1396-1404, 1985.

2. Nakayama, W. and Bergles, A.E. Cooling electronic equipment: past, present and future, in Proc. Int. Symp. on Heat Transfer in Electronic and Microelectronic Equipment, Dubrovnik, Yugoslavia, 1988, pp. 1-37.

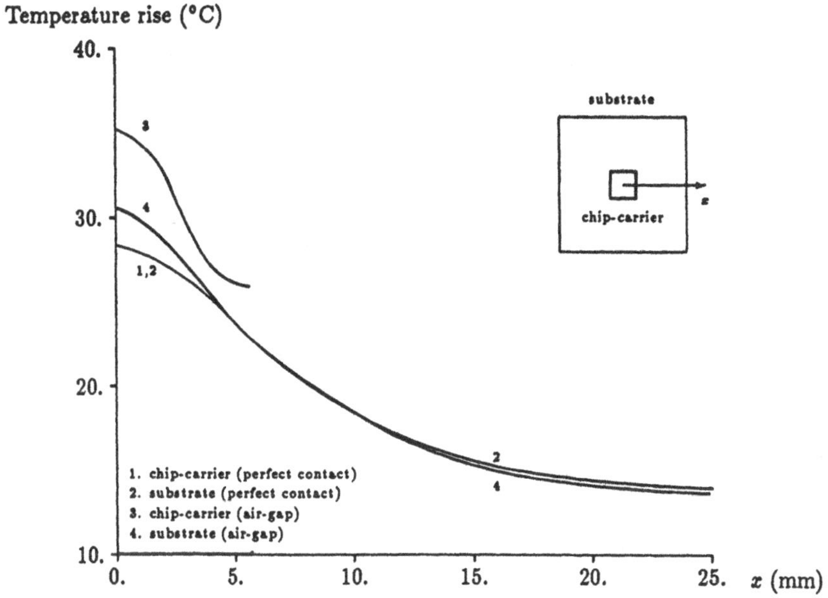

Figure 16: Influence of the contact between chip-carrier and substrate

3. Lee, C.C. and Palisoc, A.L. Thermal Analysis of Semiconductor Devices. Chapter 2, Topics in Boundary Element Research, (Ed. Brebbia, C.A.), Vol. 7, pp. 12-33, Springer-Verlag, Berlin and New York, 1990.

4. Rottiers, L. and De Mey, G. Hot Spot Effects in Hybrid Circuits, IEEE Trans. Comp., Hybr., Manufact. Techn., Vol. 11, pp. 274-278, 1988.

5. Driscart, M. and De Mey, G. A Boundary Element Method Approach to the Thermal Modelling of Microelectronic Components and Packages, in Boundary Elements in Mechanical and Electrical Engineering (Ed. Brebbia, C.A. and Chaudouet-Miranda, A.), pp. 393-404, Proceedings of the International Boundary Element Symposium, Nice, France, 1990. Computational Mechanics Publications, Springer-Verlag, 1990.

6. Driscart, M. and De Mey, G. The Boundary Element Method for Thermal Analysis of Electronic Equipment in a Multi-Dimensional Environment, in Advanced Computational Methods in Heat Transfer (Ed. Wrobel, L.C., Brebbia, C.A. and Nowak, A.J.), pp. 3-343 to 3-354, Proceedings of the First International Conference on Advanced Computational Methods in Heat Transfer, Portsmouth, England, 1990. Computational Mechanics Publications, Springer-Verlag, 1990.

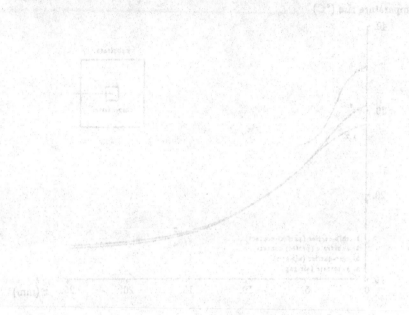

Application of the Complex Variable Boundary Element Method to the Solution of Heat Conduction Problems in Doubly and Multiply Connected Domains

A.J. Kassab, C.K. Hsieh

Department of Mechanical Engineering, University of Florida, Gainsville, Florida 32611, U.S.A.

ABSTRACT

The Complex Variable Boundary Element Method (CVBEM) was originally developed to solve two-dimensional problems in simply-connected domains based on a discretized form of the Cauchy Integral Formula for analytic functions. The present work extends the formulation to doubly-connected domains by introducing a cut in the domain, thereby rendering the stream function single-valued. The resulting CVBEM equations are shown to be a perturbed form of those derived for simply-connected domains. Cauchy-Riemann relations are used to derive boundary equations relating potentials, stream functions, and boundary fluxes when Neumann and Robin boundary conditions are imposed. The analysis is then extended to multiply-connected domains by introducing additional cuts in the domain. This results in additional perturbation terms being added to the simply-connected nodal equations. In order to facilitate computer coding, the equivalence of nodes at the cuts are established, and the codes developed for doubly-connected domains are readily extended to multiply-connected regions.

INTRODUCTION

The engineering analysis of steady heat conduction problems with negligible temperature gradients in the transverse direction leads to two dimensional thermal models governed by Laplace's equation. This equation can become analytically untractable due to the imposed boundary conditions and/or the irregular geometry of the surface bounding domain of interest. Traditionally the numerical solution of the Laplace equation has been obtained by finite difference or finite element methods. The Boundary Element Method offers a flexible alternative to the numerical solution of this problem[1-3]. Boundary Element methods are usually formulated in the real domain using real variables[1].

The Complex Variable Boundary Element Method (CVBEM) is a numerical boundary integral method that was developed by T.V. Hromadka[4-8] for the solution of two-dimensional potential problems in simply connected domains. This method is based on a discretized form of the Cauchy Integral Formula for analytic functions. The real component of the complex potential, the state variable, is identified as the temperature, while the complex component, the stream function, is related to the heat flow rate through the Cauchy-Reimann conditions. Formulation of the boundary element method in the complex plane is a recent and powerful advance, which provides advantages that are not found in the boundary element methods formulated in the real domain. For example, in the CVBEM the functions used to approximate the exact solution are analytic within the domain enclosed by the problem boundary, and they satisfy the Laplace equation throughout the domain of the problem. Further, integration along each boundary element in the CVBEM can be carried out analytically, whereas in the traditional boundary element method such integration is carried out numerically.

The extension of the CVBEM to doubly-connected domains is effected through a cut in the domain. However, the complex potential can be double-valued along the line of cut in a doubly-connected domain[9,10]. This is not encountered in simply connected domains. Analysis of multiply-connected domains is effected through further cuts and can be shown to be an extension of the developments for doubly-connected domains. The CVBEM is applied to the solution of heat conduction problems which encompass imposing Dirichlet, Neumann, and Robin conditions at the boundaries.

MATHEMATICAL BASIS OF THE CVBEM APPLIED TO CONDUCTION HEAT TRANSFER

The theoretical basis of the CVBEM is the Cauchy integral formula,

$$\omega(z_o) = \frac{1}{2\pi i} \oint_{\Gamma} \frac{\omega(\xi)}{\xi - z_o} d\xi \qquad \left\{ z_o \mid z_o \in \Omega \, ; \, z_o \notin \Gamma \right\} \tag{1}$$

which relates the value of a complex analytic function ω at a point z_o inside a domain Ω to the integral of that function along the boundary Γ of that domain; see Figure 1. The contour integral is performed so that the domain lies to the left of the contour of integration. Steady state heat conduction problems with constant thermal conductivity and no heat generation are governed by the Laplace equation. One can thus construct a complex potential, $\omega(z)=\phi(z)+i\psi(z)$, when solving these problems in a two dimensional plane. The real part of the potential $\phi(z)$, representing the state variable, is identified as the temperature; whereas the imaginary part of the potential $\psi(z)$, representing the stream function, is related to the total rate of heat flow. The two are, in turn, related by the Cauchy–Riemann conditions which yield the relation

$$\frac{\partial \phi}{\partial n} = \frac{\partial \psi}{\partial s} \tag{2}$$

where n is the outward drawn normal to Γ, and s is a tangential coordinate along Γ again in the positive direction defined earlier. Using the Fourier law of heat conduction on the left hand side of (2) enables ψ to be expressed as an integral of the heat flux.

The boundary Γ is discretized into N finite-length segments denoted by Γ_j; see Figure 1. Using isoparameteric linear boundary elements to represent the coordinate ξ and the complex potential ω in (1), a first order approximation to $\omega(z_0)$ can be derived[8] as

$$\hat{\omega}(z_o) = \frac{1}{2\pi i} \sum_{j=1}^{N} I_j(z_o) \tag{3}$$

where

$$I_j(z_o) = \frac{\bar{\omega}_{j+1}(z_o - z_j) - \bar{\omega}_j(z_o - z_{j+1})}{(z_{j+1} - z_j)} \ln \left(\frac{z_{j+1} - z_o}{z_j - z_o} \right) \tag{4}$$

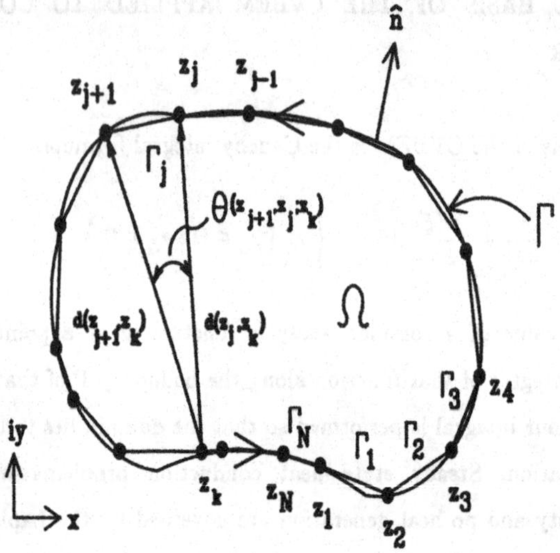

FIGURE 1. Discretization for the CVBEM in a simply connected domain.

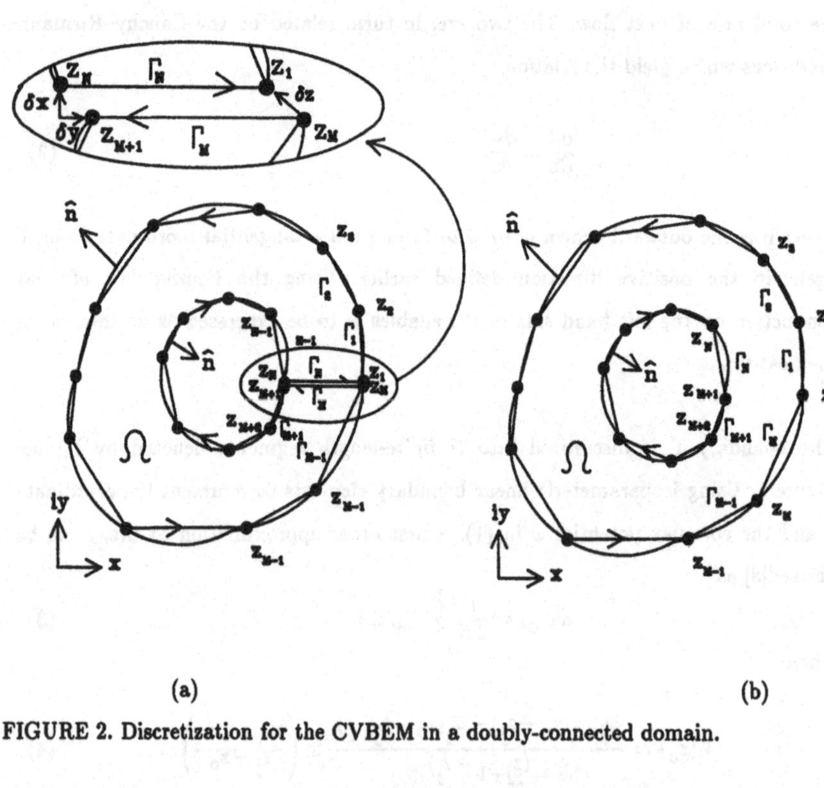

(a) (b)

FIGURE 2. Discretization for the CVBEM in a doubly-connected domain.

Here, z_0 is an interior point. The cap($\hat{}$) over a quantity refers to its being evaluated, while a bar ($\bar{}$) refers to its being specified. If $\bar{\omega}_j$ is fully specified at each node z_j , (3) can be used to evaluate $\hat{\omega}(z)$. However, in regular heat conduction problems, boundary conditions specify any $\bar{\phi}_j$, $\bar{\psi}_j$, or none of them explicitly. In order to solve for the nodal unknowns, a boundary value problem is then formed by generating nodal equations at each point z_k by moving the interior point z_0 to the boundary in a limiting process

$$\hat{\omega}(z_k) = \lim_{z_0 \to z_k} \hat{\omega}(z_0) \tag{5}$$

The resulting singularity is integrable, and evaluating the Cauchy principal value of the integral, a nodal equation can then be derived as [8]

$$\hat{\omega}(z_k) = \frac{1}{2\pi i}\,\bar{\omega}(z_k)\,\ln\left(\frac{z_{k+1}-z_k}{z_{k-1}-z_k}\right) + \frac{1}{2\pi i}\sum_{\substack{j=1 \\ j \neq k-1,k}}^{N} I_j(z_k) \tag{6}$$

Hromadka and his co-workers studied the existence, continuity, and convergence of the approximation in (6) and presented examples [4-8].

CVBEM IN DOUBLY-CONNECTED DOMAINS

The discretization scheme illustrated in Figure 2(a) is useful in formulating the CVBEM in doubly-connected domains. A cut is made in the domain, and a discrete closed circuit is defined. The outer boundary Γ_0 is discretized into M elements, which are numbered in a counterclockwise direction, while the inner boundary Γ_i is discretized into (N−M) elements, which are numbered in a clockwise direction. The cut is shared by two elements Γ_M and Γ_N ; they are equal in length but run in opposite directions as shown. For the convenience of the analysis, these two elements are separated by a small gap $\delta z = \delta x + i\delta y$, which will eventually be taken to be zero in a limiting process in order to close the gap. This approach permits the consideration of double values for the complex potentials along the line of cut. Using equation (2) and the Fourier law of conduction and recognizing that only the heat flow rates at points 1 and M and at points M+1 and N are double valued, the complex potentials at these points can be related by

$$\bar{\omega}(z_M) = \bar{\omega}(z_1) + i\Psi \tag{7}$$

$$\bar{\omega}(z_N) = \bar{\omega}(z_{M+1}) - i\Psi \tag{8}$$

where Ψ is a positive quantity defined as

$$\Psi = \frac{Q_{M,1}}{k} = -\frac{Q_{N,M+1}}{k} \tag{9}$$

In (9), k is the thermal conductivity and Q is the total heat flow rate. The subscripts for Q sequentially refer to the end- and starting-point nodal locations; therefore, $Q_{M,1}$ denotes the total rate of heat flow from point 1 to point M. Taking $\hat{n}$ to be positive pointing outward and considering a basic problem in which no heat is generated within Ω, $Q_{M,1}$ and $Q_{N,M+1}$ must be opposite in sign, and their magnitudes are related by energy conservation principles as $|Q_{M,1}| = |Q_{N,M+1}|$. In general, these Q's are not zero and Ψ does not vanish apriori. The doubly- connected domain becomes simply connected if the inner boundary vanishes, and this becomes a special case of the analysis which follows.

The boundary element problem is now formulated by using (7) and (8) and considering for each nodal point $z_k \in \Gamma$ the limit

$$\hat{\omega}(z_k) = \lim_{\substack{\delta x, \delta y \to 0 \\ z_o \to z_k}} \hat{\omega}(z) \tag{10}$$

That is, the complex integral in (4) is evaluated as each nodal point is approached from the interior, and the Cauchy integral is evaluated in the principal sense. When the limiting procedure in (10) is performed, the nodal points along the cut collapse to the same physical location, yet the contributions of elements Γ_M and Γ_N do not cancel out but are accounted for explicitly by means of the Ψ given in (9). So that the present analysis can be specialized to simply connected domains, a new numbering system is introduced as shown in Figure 2(b). The cut is effectively eliminated in the new numbering system, and in this system there is derived the nodal equation

$$\hat{\omega}(z_k) = \frac{1}{2\pi i}\,\bar{\omega}(z_k)\ln\left(\frac{z_{k+1}-z_k}{z_{k-1}-z_k}\right) + \frac{1}{2\pi i}\sum_{\substack{j=1 \\ j,j+1 \neq k}}^{N} I_{j(z_k)} + \frac{1}{2\pi}\Psi\Delta_k \tag{11}$$

where

$$\Delta_k \,;\, k \neq 1,M,M+1,N = \left[\left(\frac{z_k-z_M}{z_1-z_M}\right)\ln\left(\frac{z_1-z_k}{z_M-z_k}\right) + \ln\left(\frac{z_{M+1}-z_k}{z_1-z_k}\right) - \left(\frac{z_k-z_N}{z_{M+1}-z_N}\right)\ln\left(\frac{z_{M+1}-z_k}{z_N-z_k}\right)\right] \tag{12}$$

and for these special k's,

$$\Delta_1 = [\ln\left(\frac{z_{M+1}-z_1}{z_M-z_1}\right)+\left(\frac{z_1-z_N}{z_{M+1}-z_N}\right)\ln\left(\frac{z_{M+1}-z_1}{z_N-z_1}\right)] \quad (13)$$

$$\Delta_M = [\ln\left(\frac{z_{M+1}-z_M}{z_1-z_M}\right)+\left(\frac{z_M-z_N}{z_{M+1}-z_N}\right)\ln\left(\frac{z_{M+1}-z_M}{z_N-z_M}\right)] \quad (14)$$

$$\Delta_{M+1}= [\left(\frac{z_{M+1}-z_M}{z_1-z_M}\right)\ln\left(\frac{z_1-z_{M+1}}{z_M-z_{M+1}}\right)+\ln\left(\frac{z_N-z_{M+1}}{z_1-z_{M+1}}\right)] \quad (15)$$

$$\Delta_N = [\left(\frac{z_N-z_M}{z_1-z_M}\right)\ln\left(\frac{z_1-z_N}{z_M-z_N}\right)+\ln\left(\frac{z_{M+1}-z_N}{z_1-z_N}\right)] \quad (16)$$

It is recognized that, in (11), when k=1, k−1=M; and when k=M+1, k−1=N . The details of the derivation of (11)-(16) are given in [9,10]. Writing (11) in the format shown has the distinct advantage that it can be specialized to the analysis of simply connected domains for which the $(1/2\pi)\Psi\Delta_k$ term drops out. Indeed, we may express (11) as

$$[\widehat{\omega}(z_k)]_D=[\widehat{\omega}(z_k)]_S+\frac{1}{2\pi}\Psi\Delta_k \quad (17)$$

Here, the subscript D is used to refer to the complex potential in a doubly-connected domain and the subscript S is used to refer to the complex potential in a simply connected domain as given in (6). Performing complex algebra in (12)–(16) yields a pair of algebraic equations at each node z_k relating the evaluated temperature and stream function, $\widehat{\phi}(z_k)$ and $\widehat{\psi}(z_k)$, to the specified temperature and stream function, $\bar{\phi}(z_k)$ and $\bar{\psi}(z_k)$, as

$$\widehat{\phi}(z_k)=A^*\bar{\psi}_k+B^*\bar{\phi}_k+\sum_{\substack{j=1\\ j,j+1\neq k}}^{N}[G_{1,j}\bar{\psi}_j+G_{2,j}\bar{\psi}_{j+1}+G_{3,j}\bar{\phi}_j+G_{4,j}\bar{\phi}_{j+1}]+ST_{\phi_k} \quad (18)$$

$$\widehat{\psi}(z_k)=B^*\bar{\psi}_k-A^*\bar{\phi}_k+\sum_{\substack{j=1\\ j,j+1\neq k}}^{N}[G_{3,j}\bar{\psi}_j+G_{4,j}\bar{\psi}_{j+1}-G_{1,j}\bar{\phi}_j-G_{2,j}\bar{\phi}_{j+1}]+ST_{\psi_k} \quad (19)$$

where,

$$ST_{\phi_k}=\frac{\Psi}{2\pi}\mathrm{Re}[\Delta_k] \qquad \text{and} ST_{\psi_k}=\frac{\Psi}{2\pi}\mathrm{Im}[\Delta_k] \quad (20)$$

The coefficients A^*,B^*,G's, and Δ's are functions of geometry[9,10]. Equations (18) and (19) will be sufficient to solve most potential problems involving only ϕ and ψ. However, for heat conduction problems and especially for those involving Neumann and Robin conditions imposed at the boundaries, additional equations must be derived to relate ϕ, ψ, and the heat flux q".

Applying a Dirichlet boundary condition is equivalent to specifying the $\overline{\phi}$ values at the boundary. However, when a Neumann condition is imposed at the boundary, the Cauchy-Reimann conditions can be used to derive a relation between $\overline{\psi}_{j+l}$ and $\overline{\psi}_j$ as

$$\overline{\psi}_{j+l}=\overline{\psi}_j+\sum_{i=1}^{l}[-\left(\frac{q''}{k}\right)_{a,j+i-1}(\Delta s)_{j+i-1}] \tag{21}$$

where

$$\left(\frac{q''}{k}\right)_{a,j+i-1}=\frac{1}{2}[\left(\frac{q''}{k}\right)_{j+i}+\left(\frac{q''}{k}\right)_{j+i-1}] \tag{22}$$

$$(\Delta s)_{j+i-1}=|z_{j+1}-z_{j+i-1}| \tag{23}$$

Equations (21)-(23) approximate the heat flux q'' as linear over the element Γ_j, which is consistent with the linear approximation of ω. A Robin (or convective) condition imposed on the boundary is represented by the relation

$$\overline{\psi}_{j+l}=\overline{\psi}_j+\sum_{i=1}^{l}[\left(\frac{h}{k}\phi_\infty\right)_{a,j+i-1}(\Delta s)_{j+i-1}]+\sum_{i=1}^{l}[-\left(\frac{h}{k}\phi\right)_{a,j+i-1}(\Delta s)_{j+i-1}] \tag{24}$$

where

$$\left(\frac{h}{k}\phi\right)_{a,j+i-1}=\frac{1}{2}[\left(\frac{h}{k}\phi\right)_{j+i}+\left(\frac{h}{k}\phi\right)_{j+i-1}] \tag{25}$$

Here, h is the convective coefficient (which may also account for radiation if linearized), and ϕ_∞ is the temperature of the surroundings.

CVBEM IN MULTIPLY-CONNECTED DOMAINS

The extension of the CVBEM to multiply-connected domains can be readily accomplished once triply-connected domains have been considered. Referring to Figure 3(a), cuts are introduced in the domain, and here the cut along Γ_{ME} and Γ_{MAB} represent a new type of cut that will be used repeatedly to formulate the CVBEM in domains with connectivities higher than two. Considering the triply-connected domain in Figure 3(a), the complex potentials at points M,ME, and ME+1 are related to the complex potentials at points 1, MAB+1, and MAB by

$$\overline{\omega}(z_M)=\overline{\omega}(z_1)+i\Psi_o \qquad ,\Psi_o=-Q_{M,1}/k \tag{26}$$

$$\overline{\omega}(z_{ME})=\overline{\omega}(z_{MAB+1})-i\Psi_E \qquad ,\Psi_E=-Q_{ME,MAB+1}/k \tag{27}$$

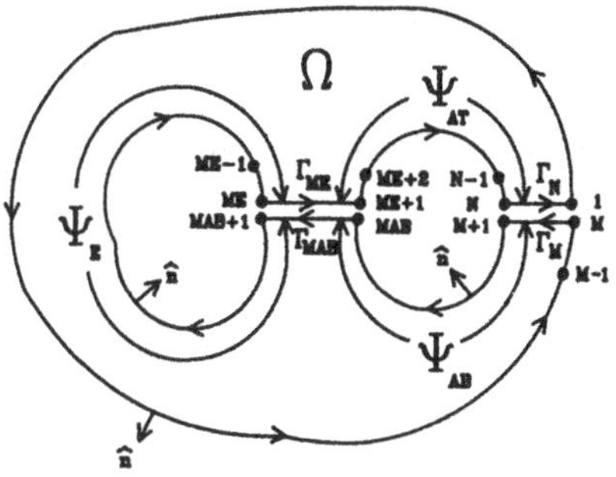

(a)

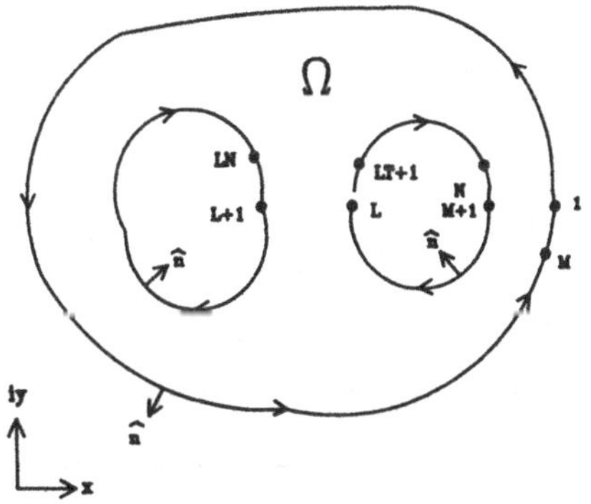

(b)

Figure 3. The CVBEM in a triply-connected domain.

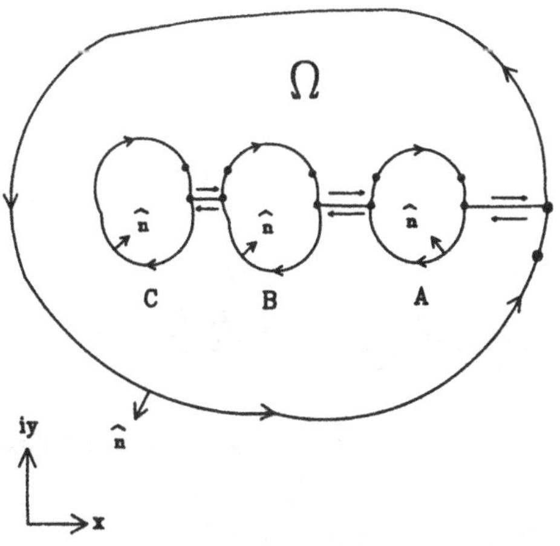

Figure 4. A multiply-connected domain with three holes.

$$\bar{\omega}(z_{ME+1})=\bar{\omega}(z_{MAB})-i\Psi_E \tag{28}$$

Additionally, the complex potentials at points N and MAB can be related to those at ME+1 and M+1 by

$$\bar{\omega}(z_N)=\bar{\omega}(z_{ME+1})-i\Psi_{AT} \qquad ,\Psi_{AT}=Q_{N,ME+1}/k \tag{29}$$

$$\bar{\omega}(z_{MAB})=\bar{\omega}(z_{M+1})-i\Psi_{AB} \qquad ,\Psi_{AB}=Q_{MAB,M+1}/k \tag{30}$$

Using the conservation of energy and performing the limiting procedure of taking an interior point to the boundary while closing the gaps across the cuts there is derived after some tedious algebra [10]

$$\hat{\omega}(z_k)=\frac{1}{2\pi i}\bar{\omega}(z_k)\ln\left(\frac{z_{k+1}-z_k}{z_{k-1}-z_k}\right)+\frac{1}{2\pi i}\sum_{\substack{j=1\\ j,j+1\neq k}}^{N} I_j(z_k)+\frac{1}{2\pi}(\Psi_o\Delta_k-\Psi_E\Xi_k) \tag{31}$$

It is noted that equation (31) now refers to the new numbering scheme in Figure 3(b), where again the cuts have been effectively eliminated. The term Δ_k has been already defined in (12)-(16). The equations for Ξ_k can be shown to be equivalent to those in (12)-(16) if the following table of correspondence between the subscripts is used:

Points in (12) – (16)	Corresponding points in Ξ equations
1	L+1
M	LN
M+1	L
N	LT+1

For example,

$$\Xi_{LN}=[\ln\left(\frac{z_L-z_{LN}}{z_{L+1}-z_{LN}}\right)\left(\frac{z_{LN}-z_{LT+1}}{z_L-z_{LT+1}}\right)\ln\left(\frac{z_L-z_{LN}}{z_{LT+1}-z_{LN}}\right)] \tag{32}$$

Here again, we may express the results in the triply-connected domain as

$$[\hat{\omega}(z_k)]_T=[\hat{\omega}(z_k)]_S+\frac{1}{2\pi}[\Psi_o\Delta_k-\Psi_E\Xi_k] \tag{33}$$

and the ST terms in the nodal equations (18)-(19) are now amended to read

$$ST_{\phi_k} = \frac{1}{2\pi}[\Psi_o \text{Re}(\Delta_k) - \Psi_E \text{Re}(\Xi_k)] \quad \text{and} \quad ST_{\psi_k} = \frac{1}{2\pi}[\Psi_o \text{Im}(\Delta_k) - \Psi_E \text{Im}(\Xi_k)] \tag{34}$$

Based on the development of the CVBEM in triply-connected domains, the CVBEM can readily be extended to multiply-connected domains with more than one hole[10]. Referring to Figure 4 the nodal boundary equation is then modified to read

$$[\hat{\omega}(z_k)]_M = [\hat{\omega}(z_k)]_S + \frac{1}{2\pi}[\Psi_o \Delta_k - \sum_{j=1}^{K} \Psi_{E,j} \Xi_{k,j}] \tag{35}$$

where the M denotes multiply-connected and K denotes the number of holes in excess of one that are enclosed in the domain. Here, a series of holes similar to the one added to extend the doubly-connected formulation to triply-connected domains are strung together to model a multiply-connected domain. The nodal equations (18) and (19) are accordingly amended to include the additional contributions to the ST_{ϕ_k} and ST_{ψ_k} terms. Equation (35) is quite general and readily reduces to the simply-connected formulation for which the Ψ-terms are zero.

ASSEMBLY AND SOLUTION OF THE BOUNDARY ELEMENT EQUATIONS

The final step in the solution by the CVBEM is the assembly of the nodal equations to solve for the unknowns. Equations (14) and (15) provide 2xN equations which can be arranged in matrix form and partitioned to give the following structure:

$$\begin{bmatrix} C_{R,\bar{\phi}} & C_{R,\bar{\psi}} \\ C_{I,\bar{\phi}} & C_{I,\bar{\psi}} \end{bmatrix} \begin{bmatrix} \bar{\phi} \\ \bar{\psi} \end{bmatrix} + \begin{bmatrix} ST_{\bar{\phi}} \\ ST_{\bar{\psi}} \end{bmatrix} = \begin{bmatrix} \hat{\phi} \\ \hat{\psi} \end{bmatrix} \tag{36}$$

Here the subscripts R and I for the partitioned matrices C refer to the fact that the set of N $\hat{\phi}$-equations are derived from the real part of (35), and the set of N $\hat{\psi}$-equations are derived from the imaginary part of (35). In the solution of heat conduction problems, the boundary conditions specify $\bar{\phi}_k, \bar{\psi}_k$, or neither explicitly. Additionally, since the Cauchy-Reimann conditions are integrated to find ψ, at least one value of ψ at the boundary is needed as a constant of integration. Three solution strategies (explicit, implicit, and hybrid) can be developed [8-10] depending on how the computed variables are related to the specified variables and on which equations are deleted from (36).

SOLUTION OF ϕ AND ψ AT INTERIOR POINTS

Once $\bar{\phi}$ and $\bar{\psi}$ are known at all boundary nodes, $\hat{\omega}(z_o)$, where $z_o \in \Omega$, can be derived by introducing cuts in the domain and accounting for the double-valued complex potentials along the cuts. For example, the complex potential for a triply-connected domain is derived as

$$\hat{\omega}(z_o) = \frac{1}{2\pi i} \sum_{j=1}^{N} I_j(z_o) + \frac{1}{2\pi} [\Psi_o \Delta - \Psi_E \Delta \Xi] \tag{36}$$

where,

$$\Delta = [\left(\frac{z_o - z_M}{z_1 - z_M}\right) \ln\left(\frac{z_1 - z_o}{z_M - z_o}\right) + \ln\left(\frac{z_{M+1} - z_o}{z_1 - z_o}\right) - \left(\frac{z_o - z_N}{z_{M+1} - z_N}\right) \ln\left(\frac{z_{M+1} - z_o}{z_N - z_o}\right)] \tag{37}$$

$$\Xi = [\left(\frac{z_o - z_{LN}}{z_{L+1} - z_{LN}}\right) \ln\left(\frac{z_{L+1} - z_o}{z_{LN} - z_o}\right) + \ln\left(\frac{z_L - z_o}{z_{L+1} - z_o}\right) - \left(\frac{z_o - z_{LT+1}}{z_L - z_{LT+1}}\right) \ln\left(\frac{z_L - z_o}{z_{LT+1} - z_o}\right)] \quad (37)$$

The extension of (36) to domains with higher connectivities follows from the above developments.

EXAMPLES

Two examples are presented to illustrate the application of the CVBEM in the solution of heat conduction problems. The first example considers the triply connected domain illustrated in Figure 5. Here, the geometry is generated using the ovals of Cassini. The inner and outer walls are imposed with Dirichlet boundary conditions using the complex potential given as $\omega(z) = (\Gamma/2\pi)\,[\ln(z - z_o) - \ln(z + z_o)]$, with $\Gamma = 4\pi$. This problem corresponds to the superposition of a source uniformly generating heat within the right cavity and a sink uniformly removing heat at an equal rate from the left cavity. This problem also arises in potential theory in the superposition of a line source and a line sink of equal strength located at $z_o = \pm 1$ within an infinite domain. This example is useful to test the CVBEM results against an analytically available solution. The CVBEM generated isotherms are shown in Figure 5. The inner and outer heat flow rates are computed as $\Psi_o = 0.05$ and $\Psi_E = 12.51$, which is in good agreement with exact values of $\Psi_o = 0$ and $\Psi_E = 12.56$. The temperature field within the triply-connected domain was scanned at 1200 points, and a maximum deviation of 2.7% from the exact solution was found.

The second example considers the heat transfer from the irregular region illustrated

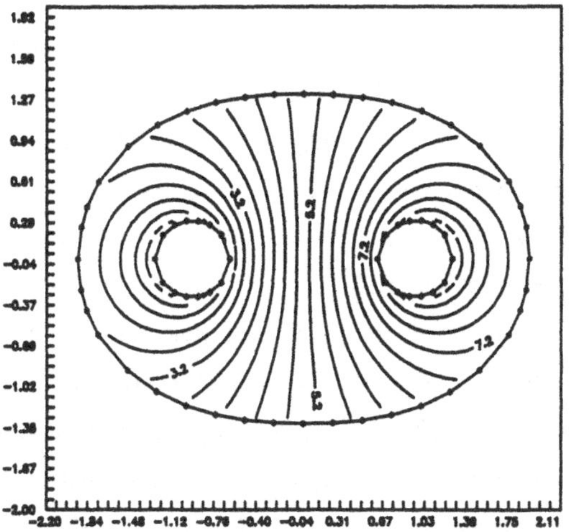

Figure 5. CVBEM generated isotherm plot for the problem with exact complex potential $\omega(z)=(\Gamma/2\pi)[\ln(z-z_0)-\ln(z+z_0)]$.

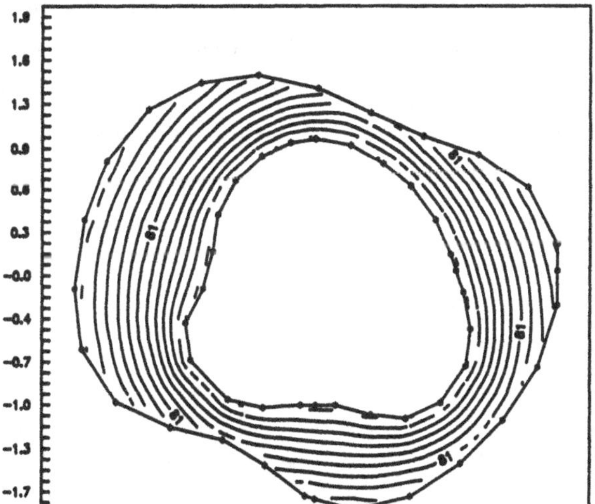

Figure 6. Isotherm plot for the heat conduction problem in an irregular region imposed with Dirichlet and Robin boundary conditions.

in Figure 6. The inner wall contains a fluid undergoing a phase change at a temperature of $\phi=100^{\circ}C$. The boundary condition at the inner cavity wall is modeled as a Dirichlet condition due to the high film coefficient there. The exterior wall is convecting heat to the surroundings at $\phi_{\infty}=20^{\circ}C$. For this problem, the ratio h/k=0.625 is taken at the exterior surface. This problem is analytically untractable. Instead, the CVBEM solution is compared with a solution obtained from a real variable boundary element (RVBEM) solution using the same discretization and linear elements. The CVBEM computed heat flow rate is $\Psi=-373.03$. This compares well with the value of $\Psi=-373.52$ computed by integrating the convective heat flux $q''/k=(h/k)(T-T_{\infty})$ over the outer boundary, using the RVBEM determined nodal temperatures. The CVBEM generated isotherms are shown in Figure 6. A comparison of the CVBEM and RVBEM nodal temperatures at the outer surface are plotted in Figure 7. There is very close agreement in both solutions, with a maximum deviation of 0.4% between the two.

CLOSING REMARKS

The CVBEM has been extended to solve heat transfer problems in multiply-connected regions. The formulation in multiply-connected domains is general, and it is readily reducible to simply-connected domains. The method is shown to be accurate, and it offers the advantage of eliminating the usual numerical quadratures required by its real variable-based counterparts. However, the method is limited to two dimension by the nature of its complex variable formulation.

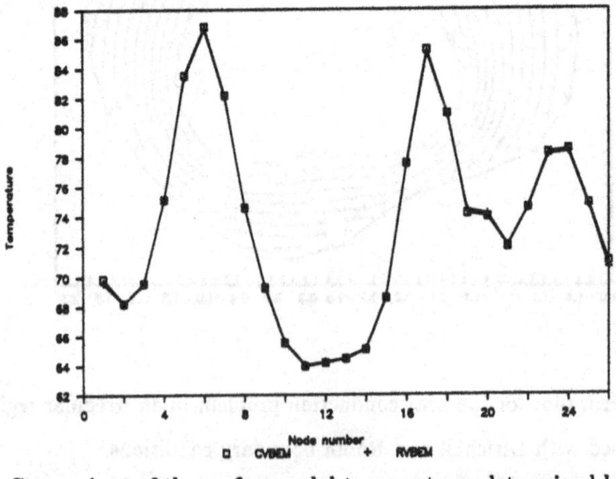

Figure 7. Comparison of the surface nodal temperatures determined by the CVBEM and the linear BEM for the conduction problem in the irregular region.

REFERENCES

1. Brebbia, C.A., Telles, J.C.F., and Wrobel, L.C., Boundary Element Techniques, Springer-Verlag, New York, 1984.
2. Wrobel, L.C. and Brebbia, C.A. Boundary Elements in thermal Problems, Numerical Methods in Heat Transfer, (Eds. Lewis, R.W., Morgan,K.W. and Zienkiewicz,O.C.), John Wiley, New York 1981.
3. Pina, H.G.L. and Fernandes, J.L.M. Applications in Transient Heat Conduction, Topics in Boundary Element Research, (Ed. Brebbia, C.A.), Springer-Verlag, New York , 1984.
4. Hromadka II and Guymon G.L. A Complex Variable Boundary Element Method: Development, International Journal of Numerical Methods in Engineering, Vol. 20, pp. 25-37, 1984.
5. Hromadka II, T.V., Guymon, G.L. and Yen, C.C. The Complex Variable Boundary Element Method: Applications, Numerical Methods in Engineering, Vol. 21, pp. 1013-1025, 1985.
6. Hromadka II, T.V. Determining Relative Error Bounds for the CVBEM, Engineering Analysis, Vol. 2, No.2, pp. 75-80, 1985.
7. Hromadka II, T.V., The Complex Variable Boundary Element Method, Lecture Notes in Engineering, (Eds. Brebbia, C.A. and Orzag, S.A.), Springer-Verlag, New York, Vol.9, 1986.
8. Hromadka II, T.V. and Lai, C., The Complex Variable Boundary Element Method in Engineering Analysis, Springer-Verlag, New York, 1987.
9. Kassab, A.J. and Hsieh, C.K. , Application of the Complex Variable Boundary Element Method to Solving Potential Problems in Doubly Connected Domains, International Journal for Numerical Methods in Engineering, Vol. 29, pp. 161-179, 1989.
10. Hsieh, C.K. and Kassab, A.J. Complex Variable Boundary Element Methods for the Solution of Potential Problems in simply and Multiply Connected Domains, Computer Methods in Mechanics and Engineering, (At Press).

SECTION 8: ELECTRICAL PROBLEMS

Comparison of Pure and Coupled BEM Methods for the Study of a Thin Plate Induction Oven

P. Dular *, A. Genon, W. Legros, J. Mauhin,
A. Nicolet, M. Umé
University of Liege, Department of Electrical Engineering, Institut Montefiore, Sart Tilman, B28-4000 Liège, Belgium
* This author is a Research Assistant with the Belgian National Fund for Scientific Research

ABSTRACT

This paper presents the use of a magnetic field computation software (LUCIE-2D) to determine eddy currents in a thin plate induction oven. The major difficulty arises from the characteristics of the plates which are magnetic, conducting, long (500 mm) and thin (1mm). Various models are compared : one boundary element model of the plate and two finite element models of the plate differing from the meshing. In the three cases, the air around the plate is modelled by the boundary element method. It appears that the skin depth, naturally involved in the BEM formulation, must be taken into account explicitly for the meshing in the FEM formulation.

INTRODUCTION

In an induction oven, the heating process is based on Joule effect (Ernst[1],Miyoshi[2]). Therefore, the determination of the eddy currents is essential for their design. The accurate computation of eddy currents involves a complete electromagnetic modelling of the oven. The particular problem studied in this paper is difficult because of the characteristics of steel plates to be heated. They are magnetic (under Curie point), conducting and with an aspect ratio of 1/500 (length 500 mm, thickness 1 mm). Moreover, at the working frequency of the oven (7400 Hz), the skin depth is smaller than the thickness and the current density cannot be considered as constant across the plate.

EDDY CURRENT

The basic equation of two-dimensional linear magnetostatics is :

$$\Delta A = - \mu J \qquad (1)$$

A potential vector which has only one non-zero component (parallel to current density)

J current density
μ magnetic permeability

Eddy currents are introduced by using the Ohm's law (Bourmanne[3]) :

$$J = \sigma E = \sigma\left(-\frac{\partial A}{\partial t} - U\right) \tag{2}$$

as the expression for the current density in equation (1). The conductor is characterized by its electrical conductivity σ.

The time derivative of A expresses the inductive effects since U = grad V is due to externally imposed electromotive force and can be interpreted as the terminal voltage of the conductor (per unit of length). Moreover, in the case of sinusoïdal excitations, the complex formalism can be adopted and the time derivative of A becomes $j\,\omega\,A$.

The equation for eddy currents is :

$$\Delta A = j\omega\sigma\mu\, A + \sigma\mu U \tag{3}$$

With $j\omega\sigma\mu = -k^2$, the standard form :

$$\Delta A + k^2 A = \sigma\,\mu\,U \tag{4}$$

is obtained, i.e. the non-homogeneous Helmholtz equation.

The fundamental parameter of the eddy current phenomena is the skin depth given by :

$$\delta = \sqrt{\frac{2}{\omega\sigma\mu}} = \frac{\sqrt{-2j}}{k} \tag{5}$$

It characterizes the depth of penetration of the electromagnetic fields inside the conductor.

One problem encountered using equation (4) is the determination of the terminal voltage U. If U is given for all the conductors, the problem can be solved. But in some cases, it is the total current which has to be imposed. This total current can be expressed as the line integral of the magnetic field (Ampere's law) on the boundary Γ of the conductor (Bamps[4]) :

$$I = \int_\Gamma \frac{1}{\mu}\frac{\partial A}{\partial n}\,d\Gamma \tag{6}$$

The practical use of this formula will be explained further.

NUMERICAL METHODS

Boundary element methods

In a non-conducting medium, there is no current and the equation (1) of magnetostatics becomes homogeneous. In this case, the boundary element method is based on the following relation (Brebbia[5]) :

$$cA = \int_{\Gamma} A \frac{\partial G}{\partial n} d\Gamma - \int_{\Gamma} G \frac{\partial A}{\partial n} d\Gamma \tag{7}$$

where :

- G is the free space Green function of the two-dimensional Laplace operator :

  $G = \frac{1}{2\pi} \ln \frac{1}{r}$ (r = distance between source point and observation point).

- c = 0.5 on a smooth boundary.

- $\partial / \partial n$ is for the normal derivative.

Integrals are taken on the boundary of the domain and involve A and $\partial A / \partial n$ (tangential induction) on this boundary. The boundary element formulation for eddy currents can be obtained by using the free space Green function of the two-dimensional Helmholtz operator :

$$H = \frac{1}{4j} H_0^{(2)} (k\, r) \tag{8}$$

where $H_0^{(2)}$ is a Hankel function (Abramowitz[6]).

As the equation (4) is non-homogeneous, the boundary element formulation involves volume integrals. To avoid them, a modified potential A' is defined by :

$$A' = A - \frac{\sigma \mu U}{k^2} \tag{9}$$

Since $\sigma\mu U/k^2$ is constant on the domain, equation (4) can be written :

$$\Delta A' + k^2 A' = 0 \tag{10}$$

The Green identity gives (Rucker[7]) :

$$c\, A' = \int_{\Gamma} A' \frac{\partial H}{\partial n} d\Gamma - \int_{\Gamma} H \frac{\partial A'}{\partial n} d\Gamma \tag{11}$$

This equation is analogous to the equation (7) but involves the Green function H and the modified potential A'.

The total current in the conductor is controlled by equation (6) which introduces a constraint on the normal gradient of A and allows the determination of the terminal voltage.

Care is necessary in numerical integration of thin plates. Self-influence coefficients involve singular kernels but because of the small thickness of the plate, several mutual coefficient computations involve nearly singular kernels. An adaptative version of the Gaussian integration rule developed by Patterson is used for both the real and the complex integrations (Patterson[8]).
In addition, a change of variable is used to "smooth" singularity or "nearly singular behaviour" (Adriaens[9]).

Finite elements
The finite element formulation is based on the Galerkin method (Silvester[10]) and we have for the domain Ω of boundary Γ:

$$\int_{\Omega} \left[\text{grad } A \,.\, \text{grad } w + \sigma\mu\left(j\omega A + U\right) w\right] d\Omega - \int_{\Gamma} w \frac{\partial A}{\partial n}\, d\Gamma = 0 \tag{12}$$

where w is a weighting function. The same function space is choosen for weighting functions w and for approximation functions of A.

The boundary term in equation (12) is usually used to apply a Neumann boundary condition. Here, it is used at the interface between the finite element and the boundary element domains to couple the methods. It allows the use of equation (6) to control the total current in the finite element domains.

INDUCTION OVEN

Figure 1 shows a thin magnetic plate induction oven. It is composed of a five turn coil fed by a current of 1800 A at a frequency of 7400 Hz. The particular case of a half-entered plate is considered. This plate is magnetic (relative permeability $\mu_r = 142$) and conducting ($\sigma = 2.5\ 10^6$ S/m). The length is 500 mm and the thickness is 1 mm (aspect ration 1/500). The total current in the plate is forced to be zero.

Nevertheless, the skin depth is $\delta = 3.1\ 10^{-4}$ m and is smaller than the thickness of the plate. Figure 2 shows the field lines around the end of the plate in the oven at a phase angle of 0 degree (maximum of excitation current). Figure 3 shows the field lines at a phase angle of 90 degrees (zero of excitation current). Three different methods have been used to determine the eddy currents in the plate.

The first two methods use the coupling between the finite elements (for the coil and the plate) and the boundary elements (for the surrounding air). The difference between those two methods lies in the discretization of the plate.

For the first one, a coarse meshing composed of a single layer of first order triangular elements has been used. The width of the elements is the one of the plate. There is no inner nodes because all the vertices of the triangles lie on the boundary of the plate. For the second method, a much finer meshing has been used which takes explicitely into account the skin depth. In order to achieve that, the meshing has been built by offset. Hence it includes two layers of elements whose width is more or less equal to the skin depth. With such a meshing, the plate appears to be composed of five slices of elements with numerous inner nodes.

In the third method, the boundary element method is used for the plate. For this method, only the boundary of the plate has to be meshed. For all those three methods, the length of the plate has been divided into about fifty pieces.

Figure 4 compares the current density modules alongside the plate for the three methods.

Figure 5 compares the phase angle of the current density alongside the plate.

Figure 6 compares the current density modulus across the plate (the section is located at the maximum of current on the curves of figure 4).

Table 1 gives the value, computed by the three methods, of the active power and of the Joule losses for the turns of the coil and for the plate. The symmetry of the results for the corresponding inductors and the energy balance give a verification of the coherence of results.

It appears clearly that the fine mesh finite element model and the boundary element model for the plate give similar results while the coarse mesh model for the plate overestimates current and losses.

CONCLUSIONS

Various methods have been used to compute the eddy currents in a thin plate induction oven. One difficulty is the shape of the plate that leads to an ill-conditioned numerical problem. That difficulty is overcome by using adaptative numerical methods.

The other difficulty comes from the small penetration depth of the field. It appears that the most accurate methods take the skin depth into account. On the one hand, the skin depth appears explicitely in the fundamental solution of the Helmholtz equation used in the boundary element formulation. On the other hand, the finite element method must take the skin depth into account at the discretization level. Therefore, the boundary element method is interesting because it avoids those meshing problems. Nevertheless, the finite element method can be naturally extended to non linear magnetic materials.

Table 1: active power and Joules losses in the induction oven (kW)

Coil	FEM (coarse mesh) Power	FEM (coarse mesh) Joule effect	FEM (fine mesh) Power	FEM (fine mesh) Joule effect	BEM Power	BEM Joule effect
1	33.55	0.0135	25.38	0.0135	24.73	0.0135
2	33.36	0.0135	25.33	0.0135	24.74	0.0135
3	30.55	0.0135	23.44	0.0135	22.90	0.0135
4	30.89	0.0135	23.47	0.0135	22.91	0.0135
5	12.84	0.0135	9.536	0.0135	9.323	0.0135
6	12.31	0.0135	9.532	0.0135	9.287	0.0135
7	1.897	0.0135	1.586	0.0135	1.535	0.0135
8	2.170	0.0135	1.560	0.0135	1.542	0.0135
9	1.022	0.0135	0.7310	0.0135	0.7260	0.0135
10	0.8372	0.0135	0.7453	0.0135	0.7213	0.0135
Plate	0.0	157.9	0.0	119.9	0.0	117.0
Total	159.4	159.3	121.3	121.2	118.4	118.3

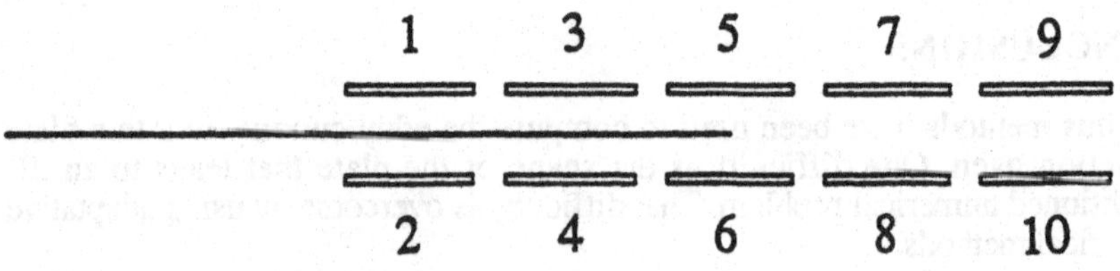

Figure 1 : Thin magnetic plate induction oven

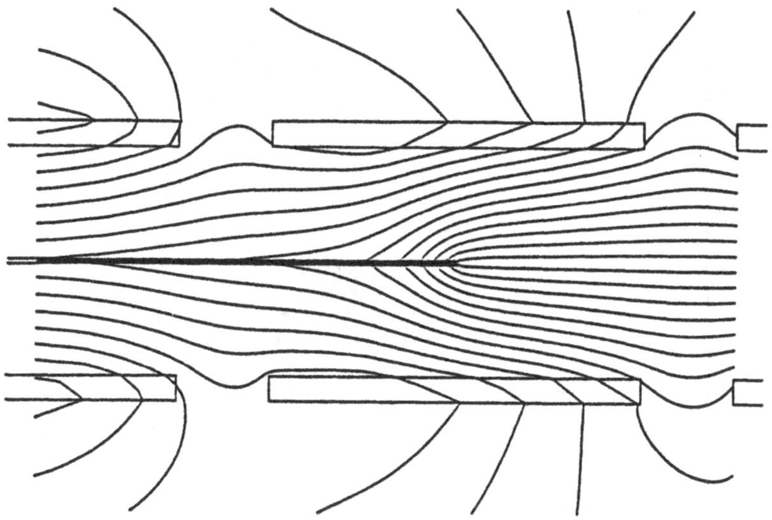

Figure 2 : Field lines around the end of the plates (zero degree phase angle)

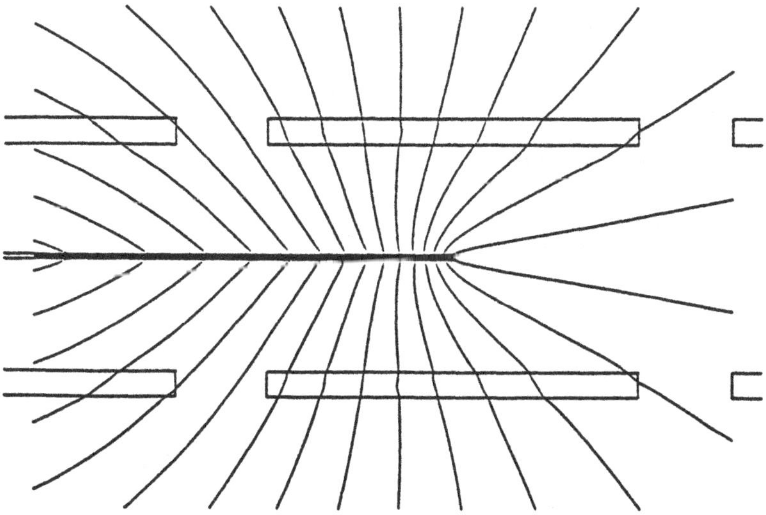

Figure 3 : Field lines around the end of the plates (90 degrees phase angle)

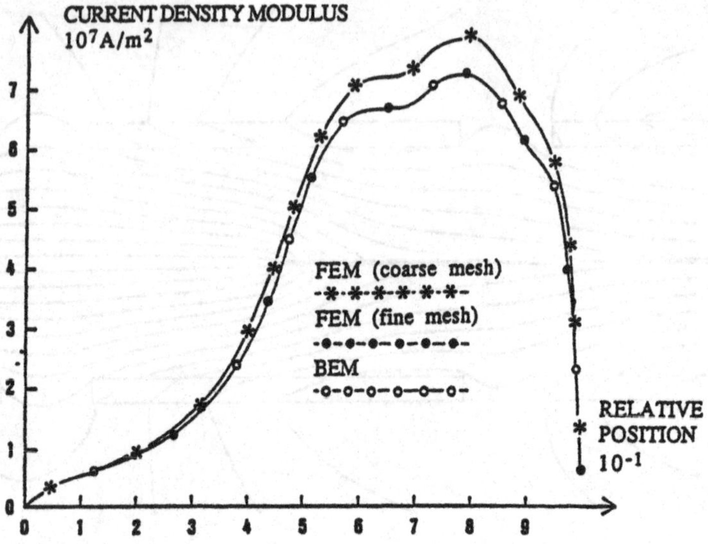

Figure 4 : Current density modules alongside the plate for the three methods

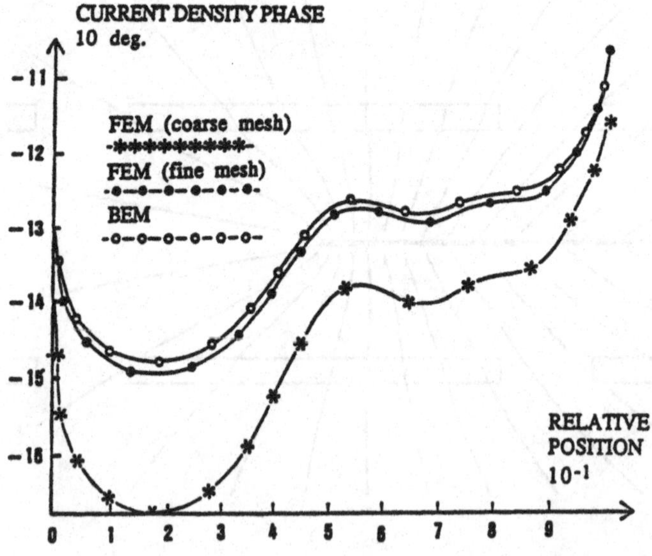

Figure 5 : Phase angle of the current density alongside the plate

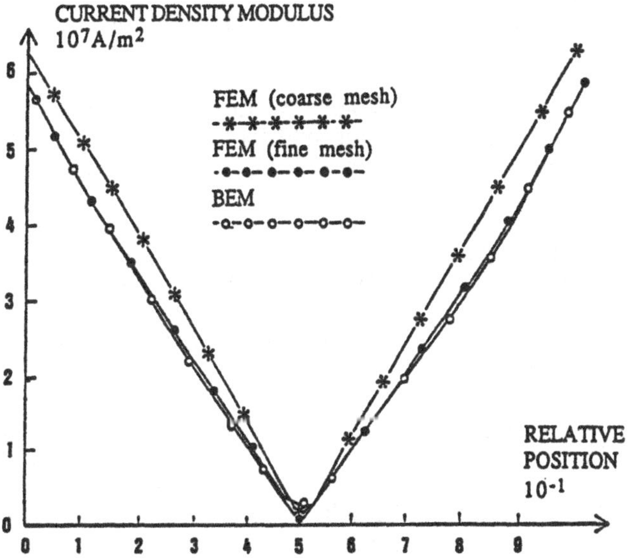

Figure 6 : Current density modulus across the plate

REFERENCES

1. Ernst, R., Gagnoud, A., Leclercq, I., Etude du comportement d'un circuit magnétique dans un système de chauffage par induction, RGE, n°9, Octobre 1987, pp.10-16.

2. Miyoshi, T., Sumiya, M., Omori, H., Analysis of induction heating system by the finite element method combined with a boundary integral equation, IEEE Transactions on Magnetics, Vol. MAG-23, n°2, March 1987, pp.1827-1832.

3. Bourmanne, P., Delincé, F., Genon, A., Legros, W., Nicolet, A., Skin effect and proximity effect in multiconductor systems with applied currents and voltages, 35th MMM Conference, San Diego, USA, October 1990.

4. Bamps, N., Bourmanne, P., Delincé, F., Genon, A., Legros, W., Nicolet, A., Comparison of various methods for the modelling of thin magnetic plates, 35th MMM Conference, San Diego, USA, October 1990.

5. Brebbia, C.A., The Boundary Element Method for Engineers, Pentech Press, London, 1984.

6. Abramowitz, M., Stegun, I.A., ed., Handbook of mathematical functions, Dover, 1972.

7. Rucker W.M., Richter, K.R., Calculation of two-dimensional eddy current problems with the boundary element method, IEEE Transactions on Magnetics, Vol. MAG-19, n°6, November 1983, pp.2429-2432.

8. Patterson, T.N.L., The optimum addition of points to quadrature formulae, Math. Comp. 22, 1968, pp.847-856.

9. Adriaens, J.P., Bourmanne, P., Delincé, F., Genon, A., Nicolet, A. , Legros, W., Numerical computations of eddy currents in thin plates, IEEE Transactions on Magnetics, Vol. MAG-26, n°5, September 1990, pp. 2376-2378.

10. Silvester, P.P., Ferrari, R.L., Finite Elements for Electrical Engineers, Cambridge University Press, Cambridge, 1990.

Hybrid Boundary Integral Equation Method for Charge Injection and Transport in Electroreceptors

M.H. Lean, P.S. Ramesh

Xerox Corporation, Webster Research Center, Mechancial Engineering Sciences Laboratory, 141 Webber Ave, N. Tarrytown, NY 10591, U.S.A.

ABSTRACT

A robust time-transient hybrid boundary integral equation method (BIEM) and method of characteristics (MOC) algorithm for simulating charge injection and transport, is described. This algorithm is used to predict space charge effects and defect sizes from empirical I-V data for axisymmetric geometries. Very rapid computation is possible in circumventing the computational bottlenecks by pre-computing all spatial integrals. Step-wise solution is then simplified to algebraic matrix operations. The speed of computation allows the temporal evolution of space charge in an electroreceptor to be dynamically captured and visualized in real-time on a highend workstation.

INTRODUCTION

Electroreceptors are used in many practical applications as receivers for charge generated using corona or other phenomena. In laser xerography, the receivers are photo-sensitive. They are precharged in the dark, and during imaging, are selectively exposed by a laser beam to leak off surface charge, resulting in a latent image to be developed with toner of the opposite polarity. During discharge, the current beam spreads in transit through the receiver thickness, and results in a wider line width than the optical beam width used in exposure. This effect leads to loss in resolution of the developed image, and becomes worse for thicker receivers and for image lines of high spatial periodicity. Another defect arises when the electroreceptor is not homogeneous, and breaks down non-uniformly under the applied electric field due to the surface charge. Typical defect sizes are hard to measure, being much smaller than the effective resolution of most electrostatic probe dimensions.

In this paper, the mechanism of charge injection and transport is modelled from empirical I-V data, and used to predict both beam spread and actual dimensions of injection sites for axisymmetric geometries. This simulation serves as a tool to quantify and infer degrees of beam spread and enables parametric studies of the problem parameters. The time-transient numerical model is constructed from a hybrid BIEM-MOC approach. Current continuity is used to ensure self-consistent calculations. Simulation of this phenomenon requires simultaneous solution of the charge conservation and current continuity equations. Given empirical I-V data from a sample in a test cell and material and geometric data, Poisson's equation for charge conservation is solved using the BIEM. Electric field in the interior is computed and tracked backwards toward the emitter, to obtain new values for space charge densities, using the MOC. Current conservation is enforced on the emitter surface by assuming uniform injection field, and new estimates for charge injection obtained. The solution proceeds toward steady-state through successive time-stepping to converge on the self-consistent space charge distribution.

PROBLEM FORMULATION

The electrostatic problem requires simultaneous treatment of both charge conservation and current continuity equations.

$$\nabla^2 \phi = -\frac{\rho}{\varepsilon} \tag{1}$$

$$\frac{\partial \rho}{\partial t} - \mu \nabla \phi \cdot \nabla \rho + \frac{\mu}{\varepsilon} \rho^2 = 0 \tag{2}$$

where ϕ, ρ, μ, and ε, are the electrostatic potential, space charge density, carrier mobility, and dielectric constant, respectively. The current density, J, is primarily conduction, and is given by:

$$\mathbf{J} = \rho \mu \mathbf{E} \tag{3}$$

assuming diffusion effects are negligible. In addition, the conservation of current

$$I = \int_0^{w/2} \mu \rho(r) \frac{\partial \phi(r)}{\partial n} 2 \pi r \, dr \tag{4}$$

is invoked at every time-step to ensure a self-consistent solution to equations(1) and (2). In this paper, the intention is to fit empirical I-V data. Therefore, the problem is to compute space charge and field distributions corresponding to the prescribed geometries and boundary and initial conditions relevant to the experiment. With inclusion of injection physics, represented by

$$\rho = \alpha \left| \frac{\partial \phi}{\partial n} \right|^{\beta} \tag{5}$$

use of the correct coefficients α and β, will allow independent prediction of current injection levels.

Integral Equation Formulation

Boundary integral formulations for (1) are derived from the use of simple-layer source distributions, σ on the problem boundary S, and γ on dielectric interface C. The respective equations to be enforced on Dirichlet, Neumann, and dielectric interfaces, respectively, are:

$$\int_{S'} G\sigma(S')\,r\,dS' + \int_{C'} G\gamma(C')\,r\,dC' = \phi(S) - \int_{R'} G\rho(R')\,r\,dR' \tag{6}$$

$$\left\{\int_{S'} \frac{\partial G}{\partial n}\sigma(S')\,r\,dS' - \frac{\sigma(S)}{2}\right\} + \int_{C'} \frac{\partial G}{\partial n}\gamma(C')\,r\,dC'$$

$$= \frac{\partial \phi(S)}{\partial n} - \int_{R'} \frac{\partial G}{\partial n}\rho(R')\,r\,dR' \tag{7}$$

$$\int_{S'} \frac{\partial G}{\partial n}\sigma(S')\,r\,dS' + \left\{\int_{C'} \frac{\partial G}{\partial n}\gamma(C')\,r\,dC' - \frac{(1+\varepsilon)}{(1-\varepsilon)}\frac{\gamma(C)}{2}\right\}$$

$$= -\int_{R'} \frac{\partial G}{\partial n}\rho(R')\,r\,dR' \tag{8}$$

where ε is the ratio of dielectric constants; and γ, the polarization charge, vanishes when $\varepsilon = 1$. The free space Green function, G, satisfies

$$-\nabla^2 G = \delta(\mathbf{r} - \mathbf{r}') \tag{9}$$

and

$$G = \frac{1}{\pi}\frac{K(m)}{[a+b]^{1/2}} \tag{10}$$

Here K(m) is the complete elliptic integral of the first kind and m, the modulus, is given by

$$m = \frac{4rr'}{(r+r')^2 + (z-z')^2} \tag{11}$$

In the denominator, parameters a and b are the standard cylindrical polar coordinate relations:

$$a = r^2 + r'^2 + (z-z')^2 \qquad b = 2rr'$$

Partial derivatives of G with respect to r and z are:

$$\frac{\partial G}{\partial r} = \frac{1}{2\pi r} \frac{[E(m) - K(m)]}{(a+b)^{1/2}} - \frac{1}{\pi} \frac{(r-r')E(m)}{(a-b)(a+b)^{1/2}} \tag{12}$$

and

$$\frac{\partial G}{\partial z} = -\frac{1}{\pi} \frac{(z-z')E(m)}{(a-b)(a+b)^{1/2}} \tag{13}$$

where E(m) is the complete elliptic integral of the second kind. Additional details on the evaluation of these integrals are contained in the Appendix.

Continuity of tangential E is *implicitly* guaranteed with the use of simple-layer sources. On dielectric interfaces, continuity of normal D is *explicitly* enforced by (8). Equations (6) to (8) are discretized and solved for boundary and interface sources which are then used to compute E-fields within the region. All integrations are performed using either the Gauss-Legendre formula for regular integrands or the specially log weighted quadrature for singular kernels.[1]

Trajectory Tracking

The method of characteristics is applied to equation (2), the current conservation equation where advection is deemed the dominant phenomenon. Along characteristics or trajectories defined by:

$$\frac{d\mathbf{l}}{dt} = -\mu \nabla \phi \tag{14}$$

the space charge density satisfies

$$\frac{d\rho}{dt} = -\frac{\mu}{\varepsilon}\rho^2 \tag{15}$$

or

$$\rho = \frac{1}{\frac{1}{\rho_0} + \frac{\mu}{\varepsilon}t} \tag{16}$$

where ρ at the emitter is recomputed at each time-step and assumed to be constant. For the steady state solution, beginning from any interior point where calculation of charge density is required, the trajectory equations are solved backward in time until the characteristic crosses a boundary. If the trajectory is found to issue from other than the emitter, the charge density at the point is taken to be zero. If the trajectory is found to issue from the emitter, then the charge density is computed using equation (16). For the transient case, the characteristic is tracked back only over one time-step, selected such that the distance traversed is less than one grid dimension. As the solution proceeds, the charge is seen to evolve from the emitter surface and spread into the interior of the receiver.

SOLUTION ALGORITHM

The solution algorithm proceeds from an initial guess at ρ. Each iteration constitutes the following steps:

1. Solve Poisson's equation (1) for a new E-field distribution.

2. Compute the new $\nabla\phi$ distribution on the volume grid.

3. Compute injected charge at the emitter surface.

4. Use the MOC to compute the new ρ distribution along trajectories defined by $\mu\nabla\phi$. The transient solution requires backward tracking over one Δt. Extension to steady state requires tracking all the way back to the emitter. No relaxation is required.

5. Perform convergence checks on ϕ and ρ to satisfy the specified exit criteria. If the solution has not converged, execution returns to step 1.

A special feature, unique to the integral method, is used to speed up re-analysis and re-calculation of field components on a fixed grid. The system matrix is inverted and stored so that subsequent solutions require only the product of $[S]^{-1}$ with the new right hand sides computed by adjusting changes to boundary conditions and/or forming the matrix product of influence coefficient matrices with the new ρ. New magnitudes for field components are computed rapidly by summing the product of influence coefficient matrices with the new source distributions. This feature can further be exploited with the use of array processors. Additional details of the algorithm are documented in the literature.[2,3,4]

RESULTS AND DISCUSSIONS

The solution algorithm is robust, and converges in less than 20 iterations for steady state problems without any relaxation. From the solution, interesting physical data, such as current beam spread, charge density fall-off, space-charge limited currents, and so forth, are obtained. Figure 1 shows a cross-section of the axisymmetric problem geometry. A high voltage bias is applied across the electroreceptor causing current to be injected from the emitter, which are them measured by the top electrode. The potential and space charge in the r-z plane are plotted in Figure 2. The potential adjacent to the emitter plate (or disc) flattens outs with increasing injection levels, whilst the space charge distribution drops off very rapidly across the thickness of the electroreceptor (Figure 3). The theoretical limit of space charge limited currents (SCLC) occurs when $\mathbf{E}\cdot\mathbf{n}$ on the emitter surface vanishes. This condition can be simulated by selecting a small value for the Neumann boundary condition. In all cases, the upper limit of the current is determined by a major re-distribution of the E-field lines such that no trajectory makes it across to the opposite face of the receiver. All E-fields begin and terminate on the emitter side. The spread factor, or ratio of current beam width at the measurement electrode to the emitter width, is shown in Figure 4. Figure 5 shows the E-field vector computed on the volume computation grid. Figure 6 shows the spread factor as functions of injection current level for various emitter widths and electroreceptor thicknesses. As expected, the spread factor is least pronouced for higher E-fields across the electroreceptor thickness and wider emitters.

This simulation is implemented within an interactive environment which graphically displays the dynamics of the computations for on-screen selectable parameters. The temporal evolution of beam spread through the receiver thickness is animated. A videotape animation of the solution dynamics will be shown at the presentation.

CONCLUDING REMARKS

This paper has introduced a novel hybrid technique for simulating charge transport through an electroreceptor. By assuming the current flow to be advection dominated, the MOC is used most suitably to compute the desired solutions. The MOC also implicitly satisfies current continuity, thus allowing for more robust convergence behavior. A planned extension to this work is to simulate the correct injection physics, as captured by equation (5), in order to independently predict current levels for any given voltage and geometric configuration. A point of note is that although domain integrals have to be evaluated, the governing equations are still enforced on the boundary. Therefore, the inherent advantage of the BIEM is retained in the use of smaller matrices compared to actual domain methods. The computational bottleneck, in evaluating the contributions of space charge to the forcing function, and the E-field on the trajectory tracking grid, is circumvented by pre-computing the spatial integrals as influence coefficient matrices. The mechanics of re-analysis and re-calculation of field parameters is then simplified to algebraic operations on matrices. With the speed-up in computation, the solution dynamics can be animated in real time to provide useful information of both physical transient effects and numerical behavior. As such, this implementation is a step in the direction of enabling computational prototyping as a parallel path to hardware development.

REFERENCES

1. Lean, M.H. and Wexler, A., Accurate numerical integration of singular boundary element kernels over boundaries with curvature, Int.J.Num.Meth.Engng. vol.21, no.2, pp. 211-228, 1985.

2. Lean, M.H. and Domoto, G.A., Charge transport in Navier-Stoke's flow, IEEE Transactions on Magnetics, vol.24, no.1, pp. 262-265, 1988.
3. Lean, M.H. and Domoto, G.A., Electrohydrodynamics of charge transport in Stoke's flow, IEEE Transactions on Magnetics, vol.23, no.5, pp. 2656-2659, 1987.
4. Lean, M.H. and Domoto, G.A., Charge transport in viscous vortex flows, Journal of Applied Physics, vol.61, no.8, pp. 3931-3933, 1987.
5. Abramowitz, M. and Stegun, I.A., Handbook of Mathematical Functions, Dover, pp. 591-592, 1965.

APPENDIX

The elliptic integrals are evaluated using polynomial approximation[5]:

$$K(m) = \sum_{i=0}^{4} a_i m_1^i + ln(\frac{1}{m_1}) \sum_{i=0}^{4} b_i m_1^i + \varepsilon(m);\ \varepsilon(m) < 2x10^{-8}$$

$$E(m) = 1 + \sum_{i=1}^{4} c_i m_1^i + ln(\frac{1}{m_1}) \sum_{i=1}^{4} d_i m_1^i + \varepsilon(m);\ \varepsilon(m) < 2x10^{-8}$$

where the complementary modulus is given by

$$m_1 = 1 - m = \frac{(r-r')^2 + (z-z')^2}{(r+r')^2 + (z-z')^2}$$

and coefficients are:

a_0=1.38629436112	b_0=0.5
a_1=0.09666344259	b_1=0.12498593597
a_2=0.03590092383	b_2=0.06880248576
a_3=0.03742563713	b_3=0.03328355346
a_4=0.01451196212	b_4=0.00441787012
c_1=0.44325141463	d_1=0.24998368310
c_2=0.06260601220	d_2=0.09200180037
c_3=0.04757383546	d_3=0.04069675260
c_4=0.01736506451	d_4=0.00526449639

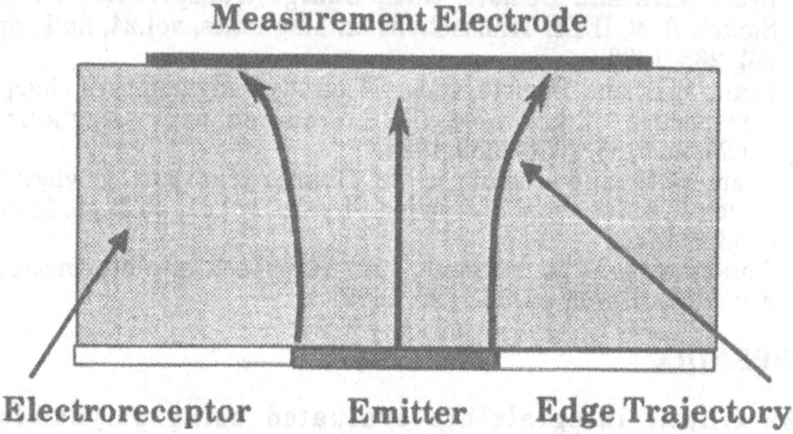

Figure 1. Electroreceptor discharge geometry

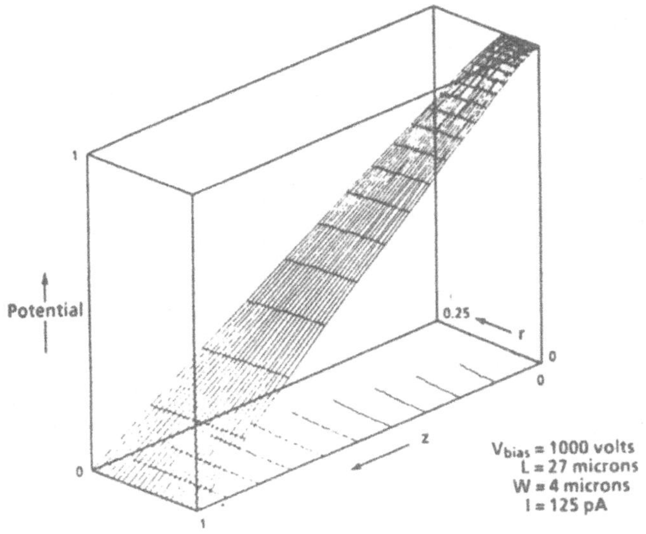

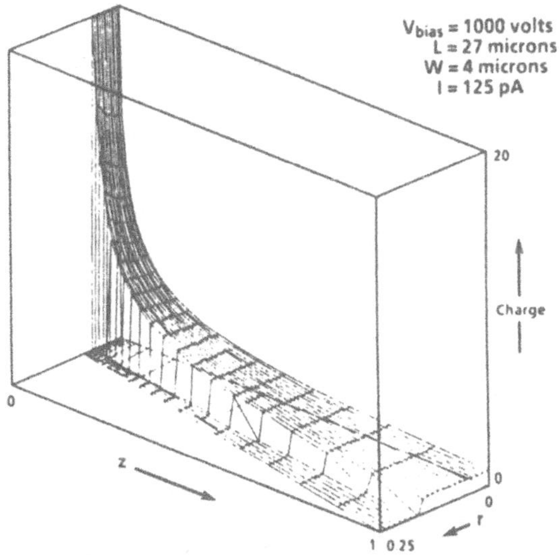

Figure 2. **Surface plots of electrostatic potential (top) and space charge density (bottom) through the electroreceptor. The actual data should be rotated through 2π about the z-axis.**

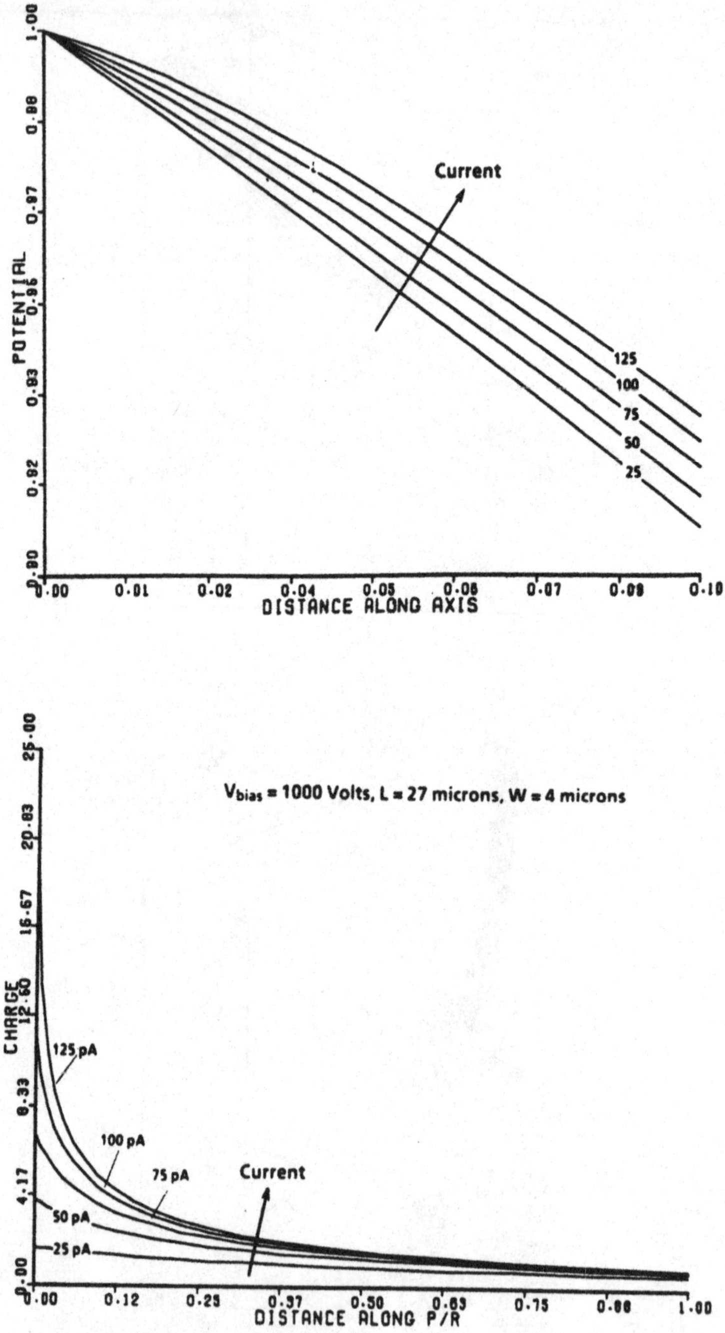

Figure 3. Electrostatic potential and space charge distribution through the electroreceptor (along the z-axis) as a function of injection level.

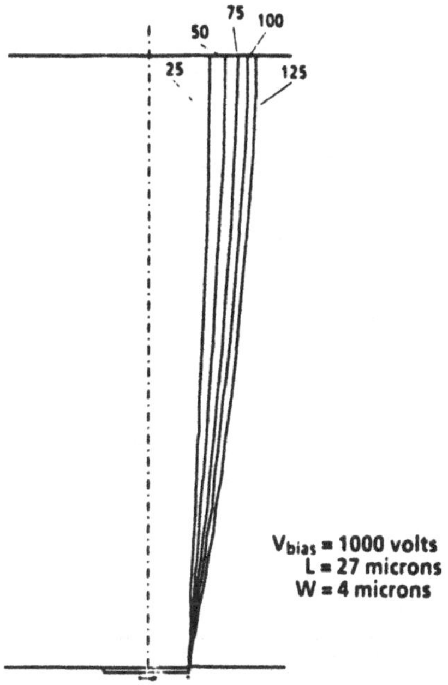

Figure 4. Current beam spread as a function of injection level. Edge trajectory shown.

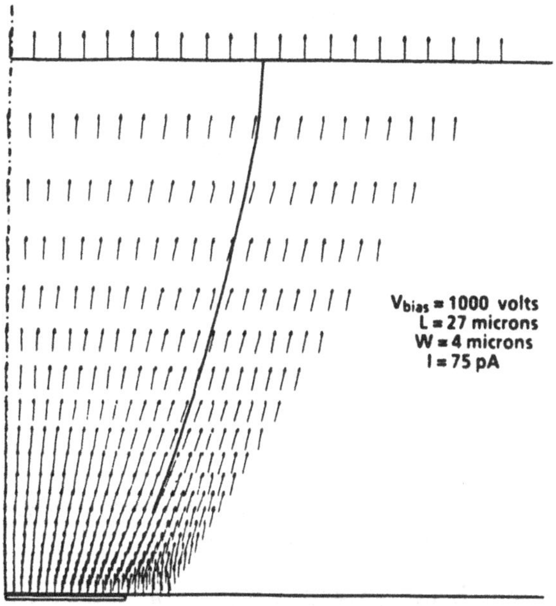

Figure 5. Electric field on computation grid with edge trajectory as indicated.

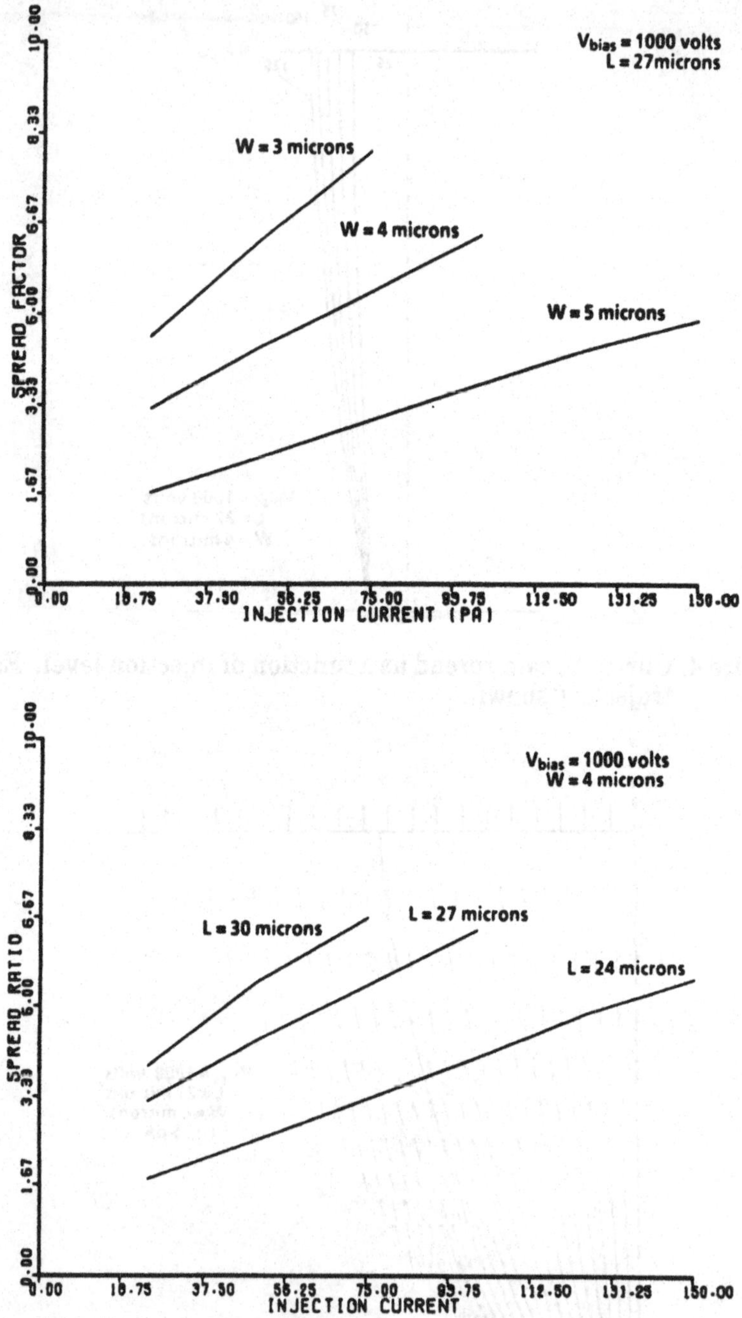

Figure 6. **Current beam spread factor as a function of injection current level for various emitter widths (top) and electroreceptor thicknesses (bottom).**

Corner Modelling for BEM in Magnetostatic

N. Bamps, P. Dular *, A. Genon, H. Hedia, W. Legros, A. Nicolet

University of Liege, Department of Electrical Engineering, Institut Montefiore, Sart Tilman, B28-4000 Liège, Belgium

* This author is a Research Assistant with the Belgian National Fund for Scientific Research

ABSTRACT

The boundary element method provides very convenient models for two-dimensional magnetostatic problems. The basic equation is a Poisson equation involving the Laplace operator. Difficulties arise at special points such as corners. A method for solving the problem of geometrical singularities at the meeting point of n domains is presented. Some degenerated cases of practical interest are studied.

INTRODUCTION

The equation of the two-dimensional magnetostatics is the Poisson equation (1) :

$$\Delta A = -\mu J \tag{1}$$

where :
A is the vector potential which has only one component
J is the current density
μ is the magnetic permeability

In the case of a current free domain, equation (1) reduces to the Laplace equation and the direct boundary element method (Brebbia[1]) is based on the following relation :

$$cA = \oint_{\Gamma} A\, \partial G/\partial n \, d\Gamma - \oint_{\Gamma} G\, \partial A/\partial n \, d\Gamma \tag{2}$$

where :
- G is the free space Green function of the two-dimensional Laplace operator;
- c = 0.5 on a smooth boundary;

- $\partial./\partial n$ is for the normal derivative.

Integrals are taken on the boundary Γ of the subdomains and the method involves only A and $\partial A/\partial n$ (tangential induction) on the boundaries.

It is known that tangential induction may be singular at some geometrical singularities such as corners. Moreover, a problem arises in defining normal and tangential components of the field when the normal vector of the boundary is not unique (Figure 1).

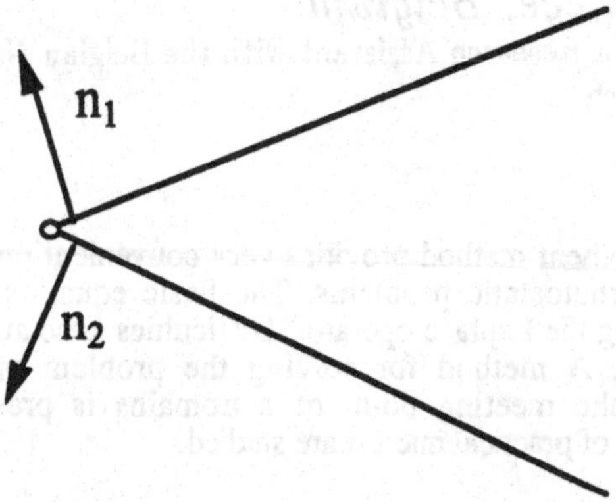

Figure 1 : non-unique normal at a corner

These problems arise in the general case of the study of meeting point of n domains or along a boundary where the boundary conditions are not continuous.

MEETING POINT OF DOMAINS

Given a node which is the meeting point of n linear magnetic and non-conducting domains, the magnetic characteristics may be considered constant in each domain, or less restrictively speaking, in the proximity of the node (Figure 2).

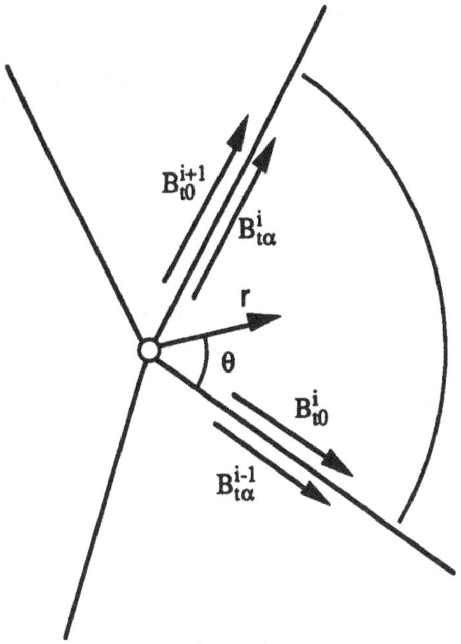

Figure 2 : meeting point of domains

The curvature of the boundaries is treated as negligible by considering a small enough neighbourhood of the node.

The angles α_i are considered to be non-zero. Mathematically, this requirement is expressed by saying that the boundary must be lipschitzian [Rektoris[2]]. Figure 3 gives a counter-example. This boundary is non-lipschitzian because of cusp points such as the tangency points (points A and B) and such as the crack tip (point C).

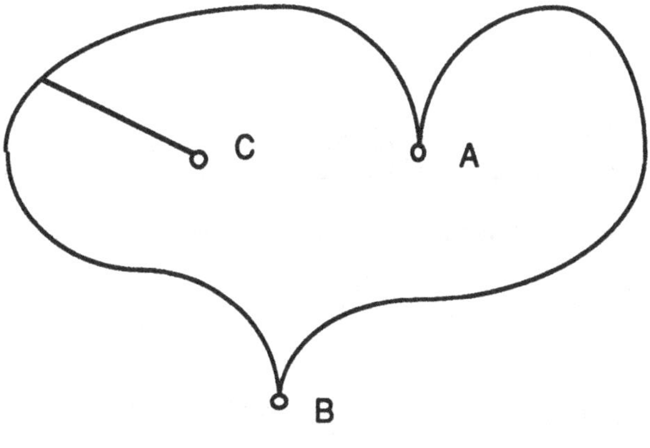

Figure 3 : non-lipschitzian boundary

The vector potential has a unique value on the node. The tangential induction is discontinuous across the boundary between two domains, and it gives two unknowns on each boundary. We have, all things considered, 2n+1 degrees of freedom for that node. The boundary conditions of the tangential induction give n equations (one per boundary), and the boundary elements give n equations (one per domain).

There is one lacking equation. The expression defining that the divergence of the induction equals zero gives this last equation. It seems paradoxical because this condition is always true when the induction derives from a vector potential but this is only true inside each domain.

In a domain touching the node, vector potential can be represented by an expression in polar coordinates centered on the node (Defourny[3]):

$$A = \sum_S r^S \left(C_S \cos S\,\theta + D_S \sin S\,\theta \right) \tag{3}$$

where S are real strictly positive coefficients.

This expression is a general solution of the Laplace equation for the vector potential. The induction is expressed by taking the curl of equation (3) :

$$B_r = \sum_S S\, r^{S-1} \left(- C_S \sin S\,\theta + D_S \cos S\,\theta \right) \tag{4}$$

$$B_\theta = \sum_S S\, r^{S-1} \left(C_S \cos S\,\theta + D_S \sin S\,\theta \right) \tag{5}$$

The minimal value of S involved in the series development determines the behaviour in the field near the node. This value is noted S*.

$S^* < 1$: B singular;
$S^* = 1$: B finite;
$S^* > 1$: B zero.

Close enough to the node ($r << 1$), the other terms of the development are negligible.

Moreover, the continuity of potential vector across two domains requires that the same exponents S are present in the series development of each domain.

The expression of the tangential induction in each domain is written as :

$$B^i_{t_0} = B^i_r(\theta = 0) = S^* r^{S^*-1} D^i \tag{6}$$

$$B^i_{t_\alpha} = B^i_r(\theta = \alpha_i) = S^* r^{S^*-1} \left(- C^i \sin S^* \alpha_i + D^i \cos S^* \alpha_i \right) \tag{7}$$

C^i and D^i constants are given by :

$$C^i = \frac{1}{S^* r^{S^*-1}} \frac{B^i_{t_0} \cos S^* \alpha_i - B^i_{t_\alpha}}{\sin S^* \alpha_i} \qquad S^* \alpha_i \neq k\pi \tag{8}$$

$$D^i = \frac{1}{S^* r^{S^*-1}} B^i_{t_0} \tag{9}$$

The non-divergence of induction is expressed by the integral formula :

$$\oint_C \vec{B} \cdot \vec{n} \, dS = 0 \tag{10}$$

A circle centered on the node is chosen as closed surface C. If C_i is the arc of C in the domain i, equation (10) becomes :

$$\sum_{i=1}^{n} \int_{C_i} \vec{B} \cdot \vec{n} \, dS = 0 \tag{11}$$

The contribution of domain i is :

$$\int_{C_i} \vec{B} \cdot \vec{n} \, dS = \int_0^{\alpha_i} B_r \, r \, d\theta = r^{S^*} [- C_i (1 - \cos S^* \alpha_i) + D_i \sin S^* \alpha_i] \tag{12}$$

When the C_i and D_i values given by equations (8) and (9) are introduced and the r^{S^*} factor is eliminated, the condition (11) becomes :

$$\sum_{i=1}^{n} \left(B^i_{t_0} + B^i_{t_\alpha}\right) \left(\frac{1 - \cos S^* \alpha_i}{\sin S^* \alpha_i}\right) = 0 \tag{13}$$

This is the missing relation, but to use this result requires the knowledge of the S^* value.

A first approach consists in considering the induction as locally constant in a domain, i.e. to take $S^* = 1$. This hypothesis is not incompatible with classical numerical methods. In fact, it is implicit when the finite element method is applied with linear variations of the vector potential. However, a more elaborated study determines the value of the critical exponant S^*.

DETERMINATION OF THE CRITICAL EXPONENT

The expressions (6) and (7) limited on the first term can be written in matrix notation as :

$$\begin{bmatrix} B_\theta^i \\ \frac{B_r^i}{\mu_i} \end{bmatrix} = S\, r^{S-1} \begin{bmatrix} \cos S^*\theta & \mu_i \sin S^*\theta \\ -\frac{\sin S^*\theta}{\mu_i} & \cos S^*\theta \end{bmatrix} \begin{bmatrix} C^i \\ \frac{D^i}{\mu_i} \end{bmatrix} \tag{14}$$

Or :

$$\underline{B}^i(r, \theta) = S\, r^{S-1} [T(\theta)]^i\ \underline{C}^i$$

The relative permeability has been introduced to ensure the continuity of the first member along a boundary :

$$\underline{B}^i(r, \alpha_i) = \underline{B}^{i+1}(r, 0) \tag{15}$$

If the matrix formulation is used and since $[T(0)]^i = [I]$ (identity matrix) :

$$[T(\alpha_i)]^i\ \underline{C}^i = \underline{C}^{i+1} \tag{16}$$

If this condition is expressed for each boundary surrounding the node, taking the boundary between the first domain and the second one, the second one and the third one, until the last one and the first one, we find :

$$\left([T(\alpha_n)]^n \cdots [T(\alpha_2)]^2\, [T(\alpha_1)]^1\right) \underline{C}_1 = \underline{C}_1 \tag{17}$$

The determinant of the matrix $[T] = \prod_{i=n}^{1} [T(\alpha_i)]^i$ is equal to 1, because this matrix is the product of unit determinant matrices. Moreover, the relation (17) shows that C_1 is an eigenvector of [T] with 1 as eigenvalue. The determinant is the product of eigenvalues, so the second one must be 1 too.

The equation used to determine the S* value is :

$$\mathrm{tr}[T] = 2 \tag{18}$$

where the trace of a matrix is the sum of its eigenvalues.

EXAMPLE

In the case of two domains, the relation (13) becomes :

$$\left(B^1_{t0} + B^1_{t\alpha}\right)\left(\frac{1 - \cos S^*\alpha_1}{\sin S^*\alpha_1}\right) + \left(B^2_{t0} + B^2_{t\alpha}\right)\left(\frac{1 - \cos S^*(2\pi - \alpha_2)}{\sin S^*(2\pi - \alpha_2)}\right) = 0 \tag{19}$$

The boundary conditions for the tangential induction are :

$$\frac{B^1_{t_0}}{\mu_1}=\frac{B^2_{t_\alpha}}{\mu_2} \quad \text{and} \quad \frac{B^1_{t_\alpha}}{\mu_1}=\frac{B^2_{t_0}}{\mu_2} \tag{20}$$

Equation (19) together with equation (20) becomes :

$$\left[\frac{1}{\mu_1}\left(\frac{1 - \cos S^*\alpha_1}{\sin S^*\alpha_1}\right)+\left(\frac{1 - \cos S^*(2\pi-\alpha_1)}{\sin S^*(2\pi-\alpha_1)}\right)\right]\left(B^1_{t_0}+B^1_{t_\alpha}\right)=0 \tag{21}$$

The following relation is obtained :

$$B^1_{t_0} = - B^1_{t_\alpha} \tag{22}$$

In this simple case, the relation depends neither on S*, nor on the characteristics of the corner such as angles or permeabilities.

The critical exponent S* is given by the smallest root of :

$$2\cos S\alpha_1 \cos S(2\pi-\alpha_1) - \left(\frac{\mu_1}{\mu_2}+\frac{\mu_2}{\mu_1}\right)\sin S\alpha_1 \sin S(2\pi-\alpha_1)-2=0 \tag{23}$$

It can be demonstrated that a root exists between 0 and 1, so S* <1 and the behaviour of the induction is always singular [Defourny[3]].

DEGENERATED CASES

In the relations between B_t, terms like sin S* α are found in the denominator. These terms must be different from zero in order to ensure the validity of the relations.

In degenerated cases, these ones are equal to zero.

For example, a colinear boundary is considered (there is no corner !) In this case, the critical exponent is given by :

$$2\cos^2 S\pi - \left(\frac{\mu_1}{\mu_2}+\frac{\mu_2}{\mu_1}\right)\sin^2 S\pi - 2 = 0 \tag{24}$$

Or :

$$-\left(\frac{\mu_1}{\mu_2}+\frac{\mu_2}{\mu_1}+2\right)\sin^2 S\pi = 0 \tag{25}$$

The smallest strictly positive S is S* = 1. Therefore, the induction has non-zero and non-singular value on that point. It is valid because no geometrical singularity exists.

As far as the relation between tangential inductions is concerned, we find sin π in the denominator. So, in this case, and in general, when the angle α_i of a domain i is equal to π, the following condition is taken :

$$B^i_{t_0} = - B^i_{t_\alpha} \tag{26}$$

Let us now consider the case where permeabilities are equal ($\mu_1 = \mu_2$).

The corner is fictive because there is no material discontinuity. So, the critical exponent is given by :

$$\cos S\, 2\pi = 1 \tag{27}$$

As expected, S* = 1.

On the other hand, the first factor of the relation of tangential induction (21) is zero, therefore, there is no relation between B_t.

When there is no discontinuity, the divergence of induction is zero and cannot be used as an additional relation.

In practice, this situation can be found in a fictive boundary. Therefore, corners which introduce geometrical singularities in fictive boundaries will be avoided. Nevertheless, the experience shows that, if the expression (26) is used for a simple corner, the solution is scarcely modified.

CONCLUSION

In practice, the relation (13) is used as an additional condition. It is computed for linear and non-conducting magnetic media in two- dimensional geometry. But, in first approximation, it can be used in saturable media (constant permeabilities are assumed), conducting media (conductivity is neglected) and also in axisymmetrical geometries.

Some authors have proposed to use special elements with a singular behaviour on the corner by means of a special function,but this method is expensive (Fix[4]).

The present tendency is to use classical functions, linear or quadratic, with adaptative meshing to approach the singular behaviour. This type of meshing is useful only if a precise knowledge of the solution near the corner is important.

In the case of magnetics, because of the nature of magnetic permeability, a phenomenon of local saturation smooths the singularities of the field and, in a way, "rounds off the angles" (Peaiyoung[5]).

Elsewhere, interesting values are global values which are scarcely modified by the local behaviour of the solution in a corner. On the other hand, the local maximum in electric field is very important.

So,in the computations of the magnetic field, a good practice is to use small elements if they touch a corner.

REFERENCES

1 C.A. BREBBIA, The Boundary Element Method for Engineers, Pentech Press, London, 1984.

2 K. REKTORYS. "Variational methods in mathematics, science and engineering", Reidel Publishing Company, 1977.

3 M. DEFOURNY, "Singular Point Theory in Laplace Field Boundary Elements" (C.A. Brebbia ed.) , Springer-Verlag, 1988, pp 165-180.

4 G.J. FIX, "Singular Finite Element Methods in Finite Elements, Theory and Application" (in "Finite Elements - Theory and Applications", D.L. Dwoyer, M.Y. Hussaini, R.G. Voigt ed.) , Springer-Verlag, 1988, pp 50-66.

5 S. PEAIYOUNG, S.J. SALON and I. MAYERGOYZ, "Some Technical Aspects of Implementing Boundary Element Equations" IEEE Transactions on Magnetics , vol 25, No 4, July 1989, pp 2998-3000.

Application of Boundary Element and Characteristic Methods for Modelling of Corona Fields

K. Adamiak

Department of Electrical Engineering, The University of Western Ontario, London, Ontario, N6A 5B9, Canada

ABSTRACT

The paper presents a new approach to the modelling of the space charge density and the electric field in monopolar dc corona discharge areas. The computational model combines the method of boundary elements for calculating the potential distribution, assuming known charge density, and the method of characteristics for estimating the charge density from a given potential distribution. Both methods are used iteratively until a self-consistent solution is obtained. The proposed numerical procedure has been applied to simulate electrical conditions in a wire-duct electrostatic precipitation device.

INTRODUCTION

An electrical corona effect occurs when high voltage is applied to a conductor with a small curvature. A very large electric field on the conductor surface results in a local electrical discharge. Free ions are generated within the thin active ionization region close to the conductor surface and move away under the influence of the electric field [1]. The rigorous physical mechanism of corona generation is rather complicated and depends on many different factors, e.g. on the polarity of the electrode and its geometry. However, a common feature in all coronas is the existence of free charges in the interelectrode space in the form of a stream of ions flowing away from the electrode.

The corona discharge is very often generated purposely. Charges produced in a corona may be used to electrify different objects, mainly small dielectric or conducting particles. A charged particle is next propelled, if placed in the electric field, so that it is possible to electrically control its movement. This idea is widely used in commercial applications. Most important are: electrostatic precipitation [2], electrostatic spraying and painting, and electrophotography.

In many situations, corona is a nuisance and may cause safety problems. The increasing voltage level for long distance high-voltage-direct-current (HVDC) transmission of electrical power has caused a great deal of interest in the investigation of the influence of corona on power losses and in the study of environmental hazard issues related to biological effects [3,4,5].

In most cases, precise modelling of the corona field is very important. In applied electrostatics it is of primary importance for determining and optimizing device performances; for HVDC transmission it is necessary in order to estimate the possible impact of a power line on the human body under a variety of conditions; in atmospheric research it provides a tool for the interpretation of different phenomena.

The electrostatic field is governed by the Poisson equation, which has been studied extensively for a long time. However, corona problems are much more difficult due to the presence of the space charge generated by coronating electrodes. Unlike in other electrostatic problems the charge density distribution is unknown in advance because it depends on the electric field. On the other hand, the field distribution is a function of charge density, so both problems are mutually coupled.

This paper presents a new technique for finding a solution to the corona discharge problem. It combines the boundary element method (BEM) and the method of characteristics (CM). Both are used iteratively until a self-consistent solution is obtained. This method has been applied to simulate the corona field in a wire-duct electrostatic precipitator. Numerical results agree very closely with the experimental data.

MATHEMATICAL MODEL

The electrostatic field in the space charge area is governed by the following subset of the Maxwell equations

$$\nabla \cdot \mathbf{E} = \frac{\rho}{\varepsilon_0} \qquad (1)$$

$$\nabla \times \mathbf{E} = 0 \qquad (2)$$

where:

ε_0 - the permittivity of free space
ρ - space charge density
$\mathbf{E}$ - electric field intensity

Electric charges in the corona field are in continual movement, therefore they create an electric current. The constitutive law relating the current density to the electric field is

$$J = \rho b E \tag{3}$$

where:

b - mobility of charges

J - current density.

From the continuity condition of current it is concluded that

$$\nabla \cdot J = 0 \tag{4}$$

Equations (1), (2), (3) and (4) form the mathematical model of the problem and should be solved as a simultaneous set of the partial differential equations. Because of its nonlinear character, the solution of the problem is far from being straightforward. All analytical techniques are practically useless, and therefore a numerical approach is necessary.

For a typical corona problem the following boundary conditions are also valid:

- conductor surfaces are equipotential with the potential level resulting from an external supply
- if a coronating electrode has a potential above a certain value, called corona onset level, the normal component of the electric field remains constant at the onset value E_0, which results from Peek derivation (Kaptsov condition) [1]. The last assumption has been validated thoroughly both analytically and experimentally with satisfactory agreement. Therefore, it is commonly adopted and will not be discussed here.

ELECTRIC FIELD CALCULATION

The corona solution process consists of two separate stages: the electric field calculation and the space charge density estimation. In the former, the electric field is determined assuming that the space density distribution is numerically available. This part is performed most conveniently after introducing a scalar quantity: the electric potential V. Its definition is

$$E = -\nabla V \tag{5}$$

V satisfies the Poisson equation

$$\nabla^2 V = -\frac{\rho}{\varepsilon_0} \tag{6}$$

In hitherto published papers, the above Poisson equation was solved most frequently by means of the finite difference method (FDM) [6], the finite element method (FEM) [3,7,12,13,14] and the charge simulation method [5,8,10,11]. It also was an attempt to apply BEM [9]; however, in the investigated model, the real distribution of charge density was not taken into account. Neither differential method - FDM or FEM - is probably the best choice in corona modelling. The electric field, used next for charge density calculation, is obtained in both approaches by numerical differentiation which may be a source of substantial error, especially at points close to the coronating conductor surface, and which also may cause instability of the iteration process.

CHARGE DENSITY PREDICTING

The second part in the solution process involves the determination of the steady flow of charges from the knowledge of the electric field intensity **E**. Several numerical approaches have been proposed during previous years. So far, the method of characteristic (MC) [7,11,12], or its specific version, assuming equipotential shells of the space charge [8,10], and the donor cell method [13,14], are the most popular. The second is more flexible in the sense that it can incorporate ions of different polarities, which is often necessary, but the algorithm is inefficient in terms of computing time. Therefore, for problems with only one polarity of charges, MC is preferred.

In MC, Equations (1), (3), (4) and (5) are combined together and form the following nonlinear partial differential equation for the unknown density of the space charge:

$$\mathbf{E}\cdot\nabla\rho = -\frac{\rho^2}{\varepsilon_0} \tag{7}$$

This partial differential equation becomes an ordinary one along the characteristic line

$$\frac{d\mathbf{r}}{dt} = b\mathbf{E} \tag{8}$$

The combination of (7) and (8) leads to

$$\frac{d\rho}{dt} = -b\frac{\rho^2}{\varepsilon_0} \tag{9}$$

Equation (9) can be integrated to yield solution

$$\rho = \frac{1}{\dfrac{1}{\rho_0} + \dfrac{bt}{\varepsilon_0}} \tag{10}$$

where ρ_0 is the charge density on the surface of the conductor.

The characteristic lines represent ion trajectories from the coronating electrode to the neutral one.

SOLUTION PROCESS

In the full model of corona, the electric field and the charge calculations should be coupled together. This is usually done on an iterative basis. The process begins with an initial estimation of the charge density. As is usual in iterative procedures, this is an extremely important step in the acceleration of the convergence of the iterations. Unfortunately, there is no reasonable and general approach for predicting the initial charge density. Usually, an estimation of the total corona current, Equation (3), and previous experience are very useful.

After the magnitude of the charges is evaluated, the Poisson equation (6) is solved to find the electric field. When differential methods (FEM, FDM) are used, the whole area must be discretized into a certain number of elements. The charge distribution, then, is updated from a given potential distribution by means of CM.

In the above iteration process, an interface between both numerical techniques is of critical importance. Commonly used linear FEM yields piece-wise constant distribution of the electric field. As a result, part of the characteristic line within one element must have a form of line segment and this can result in a very coarse representation of the characteristic line [7]. One can, of course, use higher order approximation [12] or elements with continuous derivatives, but this would only postpone the problem. Use of FEM also creates some difficulties for open problems where infinite space is to be considered [11].

In this paper, BEM has been chosen as it produces a much smoother distribution of the electric field. The solution of Equation (6) in two-dimensional space is represented in terms of a simple-layer potential [15]

$$V(P) = \int_{\Gamma} \tau(Q)\, G(P,Q)\, dl + \iint_{\Sigma} \rho(R)\, G(P,R)\, dA \qquad (11)$$

$$G(P,Q) = \frac{1}{2\pi \varepsilon_0} \ln \frac{1}{r_{PQ}}, \qquad P\in\Sigma,\, R\in\Sigma,\, Q\in\Gamma \qquad (12)$$

where:

$\tau(P)$ - unknown distribution of sources on the boundary
$\rho(R)$ - distribution of space charge determined by CM
Σ - considered domain
Γ - boundary of Σ
r_{PQ} - distance between points P and Q

Application of Equation (11) to points on the boundary, where potential or its normal derivative is known from boundary conditions, yields a set of integral equations with unknown source density $\tau(P)$. The potential V(P) is continuous across the boundary Γ, but its normal derivative is not. As a result, the integral equation of the first kind is obtained for the electrode surfaces (Dirichlet boundary conditions) and the equation of the second kind for the symmetry lines (homogeneous Neumann conditions). Numerical techniques for both equations are commonly known. In this paper, the linear boundary elements have been used [16]. The whole boundary has been discretized into some number of the line segment elements with one node located at each end. Therefore, the boundary source density has been assumed to vary linearly along the elements. Such an assumption reduces the integral equation to a set of algebraic equations with the nodal values of potential as unknowns. All entries in the system matrix were calculated by numerical contour integration, except the diagonal ones. Due to singularity, those entries must be treated analytically. The right-hand-side vector was determined by numerical evaluation of the surface integral.

After estimation of the electric field distribution, CM is used to update the initial guess of spatial density of the ionic charges. In CM, special characteristic lines are introduced. They are started at points on the surface of the corona sheath and are traced by numerical integration of equation (8). Equations (9) and (10) determine the time increment and the value of the charge at the unknown nodes. The electric field intensity **E** in (8) results from the boundary, $\tau(Q)$, and surface, $\rho(R)$, charges. Both techniques, BEM and CM, constitute the so-called "inner" iterative loop that converges to the self-consistent charge density and electric field.

However, the value of charge density on the wire surface must be assumed first. The Kaptsov condition should be used, but formally it is an inhomogeneous Neumann condition on V, so there are three boundary conditions imposed on the Poisson equation and no explicit boundary condition exist for the charge density. Therefore, iterations are necessary to converge to a value of ρ_0, which yields the proper electric field magnitude. This is an "outer" iterative loop in which charge density is adjusted at each point, where characteristic line initiates so as to produce the electric field given by Peek's formula. New estimates of ρ_0 are computed using a secant root-finding technique.

Finally, the solution process involves the following steps:

- make an initial guess for the surface charge density ρ_0 and its spatial distribution in the whole area $\rho(R)$
- apply BEM (Equation (11)) to find the boundary charge density $\tau(Q)$
- determine $\rho(R)$ from CM
- repeat steps 2 and 3 until the solution is self-consistent
- calculate the electric field **E** on the wire surface and compare it with Peek's formula
- make a new estimation for ρ_0 from the secant root-finding technique.
- repeat steps 2 through 6 until the value of the electric field intensity on the electrode surface is close enough to Peek's value.

Surprisingly, the above iteration process, although logically a little complex, appears to be very fast convergent. The "inner" loop usually requires 3-4 iterations for accuracy better than 1%. A good initial guess obviously is important for the outer loop convergence, but even with a poor initial estimation, 5-7 iterations are enough for the same accuracy level.

ANALYSIS OF CORONA DISCHARGE IN WIRE-DUCT ELECTROSTATIC PRECIPITATOR

Computational efficiency of the recommended numerical algorithm has been demonstrated in the context of a wire-duct electrostatic precipitator. It consists of two parallel plates and a set of corona wires positioned halfway between the plates and equidistant from each other. Fig.1 shows a coordinate system and geometry. Due to symmetry, it is usually enough to consider only one-quarter of a single-wire section [10] (Fig.2). The potentials on the wire and collecting plate surfaces result from supply conditions. All other parts of the boundary (wire-plate, wire-wire and midway between wires-plate lines) are lines of symmetry. The normal component of **E** across such line vanishes and this is equivalent to the homogeneous Neumann boundary condition for the scalar potential V

$$\frac{\partial V}{\partial n}\Big|_{\Gamma} = 0 \tag{13}$$

where n is normal direction.

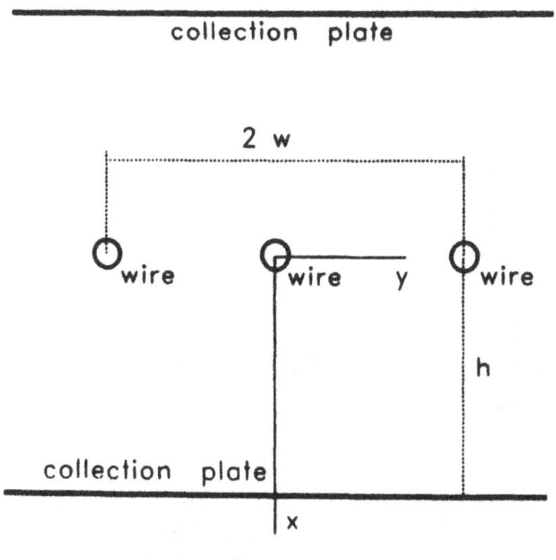

Figure 1. Model of wire-duct precipitator

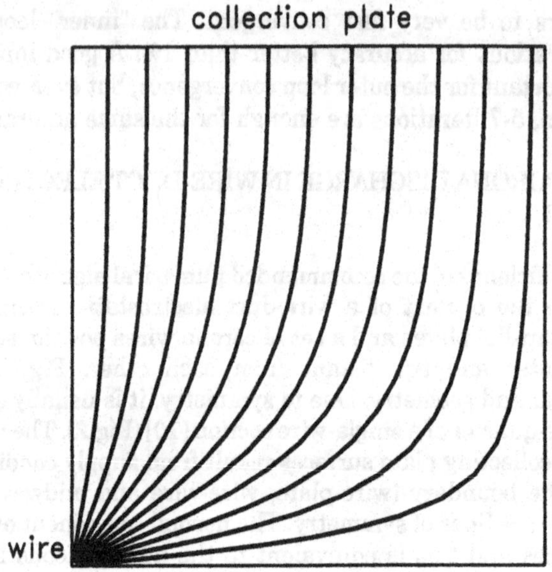

Figure 2. Distribution of the characteristic lines between the wire conductor and collection plate

The discretization density and number of characteristic lines have been found experimentally. Increasing the number of the boundary elements above 150 usually does not improve the accuracy of the solution. Similarly, 50 characteristic lines were sufficient.

The parameters used in computation were identical to those used in the experiments of Penney and Matick [17]:

- wire diameter 1.0 mm
- wire-to-wire spacing 15.24 cm
- wire-to-plate distance 11.43 cm
- ion mobility 1.82e-4 m^2/Vs

The shapes of characteristic lines are shown in Fig.2 (much larger number of lines was used in normal calculations). The lines are much smoother than those obtained by FEM [7] and the smoothness may further be improved without creating any problems. The charge density decays along each line according to Equation (10) (Fig.3).

Computational results compare favourably with experimental data. Distributions of the scalar electric potential along the wire-to-plate and midway between wires-to-plate lines are presented in Fig.4. In both cases, and at two different voltage levels, the agreement is very close.

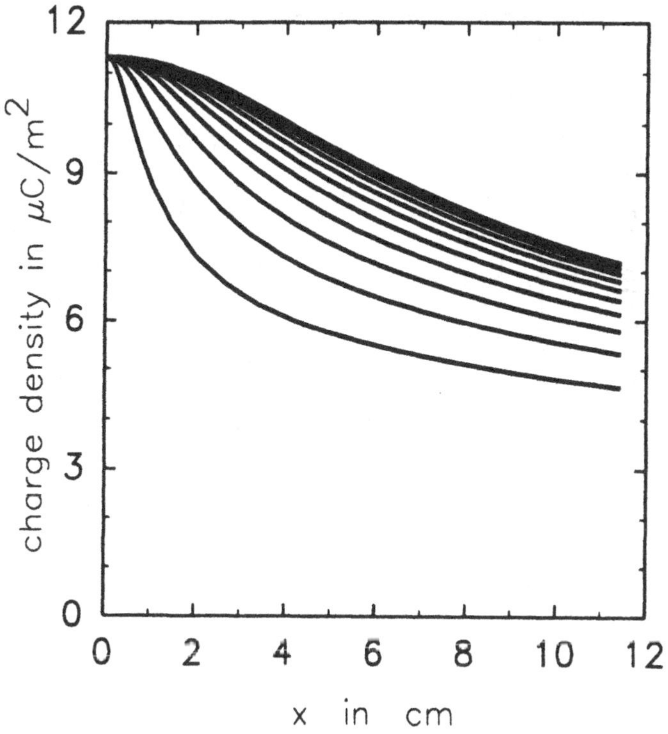

Figure 3. Spatial density distribution of the electric charge along the characteristic lines

CONCLUSIONS

This paper has described a new approach to the problem of corona modelling, based on the hybrid BEM-CM technique. This new method is capable of predicting the electric conditions in devices where the corona effect is used for particle charging. The suggested numerical algorithm is reliable and fast convergent due to the very precise prediction of the characteristic lines shape -one of the major steps in the iterative algorithm. Numerical results have been validated by the experimental data; excellent correspondence between both indicates the value of the method.

ACKNOWLEDGMENTS

The author acknowledges with thanks the financial support of the Natural Science and Engineering Council (NSERC) of Canada.

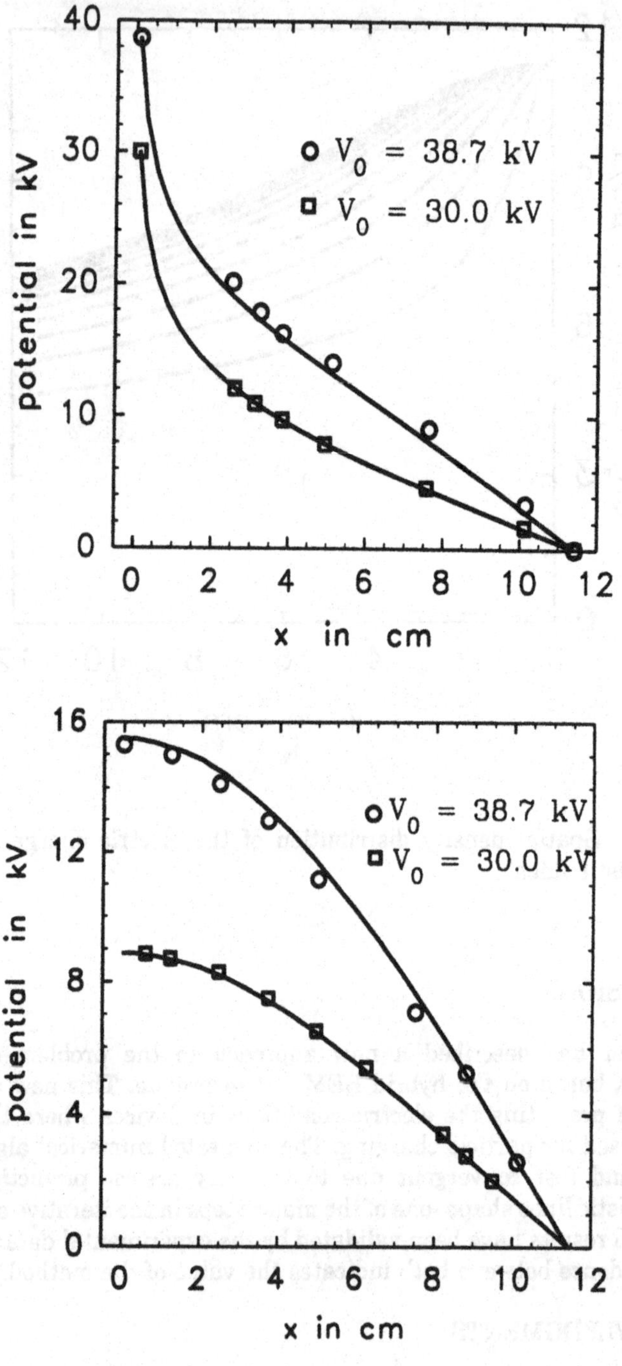

Figure 4. Experimental and numerical values of the scalar electric potential

a. wire-to-wire line
b. between wires-to-plate line

REFERENCES

1. Peek, F.W. Dielectric Phenomena in High Voltage Engineering, McGraw-Hill, 1929.

2. White, H.J. Industrial Electrostatic Precipitation, Addison-Wesley Pub., 1963.

3. Janischewskyj, W. and Gela, G. Finite Element Solution for Electric Fields of Coronating DC Transmission Lines, IEEE Trans. on Power, Apparatus and Systems, Vol.PAS-98, pp. 1000-1008, 1979.

4. Takuma, T., Ikeda, T. and Kawamoto, T. Calculation of Ion Flow Fields of HVDC Transmission Lines by the Finite Element Method, IEEE Trans. on Power, Apparatus and Systems, Vol.PAS-100, pp. 4802-4810, 1981.

5. Qin, B.L., Sheng, J.N., Yan, Z. and Gela, G. Accurate Calculation of Ion Flow Field under HVDC Bipolar Transmission Lines, IEEE Trans. on Power Delivery, Vol.3, pp. 368-376, 1988.

6. McDonald, J.R., Smith, W.B., Spencer, H.W. and Sparks, L.E. A Mathematical Model for Calculating Electrical Conditions in Wire-Duct Electrostatic Precipitation Devices, J.Appl.Phys., Vol.48, pp. 2231-2243, 1977.

7. Davis, J.L. and Hoburg, J.F. Wire-Duct Precipitator Field and Charge Computation Using Finite Element and Characteristics Method, J. of Electrostatics, Vol. 14, pp. 187-199, 1983.

8. Horenstein, M.N. Computation of Corona Space Charge, Electric Field, and V-I Characteristic Using Equipotential Charge Shell, IEEE Trans. on Industry Applications, Vol. IA-20, pp. 1607-1612, 1984.

9. Gakwaya, A., Farzaneh, M. and Prive, M. Study of the Space Charge Effect on Corona Induced Vibration by the Boundary Element Method, in Betech 86 (Ed. Connor, J.J. and Brebbia, C.A.),pp. 139-151, Proceedings of the 2nd Boundary Element Technology Conference, Massachusetts Institute of Technology. CML Publications, 1986.

10. Elmoursi, A.A. and Castle, G.S.P. Modelling of Corona Characteristics in a Wire-Duct Precipitator Using the Charge Simulation Technique, IEEE Trans. on Industry Applications, Vol. IA-23, pp. 95-102, 1987.

11. Elmoursi, A.A. and Speck, C.E. Simulation of Space Charge Fields in Unbounded Geometries, IEEE Trans. on Industry Applications, Vol. IA-24, pp. 1699-1706, 1988.

12. Butler, A.J., Cendes, Z.J. and Hoburg, J.F. Interfacing the Finite-Element Method with the Method of Characteristics in Self-Consistent Electrostatic Field Models, IEEE Trans. on Industry Applications, Vol. IA-25, pp. 533-538, 1989.

13. Ghione, G. and Graglia, R.D. Two-Dimensional Finite-Boxes Analysis of Monopolar Corona Fields Including Ion Diffusion, IEEE Trans. on Magnetics, Vol. MAG-26, pp. 567-570, 1990.

14. Levin, P.L. and Hoburg, J.F. Donor Cell-Finite Element Descriptions of Wire-Duct Precipitator Fields, Charges, and Efficiencies, IEEE Trans. on Industry Applications, Vol. IA-26, pp. 662-670, 1990.

15. Jaswon, M.A. and Symm, G.T. Integral Equation Methods in Potential Theory and Elastostatics, Academic Press, London 1984.

16. Brebbia, C.A., Telles, J.C.F. and Wrobel, L.C. Boundary Element Techniques, Springer-Verlag, Berlin 1984.

17. Penney, G.W. and Matick, R.E. Potentials in DC Corona Field, Trans. AIEE, Vol. 79, pp. 91-99, 1960.

SECTION 9: GEOMECHANICS

Boundary Integral Techniques for Stochastic Problems in Geomechanics

G.D. Manolis

Division of Structures, Department of Civil Engineering, Aristotle University, Thessaloniki 54006, Greece

ABSTRACT

A review of Boundary Element Methods (BEM) in the field of geomechanics over the last fifteen years clearly shows the substantial contributions that these methods have made in analysing geotechnical problems. There is, however, a need to expand these integral equation-based techniques to the case where the ground is represented as a stochastic medium. This is especially true in the case of time-dependent problems where, depending on the characteristics of the propagating disturbance, it is quite possible for the wavelength to be of comparable dimension as the spacing of the randomly distributed inhomogeneities in the ground. As a result of this, the propagating disturbance can be noticeably altered, primarily through reduction of its mean amplitude due to scattering from the random irregularities. In addition, the analysis of stochastic problems serves as a stepping stone for other important topics such as sensitivity and reliability of geotechnical designs to basic soil and rock material parameters. This work outlines the general formulation and solution in terms of random integral equations for problems involving a random medium and subsequently discusses approximate, perturbation-based techniques. The methodology is applied to shear wave propagation in stochastic soil.

INTRODUCTION

The BEM is a numerical analysis technique for boundary-value problems and is based on integral equation formulations, as opposed to differential equation formulations on which the finite element (FEM) and finite difference (FDM) methods are built. The tremendous advantage that the BEM exhibits is its ability to reproduce radiation type boundary conditions, a property that allows efficient modelling of infinite as well as semi-infinite media such as those commonly encountered in geomechanics. For geotechnical problems that can be approximated as linear and (piecewise) homogenous, efficient BEM solution methodologies have been generated for one-, two-, and three - dimensional cases. Also, these solutions have been extended to dynamic problems by proceeding in either the time domain or in a transformed (Fourier or Laplace) domain. Finally, the introduction of volume integrals along with incremental/iterative algorithms has extended the range of applicability of the BEM to inelastic problems, which are so prevalent in geomechanics.

A recent review of the BEM in geomechanics that lists close to two hundred references was compiled by Beskos [1]. Following a classification scheme similar to his, the areas in which the BEM has made an impact are as follows: (a) Raft and pile foundations; (b) Mining, excavation and tunnelling problems; (c) Groundwater flow; (d) Quasi-static and dynamic soil consolidation; (e) Wave propagation through ground; (f) Soil-structure interaction; and (g) Vibration isolation.

In general, the BEM is used for modelling the ground (i.e., soil or rock). The state-of-the-art today allows for proper modelling of a layered, anisotropic half-plane or half-space. The presence of a structure in the ground (i.e., piles, flexible foundations,tunnel linings, infilled trenches, and full structures) invariably results in a hybridization of the BEM and the FEM, with the latter used for modelling the structural component or structure. Inhomogeneity in soil is usually handled by introducing regions in the BEM mesh over which the material is relatively homogeneous. As far as anisotropic material behavior is concerned, recourse is done to either special fundamental solutions or to volume integrals. Rock-type materials that exhibit fissures and seams which result in cracking and slipping require the use of special boundary elements such as crack elements or elements capable of displaying displacement discontinuities, all coupled with iterative schemes since the relative motion across such imperfections is unknown.

Only a limited number of papers until now have considered representing geological media as exhibiting randomness in their material properties, despite the fact that information about soil deposits or rock structures is not known with sufficient accuracy and often variation of their properties over small distances is quite dramatic. Furthermore, laboratory measurements are conducted on samples that are small and rather homogeneous, giving the impression that the original medium is macroscopically homogeneous as well. In general, irregular changes in the form of inhomogeneities seen against the background of a rather homogeneous or piecewise homogeneous soil / rock deposit give rise to uncertainty in the description of that medium, which in turn is manifested in the medium's response.

The analysis of randomness in fields such as electromagnetic waves, atmospheric turburlence, oceanography, chamber acoustics, geophysics, biomaterials and material identification [2] is more advanced than it is in geomechanics. Among the work that has been done in the latter field is that of Cheng and Lafe in ground water flow [3,4] and Cakmak and his coworkers concerning wave propagation in ground [5,6]. It is interesting to observe that uncertainty in geomechanics is handled by using integral equation formulations, in contrast to uncertainty in structures, where finite elements are employed [7,8]. In what follows, a formal solution based on integral equations for problems involving random media will be presented and illustrated by an example drawn from wave propagation in ground. Finally, approximate solutions based on perturbation expansions will be shown and their range of validity will be discussed.

INTEGRAL FORMULATION AND FORMAL SOLUTIONS

Consider the following general stochastic differential equation:

$$L(\gamma)[u(x,\gamma)] = f(x,\gamma) \tag{1}$$

In the above, L is a differential operator of order n with random coefficients. Furthermore, u is the dependent variable, f is the forcing function and argument γ denotes a random quantity. For illustration purposes, Eqn. (1) can be identified with the harmonic scalar wave equation, in which case u is the displacement and x is a spatial variable along the direction of propagation.

The key assumption [9] is that the operator L can be decomposed into a deterministic part D and a zero-mean random part $R(\gamma)$ so that Eqn. (1) becomes

$$(D+R(\gamma))\,[u(x,\gamma)] = f(x,\gamma) \tag{2}$$

with

$$\left.\begin{aligned} D[u] &= \sum_{i=0}^{n} a_i(x)\, d^iu/dx^i \\ &\text{and} \\ R(\gamma)\,[u] &= \sum_{i=0}^{n} \beta_i(x,\gamma)\, d^iu/dx^i \end{aligned}\right\} \tag{3}$$

where coefficients a_i, β_i depend on x with $a_n=1$ and $\beta_n=0$. If a Green's function $G(x,x_1)$ exists for the deterministic operator D, then the stochastic differential equation is equivalent to the random integral equation

$$u\,(x,\gamma) + \int_{-\infty}^{\infty} N(x,x_1,\gamma)\, u(x_1,\gamma)\, dx_1 = H(x,\gamma) \tag{4}$$

where

$$H(x,\gamma) = \int_{-\infty}^{\infty} f(x_1,\gamma)\, G(x,x_1)\, dx_1 \tag{5a}$$

and

$$N(x,x_1,v) = G(x,x_1) \sum_{i=0}^{n} \beta_i(x_1,v)\, d^i/dx_1^i \tag{5b}$$

Equation (4) is a Volterra integral equation of the second kind with random kernel N and a random generalized forcing function H.

At this stage, there are a number of options available regarding the solution of Eqn. (4). Before we proceed with approximate techniques, we will first discuss the closed-form solution of the random integral equation. Following the deterministic case, the resolvent kernel Γ of N is defined through the following Neumann series [9]:

$$\Gamma(x,x_1,v) = -\sum_{k=1}^{\infty} N_k(x,x_1,v) \tag{6}$$

In the above, the iterated kernels N_k are given by the recurrence relation

$$N_k(x,x_1,v) = \int_{-\infty}^{\infty} N(x,x_2,v)\, N_{k-1}(x_2,x_1,v)\, dx_2 \qquad k=2,3,.. \tag{7}$$

with N_1=N. Thus, the formal solution of Eqn. (4) can be written as

$$u(x,v) = H(x,v)+ \int_{-\infty}^{\infty} \Gamma(x,x_1,v)\, H(x_1,v)dx_1 \tag{8}$$

where it has been assumed that the Neumann series of Eqn.(6) converges uniformly. Although the above solution methodology is quite general and applicable to the case where Eqn.(1) is a vector equation, the convergence of the Neumann series plus the construction of the resolvent kernel are difficult to establish. For the particular case where only the coefficient βo in the operator R is nonzero, convergence of the formal solution given by Eqn.(8) can be established [9].

An alternative solution to the random integral equation of Eqn.(4) is through a series expansion of the dependent variable u in the form

$$u(x,v) = \sum_{k=0}^{\infty} u_k(x,v) \tag{9}$$

Since the above decomposition is not unique, a recursive relation is chosen so that u_k is an explicit function of u_{k-1} only [10, 11]. For the linear differential operator of Eqn.(2), the decomposition can be written as follows:

$$u_0(x,v) = f(x,v) \tag{10a}$$

and

$$u_k(x,\gamma) = \int_{-\infty}^{\infty} N_k(x,x_1,\gamma)\, u_{k-1}(x_1,\gamma)\, dx_1 \quad ,k=1,2... \tag{10b}$$

If the decomposition in Eqn.(9) is written explicitly and the result is compared with Eqn.(8), it becomes obvious that the series solution is equivalent to the formal solution involving the resolvent kernel.

Consider now the first three terms in Eqn.(9), i.e.,

$$u(x,\gamma) = H(x,\gamma) - \int_{-\infty}^{\infty} N(x,x_1,\gamma)\, H(x_1,\gamma)\, dx_1 + \\ + \int_{-\infty}^{\infty}\int_{-\infty}^{\infty} N(x,x_1,\gamma)\, N(x_1,x_2,\gamma)\, H(x_2,\gamma)\, dx_2\, dx_1 \tag{11}$$

Since we are dealing with a stochastic problem, the above solution must be recast in terms of the expectation $E\{g(x,\gamma)\}=\langle g(x)\rangle$ of all the random variables $g(x,\gamma)$ involved. This operation denotes statistical averaging and its application to both sides of Eqn.(11) results in

$$\langle u(x)\rangle = \langle H(x)\rangle - \int_{-\infty}^{\infty} \langle N(x,x_1)\, H(x_1)\rangle\, dx_1 + \\ + \int_{-\infty}^{\infty}\int_{-\infty}^{\infty} \langle N(x,x_1)\, N(x_1,x_2)\, H(x_2)\rangle\, dx_1 dx_2 \tag{12}$$

If the random operator R and the forcing function f are statistically independent and if the former is a zero-mean process, then Eqn.(12) reads as

$$\langle u(x)\rangle = \langle H(x)\rangle - \int_{-\infty}^{\infty}\int_{-\infty}^{\infty} \langle N(x,x_1)\; N(x_1,x_2)\rangle\langle H(x_2)\rangle dx_1 dx_2 \tag{13}$$

where $\langle NN\rangle$ is the correlation function for the random process N and needs to be specified.

It should finally be noted that the solution methodology described so far has, despite its generality, a serious drawback in that it fails to work for the case of a zero forcing function. To overcome this difficulty, it is necessary to go back to the original random integral equation in Eqn.(4) and apply the expectation operator. The result is

$$\langle u(x)\rangle = \langle H(x)\rangle - \int_{-\infty}^{\infty} G(x,x_1)\, \langle R(x_1)\, u(x_1)\rangle\, dx_1 \tag{14}$$

Unfortunately, $\langle Ru\rangle$ is unknown and will not separate into $\langle R\rangle\langle u\rangle$. The approximation that has been used for $\langle Ru\rangle$ is to invert the differential operation in Eqn.(2), apply the random operator to both sides, and finally take the expectation of both sides of the equation [12,6]. As a result,

$$\langle R(x)u(x)\rangle = \langle R(x)D^{-1}(x)f(x)\rangle - \langle R(x)D^{-1}(x)\; R(x)u(x)\rangle \simeq -\langle R(x)D^{-1}(x)\; R(x)\rangle\langle u(x)\rangle \tag{15}$$

The only new approximation involved in Eqn.(15) is the closure approximation $\langle RD^{-1} Ru\rangle = \langle RD^{-1} R\rangle\langle u\rangle$, which is of a higher order than an approximation of the type $\langle Ru\rangle = \langle R\rangle\langle u\rangle$. As a result of the above, Eqn.(14) now reads as

$$\langle u(x)\rangle = \langle H(x)\rangle + \int_{-\infty}^{\infty} G(x,x_1)\, \langle R(x_1) \int_{-\infty}^{\infty} G(x_1,x_2)\, R(x_2)\, dx_2\rangle \langle u(x_1)\rangle\, dx_1 \tag{16}$$

where the correlation function $\langle RR\rangle$ of the random operator R needs to be specified. The above closed-form solution for the average displacement $\langle u\rangle$ invokes the additional closure approximation when compared to the formal solution given by Eqn.(13), but has the advantage that it is applicable to the case of a zero forcing function.

WAVE PROPAGATION IN A RANDOM GEOLOGICAL MEDIUM

The solution methodology given in the previous section will now be illustrated by recourse to the problem of wave propagation through ground. Under steady-state (harmonic) conditions, the governing differential equation for one-dimensional wave propagation is

$$\nabla_r.(\mu\, \nabla_r\, u(r,\gamma)) + \rho\, w^2\, u(r,\gamma) = 0 \tag{17}$$

where μ is the shear modulus of the ground, ρ is its density and w is the circular frequency of the vibration. Also, r is the radial coordinate, ∇_r is the gradient with respect to r, and $\cdot$ is the dot product. For simplicity, consider the shear modulus to be constant, while the density is random with a constant mean value ρ_0 and a space-dependent fluctuation ρ_1 about the mean, i.e.,

$$\rho(r,\gamma) = \rho_0(1+\rho_1(r,\gamma)) \tag{18}$$

We note here that ρ_1 is a zero-mean random process. Substitution of the above expression in the wave equation yields

$$\nabla_r^2\, u(r,\gamma) + k^2\, (r,\gamma)\, u(r,\gamma) = 0 \tag{19}$$

where k^2 is the random wave number given by

$$k^2(r,\gamma) = k_0^2\, (1 + \rho_1(r,\gamma)) \tag{20}$$

Also, $k_0 = w/c_s$ is the deterministic wave number and $c_s^2 = \mu/\rho_0$ is the deterministic shear wave velocity. A comparison of Eqn.(19) with the differential operator decomposition of Eqs.(2) and (3) shows that the only non-zero coefficients are

$$a_0 = k_0^2 , \quad a_1 = 2/r, \quad a_2 = 1 \quad \text{and} \quad \beta_0 = k_0^2 \, \rho_1(r,\gamma) \tag{21}$$

where variable x has been replaced by variable r.

Since there is no forcing function in this problem, the closed-form solution given by Eqn.(16) must be used. The Green's function for the deterministic harmonic wave operator is

$$G(r) = \exp\,(ik_0r) \,/\, (4\pi r) \tag{22}$$

and a common choice for the correlation function <RR> is a simple decaying exponential, i.e.,

$$\langle R(x_1)\; R(x_2)\rangle = \sigma^2 \exp\,(-r/a) \tag{23}$$

where σ^2 is the standard deviation, $r=|x_1-x_2|$, and a is the correlation length of the random inhomogeneity.

The above problem has been solved by Karal and Keller [12] using a somewhat different operator decomposition than the one described here. For a plane wave solution with amplitude A and wave number k, the solution $\langle u(r)\rangle = A \exp\,(-\,kr)$ is substituted, along with Eqs.(22) and (23), in Eqn.(16) and integration finally yields

$$k^2 = k_0^2 - i\;\sigma^2\; k_0^4\; k^{-1}\;((a^{-1} - ik_0 - ik)^{-1} + \\ +(a^{-1} - ik_0 + ik)^{-1})\;/2 \tag{24}$$

The above result indicates that the random wave number k is a complex quantity. As a consequence of the presence of the imaginary component Im(k), there is attenuation in the mean field <u> as if a damping mechanism were at work. For large values of the quantity k_0a compared to $1/\sigma$, Eqn. (24) simpifies to

$$k = k_0\;(1 + i/(2k_0a)) \tag{25}$$

The attenuation coefficient is thus proportional to 1/2a, while $R_e(k) = k_0$ and as such is independent of the correlation function chosen for <RR>.

Further details on this example can be found in Karal and Keller [12], where the additional case of elastic wave propagation through a random medium is examined. Also, shear wave propagation through one or more horizontal soil layers is developed in Chu et al [5] and Askar and Cakmak [6] using the operator decomposition of Eqn.(2). The results of these two references also show attenuation of a shear wave as it moves through homogeneous, yet stochastic soil layers.

APPROXIMATE TECHNIQUES

We will now contrast solution schemes for the differential equation given in Eqn.(1) which are based on the perturbation approach with the formal solutions given by Eqs.(13) and (16). First, consider a perturbation expansion of the dependent variable u in powers a small parameter ε as follows:

$$u(x,\gamma) = u_0(x,\gamma) + \varepsilon u_1(x,\gamma) + \varepsilon^2 u_2(x,\gamma) + \ldots. \tag{26}$$

It is important to note that success of a perturbation scheme depends on the physical properties of the problem at hand. In the wave propagation problem previously considered, for instance, it must be that the random density has a small fluctuation about the mean, i.e.,

$$\rho(r,\gamma) = \rho_0(1 + \varepsilon \rho_1(r,\gamma)) \tag{27}$$

where ε is compared to ka since the perturbation solution is the limit of the formal solution as ka→0. Then, substitution of Eqs.(26) and (27) in the wave Eqn.(17) and sorting of powers of ε yields the following system of equations:

$$\left.\begin{aligned} &\nabla_r^2 u_0 + k_0^2 u_0 = 0 \\ &\nabla_r^2 u_1 + k_0^2 u_1 = - k_0^2 \rho_1 u_o \\ &\nabla_r^2 u_2 + K_0^2 u_2 = - k_0^2 \rho_1 u_1 \\ &\ldots\ldots \end{aligned}\right\} \tag{28}$$

The problem with this perturbation expansion is that it yields a homogeneous solution for the zeroth order solution in the form $u_0 = \exp(ik_0 r)$ which, upon substitution in the equation for the first order solution, produces the secular term $r \exp(ik_0 r)$ in u_1. Thus, the solution diverges as $r \to \infty$ and, as a result, is valid for small r. The above situation can be rectified by a perturbation of the wave number as $k = k_0 + \varepsilon k_1 + \varepsilon^2 k_2 + \ldots$ and by requiring that the right-hand sides in the resulting system of equations is orthogonal to u_0 [6].

An alternative perturbation scheme can be realised [13] by going back to the original operator of Eqn.(1) and writing an integral equation statement involving the Green's function for L(γ). For the wave propagation problem of the previous section, the result is [14]

$$c(r)\, u(r,\gamma) = \int_S \{ G(r,r',\gamma)\, \partial u(r',\gamma)/\partial n' - u(r',\gamma)\, \partial G(r,r',\gamma)/\partial n' \} dS(r') \tag{29}$$

where c(r) is the jump term equal to 0.5 if r is on a smooth surface S and n is the direction along the outward pointing normal on S. By first expanding in terms of ε the Green's functions G and $\partial G/\partial n$, the dependent variables u and $\partial u/\partial n$ and the wave number k, then substituting in the wave equation and finally sorting powers of ε, the following system is obtained:

$$\left.\begin{aligned} c(r)\, u_o(r,\gamma) &= \int_S \{ G_0(r,r')\, \partial u_0(r',\gamma)/\partial n' - \\ &\quad - u_0(r',\gamma)\, \partial G_0(r,r')/\partial n \}\, dS(r') \\ c(r)\, u_1(r,\gamma) &= \int_S \{G_0(r,r')\, \partial u_1(r',\gamma))/\partial n' - \\ &\quad - u_1(r',\gamma)\, \partial G_0(r,r')/\partial n'\}\, dS(r') + \\ &\quad \int_S \{G_1(r,r',\gamma)\, \partial u_0(r',\gamma)\partial n' - \\ &\quad - u_0(r',\gamma)\, \partial G_1(r,r',\gamma)/\partial n'\}\, dS(r') \end{aligned}\right\} \tag{30}$$

....

In the above, G_0 is the Green's function for the deterministic wave operator, while G_1 is the Green's function for the wave operator with random wave number $k_1(r,\gamma)$. Once the terms in the perturbution expansion of Eqn.(26) have been evaluated, application of the expectation operator yields the desired response statistics. More details on these perturbation schemes can be found in Manolis and Shaw [13].

There are other approximate techniques available but do not necessarily involve integral equation formulations. For instance, the Taylor series expansion that still requires a small random fluctuation of the medium properties about their deterministic values is used in conjunction with the stochastic FEM [7]. A few details on the stochastic FEM are given in the Appendix so as to contrast it with integral equation formulations. Other techniques for discrete parameter systems include Fourier transforms coupled with normal mode analysis [15], exact solutions of the Fokker-Plank equation that governs the probability density function of the problem [16], and scattering expansions where the randomness is expressed in terms of a deterministic distribution of inhomogeneities in the otherwise homogeneous medium [17].

CONCLUSIONS

This paper focused on stochastic problems in geomechanics for which there is a relative paucity of work when compared with the deterministic case. In particular, it was shown that a general random integral equation formulation for the corresponding stochastic defferential operator exists and furthermore admits closed form solutions which are subject to certain conditions. This was illustrated through recourse to harmonic shear wave propagation in geological media which exhibit randomness. Finally, approximate techniques based on perturbation expansions were contrasted with the formal solutions and their inherent restrictions were discussed.

ACKNOWLEDGEMENT

The author wishes to thank Mrs. B. Baxevani for typing the manuscript.

REFERENCES

1. Beskos, D.E. Boundary Element Methods in Geomechanics, in Boundary Elements X, Vol.4, (Ed. Brebbia, C.A.), pp. 3-28, Proceedings of the 10th Int. Conf. on Boundary Element Methods, Springer-Verlag, Berlin, 1988.
2. Ishimaru, A. Wave Propagation and Scattering in Random Media, Vols. 1 and 2, Academic Press, New York, 1978.
3. Cheng, A.H.D. Heterogeneities in Flows through Porous Media by the Boundary Element Method. Chapter 6, Topics in Boundary Element Research, (Ed. Brebbia, C.A.), Vol.4, pp. 129-144, Springer-Verlag, Berlin, 1987.
4. Lafe, O.E. and Cheng, A.H.D. A Perturbation Boundary Element Code for Steady-State Groundwater Flow in Heterogeneous Aquifers, Water Resources Research, Vol.23, No.6, pp.1079-1084, 1987.
5. Chu, L.,Askar, A. and Cakmak, A.S. Earthquake Waves in a Random Medium, International Journal for Numerical and Analytical Methods in Geomechanics, Vol.5, pp. 79-96, 1981.
6. Askar, A. and Cakmak, A.S. Seismic Waves in Random Media, Probabilistic Engineering Mechanics, Vol.3, No.3, pp. 124-129, 1988.
7. Vanmarcke, E., Shinozuka, M., Nakagiri, S., Schueller, G.I., and Grigoriou, M. Random Fields and Stochastic Finite Elements, Structural Safety, Vol.3, pp. 143-166, 1986.
8. Liu, W.K., Belytschko, T. and Mani, A. Random Field Finite Elements, International Journal for Numerical Methods in Engineering, Vol.23, pp. 1831-1845, 1986.
9. Bharucha-Reid, A.T. Random Integral Equations, Academic Press, New York, 1972.
10. Adomian, G. Stochastic Systems, Academic Press, New York, 1983.
11. Benaroya, H. and Rehak, M. The Neumann Series/Born Approximation Applied to Parametrically Excited Stochastic Systems, Probabilistic Engineering Mechanics, Vol.2, No.2, pp. 74-81, 1987.

12. Karal, F.C. and Keller, J.B. Elastic, Electromagnetic, and Other Waves in a Random Medium, Journal of Mathematical Physics, Vol.5, No.4, 537-549,1964.
13. Manolis, G.D. and Shaw, R.P. Wave Motion in a Random Hydroacoustic Medium Using Boundary Integral/Element Methods, Engineering Analysis with Boundary Elements, to appear in 1991.
14. Shaw, R.P. Boundary Integral Equation Methods Applied to Wave Problems , Chapter 6, Developments in Boundary Element Methods-I (Eds. Banerjee, P.K. and Butterfield, R.), pp.121-154, Elsevier Applied Science Publishers, London, 1979.
15. Nigam, N.C. Introduction to Random Vibrations, MIT Press, Cambridge, Massachusetts, 1983.
16. Caughey, T.K. Derivation and Application of the Fokker-Planck Equation in Discrete Nonlinear Dynamic Systems Subjected to White Random Noise, Journal of the Acoustical Society of America, Vol.35, No.11, pp. 1683-1692, 1963.
17. Varadan, V.K., Ma, Y. and Varadan, V.V. Multiple Scattering Theory for Elastic Wave Propagation in Discrete Random Media, Journal of the Acoustical Society of America, Vol.77, pp.375-389, 1985.

APPENDIX

The FEM formulation for a stochastic medium whose randomness can be expressed in the general form $z = z^0(1+z^1)$, with z^0 being the deterministic term and z^1 a small random fluctation, will be briefly discussed here. Following Vanmarke et al [7], the static FEM formulation gives

$$[K(\gamma)]\ \{V(\gamma)\} = \{F(\gamma)\} \qquad (A1)$$

where [K] is the stiffness matrix and vectors {V} and {F} contain the nodal displacements and forces, respectively. The medium uncertainty is reflected in the stiffness matrix and, upon solution, on the nodal quantities. The stiffness matrix is expanded about the uncertainty using Taylor series as

$$[K(\gamma)] = [K^0] + \sum_{i=1}^{n} [K_i^1]\ z_i^1 + \frac{1}{2} \sum_{i=1}^{n} \sum_{j=1}^{n} [K_{ij}^2]\ z_i^1\ z_j^1 + \ldots \qquad (A2)$$

where n denotes the total number of random parameters z_i^1. Furthermore, superscript 0 denotes a deterministic term and superscripts 1 and 2 denote first and second order rates of change, respectively, which can be evaluated by differentiation of [K] with respect to the random variables z_i^1. A similar expansion can be used for the displacement vector. For a deterministic forcing function {F}, substitution of the aforementioned expansions in Eqn.(A1) and a subsequent perturbation-type ordering of the terms gives the following system of equations:

$$[K^0]\ \{V^0\} = \{F\}$$

$$\left.\begin{aligned}[K^0]\{V_i^1\} &= -[K_i^1]\{V^0\} \\ [K^0]\{V_{ij}^2\} &= -[K_i^1]\{V_j^1\} - [K_j^1]\{V_i^1\} - [K_{ij}^2]\{V^0\}\end{aligned}\right\} \quad (A3)$$

.....

The above system is very similar to the ones obtained for random integral equation formulations using perturbation, i.e. Eqs.(28) or (30). Once all the expansion terms for {V} are evaluated, then response statistics can be computed through application of the expectation operator. Finally, the above scheme can be extended to eigenvalue and dynamic problems. In all cases, the accuracy and convergence characteristics of the stochastic FEM formulation is dependent on the finite element mesh.

Frequency Domain BE Calculations and Interferometry Measurements in Geomechanics

L. Gaul (*), M. Plenge (**)
(*) Institute of Mechanics, Department of Mechanical Engineering, University of the Federal Armed Forces, Hamburg, Germany
(**) JAFO-Technology, Hamburg, Germany

ABSTRACT

Surface waves generated by machine foundations and diffracted by embedded structures as well as soil inhomogeneities are analyzed by improved experimental techniques and calculated by the boundary element method (BEM).

INTRODUCTION

A refined approach is developed to analyze the dynamic response of flexible or rigid foundation slabs interacting with subsoil. The study is focussed on the substructure soil and its irregularities. The field equations of subsoil are solved by boundary element method (BEM). Connecting transition elements have been formulated, such that frame-superstructures can be discretized by finite beam elements and coupled with subsoil [7]. Effects of layering, excavations for embedded foundations, trenches and obstacles in subsoil can be treated [2].

In the present paper viscoelastic constitutive equations of differential operator or hereditary integral type are generalized by implementing fractional order time derivatives as well as fractional integration into the boundary integral equations.

Small scale soil-structure-interaction experiments are performed on a lab foundation consisting of a soil box which is equipped with conventional transducers and optoelectronic wave measurement facilities [8]. Measured surface wave fields prove the necessity of layered BE-models even for soil with homogeneous compactness.

NUMERICAL TREATMENT OF SUBSOIL DYNAMICS BY BEM

The dynamics of the subsoil domain is governed by the equation of motion of a viscoelastic continuum with body forces $b_i(\mathbf{x},t)$ in terms of displacement coordinates $u_i(\mathbf{x},t)$ at location $\mathbf{x}$

$$\int_{-\infty}^{t} E_D(t-\tau)\frac{\partial u_{j,ji}}{\partial\tau}\,d\tau - e_{ijk}e_{klm}\int_{-\infty}^{t} G(t-\tau)\frac{\partial u_{m,lj}}{\partial\tau}\,d\tau + b_i = \rho\frac{\partial^2 u_i}{\partial t^2} \tag{1}$$

Steady-state elastodynamics is represented by taking the Fourier transform of Eq. (1). With the transformed variables $U_i(\mathbf{x},\omega) = F[u_i(\mathbf{x},t)]$ this yields an elliptical problem for the equations in the domain

$$c_D^{*2}\,U_{j,ji} - e_{ijk}e_{klm}c_S^{*2}\,U_{m,lj} + B_i/\rho + \omega^2 U_i = 0 \tag{2}$$

and transformed boundary conditions, whereby no initial conditions enter. The relaxation functions for plane dilatation $E_D(t)$ and shear $G(t)$ are replaced by complex moduli

$$E^*_D = \lambda^* + 2G^* = E_D(1+i\eta_D), \qquad G^* = G(1+i\eta_s) \tag{3}$$

and the complex propagation velocities of dilatational and distortional waves are

$$c_D^{*2} = E^*_D/\rho, \; c_S^{*2} = G^*/\rho\,.$$

Conventional complex moduli derived from differential operator type constitutive equations with integer time derivatives $D^k = d^k/dt^k$ $k \in N$ can be fruitfully generalized by replacing integer order time derivatives by those of fractional order D^{α_k}. E.g. the derivative of fractional order α of deviatoric stress $s_{ij}(t)$

$$D^{\alpha}s_{ij}(t) = \frac{d^{\alpha}s_{ij}(t)}{dt^{\alpha}} = \frac{1}{\Gamma(1-\alpha)}\frac{d}{dt}\int_0^t \frac{s_{ij}(t-\tau)}{\tau^{\alpha}}\,d\tau, \; 0<\alpha<1 \tag{4}$$

defined with the gamma function $\Gamma(1\text{-}\alpha) = \int_0^{\infty} e^{-x}x^{-\alpha}dx$ is the inverse operation of fractional integration attributed to Riemann and Liouville [9]. Compared to the conventional approach it has been shown, that strong frequency dependence of actual viscoelastic material over many decades can be fitted with only few parameters by adopting the fractional derivative concept. Fractional operators give rise to a richer variety of functional families and hence the possibility of improved curve fitting [10]. The defining Eq. (4) leads to the power law

$$F[D^{\alpha} s_{ij}(t)] = (i\omega)^{\alpha} F[s_{ij}(t)] = (i\omega)^{\alpha} \overset{*}{s}_{ij}(\omega) \tag{5}$$

when Fourier transform is applied. This is why integer powers $(i\omega)^k$ only have to be replaced by those of fractional order $(i\omega)^{\alpha_k}$. E.g. the complex shear modulus $G^*(\omega)$ with storage modulus $G'(\omega)$, loss modulus $G''(\omega)$ and loss factor $\eta_s(\omega)$ is generalized according to

$$G^*(\omega) = \frac{\overset{*}{s}_{ij}(\omega)}{2\overset{*}{e}_{ij}(\omega)} = \frac{\sum_{k=0}^{M} q_k (i\omega)^{\alpha_k}}{\sum_{k=0}^{N} p_k (i\omega)^{\beta_k}} = G'(\omega) + iG''(\omega) \tag{6}$$

$$G^*(\omega) = G'(\omega)[1 + i\eta_s(\omega)] \tag{6}$$

and related to the relaxation function $G(t)$ in Eq. (1) by the inverse Fourier transform

$$G(t) = F^{-1}[G^*(\omega)/(i\omega)]. \tag{7}$$

Computations with the fractional derivative concept in frequency domain require a unique selection of complex roots in the power law (5) [3].

Thermorheologically simple materials allow to introduce the influence of temperatur T in Eq. (1) by simply replacing time t by a reduced time

$$\zeta(t) = \int_0^t \phi[T(\mathbf{x}, \eta)]\, d\eta \tag{8}$$

which is based on the shift function ϕ determined from experimental data corresponding to dilatational or shear deformation. The frequency ω has to be replaced by a reduced frequency according to Eq. (8) in the complex moduli. Using the integral representation of Eq. (2), the displacement at the boundary point ξ may be written

$$c_{ij}(\xi)\, U_j(\xi) = \int_{\Gamma} [U^*_{ij}(\mathbf{x},\xi)\, t_j(\mathbf{x}) - T^*_{ij}(\mathbf{x},\xi)\, U_j(\mathbf{x})]\, d\Gamma + \int_{\Omega} \overset{*}{U}_{ij}(\mathbf{x},\xi)(B_j/\rho)\, d\Omega \tag{9}$$

where $\mathbf{x}$ are the boundary points to which the integral extends. $U_i(\mathbf{x})$ and $t_i(\mathbf{x})$ are displacement and traction components at $\mathbf{x}$, $U^*_{ij}(\mathbf{x},\xi,\omega)$ and $T^*_{ij}(\mathbf{x},\xi,\omega)$ stand for displacement and traction components of the fundamental solution at the field point $\mathbf{x}$, when a unit point load is applied at the load point ξ following the i direction, and c_{ij} is a coefficient that depends on the geometry of the boundary at ξ ($c_{ij} = \delta_{ij}/2$ for smooth boundary). One can solve the integral equation for a sufficient number of ω and numerically invert $U_i(\mathbf{x},\omega)$ to obtain the time-dependent displacement [7]. Viscoelastic material properties enter the fundamental solution

$$U^*_{ij} = \frac{1}{4\pi G^*}\left[\Psi(r, c^*_S, c^*_D)\,\delta_{ij} - X(r, c^*_S, c^*_D)\, r_{,i}\, r_{,j}\right]$$

where the potentials Ψ and X depend on the distance $r = (r_i r_i)^{1/2}$, $r_i = x_i - \xi_i$ and complex wave velocities. Only boundary integrals remain in Eq. (9) if the body forces B_i vanish. The integrals are discretized into the sum of integrals extending over the boundary elements. This yields for a boundary node a

$$c^{(a)}_{ij}\, U^a_j = U^{a\beta}_{ij}\, t^\beta_j - T^{a\beta}_{ij}\, U^\beta_j \tag{10}$$

$$\text{where} \quad U^{a\beta}_{ij} = \int_{\Gamma_{el}} U^{*a}_{ij}\, \Omega^\beta\, d\Gamma \qquad T^{a\beta}_{ij} = \int_{\Gamma_{el}} T^{*a}_{ij}\, \Omega^\beta\, d\Gamma$$

contain the shape functions $\Omega^\beta(x)$ for the boundary elements ($\Omega^\beta(x) = 1$ for constant elements), U^β_j and t^β_j are the displacement and traction nodal values. After numerical integration, the matrix formulation of Eq. (10) for all n nodes and smooth boundary yields

$$\mathbf{H}\,\mathbf{u} = \mathbf{U}\,\mathbf{t}\ ,\quad \mathbf{H} = \frac{1}{2}\,\mathbf{E} + \mathbf{T}\ , \tag{11}$$

$$\text{where} \quad \mathbf{u} = \{U^1\, U^2 \ldots U^a \ldots U^n\}^T,\quad U^a = \{U^a_1, U^a_2, U^a_3\}^T$$

$$\mathbf{t} = \{t^1\, t^2 \ldots t^a \ldots t^n\}^T,\quad t^a = \{t^a_1, t^a_2, t^a_3\}^T$$

BE-domains with different material properties such as the foundation slab Ω_a and the halfspace Ω_b in Figure 1 with associated BE-equations

$$\mathbf{H}_a \mathbf{u}_a = \mathbf{u}_a \mathbf{t}_a,\quad \mathbf{H}_b \mathbf{u}_b = \mathbf{u}_b \mathbf{t}_b \tag{12}$$

can be coupled by continuity requirements of displacements and tractions in the interface Γ^k

$$\mathbf{u}_a^k = \mathbf{u}_b^k = \mathbf{u}^k \qquad \mathbf{p}_a^k = -\mathbf{p}_b^k = \mathbf{p}^k \, . \tag{13}$$

Unknown displacements and tractions including those at the interface are obtained from the combined equations

$$\begin{bmatrix} \mathbf{H}_a^a & \mathbf{H}_a^k & \mathbf{O} & -\mathbf{U}_a^k \\ \mathbf{O} & \mathbf{H}_b^k & \mathbf{H}_b^b & \mathbf{U}_b^k \end{bmatrix} \begin{Bmatrix} \mathbf{U}_a^a \\ \mathbf{U}^k \\ \mathbf{U}_b^b \\ \mathbf{t}^k \end{Bmatrix} = \begin{bmatrix} \mathbf{U}_a^a & \mathbf{O} \\ \mathbf{O} & \mathbf{U}_b^b \end{bmatrix} \begin{Bmatrix} \mathbf{t}_a^a \\ \mathbf{t}_b^b \end{Bmatrix} \tag{14}$$

after imposing displacement and/or traction boundary conditions on parts of the boundaries Γ_a^a and Γ_b^b.

EXPERIMENTAL STUDIES ON WAVE PROPAGATION

Experimental investigations on soil-structure-interaction (SSI) are carried out at the Institute of Mechanics under the guidance of the first author. A schematic illustration and photo of the lab foundation consisting of a soil box which is equipped with conventional transducers and optoelectronic wave measurement facilities is shown in Figure 2. An experimental program has been started to measure interaction phenomena as well as surface wave propagation for homogeneous and inhomogeneous soil [8].

Figure 3 depicts the decay of vertical vibration amplitudes w related to the amplitude $\widehat{w_0}$ at the excited footing with increasing distance from the footing x for 5 excitation frequencies. The expected amplitude decay proportional to $1/\sqrt{x}$ for homogeneous soil is only fulfilled in an average.

It is interesting to note that deviations are superimposed which behave like amplitudes of damped vibrations. After proper scaling of the distance x with respect to the wavelength $\lambda_s = c_s/f$ of the generated shear wave for 5 excitation frequencies it could be shown, that the local minima and maxima of the amplitude decay coincide.

The measured amplitude oscillations cannot be explained by reflections from the wall of the soil box. This was shown by 3-d BE-calculations of the discretized entire soil box and proved by experiments as well. Even though the soil compactness is homogeneous, interference effects are created by the increase of soil-stiffness due to growing confining pressure with depth. Thus, faster running waves from greater depth interfere with slower ones closer to the surface and create local maxima and minima of the amplitude decay.

Similar effects have been measured by Haupt [6], who studied the effect by a 2-d FE-model.

The application of pulse laser holographic interferometry allows to detect soil inhomogeneities and hidden structures. The advantage of this optical contact-free measurement technique is the expanded pictorial visualization of the dynamical deformation at the soil-surface.

The interferogram in Fig. 4 shows circular interference fringes of a wavefield generated by a harmonically excited foundation on homogeneous soil. If the wave propagation is deranged by obstacles, trenches or embedded structures, the fringe pattern does change typically. In order to demonstrate this effect, a hollow quadrangular box welded from metal sheets was buried in the homogeneous soil and covered with a layer of 8 cm thickness. The shape and the location of the embedded structure can be identified very easily from the interferogram (Fig. 5). The straight front end shows up clearly and the modes of the oscillation of the cover plate becomes obvious. In the case of an impact excitation (Fig. 6) the circular shock wave front is disturbed by the differing wave velocities in sand and metal. Thus the pulse laser holographic interferometry is an appropriate method for the detection of buried structures.

CALCULATION OF SURFACE WAVES IN LAYERED VISCOELASTIC SOIL BY BEM

A first approach to simulate the measured amplitude decay (Figure 3) by 3-d BE-calculation is based on a cubic domain Ω_a describing an acceleration pickup embedded in a homogeneous viscoelastic halfspace domain Ω_b. Propagating waves are generated by harmonic vertical excitation force with complex amplitude $\hat{F}_z$ acting on a rigid square of dimensions $2a$ x $2a$ at the surface. Rigid body motion is easily implemented in Eq. (14) by coupling those displacements of u_b^b which are located in the square. The excitation force is the vertical resultant of the surface tractions. Real- and imaginary part of the complex compliance

$$\overline{u}_z aG/\overline{F}_z = f_{zz} + ig_{zz} \tag{15}$$

govern the space- and time-dependent displacement-field of the surface wave by

$$u_z(x,t) = [f_{zz}(x)\cos\omega t - g_{zz}(x)\sin\omega t]\, F_z/(aG). \tag{16}$$

Fig. 7a depicts the real- and imaginary part of the compliance (15) as well as the amplitude $(f_{zz}^2 + g_{zz}^2)^{1/2}$. From the amplitude decay it can be concluded that the homogeneous halfspace is no proper model to explain the measured results in Figure 3.

The variation of soil moduli with depth due to increasing confining pressure can be modeled by layered BE domains. Proper truncation of the surface- and interface-discretization has been analyzed by Klein [7]. Only one layer with underlying halfspace can already explain the measured amplitude decay (Figure 3). The calculated amplitudes in Figure 7b indicate that the locations of local maxima and minima fit quite well with the measured results.

The locus of surface-particle motion generated by the vertical excitation on a square base with smooth contact has been calculated by BEM as well. The path of the particle motion describes a retrograde ellipse, in contrast to the prograde-ellipse motion associated with water waves. The calculated ellipses change the orientation of their principal axes with varying distance from the center of the exciting square base (Figure 8). The measured orbits by Barkan [1] in Figure 8 underline the predicted results by BEM.

CONCLUSIONS

The BE program PROBEM developed at the Institute of Mechanics [4, 7] allows to model soil irregularities by the substructure technique. Generalized viscoelastic constitutive equations are implemented. Measured surface wave fields prove the necessity of layered BE-models even for soil with homogeneous density. Detailed 3-d models including embedded pickups lead to the deviation between the pickup-response and the response of the undisturbed soil at the pickup-location.

Displacement measurements by Laser-interferometer [5] allow to detect the difference with accelerometer measurements and lead to the important FRF of the pickup.

The information of pointwise displacements from interferometer measurements is extended by detecting surface wave-fields by adopting the technique of holographic interferometry.

REFERENCES

[1] Barkan, D. D., Dynamics of bases and foundations, translated from the Russian by L. Drashevsha, G. P. Tschebotarioff, Ed., McGraw-Hill Book Comp., New York, 1962.

[2] Gaul, L., Klein, P. & M. Plenge, Continuum-boundary element and experimental models of soil-foundation interaction. A. A. Balkema, Rotterdam, Numerical methods in geomechanics, Innsbruck, 1649 - 1661, 1988.

[3] Gaul, L., Klein, P. & S. Kempfle, Impulse response function of an oscillator with fractional derivative in damping description. Mechanics Research Communications, 16 (5), 297 - 305, 1989.

[4] Gaul, L. & B. Zastrau, Rechnergestützte (FEM, BEM, MKS) und experimentelle (CAT) Spannungs-, Verformungs- und Bewegungsanalyse statisch und dynamisch beanspruchter Strukturen. In Technologieangebote wiss. Einrichtungen Hamburgs, Ed. Arbeitskreis Technologieförderung in Hamburg , 8.04, 1989.

[5] Gaul, L. & M. Plenge, Berechnung und Messung von Körperschallausbreitungen - Identifikation bodenverlegter Störkörper, 7. Arbeitstagung "Anwendungen der Akustik in der Wehrtechnik", 180 - 201, 1990.

[6] Haupt, W. Bodendynamik. Vieweg & Sohn, Braunschweig-Wiesbaden, 1986.

[7] Klein, P., Zur Beschreibung der dynamischen Wechselwirkung von Fundamentstrukturen mit dem viskoelastischen Baugrund durch dreidimensionale Randelementformulierungen. PhD thesis, Institute of Mechanics, Univ. Fed. Armed Forces Hamburg, 1989.

[8] Plenge, M. Ein Beitrag zur Untersuchung des dynamischen Verhaltens geschichteter Baugründe. PhD thesis, Institute of Mechanics, Univ. Fed. Armed Forces Hamburg, 1990.

[9] Ross., B. Fractional calculus. Mathematics Magazine, 50 (3) , 115 - 122, 1977.

[10] Torvik, P. J. & D. L. Bagley. Fractional derivatives in the description of damping, materials and phenomena. The role of damping in vibration and noise control, ASME DE - 5 , 125 - 135, 1987.

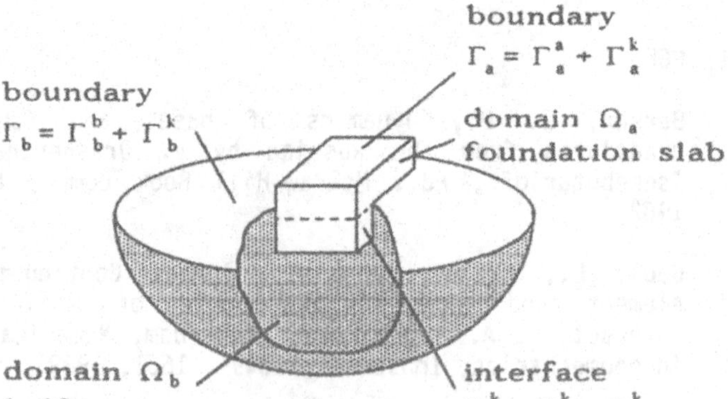

Figure 1. Notation for BE-domain-coupling

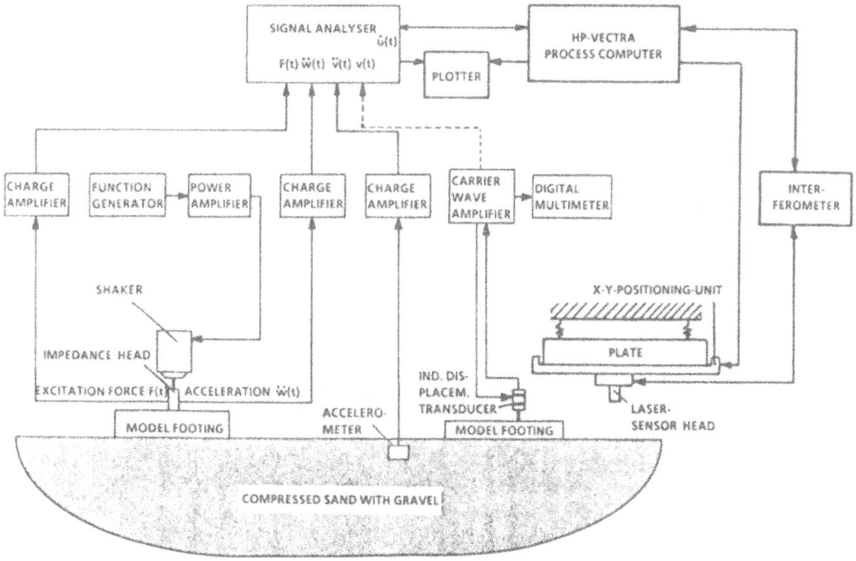

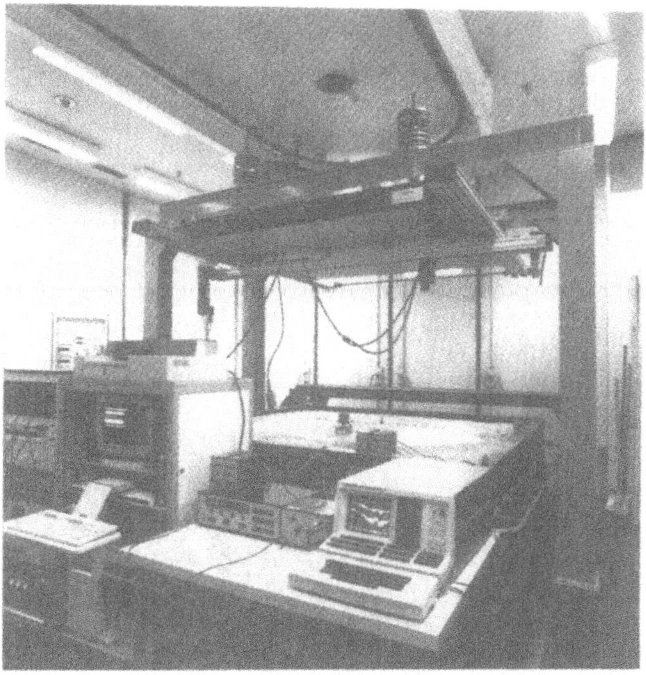

Figure 2. Lab foundation and measuring setup for experimental static and dynamic SSI investigations

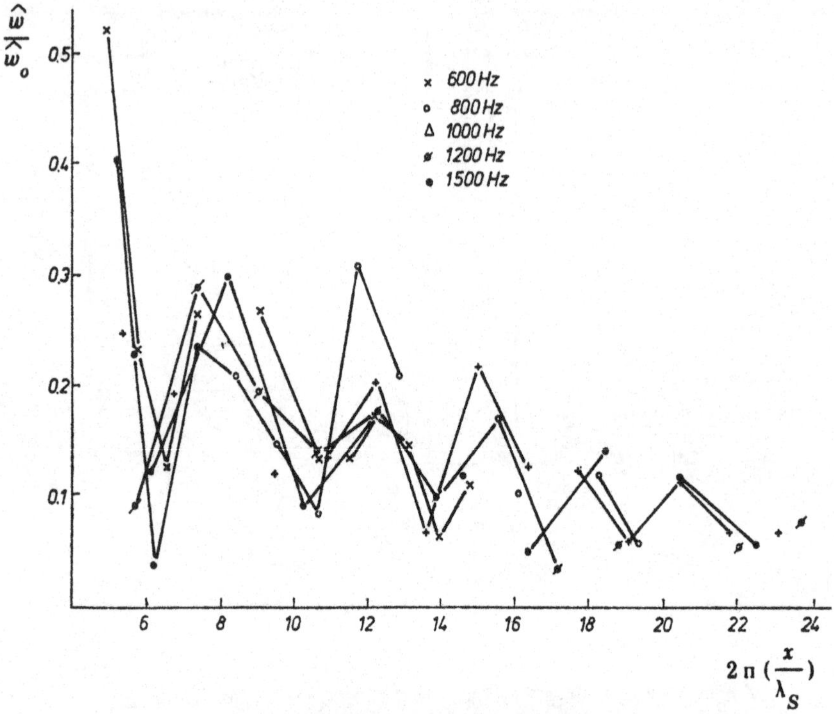

Figure 3. Measured decay of vertical vibration amplitude vs. distance from the exciting footing

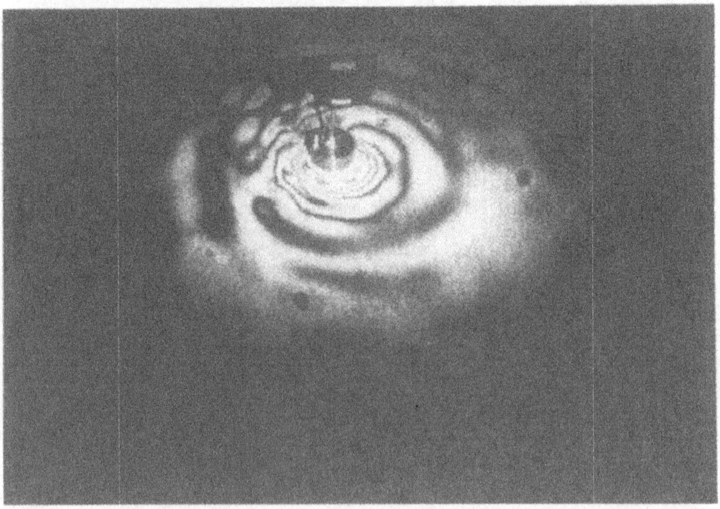

Figure 4. Interference pattern of surface wave propagation (Compacted homogeneous soil)

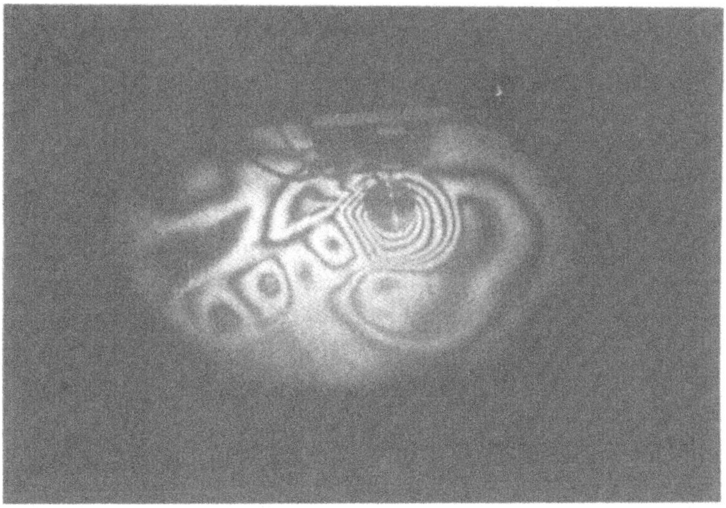

Figure 5. Interference pattern of surface wave propagation (Quadrangular box in homogeneous soil)

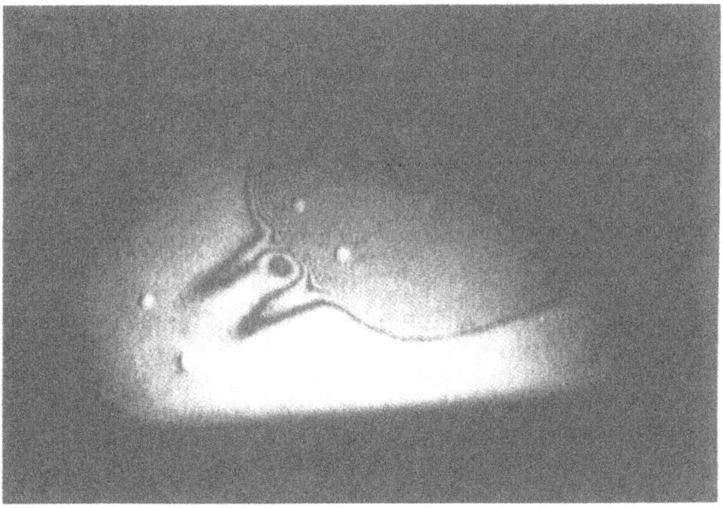

Figure 6. Interference pattern of surface wave propagation (Quadrangular box in homogeneous soil)

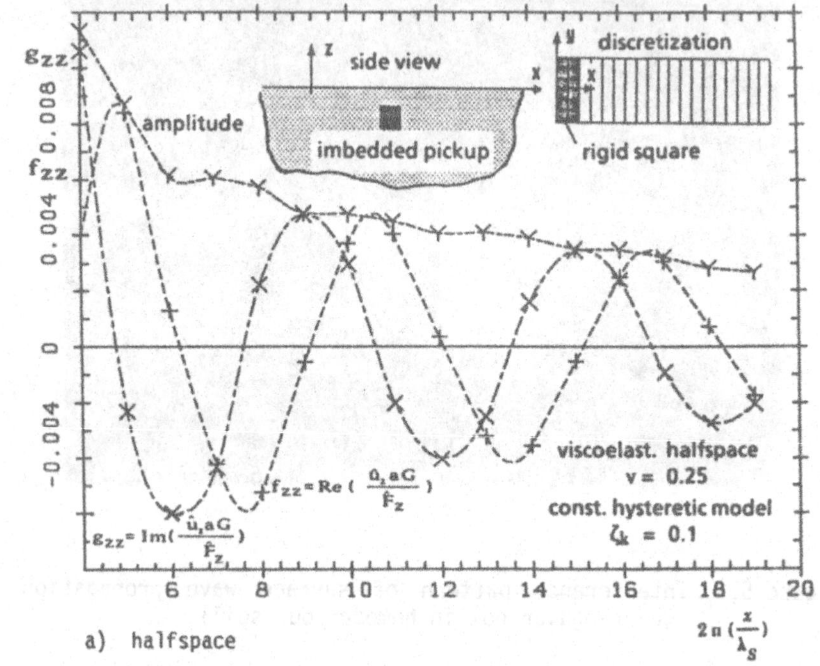

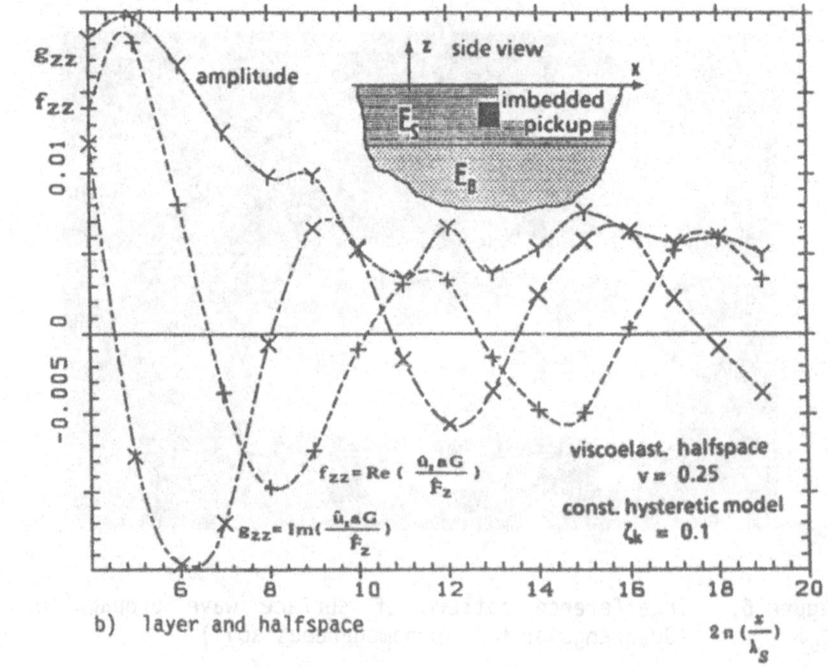

Figure 7. Vertical compliance;real-part, imaginary-part and amplitude at an embedded pickup

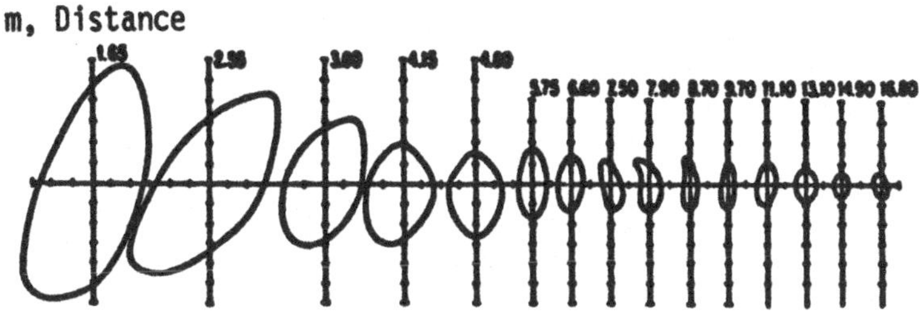

Measured by Barkan [1]

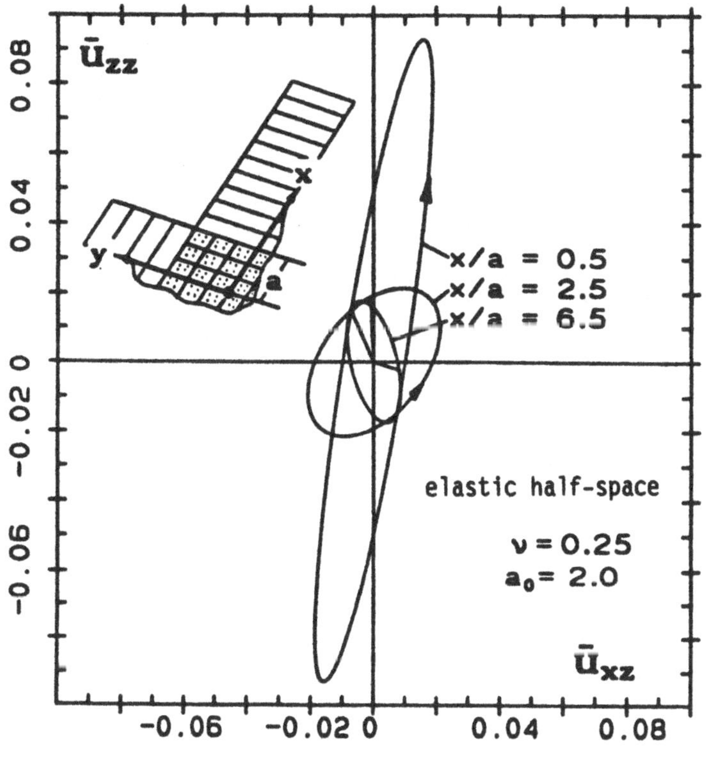

Calculated by BEM

Figure 8. Orbits of motion of surface particles at varying distances from a square base with time harmonic vertical excitation

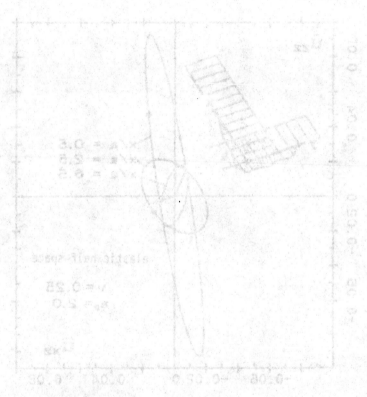

Foundation-Soil-Foundation Dynamics Using a 3-D Frequency Domain BEM

D.L. Karabalis, M. Mohammadi

Department of Civil Engineering, University of South Carolina, Columbia, SC 29208, U.S.A.

ABSTRACT

The dynamic response of a series of surface foundations including through-the-soil interaction is studied. The subgrade medium is a linear elastic half-space modelled by a 3-D BEM while the foundations are interconnected by a super-structure for which a standard FEM discretization is used. The strong through-the-soil coupling of adjacent foundations and the effect of the overlaying structure become apparent on the basis of the obtained results.

INTRODUCTION

The through-the-soil interaction of a number of adjacent foundations has received only limited attention within the general area of soil-structure interaction, in spite of several reports on the importance of the phenomenon. Foundation-soil-foundation interaction (FSFI) has been studied using, analytical [7] and Green's functions [8] approaches, as well as FEM [2,3] and BEM [1,5]. In all of the above works frequency domain analyses have been performed and the important multiple resonance phenomena associated with the presence of several foundation masses have been detected.

In this work the dynamic behavior of a group of rectangular, rigid, surface footings tied by an overlying beam structure and arranged in a way that resembles a railway track system is investigated, see Figure 1. However, the proposed methodology can routinely handle arbitrary geometries and forcing functions. The analysis is performed in the frequency domain on the basis of a 3-D BEM representation of the half-space, with externally applied loads being the only forcing functions. A similar investigation, based on an analytical 2-D model and test results, has been reported by Prange [6].

FORMULATION

The Boundary Element Method can be applied for the analysis of the through-the-soil interaction of adjacent foundations by considering either relaxed or non-relaxed

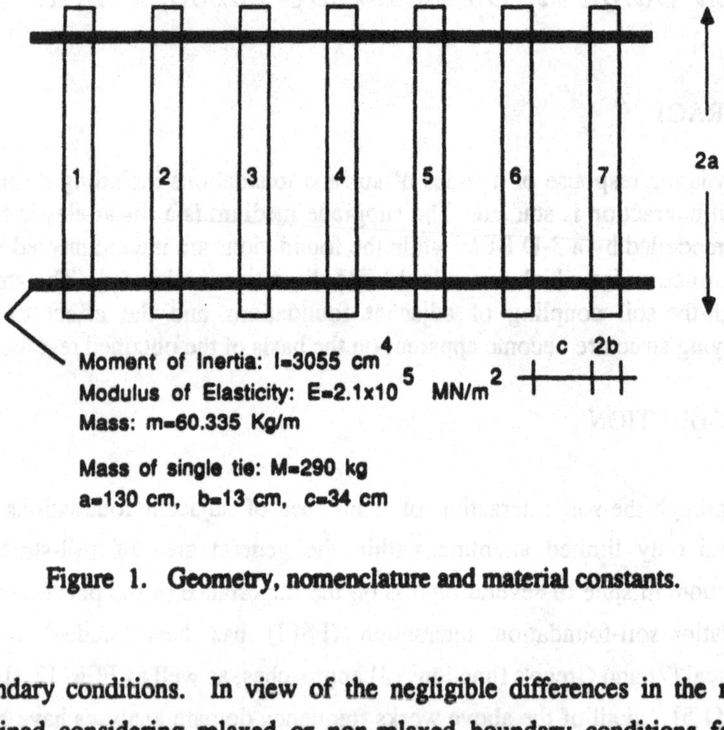

Figure 1. Geometry, nomenclature and material constants.

boundary conditions. In view of the negligible differences in the results obtained considering relaxed or non-relaxed boundary conditions for the vertical and rocking modes of vibration of a rigid surface foundation [4], and in order to minimize the computational effort, only relaxed boundary conditions have been assumed throughout this work. On the basis of this assumption, only the soil-foundation interface needs to be discretized, which for the purposes of this work has been accomplished using four-node linear, quadrilateral elements. Following standard techniques [1,4], the discretized frequency domain boundary integral equation for M nodal points on the soil-foundation interfaces can be written in a matrix form as

$$(0.5)\,\{u\}_{Mx1} = [G(\omega)]\,\{t\}_{Mx1} \tag{1}$$

where ω is the independent frequency variable, {u} and {t} comprise the vertical displacement and traction vectors, respectively, and $[G(\omega)]$ is the discretized counterpart of the fundamental solution corresponding to an infinite elastic space. The compatibility and force equilibrium equations are given by

$$\{u\}_{Mx1} = [S]\,\{D\}_{2Nx1} \tag{2}$$

$$\{P\}_{2Nx1} = -[K]\,\{t\}_{Mx1} \tag{3}$$

where N is the number rigid foundations, [S] relates nodal displacements to foundation degrees-of-freedom, [A] is a diagonal matrix defining the area corresponding to each node, the rigid displacement vector $\{D\} = \{\Delta_1......\Delta_N\ \phi_1......\phi_N\}^T$ with Δ_i and ϕ_i being the vertical and rocking degrees-of-freedom of the i-th footing, and the load vector $\{P\} = \{P_1......P_N\ M_1......M_N\}^T$ with P_i and M_i being the vertical forces and moments. Equation (1), combined with equations (2) and (3), yields a force-displacement relationship of the form

$$\begin{Bmatrix} \Delta \\ \phi \end{Bmatrix} = [C^N(\omega)] \begin{Bmatrix} P \\ M \end{Bmatrix} \tag{4}$$

where the compliance matrix $[C^N(\omega)]$ can be defined in terms of the impedance matrix $[K^N(\omega)]$ as

$$[C^N(\omega)]^{-1} = [K^N(\omega)] = (0.5)\,[K]\,[G]^{-1}\,[S] \tag{5}$$

The term C^N_{ij} (i,j=1,...,2N) of the compliance matrix represents the vertical displacement (i=1,...N) or the rotation (i=N+1,...,2N) due to a vertical force (j=1,...,N) or moment (j=N+1,...,2N) applied on the j-th degree-of-freedom of the soil-foundations system.

The inertia effects due to the mass of the foundation (and the superstructure) can also be considered using the relation

$$[K^{*N}(\omega)] = [K^{N}(\omega)] - \omega^2[M] \tag{6}$$

where [M] is a diagonal mass matrix. Similarly, the stiffness of the superstructure $[K_s]$ can be incorporated into the impedance of the soil-footings system, i.e.,

$$[K^{**N}(\omega)] = [K^{*N}(\omega)] + [K_s] \tag{7}$$

NUMERICAL RESULTS

The geometry and material properties of the railway track system studied in this work are shown in Figure 1. The soil constants are: Poisson's ratio ν = 1/3, Shear modulus G = 9500 N/cm2, and shear wave velocity c_s = 24837 cm/sec. The vertical and rocking motions of the rigid ties due to vertical and rocking loadings are the only parameters considered in this work. A (4x40) BEM discretization is used in the following examples where the magnitude of the non-dimensional compliances are plotted versus the dimensionless frequency $a_o = \omega b/c_s$, where b is specified in Figure 1.

In Figure 2(a) the vertical compliances of the loaded end-tie, in an assembly of five ties, are plotted versus frequency along with the compliance of a single tie for comparison purposes. In Figure 2(b) a similar plot is made for the loaded center-tie. Figures 3(a,b) present a similar comparison study of the rocking compliances. The effect of adjacent ties and the presence of the overlaying structure in shifting and modifying the magnitude of the resonance frequencies is clearly observed.

In an attempt to simulate the influence of adjacent ties and the rail upon the loaded tie within an "infinitely" long system, a study of a seven- and a nine-tie assembly is shown in Figures 4 and 5. In Figures 4(a,b) the vertical response of the loaded end-tie and the loaded center-tie, respectively, are plotted along with that of a single tie. Similar results for the rocking responses are shown in Figures 5(a,b). A definite convergence is observed in the compliances of the loaded end-tie. However, this is not the case for the loaded middle-tie for which case more degrees-of freedom (ties) on either side should be considered for a fair representation of an "infinitely" long system.

Similar numerical studies have been reported by the authors in reference [1]. Regretfully, due to an irreversible error the set of elastic constants and some of the results presented in reference [1] do not correspond to one another. The authors are in the process of preparing a more comprehensive report on the subject of foundation-soil-foundation interaction in a forthcoming publication.

CONCLUSION

A 3-D frequency domain BEM formulation for the dynamic analysis of a number of interacting adjacent foundations connected by a superstructure is presented. The method has been applied, as a numerical example, to a railway problem and results have been obtained in frequency domain. The amplification and shifting of certain "resonance" peaks owing to the presence of adjacent massive foundations (ties) as well as to the overlying structure (rail) are well demonstrated. The representation of an "infinitely" long track system by a truncated one seems adequate if a sequence of more than nine ties is considered.

ACKNOWLEDGEMENT

The authors wish to express their gratitude to the International Business Machines Corporation for providing remote access to their IBM 3090 supercomputer facility located at the Numerically Intensive Computing (NIC) Center, Kingston, New York. The use of the IBM 3090 vector facility, by incorporating the Engineering and Scientific Subroutine Library (ESSL), in the computer programs developed for this work provided substantial savings (up to 93% in some cases) in CPU time as compared to a "scalar" solution procedure. Sincere thanks are extended to Dr. Robert D.

Ciskowski of the IBM NIC center for his helpful instructions in using the IBM 3090 supercomputer.

REFERENCES

1. D. L. Karabalis and M. Mohammadi, "Foundation-Soil-Foundation Interaction: An Application to Railway Problems," pp. 75-87, in: C. A. Brebbia, Ed., *Boundary Elements X, Vol. 4: Geomechanics, Wave Propagation and Vibrations* (Computational Mechanics Publications, Southampton, 1988).
2. Lin, H-T, Roesset, J. M., and Tassoulas, J. T., "Dynamic Interaction Between Adjacent Foundations," Earthquake Engineering and Structural Dynamics **15**, 323-343 (1987).
3. Lysmer, J., Seed, H. B., Udaka, T., Hwang, R. N., and Tsai, C. F., "Efficient Finite Element Analysis of Seismic Soil Structure Interaction," Report No. EERC 75-34, Earthquake Engineering Research Center, University of California, Berkeley, 1975.
4. Mohammadi, M. and Karabalis, D. L., "3D Soil-Structure Interaction Analysis by the BEM: Comparison Studies and Computational Aspects," Soil Dynamics and Earthquake Engineering **9**, No. 2, 96-108 (1990).
5. Ottenstreuer, M. and Schmid, G., "Boundary Elements Applied to Soil-Foundation Interaction," in: C. A. Brebbia, ed., *Boundary Element Methods*, (Springer-Verlag, Berlin, 1981) pp. 293-309.
6. Prange, B., "Interaction of Railway Ties - Comparison With Seismic WaveData,", Proc. Int. Conf. on Recent Advances in Geotechnical Earthquake Engineering and Soil Dynamics, Vol. III, St. Louis, 1981, pp. 1053-1057.
7. Triantafyllidis, T. and Prange, B., "Dynamic Subsoil Coupling Between Rigid, Rectangular Foundations," Soil Dynamics and Earthquake Engineeing **6**, 164-179 (1987).
8. Wong, H. L. and Luco, J. E., "Dynamic Interaction Between Rigid Foundaitons in a Layered Half-Space," Soil Dynamics and Earthquake Engineering **5**, 149-158 (1986).

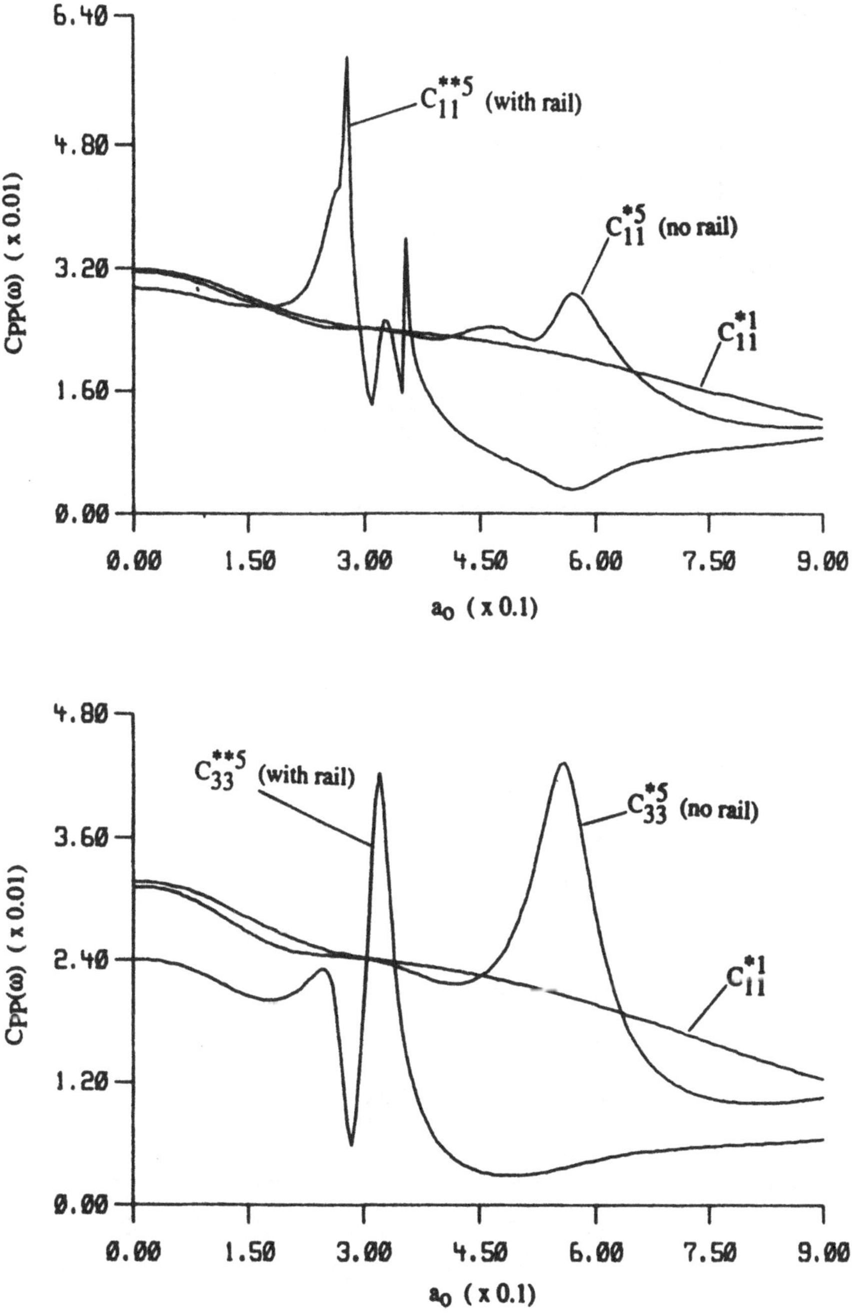

Figure 2. Vertical compliances for a five-tie system with the load applied on: (a) the end tie, and (b) the middle tie.

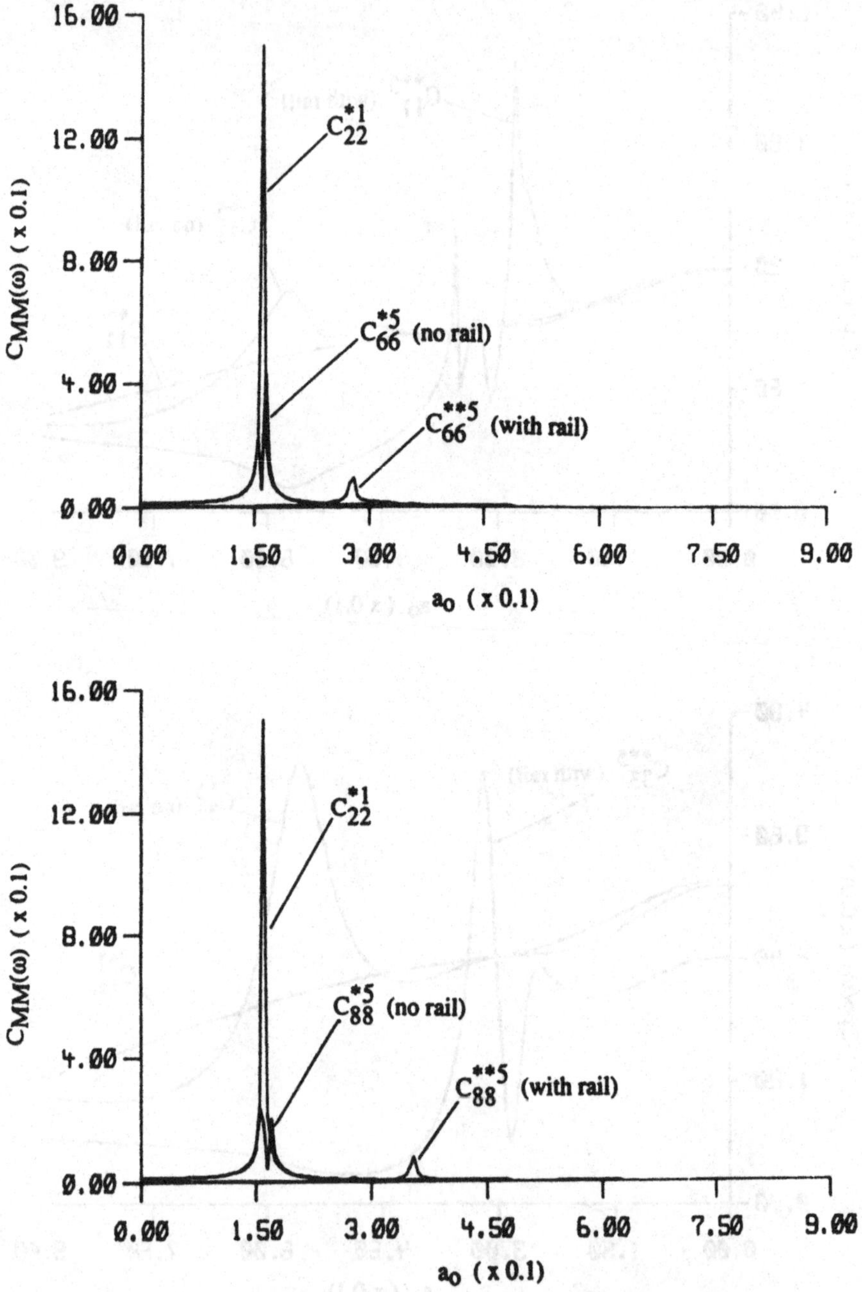

Figure 3. Rocking compliances for a five-tie system with the load applied on: (a) the end tie, and (b) the middle tie.

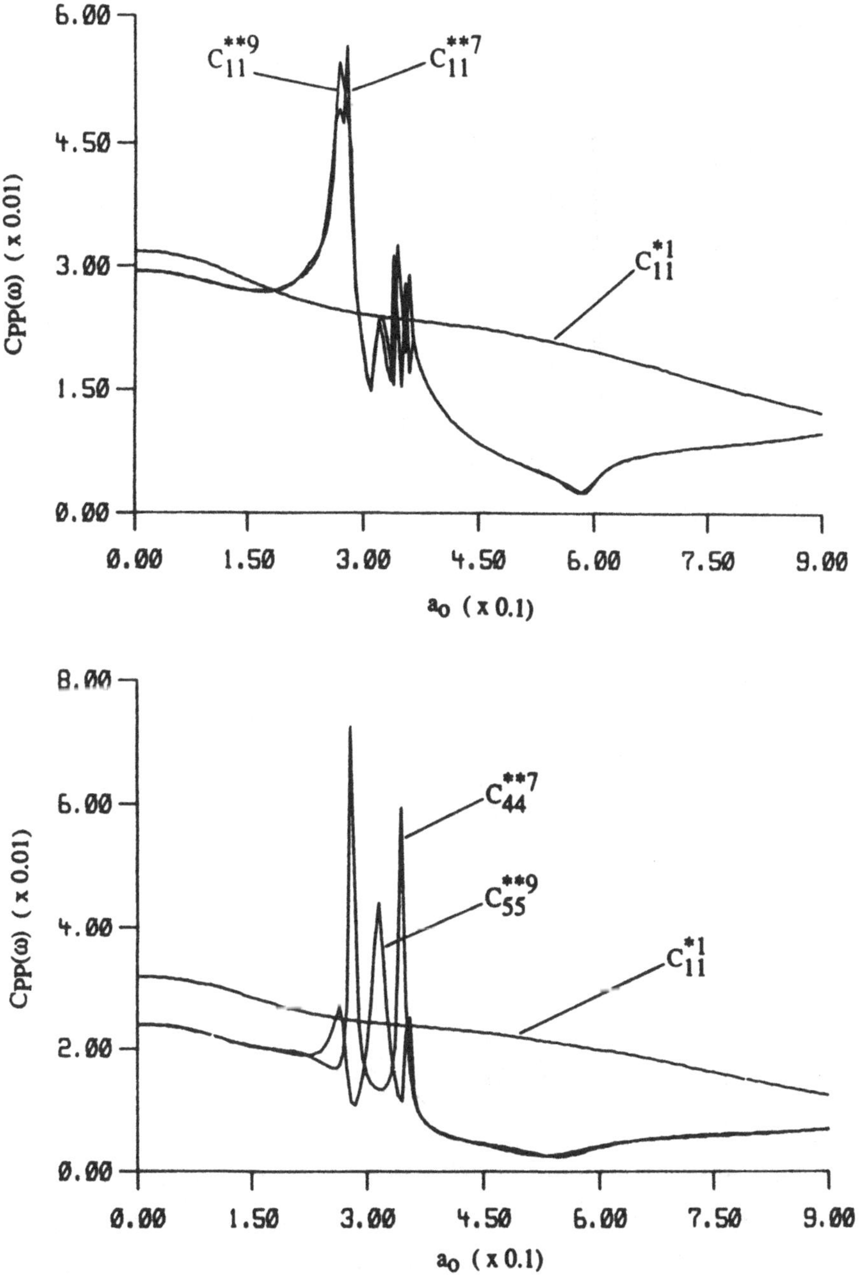

Figure 4. **Vertical compliances for a seven-tie and nine-tie system with the load applied on: (a) the end tie, and (b) the middle tie.**

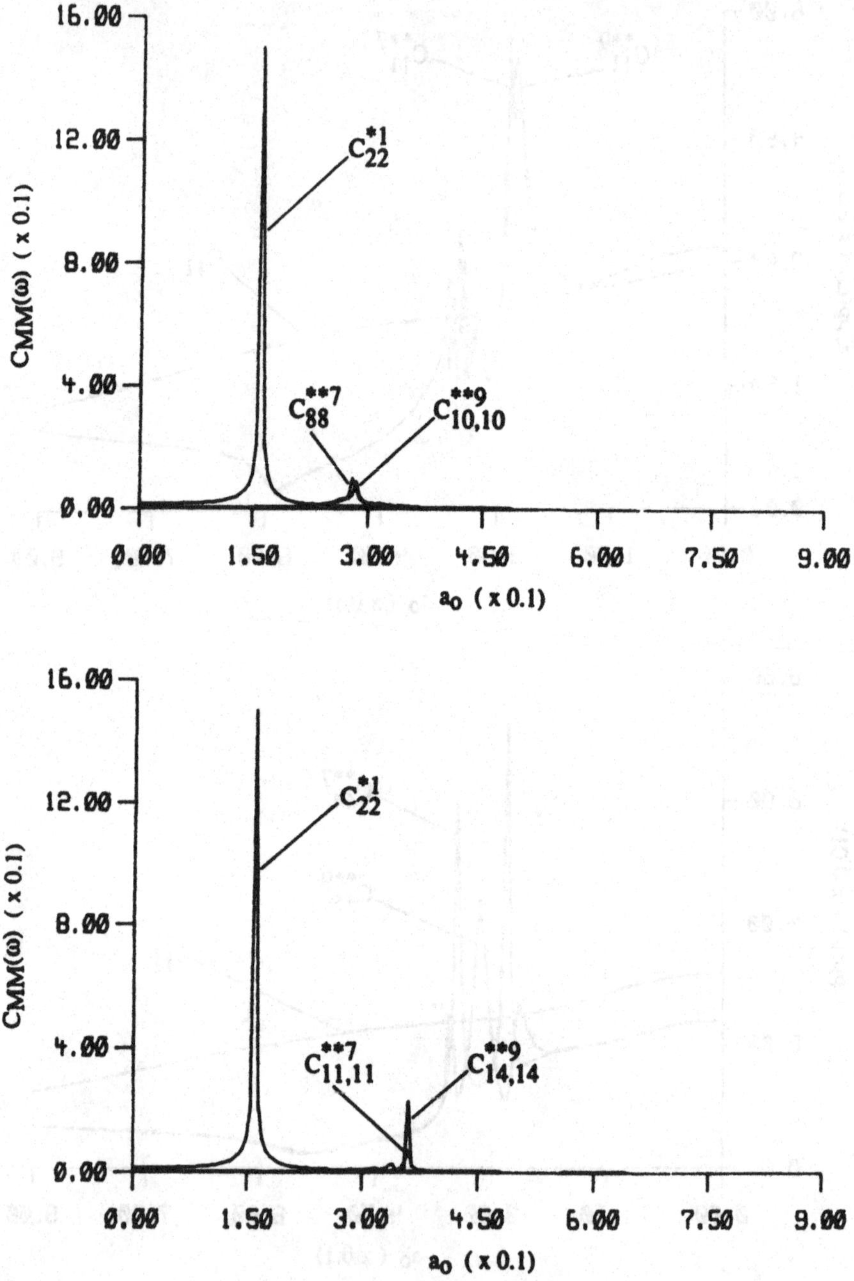

Figure 5. Rocking compliances for a seven-tie and nine-tie system with the load applied on: (a) the end tie, and (b) the middle tie.

Seismic Waves in an Impacted Half-Space: Time Domain BEM Versus the Method of Generalized Ray

H. Antes (*), P. Borejko (**), F. Ziegler (**)

(*) Institute of Applied Mechanics, Technical University of Braunschweig, D-3300 Braunschweig, Germany

(**) Civil Engineering Department, Technical University of Vienna, A-1040 Vienna, Austria

ABSTRACT

The general two-dimensional time-dependent BEM formulation accounting for the effect of initial conditions and body forces is applied to the problem of a transient line load acting normal to the surface of an elastic half-space. The complete time records of the horizontal and vertical displacements at two different receivers are numerically evaluated and compared with those obtained by the numerical evaluation of the exact solution to this problem provided by the method of generalized ray. The agreement is excellent showing the validity as well as the quality of the present numerical implementation of the time-dependent BEM.

INTRODUCTION

The new, rapidly expanding field of application of the Boundary Element Method (BEM) comprises various problems of wave propagation (for a review article see Shaw [1]).

First studies in this field were based on frequency domain formulations of the BEM. The pioneers were Cruse and Rizzo [2, 3] who have used the BEM in conjunction with the Laplace transform to study plane wave propagation and stress concentration problems. Their work has been extended and improved by Manolis and Beskos [4]. The frequency domain formulation of the BEM has also been used by Banaugh and Goldsmith [5] and Niwa et al. [6] in the study of steady-state wave propagation problems.

Recently the time domain formulations of the BEM have been applied to transient wave propagation problems. Groenenboom [7] has considered an unsteady potential flow, Mansur and Brebbia [8, 9] have studied scalar wave propagation in two and three dimensions, while elastodynamic problems with zero initial conditions and zero body forces have been examined by Mansur [10] and Mansur and Brebbia [11]. Cole et al. [12] have studied two-dimensional scalar anti-plane strain problems. Niwa et al. [13] and Manolis [14] have investigated two-dimensional wave scattering problems and Rice and Sadd [15] have considered an anti-plane strain wave scattering problem. Spyrakos and Beskos [16] have studied transient responses of rigid strip foundations subject to external forces and/or obliquely incident seismic waves. Antes [17] has provided the BEM formulation applicable to general two-dimensional elastodynamic problems including the contribution of initial conditions and body forces. Karabalis and Beskos [18] have investigated dynamic responses of three-dimensional rigid surface foundations. The three-dimensional transient dynamic problems have also been studied by Banerjee and Ahmad [19], Banerjee et al. [20], Ahmad and Banerjee [21] and Ahmad [22]. The axisymmetric transient dynamic problems have been considered by Wang and Banerjee [23].

In the present paper the BEM formulation due to Antes [17] is applied to the two-dimensional problem of a transient line load acting normal to the surface of the elastic half-space. The time-dependent integral equations (see, e.g., Eringen and Suhubi [24]) appearing in this formulation have strongly singular kernels. These equations are transformed through partial integrations into equations with kernels which are only logarithmically singular and can be integrated as Cauchy principal values. The complete theoretical seismograms (i.e., time records) for the horizontal and vertical components of the displacement vector at different receiver locations are obtained. These seismograms are then compared with those found by the numerical evaluation of the exact analytical solution to this particular problem provided by the method of generalized ray [25, 26]. The agreement is excellent which proves the validity as well as the quality of the present numerical implementation of the time domain formulation of the BEM.

TIME DOMAIN BEM: FORMULATION AND NUMERICAL IMPLEMENTATION

The component form of the equations governing in-plane motions of a homogeneous, isotropic, linear elastic medium is

$$(c_1^2 - c_2^2)\, u_{i,ij} - c_2^2\, u_{j,ii} - \ddot{u}_j = -\rho^{-1}\, b_j\,, \qquad i, j = 1, 2, \tag{1}$$

(cf. Eringen and Suhubi [24]). Here u_i are components of the (small) displacement vector **u**, the comma notation signifies partial differentiation with respect to reference coordinates x_i, a dot signifies material time differentiation, b_j are components of the body force **b** per unit mass and ρ is the mass density. The velocities c_1 and c_2 of the longitudinal (P) and shear (S) waves, respectively, for the case of plane strain, are given by

$$c_1 = [(\lambda + 2\mu)/\rho]^{1/2}\,, \qquad c_2 = (\mu/\rho)^{1/2}\,, \tag{2}$$

where λ and μ are Lamé constants. For the case of plane stress, λ has to be replaced by $\lambda' = 2\lambda\mu/(\lambda + 2\mu)$. Equations (1) are subject to initial and boundary conditions of the form

$$\left.\begin{array}{ll} u_i(\mathbf{x},t)=\bar{u}_{io}(\mathbf{x}), \quad \dot{u}_i(\mathbf{x},t)=\bar{v}_{io}(\mathbf{x}) & \text{for } t=t_o \text{ in } \Omega+\Gamma, \\ u_i(\mathbf{x},t)=\bar{u}_i(\mathbf{x},t) & \text{for } t>t_o \text{ on } \Gamma_1, \\ T_i(\mathbf{x},t)=\rho[\delta_{ik}(c_1^2-2c_2^2)u_{j,j}+c_2^2(u_{i,k}+u_{k,i})]n_k=\bar{T}_i(\mathbf{x},t) & \text{for } t>t_o \text{ on } \Gamma_2, \end{array}\right\} \tag{3}$$

where the overbar designates known quantities, $\Gamma=\Gamma_1+\Gamma_2$ and Ω denote, in turn, the boundary and the interior domain, $\bar{v}_{io}$ are components of the velocity $\mathbf{v}_o$, T_i are the components of the traction **T**, δ_{ik} is the Kronecker delta and n_k are components of the unit normal **n** to the boundary Γ_2.

The fundamental solution of equations (1), i.e., the displacement response at point **x** (receiver) of an infinite medium to a unit impulse force uniformly distributed along the line perpendicular to the plane Ω acting at point $\boldsymbol{\xi}$ (source) at time τ is given by

$$\begin{aligned} U_j^{(i)}(x,\xi;t') &= (1/2\pi\rho)\Big\{c_1^{-1}r^{-2}H(c_1t'-r)\big[R_1^{-1}(2c_1^2t'^2-r^2)r_{,i}r_{,j}-R_1\delta_{ij}\big] \\ &\quad -c_2^{-1}r^{-2}H(c_2t'-r)\big[R_2^{-1}(2c_2^2t'^2-r^2)r_{,i}r_{,j}-(R_2+R_2^{-1}r^2)\delta_{ij}\big]\Big\} \\ &= H(c_1t'-r)\overset{1}{g}{}_j^{(i)}+H(c_2t'-r)\overset{2}{g}{}_j^{(i)} \\ &= \overset{1}{U}{}_j^{(i)}+\overset{2}{U}{}_j^{(i)}, \end{aligned} \tag{4}$$

(see Antes [17]) where $R_\alpha=(c_\alpha^2t'^2-r^2)^{1/2}$ with $\alpha=1,\ 2,\ t'=t-\tau$, r denotes the distance between the source and the receiver, i.e., $r=|x-\xi|$, and H is the Heaviside step function. Note that the fundamental solution obeys the conditions of reciprocity, time translation and causality [8]. On using weighted residuals in time and space, one can obtain the following boundary integral equation ($\xi\in\Omega+\Gamma$)

$$\begin{aligned} c_{ij}u_j(\xi,t) &= \int_o^{t^+}\left[\oint_\Gamma(U_j^{(i)}T_j-T_j^{(i)}u_j)d\Gamma+\int_\Omega U_j^{(i)}b_jd\Omega\right]d\tau \\ &\quad +\rho\int_\Omega\left\langle\bar{v}_{jo}(U_j^{(i)})_o-\bar{u}_{jo}(\dot{U}_j^{(i)})_o\right\rangle d\Omega, \end{aligned} \tag{5}$$

replacing equations (1) and (3) (for details see Antes [17]). Here $c_{ij}=\delta_{ij}/2$ on a smooth surface ($\xi\in\Gamma$), $c_{ij}=\delta_{ij}$ for interior points ($\xi\in\Omega$) and the upper limit t^+ pertains to $t+\varepsilon$ with ε being arbitrarily small. In equations (5) the terms $T_j^{(i)}u_j$ and $(\dot{U}_j^{(i)})_o$ contain singularities. On integrating by parts with respect to time and applying a transformation in polar coordinates, Antes [17] derived from equations (5) the time-dependent boundary integral equation accounting for the effect of initial conditions and body forces which is suitable for the numerical integration.

Next, the spatial and time variation of the unknown displacements and tractions are approximated by the set of interpolation functions

$$\left.\begin{aligned} u_j(r,\tau) &= \sum_p\sum_m \varphi^p(r)\eta_m(\tau)u_{jp}^m, \\ t_j(r,\tau) &= \sum_p\sum_m \psi^p(r)\mu_m(\tau)T_{jp}^m, \end{aligned}\right\} \tag{6}$$

where p and m pertain to space and time, respectively, the interpolation functions $\varphi^p(r)$, $\psi^p(r)$ and $\mu_m(\tau)$ are assumed to be piecewise constant and $\eta_m(\tau)$ piecewise linear. Then, according to the adapted approximations ($N:=T_n$, $S:=T_s$), the obtained boundary integral equation is put into the form convenient for the numerical computations. The two-dimensional kernels that appear in the resultant equations are listed in reference [17]. Finally, these equations are incorporated in a time stepping algorithm together with appropriate boundary and equilibrium conditions to provide numerical solutions of two-dimensional elastodynamic problems.

METHOD OF GENERALIZED RAY

With the use of the method of generalized ray [25] one can obtain the exact solution to the plane strain problem of a transient line load acting normal to the surface of a homogeneous, isotropic, linear elastic half-space (see Fig. 1). For

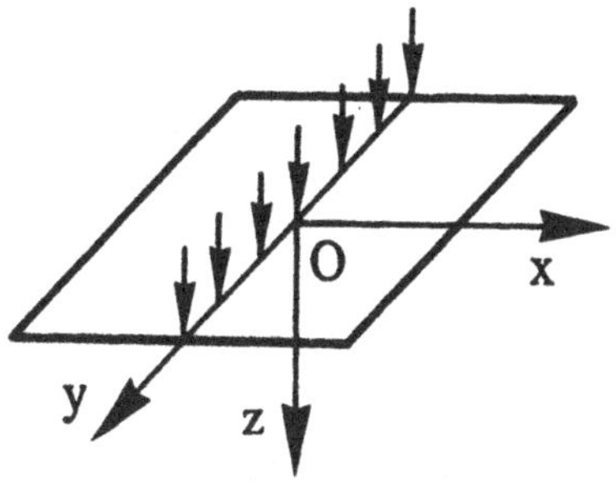

Figure 1. Transient line load acting normal to the surface of an elastic half-space.

this problem, the boundary conditions are $\sigma_{zz}=-f(t)\delta(x)$ and $\sigma_{zx}=0$ at $z=0$, where σ_{zz} and σ_{zx} are the components of the stress tensor $\boldsymbol{\sigma}$, $\delta(x)$ is the Dirac delta function and f(t) with properties $f(0^+)=0=\dot{f}(0^+)$ specifies the time behaviour of the line load. In particular, the horizontal u_x and vertical u_z displacement components at a surface receiver $\mathbf{x}=[x, 0, 0]$ are given by

$$u_i = u_{iP} + u_{iV} \qquad i = x, z,$$

where

$$\left.\begin{aligned} u_{x\alpha} &= AH(t-t_{A\alpha})\int_{t_{A\alpha}}^{t}\dot{f}(t-\tau)I_{x\alpha}(\tau)d\tau \\ &\text{and} \\ u_{z\alpha} &= -AH(t-t_{A\alpha})\int_{t_{A\alpha}}^{t}\dot{f}(t-\tau)I_{z\alpha}(\tau)d\tau, \qquad \alpha=P,SV. \end{aligned}\right\} \tag{7}$$

Here $A=(2\pi\mu)^{-1}$, $t_{A\alpha}$ is the arrival time at the receiver of the longitudinal (P) or the shear (S) wave, the upper limit t is the actual observation time and $I_{x\alpha}(\tau)$ and $I_{z\alpha}(\tau)$ are the ray integrals which are given by

$$\left.\begin{aligned} I_{x\alpha}(\tau) &= 2\mathrm{Im}\int_{0}^{\xi_1}[-iS_{\alpha}(\xi)D_{x\alpha}(\xi)]d\xi \\ &\text{and} \\ I_{z\alpha}(\tau) &= 2\mathrm{Re}\int_{0}^{\xi_1}S_{\alpha}(\xi)D_{z\alpha}(\xi)d\xi, \qquad \alpha=P,SV. \end{aligned}\right\} \tag{8}$$

where Re and Im denote, in turn, the real and imaginary parts, $i^2=-1$, the integration in the complex ξ-plane is along the Cagniard path, the source functions $S_{\alpha}(\xi)$ are defined by

$$S_P(\xi)=(\xi^2+\zeta^2)/\Delta_r(\xi) \qquad \text{and} \qquad S_{SV}(\xi)=(-i2\xi\eta)/\Delta_r(\xi), \tag{9}$$

(see Pao and Gajewski [25]) and the receiver functions $D_{x\alpha}(\xi)$ and $D_{z\alpha}(\xi)$ read

$$\left.\begin{aligned} D_{xP}(\xi)&=i\xi, \qquad & D_{xSV}(\xi)&=\zeta, \\ D_{zP}(\xi)&=-\eta, \qquad & D_{zSV}(\xi)&=i\xi. \end{aligned}\right\} \tag{10}$$

In equations (9) and (10) $\eta=(c_1^{-2}+\xi^2)^{1/2}$, $\zeta=(c_2^{-2}+\xi^2)^{1/2}$ and $\Delta_r(\xi)=4\eta\zeta\xi^2-(\xi^2+\zeta^2)^2$. The numerical evaluation of the ray integrals $I_{x\alpha}(\tau)$ and $I_{z\alpha}(\tau)$ is performed by using the Gaussian quadrature and the convolution integrals in equations (7) are calculated numerically by using Simpson`s rule (for details of the numerical implementation see Borejko and Ziegler [26]).

NUMERICAL RESULTS

For the numerical calculations the triangular time function of the source (see Fig. 8) is used·

$$f(t) = \begin{cases} 0 & t\le 0, \\ t & 0\le t\le\Delta, \\ -t+2\Delta & \Delta\le t\le 2\Delta, \\ 0 & t\ge 2\Delta, \end{cases}$$

where 2Δ is the duration of the load and it is taken as Δ=0.1. The Lamé constants of the half-space and its mass density are chosen as $\mu=\lambda=1.0$ and $\rho=1.0$. Hence, the P- and S-wave speeds are $c_1=\sqrt{3}$ and $c_2=1.0$, respectively, and the constant A in equations (7) is $A=(2\pi)^{-1}$. The characteristic time $t_o=2x_o/c_2$ is introduced , where the characteristic length x_o is selected to be $x_o=0.5$. The dimensionless time $\tau=t/t_o$ is plotted along the abscissae in Figs. 2 to 7. The rise time Δ of f(t) is also normalized by t_o. The graphs presented in Figs. 2 to 5 are calculated by menas of the time-dependent BEM formulation for three different discretizations D_1, D_2 and D_3 (see Table 1). The curves shown in Figs. 6 and 7 are evaluated by the method of generalized ray.

Table 1

Discretization	Symbol in Figs. 2 to 5	Number of elements per x_o	Δl [m]	Number of time-steps per Δ	Δt [s]
D_1	------	6	0.0833	2	0.05
D_2	– – –	12	0.0416	4	0.025
D_3	———	18	0.0278	6	0.0166

Figures 6 and 7 show the exact time records of $A^{-1}u_x$ and $A^{-1}u_z$ at two different receiver locations at the free surface of the elastic half-space (see also Borejko and Ziegler [27]). The surface response consists of the small (prograde linearly polarised) pulse followed by the large (retrograde elliptically polarised) pulse representing the fully developed Rayleigh wave. The time records of $A^{-1}u_x$ and $A^{-1}u_z$ presented in Figs. 2 to 5 for the discretizations D_2 and D_3 demonstrate an excellent agreement with those shown in Figs. 6 and 7. The records of $A^{-1}u_x$ and $A^{-1}u_z$ shown in Figs. 2 to 5 for the discretization D_1 have essentially smaller amplitudes than those shown in Figs. 6 and 7. This is due to the insufficient time discretization of f(t) which employs piecewise constant functions so that f(t) is approximated by step functions (see Fig. 8).

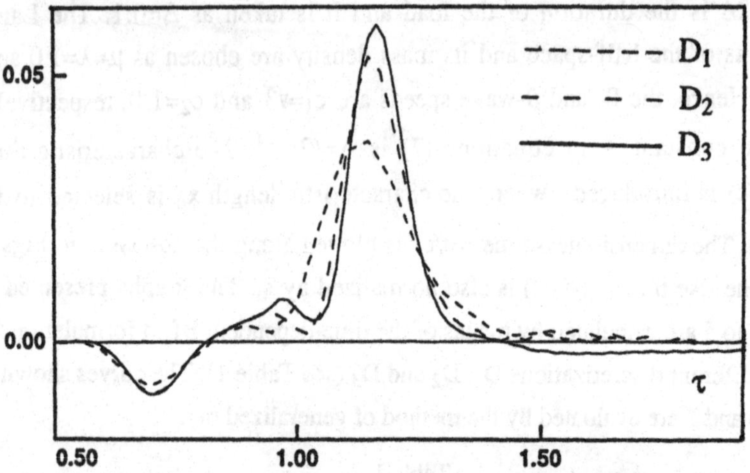

Figure 2. Time records of $A^{-1}u_x$ at the surface receiver $x=2x_o$ and $z=0$.

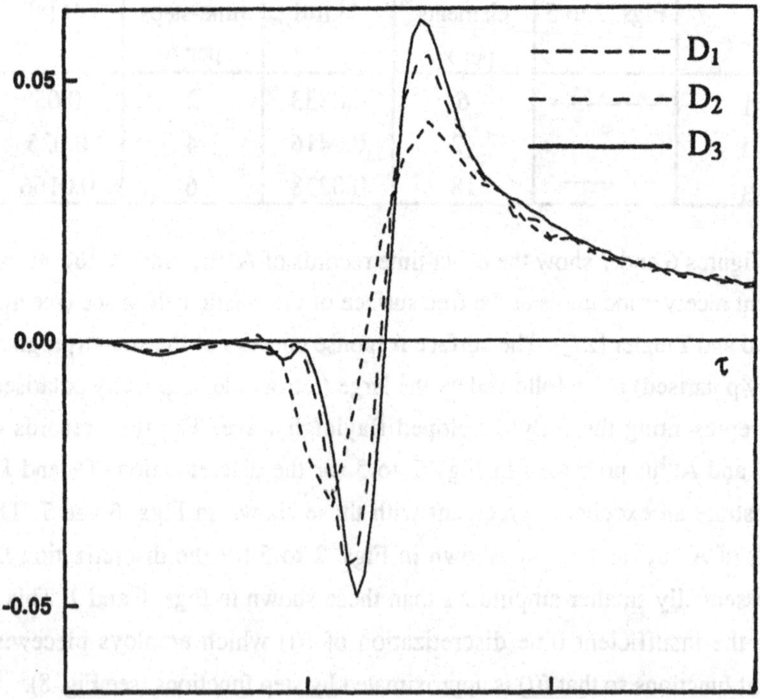

Figure 3. Time records of $A^{-1}u_z$ at the surface receiver $x=2x_o$ and $z=0$.

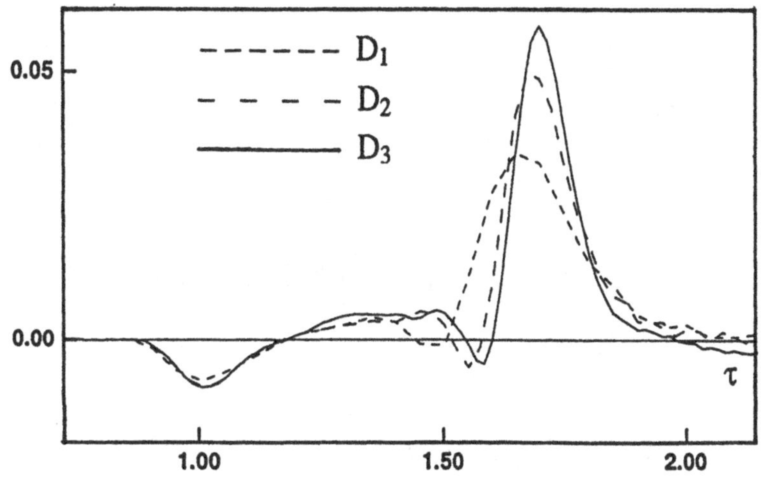

Figure 4. Time records of $A^{-1}u_x$ at the surface receiver $x=3x_o$ and $z=0$.

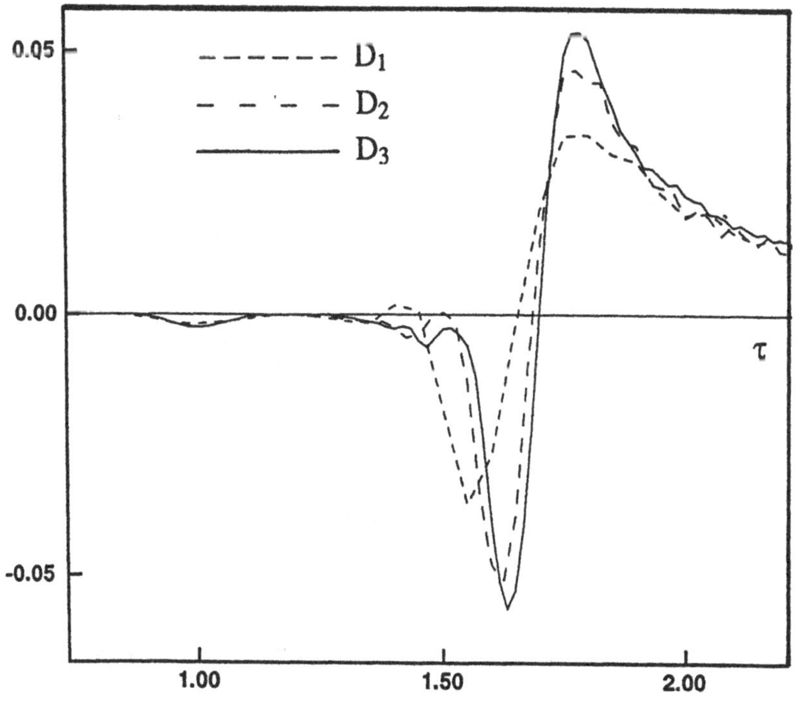

Figure 5. Time records of $A^{-1}u_z$ at the surface receiver $x=3x_o$ and $z=0$.

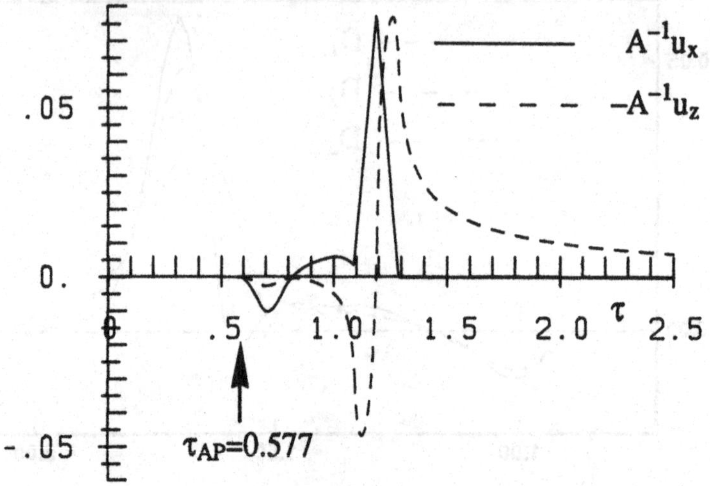

Figure 6. Time records of $A^{-1}u_x$ and $A^{-1}u_z$ at the surface receiver $x=2x_o$ and $z=0$ (τ_{AP} is the arrival time of the P-wave).

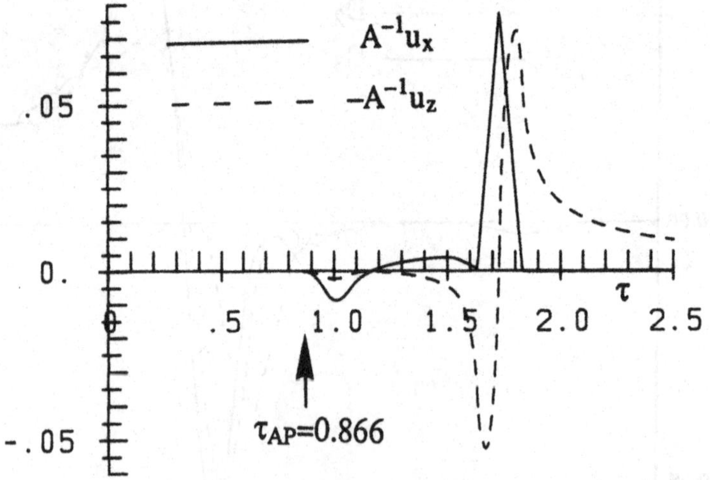

Figure 7. Time records of $A^{-1}u_x$ and $A^{-1}u_z$ at the surface receiver $x=3x_o$ and $z=0$ (τ_{AP} is the arrival time of the P-wave).

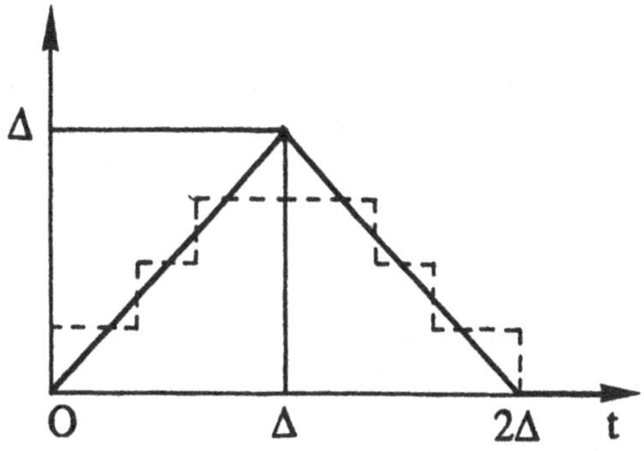

Figure 8. Piecewise constant function (– – –) approximating f(t) (——).

CONCLUSIONS

In this article, the general two-dimensional time-dependent BEM formulation accounting for the effect of initial conditions and body forces is presented and its numerical implementation is discussed. Then, it is applied to the problem of a transient line load acting normal to the surface of the elastic half-space. The complete time records of the horizontal and vertical displacements at two different surface receivers are numerically evaluated for three different discretizations. Finally, the resulting records are compared with those generated by the numerical evaluation of the exact solution to this particular problem provided by the method of generalized ray. The agreement is excellent which proves the validity as well as the quality of the present numerical implementation of the time-dependent BEM. Note that several previous implementations of the time domain BEM formulation for transient dynamic analysis [9, 12, 13, 15, 28 to 32] are known but, to our knowledge, only in the present paper and in reference [23] comparisons to exact solutions are given.

Acknowledgement P.Borejko acknowledges the financial support provided by the Austrian Science Foundation FWF under Project P6998-TEC.

REFERENCES

1. Shaw, R.P. Boundary Integral Equation Methods Applied to Wave Problems, in Developments in Boundary Element Methods (Eds. Banerjee, P.K. and Butterfield, R.), Vol.1, pp. 121-153, Applied Science Publishers, London, 1979.

2. Cruse, T.A. and Rizzo, F.J. A Direct Formulation and Numerical Solution of the General Transient Elastodynamic Problem - I, J. Math. Anal. Appl., Vol.22, pp. 244-259, 1968.

3. Cruse, T.A., A Direct Formulation and Numerical Solution of the General Transient Elastodynamic Problem - II, J. Math. Anal. Appl., Vol.22, pp. 341-355, 1968.

4. Manolis, G.D. and Beskos, D.E. Dynamic Stress Concentration Studies by Boundary Integrals and Laplace Transform, Int. J. Num. Meth. Engng., Vol.17, pp. 573-599, 1981.

5. Banaugh, R.P. and Goldsmith, W. Diffraction of Steady Elastic Waves by Surfaces of Arbitrary Shape, ASME J. Appl. Mech., Vol.30, pp. 589-597, 1963.

6. Niwa, Y., Kobayashi, S. and Fukui, T. Applications of Integral Equation Method to Some Geomechanical Problems, in Numerical Methods in Geomechanics (Ed. Desai, C.S.), pp. 120-131, American Society of Civil Engineers, New York, 1976.

7. Groenenboom, P.H.L. The Application of Boundary Elements to Steady and Unsteady Potential Fluid Flow Problems in Two and Three Dimensions, in Boundary Element Methods, Proc. 3rd Int. Seminar, Irvine (Ed. Brebbia, C.A.), pp. 37-52, Springer-Verlag, Berlin, 1981.

8. Mansur, W.J. and Brebbia, C.A. Formulation of the Boundary Element Method for Transient Problems Governed by the Scalar Wave Equation, Appl. Math. Modelling, Vol.6, pp. 307-311, 1982.

9. Mansur, W.J. and Brebbia, C.A. Numerical Implementation of the Boundary Element Method for Two-Dimensional Transient Scalar Wave Propagation Problems, Appl. Math. Modelling, Vol.6, pp. 299-306, 1982.

10. Mansur, W.J. A Time-Stepping Technique to Solve Wave Propagation Problems Using the Boundary Element Method, Ph.D. Thesis, University of Southampton, 1983.

11. Mansur, W.J. and Brebbia, C.A. Transient Elastodynamics Using a Time-Stepping Technique, in Boundary Elements (Eds. Brebbia, C.A., Futagami, T. and Tanaka, M.), pp. 677-698, Springer-Verlag, Berlin, 1983.

12. Cole, D.M., Kosloff, D.D. and Minster, J.B. A Numerical Boundary Integral Equation Method for Elastodynamics - I, Bull. Seismol. Soc. Am., Vol.68, pp. 1331-1357, 1978.

13. Niwa, Y., Fukui, T., Kato, S. and Fujiki, K. An Application of the Integral Equation Method to Two-Dimensional Elastodynamics, Theor. Appl. Mech., Vol.28, pp. 281-290, 1980, Univ of Tokyo Press.

14. Manolis, G.D. A Comparative Study on Three Boundary Element Method Approaches to Problems in Elastodynamics, Int. J. Num. Meth. Engng., Vol.19, pp. 73-91, 1983.

15. Rice, J.M. and Sadd, M.H. Propagation and Scattering of SH-Waves in Semi-Infinite Domain Using a Time-Dependent Boundary Element Method, ASME J. Appl. Mech., Vol.51, pp. 641-645, 1984.

16. Spyrakos, C.C. and Beskos, D.E. Dynamic Analysis of Embedded Rigid Strip Footings by Time Domain Boundary Element Method, in Boundary Elements VII (Eds. Brebbia, C.A. and Maier, G.), pp. 6.71-6.78, Springer-Verlag, Berlin, 1985.

17. Antes, H. A Boundary Element Procedure for Transient Wave Propagations in Two-Dimensional Isotropic Elastic Media, Finite Elements Anal. Des., Vol.1, pp. 313-322, 1985.

18. Karabalis, D.L. and Beskos, D.E. Dynamic Response of 3-D Rigid Surface Foundations by Time Domain Boundary Element Method, Earthquake Engng. Struct. Dyn., Vol.12, pp. 73-93, 1984.

19. Banerjee, P.K. and Ahmad, S. Advanced Three-Dimensional Dynamic Analysis by Boundary Element, in Proc. ASME Conf. on Advanced Topics in Boundary Element Analysis, Florida, Nov.1985, AMD, Vol.72, pp. 65-81, 1985.

20. Banerjee, P.K., Ahmad, S. and Manolis, G.D. A Time Domain BEM for Three-Dimensional Problems of Transient Elastodynamics, Earthquake Engng. Struct. Dyn., Vol.14, pp. 933-949, 1986.

21. Ahmad, S. and Banerjee, P.K. Time-Domain Transient Elastodynamic Analysis of 3-D Solids by BEM, Int. J. Num. Meth. Engng., Vol.26, pp. 1709-1728, 1988.

22. Ahmad, S. Linear and Nonlinear Dynamic Analysis by Boundary Element Method, Ph.D. Thesis, State University of New York at Buffalo, Buffalo, 1986.

23. Wang, H.-Ch. and Banerjee, P.K. Axisymmetric Transient Elastodynamic Analysis by Boundary Element Method, Int. J. Solids Structures, Vol.26, pp. 401-415, 1990.

24. Eringen, A.C. and Suhubi, E.S. Elastodynamics, Vol. II, Academic Press, New York, 1975.

25. Pao, Y.-H. and Gajewski, R.R. The Generalized Ray Theory and Transient Responses of Layered Elastic Solids, in Physical Acoustics (Eds. Mason, W.P. and Thurston, R.N.), Vol.13, pp. 183-265, Academic Press, New York, 1977.

26. Borejko, P. and Ziegler, F. Seismic Waves in Layered Soil: The Generalized Ray Theory, in Structural Dynamics - Recent Advances, Springer-Verlag, Berlin (in press).

27. Borejko, P. and Ziegler, F. Transient Elastic and Viscoelastic Responses of a Half-Space, in Proc. IUTAM-Symp. on Elastic Wave Propagation and Ultrasonic Nondestructive Evaluation (Eds. Datta, S.K., Achenbach, J.D. and Rajapakse, Y.S.), pp. 401-402, North Holland, Amsterdam, 1990.

28. Mansur, W.J. and Brebbia, C.A. Transient Elastodynamics, in Topics in Boundary Element Research (Ed. Brebbia, C.A.), Vol.2, Chap.5, pp. 124-155, Springer-Verlag, Berlin, 1985.

29. Spyrakos, C.C. and Beskos, D.E. Dynamic Response of Rigid Strip Foundations by Time Domain Boundary Element Method, Int. J. Num. Meth. Engng., Vol.23, pp. 1547-1565, 1986.

30. Spyrakos, C.C. and Antes, H. Time Domain Boundary Element Approaches in Elastodynamics: A Comparative Study, Comput. Struct., Vol.24, pp. 529-535, 1986.

31. Antes, H. and von Estorff, O. Dynamic Response Analysis of Rigid Foundations and of Elastic Structures by Boundary Element Procedures, Soil Dyn. Earthqu. Engng., Vol.8, pp. 68-74, 1989.

32. Antes, H. and Tröndle, G. Analysis of Stress Waves by Indirect BEM, in Boundary Elements in Mechanical and Electrical Engineering, Proc. Int. Boundary Element Symp., Nice, France (Ed. Brebbia, C.A. and Chaudouet-Miranda, A.), pp. 179-191, Springer-Verlag, Berlin, 1990.

Seismic Response of Soil-Structure-Interaction Problems Under Unilateral Contact Conditions

P.N. Patel (*), C.C. Spyrakos (**)
(*) Intergraph Corporation, Mail Stop LR24A1, One Madison Industrial Park, Hunstville, AL 35824, U.S.A.
(**) Department of Civil Engineering, West Virginia University, Morgantown, WV 26505, U.S.A.

ABSTRACT

Strong earthquake lateral forces can induce base overturning moments that exceed the available overturning resistance due to gravity loads causing uplift of the basemat. To address the problem of uplift, a time domain BEM-FEM methodology for a two-dimensional plane strain soil-structure interaction problem is presented. The Boundary Element Method (BEM) is applied to the soil medium and the Finite Element Method (FEM) is applied to the foundation and the super-structure. FEM interface elements are utilized to effectively simulate foundation uplift from the underlying soil medium. The two methods are combined through displacement compatibility and force equilibrium considerations at the soil-foundation interface. A representative problem of a nuclear containment structure subjected to the El Centro earthquake of 1940 is analyzed using the rigorous BEM-FEM. The results indicate that the base shear at the foundation-structure interface can cause malevolent or benevolent effects depending on various structural and soil parameters. Further, it is concluded that the effects of uplift on the structure response can not be predicted from the linear behavior under bilateral contact conditions.

INTRODUCTION

Strong shaking of structures during earthquakes may result in a partial separation of the foundation basemat from the soil. The evidence of foundation uplift has been recorded by Hanson[1] after the Alaskan earthquake of March 1964, where ice was discovered under some oil tanks. Housner[2] reported the stretching of anchor bolts of a number of tall petroleum towers during the Arvin-Tehachapi earthquake, suggesting uplift of the base of the tanks from their foundations. A large number of researchers have successfully attempted to study the separation phenomena in soil-foundation interaction. A thorough literature review on the various techniques employed to model and analyze soil-foundation separation appears in References [3] and [4].

The approach considered in this work extends the hybrid time domain BEM-FEM methodology developed by Spyrakos and Beskos [5] and Karabalis and Beskos [6], who studied the dynamic response of two- and three-dimensional foundations in complete bond with the soil. In their works the BEM-FEM formulation employed the BEM to model semi-infinite soil media and the Finite Element Method (FEM) to model the finite domain of the structure and the foundation. The BEM is particularly well suited to model soil domains because of its ability to automatically account for the radiation conditions at infinity, and reduces the spatial dimensions of the problem by one [7,8]. The expressions presented in this work allows the treatment of elastodynamic problems involving partial loss of contact between elastic bodies as is exemplified through a representative Soil Structure Interaction (SSI) problem. Further, the methodology incorporates FEM interface elements to model the foundation-soil contact area to investigate the effects of the nonlinearities arising from the soil-foundation separation. The BEM-FEM model is employed to determine the response of a representative nuclear containment building subjected to El Centro earthquake of 1940 under unilateral contact (uplift permitted) with the soil media. A parametric study is conducted for the parameters that mostly characterize structural behavior, i.e., mass, stiffness, and height of the foundation-structure system.

BEM-FEM FORMULATION

BEM Formulation:

Under the assumption of small displacement theory and homogeneous, isotropic, linear elastic material behavior, the elastodynamic displacement field of a body under conditions of plane strain is governed by Navier's equation. For zero body forces and zero initial conditions, the initial-boundary value problem defined by Navier's equation and the associated boundary and initial conditions can be expressed by the integral equations [9]

$$\frac{1}{2}u_j(\xi,t) = \oint_T \left(U_j^{(i)} * t_{(\eta)j} - T_j^{(i)} * u_j\right) dT \tag{1}$$

where $t_{(\eta)j}$ are the surface tractions, the operation * denotes time convolution and the fundamental tractions $T_j^{(i)}$, corresponding to the fundamental displacements

$$U_j^{(i)}(x,\xi;t') = \frac{1}{2\pi\rho}\Big\{\frac{1}{c_1}\frac{H(c_1t'-r)}{2}\left[\frac{2c_1^2t'^2-r^2}{R_1}r_{,i}r_{,j} - R_1\delta_{ij}\right] - \frac{1}{c_2}\frac{H(c_2t'-r)}{r^2}\times$$

$$\times\left[\frac{2c_2^2t'^2-r^2}{R_2}r_{,i}r_{,j} - \left(R_2+\frac{r^2}{R_2}\right)\delta_{ij}\right]\Big\} \tag{2}$$

$$R_\alpha = \sqrt{(c_\alpha^2t'^2-r^2)}, \quad \alpha = 1,2 \tag{3}$$

$$t' = t-\tau. \tag{4}$$

are given in Patel [4]. In the fundamental solution, H denotes the Heaviside function, and r is the distance $|x-\xi|$ between a field point x and a source point ξ.

Equation (1) can be applied to obtain the displacement response at the boundary of an elastic body. Once the boundary value problem is solved, the displacement and traction fields in the interior of the domain Ω can be easily evaluated in terms of the response obtained on the surface19,20. A solution of the integral equation (1) in closed form for general boundary conditions is not possible; thus, resort is made to a numerical solution. The numerical treatment is accomplished through spatial and time interpolations of the displacement and traction fields. More specifically, the boundary T is discretized into Q line elements, while the time variation of the displacements and tractions over each element are assumed to be constant during each time step. The displacement at the center of the element p at the time step N is given by the following discretized form of equation (1) [5]:

$$c_{ij}u^{Np} = \sum_{q=1}^{Q}\sum_{n=1}^{N}\left(\left[\int_{\Delta s} G_{ij}^{nq}ds\right]\{t^{N-l+1}\} - \left[\int_{\Delta s} F_{ij}^{nq}ds\right]\{u^{N-l+1}\}\right), \quad (5)$$

where $l=n-m+1$, G_{ij} and F_{ij} are the discretized fundamental displacements and tractions, respectively. In evaluating the line integrals of the fundamental solution pair, singularities appear when p is equal to q in equation (5) for every time step N. A detailed evaluation of these singularities is presented in Reference [4]. In the next sub-section, the expressions resulting from the application of equation (5) to every boundary element is combined through appropriate compatibility considerations with the FEM interface element equations to perform an iterative numerical treatment of the nonlinearities due to the soil-foundation uplift.

FEM Formulation:

The foundation and structure are modeled with the aid of FEM. The four node plane strain bilinear rectangular elements with two degrees of freedom per node shown in Figure 1 are used in this study. Through standard FEM procedures, the following equations of motion can be obtained by an assemblage of the individual finite elements:

$$[M]\{\ddot{\delta}\} + [C]\{\dot{\delta}\} + [K]\{\delta\} = \{f(t)\}, \quad (6)$$

where δ indicates nodal displacements. [M], [C] and [K] are the consistent mass, damping, and stiffness matrices of the foundation-structure system, respectively, and {f(t)} is the nodal force vector.

The geometric nonlinearities considered herein result from the time dependent nature of the contact area between the soil and the foundation when uplift occurs. To effectively treat these nonlinearities, thin layer four node bilinear rectangular interface elements are used to discretize the soil-foundation interface. The three modes of deformations that accurately simulate the uplift of the foundation basemat are the stick, debonding and rebonding modes. An interface element is in the stick mode when there is no relative motion between the adjoining bodies and no tensile stresses are developed due to the external disturbances.

Separation or debonding takes place when the bodies open up due to constraints of unilateral contact conditions, prohibiting the development of tensile stresses as they are incompatible with constitutive properties of the geologic materials. If the interface element in the debonding mode returns to the stick mode during subsequent loading, rebonding takes place. Interface elements similar to the one described above have been successfully used to solve static as well as dynamic two-dimensional problems, where all domains have been discretized with the aid of FEM. In this study, the equations of the interface elements are derived separately and then added to those of the foundation-structure system prior to establishing the compatibility and equilibrium criteria with soil BEM modeling.

The aspect in which the interface elements differ from regular elements, as the ones used to model the super-structure is their thickness ratio,

$$t_r = \frac{t_{int}}{t_{neigh}}, \tag{7}$$

where subscripts int and neigh pertains to the interface and the neighboring regular FEM element, respectively. The small value of thickness of the interface element is liable to cause numerical problems; however, Pande and Sharma [11] suggests guidelines that can be followed to circumvent the numerical problems. In this study, a thickness ratio of 0.1 has lead to convergent results.

The stick mode of deformation is treated like any other regular FEM element with material pròperties which are identical to those of the underlying soil media. The additional stiffness and inertia introduced in the system are negligible considering the semi-infinite soil media laying underneath the interface elements. The Young's Modulus of elasticity is reduced for the interface element in the debonding mode by

$$[D]_{debonding} = 0.001 \times [D]_{stick}, \tag{8}$$

where [D]stick is the material property matrix in the stick mode. This, in essence, creates a void element with very little stiffness. In the rebonding mode, the material properties corresponding to the soil media are reassigned, which brings back the interface element to the stick mode. More information on the complete implementation of the interface elements in the BEM-FEM methodology can be found in Reference [4].

Time Stepping Algorithm

The two approaches, BEM and FEM, are coupled appropriately by displacement compatibility and force equilibrium considerations at the soil-foundation interface. The displacements of the BEM soil elements are evaluated at the midpoint, while the FEM displacements correspond to the FEM nodes as shown in Figure 1. In order to introduce compatibility between the deformations of the interface element nodes and the soil motion, the average displacement of FEM node q is approximated by the mean value of the nodal displacements at the ends of the BEM element p. Similarly, compatibility of forces can be established if each contact force fq applied at node q is approximated by the mean value of the two resultant forces P associated with the contact stresses that develop over two successive elements joined at their common node. Thus, for the whole interface region, the compatibility relationships can be expressed as

$$\{\delta\} = [T]\{u\} \tag{9}$$

$$\{F\} = [T]\{f\} \tag{10}$$

where [T] is the transformation matrix composed of zero and 1/2 entries. In view of equations (9) and (10), equation (5) of BEM formulation can be expressed as:

$$[T]^T[\bar{B}] = [T]^T[B][T]\{\delta\}, \tag{11}$$

where

$$[B] = \frac{l}{2}[G^1]^{-1} \tag{12}$$

and

$$[\bar{B}] = l[G^1]^{-1}\left([G^2]\{t^{N-1}\} + [G^3]\{t^{N-2}\} + \ldots + [G^N]\{t^1\}\right) -$$

$$-\left([H^2]\{u^{N-1}\} + [H^3]\{u^{N-2}\} + \ldots + [H^N]\{u^1\}\right), \tag{13}$$

in which superscripts denote the time steps at which the quantities are evaluated and l represents the length of the element. It should be noted that equation (13) represents the time convolution process stipulated in equation (5), and indicates that the matrix $[\bar{B}]$ depends on the complete time history prior to the time step at which it is evaluated. Combining the BEM and FEM formulations results in the set of non-linear equations governing the response of the soil-structure system:

$$[M_t]\{\ddot{\delta}\} + [C_t]\{\dot{\delta}\} + [\bar{K}_t]\{\delta\} = \{F(t)\} \tag{14}$$

where the subscript t denotes the time dependent nature of the matrices arising from the changing contact area at the soil-foundation interface; [Mt] and [Ct] are the mass and damping matrices, respectively, $[\bar{K}_t]$ is the equivalent stiffness matrix given by

$$[\bar{K}_t] = \begin{bmatrix} [K] + [T]^T[B_{cc}][T] & [T]^T[B_{ce}][T] \\ [T]^T[B_{ec}][T] & [T]^T[B_{ee}][T] \end{bmatrix} \tag{15}$$

and the time dependent forcing function F(t) can be evaluated from

$$F(t) = \begin{bmatrix} [T]^T[\bar{B}_{cc}][T] + [K] & [T]^T[\bar{B}_{ce}] \\ [T]^T[\bar{B}_{ec}][T] & [\bar{B}_{ee}] \end{bmatrix} \begin{Bmatrix} \{\delta_c^t\} \\ \{u_e^t\} \end{Bmatrix} \tag{16}$$

where the superscript f denotes the free-field, and subscripts c and e represent the degrees of freedom corresponding to the foundation-soil contact area and the external discretization on either side of the contact area, respectively. The nonlinear system of equations is solved with the aid of the direct integration iterative scheme. Initially the foundation and the soil are assumed to be in complete bond. For each time step an iterative procedure is employed to determine the size of the correct contact area. More specifically, if at the end of a given iteration the contact area is not the same as the one assumed at the beginning of the iteration, a new contact area is computed based on the displacements found in the current iteration. The procedure is repeated until convergence that renders the contact area.

NUMERICAL EXAMPLE

The time domain BEM-FEM methodology and the approximate models are employed to obtain the response of a nuclear containment structure subjected to the horizontal component of the El Centro earthquake of 1940. The containment structure has been studied by Weidlinger Associates (1982) in their study for EPRI, in which they conducted an experimental and analytical FEM Soil-Island analyses for blast loading arising from buried explosives. The soil is characterized by a shear wave velocity of 650 ft./sec and a pressure wave velocity of 1300 ft./sec, while a modulus of elasticity 106560369 lb/ft./ft., Poisson's ratio 1/3 and mass density 3.04348 lb-sec-sec/ ft./ft. are taken to be material properties of the foundation elements. The super-structure is characterized by a modulus of elasticity E 106577648 lb/ft./ft., Poisson ratio 1/3 and mass density 0.59006 lb-sec-sec/ft./ft. These values are varied by 50 percent more and less for parametric study. The equivalent two-dimensional plane strain model and its BEM-FEM discretization are shown in Figure 1. The soil surface is discretized into 10 BEM elements at the soil-foundation contact area and 5 BEM elements on either side of the foundation. This discretization has been selected so that the faster c1 wave travels to half the length of the BEM element during a time step, a requirement necessary to secure acceptable solution accuracy30. In order to further enhance the solution accuracy, the BEM kernel matrices [G] and [H] have been evaluated on the basis of a further discretization of the twenty elements into five subelements per element. The structure-foundation system is discretized with the aid of FEM into 13 layers of 10 elements across the length of the foundation, hence the soil-foundation interface is also discretized with the aid of 10 interface elements as shown in Figure 6. No significant differences in the response time history were observed between the above discretization and a discretization consisting of 15 interface elements with a compatible super-structure FEM mesh, establishing a high level of confidence in the ability of the discretization to accurately predict the correct contact area at the soil-foundation interface.

The horizontal relative response amplitudes at point A are presented in Table 1 for the case of complete bond between the soil and foundation. The results are compared with those obtained by approximate models in time domain and frequency domain [4]. Clearly the approximate methods do not necessarily lead to conservative results.

The absolute horizontal displacement at point A is shown in Figure 2 for the first six seconds of the earthquake excitation, as it pertains to the strong motion portion of the El Centro earthquake time history. The time domain approximate model appears to give higher response amplitudes than that of the BEM-FEM under unilateral contact conditions. These differences are expected as the contact area in the approximate model is always overestimated. Also, significant differences are observed between the BEM-FEM response amplitudes due to bilateral and unilateral contact conditions. These differences are better explained from the results shown in Figure 3, where the relative displacements at point B are plotted, which shows that the unilateral contact conditions leads to higher amplitudes initially; however, the bilateral contact conditions reveals greater amplitudes at later time steps.

Under bilateral contact conditions the vertical displacement response at point A is shown in Figure 3 for various values of mass. The displacement amplitude is found to behave non-monotonically with an increase in the value of the mass parameter. Shown in Figure 4 is the vertical displacement response due to unilateral contact conditions at point A for various values of the mass parameter, a behavior that significantly differs from the one observed under bilateral contact conditions. It is observed that as the value of the mass increases, the displacement also increases monotonically. This behavior clearly indicates that one can not draw conclusions for the non-linear behavior under unilateral contact conditions based on the structural behavior under bilateral contact conditions.

A critical parameter that controls a seismic design is the base shear that develops at the foundation-structure interface. The base shear responses for bilateral and unilateral contact conditions are presented in Figures 5 and 6, and the base shear amplitudes are summarized in Table 2 for all parameter variation. The base shear amplitudes are found to behave non-monotonically with increasing mass for bilateral contact conditions, while their variation is monotonic under unilateral contact conditions. Further, the base shear amplitude is found to substantially increase for certain values of the parameters under unilateral contact conditions. This behavior is in agreement with the observations reported by Yim and Chopra [3] for structures with low slenderness ratios.

Table 1: Comparison of the horizontal relative response amplitudes.

Mass	Method 1	Method 2	BEM-FEM
50%	1.24in	1.55in	1.76in
100%	1.56in	2.32in	1.86in
150%	2.32in	2.99in	1.26in

Method 1: Approximate Time Domain Reference [3].
Method 2: Approximate Frequency Domain Reference [3].

CONCLUSIONS

Assuming that only the normal stresses in compression can occur in the area of contact between the foundation and the soil, a hybrid time domain BEM-FEM methodology is developed for the analysis of soil-structure interaction problems in the case of plane strain. The boundary element method is applied to the soil, and the finite element method is applied to the foundation-structure system. FEM interface elements are used to simulate the basemat lift-off during seismic excitation. The methodology leads to non-linear equations of motion which are solved iteratively utilizing a linear acceleration scheme.

Table 2: Base shear for various parameters-bilteral and unilateral contact conditions.

Constant parameter	Variable parameter	Base shear(1b) Bilateral contact	Unilateral contact
150%E	50%M	27310	14090
100%H	100%M	28970	20130
	150%M	17090	31210
100%M	50%E	25170	31330
100%H	100%E	26720	27630
	150%E	28970	20130
150%E	50%H	23380	45164
100%M	100%H	28970	20130
	150%H	30550	28420

Utilizing the BEM-FEM approach, a parametric study is presented for the parameters that characterize the structural behavior. The study concludes that for a specific ground excitation, uplift effects on the system response may be malevolent or benevolent depending on the structural parameters. In addition, the effects of uplift on the system response can not be predicted from the linear behavior under bilateral contact conditions.

REFERENCES

1. Hanson, R.D., "Behavior of Liquid-Storage Tanks," The Great Alaska Earthquake of 1964, Engineering, National Academy of Sciences, Washington, D.C., 1973, p.334.

2. Housner, G.W., "The Behavior of Inverted Pendulum Structures During Earthquakes," Bulletin of the Seismological Society of America, Vol. 53, No. 2, Feb. 1963, pp. 403-417.

3. Yim, C.S. and Chopra, A.K., "Earthquake Response of Structures with Partial Uplift on Winkler Foundation," Earthquake Engineering and Structural Dynamics, Vol. 12, pp. 263-281, 1984.

4. Patel, P.N., "Localized Non-Linearities due to Soil Foundation Separation in Dynamic Soil-Structure Interaction," dissertation submitted to the college of engineering of West Virginia University, Morgantown, WV, 1989.

5. Spyrakos, C.C. and Beskos, D.E., "Dynamic Response of Flexible Strip-Foundation by Boundary and Finite Elements," Soil Dynamics and Earthquake Engineering, 1986, Vol. 5, No.2, pp. 84-96.

6. Karabalis, D.L. and Beskos, D.E., "Dynamic Response of 3-D Flexible Foundations by Time Domain BEM and FEM," Soil Dynamics and Earthquake Engineering, 1985, Vol.4, No.2, pp. 91-101.

7. Banerjee, P.K. and Butterfield, R., "Boundary Elements Methods in Engineering Science," McGraw-Hill Book Company (UK) Limited, 1981.

8. Brebbia, C.A., Telles, J.C. and Wrobel, L.C., "Boundary Element Techniques, Theory and Applications in Engineering," Springer Verlag, Berlin, 1984.

9. Erigen, A.C. and Suhubi, E.S., "Elstodynamics Vol. II Linear Theory," Academic Press, N.Y., 1975.

10. Zaman, M.M., Desai, C.S. and Drumm, E.C., "Interface Model for Dynamic Soil-Structure Interaction," Journal of Geotechnical Engineering (ASCE), Vol. 110, pp. 1257-1273, 1984.

11. Pande, G. N. and Sharma, K.G., "On Joint/Interface Elements and Associated Problems of Numerical Ill-Conditioning," Short Comm., International Journal for Numerical and Analytical Methods in Geomechanics, Vol. 3, 1979, pp. 451-457.

12. Wilson, E.L., Farhoomand I. and Bathe, K-J., "Nonlinear Dynamic Analysis of Complex Structures," Earthquake Engineering and Structural Dynamics, Vol. 1, pp. 241-252, 1973.

13. Wolf, J.P., "Dynamic Soil-Structure Interaction," Prentice-Hall, Inc., Englewood Cliffs, New Jersey, 1985.

14. Gazetas, G., "Analysis of Machine Foundation Vibrations: State-of-the-Art," Soil Dynamics and Earthquake Engineering, Vol. 2, pp. 1-42, 1983.

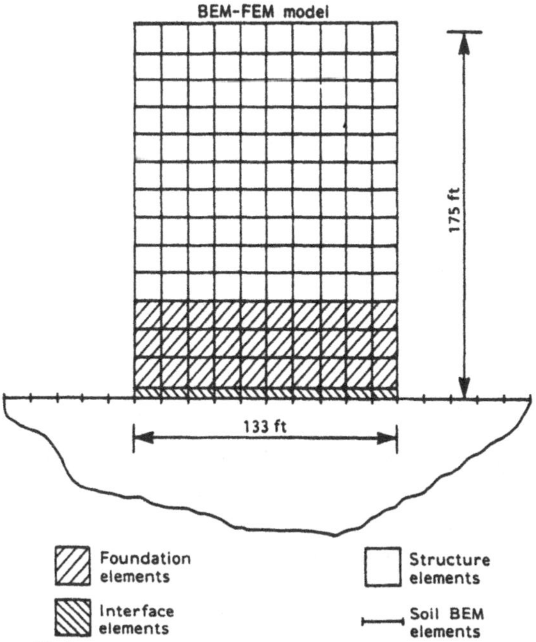

Figure 1: Plane strain BEM-FEM model of nuclear containment structure

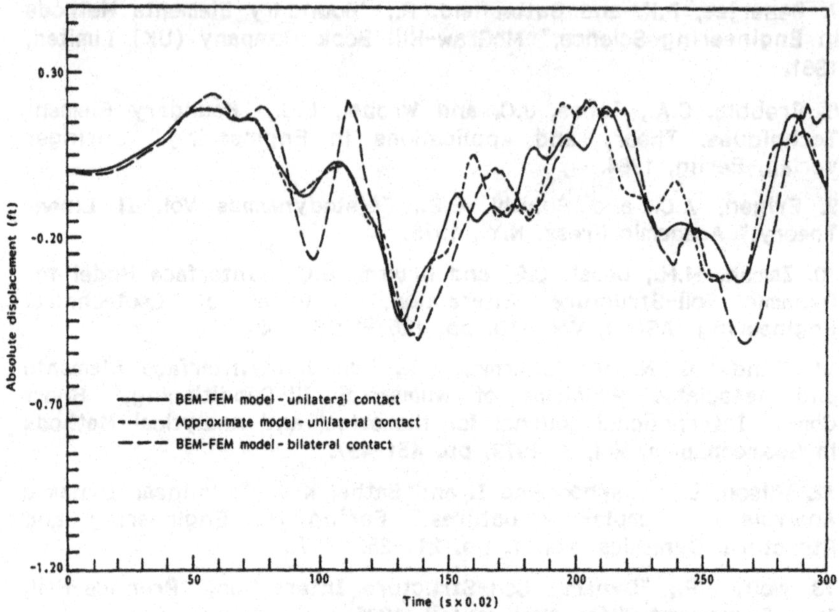

Figure 2: Comparison of the horizontal absolute response at point A

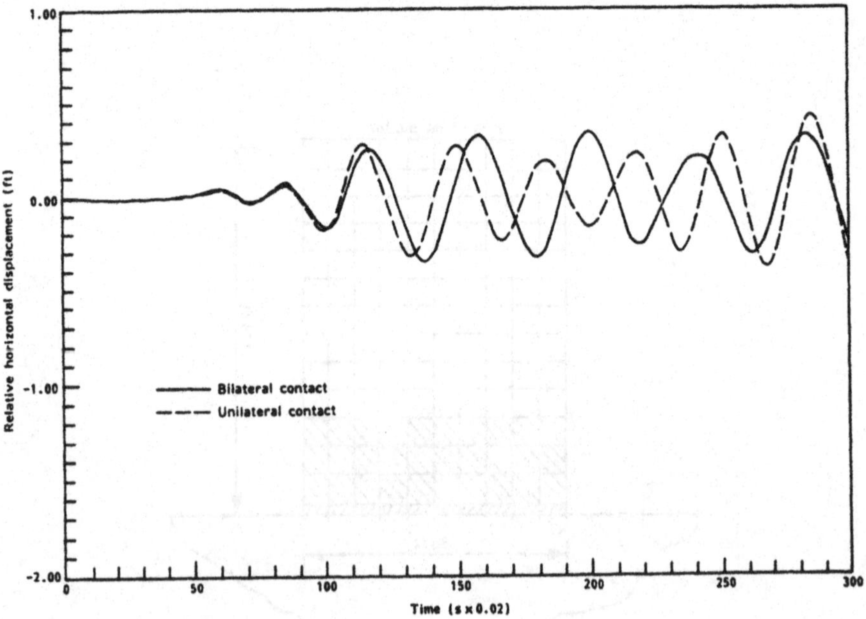

Figure 3: Relative response at point B

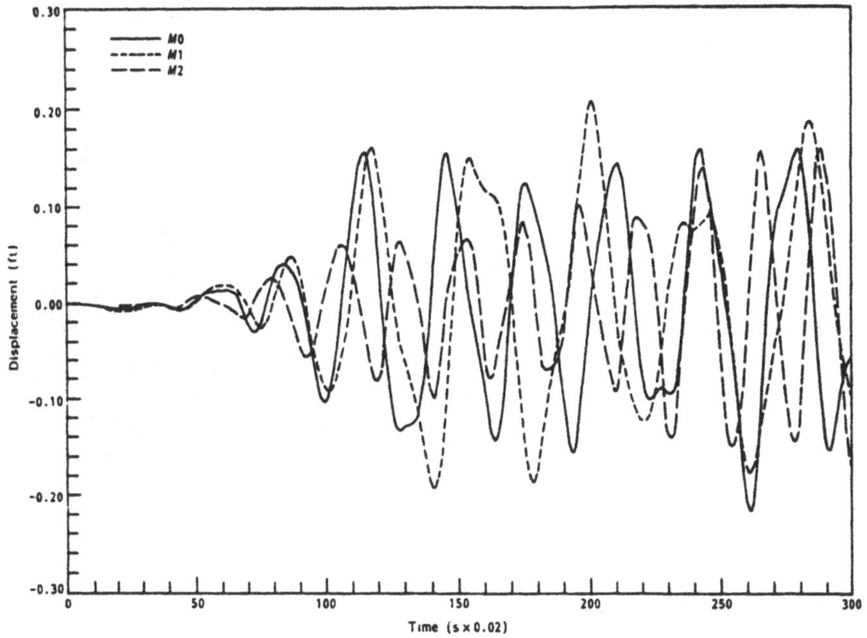

Figure 4: Vertical response at point A-bilateral contact conditions

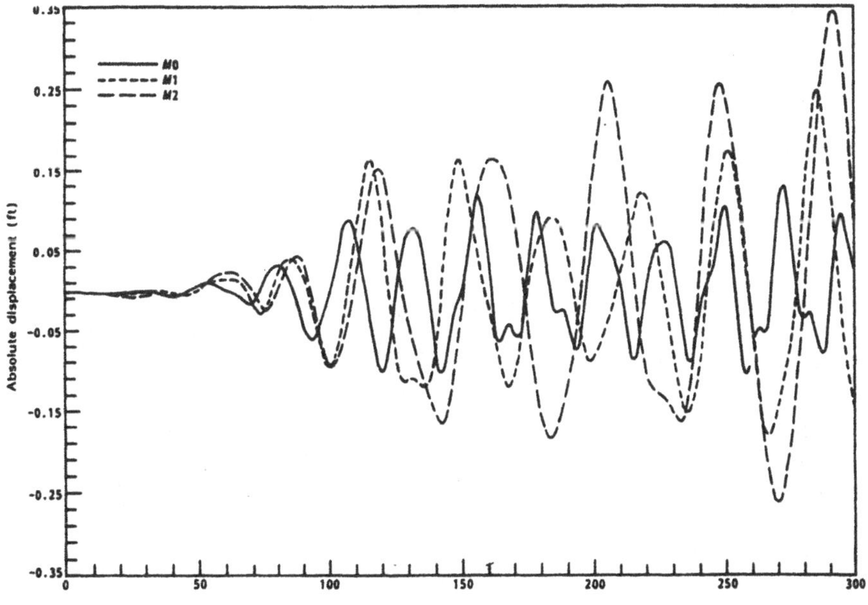

Figure 5: Vertical response at point A-unilateral contact conditions

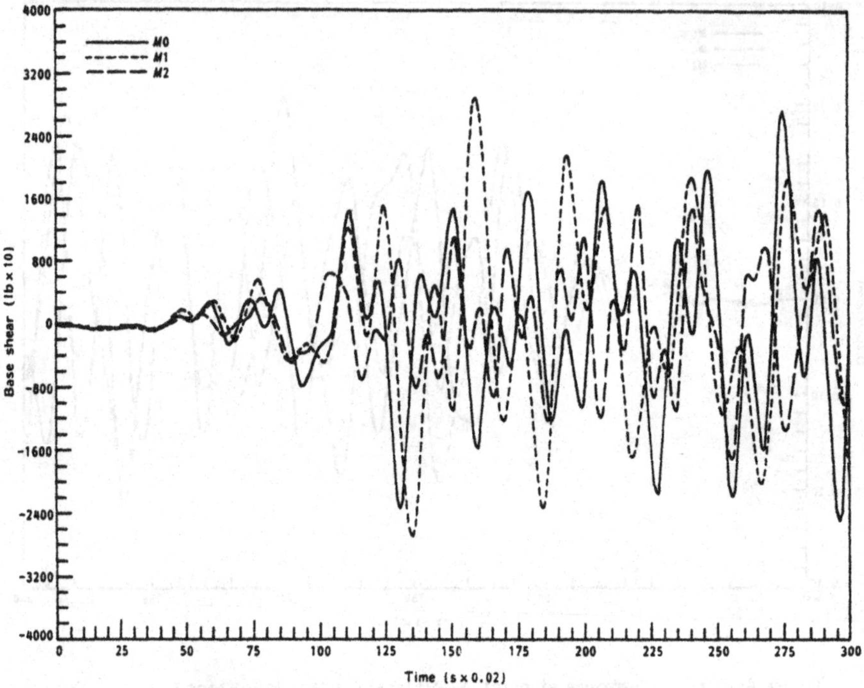

Figure 6: Base shear response-bilateral contact conditions

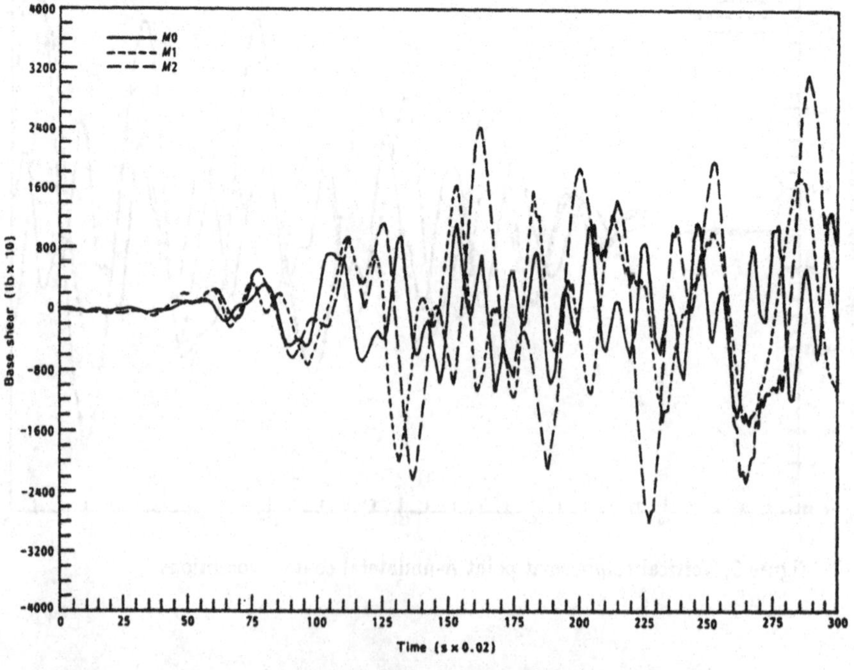

Figure 7: Base shear response-unilateral contact conditions

The Three-Dimensional Dynamic Displacement Discontinuity Method

M.G. Mack (*), S.L. Crouch (**)

(*) Dowell Schlumberger Inc., P.O. Box 2710, Tulsa, OK 74101, U.S.A.

(**) Department of Civil and Mineral Engineering, University of Minnesota, Minneapolis, MN 55414, U.S.A.

ABSTRACT

The development of a new boundary element technique applicable to dynamic problems in rock mechanics, such as joint slip and the sudden advance of excavations, is described. Both implicit and explicit versions of the technique, known as the three-dimensional dynamic displacement discontinuity method, have been developed. Comparisons of accuracy and computational efficiency are made for a simple verification problem.

INTRODUCTION

The design and analysis of mined excavations is, in some respects, an ideal application of boundary element techniques. The problems typically involve relatively simple boundaries (compared to some mechanical engineering applications) in infinite bodies, and often the global effects of changes in one part of an excavation are required either at a different part of the same excavation or on the surface of another excavation. Boundary element methods are thus particularly useful because the effect of distant, but significant, events can be captured without excessive discretization.

This paper describes the development of a boundary element method suitable for computing the dynamic response of a mined excavation to rapid enlargement of the excavation or to slip on a joint or fault plane. The excavations considered are all in *tabular* orebodies, which are extensive (hundreds or thousands of meters) in two directions, and thin (less than a few meters) in the other. Both the natural features and the mined excavations can thus be represented as crack-like features. The static displacement discontinuity method [1-4] has become widely used as a standard design tool for static analyses in this type of orebody. A dynamic displacement discontinuity method has been developed in the time domain, which allows the solution of dynamic problems with similar geometries.

THE STATIC DISPLACEMENT DISCONTINUITY METHOD

The unique feature of both natural faults and excavations in tabular orebodies is that the boundary consists of two parts in very close proximity to one another. Standard boundary element methods are poorly suited to solving such problems because nodes on opposing boundaries are so close to one another (or, in the limit, actually superimposed) that their influences are not independent. This results in ill-conditioned sets of equations which cannot be solved although the physical problem (e.g., a pressurized crack) is perfectly well-behaved.

The displacement discontinuity method has been developed to solve such problems. The method was not originally developed in the context of boundary elements but for application to specific types of mining problems [1,5]. In boundary element terminology, the method may be developed either as an indirect method (for example, [6]) or as a special case of the direct boundary element method. Forms of this derivation have been published by Weaver [7] and Cruse [8]. A brief summary of Cruse's derivation follows.

Using the Somigliana identity, we can write

$$u_i(\mathbf{x}) = \int_S U_{ij}(\mathbf{x}, \boldsymbol{\xi}) t_j(\boldsymbol{\xi}) dS(\boldsymbol{\xi}) - \int_S T_{ij}(\mathbf{x}, \boldsymbol{\xi}) u_j(\boldsymbol{\xi}) dS(\boldsymbol{\xi}) \tag{1}$$

where

$\mathbf{x}$ = any point in the body
$\boldsymbol{\xi}$ = any point on the surface
u_i = i-component of displacement
t_i = i-component of traction
U_{ij} = point force fundamental solution for displacement
T_{ij} = point force fundamental solution for traction

Equation (1) can obviously be differentiated to obtain tractions or stresses at $\mathbf{x}$. Now consider a surface S of a special shape, such that it has two parts lying on top of one another with opposite normals, such as the two surfaces of a crack. Equation (1) then becomes

$$u_i(\mathbf{x}) = \int_\Gamma U_{ij}(\mathbf{x}, \boldsymbol{\xi})[t_j(\boldsymbol{\xi}^+) + t_j(\boldsymbol{\xi}^-)]dS - \int_\Gamma T_{ij}(\mathbf{x}, \boldsymbol{\xi})[u_j(\boldsymbol{\xi}^+) - u_j(\boldsymbol{\xi}^-)]dS \tag{2}$$

taking account of the symmetry in U_{ij} and T_{ij}. If, as in the problems of interest here, the tractions on the two faces are equal and opposite, then the first integral vanishes, and we are left with

$$u_i(\mathbf{x}) = -\int_\Gamma T_{ij}(\mathbf{x}, \boldsymbol{\xi}) D_j(\boldsymbol{\xi}) dS \tag{3}$$

where D_j represents the j-th component of displacement discontinuity ($[u_j(\boldsymbol{\xi}^+) - u_j(\boldsymbol{\xi}^-)]$). Differentiation of equation (3) provides the stresses at interior points, i.e.,

$$\sigma_{ij}(\mathbf{x}) = \int_\Gamma T^{\sigma}_{kij}(\mathbf{x}, \boldsymbol{\xi}) D_k(\boldsymbol{\xi}) dS \tag{4}$$

If $\mathbf{x}$ is taken to the boundary point $\mathbf{X}$, the right-hand side of equation (4) cannot be calculated. Cruse uses integration by parts to remove the singularity. In the derivation as an indirect method [6], the integration is performed with $\mathbf{x}$ in the interior, and the limit is then taken as $\mathbf{x} \to \mathbf{X}$. The two approaches are equivalent and result in the following equations for the tractions:

$$t_i(\mathbf{X}) = \int_\Gamma T^{tr}_{ki}(\mathbf{X}, \boldsymbol{\xi}) D_k(\boldsymbol{\xi}) dS \tag{5}$$

where

$$T^{tr}_{ki}(\mathbf{X}, \boldsymbol{\xi}) = T^{\sigma}_{kij}(\mathbf{X}, \boldsymbol{\xi}) n_j(\mathbf{X})$$

and n_j is the normal to the surface at $\mathbf{X}$. The rest of the derivation follows that of most other boundary element approaches. The surface is divided into N elements, and equation (5) is written as

$$t_i(\mathbf{X}) = \sum_{n=1}^{N} \int_{\Gamma(n)} T_{ki}^{tr}(\mathbf{X}, \boldsymbol{\xi}) D_k(\boldsymbol{\xi}) dS \tag{6}$$

The form of the function D_j over the elements is specified, and the integrals are rewritten with the nodal values of D_j removed from the integral, which changes equation (6) to

$$t_i(\mathbf{X}) = \sum_{n=1}^{N} D_k^n \int_{\Gamma(n)} T_{ki}^{tr}(\mathbf{X}, \boldsymbol{\xi}) dS \tag{7}$$

if D_k is assumed constant over the elements. Similar equations are obtained for higher-order variations of D_k. Finally, with the integrals evaluated either numerically or analytically, a set of equations is written for the tractions at the nodes, which must be solved for the magnitudes of the displacement discontinuities i.e.,

$$t_i^m = \sum_{n=1}^{N} T_{ki}^{mn} D_k^n \qquad m = 1, 2, \ldots N \qquad i = 1, 2, 3 \tag{8}$$

Crouch and Starfield [3] have presented the theory and some applications of this method to three-dimensional mining problems, and it is used in the mining industry for excavation design (e.g., [4,9]).

THE DYNAMIC DISPLACEMENT DISCONTINUITY METHOD

Dynamic boundary element methods have been developed in both the frequency domain (e.g., [10,11]) and the time domain (e.g., [12]). These methods, and others developed since, suffer from the same drawbacks as typical static methods when crack-like geometries are modeled, i.e., nodes fall on top of one another and the equations become ill-conditioned.

Considering the usefulness of the displacement discontinuity method for static problems, it is natural to develop a dynamic displacement discontinuity method. Such a method would be useful in the analysis of earthquakes and fracture mechanics in addition to the mining problems considered here. Seismologists (for instance [13-18]) have modelled earthquakes as *prescribed* time-dependent displacement discontinuities across faults ("slip-functions"). These authors were mainly interested in obtaining velocities and accelerations whereas we are interested in displacements, velocities and stresses. Also, our interest is in *solving* for the displacement discontinuities when a sudden change is made to the boundary conditions rather than examining the effects of specified displacement discontinuities.

THEORETICAL DEVELOPMENT OF THE DYNAMIC METHOD

The dynamic version of the displacement discontinuity method can be derived in much the same way as the static method. The following derivation follows the work of Aki and Richards [19] who developed the solution for prescribed slip on a fault. For the method developed here, normal displacement discontinuities are also considered.

Betti's dynamic reciprocal theorem, without body forces, and with a quiescent past, can be written as

$$u_n(\mathbf{x},t) = \int_{-\infty}^{\infty} d\tau \left[\int_S G_{in}(\boldsymbol{\xi}, t-\tau; \mathbf{x}, 0) T_i(\boldsymbol{\xi}, \tau) - \int_S u_i(\boldsymbol{\xi}, \tau) c_{ijkl} n_j G_{kn,l}(\boldsymbol{\xi}, t-\tau; \mathbf{x}, 0) dS \right] \quad (9)$$

where G_{in} and $c_{ijkl}n_jG_{kn,l}$ are equivalent to U_{ij} and T_{ij} in equation (1). T_n is now used for traction to avoid confusion with t, the time. The function $G_{nk}(\mathbf{x}, t-\tau; \boldsymbol{\xi}, 0)$ is the Green's function for an instantaneously applied point load in an infinite medium. The combination $C_{ijkl}n_jG_{nk,l}$ represents the traction caused by the point load, and includes the spatial derivative of G_{nk}, the elastic properties c_{ijkl}, and the normal to the surface n_j.

Once again taking the limit as the two sides of the surface approach one another to form a crack, Σ, the term in T_n vanishes, and we are left with

$$u_n(\mathbf{x},t) = -\int_{-\infty}^{\infty} d\tau \int_{\Sigma} D_i(\boldsymbol{\xi}, \tau) C_{ijkl} n_j G_{nk,l}(\mathbf{x}, t-\tau; \boldsymbol{\xi}, 0) d\Sigma. \quad (10)$$

The significant differences between this equation and the static case are that integration must be performed with respect to time, and the Green's function is time-dependent. The Green's function for a homogeneous, isotropic elastic material in equation (10) is [19]

$$G_{ij} = \frac{1}{4\pi\rho} \{ \frac{(3\gamma_i\gamma_j - \delta_{ij})}{r^3} \tau [H(\tau - r/\beta) - H(\tau - r/\alpha)] + \frac{\gamma_i\gamma_j}{\alpha^2 r} \delta(\tau - r/\alpha) - \frac{(\gamma_i\gamma_j - \delta_{ij})}{\beta^2 r} \delta(\tau - r/\beta) \} \quad (11)$$

Substituting the Green's function into equation (10), and simplifying, we obtain

$$u_j(\mathbf{x},t) = \frac{-1}{4\pi\rho} \int_{\Sigma} [T_1(\boldsymbol{\xi}, \mathbf{x}, t) + T_2(\boldsymbol{\xi}, \mathbf{x}, t) + T_3(\boldsymbol{\xi}, \mathbf{x}, t) + T_4(\boldsymbol{\xi}, \mathbf{x}, t) + T_5(\boldsymbol{\xi}, \mathbf{x}, t)] d\Sigma(\boldsymbol{\xi}) \quad (12)$$

where ρ = density

$$T_1 = \frac{\mu}{r^4}[30\gamma_p\gamma_j\gamma_3 - 6(\gamma_j\delta_{p3} + \gamma_p\delta_{j3} + \gamma_3\delta_{pj})] \int_{r/\alpha}^{r/\beta} \tau D_p(t-\tau)d\tau$$

$$T_2 = \frac{1}{\alpha^2 r^2}[\lambda\gamma_j\delta_{p3} + \mu(12\gamma_p\gamma_j\gamma_3 - 2(\gamma_j\delta_{p3} + \gamma_p\delta_{j3} + \gamma_3\delta_{pj}))]D_p(t - r/\alpha)$$

$$T_3 = \frac{-\mu}{\beta^2 r^2}[12\gamma_p\gamma_j\gamma_3 - 2\gamma_j\delta_{p3} - 3\gamma_p\delta_{j3} - 3\gamma_3\delta_{pj}]D_p(t - r/\beta)$$

$$T_4 = \frac{1}{\alpha^3 r}[\lambda\gamma_j\delta_{p3} + 2\mu\gamma_j\gamma_p\gamma_3]\dot{D}_p(t - r/\alpha)$$

$$T_5 = \frac{-\mu}{\beta^3 r}[2\gamma_j\gamma_p\gamma_3 - \gamma_3\delta_{jp} - \gamma_p\delta_{j3}]\dot{D}_p(t - r/\beta)$$

α = P-wave velocity
β = S-wave velocity
λ, μ = Lamé constants
$\mathbf{r} = |\mathbf{x} - \boldsymbol{\xi}|$
$\gamma_i = (x_i - \xi_i)/r$ (i.e., the direction cosines)
δ_{ij} = Kronecker delta (1 for $i = j$; 0 for $i \neq j$)

The stress at any point can be obtained by differentiation of this expression.

NUMERICAL IMPLEMENTATION

The properties of the Green's function allow us to write equation (10) as

$$u_n(\mathbf{x}, t) = -\int_0^t d\tau \iint_\Sigma D_i(\boldsymbol{\xi}, \tau)T_{in}(\mathbf{x} - \boldsymbol{\xi}, t - \tau)d\Sigma. \tag{13}$$

If the D_i are assumed to be piecewise constant in both space and time, and the integrals over the elements are evaluated for discrete time intervals, we obtain (after differentiation)

$$t_i^p(\mathbf{X}) = \sum_{j=1}^{p}\sum_{n=1}^{N} D_k^{nj} \int_{\delta t^j} \int_{\Sigma(n)} T_{ki}^{tr}(\mathbf{X}, \boldsymbol{\xi})dS \tag{14}$$

Thus, a set of equations is formed relating the tractions at the element centers at times $p\Delta t$, with i ranging from 1 to N. In matrix form, this is written

$$\mathbf{t}^p = \sum_{j=1}^{p} \mathbf{B}^{jp}\mathbf{D}^j \tag{15}$$

which simply states that there is an influence matrix ($\mathbf{B}^{jp}$) defining the effect of the solution quantities at timestep j on the tractions at timestep p. The actual set of equations solved at the p-th timestep is

$$\mathbf{B}^{pp}\mathbf{D}^p = \mathbf{t}^p - \sum_{j=1}^{p-1} \mathbf{B}^{jp}\mathbf{D}^j \tag{16}$$

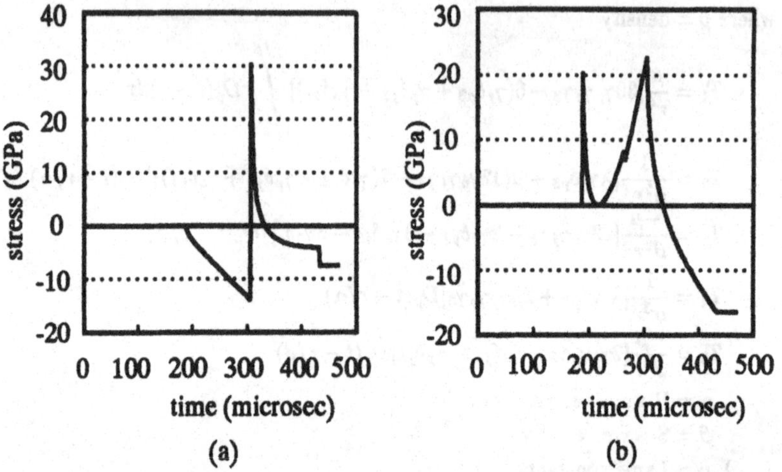

Figure 1 (a) σ_{13} due to D_1, and (b) σ_{33} due to D_3 at element center

where the right-hand side is known from previous timesteps. If a constant timestep is used, then the properties of the Green's function allow us to write the matrix $\mathbf{B}^{jp}$ as $\mathbf{B}^{p-j}$, so equation (16) becomes

$$\mathbf{B}^0\mathbf{D}^p = \mathbf{t}^p - \sum_{j=1}^{p-1} \mathbf{B}^{p-j}\mathbf{D}^j \tag{17}$$

Explicitly, for the first timestep, $\mathbf{B}^0\mathbf{D}^1 = \mathbf{t}^1$,
for the second timestep, $\mathbf{B}^0\mathbf{D}^2 = \mathbf{t}^2 - \mathbf{B}^1\mathbf{D}^1$,
for the third timestep, $\mathbf{B}^0\mathbf{D}^3 = \mathbf{t}^3 - \mathbf{B}^1\mathbf{D}^2 - \mathbf{B}^2\mathbf{D}^1$, and so forth.

At each timestep, a new matrix must be formed which accounts for the effects of the solutions from the first timestep, and the right-hand side must be recalculated.

Single-Element Behavior

The static displacement discontinuity method has been used successfully with rectangular constant-strength elements i.e., the effect of each element is represented by a single vector $\mathbf{D}$. The boundary conditions are satisfied at the centers of the elements, and the effects of corners are not important. The dynamic method has therefore been implemented with a similar spatial variation. At first, elements with piecewise constant behavior in time were considered, by substituting into equation (12) the expression

$$D_p(t) = \overline{D}_p H(t)$$

where $\overline{D}_p$ is a constant, and $H(t)$ is the Heaviside step function. The traction due to a unit step function occuring over a square was calculated at the center of the square and other points representing the centers of surrounding elements in a regular grid. A typical response at the element center is shown in Figure 1. It can be seen that there are singularities and that the solution changes sign. Also, the effect of an element at its center (the diagonal term of $\mathbf{B}^0$ for a given value of Δt) is a Dirac delta function at $t = 0$, and is then zero until signals begin to arrive from the element edges. This suggests that the behavior of the coefficient matrices ($\mathbf{B}^n$) will be very poor. It is highly unlikely that a method based on piecewise-constant elements in time would be capable of providing useful solutions.

Mathematical investigation of the behavior of piecewise-constant elements indicated that the inverse square-root singularities could be eliminated by using higher-order elements. Elements with piecewise linear behavior in time were therefore developed. The effect of a ramp-function rise in diplacement discontinuity magnitude is shown in Figure 2. This solution is much better behaved, and can be used to develop a boundary element method. One disadvantage of the dynamic method is the necessity to form the sums on the right-hand side of equation (17). As the number of timesteps increases, the time required to form the sum gets greater. To avoid this, "hat" functions (shown in Figure 3) are used as the unit solutions. Such a function has the important property that its effect at a point ends once the S-wave from the end of the "hat" has passed, and it is no longer needed in the sum. This implies that the **B** matrices have zeroes not only before the arrival of the P-wave, but also after the passing of the S-wave. This significantly improves the computational time efficiency over the use of ramp-functions.

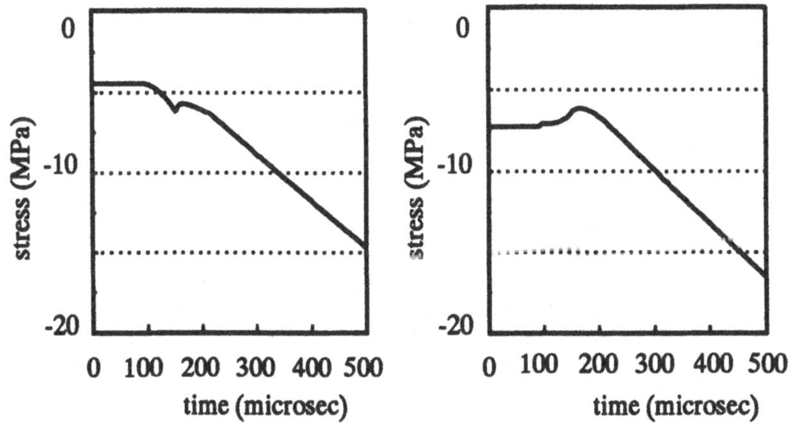

(a) (b)

Figure 2 (a) σ_{13} due to D_1, and (b) σ_{33} due to D_3 at element center - Linear elements

The Explicit Method

In the static displacement discontinuity method, there are two parts to the solution of any problem, viz., the formation of the coefficient matrix, and the solution of the resulting set of equations. When the integrals are obtained analytically, the solution of the equations can be the most time-consuming part of an analysis. When the dynamic method was first developed [20] it was designed as an *explicit* method, in which no set of equations was formed or solved.

The fundamental solutions are such that a disturbance at an element has no effect at a point in the body until the P-wave arrives at the point i.e., there is a non-zero coefficient for an interaction between two elements only if the shortest distance between them is less than αt. Thus the later **B** matrices include effects from more distant elements than the early matrices. As the magnitude of the timestep for the analysis decreases, the $\mathbf{B}^0$ matrix becomes sparser. Provided the timestep is less than a critical value ($\Delta x/2\alpha$, where Δx is the shortest element dimension) any element affects only itself in the first timestep, which means that the matrix $\mathbf{B}^0$ consists of independent 3 by 3 submatrices on the diagonal. Furthermore, on a flat element, the normal component of the solution is completely independent of the shear components, so that in fact there are N independent equations plus N sets of 2 equations. The solution of a pair of linear equations is trivial, so

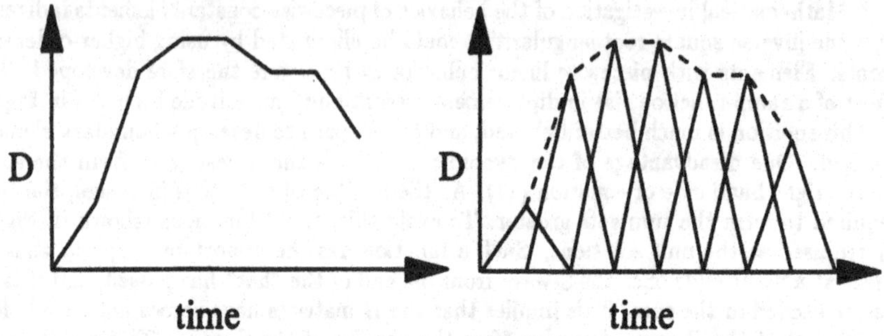

Figure 3 Superposition of "hat" functions to represent a function

the solution reduces to the solution of $3N$ independent equations. This is obviously much less time-consuming than the solution of $3N$ simultaneous equations in $3N$ unknowns.

As an example of the use of this method, consider the problem of a small area on the surface of a halfspace which is suddenly loaded. The normal displacement at a point some distance away is required. Although the current method is designed for modeling cracks in infinite domains, the halfspace problem can be solved by modeling a "large enough" crack, applying a traction over a small area, and observing the displacement discontinuity (which represents twice the displacement) at a suitable observation point. Provided displacements are not required after disturbances begin to arrive from the edges of the grid, this is a good representation of a halfspace. Figure 4 shows the representation used for this problem, and Figure 5 shows the solution, compared with the analytical solution (obtained by integrating Pekeris' [21] solution for a point load on a halfspace). It should be pointed out that the numerical solution was obtained by averaging the solutions for the four elements surrounding the observation point (which is at a corner). When the four solutions are considered separately (Figure 6), the initial peak is larger (closer to the analytical value) but some oscillations are noted in the individual solutions. In some analyses, these oscillations have been observed to become unstable (i.e., to grow in magnitude) when a timestep just less than the critical value was used. Reducing the timestep in these cases made the method stable, but no theoretical reason could be found for the oscillatory behavior.

In the explicit method, the maximum timestep is limited by the element dimension or sometimes (for stability) an even lower value, so that a large number of steps must be taken to represent a fixed time interval. A large number of matrices must be formed, and the calculation of the right-hand side must be repeated frequently (once per step). It turns out that the formation of the right-hand side is the most computationally expensive part of the solution procedure.

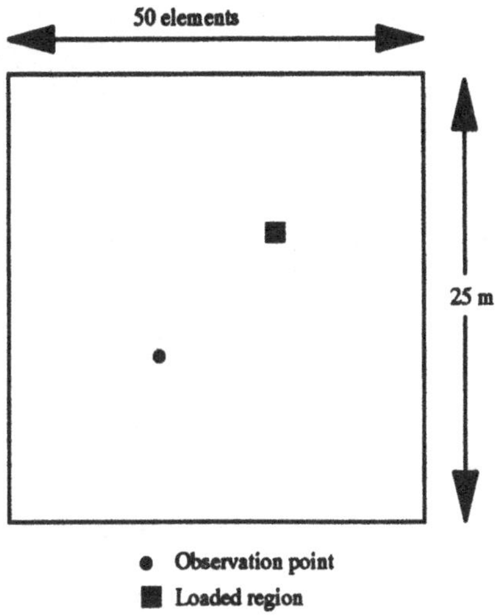

Figure 4 Halfspace model for point load

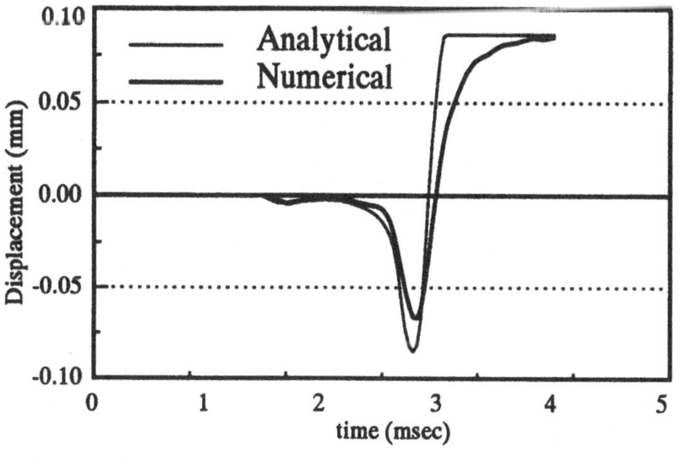

Figure 5 Normal displacement at observation point in Figure 4

The Implicit Method

Because of the expense of forming the large number of right-hand sides necessary for the explicit method, an implicit method was developed. The only difference from the explicit method is that the timestep may be increased so that an element affects its neighbors during the first timestep. Thus a set of equations must be formed and solved. However, the $\mathbf{B}^0$ matrix may remain sparse if the timestep is not much greater than the critical value, so that the strong diagonal dominance makes an iterative solution technique efficient. In the limit as the timestep reduces to the explicit value, one iteration is sufficient (but many steps must be taken). On the other hand, if the timestep is very large, a single timestep produces the final static solution. Although many iterations may be needed, only one right-hand side is formed, and only one solution is needed. This is

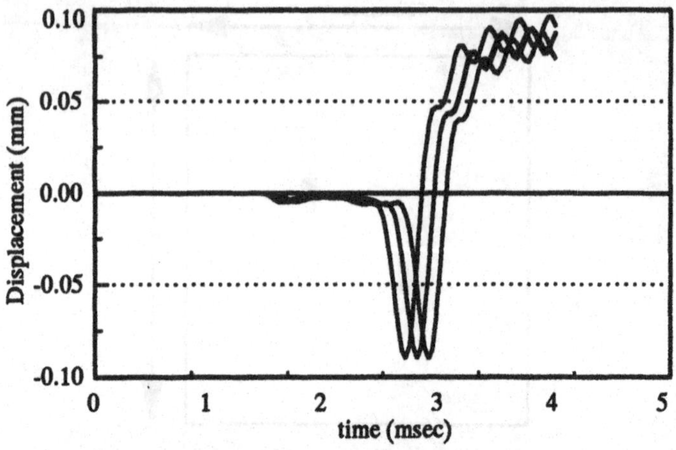

Figure 6 Individual element solutions around observation point

not very useful, however, since all the dynamic behavior is lost. At intermediate values of the timestep most of the dynamic behavior may be captured with significantly less computation. In the following section, a comparison is made between the results from the explicit method and the implicit method with several different timesteps.

COMPARISON OF EXPLICIT AND IMPLICIT METHODS

The problem of the load on a halfspace solved previously was also solved with the implicit method. Timesteps of 2, 4, 8, 24, and 120 times the magnitude used for the explicit analysis were used. Figure 7 shows how the CPU-time (on a 25 Mhz 80486 computer) varies with the number of timesteps required to represent the time interval. Apart from the first point (in which a static solution is obtained immediately), the CPU time rises rapidly with the number of steps. The storage required rises linearly to about 6 MB for the largest problem. The last point on Figure 7 represents the previously-discussed explicit solution, with a timestep of about 85% of the critical value.

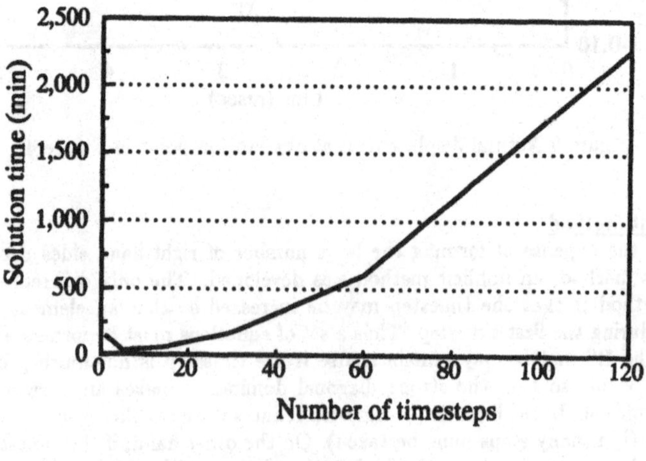

Figure 7 CPU-time required for implicit analyses

Recall that the average of the displacements at four elements was used to obtain the solutions presented previously. The same average solution, obtained using the implicit method with the different timesteps listed above, is compared with the explicit solution in Figure 8. It can be seen that even with 5 steps some sense of the solution is obtained, and that this improves with the number of steps. However, there is a some loss of accuracy even when the timestep is only doubled. The solutions at the four separate points are plotted in Figure 9 for a timestep double that used for the explicit method. The oscillations are significantly diminished compared to Figure 6. In some analyses, it has been observed that the implicit method is stable, while the oscillations cause the explicit method to become unstable.

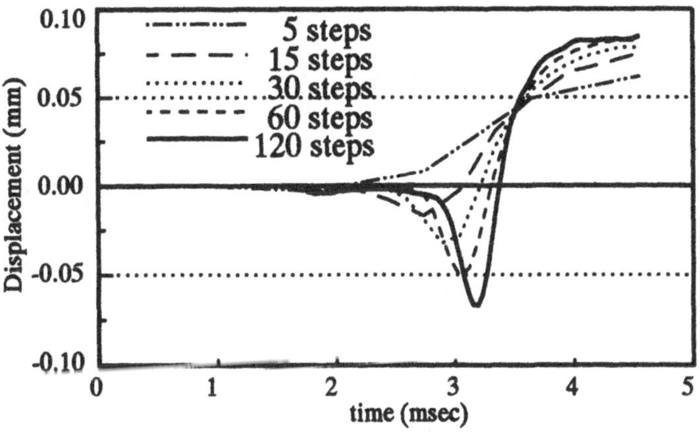

Figure 8 Comparison of solutions obtained with different timesteps

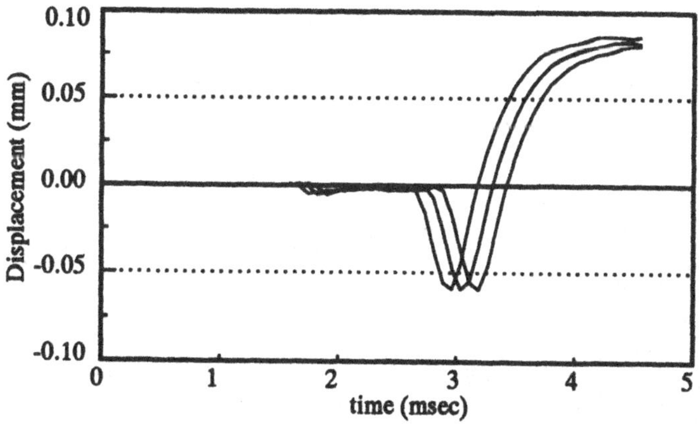

Figure 9 Individual element solutions with increased timestep

CONCLUSIONS

A new method has been developed for investigating velocities, stresses and displacements induced in excavations in tabular orebodies by dynamic loading processes such as blasting and fault slip. Both explicit and implicit implementations have been demonstrated. The explicit method is generally more accurate but requires substantially more CPU-time. It occasionally exhibits unstable oscillations which can only be corrected by reducing the timestep. No theoretical basis has been found for the timestep limitation. The implicit method, while faster and more stable, cannot always resolve the details of dynamic behavior, which may result in a loss of accuracy.

ACKNOWLEDGEMENTS

The original development of the explicit method was supported financially by Inter-Mine Services OFS (Pty) Ltd of South Africa and a Doctoral Dissertation Fellowship from the Graduate School of the University of Minnesota.

REFERENCES

1. Plewman, R. P., Diest, F. H. and Ortlepp, W. D. The development and application of a digital computer method for the solution of strata control problems, J. S. Afr. Inst. Mining Metall., Vol. 70, pp. 33-44, 1969.
2. Starfield, A. M. and Crouch, S. L. Elastic analysis of single seam extraction, in New Horizons in Rock Mechanics (Ed. Hardy, Jr., H.R. and Stefanko,R.) pp. 421-39. New York: Am. Soc. Civ. Engrs. 1973.
3. Crouch, S. L. and Starfield, A. M. Boundary element methods in solid mechanics, George Allen and Unwin,London, 1983.
4. Ryder, J. A. Excess shear stress in the assessment of geologically hazardous situations, J. S. Afr. Inst. Min. Metall., Vol. 88, pp. 27-40, 1988.
5. Salamon, M. D. G. Elastic analysis of displacements and stresses induced by mining of seam or reef deposits, Part IV, J. S. Afr. Inst. Min. Metall., Vol. 65, pp. 319-38, 1964.
6. Crouch, S.L. Solution of plane elasticity problems by the displacement discontinuity method, Int. J. Num. Methods. Engng, Vol 10, pp. 301-43, 1976.
7. Weaver, J. Three-dimensional Crack Analysis, Int. J. Solids Struct., Vol. 13, pp. 321-330, 1977.
8. Cruse, T.A. Boundary element analysis in computational fracture mechanics, Kluwer, Dordrecht, Boston, and London, 1988.
9. Spottiswoode, S.M. Volume excess shear stress and cumulative seismic moments, Proc. 2nd Intl. Symposium on rockbursts and seismicity in mines, Minneapolis, 1988, pp. 93-99, Balkema, Rotterdam, 1990.
10. Cruse, T. A. and Rizzo, F. J. A direct formulation and numerical solution of the general transient elastodynamic problem, I, J. Math. Anal. Appl., Vol. 22, pp.244-259, 1968.
11. Cruse, T. A. A direct formulation and numerical solution of the general transient elastodynamic problem, II, J. Math. Anal. Appl. Vol. 22, pp.341-355, 1968.
12. Cole, D. M., Kosloff, D. D. and Minster, J. B. A numerical boundary integral equation method for elastodynamics, I, Bull. Seismol. Soc. Am., Vol. 68, pp. 1331-1357, 1978.
13. Ben-Menahem, A. and Toksöz, M. N. Source mechanism from spectra of long period surface waves, Jnl. Geophys. Res., Vol. 68, pp. 5207-5222, 1963.
14. Haskell, N. A. Radiation pattern of surface waves from point sources in a multi-layered medium, Bull. Seismol. Soc. Am., Vol. 54, pp. 377-394, 1964.
15. Haskell N. A. Total energy and energy spectral density of elastic wave radiation from propagating faults, II, Bull. Seismol. Soc. Am., Vol. 56, pp. 125-140, 1966.
16. Haskell N. A. Elastic displacements in the near-field of a propagating fault, Bull. Seismol. Soc. Am., Vol. 59, pp. 865-908, 1969.

17. Savage, J. C. Radiation from a realistic model of faulting, Bull. Seismol. Soc. Am., Vol. 56, pp.577-592, 1966.
18. Anderson, J. G. and Richards, P. G. Comparison of strong ground motion from several dislocation models, Geophys. J. Roy. Astr. Soc., Vol. 42, pp. 347-373, 1975.
19. Aki, K. and Richards, P. G. Quantitative seismology - theory and methods, W. H. Freeman and Co.,San Francisco, 1980.
20. Mack, M.G. and Crouch, S.L. A dynamic boundary element method for modeling rockbursts, Proc. 2nd Intl. Symposium on rockbursts and seismicity in mines, Minneapolis, 1988, pp. 93-99, Balkema, Rotterdam, 1990.
21. Pekeris, C. L. The seismic surface pulse, Proc. of the National Academy of Science, Vol. 41, pp. 469-480, 1955.

SECTION 10: PLATES AND SHELLS

A Boundary Element Technique on the Solution of Cylindrical Shell Bending Problems

D. Ren, K.-C. Fu

Department of Civil Engineering, The University of Toledo, Toledo, Ohio 43606, U.S.A.

ABSTRACT

The object of this paper is to investigate the possibility of utilizing fundamental solutions in series form along with the plate singularity to implement the boundary element technique in non-shallow cylindrical shell problems with arbitrary loads. The boundary integral equations are set up by direct formulation and the series kernels of boundary integrals are derived from the more exact set of differential equations, namely, Morley-Koiter equations.

INTRODUCTION

The boundary element technique has been recognized as an efficient approximate solution method in various boundary value problems. There are also some boundary integral formulations proposed for shell bending problems [1] [2] [3] [4]. However, few of them are concerned with non-shallow shells. The difficulty is not on the formulation of the method for shells but on the finding of the fundamental solutions.

The effort of finding solutions of shells under concentrated load has been extended for a long period [5] [6] [7] [8]. Various solutions of shallow shells were given in series form [7] or in terms of special functions [8]. It is difficult to obtain numerical results using series solutions as the kernels of integral equations because of poor convergence. Chernyshev [6] showed that singularities of the solutions to a general shell under concentrated load have the same nature as those of plates. Based on this fact some authors [1] [2] employed the biharmonic potential in their boundary integral solutions to shallow shells. However, the results of using the singular solutions of plates in non-shallow shells remains doubtful in its accuracy.

In this paper, a boundary integration formulation for a circular cylindrical shell is given. The singular solutions of a thin plate under inplane point load and transverse point load are taken as the kernels of the integrals over the boundary element containing the collocation point. For other numerical integrations, a series solution to Morley-Koiter [9] equations is derived and used as the kernels of the integrals.

MATHEMATICAL STATEMENT

The equations of equilibrium of a cylindrical shell with middle surface radius R, thickness h, Young's modulus E, and Poisson's ratio ν are reduced to a set of coupled differential equations in terms of the axial membrane displacement u, circumferential membrane displacement υ and transverse displacement w in the normal coordinate system, which was first proposed by Morley [10] as a substantial improvement of Donnell's equation and confirmed by Koiter and Niordson [9] based on more rigorous derivations. It seems to be the most accurate and have the simplest form in the sense that they are consistent with Love-Kirchhoff assumption. It may be expressed in the matrix form

$$A_{ij} V_j + F_i = 0 \tag{1}$$

where the summation notation is used, both indices, i and j range from one to three. And

$$(A_{ij}) = \begin{bmatrix} \Delta - \frac{1}{2}(1+\nu)\partial_\phi^2 & \frac{1+\nu}{2}\partial_x\partial_\phi & \frac{\nu}{R}\partial_x \\ & \Delta - \frac{1+\nu}{2}\partial_x^2 & \frac{1}{R}\partial_\phi \\ \text{symmetrical} & & \frac{1}{R^2} + k\left(R\Delta + \frac{1}{2}\right)^2 \end{bmatrix} \tag{2}$$

$$\Delta = \frac{\partial^2}{\partial x^2} + \frac{\partial^2}{\partial \phi^2} \quad , \quad \partial_x = \frac{\partial}{\partial x} \quad , \quad \partial_\phi = \frac{\partial}{\partial \phi}$$

$$k = \frac{h^2}{12R^2}$$

$$(V_j) = (u, v, w)^T$$

$$(F_j) = \frac{1-\nu^2}{Eh}(F_x, F_\phi, P)^T$$

Coordinates x and ϕ are shown in Fig. 1

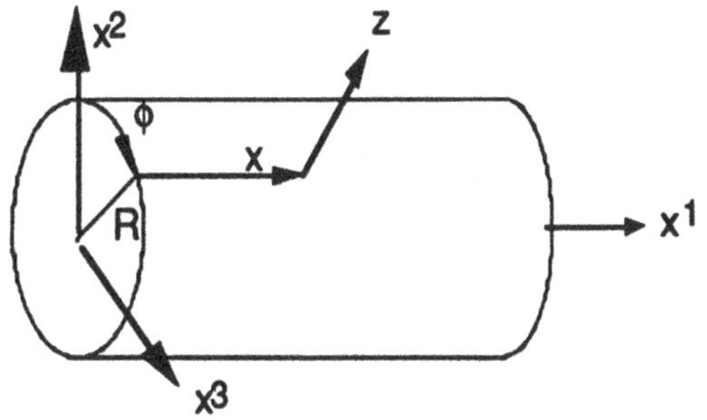

Fig. 1 Coordinates of a Cylindrical Shell

F_x, F_ϕ and P are effective loads in axial, circumferential and transverse directions, respectively.

The corresponding boundary quantities are also given in terms of displacements. For simplicity, only the boundaries of x=constant and ϕ=constant are considered. For the boundary x=constant, n=(1,0), t=(0,1):

$$M_x = D\left[\frac{\partial^2 w}{\partial x^2} + \nu \frac{\partial^2 w}{\partial \phi^2} + \frac{1-\nu}{R}\frac{\partial v}{\partial \phi} + \frac{w}{R^2}\right]$$

$$Q_x = -D\left[\frac{\partial^3 w}{\partial x^3} + (2-\nu)\frac{\partial^3 w}{\partial x \partial \phi^2} + \frac{1-\nu}{R}\frac{\partial^2 u}{\partial \phi^2} + \frac{1}{R^2}\frac{\partial w}{\partial x}\right]$$

$$N_x = \frac{Eh}{1-\nu^2}\left[\frac{\partial u}{\partial x} + \nu\left(\frac{\partial v}{\partial \phi} + \frac{w}{R}\right) - kR(1-\nu)\frac{\partial^2 w}{\partial \phi^2}\right] \tag{3}$$

$$S_x = \frac{Eh}{2(1+\nu)}\left[\frac{\partial v}{\partial x} + \frac{\partial u}{\partial \phi} - 2kR\frac{\partial^2 w}{\partial x \partial \phi}\right]$$

$$\psi = \frac{\partial w}{\partial x}$$

where $D = \dfrac{Eh^3}{12(1-\nu^2)}$, n and t are vectors in normal and tangent directions, respectively.

For the boundary ϕ = constant , n=(0,1), t=(1,0)

$$M_\phi = D\left[\frac{\partial^2 w}{\partial \phi^2} + \nu \frac{\partial^2 w}{\partial x^2} - \frac{1-\nu}{R}\frac{\partial u}{\partial x} + \frac{w}{R^2}\right]$$

$$Q_\phi = -D\left[\frac{\partial^3 w}{\partial \phi^3} + (2-\nu)\frac{\partial^3 w}{\partial x^2 \partial \phi} - \frac{1-\nu}{R}\frac{\partial^2 v}{\partial x^2} + \frac{1}{R^2}\frac{\partial w}{\partial \phi}\right]$$

$$N_\phi = \frac{Eh}{1-\nu^2}\left[\frac{\partial v}{\partial \phi} + \frac{w}{R} + \nu\,\frac{\partial u}{\partial x} + kR\Delta\,\omega\right] \tag{3a}$$

$$S_\phi = \frac{Eh}{2(1+\nu)}\left[\frac{\partial v}{\partial x} + \frac{\partial u}{\partial \phi} + 2kR\,\frac{\partial^2 w}{\partial x \partial \phi}\right]$$

$$\psi = \frac{\partial w}{\partial \phi} - \frac{v}{R}$$

All boundary quantities are corresponding boundary forces except that ψ denotes the rotations about the tangential. The meanings of them are shown in Fig. 2.

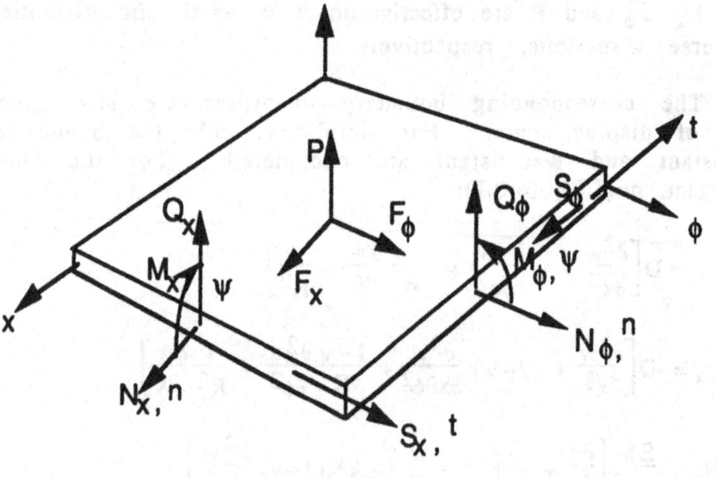

Fig. 2 Stress resultants

At a regular boundary point, there are four quantities given, say on an x=const boundary, either u or N_x, v or S_x, w or Q_x and ψ or M_x are given. The corner points and points at which boundary conditions change need to be considered in special care.

FORMULATION OF BOUNDARY INTEGRAL EQUATIONS

The direct method of boundary integral formulation invokes the generalized Green's theorem [11] [12], which is the same as the Maxwell-Betti reciprocal theorem in the case of elastic structures. Suppose that the solutions to a given cylindrical shell problem are characterized by F_i, V_i and corresponding boundary quantities ψ_n, M_n, Q_n S_n and N_n, where n denotes the normal to the boundary. And suppose that an auxiliary equilibrium state of the same shell with the same boundary but different load and/or boundary quantities is given by F_i^*, V_i^*, ψ_n^*, M_n^*, Q_n^*, and N_n^*. The Maxwell-Betti theorem states

$$\iint F_i^* V_i^* dA + \oint_s \left(M_n^* \psi_n + Q_n^* w + T_1^* u + T_2^* v\right) dS + \left[M_{st}\right]_k^* w_k$$

$$= \iint F_i V_i^* dA + \oint_s \left(M_n \psi_n^* + Q_n w^* + T_1 u^* + T_2 v^*\right) dS + [Mst]_k w_k^* \tag{4}$$

where k ranges from 1 to the number of corners, $[M_{st}]_k$ means the jump of the twist moment in the tangential plane over a corner k, which is equivalent to the reaction force given by the support. T_1 and T_2 are membrane forces in x and ϕ directions respectively. Let F_j^* in the auxiliary state be a concentrated load at (ξ, η), which corresponds to a generalized delta function. and all loads in other directions are zero. Equation (4) will give

$$CV_j = \iint F_i V_{ji}^* dA + \oint_s \left(M_n \psi_j^* + Q_n w_j^* + T_1 u_j^* + T_2 v_j^*\right) dS$$

$$- \oint_s \left(M_{n_j}^* \psi + Q_{n_j}^* w + T_{1_j}^* u + T_{2_j}^* v\right) dS$$

$$+ \left[M_{st}\right]_k w_{kj}^* - \left[M_{st}\right]_{kj}^* w_k \tag{5}$$

where C is a constant,

$$C = \begin{cases} 1 & (\xi, \eta) \text{ inside A} \\ ½ & (\xi, \eta) \text{ on smooth boundary} \\ \dfrac{\Delta\theta}{2\pi} & (\xi, \eta) \text{ is a corner} \end{cases}$$

$\Delta\theta$ is the angle spanned by the two adjacent sides of the corner inside the region. Note that the values of the constant C are taken as the same values of plates [12], because the singularities at the source points are of the same nature in shells and plates [6].

Above integral equations have two meanings. First, when the boundary quantities of a given problem are known, they supply all displacements inside the region through the boundary integration, as well as all desired stress resultant which can be obtained by differentiating the displacements expressed by Equation (5), because the displacements are continuous with respect to collocation variables (ξ, η).

Second, they offer a set of boundary integral equations which can be numerically solved for all necessary boundary quantities themselves. However, there are four unknown boundary quantities at a regular boundary point and only three equations are given in (5). So one more equation is needed. It can be obtained either by finding the solutions of an auxiliary states which is under a concentrated couple acting in desired directions or by differentiating and combining appropriate equations of (5). The later is employed here.

For the x=constant boundary, n=(1,0), t=(0,1):

$$C\psi = \iint F_i \frac{\partial V^*_{3i}}{\partial \xi} dA + \oint_s \left(M_n \frac{\partial \psi^*_3}{\partial \xi} + Q_n \frac{\partial w^*_3}{\partial \xi} + T_1 \frac{\partial u^*_3}{\partial \xi} + T_2 \frac{\partial v^*_3}{\partial \xi} \right) dS$$

$$-\oint_s \left(\frac{\partial M^*_{3n}}{\partial \xi}\psi + \frac{\partial Q^*_{3n}}{\partial \xi} w + \frac{\partial T^*_{13}}{\partial \xi} u + \frac{\partial T^*_{23}}{\partial \xi} v \right) dS$$

$$+ \left[M_{st}\right] \frac{\partial w^*_3}{\partial \xi}\bigg|_k - \frac{\partial \left[M_{st}\right]^*_3}{\partial \xi}\bigg|_k w_k \tag{6}$$

For the ϕ = constant boundary, n=(0,1), t=(-1,0):

$$C\psi = \iint Fi \left(\frac{\partial V^*_{3i}}{\partial \eta} - \frac{V^*_{2i}}{R} \right) dA$$

$$+\oint_s \left[M_n \left(\frac{\partial \psi^*_{3n}}{\partial \eta} - \frac{\psi^*_{2n}}{R} \right) + Q_n \left(\frac{\partial w^*_3}{\partial \eta} - \frac{w^*_2}{R} \right) \right.$$

$$\left. + T_1 \left(\frac{\partial u^*_3}{\partial \eta} - \frac{u^*_2}{R} \right) + T_2 \left(\frac{\partial v^*_3}{\partial \eta} - \frac{v^*_2}{R} \right) \right] dS$$

$$-\oint_s \left[M_n \left(\frac{\partial M^*_{3n}}{\partial \eta} - \frac{M^*_{2n}}{R} \right) + \psi_n \left(\frac{\partial Q^*_{3n}}{\partial \eta} - \frac{Q^*_{2n}}{R} \right) W \right.$$

$$\left. + \left(\frac{\partial T^*_{13}}{\partial \eta} - \frac{T_{12}}{R} \right) U + \left(\frac{\partial T_{23}}{\partial \eta} - \frac{T_{22}}{R} \right) V \right] dS$$

$$+ \left[M_{st}\right]_k \left(\frac{\partial w^*_3}{\partial \eta} - \frac{w^*_2}{k} \right)_k - \left(\frac{\partial \left[M_{st}\right]^*_3}{\partial \eta} - \frac{\left[M_{st}\right]^*_2}{R} \right)_k w_k \tag{6a}$$

Equation 6 or (6a), along with (5) supply enough information to solve a given boundary problem of shells.

KERNELS OF THE INTEGRAL EQUATIONS

It is obviously desirable to find the closed singular solutions to the auxiliary state with the point load along each coordinate line so that the kernels of the integral equation are highly localized which in turn give accurate numerical results. Unfortunately, such a solutions has not been found due to the coupled nature and high order of the differential equations.

On the other hand, the series solutions to shells under point load can always be found. They may be used as supplements of kernels of integral equations. Actually, most of the solutions to shallow shells or spherical shells under point load involve certain kinds of special functions which are also in series form. The key here is the convergence speed of the series. However, if the series solutions are used in regions other than the vicinity of the source point the effect of convergence speed will be reduced. Here we introduce a series solution to Morley-Koiter equations (1). The details will not be shown here since the idea was presented elsewhere [9].

Let B_{ij} = cofactor (A_{ij})

$$V_j = B_{\ell j} K_\ell (x, \phi) \tag{7}$$

where K_ℓ (ℓ = 1, 2, 3) are three undermined functions.

Substituting (7) into (1) yields

$$\mathfrak{D} K_\ell + F_\ell = 0$$

where

$$\mathfrak{D} = \det (A_{ij})$$

$$= \frac{1}{2}(1-\nu)k \left\{ \frac{1-\nu^2}{k} \partial_x^4 + \Delta^2 (R^2 \Delta + 1)^2 \right\} \mathcal{H}_\ell = 0 \tag{8}$$

where $\mathcal{H}_\ell$ is different from K_ℓ by an arbitrary constant factor. The solutions to the shell under a point load may be obtained by solving (8) with proper boundary conditions. As an example, consider an infinitely long cylindrical shell, loaded with a transverse concentrated load $\frac{1}{R}$. Set the origin of the coordinate system at the point of action and consider right half of the shell. Due to symmetry, it will carry half of the load and the solution to the homogeneous equation (8) may be written as

$$\mathcal{H}_\ell = \sum_{n=0}^{\infty} Cn(x) \cos \frac{n\phi}{R} \tag{9}$$

Substituting (9) into (8) and equating the coefficient of each trigonometric function to zero give the following set of ordinary differential equations for n=0,1,2,···.

$$\left[\frac{1-\nu^2}{k}\frac{d^4}{dx^4} + \left(\frac{d^2}{dx^2} - \frac{n^2}{R^2}\right)^2 \left(R^2 \frac{d^2}{dx^2} - n^2 + 1\right)^2\right] Cn = 0 \tag{10}$$

The characteristic equation of (10) is

$$(\alpha^2 - n^2)^2 (\alpha^2 - n^2 + 1)^2 + \frac{1-\nu^2}{k}\alpha^4 = 0 \tag{11}$$

For case $n \geq 2$, equation (11) has eight distinct roots. $\pm(p \pm iq)$ and $\pm(r \pm is)$, thus, the solution to Cn is

$$Cn = \left(A_1 \cos\frac{qx}{R} + A_2 \sin\frac{qx}{R}\right)e^{\frac{px}{R}} + \left(A_3 \cos\frac{qx}{R} + A_4 \sin\frac{qx}{R}\right)e^{-\frac{px}{R}}$$

$$+\left(A_5 \cos\frac{sx}{R} + A_6 \sin\frac{sx}{R}\right)e^{\frac{rx}{R}} + \left(A_7 \cos\frac{sx}{R} + A_8 \sin\frac{sx}{R}\right)e^{-\frac{rx}{R}}$$

The regularity at infinite requires A_1, A_2, A_5 and A_6 to be zero. The remaining four constants may be found by using four boundary conditions. This means that for case $n \geq 2$, one of $\mathcal{H}_\ell$'s is sufficient to the solutions. Let $\mathcal{H}_1 = \mathcal{H}_2 = 0$, the nth item of $\mathcal{H}_3$ ($n \geq 2$)

$$\mathcal{H}_3^n = \left[\left(A_3 \cos\frac{qx}{R} + A_4 \sin\frac{qx}{R}\right)e^{\frac{-px}{R}}\right.$$

$$\left. + \left(A_7 \cos\frac{sx}{R} + A_8 \sin\frac{sx}{R}\right)e^{\frac{-rx}{R}}\right] \cos\frac{n\phi}{R} \tag{12}$$

and $K_3 = \frac{2R^4}{1-\nu}\mathcal{H}_3$

then, from equation (7)

$$U = \frac{2R^4}{1-\nu} B_{31}\mathcal{H}_3 \;,\quad V = \frac{2R^4}{1-\nu} B_{32}\mathcal{H}_3 \;,\quad W = \frac{2R^4}{1-\nu} B_{33}\mathcal{H}_3 \tag{13}$$

The required boundary quantities may be found from equation (3) and (13). Decomposite the point load, now acting on the boundary of the half infinitely long shell, also into a Fourier Series in ϕ

$$\frac{\delta}{2R} = \sum_{m=0}^{\infty} \frac{1}{4pR^2} \cos \frac{m\phi}{R}$$

For the m-th component, the corresponding boundary conditions are

$$u=0, \quad \psi=0, \quad Q_x = \frac{1}{4\pi R^2} \cos \frac{m\phi}{R}, \quad S_x=0 \qquad \text{at } x=0$$

They offer the conditions for the determination of four constants A_3, A_4, A_7, and A_8 in Equation (12). Obviously only n=m in Equation (12) is involved in solutions to the m-th component. For the case n=0 and n=1, because four of eight roots of the characteristic equation are identically zero and rigid body movement may present, one of $\mathcal{H}_\ell$'s is not enough for the solutions. However, by picking up appropriate items of another $\mathcal{H}_\ell$, one always can find the solution. Say, for the case n=1 of above example,

Set $$\mathcal{H}_1 = A_3 x^2 \cos\frac{\phi}{R}$$

$$\mathcal{H}_3 = A_{11} x^3 \cos\frac{\phi}{R} + \left(A_7 \cos\frac{rx}{R} + A_8 \sin\frac{rx}{R}\right) e^{\frac{-sx}{R}} \cos\frac{\phi}{R}$$

Then $$u = \frac{2R^4}{1-v}\left[B_{11}\,\mathcal{H}_1 + B_{31}\,\mathcal{H}_3\right]$$

$$v - \frac{2R^4}{1-v}\left[B_{12}\,\mathcal{H}_1 + B_{32}\,\mathcal{H}_3\right]$$

$$w = \frac{2R^4}{1-v}\left[B_{13}\,\mathcal{H}_1 + B_{33}\,\mathcal{H}_3\right]$$

The remaining steps are the same as the case n≥2

In the vicinity of the collocation point, the kernels of boundary integral equations are substituted by fundamental solutions of plane stress problem and plate bending problem discussed in the Theory of Elasticity. The kernels in equation (5) and (6) take the form

$$V_{11}^* = \frac{1}{\mu}\left(\ell n\, r - \gamma\, r_x^2\right)$$

$$V_{12}^{*} = -\frac{1}{\mu}\gamma\, r_x r_y = V_{21}^{*}$$

$$V_{22}^{*} = \frac{1}{\mu_1}\left(\ell n\ r - \gamma\, r_y^2\right)$$

$$V_{33}^{*} = \frac{1}{8\pi D}\gamma^2\ \ell n\ r$$

$$V_{13}^{*} = V_{23}^{*} = V_{32}^{*} = V_{31}^{*} = 0$$

where $\mu_1 = \dfrac{4\pi E h}{(1+v)(3-v)}$

$$\gamma = \frac{1+v}{3-v}$$

$$\gamma = \left[(x - x_o)^2 + (\phi - \phi_o)^2\right]^{1/2}$$

(x_o, y_o) is the source point

$$r_x = \frac{\partial r}{\partial x} \qquad r_\phi = \frac{\partial r}{\partial \phi}$$

other kernels ψ^*, M_n^*, Q_n^*, T_1^* and T_2^* can be found through equation (3) and (3a).

NUMERICAL TREATMENT

Equation (5) and (6) or (6a) together with prescribed boundary data can be solved numerically for the remaining boundary data. The boundary is partitioned with a number of nodal points, one of which is placed at each corner. All boundary quantities are defined in terms of their nodal values. Quadratic interpolation is used for all boundary quantities and linear interpolation for geometry. At each corner equations (6) and (6a) are both formulated so the total number of system equations is four times the number of boundary nodes plus the number of corners. During the numerical integration, if a boundary element contains the collocation point, the kernels of the integral over this element are taken as those of plates, and the integrations are done analytically or by using special quadrature formula [11]. Otherwise, the series kernels and standard Gauss quadrature are used. A wide class of various boundary geometries, corners and points at which discontinuity of boundary condition occurs is currently under study. Relative unsophisticated examples are solved and comparisons with the finite element method and analytical solutions are made. They will be presented at the conference.

CONCLUSIONS

The boundary integral equations for a cylindrical shell were formulated based on Maxwell-Betti's reciprocal theorem. Solutions to Morley-Koiter equations of cylindrical shells under various point loads were found in series form. They were used as the main kernels of the boundary integral equations. As a supplement, the singular solutions to plate stretching and bending problems were taken as the kernels of integral over elements containing source points. The computational work are undertaken with considerations of various boundary discontinuities. Thus, the boundary element method is extended further for the solution of non-shallow cylindrical shell problems.

REFERENCES

1. Tottenham, H. (1979) "Boundary Element Method for Plates and Shells," Developments in Boundary Element Method-1, pp. 173-205, Applied Science Publisher.

2. Forbes, D. J. (1969) "Numerical Analysis of Elastic Plates and Shallow Shells by an Integral Equation Method," Ph.D. Thesis, University of Illinois.

3. Antes, H, (1980) "On boundary Integral Equations for Circular Cylindrical Shells," Boundary Element I (ed. Brebbia, C. A.) pp. 224-238.

4. Tosaka, N. and Miyake, S. (1983) "A Boundary Integral Equation Formulation for Elastic Shallow Shell Bending Problems," Boundary Element X (ed. Brebbia, C. A.) pp. 527-538.

5. Jahanshahi, A. (1963) "Force Singularities of Shallow Cylindrical Shells," Journal of Applied Mechanics, Vol. 30, Trans. ASME, Vol. 88, Series E, pp. 342-346.

6. Chernyshev, G. N. (11963), "On the Action of Concentrated Forces and Moments on an Elastic Thin Shell of Arbitrary Shape," Journal of Applied Mathematics and Mechanics, Vol. 27, pp. 172-184.

7. Flügge, W. and Elling, R. E. (1972) "Singular Solutions for Shallow Shells" Int. J. Solids Structures, Vol. 8, pp. 227-247.

8. Sanders, J. L. Jr. and Simmonds, J. G. (1970) "Concentrated Forces on Shallow Cylindrical Shells," Journal of Applied Mechanics, Trans., ASME, Vol. 92, pp. 367-373.

9. Niordson, F. I. (1985) "Shell Theory," North-Holland Series in Applied Mathematics and Mechanics, Elsevier Science Publishers.

10. Morley, L. S. D. (1959) "An Improvement on Donnell's Approximation for Thin-Walled Circular Cylinders," Quart. J. Mech. Appl. Math, Vol. 12, pp. 89-99.

11. Brebbia, C. A. and Walker (1980) "Boundary Element Techniques in Engineering," Newnes Butterworths.

12. Hartman, F. (1989) "Introduction to Boundary Elements," Springer-Verlag.

An Indirect Boundary Element Method for Plate Bending Problems

M. Vable (*), Y. Zhang (**)
(*) Mechancial Engineering and Engineering Mechanics, Michigan Technological University, Houghton, MI 49931, U.S.A.
(**) Walker Engineering Research Center, Gross Lake, MI 48118, U.S.A.

INTRODUCTION

The work presented here is a synopsis of the work to appear in reference [1].

An indirect boundary element formulation is constructed by using a fundamental solution of the biharmonic operator as was done by Tottenham [2]. The fundamental solution and its derivatives can be represented by four functions with two integers. The new representation is not only a simpler representation but also reveals a structure that is exploited in the development of an algorithm based on the analytical integration of line and area integrals. The conditions the unknown fictitious densities must satisfy are discussed. A square under a uniform transverse load with a variety of boundary conditions is used as a numerical example to demonstrate the effectiveness of the algorithm and to study the effect of satisfying and violating the conditions on the unknown fictitious density.

FORMULATION

Let $V^*(P)$ represent a fictitious transverse shear force at point p on the boundary. Let $M_n^*(P)$ represent a fictitious bending moment on the boundary. Let p(P) represent the actual distributed load. Let G(Q,P) represent the deflection at point Q due to a unit transverse load at point P. By superposition the displacement w(Q) at point Q can be written as

$$w(Q) = \oint_B G(Q,P)\ V^*(P)ds(P) + \oint_B G,_m(Q,P)n_m(P)M_n^*(P)ds(P)$$

$$+ \iint_R G(Q,P)\ p(P\ dx(P)\ dy(P) \qquad (1)$$

where $n_m(P)$ represents the direction cosines of the unit normal at point P on the boundary B. The comma implies the derivative in the m direction at point Q. The slope, the moment and shear force expressions can be obtained by repetitive differentiation of equation (1). This repetitive differentiation leads to long messy expressions for the derivatives of the fundamental solutions. The problem is further execrated if one seeks to evaluate the integrals of these fundamental solutions analytically. In an earlier work Vable [3] pointed out that most fundamental solutions were linear combinations of four singular functions. In plate bending problems there is a more defined structure to the fundamental solutions as reported in Zhang [4]. The structure is emphasized by defining the following functions:

$$LG^{[p,q]} = r_x^{\ p}\, r_y^{\ q} \ln(r) \tag{2}$$

$$JG^{[p,q]} = \frac{1}{2}\left[\frac{(r_x + \bar{\iota} r_y)^p}{(r_x - \bar{\iota} r_y)^q} + \frac{(r_x - \bar{\iota} r_y)^p}{(r_x + \bar{\iota} r_y)^q}\right] \tag{3a}$$

$$KG^{[p,q]} = \frac{1}{2\bar{\iota}}\left[\frac{(r_x + \bar{\iota} r_y)^p}{(r_x - \bar{\iota} r_y)^q} - \frac{(r_x + \bar{\iota} r_y)^p}{(r_x - \bar{\iota} r_y)^q}\right] \tag{3b}$$

$$PG^{[p,q]} = r_x^p\, r_y^q \tag{4}$$

where p and q are integers and $\bar{i} = \sqrt{-1}$. The complex variables used in the equation (3) is to emphasize the structure of the fundamental solution that is exploited in the development of the algorithm. The algorithm uses only real variables. In table 1 the fundamental solution and its derivative are defined in terms of the four functions defined in equations (2) through (4).

Table 1. Fundamental Solution and its derivatives

$$8\pi D\ G = LG^{[2,0]} + LG^{[0,2]}$$

$$8\pi D\ G,_{x} = 2LG^{[1,0]} + JG^{[1,0]} + PG^{[1,0]}$$

$$8\pi D\ G,_{y} = 2LG^{[0,1]} + KG^{[1,0]} + PG^{[0,1]}$$

$$8\pi D\ G,_{xx} = 2LG^{[0,0]} + JG^{[1,1]} + 2$$

$$8\pi D\ G,_{xy} = KG^{[1,1]}$$

$$8\pi D\ G,_{yy} = 2LG^{[0,0]} - JG^{[1,1]} + 2$$

$$8\pi D\ G,_{xxx} = 3JG^{[0,1]} - JG^{[1,2]}$$

$$8\pi D\ G,_{xxy} = KG^{[0,1]} - KG^{[1,2]}$$

$$8\pi D\ G,_{xyy} = JG^{[0,1]} + JG^{[1,2]}$$

$$8\pi D\ G,_{yyy} = 3KG^{[0,1]} + KG^{[1,2]}$$

$$8\pi D\ G,_{xxxx} = 2JG^{[1,3]} - 4JG^{[0,2]}$$

$$8\pi D\ G,_{xxxy} = 2KG^{[1,3]} - 2KG^{[0,2]}$$

$$8\pi D\ G,_{xxyy} = -2JG^{[1,3]}$$

$$8\pi D\ G,_{xyyy} = -2KG^{[1,3]} - 2KG^{[0,2]}$$

$$8\pi D\ G,_{yyyy} = 2JG^{[1,3]} + 4JG^{[0,2]}$$

Evaluation of Line Integrals:

Three assumptions are made for the evaluation of the line integral. (1) Assume the unknown fictitious shear force and the unknown fictitious moment can be represented by a linear combination of M piecewise continuous functions g_m. (2) Assume that each of the mth segment can be represented by N_m straight line segments. (3) Assume g_m can be expanded about the mid-point by Taylor series. Noting that the influence functions in table 1 are a linear combination of the functions defined in equation (2) through (4) the problem reduces to the evaluation of the following four integrals:

$$IL^{[k,i,j]} = \int_{Tp}^{Tp+1} t_p^k \, LG^{[i,j]} \, dt_p \tag{5}$$

$$IJ^{[k,i,j]} = \int_{Tp}^{Tp+1} t_p^k \, JG^{[i,j]} \, dt_p \tag{6a}$$

$$IK^{[k,i,j]} = \int_{Tp}^{Tp+1} t_p^k \, KG^{[i,j]} \, dt_p \tag{6b}$$

$$IP^{[k,i,j]} = \int_{Tp}^{Tp+1} t_p^k \, PG^{[i,j]} \, dt_p \tag{7}$$

where $T_p = -L/2$ and $T_{p+1} = L/2$

By Geometry shown in figure 1, the following can be derived

$$r_x + \bar{\iota}\, r_y = - e^{\bar{\iota}\theta p} (t_p - B_p) \tag{8}$$

where

$$B_p = C_p + \bar{\iota}\, D_p \tag{9}$$

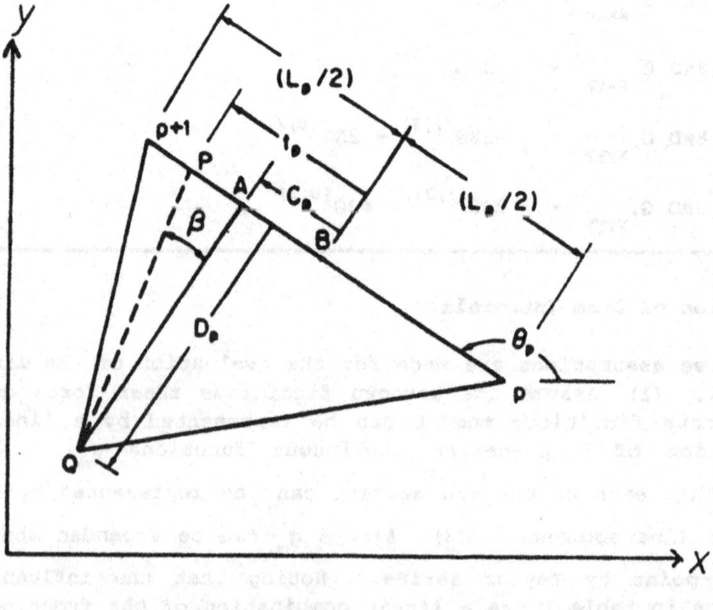

Figure 1. Geometry of the pth segment

Using equations (3), (6), and (2) we obtain

$$IJ^{[k,i,j]} = (-1)^{i+j}\left(\cos(i+j)\theta_p \; J_p^{[k,i,j]} - \sin(i+j)\theta_p \; K_p^{[k,i,j]}\right) \tag{10a}$$

$$IK^{[k,i,j]} = (-1)^{i+j}\left(\sin(i+j)\theta_p \; J_p^{[k,i,j]} + \cos(i+j)\theta_p \; K_p^{[k,i,j]}\right) \tag{10b}$$

where

$$J_p^{[k,i,j]} = \frac{1}{2}\left(I_p^{[k,i,j]} + \bar{I}_p^{[k,i,j]}\right) \tag{11a}$$

$$K_p^{[k,i,j]} = \frac{1}{2\iota}\left(I_p^{[k,i,j]} - \bar{I}_p^{[k,i,j]}\right) \tag{11b}$$

$$I_p^{[k,i,j]} = \int_{Tp}^{Tp+1} t_p^k \frac{(t_p - B_p)^p}{(t_p - \bar{B}_p)^q} dt_p \tag{12}$$

$\bar{I}$ and $\bar{B}$ are the complex conjugate of I and B of equation (9) and (12) respectively.

The integrals $J_p^{[k,i,j]}$, $K_p^{[k,i,j]}$, $IL_p^{[k,i,j]}$, and $IP_p^{[k,i,j]}$ can be evaluated by the general recursive algorithm given in reference [3]. It must be emphasized that the entire computation is carried out in the real plane and no complex arithmetic is used.

Evaluation of Area Integrals:

Two assumptions are made for the evaluation of area integrals. (1) Assume that the distributed load is piecewise constant over N subregions Rn. (2) Assume that the boundary of each subregion Rn can be represented by a set of straight line segments. It can be shown that the area integral in equation (1) and its derivatives can be found by evaluating the following integrals

$$A = -\frac{Dp}{4}\int_{Tp}^{Tp+1}\left(LG^{[2,0]} + LG^{[0,2]} - 0.25\,(PG^{[2,0]} + PG^{[0,2]})\right)ds \tag{13}$$

$$A,_{ijk} = n_k(P_p) \int_{T_p}^{T_{p+1}} G,_{ij}(Q,P)\ ds(P) \tag{14}$$

where the integration is to be performed over the pth line segment on the boundary of the Rn subregion. These integrals can be evaluated in exactly the same manner as the line integrals described earlier.

Conditions on Fictitious Densities:

In reference [1] it was shown that for moments and shear forces to be bounded the fictitious shear force and fictitious bending moment must satisfy Holder's and Hadamard conditions respectively. It is not known how these conditions can be enforced in BEM. So slightly more restrictive conditions are used, namely - the fictitious shear force must be continuous at all points on the boundary and the fictitious moment and its first derivative must be continuous at all points on the boundary. This implies Lagrange polynomials should be used for fictitious shear force approximation and a Hermite polynomial should be used for fictitious bending moment approximation.

In reference [1] it was shown that if the numerical results are to be unaffected by the choice of non-dimensionalizing parameters then the following conditions must be satisfied

$$R_i = \oint [f_i(P)\ V^*(P) - \frac{\partial f_i(P)}{\partial n\ (P)}\ M_n^*(P)]ds = 0$$

$$i = 1, \text{ to } 4. \tag{15}$$

where $f_1 = 1$; $f_2 = x$; $f_3 = y$; $f_4 = (x^2 + y^2)$

The first condition (R_1) implies that force equilibrium in the z-direction must be met by the fictitious shear force distribution. The second (R_2) and the third (R_3) condition imply that the moment equilibrium in the y and x-direction respectively must be satisfied by the unknown distribution. The last condition (R_4) can be interpreted as a second moment of some kind. The conditions of equation (15) are not explicitly enforced and may not be satisfied by the computed solution as the numerical results demonstrate. Work is under progress to overcome this problem as was done for elastostatics in Vable [5]. In this work the conditioning of the matrix and the value of the resultants R_i in equation (15) will be monitored.

Numerical Results

Extensive numerical testing was conducted for all types of boundary conditions. Boundary data was generated from a known analytical solution. Three kinds of boundary conditions were simulated and the computed solution was compared with the analytical solution at a number of points. The three types of boundary conditions that were simulated were:

Type 1: Displacement and slopes were specified to simulate the clamped type boundary conditions.
Type 2: Displacement and moments were specified to simulate simply supported boundary conditions.
Type 3: Equivalent shear force and moments were specified to simulate the free edge boundary conditions.

Each problem was solved using linear Lagrange polynomial and then using cubic Hermite polynomial for the approximation of the unknowns and the results are compared. In all problems the condition number of the matrix in the algebraic equation was computed. The matrix condition number was computed using the following definition.

Matrix condition number = $\|A\| * \|A^{-1}\|$ where $\|A\|$ and $\|A^{-1}\|$ are the norm of the matrix and its inverse respectively. The following definition for the norm of the matrix is used:

$$\|A\| = \max_{i} \sum_{j=1} |A_{ij}|$$

The four parameters R_1 though R_4 defined by equations were computed and are reported for the problem discussed below.

The displacement solution for a simply supported 1 x 1 square plate under a uniform transverse load ,with the origin at the centroid is [6]

$$w = \sum_{m=1,3,5}^{\infty} A_m \operatorname{Cosh}(m\pi\ y) + B_m(m\pi y)\sinh(m\pi y) + C_m)\ \sin(m\pi(x-0.5))$$

$$A_m = -(\frac{m\pi}{2}\tanh\frac{m\pi}{2} + 2)\ B_m \qquad (16)$$

$$B_m = 2/(\pi^5 m^5 \cosh\frac{m\pi}{2})$$

$$C_m = 4/\pi^5 m^5$$

The boundary data for slope, moment and shear force can be easily generated by repeated differentiation. A computer program was written in which the series was truncated at $m = 91$. Boundary data was generated to simulate various boundary conditions. At each point on the boundary the same kind of boundary condition was specified for type 1 and type 2. Type 3 boundary condition was not specified at all points because it would have resulted in a rigid body mode. Type 3 was specified on $y = 0.5$ and on the remaining 3 sides type 2 boundary condition was specified.

The problems were solved using linear Lagrange and cubic Hermite approximation. The mesh discretization was the same except that the length of a cubic element was constructed by combining two linear elements. The total number of unknowns for both approximations was 96. The boundary element solution was compared with the series solution and the percentage difference is reported in table 2a for the linear approximation and table 2b for the cubic approximation. Results are reported along the diagonal and $y = 0$, for the displacement w, moment M_x and shear force Q_x.

The cubic approximation yields better results for all types of boundary conditions. The results for cubic are an order of magnitude better than the linear approximation when type 3 boundary conditions are imposed on the edge $y = 0.5$. For both approximations the results deteriorate as one approaches the corner. However the accuracy of the linear approximation deteriorates more rapidly than for the cubic approximation. Near the corner $x = y = 0.49$ both approximations yield non-sensical results. It should be noted however that all three quantities(w, M_x, and Q_x) approach a zero value near the corner. Thus small differences result in very large percentage errors. As one moves from the center towards the edge along $y = 0$ the error increases as expected. But in spite of the result at $x = 0.49$ are not unreasonable for the cubic approximation.

Table 3 shows the matrix condition number for linear and cubic approximation. The cubic approximation results in a higher condition number for all types of boundary conditions.

Table 2a. Percentage error for linear Lagrange approximation.

Coordinates		Boundary Condition Type 1			Boundary Condition Type 2			Boundary Condition Type 3 on y = 0.5 Type 2 elsewhere		
x	y	w	M_x	Q_x	w_1	M_x	Q_x	w	M_x	Q_x
0.00	0.00	0.14	0.06	-	0.51	0.36	-	9.47	6.67	-
0.20	0.20	0.16	0.07	0.26	0.33	0.01	1.09	8.62	4.11	30.9
0.40	0.40	1.48	2.46	109.0	6.46	76.3	192	23.1	185	9.13
0.45	0.45	1.04	60.5	1715	34.1	504	885	107	1160	180
0.00	0.20	0.16	0.18	0.30	0.41	0.07	0.20	6.26	0.60	8.4
0.00	0.40	0.13	0.38	1.11	0.51	0.93	1.59	4.81	3.17	0.44
0.00	0.45	0.10	0.25	2.19	0.63	1.39	5.03	4.28	3.28	4.06
0.00	0.49	0.22	30.9	3.82	1.20	18.4	11.4	1.62	18.2	10.7
R_1			-0.97			-0.94			-0.92	
R_4			-0.19			-0.18			-.17	

Table 2b. Percentage error for cubic Hermite approximation.

Coordinates		Boundary Condition Type 1			Boundary Condition Type 2			Boundary Condition Type 3 on y = 0.5 Type 2 elsewhere		
x	y	w	M_x	Q_x	w	M_x	Q_x	w	M_x	Q_x
0.00	0.00	0.01	0.01	-	0.14	0.10	-	0.64	0.39	-
0.20	0.20	0.01	0.01	0.04	0.16	0.14	0.00	1.06	0.85	0.46
0.40	0.40	0.12	0.33	2.80	0.44	2.07	0.45	3.10	4.83	5.97
0.45	0.45	0.40	4.80	43.4	0.91	15.4	22.2	2.58	18.6	43.0
0.00	0.20	0.01	0.01	0.04	0.14	0.10	0.08	0.69	0.55	0.01
0.00	0.40	0.01	0.08	0.60	0.16	0.22	0.21	0.82	1.19	0.89
0.00	0.45	0.06	0.39	1.56	0.16	0.25	0.14	0.89	1.43	1.95
0.00	0.49	0.49	11.7	3.45	0.13	0.26	0.17	0.14	8.54	7.52
R_1			-1.1			-1.1			-1.1	
R_4			-0.23			-0.23			-0.23	

Table 3. Comparison of matrix condition number.

Approximation	Boundary Condition Type 1 (10^6)	Boundary Condition Type 2 (10^6)	Boundary Condition Type 3 on y = 0.5 Type 2 elsewhere (10^6)
Linear	25	73	8320
Cubic	24	479	8570

REFERENCES

1. Vable, M., and Zhang, Y., "A boundary element method for plate bending problems", Int. J. of Solids and Structures, (In print).
2. Tottenham, H., (1979), "The boundary element method for plates and shells". Development in Boundary Element MethodsEd. P.K. Banerjee and R. Butterfield, Applied Science Publishers, London.
3. Vable, M., (1987), "An algorithm based on the boundary element method for problems in engineering mechanics". Int. J. for numerical methods in engineering, 24, 1533-1540.
4. Zhang, Y., (1989), "An indirect boundary element method for isotropic and orthotropic plates", Ph.D. dissertation, Michigan Technological University .
5. Vable, M., (1990), "Importance and use of rigid body mode in boundary element method", Int. J. for numerical methods in engineering", 29, 453-472.
6. Timoshenko, S., and Woinowsky-Krieger, S, (1959), "Theory of plates and shells" McGraw Hill, New York.

A New Regularized Boundary Integral Formulation for Thin Elastic Plate Bending Analysis

T. Matsumoto, M. Tanaka, K. Hondoh

Shinshu University, 500 Wakasato, Nagano 380, Japan

ABSTRACT

The boundary integral equations conventionally used for thin elastic plate bending analysis involve the Cauchy principal value integrals and special care must be taken in evaluating them numerically when the boundary is discretized with higher order elements. In the present paper the boundary integral equations are regularized up to the integrable order by using the subtracting and adding-back technique. The obtained boundary integral equations both for the deflection and rotation are weakly singular and their discretized forms can be integrated accurately by the standard Gaussian quadrature formula. Numerical implementation of the resulting regularized integral equations is presented and effectiveness of the proposed method is discussed through some numerical demonstrations.

INTRODUCTION

In the boundary element method based on the direct method for the thin elastic plate bending problem, we usually use a pair of boundary integral equations for the deflection and rotation at a point on the boundary [1~6]. Both the above boundary integral equations have the Cauchy principal value integrals. Although the evaluation of the Cauchy principal value integral is easy for a straight element, special care must be taken for general curved elements. The use of the higher order curved elements is, however, desirable in order to obtain accurate solutions, especially for equivalent shear force.

In this paper, a regularized boundary integral equation for the rotation integrable in the standard sense is presented. All the integrals in the derived boundary integral equation can be evaluated accurately by the standard Gaussian quadrature formula. Some numerical results by the boundary element code based on the present formulation are also presented and the effectiveness of the method is demonstrated.

CONVENTIONAL FORMULATION

The basic integral equation for thin elastic plate bending problems can be written as follows :

$$w(y) = \int_\Gamma [\, w^*(x,y)V_n(x) - T_n^*(x,y)M_n(x) + M_n^*(x,y)T_n(x) - V_n^*(x,y)w(x)]d\Gamma(x)$$
$$+ \int_\Omega w^*(x,y)f(x)\, d\Omega(x) - \sum [\![\, w^*(x,y)M_{nt}(x)]\!]$$
$$+ \sum [\![\, M_{nt}^*(x,y)w(x)]\!] \,, \quad y \in \Omega \tag{1}$$

where Γ is the boundary, Ω is the domain, w is the deflection, $T_n(=\partial w/\partial n)$ is the outward normal derivative of the deflection to the boundary, M_n is the bending moment, V_n is the equivalent shear force and M_{nt} is the twisting moment. $[\![\cdot]\!]$ denotes the jump in the function inside the double brackets at a corner point and $\sum$ denotes the summation of the jumps for all corner points. w^* is the fundamental solution of the governing equation in the classical theory of thin-plate bending and satisfies the following differential equation :

$$M^*_{ij,ji} + \delta(x-y) = -D\, w^*_{,iijj} + \delta(x-y) = 0 \tag{2}$$

where M^*_{ij} is the tensor of moments corresponding to w^* and δ is the Dirac dalta function. The explicit form of w^* is given as

$$w^*(x,y) = \frac{1}{8\pi D} r^2 \ln r \tag{3}$$

where $r = |\, x-y \,|$ and D is the plate stiffness. T_n^*, M_n^*, M_{nt}^* and V_n^* are given as follows :

$$T_n^* = \frac{\partial w^*}{\partial n} = \frac{1}{8\pi D} r\,(\,2\ln r + 1\,)\frac{\partial r}{\partial n} \tag{4}$$

$$M_n^* = -\frac{1+\nu}{4\pi}(\,\ln r + 1\,) - \frac{1-\nu}{8\pi}\{2(\frac{\partial r}{\partial n})^2 - 1\} \tag{5}$$

$$M_{nt}^* = -\frac{1-\nu}{4\pi}\frac{\partial r}{\partial t}\frac{\partial r}{\partial n} \tag{6}$$

$$V_n^* = Q_n^* + \frac{\partial M_{nt}^*}{\partial s} \tag{7}$$

$$Q_n^* = -\frac{1}{2\pi r}\frac{\partial r}{\partial n} \tag{8}$$

$$\frac{\partial M_{nt}^*}{\partial s} = \frac{1-\nu}{4\pi r}(\kappa r - \frac{\partial r}{\partial n})\{2(\frac{\partial r}{\partial n})^2 - 1\} \tag{9}$$

where κ is the curvature of the boundary at x and s is the arc length along the boundary measured from a certain point on the boundary.

Taking the limit $y \in \Omega \rightarrow y \in \Gamma$ in equation (1) results in the following boundary integral equation for a point y on a smooth part of the boundary :

$$\frac{1}{2}w(y) = \int_\Gamma [w^* V_n - T_n^* M_n + M_n^* T_n - V_n^* w]\, d\Gamma + \int_\Omega w^* f\, d\Omega$$

$$- \sum [\![w^* M_{nt}]\!] + \sum [\![M_{nt}^* w]\!], \quad y \in \Gamma \tag{10}$$

Because the order of singularity of V_n^* is $1/r$, the integral for V_n^* must be evaluated in the Cauchy principal value sense.

We must consider four quantities as w, T_n, M_n and V_n on the boundary except corner points and two of them are given as the boundary conditions. Therefore, two unknowns exist at a point on the boundary and another equation is needed. For this purpose, we differentiate equation (1) with respect to y and take the limit $y \in \Omega \rightarrow y \in \Gamma$, then, multiply the both side of it by $n_i(y)$. The result is

$$\frac{1}{2}T_n(y) = \int_\Gamma [\tilde{w}^* V_n - \tilde{T}_n^* M_n + \tilde{M}_n^* T_n - \tilde{V}_n^* \{ w - w(y) \}]\, d\Gamma$$

$$+ \int_\Omega \tilde{w}^* f\, d\Omega - \sum [\![\tilde{w}^* M_{nt}]\!] + \sum [\![\tilde{M}_{nt}^* w]\!] \tag{11}$$

where $(\tilde{\cdot}) = \dfrac{\partial(\cdot)}{\partial y_i} n_i(y)$.

The order of singularity in each kernel in equation (11) increased by one and we observe $\tilde{M}_n^* \sim O(1/r)$ and $\tilde{V}_n^* \sim O(1/r^2)$. Assuming that $w(x)$ satisfies the Hölder continuity at y, we find that the integrals for $\tilde{M}_n^* T_n$ and $\tilde{V}_n^* \{w - w(y)\}$ must be evaluated in the Cauchy principal value sense. Equation (10) can be easily regularized up to the weakly singular order; however, the principal value integrals in equation (11) remain in the conventional formulation. It is cumbersome to evaluate the Cauchy principal value for higher order curved elements and we seldom find in the literature the numerical results for the higher order element discretization.

REGULARIZATION OF BOUNDARY INTEGRAL EQUATION FOR ROTATION

An alternative expression to equation (1) can be written as follows :

$$w(y) = \int_\Gamma [w^* V_n - T_n^* M_n + M_n^* T_n - Q_n^* w + M_{nt}^* \frac{\partial w}{\partial t}]\, d\Gamma$$

$$+ \int_\Omega w^* f\, d\Omega - \sum [\![w^* M_{nt}]\!] \tag{12}$$

where the following relationship is used :

$$\int_\Gamma V_n^* w\, d\Gamma = \int_\Gamma (Q_n^* + \frac{\partial M_{nt}^*}{\partial s})\, w\, d\Gamma$$

$$= \int_\Gamma Q_n^* w\, d\Gamma + \sum [\![M_{nt}^* w]\!] - \int_\Gamma M_{nt}^* \frac{\partial w}{\partial t}\, d\Gamma \tag{13}$$

Subtracting and adding back the same term, equation (12) becomes

$$w(y) = \int_\Gamma [\, w^* V_n - T_n^* M_n + M_n^* T_n - Q_n^* \{ w - w(y) \} + M_{nt}^* \frac{\partial w}{\partial t}] \, d\Gamma$$

$$- \{ \int_\Gamma Q_n^* d\Gamma \} w(y) + \int_\Omega w^* f \, d\Omega - \sum [\![w^* M_{nt}]\!] \tag{14}$$

Now, using the divergence theorem and equation (2), we obtain

$$\int_\Gamma Q_n^* d\Gamma = \int_\Gamma M_{ij,j}^* n_i d\Gamma = \int_\Omega M_{ij,ji}^* d\Omega$$

$$= - \int_\Omega \delta(x-y) \, d\Omega = -1, \quad y \in \Omega \tag{15}$$

Substituting equation (15) into equation (14) results in

$$\int_\Gamma [\, w^* V_n - T_n^* M_n + M_n^* T_n - Q_n^* \{ w - w(y) \} + M_{nt}^* \frac{\partial w}{\partial t}] \, d\Gamma$$

$$+ \int_\Omega w^* f \, d\Omega - \sum [\![w^* M_{nt}]\!] = 0 \tag{16}$$

Since Q_n^* is $O(1/r)$, we observe that equation (16) holds continuously from $y \in \Omega$ to $y \in \Gamma$ provided that $w(x)$ satisfies the Hölder continuity at y.

Now differentiating equation (16) with respect to y_i and multiplying the both sides by $n_i(y)$, we obtain

$$\int_\Gamma [\, \tilde{w}^* V_n - \tilde{T}_n^* M_n + \tilde{M}_n^* T_n - \tilde{Q}_n^* \{ w - w(y) \} + \tilde{M}_{nt}^* \frac{\partial w}{\partial t}] \, d\Gamma$$

$$+ \{ \int_\Gamma Q_n^* d\Gamma \} w_{,k}(y) n_k(y) + \int_\Omega \tilde{w}^* f \, d\Omega - \sum [\![\tilde{w}^* M_{nt}]\!] = 0 \tag{17}$$

Subtructing and adding back the same terms and using equation (15), equation (17) results in

$$\int_\Gamma [\, \tilde{w}^* V_n - \tilde{T}_n^* M_n + \tilde{M}_n^* \{ T_n - w_{,k}(y) n_k \} - \tilde{Q}_n^* \{ w - w(y) - r_k w_{,k}(y) \}$$

$$+ \tilde{M}_{nt}^* \{ \frac{\partial w}{\partial t} - w_{,k}(y) t_k \}] \, d\Gamma + \int_\Gamma (\tilde{M}_n^* n_k + \tilde{M}_{nt}^* t_k - \tilde{Q}_n^* r_k) \, d\Gamma \cdot w_{,k}(y)$$

$$- w_{,k}(y) n_k(y) + \int_\Omega \tilde{w}^* f \, d\Omega - \sum [\![\tilde{w}^* M_{nt}]\!] = 0 \tag{18}$$

In equation (18), we observe that the following relationship holds :

$$\int_\Gamma (\tilde{M}_n^* n_k + \tilde{M}_{nt}^* t_k - \tilde{Q}_n^* r_k) \, d\Gamma$$

$$= \int_\Gamma \frac{\partial}{\partial y_i} (M_n^* n_k + M_{nt}^* t_k - Q_n^* r_k) \, d\Gamma \cdot n_i(y)$$

$$= \int_\Gamma \{ (\frac{\partial}{\partial y_i} M_{jl}^*) n_j n_l n_k + (\frac{\partial}{\partial y_i} M_{jl}^*) n_j t_l t_k - (\frac{\partial}{\partial y_i} M_{jl,l}^*) n_j r_k \} \, d\Gamma \cdot n_i(y)$$

$$= \int_\Gamma \{(\frac{\partial}{\partial y_i} M^*_{jl})\delta_{lk} n_j - (\frac{\partial}{\partial y_i} M^*_{jl,l}) n_j r_k\} \, d\Gamma \cdot n_i(y)$$

$$= \int_\Omega \{\frac{\partial}{\partial y_i} M^*_{jk,j} - (\frac{\partial}{\partial y_i} M^*_{jl,lj}) r_k - (\frac{\partial}{\partial y_i} M^*_{jl,l})\delta_{kj}\} \, d\Omega \cdot n_i(y)$$

$$= \int_\Omega \{\frac{\partial}{\partial y_i} M^*_{jk,j} + \frac{\partial}{\partial y_i}\delta(x-y) r_k - \frac{\partial}{\partial y_i} M^*_{jk,j}\} \, d\Omega \cdot n_i(y)$$

$$= \int_\Omega \frac{\partial}{\partial y_i}\delta(x-y) r_k \, d\Omega \cdot n_i(y) = \delta_{ik} n_i(y) = n_k(y) \tag{19}$$

where

$$n_l n_k + t_l t_k = \delta_{lk} \tag{20}$$

$$\int_\Omega \frac{\partial}{\partial x_i}\delta(x-y) f(x) \, d\Omega_x = - \frac{\partial f(x)}{\partial x_i}\Bigg|_{x=y}, \quad y \in \Omega \tag{21}$$

are used. Using equation (19), equation (18) results in

$$\int_\Gamma [\, \tilde{w}^* V_n - \tilde{T}^*_n M_n + \tilde{M}^*_n \{T_n - w_{,k}(y) n_k\} - \tilde{Q}^*_n \{w - w(y) - r_k w_{,k}(y)\}$$

$$+ \tilde{M}^*_{nt}\{\frac{\partial w}{\partial t} - w_{,k}(y) t_k\}] \, d\Gamma + \int_\Omega \tilde{w}^* f \, d\Omega - \sum [\![\tilde{w}^* M_{nt}]\!] = 0 \tag{22}$$

In equation (22), we observe

$$\tilde{M}^*_n \sim O(1/r) \;, \quad \tilde{M}^*_{nt} \sim O(1/r) \;, \quad \tilde{Q}^*_n \sim O(1/r^2) \tag{23}$$

Therefore, equation (22) is at most weakly singular if $w(x) \in C^{1,\alpha}$ at y, i.e.,

$$| w(x) - w(y) - r_k w_{,k}(y) | < B r^{1+\alpha}, \quad 0 < \alpha \leq 1, \quad |B| < \infty \tag{24}$$

In this case, equation (22) can be continuously applied to the point y across the boundary.

The system of equations (16) and (22) can be used for the analysis of any boundary value problem of thin elastic plate bending.

By discretizing equations (16) and (22) and applying the boundary conditions, we obtain a system of liner algebraic equations for the unknown nodal values of w, T_n, M_n and V_n. In what follows, we consider the numerical treatment for equations (16) and (22). In these equations, we need to evaluate the integrals separately for Γ_S and Γ_R, where Γ_S is the set of the elements in which the source point lies and Γ_R is the rest of the elements. We can evaluate the integrals for Γ_R directly without any change, however, for Γ_S, we must cancel out explicitly the singularities of the kernels in advance.

For simplicity, consider a quadratic Lagrange element. We write here both the shape functions and interpolation functions by $\phi^p(\xi)$ (p=1,2,3), i.e.,

$$\phi^p(\xi) = \frac{(\xi - \xi^q)(\xi - \xi^r)}{(\xi^p - \xi^q)(\xi^p - \xi^r)}, \quad (p, q, r=1,2,3\,;\; q \neq r\,;\; p \neq q, r) \tag{25}$$

where ξ is the intrinsic coordinate and ξ^i (i=p, q, r) denote the coordinates of node i in the element respectively. Letting ζ be the value of the intrinsic coordinate of the source point y, the position vector of an arbitrary point x in the element from the source point becomes as follows :

$$\begin{aligned} r_i = x_i - y_i &= \sum_{p=1}^{3} \phi^p(\xi)\, x_i^p - \sum_{p=1}^{3} \phi^p(\zeta)\, x_i^p \\ &= (\xi - \zeta) \sum_{p=1}^{3} \frac{\xi + \zeta - \xi^q - \xi^r}{(\xi^p - \xi^q)(\xi^p - \xi^r)}\, x_i^p \\ &= \rho \sum_{p=1}^{3} \Phi^p(\xi, \zeta)\, x_i^p \end{aligned} \tag{26}$$

where x_i^p (p=1,2,3) denote the cartesian coordinates of the nodes with which the element is defined, and ρ and Φ^p are defined as

$$\rho = \xi - \zeta, \quad \Phi^p(\xi, \zeta) = \frac{\xi + \zeta - \xi^q - \xi^r}{(\xi^p - \xi^q)(\xi^p - \xi^r)}, \quad (p \neq q, r;\ q \neq r) \tag{27}$$

Based on the above definitions, the distance between the points x and y becomes

$$r = |r_i| = |\rho|\, \left| \sum_{p=1}^{3} \Phi^p(\xi, \zeta)\, x_i^p \right| \tag{28}$$

Similarly, the relative deflection w-$w(y)$ can be expressed by

$$w - w(y) = \rho \sum_{p=1}^{3} \Phi^p(\xi, \zeta)\, w^p \tag{29}$$

where w^p (p=1,2,3) are the nodal values of deflection.

In equation (28), only ρ tends to zero according to $r \to 0$ and the remaining part keeps non-zero. On the other hand, since the order of singularity of Q_n^* is $1/r$, it can be explicitly cancelled out in equation (16) by w-$w(y)$.

Now, we consider the treatment of $\partial w/\partial t$ and M_{nt}. For Γ_S, we can express $\partial w/\partial t$ in terms of the interpolation functions and the nodal values of deflection as follows :

$$\frac{\partial w}{\partial t} = \sum_{q=1}^{3} \frac{1}{J(\xi^q)} \left(\sum_{p=1}^{3} \frac{d\phi^p(\xi^q)}{d\xi}\, w^p \right) \tag{30}$$

where

$$J(\xi) = \left| \frac{dx_i}{d\xi} \right| = \left| \sum_{p=1}^{3} \frac{d\phi^p(\xi)}{d\xi}\, x_i^p \right| \tag{31}$$

For Γ_R, in order to avoid any approximation error, we integrate by parts the term for $\partial w/\partial t$ beforehand as follows :

$$\int_{\Gamma_R} M_{nt}^* \frac{\partial w}{\partial t}\, d\Gamma = \sum_{\Gamma_R} [\![M_{nt}^*\, w]\!] - \int_{\Gamma_R} \frac{\partial M_{nt}^*}{\partial s}\, w\, d\Gamma \tag{32}$$

The jump of M_{nt} at a corner can be calculated from the nodal values of T_n in the elements adjacent to the corner by using the following relationship :

$$M_{nt} = -D(1-\nu)\frac{\partial T_n}{\partial t} = -\frac{D(1-\nu)}{J(\xi)}\sum_{p=1}^{3}\frac{d\phi^p(\xi)}{d\xi}T_n^p \tag{33}$$

where T_n^p (p=1,2,3) are the nodal values of T_n. Besides M_{nt}, all the physical quantities but the deflection are discontinuous at corners and the treatment of them becomes cumbersome. Hence, we place the non-conforming elements adjacent to the corner points for simplicity.

Next we consider the numerical treatment of equation (22) for Γ_s. $w_{,k}(y)$ (k=1,2) do not correspond to any physical quantities given as the boundary conditions, therefore we express them in terms of $\partial w/\partial t$ and T_n as follows :

$$w_{,k}(y) = T_n(y)n_k(y) + \frac{\partial w}{\partial t}(y)t_k(y) \tag{34}$$

Then, the following relationships hold :

$$\begin{aligned}\tilde{M}_n^*\{T_n - w_{,k}(y)n_k\} + \tilde{M}_{nt}^*\{\frac{\partial w}{\partial t} - w_{,k}(y)t_k\} \\ = \tilde{M}_{ij}^*n_i\ [\ T_n\{n_j - n_j(y)\} - \{T_n - T_n(y)\}n_j(y)\ ] \\ + \tilde{M}_{ij}^*n_i\ [\ \frac{\partial w}{\partial t}\{t_j - t_j(y)\} - \{\frac{\partial w}{\partial t} - \frac{\partial w}{\partial t}(y)\}t_j(y)\ ]\end{aligned} \tag{35}$$

$$\begin{aligned}\tilde{Q}_n^*\{w - w(y) - r_k w_{,k}(y)\} \\ = \tilde{Q}_n^*\{w - w(y) - r_k t_k(y)\frac{\partial w}{\partial t}(y)\} - \tilde{Q}_n^* r_k n_k(y)T_n(y)\end{aligned} \tag{36}$$

$\frac{\partial w}{\partial t}(y)$ can also be expressed with the nodal values of w by inserting ζ into ξ in equation (30). $\tilde{M}_{ij}^*$ in equation (35) is $O(1/r)$ and the terms enclosed with { } are $O(r)$. Therefore the singularity of the kernel $\tilde{M}_{ij}^*$ is cancelled out. Also by using the interpolation functions, the related terms in equation (36) result in

$$\begin{aligned}w - w(y) - r_k t_k(y)\frac{\partial w}{\partial t}(y) \\ = \frac{\rho^2}{J(\zeta)^2}\sum_{p=1}^{3}\Phi^p(\zeta,\zeta)x_i^p\sum_{q=1}^{3}\sum_{r=1}^{3}\frac{(\xi^a+\xi^b-\xi^c-\xi^d)\,x_i^r\,w^q}{\xi^{qa}\ \xi^{qb}\ \xi^{rc}\ \xi^{rd}}, \\ (a\neq q,\ b\neq q,\ c\neq r,\ d\neq r)\end{aligned} \tag{37}$$

$$\begin{aligned}r_k n_k(y) = \frac{\rho^2}{J(\zeta)^2}\sum_{p=1}^{3}\frac{x_k^p}{\xi^{pa}\ \xi^{pb}}\sum_{q=1}^{3}\Phi^q(\zeta,\zeta)\varepsilon_{kl}x_l^q, \\ (p\neq a, b;\ a\neq b)\end{aligned} \tag{38}$$

where ε_{kl} is the permutation symbol and ξ^{qa}, ξ^{qb}, $\cdots$, ξ^{pb} are defined as

$$\xi^{qa} = \xi^q - \xi^a,\ \ \xi^{qb} = \xi^q - \xi^b,\ \cdots,\ \ \xi^{pb} = \xi^p - \xi^b \tag{39}$$

The r^{-2} order singularity of $\tilde{Q}_n^*$ is cancelled out explicitly by ρ^2 in equations (37) and (38).

As has been discussed here, all the integrals in equations (16) and (22) actually result in the integrals in the standard sense and can be evaluated accurately by using the standard Gaussian quadrature formula.

NUMERICAL EXAMPLES

In order to demonstrate the effectiveness of the present formulation, a square plate subjected to a uniform load distribution with two different boundary conditions and a circular plate subjected to a concentrated load at its center are analyzed.

In the first example, the square plate is clamped along the entire boundary. Each edge of the plate is divided into 8 quadratic elements. In Figures 1 and 2 are shown the results of the bending moment and the equivalent shear force along an edge, respectively. In these figures, x denotes the distance along the edge from the corner point and both the bending moments and the equivalent shear forces are normalized by f and L, where f is the magnitude of the distributed load and L is the length of the edge of the square plate. The domain integral terms in the integral equations originated from the uniform load distribution can be evaluated by transforming them into the equivalent boundary integrals, hence any discretization of the plate domain into internal cells is not needed here. Both results show good agreement with the analytical solutions by Timoshenko [7]. In particular, the present results of the equivalent shear force are more accurate than those obtained by Stern [2].

In the second example, the square plate is simply supported along the entire boundary. The obtained normal slopes and equivalent shear forces are shown in Figures 3 and 4. Also in this example, the present results are accurate compared with those obtained by Stern [2] and Costa-Brebbia [4].

Finally, in Table 1 are shown the results for a circular plate clamped along the entire boundary. In the table, f is the magnitude of the concentrated force and a is the radius of the circular plate. In this example, the obtained bending moment and equivalent shear force on the boundary completely agree with the analytical solutions by Timoshenko [7].

CONCLUSIONS

A new type of regularized boundary integral equation for the rotation has been presented for the thin elastic plate bending problem. By combining it with the regularized boundary integral equation for the deflection, we can avoid evaluating any Cauchy principal value integral. Therefore, higher order elements are available in the discretization of the boundary integral equations and all the integrals can be evaluated by using the standard Gaussian quadrature formula. The numerical examples have demonstrated that the analysis based on the present formulation gives us accurate solutions of the boundary quanities, including those of the equivalent shear force.

REFERENCES

(1) Niwa, Y., Kobayashi, S. and Fukui, T., An Application of the Integral Equation Method to Plate Bending, Memoirs of Faculty of Engineering, Kyoto Univ., Japan, Vol. 36-2, pp. 140-158, 1974.

(2) Stern, M., A General Boundary Integral Formulation for the Numerical Solution of Plate Bending Problems, Int. J. Solids Structures, Vol. 15, pp. 769-782, 1979.

(3) Bezine, G.P. and Gamby, D.A., A New Integral Equation Formulation for Plate Bending Problems, Recent Advances in Boundary Element Methods, (Ed. Brebbia, C.A.), pp. 327-341, Pentech Press, London, 1978.

(4) Costa, Jr., J.A. and Brebbia, C.A., Plate Bending Problems Using BEM, Proc. 6th Int. Conf. on Board the Liner the Queen Elisabeth 2, Southampton to New York, pp. 3-43 to 3-63, Springer-Verlag, 1984.

(5) Hartmann, F. and Zotemantel, R., The Direct Boundary Element Method in Plate Bending, International Journal for Numerical Methods in Engineering, Vol. 27, pp. 2049-2069, 1986.

(6) Balaš, J., Sládek, J. and Sládek, V., Stress Analysis by Boundary Element Methods, Elsevier, Amsterdam, 1989.

(7) Timoshenko, S. and Woinowsky-Krieger, S., Theory of Plates and Shells, McGraw-Hill, 1959.

(8) Kutt, H.R., The Numerical Evaluation of Principal Value Integrals by Finite-Part Integration, Numer. Math. Vol. 24, pp. 205-210, 1975.

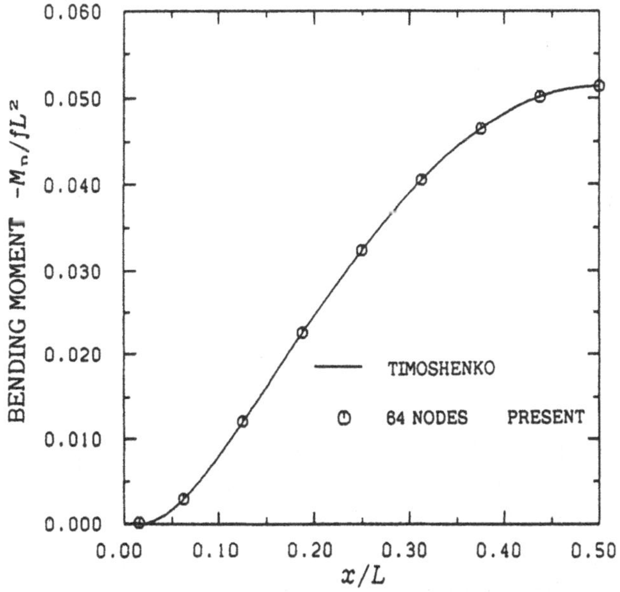

Figure 1. Results for uniformly loaded clamped square plate — bending moment.

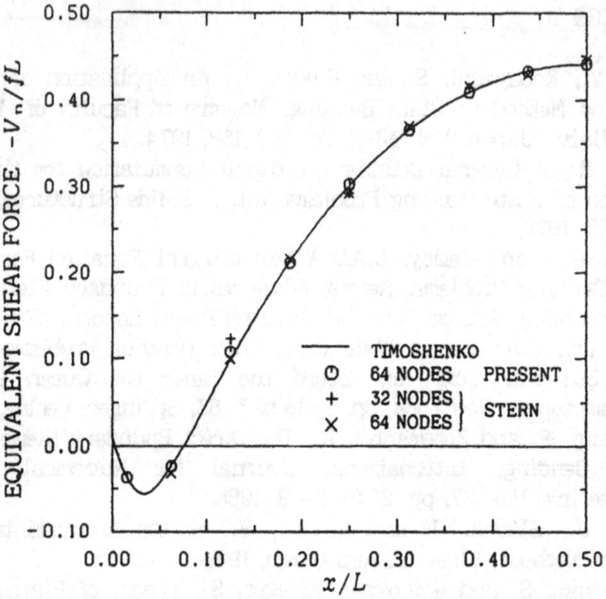

Figure 2. Results for uniformly loaded clamped square plate — equivalent shear force.

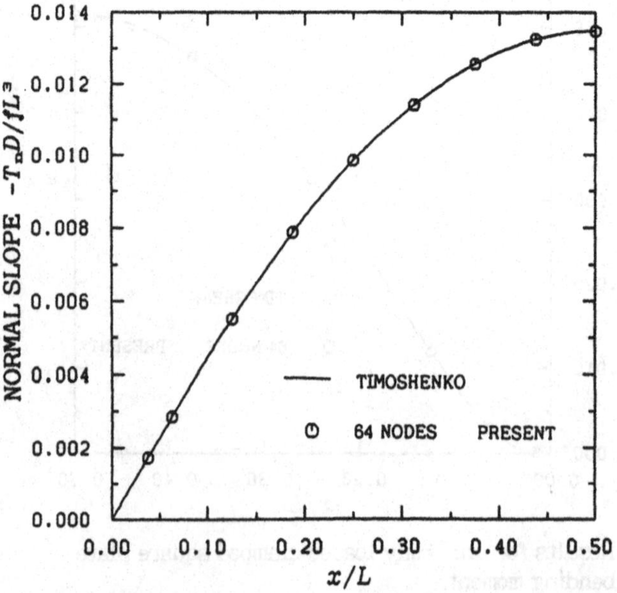

Figure 3. Results for uniformly loaded simply supported square plate — normal slope.

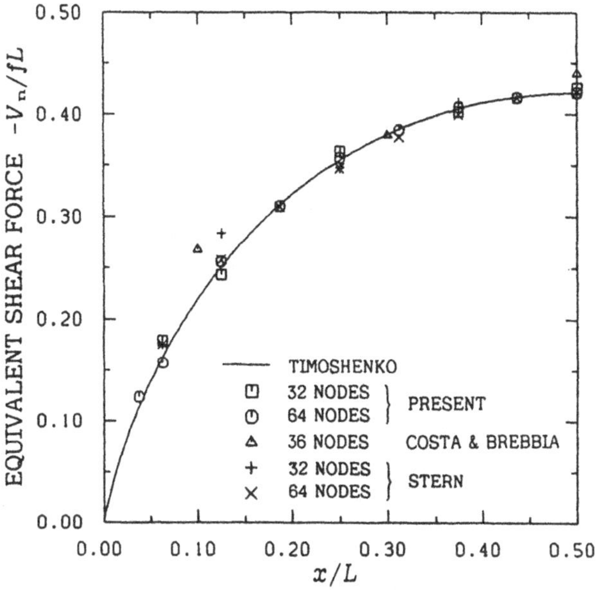

Figure 4. Results for uniformly loaded simply supported square plate — equivalent shear force.

Table 1. Results for clamped circular plate with concentrated load at its center.

PRESENT RESULTS		EXACT(TIMOSHENKO)	
M_n/f	V_na/f	M_n/f	V_na/f
-0.079577	-0.15915	-0.079577	-0.15915

An Integral Equation Formulation for the Boundary-Finite Element Model in Stiffened Plate Bending

H.-R. Shih (*), R.C. Duffield (**), J. Lin (**)
(*) Jackson State University, Jackson, Mississippi, U.S.A.
(**) University of Missouri-Columbia, Columbia, Missouri, U.S.A.

ABSTRACT

An integral equation formulation and a numerical procedure for a boundary-finite element technique are developed for the static analysis of a stiffened plate with concentric stiffeners. This formulation employs the fundamental solution associated with unstiffened plate bending and this creates not only integrals along the perimeter of the stiffened plate but additional integrals along the stiffeners and the interface between the plate and its stiffeners. Thus the domain of the plate has to be divided into zones between the stiffeners. Each zone is modeled by boundary elements and stiffeners by finite elements. The zone technique which permits coupling of the unstiffened plate boundary elements with the beam (stiffener) finite elements is presented. Numerical examples are given to demonstrate the effectiveness of this approach.

INTRODUCTION

Stiffened plates are extensively used in many important engineering structures such as vehicles, ships, aircrafts, bridges, etc. The reason that stiffened plates are used in such a wide variety of applications is due to the efficiency and economy of these types of structures. The analysis of such plates has been found to be more complex than for unstiffened plates. One of the most commonly used classical solution techniques replaces the stiffened plate with a substitute structure consisting of an equivalent orthotropic plate, which is arrived at by smearing the stiffener properties over the plate [1,2]. However, the classical approaches are generally limited in their ability to solve plate with an arbitrary shape. Two approaches generally used in the finite element analysis of stiffened plates are as follows [3-5]: (1) discretize the stiffened plate system as plates and beams; or (2) replace the stiffened plate by an equivalent isotropic/orthotropic plate. The finite element method does not restrict the type of geometric configurations or types of boundary and loading conditions which can be applied to a stiffened plate problem.

This paper presents an integral equation formulation and a numerical procedure for a boundary-finite element technique that can be used to solve bending problems of a concentrically stiffened plate. Calculus of variation is applied to the virtual work equilibrium expression for the stiffened plate to derive the governing field equations and the associated boundary conditions. The weighted residual method is then employed to formulate the integral equation for stiffened plates. This formulation employs the fundamental solution of an unstiffened plate bending problem instead of the fundamental solution to the stiffened plate problem. This creates not only integrals along the perimeter of the stiffened plate but additional integrals along the stiffeners and the plate-stiffener interfaces. Thus, the domain of the plate is divided into zones between the stiffeners. Each zone is modeled by boundary elements and stiffeners by finite elements (hence the name, boundary-finite element model). The numerical implementations of zone technique are discussed which involves employment of the equilibrium and compatibility conditions along the interface to couple the unstiffened plate boundary elements with the stiffener finite elements. The proposed method is compared against analytical solutions and finite element method solutions by using a number of examples. These numerical examples also serve to show the potential usefulness of the proposed method.

FIELD EQUATIONS AND BOUNDARY CONDITIONS

Consider a stiffened plate of arbitrary planform with a discrete number of stiffeners in both the x and y directions. Let Ω represent the plate's domain and Γ its boundary. The plate and stiffeners are assumed to be made from an elastic, homogeneous and isotropic material where the material properties of the plate may be different from those of the stiffeners. In general, stiffeners in x-direction are called ribs and those in y-direction are called stringers. The determination of the integral equation for a stiffened plate by the weighted residual method requires that the field equations and the associated boundary conditions for the structural system be known. The field equations can be obtained by first developing the equilibrium conditions for the stiffened plate in terms of virtual work. Since virtual work involves variations, the concepts of calculus of variations can be applied to the virtual work statement to achieve the field equations and the associated boundary conditions. The field equations for a stiffened plate with N ribs and M stringers can be expressed as follows

$$D \nabla^4 w + \sum_{i=1}^{N} E_{ri} I_{riy} w_{,xxxx} \delta(y-y_i) + \sum_{j=1}^{M} E_{sj} I_{sjx} w_{,yyyy} \delta(x-x_j)$$
$$= q(x,y) \qquad (1)$$

$$\sum_{i=1}^{N} G_{ri} J_{ri} \phi_{,xx} \delta(y-y_i) = 0 \qquad (2)$$

$$\sum_{j=1}^{M} G_{sj} J_{sj} \phi_{,yy} \delta(x-x_j) = 0 \qquad (3)$$

where w is the transverse deflection of the stiffened plate, $q(x,y)$ is the transversely applied load, ϕ is the twisting angle of the stiffener, and $D=Eh^3/12(1-\upsilon^2)$ is the flexural rigidity of the plate with E, h and υ being the elastic modulus, the plate thickness and Poisson's ratio, respectively. Also in above equations ∇^4 is the biharmonic differential operator, y_i and x_j are the y-location of the ith rib and the x-location of the jth stringer, respectively, E_{ri} and E_{sj} are moduli of elasticity for the ith rib and the jth stringer respectively, I_{riy} and I_{sjx} are area moments of inertia for the ith rib and the jth stringer respectively, and G_{ri}, J_{ri}, G_{sj} and J_{sj} are the shear moduli of elasticity and torsional rigidity constants for the ith rib and the jth stringer, respectively.

The associated boundary conditions are

$V_n = V_n^*$	or	w	prescribed	(4)
$M_n = M_n^*$	or	$\theta_n \left(= \frac{dw}{dn}\right)$	prescribed	(5)
$Q_{ce} = Q_{ce}^*$	or	$w\|_{\Gamma=\Gamma_{ce}}$	prescribed	(6)
$M_{rbi} = M_{rbi}^*$	or	$w_{,x}\|_{y=y_i}$	prescribed	(7)
$Q_{ri} = Q_{ri}^*$	or	$w\|_{y=y_i}$	prescribed	(8)
$M_{rti} = M_{rti}^*$	or	$\phi\|_{y=y_i}$	prescribed	(9)
$M_{sbj} = M_{sbj}^*$	or	$w_{,y}\|_{x=x_j}$	prescribed	(10)
$Q_{sj} = Q_{sj}^*$	or	$w\|_{x=x_j}$	prescribed	(11)

$M_{stj} = M_{stj}^*$ or $\phi|_{x=x_j}$ prescribed (12)

in which V_n, M_n Q_{ce}, θ_n and n represent Kirchhoff's shear force, normal bending moment, corner force, normal slope and outward normal vector on Γ, respectively, M_{rbi}, M_{rti} and Q_{ri} denote bending moment, twisting moment and shear force, respectively, on the ith rib, M_{sbj}, M_{stj} and Q_{sj} are bending moment, twisting moment and shear force, respectively, on the jth stringer, and starred quantities will denote applied boundary loads.

INTEGRAL EQUATION FORMULATION

The integral equation formulation needed for the boundary-finite element technique can be determined by the application of the weighted residual method to equations (1) through (3). For the sake of simplicity, consider the case of a stiffened plate with only one rib, as shown in Figure 1. In accordance with the procedure introduced by Brebbia [6], the weighted residual equation becomes

$$\int_\Omega \{D\nabla^4 w - q(x,y) + E_r I_{ry} w_{,xxxx} \; \delta(y-y_r)\} \overline{w}\, d\Omega -$$

$$\int_{\Gamma_w} (w - w^*) \overline{V}_n \, d\Gamma_w + \int_{\Gamma_\theta} (\theta_n - \theta_n^*) \overline{M}_n \, d\Gamma_\theta - \int_{\Gamma_M} (M_n - M_n^*) \overline{\theta}_n \, d\Gamma_M$$

$$+ \int_{\Gamma_V} (V_n - V_n^*) \overline{w} \, d\Gamma_V - \int_\Gamma (M_{rb} - M_{rb}^*) \overline{w}_{,x} \, \delta(\Gamma-\Gamma_{Mrb}) \, d\Gamma +$$

$$\int_\Gamma (Q_r - Q_r^*) \overline{w} \, \delta(\Gamma-\Gamma_{Qr}) \, d\Gamma + \sum_{e=1}^{C1} \int_\Gamma (Q_{ce} - Q_{ce}^*) \overline{w} \, \delta(\Gamma-\Gamma_{ce}) \, d\Gamma -$$

$$\sum_{e=1}^{C2} \int_\Gamma (w - w_c^*) \overline{Q}_{ce} \, \delta(\Gamma-\Gamma_{ce}) \, d\Gamma = 0 \qquad (13)$$

where $\overline{w}$ is the weighting function, $\overline{\theta}_n$, $\overline{M}_n$, $\overline{V}_n$ and $\overline{Q}_{ce}$ are the slope in normal direction, bending moment, Kirchhoff shear and corner forces associated with $\overline{w}$ function, and the starred terms denote prescribed values on the designated sections of the plate boundary indicated by Γ followed by an appropriate boundary condition parameter.

In order to obtain the inverse relationship of equation (13) from which the integral equation is obtained, it is necessary to apply Green's theorem and classical plate theory to the first term of equation (13) [7]. This gives

$$D\int_{\Omega}(\nabla^4 w)\,\overline{w}\,d\Omega = D\int_{\Omega} w\,(\nabla^4 \overline{w})\,d\Omega + \int_{\Gamma}(\overline{V}_n w - \overline{M}_n \theta_n + M_n \overline{\theta}_n -$$

$$V_n \overline{w})\,d\Gamma - \sum_{e=1}^{C}\int_{\Gamma}(Q_{ce}\overline{w} - \overline{Q}_{ce} w)\,\delta(\Gamma-\Gamma_{ce})\,d\Gamma \qquad (14)$$

In order to further reduce equation (13), it is necessary to consider the stiffener term

$$\int_{\Omega} E_r I_{ry} w_{,xxxx}\ \overline{w}\ \delta(y-y_r)\,d\Omega \qquad (15)$$

Since the finite element model is used for stiffener, and weighting function is associated with 'virtual' increment of w. The integration of equation (15) twice by parts and application of classical beam theory yield

$$\int_{\Omega} E_r I_{ry} w_{,xxxx}\ \delta w\, \delta(y-y_r)\,d\Omega = \int_{\Omega} E_r I_{ry} w_{,xx}\ \delta w_{,xx}\ \delta(y-y_r)\,d\Omega -$$

$$\int_{\Gamma} Q_r\, \delta w\, \delta(\Gamma-\Gamma_r)\,d\Gamma + \int_{\Gamma} M_{rb}\, \delta w_{,x}\ \delta(\Gamma-\Gamma_r)\,d\Gamma \qquad (16)$$

Several of terms which occur in the equation obtained from the substitution of equations (14) and (16) into equation (13) can be reduced. An example of this reduction is shown in the following group of terms

$$\int_{\Gamma} w\ \overline{V}_n\,d\Gamma\ -\ \int_{\Gamma_w}(w - w^*)\,\overline{V}_n\,d\Gamma_w\ =$$

$$\int_{\Gamma_v} w\ \overline{V}_n\,d\Gamma_v\ +\ \int_{\Gamma_w} w^*\ \overline{V}_n\ d\Gamma_w = \int_{\Gamma} w\ \overline{V}_n\ d\Gamma \qquad (17)$$

where in the final expression of equation (17), the quantity w takes on the values w^* on the boundary Γ_w. Similar types of reduction are applied to other terms and this leads to yield

$$\int_{\Omega} D\,(\nabla^4 \overline{w})\,w\,d\Omega - \int_{\Omega} q(x,y)\ \overline{w}\,d\Omega + \int_{\Gamma}(\overline{V}_n w - \overline{M}_n \theta_n + M_n \overline{\theta}_n -$$

$$V_n \overline{w})\,d\Gamma + \sum_{e=1}^{C}\int_{\Gamma}(\overline{Q}_{ce} w - Q_{ce}\overline{w})\,\delta(\Gamma-\Gamma_{ce})\,d\Gamma +$$

$$\int_{\Omega} E_r I_{ry} w_{,xx}\ \delta w_{,xx}\ \delta(y-y_r)\,d\Omega -$$

$$\int_{\Gamma} Q_r \, \delta w \, \delta(\Gamma-\Gamma_r)\, d\Gamma + \int_{\Gamma} M_{rb} \, \delta w_{,x} \, \delta(\Gamma-\Gamma_r)\, d\Gamma = 0 \qquad (18)$$

The consideration of equation (2) and the associated boundary conditions permits the weighted residual expression to be written as

$$\int_{\Omega} G_r J_r \phi_{,xx} \, \overline{w} \, \delta(y-y_r)\, d\Omega - \int_{\Gamma} (M_{rt} - M_{rt}^{*}) \, \overline{w} \, \delta(\Gamma-\Gamma_r)\, d\Gamma = 0 \qquad (19)$$

The integration by parts of the first term of equation (19) and the employment of $\overline{w} = \delta\phi$ on the boundary give

$$\int_{\Omega} G_r J_r \phi_{,x} \, \delta\phi_{,x} \, \delta(y-y_r)\, d\Omega - \int_{\Gamma} M_{rt} \, \delta\phi \, \delta(\Gamma-\Gamma_r)\, d\Gamma = 0 \qquad (20)$$

The system integral equation can be obtained by putting equations (18) and (20) together. It should be noted that equation (18) not only contain integrals along the boundary of the stiffened plate but also contain additional integrals along the stiffener. Therefore, in order to derive the integral equation formulation of boundary-finite element technique for stiffened plates, equation (18) has to be modified. The domain of the stiffened plate is divided into two zones on either side of the rib as shown in Figure 1 which shows a simplication of the structural system for explaination purposes only. Thus, the domain, Ω, consists of two zones, Ω_1 and Ω_2, and the rib (Figure 2). The interface boundary between Ω_1 and rib is called Γ_{I1} and that between Ω_2 and the rib is called Γ_{I2}. In equation (18), the third and fourth integrals which integrate around the total boundary can be replaced by integrals around each zone. Thus, integral equation for the stiffened plate system can be expressed as

$$\int_{\Omega} D(\nabla^4 \overline{w})\, w \, d\Omega - \sum_{k=1}^{2} \int_{\Omega_k} q(x,y)\, \overline{w}\, d\Omega_k + \sum_{k=1}^{2} \int_{\Gamma_k} (\overline{V}_n w - \overline{M}_n \theta_n +$$

$$M_n \overline{\theta}_n - V_n \overline{w})\, d\Gamma_k + \sum_{k=1}^{2} \int_{\Gamma_{Ik}} (\overline{V}_n w_k - \overline{M}_n \theta_{nk} + M_{nk} \overline{\theta}_n - V_{nk} \overline{w})\, d\Gamma_{Ik} +$$

$$\sum_{k=1}^{2} \sum_{e=1}^{C} \int_{\Gamma_k} (\overline{Q}_{ce} w - Q_{ce} \overline{w})\, \delta(\Gamma-\Gamma_{ce})\, d\Gamma_k +$$

$$\int_{\Omega} E_r I_{ry} w_{,xx} \, \delta w_{,xx} \, \delta(y-y_r)\, d\Omega + \int_{\Omega} G_r J_r \phi_{,x} \, \delta\phi_{,x} \, \delta(y-y_r)\, d\Omega -$$

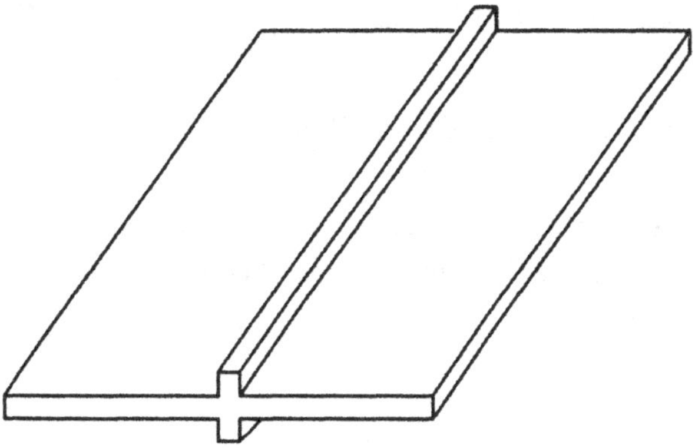

Figure 1 Stiffened Plate with one Rib

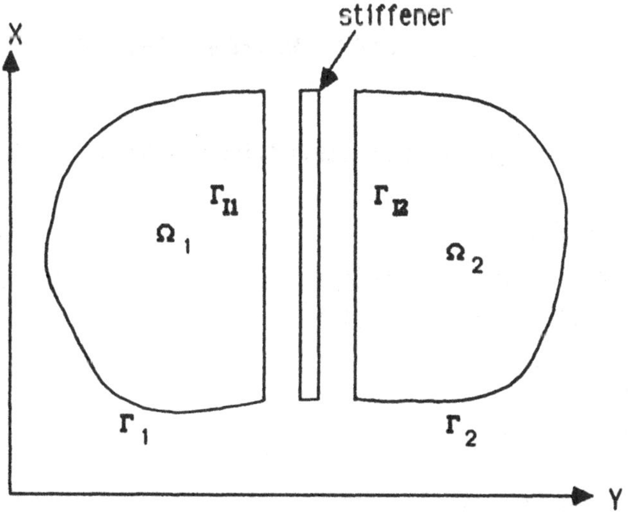

Figure 2 Domain Divided Into Two Zones and a Stiffener

$$\int_\Omega q_r\, \delta w\, \delta(y-y_r)\, d\Omega - \int_\Omega t_r\, \delta\phi\, \delta(y-y_r)\, d\Omega - \int_\Gamma Q_r\, \delta w\, \delta(\Gamma-\Gamma_r)\, d\Gamma +$$

$$\int_\Gamma M_{rb}\, \delta w_{,x}\, \delta(\Gamma-\Gamma_r)\, d\Gamma - \int_\Gamma M_{rt}\, \delta\phi\, \delta(\Gamma-\Gamma_r)\, d\Gamma = 0 \qquad (21)$$

where the terms q_r and t_r are the distributed load and torque on the rib respectively, and based on equilibrium conditions can be expressed as

$$q_r = -(V_{n1} + V_{n2}) \qquad (22)$$

$$t_r = -(M_{n1} + M_{n2}) \qquad (23)$$

Equation (21) is the governing equation for boundary-finite element solutions of stiffened plate bending problems. In the equation (21) the last seven terms are used to develop the stiffener finite elements. The rest of terms are used to produce the boundary element equation for plate. The above derivation is just for the special case of a concentrically stiffened plate with only one rib. The general equation can be acquired by extending equation (21) to include any number of ribs and stringers.

FUNDAMENTAL SOLUTIONS AND PLATE BOUNDARY INTEGRAL EQUATIONS

In order to obtain the boundary integral equations, the fundamental solution of the biharmonic equation

$$\nabla^4 \overline{w} = \delta(x-\xi)\, \delta(y-\eta) \qquad (24)$$

is obtained where $\delta(x-\xi)\,\delta(y-\eta)$ is the Dirac delta function. The fundamental solution corresponding to equation (24) [8,9] is

$$\overline{w}(A,B) = \frac{1}{8\pi D} r^2 \ln r \qquad (25)$$

where

$$r = \sqrt{(x-\xi)^2 + (y-\eta)^2} \qquad (26)$$

is the distance between the deflection point at (x,y) (point B) and load point at (ξ,η) (point A). The function $\overline{w}$ physically represents the transverse deflection at (x,y) of an infinitely extended plate due to a transverse concentrated unit load at (ξ,η). For each zone of the stiffened plate system, as shown in Figure 2, or a unstiffened plate, use of this fundamental solution leads to the following boundary integral equation

$$c(\xi,\eta)\, w(\xi,\eta) = -\int_\Gamma (\overline{V}_n w - \overline{M}_n \theta_n + M_n \overline{\theta}_n - V_n \overline{w})\, d\Gamma -$$

$$[\, w\,\overline{M}_t - M_t\,\overline{w}\,] + \int_\Omega q(x,y)\,\overline{w}\; d\Omega \qquad (27)$$

where $c(\xi,\eta)$ is a coefficient that depends on the geometry of the boundary at (ξ,η). The explicit expressions for $\overline{\theta}_n$, $\overline{M}_n$, $\overline{V}_n$ and $\overline{M}_t$ can be found elsewhere [8,9].

Equation (27) has to be complemented by an additional expression, in order to have a sufficient number of equations to correspond with the number of unknowns. Replacing $\overline{w}$ by a fundamental solution

$$\overline{w}_1 = -\frac{1}{8\pi D}(2\ln r + 1)\,[(x-\xi)\,n_\xi + (y-\eta)\,n_\eta]$$

$$= \overline{w}_{,\xi}\,n_\xi + \overline{w}_{,\eta}\,n_\eta \qquad (28)$$

the directional derivative (normal slope) of $\overline{w}$, the additional equation required becomes

$$c(\xi,\eta)\,\theta_n(\xi,\eta) = -\int_\Gamma \{\, \overline{V}_n(\overline{w}_1)[w(B)-w(A)] - \overline{M}_n(\overline{w}_1)\,\theta_n +$$

$$M_n\,\overline{\theta}_n(\overline{w}_1) - V_n\,w_1\}\,d\Gamma - [\,(w(B)-w(A))\,M_t(\overline{w}_1) - M_t\,\overline{w}_1\,] +$$

$$\int_\Omega q(x,y)\,\overline{w}_1\; d\Omega \qquad (29)$$

In equation (28), local coordinates (ξ,η) in which ξ and η are parallel to the x and y axes and origin occurs at the load point are adopted. Explicit expressions for $\overline{\theta}_n(\overline{w}_1)$, $\overline{M}_n(\overline{w}_1)$, $\overline{V}_n(\overline{w}_1)$ and $\overline{M}_t(\overline{w}_1)$, which appear in equation (29), can be found elsewhere [8,9].

NUMERICAL TREATMENT

In order to solve the above integral equations by means of boundary-finite element method, the integral along boundary, plate-stiffener interface and corresponding stiffeners must be divided into finite subintervals. In addition, the boundary and stiffener functions must be approximated by piecewise polynomials or interpolation functions. With this procedure, the problem reduces to the task of determining the nodal values of the unknown functions. The Hermite interpolation functions are selected to approximate w and $w_{,t}$, in order to make sure that the tangent slope of w is continuous. Linear interpolation functions are used to represent the remaining field variables.

Application of the solution procedure introduced above, for each zone of the stiffened plate, permits the boundary integral equations (27) and (29) to be reduced to a set of algebraic equations. If the set of "displacement terms" {deflection, slope} is denoted by {U} and the set of "force term" {moment, Kirchhoff shear, corner forces} by {T}, then the resulting boundary element equations obtainable from equations (27) and (29) can be expresssed in the following matrix form:

$$[H]\{U\} = [G]\{T\} + \{B\} \tag{30}$$

where the coefficient matrices, [H] and [G], are calculated using the fundamental solutions and the vector {B} is determined from the load distribution.

As stated in previous sections, the stiffener is modelled by finite elements. The standard finite element matrix equation can be written as

$$[K]\{U_f\} = \{F\} \tag{31}$$

where [K] is the stiffness matrix for the stiffener, {F} is the equivalent nodal force vector and $\{U_f\}$ is the vector which contains the transverse deflection, tangential slope and twisting angle nodal values.

The stiffness matrix of a stiffener element can be derived from the sixth and seventh terms of equation (21) [10]. The vector {F} is obtained by weighting the applied loads with the interpolation functions used for the displacements. Since the stiffener finite elements need to be coupled with the plate boundary element model, vector {F} has to be expressed in the form

$$\{F\} = [M]\{T_f\} \tag{32}$$

where $\{T_f\}$ is a vector composed of the nodal values of applied loads on the stiffener, and [M] is a matrix obtained from the weighting of distributed loads with the interpolation functions used for both the displacements and distributed loads [11]. Hence, equation (31) can be rewritten as

$$[K]\{U_f\} = [M]\{T_f\} \tag{33}$$

which is a form similar to the boundary element equation, resulting from equation (30). For a stiffener element, the [M] matrix can be derived from the eighth and ninth terms of equation (21) by expressing w, q_r, t_r, and ϕ within an element in term of their nodal values and corresponding interpolation functions.

In order to combine the plate boundary elements and stiffener finite elements, the compatibility and equilibrium conditions along the stiffener-plate interface must be considered. For simplicity, it is assumed that the stiffener lies along one edge of the plate, as shown in Figure 3. The continuity condition of displacement yields that the displacement of the plate along the interface is equal to the displacement of the stiffener. Also, the rotation of the plate along the interface is equal to the twisting angle of the stiffener. It should be observed that the force in the z direction transmitted from the plate to the stiffener is

$$-V_n = q_r \tag{34}$$

where q_r is distributed load on the stiffener and V_n is the Kirchhoff shear force of plate along the interface. The moment normal to the plate along interface transmits continuously distributed twisting moments to the stiffener. The magnitude of these distributed torques t_r on the stiffener is equal and opposite to the normal bending moments M_n along the interface boundary of the plate, i.e.

$$-M_n = t_r \tag{35}$$

In general boundary-finite element analysis of a stiffened plate, the entire structure is partitioned into stiffeners and plate zones. For each zone of plate, its entire boundary is characterized by noninterfaced boundaries plus interfaced boundaries where it is joined to stiffener. Hence, the equation (30) has to be rewritten as

$$[H_b \; H_I] \begin{Bmatrix} U_b \\ U_I \end{Bmatrix} = [G_b \; G_I] \begin{Bmatrix} T_b \\ T_I \end{Bmatrix} + \begin{Bmatrix} B_b \\ B_I \end{Bmatrix} \tag{36}$$

where subscripts I and b denote interface boundaries and noninterface boundaries, respectively. The application of the compatibility and equilibrium conditions along the interface and the combination of equations (36) and (33) yield the system of equations for entire stiffened plate. This system of equations then needs to be reordered according to the prescribed boundary conditions.

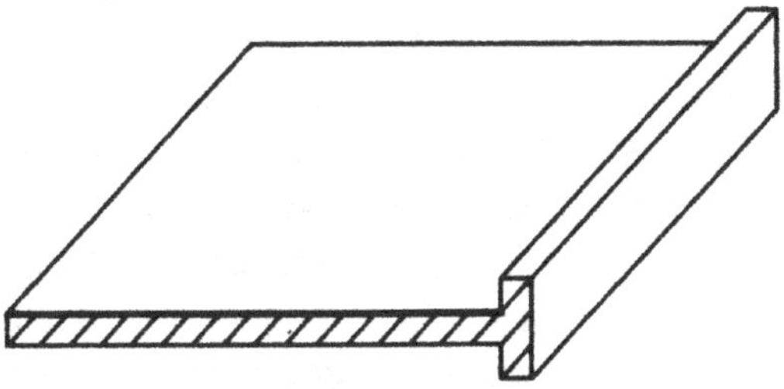

Figure 3 Plate with a Stiffener on the Edge

NUMERICAL EXAMPLES

This section presents two numerical examples to illustrate the proposed boundary-finite element model and demonstrate its accuracy by comparing the computed results against those results obtained by means of analytical methods or by means of finite element methods using the code GTSTRUDL [12].

Example 1 Simply supported square plate with one concentric stiffener
This example is a simply supported stiffened plate subjected to a uniformly distributed load of 10 psi. Figure 4 shows the detail of this plate. The plate was divided into two zones and one stiffener. The analyses have been carried out using three different element refinements of the boundary-finite element method (B-FEM). For the first analysis the boundary of each plate zone and the stiffener were discretized using 8 linear boundary elements and the 2 beam elements, respectively. The refined meshes which were described by 16 boundary and 4 beam elements, or 28 boundary and 8 beam elements. The deflection at the center of the stiffened plate as obtained from the three refined meshes along with the analytical solution [13] to the same problem are presented in Table 1. Also shown in the Table is the error between the B-FEM results and the analytical solution. The results of the boundary-finite element model of stiffened plate approach those of the analytical solution with refinement.

Example 2 A stiffened plate with a square opening
The geometry and loading for the problem are shown in Figure 5. Two opposite outer edges are simply supported and the other two outer edges are free and reinforced with stiffeners. The inner edges are free. Half of plate, which without the opening, is subjected to a uniformly distributed load of 20 psi. The elastic modulus for the plate is 30×10^6 psi, poisson's ratio is 0.3 and plate thickness is 0.4 in. The flexural and torsional rigidities of stiffeners are 34.34×10^6 lb-in^2 and 20×10^6 lb-in^2, respectively. Two analyses have been carried out. For the finite element analysis, the stiffened plate is represented by 96 four-node plate elements and 20 beam elements. For boundary-finite element model, the inner and outer edges of plate have been descretized by using 32 and 40 linear boundary elements respectively, and each stiffener has been divided into 10 beam elements. A comparison between the deflections calculated at typical points is presented in Table 2. The results obtained using boundary-finite element method are in close agreement with those obtained by GTSTRUDL.

CONCLUSION

A boundary-finite element model, which combines unstiffened plate boundary elements and stiffener finite elements, for a concentrically stiffened plate has been developed. This proposed method employs the fundamental solution of an unstiffened plate bending problem and this creates not only integrals along the perimeter of the stiffened plate but additional integrals along the stiffeners and the interface between the plate and its stiffeners. Thus, the perimeter of the stiffened plate, the stiffeners and the plate-stiffener interface have to be discretized. Although this boundary-finite element model may need more discretization than that for the boundary element model, the simple fundamental solution can reduce computational difficulties. The performance of this proposed method has been examined by solving some example problems. From the numerical results, it can be concluded that the derivation of integral equation and the development of numerical procedure are correct, and the proposed method is accurate in its ability to analyze stiffened plate structures.

REFERENCES

(1) Timoshenko, S. and Woinowsky-Krieger, S. Theory of Plates and Shells, Mcgraw-Hill, New York, 1959.

(2) Ugural, A. C. Stresses in Plates and Shells, McGraw-Hill, 1981.

(3) Rossow, M. P. and Ibrahimkhail, A. K. Constraint Method Analysis of Stiffened Plates, Comput. & Struct., Vol. 8, pp. 51-60, 1978.

(4) O'Leary, J. R. and Harari, I. Finite Element Analysis of Stiffened Plates, Comput. & Struct., Vol. 21, No. 5, pp. 973-985, 1985.

(5) Deb, A. and Booton, M. Finite Element Models for Stiffened Plates under Transverse Loading, Comput. & Struct., Vol. 28, No. 3, pp. 361-372, 1988.

(6) Brebbia, C. A. The Boundary Element Method for Engineering, John Wiley & Sons, New York, 1978.

(7) Marcano, L. MS Thesis presented to University of Missouri-Columbia, 1984.

(8) Stern, M. A General Boundary Integral Formulation for the Numerical Solution of Plate Bending Problems, Int. J. Solid Structures, Vol. 15, pp. 769-782, 1979.

(9) Hartmann, F. and Zotemantel, R. The Direct Boundary Element Method in Plate Bending, Int. J. Numer. Methods Eng., Vol. 23, pp. 2049-2069, 1986.

(10) Yang, T. Y. Finite Element Structural Analysis, Prentice-Hall, New Jersey, 1986.

(11) Brebbia, C. A. and Walker, S. Boundary Element Techniques in Engineering, Newnes-Butterworths, London, 1980.

(12) GTSTRUDL User's Manual, Georgia Tech STRUDL User's Manual, Georgia

Institute of Technology, Atlanta, Georgia, 1983.
(13) Nowacki, N. Static and Dynamics of Plate with Ribs, Archiwum Mechaniki Stosawancj, Vol. 6, No. 4, 1954.

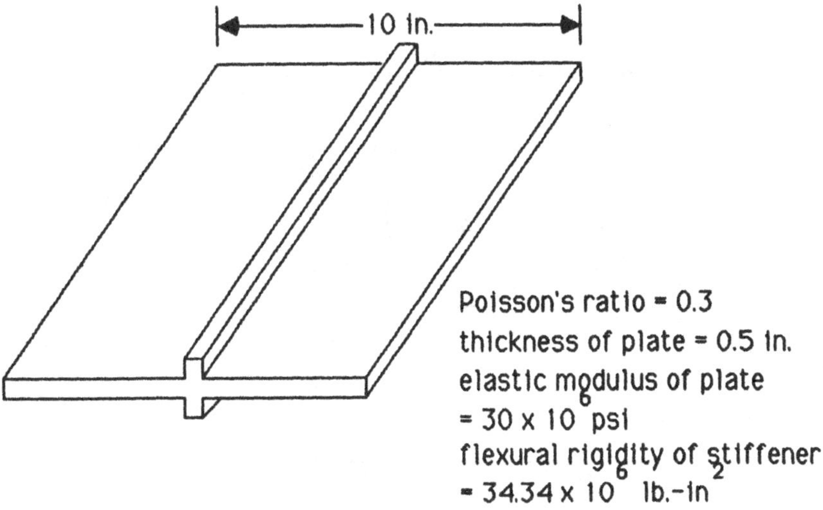

Figure 4 Simply Supported Square Plate with One Concentric Stiffener

Table 1 Central Deflection of a Simply Supported Plate with One Stiffener under Uniformly Distributed Load of 10 psi

Method	Deflection x 10^{-3} in.	% error
Ref. [13]	0.19044	---
B-FEM (8-2)	0.19621	3.030
B-FEM (16-4)	0.19150	0.557
B-FEM (28-8)	0.19111	0.352

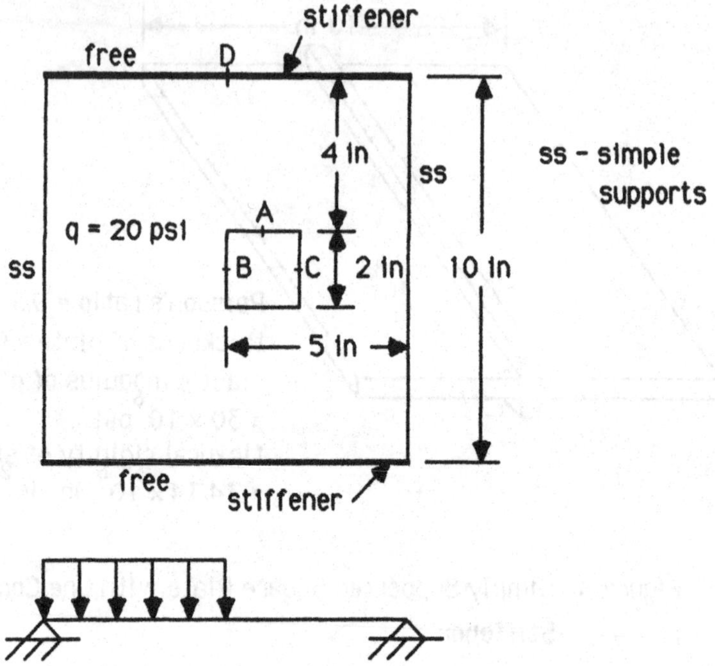

Figure 5 A stiffened plate with a square opening

Table 2 Comparison of Deflection for a Stiffened Plate With a Square Opening

Point	B-FEM (Deflect. x 10^{-3} in.)	GTSTRUDL (Deflect. x 10^{-3} in.)
A	1.2357	1.2194
B	1.6392	1.6184
C	0.9547	0.9428
D	0.1720	0.1717

A Modified Local Green's Function Technique for the Mindlin's Plate Model

R. Barbieri, C.S. de Barcellos
Universidade Federal de Santa Caterina, Departamento de Engenharia Mecanica, CP 476, 88049 Florianopolis, SC, Brasil

Abstract

The Modified Local Green's Function Method, MLGFM, was proposed as an attempt to solve non-homogeneous field problems and it is now considered to be related to the Galerkin Boundary Element Method, but without requiring the knowledge of a fundamental solution explicitly.In this paper, a MLGFM approach is presented for the Mindlin's plate model which allows the treatment of variable plate thickness without any additional effort. Finally, some numerical examples are shown and the results are compared against boundary and finite element ones.

Introduction

The Modified Local Green's Function Method (MLGFM) was originally proposed by Silva [1] and Barcellos & Silva [2] for solving potential, rod and beam problems. Recently Barbieri & Barcellos [3,4] used this method for solving singular potential problems which come out from geometrical and boundary condition discontinuities and also in the solution of non-homogeneous field potential problems. The MLGFM can now be considered as a further development of the Galerkin Boundary Element Method, since it doesn't require the explicit knowledge of the fundamental solution, but,instead, its procedure evaluates appropriate Green's function projections without the evaluation of singular integrals. Afterwards, the solution is determined very easily by the associated Boundary Element Method formalism. On the other hand, the need for treating moderately thick plates has made the Mindlin's and Reissner's plate models popular among the Finite Element Method, FEM, community. Due to difficulties encountered in modelling the transverse shear deformations, in the displacement formulation, it appeared the locking problem and many papers have been published each one proposing procedures to mitigate it.

As far as the Boundary Element Method, BEM, is concerned there has been published papers on Reissner's plate model, e.g, by Weeën [5,6] and on Mindlin's plate model, e.g.,by Barcellos & Silva [7], Silva [1], Westphal Jr. & Barcellos [8]. It has been noted that by using the BEM to the Mindlin's model, as well as the Reissner one, no locking occurs and accurate results can be obtained.

Here the MLGFM is firstly extended for treating the Mindlin's plate model and it is shown that it keeps the nice feature of BEM for not presenting locking and, moreover, that good results can be obtained, even when coarse auxiliary meshes are used, in comparison with the ones obtained by FEM and Direct Boundary Element Method, DBEM.

Mindlin's Plate Model

Let a moderately thick plate be caracterized by the middle surface as an open bounded domain, Ω, and its contour, Γ, and thickness t. The basic assumptions of this plate model are : small displacements; midsurface normals remain straight by not necessarily orthogonal to such surface during the deformation process and the through the thickness normal stress is neglected.Then, the generalized displacement field can be described by

$$u = [\ w,\ \theta x,\ \theta y\ ]^t \tag{1}$$

where its components are,respectively, the transverse displacement and the rotations about the y and x axis .

The generalized bending ε_F and the shear ε_s deformation fields can be written as

$$\varepsilon_F = \begin{bmatrix} 0 & \partial/\partial x & 0 \\ 0 & 0 & \partial/\partial y \\ 0 & \partial/\partial y & \partial/\partial x \end{bmatrix} u = \mathcal{A}_F\, u \qquad (2)$$

$$\varepsilon_S = \begin{bmatrix} \partial/\partial x & 1 & 0 \\ \partial/\partial y & 0 & 1 \end{bmatrix} u = \mathcal{A}_S\, u \qquad (3)$$

The bending and twist moments, $M = [\ M_x,\ M_y,\ M_{xy}\ ]^t$ and the transversal shear stress resultant, $Q = [\ Q_x,\ Q_y\ ]^t$, are related to ε_F and ε_S by :

$$M = \mathbb{D}_F\, \varepsilon_F \qquad \text{and} \qquad Q = \mathbb{D}_S\, \varepsilon_S \qquad (4)$$

where

$$\mathbb{D}_F = Et^3/12(1-\nu^2) \begin{bmatrix} 1 & \nu & 0 \\ \nu & 1 & 0 \\ 0 & 0 & (1-\nu)/2 \end{bmatrix} \ ; \ \mathbb{D}_S = E\ t/2\alpha(1+\nu)\ \mathbb{I} \qquad (5)$$

where E, ν, α and $\mathbb{I}$ denote, the Young modulus, Poisson's ratio, Mindlin's correction factor and the identity matrix, respectively. Hence,the total potential energy, π, for a plate under a distributed transversal load , c, can be written as

$$\pi = \int_\Omega (\varepsilon_F^{\ t}\ \mathbb{D}_F\ \varepsilon_F + \varepsilon_S^{\ t}\ \mathbb{D}_S\ \varepsilon_S - c\ w\ )\ d\Omega \qquad (6)$$

By using the principle of minimum potential energy, one obtains

$$\mathcal{A}\ u = B \qquad (7)$$

where $\mathcal{A}$ and B are,by denoting $D = E\ t^3/12(1-\nu^2)$ and $C = E\ t/2\alpha(1+\nu)$,

$$\mathcal{A} = \begin{bmatrix} -D[\partial^2/\partial x^2 + a\ \partial^2/\partial y^2]+C & -Db\ \partial^2/\partial xy & C\ \partial/\partial x \\ -Db\ \partial^2/\partial^2 xy & -D[\partial^2/\partial y^2 + a\ \partial^2/\partial x^2]+C & C\ \partial/\partial y \\ -C\ \partial/\partial x & -C\ \partial/\partial y & -C[\partial^2/\partial x^2 + \partial^2/\partial y^2] \end{bmatrix}$$

$$B = [\ c,\ 0,\ 0\ ]^t$$

where $a = (1-\nu)/2$ and $b = (1+\nu)/2$.

MLGFM Formalism for Mindlin's Plate Model

Consider the adjoint operator, $\mathcal{A}^*$, of $\mathcal{A}$ and the associated problem

$$\mathcal{A}^* \mathbb{G}(P,Q) = \delta(P,Q)\ \mathbb{I} \qquad P,Q \in \Omega \tag{8}$$

where $\delta(P,Q)$ stands for the Dirac delta function and $\mathbb{G}(P,Q)$ is the tensor fundamental solution. From now on, the capital letters "P,Q" denote domain points and the small "p,q" denote boundary points and "Q/q" stand for the observation point while "P/p" denote the load position. By pre-multiplying Equation (8) by u^t and Equation (7) by $\mathbb{G}^t(P,Q)$, one gets

$$u^t\ \mathcal{A}^*\mathbb{G}(P,Q) = u^t\ \delta(P,Q) \tag{9}$$

$$\mathbb{G}^t(P,Q)\ \mathcal{A}\ u = \mathbb{G}^t(P,Q)\ B(Q) \tag{10}$$

Transposing Equation (9), subtracting Equation (10) and integrating over Ω, keeping the observation point Q fixed and denoting the area element at the point P by $d\Omega_P$, results

$$u(Q) = \int_\Omega \{[\mathcal{A}^*\mathbb{G}(P,Q)]^t u(P) - \mathbb{G}^t(P,Q)[\mathcal{A}\ u(P)] + \mathbb{G}^t(P,Q)\ B(P)\}\ d\Omega_P \tag{11}$$

Applying the Gauss's Theorem to the first two integrals on the right hand side of Equation (11) and reminding Equation (8), yields

$$u(Q) = \int_\Omega \mathbb{G}^t(P,Q)\ B(P)\ d\Omega_P + \int_\Gamma \mathbb{G}^t(p,Q)[\mathcal{N}\ u(p)]\ d\Gamma_p\ -$$

$$\int_\Gamma u^t(p)[\mathcal{N}^*\mathbb{G}(p,Q)]\ d\Gamma_p \tag{12}$$

where $d\Gamma$ denote an arc element at the boundary point p and $\mathcal{N},\mathcal{N}^*$ are the resultant Neumann operators from $\mathcal{A}$ and $\mathcal{A}^*$,respectively . Since the knowledge of the explicit form of the Neumann operators is not required by the MLGFM, they are not detailed in here. But, by adding and subtracting

$$\mathbb{G}^t(p,Q)[\mathcal{N}'u(p)] = u(p)^t[\ \mathcal{N}'\mathbb{G}(p,Q)] \tag{13}$$

to equation (12), one obtains

$$u(Q) = \int_\Omega \mathbb{G}^t(P,Q)\ B(P)\ d\Omega_P + \int_\Gamma \mathbb{G}^t(p,Q)[(\mathcal{N}+\mathcal{N}')\ u(p)]\ d\Gamma_p\ -$$

$$\int_\Gamma {}'u^t(p)[(\mathcal{N}^*+\mathcal{N}')\mathbb{G}(p,Q)]\ d\Gamma_p \tag{14}$$

The operator $\mathcal{N}'$ can be conveniently chosen in the form

$$\mathcal{N}' = \text{diag}\ (\ k_1,\ k_2,\ k_3\ ) \tag{15}$$

and, in practice, it must be applied on the boundary part where homogeneous Dirichlet boundary conditions are specified. The scalars k_1, k_2 and k_3 can assume any non-zero value as long as the condition number of the final matrices are not compromised. Up to now, scalars ranging from 10^{-3} up to 10^6 have been tried on without any change on the generalized displacements and stress resultants. When the boundary conditions are of Neumann or Cauchy-Robin type, the k_i's values must be defined as near zero as possible since they may introduce artificial reactions if no additional treatment is made afterwards.Next, following the MLGFM formalism, one requires the Green's function to satisfy the boundary conditions

$$(\mathcal{N}^*+\mathcal{N}')\mathbb{G}(p,Q) = 0 \tag{16}$$

and one defines the subsidiary vector function

$$F(p) = (\mathcal{N} + \mathcal{N}')\ u(p) \tag{17}$$

so that the generalized displacement, Equation (14), can be written as

$$u(Q) = \int_\Omega \mathbb{G}^t(P,Q)\ B(P)\ d\Omega_P + \int_\Gamma \mathbb{G}^t(p,Q)\ F(p)\ d\Gamma_p \tag{18}$$

Note that Equation (18) does not include any derivatives of G(P,Q) or F(p).So, these integrals are much better behaved than those of Equation (12) which is similar to the Direct Boundary Element Method equation. Finally, the trace operator [9] is applied to Equation (18) resulting

$$u(q) = \int_{\Omega} G^t(P,q)\ B(P)\ d\Omega_P + \int_{\Gamma} G^t(p,q)\ F(p)\ d\Gamma_P \tag{19}$$

Equations (18) and (19) define completely the problem, without any approximation.

Interpolation Functions

In order to solve Equations (19) and then Equation (18) approximately, the generalized displacement vector is interpolated inside the domain in the same way as in the FEM, that is

$$u(P) = \begin{bmatrix} \psi_1 & & & & \psi_n & & \\ & \psi_1 & & \cdots & & \psi_n & \\ & & \psi_1 & & & & \psi_n \end{bmatrix} v = [\Psi(P)]\ v \tag{20}$$

where ψ_i are the usual interpolation functions, $v = [..\ w_i, \theta x_i, \theta y_i ..]^t$, $(w_i, \theta x_i, \theta y_i)^t$ stands for the value of the generalized displacement of node "i" and one assumes that

$$B(P) = [\Psi(P)]\ b \tag{21}$$

Similarly, at the boundary, u(p) and F(p) are approximated by using interpolation functions $\phi_i(p)$ which are traces of ψ_i.So, one may write

$$u(p) = [\Phi(p)]\ v \tag{21}$$

$$F(p) = [\Phi(p)]\ f \tag{22}$$

Substituting these interpolations into Equation (18) and using the Galerkin Method relative to the domain Ω , that is, by using ψ_j as the domain weight functions, results

$$\mathbb{A}\ \mathbf{v} = \mathbb{B}\ \mathbf{f} + \mathbb{C}\ \mathbf{b} \tag{23}$$

where

$$\mathbb{A} = \int_{\Omega} [\Psi(Q)]^t [\Psi(Q)]\ d\Omega_Q \tag{24}$$

$$\mathbb{B} = \int_{\Gamma} [G^D(p)]^t\ [\Phi(p)]\ d\Gamma_p \tag{25}$$

$$\mathbb{C} = \int_{\Omega} [G^D(P)]^t\ [\Psi(P)]\ d\Omega_P \tag{26}$$

$$[G^D(p)]^t = \int_{\Omega} [\Psi(Q)]^t\ G^t(p,Q)\ d\Omega_Q \tag{27}$$

$$[G^D(P)]^t = \int_{\Omega} [\Psi(Q)]^t\ G^t(P,Q)\ d\Omega_Q \tag{28}$$

Repeating this on equation (19), but now weighting the residues by ϕ_1, one gets

$$\mathbb{D}\ \mathbf{v} = \mathbb{E}\ \mathbf{f} + \mathbb{F}\ \mathbf{b} \tag{29}$$

where

$$\mathbb{D} = \int_{\Gamma} [\Phi(q)]^t [\Phi(q)]\ d\Gamma_q \tag{30}$$

$$\mathbb{E} = \int_{\Gamma} [G^C(p)]^t\ [\Phi(p)]\ d\Gamma_p \tag{31}$$

$$\mathbb{F} = \int_{\Omega} [G^C(P)]^t\ [\Psi(P)]\ d\Omega_P \tag{32}$$

$$[G^C(p)]^t = \int_{\Gamma} [\Phi(q)]^t\ G^t(p,q)\ d\Gamma_q \tag{33}$$

$$[G^C(P)]^t = \int_{\Gamma} [\Phi(q)]^t\ G^t(P,q)\ d\Gamma_q \tag{34}$$

Equation (29), which is similar to that which can be obtained by the Galerkin BEM , can be rewritten in terms of the prescribed boundary conditions, $\mathbf{v}_p$ and $\mathbf{f}_p$, and the unknown values, $\mathbf{v}_u$ and $\mathbf{f}_u$, as

$$\begin{bmatrix} -\mathbb{E}u & \mathbb{D}u \end{bmatrix} \begin{Bmatrix} fu \\ vu \end{Bmatrix} = \begin{bmatrix} -\mathbb{D}p & \mathbb{E}p \end{bmatrix} \begin{Bmatrix} vp \\ fp \end{Bmatrix} + \mathbb{F}\, b \tag{35}$$

The only lacking detail is how to compute the Green's function projections (27),(28),(33) and (34).

Green's Function Projections

The Green's function projections $\mathbb{G}^D(Q)$, $\mathbb{G}^D(q)$, $\mathbb{G}^C(Q)$ and $\mathbb{G}^C(q)$ constitute the crucial step of the MLGFM as originally proposed by Barcellos & Silva [2] and Silva [1]. Since the Green's tensor projection is smoother than the tensor itself and, so, its integrals, one is attempted to approximate these projections directly. For attaining this, it is enough to solve two associate problems, namely

Problem 1

$$\mathcal{A}^*(Q)\ \mathbb{G}(P,Q) = \delta(P,Q)\ \mathbb{I} \qquad P,Q \in \Omega \tag{36.1}$$

$$(\mathcal{N}^*+\mathcal{N}')(q)\mathbb{G}(p,Q) = 0 \qquad p \in \Gamma \tag{36.2}$$

Problem 2

$$\mathcal{A}^*(Q)\ \mathbb{G}(p,Q) = 0 \qquad Q \in \Omega \tag{37.1}$$

$$(\mathcal{N}^*+\mathcal{N}')(q)\mathbb{G}(p,q) = \delta(p,q)\mathbb{I} \qquad p,q \in \Gamma \tag{37.2}$$

where "Q" and "q" at the operators $\mathcal{A}^*$ and $(\mathcal{N}+\mathcal{N}')$ indicate that these operators are relative to the coordinates of the point "Q" or "q",respectively.

Now, multiply Equation (36.1) by $[\Psi(P)]$ and integrate it over the domain Ω, that is

$$\mathcal{A}^* \int_{\Omega} \mathbb{G}(P,Q)\ [\Psi(P)]\ d\Omega_P = \int_{\Omega} \delta(P,Q)\ [\Psi(P)]\ d\Omega_P \tag{38}$$

or

$$\mathcal{A}^*\ \mathbb{G}^D(Q) = [\Psi(Q)] \tag{39}$$

Since the interpolation function matrix $[\Psi(P)]$ can be writen as

$$[\Psi(P)] = [\ [\Psi_1(P)] | .. | [\Psi_j(P)] | ... | [\Psi_n(P)]\ ] \tag{40}$$

one has

$$G^D(P) = \int_\Omega G(P,Q)\ [\Psi(Q)]\ d\Omega_Q = [\ G_1^D(P),\ldots,G_j^D(P),\ldots,G_n^D(P)\ ] \quad (41)$$

where n is the number of nodes. So

$$\mathcal{A}^*\ G_j^D(P) = [\Psi_j(P)] \quad (42)$$

where the submatrices $G_j^D(P)$ are

$$G_j^D(P) = \int_\Omega G(P,Q)\ [\Psi_j(Q)]\ d\Omega_Q = [g_1^D(P),g_2^D(P),g_3^D(P)] \quad (43)$$

The tensor approximation $G_j^D(P)$, as in FEM, is achieved by minimizing the functional

$$J(G_j^D(P)) = 0.5\ \{ \sum_{l=1}^{3} \int_\Omega (\ [\mathcal{A}_F\ g_l^D(P)]^t\ D_F\ [\mathcal{A}_F\ g_l^D(P)] +$$

$$[\mathcal{A}_S\ g_l^D(P)]^t\ D_S\ [\mathcal{A}_S\ g_l^D(P)] - [g_l^D(P)]^t[\Psi_l(P)]\ )\ d\Omega +$$

$$\sum_{l=1}^{3} \int_\Gamma [\mathcal{N}' g_l^D(p)]^t g_l^D(p)\ d\Gamma\ \} \quad (44)$$

Proceeding similarly with Equation (37), the tensor approximation $G^C(P)$ is obtained. Now, the functional to be minimized is

$$J(G_j^D(P),G_j^C(P)) = 0.5\ \{ \sum_{l=1}^{3} \int_\Omega (\ [\mathcal{A}_F\ g_l(P)]^t\ D_F\ [\mathcal{A}_F\ g_l(P)] +$$

$$[\mathcal{A}_S\ g_l(P)]^t\ D_S\ [\mathcal{A}_S\ g_l(P)] - \beta\ [g_l(P)]^t[\Psi_l(P)]\ )\ d\Omega +$$

$$\sum_{l=1}^{3} \int_\Gamma (\ [\mathcal{N}' g_l^D(p)]^t g_l^D(p) - \lambda\ [g_l(p)]^t[\Phi_l(p)]\ )d\Gamma\ \} \quad (45)$$

where: (λ,β) are (0,1) for Problem 1 and (1,0) for Problem 2 ,which are to be useful in the determination of $G^D(P)$ and $G^C(P)$ respectively ; $[\Phi_l(p)] = \text{diag}\ [\phi_l(p),\phi_l(p),\phi_l(p)]$ and the vector g_l may represent either the components of $G^D_l(P),g^D_l(P)$ or $G^C_l(P),g^C_l(P)$, depending upon the desired Green's function projections and

$$G^C_j(P) = \int_\Gamma G(P,q)[\Phi_j(q)]\ d\Gamma_q = [g^C_1(P),g^C_2(P),g^C_3(P)] \tag{46}$$

By minimizing the functional (45) according the FEM technique one can write

$$(K + K_o) [G_D , G_C] = [M , m] \tag{47}$$

where **K** is the conventional finite element stiffness matrix ; K_o is the stiffness matrix which comes from the first boundary integral in (45) ; **M, m** are the matrices whose columns come from the last terms on the domain and boundary integrals in (45) and G_D, G_C are the nodal values of $G^D(Q)$ and $G^C(Q)$, respectively.

A pratical manner of defining K_o is just to assume it in the form

$$K_o = \text{diag}\ [\ k_o\ ] \tag{48}$$

where k_o is a non-zero constant which comes up only where the Dirichlet homogeneous boundary conditions are specified.

So,the determination of both g^D_l and g^C_l is performed simultaneously. The substitution of Equations (43) and (46) into Equations (27) and (28), respectively, together with Equation (24) leads us to the matrices **A**, **B** and **C**.

The matrix **F** is the transpose of **B** as it can be verified from (25) and (32). The boundary matrix **D** is evaluated and, then, the matrix **E** is obtained, since the coefficients g^C_l have already been determined and one has here adopted the set of functions ϕ_l as trace of domain interpolation functions ψ_l.

Application

A simple numerical example is include in here for illustrating the solution behavior. It concerns to uniform square plate with thickness h, clamped along its sides of length a=10, and under a uniform distributed load q=1. The domain Ω is then $\{ (x_1,x_2) \in \mathbb{R}^2 : 0 \leq x_1,x_2 \leq a \}$. Two type of elements were used for obtaining the Green's function projections; namely, the cubic and the quadratic lagrangian finite elements. The first was fully integrated using 4x4 integration points and the second was selectively integrated by the usual scheme, that is, 3x3 points for the bending portion and 2x2 points for the shear one. Due to the symmetry, only one quarter of the plate is modeled, $\{ a/2 \leq x_1,x_2 \leq a \}$ and the results for the meshes M1(3x3 equal size cubic elements) and M2 (4x4 equal size quadratic elements) are shown in Tables 1 to 3. Figures (1) and (2) show the bending and shear reactions. The results obtained by Silva [1] come from his BEM code, the Yuan's ones were evaluated by the FEM and Archer's solution comes from series expansions.

As it can be verified in the tables, one concludes that even using interpolation functions taken from displacement finite elements, satisfactory results can be obtained by the MLGFM for values of a/h up to 10E6. For higher ratios, one can see that the FEM interpolation function starts to contaminate the procedure, but in this case, the entire Mindlin's model may not be even appropriate anymore. Even using such coarse meshes the results show good agreement with Archer's solution for moderately thick plates as well as with the analytical solution for thin plates when the ratio a/h is high. The distribution of reaction moments and shear forces drawn in Figures (1) and (2) for the mesh M1 show good agreement with the analytical solution [11] and with the ones obtained by Katsikadelis & Armenakas [10] and Stern [14] using the BEM.

Table 1 - Central Deflection : $[wD/qa^4]*1000$

a/h	M1	M2	Silva[1]	Yuan [12]	Archer[13]
10	1.50441	1.50532	1.53315	1.47252	1.50458
50	1.27520	1.27603		1.27308	
10E2	1.26762	1.26846	1.271427	1.26739	
10E3	1.26510	1.26594	1.265749	1.25366	
10E4	1.26507	1.26591			
10E5	1.26508	1.26591			
10E6	1.26434	1.26615			
10E7	1.22036	1.22378			
thin plate [11] : 1.265					

Table 2 - Central Bending Moment $[M/qa^2]*100$

a/h	M1	M2	Silva [1]	Archer [13]
10	2.32035	2.33300	2.325	2.37245
50	2.30165	2.30415		
10E2	2.30434	2.30296	2.294	
10E3	2.30596	2.30280	2.291	
10E4	2.30598	2.30279		
10E5	2.30598	2.30278		
10E6	2.30415	2.30344		
10E7	2.21866	2.24366		
thin plate [11] : 2.31				

Table 3 - Bending Reaction at the Midside Node, a/2 $[M/qa^2]*100$

a/h	M1	M2	Silva [1]	Archer [13]
10	4.93807	4.95154	4.980	4.88724
50	5.12586	5.13331		
10E2	5.13303	5.13952	5.313	
10E3	5.13533	5.13413	5.160	
10E4	5.13535	5.13438		
10E5	5.13536	5.13336		
10E6	5.13237	5.13420		
10E7	5.03720	5.98997		
thin plate [11] : 5.13				

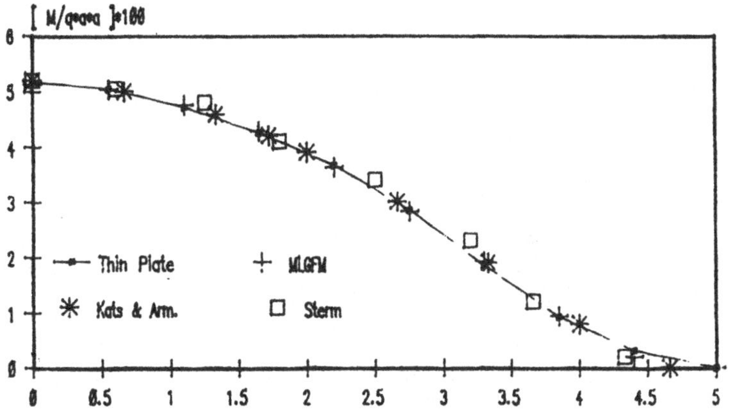

Fig.2 - Bending reaction along a half-side.

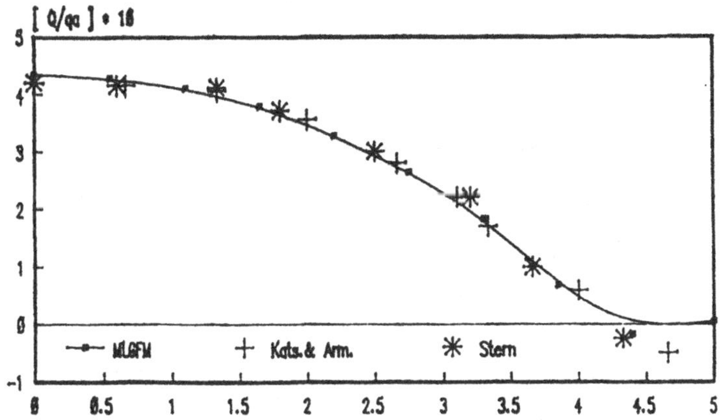

Fig.3 - Shear reaction along a half-side.

Conclusion

On the basis of the formulation derived in this paper and the numerical solution studied, it can be concluded that the Modified Local Green's Function Method can provide an effective approximate numerical technique for the solution of a wide range of plate problems since it doesn't require a fundamental solution.

In addition, the formulation presented in this paper can be extended to solve other practical plate problems, shallow shells and even shells with arbitrary shapes. In the last case,the authors believe that the MLGFM is an adequate procedure to extend the BEM methodologies into such a broad area of application.

References

1. Silva,L.H.M. Novas formulações Integrais para Problemas da Mecanica, Ph.D Thesis (in portuguese), Universidade Federal de Santa Catarina,1988.

2. Barcellos, C.S. and Silva,L.H.M. Elastic Membrane Solution by a Modified Local Green's Function Method,(Ed. Brebbia, C.A. and Venturini, W.S. ,) Proc. Int. Conf. on Boundary Element Technology, 1987, Comp. Mech. Publ. , Southampton,1987.

3. Barbieri, R. and Barcellos, C.S. Non-Homogeneous Field Potential Problems Solution by the Modified Local Green's Function Method (MLGFM),13th BEM Conf., (Ed. Brebbia, C.A.), 1991.

4. Barcellos, C.S. and Barbieri, R. Solution of Singular Potential Problems by the Modified Local Green's Function Method (MLGFM), 13th BEM Conf.,(Ed. Brebbia,C.A.), 1991.

5. Van der Weeën, F. Application of the Direct Boundary Element Method to Reissner's Plate Model, in Boundary Elements in Engineering (Ed. Brebbia, C.A.) , pp. 487-499, Proc. 4th. Int. Conf. on BEM, Southampton, 1982, Springer Verlag, Berlin and New York, 1982.

6. Van der Weeën, F. Application of the Boundary Integral Equation Method to Reissner's Plate Model, Int. J. Numerical Methods Engng., Vol 18, pp. 1-10, 1982.

7. Barcellos, C.S. and Silva, L.H.M. A Boundary Element Formulation for the Mindlin's Plate Model, (Ed. Brebbia, C.A. and Venturini, W.S.) Proc. Int. Conf. on Boundary Element Technology, 1987, Comp. Mech. Publ. , Southampton,1987.

8. Westphal Jr,T. and Barcellos, C.S. Applications of the Boundary Element Method to Reissner's and Mindlin's Plate ModelsModels, (Ed. Brebbia, C.A.) Proc. 12th BEM Conf. Comp. Mech. Publ.,1990.

9. Oden,J.T. and Reddy,J.N. An Introduction to the Mathematical Theory of Finite Elements , John Wiley & Sons, 1976.

10. Katsikadelis, J.T. and Armenakas,A.E. A New Boundary Equation Solution to the Plate Problem, J.Applied Mechanics, Vol.56,pp. 364-374 , 1989.

11. S.Timoshenko and Woinowsky-Krieger. Theory of Plates and Shell's, 2nd edn. Mac Graw-Hill, New York, 1959.

12. Yuan,Fuh-Gwo and Miller,R.E. A Cubic Triangular Finite Element for Flat Plates with Shear, Int. J. Numerical Methods in Engineering, Vol.28, pp. 109-126, 1989.

13. Archer, R.R. and Deshmukh, R.S. Numerical Solution of Moderately thick Plates, J. Engineering Mechanics Division, pp. 903-917, 1974.

14. Stern, M. A General Boundary Integral Formulation for the Numerical Solution of Plate Bending Problems, Int. J. Solids and Structures, Vol.15, pp. 769-782, 1979.

Boundary Elements and Perturbation Theory for Vibrating Plates

C.V. Camp

Department of Civil Engineering, Memphis State University, Memphis, TN 38152, U.S.A.

ABSTRACT

In this work, classical techniques from perturbation theory will be applied to develop a boundary integral formulation for low frequency forced vibrations of elastic plates. The functional form of the applied load and the plate deflection are assumed to be products of a temporal function and corresponding spatial functions. The separation of variables approach removes the difficulties associated with transient analysis. The resulting boundary element formulation requires consecutive solutions to a set of coupled non-homogeneous biharmonic equations. The domain integral normally associated with each non-homogeneous equation is transformed to a set of boundary integrals using the Rayleigh-Green identity. Numerical solutions of the perturbation-based expansion equations of forced vibrations using the boundary element method (BEM) are presented and compared with analytical analysis.

INTRODUCTION

Forced low frequency vibration of elastic plate structures is a classic problem in engineering mechanics. The dynamic behavior of a thin elastic plate may be characterized by several assumptions: the amplitude of the deflection is considered small in comparison with the plate thickness, normal stresses in the transverse direction are neglected, and the mid-plane of the plate does not undergo deformation. These conditions form the basis of elementary plate theory.

The boundary element method (BEM) is fast becoming one of the most versatile and powerful numerical methods in engineering science. One of the more appealing characteristics of BEM formulations is their compatibility with other methods. In this work, a hybrid formulation combining techniques of BEM with regular perturbation theory is presented. The resulting BEM-perturbation solution has advantages over more traditional numerical techniques and even some BEM vibration formulations.

THEORETICAL DEVELOPMENT

Consider a thin elastic plate of uniform thickness and arbitrary shape. The mid-plane of the plate lies in the x-y plane and the deflections are in the z-direction (see Figure 1). The deflections are assumed to be small in comparison to the plate thickness. The governing partial differential equation for the forced lateral vibration of the plate is [1,2]:

$$\nabla^4 w + \frac{\varrho h}{D}\frac{\partial^2 w}{\partial t^2} = \frac{q(x,y,t)}{D} \tag{1}$$

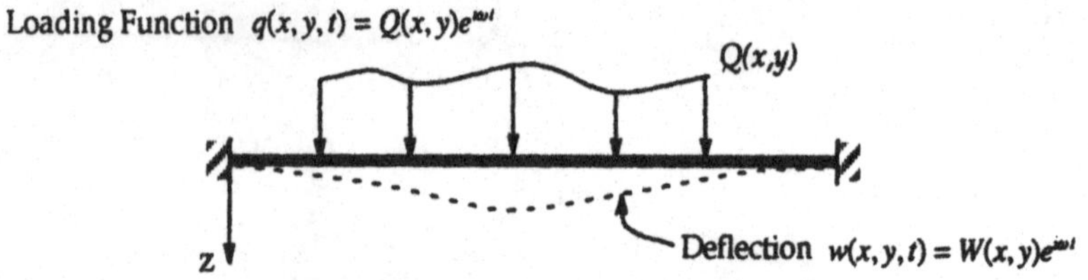

Figure 1. Vibrating Plate Definitions.

where w(x,y,t) is the deflection, h is the thickness of the plate, ρ is the mass per surface area, $D = Eh^3/[12(1-\nu^2)]$ is the flexural rigidity of the plate, and $q(x,y,t)$ is the applied transverse distributed loading function. Consider the case where the transverse loading function may be expressed as:

$$q(x,y,t) = Q(x,y)e^{i\omega t} \tag{2}$$

The expression in Equation (2) separates the loading function into a spatial part corresponding to a distributed load $Q(x,y)$ and a temporal part characterizing the oscillation of the transverse loading function at a frequency ω. In order to obtain a solution of equation (1) in terms of the distributed load $Q(x,y)$, it is convenient to express the deflection w(x,y,t) in terms of spatial and temporal components similar to the loading function. The resulting form of the deflection is :

$$w(x,y,t) = W(x,y)e^{i\omega t} \tag{3}$$

By substituting Equations (2) and (3) into Equation (1), the time component of the original governing differential equation is removed and the resulting expression is a completely spatial relationship:

$$\nabla^4 W - \left(\frac{\omega}{v}\right)^2 W = \frac{Q(x,y)}{D} \tag{4}$$

where $v = \sqrt{D/\varrho h}$

The spatial component for forced vibrations of uniform plates, given in Equation (4), is very similar in form to the governing equation for the deflection of thin plates on elastic foundations:

$$\nabla^4 z + \frac{k}{D} z = \frac{p(x,y)}{D} \tag{5}$$

where z is the deflection of the plate, k is the modulus of the foundation, and $p(x,y)$ is the distributed loading function. Several boundary element solutions to Equation (5) are available[3,4]. However, these approaches involve either time-consuming iterative techniques or have a formulation which requires repeated evaluation of complex Bessel or Kelvin functions. A BEM program employing an iterative solution procedure utilizes a simple biharmonic fundamental solution which is relatively easy to evaluate. The difficulty with this approach is that in order to obtain accurate solutions, a vast number of values of the field variables and their derivatives at internal points are required for each iteration. These intermediate calculations are very time-consuming and inefficient. In posed in Equation (1) may be accurately approximated by solving a series of simple static plate deflection problems.

BOUNDARY ELEMENT FORMULATION

Considerable work has been done in applying the boundary element method to the biharmonic equation, especially in the area of the deflection of thin elastic plates. Jawson, Maiti, and Symm [8] developed a boundary integral equation formulation for biharmonic analysis of two dimensional stress problems. Jawson and Maiti [9] extended their work to problems of clamped and simply supported elastic plates. Many other authors have developed boundary element solutions for a variety of plate problems, e.g. Altiero and Sikarskie [10], Stern [11], and Guo-Shu and Mukherjee [12].

Each of the nonhomogeneous biharmonic relationships in the sequence defined in Equation (9) may be transformed to an equivalent set of coupled Poisson-type equations by using the relationship between the field variable W and its Laplacian $\nabla^2 W = V$. Therefore Equation (9) becomes:

$$\begin{array}{ll} \nabla^2 W_0 = V_0 & \nabla^2 V_0 = \dfrac{Q(x,y)}{D} \\ \nabla^2 W_1 = V_1 & \nabla^2 V_1 = W_0 \\ \nabla^2 W_2 = V_2 & \nabla^2 V_2 = W_1 \\ \vdots & \vdots \\ \nabla^2 W_n = V_n & \nabla^2 V_n = W_{n-1} \end{array} \tag{10}$$

Since the biharmonic operator is a fourth order function and typically only two boundary conditions are prescribed for plate problems, Equation (9) and the second column of relationships in Equation (10) may be solved simultaneously for the field variable W and its Laplacian V.

Consider a general nonhomogeneous biharmonic equation and its coupled Poisson equation:

$$\nabla^4 W = f(x,y) \tag{11}$$

$$\nabla^2 V = f(x,y) \tag{12}$$

The boundary integral representation of Equations (11) and (12) may be formed by applying the Rayleigh-Green identity for biharmonic functions to Equation (11) and Green's second identity to Equation (12) (Camp and Gipson [13,14]). The resulting set of coupled integral equations for a general field point are:

$$\beta(\mathbf{p})W(\mathbf{p}) = \int_\Gamma \left(WG_1'(\mathbf{q},\mathbf{p}) - W'G_1(\mathbf{q},\mathbf{p}) + VG_2'(\mathbf{q},\mathbf{p}) - V'G_2(\mathbf{q},\mathbf{p})\right)d\Gamma + \int_\Omega f(x,y)G_2(\mathbf{q},\mathbf{p})d\Omega \tag{13}$$

$$\beta(\mathbf{p})V(\mathbf{p}) = \int_\Gamma \left(VG_1'(\mathbf{q},\mathbf{p}) - V'G_1(\mathbf{q},\mathbf{p})\right)d\Gamma + \int_\Omega f(x,y)G_1(\mathbf{q},\mathbf{p})d\Omega \tag{14}$$

addition, there are several obstacles associated with ensuring that the solution converges. A BEM formulation of Equation (5) based on the complex Kelvin fundamental solution would require complicated numerical integration procedures and repeated calls to complex functions.

The technique developed in this study is an extension of the work presented by Gipson and Reible[5]. The parameter ν, defined in Equation (4), is generally a large quantity for thin elastic plate structures. This implies that for a small value of the frequency ω a perturbation based solution procedure may be employed. Equation (4) may be rewritten as:

$$\nabla^4 W - \varepsilon W = \frac{Q(x,y)}{D} \tag{6}$$

where $\varepsilon = (\omega/\nu)^2$ is considered a small parameter. The spatial function W(x,y) may be expanded in an asymptotic series when $\varepsilon << 1$ by following the fundamentals of regular perturbation theory (Nayfeh[6,7]):

$$W(x,y) = W_0 + \varepsilon W_1 + \varepsilon^2 W_2 + \ . \ . \ . + \varepsilon^n W_n \tag{7}$$

Substituting the series approximation for W(x,y), defined in Equation (7), into the spatial component of the vibration equation, given in Equation (6), results in the following approximation:

$$\nabla^4(W_0 + \varepsilon W_1 + \varepsilon^2 W_2 + . . + \varepsilon^n W_n) - \varepsilon(W_0 + \varepsilon W_1 + \varepsilon^2 W_2 + . . + \varepsilon^n W_n) = \frac{Q(x,y)}{D} \tag{8}$$

By collecting terms of like powers of ε and acknowledging the fact that each set of coefficients must vanish independently of ε, Equation (8) is converted into an infinite set of coupled equations:

$$\begin{aligned} \nabla^4 W_0 &= \frac{Q(x,y)}{D} \\ \nabla^4 W_1 &= W_0 \\ \nabla^4 W_2 &= W_1 \\ &\ \ \vdots \\ \nabla^4 W_n &= W_{n-1} \end{aligned} \tag{9}$$

The objective of the perturbation based technique is to obtain an accurate solution for the deflection $W(x,y)$ by using just the first few terms of the infinite series of equations given in Equation (9). As the parameter ε becomes smaller, fewer terms in the expansion are required to successfully obtain an accurate approximation.

The spatial component of the original forced vibration problem has been transformed into a series of coupled nonhomogeneous biharmonic equations. Physically, each expression in the sequence defined in Equation (9) describes the deflection, W_n, of a thin elastic plate subjected to a transverse loading function W_{n-1}. The loading function for each new expression is the solution of the preceding equation. The complicated vibration problem where β is a generalized function with a value of 1.0 for a point inside the domain, some fractional value on the boundary, and zero outside the domain, and the primes denote differentiation with respect to the unit outward normal [15]. The vector **p** denotes a point on the boundary Γ or in the domain Ω, the vector **q** locates a point on the boundary Γ, and the fundamental solutions G_1 and G_2 are:

$$G_1(\mathbf{q},\mathbf{p}) = \frac{1}{2\pi}\ln|\mathbf{q}-\mathbf{p}| \tag{15}$$

$$G_2(\mathbf{q},\mathbf{p}) = \frac{1}{8\pi}\ln|\mathbf{q}-\mathbf{p}|^2(\ln|\mathbf{q}-\mathbf{p}|-1) \tag{16}$$

The functions in Equations (15) and (16) are the Laplacian and the biharmonic fundamental solutions, respectively. Both of these functions are much simpler than the Kelvin or Bessel functions required for the boundary element solution of Equation (5).

There are four types of boundary conditions for the integral formulation of the biharmonic problem in Equations (13) and (14), W, W', V, and V', where the primes denote differentiation with respect to outward normal. For a general well-posed boundary value problem, two of the four boundary conditions are prescribed at each point. The remaining two boundary quantities are determined from the simultaneous solution of Equations (13) and (14). The actual boundary conditions on the elastic plate for the forced vibration problem will be strictly applied to the equations defining the functions W_0 and V_0. The boundary conditions for the remaining correction terms, W_i and V_i, will be zero. For example, consider a plate with all edges clamped. The boundary conditions for the first correction term are: $W = W_{fixed}$ and $\partial W/\partial n = 0$. The boundary conditions on the remaining terms are $W_i = 0$ and $\partial W_i/\partial n = 0$.

A set of discrete expressions for Equations (13) and (14) may be formed by approximating the boundary Γ by a series of boundary elements. The variation of the field variables and the geometry over each element is represented by the product of a set of shape functions and a consecutive sequence of discrete boundary values. The details of the discretization and the general boundary element solution procedure for the nonhomogeneous biharmonic equation are given by Camp and Gipson [13,14].

BEM ALGORITHM

The first step in the algorithm is to solve the first equation of the infinite sequence of expressions given in Equations (9) and (10)

$$\beta(\mathbf{p})W_0(\mathbf{p}) = \int_\Gamma \left(W_0G_1'(\mathbf{q},\mathbf{p}) - W_0'G_1(\mathbf{q},\mathbf{p}) + V_0G_2'(\mathbf{q},\mathbf{p}) - V_0'G_2(\mathbf{q},\mathbf{p})\right)d\Gamma + \int_\Omega \frac{Q(x,y)}{D}G_2(\mathbf{q},\mathbf{p})d\Omega \tag{17}$$

$$\beta(\mathbf{p})V_0(\mathbf{p}) = \int_\Gamma \left(V_0G_1'(\mathbf{q},\mathbf{p}) - V_0'G_1(\mathbf{q},\mathbf{p})\right)d\Gamma + \int_\Omega \frac{Q(x,y)}{D}d\Omega \tag{18}$$

Since the spatial component of the loading function $Q(x,y)$ is a known function, the domain integration in Equations (17) and (18) may be evaluated. Equations (17) and (18) may be solved simultaneously for the values of W_0 and V_0 at the boundary. The domain integrals containing the loading function may be transformed to an equivalent set of boundary integrals when the function $Q(x,y)$ is harmonic or biharmonic in the domain Ω.

Consider the case when $Q(x,y)$ is a biharmonic function in the domain Ω, $\nabla^4Q(x,y) = 0$. The domain integrals in Equations (17) and (18) may be converted to a series of boundary integrals using the Rayleigh–Green identity:

$$\int_{\Omega}\left(Q(x,y)G_k(\mathbf{q},\mathbf{p})-G_{k+2}(\mathbf{q},\mathbf{p})\nabla^4 Q(x,y)\right)d\Omega$$

$$=\int_{\Gamma}\left(QG'_{k+1}(\mathbf{q},\mathbf{p})-Q'G_{k+1}(\mathbf{q},\mathbf{p})+(\nabla^2 Q)G'_{k+2}(\mathbf{q},\mathbf{p})-(\nabla^2 Q)'G_{k+2}(\mathbf{q},\mathbf{p})\right)d\Gamma \quad (19)$$

where G_n and $G_n{}'$ are:

$$G_n(\mathbf{q},\mathbf{p})=\frac{|\mathbf{q}-\mathbf{p}|^{2n-2}}{2^{2n-1}((n-1)!)^2\pi}\left[\sum_{j=1}^{n-1}\frac{1}{j}-\ln|\mathbf{q}-\mathbf{p}|\right] \quad (20)$$

$$G'_n(\mathbf{q},\mathbf{p})=\frac{2(n-1)|\mathbf{q}-\mathbf{p}|^{2(n-2)}}{2^{2n-1}((n-1)!)^2\pi}\left[\ln|\mathbf{q}-\mathbf{p}|+\frac{1}{2(n-1)}-\sum_{j=1}^{n-1}\frac{1}{j}\right]\left[(x_q-x_p)n_x+(y_q-y_p)n_y\right] \quad (21)$$

where n_x and n_y are the x and y direction cosines of the the unit outward normal, respectively. The second term of the domain integral in Equation (19) is identically zero if $Q(x,y)$ is biharmonic. The resulting BEM solution for the boundary values of the functions W_0 and V_0 may be obtained solely from the evaluation of surface boundary integrals.

Next, consider the second set of relationships given in Equations (9) and (10). The coupled set of equations are:

$$\beta(\mathbf{p})W_1(\mathbf{p})=\int_{\Gamma}\left(W_1G'_1(\mathbf{q},\mathbf{p})-W'_1G_1(\mathbf{q},\mathbf{p})+V_1G'_2(\mathbf{q},\mathbf{p})-V'_1G_2(\mathbf{q},\mathbf{p})\right)d\Gamma + \int_{\Omega}W_0G_2(\mathbf{q},\mathbf{p})d\Omega \quad (22)$$

$$\beta(\mathbf{p})V_1(\mathbf{p})=\int_{\Gamma}\left(V_1G'_1(\mathbf{q},\mathbf{p})-V'_1G_1(\mathbf{q},\mathbf{p})\right)d\Gamma+\int_{\Omega}W_0G_1(\mathbf{q},\mathbf{p})d\Omega \quad (23)$$

On first inspection of Equations (22) and (23), the domain integral terms involving W_0 seem to pose some difficulty in the boundary element solution procedure. Typically, many interior point calculations are required to accurately calculate the domain integral. However, the function W_0 is a biharmonic function by definition, see Equation (9). Using the Rayleigh–Green identity, given in Equation (19), the second set of equations in the sequence, Equations (22) and (23), may be transformed into:

$$\beta(\mathbf{p})W_1(\mathbf{p})=\int_{\Gamma}\left(W_1G'_1(\mathbf{q},\mathbf{p})-W'_1G_1(\mathbf{q},\mathbf{p})+V_1G'_2(\mathbf{q},\mathbf{p})-V'_1G_2(\mathbf{q},\mathbf{p})\right)d\Gamma$$

$$+\int_{\Gamma}\left(W_0G'_3(\mathbf{q},\mathbf{p})-W'_0G_3(\mathbf{q},\mathbf{p})+V_0G'_4(\mathbf{q},\mathbf{p})-V'_0G_4(\mathbf{q},\mathbf{p})\right)d\Gamma$$

$$+\int_{\Omega}\frac{Q(x,y)}{D}G_4(\mathbf{q},\mathbf{p})d\Omega \quad (24)$$

$$\beta(\mathbf{p})V_1(\mathbf{p}) = \int_\Gamma \left(V_1G_1'(\mathbf{q},\mathbf{p}) - V_1'G_1(\mathbf{q},\mathbf{p})\right)d\Gamma$$

$$+ \int_\Gamma \left(V_0G_2'(\mathbf{q},\mathbf{p}) - V_0'G_2(\mathbf{q},\mathbf{p})\right)d\Gamma + \int_\Omega \frac{Q(x,y)}{D}G_2(\mathbf{q},\mathbf{p})d\Omega \tag{25}$$

The original domain integrals of Equations (22) and (23) have been transformed into a much more convenient set of boundary integrals and a domain integral involving the loading function $Q(x,y)$. The values of W_0, W_0', V_0, and V_0' are known at points along the boundary from the solution of the previous set of Equations (17) and (18). If the loading function is also biharmonic, the remaining domain integrals in Equations (24) and (25) may be transformed into a set of boundary integrals using the relationship in Equation (19).

To determine the solution of the next correction term, the third equation in the sequence of expressions given in Equations (9) and (10) should be solved. After transforming the domain integrals using the identity given in Equation (19), the converted integral equations are:

$$\beta(\mathbf{p})W_2(\mathbf{p}) = \int_\Gamma \left(W_2G_1'(\mathbf{q},\mathbf{p}) - W_2'G_1(\mathbf{q},\mathbf{p}) + V_2G_2'(\mathbf{q},\mathbf{p}) - V_2'G_2(\mathbf{q},\mathbf{p})\right)d\Gamma$$

$$+ \int_\Gamma \left(W_1G_3'(\mathbf{q},\mathbf{p}) - W_1'G_3(\mathbf{q},\mathbf{p}) + V_1G_4'(\mathbf{q},\mathbf{p}) - V_1'G_4(\mathbf{q},\mathbf{p})\right)d\Gamma$$

$$+ \int_\Gamma \left(W_0G_5'(\mathbf{q},\mathbf{p}) - W_0'G_5(\mathbf{q},\mathbf{p}) + V_0G_6'(\mathbf{q},\mathbf{p}) - V_0'G_6(\mathbf{q},\mathbf{p})\right)d\Gamma$$

$$+ \int_\Omega \frac{Q(x,y)}{D}G_6(\mathbf{q},\mathbf{p})d\Omega \tag{26}$$

$$\beta(\mathbf{p})V_2(\mathbf{p}) = \int_\Gamma \left(V_2G_1'(\mathbf{q},\mathbf{p}) - V_2'G_1(\mathbf{q},\mathbf{p})\right)d\Gamma$$

$$+ \int_\Gamma \left(V_1G_2'(\mathbf{q},\mathbf{p}) - V_1'G_2(\mathbf{q},\mathbf{p})\right)d\Gamma$$

$$+ \int_\Gamma \left(V_0G_3'(\mathbf{q},\mathbf{p}) - V_0'G_3(\mathbf{q},\mathbf{p})\right)d\Gamma + \int_\Omega \frac{Q(x,y)}{D}G_3(\mathbf{q},\mathbf{p})d\Omega \tag{27}$$

The remaining correction terms of Equation (8) may be calculated by repeating the solution procedure described in Equations (22) – (27). The kth correction term ($k \geq 0$) may be evaluated by solving the following set of equations:

$$\beta(\mathbf{p})W_k(\mathbf{p}) = \int_\Gamma \left(W_kG_1'(\mathbf{q},\mathbf{p}) - W_k'G_1(\mathbf{q},\mathbf{p}) + V_kG_2'(\mathbf{q},\mathbf{p}) - V_k'G_2(\mathbf{q},\mathbf{p})\right)d\Gamma + B_k^1 \tag{28}$$

$$\beta(\mathbf{p})V_k(\mathbf{p}) = \int_\Gamma \left(V_kG_1'(\mathbf{q},\mathbf{p}) - V_k'G_1(\mathbf{q},\mathbf{p})\right)d\Gamma + B_k^2 \tag{29}$$

where B_k^1 and B_k^2 have the form:

$$B_k^1 = \sum_{i=0}^{k-1} \int_\Gamma \left(W_i G'_{2(k-i)+1} - W'_i G_{2(k-i)+1} + V_i G'_{2(k-i+1)} - V'_i G_{2(k-i+1)} \right) d\Gamma + \int_\Omega \frac{Q(x,y)}{D} G_{2(k+1)} d\Omega \tag{30}$$

$$B_k^2 = \sum_{i=0}^{k-1} \int_\Gamma \left(V_i G'_{k-i+1} - V'_i G_{k-i+1} \right) d\Gamma + \int \Omega \frac{Q(x,y)}{D} G_{k+1} d\Omega \tag{31}$$

The general form of the boundary integral equations defined in Equations (28) and (29) may be used to evaluate as many correction terms as desired. Once boundary values W_i are known, the general solution to the forced vibration problem may be determined from Equation (7):

$$w(x,y,t) = e^{i\omega t} \sum_{k=0}^{n} \varepsilon^k W_k(x,y) \tag{32}$$

where n is the number of expansion correction terms desired. The complete solution for forced vibrations of elastic plates may be obtained by superimposing the forced vibration solution, Equation (32), with the free vibration solution.

NUMERICAL ANALYSIS

To demonstrate the validity of the BEM solution of the perturbation expansion, the vibration of an elastic plate subjected to two types of forcing functions will be presented. The examples consider the response of a clamped circular plate of radius a subjected to a point load, P_0, and a uniform loading function, q_0, at a frequency ω. In both cases, the circular plate is discretized by 16 equally spaced nodes defining 8 quadratic elements, as shown in Figure 2. The boundary conditions are zero deflections and no slope of deflected surface, $\partial W/\partial n = 0$, at the radius of the circular plate. The forcing frequency is assumed to be a small quantity, $\omega/\nu << 1$. For both cases, analytical solutions for the small frequency perturbation problem are available for the first few terms in the expansion.

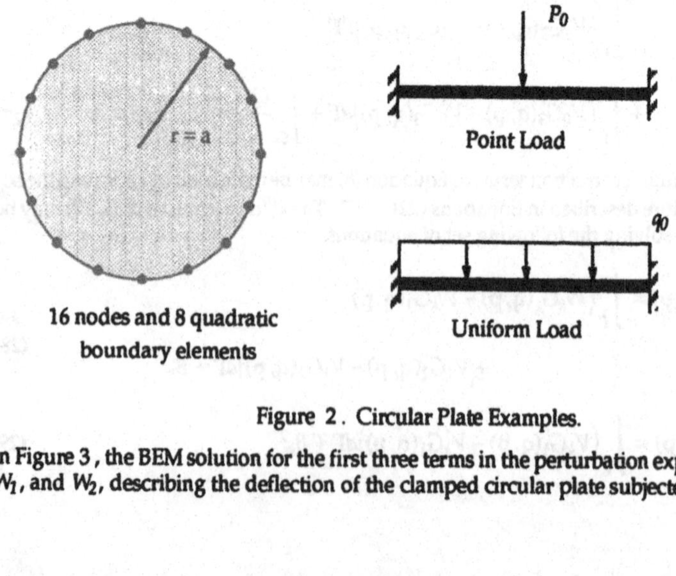

Figure 2. Circular Plate Examples.

In Figure 3, the BEM solution for the first three terms in the perturbation expansion, W_0, W_1, and W_2, describing the deflection of the clamped circular plate subjected to a point

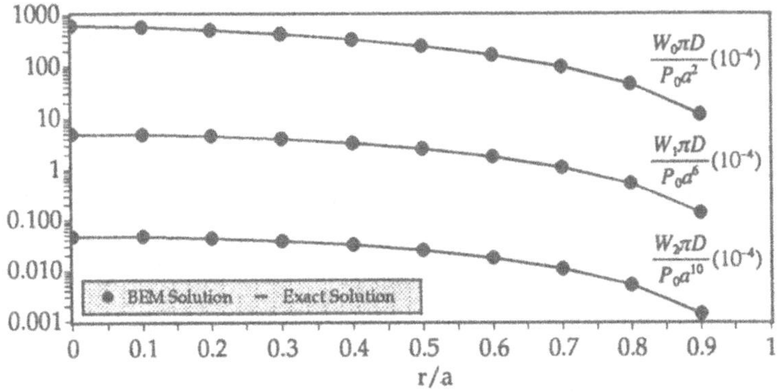

Figure 3. Perturbation Expansion Terms for a Clamped Circular Plate Subjected to a Point Load, P_0.

load vibrating at a small frequency are compared with the exact solution. The results indicate that even for a relatively unrefined model, (only 8 quadratic elements), the solution for each term is very accurate. The BEM results for a circular clamped plate subjected to a uniform load function, q_0, are compared to the exact solution and shown in Figure 4. Combining the terms defined by the regular perturbation, given in Equation (7), the center deflection for a clamped circular plate subjected to a uniform loading function is shown in Figure 5 for various values of ε. Only a few terms were required to obtain an accurate solution.

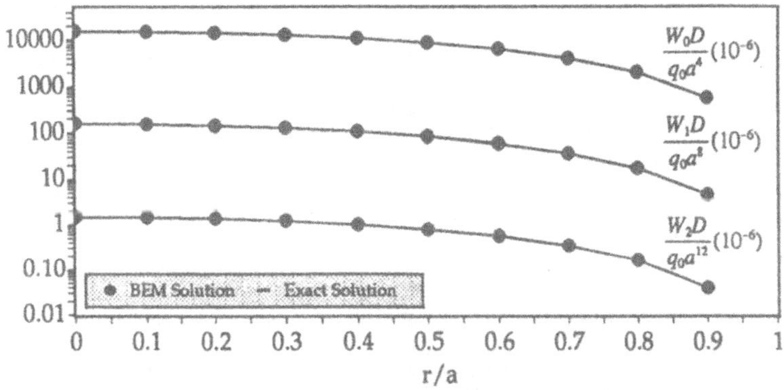

Figure 4. Perturbation Expansion Terms for a Clamped Circular Plate Subjected to a Uniform Load, q_0

The original hypothesis for the regular perturbation expansion, defined in Equation (7), was that the quantity $\varepsilon = (\omega/\nu)^2$ is considered small. As shown in Figure 5, the BEM solution is accurate when compared to an analytical solution [16] at values of ε larger than one. Some of the error associated with the BEM solution must be attributed to the coarse discretization used to represent the circular plate. The time-dependent oscillations of the elastic plate may be determined by multiplying the spatial solution, given in Figure 5, by the assumed form of the temporal component, defined in Equation (3).

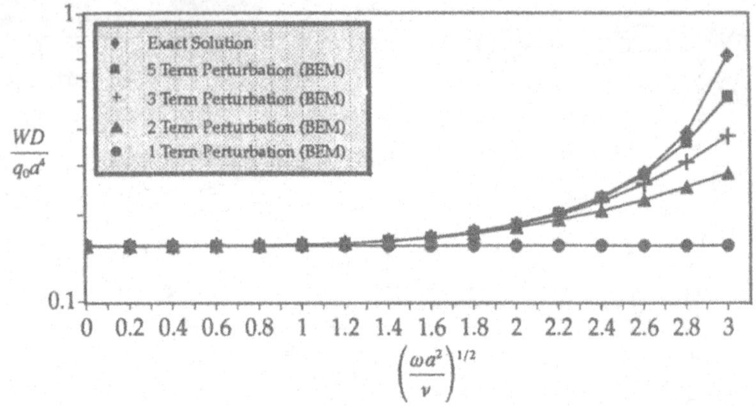

Figure 5. Center Deflection for a Clamped Circular Plate Subjected to a Uniform Load, q_0.

SUMMARY

A BEM formulation based on a regular perturbation expansion for determining the forced low frequency vibrations of elastic plates is developed. For small values of the parameter, ε, the original dynamic behavior of a vibrating plate is converted into an infinite sequence of coupled static plate problems. Several examples are presented to demonstrate the accuracy of the BEM–perturbation technique. In many cases, excellent solutions may be obtained by calculating only the first few terms in the expansion.

REFERENCES

1 W. Weaver, Jr., S. P. Timoshenko, and D.H. Young, *Vibration Problems in Engineering*, John Wiley & Sons, New York, NY, 1990.

2 J. G. Eisley, *Mechanics of Elastic Structures*, Prentice–Hall, Englewood Cliffs, NJ, 1989.

3 J. T. Katsikadelis and A. E. Armenakas, Analysis of Clamped Plates on Elastic Foundation by the Boundary Integral Equation Method, J. App. Mech., 51, 574–580, 1984.

4 J. A. Costa and C. A. Brebbia, The Boundary Element Method Applied to Plates on Elastic Foundations, Eng. Analy., 2, 174–183, 1985.

5 G. S. Gipson and D. Reible, A BEM–Perturbation Technique for Membrane Structures, Recent Applications in Computational Mechanics, ASCE, New York, NY.

6 A. H. Nayfeh, *Perturbation Methods*, John Wiley & Sons, New York NY, 1973.

7 A. H. Nayfeh, *Problems in Perturbation*, John Wiley & Sons, New York NY, 1985.

8 M. A. Jaswon, M. Maiti, and G. T. Symm, Numerical Biharmonic Analysis and Some Applications, Int. J. Solids Struct., 3, 309–332, 1967.

9 M. A. Jaswon and M. Maiti, An Integral Equation Formulation of Plate Bending Problems, J. Eng. Math., 2, 83–93, 1968.

10 N. J. Altiero and D. L. Sikarskie, A Boundary Integral Method Applied to Plates of Arbitrary Plan Form, Comp. Struct., 9, 163–168, 1978.

11 M. Stern, A General Boundary Integral Formulation for the Numerical Solution of Plate Bending Problems, Int. J. Solids Struct., 15, 769–782, 1979.

12 S. Guo–Shu and S. Mukherjee, Boundary Element Analysis of Bending of Elastic Plates of Arbitrary Shape with General Boundary Conditions, Eng. Analy., 3, 36–44, 1986.

13 C. V. Camp and G. S. Gipson, *Boundary Element Analysis of Nonhomogeneous Biharmonic Phenomena*, accepted for publication by Computational Mechanics, Southampton, UK.

14 C. V. Camp and G. S. Gipson, Biharmonic Analysis of Rectilinear Plates by the Boundary Element Method, Int. J. Numer. Methods Eng., 30, 517–539, 1990.

15 C. A. Brebbia (ed.), *The Boundary Element Method for Engineers*, Pentech Press, London, 1978.

16 Lebedev, N. N., Skalskaya, I. P., and Uflyand, Y. S., *Worked Problems in Applied Mathematics*, Dover, New York, 1965.

SECTION 11: INELASTIC PROBLEMS

DBEM Analysis of Beams on Modified Elastoplastic Winkler Foundation

P. de Simone, A.Ghersi, R. Mauro

Naples University, Italy

ABSTRACT

The problem of a beam on an elastoplastic Winkler medium is handled using a general Direct Boundary Element Method approach. The case of concentrated internal loading is considered, and the governing equations are obtained both following the classic beam subdivision approach and considering the beam loaded by a family of Dirac delta distributions and by its first generalized derivative. The two different approaches are compared and their formal equivalence is finally shown.

INTRODUCTION

Application of Boundary Element Method to the problem of a beam on a Winkler foundation is far from new. As a matter of fact, the well known method of superposition by Hetényi[6] is nothing more but the application of the Indirect Boundary Element Method to the problem (e.g. see Butterfield[3]). The more general direct approach has been subsequently applied by Butterfield[3] in the case of absence of concentrated loading at internal cross sections. In a previous paper[5] the authors have followed the DBEM approach to handle heterogeneous Winkler medium, using an iterating technique to take into account the variable part of k.

The case of a beam subjected to internal concentrated loading (forces and moments) can be handled, within the context of the Direct Boundary Element Method, following the classic approach, i.e. subdividing the beam into single portions without internal concentrated loading, then imposing satisfaction of continuity/discontinuity requirements at all the interfaces (e.g. see Hetényi[6]); a similar technique has been largely used to handle also other chain-like problems such as stratified elastic media; the weakness of some seemingly attractive solution technique related to this approach, such as the transfer-matrix (e.g. see Hetényi[6], Banerjee, Butterfield[1]), has been pointed out by Maier, Novati[7].

In the present paper the authors extend the DBEM approach to the case of a beam on a non-linear elastoplastic Winkler medium, subjected to internal concentrated loading. The classic approach of the subdivision is first followed, and a more compact approach is subsequently adopted by the use of the Dirac distribution and of its first generalized derivative. The governing equations of each approach are finally compared, and the formal equivalence of the two approaches pointed out.

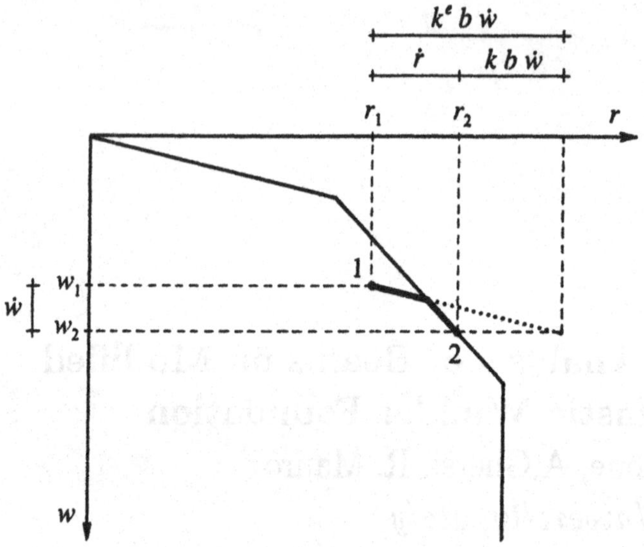

Figure 1 - displacement-reaction relationship for an elastoplastic Winkler medium.

ELASTOPLASTIC WINKLER MEDIUM

Within the framework of a discretized model, non-linear soil reaction may be represented by means of an elastoplastic model generalizing the well known linear elastic Winkler medium. In this model, the soil reaction r at a given point y depends on the deflection w of the point according to a trilinear law (fig.1). For small values of w, and in case of unloading and reloading, r is related to w by means of the usual elastic coefficient of subgrade reaction k^e

$$r = k^e \, b \, w$$

where b is the breadth of the beam.
In a similar way the elastoplastic behaviour may be expressed with the relation

$$r = (k^e - k) \, b \, w$$

In an incremental loading process, an analogous relation exists between deflection and reaction increments, $\dot{w}$ and $\dot{r}$

$$\dot{r} = (k^e - k) \, b \, \dot{w}$$

In the following sections dot will be omitted for simplicity, but all relations are to be considered incremental.

DIRECT BOUNDARY ELEMENT METHOD

The problem of a beam foundation resting on a Winkler elastoplastic medium (fig.2) can be handled very elegantly using the Direct Boundary Element Method. The method, illustrated by Butterfield[3] for the case of a linear elastic medium (constant k coefficient), can be easily extended to the case of an elastoplastic medium, as outlined below.

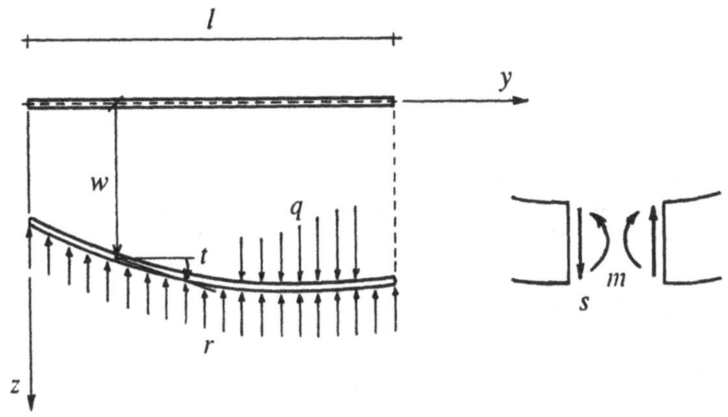

Figure 2 - sign convention for w, t, m, s, q, r.

Being l the length of the beam, EI the flexural rigidity, t, m, s the rotation, bending moment and shear, q the distributed transverse load and λ the "elastic" characteristic (see Hetényi[6]) defined as

$$\lambda = \sqrt[4]{\frac{k^{*}b}{4EI}}$$

and introducing the non-dimensional quantities

$$Y = y\lambda \qquad L = l\lambda$$

$$W(Y) = w(y)\,\lambda \qquad M(Y) = \frac{m(y)\,\lambda^3}{k^{*}b}$$

$$T(Y) = t(y) \qquad S(Y) = \frac{s(y)\,\lambda^2}{k^{*}b}$$

$$Q(Y) = \frac{q(y)\,\lambda}{k^{*}b} \qquad K(W) = \frac{k(w)}{k^{*}} \qquad R(W) = \frac{r(w)\,\lambda}{k^{*}b} = (1-K)\,W$$

the well-known relations among deflection, rotation, bending moment, shear and transverse load may be written in non-dimensional form as

$$\frac{dW}{dY} = T \qquad \frac{dM}{dY} = S$$

$$\frac{dT}{dY} = -4M \qquad \frac{dS}{dY} = -(Q-R)$$

The relation among deflection, soil reaction and transverse load thus become

$$\frac{1}{4}\frac{d^4W}{dY^4}=Q-(1-K)W \tag{1}$$

Following the general pattern of the DBEM (e.g. see Brebbia, Dominguez[2], De Simone[4]), equation (1) is multiplied by a function W^* and then integrated by parts four times over the length L of the beam, finally obtaining the relation

$$\int_L \frac{d}{dY}\left[W^*S-\frac{dW^*}{dY}M+\left(-\frac{1}{4}\frac{d^2W^*}{dY^2}\right)T-\left(-\frac{1}{4}\frac{d^3W^*}{dY^3}\right)W\right]dY+$$
$$+\int_L\left[-\frac{1}{4}\frac{d^4W^*}{dY^4}W+W^*Q-W^*(1-K)W\right]dY=0 \tag{2}$$

In this paper it is chosen as function W^* the Green's function $W^*(Y,X)$ of the infinitely long beam resting on a k^*-constant medium and loaded by a unit force at the source point $X=x\,\lambda$, i.e. the function satisfying the differential equation

$$\frac{1}{4}\frac{d^4W^*(Y,X)}{dY^4})=\delta(Y,X)-W^*(Y,X) \tag{3}$$

with distributed load equal to the Dirac delta function $\delta(Y,X)$, and with homogeneous boundary conditions at infinity ($y=\pm\infty$). Indicating with $T^*(Y,X)$, $M^*(Y,X)$ and $S^*(Y,X)$ non-dimensional rotation, bending moment and shear force associated to the displacement W^*, we have the following expressions which generalize the ones by Hetényi[6], considering that the generalized derivative of abs(Y-X) and sign(Y-X) are respectively sign(Y-X) and 2 $\delta(Y,X)$

$W^*(Y,X)=\frac{1}{2}A(Z)$	$M^*(Y,X)=\frac{1}{4}C(Z)$
$T^*(Y,X)=-\operatorname{sgn}(Y-X)\,B(Z)$	$S^*(Y,X)=-\frac{1}{2}\operatorname{sgn}(Y-X)\,D(Z)$

Z being equal to abs(Y-X) and A, B, C, D the functions by Hetényi[6]

$A(Z)=e^{-Z}(\cos Z+\sin Z)$	$C(Z)=e^{-Z}(\cos Z-\sin Z)$
$B(Z)=e^{-Z}\sin Z$	$D(Z)=e^{-Z}\cos Z$

In this way, taking into account the sifting property of δ and using Gauss' theorem we finally get the Somigliana-like identity for the displacement of the problem. Similar identities for the rotation, bending moment and shear, T, M, S, can be obtained differentiating the displacement expression with respect to X

$$W(X)=[-S^*(Y,X)\,W(Y)+M^*(Y,X)\,T(Y)-T^*(Y,X)\,M(Y)+W^*(Y,X)\,S(Y)]_0^L+$$
$$+\int_L W^*(Y,X)\,Q(Y)\,dY+\int_L K(W)\,W^*(Y,X)\,W(Y)\,dY \tag{4}$$

$$T(X)=\frac{dW(X)}{dX}=[W^*W-S^*T-4M^*M-T^*S]_0^L-\int_L T^*Q\,dY-\int_L T^*KW\,dY \tag{5}$$

$$M(X)=-\frac{1}{4}\frac{dT(X)}{dX}=\left[\frac{1}{4}T^*W-\frac{1}{4}W^*T-S^*M+M^*S\right]_0^L+\int_L M^*Q\,dY-\int_L M^*KW\,dY \tag{6}$$

$$S(X)=\frac{dM(X)}{dX}=\left[M^*W+\frac{1}{4}T^*T+W^*M-S^*S\right]_0^L-\int_L S^*Q\,dY-\int_L S^*KW\,dY \tag{7}$$

Defining vectors $\{U\}, \{P\}, \{V_Q\}, \{V_K\}, \{N_Q\}, \{N_K\}$

$$\{U(X)\}=\begin{Bmatrix} W(X) \\ T(X) \end{Bmatrix} \qquad \{P(X)\}=\begin{Bmatrix} M(X) \\ S(X) \end{Bmatrix}$$

$$\{V_Q(X)\}=\begin{Bmatrix} \int_L W^*(Y,X)\,Q(Y)\,dY \\ -\int_L T^*(Y,X)\,Q(Y)\,dY \end{Bmatrix} \qquad \{V_K(X)\}=\begin{Bmatrix} \int_L W^*(Y,X)\,K(W)\,W(Y)\,dY \\ -\int_L T^*(Y,X)\,K(W)\,W(Y)\,dY \end{Bmatrix}$$

$$\{N_Q(X)\}=\begin{Bmatrix} \int_L M^*(Y,X)\,Q(Y)\,dY \\ -\int_L S^*(Y,X)\,Q(Y)\,dY \end{Bmatrix} \qquad \{N_K(X)\}=\begin{Bmatrix} \int_L M^*(Y,X)\,K(W)\,W(Y)\,dY \\ -\int_L S^*(Y,X)\,K(W)\,W(Y)\,dY \end{Bmatrix}$$

and matrices $[H], [G]$

$$[H(Y,X)]=\begin{bmatrix} S^*(Y,X) & -M^*(Y,X) \\ -W^*(Y,X) & S^*(Y,X) \end{bmatrix} \qquad [G(Y,X)]=\begin{bmatrix} T^*(Y,X) & -W^*(Y,X) \\ 4M^*(Y,X) & T^*(Y,X) \end{bmatrix}$$

identities (4) (5) (6) and (7) can be written in the compact matrix form

$$\{U(X)\}=[H(0,X)]\,\{U(0)\}-[H(L,X)]\,\{U(L)\}+$$
$$+[G(0,X)]\,\{P(0)\}-[G(L,X)]\,\{P(L)\}+\{V_Q(X)\}+\{V_K(X)\} \tag{8}$$

$$\{P(X)\}=-\frac{1}{4}[G(0,X)]\,\{U(0)\}+\frac{1}{4}[G(L,X)]\,\{U(L)\}+$$
$$+[H(0,X)]\,\{P(0)\}-[H(L,X)]\,\{P(L)\}+\{N_Q(X)\}+\{N_K(X)\} \tag{9}$$

Collocating these expressions at the ends of the beam (or at the ends of each portion in which the beam is subdivided) we obtain a system of integral equations, the solution of which allows the evaluation of deflection, slope, bending moment and shear force at a generic point of the beam.

BEAM WITHOUT INTERNAL CONCENTRATED LOADING

If the beam is not subjected to internal concentrated loading, collocating (8) respectively at the left and right ends of the beam, the system of four integral equations is obtained

$$\left([I]-[H_{l^+l}]\right)\{U_l\}+\qquad[H_{lr}]\{U_r\}=[G_{ll}]\{P_l\}-[G_{lr}]\{P_r\}+\{V_{Ql}\}+\{V_{Kl}\}$$

$$-[H_{rl}]\{U_l\}+\left([I]+[H_{r^-r}]\right)\{U_r\}=[G_{rl}]\{P_l\}-[G_{rr}]\{P_r\}+\{V_{Qr}\}+\{V_{Kr}\}$$

which may be written

$$([I]+[H])\{U\}=[G]\{P\}+\{V_Q\}+\{V_K\} \tag{10}$$

with $[I]$ the identity matrix and matrices $[H]$, $[G]$ and vectors $\{U\}$, $\{P\}$, $\{V_Q\}$, $\{V_K\}$ being respectively

$$[H]=\begin{bmatrix}-H_{l^+l} & H_{lr}\\ -H_{rl} & H_{r^-r}\end{bmatrix} \qquad [G]=\begin{bmatrix}G_{ll} & -G_{lr}\\ G_{rl} & -G_{rr}\end{bmatrix}$$

$$\{U\}=\begin{Bmatrix}\{U_l\}\\ \{U_r\}\end{Bmatrix} \qquad \{V_Q\}=\begin{Bmatrix}\{V_{Ql}\}\\ \{V_{Qr}\}\end{Bmatrix}$$

$$\{P\}=\begin{Bmatrix}\{P_l\}\\ \{P_r\}\end{Bmatrix} \qquad \{V_K\}=\begin{Bmatrix}\{V_{Kl}\}\\ \{V_{Kr}\}\end{Bmatrix}$$

where $[H_{ij}]$, $[U_i]$ and similar terms stand for $[H(Y,X)]$, $[U(X)]$, l and r being respectively left (X,Y=0) and right (X,Y=L) ends of the beam, and l^+, r^- pointing (only when necessary) at the section immediately after or before the left and right ends. It has to be remarked that in matrix terms as $[H_{ij}]$ the first subscript i refers to the second variable X, e.g. $[H_{lr}]=[H(L,0)]$.
In the case of a linear elastic Winkler medium, K=0 and the vector $\{V_K\}$ vanishes, system (10) then degenerating into an algebraic linear system of four equations whose unknowns are only four of the eight components of $\{U\}$, $\{P\}$, the other four being given in order the problem to be well posed. Having solved (10), Somigliana identities (8) and (9) can be used to obtain deflection, slope, bending moment and shear force at a generic point.
In the case of a non linear medium, as it is in the present analysis, system (10) remains an integral equation system. The solution can be obtained treating the integral term $\{V_K\}$ as a nonlinear term, handling equations (10) by means of an iterative technique, $\{V_K\}$ being adjusted until convergence is reached.

BEAM SUBJECTED TO INTERNAL CONCENTRATED LOADING

Beam splitting approach

When a beam is subjected to concentrated loading (forces and/or moments) at n internal cross sections, a way to handle the problem which seems to follow naturally from the method of integration of differential equations consists in subdividing the beam into n+1 portions, then considering each portion as an isolated beam subjected to its own loads together with some end loading; writing the governing equations for each beam, and enforcing satisfaction of continuity/discontinuity requirements at each interface between adjacent beams, a banded linear system is finally obtained. In this way, eight more unknowns are added for each subdivision (displacement, rotation, shear, bending moment at each end), which are balanced by eight new equations (four equations for each beam together with four continuity/discontinuity relations). Following this classical approach (e.g. see Hetényi[6]) in the case of the DBEM, collocating (8) respectively at left and right ends of the generic portion i of the beam, we obtain for each portion the four equations

$$\left([I]-\left[H^i_{ll}\right]\right)\{U^i_l\}+\qquad [H^i_{lr}]\{U^i_r\}=[G^i_{ll}]\{P^i_l\}-[G^i_{lr}]\{P^i_r\}+\{V^i_{Ql}\}+\{V^i_{Kl}\} \tag{11a}$$

$$-[H^i_{rl}]\{U^i_l\}+\left([I]+\left[H^i_{rr}\right]\right)\{U^i_r\}=[G^i_{rl}]\{P^i_l\}-[G^i_{rr}]\{P^i_r\}+\{V^i_{Qr}\}+\{V^i_{Kr}\} \tag{11b}$$

At each of the n points of internal subdivision we have to write four more equations expressing compatibility and equilibrium at the interfaces

$$\{U^i_r\}=\{U^{i+1}_l\}$$

$$\{P^i_r\}-\{P^{i+1}_l\}+\{F^i\}=0 \tag{12}$$

$\{F^i\}$ being a vector of concentrated loading (moment and force) at the interface i.
Equation (11) and (12) constitute a system of $4(n+1)+4n$ equations, balancing the $8(n+1)-4$ unknowns (the eight components of $\{U\}$, $\{P\}$ in each portion, minus four end values given by boundary conditions).
As stated before, in the case of a linear elastic medium the system degenerates to an algebraic linear system containing a number of unknowns equal to the number of equations, while in the case of a non linear medium the system remains an integral one which has to be solved by means of an iterative technique.

Generalized functions loading approach

Within the framework of the Direct Boundary Element Method, the problem of a beam subjected to internal concentrated loading can be handled in an elegant and extremely concise way introducing the Dirac delta distribution δ and its first derivative δ'in the governing differential equation. As a matter of fact, the δ Dirac distribution, or simple source distribution, is equivalent to a concentrated action and can be used to express a force as a distributed load; in a similar way, the generalized first derivative δ' of the Dirac δ distribution (e.g. see Sneddon[9]), or dipole source distribution, equivalent to a concentrated dipole action, can be used to express a concentrated moment by means of some distributed (generalized) load. The mathematics of distributions, or generalized functions, is quite difficult; some reference can be made for instance in Sneddon[9], Stakgold[10], Schwartz[8]. For the present purpose it is sufficient to remember some integral properties of δ and δ'; the first one is the sifting property of δ

$$\int_{-\infty}^{+\infty}\delta(y,x)\,\phi(y)\,dy=\phi(x) \tag{13}$$

a property well-known to everyone engaged with DBEM; the second one refers to δ' and constitutes in some way a generalization of the sifting property to the generalized derivative of δ according to the relation

$$\int_{-\infty}^{+\infty} \delta'(y,x)\,\phi(y)\,dy = -\phi'(x) \tag{14}$$

The use of δ and δ' to introduce concentrated loading as distributed (generalized) ones in the governing differential equation allows to treat the beam as a unique structure, thus avoiding any need for subdivisions and additional unknowns. Following the usual steps in DBEM, taking account of properties (13) and (14), we finally obtain the Somigliana identities for a beam with internal concentrated loading, just as in the case of a beam without any internal concentrated loading. With reference to a beam with m concentrated forces and n concentrated moments acting at $m+n$ given cross sections, the governing differential equation is written, in terms of the usual non-dimensional variables

$$\frac{1}{4}\frac{d^4W}{dY^4} = Q - (1-K)\,W + F_i\delta_i + M_j\delta'_j \tag{15}$$

where use has been made of the summation convention, with i=1,..,m and j=1,...,n. Multiplying eq.(15) by the function W^*, adjoint for the linear part of the problem, and integrating by parts four times, we finally get the Somigliana identity for W

$$W(X) = [-S^*(Y,X)\,W(Y) + M^*(Y,X)\,T(Y) - T^*(Y,X)\,M(Y) + W^*(Y,X)\,S(Y)]_0^L +$$

$$+\int_L W^*(Y,X)\,Q(Y)\,dY + \int_L K(W)\,W^*(Y,X)\,W(Y)\,dY + F_iW^*(X_i) - M_jW^{*\prime}(X_j) \tag{16}$$

the subsequent derivatives of W give similar expressions for rotation, moment and shear. Collocation of the relevant Somigliana identities at the beam ends gives a non-linear system of four equations, as in the case of a beam without concentrated internal loading. The same result could have been obtained by writing Betti's theorem in a suitable form.

EQUIVALENCE OF THE TWO APPROACHES

In order to show the complete equivalence of the two approaches described in the previous section, we can write equation (11*a*) for portions i and i+1 of the beam, getting

$$\left([I]-[H^i_{l'l}]\right)\{U^i_l\} + [H^i_{lr}]\{U^i_r\} = [G^i_{ll}]\{P^i_l\} - [G^i_{lr}]\{P^i_r\} + \{U^i_{Pl}\} + \{U^i_{Kl}\} \tag{17a}$$

$$\left([I]-[H^{i+1}_{l'l}]\right)\{U^{i+1}_l\} + [H^{i+1}_{lr}]\{U^{i+1}_r\} = [G^{i+1}_{ll}]\{P^{i+1}_l\} - [G^{i+1}_{lr}]\{P^{i+1}_r\} + \{U^{i+1}_{Pl}\} + \{U^{i+1}_{Kl}\} \tag{17b}$$

where

$$[I]-[H^{i+1}_{l'l}] = \begin{bmatrix} 1/2 & 1/4 \\ 1/2 & 1/4 \end{bmatrix} \qquad \left([I]-[H^{i+1}_{l'l}]\right)^{-1} = \begin{bmatrix} 4 & -2 \\ -4 & 4 \end{bmatrix}$$

Introducing the matrix

$$[O_{lr}^{i}] = -[H_{lr}^{i}]\left([I]-\left[H_{l'l}^{i+1}\right]\right)^{-1} = \begin{bmatrix} 2W^{*}(L^{i},0) & T^{*}(L^{i},0) \\ -2T^{*}(L^{i},0) & 4M^{*}(L^{i},0) \end{bmatrix}$$

and pre-multiplying by $[O_{lr}^{i}]$ all terms of equation (17*b*) we obtain

$$[O_{lr}^{i}]\left([I]-\left[H_{l'l}^{i+1}\right]\right) = -[H_{lr}^{i}]$$

$$[O_{lr}^{i}]\,[H_{lr}^{i+1}] = [H_{lr}^{j}]$$

$$[O_{lr}^{i}]\,[G_{ll}^{i+1}] = [G_{lr}^{i}]$$

$$[O_{lr}^{i}]\,[G_{lr}^{i+1}] = [G_{lr}^{j}]$$

$$[O_{lr}^{i}]\,[U_{Pl}^{i+1}] = \begin{Bmatrix} \int_{L^{i+1}} W^{*}(L^{i}+Y,0)\,Q\,dY \\ -\int_{L^{i+1}} T^{*}(L^{i}+Y,0)\,Q\,dY \end{Bmatrix}$$

$$[O_{lr}^{i}]\,[U_{Kl}^{i+1}] = \begin{Bmatrix} \int_{L^{i+1}} W^{*}(L^{i}+Y,0)\,K\,W\,dY \\ -\int_{L^{i+1}} T^{*}(L^{i}+Y,0)\,K\,W\,dY \end{Bmatrix}$$

where *j* indicates values referred to the beam obtained joining *i* and *i*+1 portions. Adding equation (17*a*) to equation (17*b*) so pre-multiplied, and noting that

$$\left[H_{l'l}^{i}\right] = \left[H_{l'l}^{j}\right]$$

$$[G_{ll}^{i}] = [G_{ll}^{j}]$$

$$\int_{L^{i}} W^{*}(Y,0)\,Q\,dY + \int_{L^{i+1}} W^{*}(L^{i}+Y,0)\,Q\,dY = \int_{L^{i}+L^{i+1}} W^{*}(Y,0)\,Q\,dY$$

we have

$$\left([I]-\left[H_{l'l}^{j}\right]\right)\{U_{l}^{i}\} + [H_{lr}^{i}](\{U_{r}^{i}\}-\{U_{l}^{i+1}\}) + [H_{lr}^{j}]\{U_{r}^{i+1}\} =$$

$$= [G_{ll}^{j}]\{P_{l}^{i}\} + [G_{lr}^{i}](\{P_{l}^{i+1}\}-\{P_{r}^{i}\}) - [G_{lr}^{j}]\{P_{r}^{i+1}\} + \{U_{Pl}^{j}\} + \{U_{Kl}^{j}\}$$

Using compatibility and equilibrium conditions (12) and remarking that $[G_{lr}^{i}][F]$ are the displacement and rotation of the left end of beam *j* caused by $[F]$ forces, i.e. the effect of $[F]$ δ distributed load, we obtain the equation (11*a*) for beam *j*. In the same way, writing equation (11*b*) for portions *i* and *i*+1, pre-multiplying the first equation by

$$[O_{rl}^{i+1}] = [H_{rl}^{i+1}]\left([I]+\left[H_{r^-r}^{i}\right]\right)^{-1} = \begin{bmatrix} 2W^*(0,L^{i+1}) & T^*(0,L^{i+1}) \\ -2T^*(0,L^{i+1}) & 4M^*(0,L^{i+1}) \end{bmatrix}$$

and adding it to the second one we obtain (11*b*) for beam *j*. Repeating these operations for each internal subdivision we once again arrive at the same equations obtained using the generalized loading approach.

CONCLUSIONS

The Direct Boundary Element Method has been applied to a beam on a modified elastoplastic Winkler medium in the general case of concentrated internal loading following both the classical beam splitting approach and a new global one based on a systematic use of Dirac δ distribution. The formal equivalence of the two approaches has been shown and it is concluded that the more compact Dirac distribution approach should be used in applications.

REFERENCES

1. Banerjee, P.K., Butterfield, R. Boundary element methods in geomechanics. In G.Gudehus (ed.), Finite Element in Geomechanics, pp.529-570, London, Wiley, 1977.
2. Brebbia, C.A., Dominguez, J., Boundary Elements, an Introductory Course, Computational Mechanics Publications, Southampton, 1989.
3. Butterfield, R. New concepts illustrated by old problems. In P.K.Banerjee & R.Butterfield (eds.), Developments in Boundary Element Methods - 1, pp. 1-20. London, Applied Science publishers, 1979.
4. De Simone, P. The DBEM in elastostatics using Green's generalized theorem. Proc. 5th Int. Conf. Bound. Elem., Hiroshima, C.A.Brebbia, T.Futagami, M.Tanaka (eds.), pp.449-458, 1983.
5. De Simone, P., Ghersi, A., Mauro, R. Statistical approach to beams on Winkler foundation. Proc. X ECSMFE, Florence, Italy, 1991.
6. Hetényi, M. Beams on Elastic Foundation. Ann Arbor, The University of Michigan Press, 1946.
7. Maier, G., Novati, G. On boundary element-transfer matrix analysis of layered elastic systems. Engineering Analysis, 3, pp. 208-216, 1986.
8. Schwartz, L., Mathematics for the Physical Sciences, Hermann, Paris, 1966.
9. Sneddon, I.N., The Use of Integral Transforms, McGraw-Hill Book Company, 1972.
10. Stakgold, I., Boundary Value Problems of Mathematical Physics, vol.I, The MacMillan Company, New York, 1967.

A Simple Integral Equation Method for the Viscoelastic Quasistatic Analysis of Vertically Loaded Pile

X.-Y. Liu (*), J.-J. Zheng (*), W.E. Saul (**)
(*) Department of Civil Engineering, Tianjin University, Tianjin, China
(**) Department of Civil Engineering, Michigan State University, U.S.A.

Abstract

In this paper, an integral equation method for the viscoelastic analysis of vertically loaded pile is presented. By this method, applying Betti's reciprocal theorem in Laplace space, two-set of nonsingular integral equations of one dimension are established for the pile and the soil respectively. According to the condition of displacement compatibility and stress equilibrium, the simultaneous equations are solved, then the stresses and the displacements are obtained. By use of Koizumi's inversion technique the solutions in time domain are achieved. In this paper, for the interpolation approach of constant element the analytic formulas of all integrals are derived. Compared with other methods, this method has the advantages of less calculating time and higher precision, therefore it is an effective method for analysing the problem of pile-soil interaction.

INTRODUCTION

Pile is one of the major structures in Civil and Ocean Engineering. The interaction and coupled working of pile and soil make a reasonable mathematical model. As we know the work of pile-soil interaction is one of the important problems in structure engineering. This kind of problem can be solved by different numerical methds. The Finite Element Method and Integral Equation Method occupy an important position in the various different numerical methods for the pile-soil interaction analysis. The finite element method is inconvenient to represent large or infinite domain. But the boundary integral equation method can be used to solve many problems where the domain is infinite or has large volume -surface area ratio. And the main advantage of boundary integral equation method is in the reduction of the dimensionality of the problem. Therefore a coupling of two methods is a very efficient mothodlogy in geotechnical engineering. The coupling technique of the pile-soil interaction problems could be effectively employed. And the advantages of both methods can be effectively utilized.

Developing this concept further along to solve pile-soil interaction problems is the predecessor of Banerjee [1] who based on the Mindlin fundamental solution, taking a fictitious stress density and established 2-dimensional singular displacement integral equations on the interface between pile and soil. The pile shaft was assumed to be elastic. By applying the compatibility and equilibrium of pile and soil, the simulateneous equations were established. Also Yun[2] presented an analogy method, assuming one fictitious line loads distributed along the axis, then a set of singular integral equations to be established. But this method taking a pile as a line is different from the actual situation. The shortcoming of both above methods is that do not taking into account the continuous condition of stress at the contact surface. Therefore they are not perfect methods. Recently, Gi[3] et. al. presented a coupled method of singular integral equation method and difference method, alghough this method' s thoughts are explicit and the equilibrium condition on the contact surface was taken into account but there are computing works in large amounts. In view of this situation that the analytic methods of the pile-soil interaction are not perfect and simple. Therefore this paper presents a simple integral equation method for the viscoelastic quasistatic analysis of pile.

THE BASIC EQUATION OF VISCOELASTICITY

For the isotropic linear viscoelastic material, the relation of stress deviatoric components and strain deviatoric components can be used the differential operator expressing[5] as

$$P_1(D)\sigma_{ij}' = Q_1(D)e_{ij}' \tag{1}$$

where

σ'_{ij}-stress deviatoric components

e'_{ij}-strain deviatoric components

$$P_1(D) = \sum_{k=0}^{m} a_k \cdot \frac{d^k}{dt^k}$$

$$Q_1(D) = \sum_{k=0}^{m} b_k \cdot \frac{d^k}{dt^k} \qquad (2)$$

And the stress、strain deviatoric components satisfy the initial conditions, i. e. ,

$$\sum_{s=1}^{m} a_k \cdot [p^{k-1} \cdot \sigma'_{ij}(o+) + p^{k-2} \cdot \frac{\partial}{\partial t}\sigma'_{ij}(O+) + \cdots + p\frac{\partial^{k-2}}{\partial t^{k-2}}\sigma'_{ij}(O+)$$

$$+ \frac{\partial^{k-1}}{\partial t^{k-1}}\sigma'_{ij}(O+)] = \sum_{k=1}^{n} b_k \cdot [p^{k-1} \cdot e'_{ij}(o+) + p^{k-2} \cdot \frac{\partial}{\partial t}e'_{ij}(O+)$$

$$+ \cdots + p\frac{\partial^{k-2}}{\partial t^{k-2}}e'_{ij}(O+) + \frac{\partial^{k-1}}{\partial t^{k-1}}e'_{ij}(O+)] \qquad (3)$$

where

p-the parameter of Laplace transformation then assuming that elastic volume deformation[5] it gives

$$\sigma_{ii} = 3Ke_{ii} \qquad (4)$$

To make the Laplace trasformation for Eqs. (1) and (4) and noticing the initial condition Eqs. (3), we find that

$$\bar{P}_1(p)\bar{\sigma}'_{ij} = \bar{Q}_1(p)\bar{e}'_{ij} \qquad (5)$$

where

$$\bar{P}_1(p) = \sum_{k=0}^{m} a_k p^k$$

$$\bar{Q}_1(p) = \sum_{k=0}^{n} b_k p^k \qquad (6)$$

and

$$\bar{\sigma}_{ii} = 3K\bar{e}_{ii} \tag{7}$$

Let

$$\bar{Q}_1(p) \cdot / \bar{P}_1(p) = 2G_1 \tag{8}$$

thus, the constitute relation of viscoelastic problems equal to the constitute relation of elastic static problems in the Laplace space. The subscript 1 in Eqs. (8) expresses viscoelastic body.

How to calculate the viscoelastic coeficients G_1 and λ_1 in space for two kinds of viscoelastic model, they are as following

1. Maxwell model

$$a_1 = 1/(2G), \; a_0 = 1/(2\eta), b_1 = 1$$

then

$$G_1 = Gp\tau/(1 + p\tau), \; \lambda_1 = \lambda + 2G/[3(1 + p\lambda)] \tag{9}$$

2. Kelvin model

$$a_0 = 1, \; b_1 = 2\eta, \; b_0 = 2G$$

then

$$G_1 = G(1 + p\tau) \;, \lambda_1 = \lambda\text{-}2Gp\tau/3 \tag{10}$$

where (in Eqs. (9), (10))

$$\tau = \eta/G \tag{11}$$

TO ESTABLISH THE INTEGRAL EQUATIONS

This analysis was based on the following assumptions; the pile was assumed to be elastic; the soil medium sourounded the pile was assumed to be viscoelastic. As following mentioned the stress, strain, displacement, force etc mean in Laplace space, for simplicity we now cancel the bar of corresponding symbol.

1. To Establish the Integral Equations of Pile

The length of pile is l, the radius of pile is a, the distributive shear along the z axis direction is $q(z)$ shown in Fig. (1), and noticing the pile is axisymmetric. The displacement of pile shaft in z direction is $w(z)$. Assume that the bottom pressure of pile is homogeneous distributive, the total pressure is P_b, the bottom displacement of shaft is W_b, And the top displacement of shaft is W_t, the top pressure is P_t. Through the Laplace trasformation we know

$$P_t = P/p \tag{12}$$

where

P -the force in time domain

p -the force in Laplace space

The differential equation of pile under the action of axis forces can be obtained.

$$E_p(A)\frac{d^2\omega}{d\xi^2} = q(\xi) \tag{13}$$

The corresponding displacement and force fundamental solutions of Eqs. (13) are;

$$\omega^*(z,\xi) = |\xi-z|/2(E_pA)$$

$$F^*(z,\xi) = sign(\xi-z)/2 \tag{14}$$

where sign (x) means sign function i. e.

$$sign(x) = \begin{cases} 1, & x>0 \\ -1, & x<0 \end{cases} \quad (15)$$

Applying the Betti reciprocal theory, we have

$$W(z) = \int_0^l w^*(z,\xi)q(\xi)d\xi + W_bF^*(z,l) - W_tF^*(z,o)$$

$$+ P_b \cdot W^*(z,l) - P_t \cdot W^*(z,o) \quad (16)$$

which is the integral equation of pile.

2. To Establish the Integral Equation of soil

It is evident that only changng the G, ν of Mindlin fundamental solution into the G_1, ν_1, the fundamental solution of viscoelastic semi—infinitive space can be obtained. It is based on the Mindlin fundamental solution and the analogous relation of elasticity and viscoelasticity. Also applying the reciprocal theory of work, the integral equation of viscoelastic soil can be obtained

$$C.w(r,z) = \int_{\Gamma_1} q/(2\pi a) \cdot W^* d\Gamma_1 + \int_{\Gamma_1} W \cdot \sigma^*_{rz} d\Gamma_1$$

$$+ \int_{\Gamma_2} P_b/(\pi a^2) \cdot W^* d\Gamma_2 + \int_{\Gamma_2} W_b \cdot \sigma^*_{zz} d\Gamma_2 \quad (17)$$

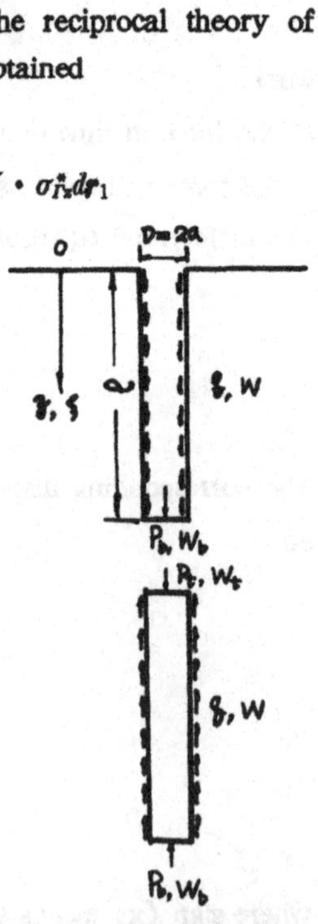

Fig. 1.

where Γ_1-expresses the side of pile contact with soil

Γ_2-expresses the botom of pile contact with soil

If we let $r = 0$, then $c = 0$ in Eqs, (17), by taking symmetry into account the integral equation can be changed into

$$\int_0^l q(\xi)W^*(0,z,a,\xi)d\xi + 2\pi a\int_0^l W(\xi)\sigma^*_{rz}(0,z,a,\xi)d\xi$$

$$+ 2P_b/a^2 \cdot \int_0^a W^*(0,z,\xi,l)\xi d\xi$$

$$+ 2\pi W_b\int_0^a \sigma^*_{zz}(0,z,\xi,l)\xi d\xi = 0 \quad (18)$$

which is the integral equation of viscoelastic soil.

THE ANALYTIC CALCULATION OF INTEGRAL COEFICIENTS AND ESTABLISHING THE MATRIX EQUATIONS

First, the pile is divided by N elements of equal length along the longitudinal direction, using the interpolation of constant element. The unknown displacements and stresses are denoted by W and F respectively i. e.

$$W = [w_1, w_2, \cdots, w_N]^T$$

$$F = [q_1, q_2, \cdots, q_N, p_b]^T \tag{19}$$

Now we know $W_t = W_1$ and $W_b = W_N$, then the integral equation (16) can be discritized and make $z=z_i(i=1,2,\cdots N)$ thus

$$W_i = \sum_{j=1}^{N} a_{ij}q_j + W_N \cdot F^*(z_i,l) - W_1 \cdot F^*(z_i,0)$$
$$+ P_b \cdot W^*(z_i,l) - p_t \cdot W^*(z_i,0) \quad (i = 1,2,\cdots\cdots,N) \tag{21}$$

where $a_{ij} = \int_{z_{j1}}^{z_{j2}} . W^*(z_i,\xi)d\xi$ (22)

The analytic expression of a_{ij} can be found in appendix.

Taking the matrix form for Eqs. (21), we have

$$A1 \cdot F + A2 \cdot W = A3 \cdot P_t \tag{23}$$

where $A1$ is $N \times (N + 1)$ matrix.

$A2$ is $N \times N$ matrix.

$A3$ is $N \times 1$ column vector

Second, the expression (18) can be discretized and taking $z=z_i(i=1,2,\cdots,N, N+1)$ thus

$$\sum_{j=1}^{N} b_{ij}q_j + \sum_{j=1}^{N} c_{ij}W_j + P_b \cdot d_i + W_N \cdot l_i = 0 \quad (i = 1,2,\cdots,N + 1) \tag{24}$$

where

$$b_{ij} = \int_{z_{j1}}^{z_{j2}} W^{*}(0, z_i, a, \xi) d\xi$$

$$c_{ij} = 2\pi a \int_{z_{j1}}^{z_{j2}} \sigma_{rz}^{*}(0, z_i, a, \xi)$$

$$d_i = (2/a^2) \cdot \int_{0}^{a} W^{*}(0, z_i, \xi, l) \xi d\xi$$

$$e_i = 2\pi \int_{0}^{a} \sigma_{zz}^{*}(o, z_i, \xi, l) \xi d\xi \tag{25}$$

The analytic expression of b_{ij}, c_{ij}, d_i, e_i can be found in appendix.

Writting equation (24) in matrix form, we have

$$B1 \cdot F + B2 \cdot W = 0 \tag{26}$$

where B_1 is $(N+1)(N+1)$ matrix.

B_2 is $(N+1) \cdot N$ matrix.

Third, Solving the simulataneous equations Eqs. (23). Eqs. (26) the displacement, the shear on side and the reaction on bottom of pile in Laplace space can be obtained. Then applying the mathematic inversion, the displacements and stresses in time domain also can be obtained. Here we employ the Laplace numerical inversion that is presented by Koizuim, this method is possessed of advantage in high precision and the shorter computing time.

If let $G_1 = G$、$\nu_1 = \nu$、$\lambda_1 = \lambda$ *and* $P = p$ then the viscoelastic problems can be changed into elastic static problems.

EXAMPLES

Example one

There is one elastic static problem, if $\mu = Ep/G$, $\nu = 0.5$. The results are shown in Fig. 2、3、4、5 for different l/D.

Where D is the diameter of pile, l is the length of pile.

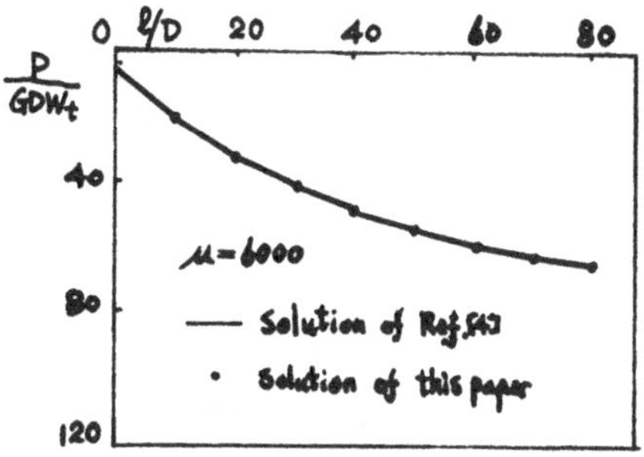

Fig. 2

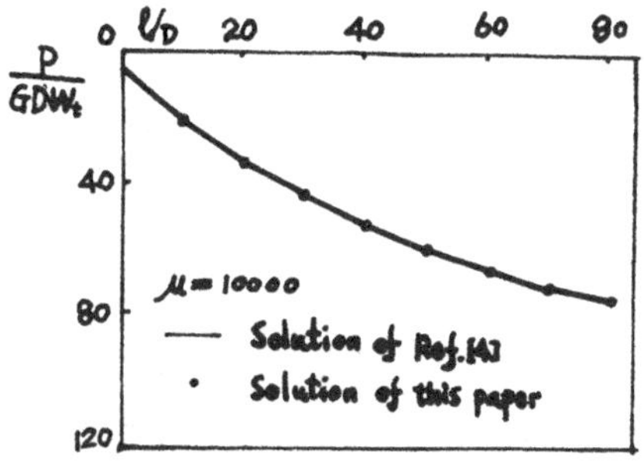

Fig. 3

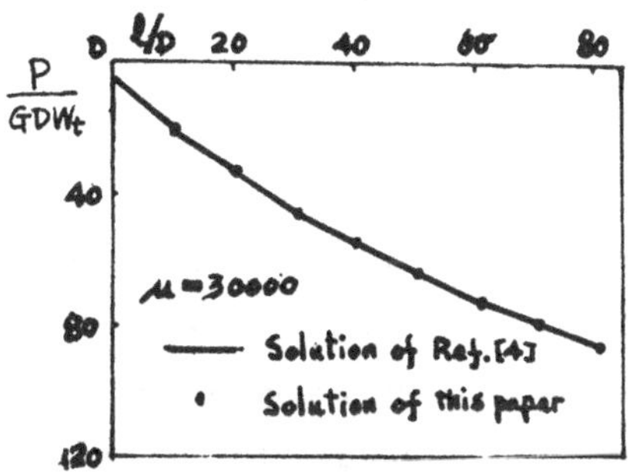

Fig. 4

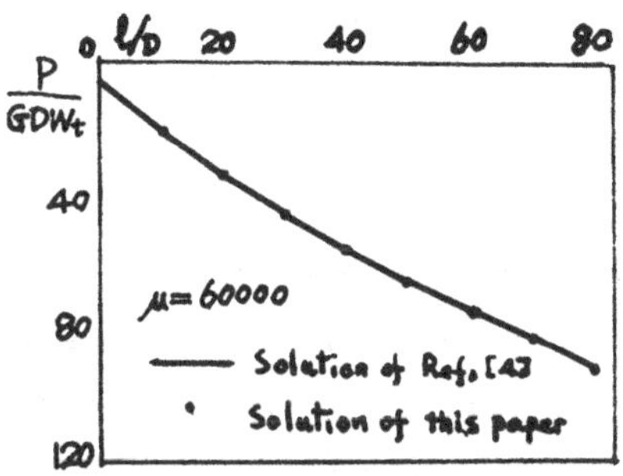

Fig. 5

As mentioned in Fig. 2、3、4、5 the solutions of this paper concide with the solution of reference[4]

Example two

There is a viscoelastic problem, the ν for soil equal 0. 3. The results are shown in Fig. 6、7. It is depicted in Fig. 6、7, the solutions in this paper are consistent with general viscoelastic pattern, for the Kelvin model, when $t/\tau \rightarrow \infty$ the viscoelastic solutions tends to elastic solutions, for the Maxwell model when $t/\tau = 0$ the viscoelastic solutions equal to the elastic solutions.

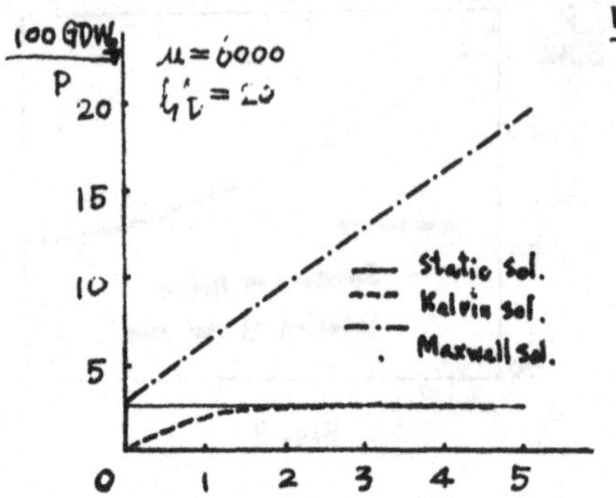

Fig. 6

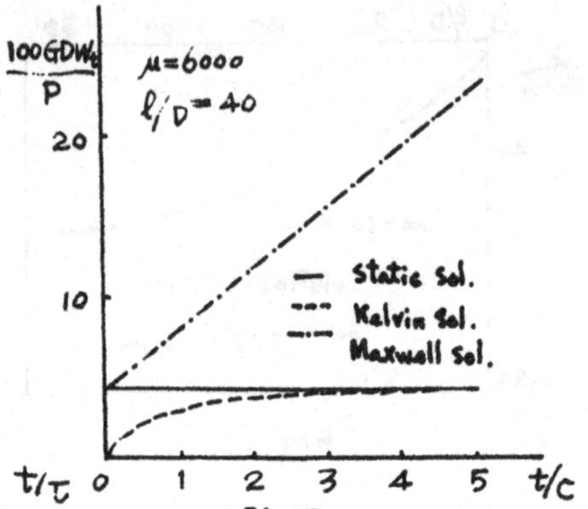

Fig. 7

CONCLUSION

This paper presents a new method, using this method the interaction problems of pile and viscoelastic medium can be solved. Through the examples it was clearly demonstrated that the solutions are reasonable. Also using the integral method in this work is different from the usually 2-dimension integral in the general references. Especially this paper presents a one dimension integral method and it' s application. By using this method the pile-soil interaction problem can be solved easily. The proposed procedure has the advantages of avoiding the singular integral, all of the coeficients are expressesed by analytic equations. Hence the calculations are more simple. Therefore the method presented in this paper is one of the simple integral methods and it makes further development for the Integral Equation Method of pile and soil interaction.

REFERENCES

1. R. Butterfied, P. K. Banerjee
   The elastic analysis of compressible pile and pile groups. Geotechnique 1971, 21(1) P43-60
2. P. K. Banerjee, T. G, Davies
   The behaviour of axially and laterally loaded single piles embeded in nonhomogeneous soils. Geotechnique, 1978. 28(3), P309-326.
3. Yun Tian-quan
   A simple integral equation method for analysis of pile. Applied Mathemat-

ics and Mechanics (in Chinese) 1981 2(3) P307-319

4. Ji Hai-po, Wang Ming-zhong, Wu Ji-ka
   The analysis of axial loaded pile with integral equation method. Acta Scientiarum Naturalium, Universitatis Pekinensis 1988. 24(3) P309-317
5. W. Flügge. Viscoelasticity. 2nd ed. Spring-Verlag 1975
6. S. Koizumi. A new method of evaluation of the Heaviside operational expression by Fourier sines. Phil. Mag. 1935. 19. P1061-1076.
7. S. William. The numerical treatment Laplace transforms; the Koizumi inversion method. Int. J. for Num. Meth. in Eng. 1984. 20 P1697-1702

APPENDIX

The analytic expression of a_{ij}, b_{ij}, c_{ij}, d_i and e_i:

$$a_{ij}=\begin{cases}(x_{j2}-x_{j1})(x_{j2}+x_{j1}-2z_i)/(4E_pA), & x_{j1}>z_i\\ [x_{j1}^2+x_{j2}^2+2z_i^2-2z_i(x_{j1}+x_{j2})]/(4E_pA), & x_{j1}<z_i<x_{j2}\\ (x_{j2}-x_{j1})(2z_i-x_{j1}-x_{j2})/(4E_pA), & z_i>x_{j2}\end{cases} \tag{27}$$

$$\begin{aligned}
e_1(x,y) &= ln[(x+\sqrt{x^2+a^2})/(y+\sqrt{y^2+a^2})]\\
e_2(x,y) &= (x/\sqrt{x^2+a^2}-y/\sqrt{y^2+a^2})/a^2\\
e_3(x,y) &= [x/(x^2+a^2)^{3/2}-y/(y^2+a^2)^{3/2}]/(3a^2)+2e_2(x,y)/(3a^2)\\
e_4(x,y) &= [x/(x^2+a^2)^{5/2}-y/(y^2+a^2)^{5/2}]/(5a^2)+4e_3(x,y)/(5a^2)
\end{aligned} \tag{28}$$

$$\begin{aligned}
b_{ij} = \{&4(1-v_1)e_1(x_{j2}-z_i,x_{j1}-z_i)+8(1-v_1)^2e_1(x_{j2}+z_i,x_{j1}+z_i)\\
&-a^2e_2(x_{j2}-z_i,x_{j1}-z_i)-[(3-4v_1)a^2+4z_i^2]e_2(x_{j2}+z_i,x_{j1}+z_i)\\
&+6a^2z_ie_3(x_{j2}+z_i,x_{j1}+z_i)-4z_i(1/\sqrt{(x_{j2}+z_i)^2+a^2}-1/\sqrt{(x_{j1}+z_i)^2+a^2})\\
&+2z_ia^2[1/(x_{j2}+z_i)^2+a^2)^{3/2}-1/((x_{j1}+z_i)^2+a^2)^{3/2}]\}/[16\pi G_1(1-v_1)]
\end{aligned} \tag{29}$$

$$\begin{aligned}
c_{ij} = \{&-(4-2v_1)e_2(x_{j2}-z_i,x_{j1}-z_i)+3a^2e_3(x_{j2}-z_i,x_{j1}-z_i)\\
&-(8-10v_1)e_2(x_{j2}+z_i,x_{j1}+z_i)+3[(3-4v_1)a^2+8z_i^2]e_3(x_{j2}+z_i,x_{j1}+z_i)\\
&-30z_i^2a^2e_4(x_{j2}+z_i,x_{j1}+z_i)+4z_i(1+v_1)[1/((x_{j2}+z_i)^2+a^2)^{3/2}\\
&-1/((x_{j1}+z_i)^2+a^2)^{3/2}]-6z_ia^2[1/((x_{j2}+z_i)^2+a^2)^{5/2}\\
&-1/((x_{j1}+z_i)^2+a^2)^{5/2}]\}\cdot a^2/[4(1-v_1)]
\end{aligned} \tag{30}$$

$$d_i = \{(3-4v_1)(\sqrt{(l-z_i)^2+a^2}-|l-z_i|)+[8(1-v_1)^2-(3-4v_1)]$$

$$\cdot\,(\sqrt{(l+z_i)^2+a^2}-|l+z_i|)-(l-z_i)^2(1/\sqrt{(l-z_i)^2+a^2}-1/|l-z^i|)$$
$$-[(3-4v_1)(l+z_i)^2-2lz_i]\cdot(1/\sqrt{(l+z_i)^2+a^2}-1/|l+z_i|)$$
$$-2z_il(l+z_i)^2\cdot[1/((l+z_i)^2+a^2)^{3/2}-1/|l+z_i|^3)\}\cdot/[8\pi G_1a^2(1-v_1)] \quad (31)$$

$$e_i=\{(1-2v_1)(l-z_i)\cdot(1/\sqrt{(l-z_i)^2+a^2}-1/|l-z_i|)-(1-2v_1)(l-z_i)$$
$$\cdot\,(1/\sqrt{(l+z_i)^2+a^2}-1/|l+z_i|)+(l-z_i)^2\cdot[1/((l-z_i)^2+a^2)^{3/2}-1/|l-z_i|^3]$$
$$+[(3-4v_1)l(l+z_i)^2-z_i(l+z_i)(5l-z_i)[1/(l+z_i)^2+a^2)^{3/2}-1/|l+z_i|^3]$$
$$+6lz_i(l+z_i)^3\cdot[l/((l+z_i)^2+a^2)^{5/2}-1/|l+z_i|^5]\}/[4(1-v_1)] \quad (32)$$

SECTION 12: DAMAGE TOLERANCE

Parametric Continuous Crack Tip Boundary Elements in Two-Dimensional Linear Elastic Fracture Mechanics

H.G. Walters (*), G.S. Gipson (**)

(*) Stimulation Research and Engineering Department, Halliburton Services, P.O. Drawer 1431, Duncan, OK 73536-0440, U.S.A.

(**) School of Civil Engineering, Oklahoma State University, Stillwater, OK 74078-0327, U.S.A.

ABSTRACT

The two-dimensional formulation of the boundary element method for linear elastic fracture mechanics is investigated. Parametric continuous Overhauser standard and crack tip boundary elements are developed along with formulas for calculation of the stress intensity factors directly from the traction solution. Results comparing the new boundary elements to more common boundary elements and other solutions are presented for a well known problem.

INTRODUCTION

The objective of this work is to better represent the geometry of a crack through the use of parametric continuous boundary elements in two dimensions and thus enhance the accuracy of the boundary element method in the solution of linear elastic fracture mechanics problems. The standard elastic fundamental solution will be employed, allowing the new elements to be implemented in existing boundary element codes easily. A method will be used for evaluating the stress intensity factors that is most appropriate for the boundary element method.

Lagrangian linear, quadratic, and cubic elements are commonly used in elastostatics. They provide only parametric positional (C^0) continuity between elements. When the boundary geometry is linear, parametric derivative continuity (C^1) between elements is satisfied; however, this is not the case when the boundary geometry is curved. The approximation of the displacements and tractions may also require derivative continuity between elements for accuracy. This is especially important in elastostatics because of its inherent dependence on a continuous strain field. C^1 continuity is also desirable since the normal direction, hence the boundary stresses, are the same as a node is approached from two different elements.

C^1 continuous curves have long been a subject of interest for many researchers, often in the context of computer-aided design. However, much of the research has gone into developing C^1 approximating curves [1] which produced surfaces for design. Watson [12] developed a hermitian cubic boundary element which requires the specification of positional and derivative data at each node. This requires the recasting of the boundary integrals to include the derivative data. Most conveniently, the C^1 continuous curve should be an interpolating curve that requires only positional data. Overhauser [8] developed such a curve based on a parabolic blending technique which forms an interpolating parametric cubic by blending parametric quadratics. Brewer and Anderson [4] developed a formulation for rapid computation which is described below. The use of the Overhauser element is described more extensively in [9, 10].

The integration procedures for the various two-dimensional elements are too lengthy to be covered in this paper; the reader is referred to [10]. One important point to remember is that the variation of the geometry, displacements, and tractions may be *independently* approximated in the following manner:

$$\begin{aligned} x_j(t) &= N_l^g(t)\, x_{lj} \\ u_j(t) &= N_l^u(t)\, u_{lj} \\ p_j(t) &= N_l^p(t)\, p_{lj} \end{aligned} \tag{1}$$

where N_l are the geometry shape functions corresponding to node l on the element, x_{lj} is the coordinate in direction j of node l of the element, and t is the parameter along the element. The superscripts on the shape functions, g, u, and p, refer to geometry, displacement, and traction, respectively.

OVERHAUSER 2D ELEMENTS

Overhauser (OVER) Element

The Overhauser curve

$$c_j(t) = (1-t)p_j(r) + tq_j(s) \tag{2}$$

shown in Figure 1, is a linear blend of two overlapping parametric parabolas $p_j(r)$ and $q_j(s)$. Reparamatizing $p_j(r)$ and $q_j(s)$ in terms of t and simplifying yields the OVER

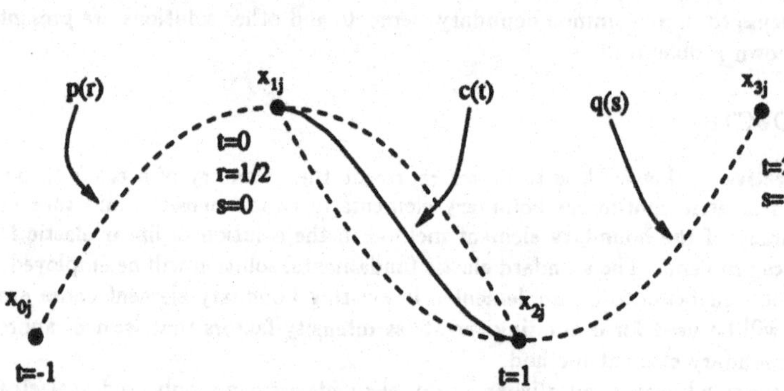

Figure 1. OVER Element Geometry and Parametric Mapping

element written in terms of shape functions

$$x_j(t) = N_0(t)x_{0j} + N_1(t)x_{1j} + N_2(t)x_{2j} + N_3(t)x_{3j} \tag{3}$$

where

$$
\begin{aligned}
N_0(t) &= \frac{-(t-1)^2 t}{2} \\
N_1(t) &= \frac{(t-1)(3t^2-2t-2)}{2} \\
N_2(t) &= \frac{-t(3t^2-4t-1)}{2} \\
N_3(t) &= \frac{(t-1)t^2}{2}
\end{aligned}
\tag{4}
$$

Note that the OVER element is only defined between x_{1j} and x_{2j} as indicated by the solid line in Figure 1. However, x_{0j} and x_{3j} do give contributions to the element assembly matrix when the element is integrated.

Right Corner Overhauser (OVRR) Element

Since the OVER element is designed to have first derivative continuity between elements, problems occur when attempting to model the region near a corner (a geometric singularity). One method of modeling a corner is to make the nodes on one side of the element coincident. Unfortunately, excessive error is introduced into the element integration by this approach, since the element integration is nearly singular for the last node. Another approach taken in References [9, 11] is to use Lagrangian cubic elements in the corners. However, C^1 continuity is lost at the point where the OVER and the Lagrangian cubic elements meet.

The approach implemented here was originated by Hibbs [5]. Hibbs developed a corner element that has C^1 continuity on one side and C^0 continuity on the other. This can be achieved by performing a quadratic extrapolation for the point x_{3j} which is "missing" (see Figure 2) when compared to the OVER element. By constructing a difference table and assuming a zero difference, x_{3j} becomes

$$
x_{3j} = 3x_{2j} - 3x_{1j} + x_{0j} \tag{5}
$$

Substituting into Equations 3 and 4 and simplifying yields the new shape functions

$$
\begin{aligned}
N_0(t) &= \frac{(t-1)t}{2} \\
N_1(t) &= -(t-1)(t+1) \\
N_2(t) &= \frac{t(t+1)}{2}
\end{aligned}
\tag{6}
$$

Left Corner Overhauser (OVRL) Element

The OVRL element is derived in a manner similar to that of the OVRR element. The shape functions become

$$
\begin{aligned}
N_0(t) &= \frac{(t-2)(t-1)}{2} \\
N_1(t) &= -(t-2)t \\
N_2(t) &= \frac{(t-1)t}{2}
\end{aligned}
\tag{7}
$$

Derivative Continuity

Derivative continuity between elements can be shown as follows. Given two overlapping OVER elements, $A_j(t)$ defined by nodes x_{0j}, x_{1j}, x_{2j}, and x_{3j}, and $B_j(t)$ defined by nodes x_{1j}, x_{2j}, x_{3j}, and x_{4j}, we find that

$$
\left.\frac{dA_j(t)}{dt}\right|_{t=1} = \left.\frac{dB_j(t)}{dt}\right|_{t=0} = -\frac{1}{2}x_{1j} + \frac{1}{2}x_{3j} \tag{8}
$$

showing C^1 continuity at node x_{2j} between elements $A_j(t)$ and $B_j(t)$. Derivative continuity between the other combinations of OVER, OVRR, and, OVRL elements may be shown in a similar manner.

OVERHAUSER 2D CRACK TIP ELEMENTS

Overhauser Left Crack Tip (CTOVRL) Element

The idea was to build the proper modeling of the $\sqrt{r}$ singularity into the displacement shape functions and $1/\sqrt{r}$ singularity into the traction shape functions at the crack tip end of the element. At the same time, the CTOVRL element leaves the middle node in its usual position and provides C^1 continuity at the other end with all of the other Overhauser element types. The CTOVRL element also lays the foundation for the three-dimensional elements derived from this type.

The element is defined by three nodes as shown in Figure 2 where the crack tip is at the x_{0j} node. The geometry shape functions are the same as the ones for the OVRL

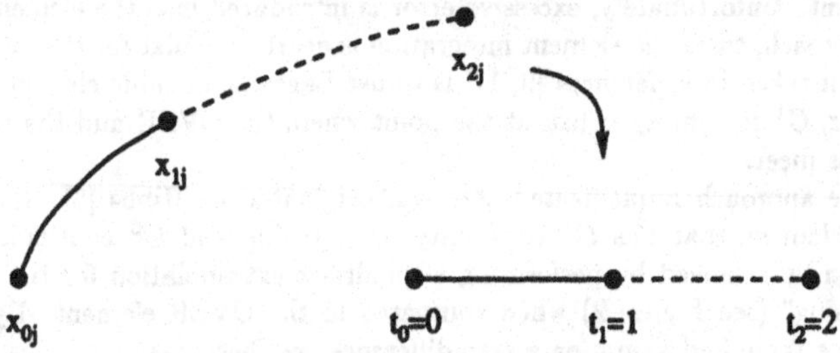

Figure 2. CTOVRL Element Geometry and Parametric Mapping

element which are given in Equation 7 and plotted in Figure 3.

The CTOVRL displacement shape functions can be derived using a simultaneous equation approach by evaluating Equation 9 at the desired values of t.

$$N_i^u(t) = a_3 t^{(\frac{3}{2})} + a_2 t + a_1\sqrt{t} + a_0$$
$$\frac{dN_i^u(t)}{dt} = 3\frac{a_3\sqrt{t}}{2} + \frac{a_1}{(2\sqrt{t})} + a_2 \qquad (9)$$

However, the derivative equation is undefined at $t = 0$; therefore, the coefficients will be solved for in the range of $1 \le t \le 2$ as shown in Equation 10 and then the shape functions will be mapped back into the desired range of $0 \le t \le 1$ by letting $t = t + 1$.

$$\begin{array}{lcc lcc}
t = & 1 & 2 & t = & 1 & 2 \\
N_0^u(t) = & 1 & 0 & dN_0^u(t)/dt = & -3/2 & -1/2 \\
N_1^u(t) = & 0 & 1 & dN_1^u(t)/dt = & 2 & 0 \\
N_2^u(t) = & 0 & 0 & dN_2^u(t)/dt = & -1/2 & 1/2
\end{array} \qquad (10)$$

The resulting CTOVRL displacement shape functions are given in Equation 11 and plotted in Figure 4.

$$N_0^u(t) = \sqrt{t+1}((\sqrt{2}+1)(t+1) + \sqrt{2}+2) - 2(\sqrt{2}+2)(t+1) + 2$$
$$N_1^u(t) = \sqrt{t+1}(-2(\sqrt{2}+1)(t+1) - 2(\sqrt{2}+2)) + (4\sqrt{2}+7)(t+1) - 1 \qquad (11)$$
$$N_2^u(t) = \sqrt{t+1}((\sqrt{2}+1)(t+1) + \sqrt{2}+2) - (2\sqrt{2}+3)(t+1)$$

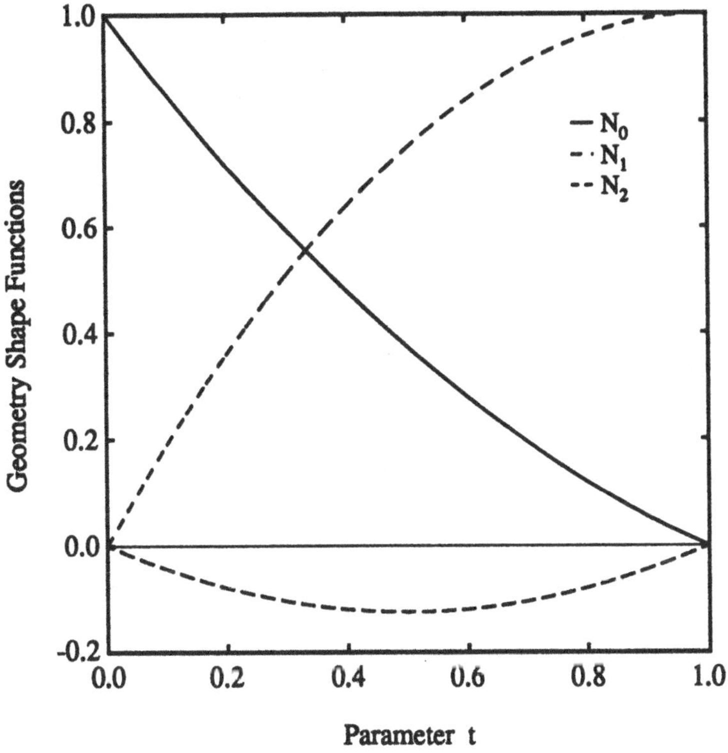

Figure 3. CTOVRL Geometry Shape Functions

The CTOVRL traction shape functions may be derived in a manner similar to that used for derivation of the CTOVRL displacement shape functions as shown in Equation 12.

$$\begin{array}{lcc lcc} t = & 1 & 2 & t = & 1 & 2 \\ N_0^u(t) = & 1 & 0 & dN_0^u(t)/dt = & -3/2 & -1/2 \\ N_1^u(t) = & 0 & 1 & dN_1^u(t)/dt = & 2 & 1/2 \\ N_2^u(t) = & 0 & 0 & dN_2^u(t)/dt = & -1/2 & 1/2 \end{array} \tag{12}$$

After solution and substitution of $t = t + 1$, the displacement shape functions yield the CTOVRL traction shape functions, given in Equation 13 and plotted in Figure 5.

$$\begin{aligned} N_0^p(t) &= \frac{(\sqrt{t+1}((\sqrt{2}+1)(t+1)+\sqrt{2}+2)-2(\sqrt{2}+2)(t+1)+2)}{\sqrt{t}} \\ N_1^p(t) &= \frac{(\sqrt{t+1}((\sqrt{2}+2)(t+1)+3(3\sqrt{2}+4))-(6\sqrt{2}+7)(t+1)-4\sqrt{2}-7)}{\sqrt{t}} \\ N_2^p(t) &= \frac{(\sqrt{t+1}((\sqrt{2}+1)(t+1)+\sqrt{2}+2)-(2\sqrt{2}+3)(t+1))}{\sqrt{t}} \end{aligned} \tag{13}$$

Overhauser Right Crack Tip (CTOVRR) Element

The CTOVRR element may be found by a global mapping of the CTOVRL nodes and appropriate changes in the sign of the element normal.

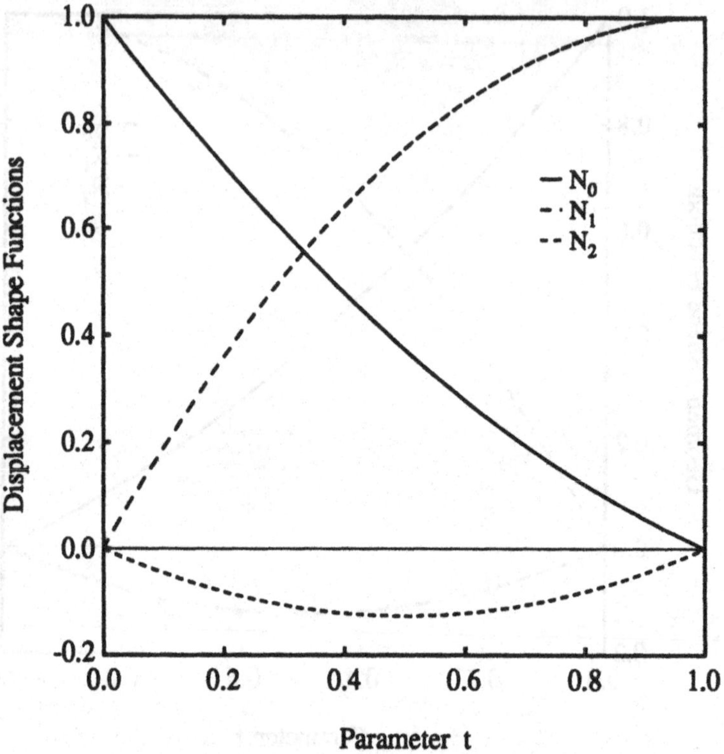

Figure 4. CTOVRL Displacement Shape Functions

Derivative Continuity

Differentiating Equations 11 and 13 and evaluating at several values of t, it can be shown that the CTOVR displacement and traction shape functions have first derivative continuity with the Overhauser family of elements.

STRESS INTENSITY FACTOR CALCULATION

The two-dimensional stress intensity factors (SIF) based on crack tip element values that may be derived for use in boundary elements fall into two categories, those based on displacements and those based on tractions. Since the CTOVR element properly models the $1/\sqrt{r}$ singularity in the tractions and the SIF based on tractions are generally more accurate [6, 10], only those will be considered here.

The traction values can be directly employed to find the SIF as follows and the crack configuration is shown in Figure 6.

Williams' [13] eigenfunction expansion for stresses of a traction-free crack in an infinite domain yields stresses defined in terms of the stress intensity factors. For r much smaller than the size of the crack and neglecting higher order terms, the expansions are

$$\sigma_{00} = \frac{K_I}{\sqrt{2\pi r}} \cos\frac{\theta}{2}\left(1 - \sin\frac{\theta}{2}\sin\frac{3\theta}{2}\right) - \frac{K_{II}}{\sqrt{2\pi r}} \sin\frac{\theta}{2}\left(2 + \cos\frac{\theta}{2}\cos\frac{3\theta}{2}\right) \quad (14)$$

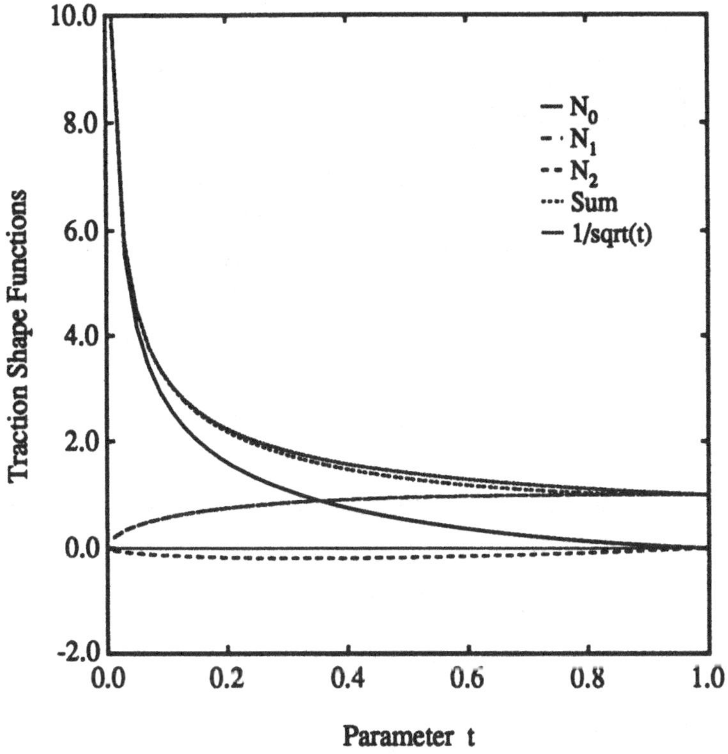

Figure 5. CTOVRL Traction Shape Functions

$$\sigma_{11} = \frac{K_I}{\sqrt{2\pi r}} \cos\frac{\theta}{2}\left(1 + \sin\frac{\theta}{2}\sin\frac{3\theta}{2}\right) + \frac{K_{II}}{\sqrt{2\pi r}}\sin\frac{\theta}{2}\cos\frac{\theta}{2}\cos\frac{3\theta}{2} \tag{15}$$

$$\sigma_{01} = \frac{K_I}{\sqrt{2\pi r}} \cos\frac{\theta}{2}\sin\frac{\theta}{2}\cos\frac{3\theta}{2} + \frac{K_{II}}{\sqrt{2\pi r}}\cos\frac{\theta}{2}\left(1 - \sin\frac{\theta}{2}\sin\frac{3\theta}{2}\right) \tag{16}$$

Boundary discretization can always be done such that $\theta = \pi$. This yields, from Equations 14, 15 and 16

$$\begin{aligned} \sigma_{00}(t) = \sigma_{11}(t) &= \frac{K_I}{\sqrt{2\pi r(t)}} \\ \sigma_{01}(t) &= \frac{K_{II}}{\sqrt{2\pi r(t)}} \end{aligned} \tag{17}$$

where the radius vector for two dimensions is

$$\begin{aligned} r(t) &= \left[(x_0(t) - x_{00})^2 + (x_1(t) - x_{01})^2\right]^{\frac{1}{2}} \\ &= \left[(N_l^g(t)x_{l0} - x_{00})^2 + (N_l^g(t)x_{l1} - x_{01})^2\right]^{\frac{1}{2}} \end{aligned} \tag{18}$$

Let $\bar{p}_j$ be defined as follows

$$\bar{p}_j = \lim_{t\to 0} p_j(t) = \lim_{t\to 0} N_l^p(t)\, p_{lj} \tag{19}$$

where

$$p_j(t) = \sigma_{ji}(t)\, n_i \tag{20}$$

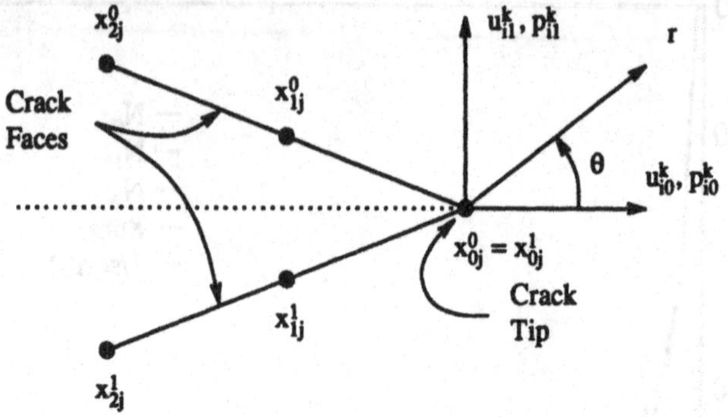

Figure 6. 2D SIF Configuration

and n_j are the outward unit normals. Combining these equations with $n_0 = 0$ and $n_1 = 1$ for a symmetric crack yields

$$\begin{aligned} \bar{p}_1 &= \lim_{t\to 0} N_l^p(t)\, p_{l1} = \lim_{t\to 0} \frac{K_I}{\sqrt{2\pi r(t)}} \\ \bar{p}_0 &= \lim_{t\to 0} N_l^p(t)\, p_{l0} = \lim_{t\to 0} \frac{K_{II}}{\sqrt{2\pi r(t)}} \end{aligned} \tag{21}$$

Rearranging

$$\begin{aligned} K_I &= \sqrt{2\pi} \lim_{t\to 0} \sqrt{r(t)} N_l^p(t)\, p_{l1} \\ K_{II} &= \sqrt{2\pi} \lim_{t\to 0} \sqrt{r(t)} N_l^p(t)\, p_{l0} \end{aligned} \tag{22}$$

Evaluating the limit portion of Equation 22 for the CTOVR element type, where $N_l^g(t)$ is given by Equation 7 and $N_l^p(t)$ is given by Equation 13 yields

$$\begin{aligned} &\lim_{t\to 0} \sqrt{r(t)} N_0^p(t) = \frac{M}{\sqrt{2}} \\ &\lim_{t\to 0} \sqrt{r(t)} N_1^p(t) = 0 \\ &\lim_{t\to 0} \sqrt{r(t)} N_2^p(t) = 0 \end{aligned} \tag{23}$$

where

$$M = \left[(-3x_{00} + 4x_{10} - x_{20})^2 + (-3x_{01} + 4x_{11} - x_{21})^2\right]^{\frac{1}{4}} \tag{24}$$

Note that for evenly spaced nodes M reduces to $\sqrt{L}$ where L is the length of the element. The traction SIF for the CTOVRL element type is therefore

$$\begin{aligned} K_I &= \sqrt{\pi} M\, p_{01} \\ K_{II} &= \sqrt{\pi} M\, p_{00} \end{aligned} \tag{25}$$

DOUBLE EDGE CRACKED PLATE

This example, the finite width plate with double edge cracks, has been used by a number of boundary element researchers [2, 7, 6]. The plate is shown in Figure 7. The model

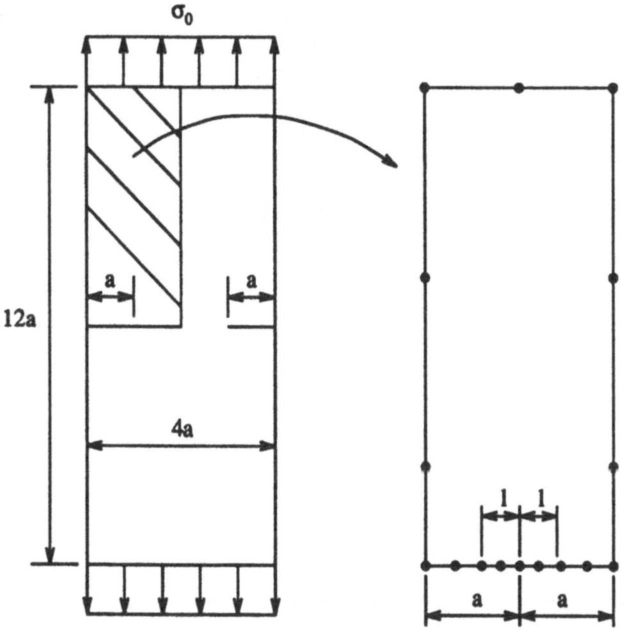

Figure 7. Double Edge Cracked Plate and BEM Mesh

dimensions and material properties are crack length $a = 1.8$ in, elastic modulus $E = 5250.0$ ksi, Poisson's ratio $\nu = 0.20$, and load stress $\sigma_0 = 1$ ksi. An approximate analytical solution accurate to 1% has been reported by Bowie [3] as $K_I = 2.737 \text{ ksi} \cdot \text{in}^{1/2}$.

Taking advantage of the symmetry of the problem, only one-quarter of the plate needs to be discretized as shown in Figure 7. The boundary element mesh consists of a combination of Lagrangian quadratic elements along the top and sides and Overhauser-type elements along the crack. The arrangement of Overhauser-type elements from left to right is OVRL, OVER, OVER, CTOVRR, CTOVRL, OVER, OVER, and OVRR.

The ratio of the length of the crack tip boundary element to the length of the crack l/a is an important parameter in analyzing the performance of crack tip boundary elements. It is desirable that the SIF values be relatively insensitive to changes in l/a. It should be noted that, as the form of Equation 24 indicates, the effective length for the CTOVR boundary elements is the distance from node x_{0j} to x_{2j}. With this in mind, values for K_I may be obtained for ratios of l/a in the range of 0.05 to 0.9.

A comparison of the SIF calculated by the traction method for the traction singularity corrected quarter point quadratic element (CTQT) [7], traction singular mid point quadratic element (CTQUA) [6], and CTOVR elements is given in Figure 8. For $l/a < 0.2$ the stress intensity factors for the CTOVR element have large percentage errors. This is due to the parameter in the Overhauser-type element becoming nonmonotonic when two elements that are very different in length are placed next to each other. As shown, for $l/a \geq 0.2$, the percent errors are similar; however, the CTOVR element is clearly less sensitive to the l/a ratio.

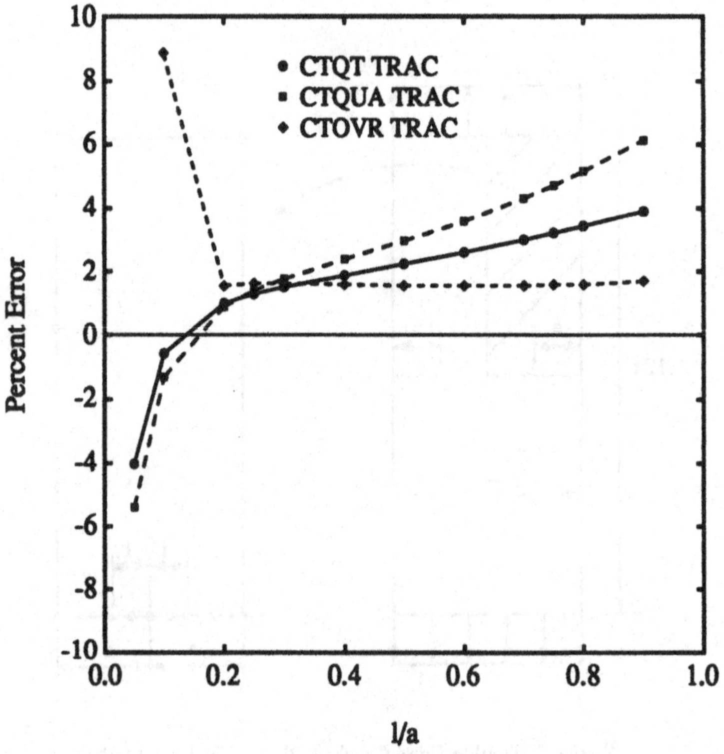

Figure 8. Comparison of Percent Errors in K_I Calculated by the Traction Method for Double Edge Cracked Plate

CONCLUSIONS

The objective of this work was to better represent the geometry of a crack through the use of C^1 continuous elements and thus enhance the accuracy of the boundary element method in the solution of linear elastic fracture mechanics problems. The Overhauser-type crack tip elements were more accurate than the other element types for the problem examined. The two-dimensional Overhauser crack tip element showed a nearly complete insensitivity to the ratio of the length of the crack tip element to the length of the crack except for extremely small values where the parameter became nonmonotonic. It is believed the more accurate results of the Overhauser crack tip elements are a result of both the C^1 continuity and a better modeling of the near field crack tip stresses. However, percentage errors for all boundary elements that properly model the tractions are fairly close, indicating that the quadratic behavior of these shape functions may be the dominant factor. While all of the Overhauser elements give excellent results, they have a disadvantage in that the meshes are slightly more difficult to assemble by hand.

ACKNOWLEDGMENTS

The author would like thank the School of Civil Engineering at Oklahoma State University for support while a student there and Halliburton Services for permission to prepare and present this paper.

REFERENCES

[1] P. Bezier. Mathematical and Practical Possibilities of UNISURF. In R. E. Barnhill and R. F. Riesenfeld, editors, *Computer Aided Geometric Design*, pages 127–152. Academic Press, 1974.

[2] G. E. Blandford, A. R. Ingraffea, and J. A. Liggett. Two-Dimensional Stress Intensity Factor Computations Using the Boundary Element Method. *International Journal for Numerical Methods in Engineering*, 17:387–404, 1981.

[3] O. L. Bowie. Rectangular Tensile Sheet with Symmetric Edge Cracks. *ASME Journal of Applied Mechanics*, 31:208–212, 1964.

[4] J. A. Brewer and D. C. Anderson. Visual Interaction with Overhauser Curves and Surfaces. *Computer Graphics*, 11(2):132–137, 1977.

[5] Terence T. Hibbs. *C^1 Continuous Representations and Advanced Singular Kernal Integrations in the Three Dimensional Boundary Element Method.* PhD thesis, Teesside Polytechnic, Middlesbrough, Cleveland UK, 1988.

[6] Z. H. Jia, D. J. Shippy, and F. J. Rizzo. On the Computation of Two-Dimensional Stress Intensity Factors Using the Boundary Element Method. *International Journal for Numerical Methods in Engineering*, 26:2739–2753, 1988.

[7] José Martínez and José Domínguez. Short Communication on the Use of Quarter-Point Boundary Elements for Stress Intensity Factor Computations. *International Journal for Numerical Methods in Engineering*, 20:1941–1950, 1984.

[8] A. W. Overhauser. Analytic Definition of Curves and Surfaces by Parabolic Blending. Technical Report SL68-40, Scientific Research Staff Publication, Ford Motor Company, May 1968.

[9] H. G. Walters. Techniques for Boundary Element Analysis in Elastostatic Influenced by Geometric Modeling. Master's thesis, Louisiana State University, Baton Rouge, Louisiana, 1986.

[10] H. G. Walters. *Parametric Continuous Crack Tip Boundary Elements in Linear Elastic Fracture Mechanics.* PhD thesis, Oklahoma State University, Stillwater, Oklahoma, 1990.

[11] H. G. Walters, J. C. Ortiz, G. S. Gipson, and J. A. Brewer III. Overhauser Boundary Elements in Potential Theory and Linear Elastostatics. In T. A. Cruse, editor, *Advanced Boundary Element Methods*, pages 459–464, New York, 1987. Springer-Verlag.

[12] J. O. Watson. Hermitian Cubic and Singular Elements for Plane Strain. In P. K. Banerjee and J. O. Watson, editors, *Developments in Boundary Element Methods 4*, pages 1–28. Elsevier Applied Science Publishers, London, 1984.

[13] M. L. Williams. On the Stress Distribution at the Base of a Stationary Crack. *ASME Journal of Applied Mechanics*, 24:109–114, 1957.

First and Second Order Weight Functions for Mixed-Mode Crack Problems

M.H. Aliabadi (*), D.P. Rooke (**)
(*) Wessex Institute of Technology, Computational Mechanics Institute, Ashurst, Hants, SO4 2AA, U.K.
(**) Materials and Structures, Royal Aerospace Establishment, Farnborough, Hants, GU14 6TD, U.K.

ABSTRACT

A new procedure, based on the decomposition of crack-tip fields, is developed to evaluate first and second order fracture mechanics weight functions. The procedure which employs the standard boundary element formulation is applied to the problems of slant edge and central cracks in rectangular sheets. The resulting stress intensity factors are found to be within 1% of alternative accurate solutions.

INTRODUCTION

In many practical situations engineers have to predict the safe life times of cracked components under different loading conditions. This often requires repeated evaluation of stress intensity factors for the given crack configuration. The use of weight functions [1] can avoid this repeated numerical analysis for different loading conditions [2]. Because of the load-independent characteristic of weight functions, they can serve as fundamental or universal functions for determining the stress intensity factors.

It can be shown [2] that the stress intensity factor for mode I and mode II deformation may be obtained from the application of Betti's reciprocal theorem. If t is the applied traction vector acting on the boundary Γ, then

$$K_N = \frac{E'}{4\sqrt{(2\pi)}B^N} \int_\Gamma t.u^N d\Gamma \quad N=I,II \quad , \tag{1}$$

where $E'=E$ for plane stress and $E'=E/(1-\nu^2)$ for plane strain, E and ν are

Young's modulus and Possion's ratio respectively. The vector u^N is the displacement field on the boundary Γ, which results from the two equal and opposite forces acting on the crack at the tip (see figure 1); B_N is the strength of the so-called Bueckner singular field [2] due to the localized forces approaching the crack tip (i.e. $c \to 0$). From equation (1) it can be seen that u^N acts as a weight function for the stress intensity factor.

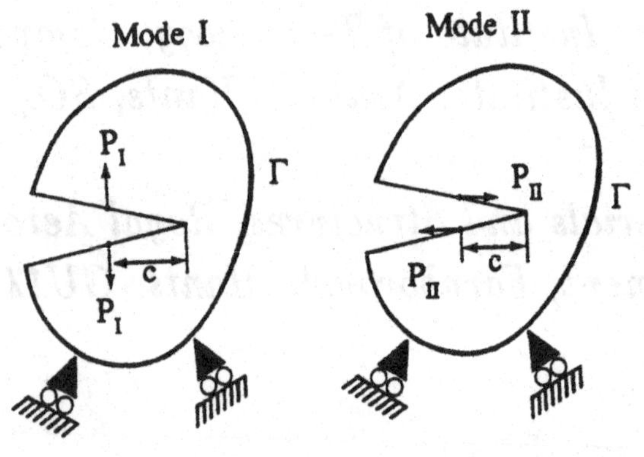

Figure 1 *Arbitrary two-dimensional body containing a crack*

In this paper a simple yet accurate procedure for the evaluation of the mixed-mode weight functions using standard boundary element method is presented. The accuracy of the technique is assessed by analysing a slant edge crack and a central crack in a rectangular sheet, for which alternative accurate solutions are available.

SUBTRACTION OF FUNDAMENTAL FIELD TECHNIQUE

Here, a procedure originally developed by Aliabadi, Cartwright and Rooke[3] is further extended to deal with mixed-mode weight function problems. This new procedure is based on the decomposition of displacement and stress fields in the finite body into singular components (u^S_j, t^S_j) due to a pair of point forces and the remaining fields (u^R_j, t^R_j) which are geometry dependent. Thus the required field (u_j, t_j) is written as:

$$u_j = (u_j - u^S_j) + u^S_j = u^R_j + u^S_j ,$$

and
$$t_j = (t_j - t^S_j) + t^S_j = t^R_j + t^S_j . \tag{2}$$

a) First-Order Weight Functions

The singular displacement and stress fields which occur in the vicinity of the crack tip as a result of the point forces, can be derived from the Westergaard complex stress function $B_N/z^{3/2}$, and are given [2] by

$$u_j^N = u_j^{N0} + \text{higher order terms}$$

and

$$\sigma_{ij}^N = \sigma_{ij}^{N0} + \text{higher order terms} \tag{3}$$

where

$$u_1^{I0} = B_0^I \mu r^{-1/2} \cos\frac{\theta}{2}\left[\frac{1}{2}(1-\kappa) + \sin\frac{\theta}{2}\sin 3\frac{\theta}{2}\right]$$

$$u_2^{I0} = B_0^I \mu r^{-1/2} \sin\frac{\theta}{2}\left[\frac{1}{2}(1+\kappa) - \cos\frac{\theta}{2}\cos 3\frac{\theta}{2}\right]$$

$$\sigma_{11}^{I0} = B_0^I r^{-3/2}\left[\cos 3\frac{\theta}{2} - \frac{3}{2}\sin\theta \sin 5\frac{\theta}{2}\right]$$

$$\sigma_{22}^{I0} = B_0^I r^{-3/2}\left[\cos 3\frac{\theta}{2} + \frac{3}{2}\sin\theta \sin 5\frac{\theta}{2}\right]$$

$$\sigma_{12}^{I0} = B_0^I r^{-3/2} \sin\theta \cos 5\frac{\theta}{2}$$

for mode I, and

$$u_1^{II0} = B_0^{II} \mu r^{-1/2} \sin\frac{\theta}{2}\left[\frac{1}{2}(1+\kappa) + \cos\frac{\theta}{2}\cos 3\frac{\theta}{2}\right]$$

$$u_2^{II0} = B_0^{II} \mu r^{-1/2} \cos\frac{\theta}{2}\left[\frac{1}{2}(\kappa-1) + \sin\frac{\theta}{2}\sin 3\frac{\theta}{2}\right]$$

$$\sigma_{11}^{II0} = B_0^N r^{-3/2}\left[-2\sin 3\frac{\theta}{2} - \frac{3}{2}\sin\theta \sin 5\frac{\theta}{2}\right]$$

$$\sigma_{22}^{II0} = B_0^I r^{-3/2} \frac{3}{2}\sin\theta \sin 5\frac{\theta}{2}$$

$$\sigma_{12}^{II0} = B_0^I r^{-3/2}\left[\cos 3\frac{\theta}{2} - \frac{3}{2}\sin\theta \cos 5\frac{\theta}{2}\right]$$

for mode II. It can be seen from (3) that the displacement fields u_j^{N0} become singular $O(r^{-1/2})$ at the crack tip $(r \to 0)$, and from σ_{ij}^{N0} that the stress fields become singular at the tip $O(r^{-3/2})$. It is these fields that are referred to as u_j^S and t_j^S in equation (2).

The weight functions to first-order accuracy can be obtained by solving the standard boundary element equations for u_j^R and t_j^R, subjected to the following boundary conditions

$$\bar{t}_j^{RN} = -\bar{t}_j^{N0} \; (= -\bar{\sigma}_{ij}^{N0} n_i) \qquad \text{since } \bar{t}_j^N = 0 \text{ everywhere}$$

and

$$\bar{u}_j^{RN} = \bar{u}_j^N - \bar{u}_j^{N0} \tag{4}$$

where n is the outward normal to the boundary. Since K_N is independent of the term B_0^N it can be set to unity without loss of generality [2].

The weight functions for either the opening mode or the sliding mode are subsequently obtained by adding the displacement field, resulting from the boundary element solution subject to the boundary conditions in (4), to the singular fields u_j^{N0}. Once the weight functions u_j^N are obtained, the stress intensity factor for any equilibrated loading is obtained from equation (1) using a simple integration. A schematic representation is shown in figure 2.

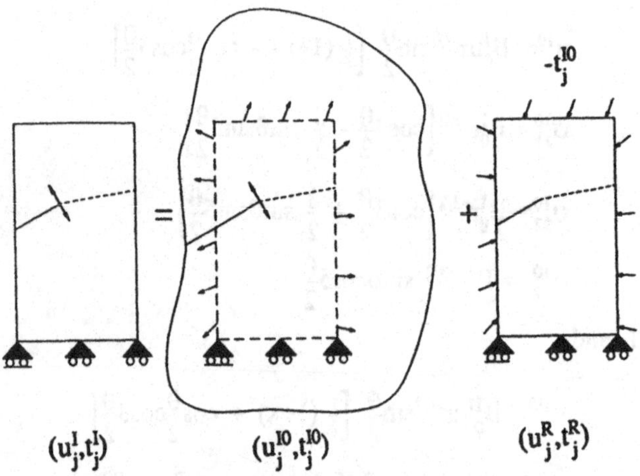

Figure 2. *A schematic representation for first order weight functions (mode I shown, mode II similar).*

b) Second-Order Weight Function

By imposing the boundary conditions due to the point forces as given in equation (4), the weight function problem is reduced to that of a traction-free crack. As such the resulting problem has the standard singularity of $O(r^{-1/2})$ in the stress field.

The singular displacement and stress fields which occur in the vicinity of the crack tip as a result of these forces, are given [2] as

$$u_j^N = u_j^{N0} + (u_j^{I1} + u_j^{II1}) + \ldots .$$

and
$$\sigma_{ij}^N = \sigma_{ij}^{N0} + (\sigma_{ij}^{I1} + \sigma_{ij}^{II1}) + \ldots . \quad N=I,II \qquad (5)$$

where

$$u_1^{I1}+u_1^{II1} = B_0^N \mu r^{1/2}\left\{b_1^I \cos\frac{\theta}{2}\left[\frac{1}{2}(\kappa-1) + \sin^2\frac{\theta}{2}\right] + b_1^{II}\sin\frac{\theta}{2}\left[\frac{1}{2}(1+\kappa) + \cos^2\frac{\theta}{2}\right]\right\}$$

$$u_2^{I1}+u_2^{II1} = B_0^N \mu r^{1/2}\left\{b_1^I \sin\frac{\theta}{2}\left[\frac{1}{2}(1+\kappa) - \cos^2\frac{\theta}{2}\right] + b_1^{II}\cos\frac{\theta}{2}\left[\frac{1}{2}(1-\kappa) + \sin^2\frac{\theta}{2}\right]\right\}$$

$$\sigma^{I1}_{11}+\sigma^{III1}_{11}= B^{N}_{0}r^{-1/2}\left\{b^{I}_{1}\cos\frac{\theta}{2}\left[1-\sin\frac{\theta}{2}\sin3\frac{\theta}{2}\right]-b^{II}_{1}\sin\frac{\theta}{2}\left[2+\cos\frac{\theta}{2}\cos3\frac{\theta}{2}\right]\right\}$$

$$\sigma^{I1}_{22}+\sigma^{III1}_{22}= B^{N}_{0}r^{-1/2}\left\{b^{I}_{1}\cos\frac{\theta}{2}\left[1+\sin\frac{\theta}{2}\sin3\frac{\theta}{2}\right]+\frac{1}{2}b^{II}_{1}\sin\theta\cos3\frac{\theta}{2}\right\}$$

$$\sigma^{I1}_{12}+\sigma^{III1}_{12}= B^{N}_{0}r^{-1/2}\left\{\frac{1}{2}b^{I}_{1}\sin\theta\cos3\frac{\theta}{2}-b^{II}_{1}\cos\frac{\theta}{2}\left[1-\sin\frac{\theta}{2}\sin3\frac{\theta}{2}\right]\right\}.$$

It can be seen from equation (5) that the stress fields $(\sigma^{I1}_{jj}+\sigma^{II1}_{jj})$ become singular at the tip $O(r^{-1/2})$. In order to deal with this singularity and hence enhance the accuracy of the solution using the standard boundary element method, these lower order singularities may also be removed by imposing the following boundary conditions:

$$\bar{t}^{RN}_{j} = -\bar{t}^{N0}_{j} - (\bar{t}^{I1}_{j} + \bar{t}^{II1}_{j})$$

and
$$\bar{u}^{RN}_{j} = \bar{u}^{N}_{j} - \bar{u}^{N0}_{j} - (\bar{u}^{I1}_{j} + \bar{u}^{II1}_{j}) \tag{6}$$

A schematic representation of the above procedure is shown in figure 3. In the above expressions the approximate values of b^{I}_{1} and b^{II}_{1} (say $b^{I(0)}_{1}$ and $b^{II(0)}_{1}$) are obtained from the solution of first-order weight functions using for example the extrapolation of near-tip fields. Using these values in the boundary conditions, a second approximation ($b^{I(1)}_{1}$ and $b^{II(1)}_{1}$ say) may be obtained by the same extrapolation procedure. This iterative process may be continued until the values of $b^{I(m)}_{1}$ and $b^{II(m)}_{1}$ are considered insignificant.

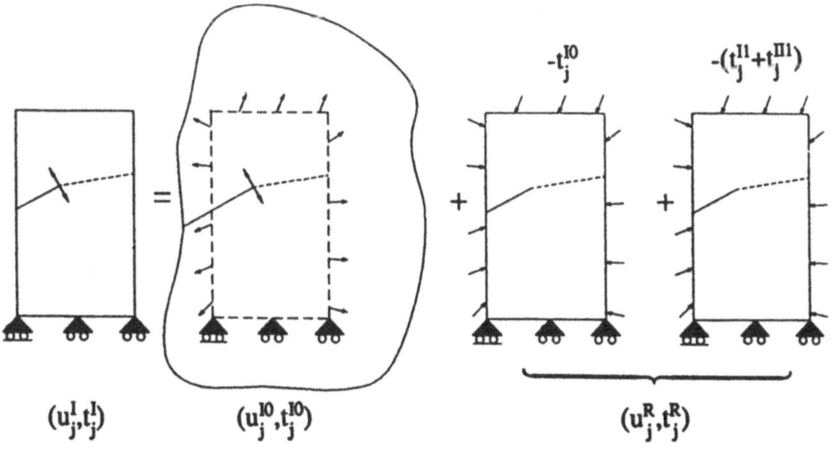

Figure 3. *A schematic representation for second order weight functions (mode I shown, mode II similar).*

TEST PROBLEMS

The accuracy of the procedures, outlined in the previous section, was checked by studying configurations for which alternative accurate results are available. The first problem studied is a slant edge crack of length a in a rectangular sheet of width w and height h, with the angle of inclination ϕ=45° (see figure 4). The ratio of crack length to width and height to width are taken as a/w=0.1 and h/w=3 respectively to simulate a semi-infinite sheet.

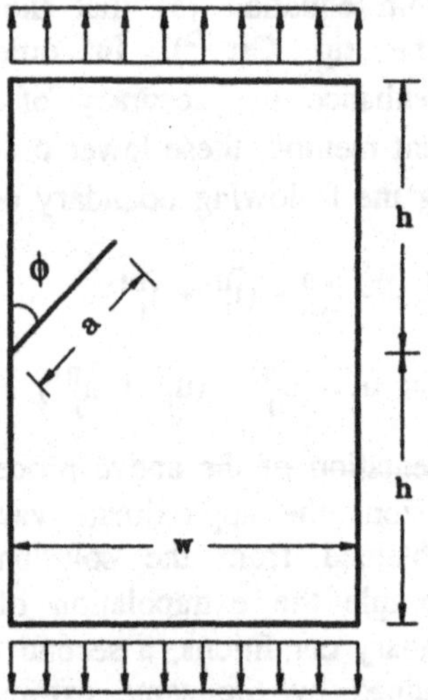

Figure 4. *Slant edge crack in a rectangular sheet.*

Table 1 shows the values of the stress intensity factors resulting from remote uniaxial tensile stress σ at the ends of the sheet, obtained by using the first and second order weight functions.

	1st order	2nd order $b_1^{I(0)}, b_1^{II(0)}$	2nd order $b_1^{I(1)}, b_1^{II(1)}$	ref[4]
$\frac{K_I}{K_0}$	0.726	0.716	0.712	0.705
$\frac{K_{II}}{K_0}$	0.375	0.369	0.368	0.364

Table 1 *Stress intensity factors for the slant edge crack configuration using 52 quadratic boundary elements* ($K_0=\sigma\sqrt{(\pi a)}$).

As can be seen from Table 1, the stress intensity factors, obtained using the first-order weight functions differ by 2-3% from the accurate values reported in ref[4]. The difference is reduced to <1% using the second-order weight functions.

The second configuration studied is that of a central slant crack of length 2a in a rectangular sheet of height to width ratio h/w=2 and the angle of inclination $\phi=30^0$ (as shown in figure 5). The calculated values of stress intensity factors due to a uniform tensile stress σ at the ends of the sheet for different ratios of a/W are shown in Table 2 and 3 for mode I and mode II respectively.

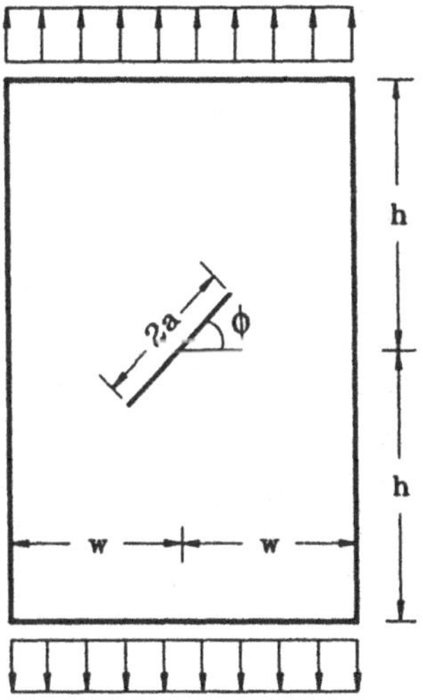

Figure 5. *Slant central crack in rectangular sheet.*

a/W	K_I/K_0 1st Order	K_I/K_0 2nd Order	K_I/K_0 Ref[4]
0.1	0.761	0.757	0.756
0.2	0.775	0.774	0.773
0.3	0.804	0.803	0.802
0.4	0.846	0.846	0.846
0.5	0.905	0.905	0.905
0.6	0.985	0.984	0.984
0.7	1.093	1.091	1.091
0.8	1.249	1.248	1.245

Table 2. *Stress intensity factor for mode I deformation. ($K_0=\sigma\sqrt{(\pi a)}$)*

a/W	K_{II}/K_0 1st order	K_{II}/K_0 2nd order	K_{II}/K_0 Ref[4]
0.1	0.436	0.434	0.434
0.2	0.438	0.437	0.437
0.3	0.443	0.442	0.442
0.4	0.451	0.450	0.450
0.5	0.463	0.462	0.462
0.6	0.480	0.480	0.480
0.7	0.510	0.508	0.508
0.8	0.551	0.550	0.550

Table 3. *Stress intensity factor for mode II deformation.* $(K_0=\sigma\sqrt{(\pi a)})$

As can be seen from Tables 2 and 3, the stress intensity factors obtained using the first- and second-order weight function are both within 1% of the accurate values reported in [4].

CONCLUSIONS

A new procedure for evaluation of the first- and second-order weight functions has been presented based on the decomposition of elastic fields. The new procedure which employs a standard boundary element formulation has been shown to provide accurate values of stress intensity factors for both mode I and mode II deformations.

ACKNOWLEDGEMENT

This work has been carried out with the support of Procurement Executive, Ministry of Defence, UK.

REFERENCES

[1] Beuckner,H.F. A novel principle for the computation of stress intensity factors. Z.Angew Meth.Mech.,**50**, 529-546 (1970).

[2] Aliabadi,M.H. and Rooke,D.P. *Numerical Fracture Mechanics.* Computational Mechanics Publication, Southampton and Kluwer Academic Publishers, The Netherlands (1991).

[3] Aliabadi,M.H. and Cartwright,D.J. and Rooke,D.P. Fracture-Mechanics Weight-Functions by the removal of singular fields using boundary element analysis. Int.J.Fracture, **40**, 271-284 (1989).

[4] Murakami,Y. Stress intensity factors handbook. Pergamon, Oxford (1987).

The Calculation of Energy Release Rates in Composite Laminates using the Boundary Element Method

D. Klingbeil, V. Karakaya

Federal Material Research and Testing Institute, D-1000 Berlin 45, Germany

Abstract

A particular failure mode of laminated composites is the delamination of layers from the free boundaries. A significant quantity to describe this failure is the energy release of the appearing cracks. The energy release rate is calculated from the displacements and stresses of both, the non-delaminated and delaminated case. All calculations are done by the BEM which turned out to be an excellent instrument for the solution of the non-delaminated as well as for the delaminated case.

1. Introduction

Usually laminated constructions are thin in one direction and they can be treated by the theory of plates and shells. This treatment does not take into consideration that interlaminar stresses occur because the direction of the unidirectional (UD) fibres and therefore the stiffness varies from layer to layer. Because the traction free boundary condition has to be satisfied the interlaminar stresses can be very large near the free boundaries or become infinite within a linear theory and may cause a delamination of layers before the limit load of the structure is reached. The is the aim to calculate the energy release rates which are significant quantities for the possibility of a delamination. Therefore it is necessary to know results for the displacements and tractions in non-delaminated and delaminated cases. Such calculations have been done by several authors with different methods, e. g. PIPES and PAGANO [7], WANG and CROSSMAN [9], DELALE [3] and with the BEM by KLINGBEIL [5, 6].

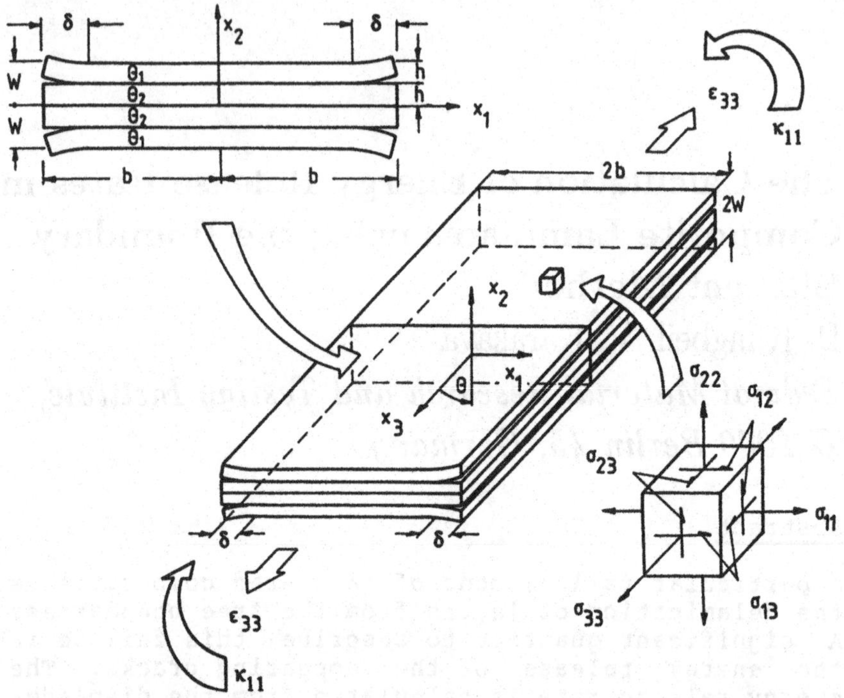

Fig. 1: Geometry and loading of a symmetrical unidirectional four layer laminate $[\Theta_1/\Theta_2/\Theta_2/\Theta_1]$ under uniform axial strain ε_{33} and curvature κ_{11}

2. Stress State in Laminated Composites

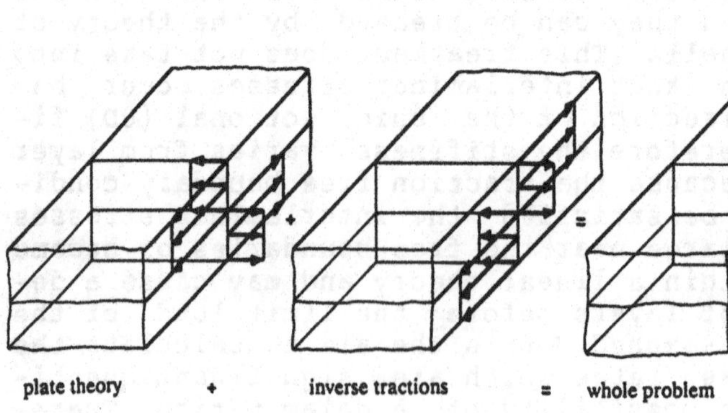

Fig.2: Decomposition of the solution

The stress state in laminates under uniaxial strain and curvature (fig. 1) consists of two parts. The first part is due to the plate theory which does not satisfy the boundary free condition, because the cross-section has to keep its form and remains perpendicular to the neutral plane. The tractions resulting from such a calculation are applied to the free boundary with the opposite sign which leads to a problem, where the stresses decrease rapidly inside the laminate from the free boundary. This is the second part of the solution. Here the second problem is solved by the BEM and the energy release rates are calculated.

3. Governing equations

The following assumptions are introduced: small displacements, linear constitutive law, homogeneous layers, and every layer is in holohedral contact with its neighbour layers. The equations describing the problem for one layer are those of equilibrium, geometrical compatibility and the constitutive law for the three dimensional stress state:

$$\sigma_{ji,j} + p_i = 0 \qquad (1)$$

$$\varepsilon_{ij} = \tfrac{1}{2}\,(u_{i,j} + u_{j,i}) \qquad (2)$$

$$\sigma_{ij} = E_{ijkl}\,\varepsilon_{kl} = E_{ijkl}\,u_{k,l} \qquad (3).$$

Latin subscripts can take the values 1 to 3. The assumption is introduced that the structure is long in x_3-direction so that any variation of all mechanical quantities are negligible in this direction. Therefore all derivatives of u_i and σ_{ij} with respect to x_3 vanish and the equations (1) to (3) are rewritten:

$$\sigma_{\alpha i,\alpha} + p_i = 0 \qquad (4)$$

$$\varepsilon_{ij} = \tfrac{1}{2}\,(u_{i,j} + u_{j,i}) \qquad (5)$$

$$\sigma_{i\beta} = E_{ijk\beta}\,u_{k,\beta} \qquad (6),$$

where the greek indices can take the values 1 and 2. It follows that only the (x_1, x_2)-plane of the laminate has to be taken into consideration. In general the constitutive law is anisotropic.

In the next step the geometrical and statical compatibility has to be postulated on those parts of the boundaries of each layer, where layers or subdomains contact each other

$$u_i(x_\alpha)_{boundary\ I} = u_i(x_\alpha)_{boundary\ I+1} \quad (7)$$

$$t_i(x_\alpha)_{boundary\ I} = -t_i(x_\alpha)_{boundary\ I+1} \quad (8),$$

for $I = 1, \ldots, N-1$, where N is the total number of layers or subdomains, $t_i = \sigma_{i\alpha} n_\alpha$ is the stress vector and n_α the outward normal vector of the boundary. The problem is formulated completely, when the geometrical and/or statical conditions on the remaining parts of the boundaries are prescribed.

4. Boundary Integral Equation

SOMIGLIANA's well known identity [1]

$$u_i(x_\alpha) = -\oint_\Gamma T_{ij} u_j d\Gamma + \oint_\Gamma U_{ij} t_j d\Gamma \quad (9)$$

gives the relation between the displacements u_i of an interior point x_α of the domain Ω and the displacement u_i and tractions t_i on its boundary Γ. U_{ij} is the fundamental solution and T_{ij} is the stress vector related to a normal vector n_α at any field point x_α. The analytical calculation of U_{ij} is given in [3] and those of T_{ij} in [5].
The boundary integral equation follows when the load point x_α lies on the boundary. This procedure causes singular integrands in both integrals of equation (9) so that CAUCHY's principle values have to be taken:

$$C_{ij} u_j(x_\alpha) + \oint_\Gamma T_{ij} u_j d\Gamma = \oint_\Gamma U_{ij} t_j d\Gamma \quad (10).$$

The values C_{ij} are the integral free terms. This boundary integral equation describes the behaviour of one subdomain or layer. The boundary integral equation is discretized by an elementation of the boundary and by introducing polynomials for u_j and t_j in each element. Crack tip elements can be used in the case of a singular behaviour of the traction t_j [4, 5, 6]. The result is a linear system of equations for each subdomain. The linear system of equations for the whole problem is formulated by using equations (7) and (8) to couple the subdomains. The right hand side results from the loads of the laminate. The coefficient matrix has a band scheme, because one layer is only in contact with its two neighbouring layers. So all advantages of this band scheme can be utilized during the solution [1].

5. Energy Release Rate

The total energy release rate G_{tot} of a crack consists of three parts

$$G_{tot} = G_{I} + G_{II} + G_{III} \qquad (11),$$

each related to the fracture modii I, II and III, where modus I represents in-plane tension, modus II in-plane shear and modus III out-of-plane shear.
E.g. the energy release rate or crack extension force G_I in linear elasticity is defined as the work which has to be done when a crack is closed by a virtual crack closure length δ and δ takes the limiting value of zero:

$$G_I = \lim_{\delta \to 0} \frac{1}{\delta} \int_0^{\delta} \sigma_{yy}(x)\ u_y(x-\delta)dx \qquad (12)$$

where $\sigma_{yy}(x)$ is the stress for the closed crack and $u_y(x-\delta)$ is the displacement for the opened crack, where all notations refer to fig. 3.

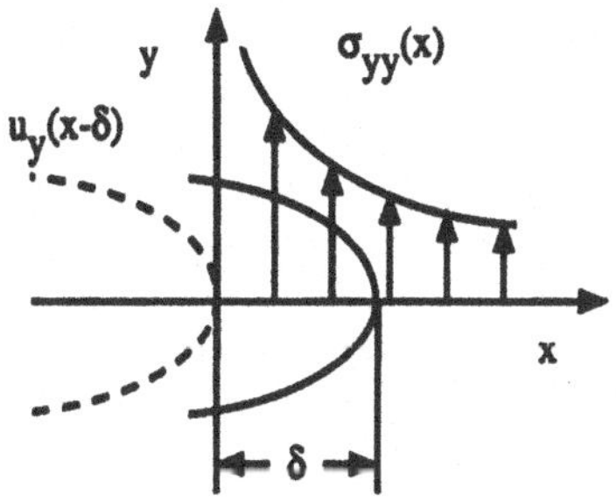

Fig.3: Basic principle of the energy release rate

It is significant that in linear isotropic elasticity the energy release rate has always a fixed, but nonzero value, because the stresses behave with $1/\sqrt{x}$ and displacements with $\sqrt{x}$ so that they cancel each other in the integrand of (12).
This conventional concept of the definition of the energy release rate has to be given up in inhomogeneous, anisotropic elasticity, because:

- in inhomogeneous, anisotropic elasticity the tractions and displacements from the two crack surfaces are different, so that each of them has to be taken into consideration with different terms.

- the stress states in the non-delaminated case and delaminated case are totally different [2]. Therefore not only the strength but also its exponent or even the function of the singularity of the tractions and the function of the displacements change, when a virtual crack closure is applied. Because the singularity of the delaminated case is always stronger than that of the non-delaminated case, it can be expected that the energy release rate becomes zero.

Therefore the following modified definitions for the energy release rates are introduced between the layers I and I+1 with a fixed, but small virtual crack closure length δ:

$$G_{I,II,III} = \frac{1}{\delta}\left(\frac{1}{2}\int_{x=-b}^{-b+\delta} t_{(2,1,3)}(x_1)\,u_{(2,1,3)}(x_1-\delta)\,dx_1 \Big|_{\text{layer I}} \right.$$

$$\left. + \frac{1}{2}\int_{x=-b}^{-b+\delta} -t_{(2,1,3)}(x_1)\,u_{(2,1,3)}(x_1-\delta)\,dx_1 \Big|_{\text{layer I+1}}\right), \qquad (13)$$

where the notation refers to figure 1 and the indices I, II, III refer to the directions 2, 1, 3.

For the calculation of energy release ratestwo BEM calculations are always necessary: one for the tractions in the non-delaminated case and another for the displacements in the delaminated case.

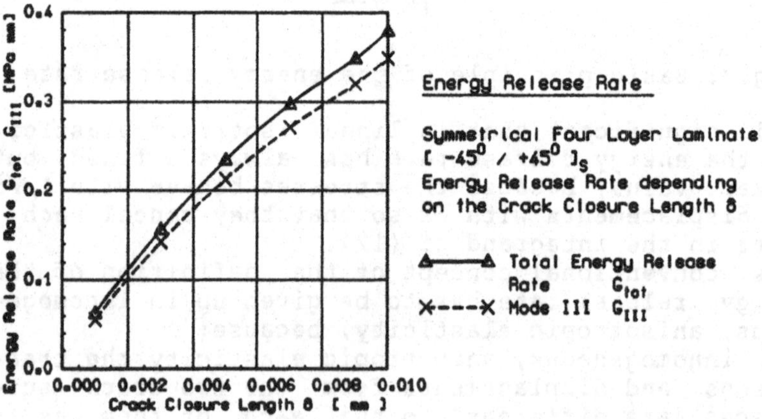

Fig. 4: Influence of the virtual crack closure length δ on the energy release rate

The first example deals with a symmetrical four layer laminate (fig. 1) with a fibre orientation of $[-45°/45°]_s$, where the index S denotes the symmetry relative to the neutral plane under uniaxial strain $\varepsilon_{33}=1$. Fig. 4 shows the influence of the virtual crack closure length δ on the energy release rate. It is seen, that the influence is significant and that for small lengths δ the energy release rate decreases rapidly. This numerical result confirms the consideration made above.

6. Examples

The designer of a laminate has to take care of three main features: the laminate must be stiff, its global limit load has to hold a certain value and the danger of a delamination before reaching the limit load must be minimized. Only the last point is discussed here and the proposed method could be an instrument to solve this problem. The two applied load cases, i. e. uniaxial strain and constant curvature, are the most important in practice. The material is an unidirectional carbon-epoxy-laminate, whose material constants are often used by other authors [7, 9].
A $[\Theta/\Theta-90°]_s$-laminate and a $[\Theta/\Theta-90°/-45°/+45°]_s$-laminate are presented to demonstrate the capabilities of the method. Because the cross-section of the laminates has two symmetry lines only a quarter has to be discretized (fig. 5).

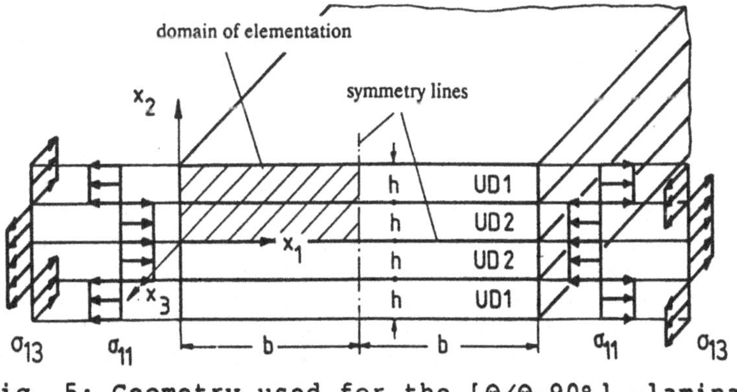

Fig. 5: Geometry used for the $[\Theta/\Theta-90°]_s$-laminate

The crack closure length δ is chosen to 0,01 mm for all calculations, and the ratio b/h is set to 1/5.
Fig. 7 and 8 show the results for the $[\Theta/\Theta-90°]_s$-laminate under uniaxial strain. It is significant that the energy release rate has marked minimums and maximums depending on the orientation angle Θ of the

fibres. The part G_{III} is dominant, because the applied load ε_{33} causes mainly shear stresses σ_{23} between the layers. The smaller parts G_I and G_{II} appear because of the anisotropic constitutive law (3). The total energy release rate G_{tot} is a significant quantity for the design. For $\Theta = 0°$ and $\Theta = 90°$ G_{tot} has its absolute minimum. In this case the constitutive law (3) is only orthotropic and no interlaminar shear stress σ_{23} occurs. The lack of this orientation is the small shear stiffness, so that it is more convenient to choose Θ in the range from 30° to 60°.
A more realistic case is the following eight layer laminate $[\Theta/\Theta-90°/-45°/+45°]_s$ under uniaxial strain and constant curvature. All results in the figures 9 to 12 refer to the crack 3 (fig. 6).

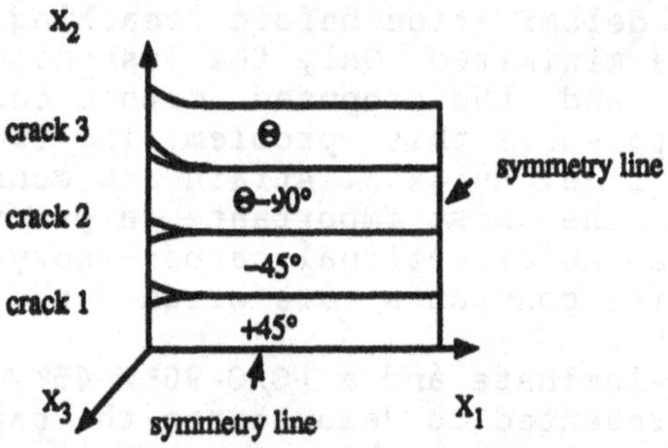

Fig. 6: Numbering of cracks for the $[\Theta/\Theta-90°/-45°/+45°]_s$-laminate

The two examples show that the anisotropy has an essential influence on the results, because the energy release rates G_I and G_{II} supply significant parts. The energy release rates resulting from the constant curvature are very much higher than those under uniaxial strain, because the distance from the neutral plane of the laminate to crack 3 (fig. 6) is large which leads to high stresses in the case of bending. In both cases it is quite simple to choose the ranges that should be avoided and that should be used for the design. The BEM solved all problems very well, no difficulties occurred related to the method. The BEM is easy to use and supplies always accurate results in a short time.

7. Conclusions

It has been worked out that the delamination of layers from the free boundary is an important feature in the design of unidirectional composite laminates. The energy release rate is a significant quan-

tity to describe the possibility of a delamination. The calculations of the displacements and tractions, which are necessary for the knowledge of the energy release rate, are done by the BEM, which turned out to be very well suited for this problem, where the stresses increase rapidly near the free boundary.
The energy release rate is given for various examples of four and eight layer laminates. The optimum design of a laminate follows, when the energy release rate has its minimum depending on the fibre orientation.

8. References

[1] Brebbia, C.A.; Telles, J.C.F.; Wrobel, L.C.: Boundary element techniques. Springer (1984).

[2] Delale, E.: Stress singularities in bonded anisotropic materials. Int. J. Solids & Structures 20 (1984), 31-40.

[3] Eshelby, J.D.; Read, W.T.; Shockley, W.: Anisotropic elasticity with application to dislocation theory. Acta Metallurgica 1 (1953), 251-259.

[4] Ezawa, Y.; Okamoto, N.: Singularity modelling in two- and three-dimensional stress intensity factor computation using boundary elements. Boundary Elements VII (eds. C.A. Brebbia, M. Tanaka), Proc. of the 7th Int. Conf. (1985), 7-3 - 7-12.

[5] Klingbeil, D.: Zur Berechnung von räumlich beanspruchten, geschichteten anisotropen Bauteilen mit starker Spannungsvariation in Querrichtung mit einem Randelementverfahren. VDI-Bericht Reihe 18, Nr.69 (1989).

[6] Klingbeil, D.: The calculation of interlaminar stresses and stress intensity factors in non-delaminated and delaminated composite laminates by a boundary element method (BEM). Proc. 5th Int. Conf. Num. Meth. Fract. Mech. (1990), 367-378.

[7] Pipes, R.B.; Pagano, N.J.: Interlaminar stresses in composite laminates under uniform axial extension. J. Comp. Mat. 4 (1970), 538-548.

[8] Rizzo, F.J.; Shippy, D.J.: A method for stress determination in plane, anisotropic bodies. Int. J. Comp. Mat. 4 (1970), 33-61.

[9] Wang, A.S.D.; Crossman, F. W.: Some new results on edge effects in symmetric composite laminates. J. Comp. Mat. 11 (1977), 92-106.

[10] Wang, S.S.: Choi, I.: Boundary layer effects in composite laminates, part I + II. J. Appl. Mech. 49 (1982), 541-560.

9. Appendix

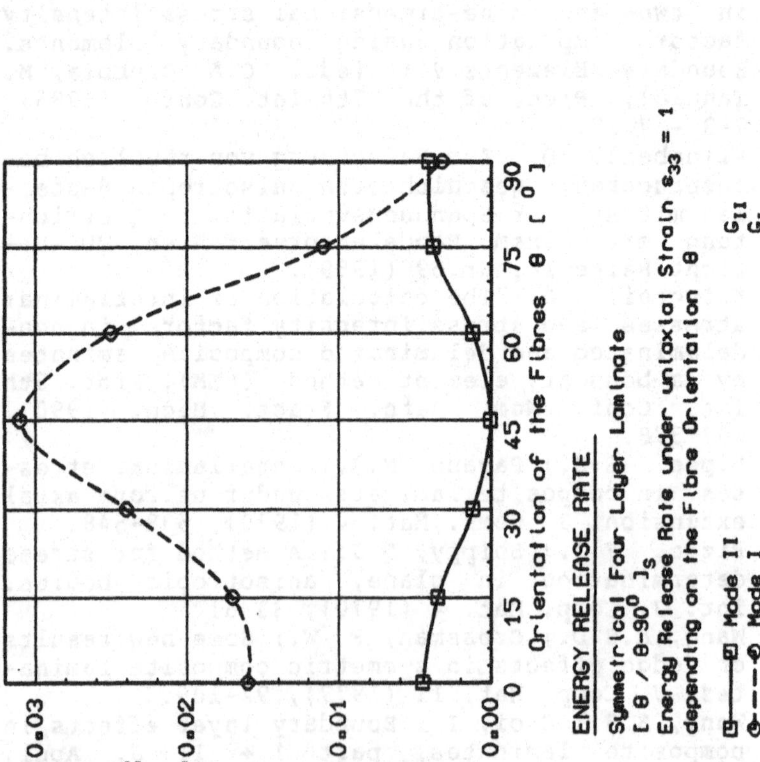

Energy Release Rate G_I G_{II} [MPa mm]
0.03
0.02
0.01
0.00
0
15
30
45
60
75
90
Orientation of the Fibres θ [°]
ENERGY RELEASE RATE
Symmetrical Four Layer Laminate
[θ / θ-90°]s
Energy Release Rate under uniaxial Strain s33 = 1
depending on the Fibre Orientation θ
Mode II G_{II}
Mode I G_I

Fig. 7

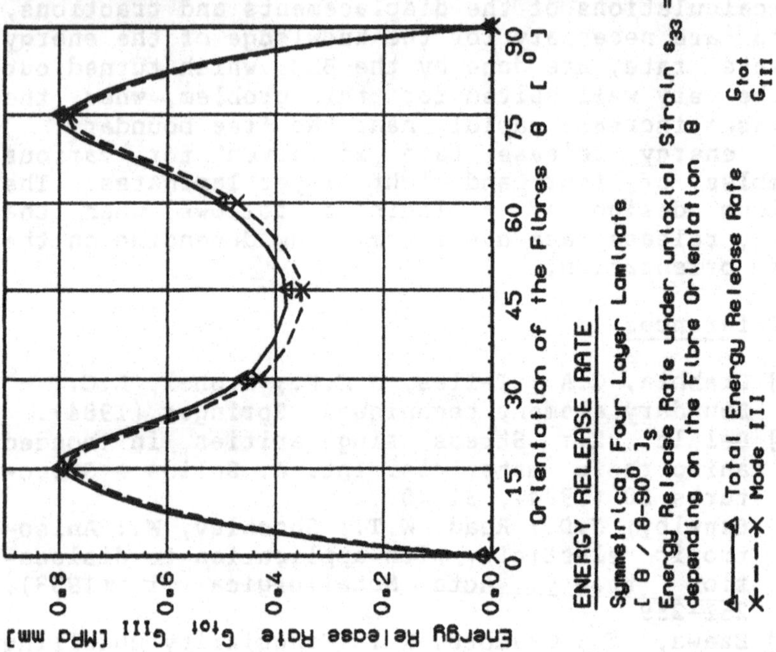

Energy Release Rate G_{tot} G_{III} [MPa mm]
0.8
0.6
0.4
0.2
0.0
0
15
30
45
60
75
90
Orientation of the Fibres θ [°]
ENERGY RELEASE RATE
Symmetrical Four Layer Laminate
[θ / θ-90°]s
Energy Release Rate under uniaxial Strain s33 = 1
depending on the Fibre Orientation θ
Total Energy Release Rate G_{tot}
Mode III G_{III}

Fig. 8

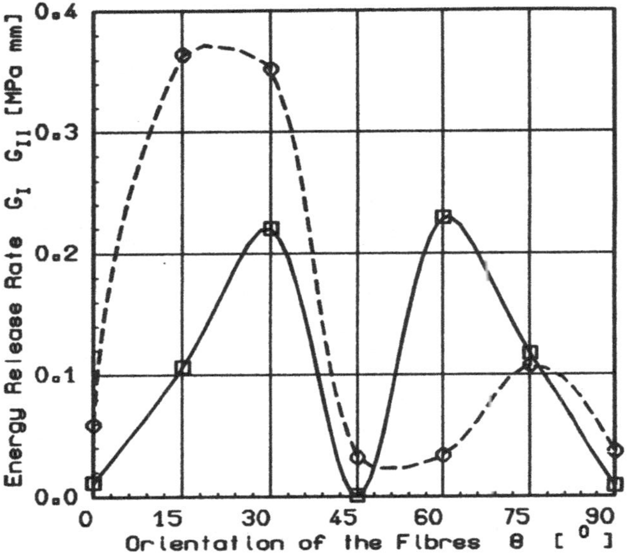

ENERGY RELEASE RATE (Crack Nr.:3)

Symmetrical Eight Layer Laminate
[θ / θ - 90^0 / -45^0 / +45^0]$_S$
Energy Release Rate under uniaxial Strain ε_{33} = 1
depending on the Fibre Orientation θ

□——□	Mode II	G_{II}
○- - -○	Mode I	G_I

Fig. 9

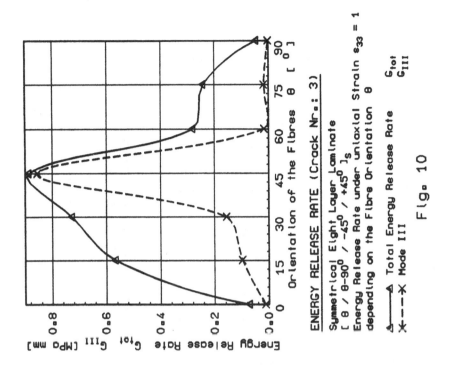

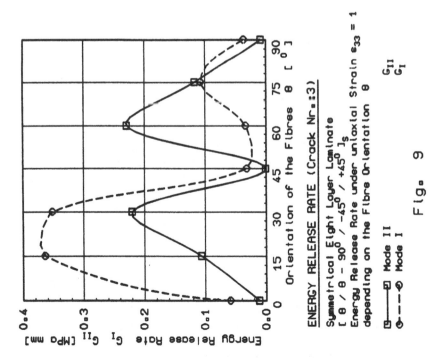

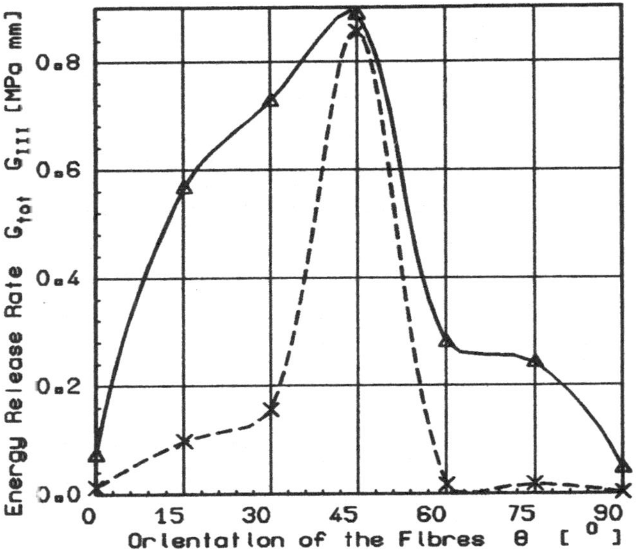

ENERGY RELEASE RATE (Crack Nr.: 3)

Symmetrical Eight Layer Laminate
[θ / θ-90^0 / -45^0 / +45^0]$_S$
Energy Release Rate under uniaxial Strain ε_{33} = 1
depending on the Fibre Orientation θ

△——△	Total Energy Release Rate	G_{tot}
×- - -×	Mode III	G_{III}

Fig. 10

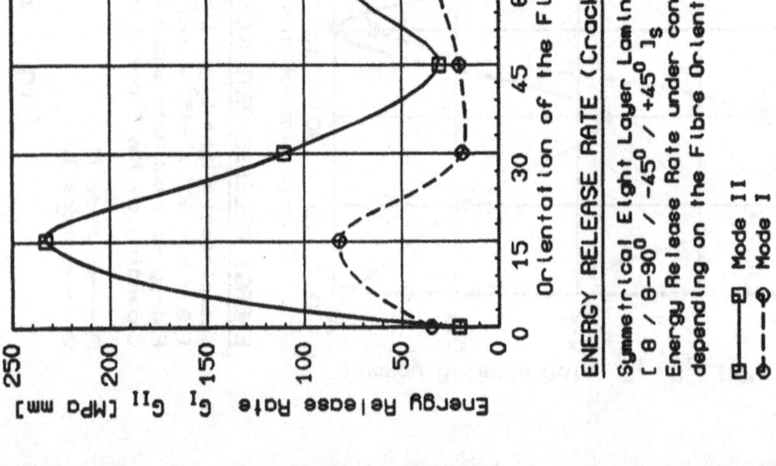
Energy Release Rate G_I G_II [MPa mm]
250
200
150
100
50
0
0 15 30 45 60 75 90
Orientation of the Fibres θ [°]
ENERGY RELEASE RATE (Crack Nr.: 3)
Symmetrical Eight Layer Laminate
[θ / θ-90° / -45° / +45°]s
Energy Release Rate under constant Curvature κ11 = 1
depending on the Fibre Orientation θ
Mode II G_II
Mode I G_I

Fig. 11

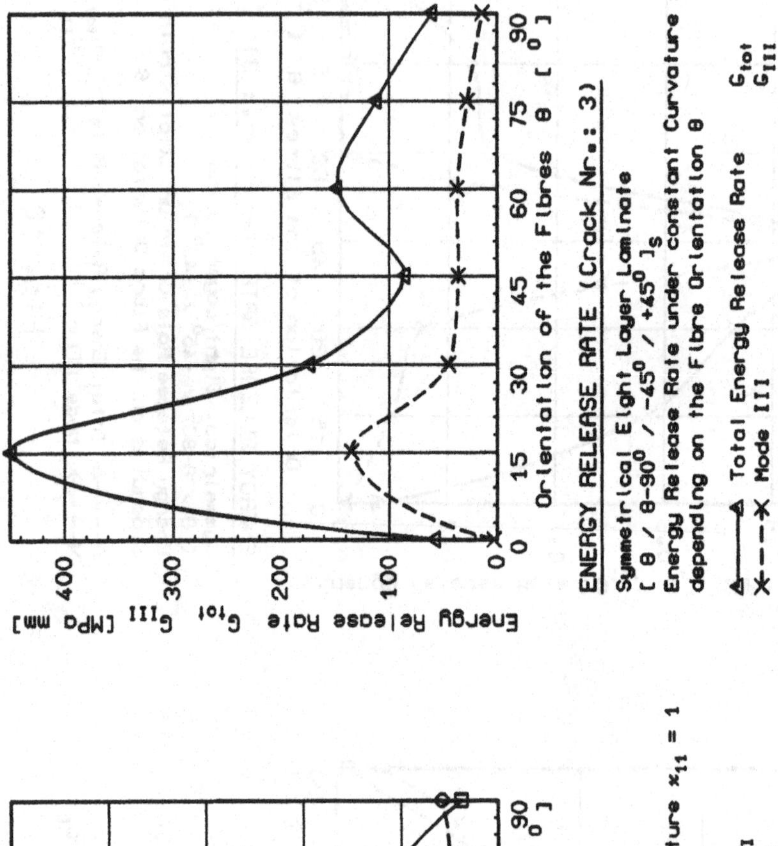
Energy Release Rate G_tot G_III [MPa mm]
400
300
200
100
0
0 15 30 45 60 75 90
Orientation of the Fibres θ [°]
ENERGY RELEASE RATE (Crack Nr.: 3)
Symmetrical Eight Layer Laminate
[θ / θ-90° / -45° / +45°]s
Energy Release Rate under constant Curvature κ11 = 1
depending on the Fibre Orientation θ
Total Energy Release Rate G_tot
Mode III G_III

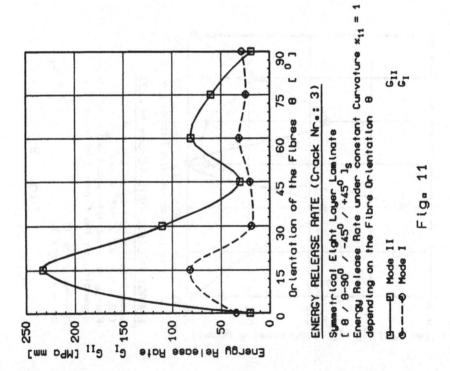

Fig. 12

SECTION 13: CONTACT MECHANICS

Numerical Solution of Thermoelastohydrodynamic Contact Problems by the Boundary Element Method

A. Gakwaya, D. Després

Department of Mechanical Engineering, Laval University, Quebec, G1K 7P4, Canada

ABSTRACT

This paper presents a boundary element formulation and numerical solution of lubricated contact problems. It uses the full Navier Stokes equation for thermoviscous incompressible lubricant fluid and the thermoelastic equations for the contacting solids. Both the solids and the lubricant are studied by the BEM.

INTRODUCTION

Contact problems between deformable solids constitute an important class of technological problems encountered in engineering industries and friction and wear are always present in most applications. Since dry friction results necessarily in undesirable surface damage, a fluid lubricant is usually inserted between the two rubbing surfaces and form a thin film carrying part of the load thus reducing the frictional effects. In order to understand the physical process taking place during the operation of a mechanical system involving such thermoelasto- hydrodynamic contact (TEHD), a combined study of fluid flow, deformation and heat transfer is required. It is rather a complex solid-fluid interaction problem [1,6]. Irreversible thermodynamic processes take place and their full understanding is not yet satisfactory (either numerically or experimentally).

Most of the works on TEHD contact problems are based on the Reynolds equations for lubrication. Also, the study of solid deformation is introduced using simplified local equations that prevent one to easily perform a global analysis [4,5,6]. Very often, a finite difference or a finite element method is applied for computer simulation. This paper presents an alternative formulation based on the boundary element techniques that employ the full field equations consistent with the assumption of linear thermoelastic solids and linear thermoviscous incompressible fluid. These equations form a coupled system that is written into equivalent boundary-domain integral equations which are then discretized using both conforming and nonconforming boundary elements[2]. The resulting set of nonlinear algebraic equations is solved iteratively using a fixed point algorithm. The corresponding computer model is then applied to study the local as well as the global behavior of some TEHD contact problems that include a slider roller and a pair of spur gear teeth[3,5].

MATHEMATICAL MODELING

We consider a thermodynamic system consisting of two thermoelastic solids A and B in a lubricated contact by means of an interfacial thin layer assumed to be a thermoviscous incompressible fluid. Outside the contact zone, the boundaries of the solids as well as their outward normal vector are denoted by S^A, n^A, S^B, n^B, and in their interfacial zone, they are denoted by S^+, n^+ for body A and by S^-, n^- for body B. The fluid layer occupies a volume V with a boundary defined in term of the interfacial boundary S_c ($=S^+ \cup S^-$) and the lateral boundary S_f(fig.1).

CONTACTING SOLIDS FIELD EQUATIONS:

We assume that the thermodynamic system is under quasi-static conditions so that the equations governing the behavior of the solids are the classical thermoelastic equations with the usual mixed thermomechanical boundary conditions.

$$\sigma_{ij,j} + f_i = 0 \tag{1}$$

$$\sigma_{ij} = \lambda\, \epsilon_{kk}\, \delta_{ij} + \mu\, \epsilon_{ij} - \beta\, \theta\, \delta_{ij} \tag{2}$$

$$\epsilon_{ij} = \tfrac{1}{2}\,(u_{i,j} + u_{j,i}) \tag{3}$$

$$k_s\, \theta_{,ii} = 0 \tag{4}$$

An associated boundary integral equation formulation is well known and may be written in operator form as[2]:

$$C\theta(x)=G(x,y)q(y)-H(x,y)\theta(y) \quad (5)$$

$$Cu(x)=K(x,y)t(y)-L(x,y)u(y)+T\theta(x,y) \quad (6)$$

where G, H are surface integral operator for the Laplace equation given e.g. by

$$Gq(y)=\int G(x,y)q(y)dSy \quad (7)$$

and K, L are corresponding operators for the thermoelastic equations defined by

$$K(.,.)t(y) = \int_S K(x,y)t(y)\, dSy \quad (8)$$

$$L(.,.)u(y) = \int_S L(x,y)\, u(y)\, dSy$$

θ, q are the temperature and heat flux while u, t are the displacement and traction vectors. $T\theta$ is the thermal body force reduced to a boundary integral[2] given by

$$T\,\theta(y) = \int_S [P(x,y)\,\theta(y) - Q(x,y)\,q(y)]\, dSy \quad (9)$$

System (5,6) is a linear decoupled thermoelastic system that can be discretized and solved following standard boundary element technique once appropriate boundary conditions are defined. These are given by

a) mechanical boundary conditions:

$$\text{on part } S_u : u_i = \bar{u}_i \qquad S_u \cup S_t = S \quad (10)$$

$$\text{on part } S_t : \sigma_{ij} n_j = t_i = \bar{t}_i$$

b) thermal boundary conditions

$$\text{on } S_\theta : \theta = \bar{\theta} \qquad S_\theta \cup S_q = S \quad (11)$$

$$\text{on } S_q : k_s \frac{\partial\theta}{\partial n} = q = \bar{q}$$

THERMOVISCOUS FLUID LAYER

For the thermoviscous fluid layer, the governing field equations are the continuity equation:

$$\nabla . v = 0, \quad (12)$$

and the momentum equation reduced to Navier-Stokes form for a viscous Newtonian fluid:

$$\mu \nabla^2 v + \mu \nabla(\nabla . v) - \nabla p = -f + \rho(v . \nabla v) \qquad (13)$$

where v, p, f are respectively the velocity, pressure and external force (assumed zero), μ is the viscosity and ρ is the density. The term $v.\nabla v = a$ is an inertia force due to convective acceleration. The energy equation is

$$\rho C_p v . \nabla \theta - \alpha \theta v . \nabla p - \mu \Phi_{inc} + k \nabla^2 \theta = 0 \qquad (14)$$

where Φinc is the viscous dissipation term, α, Cp, k are materials constants. The usual mixed boundary conditions on imposed velocity v and surface traction t or imposed temperature θ and surface heat fluxes q are assumed to be given. An integral equation formulation for these equations has been performed in [2] and can be written in operator form as:

$$Cv(x) = K^* t(y) - L^* v(y) + K_v a(y) \qquad (15)$$

for the velocity field where for example K t(y) is a surface integral operator defined by

$$K^* t = \int_s K^*(x,y) t(y) dSy \qquad (16)$$

and K^*, L^* are an appropriate fundamental solutions of the corresponding differential system [2]. K_v a is the non linear inertia term given by $K_v a = \int_v K(x,y)\, a(y)\, dV_y$. Similarily, the energy equation becomes

$$C\theta(x) = G^* q(y) - H^* \theta(y) + G_s \theta(y) + \nabla G_v \theta(y) + G_v \Phi(y) \qquad (17)$$

where the last three terms in (17) are volume integrals expressing the effects of convected heat flow and viscous dissipation (pessure effects are neglected). The boundary conditions for the thermoviscous layer are given by:

a) mechanical boundary conditions:

$$\text{on } S_v : v_i = \bar{v}_i \qquad (18)$$

$$\text{on } S_t : \sigma_{ij} n_j = t_i = \bar{t}_i \qquad (19)$$

b) thermal boundary conditions:

$$\text{on } S_\theta : \theta = \bar{\theta} \tag{20}$$

$$\text{on } S_q : k_f \frac{\partial \theta}{\partial n} = q = \bar{q}$$

SOLID-FLUID INTERFACE CONDITIONS[2,3,8,12]

In order to solve for the coupled solid-fluid problem, interfacial boundary conditions must be supplied. These include:

(i) the kinematic constraints (non interpenetration non-slip conditions:

$$\left(v_i^f - v_i^s\right) n_i = 0 \text{ on } S_c, \quad v_{it}^f = v_{it}^s \text{ on } S_c \tag{21}$$

(ii) the non negative pressure condition:

$$t_i^s n_i \leq 0 \quad \text{on } S_c \tag{22}$$

(iii) the layer load constraint:

$$t_f = -t_s \tag{23}$$

(iv) equality of fluxes at the fluid-solids interface

$$k_s \partial\theta_s/\partial n = k_f \partial\theta_f/\partial n. \tag{24}$$

Therefore the final mathematical model based on integral equation formulation is composed of:

a) a boundary integral equation for each thermoelastic solid supplemented by appropriate mixed boundary conditions for the thermomechanical equations (10,11);

b) a set of coupled boundary-domain integral equation for the velocity, pressure and temperature distribution in the thermoviscous fluid supplemented by mixed boundary conditions (19,20);

c) and the above interfacial conditions (21-24).

DISCRETIZED BOUNDARY INTEGRAL EQUATIONS

The above set of equations forms a non linear coupled system for which only approximate solutios can be envisaged. Here, we use a boundary element discretisation technique to transform them into algebraic equations. Quadratic isoparametric conforming boundary elements are generally used but, on corners, non conforming elements are also considered. Special integration techniques tailored to handling the different kinds of singular integrals encountered during the evaluation of the matrix coefficients are utilized[2,10]. The resulting discrete algebraic system of equations is (following the order of computer processing):

(i) mechanical equation for the fluid:

$$H_f\, v = G_f\, t + B_f(v) \tag{25}$$

(ii) energy equation for the fluid:

$$H_t\, \theta = G_t\, q + B_t\,(\theta) + \{\Phi\} \tag{26}$$

(iii) energy equation for the solids:

$$H\, \theta = G\, q \tag{27}$$

(iv) mechanical equation for the solids:

$$H_s\, u = G_s\, t + \{B(\theta)\} \tag{28}$$

A computer code has been developed that solve iteratively the above system.Its general flowchart is given in the algorithm given below. First, the thermoviscous fluid equations (25) and (26) are solved iteratively until convergence to a given Reynolds number. At the end of the step, boundary velocities and tractions, interior fluid velocities and their gradients and pressure distribution, boundary temperature and heat fluxes as well as interior temperature are known. Then, energy information on the solid-fluid interface is combined with other solid thermal boundary conditions and used to solve for the temperature distribution in the solids (eq. 27) whose knowledge enables one to solve for the surface displacements and traction using (eq. 28). This information on the solid deformation is then used to update the solid geometry. Once this is done, the gap shape is updated and a new BEM mesh is generated. A restart of the computational loop from the initial Re number is then performed. The iterative process stops when we have convergence of elastic displacements (within a specified tolerance).

Algorithm:

Step 1 - Initial step: data preparation: read in nodal coordinates, element connectivities, physical properties (material), boundary conditions for the fluid and the solids.

Step 2 - Compute elementary properties at gauss points: cartesian coordinates at GP, tangent vectors, jacobians and direction cosines, interpolation functions.

Step 3 - Start Reynolds number loop:

3.1 Solve for the fluid: A) the Navier stokes equation; B) the energy equation; C) if convergence GO TO D; else GO To A; D) increase Reynolds number, if Re = desired Re GO TO 3.2, else GO To 3.1 A.

3.2 Solve for the solid: A) transfer energy boundary condition at fluid-solid interface and solve the heat transfer equation; B) solve the thermo-elastic solids equations.

3.3 Update the solid geometry: convergence test: if yes STOP, END; else GO TO step 4.

Step 4 - Prepare data for next iteration: update nodal coordinates and elementary properties, go to step 3.

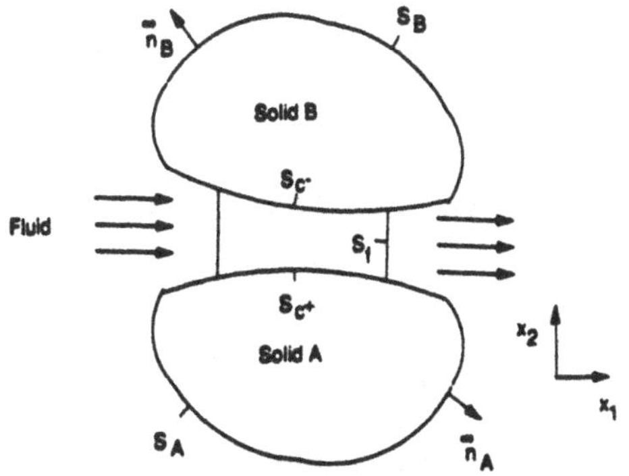

Figure 1: Thermomechanical system for fluid-solid interaction in lubricated contact

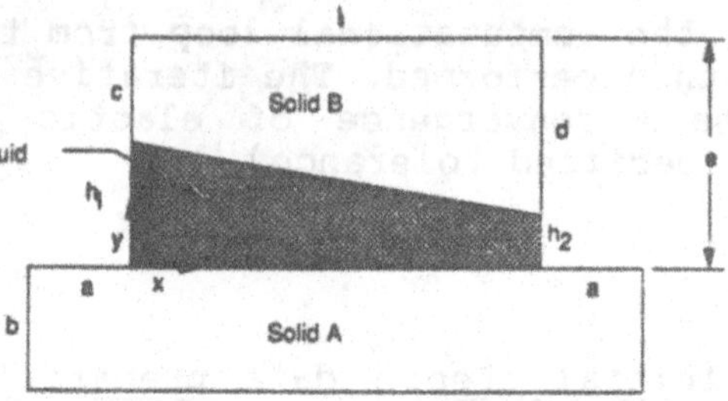

Figure 2 a: Hydrodynamic slider bearing geometry

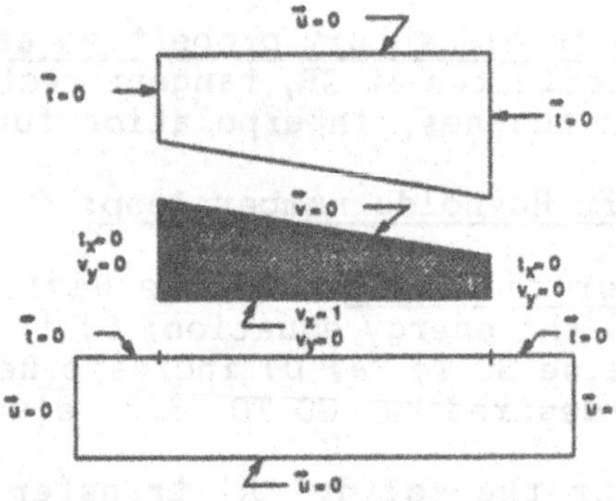

Figure 2 b: Hydrodynamic slider bearing: boundary conditions

RESULTS AND DISCUSSIONS

In order to assess the validity and usefulness of the above computational scheme, several examples problems have been solved that include a hydrodynamic slider bearing, and lubrication of a pair of spur gear teeth under thermomechanical loading.

1. Hydrodynamic slider bearing[7]

This first example is aimed at validating our computational scheme by comparing the computed results with analytical ones[5,6,12]. Figure 2 shows the data used in the computer simulation.

The geometric dimensions used in the model are:

a = .5 m, b = .1 m, c = .1 m, d = .15 m,
ϵ = .2 m, h_1 = .1 m, h_2 = .05 m, l = 1.0 m.

The physical parameters for the fluid are:

viscosity μ = .14 Pa.s, density ρ_f = 875 kg/m^3, coefficient of thermal expansion $\alpha_f = 6.48 \times 10^{-4}$ $^oK^{-1}$, thermal conducting K_f =.114 w/m.^{o}K, pressure coefficient $C_p = 0$ Reynold's number Re = 3.0. The corresponding data for the solids are Young's modulus E = 2.1 x 10^{11} Pa Poisson's ratio ν = .3, $\alpha_s = 1.782 \times 10^{-5}$ $^oK^{-1}$, K_s = 52 W/m.^{o}K. The mechanical and thermal boundary conditions are shown on figure 2 (b).

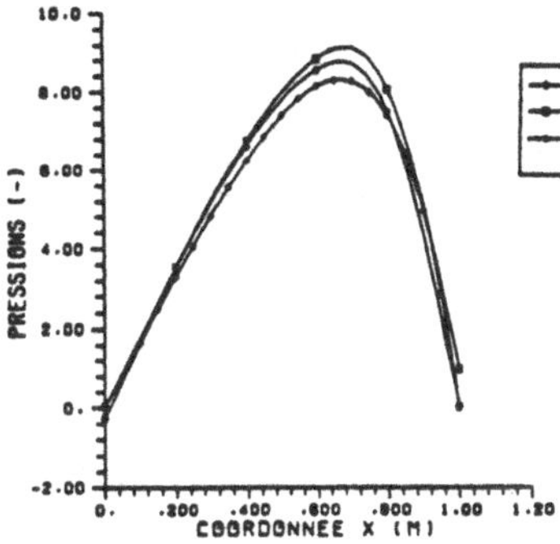

Figure 3: Slider bearing pressure distribution

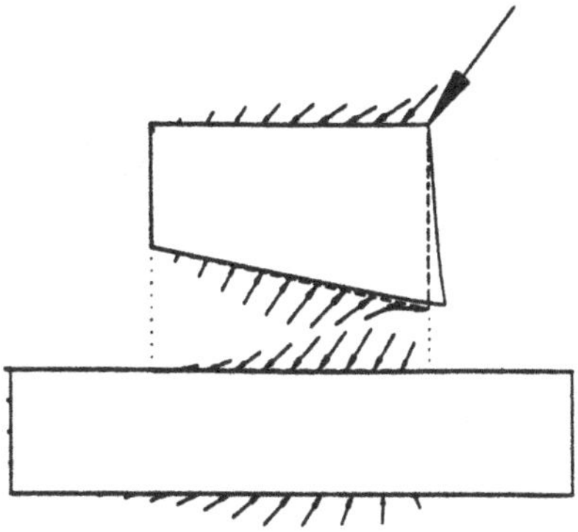

Figure 4: Slider bearing deformed shape

Figure (3) shows the pressure distribution at the fluid-solid interface as computed by BEM code and compared to analytical solution. Figure 4 gives the deformed solid shape and surface tractions distribution. Note that the high value at the upper corner is due to a bad modeling of corner discontinuity and can be eliminated easily. When thermal effects are taken into account, we essentially obtain the same kind of distribution.

2. Ball or roller bearing lubrication

We consider the case of an elastic ball (cylinder in 2D) in rolling contact with an elastic halfspace under lubricated contact conditions: simulating the case of a ball or roller bearing in a housing with a small curvature. Two cases have been studied (a) the case where the ball is under translating motion (rolling without friction) and (b) the case where the ball is under rotating motion (to simulate friction).

Figure (5) gives the geometric model and geometric parameters used in the study:

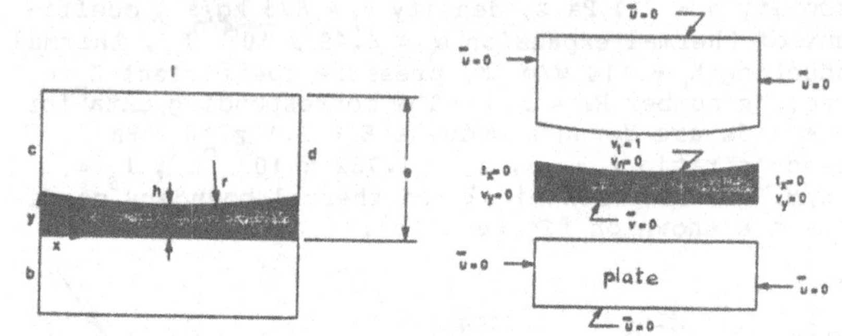

Figure 5: Ball in rolling contact with elastic plate, a) geometry, b) boundary condition.

The physical parameters employed in the simulation are:

for the fluid:
viscosity μ = .14 Pa.s, density ρ_f = 875 kg/m^3, thermal expansion coefficient α_f = 6.48 x 10^{-4} K^{-1}, thermal conductivity K_f = .114 W/m.k

and for the solids:
Young's modulus E = 2.1 x 10^5 MPa, Poisson's coefficient ν = .3, thermal expansion α_s = 52 W/m.k, thermal conductivity K_s = 1.782 x 10^{-5} K^{-1}

and we considered several loading cases:

Angular velocity	Tangential velocity	Re
1 310 RPM	2.60 m/sec	20
1 964 RPM	3.91 m/sec	30
3 280 RPM	6.51 m/sec	50

The geometric dimensions used were:
a = .002 m, b = .002 m, c = .002 m,
h = .001 m, l = .01 m, and r = .019 m (fig. 5.a).
A non slip boundary condition was imposed at the fluid-solid (ball) interface and corresponds to a normalized tangential velocity v_t = 1.0 and normal velocity v_n = 0.

Figure 6 gives the y-displacement at the fluid solid-interface while figure 7 gives the distribution of the tangential tractions. Fig. 8 and 9 show respectively the deformed shapes of the solid and the fluid velocity distribution.

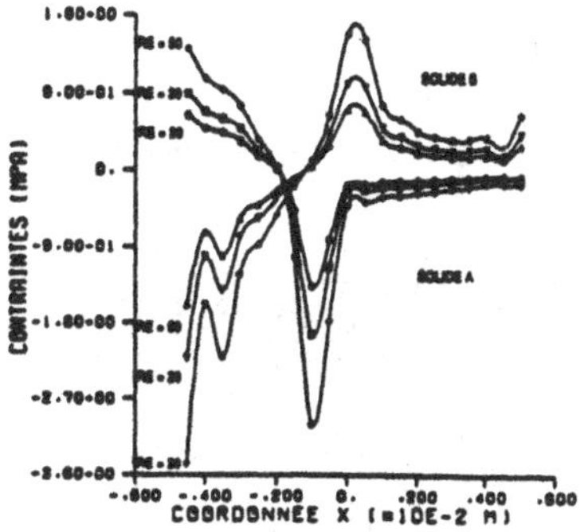

Figure 6: Tangential stress at solid-fluid interface

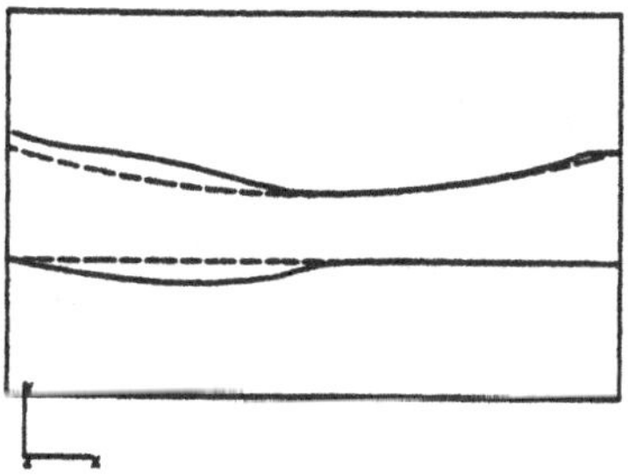

Fig. 7: Deformed solids shape

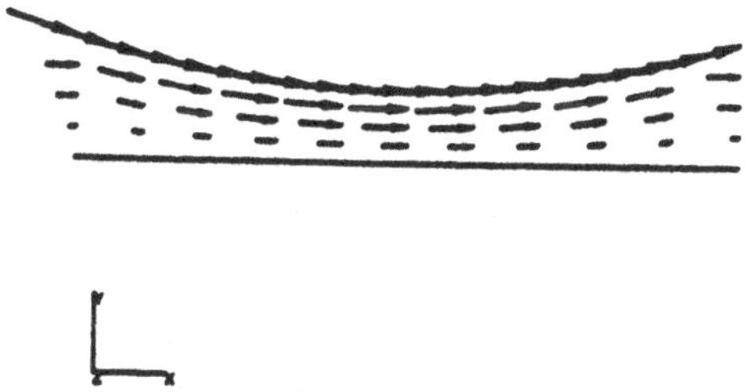

Fig. 8: Fluid velocity distribution

3. Elastic spur gear teeth in lubricated contact

To test the ability of our algorithm in practical situations , we presents in figure 10, a pair of spur gear teeth in lubricated contact. Several loading cases were studied in order to study the tooth deformation and stresses distributions and to assess the fluid load carrying capacity. These distribution compare with those obtained in [3] in 3D analysis using the FDM and a simplified integral equations for the elastic deformation.

The gears have similar dimensions and physical properties:

number of teeth N = 30, diametral pitch P = .5 mm^{-1} (module), pitch radius R = 30 mm, addendum a = 2 mm, dedendum b = 2.5 mm, pressure angle $\phi = 20^{\circ}$, axis distance C = 60 mm.

We use the same fluid and solid properties as in the preceeding example.

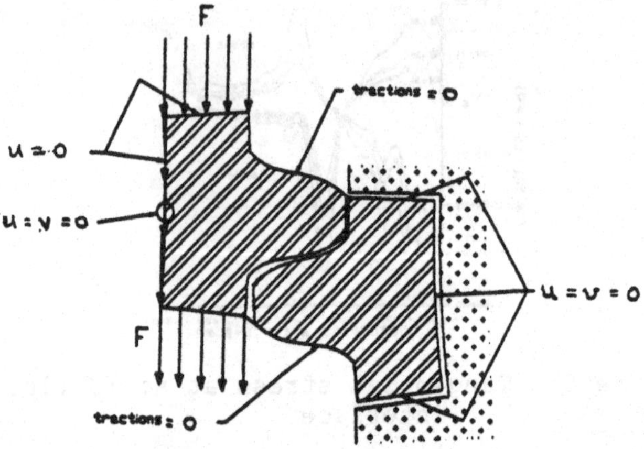

Figure 9: Gear Leeth Configuration:
a) geometry, b) boundary conditions

We assume the teeth have built-in boundary conditions at the base (other boundary conditions can of course be considered) to simulate a torque load transmission or the fluid load carrying capacity.

The fluid medium boundary conditions are as follow. At the inlet and outlet we assume zero velocity and zero pressure; on tooth face A and B, surface velocity correspond to that induced by the rotating gear; at tooth face B, the tangential velocity is that induced by the rotating tooth. The fluid Reynold's number is taken as 1 and correspond to gear angular velocity of 1 000 RPM. The termal conditions are as follows. For the solids: tooth A and tooth B basis are at zero temperature; at free teeth surface we impose zero fluxes. For the fluid, we assume zero flux at the inlet and outlet zone. On interface with tooth A, we impose $T = 0^o$ while at the interface with tooth B we impose $T = 50^o$. The computed results are given in figure 11 to 12. Figure 10 shows the pressure at fluid-solids interfaces while fig. 11 gives the tangential stresses.

Since we do not have at our disposal similar computer codes based on FEM or FDM method, we cannot assess for the moment the relative efficiency of this powerful alternative. We believe, however, that our procedure is more efficient. The limitations encountered in this study are purely of numerical type and are related to the accurate numerical evaluation of singular integrals in very thin geometries(fluid layer). Therefore special integration techniques should be developed in order to improve the accuracy and to extend the range of practical problems that can be handled by the computer model. Extension of this approach to 3D situations should consider the lubricant fluid as curved thin layer and associated Navier-Stokes equations should be written on a curved surface(oriented medium) deve-loped using a two dimensional parametric formulation.

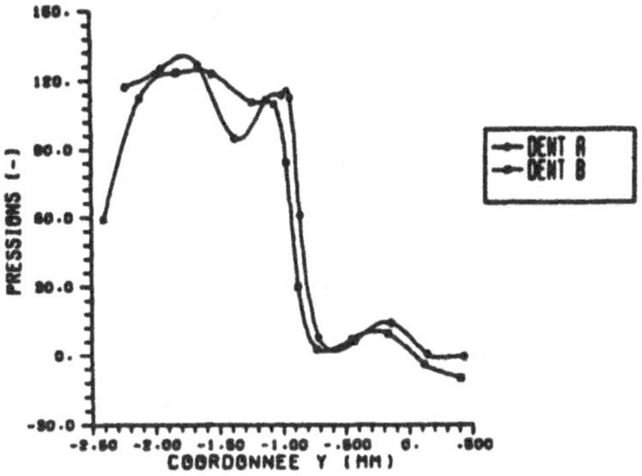

Figure 10: Normal pressure at teeth interface

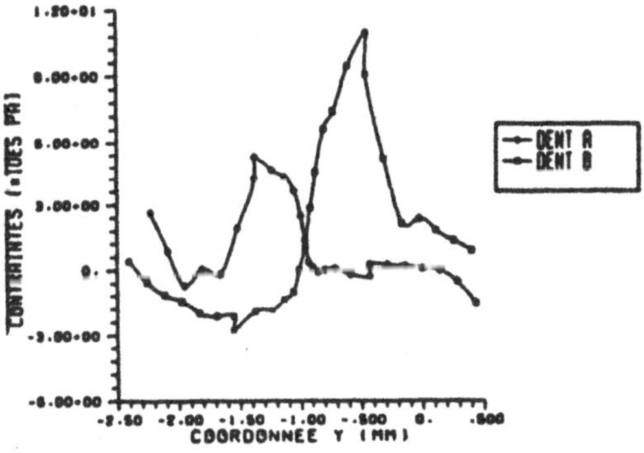

Figure 11: Tangential stresses at teeth interfaces

BIBLIOGRAPHY

1. Dowson, D."A generalized Reynolds Equation for fluid-film lubrication", Int. Jour. of Mech. Sci., vol. 4 ,pp.159-170, 1962.

2. Gakwaya, A." A Boundary element formulation for thermo-elastohydrodynamic contact problems, part I: theory " to be published in Eng. Analysis with BEM, 1991.

3. Simon, V. "Elastohydrodynamics Lubrication of hypoid gears ", ASME J. of Mech. Des., vol.114, pp195-203, 1981.

4. Blahey, A.G., Schneider, G.E.,"A numerical solution of the elasto hydrodynamic lubrication of elliptical contacts with thermal effects", Proc. O. Reynolds Centenary.,Dowson, D et al. (ed), Elsevier, pp. 219-231, 1982.

5. Curnier,A., Taylor, R.L. "A thermomechanical formulation and solution of lubricated contact between deformable solids " , ASME J. of Lub. Tech.", vol.104, pp. 109-117, 1982.

6. Cameron, A."Principles of Lubrication", Longmans, London, 1986.

7. Ezzat, H.A., Rhode, S.M., "A study of the thermo-hydrodynamic performance of finite slider bearings", Trans. ASME, J. of Lubric. Tech., pp. 298-307, 1973.

8. Brüggermann, H., Kollmann, F.G., "A numerical solution of elastohydrodynamic lubrication in an elliptical contact", Trans. ASME, J. of Lubric. Tech., pp. 392-400, 1982.

9. Prabhakaran, N., Sinhasan, R., Singh, V.D., "Elastohydrodynamic effects in elliptical bearing", Journal of Wear, vol. 118, pp. 129-145, 1987.

10. Brebbia, C.A., Wrobel, L.C., "Boundary elements techniques", Springer Verlag, 1984.

11. Brebbia, C.A. (ed), "Advances in boundary elements: vol. I, II, III", Springer Verlag-Computal. Mech., 1989.

12. Gross, W.A., "Fluid film lubrication", J. Wiley, New York, 1980.

Development of Contact Stress Analysis Programs Using the Hybrid Method of FEM and BEM

Y. Ezawa, N. Okamoto

Mechanical Engineering Research Laboratory, Hitachi Ltd., Japan

ABSTRACT

This paper deals with the development of two- and three-dimensional contact stress analysis programs using a hybrid method of the finite element method and the boundary element method. In order to improve the accuracy of analysis using the hybrid method, a new method for computing the equivalent stiffness matrix is described. Furthermore, the penalty function is applied to introduce the contact conditions. The 3-node contact element was used for two-dimensional analysis and the 5-node contact element for three-dimensional analysis. Considering the application to general slide movements, the new 8-node contact element is developed. The programs are applied to the problem of contact between a magnetic disc and a slider. A thin-film of the disc and a slider are modeled using finite elements and the substrate of the disc is modeled using boundary elements. The results show the validity and the accuracy of this procedure.

1. Introduction

The problem of contact between discs and head sliders in a magnetic head-disc-assembly has been of interest to many engineers. A disc is formed of thin-films and a substrate. Contact stress analysis is one of the key technologies for studying the tribological phenomena. The finite element method is valid for slender bodies, while the boundary element method is valid for massive bodies. Therefore, a hybrid of FEM and BEM is useful in solving contact problems of a magnetic head-disc-interface.

There are two methods of connecting FEM with BEM.[4] One is translating BEM elements into the equivalent FEM elements. The other is translating FEM elements into the equivalent BEM elements. Considering general non-linear analysis, the equivalent FEM method is more useful. However, there are some cases in which the hybrid analysis is insufficiently accurate. Therefore, it is necessary to improve the accuracy of equivalent FEM-type super elements.[5] It is insufficient to study the formulation to improve the accuracy of super elements.

Contact stress problems have non-linearity caused by the non-reversibility of slide movements between the contact surfaces, which explains the iteration procedure required in each step of loading. [1)] These many iteration procedures make it important to reduce the computing time.

The contact conditions can be described using the different contact elements formulated by each procedure. [1,2,3)] The penalty function method has the advantage which is no increase in the number of simultaneous equations.

2. Formulation of the equivalent FEM elements

2.1 Equivalent FEM elements

The boundary integral equation for an elastic solid is

$$c^i u^i_l + \int_S u_k p^*_{lk} dS = \int_S p_k u^*_{lk} dS \qquad (1)$$

where c^i is a constant depending upon the geometry at point i, u_k and p_k denote the surface displacement vector and the traction vector respectively, and $u_{lk}{}^*$ and $p_{lk}{}^*$ are kernel functions. S is the surface of the solid.[4)]

Discretization using boundary elements produces

$$[H]\{u\}=[G]\{p\} \qquad (2)$$

From this equation

$$\{p\}=[G]^{-1}[H]\{u\} \qquad (3)$$

Considering the virtual work

$$\delta\{u\}^T\{f\}=\int_S p_i \delta u_i dS \qquad (4)$$

where $\delta\{u\}$ is a virtual displacement vector, $\{f\}$ is an equivalent nodal force vector. As this relation must be valid for any value of the virtual displacement,

$$\{f\}=[M]\{p\} \qquad (5)$$

where [M] is a translation matrix calculated from eq. (4).

Substituting eq.(3) for eq.(5)

$$\{f\}=[K^*]\{u\}=[M][G]^{-1}[H]\{u\} \qquad (6)$$

In this equation, $[K^*]$ corresponds the stiffness matrix in FEM analysis. However,$[K^*]$ does not become a symmetric matrix because of the error caused by discretization. $[K^*]$ can be made a symmetric matrix using the following equation.

$$[K]=([K^*]+[K^*]^T)/2 \qquad (7)$$

This $[K]$ is the equivalent FEM-type super element. However, there are some cases in which the accuracy of the hybrid analysis using this equivalent FEM-type super element is insufficient.

2.2 Self-equilibrium condition

The element stiffness matrix in FEM satisfies the self-equilibrium condition explicitly. The equivalent FEM-type element stiffness matrix translated from BEM however, does not satisfy it explicitly. If the equivalent stiffness matrix does not satisfy the self-equilibrium condition, the external forces on the element do not balance the internal forces formed by stresses in the element. Therefore, for the purpose of improving the accuracy of the equivalent FEM-type element it is effective to satisfy explicitly the equilibrium condition in the element.[5] Two methods, the Lagrange's method of undetermined multipliers and the penalty function method are considered.

Considering the summation of surface traction on the boundary of BEM regions is zero, equation (8) is produced.

$$\int_S p_i dS = \{0\} \tag{8}$$

Discretization using boundary elements produces

$$[Q]\{p\} = \{0\} \tag{9}$$

We can introduce this condition by the Lagrange's method of undetermined multipliers. The variation of the functional is now

$$\delta\{p\}^T([G]\{p\} - [H]\{u\}) + \delta(\{\lambda\}^T[Q]\{p\}) = 0 \tag{10}$$

in which λ is a Lagrange's multiplier and $\delta\{p\}$ is a virtual traction.

Using the penalty method the governing equation is expressed as follows.

$$\delta\{p\}^T([G]\{p\} - [H]\{u\}) + \alpha\delta\{p\}^T[Q]\{p\} = 0 \tag{11}$$

in which α is a penalty number.

As this relation is valid for any value of the virtual traction, the equality of the multipliers must exist. Thus

$$\{p\} = ([G] + \alpha[Q])^{-1}[H]\{u\} \tag{12}$$

Using the above methods, the accuracy of the equivalent FEM-type element stiffness can be improved.

2.3 Discontinuity of traction

One of the problems in forming the equivalent FEM-type element is treating the discontinuity of traction. BEM elements treat surface tractions which often have discontinuity on the surfaces of which BEM elements must take account. On the other hand, the nodal forces in FEM elements do not take account of the discontinuity of the tractions.

Consider the cubic model, as illustrated in Fig. 1. The base surface is fixed and forces equivalent to unit surface traction are loaded on the upper surface. Let the cube be modeled by the equivalent FEM-type super element.

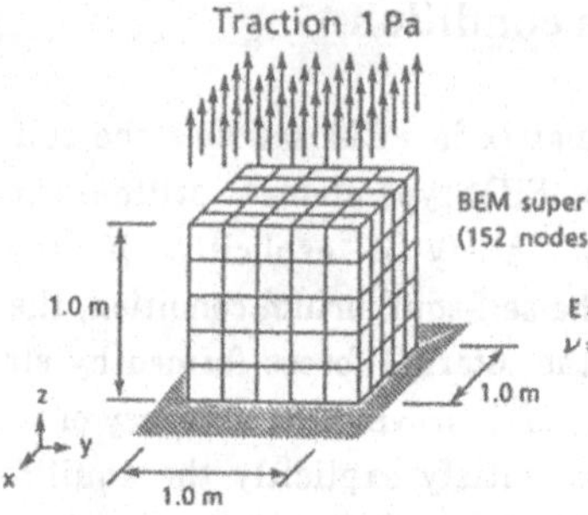

Fig. 1 Cube model (fine mesh)

The traction forces can be calculated by equation (3).
Of course, the exact solution of the {p} is {1}.

0.68	0.83	0.82	0.83	0.68
0.83	1.0	1.0	1.0	0.83
0.82	1.0	1.0	1.0	0.82
0.83	1.0	1.0	1.0	0.83
0.68	0.83	0.82	0.83	0.68

(a) Conventional

1.0	1.0	1.0	1.0	1.0
1.0	1.0	1.0	1.0	1.0
1.0	1.0	1.0	1.0	1.0
1.0	1.0	1.0	1.0	1.0
1.0	1.0	1.0	1.0	1.0

(b) Modified

Fig.2 Tractions on the upper surface

Fig. 2(a) is the calculated result of surface traction on the upper surface using the equivalent finite element method. The error around the edge, especially on the four corners, is relatively large as a result of ignoring the discontinuity of the tractions. Discontinuity was taken into consideration when calculating the matrix [G], [M] and [Q] related to the surface traction. The integration over the free surface is omitted in the calculation of each matrix. The result, in Fig. 2(b), shows no deterioration of the accuracy around the corners.

The accuracy of displacements is also checked. The difference of the displacement in the model shown in Fig. 1 is small. The coarsest mesh shown in Fig. 3 is also checked.

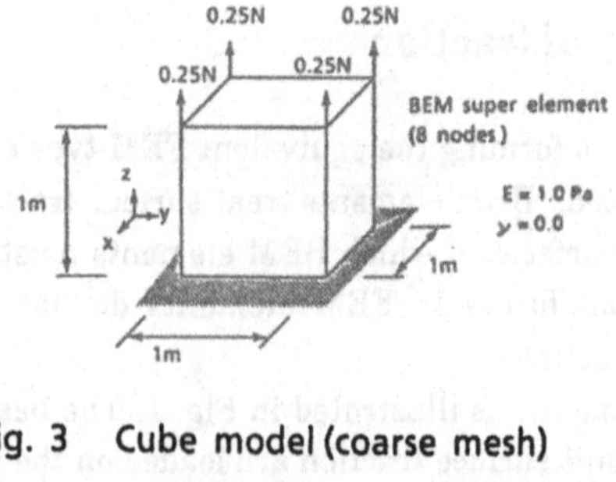

Fig. 3 Cube model (coarse mesh)

In the developed three-dimensional hybrid analysis program, the region of integration is divided into two triangles, shown in Fig. 4,

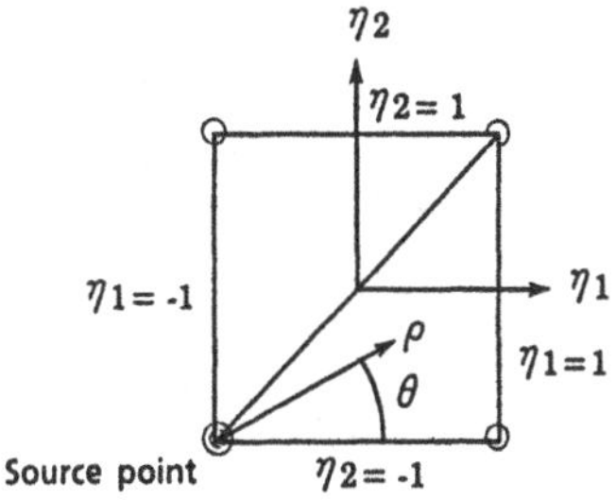

Fig. 4 Subdivided regions for integration

When the source point is located in an integral area, the coordinate transformation takes place from the rectangular system (η_1,η_2) to the polar system (ρ,θ). Singularities in the boundary integral equations can be removed by introducing this new polar coordinate system.[6] This technique allows the integral to be evaluated accurately by using the Legendre-Gauss formulas. The effect of these procedures, the coordinate transformation and the consideration of the discontinuity of the surface traction, is then checked and the results shown in Table 1. The exact value of the solution is 1. Table 1 shows that the displacement when using coordinate transformation and the consideration of the discontinuity is most accurate. The next most accurate option case is using neither method. The results show the validity of these methods used in our program.

Table 1 Displacements

	Discontinuity of surface tractions considered	Discontinuity of surface tractions not considered
Polar coordinates	1.06	1.383
No polar coordinates	0.875	1.109

3. Formulating contact analysis using penalty function method

3.1 Basic equations

Consider the contact problem of two bodies A and B, as shown in Fig. 5. The global coordinate is (x,y,z). The displacements and forces on the contact surfaces are represented by local coordinate ($\underline{x}$,$\underline{y}$,$\underline{z}$). The local coordinates $\underline{x}$ and $\underline{y}$ are defined as tangent lines on the contact point of the body A, the $\underline{y}$ coordinate being a sliding direction. The local coordinate $\underline{z}$ is defined as the inner normal line on the body A. There are three contact conditions, opening state, sticking state and sliding state, as shown in Fig. 6.

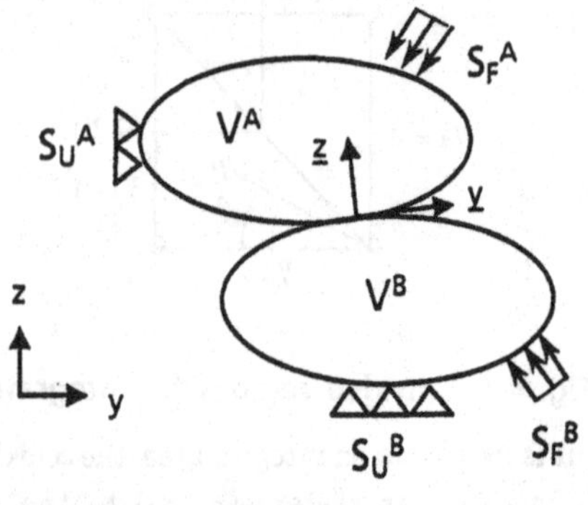

Fig. 5 Contact of two bodies

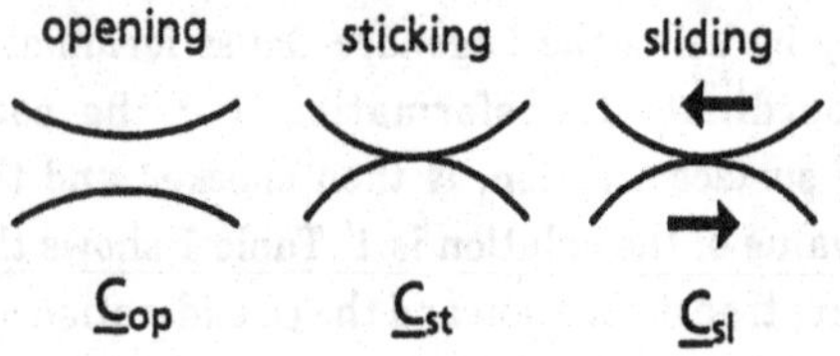

Fig. 6 Three kinds of contact state

Where the increments of loads are given and which the contact states are known, the equations for contact states are given as follows

(1) opening state

$$\underline{p_{i0}}+\underline{\Delta p_i}=0 \qquad on \quad \underline{C}_{op} \qquad (i=\underline{x},\underline{y},\underline{z}) \tag{13}$$

(2) sticking state

$$\underline{\Delta u_i^A}-\underline{\Delta u_i^B}+\underline{d_{0i}}=0 \qquad on \quad \underline{C}_{st} \tag{14}$$

(3) sliding state

$$\underline{\Delta u_z^A}-\underline{\Delta u_z^B}+\underline{d_{0z}}=0$$

$$\underline{p_{y0}}+\underline{\Delta p_y}=\pm\mu(\underline{p_{z0}}+\underline{\Delta p_z}) \qquad on \quad \underline{C}_{sl} \tag{15}$$

where "_" means the local coordinate, Δ is an incremental, $\underline{u_i}$ is a displacement vector and $\underline{p_i}$ is a pressure vector. The equilibrium conditions on contact surfaces are satisfied because $\underline{p_i}=\underline{p_{iA}}=-\underline{p_{iB}}$. $\underline{d_{0i}}$ is a distance between two surfaces. The contact conditions are determined by the following equations.

(1) open state

$$\underline{\Delta u_z^A}-\underline{\Delta u_z^B}+\underline{d_{0z}}>0 \qquad on \quad \underline{C}_{op} \tag{16}$$

(2) sticking state

$$\underline{p_{z0}}+\underline{\Delta p_z}>0$$

$$|\underline{p_{y0}}+\underline{\Delta p_y}|<|\mu(\underline{p_{z0}}+\underline{\Delta p_z})| \qquad on \quad \underline{C}_{st} \tag{17}$$

(3) sliding state

$$\underline{p_{z0}} + \underline{\Delta p_z} > 0$$

$$(\underline{p_{y0}} + \underline{\Delta p_y})(\underline{\Delta u_y^A} - \underline{\Delta u_y^B}) < 0 \qquad on\ \underline{C_{sl}} \tag{18}$$

Using the penalty function method, the variation of the approximate functional in elastic contact stress analysis is [7]

$$\begin{aligned}
\delta\Pi = \sum_{\beta}^{AB} \{ & \int_{V^\beta} (\sigma_{ij0}{}^{\beta} + \Delta\sigma_{ij}{}^{\beta})\delta\Delta\varepsilon_{ij}{}^{\beta}\, dV \\
& - \int_{S_F^\beta} (\overline{p}_{i0}{}^{\beta} + \Delta p_i{}^{\beta})\delta\Delta u_i{}^{\beta} dS + \int_{S_U^\beta} \alpha(\Delta u_i{}^{\beta} - \overline{\Delta u}_i{}^{\beta})\,\delta\Delta u_i{}^{\beta} dS \} \\
& - \int_{\underline{C_{st}} + \underline{C_{sl}}} p_{i0}\,\delta(\Delta u_i^A - \Delta u_i^B) dS \\
& + \int_{\underline{C_{st}} + \underline{C_{sl}}} \alpha(\Delta u_z^A - \Delta u_z^B + d_{0z})\delta(\Delta u_z^A - \Delta u_z^B) dS \\
& + \int_{\underline{C_{st}}} \alpha(\Delta u_x^A - \Delta u_x^B + d_{0x})\delta(\Delta u_x^A - \Delta u_x^B) dS \\
& + \int_{\underline{C_{st}}} \alpha(\Delta u_y^A - \Delta u_y^B + d_{0y})\delta(\Delta u_y^A - \Delta u_y^B) dS \\
& - \int_{\underline{C_{sl}}} \Delta p_y\,\delta(\Delta u_y^A - \Delta u_y^B) dS = 0 \\
& \qquad\qquad (\beta = A,B)
\end{aligned} \tag{19}$$

where

$$\Delta\varepsilon_{ij}{}^{\beta} = (\Delta u_{i,j}{}^{\beta} + \Delta u_{j,i}{}^{\beta})/2 \qquad in\ V^{\beta} \tag{20}$$

$$\underline{\Delta p_y} = \pm\mu(\underline{p_{z0}} + \underline{\Delta p_z}) - \underline{p_{y0}} \qquad on\ C_{sl} \tag{21}$$

and "‾" means the known value on the boundary. α is a penalty number which is larger than the Young's modulus of the material.

3.2 Contact elements

For two-dimensional contact stress analysis a 3-node triangular contact element, as shown in Fig. 7 was used. The contact elements are formulated by the penalty method mentioned in the previous paragraph.

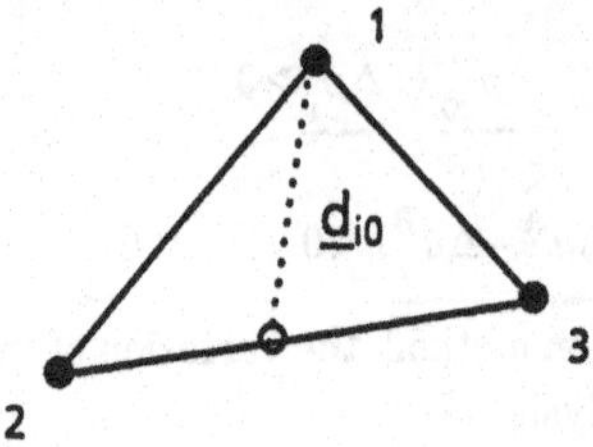

Fig. 7 Contact element for two-dimensional problems

For three-dimensional contact stress analysis, a 5-node contact element based on the penalty method, as shown in Fig. 8, was used.

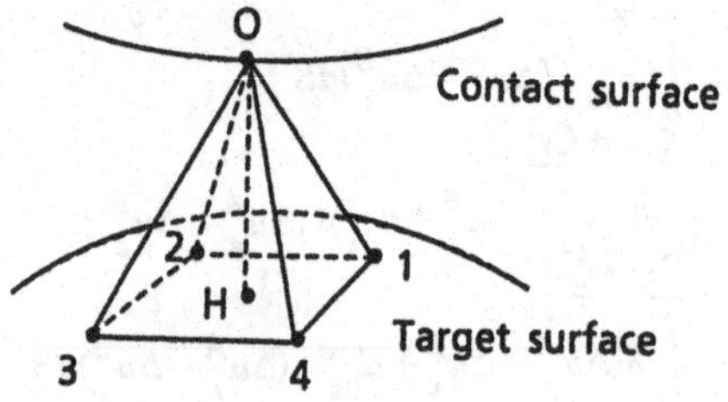

Fig. 8 Contact element for three-dimensional problems

The basic face, 1-2-3-4, is placed on the target surface and the apex, 0, is placed on the other surface, a contact surface. The foot, H, of the perpendicular line must be in the basic face 1-2-3-4, although there is a possibility that the foot H will move out of the the basic face after loading. An 8-node contact element was developed to deal with this problem. Figure 9 shows the 8-node contact element which is composed of eight 5-node contact elements.

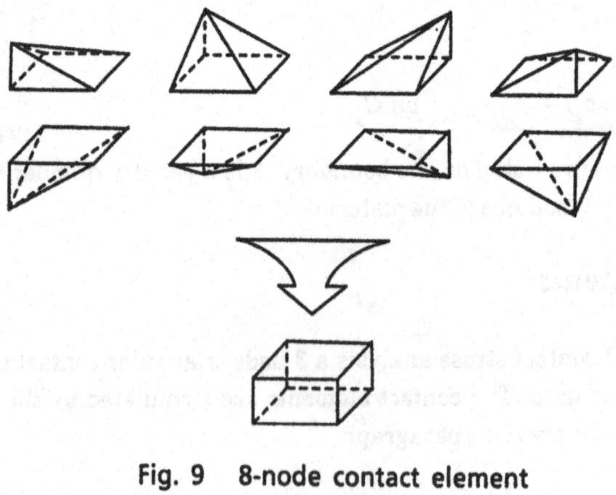

Fig. 9 8-node contact element

One feature of this 8-node contact element is the inclusion of two types of elements each with target surfaces placed on the different bodies. Therefore, even if the contact surface moves in any direction, some contact 5-node elements remain effective. (Fig. 10)

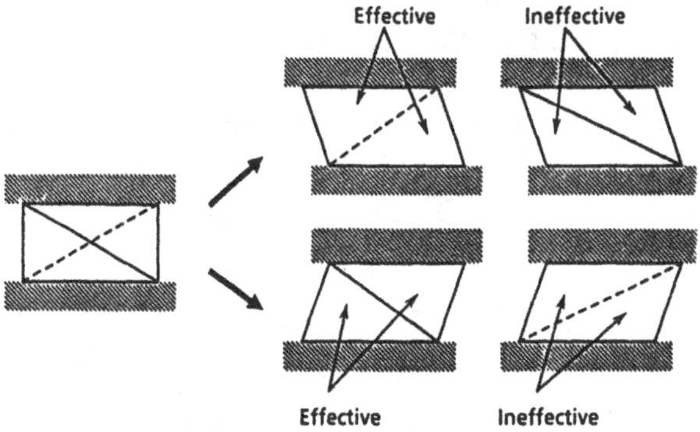

Fig. 10 Change of contact elements

4. Numerical examples

4.1 Comparison of the equivalent stiffness matrix

A rectangle is used as an illustrative example shown in Fig. 11 to compare the equivalent stiffness matrix of BEM with that of the finite element method in two-dimensional stress analysis. The area was modeled by either one finite or four boundary elements. The results are shown in Fig.12.

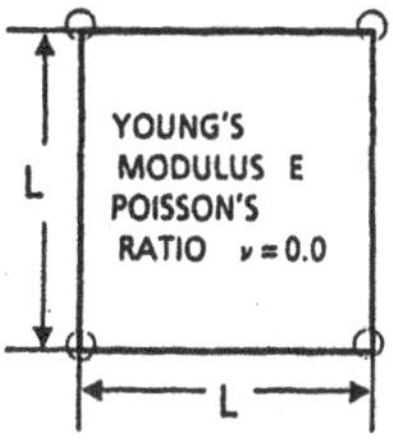

Fig. 11. A rectangle

.433	.109	-.202	-.122	-.211	-.109	-.020	.122
	.433	..122	-.020	-.109	-.211	-.122	-.202
		.433	-.109	-.020	-.122	-.211	.109
			.433	.122	-.202	.109	-.211
				.433	.109	-.202	-.122
					.433	.122	-.020
						.433	-.109
							.433

(a) Modified equivalent stiffness matrix of BEM

.454	.136	-.215	-.125	-.218	-.092	-.021	.125
	.454	.125	-.021	-.092	-.218	-.125	-.215
		.454	-.136	-.021	-.125	-.218	.092
			.454	.125	-.215	.092	-.218
				.454	.136	-.215	-.125
					.454	.125	-.021
						.454	-.136
							.454

(b) Conventional equivalent stiffness matrix of BEM

.500	.125	-.250	-.125	-.250	-.125	.000	.125
	.500	.125	.000	-.125	-.250	-.125	-.250
		.500	-.125	.000	-.125	-.250	.125
			.500	.125	-.250	.125	-.250
				.500	.125	-.250	-.125
					.500	.125	.000
						.500	-.125
							.500

(c) Stiffness matrix of FEM

Fig. 12. Comparison of the stiffness matrix (X E/L^2)

Figure 12 shows that the equivalent stiffness matrix of the conventional boundary element method does not agree with that of the finite element method nor does it include the rigid-body movement. On the other hand the stiffness matrix of the boundary element calculated by eq. (10) satisfies the rigid-body movement.

4.2 Contact stress analysis of the thin film magnetic disc regarding surface roughness

Consider the contact problem between the head slider and the magnetic disc, as shown in Fig.13.

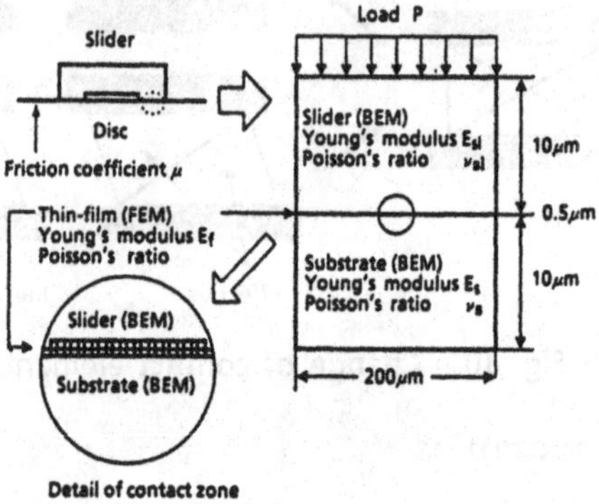

Fig. 13 Contact stress analysis of the magnetic disc

The magnetic disc consists of a substrate and thin film. Only the thin film is modeled by the finite elements. The other parts are modeled by the boundary elements. The surface of the disc has roughness as shown in Fig. 14(a). The Young's moduli of the slider, substrate and film respectively are 210 GPa, 70 GPa and 350 GPa. The Poisson's ratio for all three is 0.3. Table 2 shows the size of this problem. Fig. 14 (b) shows the deformation of a part of contact surfaces after loading.

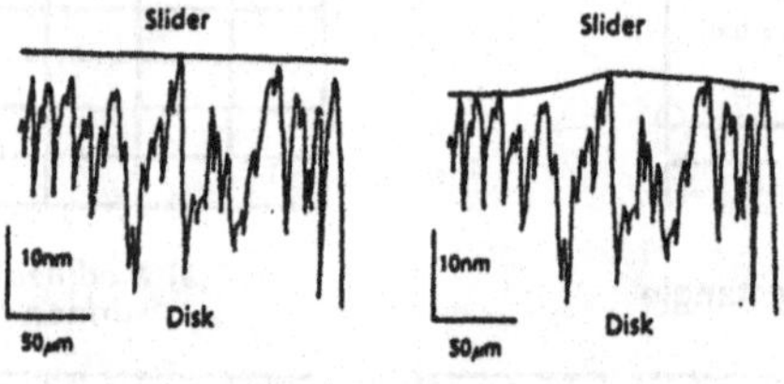

(a) Initial state (b) After loading

Fig. 14 Defomation of contact surface

Table 2 Size of the problem

	FEM/BEM
number of nodes	863
number of finite elements	400
number of boundary elements	461
number of contact elements	201
number of profiles	221256
number of incremental steps	5
CPU-time (sec) / VPU-time (sec)	125 / 35

Fig. 15 shows the relation between the average pressure and number of contact points as well as the relation between the average pressure and the maximum contact pressure.

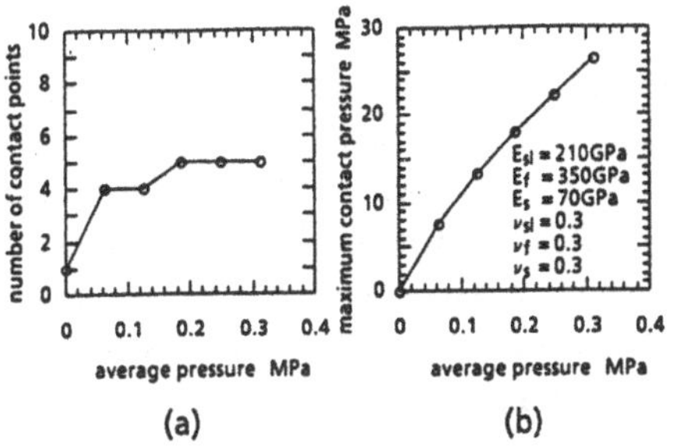

Fig. 15 Relation between average pressure and other values

4.3 Three-dimensional contact problem between a slider and disc

Consider the three-dimensional problem between a slider and disc as shown in Fig. 16. The lower block in Fig. 16 is a part of the disc comprising a substrate and a thin film. The upper block is half the slider. The lower part of the disc is modeled by the boundary elements. The other blocks are modeled by the finite elements. The slider is loaded at the center of the upper surface. Fig. 17 shows the deformation of the disc.

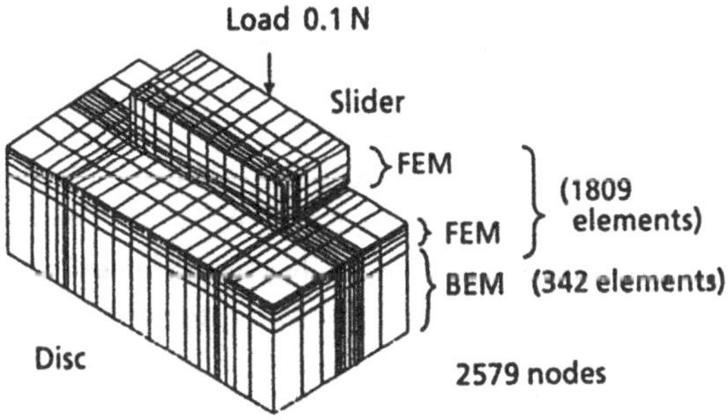

Fig. 16 Contact of a disc and slider

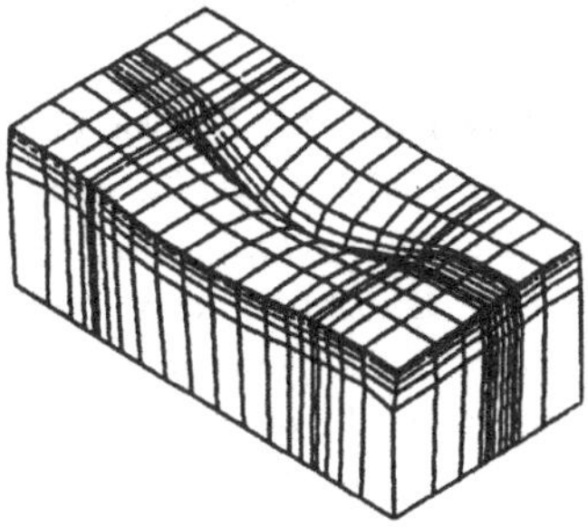

Fig. 17. Deformation of the disc

5.Conclusion

This paper presented the formulation of two- and three-dimensional contact stress analysis using a hybrid method of the finite element method and the boundary element method.

It is possible to improve the accuracy of the equivalent stiffness, by satisfying the equilibrium condition using Lagrange's method of undetermined multipliers or the penalty function method. The numerical result considering the discontinuity of traction in the equivalent stiffness matrix shows high accuracy.

The 3-node contact element for two-dimensional analysis and the 5-node contact element for the three-dimensional analysis were developed using the penalty function. Furthermore, a new 8-node contact element was developed for general slide movements. The numerical results show the validity and the accuracy of this procedure.

Acknowledgements

The authors would like to thank our former colleague, Mr. T. Nagashima of MRI Ltd., as well as Mr. M. Nomura and Mr. T. Hirase of Hitachi Information Systems, Ltd. for their valuable comments and support in developing the computer program.

References

1. Ezawa, Y. and Okamoto, N. ,"High-speed boundary element contact stress analysis using a super computer", Proceedings of the Fourth International Conference on Boundary Element Technology, Windser, 1989, pp.405 - 416.
2. Goodman, R.E., Taylor, R.L. and Brekke,T.L., "A model for the mechanics of jointed rock", Proc. ASCE, Vol.94, No.SM 3, 1968, pp.637 - 659.
3. Yamada, Y., Ezawa, Y., Nishiguchi, I. and Okabe, M., "Handy incorporation of bond and singularity elements in the finite element solution routine", Paper M9/8 in Transactions of the 5th International Conference on Structural Mechanics in Reactor Technology, Berlin, 1979.
4. Brebbia, C. A., The Boundary Element Method for Engineers, Pentech Press, London, 1978.
5. Mostoe, G. G. W., Volait, F. and Zienkiewiez, O. C., "A symmetric direct boundary integral equation method for two dimensional elastostatics, Res Mechanica, Vol.4, 1982, pp.57-82.
6. Higashimachi, T., Okamoto, N., Ezawa,Y., Aizawa T. and Ito, A. "Interractive structural analysis system using the advanced boundary element method", Boundary elements, Proceedings of the Fifth International Conference, Hiroshima, Japan, 1983, pp.847-856.
7. Kantou, Y. and Yagawa, G. "A dynamic contact buckling analysis by the penalty finite element method", Int. J. Numer. Methods Eng., Vol.29,1990, pp.755-774.

Three Dimensional Frictional Conforming Contact Using B.E.M.

A. Foces (*), J.A. Garrido (*), F. Paris (**)
(*) Escuela Técnica Superior de Ingenieros Industrales de Valladolid, Paseo del Cauce s/n, 47011 Valladolid, Spain
(**) Escuela Técnica Superior de Ingenieros Industriales de Sevilla, Avenida Reina Mercedes s/n, 410212 Sevilla, Spain

Abstract

The friction conforming contact problem in three dimensions is formulated using the Boundary Equation Method. Two sources of difficulties arise. The lack of knowledge of the correct partition of the contact zone (sliding and adhesion zones), and the nonlinearity associated to the sliding direction. A procedure to find the correct partition, starting from the adhesion situation, is proposed. The non-linear system of equations is solved applying a Newton-Raphson technique with relaxation. The applicability of the proposed procedure is shown by solving the problem of a punch on an elastic foundation.

1.- Introduction

The contact between deformable bodies is always, to greater or lesser extent, with friction. The difficulty in modelling and, in particular, of solving this problem has meant that the attempts to solve contact problems analytically have been carried out in the frictionless case, Gladwell [1], Johnson [2].

Even numerically and with reference to the three dimensional case, the problem cannot be considered completely solved. The papers of Okamoto [3], Fredrikson [4] and Torstenfelt [5], constitute some of the first references to the use of the Finite Element Method, with several approximations and hypotheses, as a tool to study this type of problem.

The application of the B.E.M. was initially carried out by Andersson on frictionless [6] and friction cases [7], always for the two dimensional situation. París and Garrido [8], [9], [10] and [11], developed a direct formulation, applying it to several situations in two dimensional configurations. More recently Garrido, Foces and Paris [12] and [13], studied the three dimensional frictionless case.

The presence of friction involves additional difficulties deriving from a lack of knowledge of the direction along which the sliding is to take place. Several resolution techniques have been used to deal with this problem, the mathematical programming and incremental techniques being the two most successfully employed. Klarbring [14] using F.E.M., and Gakwaya and Lambert [15] using B.E.M. have employed the first option, requiring the approximation of the friction law. References [3] to [5] are representative of the incremental technique.

In this paper, due to the type of problem under study (conforming case), no incremental procedure is required, the load being applied in one step once the correct partition has been found. The problem is formulated directly establishing the contact conditions (section 2) without simplifications and solving the nonlinear system of equations derived from the application of B.E.M. (sections 3 and 4), using a Newton-Raphson procedure described in section 5. A representative example (punch on elastic foundation) is solved in section 6.

2.- The Problem

The problem involves two linear elastic solids, A and B, occupying respectively the domains D^A and D^B in $\mathbb{R}^3$. An orthonormal coordinate system Ox_1,x_2,x_3, like the one represented in Figure 1-(a), will be used to refer geometrically to both domains (D^K; K=A,B). Displacements and stress vectors associated to the points of both boundaries (S^K; K=A,B), will, however, be represented in a local coordinate system -0123- also show in Figure 1-(a) (Body B). Axis "1" taken as positive following the outward normal to the boundary. The other two directions can be taken in any position inside the normal plane to direction "1", although the two systems will be taken relatively, for a point belonging to both boundaries, as is suggested in Figure 1-(b), in order to facilitate the imposition of the contact conditions.

The boundaries will be considered divided into complementary zones S_C and S^K_L, S_C being the contact surface and S^K_L being the part of the boundary free of contact. Therefore :

$$S^K = S_C \cup S^K_L \qquad (K=A,B) \qquad (1)$$

The zone free of contact will be considered subdivided, in general, into three parts : S^K_{LU} , S^K_{LT} and S^K_{LUT}. It is required that :

$$S^K_L = S^K_{LU} \cup S^K_{LT} \cup S^K_{LUT} \qquad (K=A,B) \qquad (2)$$

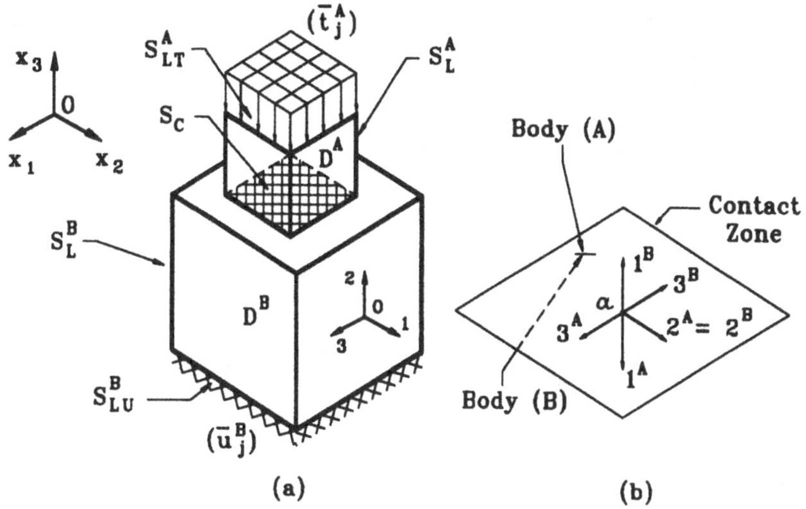

Figure 1 : Problem definition

The displacement vector is known at points belonging to S^K_{LU}, while the stress vector is known at points belonging to S^K_{LT}. S^K_{LUT} includes points with mixed boundary conditions. It is assumed that the classical equations of the Linear Elasticity Theory are satisfied at D^K, S^K_{LU}, S^K_{LT} and S^K_{LUT}.

If the local Coulomb friction law is assumed, the contact zone S_C will include points (λ) in adhesion, and points (π) where a relative displacement between both bodies can take place. It is necessary, therefore, to consider S_C partitioned into two zones, one of adhesion (S_{CA}) and the other sliding (S_{CD}) :

$$S_C = S_{CA} \cup S_{CD} \tag{3}$$

The friction law, in the stress space, is shown in Figure 2-(a), were λ represents the stress state of a pair of points in adhesion, while π represents that of a pair of sliding points.

At a pair of sliding points the following identity is satisfied :

$$|t_\tau(\pi)| = \left(t_2^2(\pi) + t_3^2(\pi) \right)^{\frac{1}{2}} = \mu \ |t_1(\pi)| \tag{4}$$

At points in adhesion, the components of stress vector must satisfy :

$$|t_\tau(\pi)| = \left(t_2^2(\lambda) + t_3^2(\lambda) \right)^{\frac{1}{2}} < |\mu \ t_1(\lambda)| \tag{5}$$

μ being the friction coefficient, t_1 the contact pressure and t_2 and t_3 the local components of the tangential stress.

The tangential stress has, at pair of points in S_{CD}, the same orientation as the relative displacement although opposite direction. Thus, with reference to Figure 2-(b), it can be established that :

$$\cos w(\pi) = -\frac{V_2(\pi)}{V_e(\pi)} = \frac{-\left(u_2^A(\pi) - u_2^B(\pi) \right)}{\left(\left(u_2^A(\pi) - u_2^B(\pi) \right)^2 + \left(u_3^A(\pi) + u_3^B(\pi) \right)^2 \right)^{1/2}} \tag{6}$$

$$\text{sen } w(\pi) = -\frac{V_3(\pi)}{V_e(\pi)} = \frac{-\left(u_3^A(\pi) + u_3^B(\pi) \right)}{\left(\left(u_2^A(\pi) - u_2^B(\pi) \right)^2 + \left(u_3^A(\pi) + u_3^B(\pi) \right)^2 \right)^{1/2}} \tag{7}$$

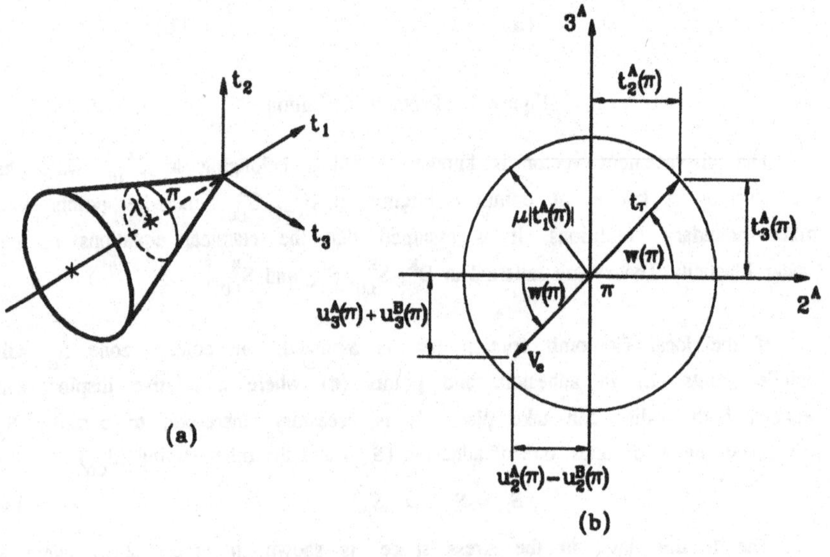

Figure 2 : Local friction law.

The following contact conditions appear at all points of the contact zone :

$$u_1^A(\alpha) + u_1^B(\alpha) = 0 \tag{8}$$

$$t_1^K(\alpha) \leq 0 \qquad (K = A,B) \tag{9}$$

$$t_1^A(\alpha) - t_1^B(\alpha) = 0 \tag{10}$$

$$t_2^A(\alpha) + t_2^B(\alpha) = 0 \tag{11}$$

$$t_3^A(\alpha) - t_3^B(\alpha) = 0 \tag{12}$$

Equation (8) expresses the absence of interpenetrations and separations at the points of the contact zone. Equation (9) states that the normal components of the stress vector must be pressures. Finally, Equations (10), (11) and (12) express the equilibrium of the stresses of both bodies.

Additionally, at points of the adhesion zone :

$$u_2^A(\lambda) - u_2^B(\lambda) = 0 \tag{13}$$

$$u_3^A(\lambda) + u_3^B(\lambda) = 0 \tag{14}$$

Finally, the following two equations, representing the law of friction adopted, must be satisfied at the sliding zone.

$$t_2^A(\pi) = - \mu\, t_1^A(\pi) \cos w(\pi) \tag{15}$$

$$t_3^A(\pi) = - \mu\, t_1^A(\pi) \operatorname{sen} w(\pi) \tag{16}$$

Although partitions (1), for the kind of problem studied, and (2) can be established a priori, this cannot be done for partition (3), because the adhesion and sliding zones are initially unknown.

3.- Boundary Integral Formulation

As is well known, working from the Reciprocity Theorem and using the Kelvin fundamental solution of the Navier equation, the Somigliana Identity is reached, the displacements at any point having the following expression :

$$C_{ij}^K(\xi)\, u_j^K(\xi) = \int_{S^K} U_{ij}^K(\xi,\gamma)\, t_j^K(\gamma)\, dS^K(\gamma) - \int_{S^K} T_{ij}^K(\xi,\gamma)\, u_j^K(\gamma)\, dS^K(\gamma) + $$

$$+ \int_{D^K} U_{ij}^K(\xi,\eta)\, X_j^K(\eta)\, dD^K(\eta) \tag{17}$$

$$i,j = x_1, x_2, x_3 \;;\; \eta \in D^K \;;\; \gamma \in S^K \;;\; \xi \in S^K \;;\; K = A,B$$

The tensor $C_{ij}^K(\xi)$ is usually called the free term of the integral equation. $T_{ij}^K(\xi,\gamma)$ and $U_{ij}^K(\xi,\eta)$ represent, respectively, the stresses and displacements corresponding to the Kelvin fundamental solution. Their general expressions can be found, for instance in Brebbia, Telles and Wrobel [16].

If the effect of the body forces is negligible compared with the effect of the boundary actions, which is a common situation, especially in contact problems, Equation (17) represents an integral expression extended only on the boundary.

It is advisable, in order to facilitate the application of the boundary and contact conditions, to transform Equation (17) into the local coordinate system previously established. Thus, if $\beta_{mj}^K(\gamma)$ is the transformation tensor between local and global coordinates, two new tensors can be defined :

$$G_{ij}^K(\xi,\gamma) = U_{im}^K(\xi,\gamma)\, \beta_{mj}^K(\gamma) \tag{18}$$

$$H_{ij}^K(\xi,\gamma) = T_{im}^K(\xi,\gamma)\, \beta_{mj}^K(\gamma) \tag{19}$$

now expressing symbolically the boundary integral Equation (17) in the form :

$$\int_{S^K} G^K_{ij}(\xi,\gamma)\, t^K_j(\gamma)\, dS^K(\gamma) - \int_{S^K} H^{*K}_{ij}(\xi,\gamma)\, u^K_j(\gamma)\, dS^K(\gamma) = 0 \tag{20}$$

$i,j = 1,2,3$ (Local coordinate system at point γ) ; $\xi,\gamma \in S^K$; $K = A,B$

The free term has been implicitly considered in Equation (20), in order to reduce the remaining equations, in the term $H^{*K}_{ij}(\xi,\gamma)$.

Assuming the correct partitions of the boundaries given by Equations (2), (3) and (7), and considering the contact conditions (Equations (8) to (16)), Equation (20) applied to a point (ξ) of the boundary S^A is :

$$\begin{aligned}
&\int_{S^A_L} G^A_{ij}(\xi,\gamma)\, t^A_j(\gamma)\, dS^A(\gamma) - \int_{S^A_L} H^{*A}_{ij}(\xi,\gamma)\, u^A_j(\gamma)\, dS^A(\gamma) + \int_{S_{CA}} G^A_{ij}(\xi,\lambda)\, t^A_j(\lambda)\, dS^A(\lambda) - \\
&- \int_{S_{CA}} H^{*A}_{ij}(\xi,\lambda)\, u^A_j(\lambda)\, dS^A(\lambda) + \int_{S_{CD}} G^A_{i1}(\xi,\pi)\, t^A_1(\pi)\, dS^A(\pi) - \int_{S_{CD}} H^{*A}_{ij}(\xi,\pi)\, u^A_j(\pi)\, dS^A(\pi) - \\
&- \int_{S_{CD}} G^A_{i2}(\xi,\pi)\, \mu\, t^A_1(\pi) \cos w(\pi)\, dS^A(\pi) - \int_{S_{CD}} G^A_{i3}(\xi,\pi)\, \mu\, t^A_1(\pi) \operatorname{sen} w(\pi)\, dS^A(\pi) = 0
\end{aligned} \tag{21}$$

$i,j = 1,2,3$; $\gamma \in S^A_L$; $\xi \in S^A$; $\lambda \in S_{CA}$; $\pi \in S_{CD}$

and applied to a point (ξ) of the boundary S^B is :

$$\begin{aligned}
&\int_{S^B_L} G^B_{ij}(\xi,\gamma)\, t^B_j(\gamma)\, dS^B(\gamma) - \int_{S^A_L} H^{*B}_{ij}(\xi,\gamma)\, u^B_j(\gamma)\, dS^B(\gamma) + \int_{S_{CA}} G^B_{i1}(\xi,\lambda)\, t^A_1(\lambda)\, dS^A(\lambda) - \\
&- \int_{S_{CA}} G^B_{i2}(\xi,\lambda)\, t^A_2(\lambda)\, dS^B(\lambda) + \int_{S_{CA}} G^B_{i3}(\xi,\lambda)\, t^A_3(\lambda)\, dS^B(\lambda) + \int_{S_{CA}} H^{*B}_{i1}(\xi,\lambda)\, u^A_1(\lambda)\, dS^B(\lambda) - \\
&- \int_{S_{CA}} H^{*B}_{i2}(\xi,\lambda)\, u^A_2(\lambda)\, dS^B(\lambda) + \int_{S_{CA}} H^{*B}_{i3}(\xi,\lambda)\, u^A_3(\lambda)\, dS^B(\lambda) - \int_{S_{CD}} H^{*B}_{i2}(\xi,\pi)\, u^B_2(\pi)\, dS^B(\pi) - \\
&- \int_{S_{CD}} H^{*B}_{i3}(\xi,\pi)\, u^B_3(\pi)\, dS^B(\pi) + \int_{S_{CD}} H^{*B}_{i1}(\xi,\pi)\, u^A_1(\pi)\, dS^B(\pi) + \int_{S_{CD}} G^B_{i1}(\xi,\pi)\, t^A_1(\pi)\, dS^B(\pi) + \\
&+ \int_{S_{CD}} G^B_{i2}(\xi,\pi)\, \mu\, t^A_1(\pi) \cos w(\pi)\, dS^B(\pi) - \int_{S_{CD}} G^B_{i3}(\xi,\pi)\, \mu\, t^A_1\, \operatorname{sen} w(\pi)\, dS^B(\pi) = 0
\end{aligned} \tag{22}$$

$i,j = 1,2,3$; $\gamma \in S^B_L$; $\xi \in S^B$; $\lambda \in S_{CA}$; $\pi \in S_{CD}$

There are three unknowns at each point of S^K_L, it being possible to write three integral equations for them. Six unknowns are associated to each point of S^A_C ($u^A_1(\lambda)$, $u^A_2(\lambda)$, $u^A_3(\lambda)$, $t^A_1(\lambda)$, $t^A_2(\lambda)$, $t^A_3(\lambda)$), six integral equations now being applicable for them ((21) and (22)). However, although each point of S_{CD} also has six unknowns associated to it ($u^A_1(\pi)$, $u^A_2(\pi)$, $u^A_3(\pi)$, $t^A_1(\pi)$, $u^B_2(\pi)$, $u^B_3(\pi)$), the sliding direction is also unknown. The same six equations as for a point of S_{CA} can now also be applied. Consequently, if partition (3) and the sliding direction at S_{CD} are correctly imposed the problem will be determined and linear.

4.- Discretization of the Problem

The boundaries will be approximated by planar triangles. Displacements and stresses are assumed to be constant over them, their value being associated to the baricenter.

N^A and N^B will represent the number of elements used to model S^A and S^B respectively. Accordingly N^A_L, N^B_L, N_C, N_{CA} and N_{CD} will represent the number of elements associated to S^A_L, S^B_L, S_C, S_{CA} and S_{CD} ($N_C = N_{CA} + N_{CD}$). The contact zone will be discretized in exactly the same manner for both bodies (same number of elements and same size for the elements to be in contact). This simplifies the application of the contact condition and is in accordance with the hypothesis of small displacements.

The integrations that appear in Equations (21) and (22) are, after the discretization, extended over the triangles. They are performed analytically (Cruse [17]) or numerically (Gauss quadrature), depending on whether or not the node where the equation is applied belongs to the element where the integration is being performed.

If Equations (21) and (22) are respectively applied to the baricenters of the triangles of the discretized bodies A and B, a system of $3N^A+3N^B$ equations is obtained. The problem, as has previously been commented presents, $3N^A$ unknowns associated to each node of S^K_L and six to the nodes of S_{CA} and S_{CD}. The total number of unknowns will be $3(N^A+N^B)+6N_C$, which is the same as the number of equations. However, the system is nonlinear if sliding zones arise, because G^K_{i2} and G^K_{i3} are nonlinearly dependent on other unknowns associated to the point ($u^A_2(\pi)$, $u^A_3(\pi)$, $t^A_1(\pi)$, $u^B_2(\pi)$, $u^B_3(\pi)$). The system is denoted by :

$$F_i(\mathbf{x}) = 0 \qquad (i = 1, \ldots , 3xN^A + 3xN^B) \tag{23}$$

where vector **x** includes the unknowns of the problem.

5.- Numerical Solution

The steps in the numerical solution are the following :

- Initial definition of the adhesion and sliding zones, assuming a sliding direction for the nodes of this zone.
- Calculation and resolution of the system of equations represented by (23)
- Checking of condition (5) for every node of S_{CA}. If all the nodes of S_{CA} satisfy this condition the final and correct solution has been reached. If the condition is not satisfied at several nodes of S_{CA} the problem must be recalculated considering these nodes as belonging to S_{CD}.

The scheme followed requires the following explanations. It is advisable to estimate S_{CD} in excess, since there is no way to detect incompatibilities in the sliding zone, due to the fact that the formulation implies that the tangential stress is opposed to the relative displacement, energy always being dissipated. Although the computational efficiency depends on the initial estimations, reasonably quick convergence is obtained taking $S_{CA} = S_C$ and taking as possible sliding directions those obtained after the solution of a frictionless case once a point has been detected to belong to the sliding zone.

The system of equations has been solved using a Newton-Raphson procedure with relaxation. Each of the equations of the system (23) can be expanded in a Taylor series :

$$F_i(x+\delta x) = F_i(x) + \sum_{j=1}^{N} \left[\frac{\partial F_i}{\partial x_j} \right]_X \delta x_j + O(\delta x^2) = 0 \qquad (24)$$

If the terms δx^2 and higher are not considered, the correction of the solution for each iteration is obtained from (24), solving the following linear system of equations :

$$\sum_{j=1}^{N} \left[\frac{\partial F_i}{\partial x_j} \right]_X \delta x_j = - F_i(x) \qquad (25)$$

Once system (25) has been solved, a new solution is estimated in the form :

$$x^{NEW} = x^{OLD} + a\ \delta x \qquad (26)$$

where a is a relaxation parameter ($a > 0.0$). Initially it is taken as 1.0, taking smaller values (always positive) if convergence is not reached. The error parameter used in this paper to define the convergence is :

$$E_F = \sum_{i=1}^{N} | F_i(x^{NEW}) | \qquad (27)$$

The convergence of the procedure has been found to be strongly dependent on the number of unknowns in the problem. This being so, the reduction of this number using a condensation technique, especially applicable to contact problems (París and Garrido [8]), produces a drastic reduction in computation time. Other factors that affect convergence, which will be commented on in the example solved, are the value of the friction coefficient and the size and discretization of the adhesion and sliding zones.

6.- Punch on Elastic Foundation

The problem shown in Figure 1 has been solved, as the classic example of conforming contact, varying the friction coefficient. The punch and the foundation have been assumed to be of the same material ($E = 2.1 \times 10^5$ MPa and $\nu = 0.2$). Due to the symmetry, only a quarter of the geometry need be considered, the dimensions being represented in Figure 3-(a). The load acting over the top of the punch is $p_o = 0.1 \times 10^5$ MPa.

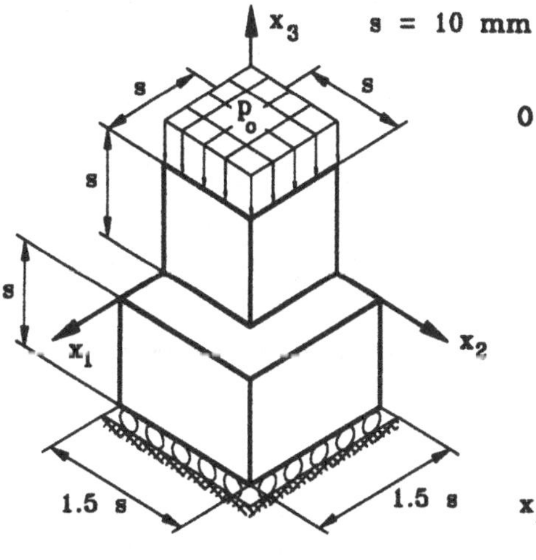

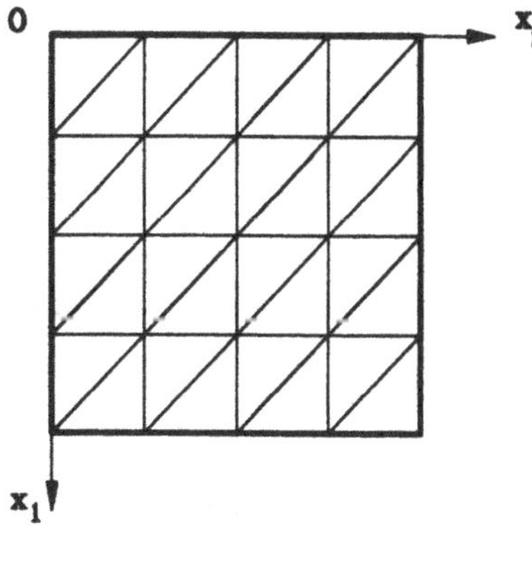

a) Geometry. b) Discretization of the contact zone.

Figure 3 : Punch on elastic foundation.

All the results presented in this paper have been obtained using 56 elements (168 d.o.f.) to discretize the boundary of the punch and 84 (254 d.o.f.) to discretize the foundation, 32 of them corresponding to the contact zone, Figure 3-(b) (the planes of symmetry do not need to be discretized, Watson [18]). The problem is modelled with 420 d.o.f., it being possible to reduce it to 192 using the condensation procedure previously referred to.

All the nodes of the contact zone are initially assumed to be in adhesion. Less than five iterations, in the cases run, have been required to reach the final solution. The nodes where Equation (5) is not satisfied are passed to the sliding zone, after each iteration. Figures 4, 5 and 6 represent the distribution of normal and tangential stresses as well as the relative displacements and their direction for three different values of μ (0.09, 0.15 and 0.20). Full adhesion appears for greater values of μ = 0.22.

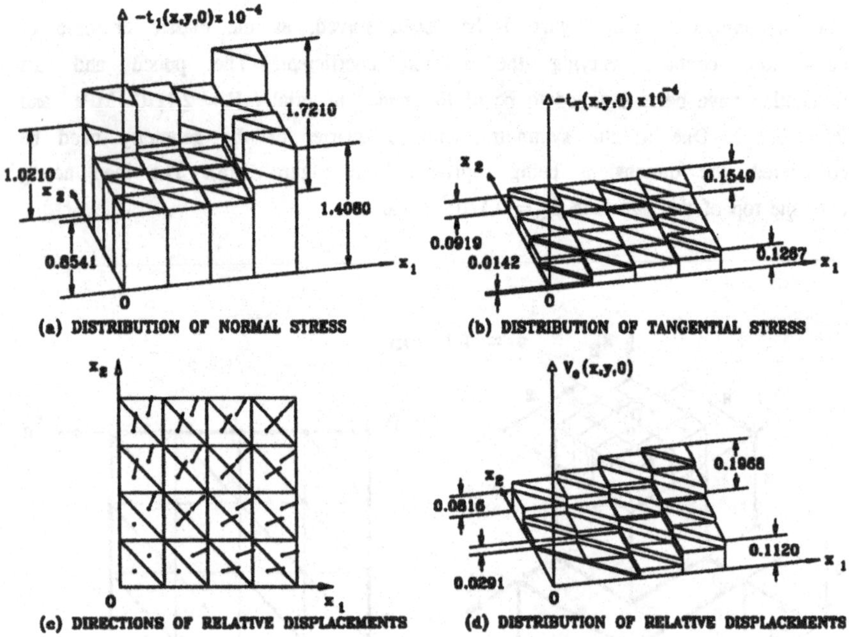

Figure 4 : Results obtained for μ = 0.09

As was expected, the size of the adhesion zone increases with the friction coefficient, whereas the relative displacements are smaller. The distribution of the normal stresses remains quite unaffected by a variation in μ.

Table 1 gives several values of interest. The first column represents the number of elements that remain in adhesion for different values of μ when the condensed system is solved. The second column represents the number of iterations required in the Newton-Raphson procedure (on the final compatible division of the contact zone), to reach convergence (in all the cases $E_F \leq 1.0\times10^{-6}$ has been taken. Finally, the third column represents the minimum value taken by the relaxation parameter throughout the previous iterations.

μ	N_{CA}	Newton-Raphson Num. of iterations	a_{MIN}
0.20	28	27	0.5
0.15	14	27	0.5
0.09	4	3	1
0.05	0	2	1

Table 1 : Convergence of the Newton-Raphson procedure

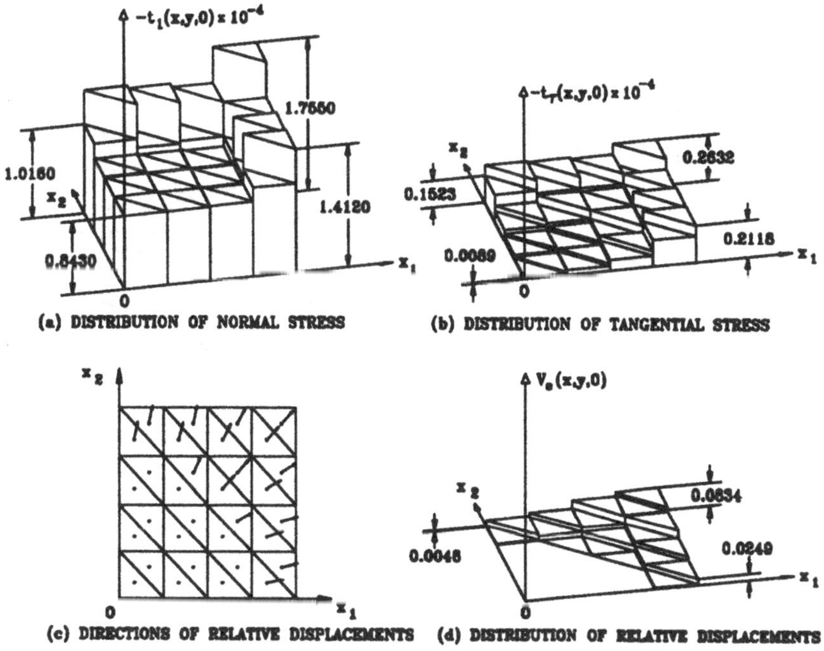

Figure 5 : Results obtained for μ = 0.15

It can be observed that the procedure requires more iterations as the friction coefficient increases, which in turn procedures a greater size of the adhesion zone. This result, at first sight contradictory because the nonlinearity is associated to the sliding zone, may be explained by taking into account that the relative displacements are smaller when the friction coefficient increases.

When the problems are solved without condensation, the values given in Table 1 increase significantly. Thus, 50 iterations and minimum relaxation parameter of 0.333 are required for μ = 0.09.

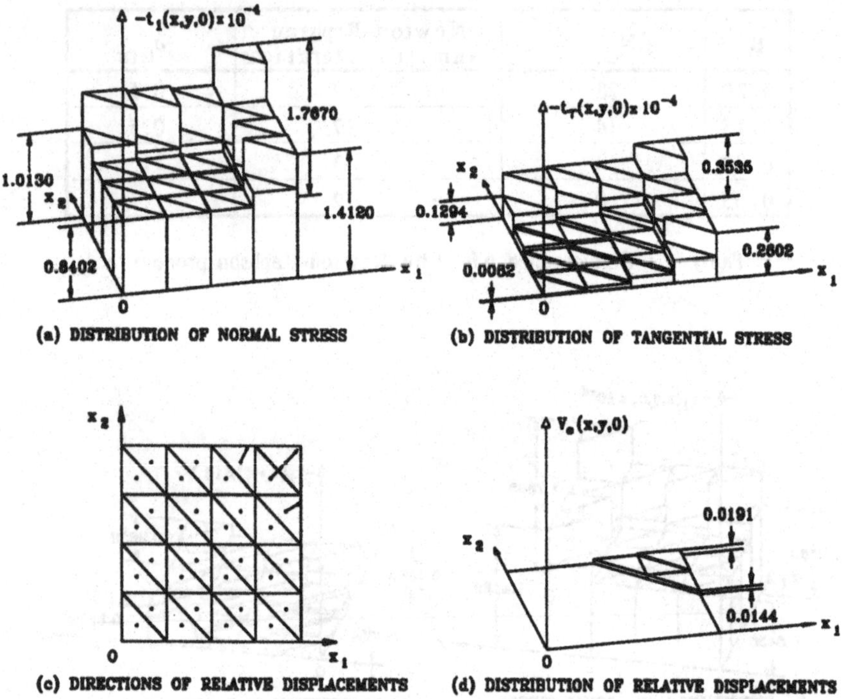

Figure 6 : Results obtained for $\mu = 0.20$

Finally, Figure 7 shows the evolution of the final contact zone in adhesion versus the friction coefficient.

7.- Conclusions

The problem of conforming contact for the three dimensional case with Coulomb friction has been formulated. The B.E.M., which again appears to be the best option to deal with contact problems, has been used to establish a numerical formulation of the configuration studied.

The determination of the correct partition of the contact zone has been performed starting from a full adhesion situation and establishing whether the stresses state of a given point continues inside the friction cone. If this is not the case the point is passed to the sliding zone. This procedure has led to a compatible solution in all the cases studied.

As regards the solution procedure of the non-linear system of equations, the Newton-Raphson method employed has led, in all the cases tested, to a solution considered satisfactory if a relaxation parameter is used. When this parameter is not used convergence is not reached for some combinations of discretization and value of the friction coefficient.

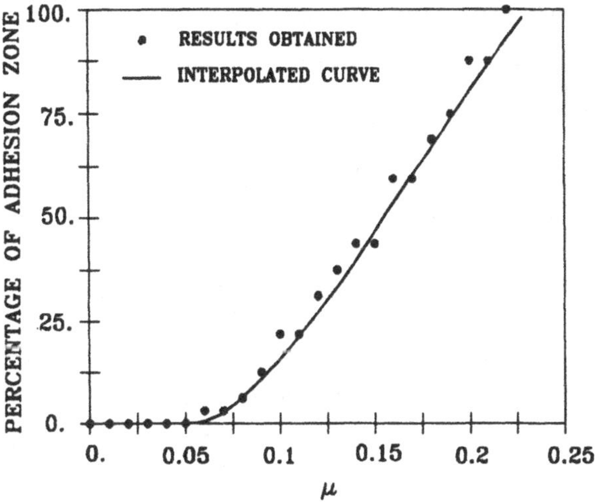

Figure 7 : Percentage of adhesion zone versus friction coefficient

The condensation of the degrees of freedom not involved in the contact appears to be a significant factor in obtaining good ratios of convergence. It has also been detected that when the sliding zone is small, and consequently the relative displacements between the two bodies tend to zero, convergence is very slow.

The solution procedure employed does not require the discretization of the law friction, which is necessary when a mathematical programming technique is applied.

The solution procedure presented requires to be extended to more general problems where receding or advancing situation can occur.

References

1.- G.M.L. Gladwell, *Contact Problems in the Classical Theory of Elasticity*, Sijthoff and Noordhoof, 1980.

2.- K. Johnson, *Contact Mechanics*, Cambridge University Press, 1985.

3.- N. Okamoto and M. Nakazawa, Finite Element Incremental Contact Analysis with Various Frictional Conditions, *International Journal of Num. M. Eng.*, **14**, 337-357, 1979.

4.- B. Fredriksson, Finite Element Solution of Surface Non - Linearities in Structural Mechanics with Special Emphasis to Contact and Fracture Mechanics Problems, *Computers & Structures*, **6**, 281-290, 1976.

5.- B. Torstenfelt, Contact Problems with Friction in General Purpose Finite Element Computer Programs, *Computers & Structures*, **16**, 487-493, 1983.

6.- T. Anderson, B. Fredrikson and B.J. Persson, The Boundary Element Method Applied to Two-Dimensional Contact Problems, *New Developments in Boundary Element Methods*, Computational Mechanics Publications-Springer Verlag, 1980.

7.- T. Anderson, The Boundary Element Method Applied to Two-Dimensional Contact Problems with Friction, *Boundary Element Method*, Computational Mechanics Publications-Springer Verlag, 1981.

8.- F. París and J.A. Garrido, On the Use of Discontinous Elements in Two-Dimensional Contact Problems, *Boundary Elements VII*, Computational Mechanics Publications-Springer Verlag, 1985.

9.- F. París and J.A. Garrido, An Incremental Procedure for Friction Contact Problems with the Boundary Element Method, *Engineering Analysis with Boundary Elements*, **6**, 202-213, 1989.

10.- F. París and J.A. Garrido, Friction Multicontact Problems with B.E.M., *Boundary Elements X*, Computational Mechanics Publications-Springer Verlag, 1988.

11.- J.A. Garrido, A. Foces and F. París, B.E.M. applied to Receding Contact Problems with Friction, *Mathematical and Computer Modelling*, **15**, 143-153, 1991.

12.- J.A. Garrido, A. Foces and F. París, Three Dimensional Conforming Frictionless Contact using BEM, *Advances in Boundary Elements*, **3**, 135-149, Computational Mechanics Publications-Springer Verlag, 1989.

13.- F. París, J.A. Garrido and A. Foces Application of Boundary Element Method to Solve Three Dimensional Elastic Contact Problems. Submitted to *Computers & Structures*.

14.- A. Klarbring, A Mathematical Programming Approach to Three-Dimensional Contact Problems With Friction, *Computers Methods in Applied Mechanics and Engineering*, **59**, 175-200, 1986.

15.- A. Gakwaya and D. Lambert, A Boundary Element and Mathematical Programming Approach for Firctional Contact Problems, *Advances in Boundary Elements*, **3**, 163-179, Computational Mechanics Publications-Springer Verlag, 1989.

16.- C.A. Brebbia, J.C.F. Telles and L.C. Wrobel, *Boundary Element Techniques*, Springer-Verlag, Berlin, Heideberg, 1984.

17.- T.A. Cruse, Application of the Boundary Integral Equation Method to Three Dimensional Stress Analysis, *Computers and Structures*, **3**, 509-527, 1973.

18.- J. Watson , Advanced Implementation of the Boundary Element Method for Two and Three - Dimensional Elastostatics, *Developments in Boundary Element Methods*, **1**, Applied Science Publications, 1979.

A Direct Solution to the BEM and Mathematical Programming Approach for Frictional Constant Contact Problems

X.-A. Kong, A. Gakwaya, A. Cardou
Department of Mechanical Engineering,
Laval University, Quebec, G1K 7P4, Canada

Abstract

A direct solution procedure for the parametric linear complementarity problem (PLCP) deduced from quasi-static frictional contact studies is presented. This direct method is suitable for constant contact area, both in two dimensional and three dimensional cases. The Coulomb's friction law is used, and for both two and three dimensional analysis, it is represented by a piecewise linear approximation polytope. Because of the direct solution of the PLCP, no iteration loop is needed, and the convergence of the algorithm can be guaranteed.

Introduction

The present article is a continuation of our previous work [1]. As has been pointed out, the frictional contact problem is highly non linear not only because of the friction phenomenon itself, but also because of the changing boundary conditions. Thus the state of contacting system is inherently path-dependent due to the unilateral nature of the frictional contact and appropriate solution algorithms have to be used. According to Kalker [5], one can classify contact problems into three major categories: (1) The normal contact problem; (2) The tangential contact problem with Coulomb like friction; (3) The combined frictional problem in which the tangential stress bounds depend on the normal pressure. This problem is still open according to Kalker [5]. However, this was solved recently in [3] for two-dimensional case. The details of the implementation in [3] are not however clear. In our previous paper [1], we have presented an extension of the work described in [3] to a more general three dimensional situation using a linear complementarity formulation (LCP) based on the BEM discretization approach of the elastic contact problem with Coulomb like friction law. This paper is aimed at presenting some preliminary results of latter formulation for constant contact zone problems.

Generally, we have to solve a whole PLCP problem for every load increment to determine the changing state of contact. However, if the contact surface is constant during the loading history (so called 'constant contact'), the resolution procedure to the PLCP becomes rather simple. And it is the purpose of this work to present an efficient algorithm for solving this PLCP by a "direct method".

The layout of the paper is then as follows. In sections 2, we summarize the main points of BEM-PLCP frictional contact approach given in [1]. In sections 3, the direct method for constant contact cases is studied. Applications of this algorithm to some typical numerical examples are presented in section 4.

BEM-PLCP formulations for frictional elastic contact problems

In this part, we will summarize some main points of BEM-PLCP approach for frictional elastic contact problems, especially the solution algorithm. For detailed description, the reader is referred to [1].

1. Boundary element formulation for elastic contact problems

Let us consider a potential contact zone of two linear elastic bodies A and B. Parts S_c^A and S_c^B of their respective boundaries S^A and S^B qualify as the potential contact surface S_c. The boundary S^A and S^B consist of three disjoint parts $S_u^{A(B)}$, $S_t^{A(B)}$ and $S_c^{A(B)}$ such that on S_u, the displacements u_i have prescribed values $u_i^{(h)}=\bar{u}_i^{(h)}$ (h=A or B) and on S_t, the surface tractions t_i have their prescribed values $\bar{t}_i$. In the potential contact surfaces, we assume their normal vectors to be almost parallel and the distance between the surfaces is of the order of the displacements. As load is applied, parts of the potential contact surfaces come into real contact and contact stresses develop. The separation function g is defined by

$$\gamma(\xi,\eta)=\zeta^B(\xi,\eta)-\zeta^A(\xi,\eta) \quad \text{(normal gap)} \tag{1}$$

where $(\xi-\eta-\zeta)$ is the local coordinates system, ζ^h(h=A,B) the local shape of body A or B.

The impenetrability condition

$$u^B n^B - (u^A n^A) n^B + \gamma(\xi,\eta) n^B \geq 0 \tag{2}$$

can then be expressed as

$$D_N = (u^B - u^A)\, n + q \geq 0 \tag{3}$$

where q is the resultant rigid body displacement in normal direction.

The boundary element approach will provide equations relating contact boundary displacements and tractions with prescribed surface tractions and displacements. Standard boundary element formulation including contact conditions then lead to the following linear BEM system for each body (after BE discretization):

$$\begin{bmatrix} H_u^{(h)} & H_v^{(h)} \end{bmatrix} \begin{Bmatrix} u^{(h)} \\ v^{(h)} \end{Bmatrix} = \begin{bmatrix} G_t^{(h)} & G_p^{(h)} \end{bmatrix} \begin{Bmatrix} t^{(h)} \\ p^{(h)} \end{Bmatrix} \qquad h = A,B \tag{4}$$

where v and p represent the contact displacements and tractions respectively. The flexibility matrix is obtained by premultiplying the inverse of matrix $H^{(h)}$ with matrix $G^{(h)}$. This leads to:

$$\begin{Bmatrix} x \\ v \end{Bmatrix}^{(h)} = \begin{bmatrix} F_{\bar{t}\bar{t}} & F_{\bar{t}p} \\ F_{p\bar{t}} & F_{pp} \end{bmatrix}^{(h)} \begin{Bmatrix} \bar{t} \\ p \end{Bmatrix}^{(h)} \tag{5}$$

where $[F]^{(h)}$ is flexibility matrix for body h and has been partitioned according to the following scheme: $x^{(h)}$ and $\bar{t}^{(h)}$ represent unknown and known data outside the potential contact zone respectively while $v^{(h)}$ and $p^{(h)}$ represent, as before, the unknown contact displacements and tractions respectively [1]. Equation (5) provides immediately a desired relation between the contact displacements and the contact tractions and external loading $\bar{t}^{(h)}$. For practical contact analysis, the global contact variables are projected onto local normal and tangential space. After this transformation and considering global equilibrium conditions, we arrive at the following fundamental equations of boundary element contact analysis [1]:

$$\begin{Bmatrix} W_N \\ W_T \end{Bmatrix} = \begin{bmatrix} F_{nn} & F_{nt} \\ F_{tn} & F_{tt} \end{bmatrix}^{(h)} \begin{Bmatrix} P_N \\ P_T \end{Bmatrix} + \begin{bmatrix} R_{ND} \\ R_{TD} \end{bmatrix}^{(B)} t^{(B)} - \begin{bmatrix} R_{ND} \\ R_{TD} \end{bmatrix}^{(A)} t^{(A)} \tag{6}$$

where W_N and W_T are relative normal and tangential contact displacements between opposite node pairs. The explicit forms of matrix [F], $[R_{ND}]$ and $[R_{TD}]$ can be found in [1].

2. Constitutive equations for contact and friction laws

In the potential contact boundary S_c, we assume the classical unilateral contact condition (Signorini's type) and Coulomb's law of friction to hold. The contact criterion (equation 2) states that: if the final normal gap $D_N>0$, then $P_N=0$, or either if $D_N=0$, then $P_N>0$ at each point of S_c. This means:

$$N_{cN}(P_N) : \{v \in R,\ v=D_N,\ D_N \geq 0,\ P_N \geq 0,\ D_N^T P_N = 0\} \tag{7}$$

and satisfies together with P_N the orthogonality property or complementary condition of mathematical programming. In the above, compressive contact pressures are taken as positive and are zero only when the bodies loose contact.

In three dimension, the isotropic Coulomb's law of friction is defined locally by the set:

$$C_T(P_N) = \{ P_N \in R,\ P_T \in R^2 : \Phi = \| P_T \| - \mu P_N \leq 0 \} \tag{8}$$

The friction conditions are: if $|P_T|<\mu P_N$ then $W_{Ti}=0$, and if $|P_T|=\mu P_N$ then $W_{Ti}=-\lambda\partial\Phi/\partial P_{Ti}$, $\lambda \geq 0$, $\Phi \leq 0$, $\lambda^T\Phi=0$. We introduce the following piecewise linear approximation of the friction cone:

$$W_{T_i}^j = \Sigma(\lambda_k^j \frac{\partial \phi_k^j}{\partial P_{T_i}}) + g_{T_i}, \qquad i=1,2 \tag{9}$$

$$\Phi_k(P_T,P_N) = [\cos\alpha_k,\ \sin\alpha_k,\ \mu]\ (P_{T1}\ \ P_{T2}\ \ P_N)^T \tag{10}$$

with the complementary conditions:

$$\lambda_k \geq 0,\ \ \Phi_k \leq 0,\ \ \lambda_k\, \Phi_k = 0 \tag{11}$$

where Φ_k, k=1,...,n are affine functions of $P_T{}^k$ and $P_N{}^k$, n the number of approximating planes and α_k, k=1,2,...,n represent the orientation of each plane in the piecewise linear approximation of the normal cone $C_T(P_N)$. We take n=8 and the orientations $\alpha_k = k\pi/4$, k=1,...,8. For two dimensional case, n=2, and:

$$\Phi_1 = P_T + \mu P_N,\quad \Phi_2 = P_T - \mu P_N \tag{12}$$

$$W_T = \lambda_1 - \lambda_2 \tag{13}$$

The global description of friction law valid simultaneously for all contact node pairs can now be deduced. Introducing following vectors: $\{W_T\}$, $\{P_T\}$, $\{g_t\}$ of dimension 2N (N: number of contact node pairs), $\{\lambda\}$, $\{\Phi\}$ of dimension nN where n is the number of planes approximating the Coulomb's friction cone as before, equations (9) and (10) then becomes:

$$\{W_T\} = [G_2]^T\{\lambda\} \tag{14}$$

$$\{\Phi\} = [G_1]\ \{P_N\} + [G_2]\ \{P_T\} \tag{15}$$

$$\{\Phi\} \leq 0,\ \ \{\lambda\} \geq 0,\ \ \{\Phi\}\{\lambda\}^T = 0 \tag{16}$$

where

$$[G_1] = \text{diag}\ \{\ [\partial_N\Phi_1{}^{(1)}, \dots, \partial_N\Phi_n{}^{(1)}], \dots, [\partial_N\Phi_1{}^{(N)}, \dots, \partial_N\Phi_n{}^{(N)}]\} \tag{17}$$
$$[G_2] = \text{diag}\ \{\ [\partial_T\Phi_1{}^{(1)}, \dots, \partial_T\Phi_n{}^{(1)}], \dots, [\partial_T\Phi_1{}^{(N)}, \dots, \partial_T\Phi_n{}^{(N)}]\} \tag{18}$$

and $\partial_N\Phi_\alpha{}^{(k)}$, $\partial_T\Phi_\alpha{}^{(k)}$, α=1,...,n, k= 1,... N represent the partial derivatives of slip function $\Phi_\alpha{}^{(k)}(P_T{}^{(k)}, P_N{}^{(k)})$ with respect to $P_N{}^{(k)}$ and $P_{Ti}{}^{(k)}$ respectively.

If we restrict ourselves to the case where the rigid body displacements are specified or can be so, the PLCP then takes the form [1]:

$$\{S\}_m = \{S\}_{m-1} + \tau\{R\}t' + [M]\ \{\Delta\Lambda\} \tag{19}$$
$$\{D_N\} \geq 0,\ \{P_N\}\geq 0,\ \{D_N\}^T\ \{P_N\} = 0 \tag{20}$$
$$\{\lambda\}\geq 0,\ \{\Phi\}\leq 0,\ \{\lambda\}^T\{\Phi\} = 0 \tag{21}$$

where $\{S\} = \{D_N, \Phi\}^T$, $\{\Delta\Lambda\} = \{\Delta P_N, \Delta\lambda\}^T$ and the loading history has been presented by a stepwise proportional approximation. For example, for a loading step during a time interval [0,T], the loading law is defined by:

$$t(\tau) = t_{m-1} + \tau t',\quad \tau \in [0,T] \tag{22}$$

where t_{m-1} is previous load level and t' is load increment.

The detailed description of general solution process to the PLCP of equations (19) - (21) can be found in [1]. Two important features of the general solution algorithm are: (1) For every load step, we have to check the variation in contact state and the exchange of some basic and non-basic variables may occur; (2) Matrix [M] of the LCP is neither a definite-positive nor a semi definite-positive one. Because we use a direct BEM approach, [M] is not even a symmetric matrix either. This brings some difficulties for the LCP solution.

However, in constant contact situation, a direct solution to the LCP can be found. This direct method does not demand any iteration operation, and the convergence of the algorithm can be guaranteed (refer to the solution process explained below).

Direct solution to constant frictional contact problem

Following standard practice in mathematical programming [1], the system of linear equations (19) can be written in tableau form (initial set up tableau):

		1	t	$\Delta\Lambda_I$	$\Delta\Lambda_J$
S_I	=	q_I	b_I	M_{II}	M_{IJ}
S_J		q_J	b_J	M_{JI}	M_{JJ}

(23)

The index sets I and J are defined in such a way that elements whose index is I or J are entries of vector S corresponding respectively to D_N (normal contact) and Φ (tangential contact). For constant contact, ΔS_I=0, and and we assume that $\Delta\Lambda_I = \Delta P_N$ is constant (for a given load step). Thus for one load step, the LCP becomes:

$$\{\Delta S_J\} = \{\ b_J\ \}\ \tau\ + [\ M_{JJ}]\ \{\Delta\Lambda_J\} \tag{24}$$

For convenience, we write equation (24) in standard LCP form:

$$w = q + M\,z \tag{25}$$
$$w,z > 0,\ \ w^T z = 0 \tag{26}$$

where $w=\{\Delta S_J\}$, $q=\{\ b_J\ \}\ \tau$, $M=[\ M_{JJ}]$, $z=\{\Delta\Lambda_J\}$. The explicit form of matrix M and vector q in equation (25) are [1]:

$$M = [M_{JJ}] = -[G_2][F_{TT}]^{-1}[G_2]^T \tag{27}$$

$$q = \{b_J\}\tau = [G_2][F_{TT}]^{-1}[R_{TD}]^{(h)}\{t\}^{(h)} \tag{28}$$

where matrix $[G_2]$ is defined by equation (8), $[F_{TT}]$ is a submatrix of the flexibility matrix $[F]$ (equation 6), $[R_{TD}]^{(h)}$ is either $[R_{TD}]^{(A)}$ or $[R_{TD}]^{(B)}$, depending on what body is loaded. Load vector $\{t\}^{(h)}$ is determined in the same way.

In the following, we discuss the direct solution procedure to the LCP equations (25) and (26) for two and three dimensional situations respectively.

1. Two dimensional analysis

The orientations α_k of equation(11) in two dimension are:

$$\alpha_1 = 0, \quad \alpha_2 = \pi \tag{29}$$

thus matrix $[G_2]$ of Coulomb's friction law is:

$$[G_2] = \text{diag}\ \{\ [G0], \ldots, [G0]\} \tag{30.1}$$

with

$$[G0] = \{1 \quad -1\}^T \tag{30.2}$$

where N is the number of contact node pairs, matrix M and vector q of equations (27) and (28) have dimensions 2N x 2N and 2N respectively. Using special structure of matrix $[G_2]$, equation (25) can be greatly simplified. Substituting equation (30) into equation (27), we have:

$$-M_{2i-1,2j} = M_{2i-1,2j-1} = -M_{2i,2j-1} = M_{2i,2j} = ([F_{TT}]^{-1})_{ij} \qquad i,j = 1,2, \ldots, N \tag{31}$$

Such a structure is illustrated below.

$$M = \begin{array}{c} \\ 1 \\ \vdots \\ i \\ \vdots \\ N \end{array} \begin{array}{c} \begin{array}{ccc} 1 & \quad j \quad & N \end{array} \\ \left[\begin{array}{ccc} \square & \cdots & \square \\ & \cdots & \\ \cdot & \boxed{\begin{array}{cc} [F_{TT}]^{-1}_{ij} & -[F_{TT}]^{-1}_{ij} \\ -[F_{TT}]^{-1}_{ij} & [F_{TT}]^{-1}_{ij} \end{array}} & \cdot \\ & \cdots & \\ \square & \cdots & \square \end{array} \right] \end{array} \tag{32}$$

With the properties of expressions (30) and (31), the first two equations of (25) become:

$$w_1 = q_1 + M_1 z \tag{33}$$

$$w_2 = -q_1 - M_1 z \tag{34}$$

where M_1 is the first line of matrix M. These relations give:

$$w_1 = -w_2 \tag{35}$$

but $w_1, w_2 \geq 0$, so

$$w_1 = w_2 = 0 \tag{36}$$

Similarly, for all components of vector w:

$$w_3 = w_4 = 0, \ldots, w_{2N-1} = w_{2N} = 0 \tag{37}$$

We arrive finally:

$$w = (0,0, \ldots, 0)^T \tag{38}$$

Equation (25) now takes the form

$$q + M z = 0 \tag{39}$$
$$z \geq 0 \tag{40}$$

Substituting equation (32) into equation (39), we have:

$$q' - [F_{TT}]^{-1} z' = 0 \tag{41}$$

where

$$z' = (z_1 - z_2, z_3 - z_4, \ldots, z_{2N-1} - z_{2N})^T \tag{42.1}$$
$$q' = (q_1, q_3, \ldots, q_{2N-1})^T \tag{42.2}$$

Note that the components of vector z' have very clear physical meaning:

$$z'_i = z_{2i} - z_{2i-1} = \lambda_1^{(i)} - \lambda_2^{(i)} \tag{43}$$

where $\lambda_1^{(i)}$, $\lambda_2^{(i)}$ are defined in equation (13) and z'_i is the relative slip at node pair i. With this physical explanation, we obtain from equation (41) the following solution for relative tangential displacements in contact region:

$$W_T = z' = [F_{TT}]q \tag{44}$$

we have until now described our direct solution procedure to the LCP - BEM frictional constant contact approach. After obtaining the solution of LCP by equation (44), the next step is to perform the block pivots on entire tableau (23) and prepare data for new load level [1].

2. Three dimensional analysis

For three dimensional constant contact problem, if we take the number of planes in the piecewise linear approximation of the normal cone equal to n=8, then the orientation matrix $[G_2]$ of equation (18) is:

$$[G_2] = \text{diag}\{[G0], \ldots, [G0]\} \tag{45}$$

where N is the number of contact node pairs, and

$$[G0] = \begin{pmatrix} \cos a1 & \sin a1 \\ \cos a2 & \sin a2 \\ \cdot & \cdot \\ \cdot & \cdot \\ \cdot & \cdot \\ \cos a8 & \sin a8 \end{pmatrix} = \begin{pmatrix} 1 & \frac{1}{\sqrt{2}} & 0 & \frac{1}{\sqrt{2}} & -1 & \frac{1}{\sqrt{2}} & 0 & \frac{1}{\sqrt{2}} \\ 0 & \frac{1}{\sqrt{2}} & 1 & \frac{1}{\sqrt{2}} & 0 & \frac{1}{\sqrt{2}} & -1 & \frac{1}{\sqrt{2}} \end{pmatrix}^T \tag{46}$$

To simplify the expression, let

$$[A] = -[F_{TT}]^{-1} \tag{47}$$

then matrix M in equation (27) can be written as:

$$M = [G_2]\,[A]\,[G_2]^T \tag{48}$$

or in submatrix form:

$$M = \begin{bmatrix} [G0] & & \\ & \cdots & \\ & & [G0] \end{bmatrix} \begin{bmatrix} A_{11} & A_{12} & \dots & A_{1N} \\ & \cdots & & \\ A_{N1} & A_{N2} & \dots & A_{NN} \end{bmatrix} \begin{bmatrix} [G0]^T & & \\ & \cdots & \\ & & [G0]^T \end{bmatrix} \tag{49}$$

where Aij are 2x2 matrices:

$$A_{ij} = \begin{bmatrix} a_{2i-1,2j-1} & a_{2i-1,2j} \\ a_{2i,2j-1} & a_{2i,2j} \end{bmatrix} \tag{50}$$

From equation (49):

$$M = \begin{bmatrix} [G0][A_{11}][G0]^T \dots [G0][A_{1N}][G0]^T \\ \cdots \\ [G0][A_{N1}][G0]^T \dots [G0][A_{NN}][G0]^T \end{bmatrix} \tag{51}$$

Let us analyze the first submatrix of the expression (51), we have:

$$[G0][A_{11}][G0]^T = \begin{bmatrix} A'_{11} & -A'_{11} \\ -A'_{11} & A'_{11} \end{bmatrix} \tag{52}$$

where

$$A'_{11} = \begin{pmatrix} a11 & \frac{1}{\sqrt{2}}(a11+a12) & a12 & \frac{1}{\sqrt{2}}(-a11+a12) \\ \frac{1}{\sqrt{2}}(a11+a21) & \frac{1}{2}(a11+a21+a12+a22) & \frac{1}{\sqrt{2}}(a12+a22) & \frac{1}{2}(-a11-a21+a12+a22) \\ a21 & \frac{1}{\sqrt{2}}(a21+a12) & a22 & \frac{1}{\sqrt{2}}(-a21+a22) \\ \frac{1}{\sqrt{2}}(-a11+a21) & \frac{1}{2}(-a11+a21-a12+a22) & \frac{1}{\sqrt{2}}(-a12+a22) & \frac{1}{2}(a11-a21-a12+a22 \end{pmatrix} \tag{53}$$

From the structure of equations (52) and (53), matrix M can be expressed as:

$$M = \begin{bmatrix} \begin{bmatrix} M1 \\ M2 \\ M3 \\ M4 \\ -M1 \\ -M2 \\ -M3 \\ -M4 \end{bmatrix} & 1 \\ \cdots & \cdots \\ [\quad] & N \end{bmatrix} = \begin{bmatrix} \begin{bmatrix} M1 \\ 1/\sqrt{2}\,(M1+M3) \\ M3 \\ 1/\sqrt{2}(-M1+M3) \\ -M1 \\ -1/\sqrt{2}(M1+M3) \\ -M3 \\ -1/\sqrt{2}(-M1+M3) \end{bmatrix} & 1 \\ \cdots & \cdots \\ [\quad] & N \end{bmatrix} \tag{54}$$

in which

$$Mi = (\overset{1}{(Mi1\ Mi2\ Mi3\ Mi4\ -Mi1\ -Mi2\ -Mi3\ -Mi4)}, \dots, \overset{N}{(\dots)}) \tag{55}$$

Let us now consider the load vector. Setting

$$[C] = [F_{TT}]^{-1}\ [R_{TD}]\ \{t\} \tag{56}$$

the vector of q in equation (28) then becomes:

$$q = [G_2]\ [C] \tag{57}$$

i.e.

$$q = \begin{bmatrix} [G0] & & \\ & \cdots & \\ & & [G0] \end{bmatrix} \begin{bmatrix} \begin{Bmatrix} c1 \\ c2 \end{Bmatrix} \\ \cdots \\ \begin{Bmatrix} c(2N-1) \\ c(2N) \end{Bmatrix} \end{bmatrix} = \begin{bmatrix} [G0] \begin{Bmatrix} c1 \\ c2 \end{Bmatrix} \\ \cdots \\ [G0] \begin{Bmatrix} c(2N-1) \\ c(2N) \end{Bmatrix} \end{bmatrix} \tag{58}$$

Investigating the first subvector of the right hand of equation (58), we have for the first contact node:

$$[G0] \begin{pmatrix} C1 \\ C2 \end{pmatrix} = \begin{pmatrix} c1 \\ \frac{1}{\sqrt{2}}(c1+c2) \\ c2 \\ \frac{1}{\sqrt{2}}(-c1+c2) \\ -c1 \\ -\frac{1}{\sqrt{2}}(c1+c2) \\ -c2 \\ -\frac{1}{\sqrt{2}}(-c1+c2) \end{pmatrix} = \begin{pmatrix} q1 \\ q2 \\ q3 \\ q4 \\ q5 \\ q6 \\ q7 \\ q8 \end{pmatrix} = \begin{pmatrix} q1 \\ \frac{1}{\sqrt{2}}(q1+q3) \\ q3 \\ \frac{1}{\sqrt{2}}(-q1+q3) \\ -q1 \\ -\frac{1}{\sqrt{2}}(q1+q3) \\ -q3 \\ -\frac{1}{\sqrt{2}}(-q1+q3) \end{pmatrix} \tag{59}$$

According to the above special structure of vector q and matrix M (equations (54), (55)), the LCP of equation (25) now take the form (again for the first contact node):

$$w_1 = -w_5 = q_1 + M_1 z \tag{60.1}$$
$$w_2 = -w_6 = 1/\sqrt{2}\ (q_1+q_3) + M_2 z \tag{60.2}$$
$$w_3 = -w_7 = q_3 + M_3 z \tag{60.3}$$
$$w_4 = -w_8 = 1/\sqrt{2}\ (-q_1+q_3) + M_4 z \tag{60.4}$$

From (60.1): $w_1 = -w_5$, and non-negative condition equation (26): $w_1, w_5 \geq 0$, so

$$w_1 = w_5 = 0$$

Similarly, from (60.2): $w_2 = w_6 = 0$, from (60.3): $w_3 = w_7 = 0$, from (60.4): $w_4 = w_8 = 0$. This conclusion stands for all other contact nodes, finally:

$$w = 0 \tag{61}$$

From equations (54), (59) and (61), we can conclude that in the first 8 equations of equation (25), there are only 2 independent ones:

$$q_1 + M_1 z = 0, \quad q_3 + M_3 z = 0 \tag{62}$$

Equation (62) can further be written as (see equations (52) and (53)):

$$q_1 + (a11\ a12\ \ldots\ \ldots\ )z' = 0, \qquad q_3 + (a21\ a22\ \ldots\ \ldots\ )z' = 0 \tag{63}$$

where z' is a vector of dimension 2N. The first two components of z' are:

$$z'_1 = z_1+(1\sqrt{2})z_2-(1\sqrt{2})z_4-z_5-(1\sqrt{2})z_6+(1\sqrt{2})z_8 \qquad (64.1)$$
$$z'_2 = (1\sqrt{2})z_2+z_3+(1\sqrt{2})z_4-(1\sqrt{2})z_6-z_7-(1\sqrt{2})z_8 \qquad (64.2)$$

i.e.

$$(z'_1 \;\; z'_2)^T = [G0]^T (\lambda_1 \; \lambda_2 \; \lambda_3 \; \lambda_4 \; \lambda_5 \; \lambda_6 \; \lambda_7 \; \lambda_8)^T \qquad (65)$$

where [G0] is defined by equation (46), vector z is identical to vector λ of equation (14). We can see from equations (45), (46) and (14) that z'_1 and z'_2 are the relative tangential displacements of the first contact node pair:

$$(W_{T1}{}^1 \;\; W_{T2}{}^1)^T = (z'_1 \;\; z'_2)^T \qquad (66)$$

Equation (63) now becomes:

$$q_1 + (a11 \; a12 \; ...)\begin{pmatrix} \begin{pmatrix} w_{T1} \\ w_{T2} \end{pmatrix} 1 \\ \cdots \\ \begin{pmatrix} w_{T1} \\ w_{T2} \end{pmatrix} N \end{pmatrix} = 0, \;\; q_3 + (a21 \; a22 \; ...)\begin{pmatrix} \begin{pmatrix} w_{T1} \\ w_{T2} \end{pmatrix} 1 \\ \cdots \\ \begin{pmatrix} w_{T1} \\ w_{T2} \end{pmatrix} N \end{pmatrix} = 0 \qquad (67)$$

Writing equation (67) for all contact node pairs, we arrive:

$$\begin{pmatrix} \begin{pmatrix} q1 \\ q3 \end{pmatrix} 1 \\ \begin{pmatrix} q9 \\ q11 \end{pmatrix} 2 \\ \cdots \\ \begin{pmatrix} q_{8(i-1)+1} \\ q_{8(i-1)+3} \end{pmatrix} i \\ \cdots \\ \begin{pmatrix} q_{8(N-1)+1} \\ q_{8(N-1)+3} \end{pmatrix} N \end{pmatrix} + [A]\begin{pmatrix} \begin{pmatrix} w_{T1} \\ w_{T2} \end{pmatrix} 1 \\ \begin{pmatrix} w_{T1} \\ w_{T2} \end{pmatrix} 2 \\ \cdots \\ \begin{pmatrix} w_{T1} \\ w_{T2} \end{pmatrix} i \\ \cdots \\ \begin{pmatrix} w_{T1} \\ w_{T2} \end{pmatrix} N \end{pmatrix} = 0 \qquad (68)$$

Let

$$q' = (\,(q1 \; q3) \; ... \; (q_{8(N-1)+1} \; q_{8(N-1)+3}))^T \qquad (69)$$
$$W_T = ((w_{T1} \; w_{T2})^1 \; ... \; (w_{T1} \; w_{T2})^N)^T \qquad (70)$$

and denote $[A] = -[F_{TT}]^{-1}$ (equation 47), then equation (68) has the form:

$$q' - [F_{TT}]^{-1}W_T = 0 \qquad (71)$$

form which the solution is:

$$W_T = [F_{TT}]q' \qquad (72)$$

After obtaining this solution, the same procedure as described in two dimensional analysis has to be followed to complete the whole PLCP and go to next load level.

Numerical examples

To illustrate the suggested direct solution method for the BEM - PLCP frictional constant contact approach, a contact problem between flat surfaces has been calculated.

Elastic punch on elastic foundation

This is a typical bi-dimensional 'conforming contact' problem which has yet been calculated by some authors [7,8,9]. The geometry of the structure is shown in figure 1, which gives also the boundary element discretization along the contact line. The punch and the foundation boundaries are divided into 42 and 56 quadratic elements respectively. As in [9], a Poisson's ratio $\nu = 0.35$ and three combinations of elastic constants E_p (elastic modulus of the punch) and E_f (elastic modulus of the foundation) are taken:

1. $E_p = 4\,000$ N/mm^2, $E_f = 40\,000$ N/mm^2, $E_p/E_f = 0.1$
2. $E_p = 4\,000$ N/mm^2, $E_f = 4\,000$ N/mm^2, $E_p/E_f = 1$
3. $E_p = 40\,000$ N/mm^2, $E_f = 4\,000$ N/mm^2, $E_p/E_f = 10$

Figures 2 and 3 show for $E_p/E_f = 1$ the relative tangential displacements W_T and tangential tractions P_T along the contact line. Different values of friction coefficient μ are taken. The results have a good agreement with that of Paris [7] (discontinuous linear boundary elements) and Takahashi and Brebbia [9] (BEM flexibility approach). Some slight differences are found near the end of the contact line. To see the influence of the ratio E_p/E_f to contact tractions, Figure 4 gives the distribution of tangential traction P_T against the contact length. As pointed out in [9], the same phenomenon can be observed: high ratio of elastic moduli produces larger sticking regions.

Conclusion remarks

We have suggested in this article a direct solution algorithm to BEM - PLCP frictional constant contact problem. The conditions for applying this method are: (1) The contact surface is constant during the whole contact process, so the relative normal contact displacements are zero; (2) Uniform normal contact tractions or some other pre-determined distributions have to be supposed to separate tangential contact variables from the whole PLCP equations. The second condition is commonly used in many elastic contact analysis, especially when the contact surface is small.

For general frictional contact problem with varying contact surface, matrix M of LCP (equation 23) is neither a definite-positive nor a semi definite-positive one. In fact, in equation (27), the dimension of square matrix M_{JJ} is 2N in two dimension and 8N in three dimension, while from matrix theory, the rank of M_{JJ} must satisfy:

$$\mathrm{rank}(M_{JJ}) \le \mathrm{Min}(\mathrm{rank}(G_2), \mathrm{rank}(F_{TT})) = \mathrm{rank}(F_{TT}) \le \begin{Bmatrix} N \text{ in two dimension} \\ 2N \text{ in three diemnsion} \end{Bmatrix}$$

thus the rank of M_{JJ} is less than its dimension. In other words, M_{JJ} is singular. M_{JJ} is a principal sub-matrix of the LCP matrix M, so the singularity of M_{JJ} confirms our conclusion: M is neither a definite-positive nor a semi definite-positive matrix. Because we have to treat a LCP with an ill-conditioned matrix, some special solution algorithms must be used and are the subject of our going research.

References

[1] Gakwaya, A., Lambert, D., Cardou, A. : "A boundary element and mathematical programming approach for frictional contact problems", submitted to Computers and Structures, September 1990.

[2] Johnson, K.L. : "Contact Mechanics", Cambridge University Press, 1985.

[3] Kwak, B. M., Lee, S. S. :" A complementary problem formulation for two dimensional frictional contact problems", Computers and structures, Vol. 28, No 4, pp. 469-480, 1988.

[4] Klarbring, A. : "A mathematical programming approach to three dimensional contact problems with friction", Computer Methods in Applied Mechanics and Engineering, Vol. 58, pp. 175-200, 1986.

[5] Kalker, J. J. : "Contact mechanical algorithms", Communication in Applied Numerical Methods, Vol. 4, pp. 25-32, 1988.

[6] Kaneko, I. : "A parametric linear complementarity problem involving derivatives", Mathematical programming, Vol. 15, pp. 146-154, 1978.

[7] Paris, F., Garrido, J. A. : "On the use of discontinuous elements in two-dimensional contact problems", in Boundary Element 7, Proc. of the 7th Int. Conf., C. A. Brebbia and G. Maier (ed), Computational Mechanics Publications, pp. (13-17) -(13-39), September,1985.

[8] Fredriksson, B. : "Finite element solution of surface nonlinearities in structural mechanics with special emphasis to contact and fracture mechanics problems", Computers and Structures, Vol. 6, pp.281-290,1976.

[9] Takahashi, S., Brebbia, C. A. : "Analysis of contact problems in elastic bodies using a B.E.M. flexibility approach", BEM X, C. A. Brebbia edit, Springer Verlag, pp. 353-379, September 1988.

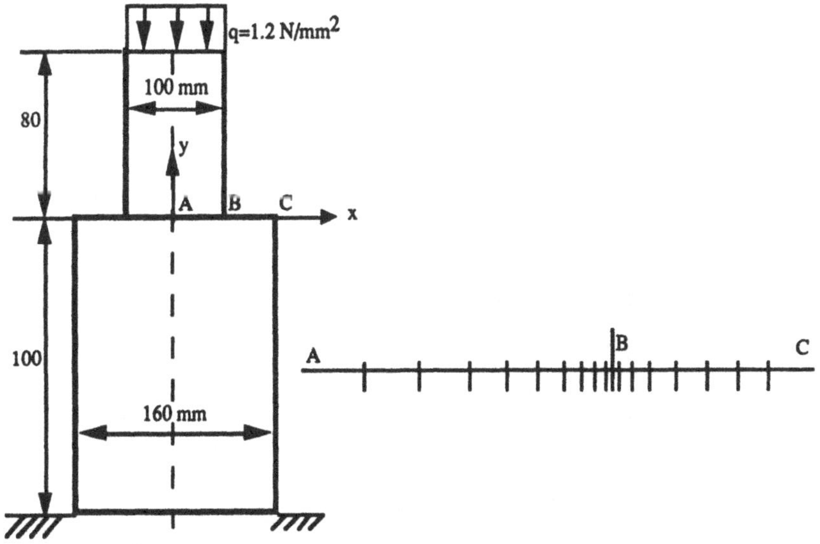

Figure 1 Structure geometry and BE discretization

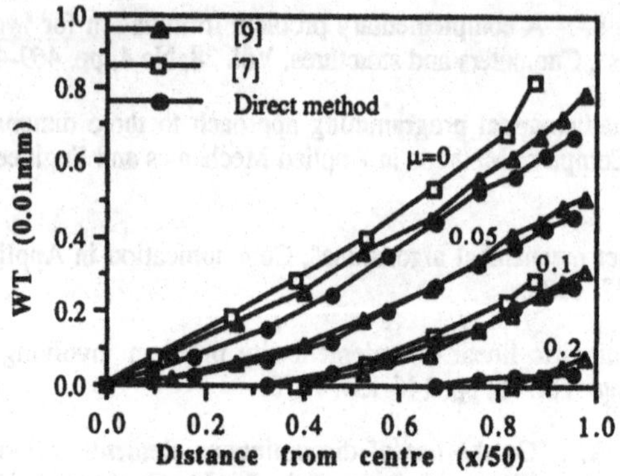

Figure 2 W_T for different values of μ

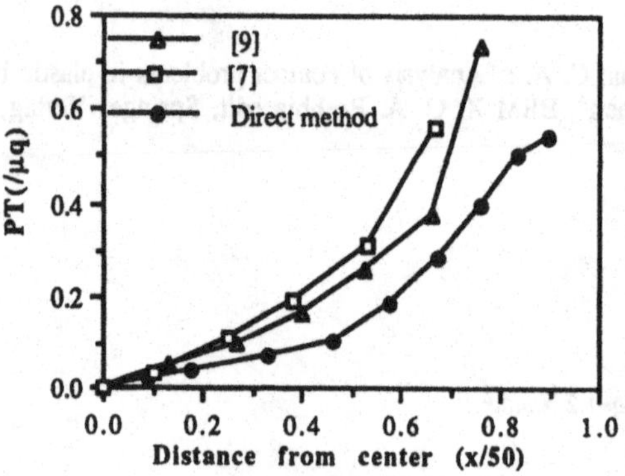

Figure 3 P_T for $\mu = 0.2$

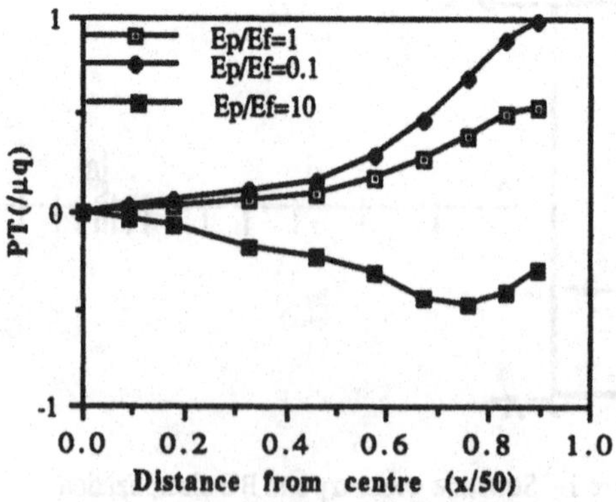

Figure 4 P_T for different values of E_P/E_F (μ=0.2)

SECTION 14: INDUSTRIAL APPLICATIONS

New Directions in the BEASY Boundary Element Method Software

J. Trevelyan, C.J. Gibson
Computional Mechanics, Inc., Billerica, Massachusetts, U.S.A.

ABSTRACT

The BEASY software has been used for a decade in industry and academic organizations for the boundary element analysis of potential flow and stress problems. During the late 1980's, the developments and enhancements made to the code responded to requests for improved performance. These requests having largely been satisfied, the direction of development has changed towards the introduction of new analysis capabilities. This paper discusses the recent developments and gives examples of their application to engineering problems.

INTRODUCTION

BEASY[1] is a commercial software package which uses the boundary element method (BEM) to solve a variety of engineering problems. It has been used since 1981 by both industrial companies and academic departments. Updates and new versions of the software are released periodically to make more features available to these organizations. As far as possible, the developments and enhancements which are implemented are driven by the requests of the users, particularly those in the commercial world.

During the mid-1980's, these users showed a clear desire to use the boundary element method, as the benefits in data preparation over traditional finite element modeling were usually very apparent. However, the level of performance (i.e. the usage of computer resources) needed to be improved. So in spite of the obvious man-time benefits, the computational difficulties associated with the more complicated integration and the full, unsymmetric BEM matrices were causing some concern.

The introduction of hybrid models containing a mixture of continuous and discontinuous elements, improved integration schemes, improved solvers and memory management algorithms have, since then, addressed these concerns. Figure 1 shows the evolution of increasingly efficient versions of the software. This chart has been produced by taking an identical input data file and recording the cpu time taken by different versions of the BEASY code in solving the problem. While the figure shows a clear improvement, it underestimates the real improvements which have been made as some more advanced elements have been added. These elements would further reduce the cpu time by allowing a reduction in the number of degrees of freedom without noticeable loss in accuracy.

Now the performance has been improved so substantially, the developments have turned to the implementation of new types of analysis, new boundary conditions and various other new capabilities. In this respect the policy has been to implement first those types of analysis which naturally lend themselves to analysis by the BEM. Contact analysis with gap elements, fracture mechanics, interference problems, etc. are all essentially boundary phenomena in that they all relate to analysis of behavior at the surface of a component. These have been implemented before more volumetric type phenomena such as material nonlinearity have been tackled.

NEW FEATURES

Contact analysis

One of the most consistently requested areas of development has been contact analysis with gap elements. This has been implemented in BEASY using a multipoint constraint algorithm which connects degrees of freedom internally in the matrix solution. The load is applied incrementally to accelerate the convergence of the iteration.

Gaps can be applied between elements and external rigid surfaces, or between elements on either side of a zone interface, enabling contact between two regions to be analysed. For this latter case, a dummy set of elements is created automatically, and their degrees of freedom linked to the original elements' degrees of freedom by the particular gap geometry imposed by the user.

Starting with a small percentage of the applied loading, the program performs an iteration procedure; first analysing, then readjusting the boundary conditions to prevent overlapping or to allow lift-off, reanalysing, etc. until the contact area has been found. The load is increased and the iteration continues. A solution is obtained when the iteration converges for 100% of the load applied by the user.

The first example of contact analysis is in 2D, in which a connecting rod from an automobile engine is subjected to a load from the crank pin (figure 2). The connecting rod has been manufactured in two parts and bolted together around the crank pin. The mating surface will open up if the load from the crank pin is large enough to overcome the prestress supplied by the bolt. This opening, which is to be prevented in this design, is inhibited by the presence of the pin itself.

The load from the crank pin causes an ovalling of the originally circular hole, and this too is handled using gap elements. The displaced shape (figure 3) shows the crank pin pulling away from the connecting rod at the left, and the mating surface also starting to open slightly.

The Von Mises stress contours are shown in figure 4. The true stress distribution can be predicted only by using a non-linear gap element in the analysis to ensure the true contact area is used in the calculation.

The second example of a contact analysis is of essentially the same component - an automobile engine connecting rod with the crank pin. This time the analysis is done in 3D, which will clearly give a more accurate stress picture around the high stress region, as the component is far from 2D in reality. In this case, the contact surface is restricted to the surface between the crank pin and the connecting rod (figure 5).

The deformed shape (figure 6) clearly shows the same ovalling effect as in the 2D example, and the gap opening up between the connecting rod and the crank pin. However, the Von Mises stress contours (figure 7) show the peak stress to build up in an area in the 'shoulder' of the connecting rod in a fashion which could not have been predicted by a 2D analysis.

The contact analysis will be useful in improving the quality of results of the stress analysis of threaded joints, which has been a popular application of boundary elements due to the ease of modeling the true geometry (and the corresponding difficulty with the finite element method)[2].

New boundary conditions

A number of new boundary conditions have been added to the BEASY analysis. These include press fit, shrink fit, prestress and a number of thermal analysis conditions. The press fit and shrink fit conditions simply relate the displacements on either side of a zone interface (u_1 and u_2) using an equation of the form

$$u_1 + u_2 = \bar{u}$$

The sign of the value the user specifies for $\bar{u}$ determines whether the condition is a press fit (interference fit) or shrink fit.

The prestress condition is a simple spring (either between zones or to an external rigid boundary) with an added traction term modeling the prestress. This is useful for local analysis around bolted joints.

The example shown in this section is a 3D analysis of a ball joint. The ball and the socket are initially unable to fit, as there is a small interference between the two components. They are then put together either by forcing or by heating the socket before putting them together.

The model uses two planes of symmetry, and figure 8 shows how this immensely reduces the data preparation effort. No elements need to be used on a plane of symmetry, making the model look hollow. (This effect was also seen in the 2D crankshaft analysis described above and shown in figure 2).

Figures 9 and 10 show the stress distribution on the mating surfaces of the ball and the socket respectively. (The high stresses on the shaft attached to the ball are due to an axial load applied at the end of the shaft).

Mixed 2D/axisymmetric analysis

This is a feature which has been requested consistently by BEASY users in the aerospace industry, most notably in the analysis of turbine jet engines. The rotating machinery contains many components which are either truly axisymmetric or close enough to be considered axisymmetric in a stress analysis. There are also a number of blades which may have complex curved geometries, but which could be considered to act in a plane stress way.

This feature has been implemented by replacing the standard boundary element primary variables (displacement and traction) by displacement and force. This is of course a simple transformation, but some care is necessary to choose the correct surface area to consider where two zones of different thickness meet. Once this transformation of variables is made, equilibrium is maintained between the two zones, and the remainder of the BEASY code (or any other boundary element code, for that matter) can be used in the normal way to solve the problem.

As an example of this type of analysis, figure 11 shows a turbine disk with a blade fitted under a simple rotational load. The displaced shape (figure 12) shows the general elongation of the component in the radial direction, with some bending taking place in the drum as the more massive disk tries to displace further under the centrifugal loading. The stress contour plot (figure 13) shows the peak stress to occur at the bore (inner diameter) as expected. Stress discontinuities are seen at the interface between the plane stress blade and the axisymmetric disk.

Fracture mechanics analysis

Boundary elements have long been known to give highly accurate stress solutions on the boundary[3,4,5]. Furthermore the formulation of the boundary integral equation, and in some cases its expression in boundary element software codes, allow for discontinuity in stress values. The combination of these factors, quite apart from the other clear benfits to be gained from boundary elements, make the technique a natural choice for fracture mechanics analysis.

BEASY will automatically use a discontinuous element[1] where there is a sharp angle in the geometry, or where there is a step change in boundary conditions. These two cases will cover crack problems automatically, allowing the sharp dicontinuitites in stress which are found to be predicted accurately. So from the users perspective a fracture mechanics model is identical to a model for a more conventional stress analysis. It is necessary only to make the elements small enough in the vicinity of the crack tip/crack front to model the sharp variation in stress.

BEASY's pre and postprocessor BEASY-IMS (Interactive Modeling System) has been developed to display tables and graphs of stress intensity factors. The user selects the mesh point at the crack tip and a range of points at which the stress intensity factors at the crack tip should be evaluated. The graphical display allows the user to predict the true stress intensity factors as the limit of the calculation as the point of evaluation approaches the crack tip itself.

Improved 3D stress analysis

Some improvements have been made to the 3D stress analysis module in terms of performance and accuracy. The previous restrictions of 3:1 on both element aspect ratios and mesh grading have been relaxed to 8:1, both by the implementation of more accurate integration schemes. This improvement carries no extra cpu cost for most problems. This often means that far fewer elements need to be used for 3D problems which include detailed areas such a fillets or other features requiring small or thin elements. So larger models can be considered.

In addition to the new integration schemes, there has also been an improvement in both the algorithm which is used to determine which elements should be continuous and which discontinuous, and in the use of memory during the equation solution phase. The first of these was implemented primarily to improve reliability for distorted and complicated meshing, but does in fact reduce the number of degrees of freedom for many problems, and therefore reduces the cpu time.

The memory management scheme uses the larger memory capacities of today's computers to store parts (or all) of the matrix equation during reduction. This drastically reduces the disk i/o (input/output) required. This has given rise to analysis runs in which the total elapsed (or "wall-clock") time exceeds the cpu time by only a small percentage, even for substantial 3D stress models.

For example, the 3D stress analysis of a crankshaft (including the fillets) has been made possible. The model shown in figure 14 consists of nearly 2000 elements in 8 zones. This model ran on a CRAY X-MP using 4MW of memory, taking 37:16 minutes cpu time and only 42:04 minutes elapsed time. On an IBM RS6000 model 530, the elapsed time was only 83 minutes, showing the rapid advances being made in hardware and software performance on workstations.

Figure 15 shows a detail of a stress contour plot, showing the stress to build up to a peak in the fillets as expected.

CONCLUSIONS

The philosophy of the development of BEASY has always been to trust that the direction suggested by the users is sound, and this has turned out to be a successful policy. Of course, not everybody's requirements are the same, and with limited resources it is impossible to satisfy all needs. However, there is a surprisingly large agreement between the requirements of the different users in different disciplines.

The developments in performance that culminated in version 3.3 were necessary to make a technology which was just emerging (in practical terms, if not the underlying theory) into a practical option for users in companies who charged for computer resources as well as for engineers time. Now with a solid, reliable and efficient linear analysis 'core', the development of various new features has begun, and the release of version 4.0 brings these features to the users of the software in industry and academic departments.

The software continues to develop, and further updates are planned for the user community in later releases.

REFERENCES

1. **BEASY User Guide**, Version 4.0, Computational Mechanics, Southampton, UK and Billerica, MA, USA.
2. A.I. Wanderlingh, **Determination of the Maximum Thread Stress in Threaded Structures using the Boundary Element Method**, Proc. 11th International Conference on Boundary Element Methods, Vol.3., Computational Mechanics Publications, 1989.
3. C.A. Brebbia, **The Boundary Element Method for Engineers**, Pentech, 1978.
4. C.A. Brebbia, J.C.F. Telles, L.C. Wrobel, **The Boundary Element Method : Theory and Applications**, Springer Verlag, 1984.
5. C.A. Brebbia, J. Dominguez, **Boundary Elements : An Introductory Course**, Computational Mechanics Publications, 1988.

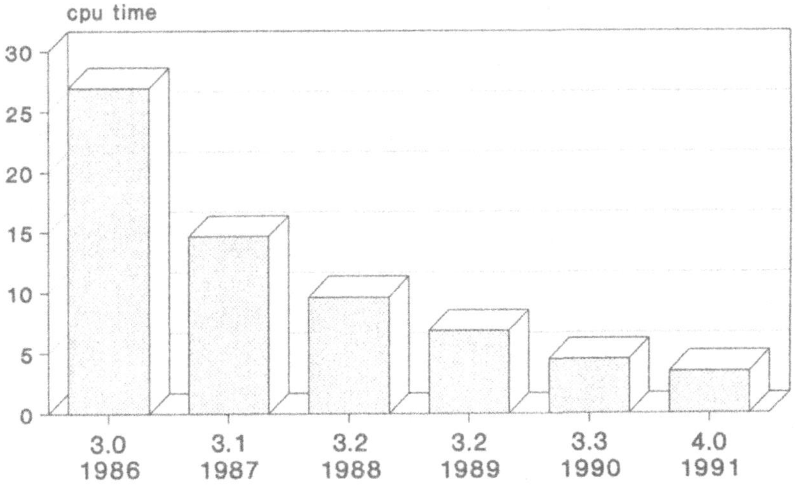

Fig. 1. BEASY performance improvements

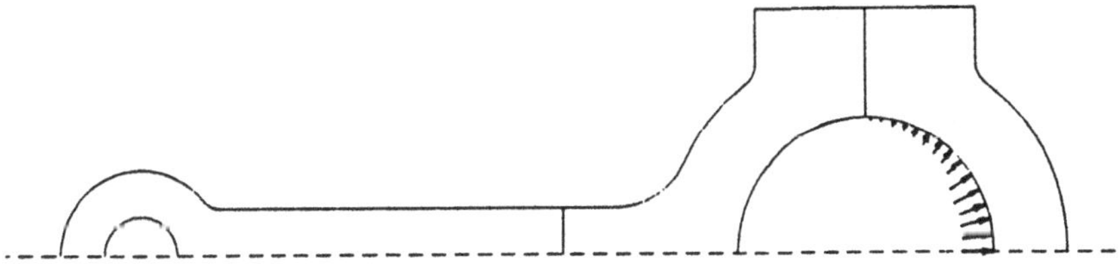

Fig. 2. 2D connecting rod/pin contact problem

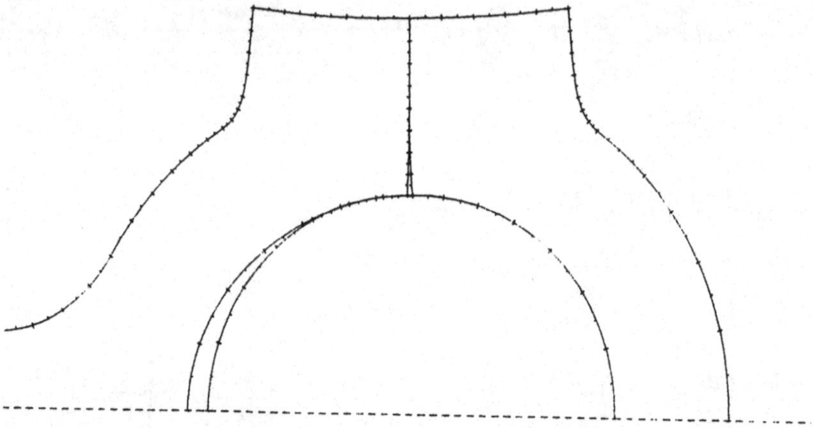

Fig. 3. Connecting rod/pin displaced shape
Note gaps opening

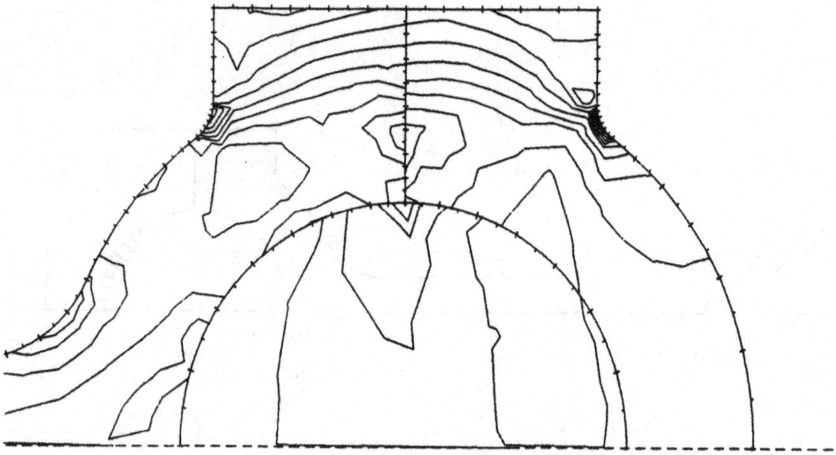

Fig. 4. Contours of stress results

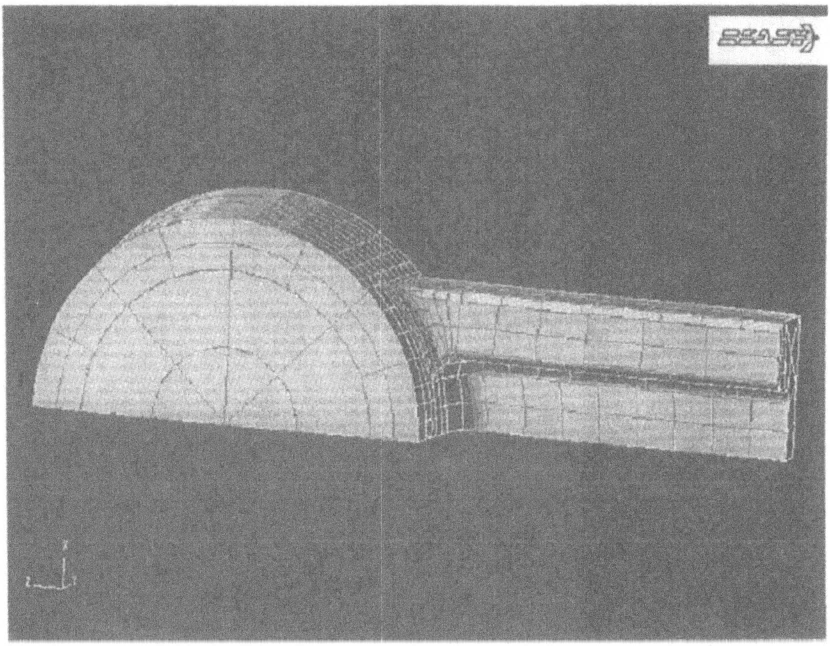

Fig. 5. 3D connecting rod/pin contact problem

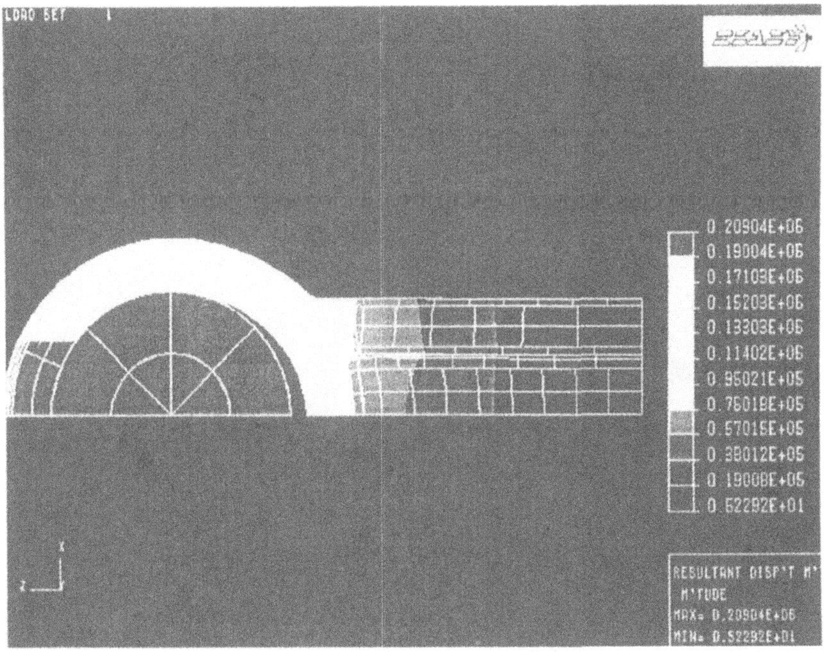

Fig. 6. 3D connecting rod/pin contact problem
Contour plot on deformed shape

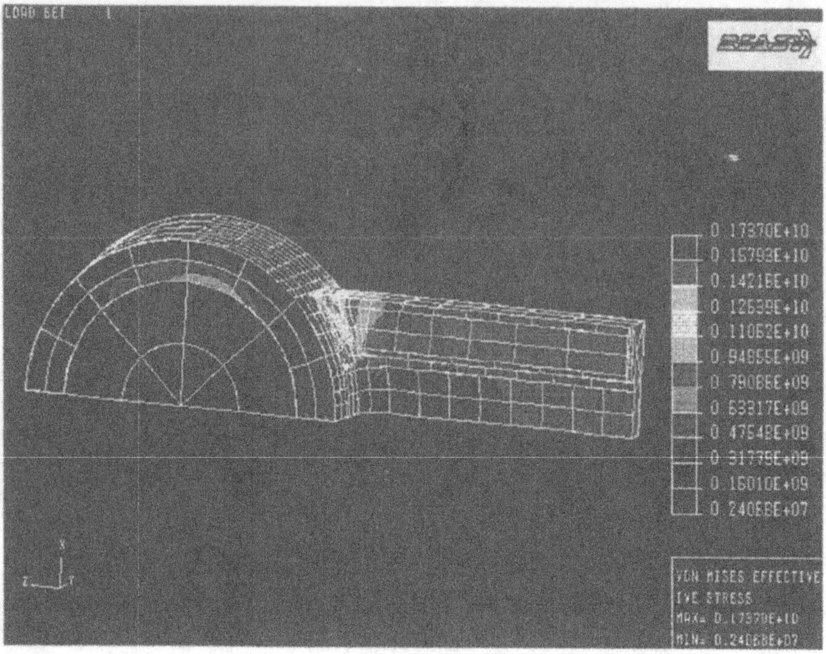

Fig. 7. Von Mises stress contours on connecting rod

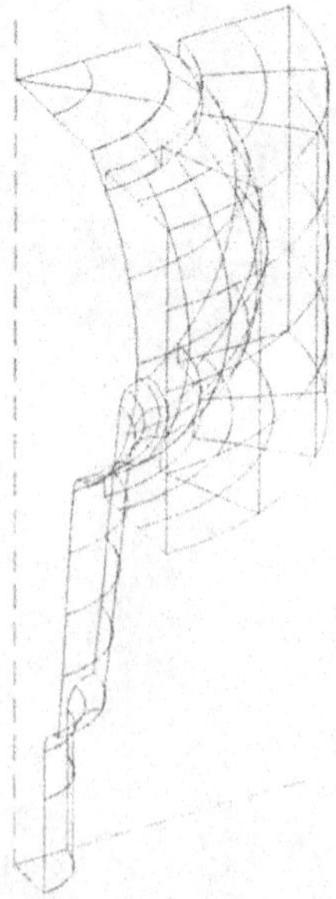

Fig. 8. 3D ball joint interference fit model

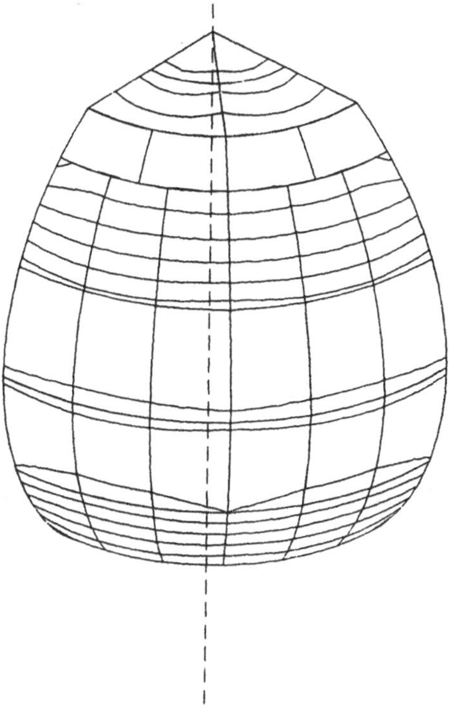

Fig. 9. Ball joint model - stresses on ball surface

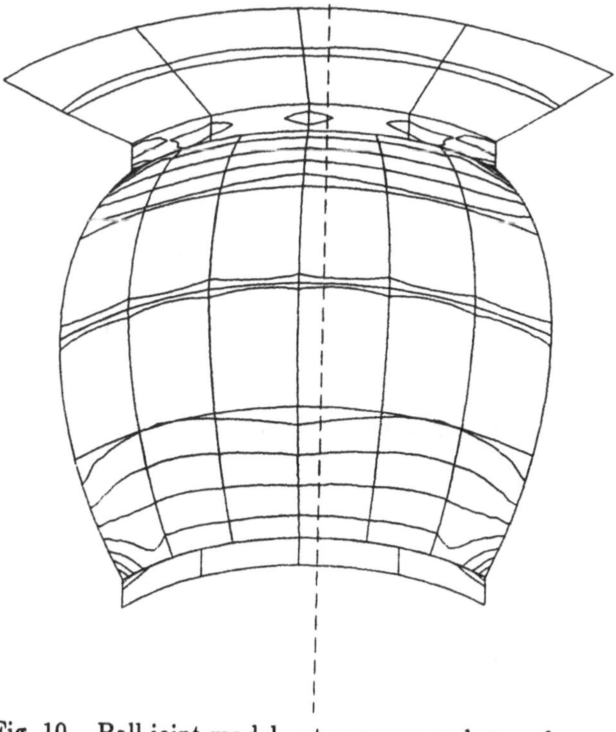

Fig. 10. Ball joint model - stresses on socket surface

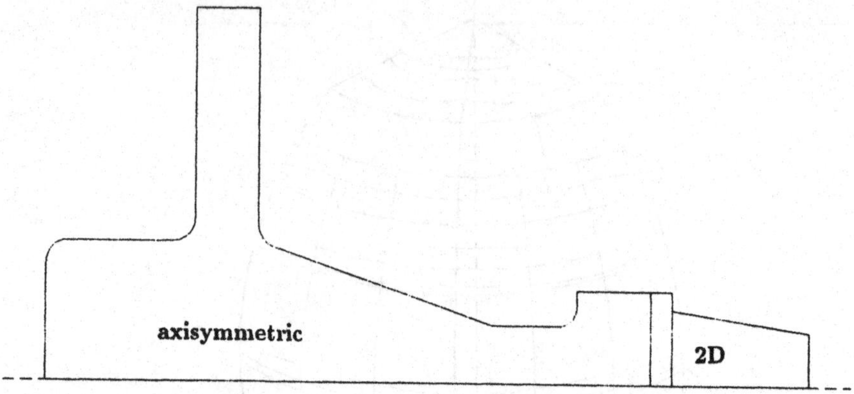

Fig. 11. Mixed 2D/axisymmetric blade/disk model

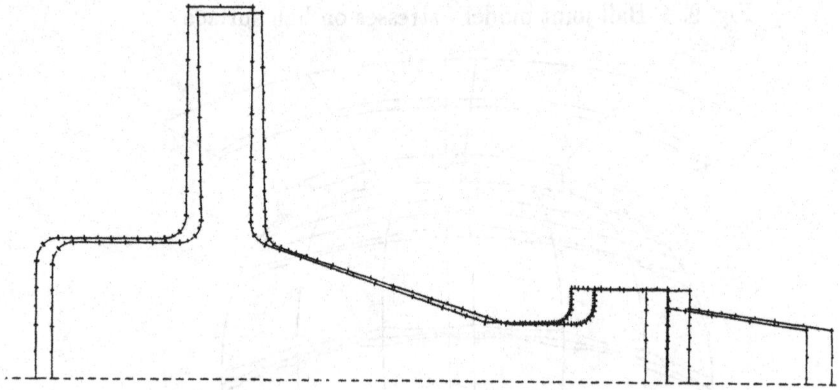

Fig. 12. Deformed shape of blade/disk model

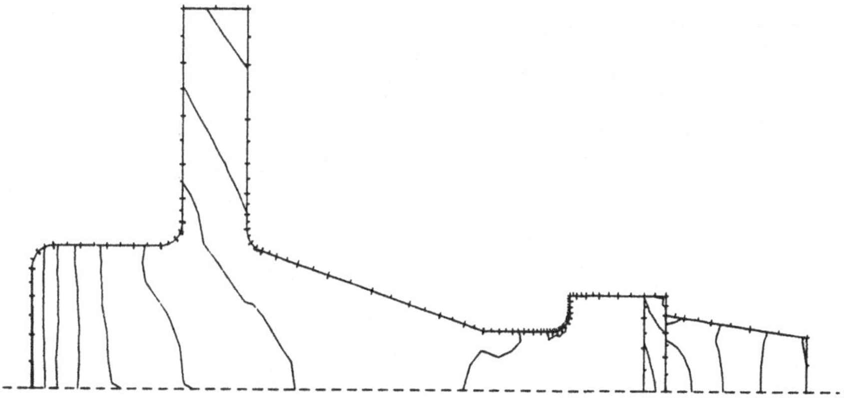

Fig. 13. Stress contours on blade/disk model
Note discontinuities between 2D and axisymmetric zones

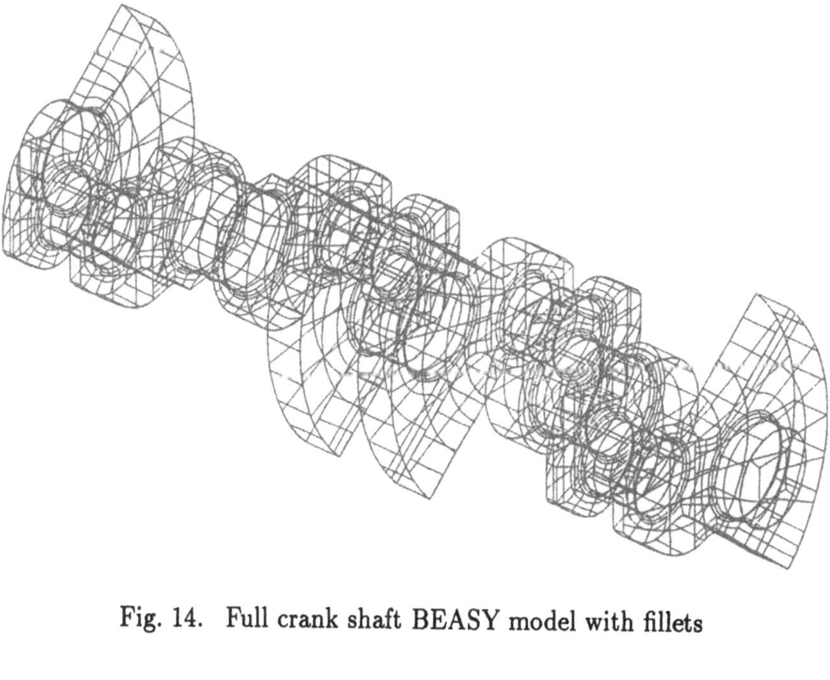

Fig. 14. Full crank shaft BEASY model with fillets

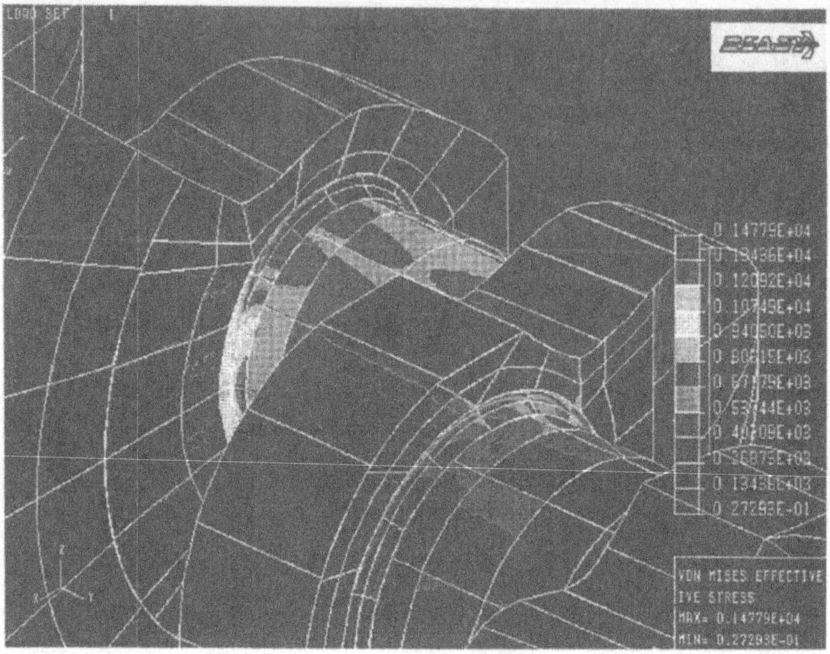

Fig. 15. Detail of stress contour plot

Integrated Computer Aided Surface Modeling and BEM Analysis

L.M. Kamoso, A. Gakwaya

Department of Mechanical Engineering,

Laval University, Quebec, G1K 7P4, Canada

ABSTRACT

An integration of Computer aided free form surface modelling and Boundary Element Analysis in the design process of mechanical systems is presented using an new non conforming boundary element. A periodic bicubic B-spline patches representation is introduced to define the surface bounding the parts to be designed. For BEM analysis of mixed elastostatic problems we use a new element having geometric shape functions different from the usual Lagrangian interpolation functions used in isoparametric formulation, and which satisfy CAD requirements. Elements nodes are automatically generated and depend on geometric design criteria.

1)- INTRODUCTION

In computer aided design, the representation of shapes of mechanical parts requires the manipulation of surfaces or volumes elements. Two important approaches are generally used in mechanical computer aided design (MCAD). The first approach is the constructive solid geometry (CSG), which uses boolean operations to combine simple primitives, like cylinder, spheres, prisms...., to built a complex solid object but require a boundary evaluation[2,3] for its graphical display. The second approach is the boundary representation (Brep) in which a body is characterised by its exterior shape that is represented by various surface representation models, the most recent one being the non-uniform rational B spline scheme[3]. In the Brep approach, two classes can be distinguished: plane facet polyhedral model and free form surface modelling. In the polyhedral representation, only plane equation, edges and vertices defining the connection between adjacent faces are represented.The modelling of complex 3D shapes is achieved by decomposition of the complicated form into a number of more simple forms that are then patched together so that they accurately represent the original form [3,12]. For free form surface modelling the boundary is represented by elementary curved surfaces; these elements are called patches. The reason for partitioning surface into a finite number of patches is to ensure easier mathematical analysis and efficient computer implementation, because a lower continuity requirement can be used with small patches that are cheap to evaluate and display. From the point of view of a designer, the surfaces to be modelled are unkown and little information is available. In this case, interpolation or approximation techniques can be used to define it. If the interpolation technique is used, one must provide enough points on the surface[1,5], and some smothness or fairness properties must be satisfied. However, with the approximation technique[2,3], some control vertices defining a polyhedral approximation of the desired surface are usually sufficient to represent a new smooth and fair shape. With the recent powerful non uniform rational B-spline modelling, it is possible to represent both the cubic and the quadratic surfaces[2,3] using the same mathematical settings and both interpolating and approximating surfaces are easily generated. In this paper, a bicubic B-spline representation is used to define the surface bounding the body whose shape is to be optimized under some geometric and mechanical constraints. Boundary elements automatically generated on these surfaces, and using the geometric B-spline approximation coupled with lagrangian interpolation of field variables can be readily used in stress analysis computations.

B-SPLINE SURFACE REPRESENTATION

Let u,v be the parametric variables in a 3D surface representation. A B-spline patch of order k in direction u and order l in direction v is defined by :

$$\Gamma(u,v) = \sum_{j=0}^{R}\sum_{i=0}^{S} N_{i,k}(u) M_{j,l}(v) p_{ij} \tag{1}$$

where R+1,S+1 are the number of control vertices in u,v direction, $N_{i,k}(u)$, $M_{j,l}(v)$ are the B-spline basis functions of order k and l respectively and P_{ij} are the control vertices defining a polyhedral approximation of the sought surface[3], but on which the surface does not necessarily pass. One can choose the degree of the basis functions according to the continuity and smoothness level needed by the application.

For example, with a given set of R by S vertices, a periodic bicubic subpatch is defined with sixteen specific vertices by[2]:

$$\Gamma_{ij}(u,v) = \begin{bmatrix} u^3 & u^2 & u & 1 \end{bmatrix}.[M].\begin{bmatrix} P_{ij} & P_{(i+1)j} & P_{(i+2)j} & P_{(i+3)j} \\ P_{i(j+1)} & P_{(i+1)(j+1)} & P_{(i+2)(j+1)} & P_{(i+3)(j+1)} \\ P_{i(j+2)} & P_{(i+1)(j+2)} & P_{(i+2)(j+2)} & P_{(i+3)(j+2)} \\ P_{i(j+3)} & P_{(i+1)(j+3)} & P_{(i+2)(j+3)} & P_{(i+3)(j+3)} \end{bmatrix}.\begin{bmatrix} M^T \end{bmatrix}.\begin{Bmatrix} v^3 \\ v^2 \\ v \\ 1 \end{Bmatrix} \tag{2}$$

where $u, v \in [0,1]$ are parametric variables on the surface

$i = 1,2,\cdots,R-3 \quad ; \quad j = 1,2,\cdots,S-3$

$$[M]=\frac{1}{6}\begin{bmatrix}-1 & 3 & -3 & 1\\ 3 & -6 & 3 & 0\\ -3 & 0 & 3 & 0\\ 1 & 4 & 1 & 0\end{bmatrix} \tag{3}$$

In general, in equation(1), $x_i \le u \le x_{i+1}, y_j \le v \le y_{j+1}$, where x_i, y_j are elements of the knot vectors for direction u and v respectively[2,3]. The influence of a single control vertex is limited to $\left(\pm\frac{k}{2},\pm\frac{l}{2}\right)$ spans in each direction. This local effect, combined with the possibility of modifying the degree of the shape functions independently of the number of control vertices, confers a great flexibility to the B-spline representation relative to the other approximating or interpolating representations [3,10]. If the number of control vertices is equal to order in each parametric direction and there are no interior knot values, then the B-spline surface reduces to a Bezier surface[2]. Greater flexibility could be obtained if a rational scheme was used in place of equation (1) [3].
Using the B-spline parametric equation, points on any subpatch can be found by choosing particular values of the parameters u and v in the bounded interval. Thus it is possible to define all the required points on the surface on which functions like displacements and tractions can be interpolated for analysis either by the FEM or by the BEM. Also, if these points are considered as nodes, it is possible to subdivide the patch into boundary elements and thus generating all the FE mesh information[15].

2)- SUPERPARAMETRIC B-SPLINE ELEMENTS

Very often, isoparametric formulations are used in both the FEM and the BEM methods and give very good results[1]. However, in the context of shape optimization, interpolated domain may converge to unrealistic shape when nodal coordinates are used as design variables[4]. For this reason and considering the advantages of B-spline representation evoked above, we propose, in this paper, to separate the surface shape representation and the interpolation of the field variables, and thus introduce a superparametric boundary element. The boundary nodes and elements shape is defined using the B-spline patch functions while the field variables are approximated using standard Lagrangian shape functions[5,15]. This means that all geometric properties are calculated on the real design shape, and all other physical variables will be interpolated, as usual done with quadratic Lagrangian functions[1,5,7].

SURFACE MESHING

Let us consider fig. 1, which represents a B-spline surface subdivided into NU subpatches in u direction and NV subpatches in v direction. If we need to mesh this surface, first we define a uniform subdivision of the surface into r by s elements of given nodes, and then we calculate the position of these nodes using the parametric equation (4) for each subpatch, number them and generate the elements connectivity.

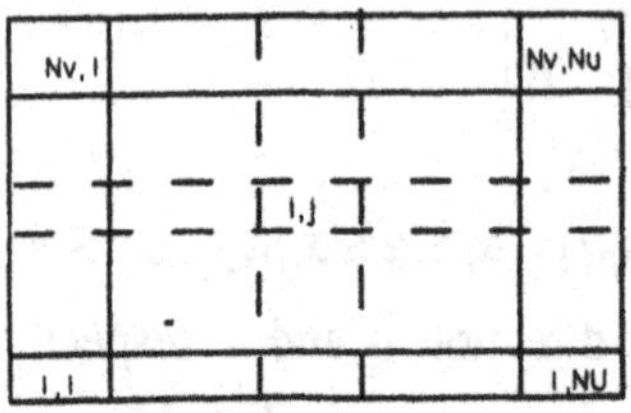

Fig.1 Surface subdivided into subpatches

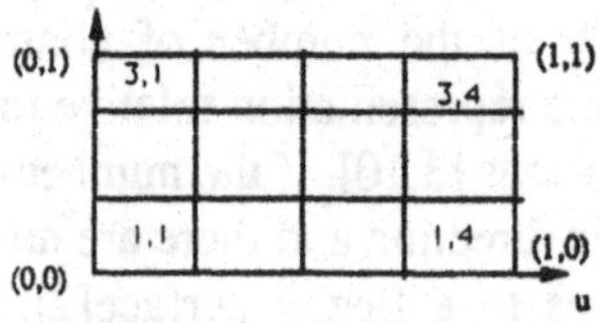

Fig.2 A subpatch subdivided into 3 by 4 elements

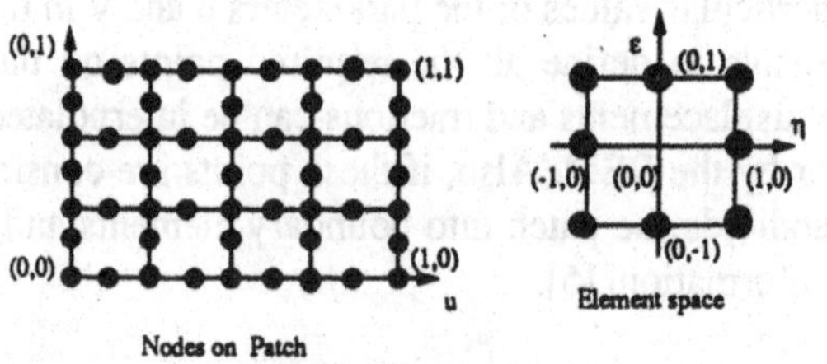

Fig.3 Subpatch meshed into quadrilateral elements (8 nodes)

For a B-spline surface represented by R subpatches in u direction and S subpatches in v direction, each subpatch is a B-spline of order k,l and can be divided into r elements in the u direction and s elements in the v direction. If we consider quadratic quadrilateral elements of type Q8, we obtain for a bicubic subpatch of fig. 2, a mesh representation with typical element in fig.3, triangular elements can also be obtained. To compute the nodal coordinates of each element, equation (1) can be used. Then, the ith subpatch $\Gamma_i(u,v)$ can be written in terms of B-spline basis shape functions and the defining 16 control vertices P_{ij}, j=1,...,16

$$\vec{\Gamma}_i(u,v) = \left[N_1(u,v) \cdots N_{16}(u,v)\right] \bullet \begin{Bmatrix} p_{i1} \\ \vdots \\ p_{i16} \end{Bmatrix} \tag{4}$$

Where $N_i(u,v)$ are shape functions easily derived from equation (1)[1,2].

Introducing local coordinates ξ,η on each boundary element, a coordinate transformation equation must be defined to relate the parametric space (u,v), to the element variable interpolation space (ξ,η). The (u,v) parametric bounds on each subpatch are defined in terms of mesh parameters by:

$$u \in \left[\frac{i-1}{R} \quad \frac{i}{R}\right] \qquad v \in \left[\frac{j-1}{S} \quad \frac{j}{S}\right] \tag{5}$$

with i and j defining the element position on the subpatch in u and v direction. For Gauss integration schemes, needed in subsequent treatment, we require a transformation from (u,v) space defined in interval [0,1] to space (ξ,η) defined on interval [-1,1]. This transformation between the parametric patch space (u,v) to the local element coordinate space (ξ,η) is given by:

$$u, v \in [0,1] \quad \Rightarrow \quad \varepsilon, \eta \in [-1 \quad 1] \tag{6}$$

$$u = \frac{2i-1}{2r} + \frac{1}{2r}\xi \quad \Rightarrow du = \frac{1}{2r}d\xi$$

$$v = \frac{2j-1}{2s} + \frac{1}{2s}\eta \quad \Rightarrow dv = \frac{1}{2s}d\eta \tag{7}$$

It allows integrals on the subpatch to be mapped into integrals on the reference boundary element and is given by:

$$\int_{\frac{2i-1}{2r}}^{\frac{i}{2r}} \int_{\frac{2j-1}{2s}}^{\frac{j}{2s}} dv \bullet du \quad \Rightarrow \int_{-1}^{1}\int_{-1}^{1} \frac{1}{2s} \bullet \frac{1}{2R} d\eta \bullet d\xi \tag{8}$$

GEOMETRIC FEATURES OF THE SUPERPARAMETRIC ELEMENT.

The geometric characteristics of the new element are defined in terms of the B-spline basis shape functions and parameters.

* Tangent vectors and unit vector on a subpatch are calculated by

$$\vec{T}_u(u,v) = \left[\frac{\partial N(u,v)}{\partial u}\right]\{p\} \quad \Rightarrow \quad \vec{t}_u = \frac{\vec{T}_u}{\left\|\vec{T}_u\right\|} \tag{9}$$

$$\vec{T}_v(u,v) = \left[\frac{\partial N(u,v)}{\partial v}\right]\{p\} \quad \Rightarrow \quad \vec{t}_v = \frac{\vec{T}_v}{\left\|\vec{T}_v\right\|} \tag{10}$$

* The unit normal vector is calculated by :

$$\vec{N}(u,v) = \vec{T}_u \times \vec{T}_v \quad \Rightarrow \quad \vec{n} = \frac{\vec{N}}{\left\|\vec{N}\right\|} \tag{11}$$

* The surface jacobian is given by:

$$J_s(u,v) = \left\|N(u,v)\right\| \tag{12}$$

so that $\int_s f(x,y,z)ds = \int_s f'(u,v)J_s(u,v)du.dv$

FIELD VARIABLES APPROXIMATION

Once we have determined these geometric features, we need to specify how physical variables such as displacement and tractions will be approximated. In this case, we employ the usual Lagrangian interpolation scheme[15].

$$u(x) = \left[\tilde{N}(\xi,\eta)\right]\left\{u_n^e\right\}, \tag{13}$$

$$t(x) = \left[\tilde{N}(\xi,\eta)\right]\left\{t_n^e\right\} \tag{14}$$

where $\tilde{N}(\xi,\eta)$ are standard boundary finite element Lagrangian interpolation functions that can be found in [15], and $\{u_n^e\},\{t_n^e\}$ are element nodal variables. The interpolation functions depend, of course, on the type of the element used.

The advantage of using such a superparametric element is that, with the geometry being well represented and decoupled from the physical variables interpolation, we can easily control the shape variation and its up dating. Also, if rational B-spline representation is used, we have additional flexibility of being able to represent cubic and quadratic geometries and to mix them with quadratic interpolation for the variables. However, in the finite element literature, superparametric elements are not popular, because they apparently do not satisfy some convergence requirement[15]. It is the purpose of this paper to study the accuracy of this new element in the context of coupled BEM analysis and shape optimization.

3)-BEM ANALYSIS USING THE B-SPLINE ELEMENT

Assuming we have an elastic solid body whose shape, or part of it ,is to be optimized, the governing equation for the displacement and the traction fields is given by the usual boundary integral equation (9) [1,5] (with body force neglected for simplicity)

$$C_i\, u_{ij}(x^\alpha) + \iint_\Gamma T_{ij}^*(x^\alpha,y)\, u_j\, ds_y = \iint_\Gamma U_{ij}^*(x^\alpha,y)\, t_j\, ds_y \tag{15}$$

where x^α is the obsevation point and y is the load point,

T_{ij}^*, U_{ij}^* are traction and displacement fundamental solutions for elastostatic problems given in 3D by[1].

$$T_{ij}^{*}(x^{\alpha},y)=\frac{-1}{8\pi(1-\nu)r^{2}}\left\{\left[(1-2\nu)\delta_{ij}-3r_{,i}r_{,j}\right]\frac{\partial r}{\partial n}-(1-2\nu)(r_{,i}n_{j}-r_{,j}n_{i})\right\} \tag{16}$$

$$U_{ij}^{*}(x^{\alpha},y)=\frac{-1}{16\pi(1-\nu)Gr}\left\{\left[(3-4\nu)\delta_{ij}+r_{,i}r_{,j}\right]\right\} \tag{17}$$

where $r(x^{\alpha},y)$ is the distance between field point x and load point y on the boundary. Morever,

$$r_{i}=x_{i}-y_{i}\quad\Rightarrow r=(r_{i}\bullet r_{i})^{1/2} \tag{18}$$

$$r_{,i}=\frac{\partial r}{\partial x_{i}}=\frac{r_{i}}{r}\quad\Rightarrow\frac{\partial r}{\partial n}=\left[grad_{i}(r)\right]\bullet\{n_{i}\} \tag{19}$$

With the B-spline element, the distance between points x and y is given by:

$$\begin{cases} y_{i}=\left[N_{i}^{y}\right]\{p_{i}^{y}\} \\ x_{i}=\left[N_{i}^{x}\right]\{p_{i}^{x}\}\end{cases}\Rightarrow\begin{bmatrix} r_{i}=\left[N_{i}^{x}\right]\{p_{i}^{x}\}-\left[N_{i}^{y}\right]\{p_{i}^{y}\} \\ r=(r_{i}\bullet r_{i})^{1/2}\end{bmatrix} \tag{20}$$

Where $\{p_{i}^{x}\},\{p_{i}^{y}\}$ are the vector of control vertices influencing point x and point y respectively. The above equation shows the dependency of the distance function and hence of the kernels T* and U* on the control point P_{ij}.

The surface element dS thus becomes, after using equation (6) for surface jacobian.

$$dS=J_{s}(u,v)du.dv=J_{s}(u(\xi).v(\eta))J_{\xi\eta}d\xi d\eta \tag{21}$$

Introducing the B-spline surface representation and meshing (eq ..15) we partition the boundary into a varying part Γ_{v} and a fixed part Γ_{f} such that:

$$\Gamma=\Gamma_{v}\cup\Gamma_{f} \tag{22}$$

Also we assume that the boundary is subdivided into NELT elements among which, M elements have a fixed shape and (NELT-M) elements have a varying shape. Applying standard BEM discretisation and approximation technique[1,5], the discretized integral equation becomes :

$$\begin{aligned} Cu_{i}(x^{\alpha}) &= \int_{\Gamma=\Gamma_{f}\cup\Gamma_{v}}U_{ij}^{*}t_{j}ds-\int_{\Gamma=\Gamma_{f}\cup\Gamma_{v}}T_{ij}^{*}u_{j}ds \\ &=\sum_{e=1}^{M}\left(\int_{\Gamma_{f}^{e}}U_{ij}^{*}t_{j}ds-\int_{\Gamma_{f}^{e}}T_{ij}^{*}u_{j}ds\right)+\sum_{e=M+1}^{NELT}\left(\int_{\Gamma_{v}^{e}}U_{ij}^{*}t_{j}ds-\int_{\Gamma_{v}^{e}}T_{ij}^{*}u_{j}ds\right)\end{aligned} \tag{23}$$

which, upon applying the numerical integration and assembling process[5], yields the following matrix equation:

$$\begin{bmatrix} H_{ff} & H_{fv}(p) \\ H_{vf}(p) & H_{vv}(p) \end{bmatrix} \begin{Bmatrix} u_f \\ u_v \end{Bmatrix} = \begin{bmatrix} G_{ff} & G_{fv}(p) \\ G_{vf}(p) & G_{vv}(p) \end{bmatrix} \begin{Bmatrix} t_f \\ t_v \end{Bmatrix} \tag{24}$$

where the variables associated with the fixed shape (index f) and those that depend on the design variable (index v), have been separated.

After introducing the boundary conditions [1,5], we obtain:

$$[A]\{x\} = [B]\{d\} = \{F\} \tag{25}$$

where A is the state matrix and {x} is the unknown state variables vector, while {d} is the known vector of boundary conditions and {F} is the loading vector. Solving this equation, we obtain the unknown boundary displacements and tractions, from which we can calculate boundary stresses as explained in [9].

SURFACE STRESSES REPRESENTATION

Since the stresses will be used in sensitivity analysis, or in gradient calculation, a convenient expression for their boundary representation must be used. Consider a boundary point κ on the three dimensional surface and define a local cartesian coordinate system (τ,ρ,n) with origin at κ as shown in fig(4)

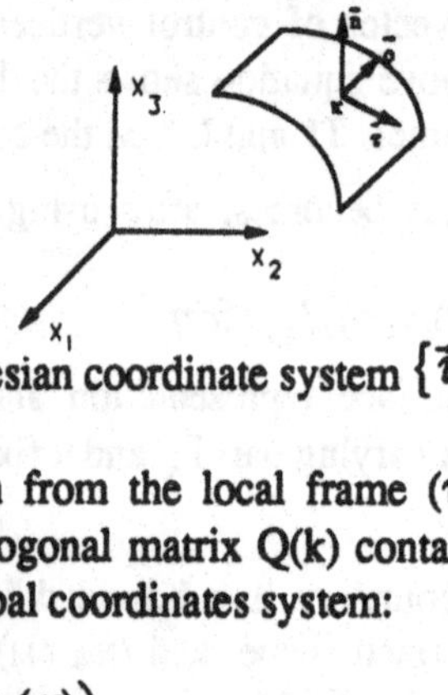

Fig.4 Local cartesian coordinate system $\{\vec{\tau}(\kappa), \vec{\rho}(\kappa), \vec{n}(\kappa)\}$

The matrix transformation from the local frame (τ,ρ,n) to the global frame (x,y,z) is given by an orthogonal matrix Q(k) containing the direction cosines of the local axes in the global coordinates system:

$$Q(\kappa) = \begin{pmatrix} \tau_1(\kappa) & \tau_2(\kappa) & \tau_3(\kappa) \\ \rho_1(\kappa) & \rho_2(\kappa) & \rho_3(\kappa) \\ n_1(\kappa) & n_2(\kappa) & n_3(\kappa) \end{pmatrix} \tag{26}$$

and such that any global vector $\vec{v}$ is transformed into its local components v_i' by:

$$v_i' = Q_{ij}(\kappa) v_j \tag{27}$$

From the definition (5,7) of the tangent vectors, jacobian and outward normal vector, we can define the local cartesian coordinate system as follows[5,9]:

$$\vec{\tau}=\vec{t}_u \;;\; \vec{n}=\vec{n} \;;\; \vec{\rho}=\vec{n}\times\vec{\tau} \tag{28}$$

Then the local tangential displacement and tractions on the boundary are found by projection onto the tangent plane of the global quantities :

$$u_i'(\kappa)=\left[\tilde{N}^e(\kappa)\right]\left\{u_s^e\right\}m_{is} \tag{29}$$

$$t_i'(\kappa)=\left[\tilde{N}^e(\kappa)\right]\left\{t_s^e\right\}m_{is} \tag{30}$$

where i=1,2 ; s=1,2,3 and

$$\vec{m}_{1s}=\vec{\tau}_s \;;\; \vec{m}_{2s}=\vec{\rho}_s \tag{31}$$

The corresponding tangential strains are[5,9]:

$$\varepsilon_{11}'(\kappa)=\sum_{a=1}^{nnel}u_j(\eta^{aq})E_{11j}^{aq} \tag{32}$$

$$\varepsilon_{12}'(\kappa)=\sum_{a=1}^{nnel}u_j(\eta^{aq})E_{12j}^{aq} \tag{33}$$

$$\varepsilon_{22}'(\kappa)=\sum_{a=1}^{nnel}u_j(\eta^{aq})E_{22j}^{aq} \tag{34}$$

where η^{aq} is the interpolation point on the element;

$$E_{11j}^{q}=\frac{1}{h^q(\xi_1,\xi_2)}\tau_j(\kappa)\left[\frac{\partial\tilde{N}(\xi_1,\xi_2)}{\partial\xi_1}\right] \tag{35}$$

$$E_{12j}^{q}=\frac{1}{2g^q(\xi_1,\xi_2)}\left(\tau_j(\kappa)h^q(\xi_1,\xi_2)\left[\frac{\partial\tilde{N}(\xi_1,\xi_2)}{\partial\xi_2}\right]-A^q(\xi_1,\xi_2)E_{11j}^{q}\right)+\frac{1}{2h^q(\xi_1,\xi_2)}\rho_j(\kappa)\left[\frac{\partial\tilde{N}(\xi_1,\xi_2)}{\partial\xi_1}\right] \tag{36}$$

$$E_{22j}^{q}=\frac{1}{g^q(\xi_1,\xi_2)}\rho_j(\kappa)\left(h^q(\xi_1,\xi_2)\left[\frac{\partial\tilde{N}(\xi_1,\xi_2)}{\partial\xi_2}\right]-\frac{A^q(\xi_1,\xi_2)}{h^q(\xi_1,\xi_2)}\left[\frac{\partial\tilde{N}(\xi_1,\xi_2)}{\partial\xi_1}\right]\right) \tag{37}$$

$$h^q(\xi_1,\xi_2)=\left\|\vec{T}_u\right\| \;;\; k^q(\xi_1,\xi_2)=\left\|\vec{T}_v\right\| \tag{38}$$

$$A^q(\xi_1,\xi_2)=\vec{T}_u\bullet\vec{T}_v \;;\; g^q(\xi_1,\xi_2)=\left\|\vec{T}_u\times\vec{T}_v\right\|=\left\|\vec{n}(\xi_1,\xi_2)\right\| \tag{39}$$

The local cartesian components of the stress tensor at the point κ on the boundary can be obtained with Hooke's law , and written in local frame[5,9]:

$$\sigma_{ij}(\kappa)=2\mu(\varepsilon_{ij}(\kappa)+\frac{\nu}{1-2\nu}\varepsilon_{kk}\delta_{ij}) \tag{40}$$

Once we know the surface stresses on each element, it is easy to compute the equivalent Von Mises stress:

$$\sigma^2_{vm}(\kappa)=\langle\sigma\rangle[V]\{\sigma\} \tag{41}$$

where {σ} is stress tensor component and [V] a matrix of constants coefficients[17].

From the above equations(32)-(41) we see that the stress variations can be expressed in terms of tangential strain variation and normal traction variations, which are obtained by BEM sensitivity analysis.

4) SENSITIVITY ANALYSIS

Now, if one wants to perform a sensitivity analysis of the Von Mises stresses, we need to evaluate the following expression [6,7], obtained from eq (23) :

$$\frac{\partial C_{ij}}{\partial p}u+C_{ij}\frac{\partial u}{\partial p}=\sum_{e=M+1}^{NELT}\left[\int_{\Gamma_v}\left(\frac{\partial T^*_{ij}}{\partial p}u+T^*_{ij}\frac{\partial u}{\partial p}\right)ds+\int_{\Gamma_v}T^*_{ij}u\frac{\partial ds}{\partial p}\right]$$
$$-\sum_{e=M+1}^{NELT}\left[\int_{\Gamma_v}\left(\frac{\partial G^*_{ij}}{\partial p}t+G^*_{ij}\frac{\partial t}{\partial p}\right)ds+\int_{\Gamma_v}G^*_{ij}t\frac{\partial ds}{\partial p}\right] \tag{42}$$

Using the substructuring technique [6,11], integration on the varying part only will be required. Otherwise, we must integrate on the whole boundary, because the varying boundary element influences the fixed one. After the discretisation and the assembling process we obtain:

$$[H]\frac{\partial\{u\}}{\partial p}=[G]\frac{\partial\{t\}}{\partial p}+\left(\frac{\partial[G]}{\partial p}.\{t\}-\frac{\partial[H]}{\partial p}.\{u\}\right) \tag{43}$$

where [H], [G] are matrices obtained in standard BEM program so that we need to evaluate only $\frac{\partial G}{\partial p}$ and $\frac{\partial H}{\partial p}$. Introducing the boundary conditions, we obtain, after rearrangement using the standard BEM procedures [1,5], the following equation:

$$[A]\frac{\partial\{x\}}{\partial p}=\left(\frac{\partial[G]}{\partial p}.\{t\}-\frac{\partial[H]}{\partial p}.\{u\}\right)=\frac{\partial[-A]}{\partial p}.\{x\} \tag{44}$$

for the determination of the displacement and traction variation (sensitivity) with respect to the design parameters. These are required to compute the stress sensitivities on the boundary using equations(29)-(41). Matrix [A] in equation (44) is the same as the one appearing in standard BEM analysis and its derivative can be calculated by direct differentiation and numerical integration of kernels T* and U* and assembled in the same way.

The knowledge of the derivative $\left[\frac{\partial A}{\partial p}\right]$ of the state matrix requires the evaluation of the derivatives of the singular kernels with respect to p and involves the evaluation of $\frac{\partial T_{ij}^*}{\partial p}, \frac{\partial U_{ij}^*}{\partial p}$.

Since :

$$T_{ij}^*(x^\alpha, y) = \frac{\phi_1}{r^2}\left\{\left[\phi_2\delta_{ij} - 3\frac{r_i r_j}{r^2}\right]\frac{r_k . n_k}{r} - \phi_3\left(\frac{r_i n_j}{r} - \frac{r_j n_i}{r}\right)\right\} \tag{45}$$

$$U_{ij}^*(x^\alpha, y) = \frac{\phi_4}{r}\left\{\phi_5\delta_{ij} - \frac{r_i r_j}{r^2}\right\} \tag{46}$$

where $\phi_1, \phi_2, \phi_3, \phi_4, \phi_5$ are material constants [1].

One can show that for the points neighbouring the source point, the distance r and its components r_i are of the same order of magnitude. Then the derivative of r_i with respect to P is of order ε and the kernels derivatives become $\frac{\partial T_{ij}}{\partial p} = O(\frac{1}{\varepsilon^3}), \frac{\partial U_{ij}}{\partial p} = O(\frac{1}{\varepsilon^2})$ hence increasing the order of their singularity. Therefore, special numerical integration routines must be used for the accurate evaluation of integrals containing such kernels.

To assess the validity of the developed formulation we applied it to benchmark problem treated in reference[5,15]. The numerical results will be presented later.

5)- CONCLUSION

It is shown that the integration of free form surface modelling technique and BEM analysis is suited for sensitivity analysis hence it is a very useful tool in design optimization, since one can work with a real surface and the shape optimisation problem uses natural CAD parameters used daily in the interactive computer aided design and integrated in the boundary element analysis through superparametric element. The B-spline properties ensure that one obtains well behaved shapes.

REFERENCES

1. C.A Brebbia, "Boundary Elements Techniques", Springer Verlag, (1984).
2. M.E Mortenson, "Geometric Modeling", Wiley,New York,(1985).
3. D.F Rogers and J.A Adams,"Mathematical Elements for Computer Graphics", McGraw Hill, (1990).
4. Tse-Min Yao and K.K Choi, "3D Shape optimal design and automatic finite Element regriding", Int. J. Num. Meth. Eng,Vol.28,369-384(1989).

5. A.Gakwaya, "Développement et application de la méthode des éléments finis de frontières à la mecanique des solides", Ph. D. Thesis, Université Laval, (1983).
6. E. Sandgren and S.J. Wu, "Shape optimization using the boundary element method with substructuring", Int. J. Num. Meth. Eng., Vol.26, pp 1913-1924,(1988).
7. S.SAigal,R.Aithal and J.H. Kane, "Conforming Boundary element plane elasticity for shape design sensitivity", Int. J. Num. Meth. Eng., vol.28, 2795-2811 (1989).
8. A.D. Belegundu and S.D. Rajan, "A shape optimization approach based on natural design variables and shape functions", Comput.Meth.Appl.Mec.Eng. vol 66,pp87-106, (1988).
9. V.Sladek and J.Sladek "Improved compution of stress using the boundary element method", Appl. Math. Modelling, vol.10, August 249-255,(1986)
10. G. J. Peter,"Interactive computer graphics application of the parametric Bi-cubic surface to engineering design problems",in Computer Aided Geometric Design. Academic press (1974).
11. J.H. Kane, "Optimization of continuum structures using a boundary element formulation", Ph.D. Thesis, University of connecticut, (1986).
12. V. Braibant, " Optimisation de forme des structures en vue de la conception assistée par ordinateur", Ph.D. thesis, Université de Liège (1985).
13. M Hassan Imam, " Three dimensional shape optimisation ", Int. J. Num Meth. Eng., Vol 18,661-673,(1982).
14 M. Defourny, " OPtimization techniques and BEM", BEM IX, vol.3, PP 462-472, Computational Mechanics Publications, Springer-Verlag,1987.
15. G. Dhatt, G. Touzot,"Une presentation de la Methode des Elements Finis" , Presses de l'Universite Laval, Maloine, (1981).
16. A. GAKWAYA, G.DHATT, A. CARDOU, " An implementation of stress discontinuity in the boundary element method and application to gear teath" Applied Math. Modelling, vol. 8, pp 319-327, (1984).
17. M. N6 and J.M Aguinagald,"Finite element method and optimality criteria based structural optimization", Computer & structures, vol.27 no.2, pp 287-295,(1987).

Stress Analysis of the Turbine Rotor Disk by the Axisymmetric Boundary Element Method

J.W. Kim

Department of Mechanical Engineering, Korea Military Academy, Seoul, 139-799, Korea

ABSTRACT

The BEM for linear elastic stress analysis is applied to the rotating axisymmetric body problem which also involves the thermoelastic effects due to steady-state thermal conduction. The axisymmetric BEM formulation is briefly summarized and an alternative approach for transforming the volume integrals associated with such body force kernels into equivalent boundary integrals is described in a way of using the concept of inner product and vector identity. A discretization scheme is outlined for numerical treatment of the resulting boundary integral equations, and it is consequently illustrated by determining the stress distributions of the turbine rotor disk for which a FEM stress solution has been furnished by author.

INTRODUCTION

In a class of stress analysis problems, axisymmetric elastic bodies are subjected to boundary tractions and displacements as well as thermal and centrifugal body forces. The strength design of gas turbojet engines involves a large number of axisymmetric parts such as the compressor and turbine disks. It is clear in stress analysis of such a turbine rotor disk that inclusion of rotational inertia body force loading and steady-state thermoelastic effects are major importance.

Application of the Boundary Element Method(BEM) to thermoelastic and centrifugal axisymmetric loading problems was first formulated by Cruse et al.[1] but limited numerical implementation to linear elements sometimes circular arcs in shape. Recently further developments included the use of isoparametric quadratic elements for thermoelastic problems (Bakr and Fenner [2]) and rotating axisymmetric body problems (Abdul-Mihsein et al.[3]). Numerical examples are however confined to simple axisymmetric geometries such as thick-walled cylinder, hollow sphere, and rotating disk of uniform thickness, and separately treated in centrifugal body force and thermal loading problems.

The main purpose of present study is to determine the stresses of the turbine rotor disk subjected to both thermal and rotational inertia force by the axisymmetric BEM. While the formulation of the method follows closely in spirit the ideas outlined for thermoelastic analysis by Bakr and Fenner [2] and centrifugal loading by Abdul-Mihsein et al.[3], combination of two loadings is magnified by numerical treatment procedure. The development of the axisymmetric BEM is summarize in order to motivate the subsequent numerical extension to thermoelastic rotating axisymmetric body problems and new procedure for deriving the fundamental thermal and centrifugal body force kernels on equivalent boundary integral are presented in a efficient way of using the concept of inner product and invariant vector operation. Finally axisymmetric BEM solutions to the model problem of the turbine rotor disk in small turbojet engines are presented and compared with the axisymmetric FEM solutions.

AXISYMMETRIC BEM FORMULATIONS

Consider an axisymmetric body of arbitrary cross-sectional shape with an interior point p in Ω and a boundary point Q on $\partial\Omega$ as shown in Fig. 1. Following the analysis of the weighted residual statements or the Betti's reciprocal theorem, the interior displacement solution formula can be obtained as known the Somigliana integral for centrifugal and thermoelastic loading (Cruse et al.[1]) which is given by

$$u_i(p) = -\int_{\partial\Omega} T_{ji}(p,Q)u_j(Q)ds + \int_{\partial\Omega} U_{ji}(p,Q)t_j(Q)ds$$

$$+\rho\omega^2\int_{\Omega} U_{ji}(p,q)x_i(q)dV + \frac{\alpha E}{1-2\nu}\int_{\Omega} U_{ji,j}(p,q)\Phi(q)dV \qquad (1)$$

where u_i is the displacements, t_i is the tractions, Φ is the temperature and x_i is the location relative to the center of rotation of q in Ω. The constants are mass density ρ, angular velocity ω, Poisson's ratio ν, Young's modulus E and coefficient of thermal expansion α. T_{ji} and U_{ji} are the traction and displacement kernel functions associated with the three dimensional fundamental solutions to the Navier's equation of equilibrium.

For the axisymmetric BEM formulation two alternative approaches have been used to derive the kernel functions. The first (Rizzo and Shippy [4]) is to integrate the three dimensional Kelvin's solutions along the circular path around the axis of rotational symmetry. The second approach is to obtain the axisymmetric component form of Galerkin vector directly (Kermanidis [5]; Cruse et al.[1]) and is used the one here. As described in the second approach the complementary displacement vector $\underline{u}^*$ associated with the displacement kernel can be represented in terms of Galerkin vector $\underline{G}$ as

$$\underline{u}^* = \nabla^2\underline{G} - \frac{1}{2(1-\nu)}\underline{\nabla}(\underline{\nabla}\cdot\underline{G}) \qquad (2)$$

Then substitution of this expression into the Navier's equation gives the following biharmonic equations in radial $-r$ and axial $-z$ direction, respectively

$$\left(\nabla^2-\frac{1}{r^2}\right)\left(\nabla^2-\frac{1}{r^2}\right)G_r = -\frac{F_r}{\mu}$$

$$\nabla^2\nabla^2 G_z = -\frac{F_z}{\mu} \qquad (3)$$

where (G_r, G_z) are the Galerkin vector components, (F_r, F_z) are the fundamental body force components which can be represented through the use of Dirac delta function, and μ is shear modulus. Integral transform methods (ex. Hankel transform by Bakr and Fenner [2]) can be then used to obtain solutions for (G_r, G_z) from Eq.(3) and the results are given in terms of Legendre functions of the second kind

$$
\begin{aligned}
G_r &= \sqrt{(Rr)}\sqrt{(\gamma^2-1)}\frac{Q^{-1}_{+1/2}(\gamma)}{8\pi^2\mu} \\
G_z &= \sqrt{(Rr)}\sqrt{(\gamma^2-1)}\frac{Q^{-1}_{-1/2}(\gamma)}{8\pi^2\mu} \\
\gamma &= 1+\frac{[(Z-z)^2+(R-r)^2]}{2Rr}
\end{aligned}
\tag{4}
$$

where (R, Z) represent the components of load point p and (r, z) the field point q in Ω. The displacement kernels U_{ij} are then determined by substitution of Eq.(4) into the relations

$$
\begin{aligned}
2(1-\nu)U_{rr} &= (1-2\nu)\left(\nabla^2-\frac{1}{r^2}\right)G_r+\frac{\partial^2 G_r}{\partial z^2} \\
2(1-\nu)U_{zr} &= -\frac{\partial^2 G_r}{\partial r\partial z}\left(\frac{1}{r}\right)\frac{\partial G_r}{\partial z} \\
2(1-\nu)U_{rz} &= -\frac{\partial^2 G_z}{\partial r\partial z} \\
2(1-\nu)U_{zz} &= (1-2\nu)\nabla^2 G_z+\frac{\partial^2 G_z}{\partial r^2}+\left(\frac{1}{r}\right)\frac{\partial G_z}{\partial r}
\end{aligned}
\tag{5}
$$

Now the volume integrals of last two terms in Eq.(1) limits the potential benefits of BEM formulation. In order to reduce the volume integrals to equivalent boundary integrals in Eq.(1) it is convenient to use the concept of inner product of body force and fundamental displacement. The first volume integral in Eq.(1) associated with the body force vector for rotational inertia loading has the following inner product of body force $\underline{b}$ and fundamental displacement $\underline{u}^*$

$$
\begin{aligned}
\langle \underline{b}, \underline{u}^* \rangle_\Omega &= \rho\omega^2 \int_\Omega (\underline{x}\cdot\underline{u}^*)dV \\
&= \rho\omega^2 \int_\Omega \left[\nabla^2\underline{G}-\frac{1}{2(1-\nu)}\underline{\nabla}(\underline{\nabla}\cdot\underline{G})\right]\cdot\underline{x}dV
\end{aligned}
\tag{6}
$$

Substitution of the following vector identity

$$
\underline{\nabla}(\underline{\nabla}\cdot\underline{G}) = \nabla^2\underline{G} + \underline{\nabla}\times(\underline{\nabla}\times\underline{G}) \tag{7}
$$

and Green theorem for volume integrals to surface integrals

$$\int_{\Omega} \underline{x}\cdot[\underline{\nabla}\times(\underline{\nabla}\times\underline{G})]dV = \int_{\partial\Omega} \underline{x}\cdot(\underline{n}\times\underline{\nabla}\times\underline{G})dS$$

$$\int_{\Omega} \underline{x}\cdot\nabla^2\underline{G}dV = \int_{\partial\Omega}[\underline{x}(\underline{\nabla}\cdot\underline{G}-\underline{G}(\underline{\nabla}\cdot\underline{x})]\cdot\underline{n}dS - \int_{\Omega}[\underline{x}\cdot\underline{\nabla}\times(\underline{\nabla}\times\underline{G})]dV \qquad (8)$$

Eq.(6) can be then directed to the equivalent boundary integral

$$\rho\omega^2\int_{\Omega} \underline{x}\cdot\underline{u}^{*}dV = \rho\omega^2\frac{(1-2\nu)}{2(1-\nu)}\int_{\partial\Omega}[\underline{x}(\underline{\nabla}\cdot\underline{G}-\underline{G}(\underline{\nabla}\cdot\underline{x})]\cdot\underline{n}dS$$

$$-\rho\omega^2\int_{\partial\Omega} \underline{x}\cdot(\underline{n}\times\underline{\nabla}\times\underline{G})dS \qquad (9)$$

In order to reduce the second volume integral in Eq.(1) consider thermal body force and boundary traction of heat flux

$$\underline{b} = -\frac{\alpha E}{(1-2\nu)}\underline{\nabla}\Phi, \qquad \underline{t} = \frac{\alpha E}{(1-2\nu)}\Phi\underline{n} \qquad (10)$$

Then application of two inner product of above thermoelastic effects and fundamental displacement $<\underline{b},\underline{u}^{*}>_{\Omega}, <\underline{t},\underline{u}^{*}>_{\partial\Omega}$ to the Betti's reciprocal theorem being the basis for derivation of Somigliana identity yields an equivalent representation to the second volume integral in terms of invarient vector form as

$$\frac{\alpha E}{(1-2\nu)}\int_{\Omega}\Phi(\underline{\nabla}\cdot\underline{u}^{*})dV = \frac{\alpha E}{(1-2\nu)}\int_{\Omega}\Phi\left\{\underline{\nabla}\cdot\left[\nabla^2\underline{G}-\frac{1}{2(1-\nu)}\underline{\nabla}(\underline{\nabla}\cdot\underline{G})\right]\right\}dV \qquad (11)$$

Use of the vector identity such as $\underline{\nabla}\cdot(\nabla^2\underline{G}) = \nabla^2(\underline{\nabla}\cdot\underline{G})$ and condition of steady-state heat conduction $\nabla^2\Phi = 0$ in Ω. Proceeding the same argument to Green theorem, Eq.(11) can be finally reduced to the following boundary integral equation

$$\frac{\alpha E}{(1-2\nu)}\int_{\Omega}\Phi(\underline{\nabla}\cdot\underline{u}^{*})dV = \frac{\alpha E}{2(1-\nu)}\int_{\partial\Omega}[\Phi\underline{\nabla}(\underline{\nabla}\cdot\underline{G}-(\underline{\nabla}\cdot\underline{G})\underline{\nabla}\Phi]\cdot\underline{n}dS \qquad (12)$$

These two transformed Eqs.(9) and (12) can be utilized to construct the corresponding boundary integral formula for the displacement.

Reduction of Somigliana identity of Eq.(1) to the boundary integral representation is accomplished by taking the limit of each term as the interior point p approaches a boundary point P. The resulting integral equations for the boundary displacement can be expressed as

$$C_{ij}(P)u_j(P) = \int_{\partial\Omega}[U_{ij}(P,Q)t_j(Q) - T_{ij}(P,Q)u_j(Q) + \alpha V_{ij}(P,Q)\Phi_j(Q)$$

$$+\rho\omega^2 W_i(P,Q)]dS \qquad (13)$$

where the subscripts i and j range over r and z except temperature Φ_1, normal derivative of $\Phi_2 = \underline{\nabla}\Phi \cdot \underline{n}$. The thermal loading kernels V_{ij} and rotational inertia force kernels W_i are defined from Eqs.(12) and (9), respectively. The coefficients C_{ij} can be evaluated in the limiting process used in deriving boundary integral equation. This procedure is described in detail by Cruse et al.[1] and Hartmann [7].

NUMERICAL TREATMENT AND RESULTS

The boundary contour $\partial\Omega$ of the typical Θ-plane in axisymmetric body is divided into elements and the integration of Eq.(13) performed over each element. The coordinates of any point on the element can be expressed in terms of the usual quadratic interpolation functions as

$$r(\xi) = N_i(\xi) r_i \qquad ; \qquad z(\xi) = N_i(\xi) z_i$$

where

$$N_1(\xi) = \frac{1}{2}(1-\xi) - \frac{1}{2}(1-\xi^2) \tag{14}$$

$$N_2(\xi) = \frac{1}{2}(1+\xi) - \frac{1}{2}(1-\xi^2)$$

$$N_3(\xi) = (1-\xi^2)$$

Similarly the displacement and traction can be also represented as

$$\begin{aligned} u_r(\xi) &= N_i(\xi) u_{ri} \qquad ; \qquad u_z(\xi) = N_i(\xi) u_{zi} \\ t_r(\xi) &= N_i(\xi) t_{ri} \qquad ; \qquad t_z(\xi) = N_i(\xi) t_{zi} \end{aligned} \tag{15}$$

The discretization of the boundary integral equation of (13) can consequently yields in axisymmetric form

$$C_{rr}(P)u_r(P) + C_{rz}(P)u_z(P) =$$

$$2\pi \sum_{m=1}^{M} \sum_{n=1}^{3} \left[\begin{array}{c} -u_{rn}^{m}(Q) \int_{-1}^{+1} T_{rr}(P,Q(\xi)) \\ -u_{zn}^{m}(Q) \int_{-1}^{+1} T_{rz}(P,Q(\xi)) \\ +t_{rn}^{m}(Q) \int_{-1}^{+1} U_{rr}(P,Q(\xi)) \\ +t_{zn}^{m}(Q) \int_{-1}^{+1} U_{rz}(P,Q(\xi)) \\ +\Phi_{n}^{m}(Q) \int_{-1}^{+1} V_{r1}(P,Q(\xi)) \\ +\frac{d\Phi}{dn}{}_{n}^{m}(Q) \int_{-1}^{+1} V_{r2}(P,Q(\xi)) \\ +\rho\omega^2 \frac{(1-2\nu)}{2(1-\nu)} \int_{-1}^{+1} W_1(P,Q(\xi)) \\ +\rho\omega^2 \int_{-1}^{+1} W_2(P,Q(\xi)) \end{array} \right] N_m(\xi) r(\xi) J(\xi) d\xi \tag{16}$$

$$C_{zr}(P)u_r(P)+C_{zz}(P)u_z(P) = 2\pi\sum_{m=1}^{M}\sum_{n=1}^{3}\begin{bmatrix} -u_{rn}^{m}(Q)\int_{-1}^{+1} T_{zr}(P,Q(\xi)) \\ -u_{zn}^{m}(Q)\int_{-1}^{+1} T_{zz}(P,Q(\xi)) \\ +t_{rn}^{m}(Q)\int_{-1}^{+1} U_{zr}(P,Q(\xi)) \\ +t_{zn}^{m}(Q)\int_{-1}^{+1} U_{zz}(P,Q(\xi)) \\ +\Phi_{n}^{m}(Q)\int_{-1}^{+1} V_{z1}(P,Q(\xi)) \\ +\frac{d\Phi}{dn}{}_{n}^{m}(Q)\int_{-1}^{+1} V_{z2}(P,Q(\xi)) \end{bmatrix} N_m(\xi)r(\xi)J(\xi)d\xi \qquad (17)$$

where M is the total number of elements, and $J(\xi)$ is the Jacobian of transformation. u_{rn}^{m} denotes the radial displacement component of n node in the m_{th} element. Each node on the boundary is taken in turn as the load point and the indicated numerical integration performed (ex. Gaussian quadrature) over the entire boundary, leading to a set of linear algebraic equations which can be finally given by

$$[H]\{u\} = [G]\{t\}+[V_1]\{\Phi\}+[V_2]\left\{\frac{d\Phi}{dn}\right\} + \rho\omega^2\frac{(1-2\nu)}{2(1-\nu)}[W_1]+\rho\omega^2[W_2] \qquad (18)$$

Matrices $[H]$ and $[G]$ contain the integrals of traction and displacement kernels, respectively, while $[V_1]$ and $[V_2]$ involve the integrals of thermoelastic kernels defined from Eq.(12), and $[W_1]$ and $[W_2]$ contain the integrals of the rotational body force kernels from Eq.(9). Before solving Eq.(18), the boundary conditions are applied. These take the form of either prescribed displacement or tractions over each element. The equation can be rearranged such that all the unknown displacements and tractions are on the left hand side and all the known quantities including centrifugal and thermal loading terms are on the right and reduced the final stiffness equation as

$$[A]\{x\} = \{y\} \qquad (19)$$

The stiffness $[A]$ is in general fully populated with non-zero coefficients, and is not symmetric. The equation are best solved by direct elimination procedure like Gauss-Jordan technique.

Fig. 2 shows the three dimensional CAD configuration of the turbine rotor disk. Due to symmetry only half the disk is modelled with 46 quadratic elements and 92 nodal points as shown in Fig. 3. The same disk is analyzed using the FEM with 104 quadrilateral elements and 135 nodal points as also shown in Fig. 4. The material properties and loading data are as follows ;

Young's modulus, E	= 197.2 GPa,
Poisson's ratio	= 0.3
mass density	= 8000 kg/m^3

rim loading	= 47.29 MPa
angular velocity, ω	= 4241.15 rad/s
Coefficient of thermal expansion	$= 9.88 \times 10^{-6}$ m/m°C
temperature at inner surface, T_i	= 300 °C
temperature at outer surface, T_o	= 900 °C

The temperature distribution is prerequisite for such a boundary value problem. This can be obtained from an independent routine by solving the Laplace equation of steady-state heat conduction

Table 1 shows the results of the BEM and FEM hoop stress solutions together with the corresponding temperature distribution, and Fig. 5 depicts the comparison of these two stress distributions. Turbine disk is usually made thicker near its hub and taper down to a smaller thickness towards the periphery. The reason for this is the hoop stress concentration near center of rotation as clearly illustrated in Fig. 5. The BEM results of the radial stress distribution are also compared with the FEM one in Fig. 6. The close agreement between the BEM and FEM results confirms the accuracy of both model.

CONCLUSIONS

The present study is an application of the BEM to the stress analysis problem for which axisymmetric body is subjected to thermal loading and rotational inertia body forces. An extension of the axisymmetric BEM formulation has been demonstrated to construct such kernel functions over equivalent boundary integrals. A discretization scheme is outlined for numerical treatment of the resulting boundary integral equations, and it is finally illustrated by solving the model problem of the turbine rotor disk for which a FEM solution has been furnished by author.

ACKNOWLEDGEMENTS

The author is grateful for the support provided by a grant from the Korea Science and Engineering Foundation.

REFERENCES

1. Cruse, T. A., Snow, D. W. and Wilson, R. B., Numerical Solutions in Axisymmetrc Elasticity, Comp.& Struct., Vol.7, pp. 445 - 451, 1977.

2. Bakr, A. A. and Fenner, R. T., Boundary integral Equation Analysis of Axisymmetric Thermoelastic Problems, J. Strain Analysis, Vol. 18, pp. 239 - 251. 1983.

3. Abdul-Mihsein, M. T., Bakr, A. A. and Parker, A. P., Stresses in Axisymmetric Rotating Bodies Determined by the by the Boundary Integral Equation Method, J. Strain Analysis, Vol. 20, pp. 79 - 86, 1985.

4. Rizzo, F. J. and Shippy, D. J., A Boundary Element Method for Axisymmetric Elastic Bodies, Developments on BEM, Vol. 4(Banerjee and Watson Eds.) Elsevier Applied Sc. Pub., pp. 67 - 90, 1986.

5. Kermanidis, T., A Numerical Solutions for Axially Symmetrical Elasticity Problems, Int. J. Solids. Struct., Vol. 11, pp.493 - 500, 1975.

6. Bakr, A. A. and Fenner, R. T., Use of the Hankel Transform in Boundary Integral Methods for Axisymmetric Problems, Int. J. Numer. Meths. Engng., Vol. 19, pp. 1765 - 1769, 1983.

7. Hartmann, F., Computing the C - Matrix in Non-smooth Boundary points, New Developments in BEM(Brrebia, C. A., Eds.) Butterworths, pp. 367 - 379, 1980.

Table 1. Comparison of BEM and FEM solution for the hoop stress in the turbine rotor disk

Radius $R\times10$ (*mm*)	Temperature $\Phi\times10^2$ (°C)	Temp. Gradient $d\Phi/dn\times10$ (°C/*mm*)	BEM Sol. $\sigma_\theta\times10^2$ (MPa)	FEM Sol. $\sigma_\theta\times10^2$ (MPa)
0.749	3.000	-1.676	9.584	-
0.940	3.292	0.0	7.936	8.550
1.295	3.711	0.0	5.764	6.371
1.422	4.037	0.0	4.751	4.992
2.007	3.213	0.0	4.206	4.399
2.362	4.562	0.0	3.882	4.109
2.692	4.784	0.0	3.696	3.896
3.073	5.040	0.0	3.509	3.647
3.581	5.390	0.0	3.254	3.482
4.013	5.699	0.0	2.965	3.185
4.394	5.976	0.0	2.641	2.675
4.750	6.232	0.0	2.468	2.324
5.080	6.465	0.0	2.144	1.931
5.385	6.671	0.0	1.779	1.627
5.690	6.862	0.0	1.393	1.317
6.020	7.052	0.0	1.041	0.965
6.350	7.236	0.0	0.896	0.820
6.655	7.416	0.0	0.800	0.958
6.934	7.601	0.0	0.669	0.869
7.163	7.771	0.0	0.469	0.607
8.052	8.545	0.0	-0.820	-0.586
8.306	8.760	0.0	-1.538	-1.558
8.463	8.876	0.0	-2.103	-2.310
8.651	9.000	0.632	-3.199	-

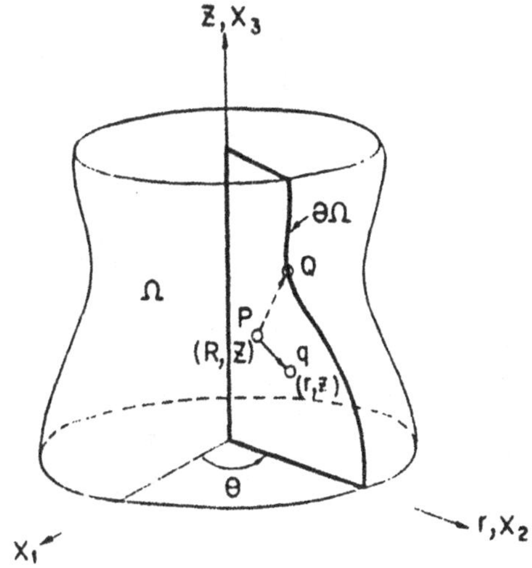

Fig.1 Geometry of the axisymmetric solution domain

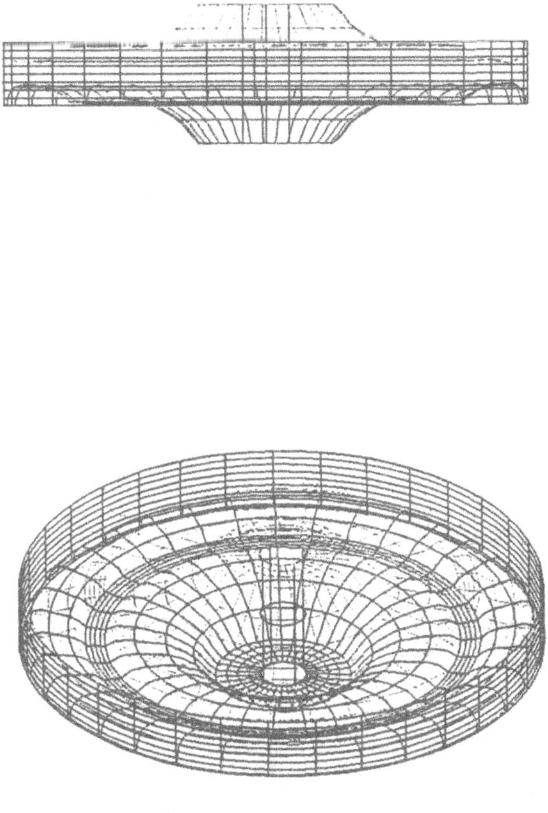

Fig.2 3-D Configuration of the turbine rotor disk

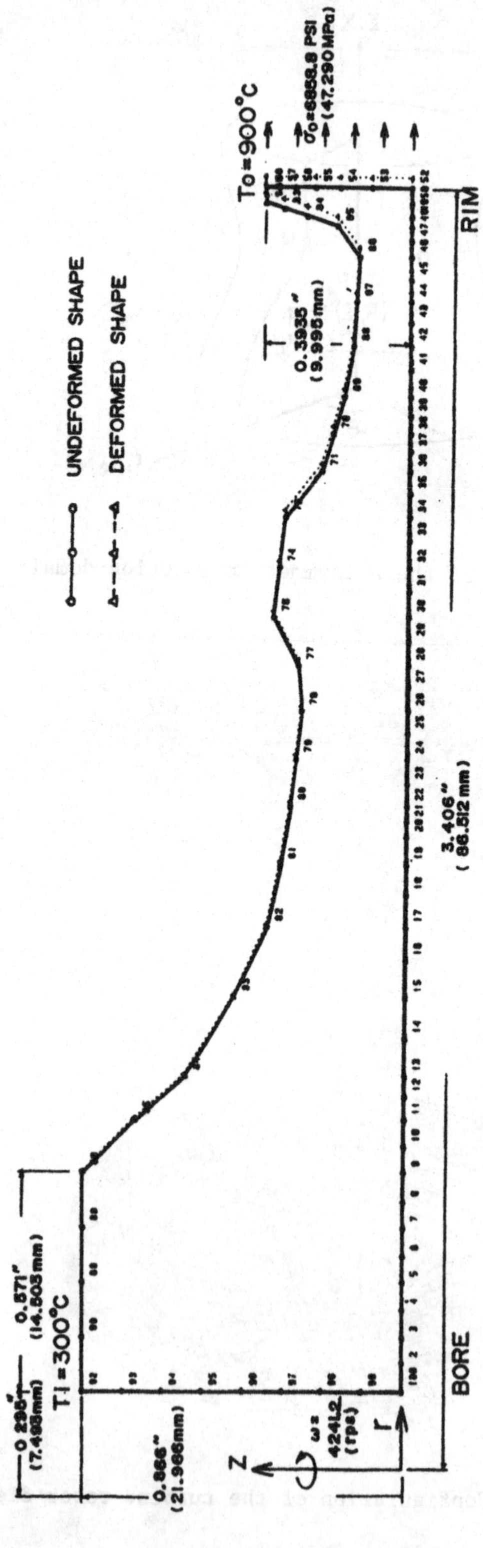

Fig.3 BEM meshes for the turbine rotor disk

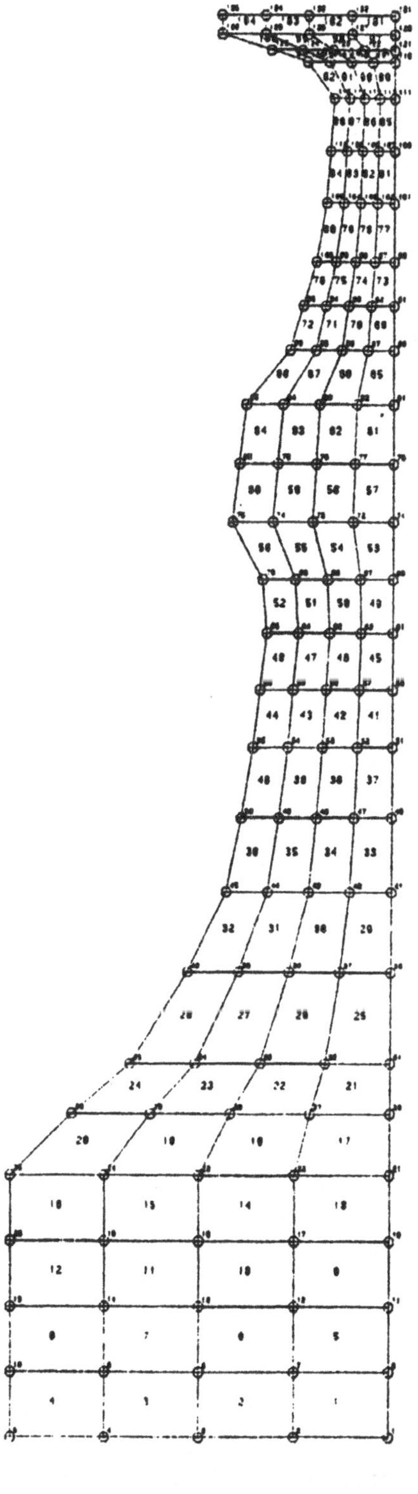

Fig.4 FEM meshes for the turbine rotor disk

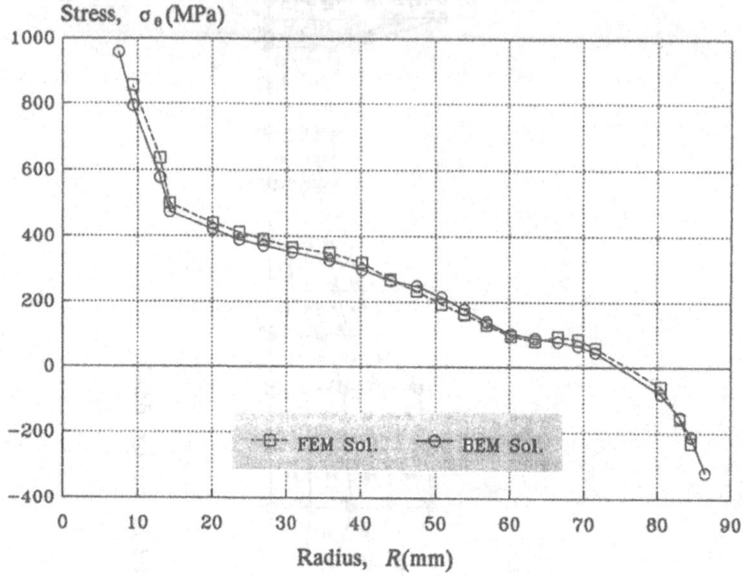

Fig.5 Hoop stress distributions in the turbine rotor disk

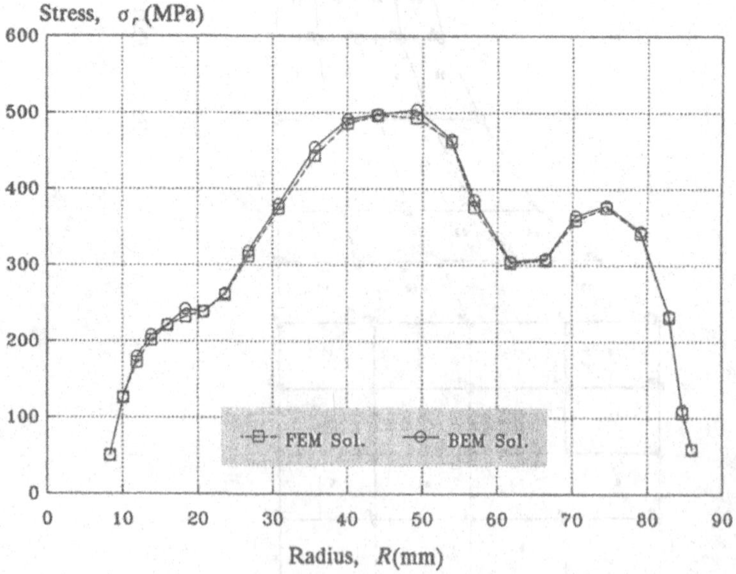

Fig.6 Radial stress distribution in the turbine rotor disk

3-D Spline Boundary Element Method for the Analysis of Sluice*

Y. Wang, Y. Zhang, G. Zhang, T. Dun

Department of Civil Engineering, Hefei University of Technology, 230009 Hefei, P.R. China

ABSTRACT

We analyse the sluice chamber by 3-D spline boundary element method. The floor, piers and intercepting walls are its substructures, the traffic bridge, operating bridge and altar are treated as internal supports. The foundation and neighbouring loads may be arbitrary, but it needs the known settlement field of earth's surface under unit forces. Even with coarse division, we can obtain the displacement field, stress field and reaction field of chamber in any operating modes with fine precision, whether taking the planar gate scheme or the radial one.

INTRODUCTION

The sluice chamber structure is very complicated and its operating modes are variable. In engineering, the analyses used , such as Zemochkin's method and Gorbunov-Posadov method etc, all approximately simplify the chamber as finite beam or beam system, it is very coarse. The variations analysis of chamber structure, which is put forward by Fu Zuoxing in the 1980s, remedies this defect to some extent, but the displacement function and the number of series terms are all varied as chamber construction and operating modes, and short of strict proof on calculating precision too, so it has some limitations for employment. However, for 3-D finite element method for the analysis of chamber,it is difficultly developed to engineering calculation with a massive scale of solution and lower stress precision. In this paper, we adopt a new method, i.e. we analyse chamber as a whole by using 3-D spline boundary element method. Floor, piers and intercepting walls are treated as its substructures, and traffic bridge, operating bridge and altar as internal supports. In this method, due to adopting fine spline interpolation and limiting unknown quantites only on boundary,so we can obtain fine solutions of stress and displacement for any operating modes even with coarse division. The foundation and neighbouring loads may be arbitrary, but it needs the known settlement field of the earth's surface under unit forces. Meanwhile, some new techniques[4].[5] are adopted for calculating boundary stress and fast forming coefficient matrix, it is further favorable to develop the present method into engineering or into the good evaluation of simplifed methods.

SPLINE BOUNDARY ELEMENT METHOD

Betti theorem is the theoretical foundation of Boundary Element Method. According to it, we write easily the boundary integral equation as

$$c_{ij} u_j(\xi) + \int_\Gamma p_{ij}^* u_j \, d\Gamma = \int_\Gamma u_{ij}^* p_j \, d\Gamma \qquad (1)$$

in which Γ is the boundary of structure(Fig.1), i,j=1,2,3, u_j, p_j are the components of boundary displacement and traction respectively. u_{ij}^*, p_{ij}^* are Kelvin solutions and represent the components of displacement and traction at field point X in direction j respectively, when the unit force in direction i acts on the singular point ξ of the infinite domain.

$$u_{ij}^* = \frac{1}{16\pi G(1-\nu) r}\left[(3-4\nu)\delta_{ij} + \frac{r_i}{r}\frac{r_j}{r}\right]$$

$$p_{ij}^* = \frac{1}{8\pi(1-\nu) r^2}\left[(1-2\nu)\left(\frac{r_i}{r}n_j - \frac{r_j}{r}n_i\right) - \left((1-2\nu)\delta_{ij} + \frac{r_i}{r}\frac{r_j}{r}\right)\frac{\partial r}{\partial n}\right]$$

where, $r = (r_i r_i)^{1/2}$, $\frac{\partial r}{\partial n} = \cos(r,n) = \frac{r_i}{r} n_i$, $r_i = x_i(X) - x_i(\xi)$

n_j is the direction cosines of the exterior normal to the point X. c_{ij} are singular coefficients created by the singularity of p_{ij}^*, if ξ is internal point, $c_{ij} = \delta_{ij} = \begin{cases} 1, & i=j \\ 0, & i\neq j \end{cases}$.

In formula (1), the summation convention is adopted that whenever a subscript is repeated in a given term we are to sum over that index. For the body force terms such as dead weight, a steady-state thermal loading, etc. we can transform them to the boundary integral term and add to the right-hand side of the formula(1)[6].

In view of the unknown quantities being purely on the Γ-surface we only need to divide Γ into several boundary elements. If regarding each boundary node as singular point and interpolating the variables by spline functions, we can write the boundary element equation in matrix form

$$HU = GP \qquad (2)$$

where, U,P stand for the column matrix of full displacement components and traction components on the boundary respectively. Singular coefficient terms have been contained in the main submatrix $h_{\underline{ll}}$ of coefficient matrix.

Solving formula(2) and the reaction-settlement equations, we can obtain the displacement field and reaction field.

Using the method of rigid-body movement, we can deduce:

$$\underset{(3\times3)}{h_{\underline{ll}}} = -\sum_{\substack{m=1 \\ m\neq l}}^{n} h_{lm} \qquad (3)$$

that is to say, the main submatrix may be expressed as the

negative algebraic sum of pertinent non-diagonal submatrices, so we can avoid the calculation of singular coefficients and principal values, and N is the total numbers of nodes.

In this paper, unknowns are interpolated by cubic B-spline functions, that is

$$f(\zeta)=\lfloor b_1(\zeta)\rceil [A]^{-1}\{f_1\}$$

in which $f(\zeta)$ is boundary function, $\zeta=S/\Delta S$ (Fig.2), $\{f_1\}$ is a column matrix of nodal parameters, $l=-1,0,1,\ldots, n, n+1$, $b_1(\zeta)$ is cubic B-spline functions[7]. $\lfloor\ \rceil$ represents row matrix, [A] is the interpolating matrix, -1 means inverse.

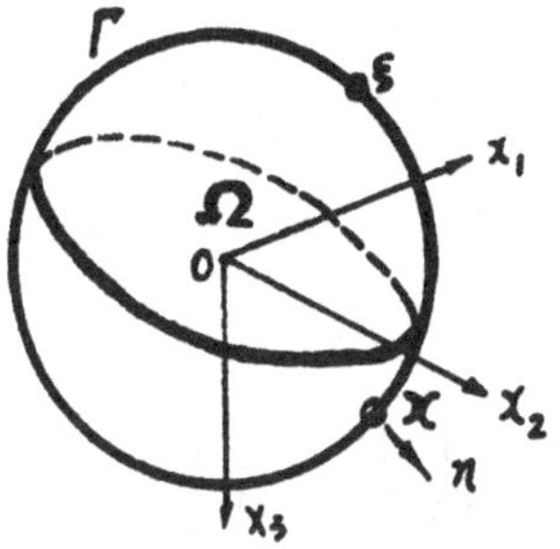

Fig.1

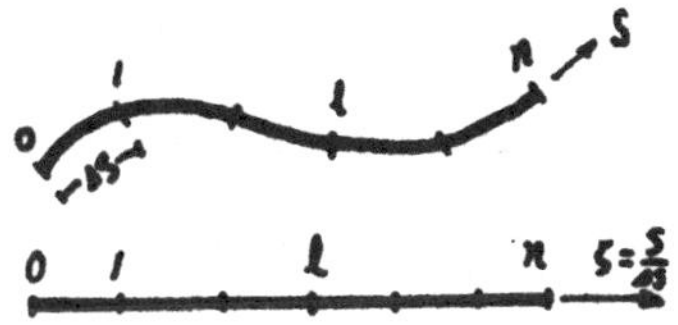

Fig.2

In the case of controlling the first derivative of endpoints, we have

$$\underset{(n+3)(n+3)}{[A]^{-1}}=\begin{bmatrix} -3 & 0 & 3 & & & \\ 1 & 4 & 1 & & & \\ & 1 & 4 & 1 & & \\ & & \cdot & \cdot & \cdot & \\ & & & 1 & 4 & 1 \\ & & & -3 & 0 & 3 \end{bmatrix}^{-1}, \quad \{f_1\}=[f,_\zeta(0)\ f_0 \ldots f_n\ f,_\zeta(n)]^t.$$

The first derivative of end point may be interpolated by Lagrangian functions too, for cubic interpolation, we arrive at

$$\underset{(n+3)(n+3)}{[A]^{-1}}=\begin{bmatrix} -7 & 26 & -34 & 16 & 1 & -2 & \\ 1 & 4 & 1 & & & & \\ & 1 & 4 & 1 & & & \\ & & \cdot & \cdot & \cdot & & \\ & & & 1 & 4 & 1 & \\ & & & & 1 & 4 & 1 \\ & -2 & 1 & 16 & -34 & 26 & -7 \end{bmatrix}^{-1}, \quad f_1 = [0\ f_0\ \ldots\ f_n\ 0]^t$$

In this algorithm, we have still got fine calculating precision even in the non-symmetric case.

In sluice the boundary is 2-D surface. By adopting Kronecker product we can write easily the 2-D spline interpolation formulae of variables, too. [8]

STRESS CALCULATION

After solving for boundary displacement field and reaction field, we can calculate the stress of internal points one by one.

$$\sigma_{ij}(\xi) = \int_\Gamma (u^*_{ijk} p_k - p^*_{ijk} u_k) d\Gamma \qquad \xi \in \Omega \tag{4}$$

in which,

$$u^*_{ijk} = \frac{1}{8\pi(1-\nu)r^2}[3\frac{r_i r_j r_k}{r^3} + (1-2\nu)(\frac{r_i}{r}\delta_{jk} + \frac{r_j}{r}\delta_{ki} - \frac{r_k}{r}\delta_{ij})]$$

$$p^*_{ijk} = \frac{G}{4\pi(1-\nu)r^3}\{3\frac{\partial r}{\partial n}[(1-2\nu)\frac{r_k}{r}\delta_{ij} + \nu(\frac{r_i}{r}\delta_{jk} + \frac{r_j}{r}\delta_{ki}) - 5\frac{r_i r_j r_k}{r^3}]$$
$$+3\nu(\frac{r_i r_k}{r^2} n_j + \frac{r_j r_k}{r^2} n_i) + (1-2\nu)(3\frac{r_i r_j}{r^2} n_k + n_i \delta_{jk} + n_j \delta_{ki}) - (1-4\nu) n_k \delta_{ij}\}$$

The discretization model about formula (4) may be written as

$$\boldsymbol{\sigma}(\xi) = G'P - H'U \qquad \xi \in \Omega \tag{5}$$

in which, $\boldsymbol{\sigma} = [\sigma_{11} \ \sigma_{22} \ \sigma_{33} \ \sigma_{23} \ \sigma_{31} \ \sigma_{12}]^t$. Similarly, as to dead weight term and steady-state thermal loading, we can transform them to the equivalent boundary integral terms and add to the right-hand side of formula(4) or (5).

However, using formulae (4),(5), we cannot calculate boundary stress directly, because p^*_{ijk} and u^*_{ijk} have high order singularity as $\xi \in \Gamma$ and the results will be incorrect. While using common method, that is to say, calculating boundary stress according to the tangential derivative of boundary displacement or its difference approximation, we will obtain the results with bad precision, so we have created a new method[4], in which boundary stress is expresses as

$$C'\boldsymbol{\sigma}(\xi) = G'P - H'U \tag{6}$$

where, C' is 6x6 matrix of stress singular coefficient, if I_6 stand for unit matrix, then

for smooth boundary point, $C' = \frac{1}{2} I_6$

for internal point, $C' = I_6$

while in the H' and G' matrices, using the way of rigid-body movement and unit-stress field, we can express the main submatrix by corresponding non-diagonal ones.

$$\mathbf{h}'_{\underline{ll}} = -\sum_{\substack{m=1 \\ m \neq l}}^{N} \mathbf{h}'_{lm} \tag{7}$$

$$\mathbf{g}'_{\underline{ll}} = C'\hat{\boldsymbol{\sigma}} - \sum_{\substack{m=1 \\ m \neq l}}^{N} (\mathbf{g}'_{lm}\hat{P}_m - \mathbf{h}'_{lm}\hat{U}_m) \tag{8}$$

in which,

$$\hat{\sigma}=\begin{bmatrix}0&0&0\\0&0&0\\0&0&1\\0&1&0\\1&0&0\\0&0&0\end{bmatrix}, \qquad \hat{P}_m=\begin{bmatrix}n_3&0&0\\0&n_3&0\\n_1&n_2&n_3\end{bmatrix}_m$$

$$\hat{U}_m=\begin{bmatrix}r_3/2G&0&-\nu r_1/E\\0&r_3/2G&-\nu r_2/E\\r_1/2G&r_2/2G&r_3/E\end{bmatrix}_m$$

For non-diagonal submatrices, we can further eliminate the singularity effect by the techniques of the polar coordinate transformation and Kutt quadrature formula in the neighbourhood of the singular point, so we can avoid the obstacle of strong singularity in the integrand. This method can be used to calculate the stress at the arris and corners too, for details, see reference [4].

INTERACTION BETWEEN FOUNDATION AND STRUCTURES

Under the action of floor pressure R_j and neighbouring-loads $\bar{R}_j$ the displacement of point ξ on the earth's surface is:

$$u_i(\xi)=\int_{\Gamma c} R_j\,\eta_{ij}\,d\Gamma+\bar{u}_i \tag{9}$$

in which, Γ_c is the contact domain between the bottom of structure and the foundation.(Fig.3) η_{ij} is the influence function, ordinarily, it varies with r_i and foundation parameters, $r_i=x_i(X)-x_i(\xi)$ e.g. for half space foundation, they are given by Boussinesq formula and Cerruti formula.[9]

$$\bar{u}_i=\int_{\Gamma n}\bar{R}_j\,\eta_{ij}\,d\Gamma$$

Γn is the neighbouring domain.

By formulating for all nodes on Γ_c, we can arrive at the equations of the earth's surface displacement in the discretization form.

$$U_c=G_c\,P_c+\bar{U}_c \tag{10}$$

in which U_c and P_c are the column matrix for the displacement components and the reaction components of foundation at nodes on Γ_c respectively, $\bar{U}_c$ is produced by neighbouring loads.

Solving equations (2) and (10), we can obtain the boundary displacement field of structure and the reaction field of foundation, where, it is obvious that we have used the continuity condition of displacement and the condition $P_j=-R_j$.

NUMERICAL EXAMPLE

Example 1. A square sheet subjected to unit traction in the direction of x_2, its measurements, division and coordinate system are all shown as Fig.4. Variables are interpolated by cubic B-spline functions in the direction of x_1, x_2, while in the direction of x_3 by linear function, and the derivatives of endpoints are determined by cubic Lagrangian function. The results of stress are shown in table 1, while the calculated errors of displacement do not exceed 0.5%. This example is used to illustrate the effectiveness and reliability for 3-D spline boundary element method, and shows that it is not only the same order of accuracy but also appropriate effort for calculating boundary stress and internal stress.

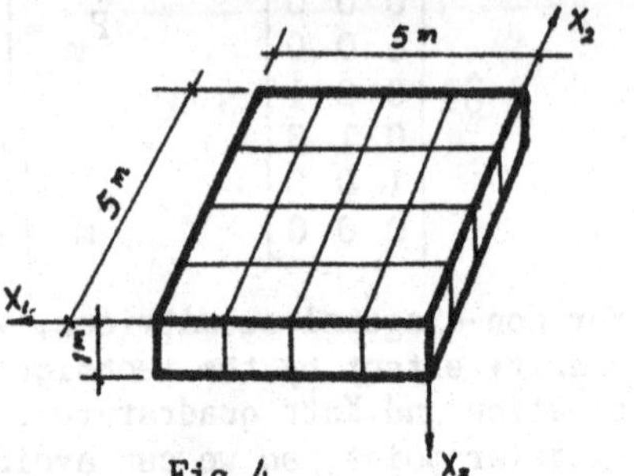

Fig.4

Table 1.

	nodes coordintes (x_1 , x_2 , x_3)	σ_{11}	σ_{22}	σ_{33}
Boundary points	(1.25, 1.25, 0)	0.2954×10^{-3}	0.9817	-0.2307×10^{-3}
	(1.25, 2.5, 0)	-0.7520×10^{-3}	0.9833	0.2731×10^{-3}
	(2.5, 2.5, 0)	0.1144×10^{-4}	0.9840	0.1940×10^{-3}
	(1.25, 1.25, 1.0)	0.4298×10^{-3}	0.9825	-0.1701×10^{-3}
	(1.25, 2.5, 1.0)	-0.8802×10^{-3}	0.9839	0.9408×10^{-3}
	(2.5, 2.5, 1.0)	0.3127×10^{-4}	0.9841	0.3821×10^{-3}
internal points	(1.25, 1.25.0.1)	0.1912×10^{-3}	0.9879	-0.7745×10^{-3}
	(2.5, 2.5, 0.1)	0.7841×10^{-3}	0.9807	-0.9920×10^{-3}
	(1.25, 1.25, 0.5)	0.7504×10^{-3}	0.9843	0.4055×10^{-3}
	(2.5, 2.5, 0.5)	0.7139×10^{-3}	0.9865	-0.4105×10^{-3}
	(1.25, 1.25, 0.9)	0.1203×10^{-3}	0.9853	-0.8790×10^{-3}
	(2.5, 2.5, 0.9)	0.4397×10^{-3}	0.9851	-0.5836×10^{-3}
exact solutions		0	1.0	0

Example 2. A square plate resting on elastic half space foundation is subjected to uniform vertical loads $P_3 = 1 \times 10^{-2}$ MPa, for plate $E = 2 \times 10^4$ MPa, $\nu = 1/6$, for foundation Es=20 MPa, $\nu_s = 0.35$. Its measurements, division and coordinate system are also shown in Fig.4. The results of vertical displacements are shown in table 2, the reaction force (R_3) of foundation in table 3.

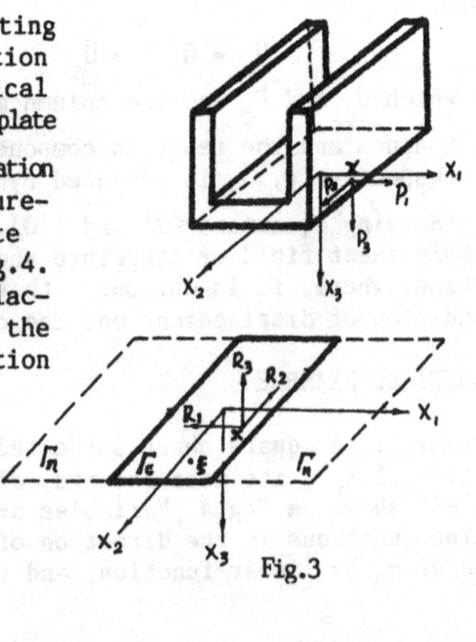

Fig.3

Table 2. Vertical displacements of foundation plate (10^{-2}m)

x_2 (m) \ x_1(m)	0	1.25	2.5	3.75	5
0	0.1756 0.1806	0.1765 0.1816	0.1769 0.1819	0.1765 0.1816	0.1756 0.1806
1.25		0.1774 0.1823	0.1777 0.1827	0.1774 0.1823	0.1766 0.1816
2.5			0.1781 0.1830	0.1778 0.1827	0.1770 0.1819
3.75				0.1774 0.1823	0.1766 0.1816
5					0.1756 0.1806

Notes : In Tab.2, for all points, the values in the first row are the present solution and in the second row the solution using spline boundary element method for Reissner's plate by [10].

Table 3. The reaction of foundation (10^{-2} MPa)

x_2(m) \ x_1 (m)	0	1.25	2.5	3.75	5
0	8.125	2.037	2.232	2.137	8.125
1.25		0.270	0.365	0.270	2.034
2.5			0.480	0.369	2.226
3.75				0.270	2.034
5					8.125

From Tables 2,3 we know the symmetry of data is very good. The results of vertical displacements coincide with each other, and the reaction of foundation in the present solution satisfies the equilibrum condition, the difference between the resultant of reaction and the sum to vertical loads is less than 1.7% of the latter. this is far better than the results by [10]. It is evident that 3-D spline bounary element method for calculating foundation plate is effective and reliable too.

Example 3. Use 3-D spline boundary element method to analyse the sluice chamber in the period of construction, whose constructional slits are on the piers (Fig.5). We know : for structure, $E=2 \times 10^4$ MPa $\nu = 1/6$, for half space foundation, $E_s = 20$ MPa $\nu_s = 0.35$, the unit weight of concrete is 24 KN/M^3.

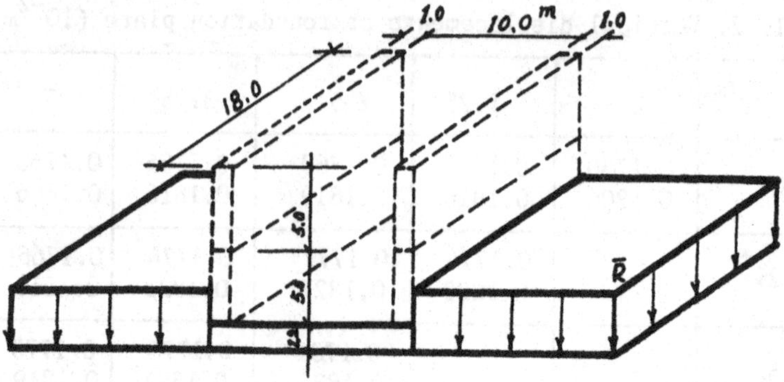

Fig.5

Construction proceduce is as follows, first, we pour the floor, second, two layers of piers successively, each layer is five metres high, and the construction for neibourhood chambers will delay to one period. We suppose that the loads of floor in the period of construction do not exceed the primary foundation pressure before dig, i.e. the foundation strain is elastic. So we can analyze the chamber in construction in three steps as follows:

Step 1, when floor being poured, its weights are passed on foundation directly and supposing the dead weights do not produce any internal forces in floor after hardening (on the safe side).

Step 2, we analyse the displacement and stress of floor under the weight of five metres high pier blocks. In this case, the neighbouring loads are produced by the weight of floor, one chamber width each side.

Step 3, composing the floor and the five metres high pier blocks as a whole and analysing its displacement and stress under the action of the weight of another 5-metres high pier blocks. In this case, neighbouring loads are the average reaction calculated in second step.

In the calculation of construction period, we omit the tangential reaction and only require to satisfy the continuity-condition of vertical displacement, and make use of symmetry to solve it. The division is shown as Fig.6. After combining the three steps above, we can obtain the displacement field, reaction field and internal forces distribution along the axis x_1, x_2.(tables 4,5 and Figs 7,8)

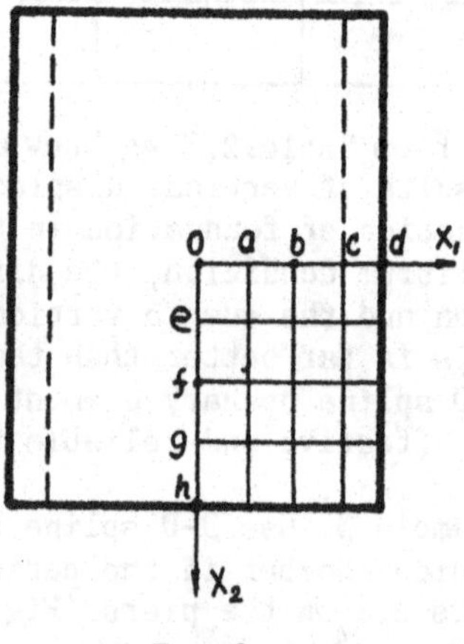

Fig.6

According to the stress distribution shown in Figs 7,8, and Lagrange interpolation, we can calculate easily the bending moments in each section (table 5) for the design of reinforcement. Here

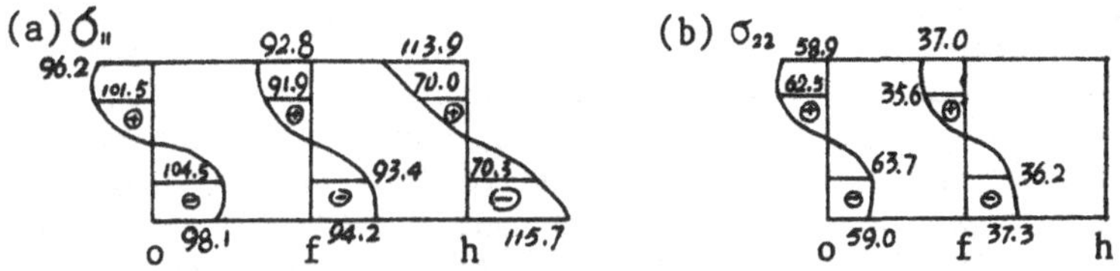

Fig.7 Stress distribution for the sections of o, f, h on the x_2 axis (unit: 10^{-2}MPa)

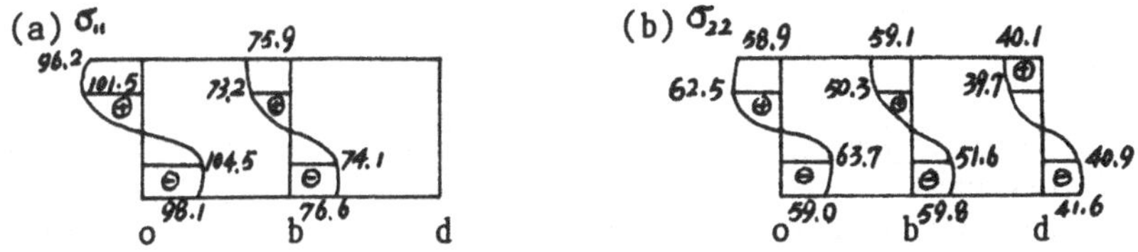

Fig.8 Stress distribution for the sections of o, b, d on the x_1 axis (unit: 10^{-2} MPa)

Table 4. The reaction forces of foundation(unit:10^{-2}MPa)

x_2-coordinate symbol \ x_1	o	a	b	c	d
o	3.1696 (3.5037)	2.9867 (3.3043)	2.8926 (2.8457)	1.9996 (1.7576)	6.1630 (6.9026)
e	3.2673 (3.5672)	3.1017 (3.4608)	2.8680 (2.9488)	2.0579 (1.8584)	5.9147 (6.6836)
f	3.7137 (3.9486)	3.5426 (3.8120)	3.2194 (3.1661)	2.5062 (2.2807)	5.0556 (5.7128)
g	3.8252 (4.1359)	3.7236 (3.9178)	3.3930 (3.3510)	2.0126 (2.0995)	3.1297 (3.5365)
h	19.382 (20.155)	20.148 (19.719)	20.418 (20.365)	20.575 (19.029)	23.247 (24.547)

Notes: The results in () are corresponding to 50% neighbouring loads.

Table 5. Bending moment on the section of floor(unit:10 KN·m/m)

section		o	b	d	f	h
bending moment	M_{11}	36.4 (28.7)	26.8	0.0	33.5 (27.4)	30.6 (33.3)
	M_{22}	22.2 (19.4)	19.3 (14.8)	14.6 (11.9)	13.1	0

Notes: The results in () are corresponding to 50% neighbouring loads.

stresses of four points are calculated for each section. In fact, due to the bending characteristic of floor , for each section, it is enough only to calculate one boundary point stress and one internal point stress.

For comparison, we also suppose that only 50% settlement is produced in the period of construction, the values in ( ) shown in tables 4,5 are the corresponding results. We can see that the consolidation extent of foundation has a light influence on reaction but serious influence on internal forces, even the sections of the maximun bending moment are different from each other.

In contrast, we also calculate it by Gorbunov-Posadov method in same conditions. The maximum moment of floor is 490 KN·m/m if neighbouring loads are wholly considered, and 432 KN m/m for considering only 50%. These values are far more than those shown in table 5. Obviously, 3-D spline boundary element method, fully considering the space resistance of the floor, will be more reasonable and economic.

DISCUSSION

(1). 3-D spline boundary element method for the analysis of sluice has fairly great attraction. The cell division of structure may be coarse, and unknowns are limited on the boundary, so the degrees of freedom are fewer, but the calculating precision is quite fine. On the other hand, if using fast quadrature technique (FQT), the computational efficiency for forming the coefficient matrices of equations can almost compare with Finite Element Method, while the idea of FQT is much simpler, it is based on fully eliminating the singularity of the boundary integral equations and utilizing spline functions to interpolate fundamental solutions and Jacobians, too.

(2). The present method can also be used to analyse the sluice whose floor is allowed to crack, i.e. it is designed to constrain the crack width. In this case, we may approximately discount the elastic modulus of floor according to the standard or by experience, such as taking a multiplier 0.65. According to calculation, as reducing the rigidity of floor, the distribution of internal forces tend to uniform,and the crest values will be reduced too.

(3). As space is limited, in this paper we only calculate the chamber structure in the construction period, in fact, for the completion period and operation period, the analysis is similar. According to calculating experiences, for completion period, we can purely consider normal reaction and the continuity conditions of vertical displacement, and omit the effect of the internal supports caused by upper structures. For operation period, we may determine the tangential reactions by the horizontal equilibrium conditions if supposing their distributions previously (e.g. uniform distribution) and still only require to satisfy the continuity condition of vertical displacements. It follows that the precision is still fine, not only reducing the degrees of freedom but also avoiding the serious ill-condition of the determining equations. But in the operation period we must consider the effect of the internal supports and unoverlap-conditions of the piers. These factors have little interference to the internal forces of floor but notable influence on the piers.

(4). If it is possible that the floor may be separated locally from the foundation, we can search for the contact domain by iteration method. Further details are given in references [10],[11].

(5). The present method may be fitted to analyse the chambers of any constructional slit scheme and convenient to calculate the chamber either with raft floor or with inverted arched floor, both taking the planar gate and the radial one. For the latter, we can also determine the stress concentration of piers caused by the gate arms.

(6). 3-D spline boundary element method can also be used to analyse gravity dam, arch dam and the lower structures of the hydropower station, whether in static and dynamic analysis or in optimum structural design. For them, we will introduce in other papers in succession.

REFERENCES

Book

1. Tan Songxi, Design of sluices, Hydraulic and Electric Press, 1986, in chinese.

6. Brebbia, C.A. et al, Boundary Element Techniques, Springer-Verlag, 1984.

9. Xu Zhilong, Elasticity (second edition), Vol.1, Advance Edcation Press, 1982, in chinese.

Paper in a journal

2. Fu Zuoxin, Zhou Ti. The Integral Analysis of Sluice Floor, Journal of Hydraulic Engineering, No.5. 1986, page17-23, in chinese.

3. Fu Zuoxin, The Internal Forces Analysis of Sluice floors, Journal of East China Institute of Hydraulic Engineering, No.1, 1981, page 1-13, in chinese.

4. Wang Youcheng et al. Particular Solutions Method Adjusted by Singularity For Calculating Boundary Stress, Proc. 4th China-Japan Symposium on BEM, Beijing, 1991.

5. Wang Youcheng, Jiang Longzhi. Singular Integral and Fast Calculation of Integral Principal Values in Boundary Element Method, The Theory of The Combination Method of Analytic and Numerical Solutions and Applications in Engineering, Hunan University Press, 1989, page 248-253. in chinese.

7. Wang Youcheng. Spline Boundary Element Method For Kirchhoff's plate, Computational Structural Mechanics and Applications Vol.3 (1986) No.1, Page41-50. in chinese.

8. Wang Youcheng et al. Kirchhoff Type Spline Boundary Element Method for Problem of Shallow Shell, Computational Structural Mechanics and Applications Vol.5(1988). No.2, page 1-10. in chinese.

10. Wang Youcheng et al. Spline Boundary Element Method for Reissner's Plate and Its Application to Foundation Plates, Boundary Elements IX (Ed. Brebbia C.A et al). Springer-Verlag, 1987, Vol.2, 111-125.

11. Su Ganlong, Wang Youcheng. The Finite Element Analysis of Structures on Generalized Foundation, Journal of Hydraulic Engineering, No.10. 1984. in chinese.

* The project was supported by the National Natural Science Foundation of China.

SECTION 15: DESIGN SENSITIVITY AND OPTIMIZATION

Local Shape Optimization Using Boundary Element Zooming Method

N. Kamiya, E. Kita

Department of Mechanical Engineering, Nagoya University, Nagoya, 464-01 Japan

ABSTRACT

This paper is concerned with a new practical shape optimization method, especially for obtaining local optimized shape.

In this method, firstly, initial assumed shape is analyzed by finite element method (FEM) and a subregion undergoing boundary modification is separated from the original entire region. The subregion referred to as the zooming region is re-discretized by boundary elements for the shape optimization. In order to decide the zooming region boundary, a parameter "Region Partition Index (PRI)" is defined. This method has the following advantages:

1. Since the zooming region is discretized by boundary elements, shape modification is easier.
2. Only small part of the entire region is successively modified, and high computational efficiency can be obtained.
3. The global matrix formulation is simple and straightforward and therefore, commercial softwares of FEM and BEM can be employed without essential modification.

The present method is applied to the local shape optimization of two-dimensional elastic problem to confirm its effectiveness.

KEYWORDS : Shape optimization; Boundary element method; Zooming analysis; 2D-elastic problem.

INTRODUCTION

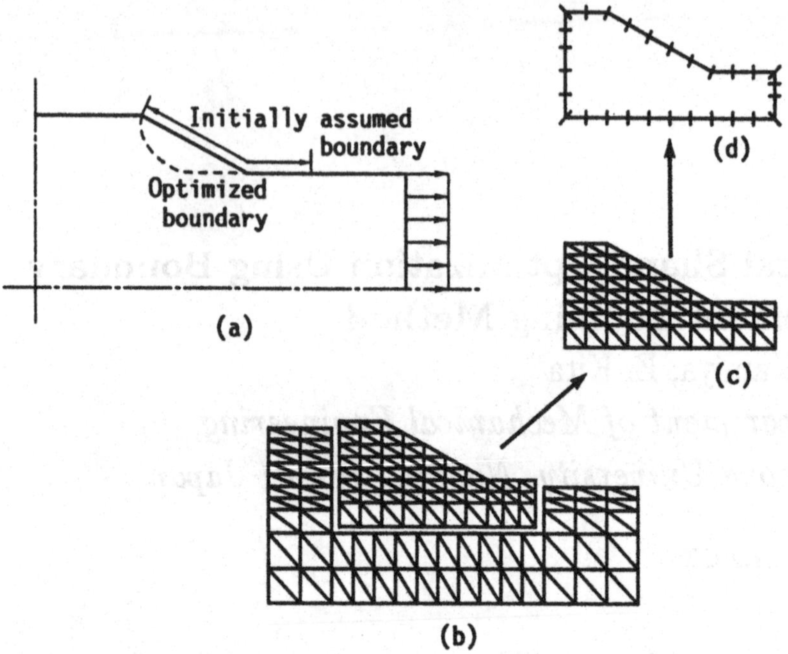

Figure 1: Shape optimization by BE zooming method

Several papers are concerned with shape optimization methods employing finite element method (FEM) and boundary element method (BEM). In shape optimization method, initially assumed shape is modified iteratively so as to satisfy design object, thus easy shape modification and high computational efficiency are necessary. Because BEM needs only boundary discretization of the object under consideration governed by linear and homogeneous differential equations and therefore, shape modification is very easy and remeshing is less necessary than FEM. BEM has a shortcoming, however; its coefficient matrices are fully-populated and matrix formulation consumes a lot of time. Several papers presented schemes to effectively formulate the coefficient matrices at each shape modification [1, 2, 3]. In these methods, the coefficient matrices of the entire region are reduced to the submatrix related to the nodes which are moved during successive shape modification. These methods, however, need a complicated matrix formulation and therefore, one cannot directly employ the commercial softwares of FEM and BEM.

In this paper, we present a new scheme (Figure 1); firstly, the object under consideration is analyzed by FEM and then, a subregion undergoing boundary modification is separated from the entire region. The subregion referred to as the zooming region is re-discretized by boundary elements for the shape optimization. The process to decide the zooming region is similar to the ordinary zooming method of FEM [4], except that the FE zooming zone is not changed but is analyzed. Matrix formulation for the zooming region boundary is so simple that commercial softwares of FEM and BEM can be employed without essential modification.

By the way, the present method has two key-points; how to determine the zooming region and how to generate its boundary element mesh. For the first, as the measure to enable to decide the zooming region, we will employ a new index named "Region partition index (RPI)" and, for the second, one develops a mesh data converter which automatically converts finite element analysis data for the zooming region into boundary element analysis data. The present method is applied to the local shape optimization of two-dimensional linear elastic body in order to examine its effectiveness.

SHAPE OPTIMIZATION BY BE ZOOMING METHOD

Now, we consider that a relatively small part of the large object is optimized (Figure 1(a)). In the present method, the shape optimization is carried out along the following steps:

1. Carry out the initial finite element analysis of the object under consideration (Figure 1(b)).
2. Specify design condition (design variable, its maximum allowable value and so on) and determine RPI distribution.
3. Decide zooming region by RPI distribution (Figure 1(c)).
4. Generate data for boundary element analysis of zooming region (Figure 1(d)).
5. Carry out the shape optimization of zooming region.

As shown in Figure 2, all data are exchanged through the intermediate data-file between analyzer and FEM and BEM solvers, which simplifies matrix formulation and enable to employ commercial softwares.

REGION PARTITION INDEX (RPI)

Definition of RPI

We consider the numerical solution (displacement field) of initial shape and the one of slightly modified shape as $\boldsymbol{u}^0$ and $\boldsymbol{u}$, respectively. The variation of numerical solution by the shape modification, which is referred to $\boldsymbol{d}_u$, is:

$$\boldsymbol{d}_u \equiv \boldsymbol{u} - \boldsymbol{u}^0 \tag{1}$$

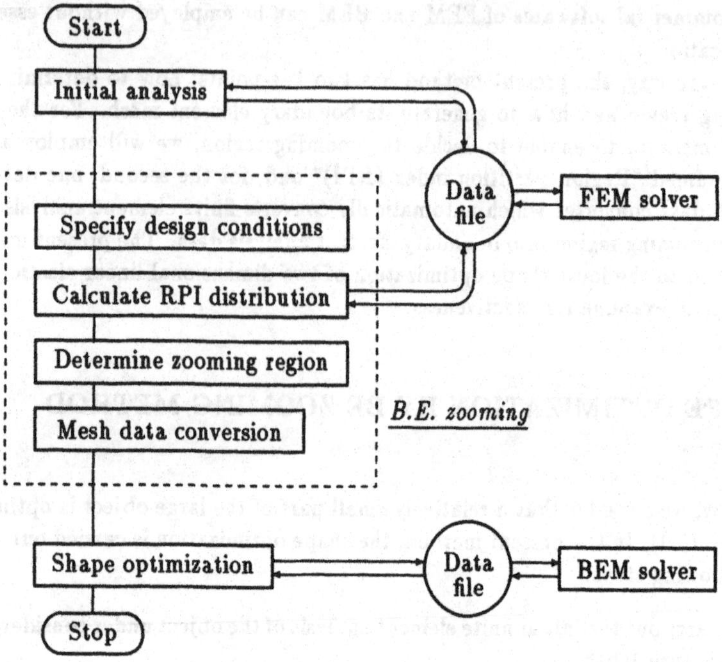

Figure 2: Flow of the procedure

Considering the displacement-strain relation and the linear elastic constitutive relation, the variation of strain and stress by the shape modification are expressed as follows:

$$\left.\begin{array}{rcl} d_\epsilon & = & Ad_u \\ d_\sigma & = & (DA)d_u \end{array}\right\} \tag{2}$$

where A are the matrix of differential operator connecting between the displacement and strain fields in the two-dimensional space and D the stiffness coefficient matrix.

The measure to decide the zooming region is defined by the energy norm:

$$\begin{aligned} ||d|| &= \left(\int_S d_\epsilon{}^T d_\sigma dS\right)^{1/2} \\ &= \left(\int_S (Ad_u)^T D(Ad_u) dS\right)^{1/2} \end{aligned} \tag{3}$$

RPI is defined by the relative percentage expression of the energy norm:

$$\eta \equiv \frac{||d||}{||u^0||} \times 100 \qquad \% \tag{4}$$

where $||u^0||$ is the energy norm of initial shape.

Calculation of RPI

Denoting design variables as z, the Taylor expansion of numerical solution can be expressed as follows:

$$\begin{aligned} u &= u^0 + \frac{\partial u}{\partial z}dz + \frac{1}{2!}\frac{\partial^2 u}{\partial z^2}dz^2 + \ldots \\ &\simeq u^0 + \frac{\partial u}{\partial z}dz \\ &\equiv u^0 + J^z dz \end{aligned}$$

The component expression of the above equation is as follows:

$$u_i = u_i^0 + \sum_{j=1}^{N_d} J_{ij}^z dz_j \tag{5}$$

where N_d is the total number of the design variables. Considering the maximum norm of the above equation, the upper bound of u_i is expressed as follows:

$$\begin{aligned} u_i &= u_i^0 + \sum_{j=1}^{N_d} J_{ij}^z dz_j \\ &< u_i^0 + \sqrt{\sum_{j=1}^{N_d} \left(J_{ij}^z dz_j^{max}\right)^2} \equiv \bar{u}_i \end{aligned}$$

where dz_j^{max} is maximum allowable value of dz_j. Substituting $\bar{u}_i$ into Eq.(1), one can calculate RPI from Eq.(4).

BOUNDARY ELEMENT MESH GENERATION

Boundary element mesh data of zoomed-up region is generated by the following three steps:

1. Determine zooming region.
2. Generate data for boundary element analysis by FE/BE mesh data converter.
3. Refine the generated boundary element mesh.

In what follows, we briefly discuss the above-mentioned steps.

Decision of Zooming Region

The process to decide the zooming region can be summarized as follows:

1. Specify the allowable value of RPI η_c.

2. Construct the zooming region by the finite elements with η_i which satisfies $\eta_c < \eta_i$.

3. Supplement some finite elements to construct the zooming region with smooth single-connected boundary.

FE/BE Mesh Converter

When triangular finite elements are employed, the process of mesh converter can be summarized as follows:

1. Consider a triangular finite element E_1, which is included in the zooming region and its three segments as L_1, L_2 and L_3, respectively.
2. Check a neighboring element close to L_1 and name it as E_2 if it exists.
3. When E_2 does not exist, i.e., L_1 is thought to be a part of the real boundary of the object under consideration,
   - restore L_1 as a boundary element,
   - restore the boundary condition data specified at L_1 and,
   - go to step 5.
4. When E_2 exists, i.e., L_1 is thought to be included in the object under consideration, the process is divided into two step.
   (a) When E_2 is included in the zooming region, go to step 5.
   (b) When E_2 is not included in the zooming region, i.e., L_1 is a part of zooming boundary, the following operations are carried out:
      - restore L_1 as a boundary element,
      - restore the boundary condition data specified at L_1 and,
      - go to step 5.
5. For segments L_2 and L_3, carry out the operations from step 2 to step 4.
6. For all elements within the zooming region, carry out the operations from step 1 to step 5.
7. Remove the duplication of the boundary element data defined through the above-mentioned process.

On the internal boundary between the zooming region and the remaining, instead of the traction solution, the displacement solution of finite element analysis is taken as the boundary condition, because the displacement solution generally has a higher accuracy than the traction solution, and given at each node.

Remeshing Process

The above-mentioned process often makes relatively rough boundary element mesh and then, boundary element mesh is refined:

1. Divide the boundary elements on the boundary to be modified.
2. Divide large elements when the length ratio of the largest to the shortest element is bigger than maximum allowable value.

OPTIMIZATION PROCEDURE

Optimization problem is formulated so as to minimize the object function under the constraint conditions by modifying the design variables. The object function W, the constraint conditions $\boldsymbol{f}$ and the design variables $\boldsymbol{z}$ can be defined as follows:

$$W = \boldsymbol{w}^T\boldsymbol{w} \tag{6}$$

$$\boldsymbol{w} = \{w_1 w_2 \dots w_{N_e}\}^T$$

$$\boldsymbol{f} = \{f_1 f_2 \dots f_{N_c}\}^T = \boldsymbol{o} \tag{7}$$

$$\boldsymbol{z} = \{z_1 z_2 \dots z_{N_d}\}^T \tag{8}$$

where w_i is the element of the object function and N_e, N_c and N_d are the total numbers of the elements, the constraint conditions and the design variables, respectively.

If the initial assumed shape is optimized so as to obtain uniform stress distribution, w_i is given as the follows:

$$w_i \equiv \sigma_i/\sigma_c - 1 \quad (\sigma_i > \sigma_c) \tag{9}$$

where σ_i and σ_c are the equivalent stresses at node i and reference stress, respectively.

The design variables are taken as the coordinates of boundary nodes which are displaced towards the outward normal direction. Besides, the upper and lower bounds of the design variables are specified as follows:

$$|z_j| \leq \Delta z_j^{max} \qquad (j = 1, \dots, N_d) \tag{10}$$

where Δz_j^{max} is the maximum allowable change for the design variables. The problem defined by Eqs.(6) to (10) is tackled by the sequential linear programming (SLQ) method together with the BEM analysis.

We may sometimes encounter a zig-zag shaped boundary during shape optimization/change process due to pile-up of calculation error or an inadequate initially assumed shape. In order to overcome this difficulty, we employ the two supplementary facilities presented in the previous paper [3]:

1. neglect of weak design sensitivities, and
2. smoothing of boundary profile by B-spline.

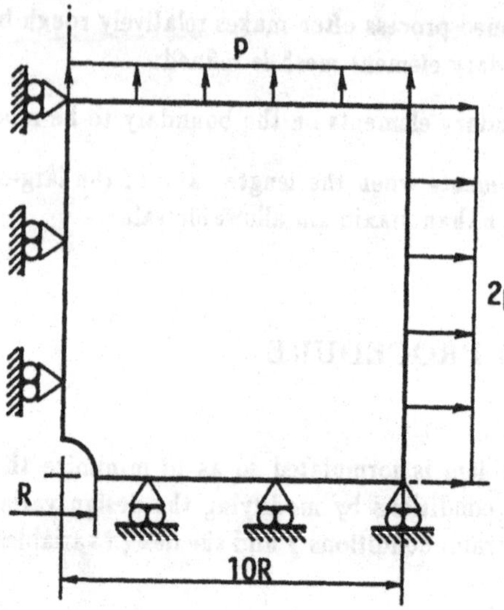

Figure 3: The problem under consideration (A quarter is shown.)

NUMERICAL EXAMPLE AND DISCUSSION

A flat square plate with a circular hole under bi-axial stretching forces is considered as a example (Figure 3). Considering the symmetry, the plane stress analysis of a quarter of the object is carried out and the zooming regions are decided for different η_c's. A hole shape is modified to obtain uniform stress distribution along its boundary. In this example, Poisson ratio is taken as $\nu = 0.3$. The optimized shapes are compared with theoretical one [5].

Zooming region: By taking initial finite element mesh shown in Figure 4, the strain energy density (SED) distribution shown in Figure 5 is obtained. In addition, by taking 4 nodes along an original circular hole as the design variables for deciding the zooming region and the radius of the hole as Δz_j^{max}, RPI distribution shown in Figure 6 is obtained. A comparison of Figures 5 and 6 shows that there is a little deviation between SED and RPI distributions and therefore, we mention that SED, which is often employed in the ordinary FE zooming, is not appropriate for deciding BE zooming region. Figure 7 indicates the zooming regions decided for η_c=3%, 5% and 7%.

Shape optimization: Along the zooming regions obtained above, shown in Figure 7, boundary element meshes are generated by the mesh data converter. In these meshes, the hole periphery is divided into 12 equal length elements and the outer boundary except for the symmetric axes is divided into 32 equal length elements and therefore, total number of nodes are 46 (Figure 8). Since BEM solver employed here automatically satisfies boundary condition on the symmetric axes and therefore, boundary elements are not taken on the axes.

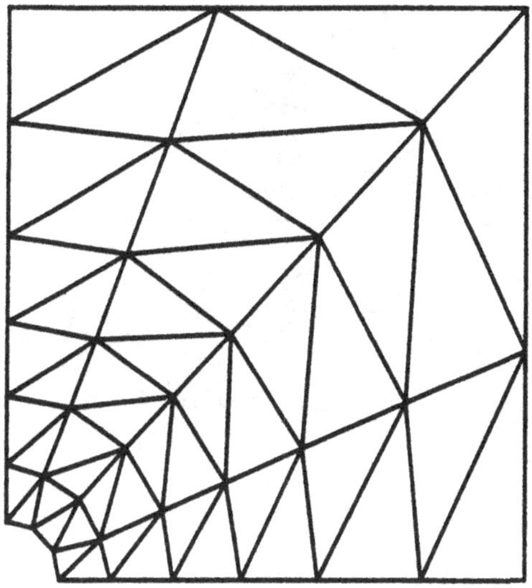

Figure 4: F.E. mesh

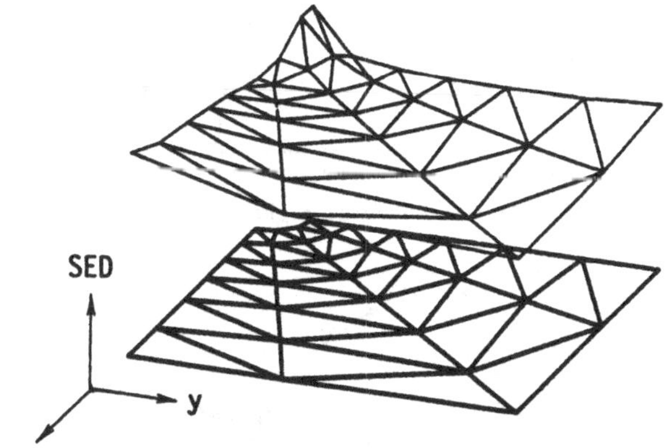

Figure 5: SED distribution

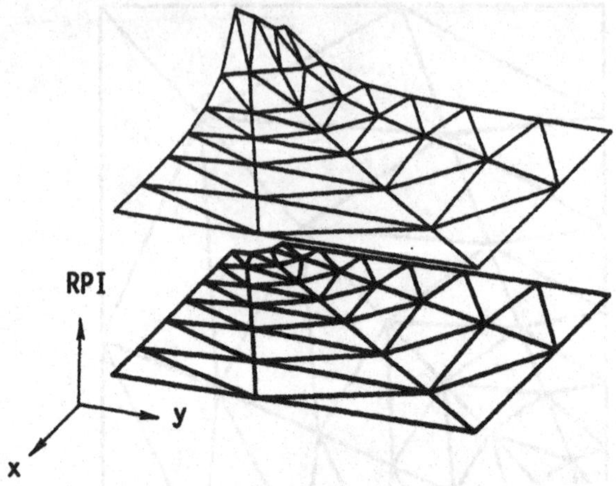

Figure 6: RPI distribution

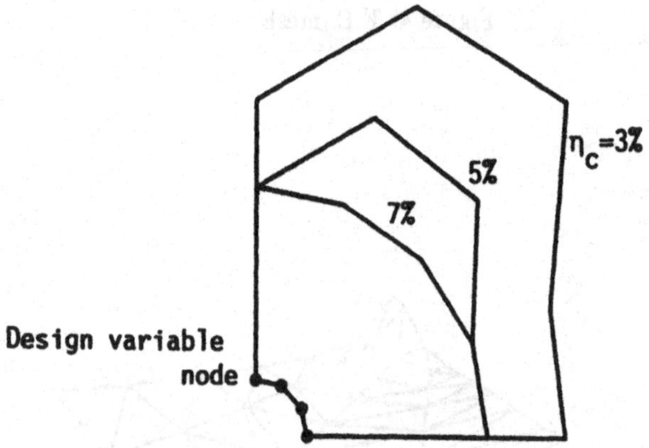

Figure 7: Zoomed-up region

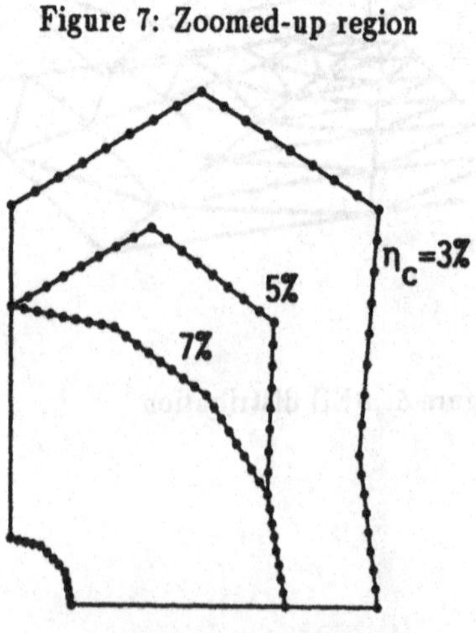

Figure 8: B.E. mesh

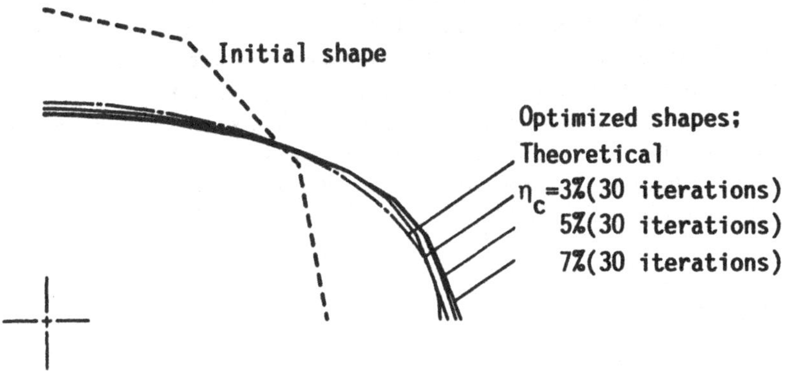

Figure 9: Optimized and theoretical shapes

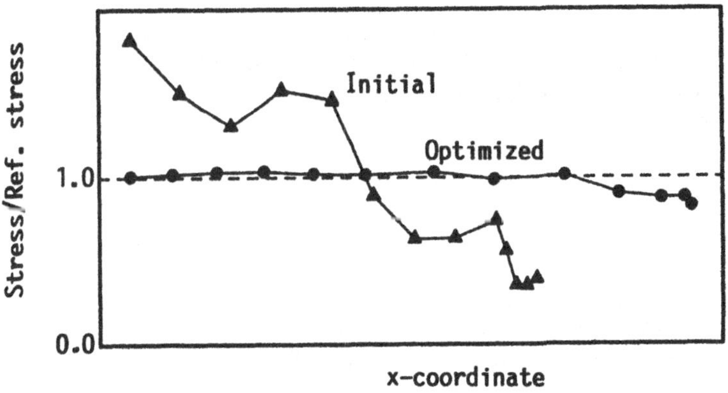

Figure 10: Stress distributions ($\eta_c = 3\%$)

By taking the nodes along the hole boundary as the design variables for shape optimization, the hole shape is optimized to obtain uniform stress distribution along its boundary. Initial and optimized shapes are compared in Figure 9 and for η_c=3%, the stress distributions of initial and optimized shapes are shown in Figure 10. Figures 9 and 10 show that the optimized shapes are nearly elliptic and their stress distributions are almost uniform. One can mention that the optimized shape decided from initial shape with smaller RPI shows better agreement with the theoretical one.

CONCLUSIONS

This paper presented local shape optimization of structural and machine elements employing the boundary element (BE) zooming method. Firstly, the entire region under consideration is analyzed by finite element method (FEM) and then, a zooming region is separated from the original entire region and re-discretized by boundary elements for the shape optimization. This process has two key-points; how to decide boundary element region and how to generate boundary element mesh from finite element mesh. For the first, "Region partition index (RPI)" was proposed as the measure to determine zooming region and for the second, mesh data converter was developed to automatically convert finite element mesh into boundary element mesh. We carried out a simple numerical experiment and could obtain sound result.

Although only simple numerical example was indicated in this paper, the present method may also be useful for local shape optimization and stress distribution of the object with more complicated shape.

REFERENCES

[1] N. Kamiya, T. Nagai, and E. Kita. Design boundary elements and shape optimization. *Software Eng. Workst.*, 4:11–15, 1988.

[2] E. Sandgren and S.-J. Wu. Shape optimization using the boundary element method with substructuring. *Int. J. Num. Meth. Eng.*, 26:1913–1924, 1988.

[3] N. Kamiya and E. Kita. Local shape optimization of a two-dimensional elastic body. *Finite Elem. Anal. Des.*, 6:207–216, 1990.

[4] I. Hirai B. P. Wang and W. D. Pilkey. An effective zooming method for finite element analysis. *Int. J. Num. Meth. Eng.*, 20:1671–1683, 1984.

[5] S. P. Timoshenko and J. N. Goodier. *Theory of Elasticity.* McGraw-Hill Ltd., Third edition, 1982.

A Modified Finite Difference Method to Shape Design Sensitivity Analysis

Z. Zhao

School of Civil and Structural Engineering,
Nanyang Technological University,
Nanyang Avenue, 2263 Singapore

ABSTRACT

A new method is presented in this paper for the calculation of shape design sensitivity with kinematical design boundary. This new method modifies the traditional finite difference approach, such that the variation of the structural response due to the change of the kinematic boundary is replaced by an equivalent problem, and the final design sensitivity is expressed as the solutions of the initial structure under the perturbation displacements on the design boundary. Two examples are used to demonstrate the proposed new formulation.

INTRODUCTION

During the last two decades, design sensitivity analysis has been widely investigated, and various approaches have been developed [1]-[11].

In traditional finite difference method (FDM), the design sensitivity is obtained by comparing the solutions of the initial design and the perturbed design [12]-[14]. The FDM has the advantage of being simple in concept and easy to implement. However, the FDM has two serious drawbacks: 1) the accuracy of the shape design sensitivity often depends on the choice of the perturbation step, 2) the computational cost is usually higher as the system matrix of the perturbed structure has to be evaluated and solved for each design variable [15].

Recently, a new approach - finite difference load method (FDLM), has been proposed by the author for the design sensitivity analysis with traction design boundary [16]. This new approach employs the concept of the traditional finite difference method, and the final design sensitivity can be evaluated from the solutions of a boundary value problem, which has the identical domain with the initial problem, but with different boundary traction. The FDLM does not require the perturbed structure, and the final design sensitivity is independent of the design perturbation.

In this paper, the FDLM is extended to the kinematical design boundary, in which the kinematical boundary is free to move, and the traction boundary is fixed. Again the same finite difference procedure as in the FDLM derivation is used to derive the design sensitivity formulation, and design sensitivity can be obtained from the solutions of the initial structure under new perturbation displacements. This new approach is named as finite difference displacement method (FDDM).

DERIVATION OF THE FDDM

The finite difference displacement method (FDDM) for 2-D continuum structures is considered as follows. Fig. 1 shows a 2-D structure with traction boundary τ_2 and kinematical boundary τ_1, where τ_1 is the design boundary. The boundary conditions are:

$$\begin{aligned} &P = P_0 && \text{on } \tau_2 \\ &\left.\begin{aligned} U_n &= U_{n0} \\ U_s &= U_{s0} \end{aligned}\right\} && \text{on } \tau_1 \end{aligned} \tag{1}$$

where P_0 is the applied load, U_{n0} and U_{s0} are the components of the displacement in normal and tangential directions respectively.

Given a design perturbation δb, the kinematical design boundary will change from τ_1 to τ' as shown in Fig. 1. The structure after design perturbation is named as the perturbed structure, which occupies the domain Ω_a. In general, the initial displacements on τ_1 will change during the design perturbation, depending on the design perturbation δb. The new displacements on τ' are noted as U_n^a and U_s^a, i.e. $U_{n0} \rightarrow U_n^a$, $U_{s0} \rightarrow U_s^a$. Fig. 2 shows the perturbed structure under the new displacement boundary conditions.

For a structural response $g(x)$, $x \in \Omega_a$, the design sensitivity due to the design perturbation δb can be written as:

$$\frac{dg}{db} = \lim_{\delta b \to 0} \frac{g_3 - g_1}{\delta b} \tag{2}$$

where g_1 is the response of the initial structure under loads P, and displacements U_{n0} and U_{s0}, and g_3 is the response of the perturbed structure under load P, displacements U_n^a and U_s^a. In the following, we name the initial problem as problem 1, and the perturbed problem as problem 3.

In what follows, a new procedure is developed so that the design sensitivity of (2) involves the original structure only.

It can be seen from (2) that the design sensitivity involves two solutions of g_1 and g_3 which are calculated on the basis of different domains (Ω in problem

1 and Ω_a in problem 3), therefore the first step is to transform problem 1 into an equivalent problem. The purpose of such a transformation is to keep the structural response of the problem 1 unchanged, and at the same time transform the domain Ω into Ω_a. This can be easily done by removing the part between τ_1 and τ' from Ω, and assigning the displacements U_n' and U_s' on τ' as a boundary condition, where U_n' and U_s' are the components of displacement in normal and tangential directions along the boundary τ' of the problem 1. This newly formed structure under load P_0 on τ_2, U_n' and U_s' on τ', is named as problem 2. It is obvious that the problems 1 and 2 have the identical solutions for $x \in \Omega_a$, therefore $g_2 = g_1$, where g_2 and g_1 are the structural responses of the problems 2 and 1 respectively. So the (2) becomes

$$\frac{dg}{db} = \lim_{\delta b \to 0} \frac{g_3 - g_2}{\delta b} \tag{3}$$

Given a design perturbation δb, the design boundary movement can be written as:

$$\begin{aligned} V_n &= f_n(\xi)\delta b \\ V_s &= f_s(\xi)\delta b \end{aligned} \tag{4}$$

where V_n and V_s are the components of the design boundary movement in normal and tangential directions respectively. f_n and f_s are the interpolation functions, and ξ is the local coordinate.

Since V_n and V_s are small, the displacements along the boundary τ' in the problem 1 can be approximated as:

$$\begin{aligned} U_n^+ &= U_{n0} + \frac{\partial U_n}{\partial n^-}V_n + \frac{\partial U_n}{\partial s}V_s \\ U_s^+ &= U_{s0} + \frac{\partial U_s}{\partial n^-}V_n + \frac{\partial U_s}{\partial s}V_s \end{aligned} \tag{5}$$

Two sets of notations are used here: U_{n0} and U_{s0} are the initial displacements on τ_1, and U_n and U_s represent the displacements on τ_1 in general sense. n^- is the inward normal direction, and s is the tangential direction.

The n and s directions in (5) correspond to the boundary τ_1. By considering the change of the normals between τ_1 and τ', we have local displacements

$$\begin{aligned} U_n' &= U_n^+ cos(\theta) + U_s^+ sin(\theta) \\ U_s' &= -U_n^+ sin(\theta) + U_s^+ cos(\theta) \end{aligned} \tag{6}$$

where θ is the difference of the normals between τ' and τ_1, and U_n' and U_s' are the displacement components in normal and tangential directions under the local coordinates of boundary τ'.

The two boundary value problems 2 and 3 have the same domain Ω_a. g_2 is the solution of problem 2 under the following boundary conditions:

$$\begin{aligned} P &= P_0 && \text{on } \tau_2 \\ \left.\begin{aligned} U_n &= U_n' \\ U_s &= U_s' \end{aligned}\right\} && \text{on } \tau' \end{aligned} \tag{7}$$

g_3 is the solution of problem 3 under the following boundary conditions:

$$\begin{aligned} P &= P_0 && \text{on } \tau_2 \\ \left.\begin{aligned} U_n &= U_n^a \\ U_s &= U_s^a \end{aligned}\right\} && \text{on } \tau' \end{aligned} \tag{8}$$

For a linear elastic structure, there are two properties frequently used in the structural analysis: 1) a structural response g is proportional to the applied loads and the applied displacements; 2) the superposition principle states that for a linear structure under two sets of boundary conditions, the sum of the solutions under individual boundary condition equals to the solution of the structure under the sum of boundary conditions. Those two properties make it possible to combine the problems 2 and 3 together so that the solution of this new problem (named as perturbation problem) equals to the design sensitivity dg/db.

The perturbation problem has the following boundary conditions:

$$\begin{aligned} P^* &= 0 && \text{on } \tau_2 \\ \left.\begin{aligned} U_n^* &= \lim_{\delta b \to 0} \frac{U_n^a - U_n'}{\delta b} \\ U_s^* &= \lim_{\delta b \to 0} \frac{U_s^a - U_s'}{\delta b} \end{aligned}\right\} && \text{on } \tau' \end{aligned} \tag{9}$$

where U_n^* and U_s^* are the components of the displacement in the normal and tangential directions, which are named as perturbation displacement. $P^* = 0$ implies that τ_2 is a traction free boundary.

The limiting process $\lim \delta b \to 0$ indicates that the perturbed structure approaches the initial structure (i.e. $\Omega_a \to \Omega$), therefore the perturbation problem becomes a problem with domain Ω under the displacement boundary conditions U_n^* and U_s^* on τ_1, and traction free on τ_2. The above procedure has transformed the finite difference method into a finite difference displacement method (FDDM), which does not involve the perturbed structure and is independent of the perturbation step.

The solution of the perturbation problem represents the design sensitivity dg/db. In order to solve the perturbation problem, we need to evaluate the perturbation displacements. Substitute (5) and (6) into (9), after some lengthy algebra derivation, finally we shall have

$$\begin{aligned} U_n^* &= -\frac{\partial U_n}{\partial n^-} f_n(\xi) - \frac{\partial U_n}{\partial s} f_s(\xi) - c_i U_{s0} + \frac{\partial U_n}{\partial b} \\ U_s^* &= -\frac{\partial U_s}{\partial n^-} f_n(\xi) - \frac{\partial U_s}{\partial s} f_s(\xi) + c_i U_{n0} + \frac{\partial U_s}{\partial b} \end{aligned} \tag{10}$$

where $c_i = \lim_{\delta b \to 0} \sin(\theta)/\delta b$, $\partial U_n/\partial b$ and $\partial U_s/\partial b$ denote the variations of the kinematical boundary conditions with respect to the change of design variable b.

(10) is derived for the case of inward design movement. In order to keep the conventional notation that V_n is positive in the n^+ direction, we can change the sign of f_n, so (10) becomes:

$$\begin{aligned} U_n^* &= \frac{\partial U_n}{\partial n^-} f_n(\xi) - \frac{\partial U_n}{\partial s} f_s(\xi) - c_i U_{s0} + \frac{\partial U_n}{\partial b} \\ U_s^* &= \frac{\partial U_s}{\partial n^-} f_n(\xi) - \frac{\partial U_s}{\partial s} f_s(\xi) + c_i U_{n0} + \frac{\partial U_s}{\partial b} \end{aligned} \tag{11}$$

If the boundary element method is employed in the structural analysis, after the analysis of the initial problem, the solutions of displacements and stresses on the kinematical boundary can be used to calculate the derivatives of the displacements as follows (plane strain):

$$\begin{aligned} \frac{\partial U_n}{\partial s} &= u_{x,s} n_x + u_{y,s} n_y \\ \frac{\partial U_s}{\partial s} &= u_{x,s} s_x + u_{y,s} s_y \\ \frac{\partial U_n}{\partial n^-} &= -\frac{1}{2G}((1-\nu)\sigma_n - \nu\sigma_s) \\ \frac{\partial U_s}{\partial n^-} &= -(\frac{\sigma_{sn}}{G} - \frac{\partial U_n}{\partial s}) \end{aligned} \tag{12}$$

where u_x and u_y are the displacements in x and y direction respectively, n_i and s_i are the components of unit normal and unit tangential in i direction, G is the shear modulus, and σ_n, σ_s and σ_{sn} are the normal, tangential and shear stresses respectively.

It should be noted that the solutions of the perturbation problem represent the design sensitivity of displacement and stress on the fixed boundary. For the case of constraints on the design boundary, there will be extra terms in the design sensitivity formulation which come from the contribution of the moving boundary. The formulation for such constraints will be reported later.

NUMERICAL EXAMPLES

A clamped thin beam

Two examples are used to demonstrate the accuracy of the FDDM. The first example is a simple thin beam with fixed ends under a point load as shown in Fig. 3, in which the analytical solution is available.

The bending moment at position x is

$$M(x) = M_A - P_A x + kp \tag{13}$$

where $k = 1$ if $x \geq a$, otherwise $k = 0$, and

$$\begin{aligned} M_A &= \frac{pab^2}{(a+b)^2} \\ P_A &= \frac{p}{(a+b)^3}(ab^2 + b(a+b)^2 - a^2b) \end{aligned}$$

Consider b as a design variable, then the design sensitivity of the bending moment $M(x)$ can be obtained analytically by differentiating (13),

$$\frac{\partial M}{\partial b} = -\frac{2a^2bp}{(a+b)^4}(3x - a - b) \tag{14}$$

It can be seen from Fig. 3 that the change of b can also be treated as the right hand fixed kinematical boundary moving to the right, so we can use the FDDM to carry out the design sensitivity calculation. In the beam, the kinematical boundary conditions at $x = a + b$ are:

$$\begin{aligned} \theta_0 &= 0 \\ v_0 &= 0 \end{aligned} \tag{15}$$

where θ_0 is the angle of the deflection, and v_0 is the vertical displacement.

If we modify the (11) for the beam, we have the perturbation displacements as:

$$\begin{aligned} \theta^* &= -\frac{d\theta}{db}|_{x=a+b} = \frac{a^2bp}{(a+b)^2} \\ v^* &= -\frac{dv}{db}|_{x=a+b} = 0 \end{aligned} \tag{16}$$

where positive θ^* indicates the right hand fixed end has a rotation in anti-clockwise direction.

The beam with displacements θ^* and v^* will result in the following bending moment

$$M = -\frac{2a^2bp}{(a+b)^4}(3x - a - b) \tag{17}$$

which is exactly the same as the analytical solution. Thus we establish the equivalency of the FDDM with the analytical solution for the beam extension problem.

A cantilever beam

The dimension and the boundary element discretization are shown in Fig. 4, in which $b = 2\,m$, $L = 10\,m$, $P = 15\,N$, and $E1$ to $E10$ denotes the element's number. The material properties are : Young's Modulus $E = 1.0 \times 10^7\,N/m^2$, and Poisson's ratio $\nu = 0.3$. The left hand fixed end is the design boundary.

First, let us assume a cosine type design perturbation as shown in Fig. 5(a), in which the design perturbations are: $V_n = \delta b \cos(\pi y/2)$, and $V_s = 0$. The stress sensitivity results (average over the element) of the lower surface of the beam are plotted in Fig. 6, in which the FDM presents the finite difference results, and the FDDM the finite difference displacement method. The FDM results are the stable results after running a series of step sizes, and the final perturbation step is taken as $\delta b = 1.0 \times 10^{-4}$. It can be seen from Fig. 6 that the agreements between the FDM and FDDM are excellent. Generally speaking the design boundary perturbation has very little influence on the stress distribution of the beam, with the exception of elements which are close to the fixed end.

Next, we will examine the case in which the fixed end has an initial displacement in x direction as shown in Fig. 5(b), where $U_{n0} = 1.2 \times 10^{-5} \times (1 - |y|)\,m$. This initial displacement is assumed to be constant during the design perturbation. Fig. 7 shows the results of the stress sensitivity by the FDM and the FDDM, and the stress distribution of the initial structure (note that the real stress value is scaled down by 0.1 in Fig. 7). The FDDM results agree with the FDM very well except for elements 1 and 2. The poor accuracy of elements 1 and 2 may be caused by the following factors: 1) as it can be seen from the initial stress distribution, the stress variation is large near the left hand end, so the mesh may not be fine enough to model the stress variation within this area; 2) the initial displacement field has little influence on the stress sensitivity of the beam surface except the area near the left hand end, in which case the stress sensitivity variation has increased substantially. This again makes the BEM solutions incapable to model the stress variation accurately.

CONCLUSIONS

A novel approach to shape design sensitivity analysis is presented for the case of the kinematical design boundary, which overcomes the two major drawbacks of the traditional finite difference method. The FDDM formulation is based on the continuum model, therefore no discretization approximations are involved during the derivation of the perturbation displacements. Since the perturbation displacements are acting on the initial structure, the solution under these displacements can be solved efficiently by using the already factorized matrices during the initial problem analysis. Subsequently, it is very easy to implement using the existing analysis programs. The 2-D formulation for perturbation displacements can be extended to 3-D problems using the same concept. FDLM can be coupled with either the FEM or the BEM for shape design sensitivity analysis.

The numerical results indicate that the FDDM is a reliable and accurate approach. Poor results may appear near the design perturbation boundary, especially where there is a large stress variation or stress sensitivity variation. Further research is under taken to investigate the accuracy of the perturbation displacements using different numerical methods.

References

[1] E. J. Haug, K. K. Choi, and V. Komkov, *Design Sensitivity Analysis of Structural Systems*, Academic Press, New York, 1985.

[2] B. R. Haber, 'A New Variational Approach to Structural Shape Design Sensitivity Analysis', *Computer Aided Optimal Design: Structural and Mechanical Systems*, Springer-Verlag, 573-588, 1987.

[3] K. Dems, Z. Mroz, 'Variational Approach by Means of Adjoint System to Structural Optimization and Sensitivity Analysis', *Int. J. Solids Structures*, 19, 677-692, 1983. 20, 527-552, 1984.

[4] K. Dems, 'Sensitivity Analysis in Thermoelasticity Problems', *Computer Aided Optimal Design: Structural and Mechanical Systems*, Springer-Verlag, 563-572, 1987.

[5] C. A. Mota Soares, K. K. Choi, 'Boundary Elements in Shape Optimal Design of Structures', *The Optimum Shape: Automated Structural Design* (ed J. A. Bennett and M. E. Botkin), Plenum, NewYork, 199-231, 1986.

[6] B. M. Kwak, J. H. Choi, ' Shape Design Sensitivity Analysis Using Boundary Integral Equation for Potential Problems', *Computer Aided Optimal Design: Structural and Mechanical Systems* (ed. C. A. Mota Soares), Springer-Verlag, 1987.

[7] Z. Zhao, R. A. Adey, 'Shape Optimization Using the Boundary Element Method', *Computer Aided Optimum Design: Recent Advance*, CMP and Springer-Verlag, Southampton, 1989.

[8] S. Saigal, R. Aithal, J. H. Kane, ' Semianalytical Sensitivity Formulation in Boundary Elements', *AIAA Journal*, Vol. 27, No. 11, 1615-1621, 1989.

[9] B. M. Barthelemy, R. T. Haftka, 'Accuracy Analysis of the Semi-analytical Method for Shape Sensitivity Calculation', *Proc. AIAA/ASME/ASCE/AHS 29th Structures, Structural Dynamics and Material Conf.*, USA, 1988.

[10] J. H. Kane, 'Optimization of Continuum Structures Using a Boundary Element Formulation', PhD Thesis, The University of Connecticut, 1986.

[11] S. J. Wu, 'Application of the Boundary Element Method for Structural Shape Optimization', PhD Thesis, The University of Missouri, Columbia, 1986.

[12] V. U. Nguyen, R. Arenicz, 'Sensitivity Analysis of Underground Excavation Using Boundary Element Methods', *Proc. BETECH 85* (eds. C. A. Brabbia, B. J. Noye), Springer-Verlag, 1985.

[13] S. Y. Wang, Y. Sun, R. H. Gallagher, 'Sensitivity Analysis in Shape Optimization of Continuum Structures', *Comput. Struct.*, 20, 1985.

[14] R. T. Haftka, D. S. Malkus, 'Calculation of Sensitivity Derivatives in Thermal Problems by Finite Differences', *Int. Numer. Meth. Engng.*, 17, 1981, pp. 1811-1821.

[15] C. J. Camarda, H. M. Adelman, 'Static and Dynamic Structural Sensitivity Derivative Calculations in the Finite Element Based Engineering Analysis Language (EAL) System', NASA TM-85743, 1984.

[16] Z. Zhao., R. A. Adey, 'A Finite Difference Based Approach to Shape Design Sensitivity Analysis' *Proc. BEM12* (eds. M. Tanaka, C. A. Brebbia, T. Honma), CMP and Springer-Verlag, 1990.

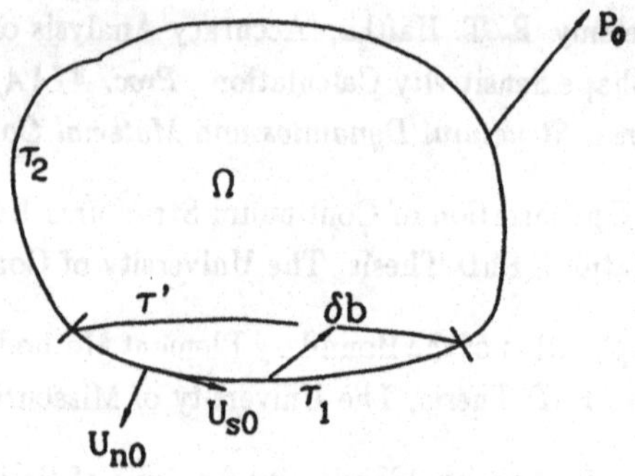

Fig. 1 The definition of the initial problem and the design boundary perturbation

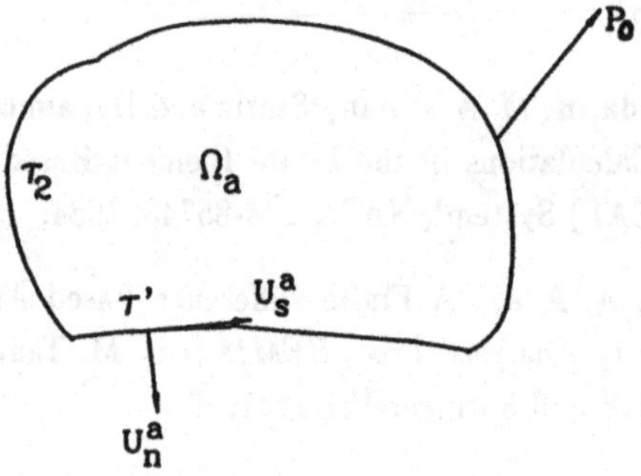

Fig. 2 The perturbed problem

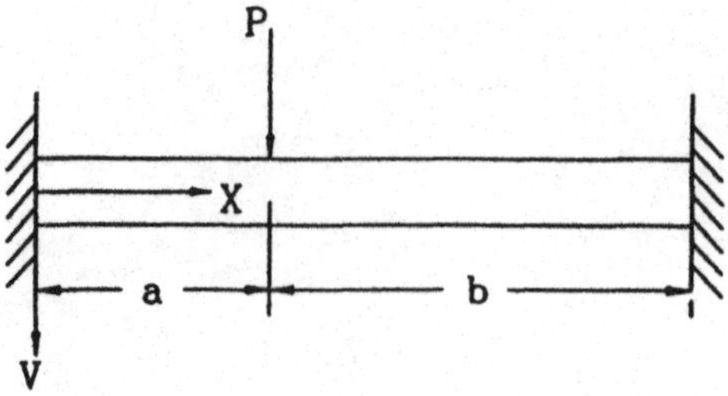

Fig. 3. A clamped beam

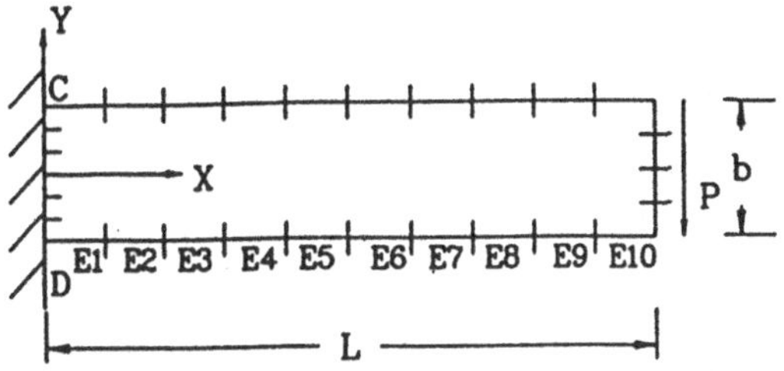

Fig. 4 The BEM modelling of the beam

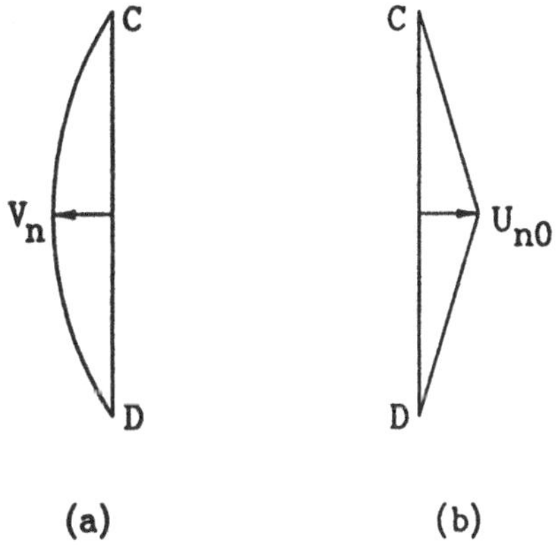

Fig. 5 (a) The design boundary perturbation

(b) The initial normal displacement

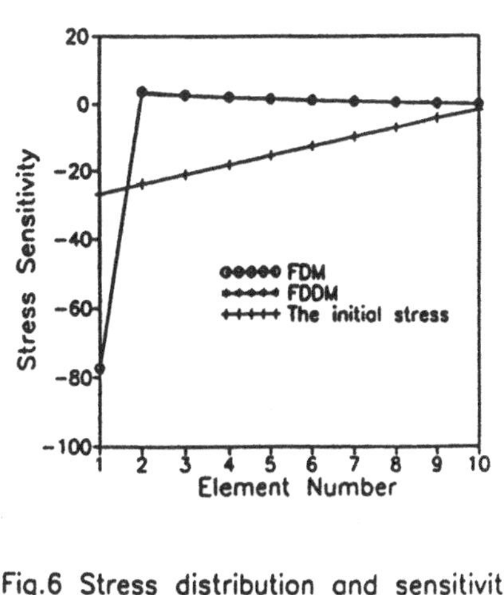

Fig.6 Stress distribution and sensitivities of the lower surface of the beam

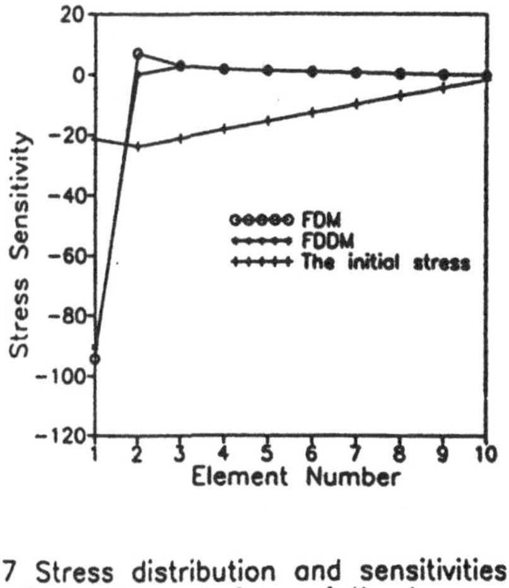

Fig.7 Stress distribution and sensitivities of the lower surface of the beam

Two Continuous Approaches with Boundary Representation to Shape Sensitivity Analysis

J. Unzueta, E. Schaeidt, S. Casado, J.A. Tárrago
Analysis & Design Department, LABEIN, Cuesta de Olabeaga, 16, 48013 BILBAO, Spain

ABSTRACT

Two continuous approaches using boundary information are presented to develop the sensitivity analysis with shape as design variable for 2D + Axisymmetric potential problems. The functions to be derived are functionals applied on Dirichlet, Neumann, Float or interfaces boundaries, involving the gradient of the state variable. Direct and adjoint methods are used to obtain the desired sensitivities with BEM as analysis technique. A numerical example is presented.

LITERATURE SURVEY

Many papers have come up over the past few years in the field of shape design sensitivity analysis (see Haftka [1] and references quoted therein). There have been two common approaches contributing to this topic, namely the discrete approach and the continuous approach.

The first approach derives the implicit discretized equilibrium equation. The accuracy of sensitivities is dependent on the mesh distortion rate, the calculation of each derivative can be more expensive than the analysis and a significant programming effort is required to develop the numerical integration of sensitivity Kernels when BEM is used as analysis tool (Deforuny [2], Kane [3], Saigal [4]).

The second approach uses a continuous model of the problem and the material derivative method of continuum mechanics to obtain computable expressions for the effect of shape variation on the functionals arising in the shape design problem. Two equivalent formulations, either on the boundary or in the domain, can be obtained for the continuous approach depending upon the way of calculating the derivatives of the functionals. The expressions reached through the boundary approach are functions of the partial shape derivatives of state variables. The domain approach, on the other hand, includes the total shape derivatives.

To evaluate the sensitivity equation derived by the boundary approach, only the design velocity field along the varied boundary is required. This represents a considerable computational saving compared to the domain approach, in which the design velocity field over the entire domain needs to be specified. As it is done in this paper, the boundary approach is usually employed for shape sensistivity analysis in conjunction with the boundary element method for analysis. Thus, inherent numerical difficulties in MEF regarding lack of accuracy for the response over the boundaries can all be avoided.

The continuous shape sensitivity is obtained in terms of state variables, their shape derivatives and geometric changes (velocity field). The evaluation of those derivatives requires the response of an associated problem, for each shape variable by directly differentiating the governing equations - direct method -, or an adjoint problem for each functional which removes the shape derivatives from the formulation - adjoint method.

Since it has been concluded that the adjoint method is more efficient than the direct method if the number of functionals is less than the number of design variables, and active constraints selection algorithms are included in most optimization codes, developments in shape sensitivities analysis were oriented towards the adjoint method. Examples of that can be found in references Choi [5], Haug [6], Meric [7-9], Dems [10], Mróz [11], Kwak [12], Choi [13].

Although the formal adaptation of the adjoint technique is conceptually straightforward, major computational difficulties arise when evaluating sensitivities of functionals at discrete boundary points, or sensitivities of boundary functionals involving spatial gradients of state variables. The problem stems from the fact that the adjoint solution cannot be expressed in terms of the boundary element formulation, because they give rise to infinite integrals (i.e., in elasticity the adjoint solution would correspond to a concentrated force and moment for displacement and stress sensitivities, respectively). To work out this singular adjoint solution, some empirical guidelines are found in reference Mota Soares [14] in which the adjoint solution is represented by a statically equivalent triangular load and in Zhao [15], using the fundamental solution of semi-plane spaces. In this paper, another very efficient alternative is proposed.

Nevertheless, the direct method, which does not present these drawbacks because it calculates sensitivities of the entire response field regardless of the functionals, has been mostly ignored and few references (Dems [16-18]) using this approach can be found. In this work, the direct method is also studied.

MATERIAL DERIVATIVE FOR SHAPE SENSITIVITY ANALYSIS

Since the domain Ω is to be varied, it is treated as a continuum moving with a "time like" parameter t in the material derivative formulation of shape design sensitivity analysis. The variations of a point $\underline{x}$ in the domain are expressed in terms of a velocity field $\underline{V}$, which defines the direction of movement of $\underline{x}$ on the nominal domain Ω to points $\underline{x}_t$ on a deformed domain Ω_t, given by the transformation,

$$\underline{x}_t = \underline{x} + t\underline{V}(\underline{x})$$

Then, the variation of a general differentiable function Ø with respect to a shape variation is expressed as the material derivative or total derivative of Ø at t = 0,

$$\dot{\phi} = \frac{d\phi_t(\underline{x}_t)}{dt}\Big|_{t=0} = \lim_{t\to 0}\frac{\phi_t(\underline{x}_t) - \phi(\underline{x})}{t} = \phi'(\underline{x}) + \underline{\nabla}\phi\underline{V}(\underline{x})$$

where,
$\phi_t(\underline{x}_t)$ is the solution of the boundary value problem on Ω_t, evaluated at a point $\underline{x}_t$ that moves with t, and

$$\phi'(\underline{x}) = \lim_{t\to 0}\frac{\phi_t(\underline{x}) - \phi(\underline{x})}{t}$$

namely local derivative, is the variation of ϕ at point $\underline{x}$.

Material derivatives of the normal and tangential unit vectors to a boundary

Let $\underline{n}$ be a normal vector to a boundary Γ. Its material derivative is (Zolésio [19]),

$$\dot{\underline{n}} = (\underline{n}D\underline{V}\underline{n})\underline{n} - D\underline{V}^T\underline{n}$$

where

$$D\underline{V} = \left(\frac{\partial V_i}{\partial x_j}\right)$$

and by simplicity, it will be understood that $\underline{a}B\underline{c} = \underline{a}^T.B.\underline{c}$

Let $\underline{N}_o$ and $\underline{S}_o$ be two unitary extensions of the fields $\underline{n}$ and $\underline{s}$ respectively, then

$\underline{V} = u\underline{N}_o + v\underline{S}_o$, $u = V_n$, $v = V_s$

$D\underline{V}^T\underline{n} = \underline{\nabla}u + vD\underline{S}_o^T\underline{n}$, because $D\underline{N}_o^T\underline{n} = 0$

$(\underline{n}D\underline{V}\underline{n})\underline{n} = (\underline{n}\underline{\nabla}u)\underline{n} + v(\underline{n}D\underline{S}_o\underline{n})\underline{n}$, because $D\underline{S}_o^T\underline{s} = 0$

Therefore,

$$\dot{\underline{n}} = -\frac{\partial V_n}{\partial s}\underline{s} + V_s D\underline{N}_o\underline{s} \qquad (1)$$

$$\underline{n}' = -\frac{\partial V_n}{\partial s}\underline{s} - V_n D\underline{N}_o\underline{n} \qquad (2)$$

Since $n_x = s_y$ and $n_y = -s_x$, then

$$\dot{\underline{s}} = \frac{\partial V_n}{\partial s}\underline{n} + V_s D\underline{S}_o\underline{s} \qquad (3)$$

$$\underline{s}' = \frac{\partial V_n}{\partial s}\underline{n} - V_n D\underline{S}_o\underline{n} \qquad (4)$$

Material derivatives of integral functionals

Let ψ_t be a functional defined over an open or closed boundary Γ_{At} piece-wise C^{k+1} regular on the varied domain Ω_t in R^2.

$$\psi_t = \int_{\Gamma_{At}} g_t(\underline{x}_t)\,d\Gamma_t \qquad (5)$$

Defining χ as a characteristic function,

$$\chi(\underline{x})|_\Gamma = \begin{cases} 1, \underline{x}\in\Gamma_A \\ 0, \underline{x}\in\Gamma_A \end{cases}$$

the expression (5) has the form

$$\psi_t = \int_{\Gamma_t} \chi_t(\underline{x}_t)\, g_t(\underline{x}_t)\,d\Gamma \qquad (6)$$

where Γ_t is $\partial\Omega_t$.

The material derivative of ψ at Γ is,

$$\psi' = \int_{\Gamma_A} [g' + (\nabla g \underline{n} + gH)V_n] d\Gamma + \sum^{corners \Gamma_A} [gV_s] + gV_s|_A^B \qquad (7)$$

H being the curvature of Γ, [ ] a jump operator and A, B, the initial and final points of Γ_A. In many cases, it will be interesting to formulate functionals weighted with the measure m_A of Γ_A,

$$\psi_t = \frac{1}{m_{At}} \int_{\Gamma_{At}} g_t(\underline{x}_t) d\Gamma_t \qquad (8)$$

also defined over an open or closed boundary Γ_{At} piece-wise C^{k+1} regular in R^2. The material derivative of ψ at Γ is,

$$\psi' = \frac{1}{m_A} [\int_{\Gamma_A} [g' + (\nabla g \underline{n} + (g - \psi)H)V_n] d\Gamma +$$

$$+ \sum^{corners \Gamma_A} [(g - \psi)V_s] + (g - \psi)V_s|_A^B] \qquad (9)$$

SHAPE SENSITIVITY IN ELECTROSTATIC PROBLEMS

Let Ω be an heterogeneous domain in R^2 over which an electrostatic phenomenon is defined, with Dirichlet, Neumann, Float (screens at unknown constant potential) and interface boundary conditions. The governing equations are:

$$\nabla^2_k \phi = 0 \qquad \text{in } \Omega_i$$

$$\phi = \phi_o \qquad \text{on } \Gamma^i_\phi \qquad (10)$$

$$\underline{\nabla}_k \phi \underline{n} = q_o \qquad \text{on } \Gamma^i_q$$

$$\phi = cte \qquad \text{on } \Gamma^l_F \qquad l = 1 \ldots m^i$$

$$\int_{\Gamma_F^l} \underline{\nabla}_k \phi \underline{n} \, d\Gamma = 0$$

$$\phi^i = \phi^j \qquad \text{on } \Gamma^{i/j}_I \qquad i, j = 1 \ldots\ldots n$$
$$q^i + q^j = 0$$

being,
n = number of subregions of Ω
m^i = number of screens in Ω^i
$\Gamma^i = \partial\Omega^i$
$\Gamma^i_\phi \cup \Gamma^i_q \cup \Gamma^i_F \cup \Gamma^i_I = \Gamma^i$
$\underline{\nabla}_k = (K_x\ \partial/\partial x,\ K_y\ \partial/\partial y)$
K_x, K_y = orthotropic permitivities in Ω^i.

The functionals determining a design in electrostatics (and in most physical phenomenons) are functions of the state variable gradient on the boundaries, that is, punctual functionals,

$$\psi = g(\underline{\nabla}\phi) \qquad \textit{on whichever point of } \Gamma \quad (11)$$

which material derivative is,

$$\psi' = g_{\underline{\nabla}\phi}\ [\underline{\nabla}\phi' + D^2\phi\underline{V}] \qquad (12)$$

and also integral functionals,

$$\psi = \frac{1}{m_A}\int_{\Gamma_A} g(\underline{\nabla}\phi)d\Gamma \qquad (13)$$

which has the form of eq.(8). Therefore, its material derivative is like eq.(9), where

$$g' = g_{\underline{\nabla}\phi}\underline{\nabla}\phi' \qquad (14)$$

<u>Direct method</u>

For calculating the material derivatives of eq. (11) and (13), that is eq.(12) and (9) with (14) the direct differentation of the governing equations is needed in order to get the response derived.

- In Ω,

$$\nabla_k^2 \phi = 0$$

Taking the material derivative at both hands of the equality,

$$\nabla_k^2 \phi' = 0 \qquad (15)$$

where the nullity of the Laplacian has been imposed.

- On $\Gamma\phi$,

$$\phi = \phi_o$$

$$\phi' = - \underline{\nabla}\phi \underline{V} \qquad (16)$$

where ϕ_o has been suposed constant.

- On Γq

$$q = q_o$$

$$q' = -\underline{\nabla} q \underline{V} \qquad \text{because } q_o = \text{constant}$$

$$\underline{\nabla}_k \phi' \underline{n} = - \underline{\nabla}_k \phi \underline{n}' - \underline{\nabla} q \underline{V}$$

Taking into account eq.(2), after operating

$$\underline{\nabla}_k \phi' \underline{n} = \frac{\partial V_n}{\partial s} \underline{\nabla}_k \phi \underline{s} - n_k D^2 \phi \underline{V} - V_s \underline{\nabla}_k \phi \frac{\partial n}{\partial s} \qquad (17)$$

where

$$D^2 = [\frac{\partial^2}{\partial x_i \partial x_j}]$$

- On Γ_F^i,

$$\partial \phi / \partial s = 0$$

Taking the material derivative and since

$$(\underline{\nabla}\phi)^{\cdot} = \underline{\nabla}\dot{\phi} - D\underline{V}^T \underline{\nabla}\phi$$

$$D\underline{V} = V_n D\underline{N}_o + \underline{n}\underline{\nabla} V_n + V_s D\underline{S}_o + \underline{s}\underline{\nabla} V_s$$

and from eq. (3), after operating

$$\partial \dot{\phi} / \partial s = 0$$

that is,

$$\phi' = \text{cte} - \underline{\nabla}\phi \underline{V} \qquad (18)$$

$$\int_{\Gamma_F} \underline{\nabla}_k \phi \underline{n} d\Gamma = 0$$

similar to the process followed on Γ_q and from eq.(2),

$$\int_{\Gamma_F} \underline{\nabla}_k \phi' \underline{n} d\Gamma = \int_{\Gamma_F} [\frac{\partial V_n}{\partial s} \underline{\nabla}_k \phi \underline{s} - (\underline{n}_k D^2 \phi \underline{n} + qH) V_n] d\Gamma$$

where corners have not been considered on Float boundaries.

- On $\Gamma_I^{i/j}$,

$$\phi^i = \phi^j$$

$$\phi^{i'} = \phi^{j'} + (\underline{\nabla}\phi^j - \underline{\nabla}\phi^i)\underline{V} \tag{19}$$

$$q^i = - q^j$$

Differentiating both hands of this equation, similar to (17),

$$\underline{\nabla}_k \phi^{i'} \underline{n}^i = - \underline{\nabla}_k \phi^{j'} \underline{n}^j - \underline{V} \sum^2 [D^2 \phi \underline{n}_k] - \sum^2 [\underline{\nabla}_k \phi \dot{\underline{n}}] \tag{20}$$

where $\dot{n}$ is given in eq.(1).

The problem governed by eq. (15-20) allows the calculus of the unknown term $\underline{\nabla}\phi'$ in eq.(12) and (9) with (14) and therefore, the desired sensitivities.

<u>Adjoint method</u>

For the adjoint method, the functionals need to be expressed in an integral form. Thus, eq. (13) will be studied as a general case since it can be used to express functionals at isolated points:

$$\psi = g(\underline{\nabla}\phi) \approx \frac{1}{m_A} \int_{\Gamma_A} g(\underline{\nabla}\phi) d\Gamma \qquad , \quad m_A \to 0$$

Thus, the smaller m_A value that is taken, the better approximation will be obtained for ψ and therefore, for ψ'. For calculating (9) with (14) an adjoint problem is created. This problem obtains the integral part, denoted by P, containing the local derivatives of the state variable,

$$P=\int_{\Gamma}\chi g_{\underline{\nabla}\phi}\underline{\nabla}\phi' d\Gamma \qquad (21)$$

It is necessary to express $\underline{\nabla}\phi'$ in terms of ϕ' and/or $\underline{\nabla}_k\phi'\underline{n}$,

$$\underline{\nabla}\phi'=(\underline{\nabla}_k\phi'\underline{n})\underline{u}+\frac{\partial\phi'}{\partial s}\underline{v} \qquad (22)$$

being,

$\underline{u} = \underline{n}/(\underline{n}_k\underline{n})$ $\qquad$ $\underline{v} = \underline{s}_{1/k} / \underline{n}_k\underline{n}$

$\underline{s}_{1/k} = (K_y s_x, K_x s_y)$ $\qquad$ $\underline{n}_k = (K_x n_x, K_y n_y)$

Substituting (22) into (21) and taken into account that,

$$f(s_o)=\int_{\Gamma}\delta(s,s_o)f(s)d\Gamma$$

the expression (21) reaches the form after manipulation,

$$P=\sum_i [\int_{\Gamma^i}\chi_A[g_{\underline{\nabla}\phi}\underline{u}(\underline{\nabla}_k\phi'\underline{n})-\phi'\frac{\partial}{\partial s}(g_{\underline{\nabla}\phi}\underline{v})]d\Gamma+$$

$$+\sum^{corners\,\Gamma_A}\int_{\Gamma^i}\delta(s,s_o)\phi'[g_{\underline{\nabla}\phi}\underline{v}]d\Gamma+\int_{\Gamma^i}\delta(s,s_o)\phi' g_{\underline{\nabla}\phi}\underline{v}|_A^B d\Gamma]$$

where s_o is the parameter of Γ at the corners of Γ_A or at the initial and final points A, B of Γ_A, and superscript i, the subregion index of Ω.

The adjoint variable is introduced adding to P the next null quantity,

$$\sum_i [\int_{\Omega^i} (\underline{\nabla}\phi^a \underline{\nabla}_k \phi' - \underline{\nabla}_k \phi^a \underline{\nabla}\phi') d\Omega] = 0$$

Applying Green formula in each subdomain Ω^i, taking into consideration that

$$\nabla_k^2 \phi' = 0 \qquad \text{in } \Omega^i$$

due to the equilibrium equation of (10) and imposing that

$$\nabla^2_k \phi^a = 0 \qquad \text{in } \Omega^i \qquad (23)$$

as equilibrium equation of the adjoint problem, and calling

$$\chi_A g_{\underline{\nabla}\phi} \underline{u} = a_\gamma \qquad ,on \ \Gamma^i_\gamma$$

$$-\chi_A \frac{\partial}{\partial s}(g_{\underline{\nabla}\phi}\underline{v}) + \sum^{corners \Gamma_A \cap \Gamma^i_\gamma} \delta(s,s_o)[\![g_{\underline{\nabla}\phi}\underline{v}]\!] + \delta(s,s_o) g_{\underline{\nabla}\phi}\underline{v}|_A^B = b\gamma \qquad ,on \ \Gamma^i_\gamma$$

where, now A, B are the initial and final points of $\Gamma_A \cap \Gamma_\gamma^i$, and doing that,

$$\phi_a = - a_\phi \qquad \text{on } \Gamma_\phi^i \quad (24)$$

$$\underline{\nabla}_k \phi^a \underline{n} = b_q \qquad \text{on } \Gamma_q^i \quad (25)$$

$$\left.\begin{array}{l} \phi^a = cte \\ \int_{\Gamma_F^i} \underline{\nabla}_k \phi^a \underline{n} \, d\Gamma = 0 \end{array}\right\} \qquad \text{on } \Gamma_F^i \quad (26)$$

$$\left.\begin{array}{l} a_I^i + \phi^{ia} = a_I^j + \phi^{ja} \\ q^{ia} + q^{ja} = b_I^i + b_I^j \end{array}\right\} \qquad \text{on } \Gamma_I^{i/j} \ (27)$$

as boundary conditions of the adjoint problem, P can be rewritten in the form,

$$P=\int_{\Gamma_A} g_{\underline{\nabla}\phi}\underline{\nabla}\phi' d\Gamma=\sum_i [\int_{\Gamma_\phi^i}(\chi_A\frac{\partial}{\partial s}(g_{\underline{\nabla}\phi}\underline{\nu})+\underline{\nabla}_k\phi^a\underline{n})\underline{\nabla}\phi\underline{V}d\Gamma-$$

$$-\sum^{corners\,\Gamma_A\cap\Gamma_\phi^i}[g_{\underline{\nabla}\phi}\underline{\nu}](\underline{\nabla}\phi\underline{V})-g_{\underline{\nabla}\phi}\underline{\nu}(\underline{\nabla}\phi\underline{V})|_A^B+$$

$$+\int_{\Gamma_q^i}(\chi_A g_{\underline{\nabla}\phi}\underline{u}+\phi^a)(\frac{\partial V_n}{\partial s}\underline{\nabla}_k\phi\underline{s}-\underline{n}_k D^2\phi\underline{V}-V_s\underline{\nabla}_k\phi\frac{\partial\underline{n}}{\partial s})d\Gamma+$$

$$+\int_{\Gamma_I^i}[(\chi_A\frac{\partial}{\partial s}(g_{\underline{\nabla}\phi}\underline{\nu})+q^a)\underline{\nabla}\phi\underline{V}-(\chi_A g_{\underline{\nabla}\phi}\underline{u}+\phi^a)(\underline{V}D^2\phi\underline{n}_k+$$

$$+\underline{\nabla}_k\phi(V_s\frac{\partial\underline{n}}{\partial s}-\frac{\partial V_n}{\partial s}\underline{s}))]d\Gamma-\sum^{corners\,\Gamma_A\cap\Gamma_I^i}[g_{\underline{\nabla}\phi}\underline{\nu}](\underline{\nabla}\phi\underline{V})-g_{\underline{\nabla}\phi}\underline{\nu}(\underline{\nabla}\phi\underline{V})|_C^D+$$

$$+\int_{\Gamma_F^i}[\underline{\nabla}_k\phi^a\underline{n}(\underline{\nabla}\phi\underline{V})+\frac{\partial V_n}{\partial s}\underline{\nabla}_k\phi\underline{s}-(\underline{n}_k D^2\phi\underline{n}+qH)V_n]d\Gamma$$

where it has been supposed that $\Gamma_A \cap \Gamma_F^1 = 0$. P allows the evaluation of (9) with (14), after having solved the adjoint problem defined by eq. (23-27).

Comparing the adjoint system with eq. (10), it can be seen that only a second member has to be computed. If a Gauss triangularization has been employed, a lot of time will be spared if the upper triangular has been kept on disk.

For axisymmetric problems, the null term to add is:

$$\sum_i [\int_{\gamma^i}\phi^a\underline{\nabla}_k\phi'\underline{n}rd\gamma-\int_{\gamma^i}\underline{\nabla}_k\phi^a\underline{n}\phi' rd\gamma]=0$$

where γ^i is the curve Γ^i (r, z, θ_o), θ_o = const. Following the same steps as before, the obtaining of P and adjoint boundary conditions is straightforward.

Numerical treatment of the singular adjoint boundary conditions

Eq. (25-27) show the singularity introduced in the adjoint problem by the b_γ boundary conditions. They have a finite part which does not cause any complication and a singular part to be tackled. The procedure proposed consists of removing from the boundary conditions the singular part of coeficients b_γ and dealing with it into the B.E.M. formulation. Thus, it is possible to calculate the exact effect of singular boundary conditions for all nodes except the one which b_γ is on

$$\int_\Gamma W(s,s_l)\delta(s,s_o)d\Gamma = W(s_o,s_l) \qquad ,s_l \neq s_o \qquad on\Gamma$$

where W is the fundamental solution, s_l the source point and s_o the functional point.

When $s_l = s_o$, an approximation of Dirac delta is used,

$$\delta(s,s_o) \approx \frac{1}{\Pi}\frac{\varepsilon}{[r-r_o]^2+\varepsilon^2}+h \qquad , \; \varepsilon,h\to 0$$

where $r_o = r(s_o)$ and h a remaining value which makes the integral

$$\int_\Gamma \delta(s,s_o)d\Gamma = 1$$

be satisfied. The election of ε has influence on the accuracy of results. Sensitivity values presented in Figure 2 correspond to

$$\varepsilon = \frac{2\log_e(m+1)}{\Pi p} \qquad , \; p=10$$

m being the measure of the boundary element on which b_γ is, and p the ratio of the δ height to m.

EXAMPLE

Figures 1 to 3 show a real life design application example. Figure 1 shows the design models, both the initial geometry (a) and final geometry (b) (regarding tidiness the boundary elements analysis models have not been presented). It is an axisymmetric insulator, of which only half has been modelled, surrounded by air. The electrode is connected to 75 Kv. Dimensions selected as design variables are displayed. The optimization problem consists of determining V1, V2, V3, V4 and V5 in order to minimize,

$$\frac{1}{m_c}\int_{\Gamma_c}(|\nabla\phi|)^2 d\Gamma$$

where Γ_c is the curved interface boundary close to ground. There are two behaviour constraints at interface points A and B,

$$|\nabla\phi| \begin{matrix} \leq 400 v/mm \text{ at } A \\ \leq 300 v/mm \text{ at } B \end{matrix} \qquad \textit{by the direct method}$$

$$|\nabla\phi| - \frac{1}{mp}\int_{\Gamma_p}|\nabla\phi| d\Gamma \begin{matrix} \leq 400 v/mm \text{ at } A \\ \leq 350 v/mm \text{ at } B \end{matrix} \qquad \textit{by the adjoint method}$$

where boundary Γ_p is the anterior and posterior boundary element to points A or B.

Figure 2 shows sensitivities at initial geometry obtained with the adjoint method (a) compared to finite differences (b).

Figure 3 shows objective function (a) and maximum constraint violation (b) history. Notice that even though one constraint (A) is violated at initial shape, there is an objective function decrease during the optimization process. Constraint A is active at the optimum.

REFERENCES

1. Haftka R.T. and Adelman H.M. Recent Developments in Structural Sensitivity Analysis. Third International Conference on CAD/CAM, Robotics and Factories of the Future, Southfield, Michigan, August 14-17, 1988.
2. Defourny M. Optimization Techniques and Boundary Element Method. Proc. Boundary Elements X. Vol 1: Mathematical and Computational Aspects. Springer-Verlag, 1988.
3. Kane J.H., Saigal S. Design Sensitivity Analysis of Solids Using BEM. Journal of Engineering Mechanics. Vol 114, pp 1703-1722, 1988.
4. Saigal S., Borggaard J.T., Kane J.H. Boundary Element Implicit Differentiation Equations for Design Sensitivity of Axisymetric Structures. International Journal for Solids and Structures. Vol 25, No 5, pp. 527-538, 1989.
5. Choi K.K., Seong H.G. A Domain Method for Shape Design Sensitivity of Built-up Structures. Computer Methods in Applied Mechanics and Engineering, 57, pp. 1-15, 1986.
6. Haug E.J., Choi K.K. Komkov V. Design Sensitivity Analysis of Structural Systems. Academic Press. New York, 1986.
7. Meric R.A. Boundary Elements in Shape Design Sensitivity Analysis of Thermoelastic Solids. Computer Aided Optimal Design: Structural and Mechanical Systems. Ed: C.A. Mota Soares. Springer-Verlag, Berlin, Heidelberg, 1987.
8. Meric R.A. Shape Design Sensitivity Analysis for Non-Linear Anisotropic Heat Conducting Solids and Shape Optimization by the BEM. International Journal for Numerical Methods in Engineering. Vol 26, pp 109-120, 1988.
9. Meric R.A. Boundary Elements for Exact and Optimal Shape Inverse Problems. Boundary Element Techniques: Applications in Engineering. Ed: C.A. Brebbia, N.G. Zamani, CMP, 1989.
10. Dems K. Sensitivity Analysis in Thermoelasticity Problems. Computer Aided Optimal Design: Structural and Mechanical Systems. Ed: C.A. Mota Soares, Springer-Verlag, Heidelberg, 1987.
11. Mróz Z. Sensitivity Analysis and Optimal Design with Account for Varying Shape and Support Conditions. Computer Aided Optimal Design: Structural and Mechanical Systems. Ed: C.A. Mota Soares, Springer-Verlag, 1987.
12. Kwak B.H., Choi J.H. Design Sensitivity Analysis Based on Boundary Integral Equation Method Considering General Shape Variations. Part I: For Self-Adjoint Elliptic Operator Problems. KSME Journal. Vol 1, No 1, pp. 70-73, 1987.
13. Choi J.H., Kwak B.H. Shape Design Sensitivity Analysis of Elliptic Problems in Boundary Integral Equation Formulation. Mech. Struct. & Mach. 16 (2), 147-165, 1988.

14. Mota Soares C.A., Choi K.K. Boundary Elements in Shape Optimal Design of Structures. The Optimum Shape: Automated Structural Design, Ed: J.A. Bennet, M.E. Botkin, Plenum, New York, pp. 199-231, 1986.
15. Zhao Z, Adey R.A. The Accuracy of the Variational Approach to Shape Design Sensitivity Analysis. Boundary Elements in Mechanical and Electrical Engineering. Ed: C.A. Brebbia, Springer-Verlag, 1990.
16. Dems K. Sensitivity Analysis in Thermal Problems-II: Structure Shape Variation. Journal of Thermal Stresses, Vol 10, pp. 1-16, 1987.
17. Dems K., Mroz Z. Variational Approach to Sensitivity Analysis in Thermoelasticity Journal of Thermal Stresses, Vol 10, pp. 283-306, 1987.
18. Dems K., Haftka R.T. Two Approaches to Sensitivity Analysis for Shape Variation of Structures. Mechanics of Structures and Machines, 16 (4), 501-522, 1989.
19. Zolesio J.P. The material Derivative (or Speed) Method for Shape Optimization. Optimization of Distributed Parameter Structures (Eds. E.J. Haug, J. Cea), Sijthoff & Noordhoff, Alphen aan den Rijn, Netherlands, pp. 1089-1153, 1981.

Acknowledgement: This research was carried out under project OCIDE 132135.

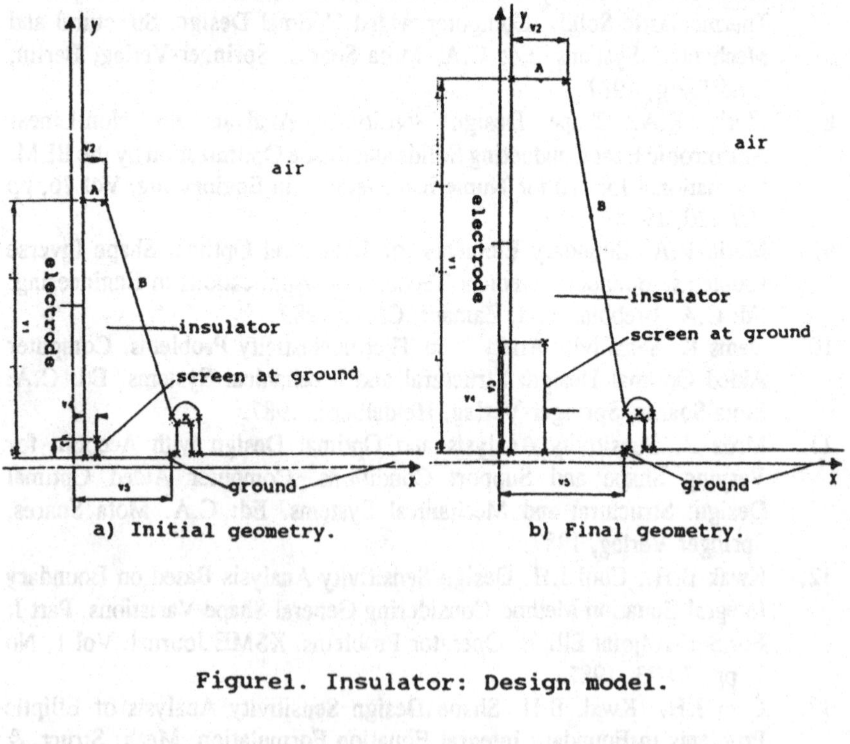

Figure1. Insulator: Design model.

	VARIABLE 1	VARIABLE 2	VARIABLE 3	VARIABLE 4	VARIABLE 5
OBJECT. FUNC.	207.342631	592.113125	-1235.7376	-817.86473	-1682.8231
CONSTRAINT 1	-0.002277	-0.020854	0.000991	0.000548	0.002543
CONSTRAINT 2	-0.004620	-0.009060	0.002243	0.001227	0.002316

a) Direct Method

	VARIABLE 1	VARIABLE 2	VARIABLE 3	VARIABLE 4	VARIABLE 5
OBJECT. FUNC.	213.595520	608.465140	-1342.4478	-891.67746	-1820.6793
CONSTRAINT 1	-0.001832	-0.020491	0.000994	0.000549	0.002596
CONSTRAINT 2	-0.004619	-0.009060	0.002245	0.001226	0.002306

b) Adjoint Method

	VARIABLE 1	VARIABLE 2	VARIABLE 3	VARIABLE 4	VARIABLE 5
OBJECT. FUNC.	213.917890	627.283961	-1209.8706	-779.07712	-1973.5038
CONSTRAINT 1	-0.003756	-0.020767	0.000993	0.000547	0.002191
CONSTRAINT 2	-0.004525	-0.009000	0.002247	0.001220	0.001726

c) Finite Differences

Figure 2. Insulator: Sensitivity values

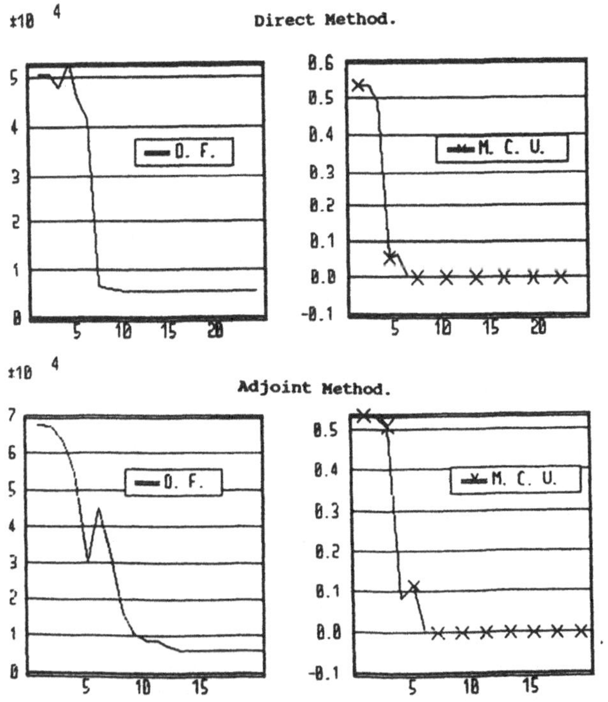

a) Objective function v. optimization iterations.

b) Maximum constraint violation v. optimization iterations.

Figure3. Insulator: Optimization history.

Application of the EBP-BEM to Optimizing the Cavity of Large Aluminum Reduction Cells

J. Lu, Z. Wu

Electrical Engineering Department, Naval Engineering Institute, Jiefang Road 339-401, Wuhan, P.R. China

ABSTRACT

The hybrid Equivalent Boundary Principle - Boundary Element Method (EBP-BEM) is developed for optimizing the current field in large aluminum reduction cells. The method of computing the inverses of revised matrices is given. Taking a typical 160 KA cell as an example, we determine the optimal ridge in the cell cavity and calculate the electromagnetic forces in the cell under the conditions of the determined bus disposition and the optimal ridge.

INTRODUCTION

It is a glaring problem to weaken the electromagnetic forces in the melt (melted metal, bath) for reducing the dissipation of electric energy and raising the current efficiency in research of large aluminum reduction cells (Fig. 1a). The vertical component of magnetic field and the horizontal current are main factors in causing the melt oscillation [1]. The vertical magnetic field in the cell can be reduced by disposing the bus rationally. In this respect, the researchers have accumulated a wealth of experience [2,3]. After the cells are built, the effective and feasible way the horizontal current in the melted metal can be checked is to obtain the optimal cavity shape of the cell by controlling some operating variables.

In the past, suitable ridge was determined by test after the cell was put into operation. However, it is impossible to observe and measure carefully the working conditions of the cell, since the upper part of the melt in the cell is covered with a layer of solid matter and the temperature in the cell is very high. Therefore, the above-mentioned method is both rough and blind to some extent. In addition, the cost of test is high. Application of computing techniques helps the development of the research of electromagnetic field in the cell [3,4].

Along with the development of computing techniques, the optimization method has been used in the design of electric devices [5,6]. The BEM is known to have superiority over other

Figure 1a. Aluminum reduction cells.

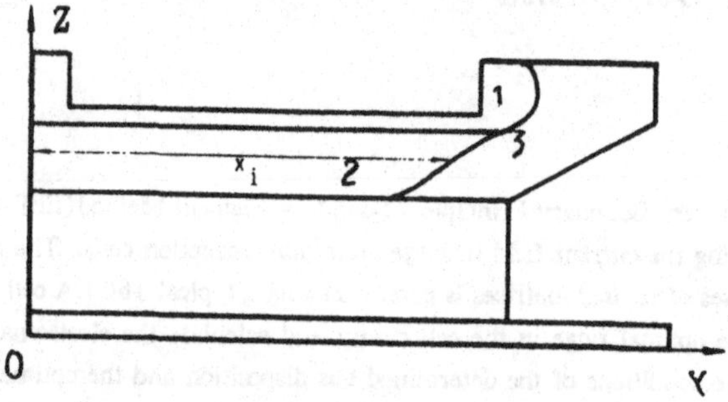

Figure 1b. Calculation model of the current field in the cell. 1. Bath. 2. Melted metal. 3. Ridge.

numerical methods in the shape optimization [7,8]. However, because of the dense coefficient matrices, a large amount of computation is still required in using the BEM directly. To simplify the computation, we suggest that the EBP-BEM which is presented in Ref. [9] and used in practical engineering [10] is used in the shape optimization. The advantages of the EBP-BEM in the shape optimization are mainly embodied in the following: the region to be solved can be equivalently reduced according to the Equivalent Boundary Principle, and the variance of the boundary can be conveniently handled by using the BEM.

In this paper, the EBP-BEM is developed for optimizing the shape of the cavity of large aluminum reduction cells. The method of computing the inverses of revised matrices is given. The practical application shows that the method is highly efficient in shape optimization.

SHAPE OPTIMIZATION BY EBP-BEM

The current field in the cell belongs to steady current field. The electric potential u and the current density $\mathbf{J}$ satisfy the following

$$\mathbf{J} = -\gamma \nabla u \tag{1}$$

$$\nabla \cdot \mathbf{J} = 0 \tag{2}$$

where γ is the electrical conductivity. We discuss the case of $\nabla^2 u = 0$ below.

The region V which must be optimized is shown in Fig. 2. A proper interface $\bar{S}$ is selected to divide the whole region V into two subregions; one (denoted by V_2) is the part having changeable boundary on which the object function is to be estimated, and the other (denoted by V_1) is the part without shape change during the whole optimization process.

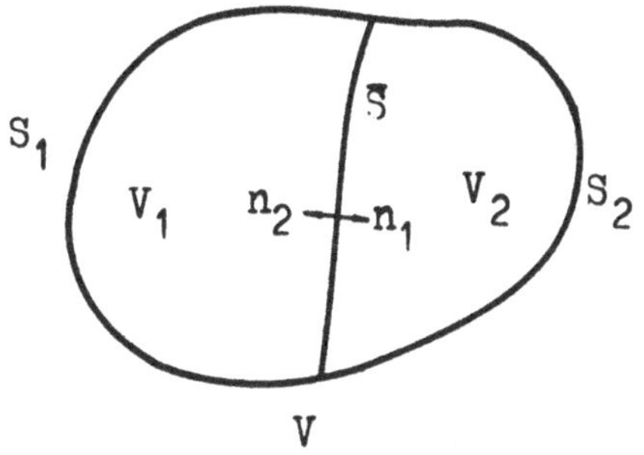

Figure 2

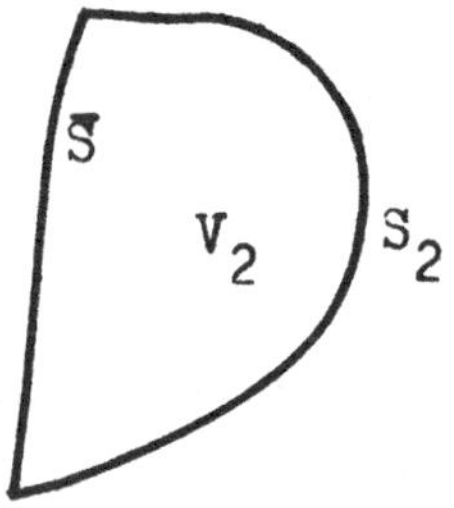

Figure 3

The potential in V_1 can be written as

$$c_i u_i = \int_{S_1} \left[u^* \frac{\partial u}{\partial n} - u \frac{\partial u^*}{\partial n} \right] ds + \int_{\bar{S}} \left[u^* q_{s1} - u_{s1} \frac{\partial u^*}{\partial n_1} \right] ds \tag{3}$$

where u^* is the fundamental solution, u_{s1} is the potential on $\bar{S}$ (consider it belongs to V_1) and $q_{s1} = \partial u_{s1} / \partial n_1$. Introducing the Equivalent Boundary Principle of electromagnetic field [9,10]

$$q_{s1} = I + T u_{s1} \tag{4}$$

and considering the interface conditions

$$u_{s2} = u_{s1} \tag{5}$$

$$q_{s2} = -q_{s1} \tag{6}$$

yield the following expression for the potential in V_2

$$c_i u_i = \int_{S_2} \left[u^* \frac{\partial u}{\partial n} - u \frac{\partial u^*}{\partial n} \right] ds - \int_{\bar{S}} \left[u^* (I + T u_{s2}) + u_{s2} \frac{\partial u^*}{\partial n_2} \right] ds \tag{7}$$

where u_{s2} is the potential on $\bar{S}$ (belonging V_2) and $q_{s2} = \partial u_{s2} / \partial n_2$. The current density in V_2 is

$$J_x = -\gamma \int_{S_2+\bar{S}} \left[\frac{\partial u}{\partial n} \frac{\partial u^*}{\partial x} - u \frac{\partial}{\partial x}\left(\frac{\partial u^*}{\partial n} \right) \right] ds \tag{8}$$

$$J_y = -\gamma \int_{S_2+\bar{S}} \left[\frac{\partial u}{\partial n} \frac{\partial u^*}{\partial y} - u \frac{\partial}{\partial y}\left(\frac{\partial u^*}{\partial n} \right) \right] ds \tag{9}$$

By using the discretizing techniques described in Ref. [11], the discretization equations corresponding to the subregion V_1 can be obtained as

$$[\mathbf{H}_{S_1} \quad \mathbf{H}_{\bar{S}1}] \begin{bmatrix} \mathbf{U}_{S_1} \\ \mathbf{U}_{\bar{S}1} \end{bmatrix} = [\mathbf{G}_{S_1} \quad \mathbf{G}_{\bar{S}1}] \begin{bmatrix} \mathbf{Q}_{S_1} \\ \mathbf{Q}_{\bar{S}1} \end{bmatrix} \tag{10}$$

where $\mathbf{U}_{S_1}$ $(\mathbf{Q}_{S_1})$ and $\mathbf{U}_{\bar{S}1}$ $(\mathbf{Q}_{\bar{S}1})$ are the potential (flux) vectors on S_1 and $\bar{S}$ (belonging to V_1) respectively. Eq. (10) may be rearranged as

$$\begin{bmatrix} \mathbf{K}_{11} & \mathbf{K}_{12} \\ \mathbf{K}_{21} & \mathbf{K}_{22} \end{bmatrix} \begin{bmatrix} \mathbf{U}_{S_1} \\ \mathbf{U}_{\bar{S}1} \end{bmatrix} = \begin{bmatrix} \mathbf{Q}_{S_1} \\ \mathbf{Q}_{\bar{S}1} \end{bmatrix} \tag{11}$$

From Eqs. (11) and (4) we have

$$\mathbf{I} = \begin{cases} \mathbf{K}_{21}\mathbf{U}_{S_1} & \text{when } S_1 \text{ is an essential boundary} \\ \mathbf{K}_{21}\mathbf{K}_{11}^{-1}\mathbf{Q}_{S_1} & \text{when } S_1 \text{ is a natural boundary} \end{cases} \tag{12}$$

$$\mathbf{T} = \begin{cases} \mathbf{K}_{22} & \text{when } S_1 \text{ is an essential boundary} \\ \mathbf{K}_{22} - \mathbf{K}_{21}\mathbf{K}_{11}^{-1}\mathbf{K}_{12} & \text{when } S_1 \text{ is a natural boundary} \end{cases} \tag{13}$$

and

$$\mathbf{Q}_{\bar{S}2} = -(\mathbf{I} + \mathbf{T}\mathbf{U}_{\bar{S}2}) \tag{14}$$

where $\mathbf{U}_{\bar{S}2}$ and $\mathbf{Q}_{\bar{S}2}$ are the potential and flux vectors on $\bar{S}$ (belonging to V_2) respectively. According to Eq. (7) the discretization equations corresponding to V_2 can be written as

$$[\mathbf{H}_{\bar{S}2} + \mathbf{G}_{\bar{S}2}\mathbf{T} \quad \mathbf{H}_{S_2}] \begin{bmatrix} \mathbf{U}_{\bar{S}2} \\ \mathbf{U}_{S_2} \end{bmatrix} = [\mathbf{G}_{\bar{S}2} \quad \mathbf{G}_{S_2}] \begin{bmatrix} \mathbf{I} \\ \mathbf{Q}_{S_2} \end{bmatrix} \tag{15}$$

in which $\mathbf{H}_{\bar{S}2}$, $\mathbf{H}_{S_2}$, $\mathbf{G}_{\bar{S}2}$ and $\mathbf{G}_{S_2}$ satisfy the following relation

$$[\mathbf{H}_{\bar{S}2} \quad \mathbf{H}_{S_2}] \begin{bmatrix} \mathbf{U}_{\bar{S}2} \\ \mathbf{U}_{S_2} \end{bmatrix} = [\mathbf{G}_{\bar{S}2} \quad \mathbf{G}_{S_2}] \begin{bmatrix} \mathbf{Q}_{\bar{S}2} \\ \mathbf{Q}_{S_2} \end{bmatrix} \tag{16}$$

where $\mathbf{U}_{S_2}$ and $\mathbf{Q}_{S_2}$ are the potential and flux vectors on S_2 respectively.

Owing to introducing the Equivalent Boundary Principle, the shape optimization of the region V shown in Fig. 2 can be changed into that of the region in Fig. 3, which reduces the size of the matrices in the global boundary elements computation. The I and T can be solved once in the optimizing process and only Eq. (15) is to be solved during iteration steps. So the memory space of computer is saved and computation efficiency is raised.

COMPUTATION OF INVERSES OF REVISED MATRICES

When the Boundary Element Method is used to optimize the boundary shape of electromagnetic field, the coefficient matrix A in the following boundary element equations

$$AX = F \tag{17}$$

has the following characteristics: In A, only the elements concerning the changeable boundary may change. And the changeable elements are blocked. To meet the need of iteration, we develop here the method of computing the inverses of revised matrices.

Corresponding to Fig. 3, the matrix A may be written as

$$A = \begin{bmatrix} A_{\bar{S}\bar{S}} & A_{\bar{S}S_2} \\ A_{S_2\bar{S}} & A_{S_2S_2} \end{bmatrix} \tag{18}$$

in which $A_{\bar{S}S_2}$, $A_{S_2\bar{S}}$ and $A_{S_2S_2}$ are changeable while $A_{\bar{S}\bar{S}}$ is not during iteration steps. Suppose the number of nodes on $\bar{S}$ and S_2 is $n_{\bar{S}}$ and n_{S_2} respectively, then, $A_{\bar{S}\bar{S}}$ and $A_{S_2S_2}$ are $n_{\bar{S}} \times n_{\bar{S}}$ and $n_{S_2} \times n_{S_2}$ sub-matrices respectively, $A_{\bar{S}S_2}$ and $A_{S_2\bar{S}}$ are $n_{\bar{S}} \times n_{S_2}$ and $n_{S_2} \times n_{\bar{S}}$ sub-matrices respectively.

A^k is used to denote the matrix A after the k-th iteration step, then

$$A^{k+1} = A^k + \Delta A^k \tag{19}$$

A^{k+1} is called the revised matrix which is modified on the basis of A^k. If $(A^k)^{-1}$ has been got, the $(A^{k+1})^{-1}$ can be calculated from $(A^k)^{-1}$[12].

We write ΔA^k in the following form

$$\Delta A^k = \underbrace{\begin{bmatrix} O & \Delta A^k_{\bar{S}S_2} \\ \Delta A^k_{S_2\bar{S}} & \Delta A^k_{S_2S_2} \end{bmatrix}}_{n_{\bar{S}} \quad n_{S_2}} \begin{matrix} \} \; n_{\bar{S}} \\ \} \; n_{S_2} \end{matrix}$$

$$= \underbrace{\begin{bmatrix} O & \Delta A^k_{\bar{S}S_2} \\ O & O \end{bmatrix}}_{\Delta A'^k} + \underbrace{\begin{bmatrix} O & O \\ \Delta A^k_{S_2\bar{S}} & \Delta A^k_{S_2S_2} \end{bmatrix}}_{\Delta A''^k} \tag{20}$$

and define that

$$B^k = (A^k)^{-1} = \underbrace{\begin{bmatrix} B^k_{\bar{S}\bar{S}} & B^k_{\bar{S}S_2} \\ B^k_{S_2\bar{S}} & B^k_{S_2S_2} \end{bmatrix}}_{n_{\bar{S}} \quad n_{S_2}} \begin{matrix} \} \; n_{\bar{S}} \\ \} \; n_{S_2} \end{matrix} \tag{21}$$

$$B'^k = (A^k + \Delta A'^k)^{-1} = \underbrace{\begin{bmatrix} B'^k_{\bar{S}\bar{S}} & B'^k_{\bar{S}S_2} \\ B'^k_{S_2\bar{S}} & B'^k_{S_2S_2} \end{bmatrix}}_{n_{\bar{S}} \quad n_{S_2}} \begin{matrix} \} \; n_{\bar{S}} \\ \} \; n_{S_2} \end{matrix} \tag{22}$$

thus

$$B'^{k} = B^{k} - B^{k}_{1c} \cdot \Delta A^{k}_{SS_2}(E_{n_{S_2}} + B^{k}_{S_2S} \cdot \Delta A^{k}_{SS_2})^{-1} B^{k}_{2r} \tag{23}$$

$$(A^{k+1})^{-1} = B^{k+1} = [E_n - B'^{k}_{2c}(E_{n_{S_2}} + \Delta A^{k}_{2} \cdot B'^{k}_{2c})^{-1} \cdot \Delta A^{k}_{2}]B'^{k} \tag{24}$$

In Eqs. (23) and (24)

$$B^{k}_{1c} = \begin{bmatrix} B^{k}_{SS} \\ B^{k}_{S_2S} \end{bmatrix}, \quad B^{k}_{2r} = [B^{k}_{S_2S} \quad B^{k}_{S_2S_2}]$$

$$B'^{k}_{2c} = \begin{bmatrix} B'^{k}_{SS_2} \\ B'^{k}_{S_2S_2} \end{bmatrix}, \quad \Delta A^{k}_{2} = [\Delta A^{k}_{S_2S} \quad \Delta A^{k}_{S_2S_2}]$$

$E_{n_{S_2}}$ and E_n are $n_{S_2} \times n_{S_2}$ and $(n_S + n_{S_2}) \times (n_S + n_{S_2})$ unity matrices respectively.

Using Eqs. (23) and (24), we can get $(A^{k+1})^{-1}$ from $(A^{k})^{-1}$ through computing the inverses of relatively low order matrices twice. In the present problem under discussion, the changeable boundary is mainly the ridge in the cell, and indispensable nodes are much fewer on the ridge than on the entire boundary of the melt. With the method of computing the inverses of revised matrices, computation efficiency in iteration is raised.

OBJECT FUNCTION

The vertical current is the normal working current but the horizontal current in the melt is unfavourable to electrolysing. Extremely harmful to the electrolysing in the cell are the electromagnetic forces caused by the interaction of the horizontal current with the vertical component of magnetic field in the melt. The magnetic field in the cell has been well studied. And once the cell has been built, the bus arrangement is fixed. The present principal problem is to reduce the horizontal current in the melt so as to further weaken the electromagnetic forces causing melt oscillation.

Fig. 4 shows the distribution of the horizontal current density on the same horizontal level of the melted metal in the cell both with and without a ridge. It is clear that the ridge affects greatly the current distribution in the cell after the cell has been built. The cell without a ridge has a larger horizontal current density in the middle part than the one with a ridge. The ridge can effectively check the horizontal current in the middle part of the cell, but it leads to an increase in the horizontal current in the area near the edge of the cell. For this reason, to get an optimal ridge, i. e., to design an optimal shape of the cell cavity is the problem to be solved.

The results in Ref. [9] show that the distribution of the horizontal current density in different levels of the melted metal in the cell with the same ridge is almost the same. Therefore, the horizontal current density J_x on the middle level of the melted metal can be used to

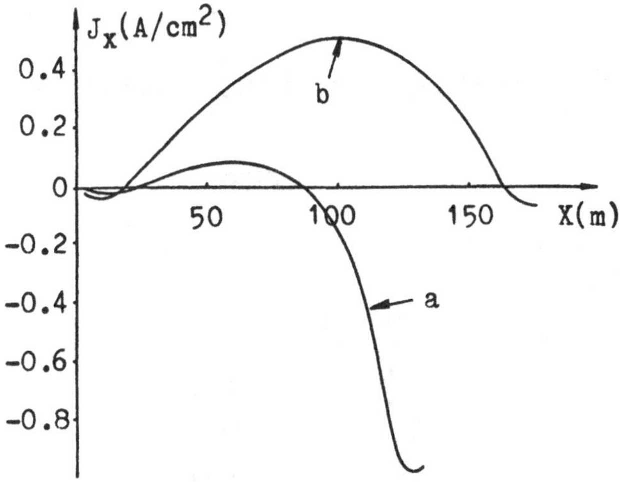

Figure 4. Horizontal current density distribution in the melted metal. a. Corresponding to the cell with a ridge. b. Corresponding to the cell without a ridge.

measure the value of the horizontal current in the whole melted metal.

From above, we can select

$$f(X) = \sum_{k=1}^{N} J_{xk}^2(X) \tag{25}$$

to be minimized as the object function. In Eq. (25), N is the number of internal nodes selected on the middle level of the melted metal and J_{xk} is the horizontal current density at the k-th node.

$$X = (x_1 \quad x_2 \quad \cdots \quad x_L) \tag{26}$$

is the design variable vector of L elements. The ridge is equally divided Z axially. L is the number of nodes on the ridge, $x_i(i=1,2,\cdots,L)$ represents the horizontal distance from the cell centre to the i-th node on the ridge(see Fig. 1b).

OPTIMIZATION

The object function described in Eq. (25) consists of terms in quadratic form. This is a typical nonlinear optimization problem. Eq. (25) can be written as

$$f(X) = J_X^T J_X \tag{27}$$

where

$$J_X = (J_{x1} \quad J_{x2} \quad \cdots \quad J_{xN})^T \tag{28}$$

Using the method of Gauss-Newton least squares [13], we get the coordinates of the ridge after the k-th iteration step

$$X^{k+1} = X^k - (2(K^k)^T K^k)^{-1} \nabla f^k \tag{29}$$

in Eq. (29)

$$K = \begin{bmatrix} \frac{\partial J_{x1}}{\partial x_1} & \frac{\partial J_{x1}}{\partial x_2} & \cdots & \frac{\partial J_{x1}}{\partial x_L} \\ \frac{\partial J_{x2}}{\partial x_1} & \frac{\partial J_{x2}}{\partial x_2} & \cdots & \frac{\partial J_{x2}}{\partial x_L} \\ \cdots & \cdots & \cdots & \cdots \\ \frac{\partial J_{xN}}{\partial x_1} & \frac{\partial J_{xN}}{\partial x_2} & \cdots & \frac{\partial J_{xN}}{\partial x_L} \end{bmatrix} \tag{30}$$

$$\nabla f(X) = \left[\frac{\partial f(X)}{\partial x_1} \quad \frac{\partial f(X)}{\partial x_2} \quad \cdots \quad \frac{\partial f(X)}{\partial x_L}\right]^T = 2K^T J_X \tag{31}$$

APPLICATION

Applying the above principle and method, and adopting the calculation model (Fig. 1b) established in Ref. [9], we obtain the optimal ridge of the 160 KA prebaked anode cell (Fig. 5a). Linear elements are used in this example. We select 315 nodes on the boundaries and interfaces of the current field of the cell. After the current field is equivalently reduced on Equivalent Boundary Principle, the discretization nodes on the boundary of the melt are 80, among which only 12 nodes are on the changeable boundary in optimizing process. According to the method of computing the inverses of revised matrices we calculate the inverses of two 12×12 matrices during iteration to find the unknown potential and flux on the boundary of the melt.

The current distribution in the cell with the optimal ridge is shown in Fig. 5a. For comparision, Fig. 5 includes the current distribution in the cell with a typical ridge which is determined by test. Figs. 6 and 7 show respectively the vertical component B_z of the magnetic field

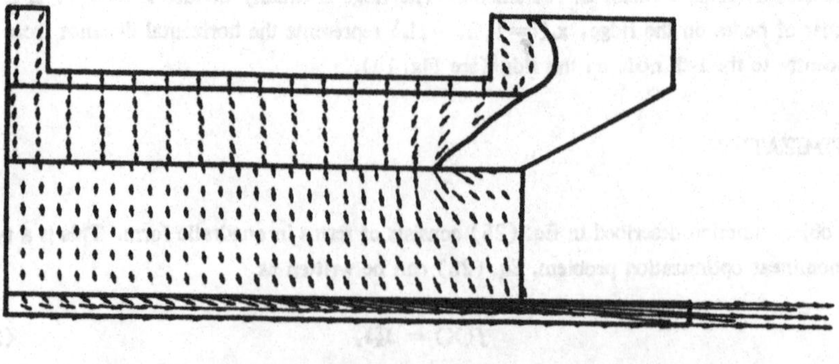

Figure 5a Current distribution in the cell with the optimal ridge.

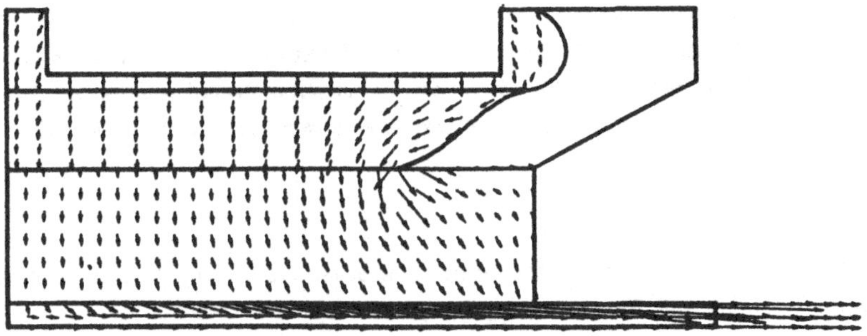

Figure 5b Current distribution in the cell with a typical ridge which is determined by test.

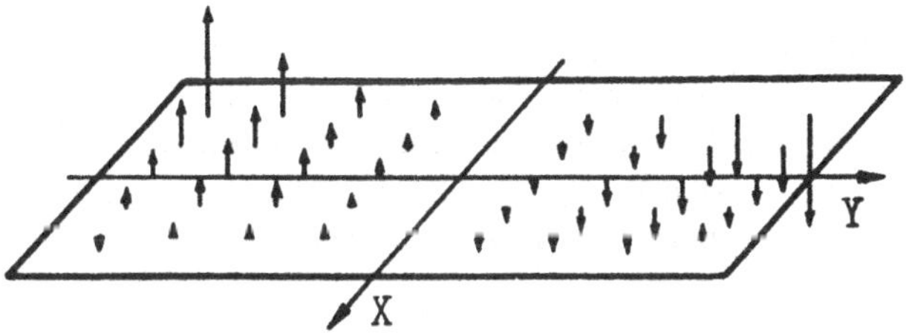

Figure 6. Vertical component B_z of the magnetic field in the melted metal.

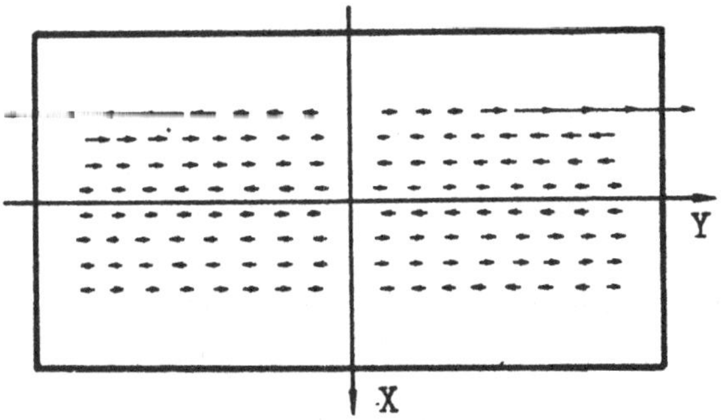

Figure 7. Electromagnetic forces caused by the interaction of B_z with the horizontal current in the melted metal.

and the electromagnetic forces caused by the interaction of B_z with the horizontal current in the melted metal in the cases of the determined bus disposition and the optimal ridge.

Figs. 5 and 7 show clearly that, in the cell with the optimal ridge, the horizontal current in the melt is distinctly reduced, the electromagnetic forces are weakened, and the distribution of the forces is very reasonable. Moreover, the current in the cell with the optimal ridge is well-distributed so that overheat in some parts is eliminated and current works efficiently.

CONCLUSIONS

The EBP-BEM is used to optimize the cavity shape of large aluminum reduction cells, which provides a powerful tool for determining the optimal shape of the ridge in the cell.

In shape optimization the advantages of the Equivalent Boundary Principle and the Boundary Element Method are brought into full play. The EBP-BEM saves greatly the computing time and memory space of computers and is to be well received in practical engineering. This method has broad prospects in the shape optimization.

Using the method of computing the inverses of revised matrices, we can obtain the inverses of revised matrices through computing the inverses of matrices of relatively low order, which raises computation efficiency in iteration steps of shape optimization. The method is very useful for computation, especially when the differences between the number of nodes both on the changeable boundary in iteration steps and the entire boundary of the region considered are great.

The optimal ridge enable the horizontal current in the melt to decrease and the current distribution to be even, thus weakening the electromagnetic forces which cause oscillation of the melt, reducing the dissipation of electrical energy and raising the current efficiency in the electrolysis.

REFERENCES

1. Mori, K., Shiota, N., Urata, N. and Ikeuchi, H. The Surface Oscillation of Liquid Metal in Aluminum Reduction Cells, Light Metals 1976, pp. 77—95, 1976.

2. Blanc, J. M. and Entner, P. Application of Computer Calculations to Improve Electromagnetic Behaviour, Light Metals 1980, pp. 285-295, 1980.

3. Evans, J. W., Zundelevigh, Y. and Sharma, D. A Mathematical Model for Prediction of Cuerrents, Magnetic Fields, Melt Velocities, Melt Topography and Current Efficiency in Hall-Heroult Cells, Metallurgical Trans. B. Vol. 12B, pp. 353-360, 1981

4. Ciancarlo, A. S. Automatic plotting as Efective Aid for Theoretical Studies in Reduction Cell Phenomena, Light Metals 1980, pp. 243-272, 1980.

5. Symm, G. T. A Problem in Magnetic Design, Boundary Elements in Mechanical and Electrical Engineering, (Ed. Brebbia, C. A. and Chaudouet-Miranda, A.), pp. 432-

440, Computational Mechanics Publications, 1990.

6. Ressenschuck, S. Mathematical Optimization Techniques for the Design of Permanent Magnet Synchronous Machines Based on Numerical Calculation, IEEE Trans. on Mag. Vol. 26, No. 2, March 1990.

7. Soares, C. A. M., Rodrigues, H. C., Faria, L. M. O. and Haug, E. J. Optimization of the Shape of Solid and Hollow shaft Using Boundary Elements, Proc. 5th Int. Conf. BEM (Ed. Brebbia, C. A. et al.), pp. 883-889, Springer-Verlag, 1983.

8. Brebbia, C. A. and Hernandez, S. (Eds.) Computer Aided Optimum Design of Structures; Recent Advances, Conf. Proc. Opti 89, Computational Mechanics Publications, 1989.

9. Lu, J. and Chang, W. Hybrid Equivalent Boundary Principle - Boundary Element Method for Calculating Electromagnetic Field, Electromagnetic Fields in Electrical Engineering, (Ed. Ding, S.), pp. 357-360, International Academic Publishers, 1989.

10. Lu, J. and Chang, W. Calculation of Current Field in Large Aluminum Reduction Cells Using the BEM, Boundary Elements X, (Ed. Brebbia, C. A.), Vol. 2, pp. 493-504, Computational Mechanics Publications, 1988.

11. Brebbia, C. A. Boundary Element Techniques in Engineering, Newnes-Butterworths, London, 1980.

12. Zhang, K. F. Inversing Revised Matrices, J. of the Naval Engineering Institute, No. 3, pp. 1-29, 1980.

13. Chai, X. S. Optimization and Optimal Control, Tsinghua University Press, 1983.

SECTION 16: INVERSE PROBLEMS

Identification of Defects by the Elastodynamic Boundary Element Method Using Noisy Additional Information

M. Tanaka (*), M. Nakamura (*), T. Nakano (**), H. Ishikawa (***)

(*) Department of Mechanical Systems, Faculty of Engineering, Shinshu University, 500 Wakasato, Nagano 380, Japan

(**) Kawasaki Heavy Industries, Ltd., 2-4-1 Hamamatsu-cho, Minato-ku, Tokyo 105, Japan

(***) Graduate School of Shinshu University, 500 Wakasato, Nagano 380, Japan

ABSTRACT

The inverse problem under consideration deals with identification of unknown defects included in structural components by means of the boundary element method using measured data as additional information. In most inverse problems, identification results are much influenced by errors included in the measured data. The present paper is concerned with a boundary element method simulation to investigate how the measurement errors influences the identification results. Numerical experiment is carried out for the case where the given displacement information includes probabilistic errors with a normal distribution.

INTRODUCTION

It is very important in engineering to estimate the safety margin of structural components by finding defects with non-destructive means. Various experimental techniques are available for such non-destructive inspection. Recently, there is a growing interest in computational approaches, that is, the non-destructive inspection or evaluation is formulated as an inverse problem, and it is analyzed by using the computational software so far developed for the corresponding direct problem[1-8].

The authors previously reported on some investigations of the elastodynamic inverse analysis for the identification of unknown defects in a structural component[9-12]. In our elastodynamic inverse problem, it is assumed that the structural component includes an internal crack or defect and is subjected to time-harmonic excitation. Supposed that the dynamic responses at some points on the boundary can be used as additional information, the inverse problem can be reduced to an optimum problem solvable by means of the standard optimization technique. The objective function of the resulting optimum problem is chosen as the square sum of residuals between the measured data and the dynamic responses computed by the boundary element software available for steady-state elastodynamics. Since in general the inverse problems are of an ill-posed nature, identification results are much influenced by errors included in the measured data. Therefore, the influence of the errors on the identification results should be investigated to develop a reliable, efficient procedure for inverse analysis.

In the present paper, computer simulation is carried out for the defect identification when the given additional information includes probabilistic errors with a normal distribution. Finally, a practical method is proposed for improvement of accuracy in the identification results.

DEFECT IDENTIFICATION PROCEDURE

It is assumed that the material of the structural component under consideration is homogeneous and isotropic, and obeys Hooke's law. The inverse problem of defect identification can be defined such that the shape and the location of the internal defect in the structural component are not known, while the dynamic responses at some selected points on the boundary are given as additional information. This inverse problem is reduced to an optimum problem which minimizes the objective function chosen as the square sum of the residuals between the boundary element solutions and the measured data (reference data).

When the displacement responses are given as additional information, the objective function is the square sum of residuals between the measured displacements $\bar{u}_i^n$ at selected points on the boundary and the corresponding displacements u_i^n computed by the boundary element method assuming the unknown parameters. In this case, the objective function can be expressed as

$$W = \sum_{n=1}^{M} \sum_{i=1}^{D} (u_i^n - \bar{u}_i^n)^2 \tag{1}$$

where M is the number of measuring points, and D is the dimension of space. W is the function of unknown parameters to be identified. We employ in this study the conjugate gradient method[13,14] to determine an optimal set of parameters.

ERRORS IN ADDITIONAL INFORMATION

We can imagine various sources for errors in the measured data. The errors to be considered can be, for example, errors in the location of measuring point, and errors in the magnitude and location of exciting force. In this study, we investigate the case where the measured displacement values include probabilistic errors with a normal distribution.

It is assumed that the displacement component $\bar{u}_i^n$ in the i-direction at a measuring point n on the boundary contains the error e_i^n with the normal distribution $N(m, \sigma^2)$, where m denotes the mean value of the errors, and σ the standard deviation. Using the mean value of the displacement $\bar{u}_{oi}^n$, we can write the reference data $\bar{u}_i^n$ with the error e_i^n as follows:

$$\bar{u}_i^n = \bar{u}_{oi}^n + e_i^n \tag{2}$$

Introducing a random number R with the normal distribution in the range [-1,1], we can express e_i^n by

$$e_i^n = \bar{u}_{oi}^n \frac{H}{100} R \tag{3}$$

where H denotes the percentage amount of dispersion in measured values. Accordingly, the reference data including the error with any amount of dispersion H can be expressed as follows:

$$\bar{u}_i^n = \bar{u}_{oi}^n \left(1 + \frac{H}{100} R \right) \tag{4}$$

In the numerical simulation for investigating the influence of measurement errors on the identification results, the reference data $\bar{u}_i^n$ with error can be produced in the above-mentioned manner from the displacement value $\bar{u}_{oi}^n$ which can be obtained by boundary element analysis using the exact parameter values.

NUMERICAL SIMULATION OF DEFECT IDENTIFICATION

In order to clarify the influence of measurement errors on identification results, we simplify the problem by dividing the defect identification process into two parts, that is, the identification of defect location and that of defect shape.

In the numerical experiment we assume that the rectangular plate model subject to a plane stress condition has a circular hollow defect as shown in Fig.1. The rectangular plate is assumed to have the material constants as shown in Table 1. The boundary Element discretization used is shown in Fig.2: the outer boundary of the plate and the boundary of the defect are divided into 50 and 16 constant elements, respectively. The same boundary element discretization is used during the iterative computation of optimization. It is noted that the exact displacement values are computed by elastodynamic boundary element analysis under the above-mentioned assumptions. The reference data $\bar{u}_i^n$ including errors with the dispersion of H% can be produced by using equation(4).

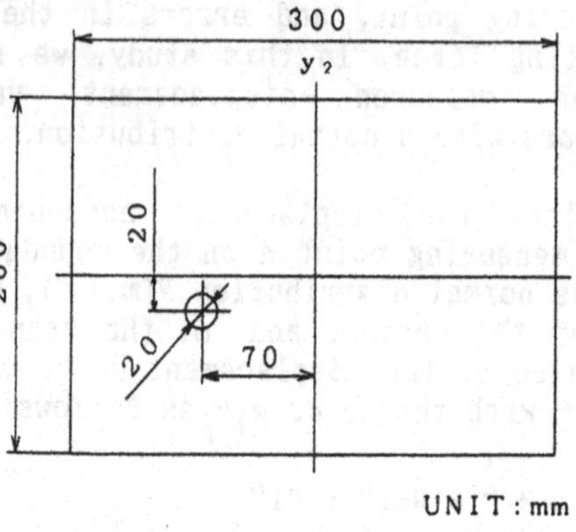

Fig.1 Rectangular plate model with a circular hollow defect

Table 1. Material constants

Young's modulus E	210 GPa
Poisson's ratio ν	0.3
Mass density ρ	7.80×10^3 kg/m^3

The time-harmonic exciting force with angular frequency $\omega = 10^4$ rad/sec(≒ 1600Hz) is applied to the rectangular plate. It is assumed that the exciting force is applied at 6 different locations as shown in Fig.3. We now investigate the cases where the numbers of measuring points are 6 and 35, while the dispersions in reference data are 1% and 5%. Numerical experiment is carried out for five sets of the reference data produced by using a different series of the random number for every case. The measuring points are assumed to be distributed almost uniformly on the boundary except for the fixed part. The case of 6 measuring points is shown in Fig.4.

<u>**Identification of defect location**</u>
In this case, optimization parameters to be identified are y_1 and y_2, the coordinates of the center point of the defect. First, a circular defect with the radius of 10mm is assumed on the same location as the real defect, then the iterative computation is carried out.

Figures 5 and 6 show the identification results for the reference data with 5% and 1% dispersions of measurement errors, respectively, in which the number of measuring points is 35. In these figures, the vertical axis denotes the identification errors, and the horizontal axis the sum of absolute values of sensitivity ϕ_j^a. It is noted that the sensitivity S_j^n is defined for the jth parameter y_j of the nth measuring point as follows:

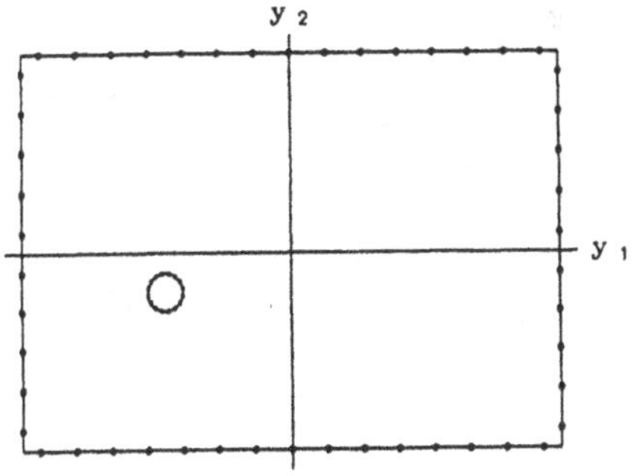

Fig.2 Discretization of rectangular plate model

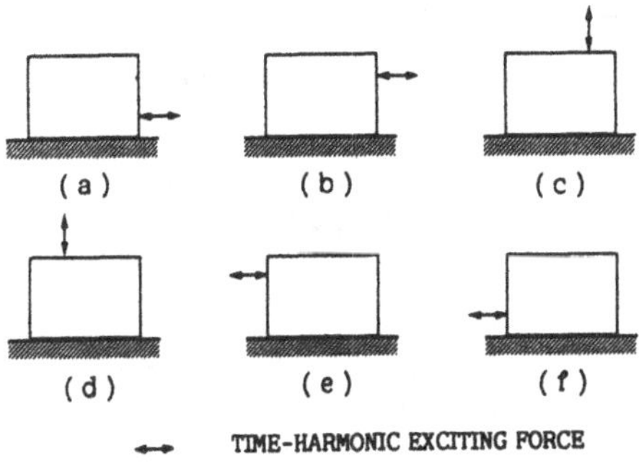

Fig.3 Locations of time-harmonic exciting force

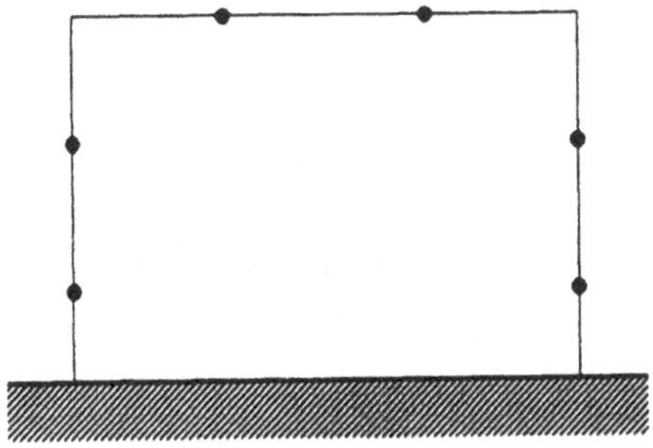

Fig.4 Measuring point locations in case of 6 points

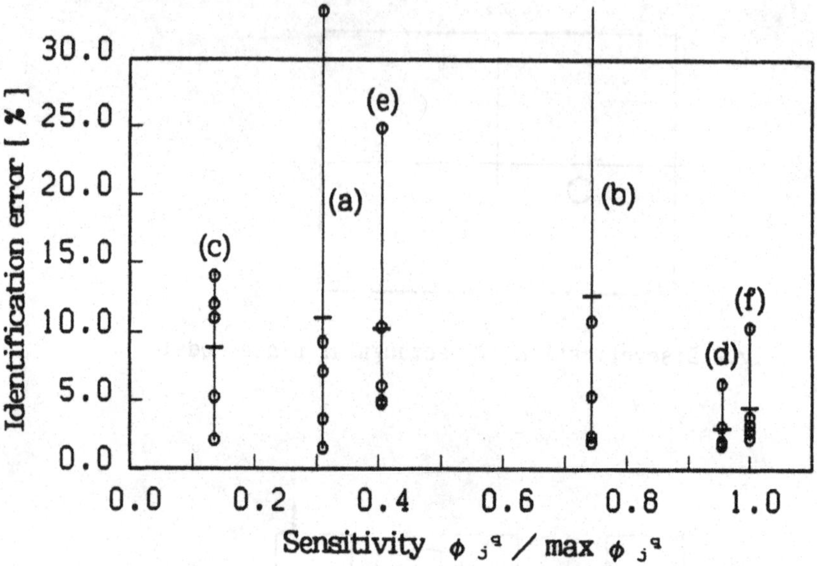

Fig.5 Identification errors of defect location
(35 measuring points, 5% dispersion of measurement error)

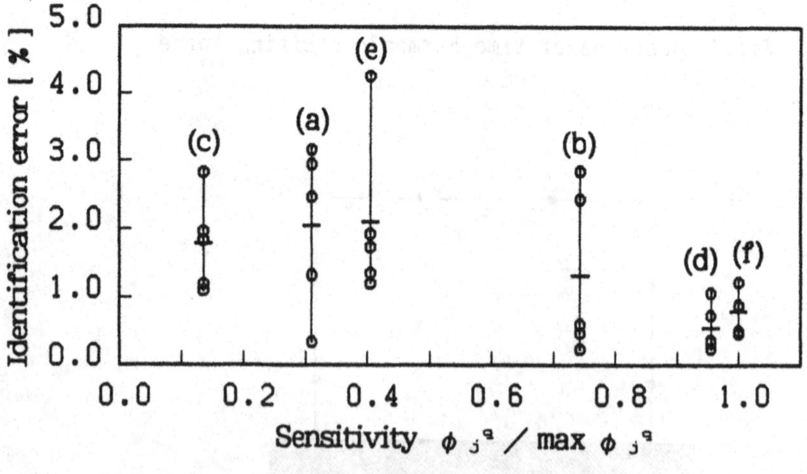

Fig.6 Identification errors of defect location
(35 measuring points, 1% dispersion of measurement error)

$$S_j^n \equiv \frac{\partial W}{\partial y_j} \tag{5}$$

and that the sum of absolute values of sensitivity is given as

$$\phi_j^q = \sum_{n=1}^{N} | S_j^n | \tag{6}$$

In the figures, we make dimensionless the absolute values' sum of sensitivity for each exciting force location by dividing the greatest sum among them.

The identification error of defect location, E_d, denotes the distance d between the real location and the identified one, which is expressed in percentage by dividing the horizontal length L of the rectangular plate. That is,

$$E_d = \frac{d}{L} \times 100 \% \tag{7}$$

The symbol o in the figures denotes the identification errors, and the symbol — the average of five numerical experiments. As shown in the figures, it can be seen that in general the dispersion and also the average of identification errors are small as the sum of absolute values of sensitivity becomes large. Also, it is found that the case of 1% errors included in the reference data has a smaller dispersion of the identification errors than that of 5%. It is interesting to note that no identification results could be obtained in the case of 6 measuring points.

<u>Identification of defect shape</u>

We now investigate the influence of errors on the identification of defect shape. A circular defect with the radius of 15mm is assumed on the same location as the real one. In this case, the optimization parameter to be identified is a radius r of the defect. The number of measuring points, the dispersion of measurement errors and the locations of exciting force are the same as the first case mentioned-above.

Figures 7 and 8 show the identification errors of defect shape in the case of 35 measuring points, while Figs. 9 and 10 show those in the case of 6 measuring points. In these figures, the vertical axis denotes the relative error E_r of the identified radius r' for r of the real defect. That is,

$$E_r = \frac{| r' - r |}{r} \times 100 \% \tag{8}$$

In the case of defect shape identification, similar results have been obtained to the case of the defect location identification.

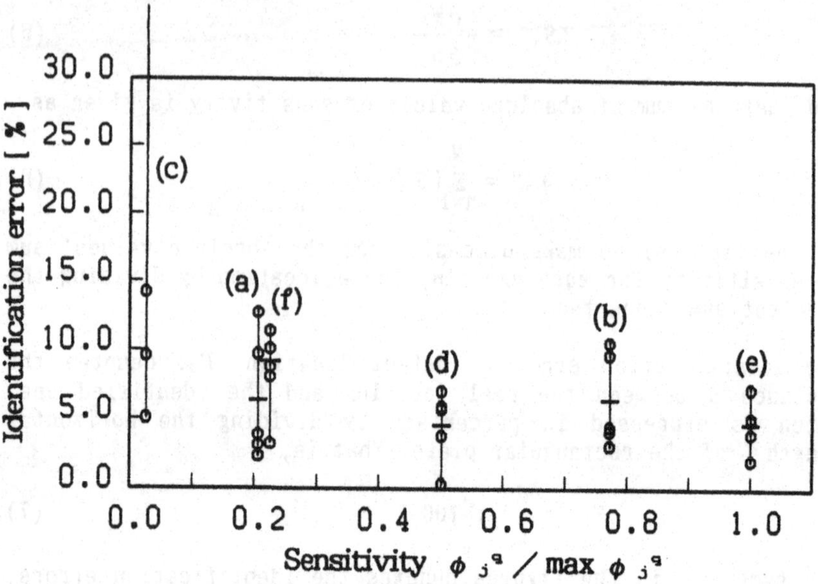

Fig.7 Identification errors of defect shape
(35 measuring points, 5% dispersion of measurement error)

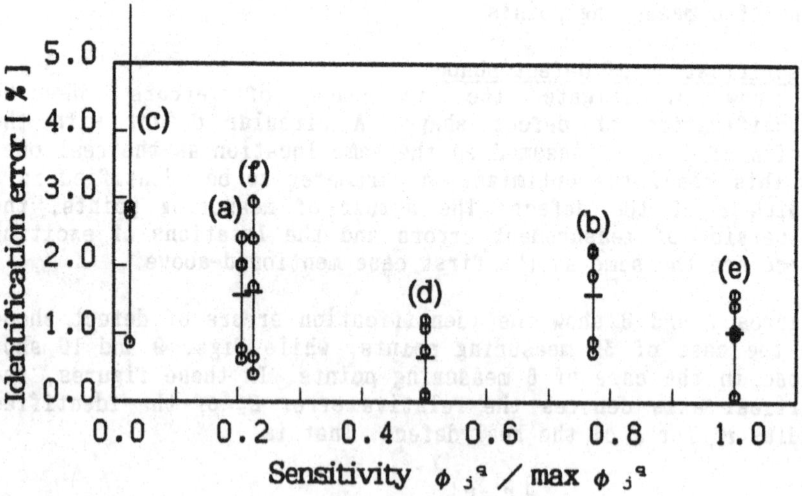

Fig.8 Identification errors of defect shape
(35 measuring points, 1% dispersion of measurement error)

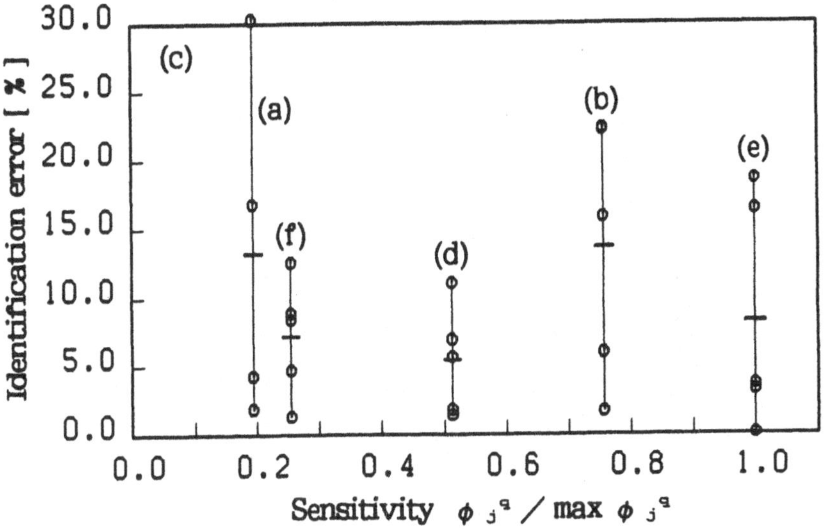

Fig.9 Identification errors of defect shape
(6 measuring points, 5% dispersion of measurement error)

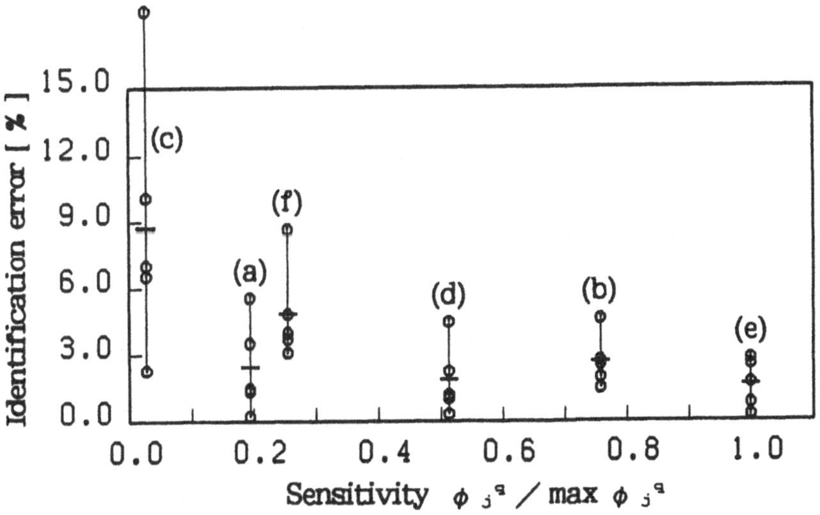

Fig.10 Identification errors of defect shape
(6 measuring points, 1% dispersion of measurement error)

Table 2. Improvement in identification results by applying method of multiple force applications to identifying defect location

Number of measuring points	Location of exciting force	Identification errors %						
		1	2	3	4	5	Average of errors	Dispersion of errors
35	(f)	3.28	2.78	3.89	10.4	2.26	4.52	8.14
	(d)	2.05	1.89	1.74	6.29	3.12	3.02	4.55
6	(f)	2.60	25.7	63.1	4.06	3.32	19.8	60.5
	(d)	4.76	0.498	20.8	10.6	11.6	9.65	20.3
	(f).(d)	0.874	4.13	4.10	0.918	2.76	2.56	3.26

Improvement of identification accuracy

In Table 2 comparison is made for influences on the identification errors of defect location when the numbers of measuring points are 6 and 35. The dispersion of measurement errors is assumed as 5%, and numerical experiment is carried out five times for each exciting force location of (f) and (d) in which the large sensitivities are obtained at measuring points. The range of errors in Table 2 expresses the difference between the maximum and minimum absolute values of identification errors. It can be concluded from this comparison that a larger number of measuring points could lead to a more accurate identification result.

In order to improve the identification accuracy, we apply the method of multiple force applications, which was previously proposed by the authors[9,10], to the above case of 6 measuring points. In this method, each exciting force of (f) and (d) is applied separately, and then the objective function is evaluated as the algebraic sum of displacement residuals for each excitation. The lowest column in Table 2 shows the results obtained by this method of multiple force applications for this case. It is revealed that the method of multiple force applications improves drastically the identification results making the average and the dispersion of identification errors fairly less than those obtained by the method of single force application.

CONCLUSIONS

The boundary element method simulation has been carried out to investigate the influence of probabilistic errors included in the given additional information on the defect identification results in the elastodynamic inverse problems. It is revealed from this numerical experiment that a successful identification is possible even in such cases, if we choose appropriately measuring points with lager sensitivities to the parameters to be identified. It is also demonstrated that the method of multiple force applications could improve drastically the identification results, and hence this method can be recommended for use in the case of noisy additional information.

REFERENCES

1. Tanaka, M. , Some Recent Advances in Boundary Element Research for Inverse Problems, Boundary Element X, (Ed. Brebbia, C.A.), Vol.1, pp.567-582, CM Publications, Southampton and Boston; Springer-Verlag, Berlin and New York, 1988.
2. Blakemore, M. and Georgiou, G.A.(Eds.), Mathematical Modeling in Non-destructive Testing, Oxford Univ. Press, Oxford and New York, 1988.

3. Proceedings of the JSME Symposium on Computational Methods and Their Applications to Inverse Problems, (in Japanese), No.890-34, 1989.
4. Ohtsu, M., Acoustic Emission Characteristics in Concrete and Diagnostic Applications, Journal of Acoustic Emission, Vol.6, pp.99-108, 1987.
5. Kubo, S., Sakagami, T. and Ohji, K., Reconstruction of a Surface Crack by Electric Potential CT Method, Computational Mechanics '88, (Eds. Atluri, S. N. and Yagawa,G.), Proc. Int. Conf. Computational Eng. Sci., Vol.1, pp.12.i.1-5, Springer-Verlag, Berlin and New York, 1988.
6. Murai, T. and Kagawa, Y., Boundary Element Iterative Techniques for Determining the Interface Boundary between Two Laplace Domains, Int. J. Num. Meth. Eng., Vol.23, pp.35-47, 1986.
7. Yagawa, G. and Fukuda, T., Applications of the Electric Potential Method to Fracture Mechanics, Science of Machine, (in Japanese), Vol.35, pp.1237-1244,1324-1330, 1983; Vol.36, pp.27-32, 1984.
8. Nishimura, N. and Kobayashi, S., Regularised BIEs for Crack Shape Determination Problems, Boundary Elements XII, (Eds. Tanaka, M., Brebbia, C.A. and Honma, T.), Vol.2, pp.425-434, CM Publications, Southampton and Boston; Springer-Verlag, Berlin and New-York, 1990.
9. Tanaka, M., Nakamura M., and Nakano T., Defect Shape Identification by Means of Elastodynamic Boundary Element Analysis and Optimization Technique, Advances in Boundary Elements, (Eds. Brebbia, C.A. and Connors, J.J.), Proc. of 11th Int. Conf. on Boundary Element Methods, Vol.3, pp.183-194, CM Publications, Southampton and Boston; Springer-Verlag, Berlin and New York, 1989.
10. Tanaka, M. , Nakamura, M. and Nakano, T., Defect Shape Identification by Elastodynamic Boundary Element Method Using Strain Responses, Advances in Boundary Element Methods in Japan and USA, (Eds. Tanaka, M., Brebbia C.A. and Shaw, R.), CM Publications, Southampton and Boston, pp.137-151, 1990.
11. Nakano. T., Tanaka, M. and Nakamura, M., Defect Identification by Elastodynamic BEM - Consideration on Selection of Additional Information, Proc. of JSME, (in Japanese), No.907-1, pp.65-66, 1990.
12. Tanaka, M., Nakamura, M. and Nakano, T., Detection of Cracks in Structural Components by the Elastodynamic Boundary Element Method,, Boundary Elements XII, (Eds. Tanaka, M., Brebbia, C.A. and Honma, T.), Vol.2, pp.413-424, CM Publications, Southampton and Boston; Springer-Verlag, Berlin and New-York, 1990.
13. Rao, S.S. Optimization, Halsted Press, New York, pp.248-330, 1979.
14. Fox, R.L., Optimization Methods for Engineering Design, Addison-Wesley Publishing Co., Massachusetts, pp.38-116, 1971.

The Solution of a Nonlinear Inverse Elliptic Problem by the Boundary Element Method

D.B. Ingham, Y. Yuan

Department of Applied Mathematical Studies, The University of Leeds, Leeds LS2 9JT, U.K.

ABSTRACT

In this paper we investigate the numerical solution of a class of nonlinear inverse elliptic equations. The Boundary Element Method, combined with a minimal energy constraint, is used and it is found that this technique gives a good, stable approximation to the solution.

INTRODUCTION

If the heat flux and/or temperature is specified at the surface of a solid body, whose thermal properties are assumed to be constant, then the steady state temperature distribution within the body is governed by the Laplace equation for which there is a unique solution. For solving such problems numerically it is very advantageous to use the Boundary Element Method (BEM), since the existence of the fundamental solution which can be used to convert the Laplace equation into an integral equation involving boundary integrals, see Jaswon and Symm [1], Brebbia [2] and Ingham and Kelmanson [3] is well known. Consequently, only the boundary data is required in order to compute the solution at any interior point of the solution domain and this contrasts with the finite difference (FD) and finite (FE) element method which require interior data. An immediate advantage of this is that the system of algebraic equations generated by a BEM is considerably smaller than that generated by an equivalent FD or FE approximation.

The solution of nonlinear problems in heat transfer using the BEM technique is well established, see for example the investigations of Khader [4], Bialecki and Nowak [5], Ingham *et al.* [6], Khader and Hanna [7] and Ingham and Kelmanson [3]. All these studies deal with nonlinear boundary conditions whilst Khader [4], Bialecki and Nowak [5], Khader and Hanna [7] and Ingham and Kelmanson [3] deal with temperature dependent thermal conductivity by first employing the Kirchhoff transformation.

Han [8] and Falk and Monk [9] studied the Cauchy problem for elliptic equations by using the minimal energy technique and Ingham, Yuan and Han [10] investigated an inverse heat transfer problem by using this technique. However, all these studies were restricted to linear problems. In this paper we consider the steady state solution of the nonlinear heat conduction equation,

$$\nabla\cdot(f(T)\ \nabla T) = 0 \qquad \text{in } \Omega \tag{1}$$

where Ω is the region enclosed by the boundary, $\partial\Omega$, of the solid body, T is the temperature of the solid and f(T) is the thermal conductivity of the body which is temperature dependent. We will assume that at every point on the surface of the body the temperature is prescribed. If f(T) is known then the techniques as described in Ingham and Kelmanson [3] may be applied. However, frequently in practice the detailed variation of the thermal conductivity with temperature is unknown but extra information in the interior of the body is known, e.g. the temperature may be measured at a number of points within the body. This phenomenon falls into the general class of problems known as an inverse heat conduction problem since extra conditions are specified on a problem than what is normally required but there is an unknown function within the governing equation. In this paper we will show how the BEM may be modified in order to solve equation (1) subject to the conditions

$$T(x,y) = \phi(x,y) \qquad \text{on } \partial\Omega \tag{2a}$$

$$T(x,y) = \psi(x,y) \qquad \text{on } \Gamma \tag{2b}$$

where ϕ and ψ are given functions and Γ is a set of interior points to the boundary $\partial\Omega$.

A transformation of equation (1) is employed such that all the nonlinear aspects of the problem are transferred to the boundary of the solution domain. The function f(T) is then represented by a piecewise quadratic function $\bar{f}(T)$ and a modified BEM developed. In this paper all the calculations have been performed in a square region with T given on the boundary $\partial\Omega$, but the extension to boundaries of arbitrary given shape is trivial.

FORMULATIONS

In order to solve the problem (1) and (2) by using the BEM we introduce the transformed variable A which satisfies

$$\nabla A = f(T)\ \nabla T \tag{3}$$

Thus the governing equation (1) becomes the Laplace equation

$$\nabla^2 A = 0 \tag{4}$$

and the application of the BEM to the solution of equation (4) is well documented. The formula for A may be expressed as, see Ingham and Kelmanson [3],

$$\eta(p)\ A(p) = \int_{\partial\Omega} A(p)\ \ln'|p-q|\ dq - \int_{\partial\Omega} A'(p)\ \ln|p-q|\ dq \tag{5}$$

where

(i) $p \in \bar{\Omega}$, $q \in \partial\Omega$.

(ii) The prime (′) denotes differentiation with respect to the outward normal to $\partial\Omega$ at q.

(iii) $\eta(p) = \begin{cases} 2\pi & \text{if } p \in \Omega \\ \theta & \text{if } p \in \partial\Omega, \text{ where } \theta \text{ is the angle included between the tangents to } \partial\Omega \text{ on either side of } p. \end{cases}$

If either A or A′ is prescribed at each point $q \in \partial\Omega$ then the solution of the boundary integral equation obtained by letting $p \in \partial\Omega$ in equation (5) determines the boundary distribution of both A and A′. Equation (5) may now be used to generate the solution A(p) at any point $p \in \bar{\Omega}$.

Let

$$g(T) = \int^{T} f(\tau)\ d\tau \tag{6}$$

and employing equation (3), we may write the transformation in the form

$$A = g(T), \qquad A' = f(T)\ T' \tag{7}$$

Combining equations (5) and (7) we obtain

$$\int_{\partial\Omega} g(T(q))\ \ln'|p-q|\ dq - \int_{\partial\Omega} T'(q) f(T(q))\ \ln|p-q|\ dq$$

$$= \eta(p)\ g(T(p)) \qquad p \in \bar{\Omega},\ q \in \partial\Omega \tag{8}$$

as the nonlinear integral equation on $\partial\Omega$.

In order to obtain the numerical solution from the integral equation (8) the boundary $\partial\Omega$ is first subdivided into N segments $\partial\Omega_j$, j= 1,2,...,N. On each segment $\partial\Omega_j$ we take constant T_j and T'_j to represent T and T′, where T_j and T'_j take the value of T and T′ at the midpoint of the segment $\partial\Omega_j$, respectively. The integral formula then becomes

$$\sum_{j=1}^{N} g(\phi_j) \int_{\partial\Omega_j} \ln'|p-q|\ dq - \sum_{j=1}^{N} \phi'_j\ f(\phi_j) \int_{\partial\Omega_j} \ln|p-q|\ dq$$

$$= \eta(p)\ g(\ T(p)) \tag{9}$$

Taking $p_i = q_i \in \partial\Omega_i$, we obtain

$$\sum_{j=1}^{N} g(\phi_j)\{ \int_{\partial\Omega_j} \ln'|p_i-q|\, dq - \delta_{ij}\eta_i \} - \sum_{j=1}^{N} \phi'_j f(\phi_j) \int_{\partial\Omega_j} \ln|p-q|\, dq$$

$$= 0 \qquad i=1,2,\ldots,N \qquad (10)$$

We now let

$$E_{ij} = \int_{\partial\Omega_j} \ln'|p_i-q|\, dq - \delta_{ij}\eta_i$$

$$G_{ij} = \int_{\partial\Omega_j} \ln|p_i-q|\, dq$$

then, from (10), we obtain the linear system of equations

$$E_{ij}\, g(\phi_j) - G_{ij}\, \phi'_j f(\phi_j) = 0 \qquad i=1,\ldots,N \qquad (11)$$

Similarly, if $p_i \in \Gamma \subset \Omega$, we let

$$EI_{ij} = \int_{\partial\Omega_j} \ln'|p_i-q|\, dq$$

$$GI_{ij} = \int_{\partial\Omega_j} \ln|p_i-q|\, dq$$

then, we have

$$EI_{ij}\, g(\phi_j) - GI_{ij}\, \phi'_j\, f(\phi_j) = g(T(p_i))$$

$$i=1,2,\ldots,k \qquad (12)$$

MINIMAL ENERGY SCHEME

In order to obtain the numerical solution of the problem (1) and (2) we first use a k-piecewise quadratic function $\bar{f}(T)$ to represent the unknown function $f(T)$. In view of the maximal principle there are points p_1 and $p_2 \in \partial\Omega$ such that

$$T(p_1) = \min_{p\in\Omega} T(p) = m$$

$$T(p_2) = \max_{p\in\Omega} T(p) = M$$

Hence the defining region of $f(T)$ is a closed interval $[m,M]$, and this is also true for the function $\bar{f}(T)$. Subdividing $[m,M]$ into k equal intervals with the mesh points at $m = t_0, t_1, \ldots, t_{k-1}, t_k = M$, then $\bar{f}(T)$ may be written in the form

$$\bar{f}(T) = \begin{cases} a_1T^2 + b_1T + c_1 & T \in [t_0, t_1] \\ \vdots & \\ a_kT^2 + b_kT + c_k & T \in [t_{k-1}, t_k] \end{cases}$$

where a_i, b_i and c_i, $i=1,\ldots k$, are constants to be determined. In the transformation (6) we let $\bar{f}(T)$ represent $f(T)$ and then we obtain a piecewise third-order polynomial, say $\bar{g}(T)$, which may be written in the form

$$\bar{g}(T) = \begin{cases} \frac{1}{3} a_1 T^3 + \frac{1}{2} b_1 T^2 + c_1 T + d_1 & T \in [t_0, t_1] \\ \vdots & \\ \frac{1}{3} a_k T^3 + \frac{1}{2} b_k T^2 + c_k T + d_k & T \in [t_{k-1}, t_k] \end{cases}$$

where d_i, $i=1,\ldots k$, are to be determined integral constants of the transformation (6).

Inserting the functions $\bar{f}$ and $\bar{g}$ into the linear system (11), we obtain

$$E_{ij}\, \bar{g}(\phi_j) - G_{ij}\, \phi'_j \bar{f}(\phi_j) = 0 \qquad i=1,\ldots,N \tag{13}$$

which includes N equations and N + 4k unknown variables. In order to solve problem (13) we need 4k extra equations. The continuity of $\bar{f}(T)$, $\bar{f}'(T)$ and $\bar{g}(T)$ at the mesh points, except the two end points T = m and T = M, may offer 3k-3 equations, and clearly $d_1 = 0$ is another equation. Since (13) is a homogeneous system of equations we need some extra condition to solve it and therefore $\bar{f}(T)$ has to be fixed at one point. However, if only one point is fixed the accuracy of solution is not good, see example 1, so we suppose that $\bar{f}(T)$ is fixed at two mesh points, for example at the two end points m and M. If the set Γ contains k points then the equation (2b) gives rise to the remaining k equations. Combining all the conditions above, and (13), we obtain a new solvable system of equations

$$\begin{cases} E_{ij}\bar{g}(\phi_j) - G_{ij}\phi'_j\bar{f}(\phi_j) = 0 & i=1,..,N \\ EI_{ij}\bar{g}(\phi_j) - GI_{ij}\phi'_j\bar{f}(\phi_j) = \bar{g}(\psi_i) & i=1,..,k \\ \left.\begin{array}{l} a_i t_i^2 + b_i t_i + c_i = a_{i+1} t_i^2 + b_{i+1} t_i + c_{i+1} \\ 2a_i t_i + b_i = 2a_{i+1} t_i + b_{i+1} \\ \frac{1}{3}a_i t_i^3 + \frac{1}{2}b_i t_i^2 + c_i t_i + d_i = \\ \qquad \frac{1}{3}a_{i+1} t_i^3 + \frac{1}{2}b_{i+1} t_i^2 + c_{i+1} t_i + d_{i+1} \end{array}\right\} & i=1,..,k-1 \\ d_1 = 0 & \\ a_1 m^2 + b_1 m + c_1 = s, \quad a_k M^2 + b_k M + c_k = S & \end{cases} \tag{14}$$

where s = f(m) and S = f(M). Unfortunately, it is not possible to obtain an accurate solution by solving this problem directly and therefore we introduce the minimal energy method.

Instead of considering the problem (1) and (2) we investigate the solution of the following problem

$$\begin{cases} \nabla(f(T)\ \nabla T) = 0 & \text{in } \Omega \\ T = \phi & \text{on } \partial\Omega \\ |\ T|_{\Gamma} - \psi\ | \le \varepsilon & \text{on } \Gamma \end{cases} \tag{15}$$

where $\varepsilon > 0$ is a preassigned small quantity. Clearly the solution of problem (15) may be considered as an approximate solution of problem (1) and (2). Inserting transformation (6) into (15) we obtain

$$\nabla^2 g(T) = 0 \qquad \text{in } \Omega \tag{16a}$$
$$g(T) = g(\phi) \qquad \text{on } \partial\Omega \tag{16b}$$
$$g(\psi-\varepsilon) \le g(T) \le g(\psi+\varepsilon) \qquad \text{on } \Gamma \tag{16c}$$

We write

$$a(u,v) = \iint_{\Omega} \nabla u \cdot \nabla v \ dxdy$$

$$J(u) = \frac{1}{2} a(u,u) = \frac{1}{2} \iint_{\Omega} |\nabla u|^2 dxdy$$

$$= \frac{1}{2} \int_{\partial\Omega} u \frac{\partial u}{\partial n} ds$$

where n is the outward normal to $\partial\Omega$ and $J(u)$ is called the energy functional of the Laplace equation

$$\begin{cases} \nabla^2 u = 0 & \text{in } \Omega \\ u(x,y) = \phi(x,y) & \text{on } \partial\Omega \end{cases} \tag{17}$$

It is well known that the Laplace equation (17) is equivalent to the following minimal problem

$$J(u) = \inf_{v \in \overset{*}{H}(\Omega)} J(v)$$

where $\overset{*}{H}(\Omega) = \{\ u \in H^1(\Omega);\ u|_{\partial\Omega} = \phi\ \}$, and $H^1(\Omega)$ is the normal Sobolev space. Similarly, considering the problem (16a) and (16b), we know that the problem (16a) and (16b) is equivalent to the minimal problem

$$J(\ g(T)) = \inf_{\tilde{g}(T) \in \overset{*}{H}(\Omega)} J(\ \tilde{g}(T)) \tag{18}$$

where $\tilde{g} \in C^2[m,M]$, and $T \in H^1(\Omega)$, which on discretization becomes

$$J(\ g(T)) = \sum_{i,j=1}^{N} \int_{\partial\Omega} g(\phi_i) \frac{\partial}{\partial n} g(\phi_j)\ ds$$

$$= \sum_{i,j=1}^{N} \int_{\partial\Omega} g(\phi_i)\ G^{-1}_{ik} E_{kj}\ g(\phi_j)\ ds \tag{19}$$

Inserting $\bar{g}(T)$ into (18) we obtain

$$J(\bar{g}(T)) = \sum_{i,j=1}^{N} \int_{\partial\Omega} \bar{g}(\phi_i)\, G^{-1}_{ik} E_{kj}\, g(\phi_j)\, ds \tag{20}$$

It is clear that $\inf \sum_{i,j=1}^{N} \int_{\partial\Omega} \bar{g}(\phi_i)\, G^{-1}_{ik} E_{kj}\, \bar{g}(\phi_j)\, ds$ is equivalent to $\inf \sum_{i,j=1}^{N} \bar{g}(\phi_i)\, G^{-1}_{ik} E_{kj}\, \bar{g}(\phi_j)$, and we set all the continuity conditions on $\bar{f}(T)$ and $\bar{g}(T)$ and the inequality (16c) as the constraining conditions and then the problem reduces to finding $\bar{g}$ which satisfies

$$\begin{cases} J(\bar{g}(T)) = \inf \sum_{i,j=1}^{N} \bar{g}(\phi_i)\, G^{-1}_{ik} E_{kj}\, \bar{g}(\phi_j) \\ \bar{g}(\psi_i - \varepsilon) \le \sum_{j=1}^{N} W_{ij}\, \bar{g}(\phi_j) \le \bar{g}(\psi_i + \varepsilon) \qquad i=1,\ldots,k \end{cases} \tag{21}$$

where $W_{ij} = EI_{ij} - GI_{il} G^{-1}_{lm} E_{mj}$, $\psi_i = T(p_i)$ and $p_i \in \Gamma$. Solving the constrained minimal problem (20), using the Nag routine E04UCF, we obtain the functions $\bar{g}(T)$ and $\bar{f}(T)$. Then we may obtain an approximate solution of the problem (1) and (2) by the use of the BEM, see Ingham and Kelmanson [3].

EXAMPLES

We have described the minimal energy scheme and we now give two examples to verify the accuracy of this technique. In both of the examples the calculations have been performed in a square region Ω, and the set Γ contains $n \times n$ points which are evenly distributed in a checker-board formation in the solution domain Ω.

EXAMPLE 1.

We consider the problem in which

$$\begin{cases} \phi(x,y) = 0 & 0 \le x \le 1,\ y = 0, \\ \phi(x,y) = \ln(1+2y) & 0 \le y \le 1,\ x = 1, \\ \phi(x,y) = \ln(1+2x) & 0 \le x \le 1,\ y = 1, \\ \phi(x,y) = 0 & 0 \le y \le 1,\ x = 0, \end{cases}$$

and

$$\psi(x,y) = \ln(1+2xy) \qquad (x,y) \in \Gamma$$

Clearly, $f(T) = \exp(T)$ and $T = \ln(1+2xy)$ is the analytical solution of the problem (1) and (2). Numerical solutions were obtained with $N = 160$, i.e. 40 segments on each side of the boundary $\partial\Omega$, and $k = 4$, 8 and 12, where k denotes the number of pieces of the piecewise quadratic function $\bar{f}(T)$. The set Γ was taken to contain 25, 64 and 100 points for $k = 4$, 8 and 12 respectively. The function $\bar{f}(T)$ was fixed at the two end points, $T = 0$ and $T = \ln(3)$.

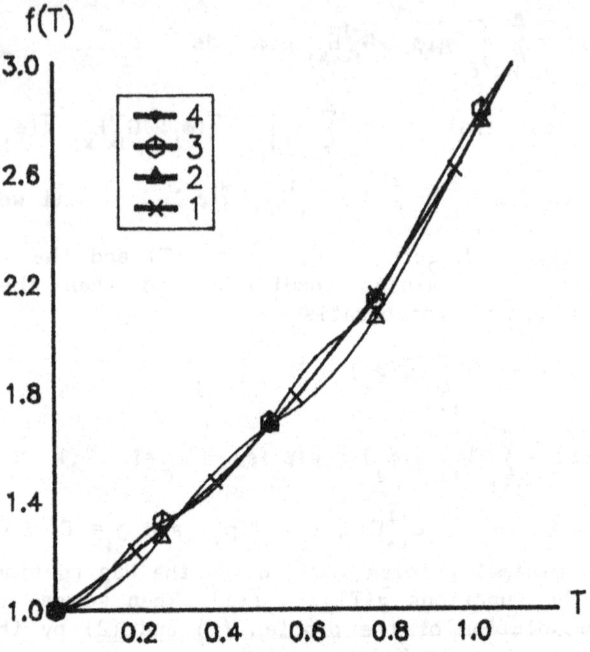

Fig. 1. The analytical solution f(T) and the numerical solution $\overline{f}(T)$, where 1 is f(T), and 2, 3 and 4 are $\overline{f}(T)$ with k = 4, 8 and 12, respectively.

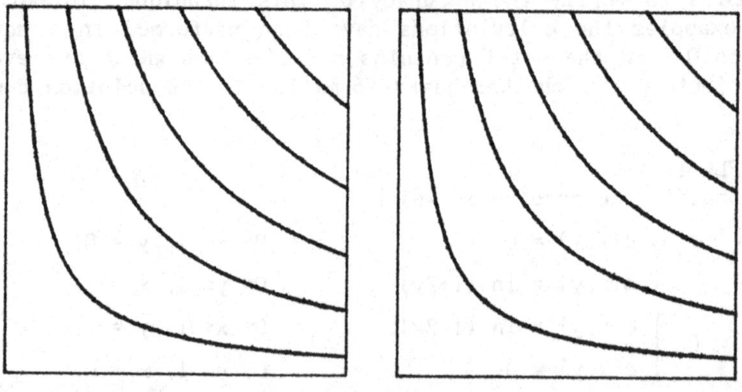

(a) (b)

Fig. 2. The lines of constant T for example 1, where (a) illustrates the analytical solution and (b) shows the numerical solution with k = 4.

In Fig. 1 we present a comparison of the the analytical solution f(T) and the numerical solutions $\overline{f}(T)$. For k = 4, 8 and 12 we obtain the values of the minimal energy, Eng, to be 2.64588, 2.65087 and 2.65369, respectively. The correct value being Eng = 8/3 and this, like Fig. 1, shows that as k increases then the solution becomes more accurate. Fig. 2 shows the lines of constant T for the analytical solution and the numerical solution with k = 4. The differences between these solutions are graphically indistinguishable and as k increases it is found that the numerical solution is convergent and stable, and the accuracy improves. A rough estimate shows that the relative error is about 0.1% for $\bar{f}(T)$ and 0.005% for T when k = 12.

In Fig. 3. we show the numerical solutions $\bar{f}(T)$ by solving the system of equations (14) directly and using the minimal energy method when only one point, T = 0, is fixed for $\bar{f}(T)$. It is deserved that in order to obtain a good accuracy then at least two points should be fixed for $\bar{f}(T)$.

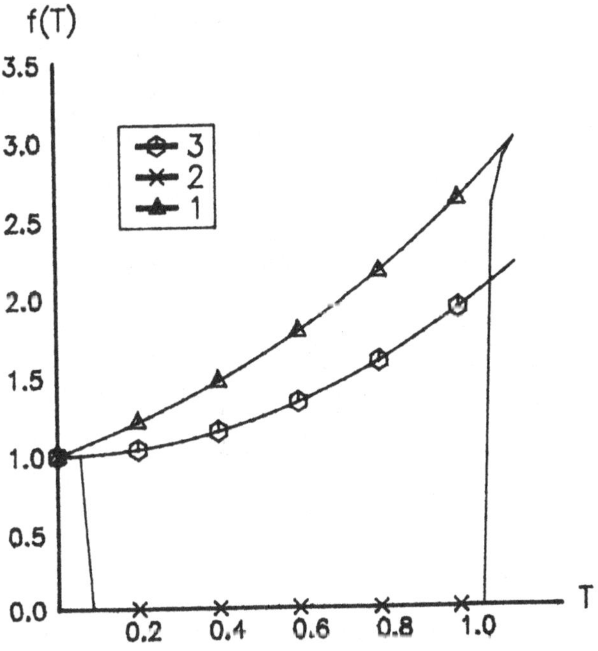

Fig. 3. The numerical solution $\bar{f}(T)$ using direct method, line 2, and minimal energy method with one point, line 3, for example 1. Line 1 is the analytical solution.

EXAMPLE 2

In this example we consider a problem in which there is no simple analytical solution for T. However, in order to test the numerical technique developed here we assume that f(T) is given, and for this case we take $f(T) = 1 + 2T + 3T^2$, and evaluate the solution using the method described by Ingham and Kelmanson [3]. Having obtained the solution we then specify T at the interior mesh points and attempt to determine f(T) and T using the minimal energy method. The boundary condition is

$$\phi(x,y) = \begin{cases} x & 0\le x\le 1, \quad y=0 \\ 1 & 0\le y\le 1, \quad x=1 \\ 1 & 0\le x\le 1, \quad y=1 \\ y & 0\le y\le 1, \quad x=0 \end{cases}$$

Solutions were obtained with N = 40, 80 and 160, and k = 4, 6, 10 and 12, respectively, and Γ containing 100 points.

Mesh points (T)	Numerical Solutions			Analytical Solution
	N = 40	N = 80	N = 160	
0.0	1.00010	1.00010	1.00010	1.0
0.1	1.29759	1.25931	1.23813	1.23
0.2	1.55102	1.53241	1.52295	1.52
0.3	1.84471	1.85511	1.86715	1.87
0.4	2.20180	2.25937	2.27138	2.28
0.5	2.69790	2.72524	2.74503	2.75
0.6	3.24347	3.26794	3.27838	3.28
0.7	3.90755	3.88622	3.87544	3.87
0.8	4.44168	4.50161	4.51483	4.52
0.9	5.26187	5.24437	5.23488	5.23
1.0	5.99970	5.99970	5.99970	6.0

Table 1. The values of $\bar{f}(T)$ for example 2 with N = 40, 80 and 160, respectively.

Table 1 shows the numerical solution $\bar{f}(T)$ with N = 40, 80 and 160, respectively, and k = 10, and a rough estimate shows that the relative error is about 0.1%. Table 2 shows the minimal energy in each case. Although the exact value for the minimal energy is not known in this case it is observed that results obtained appear to be consistent with an estimate of about 9.4. In Fig. 4 we present the lines of constant T for N= 160, and k = 4, 8 and 12, respectively. It is clear that as number of pieces of $\bar{f}(T)$ increases then the solution is converging and stable.

k \ N	40	80	160
4	9.41843	9.41694	9.41591
8	9.40172	9.39915	9.39493
12	9.39764	9.38505	9.37974

Table 2. The variation of J(T) as a function of k and N for example 2.

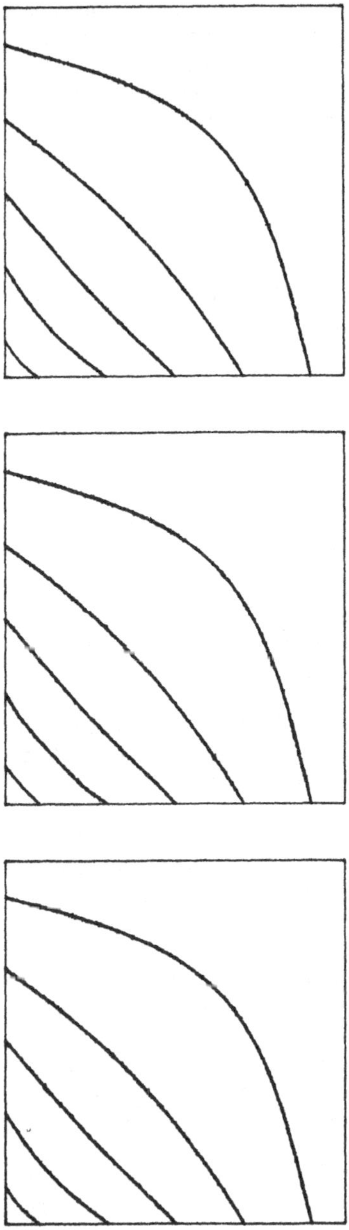

Fig. 4 The lines of constant T for example 2, where (a), (b) and (c) are the solution for N = 160 and k = 4, 8 and 12, respectively

In conclusion, in this paper a Boundary Element Method with a minimal energy constraint has been presented which enables an accurate treatment of a class of nonlinear inverse elliptic equations. In view of the numerous calculations performed we have always been able to obtain both $\overline{f}(T)$ and T to a high degree of accuracy.

ACKNOWLEDGEMENT

The authors would like to thank the Sino-British Friendship Fellowship Scheme for the support of Y. Yuan

References

1. Jaswon, M.A. and Symm, G.T. Integral Equation Methods in Potential Theory and Elastostatics, Academic Press, London, 1977.

2. Brebbia, C.A. (Editor), Proc. 12th Int. Conf. on Boundary Element Methods in Engineering, Springer-Verlag, 1989.

3. Ingham, D.B. and Kelmanson, M.A. Boundary Integral Equation Analysis of Singular, Potential and Biharmonic Problems, Lecture Notes in Engineering, Springer-Verlag, 1984.

4. Khader, M.S. Heat Conduction with Temperature Dependent Thermal Conductivity, Paper 80-HT-4, National Heat Transfer Conf., ASME, Orlando, Florida, 1980.

5. Bialecki, R. and Nowak, A.J. Boundary Value Problems in Heat Conduction with Nonlinear Material and Boundary Conditions, Appl. Math. Modelling, pp.416-421, 1981.

6. Ingham, D.B., Heggs, P.J. and Manzoor, M. Boundary Integral Equation Solution of Nonlinear Plane Potential Problem, IMA J. NUM. Anal., Vol.1, pp.416-426, 1981.

7. Khader, M.S. and Hanna, M.C. An Iterative Boundary Numerical Solution for General Steady Heat Conduction Problems, Trans. ASME J.Heat. Transfer, Vol.103, pp.26-31, 1981.

8. Han, H. The Finite Element Method in a Family of Improperly Posed Problems, Math. of Comp., Vol.38, No.157, pp.55-65, 1982.

9. Falk, R.S. and Monk, P.B. Logarithmic Convexity for Discrete Harmonic Functions and the Approximation of the Cauchy Problem for Poison's Equations, Math. of Comp., Vol.47, pp.135-149, 1986.

10. Ingham, D.B., Yuan, Y. and Han, H. The Boundary Element Method for an Improperly Posed Problem, to be published in JIMA, 1991.

SECTION 17: SPECIAL TECHNIQUES

The Method of Infinite Matrices - An Alternative to BEM?

G.T. Symm

Division of Information Technology and Computing, National Physical Laboratory, Teddington, Middlesex, TW11 OLW, U.K.

ABSTRACT

The method of infinite matrices (or the infinite-matrix method) has been used, for several years, for the solution of the wave equation in the field of acoustics. Two particular applications are outlined in this paper – computations of the free-field correction for a laboratory standard microphone mounted on a semi-infinite rod and of the transmission characteristics of an acoustic coupler. The results suggest that, for certain boundary value problems with interfaces, the method of infinite matrices may be a viable alternative to the boundary element method.

INTRODUCTION

This is an expository paper on the method of infinite matrices, as applied to two problems in acoustics – computations of the free-field correction for a laboratory standard microphone mounted on a semi-infinite rod and of the transmission characteristics of an acoustic coupler. Software for the solution of these problems has been prepared for the Division of Radiation Science and Acoustics at NPL, for applications relating to acoustic standards and calibration of microphones.

The two problems, both of which have rotational symmetry, were first solved by the infinite-matrix method by Matsui and Miura, as described in a series of publications (Matsui, 1960, 1971; Matsui and Miura, 1967; Miura and Matsui, 1974). The present paper is essentially a review of the work of these authors, clarifying some of the concepts involved and removing various ambiguities, without repeating all of the technical details, and indicating similarities between their method and the boundary element method.

In the following sections, we describe first the problems and then the method of infinite matrices, including some of the analysis on which it is based and its numerical implementation. In conclusion, we draw parallels with the boundary element method.

TWO PROBLEMS IN ACOUSTICS

A microphone mounted on a rod

We consider first the configuration illustrated in Figure 1, where a condenser-type microphone of radius a_1 is mounted on the end of a semi-infinite rod of the same radius. The microphone has a diaphragm of radius b, with a back plate of radius c, set in a recess of radius a_2 and depth l. A plane sound wave, with velocity potential e^{ikz}, where k is the wave number, approaches the microphone parallel to the (z-)axis of rotational symmetry.

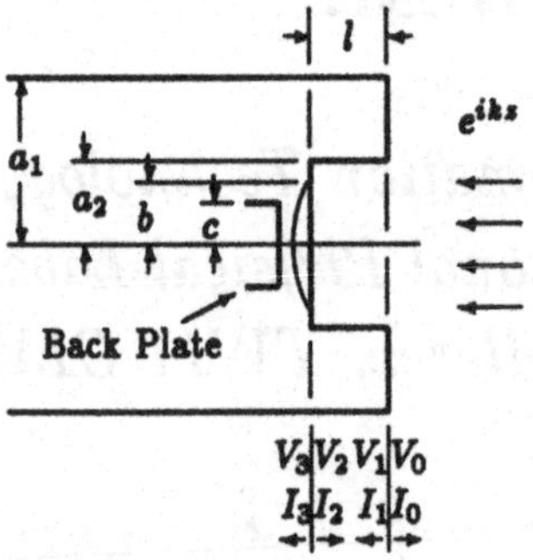

Figure 1: Sectional view of a microphone on a rod

The problem here is to compute the free-field correction $[FC]$ of the microphone, $[FC]$ being defined as 20 times the logarithm (to the base 10) of the ratio of the free-field response to the pressure response. The pressure response of this type of microphone is proportional to, and may be represented by, the volume displacement within the area adjacent to the back plate when unit sound pressure is applied uniformly over the bottom surface of the recess. The free-field response is represented by the corresponding volume displacement when a plane wave of unit sound pressure strikes the face of the microphone.

An acoustic coupler

Our second problem is illustrated in Figure 2, where two microphones, each of radius a_1, are coupled together by a cylinder of length l_1 and the same radius. Both microphones are of the condenser type described above and their internal dimensions are denoted by a, b, c and l with subscripts S and R for source and receptor microphones respectively.

In this case, we wish to compute two of the transmission characteristics of the coupler, $[TC]_{soft-soft}$ and $[TC]_{hard-hard}$, corresponding to microphones with finite and infinite diaphragm tensions respectively. Each transmission characteristic is defined as the ratio of the response of the coupler to that of an ideal coupler, i.e. one for which the effects of the wave motion and diaphragm impedance may be neglected. The response of the coupler is defined as the volume displacement of the receptor diaphragm in the area adjacent to its back plate when the source diaphragm is driven by a unit electrostatic force from its back plate.

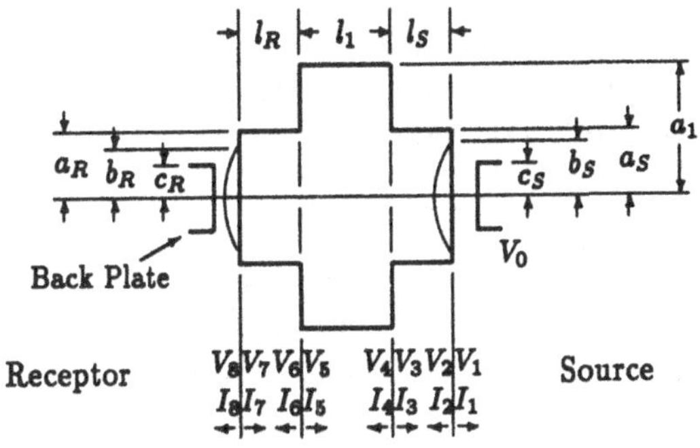

Figure 2: Sectional view of an acoustic coupler.

THE METHOD OF INFINITE MATRICES

The method of infinite matrices, as applied to the above problems, is based upon analytical analyses of the Helmholtz wave equation in cylindrical polar coordinates and of the equation of motion of a circular diaphragm with a concentric circular back plate. Each problem domain is divided into several sections (such as the three cylinders and two diaphragms in Figure 2), which are initially treated separately (much as they might be in the boundary element method). Infinite series solutions in each section are then coupled together at the interfaces between sections by means of infinite matrices.

Thus, as will be amplified later, for the first problem, the microphone mounted on a rod, the method leads (Matsui, 1971) to the equations:

$$AV_0 + BI_0 = V_{IN}, \quad (1)$$
$$V_1 = DV_0, \quad (2)$$
$$I_0 = -(a_2/a_1)^2 D^{\mathrm{T}} I_1, \quad (3)$$
$$V_1 = Z_D I_1 + Z_T I_2, \quad (4)$$
$$V_2 = Z_T I_1 + Z_D I_2, \quad (5)$$
$$V_3 = FV_2, \quad (6)$$
$$I_2 = -(b/a_2)^2 F^{\mathrm{T}} I_3, \quad (7)$$
$$V_3 = Z_M I_3, \quad (8)$$

where the various V_i and I_i are column vectors of coefficients of certain orthogonal functions in infinite series representations of acoustic pressure and particle velocity respectively, while A, B, D, F, Z_D, Z_M and Z_T are all infinite matrices and the superscript T denotes transpose.

From the solution of these equations, the free-field correction is given by:

$$\begin{aligned}[FC] = \; & 20\log_{10}|SUM\, Z_M^{-1} F\{E + [Z_D - Z_T Z_{DP}^{-1} Z_T] \\ & (b/a_2)^2 F^{\mathrm{T}} Z_M^{-1} F\}^{-1} Z_T Z_{DP}^{-1} D A^{-1} V_{IN}| \\ & -20\log_{10}|SUM\, Z_M^{-1} FG|, \end{aligned} \quad (9)$$

where E is the identity matrix and

$$Z_{DP} = Z_D + (a_2/a_1)^2 DA^{-1}BD^{\mathrm{T}}, \quad (10)$$

while SUM and G are infinite row and column vectors respectively.

Similarly, for the second problem, the acoustic coupler, the method leads (Matsui and Miura, 1974) to the equations:

$$V_1 - V_0 = Z_{MS} I_1, \tag{11}$$
$$V_1 = F_S V_2, \tag{12}$$
$$I_2 = -(b_S/a_S)^2 F_S^{\mathrm{T}} I_1, \tag{13}$$
$$V_2 = Z_{DS} I_2 + Z_{TS} I_3, \tag{14}$$
$$V_3 = Z_{TS} I_2 + Z_{DS} I_3, \tag{15}$$
$$V_3 = D_S V_4, \tag{16}$$
$$I_4 = -(a_S/a_1)^2 D_S^{\mathrm{T}} I_3, \tag{17}$$
$$V_4 = Z_{D1} I_4 + Z_{T1} I_5, \tag{18}$$
$$V_5 = Z_{T1} I_4 + Z_{D1} I_5, \tag{19}$$
$$V_6 = D_R V_5, \tag{20}$$
$$I_5 = -(a_R/a_1)^2 D_R^{\mathrm{T}} I_6, \tag{21}$$
$$V_6 = Z_{DR} I_6 + Z_{TR} I_7, \tag{22}$$
$$V_7 = Z_{TR} I_6 + Z_{DR} I_7, \tag{23}$$
$$V_8 = F_R V_7, \tag{24}$$
$$I_7 = -(b_R/a_R)^2 F_R^{\mathrm{T}} I_8, \tag{25}$$
$$V_8 = Z_{MR} I_8, \tag{26}$$

where the subscripts S and R refer to source and receptor microphones respectively and the unit subscript to the cylinder in between.

In this case, equations (14)-(23) combine to yield

$$V_2 = Z_{D22} I_2 + Z_{T27} I_7, \tag{27}$$
$$V_7 = Z_{T72} I_2 + Z_{D77} I_7, \tag{28}$$

where

$$Z_{D22} = Z_{DS} - Z_{TS} Z_{DSPD}^{-1} Z_{TS}, \tag{29}$$
$$Z_{T27} = Z_{TS} Z_{DSPD}^{-1} D_S Z_{T1} (a_R/a_1)^2 D_R^{\mathrm{T}} Z_{DRP}^{-1} Z_{TR}, \tag{30}$$
$$Z_{T72} = Z_{TR} Z_{DRPD}^{-1} D_R Z_{T1} (a_S/a_1)^2 D_S^{\mathrm{T}} Z_{DSP}^{-1} Z_{TS}, \tag{31}$$
$$Z_{D77} = Z_{DR} - Z_{TR} Z_{DRPD}^{-1} Z_{TR}, \tag{32}$$

with Z_{DRP} and Z_{DSP} given by

$$Z_{DmP} = Z_{Dm} + D_m Z_{D1} (a_m/a_1)^2 D_m^{\mathrm{T}}, \tag{33}$$

with m, for microphone, equal to R and S respectively, and Z_{DRPD} and Z_{DSPD} given by

$$Z_{DmPD} = Z_{Dm} + D_m \{Z_{D1} - Z_{T1} (a_q/a_1)^2 D_q^{\mathrm{T}} Z_{DqP}^{-1} D_q Z_{T1}\} (a_m/a_1)^2 D_m^{\mathrm{T}}, \tag{34}$$

where $q = R$ when $m = S$ and vice versa.

Further algebra reveals that the vector V_8, which represents the pressure on the receptor diaphragm, is given in terms of the vector V_0, representing the static drive force acting on the source diaphragm through its back plate, by

$$\begin{aligned} V_8 &= F_R\{E + Z(b_R/a_R)^2 F_R^T Z_{MR}^{-1} F_R\}^{-1} Z_{T72} \\ &\quad \{E + (b_S/a_S)^2 F_S^T Z_{MS}^{-1} F_S Z_{D22}\}^{-1} (b_S/a_S)^2 F_S^T Z_{MS}^{-1} V_0, \end{aligned} \tag{35}$$

where

$$\begin{aligned} Z = Z_{D77} - Z_{T72}\{E + (b_S/a_S)^2 F_S^T Z_{MS}^{-1} F_S \\ Z_{D22}\}^{-1} (b_S/a_S)^2 F_S^T Z_{MS}^{-1} F_S Z_{T27}. \end{aligned} \tag{36}$$

From this result (35), abbreviated here to

$$V_8 = F_R E_R Z_{T72} E_S (b_S/a_S)^2 F_S^T Z_{MS}^{-1} V_0, \tag{37}$$

with

$$E_R = \{E + Z(b_R/a_R)^2 F_R^T Z_{MR}^{-1} F_R\}^{-1}, \tag{38}$$

$$E_S = \{E + (b_S/a_S)^2 F_S^T Z_{MS}^{-1} F_S Z_{D22}\}^{-1}, \tag{39}$$

Miura and Matsui (1974) derived the transmission characteristic

$$[TC]_{soft-soft} = \frac{SUM_R Z_{MR}^{-1} F_R E_R Z_{T72} E_S F_S^T Z_{MS}^{-1} SUM_S^T}{\pi a_S^2 SUM_R Z_{MR}^{-1} F_R G(\gamma P_0/i\omega V_c) G^T F_S^T Z_{MS}^{-1} SUM_S^T}, \tag{40}$$

where γ is the ratio of the specific heats of the coupler medium, P_0 is the static atmospheric pressure in the coupler, V_c is the coupler volume and ω is the angular frequency of vibration. The corresponding transmission characteristic $[TC]_{hard-hard}$ is given by the right hand side of equation (40) with E_R and E_S each replaced by E alone.

There is not room in the confines of this paper to give details of all the matrices involved in this formulation but the essence of the method may be appreciated from the following analysis of wave propagation in a cylinder.

THE SOUND FIELD IN A RIGHT CIRCULAR CYLINDER

In cylindrical polar coordinates (ρ, ϕ, z), the Helmholtz wave equation

$$\nabla^2 u + k^2 u = 0, \tag{41}$$

which governs the velocity potential u, for a wave number $k \equiv \omega/c_0$ (where c_0 is the speed of sound), takes the form

$$\frac{1}{\rho}\frac{\partial}{\partial \rho}\left(\rho \frac{\partial u}{\partial \rho}\right) + \frac{1}{\rho^2}\frac{\partial^2 u}{\partial \phi^2} + \frac{\partial^2 u}{\partial z^2} + k^2 u = 0, \tag{42}$$

which, in the case of cylindrical symmetry, i.e. when u is independent of ϕ, reduces to

$$\frac{1}{\rho}\frac{\partial}{\partial \rho}\left(\rho \frac{\partial u}{\partial \rho}\right) + \frac{\partial^2 u}{\partial z^2} + k^2 u = 0. \tag{43}$$

Under the assumption that the variables may be separated, i.e. setting

$$u = R(\rho)Z(z), \tag{44}$$

in equation (43), we obtain

$$\left\{R''(\rho) + \frac{1}{\rho}R'(\rho)\right\} Z(z) + R(\rho)Z''(z) + k^2R(\rho)Z(z) = 0, \tag{45}$$

whence it follows that

$$\frac{R''(\rho)}{R(\rho)} + \frac{1}{\rho}\frac{R'(\rho)}{R(\rho)} + C = 0 \tag{46}$$

and

$$\frac{Z''(z)}{Z(z)} + k^2 - C = 0, \tag{47}$$

where C is a constant.

If $C = 0$, equation (46) reduces to

$$\rho R''(\rho) + R'(\rho) = 0, \tag{48}$$

whence $\rho R'(\rho)$ is a constant and, in a circular cylinder of radius a, the boundary condition that radial particle velocity vanishes on the cylindrical wall, viz.

$$R'(\rho) = 0 \text{ on } \rho = a, \tag{49}$$

implies that this constant is zero. It follows that $R(\rho)$ is itself a constant, which may be "normalised" to unity with no loss of generality. Meanwhile equation (47) becomes

$$Z''(z) + k^2Z(z) = 0, \tag{50}$$

whence

$$Z(z) = h_0 \cos kz + g_0 \sin kz, \tag{51}$$

where h_0 and g_0 are further constants.

If $C \neq 0$, equation (46) gives

$$\rho^2R''(\rho) + \rho R'(\rho) + C\rho^2R(\rho) = 0, \tag{52}$$

which, with the substitutions

$$C = K^2 \text{ and } R(\rho) \equiv F(x) \text{ with } x = K\rho, \tag{53}$$

becomes

$$x^2F''(x) + xF'(x) + x^2F(x) = 0, \tag{54}$$

the solution of which is the Bessel function $J_0(x)$. It follows that equation (52) has the solution $R(\rho) = J_0(K\rho)$ and, since $J_0'(x) = -J_1(x)$, the boundary condition (49) implies that $K = j_{1m}/a$, where j_{1m}, $m = 1, 2, \ldots$, are the positive zeros of the Bessel function $J_1(x)$. Thus, dividing R by a "normalising" factor $J_0(j_{1m})$, we obtain

$$R(\rho) = \frac{J_0(j_{1m}\rho/a)}{J_0(j_{1m})}. \tag{55}$$

In this case, recalling that $C = K^2 = (j_{1m}/a)^2$, we define κ_m by

$$\kappa_m^2 = \left(\frac{j_{1m}}{a}\right)^2 - k^2, \tag{56}$$

whereupon equation (47) becomes

$$Z''(z) - \kappa_m^2 Z(z) = 0, \tag{57}$$

with the solution

$$Z(z) = h_m \cosh \kappa_m z + g_m \sinh \kappa_m z, \tag{58}$$

where h_m and g_m are further constants.

Collecting the above results, we obtain the general solution of equation (43), subject to the boundary condition expressed by equation (49), in the form

$$u = h_0 \cos kz + g_0 \sin kz + \sum_{m=1}^{\infty} (h_m \cosh \kappa_m z + g_m \sinh \kappa_m z) \frac{J_0(j_{1m}\rho/a)}{J_0(j_{1m})}, \tag{59}$$

where the coefficients h_m and g_m, $m = 0, 1, 2, \ldots$, depend upon the boundary conditions on the ends of the cylinder.

It follows from equation (59) that the sound pressure $p \equiv i\omega\rho_0 u$ (where ρ_0 is the density of the medium within the cylinder) and the axial component of the particle velocity $v = -\nabla u$ may be represented on the cylinder ends, planes of constant z, by infinite series in $J_0(j_{1m}\rho/a)/J_0(j_{1m})$ (normalised Dini series). This observation provides a starting point for the derivation of several of the matrices in the infinite-matrix method.

DERIVATION OF MATRICES

Suppose first that our cylinder has length l and is terminated by the planes $z = z_1$ and $z = z_2$. Let V_1 and V_2, and I_1 and I_2, be infinite column vectors containing the ordered coefficients of the normalised Dini series representations of sound pressure, and axial particle velocity (directed into the cylinder), respectively, on the two ends. [For example, in the notation above, the first element of V_1 will be $i\omega\rho_0(h_0 \cos kz_1 + g_0 \sin kz_1)$, while, from the derivative of equation (59) with respect to z, the $(m+1)$th element of I_2 will be $\kappa_m(h_m \sinh \kappa_m z_2 + g_m \cosh \kappa_m z_2)$].

Then, it is easy, though rather tedious, to show that (whatever the values of the coefficients h_m and g_m, $m = 0, 1, 2, \ldots$):

$$V_1 = Z_D I_1 + Z_T I_2, \tag{60}$$

$$V_2 = Z_T I_1 + Z_D I_2, \tag{61}$$

where Z_D and Z_T are infinite diagonal matrices, representing driving-point and transfer impedances respectively, with non-zero elements, compactly displayed as follows:

$$Z_D = \frac{\gamma P_0}{i\omega l}\left[\frac{kl}{\tan kl}, \frac{-k^2 l}{\kappa_1 \tanh \kappa_1 l}, \frac{-k^2 l}{\kappa_2 \tanh \kappa_2 l}, \ldots\right], \tag{62}$$

$$Z_T = \frac{\gamma P_0}{i\omega l}\left[\frac{kl}{\sin kl}, \frac{-k^2 l}{\kappa_1 \sinh \kappa_1 l}, \frac{-k^2 l}{\kappa_2 \sinh \kappa_2 l}, \ldots\right]. \tag{63}$$

Equations (60) and (61) are precisely equations (4) and (5) of our formulation and equations (62) and (63), with appropriately subscripted values of l, define all the matrices in equations (14), (15), (18), (19), (22) and (23).

Suppose next that a cylinder of radius a_1 meets a cylinder of radius a_2 ($< a_1$), on a certain z–plane on which the series for the sound pressure in the two cylinders reduce to the Dini series

$$h_0 + \sum_{m=1}^{\infty} h_m \frac{J_0(j_{1m}\rho/a_1)}{J_0(j_{1m})}, \ \rho < a_1, \tag{64}$$

and

$$g_0 + \sum_{\mu=1}^{\infty} g_\mu \frac{J_0(j_{1\mu}\rho/a_1)}{J_0(j_{1\mu})}, \ \rho < a_2, \tag{65}$$

respectively, where the z-dependence and other factors have been included in the coefficients. Then, equating expressions (64) and (65), for continuity of pressure for $\rho < a_2$, and using the orthonormal properties of J_0 given by

$$\frac{2}{a^2}\int_0^a \frac{J_0(j_{1m}\rho/a)J_0(j_{1\mu}\rho/a)\rho\, d\rho}{J_0(j_{1m})J_0(j_{1\mu})} = \begin{cases} 1, & \mu = m, \\ 0, & \mu \neq m, \end{cases} \tag{66}$$

together with other properties of the Bessel functions (Gray et al., 1931), we find that

$$g_0 = h_0 + \sum_{m=1}^{\infty} h_m \frac{2}{j_{1m}a_2/a_1} \frac{J_1(j_{1m}a_2/a_1)}{J_0(j_{1m})}, \tag{67}$$

while

$$g_\mu = \sum_{m=1}^{\infty} h_m \frac{2j_{1m}a_2/a_1}{(j_{1m}a_2/a_1)^2 - j_{1\mu}^2} \frac{J_1(j_{1m}a_2/a_1)}{J_0(j_{1m})}, \ \mu = 1, 2, \ldots. \tag{68}$$

If, in this case, the vectors V_1 and V_2 hold the coefficients of the series in the two cylinders respectively, it follows that

$$V_2 = DV_1, \tag{69}$$

where

$$D = \begin{bmatrix} 1 & \cdots & \frac{2}{j_{1m}a_2/a_1} \frac{J_1(j_{1m}a_2/a_1)}{J_0(j_{1m})} & \cdots \\ \vdots & & \vdots & \\ 0 & \cdots & \frac{2j_{1m}a_2/a_1}{(j_{1m}a_2/a_1)^2 - j_{1\mu}^2} \frac{J_1(j_{1m}a_2/a_1)}{J_0(j_{1m})} & \cdots \\ \vdots & & \vdots & \end{bmatrix}. \tag{70}$$

This is precisely the matrix D which was introduced in equation (2) above and the matrices D_S and D_R, introduced in equations (16) and (20), are of the same form in terms of appropriate radii.

Similarly, if the components of particle velocity in the two cylinders, normal to the plane of the interface, take the forms (64) and (65) respectively, the continuity of particle velocity for $\rho < a_2$ and the condition of zero particle velocity for $a_2 < \rho < a_1$ together yield the result

$$I_1 = -\left(\frac{a_2}{a_1}\right)^2 D^{\mathrm{T}} I_2. \tag{71}$$

Equations (3), (17) and (21) are of this form.

SUMMARY OF OTHER EQUATIONS

For a vibrating diaphragm of radius b, with a back plate of radius c ($< b$), the displacement $\eta(\rho)$ (with time dependence $e^{i\omega t}$) satisfies the differential equation

$$-\frac{T}{\rho}\frac{d}{d\rho}\left(\rho\frac{d\eta}{d\rho}\right) + \left\{(i\omega)^2\sigma + i\omega\mu w(\rho)\right\}\eta = p(\rho), \tag{72}$$

where $p(\rho)$ is the applied sound pressure, T is the tension in the diaphragm, σ is its surface density, μ is the air damping coefficient (resistance per unit area – assumed uniform) of the air film between the diaphragm and its back plate and $w(\rho) = 1$ for $\rho < c$ and $w(\rho) = 0$ for $c < \rho < b$. The relevant boundary condition is $\eta(b) = 0$ and, in this case, expanding the sound pressure and the displacement as normalised Fourier-Bessel series:

$$p(\rho) = \sum_{n=1}^{\infty} p_n \frac{J_0(j_{0n}\rho/b)}{J_1(j_{0n})} \tag{73}$$

and

$$\eta(\rho) = \sum_{\nu=1}^{\infty} \eta_\nu \frac{J_0(j_{0\nu}\rho/b)}{J_1(j_{0\nu})}, \tag{74}$$

we find that

$$p_n = (i\omega)^2\sigma\eta_n + T(j_{0n}/b)^2\eta_n + i\omega\mu\sum_{\nu=1}^{\infty} C_{n\nu}\eta_\nu, \; n = 1, 2, \ldots, \tag{75}$$

where

$$C_{n\nu} = \frac{2}{b^2}\int_0^c \frac{J_0(j_{0n}\rho/b)J_0(j_{0\nu}\rho/b)\rho\, d\rho}{J_1(j_{0n})J_1(j_{0\nu})}. \tag{76}$$

It follows that, if V and I contain the coefficients of the series for sound pressure and velocity (orthogonal to the plane of the diaphragm), viz. p_n and $i\omega\eta_n$ respectively with $n = 1, 2, \ldots$, then

$$V = Z_M I, \tag{77}$$

where

$$Z_M = \begin{bmatrix} i\omega\sigma + \frac{T}{i\omega}\left(\frac{j_{01}}{b}\right)^2 + \mu C_{11} & \ldots & \mu C_{1n} & \ldots \\ \vdots & & \vdots & \\ \mu C_{n1} & \ldots & i\omega\sigma + \frac{T}{i\omega}\left(\frac{j_{0n}}{b}\right)^2 + \mu C_{nn} & \ldots \\ \vdots & & \vdots & \end{bmatrix}. \tag{78}$$

This is the matrix which first appears in equation (8) above and the matrices Z_{MS} and Z_{MR}, in equations (11) and (26), are of this form too.

By arguments analogous to those relating to the matrix D in the previous section, the continuity of sound pressure on a diaphragm of radius b at the end of a cylinder of radius a ($> b$) yields the equation

$$V_1 = FV_2, \tag{79}$$

relating the coefficients of the corresponding Fourier-Bessel and Dini series respectively, where

$$F = \begin{bmatrix} \frac{2}{j_{01}} & \cdots & \frac{2j_{01}}{j_{01}^2-(j_{1m}b/a)^2}\frac{J_0(j_{1m}b/a)}{J_0(j_{1m})} & \cdots \\ \vdots & & \vdots & \\ \frac{2}{j_{0n}} & \cdots & \frac{2j_{0n}}{j_{0n}^2-(j_{1m}b/a)^2}\frac{J_0(j_{1m}b/a)}{J_0(j_{1m})} & \cdots \\ \vdots & & \vdots & \end{bmatrix}. \tag{80}$$

Similarly, the axial particle velocity in the cylinder is given in terms of the velocity of the diaphragm by

$$I_2 = -\left(\frac{b}{a}\right)^2 F^{\mathrm{T}} I_1. \tag{81}$$

With a replaced by a_2, equation (80) defines the matrix F in equation (6) and equation (81) yields equation (7). Equations (12), (13), (24) and (25) and the matrices therein are similarly derived.

In the present notation, unit sound pressure on the end of a cylinder is represented by the vector of Dini series coefficients:

$$G \equiv [1, 0, 0, \ldots]^{\mathrm{T}}. \tag{82}$$

It follows from equations (77) and (79) that, when this cylinder is a microphone recess, the corresponding velocity of the diaphragm is

$$I = Z_M^{-1} F G, \tag{83}$$

whence, by integration of the terms of the Fourier-Bessel series (74) over the area of the backplate, the pressure response of the microphone is

$$(i\omega)^{-1} SUM\, Z_M^{-1} F G, \tag{84}$$

where

$$SUM = 2\pi bc\left[\frac{J_1(j_{01}c/b)}{j_{01}J_1(j_{01})}, \ldots, \frac{J_1(j_{0n}c/b)}{j_{0n}J_1(j_{0n})}, \ldots\right]. \tag{85}$$

Equation (84) is the source of the second term of expression (9) for the free-field correction of the microphone mounted on a rod. The row vector SUM enters the first term in a similar manner. The column vector G, defined by equation (82), also appears in the expression for the transmission characteristic of the acoustic coupler, equation (40), in which SUM_S and SUM_R are derived from equation (85) with appropriate source and receptor radii.

In equation (1) the vector V_{IN}, describing the incident wave, is equal to G for normal incidence, while the matrices A and B are given explicitly by Matsui (1970) in terms of certain "split" functions. These functions arise from the analysis of the scattered field associated with a semi-infinite rod of circular cross-section, carried out by Jones (1955) using the Wiener-Hopf technique. The evaluation of these functions is adequately described by Matsui (1960) and, since they are relevant to only one of our two problems, we will not discuss them further here.

NUMERICAL IMPLEMENTATION

To implement the above formulation numerically, we use the method of reduction (Kantorovich and Krylov, 1964, p.25), whereby we truncate each of the infinite (square) matrices and vectors to a finite order N and obtain the solution to each problem as the limit of the solutions to these finite systems as N tends to infinity. The matrices are computed in Fortran 77 and manipulated to give the required results using MATLAB – a high-performance interactive software package for scientific and engineering numerical computation (The MathWorks, 1989).

In practice, N does not need to be very large, as indicated by Table 1, which shows successive approximations (obtained using finite matrices) to the free-field correction (9), at a frequency of 3KHz, for a typical microphone mounted on a rod. The dimensions of the microphone are given by $a_1 = 0.01189\,\mathrm{m}$, $a_2 = 0.00932\,\mathrm{m}$,

Order of matrices N	Approximation to $[FC]$
2	1.939
3	1.985
4	2.021
5	2.039
6	2.043
7	2.041
8	2.041

Table 1: Successive approximations to the free-field correction of a microphone mounted on a rod for a frequency of 3000Hz.

$b = 0.00838\,\mathrm{m}$, $c = 0.00686\,\mathrm{m}$ and $l = 0.00196\,\mathrm{m}$. Other physical data are $c_0 = 343.59\,\mathrm{m\,s^{-1}}$, $\gamma = 1.402$, $\mu = 4510\,\mathrm{kg\,m^{-2}s^{-1}}$, $P_0 = 101325\,\mathrm{kg\,m^{-1}s^{-2}}$, $\rho_0 = 1.203\,\mathrm{kg\,m^{-3}}$, $\sigma = 0.07\,\mathrm{kg\,m^{-2}}$ and $T = 2430\,\mathrm{kg\,s^{-2}}$. Though the convergence is not monotonic, and slight oscillations in the results continue as N is increased further, the value for $N = 7$ is within 0.2% of that for $N = 40$ and $N = 7$ is accepted to obtain results for a range of frequencies as presented in Table 2. Similar results presented by Matsui (1971) agree with experimental measurements to within 0.1dB.

For the acoustic coupler, we find that $N = 10$ suffices to determine the required transmission characteristics to an acceptable accuracy, in agreement with Matsui and Miura (1967).

CONCLUSION

For problems of acoustics with rotational symmetry, involving a number of connected domains, the method of infinite matrices appears to be very effective. Though the problems discussed could have been solved by the boundary element method, by formulating the appropriate integral equations in each section of the domain of interest and coupling the solutions at the interfaces through the conti-

Frequency $\omega/2\pi$ (Hz)	Correction $[FC]$ (dB)
1000	0.256
2000	0.971
3000	2.041
4000	3.359
5000	4.800
6000	6.208
7000	7.426
8000	8.343
9000	8.937

Table 2: Computed free-field corrections of a microphone mounted on a rod for a range of frequencies.

nuity of pressure and particle velocity, the present method has certain advantages. In particular, the interface conditions are satisfied exactly by the infinite matrices and it is only at the final stage of reduction that any approximation is made. In each of the examples considered here, the solution derived from the reduced (finite) matrices converges rapidly to the required solution.

While the infinite-matrix method undoubtedly involves more theoretical analysis than the boundary element method, much of it has to be carried out only once, the same matrices being applicable to many problems as illustrated above. Though not as generally applicable as boundary element or finite element methods, the method of infinite matrices is not restricted to problems with rotational symmetry and has possibilities for the analysis of waveguides. For field problems in composite domains with uniform cross sections, the method is certainly worth consideration.

ACKNOWLEDGEMENTS

The author is grateful to Dr. D.R. Jarvis and R.E. Barham, of the Division of Radiation Science and Acoustics at NPL, who initiated this work, and also to student visitors to NPL, I. Zakiuddin and P.R. Johnson, who assisted with the computer programming. Thanks are also due to Dr. M.G. Cox for valuable comments on the draft of this paper.

REFERENCES

Gray, A., Mathews, G.B. and MacRobert, T.M. (1931) *Bessel Functions* (Second Edition, reprinted), MacMillan and Co., London.

Jones, D.S. (1955) The scattering of a scalar wave by a semi-infinite rod of circular cross section. *Phil. Trans. Roy. Soc. London*, **A247**, 499-528.

Kantorovich, L.V. and Krylov, V.I. (1964) *Approximate Methods of Higher Analysis*, P. Noordhoff, Ltd., Groningen, The Netherlands.

Matsui, E. (1960) On the free-field correction for laboratory standard microphones mounted on a semi-infinite rod. *National Bureau of Standards Report* 7038, USA.

Matsui, E. (1971) Free-field correction for laboratory standard microphones mounted on a semi-infinite rod. *J. Acoust. Soc. Amer.*, **49**, 5(2), 1475-1483.

Matsui, E. and Miura, H. (1967) On the transmission characteristics of the 20cc coupler. *Electrotechnical Laboratory Report*, Tokyo, Japan. (Summarised at the 6th International Congress on Acoustics, Tokyo, August 21-28, 1968.)

Miura, H. and Matsui, E. (1974) Analysis of the wave motion within a coupler for the pressure calibration of laboratory standard microphones. *J. Acoust. Soc. Japan*, **30**, 12, 639-646 (in Japanese).

The MathWorks (1989) *PRO-MATLAB, User's Guide*, The MathWorks, Inc., South Natick, MA, USA.

Robust Boundary Element Scheme for Helmholtz Eigenvalue Equation

N. Kamiya, E. Andoh

Department of Mechanical Engineering, Nagoya University, Nagoya, 464-01 Japan

Abstract

This paper presents a new and robust boundary element scheme for analyzing the Helmholtz eigenvalue probelm. The method is based on a combination of the multiple reciprocity boundary element method using the fundamental solutions of the Laplace and its related higher order equations, and efficient technique for the eigenvalue determination. The method completely removes domain integrals, complicated integrals of the fundamental solutions and solution of transcendental equation. Some numerical examples for two-dimensional problems demonstrate superior usefulness.

Introduction

Multiple Reciprocity Method (MRM) in the boundary element methods (BEM) was developed by Nowak and Brebbia [1, 2, 3] to convert domain integral to the corresponding boundary integral. The method was first considered for transient heat transfer and extended further to Poisson and Helmholtz equations.

Eigenvalue analysis is known as one of the essential informations in various engineering fields especially in designing. BEM has been also applied to the eigenvalue problem with acoustics, vibration among others. Although the BEM formulation with the fundamental solution to the Helmholtz equation can be carried out only with the boundary of the domain under consideration, the conventional method for obtaining eigenvalue becomes complicated, because the integral equation in terms of complex-valued variable includes the unknown wavenumber and yields transcendental equation.

In the authors' previous paper [5], a new robust scheme for the Helmholtz eigenvalue analysis was proposed using the MRM formulation in terms of the higher order fundamental solutions to the Laplace operator. This study, as an further extension, is devoted to more examination of the method. The proper equation determining the eigenvalues by the proposed method is demonstrated and compared with the conventional method in the category of BEM. Some suggestions and discussion on the convergency of the formulation are given. Numerical illustrations of the two-dimensional simple examples compare validity and efficency of several types of solution schemes.

MRM for Helmholtz eigenvalue problem

In order to examine various solution schemes for the eigenvalue determination in the framework of BEM formulation, we summarize the equations here. The Helmholtz equation in terms of the potential u (complex-valued) in two-dimensional domain Ω surrounded by the boundary Γ

$$\nabla^2 u + k^2 u = 0 \qquad (\text{in } \Omega) \tag{1}$$

where k is the wavenumber, which is unknown in the eigenvalue problem. The conventional boundary integral formulation employs the fundamental solutions to the above equation (1); i.e.,

$$\nabla^2 v + k^2 v + \delta = 0 \tag{2}$$

$$v = -\frac{i}{4} H_0^2(kr) \tag{3}$$

where δ, H_0^2, r are Dirac delta function, Hankel function and the distance between the source and integration points, respectively. i denotes the imaginary unit, $i = \sqrt{-1}$. The resulting integral equation using Eqs.(2) and (3) is

$$cu + \int_\Gamma (u\frac{\partial v}{\partial n} - qv)d\Gamma = 0 \tag{4}$$

where c is constant depending on the place where the source point is taken and q is the flux, i.e., outward normal derivative of u on the boundary; $q = \partial u/\partial n$. Equation (4) yields the discretized matrix equation using the boundary elements as follows ;

$$\mathbf{Su} = \mathbf{Tq} \tag{5}$$

in terms of the vectors $\mathbf{u}$ and $\mathbf{q}$ of their nodal values. The matrices $\mathbf{S}$ and $\mathbf{T}$ are constructed by integration of the complex-valued fundamental solution multiplied by the interpolation function taken on the boundary elements. Since the fundamental solution, Eq(3), include k, the matrices $\mathbf{S}$ and $\mathbf{T}$ are thought to be highly complicated functions of k.

The MRM formulation employs the fundamental solution u_0^* of the Laplace equation and those of related Laplace differential operator;

$$\nabla^2 u_0^* + \delta = 0 \tag{6}$$

$$u_0^* = -\frac{1}{2\pi} \ln r \tag{7}$$

$$\begin{aligned}\nabla^2 u_{j+1}^* &= u_j^* \\ q_j^* &= \partial u_j^*/\partial n \qquad\qquad (j=1,2,\ldots) \qquad (8)\\ u_j^* &= -\frac{1}{2\pi}r^{2j}\frac{1}{4^j(j!)^2}\left(\ln r - \sum_{l=1}^{j}\frac{1}{l}\right)\end{aligned}$$

(9)

The resulting integral equation is

$$cu + \sum_{j=0}^{n}(-k^2)^j \int_\Gamma (uq_j^* - qu_j^*)d\Gamma = (-1)^n(k^2)^{n+1}\int_\Omega uu_n^* d\Omega \qquad (10)$$

The right-hand-side of the above equation still remains as the domain integral but is negligible for engineering practice by the reason mentioned in Appendix.(The proof of convergency of Eq.(10) is partially stated in Ref.[3]). Therefore, sufficently huge n, Eq.(10) is read as

$$cu + \sum_{j=0}^{n}(-k^2)^j \int_\Gamma (uq_j^* - qu_j^*)d\Gamma = 0 \qquad (11)$$

and the discretized form as

$$\mathbf{Hu} = \mathbf{Gq} \qquad (12)$$

where

$$\begin{aligned}\mathbf{H} &= \mathbf{H}_0 - k^2\mathbf{H}_1 + \ldots + (-k^2)^n\mathbf{H}_n \\ \mathbf{G} &= \mathbf{G}_0 - k^2\mathbf{G}_1 + \ldots + (-k^2)^n\mathbf{G}_n\end{aligned} \qquad (13)$$

The matrices $\mathbf{H}_j$ and $\mathbf{G}_j$ are the corresponding integrals to the fundamental solutions q_j^* and u_j^* multiplied by the interpolation function on the boundary element, respectively. $\mathbf{H}$ and $\mathbf{G}$ are polynominals of k and, therefore, do not include k inside their respective terms $\mathbf{H}_j$ and $\mathbf{G}_j$.

Proper equations

Consider the Helmholtz differential equation and/or its discretized equatios (5) and (12) under the following boundary condition:

$$\begin{aligned}u &= 0 \qquad \text{on } \Gamma_u \\ q &= 0 \qquad \text{on } \Gamma_q \\ \Gamma &= \Gamma_u + \Gamma_q\end{aligned} \qquad (14)$$

The matrix equation (5) is written as

$$\begin{bmatrix}\mathbf{S}_u & \mathbf{S}_q\end{bmatrix}\begin{bmatrix}\mathbf{0}\\ \mathbf{u}_q\end{bmatrix} = \begin{bmatrix}\mathbf{T}_u & \mathbf{T}_q\end{bmatrix}\begin{bmatrix}\mathbf{q}_u\\ \mathbf{0}\end{bmatrix} \qquad (15)$$

where suffixes u and q stand for the values on the potential prescribed boundary Γ_u and on the flux prescribed boundary Γ_q, respectively. Rearranging terms on both sides of Eq.(15), one gets

$$\begin{bmatrix}-\mathbf{T}_u & \mathbf{S}_q\end{bmatrix}\begin{bmatrix}\mathbf{q}_u\\ \mathbf{u}_q\end{bmatrix} = \begin{bmatrix}-\mathbf{S}_u & \mathbf{T}_q\end{bmatrix}\begin{bmatrix}\mathbf{0}\\ \mathbf{0}\end{bmatrix} \qquad (16)$$

which is represented formally as

$$\mathbf{Au} = \mathbf{B0} \tag{17}$$

Consequently, the proper equation is

$$det\mathbf{A} = 0 \tag{18}$$

where

$$\mathbf{A} = \left[\begin{array}{cc} -\mathbf{T}_u & \mathbf{S}_q \end{array} \right] \tag{19}$$

Conventional method for solving transcendental Eq.(18) with respect to wavenumber k is based on construction of variation of $|det\mathbf{A}|$ to k, which is generally cumbersome and encounters serious difficulties, because elements of $\mathbf{A}$ is complex-valued and involving unknown k and $|det\mathbf{A}|$ has not easy-distinguishable zero-points as shown in later examples.

On the other hand, Eq.(12), for the MRM formulation, can be written corresponding to the boundary condition (15), as

$$\left[\begin{array}{cc} \mathbf{H}_u & \mathbf{H}_q \end{array} \right] \left[\begin{array}{c} \mathbf{0} \\ \mathbf{u}_q \end{array} \right] = \left[\begin{array}{cc} \mathbf{G}_u & \mathbf{G}_q \end{array} \right] \left[\begin{array}{c} \mathbf{q}_u \\ \mathbf{0} \end{array} \right] \tag{20}$$

and

$$\left[\begin{array}{cc} -\mathbf{G}_u & \mathbf{H}_q \end{array} \right] \left[\begin{array}{c} \mathbf{q}_u \\ \mathbf{u}_q \end{array} \right] = \left[\begin{array}{cc} -\mathbf{H}_u & \mathbf{G}_q \end{array} \right] \left[\begin{array}{c} \mathbf{0} \\ \mathbf{0} \end{array} \right] \tag{21}$$

Equation(21) is also written formally as Eq.(17) and,

$$det\mathbf{A} = 0 \tag{22}$$

where

$$\mathbf{A} = \left[\begin{array}{cc} -\mathbf{G}_u & \mathbf{H}_q \end{array} \right] \tag{23}$$

In this case, unlike the complex-valued formulation, the elements of $\mathbf{A}$, i.e., $\mathbf{G}_u$, $\mathbf{H}_q$, are polynominals in terms of k (k is included explicitly), which reduces computational task greatly. The variation of $det\mathbf{A}$ with respect to k can be still employed for determination of eigenvalues. It was pointed out, however, that Eq.(22) may not give the true proper equation in this real-valued formulation [5]. Instead of Eq.(22), the proper equation is given as

$$det\mathbf{A}/det\mathbf{R}(\equiv \eta) = 0 \tag{24}$$

$\mathbf{R}$ appearing in the denominator is defined by

$$\mathbf{R} \equiv \left[\begin{array}{c} \mathbf{a}_1 \ldots \mathbf{a}_{m-1}, \ -\mathbf{b}_m, \ \mathbf{a}_{m+1} \ldots \mathbf{a}_N \end{array} \right] \tag{25}$$

i.e., the mth column of $\mathbf{A}$ (Nth order matrix) is replaced by $-\mathbf{b}_m$, mth column of $\mathbf{B}$ which is the right-hand-side coefficent matrix in Eq.(21)

$$\mathbf{B} = \left[\begin{array}{cc} -\mathbf{H}_u & \mathbf{G}_q \end{array} \right] \tag{26}$$

m is taken arbitrarily between 1 and N. Equation (24) is the necessary and sufficent condition for the nontrivial eigenvector. The distribution of η with respect to k is the essential information obtaining rough approximation of the eigenvalue. Precise solution can be determined by successive approximation using the following Newton method starting with the above-mentioned rough estimation [6]. The Newton recursive relation from sth to $s+1$st iteration is

$$k_{s+1} = k_s - \frac{1}{tr(\mathbf{E}_a) - tr(\mathbf{E}_r)} \tag{27}$$

where

$$\begin{aligned} \mathbf{E}_a &= \mathbf{L}_a{}^{-1}\mathbf{P}_a \frac{d\mathbf{A}(k_s)}{dk}\mathbf{U}_a{}^{-1} \\ \mathbf{E}_r &= \mathbf{L}_r{}^{-1}\mathbf{P}_r \frac{d\mathbf{R}(k_s)}{dk}\mathbf{U}_r{}^{-1} \end{aligned} \tag{28}$$

$$\begin{aligned} \mathbf{P}_a\mathbf{A}(k_s) &= \mathbf{L}_a\mathbf{U}_a \\ \mathbf{P}_r\mathbf{R}(k_s) &= \mathbf{L}_r\mathbf{U}_r \end{aligned} \tag{29}$$

i.e., LU decomposition of $\mathbf{A}$ and $\mathbf{R}$ with the permutation matrix $\mathbf{P}$ for pivoting is only necessary computation for this scheme (suffixes a and r are for the matrices $\mathbf{A}$ and $\mathbf{R}$, respectively).

Numcrical illustrations

As an example for examination of the above-stated three different proper equations (18),(22) and (24), we here consider a rectangular region, longitudinal side $L_x = 1$, transverse side $L_y = 0.6$ as shown in Figure 1. The three types of boundary condition, the Dirichlet, the Neumann and their Mixture are thought, of which rigorous analytical solutions are obtainable.

For all these problems, the varations of $|det\mathbf{A}|$ (Eq.(19)), $det\mathbf{A}$ (Eq.(23)) and $det\mathbf{A}/det\mathbf{R}$ (Eqs.(23),(25)), denoted as $|det\mathbf{A}_1|$, $det\mathbf{A}_2$ and $det\mathbf{A}_2/det\mathbf{R}$ in Figures 2 ~ 4 (48 elements), with respect to k are first computed; these results are only one and the last informations for the method obeying first two formulations, while, for the last formulation, iteration by Eq.(27) follows.

In view of Figures 2 to 4, one can mention that the zero-points of $|det\mathbf{A}_1|$ are not clearly be indicated, because $\mathbf{A}_1$ is complex-valued and only its absolute magnitude of $det\mathbf{A}_1$ approches relatively null (not strictly). Computational approximation and error sometimes makes the situation wrong. It is difficult to distinguish eigenvalues extremely close to each other. The variation of $det\mathbf{A}_2$ seems different from that of $|det\mathbf{A}_1|$, but is not always convenient for detemination of the eigenvalues; difficuties arise for degenerate eigenvalues due to equal roots of $det\mathbf{A}_2 = 0$ as shown in, say, Figures 3 and 4. In addition, it must be pointed out that the zero-point of $det\mathbf{A}_2 = 0$ is not necessarily that of $det\mathbf{A}_2/det\mathbf{R}$ as clearly observable in Figures 3 and 4. Table 1 compares the analytical results with the zero-points of $det\mathbf{A}_2/det\mathbf{R}$. Comparison between two results $det\mathbf{A}_2/det\mathbf{R} = 0$ and $det\mathbf{A}_2 = 0$ proves that the necessary and sufficient condition for the proper equation is not $det\mathbf{A}_2 = 0$ but $det\mathbf{A}_2/det\mathbf{R} = 0$. The distribution $det\mathbf{A}_2/det\mathbf{R}$ can separate degenerate eigenvalues skilfully.

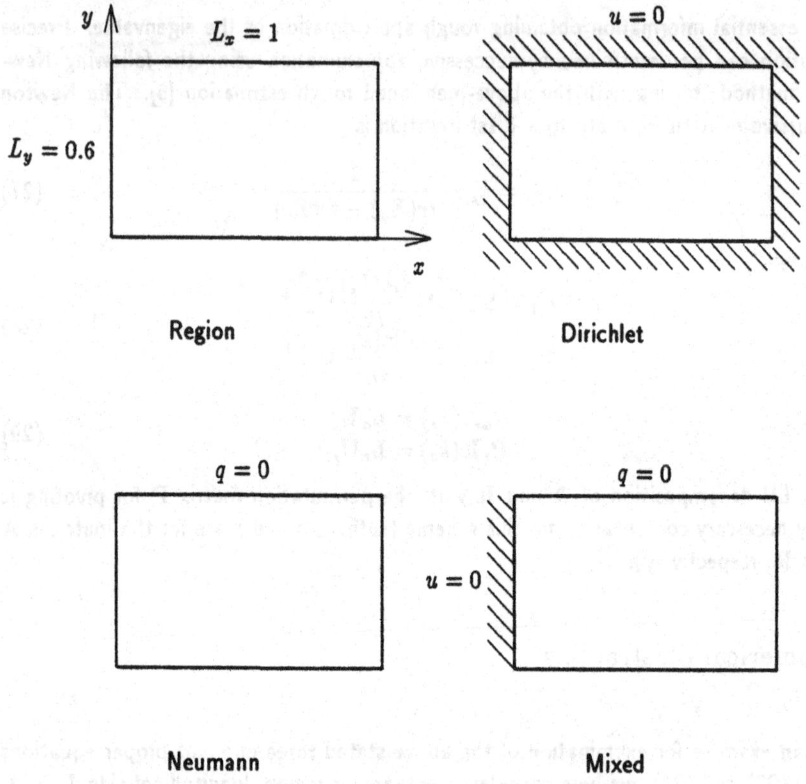

Figure 1: Coordinates and Boundary Conditions

Table 1: Analytical and Numerical solutions

Dirichlet		Neumann		Mixed	
Anal.Sol.	Present	Anal.Sol.	Present	Anal.Sol.	Present
6.10	6.10	3.14	3.14	1.57	1.57
		5.23	5.23	4.71	4.71
		6.10	6.11	5.46	5.46
		6.28	6.28	7.04	7.04
				7.85	7.85

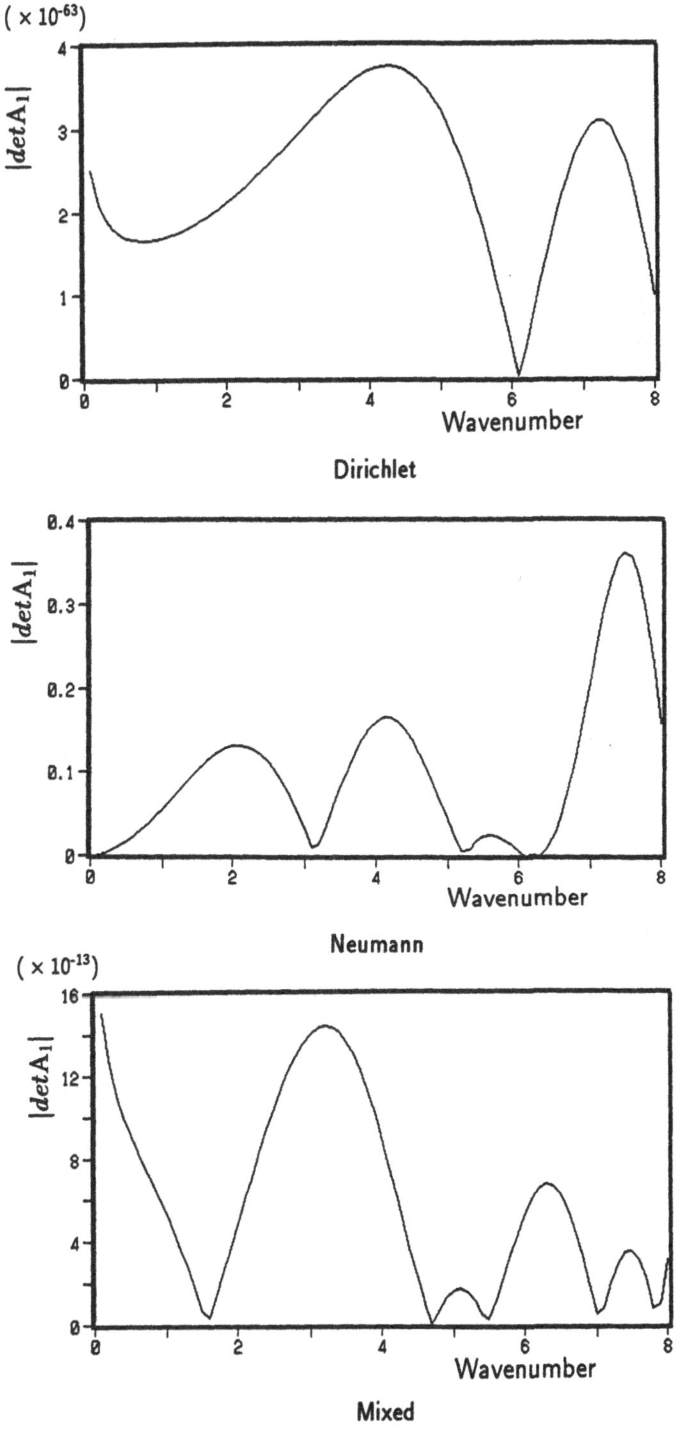

Figure 2: Variations of $|det\mathbf{A}_1|$

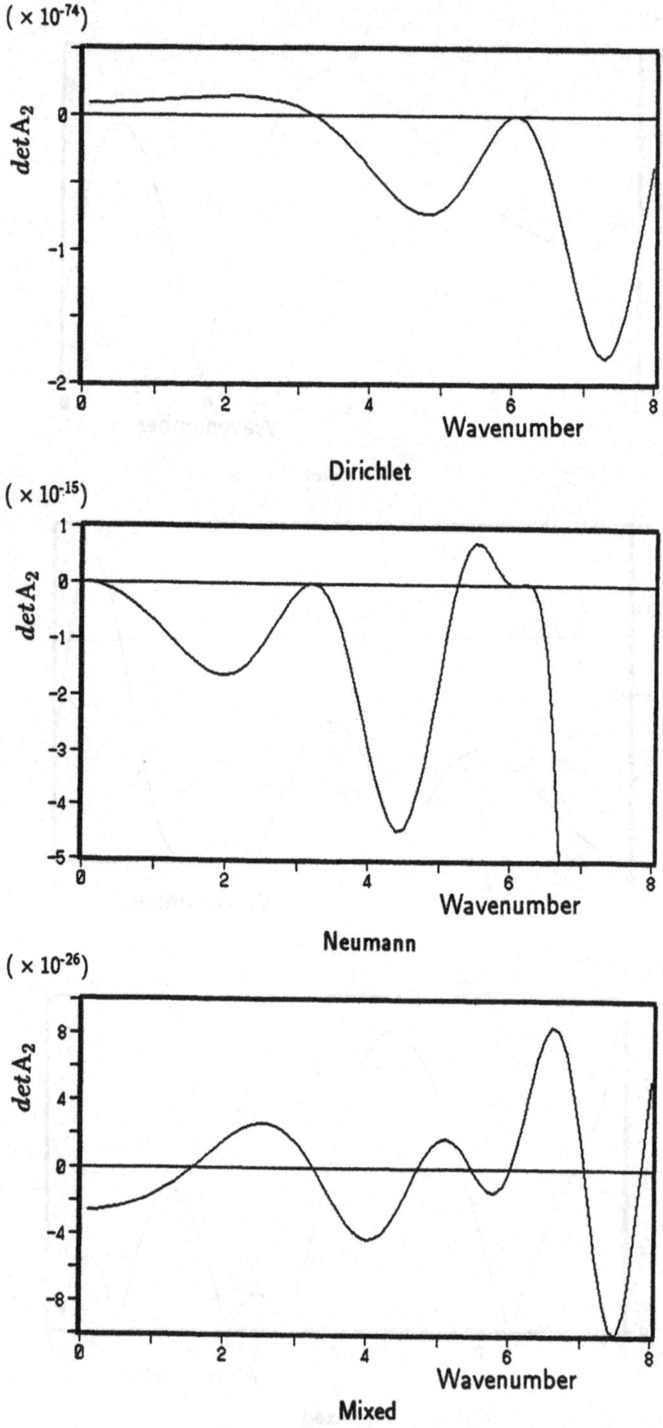

Figure 3: Variations of $det\mathbf{A}_2$

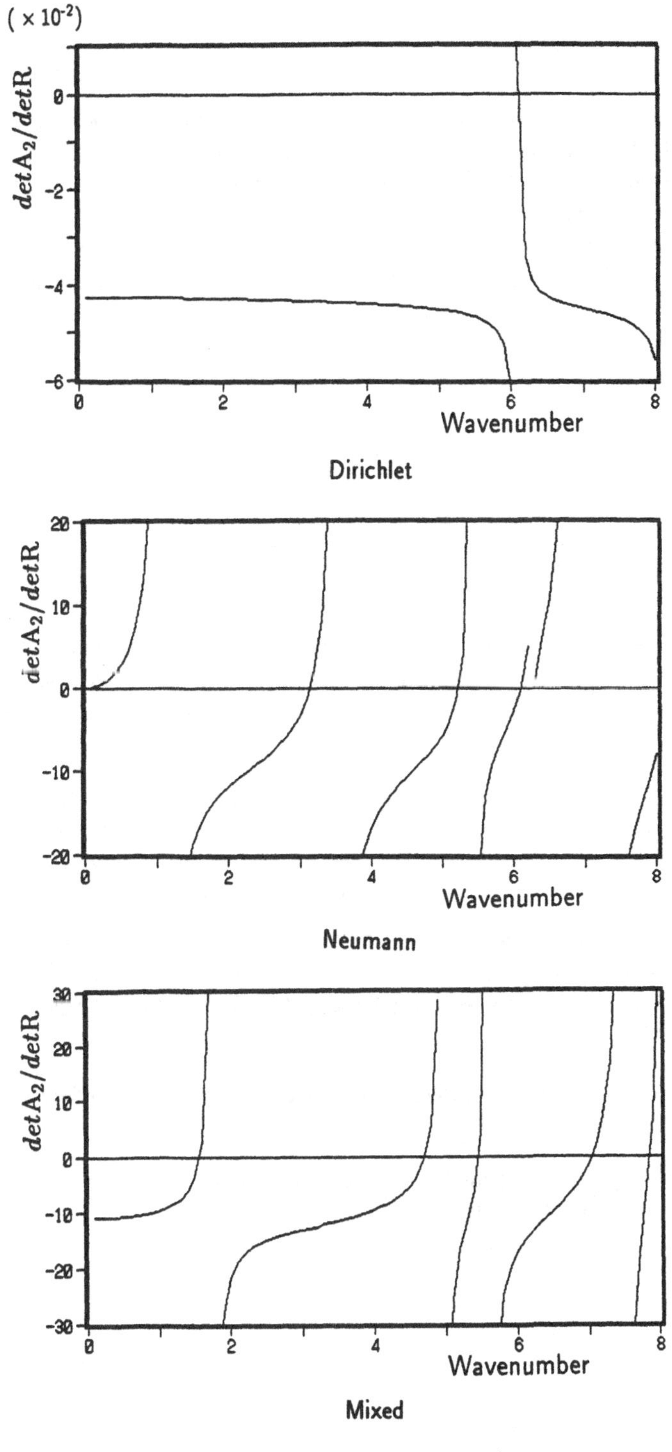

Figure 4: Variations of $det\mathbf{A}_2/det\mathbf{R}$

Conclusions

An efficent and robust eigenvalue analysis scheme for the Helmholtz equation based on the MRM boundary integral equation formulation was examined. The proper equation corresponding to the present real-valued formulation was validated by numerical examples. Precise solution could be obtained by Newton method applied to specifically but simply preconditioned (LU decomposition) matrices. Further study shuld be focused on development of efficient computation of the determination of polynominal matrix.

References

[1] Nowak, A.J., Temperature Fields in Domains with Heat Sources Using Boundary Only Formulations, Proc. 10th BEM Conference (Ed. Brebbia, C.A.), Springer - Verlag, Vol. 2, pp. 233 - 247, 1988.

[2] Brebbia, C.A., On Two Different Methods for Transforming Domain Integrals to the Boundary, Proc. 11th BEM Conference (Eds. Brebbia, C.A. and Connor, J.J.), Computational Mechanics Publications and Springer - Verlag, pp. 59 - 74, 1989.

[3] Nowak, A.J. and Brebbia, C.A., Solving Helmholtz Equation by Boundary Elements Using the Multiple Reciprocity Method, Computers and Experiments in Fluid Flow (Eds. Carlomagno, G.M. and Brebbia, C.A.), Computational Mechanics Pubications and Springer - Verlag, pp. 265 - 270, 1989.

[4] Niwa, U., Kobayashi, S. and Kitahara, M., Determination of Eigenvalues by Boundary Element Mehods, Chapter 7, Developments in Boundary Element Methods, (Eds. Banerjee, P.K. and Shaw, R.P.), Vol. 2, Applied Science Publishers, 1982.

[5] Andoh, E., and Kamiya, N., Helmholtz Eigenvalue Analysis by Bondary Element Method, (submitted).

[6] Yang, W.H., A Method for Eigenvalues of Sparse λ - Matrices, Int. J. Num. Meth. Eng., Vol. 19, pp. 943 - 948, 1983.

Appendix

Neglect of the domain integral appearing MRM boundary element formulation: Suppose that the domain Ω under consideration is bounded and, therefore, the distance between source and integration points is finite. Wavenumber k is also finite. The domain integral appeared in Eq.(10) is

$$I = (-1)^n (k^2)^{n+1} \int_\Omega u u_n^* d\Omega \tag{A1}$$

and its absolute value becomes

$$\begin{aligned} |I| &= |(-1)^n (k^2)^{n+1} \int_\Omega u u_n^* d\Omega| \\ &\leq (k^2)^{n+1} |u|_{max} |u_n^*|_{max} \int_\Omega d\Omega \end{aligned} \tag{A2}$$

where the fundamental solution is given as

$$u_n^* = -\frac{1}{2\pi} r^{2n} \frac{1}{4^n (n!)^2} (\ln r - s_n) \tag{A3}$$

$$s_n = \begin{cases} 0 & (n = 0) \\ \sum_{l=1}^{n} \frac{1}{l} & (n \geq 1) \end{cases} \tag{A4}$$

Now, we define the following quantity J;

$$J = (k^2)^{n+1} |u_n^*|_{max} \geq 0 \tag{A5}$$

For sufficiently large n,

$$|\ln r - s_n| = s_n - \ln r \tag{A6}$$

By virtue of Eqs.(A3) ~ (A6),

$$J = \frac{r^{2n} (k^2)^{n+1}}{2\pi 4^n (n!)^2} (s_n - \ln r) \tag{A7}$$

We examine the magnitude of several terms appearing in Eq.(A7). First,

$$\begin{aligned} s_n &= 1 + \frac{1}{2} + \frac{1}{3} + \ldots + \frac{1}{n} \\ &< \int_1^n \frac{dx}{x} + 1 = \ln n + 1 \end{aligned} \tag{A8}$$

Since the following inquality holds,

$$\begin{aligned} \ln n! &= \ln 1 + \ln 2 + \ln 3 + \ldots + \ln n \\ &> \int_1^n \ln x dx = \ln \frac{n^n e}{e^n} \end{aligned} \tag{A9}$$

we obtain

$$\frac{1}{n!} < \frac{1}{e} \left(\frac{e}{n} \right)^n \tag{A10}$$

Substituting Eqs.(A8) and (A10) into Eq.(A7), we have

$$\begin{aligned} J &< \frac{k^2}{2\pi e^2}\left\{\left(\frac{rke}{2}\right)^n \frac{1}{n^n}\right\}^2 (\ln n + 1 - \ln r) \\ &= \frac{r^2k^4}{8\pi}\left\{\frac{\left(\frac{rke}{2}\right)^{n-1}}{n^{n-1}}\right\}^2 \left(\frac{\ln n}{n^2} + \frac{1}{n^2} - \frac{\ln r}{n^2}\right) \\ &= \frac{r^2k^4}{8\pi}KL \end{aligned} \tag{A11}$$

where

$$C = \frac{1}{2}rke \tag{A12}$$

$$K = \left(\frac{C}{n}\right)^{n-1} \tag{A13}$$

$$L = \left(\frac{\ln n}{n^2} + \frac{1}{n^2} - \frac{\ln r}{n^2}\right) \tag{A14}$$

Considering the identity

$$\ln K = C\frac{\ln\left(\frac{C}{n}\right)}{\left(\frac{C}{n-1}\right)} \tag{A15}$$

and the following limit values as $n \to \infty$

$$\lim_{n\to\infty}\left(\frac{C}{n-1}\right) = 0 \tag{A16}$$

$$\lim_{n\to\infty}\ln\left(\frac{C}{n}\right) = -\infty \tag{A17}$$

we determine the limit of K as

$$\lim_{n\to\infty} K = 0 \tag{A18}$$

Besides, it is evident that

$$\lim_{n\to\infty} L = 0 \tag{A19}$$

Consequently, with the help of Eqs.(A18) and (A19), Eq.(A11) yields,

$$\lim_{n\to\infty} J = 0 \tag{A20}$$

and thus, finally

$$\lim_{n\to\infty} |I| = 0 \tag{A21}$$

QED.

Solution of Singular Potential Problems by the Modified Local Green's Function Method (MLGFM)

C.S. de Barcellos, R. Barbieri

Universidade Federal de Santa Catarina, Departamento de Engenharia Mecanica, CP 476, 88049 Florianopolis, SC, Brasil

Abstract

This paper illustrates the applicability of the MLGFM to solve singular potential problems. Two examples are solved for different meshes and the results, which generally show good agreement with analytical ones, are discussed. It is also shown that the convergence rate has a similar behavior than that of the Finite Element Method.

Introduction

During the last few decades, there has been an increasing interest in the solution of singular problems. The correct representation of the singularity is a rather complex numerical problem and among the most used methodologies are the quarter-points and the self-adaptive techniques for the Finite Element Method, FEM, and Boundary Element Method, BEM.

On the other hand, there exists the Modified Local Green's Function Method, MLGFM [1,2,3], which is related to the Galerkin Boundary Element Method [4] and uses the Finite Element Method, FEM, to solve auxiliary problems. Recently, the authors have reached great efficiency in solving singular problems using the MLGFM , even when conventional FEM elements and coarse meshes are used. The superconvergence property of the MLGFM can easy the adaptive procedures or may even, in some problems, be a reliable and advantageous alternative.

Here the MLGFM is used for solving Poisson's equations with singularities associated with the domain geometry and discontinuous boundary conditions.

Non-Regular Domains

As a matter of illustrating an elliptic boundary-value problem, consider one is interested in the function u(x) such that

$$-\nabla\cdot[a(x)\ \nabla\ u(x)] + b(x)\ u(x) = f(x) \qquad x \in \Omega \tag{1.1}$$

$$u(x) = 0 \qquad x \in \Gamma \tag{1.2}$$

where Ω is the domain with boundary Γ. The variational form of problem (1) consists in the look for a function $u(x) \in \mathcal{H}^1_0(\Omega)$ such that

$$\int_\Omega \{\ a(x)\nabla\ u(x)\cdot\nabla\ v(x) + b(x)\ u(x)v(x)\ \}\ d\Omega = \int_\Omega f(x)v(x)\ d\Omega \tag{2}$$

where $\mathcal{H}^k_0(\Omega)$ denotes the Hilbert space of order k defined as the completion under the norm

$$\|u(x)\|_{\mathcal{H}^m} = \left[\sum_{\alpha\le m} \|\ D^{|\alpha|}u(x)\ \|_{\mathcal{L}_2(\Omega)}\right]^{1/2} = \left[(u(x),u(x))_{\mathcal{H}^m}\right]^{1/2} \tag{3}$$

of the space of the k-times continuously differentiable functions with compact support in Ω. In the definition (3), one has :$\alpha = (\alpha_1,\alpha_2),\alpha_j \ge 0$, integers; $|\alpha| = \alpha_1 + \alpha_2$; $D^{|\alpha|}u(x) = \partial^{|\alpha|}u(x)/\partial x_1^{\alpha_1}\partial x_2^{\alpha_2}$ and $(\cdot,\cdot)_{\mathcal{H}^m}$ denotes the inner product defined on $\mathcal{H}^m(\Omega)$ [5].

If the set of data [a(x),b(x),f(x)] is sufficiently regular and the boundary is a closed almost everywhere regular curve, the solution of (1) and (2) are the same. In fact, if $f(x) \in \mathcal{H}^r(\Omega)$; $r \ge 0$ and if $a(x),b(x) \in \mathcal{C}^\infty(\bar{\Omega})$ such that $a(x) \ge a_0 > 0$; $b(x) > 0$, where $\bar{\Omega}$ is the closure of Ω and a_0 is a constant, then the solution u(x) has the derivative of order r+2 in $\mathcal{H}^0(\Omega)$, i.e.,

$$u(x) \in \mathcal{H}^{r+2}(\Omega) \cap \mathcal{H}^1_0(\Omega) \tag{4}$$

In order to discuss the effect of the singularities on Γ, one may consider, without loss of generality, the particular case where $a(x)=1$, $b(x)=0$ and $f(x) \in \mathcal{C}^{\infty}(\Omega)$, $\Omega \in \mathbb{R}^2$. In such a case, the equation (2) reduces to

$$\int_{\Omega} \nabla u(x) \cdot \nabla v(x) \, d\Omega = \int_{\Omega} f(x) v(x) \, d\Omega \qquad \forall v(x) \in \mathcal{H}_0^1(\Omega) \tag{5}$$

which is associated to the Poisson's equation

$$-\Delta u(x) = f(x) \qquad x \in \Omega \tag{6}$$

$$u(x) = 0 \qquad x \in \Gamma \tag{7}$$

where $\Delta = \partial^2/\partial x_1^2 + \partial^2/\partial x_2^2$.

Let a smooth boundary except at a vertex "P" where there is an internal angle α, Fig.(1). The loss of regularity of the solution $u(x)$ occurs due to its behaviour in the neighborhood of the point "P" for certain values of α.

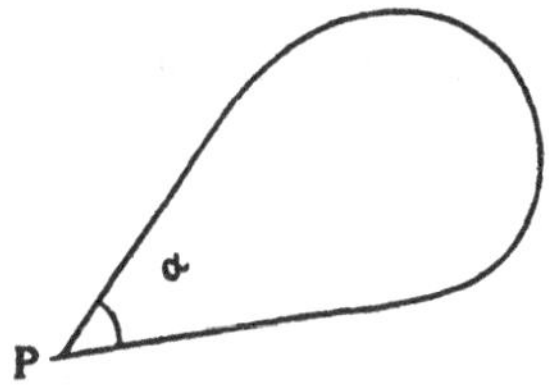

Fig.1 - Typical vertex.

The solution near the singularity can be written in the form

$$u(r,\theta) = \sum_{n=1}^{\infty} A_n \, r^{\mu_n} \, \Psi_n(\theta) + \text{" smooth terms "} \tag{7}$$

where $A_n\Psi_n$ are functions which depend upon the angular coordinate θ alone and are sufficiently well behaved, but r^{μ_n} may have a singular behavior, i.e, their derivatives may be unbounded when $r \to 0$. Moreover, if π/α is not an integer, a singular term of the series is

$$r^{\pi/\alpha} \sin(\ \pi\Theta/\alpha) \tag{8}$$

and an increase of the angle α causes a higher singularity on the function $u(r,\Theta)$. In fact, it can be shown [5] that the solution $u(x)$ of equation (5) is such that

$$u(x) \in \mathcal{H}^{1+\beta-\varepsilon}(\Omega) \tag{9}$$

where ε is an arbitrary small positive constant, β is equal to π/α_J and α_J is the greatest internal angle of all of the vertices defined by the discretization as illustrated in Fig.(2).

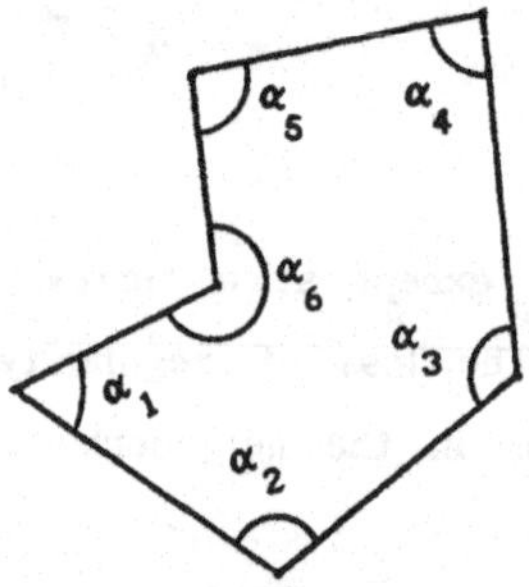

Fig.(2) Boundary Discretization and Internal Angles

Additional Irregularities

Other types of irregularities can occur in the solution of a Boundary Value Problem (BVP) even when the geometry is sufficiently regular. Some of these sources of irregularity are :

1-Singular Data. This is frequently found when the excitation is of the Dirac δ-function type, but the boundary condition $u\equiv g$ and/or $\partial u/\partial n\equiv h$ may also include jump type of singularities.

2-Mixed Boundary Conditions. If $u\equiv g$ is prescribed on Γ_1 and $\partial u/\partial n\equiv h$ is prescribed on Γ_2; $\Gamma = \Gamma_1 \cup \Gamma_2$, the solution $u(x)$ may have singularity at the junction points $\bar{\Gamma}_1 \cap \bar{\Gamma}_2$.

3-Discontinuous Coefficients. If $a(x)$ and/or $b(x)$ is discontinuous in domain points or are zero over another points, the solution may present singularities on these points.

For such singularities, each point must be analysed separately for identification of the solution regularity, i.e., to determine which is Hilbert space of maximum order which contains the solution.

Numerical Results

An adequate mathematical approach for the MLGFM is found in [1], while the formalism can be found in [2] and [3] a companion paper in this conference. So, the MLGFM formalism will not be reproduced here and the solution of two examples are shown to illustrate its accuracy.

Problem 1. Find $u(x) \in \mathcal{H}^{3/2-\varepsilon}(\Omega)$ such that

$$\Delta\, u(r,\theta) = 0 \qquad r,\theta \in \Omega$$

$$u(r,\theta) = r^{1/2}\cos(\theta/2) \qquad r,\theta \in \Gamma_1$$

$$\partial u(r,\theta)/\partial\theta = 0 \qquad r,\theta \in \Gamma_2$$

where $\Omega = \{ (x_1,x_2) \in \mathbb{R}^2,\ -1 \le x_1 \le 1\ ,\ 0 \le x_2 \le 1\}$, Γ_1 and Γ_2 are shown in Fig.(3). The exact solution is

$$u(r,\theta) = r^{1/2}\cos(\theta/2)$$

Here the domain is not partitioned as far as MLGFM is concerned. The FEM and BEM meshes for the Green's matrices aproximations are shown in Fig.(3). The elements adopted here were the lagrangian quadratic finite element and the continuous quadratic boundary line element. Note that near the singularity , four elements in a geometrical progression equal 0.50 were used. Double nodes are used where discontinuities of boundary conditions were present.

The curves of iso-potential and equal error in the potential are shown in Fig.(4). One may notice that in most of the domain the error is less then 0.10% and that even near the singularity the error is between 0.10% and 1.00% and,so, the energy error is very small. The energy norm obtained by adaptive finite element [6] is about 0.8869 which compares well with the 0.8820 obtained by MLGFM. According to the ℓ_∞ norm, the maximum error in the potential is 5.135D-3 . As far as fluxes are concerned, one has on the singular node a value about

10.0D4 and out of the singular region the errors are of the same order as those found for the potential.

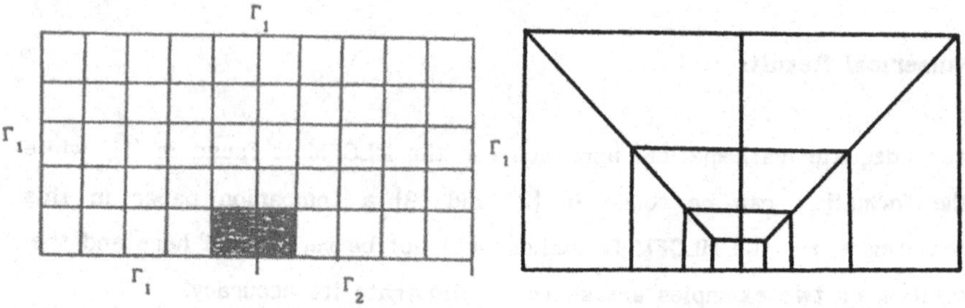

(a) (b)

Fig.3. Domain and Boundary Definition Including the FEM Mesh (a) and Detail of Mesh Near the Singular Point (b).

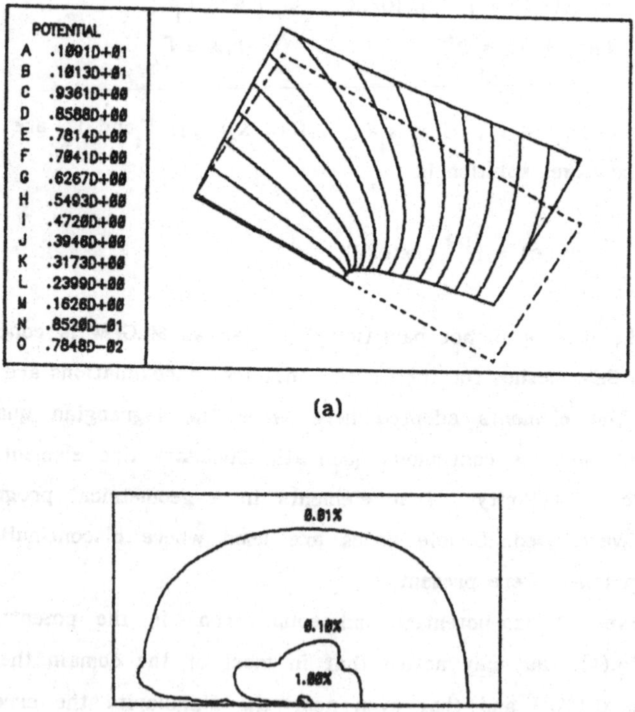

(a)

(b)

Fig.4. Isopotential Lines (a) and Error in the Potential Function (b).

Following the works of Szabo [7,8] and Babuska & Guo [9] the mesh for the Green's function projection was redrawn in such a way that the elements near the singular point have their side lengths varying according a geometric progression with ratio equal to 0.25 and seven cubic lagrangian elements were used instead of four.This new coarse mesh also showed to be efficient for this type of singularity and the potential and flux results near the singular point are shown in Figures (5),(6) and (7).

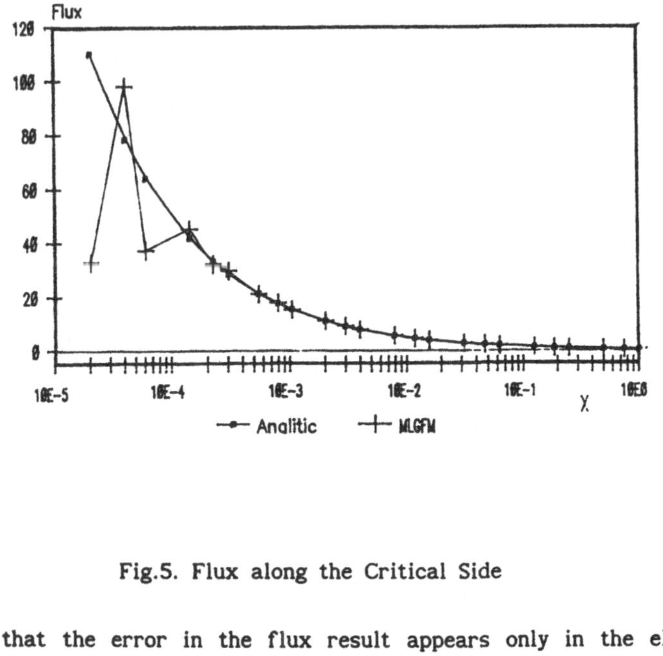

Fig.5. Flux along the Critical Side

Note that the error in the flux result appears only in the element which is nearest to the singular point and that the MLGFM solution presents the same type of oscillation around the analytical solution as those found in other applications using the BEM to study this type of singularity,[10] and [11].

When treating the h and/or p BEM adaptive techniques, some authors have used as a convergence criteria the integral of the flux along the boundary. Such integral is of the order of

$$\int_{\Gamma} r^{-1/2}\, dr = 2\, r^{1/2}$$

and, near the singularity, this value behaves like ε^2 which leads to convergence even when satisfactory results are only found away from the critical region.

As far as potential is concerned, the Figures (6) and (7) show the percentage error over the elements near the singularity and along the critical side.Since the geometric progression ratio is 0.25, the side length of these elements are of the order of $1/4^6$ and $1/4^7$, which indicate very satisfactory results with maximum error of about 4.4%.

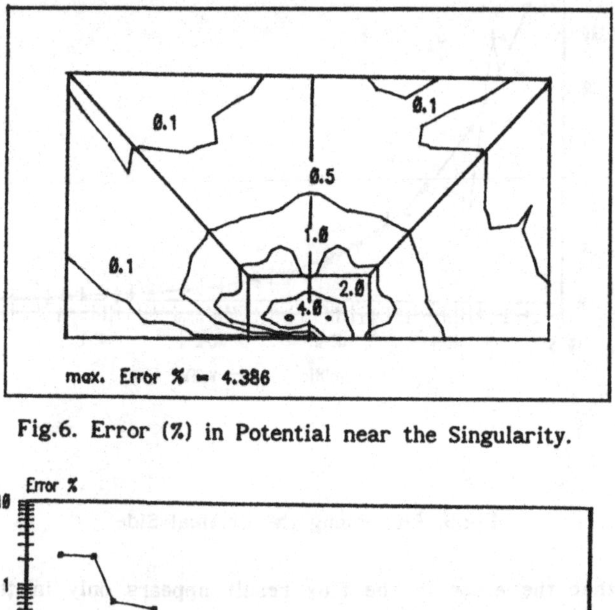

Fig.6. Error (%) in Potential near the Singularity.

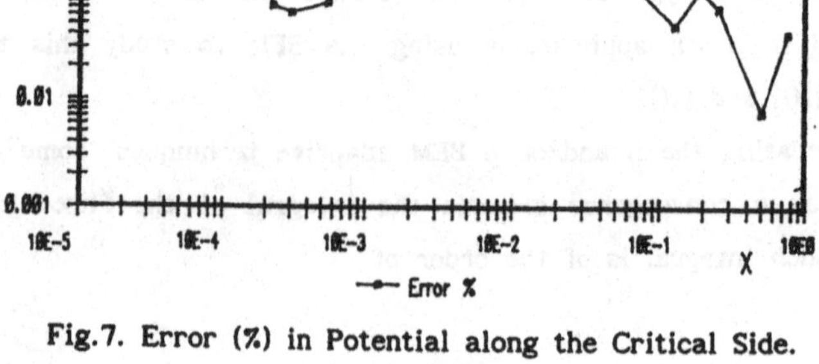

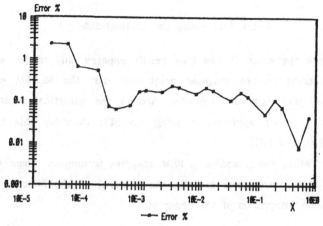

Fig.7. Error (%) in Potential along the Critical Side.

Problem 2. Find $u(x) \in \mathcal{H}^{5/3-\varepsilon}(\Omega)$ such that

$$-\Delta\, u(x) = 1 \qquad x \in \Omega$$

$$u(x) = 0 \qquad x \in \Gamma$$

where Ω is the L-shaped domain drawn in Figure (8).

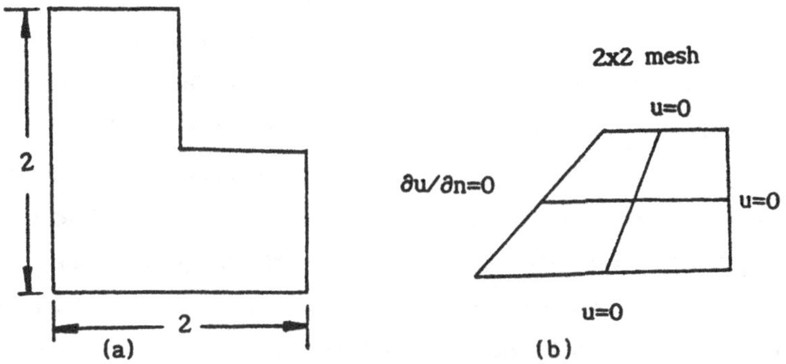

Fig.8. L-shaped Domain (a) and the Coarse Mesh Layout (b).

Here, due to the symmetry, only one half of the domain is partioned for the auxiliary FEM approximations. Along the line of symmetry, the flux is set equal to zero and double nodes are used wherever there are boundary condition discontinuities.

This problem was also solved by using p-adaptive Finite Element Method when the energy estimate 0.10703473 was obtained and which is now taken as a reference for the Figure (9).In this figure,h stands for the maximum external diameter among the mesh elements.

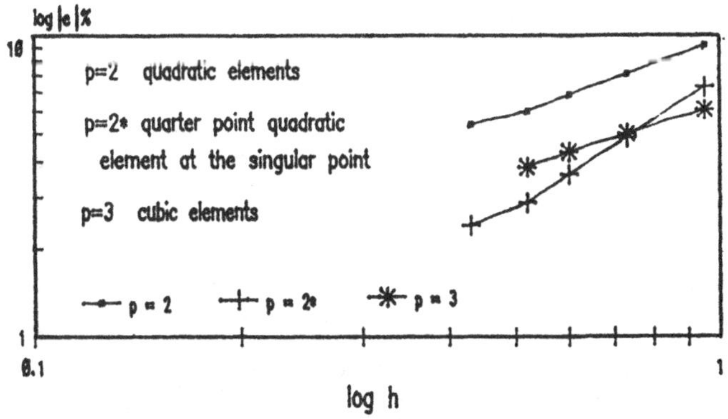

Fig.9. Error (%) in Energy Norm versus h.

The error norm is defined as

$$\|e\|\% = (|U_E - U_{MLGFM}|/U_E)^{1/2} * 100$$

where U_E is the exact energy value, which is here taken as the reference value mentioned above, and U_{MLGFM} is the MLGFM approximation.

Following again the works of Szabo [7,8] and Babuska & Guo [9] with a 6x6 mesh, the nearest element to the singular point is divided again according to a geometric progression with ratio equal to 0.20 and four elements per side. Then the results, in terms of energy,are found to be of the order of 0.106944661 ($\|e\|\%$=2.89%) and 0.10698829 ($\|e\|\%$=2,08%) for conventional quadratic elements and for a quarter point quadratic element at the singular point, respectively. These results are in agreement with the tendencies shown in Figure (9). Moreover, it can be noticed from this figure that the convergence is linear as it happens with the FEM .It must be noted that the 2.89% error in energy was attained by using uniform quadratic elements while the p-adaptive finite element estimate was found according the Szabo extrapolation procedures with interpolation functions of 6th, 7th and 8th order for obtaining the energy error of 2.59% using similar mesh.

Finally, the results for the fluxes along the critical side are illustrated in Figure (10) for the 6x6 regular mesh as well as the 6x6 modified mesh with a quarter point finite/boundary element at the singular point.

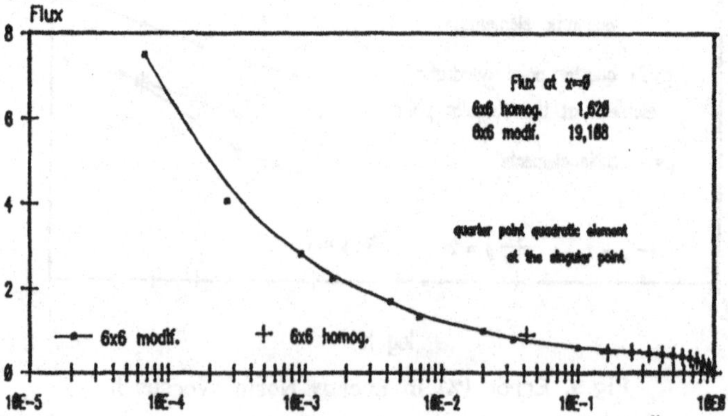

Fig.10. Flux at Critical Side

Conclusion

It's shown in Problem 1 that the MLGFM solutions presents the same behavior as the DBEM ones with flux oscillations near the singularity. Considering the Problem 2, it can be verified that the rate of convergence h follows that of the FEM and that it is possible to treat singularities even with lower order elements. Possibly, a way to pursue better results consists in the use of mixed formulations where different interpolation functions for the potential and for the boundary flux are selected.

References

1. Silva, L.H.M. Novas Formulações Integrais para Problemas da Mecanica, Ph.D. Thesis (in portuguese), Universidade Federal de Santa Catarina, 1988.

2. Barcellos, C.S. and Silva, L.H.M. Elastic Membrane Solution by a Modified Local Green's Function Method, (Ed. Brebbia, C.A. and Venturini, W.S.) Proc. Int. Conf. on Boundary Element Technology,1987, Comp. Mech. Publ., Southampton, 1987.

3. Barbieri, R. and Barcellos, C.S. Non Homogeneous Field Potential Problems Solution by the Modified Local Green's Function Method (MLGFM), 13th BEM Conf.,(Ed. Brebbia, C.A.),1991.

4. Barbieri,R. and Barcellos, C.S. A Modified Local Green's Function Technique for the Mindlin's Plate Model, 13th BEM Conf.,(Ed. Brebbia, C.A.),1991.

5. Carey, F.G. and Oden, J.T. Finite Elements : A Second Course, Vol.II Prentice-Hall,Inc., Englewood Cliffs, New Jersey,1983.

6. Kelly, D.W.; Gago, J.P.S.R. and Zienkiewicz, O.C. A Posteriori Error Analysis and Adaptive Processes in the Finite Element Method : Part I-Error Analysis,Int. J. Num. Meth. Eng., Vol.19,pp. 1593-1619,1983.

7. Szabo, B.A. Estimation and Control of Error Based on p Convergence, Chapter 3,Accuracy Estimates and Adaptive Refinements in Finite Element Computations, Ed. Babuska,I.; Zienkiewicz,O.C.; Gago,J. and Oliveira,E.R.A. ,1986, John Wyley & Sons Ltda.

8. Szabo, B.A. Mesh Design for the p-Version of the Finite Element Method , Comp. Meth. in Applied Mechanics and Engineering, Vol.55, pp.181-197, 1986.

9. Babuska,I. and Guo,B.Q. The h-p Version of the Finite Element Method for Domains with Curved Boundaries, SIAM J. Num. Anal., Vol.25, pp.837-861,1988.

10. Alarcón,E. and Reverter,A. P-Adaptive Boundary Elements, Int. J. Num. Meth. Eng., Vol.23, pp.801-829,1986.

11. Costa Jr.,J.A. Formulação h-Auto-Adaptiva de Elementos de Contorno para Problemas de Potencial, 3rd. Brazilian Thermal Science Meeting, (Ed. Maliska, C.A.; Melo,C. and Prata, A.T.), pp.421-428, 1990.

Boundary Element Analysis of Reinforced Concrete Members

M. Ameen, B.K. Raghu Prasad

Civil Engineering Department, Indian Institute of Science, Bangalore 560 012, India

ABSTRACT

A boundary integral equation solution of the linear elastic behaviour of reinforced concrete members under static loading is presented. A linear interpolation model representing the axial deformation of the steel reinforcement is employed. Bond between steel and concrete is not modeled. Potentials of the model developed are demonstrated using a singly reinforced R.C.C. beam. The results are seen to be in good agreement with the results obtained using standard finite element method.

1. Introduction

Analysis of reinforced concrete members using finite element method has been in vogue for the last two decades. Now it is of interest to try boundary element method of analysis for reinforced concrete structures [1] due to its advantages of reduction in dimensionality and requirement of lesser computer memory and time. Boundary element method does not require any elaborate description due to the availability of a good many number of text books, e.g. [2, 3]. The use of BEM for analysis of RCC structures, to the best of the knowledge of the authors, was first attempted by Gopal Rao et al [1]. Though the model had certain drawbacks it certainly did show the feasibility of the use of BEM for the analysis of RCC components. The major limitations of the model were (i) the steel had to be modeled using smeared steel elements which required discretization of the domain and thus resulted in domain integration (ii) the model depended on an iterative scheme even for an elastic analysis (iii) the model failed to give good results for certain percentages of steel.

The present work eliminates the disadvantages of the above model by using one dimensional linear interpolation model to represent steel. The elastic analysis is not iterative and makes use of a special form* of the Somigliana's identity [2, 3]. A 2-D analysis only is attempted though the extension to 3-D is quite straightforward. Inelastic analysis of RCC members using the present developed model is discussed.

Elastic Analysis of RCC Members

Consider a reinforced concrete member as shown in Fig. 1(a). The reinforcement is assumed to have only axial stiffness with zero flexural stiffness. Due to this assumption, under the action of loads, the bar will be subjected to only axial force. The problem can be seen to be a superposition of two problems as shown in Figs. 1(b) and (c). In the present analysis the problem corresponding to Fig. 1(b) is solved using boundary integral equation method and that in Fig. 1(c) is solved using FEM. The two solutions are combined using the equilibrium and compatibility requirements along AB.

For the problem shown in Fig. 1(b), the displacements $u_i(\xi)$ at any internal points 'ξ' can be written, using the Somigliana's identity, as

$$u_i(\xi) = \int_\Gamma [p_j(x)u^*_{ij}(\xi,x) - p^*_{ij}(\xi,x)u_j(x)]d\Gamma(x) + \int_{\Gamma_1} q_j(x)\, u^*_{ij}(\xi,x)\, d\Gamma_1(x) \qquad (1)$$

By bringing the field point onto a point on the boundary Γ of the domain, we obtain,

$$c_{ij}(\xi)u_j(\xi) = \int_\Gamma [p_j(x)u^*_{ij}(\xi,x) - p^*_{ij}(\xi,x)u_j(x)]d\Gamma(x) - \int_{\Gamma_1} q_j(x)\, u^*_{ij}(\xi,x)\, d\Gamma_1(x) \qquad (2)$$

where c_{ij}, u^*_{ij}, p^*_{ij} etc. are as defined and used in ref. [3].

Now the boundary Γ is discretized into N elements using a convenient interpolation model and the line AB into M linear elements. By taking the field point successively to all the nodal points on Γ and Γ_1 we arrive at a system of simultaneous linear algebraic equations which in matrix notation can be writtren as

$$[H_b\ H_i] \begin{Bmatrix} u_b \\ u_i \end{Bmatrix} = [G_b\ G_i] \begin{Bmatrix} p_b \\ p_i \end{Bmatrix} \qquad (3)$$

where the suffixes b and i denote the boundary nodes and nodes along the line AB respectively. For the problem shown in Fig. 1(c), a finite element analysis is carried out to obtain

$$[K]\ \{u^s_i\} = \{p^s_i\} \qquad (4)$$

where u_i^s, p_i^s are the displacements and tractions respectively at steel nodes. The conditions of compatibility and equilibrium at nodes along Γ_1 leads to

$$\{u_i\} = \{u_i^s\} \quad \text{and} \quad \{p_i\}+\{p_i^s\} = \{0\} \tag{5}$$

Premultiplying equation (4) by $[G_i]$ and adding to (3), we get,

$$[H_b \quad H_i + G_iK] \begin{Bmatrix} u_b \\ u_i(=u_i^s) \end{Bmatrix} = [G_b \; G_i] \begin{Bmatrix} p_b \\ p_i+p_i^s(=o) \end{Bmatrix} \tag{6}$$

The above system of equations can readily be solved to obtain the unknown displacements and tractions over the boundary as well as along the line Γ_1. Then equation (1) can be used to determine the displacement at any interior point. Consequently, the stress state at this point can be obtained by invoking the strain displacement relation, $\varepsilon_{ij} = 1/2\ (u_{i,j} + u_{j,i})$ and then the Hooke's law, $\sigma_{ij} = 2\ \mu\ \varepsilon_{ij} + \lambda \varepsilon_{kk}\ \delta_{ij}$, where μ and λ are Lame's constants. Thus,

$$\sigma_{ij}(\xi) = \int_{\Gamma} [u^*_{ijk}(\xi,x)p_k(x) - p^*_{ijk}(\xi,x)u_k(x)]d\Gamma(x)$$

$$+ \int_{\Gamma_1} q_k(x)\, u^*_{ijk}(\xi,x)\, d\Gamma_1(x)$$

where u^*_{ijk}, p^*_{ijk} are as defined in ref. [2].

Discussion and scope for future work

The formulation presented above is simple and straight-forward. The model could be used to analyse reinforced concrete members in the inelastic range also. One of the many constitutive relationships available for concrete, see e.g. [4,5,6], could be used along with the regular inelastic boundary element technique [1, 7, 8]. The accuracy of the model could be improved by considering the bond slippage by the use of proper bond link elements, see e.g. [9, 10]. The stiffness effects of these bond link elements could be accommodated in equation (4). The authors are, in fact, currently involved in extending the model to problems on applications of fracture mechanics to reinforced concrete beams.

Numerical results

A computer program RCBEM is written to carry out the above procedure of analysis of RCC member. An example problem is solved using this program. The results are compared with those obtained by standard FEM package SAPIV.

Example

An RCC beam with tension reinforcement alone having the geometry and properties of materials as shown in Fig. 2(a) is analysed using RCBEM. The BEM discretization is shown in Fig. 2(b). The same problem is solved using the SAP IV package with the FEM discretization as shown in Fig. 2(c). The results are presented in Table 1.

Conclusions

The boundary element modeling of steel reinforcement in RCC members under static loading is considered. Thus the versatility of BEM in obtaining numerical solution of complex problems of composites such as RCC is highlighted. Construction of solution beyond the linear elastic range using the above developed modeling is indicated to be straightforward. Improved solutions with more accuracy can be obtained by boundary element modeling of the bond developed over the contact area between reinforcing steel and concrete. Results are compared with those obtained using the more well established finite element packages and it can be observed that the results are in good agreement with the latter. Effective means of using the developed method to more complex problems are briefed.

Acknowledgements

The authors wish to express their thanks to Mr A R Gopalakrishnan, Senior Scientific Officer, Department of Civil Engineering, Indian Institute of Science, Bangalore, for providing the numerical results using SAP IV.

References

1. Gopala Rao, N., Channakeshava, C., Iyengar, K.T.S. and Raghu Prasad, B.K., "Elastic and Inelastic Analysis of Reinforced Concrete Structures using the Boundary Element Method", Engg. Analysis, Vol. 5, No. 3, 1988.

2. Brebbia, C.A., Telles, J.C.F. and Wrobel, L.C., "Boundary Element Techniques - Theory and Application in Engineering", Springer-Verlag, Berlin and New York, 1984.

3. Banerjee, P.K. and Butterfield, R., "Boundary Element Methods in Engineering Science", Mc-Graw Hill, 1981.

4. Bazant, Z.P. and Kim, S.S., "Plastic-Fracturing Theory for Concrete", Jl. Engg. Mechanics, Vol. 105, No. EM3, June 1979, pp. 407-428.

5. Bazant, Z.P. and Bhat, P., "Endochronic Theory of Inelasticity and Failure of Concrete", Jl. Engg. Mechanics, Vol. 102, No. EM4, Apr. 1976, pp. 701-722.

6. Han, D.J. and Chen, W.F., "Strain-Space Plasticity Formulation for Hardening-Softening Materials with Elasto-plastic Coupling", Int. Jl. Solids and Strucs., Vol. 22, No. 8, 1986, pp. 935-950.

7. Telles, J.C.F., "The Boundary Element Method Applied to Inelastic Problems", Lecture Notes in Engg., Vol. 1 (Brebbia and Orszerg eds.) Springer-Verlag, Berlin, 1983.

8. Henry Jr., D.P. and Banerjee, P.K., "A New BEM formulation for Two and Three Dimensional Elastoplasticity using Particular Integrals", Int. Jl. Numerical Methods in Engg., vol. 26, 1988, pp. 2079-2096.

9. Ngo, D. and Scordelis, A.C. "Finite Element Analysis of Reinforced Concrete Beams", ACI Journal, Mar. 1967, pp. 152-163.

10. Nam, C.H. and Salmon, C.G., "Finite Element Analysis of Concrete Beams", Jl. Structural Division, Vol. 10, No. ST12, Dec. 1974, pp. 2419-2431.

	RCBEM		SAP IV
	48 elements	110 elements	48 elements
deflection at midspan (mm)	0.5754	0.6061	0.6061
stress at bottom fibre (MPa) at midspan	167.25	176.05	171.30
stress at top fibre (MPa) at midspan	175.12	184.30	179.58

TABLE 1

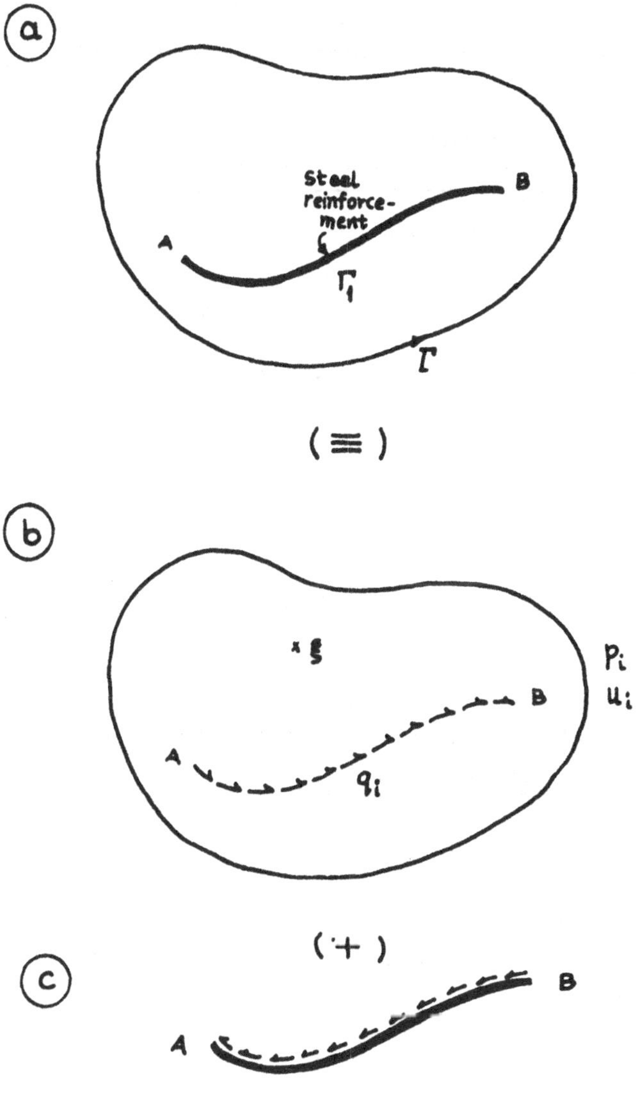

Fig. 1 Concrete member with the steel reinforcement.

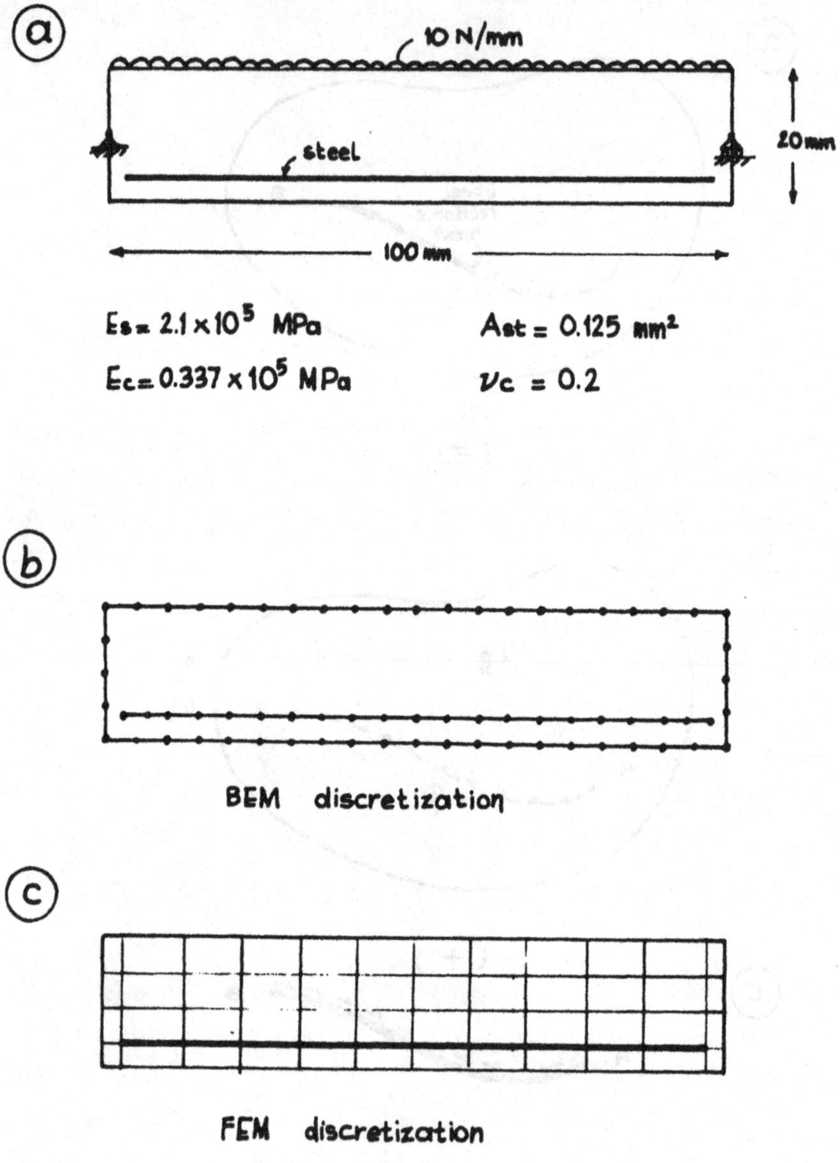

Fig.2. Example Problem

SECTION 18: NUMERICAL ASPECTS

Consistent Regularization of Both Kernels in Hypersingular Integral Equations

T.J. Rudolphi, K.H. Muci-Küchler

Department of Aerospace Engineering and Engineering Mechanics, Iowa State University, Ames, IA 50011, U.S.A.

ABSTRACT

A common method to regularize the stronger kernel that appears in derivatives of boundary integral equations, or hypersingular equations, is through subtraction of the ascending terms of the Taylor series expansion of the density function in the vicinity of the singular point. This isolates the kernel in an integral with unit density, which can generally be integrated analytically, or equivalently reduced in singularity order by the use of known solutions to the equation. In this process, terms arising from the analytically integrated term can be associated with the "flux" density and provide a regularization for the weaker kernel as well. This kind of formal development is, however, not well suited for computational purposes, since the added back flux term must be determined through derivatives of the primary variable on the boundary.

To avoid this mixing of the boundary variables we use only the tangent derivative part of the Taylor expansion to regularize the strong kernel and perform a separate regularization of the weaker kernel. This process relies on assumptions about the geometry and flux variable consistent with those required to regularize the stronger singularity. Formulations are given for two types of hypersingular equations of elasticity.

INTRODUCTION

Hypersingular integral equations have been found to be useful in many applications. Most notably, these types of equations have been used in problems involving thin, crack-like boundaries such as for fluid flow about thin airfoils [1], crack problems of elasticity [2], scattering [3], electroplating [4] and suppression of the so-called fictitious eigenfrequencies of exterior acoustics [5]. The need for these types of equations has been motivated by the various applications and the numerical treatment has been, by and large, ad hoc.

A systematic treatment of hypersingular boundary integral equations, both analytically and numerically, for scattering problems was produced recently by Krishnasamy, et al. [6]. Here the logical extension of the work of Brandao [1] was developed for acoustics and elastodynamics in the frequency domain. The basic idea therein is to remove the strong singularity of the kernel by subtraction of a sufficient number of terms of the Taylor series expansion of the density function about the singular point to regularize the kernel. The added back terms are then treated analytically by Stokes theorem or exact integration to reduce the strongly singular integrals to weakly singular, improper ones. This analytical work is all performed prior to the limit of actually forming an integral equation so contending with Hadamard finite part or Cauchy principal valued integrals is avoided.

For the case of closed domains, both interior or exterior, the strongly singular integrals of the added back terms of the regularization can be expressed in terms of weaker ones through the use of certain integral identities pertaining to the fundamental solution and its derivatives. These identities can be obtained from elementary boundary value problems [7] or directly from the governing differential equation [8]. By this approach or by integration via Stokes theorem for open regions, the integrated and most strongly singular added back term results in a singularity comparable to the weaker kernel of the representation and, by direct association of these terms, the weaker kernel is also regularized as the limit is taken to derive the integral equation.

The associated density term difference which regularizes the weaker kernel is, however, composed of the flux and derivatives of the primary variable so, although analytically regularized, is not suited for numerical treatment since it mixes the two independent boundary variable types. An alternate treatment of these terms is proposed in this paper which leads to a completely regularized integral equation readily suited for numerical approximation. We consider only the case of elastostatics, but other physical problems amenable to boundary integral formulations can be similarly treated, including frequency domain elastodynamics and acoustics.

Elastostatics

As is typical, a boundary integral equation is obtained from an integral representation. We start then with the displacement representation at a point ξ in the closed domain bounded by S,

$$u_i(\xi) = \int_S \Big[U_{ji}(\xi,x)t_j(x) - T_{ji}(\xi,x)u_j(x) \Big] dS(x) \qquad (1)$$

where U_{ij} denotes the components of the fundamental solution and T_{ij} the components of the traction vector due to U_{ij}. For ease in the formal development, we use the general anisotropic relations

$$t_i = E_{ijkl} n_j u_{k,l}$$
$$T_{im} = E_{ijkl} n_j U_{km,l} \qquad (2)$$

to relate tractions and displacements, where E_{ijkl} are the elastic constants and n_j are unit vector components normal to a plane of interest.

To establish the various integral identities pertaining to U_{ij} and T_{ij} and their derivatives which will be useful in the subsequent regularizations, we require that the representation (1) satisfy the general linear displacement field,

$$u_i(x) = a_i + b_i r_j e_j \tag{3}$$

where $r_i = x_i - \xi_i$, e_j are components of an arbitrary unit vector and a_i and b_i are arbitrary constants. The tractions due to this field on a plane with normal n_i components are

$$t_i(x) = E_{ijkl} n_j b_k e_l \tag{4}$$

so when this linear field is substituted into the representation (1), the relations

$$\int_S T_{ij} dS + \delta_{ij} = 0 \tag{5}$$

and

$$\int_S T_{li} r_m dS = \int_S E_{jklm} n_k U_{ji} dS \tag{6}$$

result when a_i and b_i are arbitrary. The first of these, equation (5), is recognized as that identity resulting from the rigid motion argument and often used to regularize the usual boundary integral equation.

The derivatives of identities (5) and (6) with respect to the coordinates at ξ, i.e., ξ_i, produce two other relations useful in the regularization of the hypersingular equation. They are

$$\int_S \frac{\partial T_{ji}}{\partial \xi_k} dS = 0 \tag{7}$$

and

$$\int_S \frac{\partial T_{mi}}{\partial \xi_k} r_n \, dS + \delta_{im}\delta_{kn} = \int_S E_{jlmn} n_l \frac{\partial U_{ji}}{\partial \xi_k} dS \qquad (8)$$

where, in the establishment of the last identity, equation (5) has been used to produce the $\delta_{im}\delta_{kn}$ term. The above identities - (5), (6), (7) and (8) - were established for the potential problem in [7] and also in [8] starting directly from the differential equation defining U_{ij}. They are now used to derive a completely regular form of derivative boundary integral equations.

The derivative of the representation (1) with respect to ξ_k is,

$$\frac{\partial u_i}{\partial \xi_k}(\xi) + \int_S \frac{\partial T_{ji}}{\partial \xi_k}(\xi,x) u_j(x) dS = \int_S \frac{\partial U_{ji}}{\partial \xi_k}(\xi,x) t_j(x) dS \qquad (9)$$

As the kernel $\partial T_{ji}/\partial \xi_k$ of the integral on the left contains the stronger, $O(1/r^2)$ singularity, we first remove it by subtraction of the first two terms of a Taylor's expansion of u_j at ξ. Thus,

$$\int_S \frac{\partial T_{ji}}{\partial \xi_k} u_j dS = \int_S \frac{\partial T_{ji}}{\partial \xi_k} \left[u_j(x) - u_j(\xi) - \frac{\partial u_j}{\partial \zeta}(\xi) r_l \zeta_l \right] dS$$

$$+ \int_S \frac{\partial T_{ji}}{\partial \xi_k} dS \, u_j(\xi) + \int_S \frac{\partial T_{ji}}{\partial \xi_k} r_l dS \, \zeta_l \frac{\partial u_j}{\partial \zeta}(\xi) \qquad (10)$$

where ζ_l are the components of the unit tangent vector to S at a point z, the eventual limit point of ξ, on the boundary. Note here that the normal derivative part of the gradient, $\partial u_j/\partial \nu$, ν_l being the unit normal components at z, is not included in the subtraction since it plays no role in the regularization. Alternately, the term in brackets of equation (10) is $O(r^2)$ for $\xi = z$ and $x \to z$ or $r \to 0$ when u_i has continuous first derivatives at ξ, which is sufficient to remove the singularity of the integral, without including the normal derivative component.

Then by identity (7), the first added back term is zero and the second can be replaced by use of identity (8). Using identities (7) and (8) with (10) and incorporating the results into equation (9), we have

$$\frac{\partial u_i}{\partial \xi_k}(\xi) + \int_S \frac{\partial T_{ji}}{\partial \xi_k} \left[u_j(x) - u_j(\xi) - \frac{\partial u_j}{\partial \zeta}(\xi) r_l \zeta_l \right] dS$$

$$+ \int_S E_{mnjl} n_n \frac{\partial U_{mi}}{\partial \xi_k} dS \; \zeta_l \frac{\partial u_j}{\partial \zeta}(\xi) - \delta_{ij} \delta_{kl} \; \zeta_l \frac{\partial u_j}{\partial \zeta}(\xi)$$

$$= \int_S \frac{\partial U_{ji}}{\partial \xi_k} t_j dS \qquad (11)$$

This is now a form of the gradient equation where, through regularization, the stronger singularity has been removed and, through the identities (7) and (8), replaced with weakly singular integrals that will exist as ordinary improper integrals as ξ is taken to z on the boundary. The singularity on the right side of the equation is also to be regularized, but the choice of terms to subtract depends upon which component (normal or tangent) of the gradient equation we wish to use.

To form the traction equation we contract the displacement gradient equation (11) on $E_{pqik} \nu_q$ to get,

$$t_p(\xi) + E_{pqik} \nu_q \int_S \frac{\partial T_{ji}}{\partial \xi_k} \left[u_j(x) - u_j(\xi) - \frac{\partial u_j}{\partial \zeta} (\xi) r_l \zeta_l \right] dS$$

$$+ E_{pqik}\nu_q \int_S E_{mnjl}n_n \frac{\partial U_{mi}}{\partial \xi_k} dS\ \zeta_l \frac{\partial u_j}{\partial \zeta}(\xi)$$

$$- E_{pqik}\nu_q\ \delta_{ij}\delta_{kl}\zeta_l \frac{\partial u_j}{\partial \zeta}(\xi) = E_{pqik}\nu_q \int_S \frac{\partial U_{ji}}{\partial \xi_k} t_j dS \tag{12}$$

On the right side we now regularize the kernel by first subtraction of n_q and then $t_j(\xi)$, i.e.,

$$\nu_q \int_S \frac{\partial U_{ji}}{\partial \xi_k} t_j dS = \int_S \frac{\partial U_{ji}}{\partial \xi_k} (\nu_q - n_q) t_j dS$$

$$+ \int_S \frac{\partial U_{ji}}{\partial \xi_k} n_q \Big[t_j(x) - t_j(\xi)\Big] dS + \int_S \frac{\partial U_{ji}}{\partial \xi_k} n_q dS\ t_j(\xi) \tag{13}$$

Putting the result of equation (13) into equation (12) while noting that with identity (5) and equation (2), the last term of equation (13) and the $t_p(\xi)$ on the left cancel to give,

$$E_{pqik}\nu_q \int_S \frac{\partial T_{ji}}{\partial \xi_k} \left[u_j(x) - u_j(\xi) - \frac{\partial u_j}{\partial \zeta}(\xi) r_l \zeta_l \right] dS$$

$$+ E_{pqik}\nu_q \left[\int_S E_{mnjl} n_n \frac{\partial U_{mi}}{\partial \xi_k} dS - \delta_{ij}\delta_{kl} \right] \zeta_l \frac{\partial u_j}{\partial \zeta}(\xi)$$

$$= E_{pqik}\int_S \frac{\partial U_{ji}}{\partial \xi_k} (\nu_q - n_q) t_j dS + E_{pqik}\int_S \frac{\partial U_{ji}}{\partial \xi_k} n_q \Big[t_j(x) - t_j(\xi)\Big] dS$$

(14)

Finally, some simplification of the second term on the left can be made by the use of identity (5) and equation (2), i.e.,

$$E_{pqik}\nu_q\left[\int_S E_{mnjl}n_n\frac{\partial U_{mi}}{\partial \xi_k}dS - \delta_{ij}\delta_{kl}\right]$$

$$= \int_S E_{mnjl}n_n E_{pqik}\nu_q \frac{\partial U_{mi}}{\partial \xi_k}dS - E_{pqjl}\nu_q$$

$$= \int_S \left[E_{mnjl}\nu_n T_{pm} - E_{mnjl}n_n T^0_{pm}\right]dS = -E_{mnjl}\int_S \left[T^0_{pm}n_n - T_{pm}\nu_n\right]dS \tag{15}$$

where

$$T^0_{pm} = -E_{pqik}\nu_q\frac{\partial U_{mi}}{\partial \xi_k} \tag{16}$$

The final form of the traction equation is then, putting equation (15) into (14) and lastly taking the limit $\xi \to z \in S$,

$$E_{pqik}\nu_q\int_S \frac{\partial T_{ji}}{\partial \xi_k}\left[u_j(x) - u_j(z) - \frac{\partial u_j}{\partial \zeta}(z)r_l\zeta_l\right]dS$$

$$- E_{mnjl}\int_S \left[T^0_{pm}n_n - T_{pm}\nu_n\right]dS\ \zeta_l\ \frac{\partial u_j}{\partial \zeta}(z)$$

$$= \int_S \left[T_{pj} - T^0_{pj}\right]t_j dS - \int_S T_{pj}\left[t_j(x) - t_j(z)\right]dS \tag{17}$$

In this form, all terms are, at most, weakly singular and can be integrated as ordinary improper integrals. We note that, for numerical purposes, all terms on the left side of the equation involve the boundary variables u_i or their tangent derivatives while all terms on the right involve only the tractions t_i. Thus the equation is well arranged for numerical approximation, with the tangential derivative determined by derivatives of shape functions on elements when the boundary is discretized.

The development of a tangent derivative equation is somewhat easier than the traction equation. One need only contract the gradient equation (11) on ζ_k to get

$$\zeta_k \frac{\partial u_i}{\partial \xi_k}(\xi) + \int_S \zeta_k \frac{\partial T_{ji}}{\partial \xi_k}\left[u_j(x) - u_j(\xi) - \frac{\partial u_j}{\partial \zeta}(\xi) r_l \zeta_l\right] dS$$

$$+ \int_S \zeta_k E_{mnjl} n_n \frac{\partial U_{mi}}{\partial \xi_k} dS \; \zeta_l \frac{\partial u_j}{\partial \zeta}(\xi) - \delta_{ij}\delta_{kl}\zeta_k\zeta_l \frac{\partial u_j}{\partial \zeta}(\xi)$$

$$= \int_S \zeta_k \frac{\partial U_{ji}}{\partial \xi_k} t_j dS \qquad (18)$$

and immediately the first and last terms on the left side cancel. On the right side, we now regularize through subtraction of the tangent components $s_k(\xi)$, i.e.,

$$\int_S \zeta_k \frac{\partial U_{ji}}{\partial \xi_k} t_j dS = \int_S \frac{\partial U_{ji}}{\partial \xi_k} (\zeta_k - s_k) t_j dS$$

$$+ \int_S \frac{\partial U_{ji}}{\partial \xi_k} s_k \left[t_j(x) - t_j(\xi)\right] dS + \int_S \frac{\partial U_{ji}}{\partial \xi_k} s_k dS \; t_j(\xi) \qquad (19)$$

but the last term is zero (on a closed boundary S). Also, the third term on the left side of equation (18) becomes,

$$\int_S \zeta_k E_{mnjl} n_n \frac{\partial U_{mi}}{\partial \xi_k} dS =$$

$$\int_S E_{mnjl} \frac{\partial U_{mi}}{\partial \xi_k} (\zeta_k n_n - s_k \nu_n) dS + E_{mnjl} \int_S \frac{\partial U_{mi}}{\partial \xi_k} s_k dS \; \nu_n \qquad (20)$$

and again the last term is zero.

Putting the results of equations (19) and (20) into (18) and proceeding to the limit we have,

$$\zeta_k \int_S \frac{\partial T_{ji}}{\partial \xi_k} \left[u_j(x) - u_j(z) - \frac{\partial u_j}{\partial \zeta}(z) r_1 \zeta_1 \right] dS +$$

$$\int_S E_{mnjl} (\zeta_k n_n - s_k \nu_n) \frac{\partial U_{ml}}{\partial \xi_k} dS \, \zeta_1 \frac{\partial u_j}{\partial \zeta}(z)$$

$$= \int_S \frac{\partial U_{ji}}{\partial \xi_k} (\zeta_k - s_k) t_j dS + \int_S \frac{\partial U_{ji}}{\partial \xi_k} s_k \left[t_j(x) - t_j(z) \right] dS \quad (21)$$

which, like the traction equation (17), is a completely regularized form of the tangent derivative equation where all terms are suitable for numerical integration.

DISCUSSION AND CONCLUSIONS

Two regularized forms of the gradient equation pertaining to two-dimensional elasticity have been developed. The regularizations used here are based on two major premises. First, in order that the regularizations by subtraction be effective, one must require a certain degree of continuity of the density functions. Second, the replacement of the stronger singularities of the added back terms by weaker ones utilizes the basic identities of the fundamental solution and its derivatives.

The degree of continuity required of the density functions for the regularization of the stronger singularity has been addressed by Krishnasamy, et al. [9]. Technically, for the integrals where both the displacement function and the tangential derivative are subtracted to remove the $O(1/r^2)$ singularities, the displacements must have Hölder continuous first derivatives. Practically, one would require continuous first derivatives of the displacements at the collocation points, presuming the collocation technique is used in the numerical solution. If regular, C^0 continuous boundary elements are used, this then requires collocation at interior points of the elements and precludes collocation at element junctions. Alternately, one could use spline or Hermitian elements to satisfy the continuity requirements. Regardless of the interpolation used for the displacements, the tangential derivative component should be continuous, so collocation at corners of the boundary is, in general, not possible.

Consistent with the requirements on the displacements, the weaker singularities of the derivative equations have been removed by subtraction of both boundary normal or tangent components and tractions. Since the stronger singularity removal requires continuous tangent derivatives, then the presumption of continuous normal and tangent vectors and continuous traction components is consistent.

The second aforementioned issue concerning the use of the various identities of the fundamental solution is clear. Since the identities were developed from the representation, then they are valid for problems where the representation is valid. Here, this implies closed regions. For regions with special boundaries, such as for cracks where two parts of the boundary overlap, then the representation degenerates and the identities may no longer apply or be useful. For problems with crack-like boundaries, the representation can be specialized so that part of the boundary is an open arc and for such cases the added back terms may have to be integrated by use of Stokes theorem as in reference [6].

Specialization of the traction and tangent derivative equation for the case of isotropy and specific development of the kernels for numerical purposes has not yet been completed. The analogous forms of these equations for potential problems have been used [7,10] and have been shown to provide stable numerical solutions. Similar results can be expected for the equations of elasticity.

The utility of the traction and tangent derivative equations is problem dependent. Whereas the traction equation and the regular boundary integral are not independent formulations of a problem and must play complementary roles in their usage, the tangent derivative equation is an independent one which can be concurrently used with regular or the traction boundary integral equations. Whatever the use, the regularized equations developed here provide the basis for computation algorithms.

REFERENCES

1. Brandao, M. P. Improper Integrals in Theoretical Aerodynamics: The Problem Revisited, AIAA Journal, 25, No. 9, pp. 1258-1260, 1987.

2. Ioakimidis, N. I. A Natural Approach to the Introduction of Finite-part Integrals into Crack Problems of Three-Dimensional Elasticity, Engineering Fracture Mechanics, 16(5), pp. 669-673, 1982.

3. Burton, A. J. and Miller, G. F. The Application of Integral Equation Methods to the Numerical Solution of Some Exterior Boundary-valve Problems, Proc. Roy. Soc., Ser. A, 323, pp. 201-210, 1971.

4. Gray, L. J. and Giles, G. E. Application of the Thin Cavity Method to Shield Calculations in Electroplating (Ed. Brebbia, C. A.), Vol. 2, pp. 441-452, Proceedings of the Boundary Elements X Conference, CMP, Springer-Verlag, Southampton, 1988.

5. Ingber, M. S. and Hickox, C. E. A Modified Burton-Miller Algorithm for Treating the Uniqueness of Representation Problem for Exterior Acoustic Radiation and Scattering Problems, Engineering Analysis with Boundary Elements, to be published.

6. Krishnasamy, G., Rudolphi, T. J., Schmerr, L. W. and Rizzo, F. J. Hypersingular Boundary Integral Equations: Some Applications in Acoustics and Elastic Wave Scattering, JAM, Vol, 57, pp. 404-414, June, 1990.

7. Rudolphi, T. J. The Use of Simple Solutions in the Regularization of Hypersingular Boundary Integral Equations, Mathl. Comput. Modelling, Vol. 15, No. 3-5, pp. 269-278, 1991.

8. Liu, Y. and Rudolphi, T. J. Some Identities for Fundamental Solutions and Their Applications to Non-singular Boundary Element Formulations, Engr. Anal. with Boundary Elements, to be published.

9. Krishnasamy, G., Rizzo, F. J., Rudolphi, T. J. Continuity Requirements for Density Functions in the Boundary Integral Equation Method, Computational Mechanics, to be published.

10. Rudolphi, T. F. Higher Order Elements and Element Enhancement by Combined Regular and Hypersingular Boundary Integral Equations (Eds. Annigeri, B. S. and Tseng, K.), pp. 448-455, Proceedings of the IABEM-89 Int. Sym. on BEM, East Hartford, CN, Springer-Verlag, Oct. 1989.

Effective Iterative Solution Methods for Boundary Element Equations

S. Rjasanow

Department of Mathematics, University of Kaiserslautern, D-6750 Kaiserslautern, Germany

ABSTRACT

The precondition techniques are developed for matrices arising from Galerkin method with different trial functions (piecewise constant, linear and quadratic) for two-dimensional and three-dimensional, simple and multiple connected domains with Dirichlet and Neumann boundary conditions.

INTRODUCTION

The boundary element method (BEM) leads to an algebraic system of equations with a full dense matrix (Brebbia, Walker [1], Wendland [2]). The Gaussian elimination is usually used for solving such linear systems. If we denote the discretization parameter in one space direction as h, we obtain matrices having the dimension $O(h^{-1})$ in two- and $O(h^{-2})$ in three-dimensional case. The number of arithmetical operations are $O(h^{-3})$ and $O(h^{-6})$ respectively. Since the given matrices are generally large, more efficient methods for the solution of BEM linear systems are necessary.

The BEM for the Dirichlet or Neumann problem in a two-dimensional circular domain and in a three-dimensional rotational domain leads to a system with a matrix having a special structure. It is a circulant or block-circulant structure. Such systems can be solved directly using Fast Fourier Transform (FFT), and the number of arithmetical operations would be only $O(h^{-1}\log h^{-1})$ in two- and $O(h^{-3}\log h^{-1})$ in three-dimensional case. Such matrices can be used for preconditioned iterative solution methods for the systems arising from general arbitrary domains. The precondition and system matrices are spectral equivalent. The rate of convergence of the Conjugate Gradient Method (CGM) is independent of h and the numerical work will be drastically decreasing to $O(h^{-2})$ in two- and $O(h^{-4})$ in three-dimensional case.

BOUNDARY INTEGRAL EQUATIONS

In this paper we consider the Dirichlet and Neumann problems for Poisson equation in two- and three-dimensional domain Ω:

$$\Delta u(x) = f(x),\ x \in \Omega, \tag{1}$$

$$(\mathcal{P}_1 u)(x) + (\mathcal{P}_2 v)(x) = g(x),\ x \in \Gamma = \partial\Omega, \tag{2}$$

where, $v(x)$ is the normal derivative of the function $u(x)$:

$$v(x) = \frac{\partial u(x)}{\partial n_x},$$

and n_x ($x \in \Gamma$) is the unit outward normal of the boundary Γ at x. The Operators $\mathcal{P}_1$ and $\mathcal{P}_2$ are

$$\mathcal{P}_1 = \mathcal{I},\ \mathcal{P}_2 = \mathcal{O}$$

for Dirichlet and

$$\mathcal{P}_1 = \mathcal{O},\ \mathcal{P}_2 = \mathcal{I}$$

for Neumann problem.

It is known (Wendland [2], Costabel [3]) that there are two relations on Γ between $u(x)$ and $v(x)$

$$(\mathcal{A}_1 v)(y) - \left(\left(\frac{1}{2}\mathcal{I} + \mathcal{B}\right) u\right)(y) = f_1(y),\ y \in \Gamma; \tag{3}$$

$$-\left(\left(\frac{1}{2}\mathcal{I} - \mathcal{B}'\right) v\right)(y) + (\mathcal{A}_2 u)(y) = f_2(y),\ y \in \Gamma. \tag{4}$$

The equation (3) is well known Green's third identity for potential theory. The operators $\mathcal{A}_1$ and $\mathcal{B}$ are the simple and double layer potentials:

$$(\mathcal{A}_1 v)(y) = \int_\Gamma u^*(x,y) v(x) ds_x;$$

$$(\mathcal{B}u)(y) = \int_\Gamma \frac{\partial u^*(x,y)}{\partial n_x} u(x) ds_x.$$

The equation (4) can be obtained (Blue [4]) from the second Green's identity. Here, $\mathcal{B}'$ is the adjoint operator of $\mathcal{B}$:

$$(\mathcal{B}' v)(y) = \int_\Gamma \frac{\partial u^*(x,y)}{\partial n_y} v(x) ds_x,$$

and $\mathcal{A}_2$ defined by

$$(\mathcal{A}_2 u)(y) = \int_\Gamma \frac{\partial^2 u^*(x,y)}{\partial n_x \partial n_y} (u(x) - u(y)) ds_x$$

is the operator with the hypersingular kernel, where $u^*(x,y)$ is the known fundamental solution of the Laplace equation:

$$u^*(x,y) = \begin{cases} -\dfrac{1}{2\pi} \log|x-y| & \textit{for } \Omega \subset R^2; \\ \dfrac{1}{4\pi} \dfrac{1}{|x-y|} & \textit{for } \Omega \subset R^3. \end{cases}$$

Any one of the integral equations (3),(4) can be used for the numerical solution of the boundary value problem (1),(2).

Now we shall consider (Costabel [3], Costabel, Wendland [5]) the properties of the operators $\mathcal{A}_1, \mathcal{A}_2$, $\frac{1}{2}\mathcal{I} + \mathcal{B}$ and $\frac{1}{2}\mathcal{I} - \mathcal{B}'$. Let Ω be a bounded simple connected domain whose boundary Γ is sufficiently smooth. All four operators are pseudo-differential operators of integer order on Γ:

$$\begin{aligned} \mathcal{A}_1 &: H^s(\Gamma) \mapsto H^{s+1}(\Gamma); \\ \mathcal{A}_2 &: H^s(\Gamma) \mapsto H^{s-1}(\Gamma); \\ \frac{1}{2}\mathcal{I} + \mathcal{B},\ \frac{1}{2}\mathcal{I} - \mathcal{B}' &: H^s(\Gamma) \mapsto H^s(\Gamma); \\ & s \in R, \end{aligned}$$

where $H^s(\Gamma)$ is the Sobolev space (Adams [6]) on boundary Γ. The operators $\mathcal{A}_1$ and $\mathcal{A}_2$ are selfadjoint, i.e.

$$< \mathcal{A}_1 u, v >_0 = < u, \mathcal{A}_1 v >_0,\ \forall u, v \in H^{-1/2}(\Gamma);$$
$$< \mathcal{A}_2 u, v >_0 = < u, \mathcal{A}_2 v >_0,\ \forall u, v \in H^{1/2}(\Gamma),$$

where $< \cdot, \cdot >_0$ denotes the duality pairing between $H^s(\Gamma)$ and $H^{-s}(\Gamma)$, i.e. $L_2(\Gamma)$ inner product

$$< u, v >_0 = \int_\Gamma u(x)v(x)ds_x .$$

The operators $\mathcal{A}_1$ and $\frac{1}{2}\mathcal{I} - \mathcal{B}'$ satisfy a Gårding inequality, which can be formulated for an operator $\mathcal{A} : H^s(\Gamma) \mapsto H^{-s}(\Gamma)$: there exists a constant $\gamma > 0$ such that

$$< \mathcal{A}u, u >_0 \geq \gamma \|u\|_s^2,\quad \forall u \in H^s(\Gamma), \tag{5}$$

where $\|\cdot\|_s$ denotes the Sobolev norm in $H^s(\Gamma)$. The operators $\mathcal{A}_2$ and $\frac{1}{2}\mathcal{I} + \mathcal{B}$ satisfy (5) for all functions u with $< u, 1 >_0 = 0$, since both are singular operators:

$$(\mathcal{A} \cdot 1)(y) = 0.$$

The most effective iterative methods (for example CGM) need symmetric and positive definite matrix. These conditions can be **satisfied** if we choose equation (3) for the numerical solution of Dirichlet problem :

$$(\mathcal{A}_1 v)(y) = f_1(y),\ y \in \Gamma, \tag{6}$$

and the equation (4) for Neumann problem

$$(\mathcal{A}_2 v)(y) = f_2(y),\ y \in \Gamma. \tag{7}$$

TWO-DIMENSIONAL CASE

We begin with a parametrization of Γ by 1-periodic representation

$$\Gamma = \left\{ x \in R^2 : x = x(t),\ 0 \leq t < 1,\ |\dot{x}(t)| \geq \kappa > 0 \right\}. \tag{8}$$

Now we can rewrite the equations (6) and (7) as follows:

$$(\mathcal{A}_1 v)(\tau) = \int_0^1 u^*(x(t), x(\tau)) v(x(t)) |\dot{x}(t)| dt = f_1(\tau),$$

$$(\mathcal{A}_2 u)(\tau) = \int_0^1 \frac{\partial^2 u^*(x(t), x(\tau))}{\partial n_{x(t)} \partial n_{x(\tau)}} (u(x(t) - u(x(\tau))) |\dot{x}(t)| dt = f_2(\tau),$$

$$0 \leq \tau < 1.$$

For the circular boundary Γ

$$\Gamma = \left\{ x \in R^2 : x = \frac{1}{2} \begin{pmatrix} \cos 2\pi t \\ \sin 2\pi t \end{pmatrix}, 0 \le t < 1 \right\}$$

this yields:

$$(A_1 v)(\tau) = -\frac{1}{2} \int_0^1 \log|\sin \pi(t-\tau)| v(t) dt = f_1(\tau), \tag{9}$$

$$(A_2 u)(\tau) = -\frac{1}{2} \int_0^1 \frac{u(t) - u(\tau)}{\sin^2(t-\tau)} dt = f_2(\tau), \tag{10}$$

$$0 \le \tau < 1.$$

The following Lemma is proved in (Rjasanow [8],[9]).

Lemma 1 *The operators A_1 and A_2 from (9),(10) have the eigenfunctions and eigenvalues:*

$$A_1 v_k = \lambda_k^{(1)} v_k,$$
$$A_2 v_k = \lambda_k^{(2)} v_k,$$
$$v_k = e^{i2\pi k t},$$
$$k \in Z,$$
$$\lambda_0^{(1)} = \frac{1}{2}\log 2, \ \lambda_0^{(2)} = 0,$$
$$\lambda_k^{(1)} = \frac{1}{4|k|}, \ \lambda_k^{(2)} = |k|,$$
$$k \in Z \setminus \{0\}.$$

DISCRETIZATION

For the discretization of (6),(7) we begin with the approximation of $v(t)$ and $u(t)$ by piecewise polynomials $\phi_l^{(\nu)}(t)$, $l = 1, \ldots, n$ of degree $\nu = 0, 1, 2$ (B-Splines). We divide the basic interval $[0,1)$ into $n > \nu + 1$ subintervals

$$[0,1) = \bigcup_{l=1}^{n} [t_l, t_{l+1}), \ t_l = (l-1)h, \ h = 1/n,$$

and introduce a n-dimensional subspace H_n of 1-periodic functions (see Rjasanow [9])

$$H_n = \{u_h : u_h = \Phi_n^\nu(t) y, \ y \in R^n\} = span(\phi_1^{(\nu)}(t), \ldots, \phi_n^{(\nu)}(t)).$$

GALERKIN METHOD

The Galerkin method for the equations (6),(7) leads to:

Find $v_h(t) \in H_n$ such that Galerkin equations

$$< A v_h, w >_0 = < f, w >_0 \tag{11}$$

are satisfied for all functions $w \in H_n$.

We obtain from (11) the following systems of linear equations:

$$\begin{aligned} A_1 y_1 &= b_1, \\ A_2 y_2 &= b_2, \\ A_1, A_2 &\in R^{n\times n}, \\ y_1, y_2, b_1, b_2 &\in R^n. \end{aligned}$$

The matrices A_1 and A_2 have a special structure (Reichel [7], Rjasanow [8]).

Lemma 2 *The matrices A_1 and A_2 are symmetric, positive definite (A_2 is semi-definite) and circulant.*

We introduce the most simple circulant matrix as

$$J = \begin{pmatrix} 0 & 1 & 0 & \cdots & 0 \\ 0 & 0 & 1 & \cdots & 0 \\ \dots & \dots & \dots & \dots & \dots \\ 0 & 0 & 0 & \cdots & 1 \\ 1 & 0 & 0 & \cdots & 0 \end{pmatrix}, \quad J \in R^{n\times n} \tag{12}$$

It is not difficult to see that each circulant matrix $A \in R^{n\times n}$ satisfies

$$A = \sum_{l=1}^{n} a_{1l} J^{l-1} = a(J), \tag{13}$$

thus A is a polynomial function of the matrix J, furthermore (Woewodin, Tyrtyshnikow [10], Davis [11]) each circulant matrix A can be written as

$$A = n^{-1} F \Lambda F^*, \tag{14}$$

where F denotes the matrix of discrete Fourier transform

$$f_{kl} = e^{\imath \frac{2\pi}{n}(k-1)(l-1)}, \quad k, l = 1, \dots, n$$

and Λ is the diogonal matrix with the eigenvalues of A:

$$\Lambda = diag(\lambda_1, \dots \lambda_n).$$

The eigenvalues of A_1 and A_2 can be obtained by the following Lemma (see Rjasanow [8],[9])

Lemma 3 *The eigenvalues of the matrices A_1 and A_2 have the form*

$$\lambda_j^{(1)} = \begin{cases} \frac{1}{2} h \log 2, \; j = 1; \\ \frac{1}{4} h^2 \left(\frac{\sin \pi s}{\pi}\right)^{2\nu+2} \sum_{k=0}^{\infty} \frac{1}{(k+s)^{2\nu+3}} + \frac{1}{(k+1-s)^{2\nu+3}}, \; j > 1; \end{cases}$$

$$\lambda_j^{(2)} = \begin{cases} 0, \; j = 1; \\ \frac{2}{h} \left(\frac{\sin \pi s}{\pi}\right)^{2\nu+2} \sum_{k=0}^{\infty} \frac{1}{(k+s)^{2\nu+1}} + \frac{1}{(k+1-s)^{2\nu+1}}, \; j > 1; \end{cases}$$

$$s = \frac{j-1}{n}.$$

It is very easy to solve a system of linear equations

$$Aw = r$$

with the circulant matrix A. The solution w can be written with the help of (14) as

$$w = A^{-1}r = n^{-1}F\Lambda^{-1}F^{*}r,$$

and w is computed using FFT (Cooley, Tukey [12], Henrici [13]) needs only $O(n\log n) = O(h^{-1}\log h^{-1})$ arithmetical operations. We use the pseudo inverse matrix A_2^{+} to solve a linear system with the matrix A_2 :

$$A_2^{+} = n^{-1}F\Lambda^{+}F^{*}, \ \Lambda^{+} = diag(0, 1/\lambda_2^{(2)}, \ldots, 1/\lambda_n^{(2)}).$$

PRECONDITIONED CONJUGATE GRADIENT METHOD

One of the most effective iterative solution methods for the symmetric, positive definite system of linear equations

$$Ay = b, \ A \in R^{n \times n}, \ A = A^{T} > 0, \ y, b, \in R^{n}$$

is the conjugate gradient method (Hestenes,Stiefel [14]):

1. $y_0 \in R^n$;
   $r_0 = Ay_0 - b, \ w_0 = B^{-1}r_0;$
   $s_0 = w_0;$
2. for $k = 0, 1, \ldots$
   $$y_{k+1} = y_k - \alpha_{k+1}s_k, \ \alpha_{k+1} = \frac{(r_k, w_k)}{(As_k, s_k)};$$
   $$r_{k+1} = r_k - \alpha_{k+1}As_k, \ w_{k+1} = B^{-1}r_{k+1};$$
   $$w_{k+1} = w_{k+1} + \beta_{k+1}s_k, \ \beta_{k+1} = \frac{(r_{k+1}, w_{k+1})}{(r_k, w_k)},$$

where $B = B^{T} > 0$ is a preconditioning matrix.

For general arbitrary domains Ω (not being circular), the Galerkin approach still yields (Wendland [2], Costabel [3], Rjasanow [9]) a symmetric, positive definite matrix A (for the Neumann problem positive semi-definite) without having the additional special circulant property. In order to solve the resulting system of equations, we are able to use the preconditioned conjugate gradient method with the preconditioning matrix $B = A_1$ or $B = A_2$. Here, one step of the iterations requires the matrix-vector multiplication ($O(n^2)$ Operations), two scalar products ($O(n)$ Operations) and the solution of the preconditioning system ($O(n\log n)$ Operations). Hence, the work per iteration is mainly dominated by the work for the matrix-vector multiplication. The total number of operations is dependent on the spectral condition number $\kappa(B^{-1}A)$, which is bounded, the bound is independent of n (Rjasanow [8],[9]). In this case we need only $O(n^2) = O(h^{-2})$ arithmetical operations for the entire process.

MULTIPLE CONNECTED DOMAIN

Now let Γ be a multiple connected boundary

$$\Gamma = \bigcup_{l=1}^{m} \Gamma_l,$$

where each boundary Γ_l, $l = 1, \dots, m$ is simple connected and can be parametized (see (8)). The Galerkin approach for the Dirichlet problem yields in this case a symmetric positive definite matrix A having the following block structure

$$A = \begin{pmatrix} A_{11} & \cdots & A_{1m} \\ \cdots & \cdots & \cdots \\ A_{m1} & \cdots & A_{mm} \end{pmatrix}, \quad A \in R^{N \times N},\ N = \sum_{l=1}^{m} n_l,$$

where the blocks $A_{kl} \in R^{n_k \times n_l}$ are Galerkin matrices for the operators

$$\int_{\Gamma_l} u^*(x, y) v(x) ds_x, \quad y \in \Gamma_k,$$

and n_k denotes the number of discretization points on Γ_k. The preconditioning matrix B can be constructed as follows

$$B = diag(B_1, \dots, B_m), \quad B \in R^{N \times N},$$

where each diagonal block $B_k \in R^{n_k \times n_k}$ is the Galerkin matrix for the operator (9). The number of iterations of CGM is independent of $h = \max h_k$, but is dependent of the geometry of the boundary Γ. This is shown in (Staude [15]) .

THREE-DIMENSIONAL CASE

Let $\Gamma = \partial\Omega$, $\Omega \in R^3$ be a bounded simple connected surface given by a parametric representation

$$\Gamma = \left\{ x \in R^3,\ x = x(t, z),\ 0 \le t < 1,\ 0 \le z \le 1 \right\}. \tag{15}$$

With the help of (15) we can write the boundary integral equation (6)

$$\frac{1}{4\pi} \int_{\Gamma} \frac{v(x)}{|x - y|} ds_x = f(y), \quad y \in \Gamma$$

as an integral equation over $[0,1) \times [0,1]$:

$$\frac{1}{4\pi} \int_0^1 \int_0^1 \frac{v(t, z)}{|x(t, z) - x(\tau, w)|} J(t, z) dt dz = f(\tau, w), \quad (\tau, w) \in [0, 1) \times [0, 1], \tag{16}$$

where $y = x(\tau, w), v(t, z) = v(x(t, z)), f(\tau, w) = f(x(\tau, w))$ and

$$J(t, z) = \sqrt{\left|\frac{\partial x}{\partial t}\right|^2 \left|\frac{\partial x}{\partial z}\right|^2 - \left(\frac{\partial x}{\partial t}, \frac{\partial x}{\partial z}\right)^2} \tag{17}$$

is the Jacobian of (15).

DISCRETIZATION

The functions on Γ are identified with functions on $[0,1) \times [0,1]$, 1-periodic in the first argument t. We devide the domain $[0,1) \times [0,1]$ into rectangles using nodes

$$\{(t_k, z_l) = (h_t(k-1), h_z(l-1)),\ k = 1,\dots,n,\ l = 1,\dots,m\},$$

where $h_t = 1/n$ and $h_z = 1/(m-1)$. For the approximation of $v(t,z)$ we introduce an N-dimensional subspace of functions H_N :

$$H_N = \left\{ v_N(t,z) = \sum_{i_1=1}^{n} \sum_{i_2=1}^{m} y_{(i_1,i_2)} \phi_{(i_1,i_2)}(t,z) = \Phi y,\ y \in R^N,\ N = n \cdot m \right\}.$$

As $\phi_{(i_1,i_2)}(t,z)$ one can use the piecewise constant, linear or quadratic functions. The Galerkin method for (16) leads to:

Find $v_N \in H_N$ such that the Galerkin equations

$$< Av_N, w >_0 = < f, w >_0, \tag{18}$$

hold for all $w \in H_N$.

The equations (18) are replaced by the algebraic system

$$Ay = b,\ A \in R^{N\times N},\ y, b \in R^N, \tag{19}$$

where the matrix A is symmetric and positive definite having a block structure

$$A = \begin{pmatrix} A_{11} & \cdots & A_{1m} \\ \dots & \dots & \dots \\ A_{m1} & \cdots & A_{mm} \end{pmatrix},\ A_{i_2 j_2} \in R^{n\times n}.$$

ROTATIONAL DOMAIN

Let Γ be given by a parametric representation

$$\Gamma = \left\{ x \in R^3,\ x = \begin{pmatrix} R(z)\cos 2\pi t \\ R(z)\sin 2\pi t \\ z \end{pmatrix},\ 0 \le t < 1,\ 0 \le z \le 1 \right\}.$$

That is, Γ originates from the rotation of the curve

$$\{R(z),\ 0 \le z \le 1\}$$

about the z-axis. in this case the matrix A has some special properties: each block $A_{i_2 j_2}$ of A is symmetric and circulant, i.e. the matrix A is symmetric, positive definite, block circulant matrix with symmetric blocks. Such matrices are investigated in (Meyer, Rjasanow [16], Rjasanow [9]), where very efficient algorithms for the numerical solution of linear systems of equations are constructed.

The matrix A has the following form

$$A = A(J) = \begin{pmatrix} a_{11}(J) & \cdots & a_{1m}(J) \\ \dots & \dots & \dots \\ a_{m1}(J) & \cdots & a_{mm}(J) \end{pmatrix},$$

where each block is a polynomial function (13) of the matrix J defined in (12). If we solve n eigenvalue problems

$$A(\omega_k)x_l(\omega_k) = \lambda_l(\omega_k)x_l(\omega_k), \quad l = 1, \dots, m \tag{20}$$
$$\omega_k = e^{i\frac{2\pi}{n}(k-1)}, \quad k = 1, \dots, n$$

with symmetric matrices $A(\omega_k) \in R^{m\times m}$ (Meyer, Rjasanow [16]), the matrix $A(J)$ can be given by the formula:

$$A = \frac{1}{n}\sum_{k,l} \frac{1}{\lambda_l(\omega_k)} (x_l(\omega_k) \otimes f_k)(x_l(\omega_k) \otimes f_k)^*, \tag{21}$$

where f_k are the eigenvectors of the matrix J:

$$Jf_k = \omega_k f_k, \; f_k = (\omega_k^0, \omega_k^1, \dots, \omega_k^{n-1})^T,$$

and $\otimes$ is the Kronecker product.

The algorithm to compute the solution y is the following:

1. $C := (F^*B)^T \in C^{m\times n}$;
2. $Y := 0$;
3. for $l = 1, \dots, m$

   3.1 $D_l = diag(d_1, \dots, d_n)$ with $d_k = \dfrac{x_l^T(\omega_k)Ce_k}{n\lambda_l(\omega_k)}$;

   3.2 $Y := Y + Re(FD_lX_l^T)$,

where

$$X_l = (x_l(\omega_1)\vdots\dots\vdots x_l(\omega_n)) \in R^{m\times n} \quad \text{– eigenvectors of (20);}$$
$$B = (b_1\vdots\dots\vdots b_m) \in R^{n\times m} \quad \text{– given right side of (19);}$$
$$Y = (y_1\vdots\dots\vdots y_m) \in R^{n\times m} \quad \text{– solution of (19);}$$
$$e_k = (0, \dots, 0, 1, 0, \dots, 0)^T \in R^m.$$

We use FFT for the steps 1. and 3.2 which requires $O(m^2n\log n) = O(h^{-3}\log h^{-1})$ arithmetical operations for the entire algorithm. Furtheremore we need only $O(nm^3) = O(h^{-4})$ arithmetical operations (Garbow, Boyle, Dongarra, Moler [17], Smith, Boyle, Garbow, Ikebe, Klema, Moler [18]) to solve all eigenvalue problems (20).

NUMERICAL EXPERIMENTS

The presented methods for the iterative solution of systems of linear equations arising from boundary element equations have been used for the Dirichlet problem in two-dimensional domain Ω with the boundary

$$\Gamma = \left\{x \in R^2, \; x = \frac{1}{2}\begin{pmatrix} \cos 2\pi t \\ \sin 2\pi t(2 - \frac{3}{2}\sin 2\pi t)\end{pmatrix}, \; 0 \le t < 1\right\}$$

and in three-dimensional domain Ω with the boundary

$$\Gamma = \left\{ x \in R^2,\ x = \frac{1}{2} \begin{pmatrix} R(z)\cos 2\pi t \\ R(z)\sin 2\pi t(2 - \frac{3}{2}\sin 2\pi t) \\ z \end{pmatrix},\ 0 \le t < 1,\ 0 \le z \le 1 \right\}$$

with

$$R(z) = \sqrt{z(1-z)}.$$

We have used the piecewise linear ($\nu = 1$) in two- and piecewise constant ($\nu = 0$) functions in three-dimensional case for approximation $v(x)$. We have used the Galerkin method for the discretization of two-dimensional problem and collocation method (panel method) for three-dimensional problem.
The tables show the number of iterations (It) required for the accuracy $\epsilon = 10^{-8}$ and the corresponding computing times (CPU) measured in CPU seconds on the INMOS T800-transputer for the conjugate gradient method in two- and gradient method in three-dimensional case. We consider both methods with and without ($B = I$) preconditioning. The following tables also show the computing times for Cholesky and Gaussian elimination.

Table 1 *Two-dimensional case*

	without		with		Cholesky
n	It	CPU	It	CPU	CPU
16	18	0.07	10	0.16	0.02
32	30	0.39	10	0.40	0.08
64	44	1.94	11	1.16	0.46
128	61	9.97	11	3.35	3.03
256	82	51.42	11	10.59	21.83
512	114	279.56	11	36.12	165.16

Table 2 *Three-dimensional case*

	without		with		Gaussian
n	It	CPU	It	CPU	CPU
32	50	0.58	14	1.03	0.11
64	69	2.02	19	3.45	0.75
128	75	11.43	19	12.98	5.50
256	140	81.76	31	54.67	42.13
512	167	382.72	32	159.73	329.55

REFERENCES

1. Brebbia C.A. and Walker S. Boundary Element Techniques in Engineering, Newnes-Butterwords, 1980.

2. Wendland W.L. Bemerkungen zu Randelementmethoden und ihren mathematischen und numerischen Aspekten, Mitteilungen der GAMM, Heft 2, pp. 3-27, 1986.

3. Costabel M. Principles of Boundary Elements Methods, Comp. Phys. Reports, 6, pp. 243-274, 1987.

4. Blue J.L. Boundary Integral Solutions of Laplace's Equation, The Bell system Technical Journal, Vol.57, 8, pp. 2797-2822, 1978.

5. Costabel M. and Wendland W.L. Strong ellepticity of boundary integral operators, J. Reine Angew. Math., 372, pp. 34-63, 1986.

6. Adams R.A. Sobolev Spaces, Academic Press, New York, 1975.

7. Reichel L. A Method for Preconditioning Matrices Arising from Linear Integral Equations for Elliptic Boundary Value Problems, Computing, 37, pp. 123-136, 1986.

8. Rjasanow S. Untersuchungen von Eigenschaften der Galerkin- Kollokationsmethode fuer die Randintegralgleichungen, Preprint 48, TU Karl-Marx-Stadt, 1987.

9. Rjasanow S. Vorkonditionierte iterative Aufloesung von Randelementgleichungen fuer die Dirichlet-Aufgabe, Wiss. Schriftenreihe der TU Chemnitz, 7, 1990.

10. Woewodin W.W. and Tyrtyshnikow E.E. Numeracal Processes with Toeplitz Matrices, Nauka, Moskow, 1987 (russian).

11. Davis P.J. Circulant Matrices, John Willy and sons, New York, Chichester, Brisbane and Toronto, 1977.

12. Cooley J.W. and Tukey J.W. An Algorithm for the Machine Calculation of Complex Fourier Series, Math. Comput., 19, pp. 297-301, 1965.

13. Henrici P. Fast Fourier Methods in Computational Complex Analysis, SIAM Rev., 21, pp. 481-527, 1979.

14. Hestenes M.R. and Stiefel E. Methods of Conugate Gradients for Solving Linear Systems, J. Res. NBS, 49, 1952.

15. Staude H. Anwendung von Randintegralmethoden auf mehrfach zusammenhaengende Gebiete, Diplomarbeit, TU Karl-Marx-Stadt, 1989.

16. Meyer A. and Rjasanow S. An Effective Direct Solution Method for Certain Boundary Integral Equations in 3D, Math. Methods in Appl. Sciences, Vol. 13, pp. , 1990.

17. Garbow B.S., Boyle J.M., Dongarra J.J. and Moler C.B. Matrix Eigensystem Routines-EISPACK Guide Extension, Lect. Notes in Comp. Science, 51, Springer Verlag, 1977.

18. Smith B.T., Garbow B.S., Boyle J.M., Ikebe Y., Klema V.C. and Moler C.B. Matrix Eigensystem Routines-EISPACK Guide, Lect. Notes in Comp. Science, 6, Springer Verlag, 1976.

A New Approach Treating Corners in Boundary Element Method

Z. Wang, Q. Wu

Department of Computer Science and Technology, Tsinghua University, Beijing, 100084, China

ABSTRACT

In order to treat corners with continuous linear element, a concept of multiple normal derivatives at a corner is presented for solution of a potential problem in the region composed of one or several mediums. The implementation of this concept improves the result accuracy and avoids introducing unnecessary unknowns. It can be used for improvement of partially discontinuous element and easily generalized to other continuous element of higher degree.

INTRODUCTION

Treating corners in the boundary element method (BEM) is an important problem because it can be met almost everywhere and is directly concerned with accuracy and efficiency of solution. Several strategies for it have been proposed in the past.

The first one chose corner nodes as unknown nodes. But it treated the normal derivatives at the corner as a scalar variable and caused poor accuracy [2]. Another approach introduced by Riccardella [3] adopted a mesh that rendered discontinuous at corner by not joining adjacent elements and leaving two spaced freedom nodes close together on either side of the corner, in 2–D case. Although satisfactory results could be obtained, strong linear dependence in the algebraic system would be caused when the double freedom nodes were too close. Afterwards fully discontinuous and partially discontinuous elements were proposed by Patterson et al [4] and have been widely used in practice. But introducing more unknowns was one defect anyway, even for the partially discontinuous element. The modified multiple–node concept [5] was proposed by Mitra et al. Up to now, it seems to be a quite successful method resolving the effect of discontinuous boundary conditions at a corner.

This paper presents a concept called multiple normal derivatives at a corner. It is similar to the modified multiple–node concept, but the treatment is quite different from the method in [5]. Besides, different kinds of corners with multiple mediums are discussed in this paper.

In the following sections, the concept of multiple normal derivative, implementation of it, improvement on the partially discontinuous element and some results will be presented one by one.

COMPUTATIONAL PROBLEM

In order to evaluate the performance of VLSI circuits, it is necessary to calculate the parasitic resistance involved in the circuits accurately [7, 8]. Two typical parasitic resistors are shown in Fig. 1. The thick lines in figures represent the forced boundary Γ_u throughout and the rest is the natural boundary Γ_q or borderline Γ_b, which is indicated as a dash line, between two mediums.

In calculation of resistance a set of 2–D Laplace's equations with the mixed boundary conditions should be solved. For simplicity, a problem involving only mediums 1 and 2 can be written as follows

$$-\sigma_k \Delta u = 0 \qquad \text{in } \Omega_k \ (k=1,2) \tag{1}$$

$$u_1 = u_2 \qquad \text{on } \Gamma_b \tag{2}$$

$$\sigma_1 \frac{\partial u_1}{\partial n_1} = \sigma_2 \frac{\partial u_2}{\partial n_1} \qquad \text{on } \Gamma_b \tag{3}$$

$$u = u_0 \qquad \text{on } \Gamma_u \tag{4}$$

$$q = \frac{\partial u}{\partial n} = q_0 = 0 \qquad \text{on } \Gamma_q \tag{5}$$

where u_k is the electrical potential on Γ_b defined from the region Ω_k $(k=1,2)$, σ_k the sheet conductivity of medium k $(k=1,2)$. Besides, n represents the normal direction on the boundary, and n_1 on the borderline dependent on Ω_1. The continuity of u and q on the borderline between Ω_1 and Ω_2 is described in equations (2) and (3). In order to get the electrical field $\partial u / \partial n$ on the forced boundary, the BEM is employed to solve this set of two dimensional potential equations defined in the region of two homogeneous mediums. Hence the resistance value can be evaluated by calculating the line integral of the electrical field on the forced boundary.

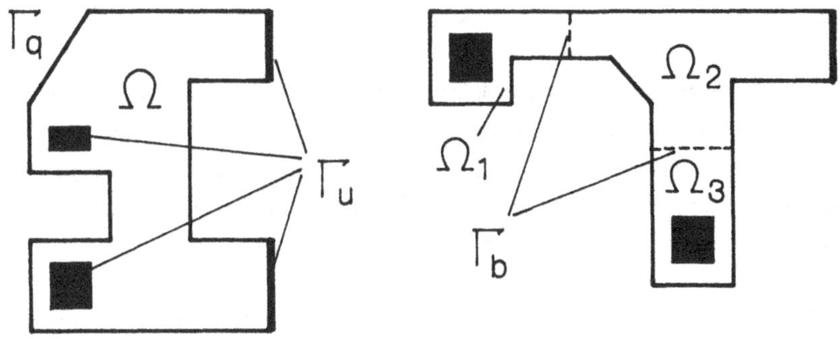

Fig. 1 Two typical parasitic resistors.

First it is transformed into integral equation as follows [1]:

$$c_s u_s + \int_{\partial\Omega_k} q^* u d\Gamma = \int_{\partial\Omega_k} u^* q d\Gamma \qquad (k = 1,2) \tag{6}$$

where u^* is the fundamental solution and $q^* = \partial u^* / \partial n$, $\partial\Omega_k$ is the entire boundary surrounding medium k. Then $\partial\Omega_k$ can be divided into a series of M_k boundary elements and the u and q can be approximated as follows:

$$u = [\varphi(x)]\{u\} \qquad (x \in \Gamma) \tag{7}$$

$$q = \frac{\partial u}{\partial n} = [\varphi(x)]\{q\} \qquad (x \in \Gamma) \tag{8}$$

here {u} and {q} are the column vectors of the nodal values which are generally defined on a set of nodes on the total boundary $\Gamma = \partial\Omega_1 + \partial\Omega_2$, $[\varphi(x)]$ is a row vector of the shape function defined on the total boundary to support the nodal values of u and q. By solving a linear algebraic system from the equation (6) the electrical field on the forced boundary can be obtained.

MULTIPLE NORMAL DERIVATIVES

The problem of treating corners is concentrated on the existence of several normal derivatives corresponding to different normal directions at a corner. From the viewpoint of mathematics or physics, it is reasonable to accept the above fact. It means that multiple normal derivatives, the number of which is always more than one, should be introduced at a corner. But, the difficulty, i.e. the number of equations formulated by using all nodes which define the nodal values of u and q is less than the number of nodal unknowns of u and q, may be caused when continuous elements are employed. Fortunately, only some corners will lead to the difficulty.

Before discussion we give some simple definitions about a node.

We call the unknown variables defined at a node as freedom nodes. The number of freedom nodes at node i is written as F_i, and the number of independent equations of node i is specified as E_i. E_i of node i is equal to k if the node belongs to k mediums simultaneously, since it can be used k times as the source point for building up the discrete boundary element equations.

Then, the classification of every node can be done. It is found that there are only two kinds of node when the concept of the multiple normal derivatives with continuous elements is employed. The first kind of node means that $F_j - E_j = 1$ at node j and the other one, $F_j = E_j$. Obviously, it is always true that $F_j = E_j$ when node j is not a corner node. Therefore, the following paragraphs will distinguish the different corners.

<u>First Kind of Corners</u>

The corner j shown in Fig. 2(a) is formed by two pieces of forced boundary on which Dirichlet condition is specified and a borderline between two mediums. Obviously, based on the concept of the multiple normal derivatives, at corner j there are three normal derivatives q_{j1}, q_{j2} and q_{j3} defined by the corresponding normal directions $\mathbf{n}_{j1}$, $\mathbf{n}_{j2}$ and $\mathbf{n}_{j3}$ of three joining adjacent elements e_{j1}, e_{j2} and e_{j3} respectively. All of them should be taken as independent variables, i.e. freedom nodes, in the continuous linear elements. On the other hand, only two independent equations can be built up by using the node j twice as the source point in the medium 1 and 2. So, $F_j - E_j = 1$ for the corner node j, i.e. the number of freedom nodes defined at node j is one more than the number of the independent equations provided by corner node j.

To keep consistency between the freedom nodes and independent equations one more source point S on the boundary or borderline should be selected for building up an extra equation in medium 1 or 2. In fact, this equation can be built up anywhere but all known nodes which are located at two endpoints of every element. For simplicity, the middle point S, which is shown in Fig. 2(a), of the element e_{j1} is taken as a new source point of the fundamental solution u^* for the extra equation.

It should be pointed out that F_j is always only one more than E_j no matter how many mediums between two pieces of the forced boundary there are. This means that it is enough to add only one extra source point and use it once in either medium involving it. The corner in Fig. 2(b) intersected by two pieces, which become a straight line, of Dirichlet boundary and a borderline can be treated as a particular case of Fig. 2(a). It belongs to one of the first kind of corners.

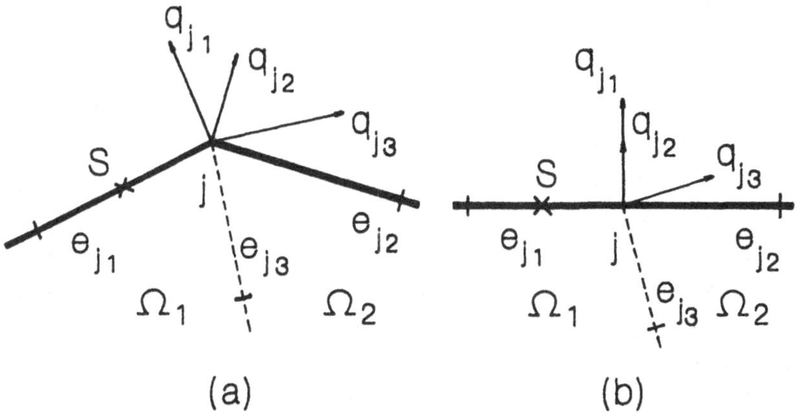

Fig. 2 The corner with Dirichlet conditions on both sides. (a) The corner composed of two mediums, (b) A particular case.

As shown in Fig. 3(a), there are two normal derivatives q_{j1} and q_{j2} defined by the corresponding n_{j1} and n_{j2} as well as the potential u_j as three freedom nodes at a corner j formed by two pieces of the borderline between medium 1 and 2. But only two independent equations can be built up by using the node j. Hence, again F_j is one more than E_j. One extra source point should be added on the borderline for one extra equation. It is selected at middle point S of the element e_{j1}, as shown in Fig. 3(a), for building up the extra equation in either medium 1 or 2.

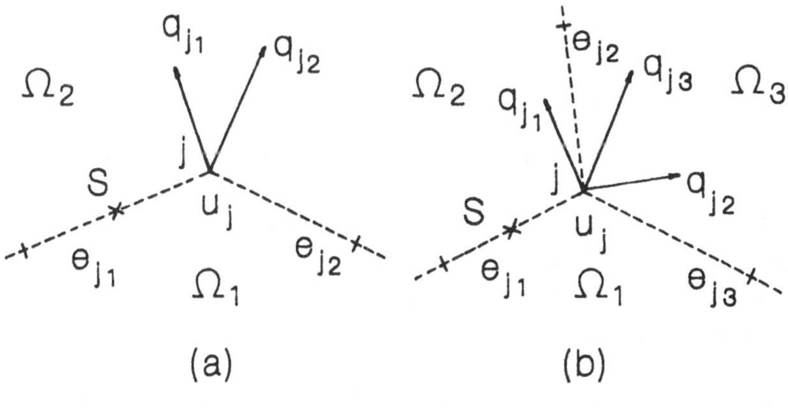

Fig. 3 The corner formed by the borderlines. (a) A corner composed of two mediums, (b) A corner composed of three mediums.

It is easy to see that the feature that F_j is one more than E_j is independent of the number of mediums forming the corner j. For example, in Fig. 3(b), $F_j - E_j$ is equal to one for corner j composed of three mediums. Therefore, increment of the freedom nodes for the first kind of corners is very small in the scheme of the multiple normal derivatives.

Second Kind of Corners

Fig. 4 and 5 give different configurations of the second kind of corners. Corner j formed by two pieces, which belong to two different mediums 1 and 2 respectively, of the natural boundary in Fig. 4 has two freedom nodes, one is the normal derivative q_{j3} defined by the the normal direction on the borderline and the other u_j. Obviously, the corner node j can be employed as the source point twice for contributing two independent equations. Therefore, no extra source point needs to be added anywhere in this case because of the consistency between F_j and E_j. Even if the corner j is composed of only one medium or many mediums the feature is still held.

A particular corner like Fig. 2(b) may be met when two pieces of Neumann boundary fall into one straight line. But we can deal with it as that of Fig. 4.

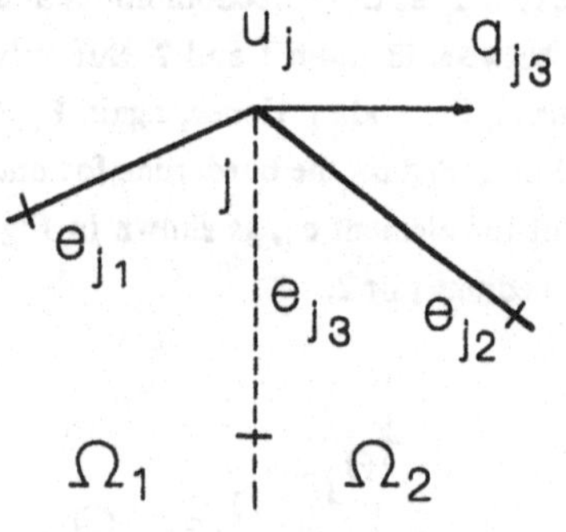

Fig. 4 The corner with Neumann conditions on both sides.

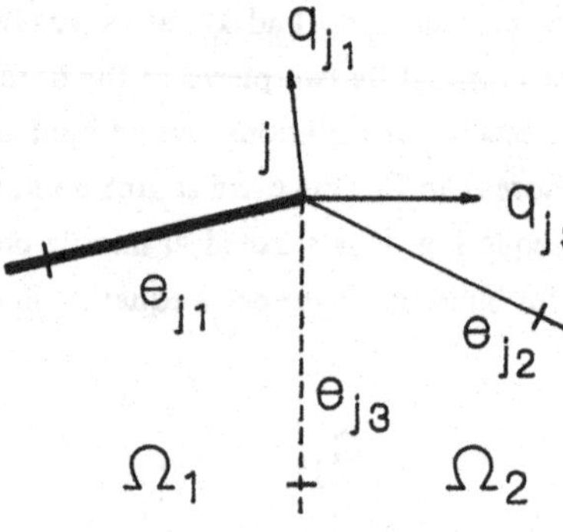

Fig. 5 The corner with mixed conditions on both sides.

The corner j of Fig. 5 is composed of two mediums between the forced boundary and the natural one. The normal derivatives q_{j1}, which is specified by n_{j1} of e_{j1}, and q_{j3} on the borderline should be introduced as two freedom nodes. F_j is equal to E_j in this case and no extra equation needs to be added.

We have discussed how many freedom nodes should be introduced in the different configurations of corner based on the concept of multiple normal derivatives at a corner. For a clear comparison between the multiple normal derivative and the partially discontinuous element, a table listing the number of freedom nodes required is given below. Suppose that all of the corners are composed of k mediums. We call the corners, as shown in Fig. 2, 3, 4 and 5, as I.a, I.b, II.a and II.b respectively.

Table 1. The comparison of freedom nodes at corner between Multiple Normal Derivative (M.N.D.) and Partially Discontinuous Element (P.D.E).

	M.N.D.	P.D.E.	P.D.E. – M.N.D.
I.a	k + 1	2k	k − 1
I.b	k + 1	2k	k − 1
II.a	k	2k	k
II.b	k	2k	k

Table 1 indicates that the freedom nodes introduced by the M.N.D. are always less than or equal to that by the P.D.E. The more mediums to form a corner there are, more advantage can be taken in reducing the size of the linear algebraic system.

NUMERICAL IMPLEMENTATION

In the multiple normal derivative we suppose that the potential u is continuous on the boundary or borderline even if at corner, but the normal derivative q at a corner is usually discontinuous and should be defined by several values which depend on the normal directions of the corresponding boundary elements. This means that more nodal values of the normal derivative will be introduced than the potential values.

Suppose that $\{u\}$ and $\{q\}$ are the column vectors of the nodal values of u and q respectively. $[\varphi(x)]$ and $[\psi(x)]$ are different row vectors of the shape function defined on the total boundary Γ involving borderline to support the nodal values of u and q respectively. Therefore, u and q can be presented as follows:

$$u = [\varphi(x)]\{u\} \qquad (x \in \Gamma) \tag{9}$$

$$q = [\psi(x)]\{q\} \qquad (x \in \Gamma) \tag{10}$$

Substitute the equations (9) and (10) into the integral equation (6) and a modification in the discrete BEM equations by using the concept of the multiple normal derivative can be formulated during the integral stage. Thus, numerical implementation of the multiple normal derivative is very simple after distinguishing each corner and giving appropriate symbols to the boundary elements which are connected with the corner node.

IMPROVEMENT OF PARTIALLY DISCONTINUOUS ELEMENT

Besides the above improvement of continuous linear elements an improved scheme of the partially discontinuous linear element is used in our simulator. It is easy to find that it is unnecessary to use the discontinuous elements at each corner like [4]. We use the discontinuous element only at the first kind of corners where the number of the freedoms is one more than that of the independent equations with the continuous element. The continuous elements can be used at the second kind of corner to avoid introducing unnecessary freedoms. For example, in Fig. 1, the boundary of the left resistor is divided into 60 linear elements. Then, 81 independent variables for freedom nodes are introduced when using the original partially discontinuous element, while only 68 variables when using improved either continuous or partially discontinuous element presented in this paper. In consideration of computational experience that total cost in the BEM is approximately proportional to (freedom nodes)$^{\gamma}$ where γ, $1.5<\gamma<3$, increases with growth of the freedoms, the improvement here can bring greater advantage, especially for complex boundary geometry with more corner nodes.

NUMERICAL RESULTS

1. Model Problem

To demonstrate the effect of the multiple derivatives at corners some results of a model problem will be shown below. This model resistor, which is similar to a steady heat conduction through a rectangular prism quoted in [1], is shown in Fig. 6. Its shape and boundary conditions are so simple that the analytic answer for potential u can be easily given as $u(x,y)=x/16$. Hence the exact resistance value is equal to 2 when $\sigma = 1$.

For a clear comparison we show four solutions using, (a) multiple normal derivatives at a corner, (b) continuous element without improvement [1], (c) original partially discontinuous element at all corners [4] and (d) discontinuous mesh with double close nodes at corners [2], in Table 2. In each solution of the BEM, there are four equal linear elements on DA and CB, two equal elements on AB and DC. In this table, Maximum error of u is defined as follows

$$\text{Max. err. of } u = \text{Max}|u_i - u_i^*| \tag{11}$$

where node i is located on boundary DA or CB, u_i is the numerical solution at point i, and u_i^* the exact value. Maximum error of q and resistor value is similar.

Table 2. Comparison of solution precision of methods (a), (b), (c) and (d).

	Max. err. of u	Max. err. of q	Max. err. of res. value
Method a	5.9×10^{-11}	2.0×10^{-12}	4.0×10^{-11}
Method b	2.9×10^{-2}	3.2×10^{-2}	5.4×10^{-1}
Method c	2.1×10^{-6}	7.4×10^{-7}	1.3×10^{-6}
Method d	8.4×10^{-5}	2.1×10^{-4}	1.4×10^{-3}

The results in Table 2 show that method (a) is the most precise one, if compared with other three methods. Astonishing agreement of its solution with the exact solution demonstrates correctness of the concept of multiple normal derivatives at a corner when using continuous linear element. In addition, it can be seen that only 12 freedom nodes are used for methods (a) and (b) in discretization mentioned above, but 16 for methods (c) and (d). Although methods (c) and (d) can give resistance value with enough precision, method (a) is the most attractive because of very high precision and less freedom nodes. A shortcoming of method (d) should be pointed out that very strong linear dependence will be induced in the algebraic system when the double nodes are too close.

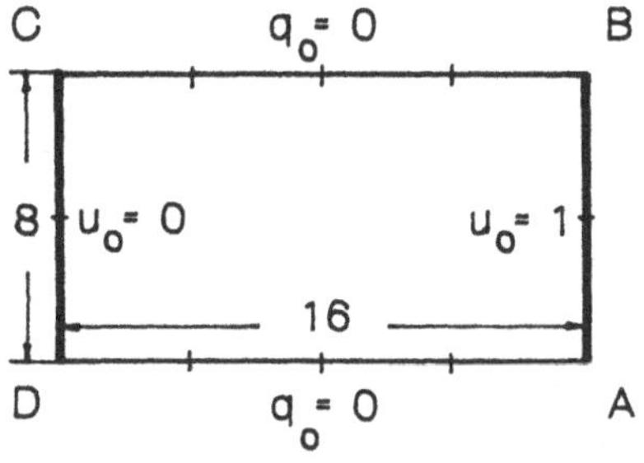

Fig. 6 A model resistor.

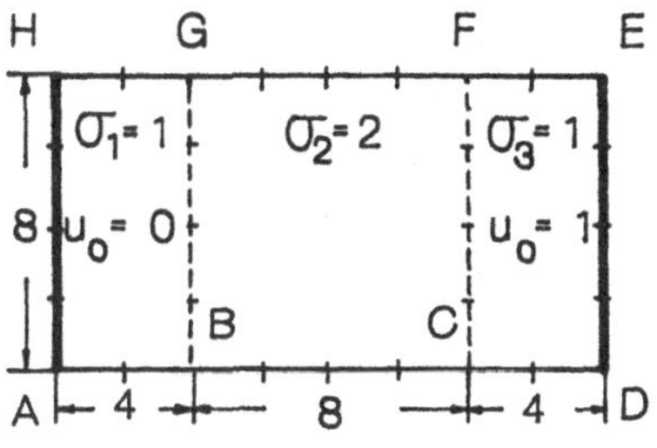

Fig. 7 A resistor with mediums.

2. Example with Mediums

A simple resistor composed of three mediums is shown in Fig. 7. The sheet conductivities and dimensions of three mediums and boundary conditions are specified in that figure. Because of its simplicity the exact answer of potential u and resistance value can be obtained. We calculate this resistance value by using the continuous linear elements with the multiple normal derivatives at a corner under two different partitions of the boundary. In the first division shown in Fig. 7, the size of each boundary element is equal to two and four elements are employed on each borderline between two mediums. Since all corners A–H, as shown in Fig. 7, belong to the second kind of corners it is unnecessary to introduce any extra source point. In the second partition, each boundary line in a medium is divided two elements. To demonstrate precision of two numerical solutions table 3 is given below. The definitions of Max. err. of u et al are similar to equation (11).

Table 3. Solution precision with the multiple normal derivative in a region composed of three mediums under two divisions..

Num. of unknowns	Max. err. of u	Max. err. of q	Max. err. of res. value
40	2.0×10^{-11}	5.2×10^{-12}	1.3×10^{-11}
24	6.3×10^{-11}	4.2×10^{-11}	7.2×10^{-11}

From Table 3 we can see that implementation of the multiple normal derivative at corners in a region of mediums is quite successful. The precision of two results is close to that in single medium shown in Fig. 6. Meanwhile, the scheme based on this concept is very economical. On the other hand, to treat 8 corners A to H by using the partially discontinuous element, it is necessary to increase 12 more freedom nodes than the multiple normal derivatives. Besides, when the first kind of corners with more than one medium is met, the number of freedom nodes introduced by the improved partially discontinuous element is usually more than that by the multiple normal derivatives.

CONCLUSIONS

A principle concept, multiple normal derivatives of a corner, for improving precision of the solution of Laplace's problem in treatment of the corners is presented. Each corner formed by one or several mediums belongs to either first or second kind of corners based on that concept. One extra source point located on the boundary or borderline for building an extra equation needs to be added only for the first kind of corners. On this basis two important improvements have been implemented in our resistance simulator. First we improve the continuous element. Second an improved partially discontinuous element scheme, in which no unnecessary normal derivative freedom is introduced, is implemented. It should be stressed that these improvements can be generalized to other higher order element, 3–D domain and adaptive technology as well.

ACKNOWLEDGMENT

This work was supported in part by the National Natural Science Foundation of China under Grant 69076430.

REFFERENCES

1. Brebbia, C.A. The Boundary Element Method for Engineers, Pentech Press,

1978.
2. Jaswon, M. and Symm, G.T. Integral Equation Methods in Potential Theory and Elastostatics, Academic Press, London, 1977.
3. Riccardella, P. An Implementation of the Boundary Integral Techniques for Plane Problems in Elasticity and Elasto–Plasticity, Ph.D. Thesis, Carnegie–Mellon University, Pittsburgh, 1973.
4. Patterson, C. and Sheikh, M. A. Interelement Continuty in the Boundary Element Method, Topics in Boundary Element Research, (Ed. Brebbia, C.A.), Vol. 1, pp. 123–141, Springer–Verlag and Berlin, 1984.
5. Mitra, A.K. and Ingber, M.S. Resolving Difficulties in the BIEM Caused by Geometric Corners and Discontinuous Boundary Conditions, Boundary Elements IX , Mathematical and Computational Aspects, (Ed. Brebbia, C.A. Wendland, W.L. and Kunh, G.), Vol. 1, pp. 519–532, CMP, Springer–Verlag, 1987.
6. Alarcon, A. Martin, A. and Paris F. Improved Boundary Elements in Torsion Problems, Proc. Int. Conf. on Recent Advances in B.E.M., Southhampton University, pp. 149–165, 1978.
7. Mitsuhashi, T. and Yoshida, K. A Resistance Calculation Algorithm and Its Application to Circuit Extraction, IEEE Trans. on Computer–Aided Design, Vol. 6, No. 3, pp. 337–345, 1987.
8. Kemp, A. J., Pretorius J. A. and Smit W. The Generation of a Mesh for Resistance Calculation in Integrated Circuits, IEEE Trans. on Computer–Aided Design, Vol. 7, No. 10, pp. 1029–1037, 1988.

An Overview of Integration Methods for Hypersingular Boundary Integrals

E. Lutz (*), L.J. Gray (**), A.R. Ingraffea (***)
(*) 486 Engineering and Theory Center, Cornell University, Ithaca, NY 14850, U.S.A.
(**) Mathematical Sciences Section, Oak Ridge National Labs, Oak Ridge, TN 37831, U.S.A.
(***) 317 Hollister Hall, Cornell University, Ithaca, NY 14850, U.S.A.

Abstract

Several methods of analyzing the hypersingular gradient BIE have been developed recently. This paper is a review highlighting the numerous common aspects and several differences among the methods. Significant common aspects include (a) a regularization of constant and linear terms, (b) analysis of integration points near rather than on the surface, and (c) analysis of the *neighborhood* of the singular point rather than of individual elements.

Introduction

Singular boundary integral equations have been used since the middle 1960's for solution of boundary value problems such as potential flow, wave scattering, and elasticity. Historically, there are three phases in the development of this technology. First, theoretical analysis of singular integrals began prior to the computer era. This style of analysis ([CZ52], [Mik65], [MP80], [Kup65]) is known as generalized potential theory and is still present today in the 'indirect' formulations of boundary integrals. Second, engineering researchers in the 1960's ([Riz69], [Cru69], [Jas63]) modified the formulation so that the abstract boundary integral would make direct use of true 'physical unknowns' (e.g. displacements and tractions in elasticity problems) rather than the elusive 'surface densities' used in the earlier indirect forms.

Third, a particular style of *impelementation* was developed ([LW76], [BB81], [BTW84], [BD88]) in the 1970's and 1980's by borrowing interpolation technology from the finite element method. This particular blend of theory and implementation is known as the 'boundary element method' (BEM). From an end-user's view, this method is competetive with (i.e. has advantages for some problems, disadvantages for others) the finite element and finite difference methods as a day-to-day engineering tool.

In the late 1980's, several methods have been developed to deal with the gradient (derivative) the boundary integral for the BEM. This integral contains 'hypersingular' terms whose behavior was not previously understood. This paper is a review highlighting the numerous common aspects and several differences among these methods. Significant common aspects include (a) a regularization of constant and linear terms, (b) analysis of integration points near rather than on the surface, and (c) analysis of the *neighborhood* of the singular point rather than of individual elements.

Organization

Section 1 introduces the hypersingular integration problem in the BEM. Section 2 gives the notation and terminology for the BIE and its gradient, and identifies several points that are in common to all methods of analyzing the hypersingular problem. Section 3 describes how the various methods differ from one another.

To simplify the presentation, mathematical examples are given only for the Laplace equation, with occasional reference to elasticity. A complete set of analogous forms for elasticity is found in [Lut91].

1 The Singular Integration Problem

From a programmer's view, the boundary element method presents challenging numerical problems associated with computation of singular integrals, i.e. integrals in which the integrand is unbounded at certain points in the domain of integration. This problem is resolved in the primary BIE by either (a) a Cauchy principal value analysis or (b) a rigid body motion argument.

Analysis of (a) the elastic field around a crack, (b) the electric field around a thin non-conducting sheet in a conducting medium, and (c) acoustic wave scattering in a medium with a thin rigid inclusion are all characterized by a single surface placed so that *both* of its (oriented) sides are incident to the same physical region. Sign changes in the integrands at mated points on opposite sides of the feature lead to cancellation of terms describing normal gradients of the solution fields (respectively, tractions, normal fluxes, and normal velocities) in the primary BIE.

An established remedy for this degeneracy is to subdivide the geometry into multiple regions so that opposite sides of the thin feature appear in different regions ([BIL81], [JSR90]). This is effective, but introduces additional geometric processing not present in pure boundary-only methods. An alternative is to seek some other boundary integral in which the solution functions on the singular surface are fully represented.

Several researchers (e.g. [Bui77], [Wea77], [PCH87], [IP88], [RB90], [CN90], [KSRR90], [GMI90], [Lut91]) have attempted to resolve this difficulty by applying some form of the the derivative (gradient) of the original BIE. Differentiation raises the degree of singularity in the integrands. The Cauchy singularity of the primary BIE becomes an apparently-more-difficult 'hypersingular' term, and new instances of the Cauchy type of singularity appear. An early analysis addressing many issues not resolved until the current time is found in ([CVB71]).

Cursory reading of the various analyses might leave one with the impression that there are significant differences among the assumptions, methods, and sometimes-complicated results. This is not the case. This paper is an overview that attempts to emphasize the common aspects of the various approaches.

2 The Boundary Integral Equation

This section gives the notation for the Laplace BIE and then identifies issues that are common to all analyses of the gradient BIE.

The primary boundary integral equation says that for any function that satisfies $\nabla^2 u = 0$ the solution values $u(\mathbf{x})$ for any point *not on the boundary S* of a region (see Fig. 1(a)) can be defined *exactly* by the following integral that involves only (a) the boundary S, (b) the solution values $u(\mathbf{y})$ and $\frac{\partial u(\mathbf{y})}{\partial \mathbf{n}(\mathbf{y})} S$, and (c) a kernel function G and its normal derivative H, in the form

$$\gamma(\mathbf{x}, S)u(\mathbf{x}) = \int_S G(\mathbf{x}, \mathbf{y})\frac{\partial u(\mathbf{y})}{\partial \mathbf{n}(\mathbf{y})} - H(\mathbf{x}, \mathbf{y}, \mathbf{n})u(\mathbf{y})\, dS(\mathbf{y})$$

or, omitting references to geometric parameters $\mathbf{x}$ and $\mathbf{y}$

$$\gamma \delta_{ik} u_k = \int_S G\frac{\partial u}{\partial \mathbf{n}} - Hu\, dS \tag{1}$$

The kernels G and H are defined by

	2D	3D
G	$\frac{1}{4\pi}\ln\frac{1}{r}$	$\frac{1}{4\pi r}$
$H = \frac{\partial G}{\partial \mathbf{n}(\mathbf{y})}$	$-\frac{1}{4\pi}\frac{\mathbf{r}\cdot\mathbf{n}}{r^2}$	$-\frac{1}{4\pi}\frac{\mathbf{r}\cdot\mathbf{n}}{r^3}$

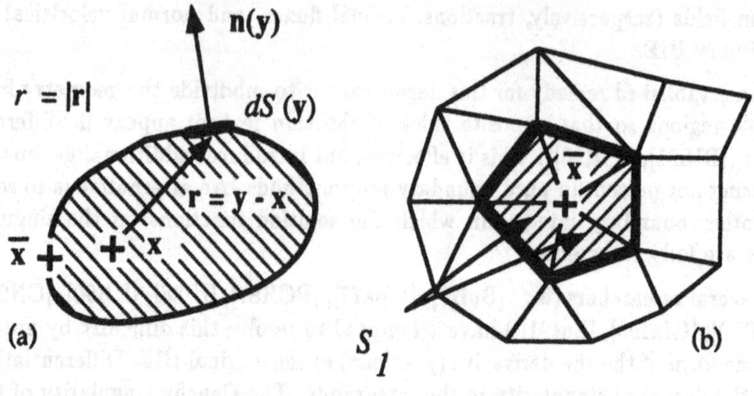

Fig. 1 (a) General (2D) body for boundary integrals.
(b) Patch of elements incident to singular point on 3D surface.

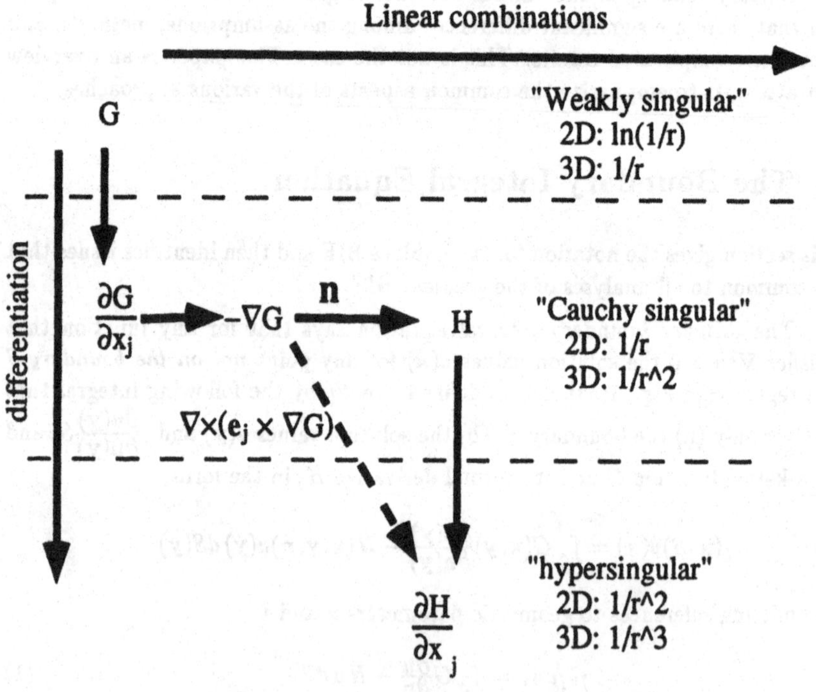

Fig. 2 Relationships and singularities of Laplace kernels

where $\mathbf{r} = \mathbf{y} - \mathbf{x}$, $r = |\mathbf{r}|$ and $\mathbf{n}$=$\mathbf{n}(\mathbf{y})$ is the outward surface normal at $\mathbf{y}$.

The coefficient $\gamma(\mathbf{x},S)$ is 1 for any interior point and 0 for any exterior point. Some types of singularity analysis define values of γ when $\mathbf{x}$ is exactly *on* the boundary, e.g. at surface point $\overline{\mathbf{x}}$. For the Laplace equation, this boundary value is exactly the including angle of the surface at the singular point. Other analyses leave γ 'undefined' on the boundary and instead produce expressions for (different) limiting values of the integrals when $\mathbf{x}$ approaches point $\overline{\mathbf{x}}$ from either the interior or exterior.

Taking partial derivatives of (1) with respect to a coordinate of $\mathbf{x}$ produces a corresponding boundary integral for partial derivatives of u:

$$\gamma \frac{\partial u(\mathbf{x})}{\partial x_j} = \int_S \frac{\partial G}{\partial x_j} \frac{\partial u}{\partial \mathbf{n}} - \frac{\partial H}{\partial x_j} u \, dS \tag{2}$$

The kernels H, $\frac{\partial G}{\partial x_j}$, and $\frac{\partial H}{\partial x_j}$ in (2) and (1) are derived from G by a sequence of differentiations, as in Fig. 2. As differentiation raises the singularity of the kernel, computations for points $\mathbf{x}$ that are approaching a surface point $\overline{\mathbf{x}}$ (see Fig. 1(a)) become increasingly difficult.

When dealing with discretized surface functions, integrations are carried out only over some small portion of the surface. In subsequent sections, S_1 always denotes this local patch (Fig. 1(b)).

2.1 Regularization

All analyses of the gradient BIE depend on isolation of the various degrees of singularity. This is done by a regularization step as follows. Write the potential function in the neighborhood of a surface point $\overline{\mathbf{x}}$ as

$$u(\mathbf{y}) = \overline{u} + \overline{a_i}(y_i - \overline{x_i}) + w(y) \tag{3}$$

where $\overline{u}$ and the $\overline{a_i}$ are the potential and fluxes at $\overline{x}$ itself and $w(\mathbf{y})$ contains residual terms of order $O(|\mathbf{y} - \overline{\mathbf{x}}|^2)$. Because (3) is a Cartesian expansion (not just a surface function), it can be differentiated to obtain

$$\frac{\partial u(\mathbf{y})}{\partial y_j} = \delta_{ij}\overline{a_i} + \frac{\partial w(\mathbf{y})}{\partial y_j} \tag{4}$$

where $\frac{\partial w(\mathbf{y})}{\partial x_j}$ is of order $O(r)$.

Applying the series expansion, the gradient BIE (2) over surface patch S_1 becomes

$$\delta_{ij}\overline{a_i} + \omega(\mathbf{x}) \quad = \quad -\overline{u} \boxed{\int_{S_1} -\frac{\partial H}{\partial x_j} \, dS} \tag{5}$$

$$+\boxed{\overline{a}_i \int_{S_1} \frac{\partial G}{\partial x_j} \delta_{ij} n_i \frac{\partial H}{\partial x_j} (y_i - \overline{x}_i)\, dS} \tag{6}$$

$$+\int_{S_1} G \frac{\partial w}{\partial y_i} n_i - \frac{\partial H}{\partial x_j} w\, dS \tag{7}$$

(5) and (6) are the difficult (more than weakly singular) terms that require special analysis. An important aspect of (6) is that the integral involves both $\frac{\partial G}{\partial x_j}$ *and* the hypersingular $\frac{\partial H}{\partial x_j}$ kernel multiplied by a linear factor. This is in contrast to the standard implementation practice of treating each kernel (multiplied by a surface function) as an independent integration problem.

The following paragraphs outline several points where integration of the gradient BIE requires revision of the standard way of organizing BIE computations.

Domain of integration Except for the Cauchy singularity, integrations of the primary BIE can proceed 1 *element* at a time. That is, in Fig. 1(b), integrals for $\overline{\mathbf{x}}$ can be done separately over the various incident elements. However, the boxed singular expressions (5) and (6) are well defined only over the entire surrounding neighborhood S_1 around $\mathbf{x}$; there is a single value for all of S_1 in the figure.

Surface Values versus Limiting Values Another feature common to all analyses is that they never allow the singular point to actually be placed on the surface. Point $\mathbf{x}$ of (1) is always considered to be *close to* some surface point $\overline{\mathbf{x}}$ (Fig. 1(a)). Each method provides some argument for existence of a finite limit.

When dealing with cracks, both directions of approach are significant. A point approaching one side of a crack will appear to be on the *inside* of (oriented) elements on that crack face and *outside* of oppositely oriented elements on the opposing face.

Continuity Requirements and the Cartesian Series Expansion The use of a series expansion for *both constant and linear terms* imposes a difficult continuity requirement, particularly at corners. The reader may consult the references for examples of the creative methods that have been applied to achieve this continuity in the absence of simple C^1 element formulations.

Use of the Cartesian (rather than parametric) series does not preclude the usual independent parametric interpolation of potential and flux. However, it does require that the implementation be able to transform the parametric expansions into Cartesian forms of the appropriate continuity at any surface point where the gradient BIE is to be computed.

3 Summary of Methods

This section summarizes the general strategies that are used in various analyses of hypersingular integrals. Fig. 3 shows a classification of approaches. At the top of the decision tree, there is an initial decision as to whether the various kernels in the integrals will be (a) expanded into their (sometimes lengthy) algebraic forms or (b) left in generalized form.

The left side of the tree leads to methods based on manipulating the algebraic expansions. The crucial aspect of all these methods is to include explicitly the distance of $\mathbf{x}$ to the surface point $\overline{\mathbf{x}}$ in the initial statement of the integral. This makes the algebra more complicated than a surface-only expansion, but the limiting values of the integrals turn out to be well defined as the distance vanishes. There is a lower level branch based on whether or not the algebraic manipulations are done in the context of an 'exclusion zone' on the surface.

Methods in the right branch of the tree do not expand the kernels. Instead, they invoke special properties that the kernels posess *because of their special relationship to the differential equation.* The ability to achieve convincing results by this generalized reasoning is an illustration of the remarkable power of Green's functions.

3.1 Modal Solution Method

The modal method is a generalized analysis that produces extremely concise results. By considering (as isolated problems) (a) a globally constant potential solution mode $u(\mathbf{x}) = \overline{u}$ and then (b) a globally linear potential solution mode $u(\mathbf{x}) = a_i(x_i - \overline{x}_i)$, it can be shown ([Lut91]) that

$$\int_{S_1} \frac{\partial H}{\partial x_j}\, dS = -\int_{S_2} \frac{\partial H}{\partial x_j}\, dS \tag{8}$$

and, for each i and j,

$$\begin{aligned}\int_{S_1} (G\delta_{ij} - \frac{\partial H}{\partial x_j}(y_i - \overline{x}_i)\, dS &= \gamma(\mathbf{x}, S_1 + S_2)\delta_{ij} \\ &\quad -\int_{S_2} (G\delta_{ij} - \frac{\partial H}{\partial x_j}(y_i - \overline{x}_i)\, dS\end{aligned} \tag{9}$$

where S_2 is any surface (*not necessarily the remainder of the actual geometry*) such that (a) $\mathbf{x}$ is not on either surface and (b) the composite surface $S_1 + S_2$ is closed (i.e. encloses volume), as in the generic 2D Fig. 4(a). If $\mathbf{x}$ is not near to S_2, this converts the difficult integral over S_1 to a non-singular one over S_2. Comparable expressions apply to elasticity, with tensor operators but no essential analytic differences.

For non-crack points, S_2 may be taken to be all elements other than those actually incident to the singular point. For crack-surface points, this would include points on the 'far' crack surface, which are geometrically near the singular point, hence making the S_2 integral just as hard as the S_1 integral. This is

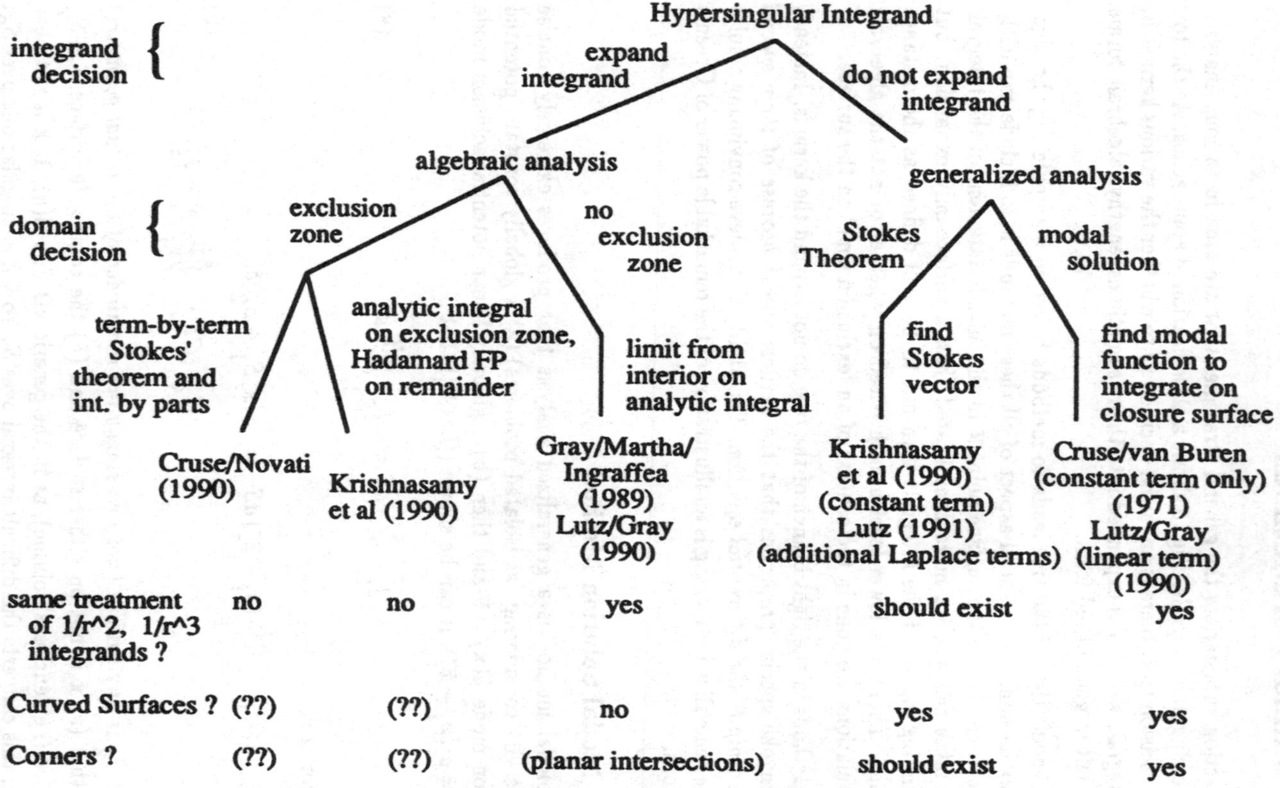

Fig. 3: Decision tree for hypersingular integration

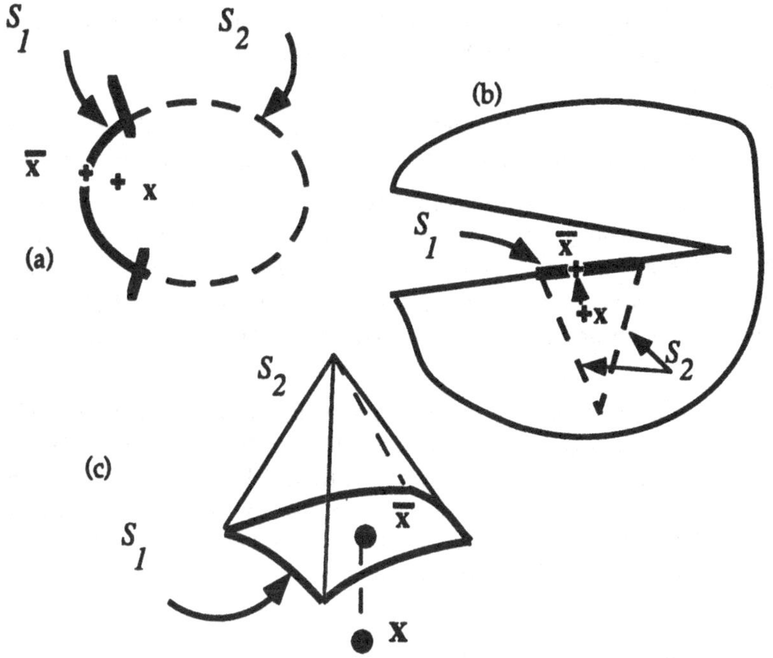

Fig. 4 Closure Surfaces for modal integration. (a) generic complete solid; (b) special construction over 2D crack surface; (c) ruled surface construction over 3D crack surface.

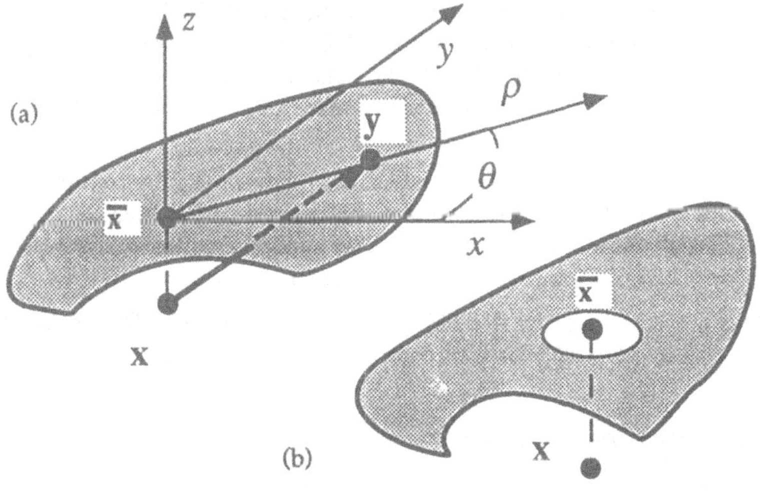

Fig. 5 Surface patches for analysis of hypersingular integral as x approaches the surface. (a) polar coordinate system; (b) typical exclusion zone.

remedied by constructing a temporary surface. Fig. 4(b) shows a construction for a 2D crack surface, and Fig. 4(c) shows a ruled-surface closure over a 3D surface patch.

These expressions are a higher order analogy of the well known 'rigid body mode' arguments used in the primary BIE, hence the name 'modal method'.

These expressions are attractive because they (a) are extremely concise, (b) can be implemented using only repeated calls to existing integration routines for the integral over S_2, (c) are independent of the geometry of S_1 (i.e. apply when S_1 is curved and has corners), (d) apply in both 2D and 3D, and (e) use only general reasoning, with no manipulation of the algebraic expressions for the kernels.

An interior-to-exterior jump occurs in (9) when $\mathbf{x}$ crosses the surface and the value of γ changes between 0 and 1. The facts that the integrals (a) are finite and (b) approach fixed limits when $\mathbf{x} \rightarrow \overline{\mathbf{x}}$ from either direction are clear because all terms on the right are well behaved as long as $\overline{\mathbf{x}}$ is not near the closure surface. One difficulty that is introduced is that as $\overline{\mathbf{x}}$ is moved closer to the boundary of S_1 the integrals over S_2 become nearly-singular for some parts of S_2.

3.2 Limit from the Interior

The 'limit from the interior' process is an algebraic analysis with no exclusion zone. As in Fig. 5(a), $\mathbf{x}$ is considered to approach a planar surface. At a (small but finite) distance from the surfaces, all kernels (hypersingular, Cauchy, and weaker, if desired) are written out in full and integrated analytically in the ρ direction. Simple limit analysis at this point cancels all distance-to-surface terms. A jump between interior and exterior values of the Cauchy singularity (6) occurs due to ratios of signed and unsigned distance-to-surface terms at the time limits are taken.

The result of this process is a contour integral over the outer contour of the patch. This is well defined as long as that contour remains far from the singular point. ([GL90]) show that the process can be applied simultaneously to planar surfaces with a common edge at the singular point.

This process is used by ([GL90]), ([RB90]), and ([Gra89]) for the Laplace equation and by ([GMI90]) for elasticity.

3.3 Exclusion Zone Methods

Krishnasamy *et al* [KSRR90], Cruse *et al* ([CN90]) and Iokimidis *et al* [IP88] use algebraic methods with an exclusion zone Fig. 5(b). They are also known as (Hadamard) finite parts approaches. These are a higher-order analogy of Cauchy principal value methods. The key elements of these analyses are that (a) an exclusion zone is placed around the singular surface point, (b) the interior or exterior point $\mathbf{x}$ approaches the surface faster than the exclusion zone vanishes. Arguments follow to show that integrals over the exclusion zone and the non-excluded part are all either finite or have infinite parts that are cancelled by matching terms from the other surface portion.

A common requirement in these methods is that the distance between $\mathbf{x}$ and $\bar{\mathbf{x}}$ vanish faster than the size of the exclusion zone. Portions of ([KSRR90]) are identical to the limit-from-the interior process, with the rate-of-vanishing requirement forcing the exclusion zone to appear to be a finite size relative to the vanishing distance-to-surface. The notation in ([CN90]) is different because it is based on an indirect (displacement discontinuity) form of the BIE. Their regularization is not done *a priori*, but results from applications of Stokes' theorem and the divergence theorem to various individual terms.

All of these approaches defer the linear term (6) of the regularized expression for treatment by Cauchy principal value analysis. The jump between interior and exterior values occurs due to the free term of the Cauchy principal value analysis.

3.4 Generalized Stokes Methods

When a surface integrand can be shown to be the curl of another vector function, Stokes' theorem can be applied to convert a surface integral to a contour integral that is completely independent of the surface geometry. Fig. 3 has a branch for *generalized* application of this conversion. This is an incomplete branch of the tree.

([KSRR90]) gives an example of such conversion for the integral (5). The resulting vector expression is shown in Fig. 2. The corresponding conversion of the hypersingular elasticity S_{ijk} kernel is given in ([Lut91]), along with additional cases for the Laplace equation. In principle, any integration done by the modal method can be done by a contour integral that is independent of the surface geometry. The only elasticity conversions known at present are the S_{ijk} case just noted and the primary BIE Cauchy case given in ([GM87]).

4 Conclusions

We have discussed the common aspects of several approaches to computing the integrals in a hypersingular BIE. It is to be emphasized that the various results are fundamentally equivalent.

The modal methods are easy to work with if (a) the data structures in use permit construction of the necessary (temporary) closure surface for crack surface points and (b) integration procedures are accurate for near-singular integrals that often occur on the closure surface. ([Lut91]) demonstrates several examples of effective application of this method, including an example analysis of cracks in the US space shuttle fuel door hinge.

The methods grouped here as 'algebraic' present grueling analysis and coding tasks, particularly for curved surfaces. However, the analytic results may be highly efficient.

Conversion to contour integrals via Stokes theorem is a very attractive operation when done in a generalized manner.

All methods are bound by the physical constraint that the integrations are well defined only when there is a C^1 interpolation of the solution function on the surface so that the *Cartesian* (rather than parametric) series expansions can be applied.

The analyses summarized in this paper have greatly improved our understanding of the behavior of the gradient BIE. These methods are now being used to model cracks and crack-like features. More broadly, the rigors of the hypersingular case have contributed to a clearer understanding of the milder singularities in the primary BIE.

Acknowledgements: Financial support was provided by a fellowship from the Unisys Corporation. and by the Applied Mathematical Sciences Program, Office of Energy Research, US Department of Energy under contract DE-AC05-84OR21400. Computer facilities were provided by the Cornell Fracture Group and Cornell University Program of Computer Graphics, funded by Dowell Schlumberger and the Digital Equipment Corporation.

References

[BB81] P.K. Banerjee and R. Butterfield. *Boundary Element Methods in Engineering Science.* McGraw Hill, 1981.

[BD88] C.A. Brebbia and X.Y. Dominguez. *Boundary Elements: An Introductory Course.* Computational Mechanics Publications (McGraw-Hill), 1988.

[BIL81] G.E. Blandford, A.R. Ingraffea, and J.A. Liggett. Two dimensional stress intensity factor computations using the boundary element method. *International Journal for Numerical Methods in Engineering*, 17:387–404, 1981.

[BTW84] C.A. Brebbia, J.C.F. Telles, and L.C. Wrobel. *Boundary Element Techniques - Theory and Applications in Engineering.* Springer-Verlag, 1984.

[Bui77] H.D. Bui. An integral equation for solving the problem of plane crack of arbitrary shape. *J. Mech. Physics and Solids*, 25:23–39, 1977.

[CN90] T.A. Cruse and G. Novati. Traction bie formulations and applications to non-planar and multiple cracks. In *22nd ASTM Conference on Fracture Mechanics.* ASTM, 1990.

[Cru69] T.A. Cruse. Numerical solutions in three dimensional elastostatics. *International Journal of Solids and Structures*, 5:1259–1274, 1969.

[CVB71] T.A. Cruse and W. Van Buren. Three-dimensional elastic stress analysis of a fracture specimen with an edge crack. *International Journal of Fracture Mechanics*, 7:1–15, 1971.

[CZ52] A.P. Calderon and Zygmond. On the existence of certain singular integrals. *Acta Mathematica*, 88:85–139, 1952.

[GL90] L.J. Gray and E.D. Lutz. On the treatment of corners in the boundary element method. *Journal of Computational and Applied Mathematics*, 32:369–386, 1990.

[GM87] N. Ghosh and S. Muhkerjee. A new boundary element method formulation for three dimensional problems in linear elasticity. *Acta Mechanica*, 67:107–119, 1987.

[GMI90] L.J. Gray, Luiz F. Martha, and A.R. Ingraffea. Hypersingular integrals in boundary element fracture analysis. *International Journal for Numerical Methods in Engineering*, 29:1135–1158, 1990.

[Gra89] L.J. Gray. Boundary element method for regions with thin internal cavities. *Engineering Analysis*, 6:180–184, 1989.

[IP88] N.I. Ioakimidis and M.S Pitta. Remarks on the gaussian quadrature rule for finite-part integrals with a second-order singularity. *Computer Methods in Applied Mechanics and Engineering*, 69:325–343, 1988.

[Jas63] M.A. Jaswon. Integral equation methods in potential theory, i. *Proceedings of the Royal Society*, 275:23–32, 1963.

[JSR90] Z.H. Jia, D.J. Shippy, and F.J. Rizzo. Boundary-element analysis of wave scattering from cracks. *Communications in Applied Numerical Methods*, 6:591–601, November 1990.

[KSRR90] L.W. Krishnasamy, L.W. Schmerr, T.J. Rudolphi, and F.J. Rizzo. Hypersingular boundary integral equations: Some applications in acoustic and elastic wave scattering. *Journal of Applied Mechanics*, 57:404–414, 1990.

[Kup65] V.D. Kupradze. *Potential Methods in the Theory of Elasticity*. D. Davey, 1965.

[Lut91] E.D. Lutz. *Numerical Methods for Hypersingular and Near-Singular Boundary Integrals in Fracture Mechanics*. PhD thesis, Cornell University, Ithaca, NY, USA, 1991.

[LW76] J.C. Lachet and J.O. Watson. Effictive numerical treatment of boundary integral equations: A formulation for three-dimensional elastostatics. *International Journal for Numerical Methods in Engineering*, 10:991–1005, 1976.

[Mik65] S.G. Mikhlin. *Multidimensional Singular Integrals and Singular Integral Equations*. Pergamon Press, 1965.

[MP80] S.G. Mikhlin and S. Proessdorf. *Singular Integral Operators*. Springer-Verlag, 1980.

[PCH87] E.Z. Polch, T.A. Cruse, and C.-J. Huang. Traction bie solutions for flat cracks. *Computational Mechanics*, 2:253–267, 1987.

[RB90] M. Rezayat and T. Burton. A boundary-integral formulation for complex three-dimensional geometries. *International Journal for Numerical Methods in Engineering*, 29:263–273, 1990.

[Riz69] F.J. Rizzo. An integral equation approach to boundary value problems of classical elastostatics. *Quarterly of Applied Mathemathics*, 25:83–95, 1969.

[Wea77] J. Weaver. Three dimensionial crack analysis. *International Journal of Solids and Structures*, 13:321–330, 1977.

SECTION 19: COMPUTATIONAL ASPECTS

Conforming and Non-Conforming Elements in a Three-Dimensional H-Adaptive Boundary Element Method

S.H. Crook, R.N.L. Smith

Applied & Computational Mathematics Group, RMCS, Shrivenham, Swindon, Wiltshire, SN6 8LA, U.K.

ABSTRACT

Self-adaptive boundary element methods use some indicator of local error on each element to redesign (and hopefully improve) an existing mesh. An h-adaptive scheme works by increasing the number of elements in the mesh. In two dimensions, a given element is only coupled to the adjacent element at one node and it may therefore be subdivided without any fundamental effects on the adjacent elements. However, in three dimensions elements are joined along edges and a subdivision of one element means that the edge representation may no longer be continuous. We consider two approaches; one, that the new nodes on the subdivided element are assumed to move isoparametrically on the smaller (subdivided) elements and two, that these new nodes are constrained to follow the same isoparametric variation as the undivided element. Clearly, the first kind of subdivision is discontinuous or non-conforming and the second, continuous or conforming. Both types of subdivision are successfully applied to a test problem and we discuss the relative merits of the two methods.

INTRODUCTION

Boundary element methods are now being applied to more and more complex problems and therefore the discretisation, or mesh, becomes more difficult to design. Elements must model the variation of geometry, traction and displacements efficiently in order to obtain an accurate solution. If any one of these three parameters cannot be modelled effectively by the approximating elements then the number of degrees of freedom must be increased. Part of the boundary solution will always be unknown so that some intuition is required in distributing elements. The quality of the mesh is therefore dependent on the knowledge of the scientist or engineer who formulates the problem. A method that can automatically refine the mesh based on results from an initial mesh overcomes these difficulties. Such methods have been developed in recent years and are known as adaptive or self-adaptive refinement methods. At present these are of three basic kinds: a polynomial or p-type, an element subdivision or h-type, and a nodal distribution or r-type method. Hybrid methods combining the merits (and demerits) of any of the methods are also possible, a combination of h- and p-adaptive methods known as an hp-adaptive method, being the most common.

Much of the work on the self-adaptive BEM parallels developments in the FEM and many of the results quoted are based on well established results obtained for the FEM. Some of these results are without explicit proof for the BEM and thus may only be implied by the numerical work. Nevertheless, the majority of the numerical work supports the assumption of a broad similarity between BE and FE adaptive behaviour. Adaptive methods typically start with a simple mesh and compute a first approximation to the solution. The accuracy of this solution can then be assessed using suitable *error estimators*. If excessive error is evident then the mesh is refined in some fashion, with local *error indicators* on each element guiding where refinement is required. A further solution is calculated using this new mesh and the procedure repeated. Successive steps are made in this way, continuing until the error estimator indicates that no further refinements are necessary.

RESIDUALS AND ERROR ESTIMATES

Adaptive BEM techniques commonly base error estimators and indicators on an approximation to the *residual function*. The integral equation statement of the BEM for elasticity (omitting body forces) may be written as,

$$c_{ij}(x)u_j(x) + \int_S T_{ij}(x,y)u_j(y)dS_y = \int_S U_{ij}(x,y)t_j(y)dS_y \tag{1}$$

for a field point x and source points y on the boundary of a discretized mesh. $T_{ij}(x,y)$ and $U_{ij}(x,y)$ are the known fundamental solutions. If boundary tractions t, displacements u and geometrical coordinates are represented by simple interpolation polynomials based on collocation nodes then, after numerical integration, we have the usual system in matrix form,

$$HU = GT \tag{2}$$

See, for example, Brebbia and Dominguez [1]. After Gauss elimination, the approximate solutions should satisfy Equation (2) at any collocation (or mesh) point. However, for non-mesh points the approximation will have some degree of error associated, with it, i.e.,

$$HU - GT = R \tag{3}$$

where the residual R is in general non-zero. The residual R is calculated by letting the field point in Equation (1) be a non-mesh point. There are no equations to be solved, however, as all the tractions and displacements are known at the nodes. The errors in traction and displacement satisfy the exact integral equation with the same residual as the approximate solution. The residual Equation (3) may thus be considered as an approximate weighted integral of the actual errors, where the weighting functions impart a local character because of their singular behaviour.

An error estimator provides an indication of the global error in the solution, and is used to determine whether further refinement of the mesh is necessary. An error estimator η should satisfy a condition of the form

$$C_1\eta \leq \|e\| \leq C_2\eta \tag{4}$$

where $e = U - u$ is the difference between the exact solution and the approximate solution, and the constants C_1, C_2 are independent of u. A similar relationship between estimators and actual errors has been observed to hold numerically for a large class of problems in the FEM, and is assumed to hold for the BEM. An integral norm of the residual function is often used as an error estimator in the BEM. Researchers such as Parreira [4], [5], have used L_2-norms,

$$\eta^2 = \|r\|_{L_2}^2 = \int_\Gamma \sum_i r_i^2 d\Gamma \tag{5}$$

and Rank [8], [9] H^1-norms,

$$\eta^2 = \|r\|_{H^1}^2 = \int_\Gamma \sum_i \left[r_i^2 + \left(\frac{\partial r_i}{\partial s}\right)^2\right] d\Gamma \tag{6}$$

where r_i is the residual in the x_i coordinate direction. In this paper we only consider the L_2-norm.

Using Equation (5), error indicators can be defined over each of the N_e element subdivisions of the mesh, where $\Gamma = \cup_e^{N_e}\Gamma_e$, so that,

$$\chi_e^2 = \int_{\Gamma_e} \sum_i r_i^2 d\Gamma = \sum_i \int_{\Gamma_e} r_i^2 d\Gamma \tag{7}$$

and for ease of computation the error estimator can be related to the error indicators,

$$\eta^2 = \sum_e \chi_e^2. \tag{8}$$

Error indicators are thus local to a given element and are used to decide if local mesh refinement is required.

<u>Indicators and mesh refinement</u>

Actual evaluation of an error indicator for a given element requires some estimate of the integral of the residual function. The aim of any integration scheme used should be to minimize the number of sample values of residual required to calculate a sufficiently accurate result. This is a most important consideration as the information used to calculate the residual is relatively expensive to obtain and is redundant at the next step in many adaptive approaches. In three dimensions, the accuracy required in estimates of residuals and the effectiveness of integration schemes for error indicators are considered by Crook and Smith [2]. Numerical integration with 2 by 2 Gauss points is used for the indicators employed here.

Note that error indicators need not be determined as accurately as the final BEM solution. An error in the indicators may cause unnecessary refinement or fail to implement required refinement. This is inefficient but does not introduce any additional error into the BEM solution.

Once error estimators and indicators are available, largely heuristic judgements are made as to which elements to refine. We must assume that if one indicator is larger than another, then the associated element has a larger error. The scheme used in our current program is that refinement occurs if the maximum error indicator on the element is within the *error indicator tolerance*, α, of the maximum of all the error indicators. That is, refinement occurs for element j if,

$$\chi_j^2 \geq \alpha \max_e \chi_e^2. \tag{9}$$

A choice of $\alpha = 1.0$ normally results in only one element being refined at each stage and $\alpha = 0.0$ gives uniform refinement where all elements are refined at each stage. The value of α may have an influence on the final solution but at present no particular choice can be advanced in favour of any other. Rank [9] chose a value of 50% of the maximum for refinement in two dimensions but comments that "the numerical results proved not to be very sensitive to the choice of this parameter". It may be true that the optimum value is problem specific and if so, then the ultimate choice should be left to the individual preference of the user. Here α is set to 0.8, and the adaptive process is continued until the maximum mesh size is reached.

h-ADAPTIVE REFINEMENT

In an *h*-adaptive scheme, refinement is made by increasing the number of elements, thereby reducing the average size of the elements. Parreira [6] has used, in two-dimensional analysis, the error indicators to define in which direction refinement is required. In three dimensions a similar scheme could be developed where the residuals are 'resolved' into components in the local coordinate directions. The element refinement would then be graded differently in each local coordinate direction. The main problem with this sort of refinement would be the large number of nodes generated, whereas if refinement is uniform in each direction (the element is divided into four quarters) then nodes on adjacent elements may coincide. Using this simple subdivision, the overall problem size should not increase too dramatically with increasing refinement.

The refinement scheme adopted here is to divide the element into quarters which are determined by the quadrants of the local coordinate system. Figure 1 shows the original nodes as (∘) and the introduced nodes as (•) where the element on the right has been refined. Using this refinement scheme one element becomes

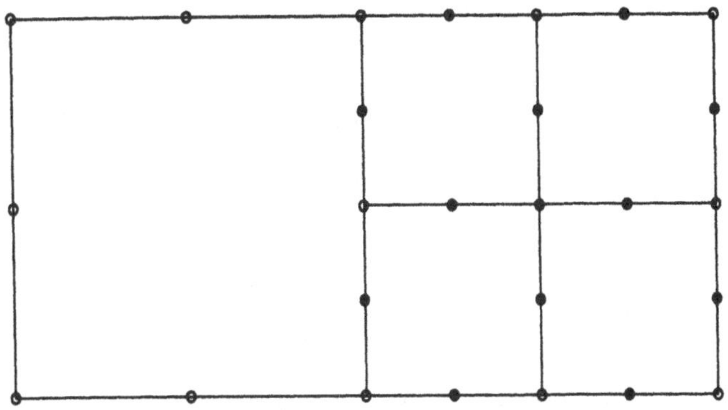

Figure 1: Refinement of an element for an h-adaptive process

four and thirteen new nodes are introduced. Continuity is maintained within the refined element but may not be across element boundaries. On adjacent refined elements (if both elements were refined) some of the introduced nodes may coincide. Such nodes are treated as one node thereby reducing the total number of new introduced nodes and most importantly, in this way a linearly dependent system of equations is avoided.

A node that is positioned on the boundary between adjacent elements and is considered to belong to one of the elements but not the other is termed *interstitial* here. Normally implementations of the BEM seek to avoid such nodes as the integrals they introduce are difficult to deal with. Continuity of geometry, displacement and traction is not, in general, maintained between elements and integration involving the interstitial nodes is complicated.

Two models have been developed to cope with the mesh topology generated by interstitial nodes. The first, where local solution continuity is enforced between elements and the second, where it is allowed to be partially discontinuous. In Figure 1 the two nodes introduced on the boundary between the standard and the subdivided elements are examples of interstitial nodes. They are given special treatment by the program whichever model is being used and integration routines have been introduced to deal specifically with them.

CONFORMING ELEMENT MODEL

The continuous element model treats all interstitial nodes as nodes where the solution is known a *priori*. The overall philosophy of this model is that at the interstitial node all displacement and traction values are either known as boundary conditions or may be described in terms of the surrounding nodal solutions. These nodes may be disregarded for the solution of the system of equations formed thus reducing the size of the problem.

Four different boundary condition pairs have to be considered for the interstitial node, u-u, u-t, t-u and t-t, where u represents a displacement boundary

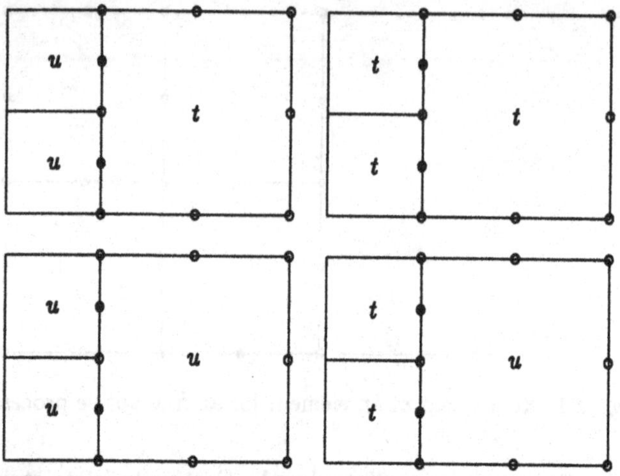

Figure 2: Boundary Condition Couplings for Refined Meshes

condition and t a traction condition. These configurations are illustrated in Figure 2 where the interstitial nodes are shown as solid circles (•) and other nodes as simple circles (∘). In each case a subdivision has been made in one element but not in the adjacent element. The interstitial nodes at the mid-edge of the smaller elements are used in the representation of the local geometry and boundary conditions and then disregarded. There is only one unknown quantity at the interface providing that the boundary conditions in the second and third cases are the same for both small and large elements. The solution at the mid-edge nodes may be expressed as a linear combination of the values at the nodes from the large element lying on the interface.

The program used here forms the H matrix in its entirety by considering each element in turn and looping through the nodes. The formation of the matrix is therefore column by column. Boundary conditions are applied in two stages. Firstly, the G integrals for the traction conditions are computed, multiplied by the boundary condition, and then put directly into the equation right hand side. Secondly, displacement conditions are applied by swapping H contributions calculated previously into the right hand side. The formation of the influence matrix is completed by computing the G integrals to replace the swapped out contributions.

For the continuous model, contributions to the influence matrix are omitted when integrating from the interstitial nodes. When the source element contains such a node, however, integrals are collected in the normal way and put into the H matrix. The matrices are thus the same as that for a conforming mesh except that rows corresponding to the interstitial nodes are empty. The contributions in the columns for the special nodes may then be apportioned into the matrix according to the expression of the solution at the node. A simple example clarifies this point (for a node with local coordinates *on the larger element* of $\xi = -1, \eta = \frac{1}{2}$). The displacement at the interstitial node u_I is represented by the shape functions and the nodal values on the large element,

$$u_I = \sum_i^n \phi_i u_i. \tag{10}$$

The shape functions may be evaluated at the node giving,

$$u_I = -\frac{1}{8}u_1 + \frac{3}{8}u_7 + \frac{3}{4}u_8. \tag{11}$$

Nodes 1, 7 and 8 are the nodes on the interface between the large and small elements. Any contributions for the interstitial node would therefore be added in proportions $-\frac{1}{8}u_1, \frac{3}{8}u_7, \frac{3}{4}u_8$ to the contributions at the interface nodes 1, 7 and 8 respectively. The column corresponding to contributions over the interstitial node is zeroed. Row sums may be calculated as normal prior to the rearrangement of the matrix, but care must be taken to distribute the values in a similar fashion to the other contributions.

The boundary conditions may then be applied. Traction conditions on an element are treated normally by computing their values over the element and directly forming the equation RHS. Displacement conditions must however take account of the rearrangement previously performed. The contributions at the interstitial node have already been added to those of other nodes and there is no need to include them at this stage. The zeroing of the column corresponding to the interstitial nodes when the H matrix is rearranged means that no special coding is required. The displacement conditions may be calculated and entered in the RHS, and the contributions in the influence matrix replaced by G integrals calculated in the normal fashion. The influence coefficients for interstitial nodes must be re-distributed as described above. The matrix as formed has rows and columns missing and must therefore be rewritten to form a normal square matrix before the equations may be solved.

NON-CONFORMING ELEMENT MODEL

The non-conforming element model retains all the nodes generated and thus results in a larger problem than in the conforming case. The model is a hybrid as it is mostly continuous but has discontinuities associated with the refined elements. These discontinuities only occur for mid-edge nodes here since a single element is divided into four sub-elements. The interstitial nodes are required for the representation of the local field variables, but are not used for the adjacent element solution. Thus the model allows the displacement across a boundary to be discontinuous at the interstitial node. No significant problems have been observed however. Indeed the only real difference between this model and conventional discontinuous element solutions, as discussed by Manolis and Banerjee [3] is that nodes are placed on the interface between the elements, rather than internally. Linear dependence of the BEM equations is avoided as the nodes are all uniquely positioned in a geometric sense. Introduction of new hybrid continuous/discontinuous elements such as those described by Patterson and Elsebai [7] is therefore not necessary.

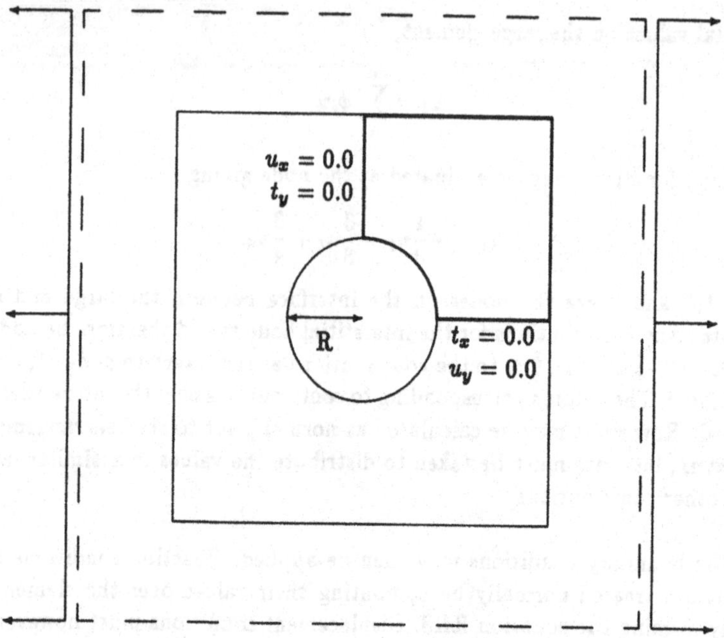

Figure 3: Infinite block weakened by central hole

The main problem encountered is with the calculation of the free term. A conventional non-conforming mesh is able to avoid this as all nodes are positioned on the internal smooth boundary of the element. A scheme has been established that permits the continued use of the row sum technique and thus allows indirect computation of the free term and the Cauchy principal value integrals.

The technique adopted is similar to that previously discussed in Crook and Smith [2] for use with residual calculation. The general principle remains the same; in order to avoid computation of many strongly singular integrals the target element is subdivided and hence the singular point is isolated. Application of this method with residuals is simpler, however, because each sub-element may be treated separately and the *HU* (or *GT*) products calculated and summed individually for each. Displacement and tractions are not known when the influence matrix is formed. All of the integrals have, therefore, to be related back to the unknown quantities at the mesh nodes on the original undivided element. The solution at all nodes that are introduced for the sub-elements may be expressed in terms of the solution at the original nodes via the shape functions. Contributions at these extra nodes may thus be apportioned between the mesh nodes and added into the matrix of coefficients.

COMPARISON OF ADAPTIVE SCHEMES

The analytical solution to the problem of an infinite two-dimensional sheet with a central hole is well-known (see, for example, Reismann and Pawlik [10]). Very few non-trivial analytical solutions exist for finite regions, but use may be made of an infinite region solution if appropriate boundary conditions are applied. The problem (see Figure 3) is modelled by considering only a 'cutout' area from the central region of the infinite plate and using symmetry to reduce the area meshed (represented by the thicker lines in the diagram). Appropriate traction boundary conditions may be calculated from the exact solution for the stresses and applied to the 'external' boundaries of the mesh. Zero displacements are specified along the x and y axis in the direction perpendicular to the face of the element and zero displacements are applied to the top and bottom surfaces of the mesh. to represent plane strain conditions. Our program incorporates a dedicated routine which automatically calculates the appropriate boundary conditions element by element.

The presence of the hole in the plate produces a non-linear solution which the quadratic elements used cannot represent exactly. The initial mesh is shown in Figure 5 with only 10 elements and 32 nodes, and the adaptive scheme was applied until the mesh size became too large for the available computer space. Examples of the adaptive meshing sequence for each element type are depicted in Figure 4 and Figure 5 for a plate with a hole of radius $R = 1.0$ and using $\alpha = 0.8$. The corresponding error estimators are compared to estimates of the actual error (calculated in the same norm) in Table 1 and Table 2. These tables also include the number of nodes and elements in each mesh. It is clear from these results that the scheme produces meshes that converge to an accurate solution although not uniformly. Both types of element show a sudden increase in the actual error between mesh 3 and mesh 4 as new elements are introduced but this is not reflected in the estimates, possibly an effect of starting with such a simple mesh and the correspondingly large errors.

Error estimators have also been computed for a sequence of meshes using $\alpha = 1$ to generate the maximum number of points and these are plotted in Figure 6 against the corresponding actual error norms. There is a good overall correlation between actual and estimated error norms.

For a given number of elements, the actual errors produced by conforming element meshes are slightly greater than those produced by the corresponding non-conforming mesh. However, when the number of nodes is considered, there are clear differences. Not all nodes on conforming elements form part of the solution matrix since the interstitial ones are discarded. This reduction in number of nodes makes a considerable difference to the time for solution of the final system of equations. In Table 1, the total number of nodes in each mesh is listed and this is the size of the final solution matrix; whereas in Table 2 the total number of nodes listed is reduced by the number of interstitial nodes in the 'inter' column. The conforming elements achieve a given accuracy with roughly three-quarters of the number of nodes needed by non-conforming elements and a corresponding reduction in solution time.

DISCUSSION

The results presented here for the single example of a hole in a plate show that a simple three-dimensional h-adaptive procedure converges towards the exact (plain strain) solution. The conforming elements used require significantly fewer nodes than the non-conforming elements to attain a given accuracy but more examples must be considered before firm conclusions may be drawn. The non-conforming integration procedure is much simpler to implement.

A more sophisticated remeshing scheme which adapts in a direction sensitive manner should also be worthy of investigation.

ACKNOWLEDGEMENT

This work has been partially supported by the Procurement Executive, Ministry of Defence, United Kingdom.

REFERENCES

[1] C. A. Brebbia and J. Dominguez, *Boundary Elements: An Introductory Course.* Computational Mechanics Publications, 1989.

[2] S. H. Crook and R. N. L. Smith, "Numerical Residual Calculation and Error Estimation for Boundary Element Methods," *Engineering Analysis*, to appear.

[3] G. D. Manolis and P. K. Banerjee, "Conforming Versus Non-Conforming Boundary Elements in Three-Dimensional Elastostatics," *International Journal for Numerical Methods in Engineering*, vol. 23, pp. 1885–1904, 1986.

[4] P. Parreira, "Self-adaptive p-Hierarchical Boundary Elements in Elastostatics," in *Proceedings of the Ninth International Conference on Boundary Element Methods*, C. A. Brebbia *et al.*, Eds., pp. 351–373, 1987.

[5] P. Parreira, "Further Developments on Error Indicators and Estimators for Adaptive Hierarchical Boundary Elements," in *Proceedings of the Eleventh International Conference on Boundary Element Methods*, C. A. Brebbia and J. J. Connor, Eds., Vol. 1, pp. 105–122, 1989.

[6] P. Parreira, "Residue Interpolation for Error Estimation in *p*-Adaptive Hierarchical Collocation Boundary Elements," in *Proceedings of the International Boundary Element Symposium*, C. A. Brebbia and A. Chaudouet-Miranda, Eds., pp. 483–498, 1990.

[7] C. Patterson and N. A. S. Elsebai, "A Regular Boundary Method using Non-Conforming Elements for Potential Problems in Three Dimensions," in *Proceedings of the Fourth International Conference on Boundary Element Methods*, C. A. Brebbia, Ed., pp. 112–126, 1982.

[8] E. Rank, "Adaptive Boundary Element Methods," in *Proceedings of the Ninth International Conference on Boundary Element Methods*, C. A. Brebbia *et al.*, Eds., pp. 259–278, 1987.

[9] E. Rank, "Adaptive *h*-, *p*- and *hp*- Versions for Boundary Integral Element Methods," *International Journal for Numerical Methods in Engineering*, vol. 28, no. 8, pp. 1335–1349, 1989.

[10] H. Reismann and P. S. Pawlik, *Elasticity—Theory and Applications*. John Wiley & Sons, 1980.

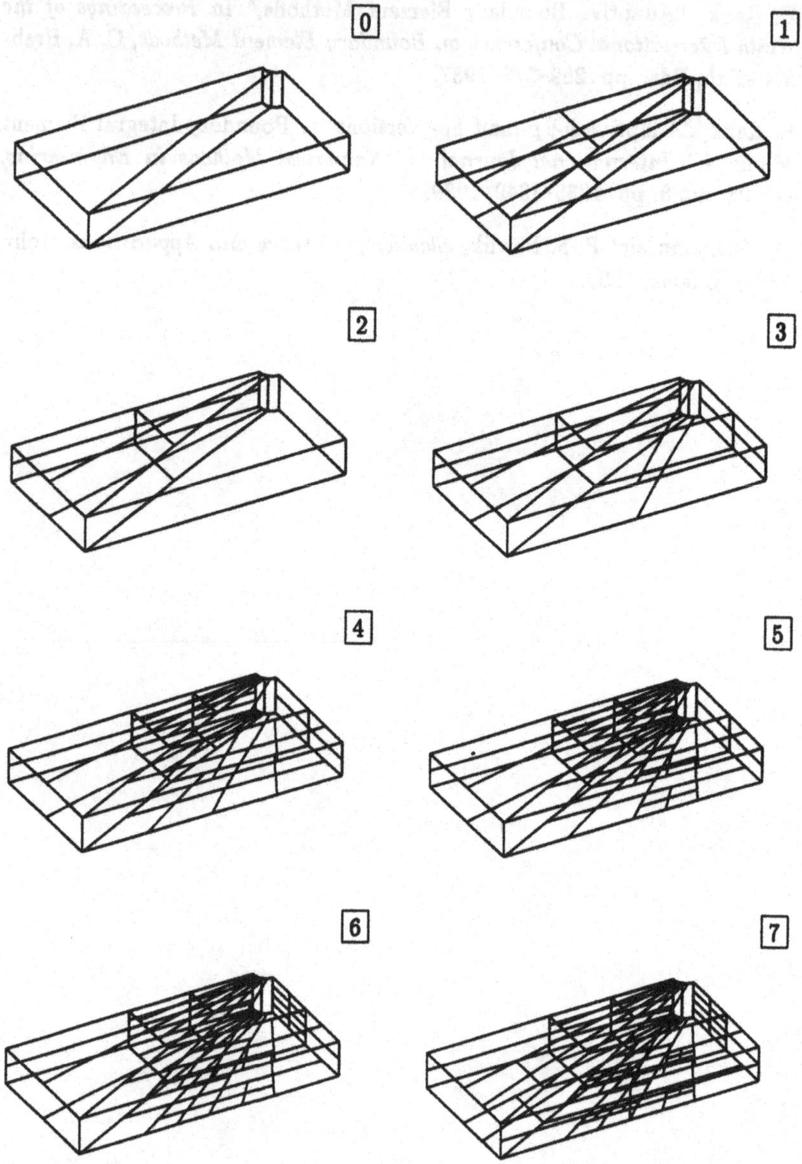

Figure 4: Adaptive mesh sequence for non-conforming elements

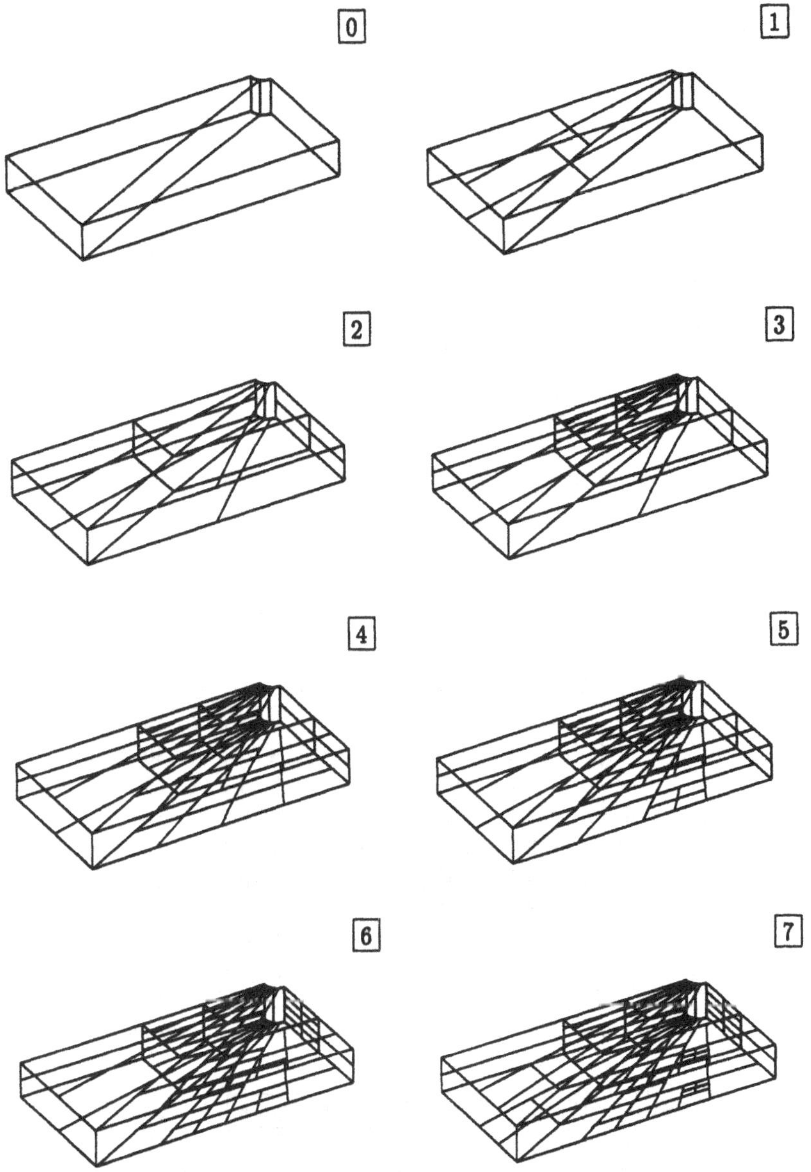

Figure 5: Adaptive mesh sequence for conforming elements

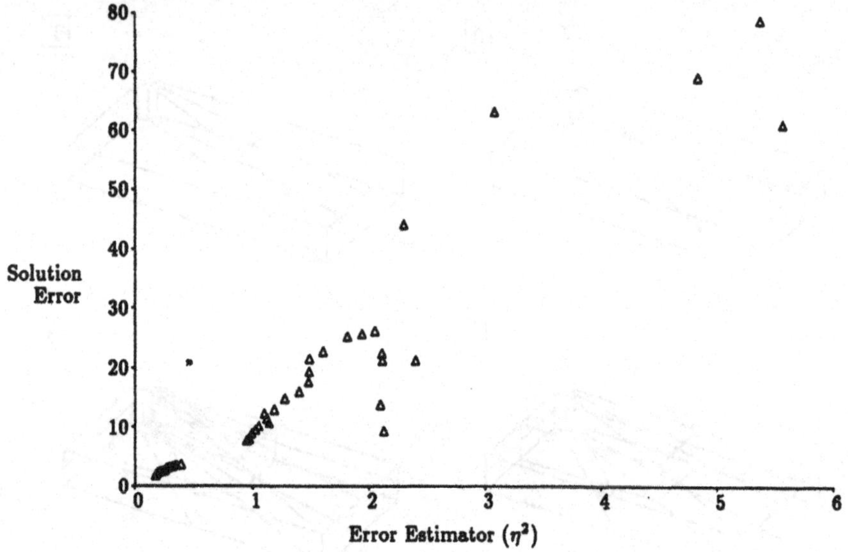

Figure 6: Non-conforming Error Estimators versus Solution Error

Iteration	Elems	Nodes	Inter	Estimate	Actual
0	10	32	0	5.56	61.44
1	16	58	16	4.83	69.30
2	19	67	16	3.08	63.29
3	28	98	24	2.13	9.36
4	58	214	76	1.29	18.12
5	70	258	92	1.10	13.43
6	88	326	120	0.95	7.68
7	115	425	156	0.40	3.47
8	157	577	208	0.20	1.83

Table 1: Infinite block with hole $R = 1.0$ using Non-Conforming model

Iteration	Elems	Nodes	Inter	Estimate	Actual
0	10	32	0	5.56	61.44
1	16	58	16	4.35	75.77
2	28	98	24	1.91	10.67
3	46	166	52	1.17	6.86
4	64	232	76	1.10	21.16
5	70	258	92	1.04	16.10
6	88	326	120	0.91	9.43
7	100	370	136	0.90	6.17
8	169	605	192	0.25	2.33

Table 2: Infinite block with hole $R = 1.0$ using conforming model

P-Version of the Boundary Element Method for Elastostatic Problems

A.W. Crim, P.K. Basu

Vanderbilt University, Nashville, TN 37235, U.S.A.

ABSTRACT

The concept of p-version boundary element modeling of problems of elastostatics of the two-dimensional continua is presented. Areas unique to the p-version boundary element formulation are discussed. These include the development of the equations of nodal collocation for the p-version, selection of higher order hierarchial shape functions, techniques for integrating the product of the kernel functions and these shape functions, strategies for selecting collocation points used in approximating the unknowns associated with the higher order shape functions, and program organization. A numerical example which demonstrates the performance of the p- version formulation is included.

INTRODUCTION

The p-version of the finite element method has been an important development in the computer based solution to the problems of the continua. In p-extension, the convergence to the true solution is approached by increasing the degree of piecewise polynomials while keeping the discretization of the physical space unchanged. This is contrary to conventional or, so called, h-version in which the convergence to true solution is approached by refining the physical space while keeping the function space, or p-level, fixed. The main advantages of the p-version of the finite element method in the case of problems of elastostatics are its faster rate of convergence, robustness, and ease of input data preparation and model refinement [1,2].

Researchers in the boundary element method are naturally interested to determine if it is possible to develop boundary element formulations which can realize the advantages that the p- version finite elements have exhibited. Whereas the concept of approximating the system variables by piecewise hierarchial polynomials is the same for both methods, the actual formulation of the system of equations is entirely different. Alarcon and Reverter [3,4] have presented a hierarchial boundary element formulation for two-dimensional potential problems based on the point collocation method along with discussions about adaptivity. Rank [5] and Postell and Stephan [6] have considered adaptive h-, p-, and h-p versions of the boundary element method for two-dimensional potential problems based on the Galerkin method. This work presents a p-version extension to the direct boundary integral equation formulation of two-dimensional plane elastostatic problems. The implementation of p-version boundary element method is presented as an extension of the existing h-version implementations.

P-VERSION BOUNDARY ELEMENT FORMULATION

In p-version formulations it is assumed that as the degree p of the approximating functions approaches a large value, the solution converges to the true solution. Key to the formulation is the appropriate selection of the shape functions to approximate the system variables [1]. In order to achieve the advantages of ease of data preparation and model refinement, shape functions must be selected that provide the order to be increased by simply augmenting the lower order shape functions. For the boundary element method this would allow retaining the prior integration data. Lagrange polynomials, which are currently used in most h- version boundary element programs, are non-hierarchial in nature, requiring a new set of shape functions for each higher order of approximation and thus recomputation of the system of matrices.

Hierarchial Shape Functions for Boundary Elements

Hierarchial shape functions for boundary elements are characterized by a set in which the shape functions of order $p-1$ is a subset of the ones of order p. Using such hierarchial shape functions the p- version formulation for the boundary element method can be implemented. Figure 1 illustrates the generation of shape functions for boundary elements to model two-dimensional regions [7]. The linear approximation (p=1) of such an element can be expressed as

$$\hat{\phi}^e = \phi_0 N_0^e + \phi_1 N_1^e \tag{1}$$

where

$$N_0^e = -\frac{1}{2}(\xi - 1),\quad N_1^e = \frac{1}{2}(\xi + 1),\quad -1 \leq \xi \leq 1 \tag{2}$$

For a hierarchial quadratic approximation (p=2) we add $a_2 N_2^e$ to the linear approximation, where N_2^e is a quadratic function of the form

$$N_2^e \alpha_0 + \alpha_1 \xi + \alpha_2 \xi^2 \tag{3}$$

In order to satisfy the C^0 continuity of the approximation ϕ between segments, the coefficients of N_2^e are chosen to yield $N_2^e = 0$ at $\xi = \pm 1$. A suitable choice

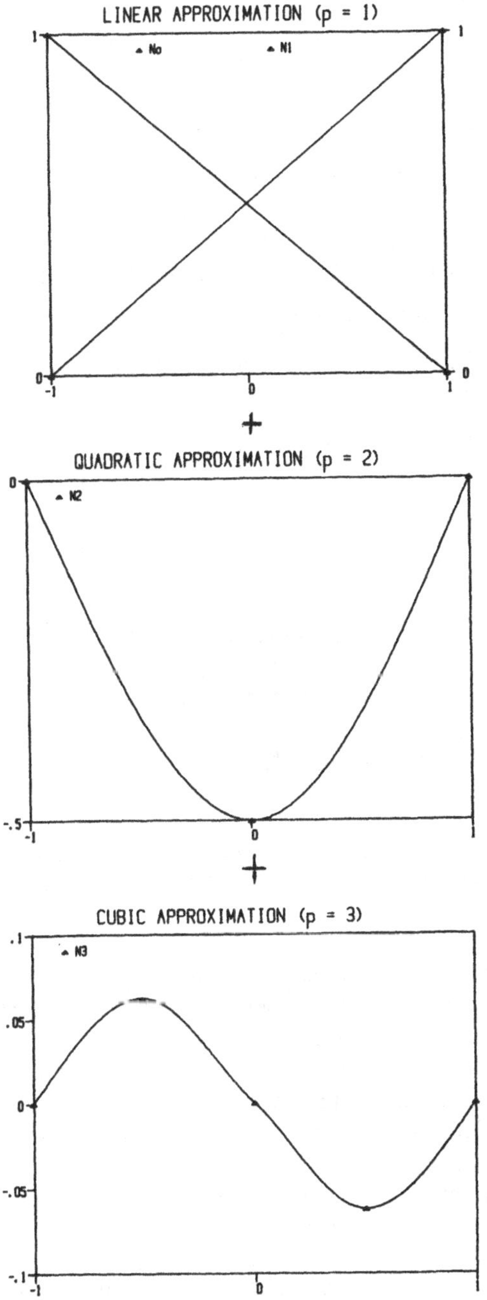

Figure 1. Hierarchial Shape Functions

is the symmetric parabolic function

$$N_2^e = \frac{1}{2}(\xi^2 - 1) \tag{4}$$

The hierarchial quadratic approximation is then

$$\hat{\phi}^e = \phi_0 N_0^e + \phi_1 N_1^e + a_2 N_2^e \tag{5}$$

For a hierarchial cubic approximation (p=3) we add $a_3 N_3^e$ to the hierarchial quadratic approximation where N_3^e is a cubic function of the general form

$$N_3^e = \alpha_0 + \alpha_1 \xi + \alpha_2 \xi^2 + \alpha_3 \xi^3 \tag{6}$$

with zero values at $\xi = \pm 1$ and one or more of the coefficients equal to zero. A suitable choice is the cubic function

$$N_3^e = \frac{1}{6}(\xi^3 - \xi) \tag{7}$$

The hierarchial cubic approximation is then

$$\hat{\phi}^e = \phi_0 N_0^e + \phi_1 N_1^e + a_2 N_2^e + a_3 N_3^e \tag{8}$$

For each increasing order of approximation p, a term $a_p N_p^e$ is added to the previous approximation. Coefficients of N_p^e are selected so that $N_p^e = 0$ at $\xi = \pm 1$. The hierarchial approximation of order p is then

$$\hat{\phi}^e = \phi_0 N_0^e + \phi_1 N_1^e + \sum_{j=2}^{p} a_j N_j^e \tag{9}$$

It may be noted here that the coefficients (ϕ_0, ϕ_1) associated with the first two shape functions represent values of the unknown function at the end points.

Discretized Formulation for Hierarchial Boundary Elements

The computation of the integrals and development of the system of equations for p-level of one is identical to that for the h-version of the boundary integral equation method. That is because the collocation points coincide with the nodes defining the segment and the only unknowns are the displacements and tractions at the nodal points. When the p-level is increased, new unknowns associated with the hierarchial higher order shape functions are appended, requiring the introduction of new collocation points to setup the additional equations needed to solve the problem.

The direct formulation for the elastostatic boundary integral equation can be expressed as follows [8]

$$C_{ij}(x)\, u_j(x) + \int_s T_{ij}(x,y)\, u_i(y)\, d\,Sy = \int_s U_{ij}(x,y)\, t_i(y)\, d\,Sy \tag{10}$$

where U_{ij} and T_{ij} represent the fundamental solutions for displacements u_i and tractions t_i at point y due to a unit load applied at point x in the j direction. This equation is solved by dividing the boundary S into elements b and evaluating the integrals on the elements using the local coordinate system ξ. The first integral can be expressed as

$$\int_{s_b} T(x^n, y(\xi))\, N_j(\xi)\, J(\xi)\, d\xi \tag{11}$$

Evaluating this integral for each collocation point x^n over all elements for each shape function $N_j(\xi)$ yields coefficients h_{ij} for the displacement matrix H. Each row corresponds to an unknown displacement associated with x^n and each column corresponds to a shape function.

The second integral can be expressed as

$$\int_{s_b} U(x^n, y(\xi))\, N_j(\xi)\, J(\xi)\, d\xi \tag{12}$$

Likewise, evaluating this integral for each collocation point x^n over all the elements corresponding to each shape function $N_j(\xi)$ yields the coefficients g_{ij} for the traction matrix G.

Adding the contribution of the free term $c_{ij}(x^n)$ to H, equation (10) takes the following matrix form

$$H\ u = G\ t \tag{13}$$

Suppose that for the next iteration the degree of the approximating functions is increased. For each increasing order of approximation per element, a term $a_p N_p$ is added to the previous approximation. Evaluating the integral

$$\int_{s_b} T(x^n, y(\xi))\, N_p(\xi)\, J(\xi)\, d\xi \tag{14}$$

for each collocation point x^n over the element b for the added shape function $N_p(\xi)$ yields a new column of coefficients h_{ij} corresponding to the newly added term a_p.

In order to compute the remaining coefficients needed a new collocation point x^p must be selected on element b. Thus evaluating the integral

$$\int_{s_b} T(x^p, y(\xi))\, N_{j+1}(\xi)\, J(\xi)\, d\xi \tag{15}$$

for the new collocation point x^p over all elements for each shape function $N_{j+1}(\xi)$, which includes the new shape function $N_p(\xi)$, yields a new row of coefficients h_{ij} corresponding to the newly added term a_p.

The coefficients for the G matrix can be computed in the same manner.

It is also necessary to compute the contribution of the free term $c_{ij}(x^P)\, u_j(x^P)$. Since the variable $u_j(x^P)$ will not be solved for directly it is necessary to make use of the expression [9]

$$u_j(x^P) = u_0 N_0(x^P) + u_1 N_1(x^P) + a_p N_p(x^P) \tag{16}$$

Thus,

$$c_{ij}(x^P)\, u_j(x^P) = c_{ij}(x^P)\, [u_0 N_0(x^P) + u_1 N_1(x^P) + a_p N_p(x^P)] \tag{17}$$

which adds the coefficients corresponding to all the shape functions approximating u over element b.

The augmented matrix equation is then made up of new columns and rows where the added columns represent the contribution of the new shape functions with the source at the previous collocation points, and the added rows represent the contribution of all shape functions with the source located at each new collocation point [4]. By then applying the appropriate boundary conditions a system of equations can be formulated as in the h-version.

COMPUTER IMPLEMENTATION

The computer implementation of the p-version boundary element method for modeling problems of elastostatics of two-dimensional continua builds on h-version implementations. Several points unique to the p-version formulation by the point collocation method need special attention. These include selection of proper shape functions for the higher order representations, techniques for integration of the product of the kernel functions and these shape functions, strategies for selecting collocation points used in approximating the unknowns associated with the higher order shape functions, and program organization.

Hierarchial Shape Functions

There are two constraints that must be met in the selection of shape functions of order higher than one for two dimensional boundary elements. These constraints are that the value of the shape functions be zero at $\xi = \pm 1$, and that they be independent. There exists a great many sets of functions that satisfy these constraints, and if the integrals could be computed with unlimited precision then in theory there would be no difference in the solution for a given p-level with any set of shape functions chosen. Unfortunately, however, due to the limited precision of the computations carried out on a digital computer, round-off errors do not allow this to hold good. The question then is how to select the best set of shape functions to be used in the computer implementation of the p-version boundary element formulation.

Since the shape functions used in the h-version boundary element formulation are the same as used for the h-version finite element formulation [10], it follows that p-version finite element research could supply the p-version boundary element formulation with like shape functions. P-version finite element research has found that round-off error can be reduced significantly if the off-diagonal terms of the matrices are small. This can be achieved if the integrands appearing in the off-diagonal terms consist of product of orthogonal functions [1]. Shape functions of this type were used by Basu [9]. The general form of these shape functions defined over the region $0 \le \xi \le +1$ for $p > 1$ is

$$N_p(\xi) = \xi^{p-1}(\xi - 1) \tag{18}$$

These shape functions are shifted (or biased) toward $\xi = +1$. In an attempt to make these shape functions symmetric, shape functions based on Legendre polynomials were developed [1]. The general form of the Legendre shape functions defined over the region $-1 \le \xi \le +1$ for $p > 1$ is

$$N_p(\xi) = \frac{1}{p!}(\xi^p - 1) \quad p \text{ is even} \tag{19}$$

$$N_p(\xi) = \frac{1}{p!}(\xi^p - \xi) \quad p \text{ is odd} \tag{20}$$

The Legendre shape functions are said to be orthogonal in the functional norm. The latest p-version finite element research has led to the development of a set of shape functions that are orthogonal in the energy norm [11]. These shape functions are based on the integrals of the Legendre polynomials. The Rodriguezes formula [11] for Legendre polynomials is given by

$$p_i(t) = \frac{1}{2^i\, i!}\frac{di}{dt^i}(t^2 - 1)^i \quad \text{where } i = 0, 1, 2, \ldots \tag{21}$$

The general form of the shape functions derived from the integrals of the Legendre polynomials defined over the region $-1 \le \xi \le +1$ for $p > 1$ is

$$N_p(\xi) = \frac{\sqrt{2_{p-1} - 1}}{2}\int_{-1}^{\xi} p_{p-1}(t)\, dt \tag{22}$$

The shape functions typically used in p-version boundary element formulations are based on Legendre polynomials as given in Equation 19 [1,2]. It is important to note that experimentation with these three sets of shape functions in p-version boundary element programs has resulted in nearly identical results. Therefore any one of these sets of orthogonal shape functions are suitable for p-version boundary element computer implementation.

Another consideration in selecting the set of shape functions to use would be in coupling a p-version boundary element region with a p-version finite element region [9]. In this case both methods would have to use the same set of shape functions, and typically the finite element formulation would control.

Integration of the Kernel and p-level Shape Function Products

Typical h-version boundary element programs follow one of two methods [12] for computing the integration coefficients which make up the system of equations approximating the integral equation. The first is to move the source or pole from collocation point to collocation point, integrating all the shape functions for each element. This method computes the H and G matrices in a row by row fashion. The second method is to move from element to element, locating the source at all the collocation points and integrating the shape functions for each element. This method computes the H and G matrices in a column by column fashion. Either method may be extended to the p-version. The row by row method was chosen for the present study. For this case the integration coeffici nts of H and G could be computed in one program module and stored on the disk by rows. A second module could read each row, which corresponds to a collocation point, and build the system matrix by applying the boundary conditions at that collocation point.

Like the h-version, there are three different types of integrals that need to be evaluated. Techniques developed for h-version integration can be applied or modified for each type. In the first case, the source does not lie on the element being integrated over and the integrals may be computed by standard Gaussian quadrature as in the h-version. The other two cases occur when the source does lie on the element being integrated over. Here the integrals are singular and special attention must be given to the integration procedures. Also, unlike the h-version, shape functions for the p-version are not zero at the collocation points corresponding to the higher order terms.

The weakly singular integrals of $U(x,y(\xi))N_j(\xi)$ may be evaluated in the same manner as the h-version. These integrals have integrable singularities that can be evaluated by dividing the element into subelements and applying the Gaussian quadrature formula for the product of a polynomial and log $1/r$ [8].

The strongly singular integrals of $T(x,y(\xi))N_j(\xi)$ exist only in the Cauchy principal value sense [8] and must be evaluated by special numerical techniques. The popular h-version method of evaluating the sum of the integrals of $T(x,y(\xi))N_j(\xi)$ and the coefficient of the free term $c(x)$ by considering rigid body translations is valid only for p-level of one. Only the off diagonal terms corresponding to the p-level one shape functions are valid in this summation.

A numerical technique that works well for integrating shape functions corresponding to p-levels greater than one is that of subtracting out the singularity [9,10,12]. For curved elements this integral must be split into a flat portion which is evaluated analytically and a curved portion which can be evaluated numerically [10].

<u>Selection of the p-level Collocation Points</u>
The collocation points for p-level of one are identical to that of the h-version linear boundary element. For this case the collocation points are located at the geometric nodes at each end of the element and the only unknowns are the displacements and tractions at these nodes. When the p-level is increased, new unknowns associated with the higher order shape functions are introduced thus

requiring the selection of a new unique collocation point for each additional higher order term. These new p-level collocation points are used to compute the coefficients of the system of equations approximating the integral equation corresponding to the new p-level unknowns.

While these p-level collocation points may be located anywhere along the element, two patterns will be discussed here. The first pattern is the symmetric selection [9]. Here the collocation points are located symmetrically along the element. For p=2 the p-level collocation point is located at the center of the element. For p=3 the two p-level collocation points are located at the third points of the element. For p=4 the three p-level collocation points are located at the quarter points of the element, and so on.

The second pattern is the true hierarchial selection [9]. Here the p-level collocation points do not change with each increment p-level approximation. This pattern then facilitates the hierarchial nature of the shape functions requiring only integral computations for each new p-level shape function. For p=2 the p- level collocation point is located at the center of the element. For p=3 the p-level collocation point can be located at either quarter point of the element. For p=4 the p-level collocation point is located at the other quarter point. As the p-level increases the location of the corresponding p-level collocation point alternates on either side of the center of the element in a symmetric pattern.

The drawback to this pattern is that only the even p-levels produce a symmetric distribution of collocation points and thus symmetrical results. Indeed, experimentation has shown that the results from odd p-levels for the non-symmetric true hierarchial patterns are generally not as good as those from symmetric patterns, and can be influenced by the location of the collocation points.

Program Organization

From the two patterns of collocation point distributions presented here two different approaches to the program organization can be implemented [9]. The first would be a semi-hierarchial implementation using symmetrical shape functions. The system of equations for p=1 would be retained. With each increasing p-level approximation, the coefficients for all the p-level shape functions would have to be computed since the location of the collocation points for p levels greater than one will have changed. This would result in the addition of rows and columns to the original system of equations for p-level of one for each approximation.

The second approach is a true hierarchial implementation based on a hierarchial selection of collocation points. The system of equations for each approximation would be saved. As the p-levels for various elements are increased for a new approximation, new columns and rows can be added to the system of equations from the prior approximation producing a new system of equations which would then be saved for use by successive approximations.

For each implementation the same amount of data entry would be required. While the semi-hierarchial implementation would require greater computation for each successive approximation, the results for odd p-level combinations would be better than that for the true hierarchial implementation. The selection of which implementation to use would depend on the number of refinements required before an acceptable solution is obtained.

EXAMPLE PROBLEM

The following cantilever beam with shear loading ($E = 2.3,\ \nu = .3$) as shown in Figure 2 is presented here to compare the h-version and p-version boundary element results. This cantilever problem was used [10] to demonstrate a h-version boundary element program. Tables 1 and 2 show the results based on four and eight element discretizations for p levels ranging from 2 to 8. The four element discretization is far too crude to produce reasonable results for linear h-version or $p = 1$. The true hierarchial collocation point pattern is presented along with the symmetric pattern for even p-levels. Displacements are shown for the top and bottom of the free end to demonstrate how the results are affected by the location of the collocation points.

The results for p-level of two are very close to the quadratic h-version for both the four and eight node discretizations. Results compared to the p-version finite element method [13] using four triangular elements illustrate that by increasing the p-level monotonic convergence to the true solution is achieved.

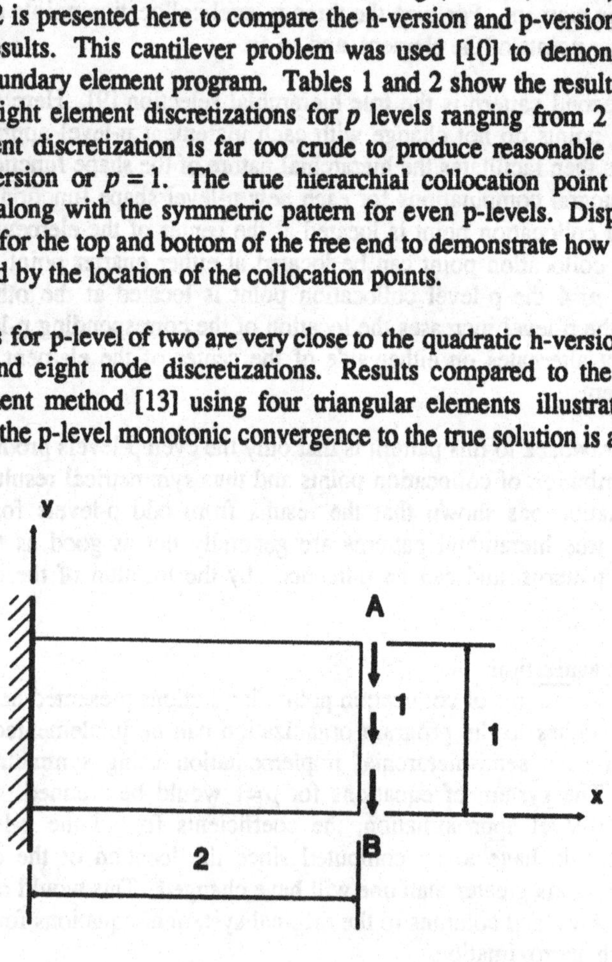

Figure 2. Cantilever Beam Problem

Table 1. Cantilever Beam (Four Element Model)				
	Deflection at A		Deflection at B	
Comparison	Horizontal	Vertical	Horizontal	Vertical
h-Version quadratic	4.0841	-12.4934	-4.0841	-12.4934
p-Version Finite Element [13] 4 elements, p=8	5.382	-16.61	-5.379	-16.61
Present (True Hierarchial)				
p=2	4.0862	-12.4446	-4.0862	-12.4446
p=3	4.5358	-14.1643	-4.5348	-14.1831
p=4	4.6188	-14.5494	-4.6188	-14.5494
p=5	4.4690	-14.4127	-4.7793	-14.6340
p=6	5.3152	-16.2264	-5.3152	-16.2264
p=7	5.2641	-16.1324	-5.3290	-16.1814
p=8	5.6961	-17.1058	-5.6961	-17.1058
Present (Symmetric)				
p=3	4.5620	-14.3398	-4.5620	-14.3398
p=5	4.6232	-14.5343	-4.6232	-14.5343
p=6	4.7682	-14.9271	-4.7682	-14.9271
p=7	5.2076	-16.0385	-5.1587	-16.1343

Table 2. Cantilever Beam (Eight Element Model)

Comparison	Deflection at A Horizontal	Deflection at A Vertical	Deflection at B Horizontal	Deflection at B Vertical
h-Version [10]	4.681	-14.67	-4.681	-14.67
p-Version Finite Element [13] 4 elements,				
p=8	5.382	-16.61	-5.379	-16.61
Present (True Hierarchial)				
p=2	4.6592	-14.6105	-4.6592	-14.6105
p=3	4.7320	-14.7834	-4.7079	-14.7705
p=4	4.7311	-14.8504	-4.7311	-14.8504
p=5	4.7984	-15.0315	-4.9560	-15.1797
p=6	5.1946	-15.8540	-5.1946	-15.8540
p=7	5.2475	-15.9458	-5.2807	-15.9754
p=8	5.4423	-16.3647	-5.4423	-16.3647
Present (Symmetric)				
p=3	4.7431	-14.8585	-4.7431	-14.8585
p=5	4.7184	-14.8295	-4.7184	-14.8295
p=6	4.8085	-15.0494	-4.8085	-15.0494
p=7	5.0352	-15.5268	-5.0364	-15.5275

SUMMARY AND CONCLUSIONS

P-version of the boundary element method for two dimensional plane elastostatic problems was developed and implemented. A number of hierarchial shape functions were presented and two alternative schemes were evaluated and implemented. An example problem illustrated that monotonic convergence to the true solution was achieved by increasing the p-level. In the case of the true hierarchial implementation, which required less computational effort, the best performance was observed for even p- levels. In the semi-hierarchial implementation scheme, no such restriction was found to be necessary but involved more computational effort.

The results here prove that it is possible to develop p- version boundary element formulations that can realize the advantages that the p-version finite element formulations have exhibited. The convergence pattern, although always monotonic, with increasing p-level was found to be dependent upon the location of collocation points used to compute the coefficients and more research is necessary to arrive at the optimal location of these points. Further development with the two methods naturally lead to the combination of the p-versions [9] just as the combination of the h-versions have successfully been done. Research work in this area is in progress.

REFERENCES

1. Basu, P.K. and Lamprecht, R.M. Some Trends in Computerized Stress Analysis, Proceedings of the Seventh ASCE Conference in Electronic Computation, Washington University, St. Louis, MO, August 1979.

2. Basu, P.K. and Peano, A. Adaptivity in P-version Finite Element Analysis, Journal of Structural Engineering, ASCE, Vol. 109, pp. 2310-2323, 1983.

3. Alarcon, E., Reverter, A., and Molina, J. Hierarchial Boundary Elements, Computers and Structures, Vol. 20, pp. 151-156, 1985.

4. Alarcon, E., and Reverter, A. p-Adaptive Boundary Elements, Int. J. Numer. Meth. Eng., Vol. 23, pp. 801-829, 1986.

5. Rank, E. Adaptive h-, p-, and hp-Versions for Boundary Integral Element Methods, Int. J. Numer. Meth. Eng., Vol 28, pp. 1335-1349, 1989.

6. Postell, F.V., and Stephan, E.P. On the h-, p-, and h-p Versions of the Boundary Element Method Numerical Results, Computer Methods in Applied Mechanics and Engineering, Vol. 83, pp. 69-89, 1990.

7. Zienkiewicz, O.C. and Morgan, K. Finite Elements and Approximation, John Wiley and Sons, New York, 1983.

8. Watson, J.O. Advanced Implementation of the Boundary Element Method for Two- and Three- Dimensional Elastostatics, Developments in Boundary Element Methods-1, (Ed. P.K. Banerjee and R. Butterfield), pp. 31-63, Applied Science Publishers, London, 1979.

9. Crim, A.W. Development of Boundary Element Method and Combination with Finite Element Method in p-Version, Doctoral Dissertation, Notes, Vanderbilt University.

10. Banerjee, P.K., and Butterfield, R. Boundary Element Methods in Engineering Science, McGraw-Hill, London, 1981.

11. Basu, P. K. Dimensional Reduction of Structural Plates and Shells, NSF Research Report, Grant No. CEE-84115675, Vanderbilt University, Nashville, Tennessee 4, 1986.

12. Doblare, M. Computational Aspects of the Boundary Element Method, Topics in Boundary Element Research, Volume 3: Computational Aspects, (Ed. C. A. Brebbia), pp. 51-131, Springer-Verlag, Berlin, 1987.

13. Basu, P.K., Rossow, M.P. and Szabo, B.A. Technical Documentation and User's Manual: Comet-x, Report No R-340, Federal Railroad Administration, 1977.

Application of Overhauser C^1 Continuous Boundary Elements to "Hypersingular" BIE for 3-D Acoustic Wave Problems

Y.Liu, F.J. Rizzo

Department of Theoretical and Applied Mechanics, University of Illinois at Urbana-Champaign, 104 S. Wright Street, Urbana, Illinois 61801, U.S.A.

1. INTRODUCTION

The C^1 continuous representations of boundary geometry and variables in the BIE/BEM not only provide more accurate results, but are also demanded by some BIE formulations, such as the "hypersingular" BIE formulations of certain problems [1, 2]. There are several types of C^1, or even C^2, boundary elements in the literature. Liggett and Salmon [3] introduced cubic spline interpolation in the discretization of BIE formulation. In this interpolation, nodal values of the function and second derivative of the function are used. The nodal values of the second derivative are obtained in terms of the nodal values of the function by solving a separate set of equations. Thus, this method has a global character in the sense that the function inside one element is actually determined by all the nodal values of the function on the entire curve. Watson [4] introduced Hermitian cubic elements in BEM for 2-D problems. The problem with the Hermitian elements is that the tangential derivatives of the function and even cross-derivatives (for 3-D problems) need to be introduced at the nodes. These derivatives are usually not the boundary variables in the conventional BIE formulation and are troublesome to deal with. B-splines have been used in BEM for some time and a more recent work is given by Cabral *et al* [5]. The important character of the B-splines is that the spline curves do not pass through the specified nodes (for geometry) or the nodal values (for function). The geometry (or function) within an element is defined by four control points (or coefficients) multiplied by blending functions. Positions of the control points or values of the coefficients are unknown in advance and need to be determined by solving a system of linear equations (of the same size as that of the BEM system) relating the control points to the nodes or the coefficients to nodal values of the function. Although the matrix of this system has some special features, the solution of this additional system will reduce the efficiency of the method, especially for large,real engineering problems. All the above mentioned C^1 or C^2 boundary elements have been applied only to 2-D problems.

Overhauser C^1 continuous line elements for 2-D problems were developed by Ortiz *et al* [6], and surface elements for 3-D problems by Hall and Hibbs [7-9]. The main advantage of the Overhauser elements is that the nodes or nodal values of function are used directly in representing the geometry or function on the elements. No derivatives of the function (as in the cubic splines and Hermitian elements) or some intermediate quantities (such as the control points or coefficients in B-splines) are involved. Thus the definitions of the Overhauser elements are straightforward and easy to program. Numerical results obtained by using Overhauser line elements for 2-D problems [6, 10] clearly show greater accuracy and efficiency compared with the commonly applied C^0 elements and other cubic spline elements. Numerical examples of the Overhauser surface (quadrilateral and triangular) elements for 3-D problems, though limited in numbers [7-9], show similar advantages of the Overhauser elements over the usual surface elements regarding accuracy and efficiency.

In this paper, the Overhauser surface elements are applied to 3-D acoustic wave problems for which the "hypersingular" BIE [11] is employed to overcome the fictitious eigenfrequency difficulty of the conventional BIE. This "hypersingular" BIE formulation requires, theoretically, C^1 continuity of the density functions [2] in the neighborhood of the source point. Data from numerical experiments involving acoustic scattering problems show that the Overhauser surface elements, in general, can give comparably accurate results with much fewer nodes on the boundary and less computer running time, compared with two types of quadratic elements, namely, conforming quadratic elements and non-conforming quadratic elements. Thus in addition to their high accuracy, the Overhauser surface elements can produce a much smaller system of linear algebraic equations to solve. This is an important feature for the applications of BEM on microcomputers or workstations.

2. THE OVERHAUSER ELEMENTS

The Overhauser line element [6, 7] for 2-D problems is defined by four nodes along a curve, where the two inner nodes are at the ends of the element considered and the two outer nodes are on the adjacent elements. Two parabolas are constructed by the first three and last three nodes, respectively. The Overhauser curve is thus formed by a linear blending of the two parabolas. The curve is guaranteed to have inter-element continuous slopes at the nodes. Functions defined on the curve are interpolated in a similar way.

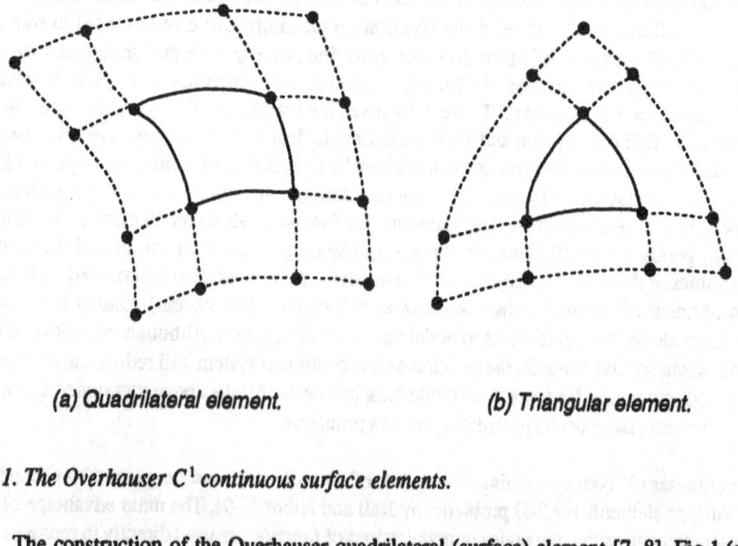

(a) Quadrilateral element. (b) Triangular element.

Fig. 1. The Overhauser C^1 continuous surface elements.

The construction of the Overhauser quadrilateral (surface) element [7, 8], Fig.1.(a), is a straightforward generalization of the line element. Sixteen nodes are used in the definition of the quadrilateral element, where four nodes are placed at the corners of the element and twelve others are on the surrounding elements. There are altogether sixteen shape functions employed in the interpolations of the surface and functions defined on the surface. Construction of the Overhauser triangular element [8, 9], Fig.1.(b), on the other hand, is much more complicated. The triangular elements are necessary because the quadrilateral elements alone are not sufficient to produce a mesh which can ensure C^1 continuity for all kinds of closed surfaces, [8, 9]. Twelve nodes, three on the corners of the element and nine others on the surrounding (quadrilateral) elements, are used in the definition of the triangular elements. In contrast to the quadrilateral element, the twelve shape functions for the triangular element are lengthy and the derivatives of the shape functions are even more tedious. However, the efficiency in computation of the Overhauser elements will not be hindered by their larger number and lengthy expressions of the shape functions, as will be discussed in the last section. A complete set of the Overhauser C^1 continuous surface elements, including the reduced version of the quadrilateral elements for surfaces with corners or edges and the transition elements for imbedding the triangular element in an otherwise quadrilateral mesh, is presented in detail in [8].

The most important feature of the Overhauser elements is that unlike all other C^1 or C^2 boundary elements, only the nodes or nodal values of functions are employed in the definition of the elements. This same feature is used in the definitions of lower order elements (e.g. linear or quadratic elements). The computer program using lower order elements would keep the same structure if the lower order elements were replaced with the Overhauser elements. No additional work, such as finding the nodal values of the second derivatives for cubic splines [3] and the control points or coefficients for B-splines [5], would be needed except for handling the shape functions. Thus upgrading existing codes with the Overhauser elements is quite straightforward.

The Overhauser quadrilateral and triangular elements developed by Hall and Hibbs [7-9] are applied in this paper. Little modification is made for the triangular element. A new set of side parameters used in the definition of the triangular elements is introduced, which can significantly simplify the expressions of the derivatives of the shape functions. The transition (quadrilateral) elements introduced by Hall and Hibbs are not used. Instead, they are simulated by using the original quadrilateral element, which is done by specifying same coordinates for associated nodes, i.e. treating the transition elements as special cases of the original quadrilateral elements. In this way, one will not need to evaluate and store additional sets of shape functions for the transitional elements. It is tempting for one to simulate the triangular element by the quadrilateral element. However, careful consideration reveals that the inter-element smoothness will be violated by the simulated triangular element.

3. THE COMPOSITE BIE FORMULATION

The conventional boundary integral equation for exterior acoustic problems is

$$C(P_o)\phi(P_o) = \int_S \left[G(P,P_o)\frac{\partial\phi(P)}{\partial n} - \frac{\partial G(P,P_o)}{\partial n}\phi(P) \right] dS(P) + \phi^I(P_o), \qquad \forall P_o \in S, \quad (1)$$

where ϕ is the total wave, ϕ^I the incident wave (for scattering problems), $G(P,P_o)$ the Green's function for the Helmholtz equation, n the outward normal to the boundary S of the exterior domain E, and the coefficient $C(P_o)$ depends on the smoothness of S.

The solution of Eq. (1) suffers a nonuniqueness problem when the wavenumber is near or equal to one of the so called fictitious eigenfrequencies [12]. How to circumvent this fictitious eigenfrequency difficulty (FED) for the exterior problems has been a major research area in the applications of BIE to acoustics. The most effective method to deal with the FED is due to Burton and Miller [13]. They proved that the composite BIE formulation, using a linear combination of the conventional BIE (Eq. (1)) and the following "hypersingular" BIE

$$\frac{\partial\phi(P_o)}{\partial n_o} = \int_S \left[\frac{\partial G(P,P_o)}{\partial n_o}\frac{\partial\phi(P)}{\partial n} - \frac{\partial^2 G(P,P_o)}{\partial n \partial n_o}\phi(P) \right] dS(P) + \frac{\partial\phi^I(P_o)}{\partial n_o}, \qquad P_o \to S, \quad (2)$$

where n_o is the outward normal at $P_o \in S$, can provide unique solutions at all wavenumbers. The major difficulty in implementing this composite BIE formulation has been the treatment of the hypersingular integral in Eq. (2). A recently proposed approach [11] to deal with this hypersingular integral is to transform Eq. (2), by employing some integral identities established in [14] for the static Green function $\overline{G}(P,P_o)$, into the following weakly-singular form,

$$\frac{\partial\phi(P_o)}{\partial n_o} + \int_S \frac{\partial^2 \overline{G}(P,P_o)}{\partial n \partial n_o}[\phi(P) - \phi(P_o) - \phi_{,k}(P_o)(x_k - x_{ok})]dS(P)$$
$$+ \int_S \frac{\partial^2}{\partial n \partial n_o}[G(P,P_o) - \overline{G}(P,P_o)]\phi(P)dS(P)$$
$$= \int_S \frac{\partial G(P,P_o)}{\partial n_o}[\phi_{,k}(P) - \phi_{,k}(P_o)]n_k(P)dS(P)$$
$$+ \int_S \frac{\partial}{\partial n_o}[G(P,P_o) - \overline{G}(P,P_o)][\phi_{,k}(P_o)n_k(P)]dS(P)$$
$$+ \frac{\partial\phi^I(P_o)}{\partial n_o}, \qquad \forall P_o \in S, \quad (3)$$

where x_k and x_{ok} are coordinates of P and P_o, respectively. All the integrals in Eq. (3) are at most weakly singular and the commonly used quadratures for conventional BIE are sufficient to compute them. However, there is a theoretical restriction on the density function in either Eq. (2) or Eq. (3). For the hypersingular integral in Eq. (2) to exist or for the procedures leading to Eq. (3) to be valid, the density function ϕ must be C^1 continuous, at least at the source point P_o, see e.g. [1,2,11]. This restriction on the "hypersingular" BIE will demand, theoretically, C^1 boundary elements in the discretization of Eq. (3).

The Overhauser elements will be applied to the composite BIE formulation, i.e. the linear combination of Eq. (1) and Eq. (3), in the next section. For comparison, two types of quadratic boundary elements are also applied, namely, conforming and non-conforming quadratic elements (quadrilateral and triangular). The conforming quadratic elements are the commonly used eight node quadrilateral and six node triangular elements in the BEM literature. These elements are in the C^0 element category and violate the smoothness requirement for Eq. (6). Nevertheless, if the smoothness requirement is relaxed in some sense and certain techniques are employed to handle the nonuniqueness of the gradient of ϕ at the source point, good results can be obtained by using these conforming elements [11]. However, the validity of applying the conforming quadratic elements to "hypersingular" BIE's is still an open question. The non-conforming quadratic elements are obtained by simply moving the nodes some distance inside the elements. Hence the C^1 continuous requirement on the density function is satisfied in the neighborhood of the source point.

4. NUMERICAL RESULTS

The scattering problem of a plane incident wave ϕ^I from a rigid sphere ($\partial\phi/\partial n = 0$ on the boundary) of radius a is considered here. The magnitudes of the ratios of the scattering wave ϕ^S to ϕ^I at a radius $R=5a$ are plotted versus the angle θ between the direction of the incident wave and $\mathbf{R}$. In all the cases, M is the total number of elements on the sphere and N the number of nodes. Two plots of the Overhauser element meshes for the sphere are shown in Fig. 2.

All the BEM results reported here were obtained by applying the composite BIE formulation mentioned in the previous section and the wavenumbers studied are fictitious eigenfrequencies at which the conventional BIE formulation cannot provide unique solutions (indicated by a large condition number of the coefficient matrix). The computation was performed on an Apollo 10000 machine.

Figure 3 shows the convergence of the Overhauser elements at wavenumber $ka = 2\pi$. The convergence of the results is observed as the number of elements increases. It is noticed that results on the shadow side ($\theta = 0$ degree) converge at a slower rate than on the illuminated side ($\theta = 180$ degree, backscattering direction). This is a typical phenomenon for the BEM solutions of this problem.

Figure 4 is a comparison of the conforming quadratic, non-conforming quadratic and Overhauser elements, at $ka = \pi$, where the numbers of elements are fixed at $M = 56$. The numerical values at the shadow side by the Overhauser elements are not as good as the conforming and non-conforming quadratic elements, but the computer running time for the Overhauser elements is much less than those for the latter two. Notice that the size of the system of equations (N by N) for the Overhauser elements is only about *1/3* of that for the conforming elements and *1/8* of that for non-conforming elements.

Figure 5 is a comparison of the three types of elements, at $ka = \pi$, where about the same numbers of nodes are used for the three elements. This is probably a more important comparison since number of nodes used is closely related to the size of the system of equations and usually is chosen as the parameter for comparison of different elements in BEM literature, e.g. [3]. It is shown that the best results are achieved by the Overhauser elements in this case. The results by the non-conforming elements are unacceptable due to the small number of elements which can be generated from the given number of nodes. However, the running time for Overhauser elements is the longest in this case, while for the non-conforming elements,the shortest. The running time (mainly the system formation time plus the solution time) for the three types of elements will be discussed in more detail in the next section.

Figure 6 is a more realistic comparison of the three elements, at $ka = 2\pi$. First, the non-conforming elements were tested by using meshes of increasing numbers of elements (or nodes) until reasonably good results were obtained. Then, the conforming and Overhauser elements were tested in the same way until the results of about the same accuracy as that by the non-conforming elements were achieved. The results by the final meshes of the three elements were plotted and the running time recorded. It is observed that for the conforming elements, about the same number of nodes as that for the non-conforming elements is needed to achieve about the same accuracy, while for the Overhauser elements, the number of nodes needed is only nearly *1/3* of those for the other two types of elements. More importantly, the running time of the Overhauser elements is the shortest, about *3/4* of that of the non-conforming elements and *1/3* of that of the conforming elements.

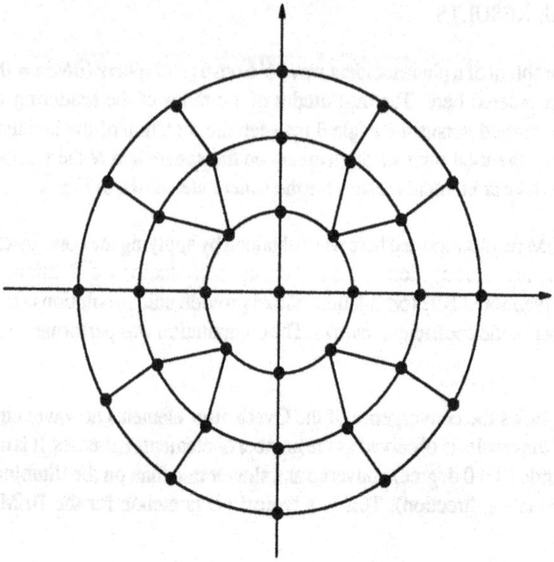

(M = 56, N = 54)

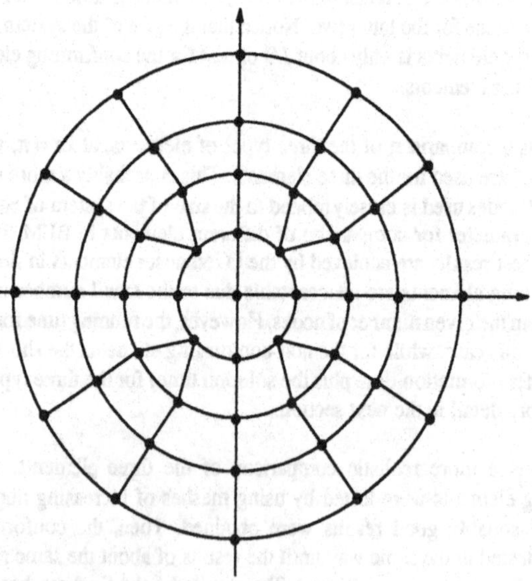

(M = 80, N = 78)

Fig. 2. Overhauser element meshes for the sphere.

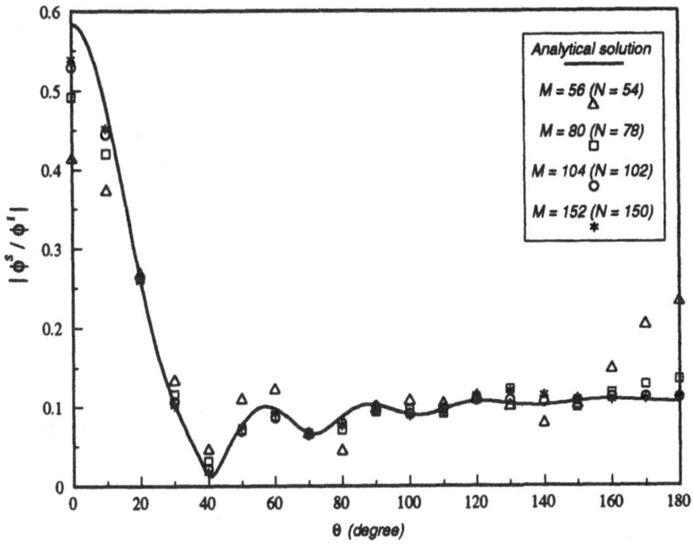

Fig. 3. Convergence of the Overhauser elements at wavenumber $ka = 2\pi$.

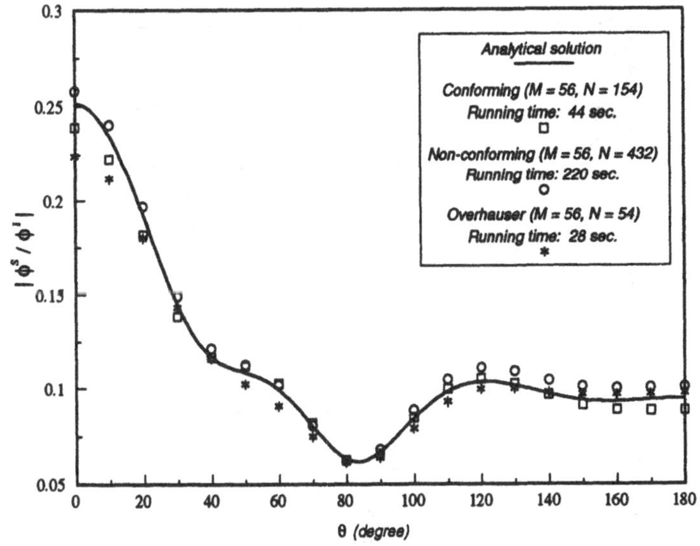

Fig. 4. Comparison at wavenumber $ka = \pi$, for a fixed number of elements.

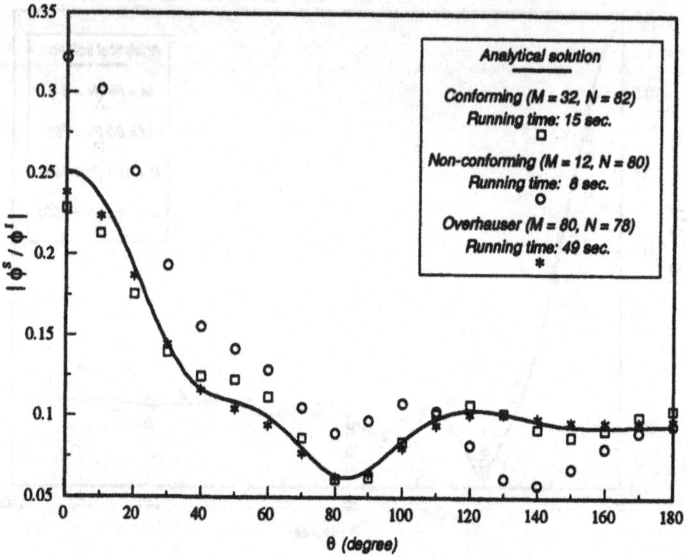

Fig. 5. Comparison at wavenumber ka = π, for a 'fixed' number of nodes.

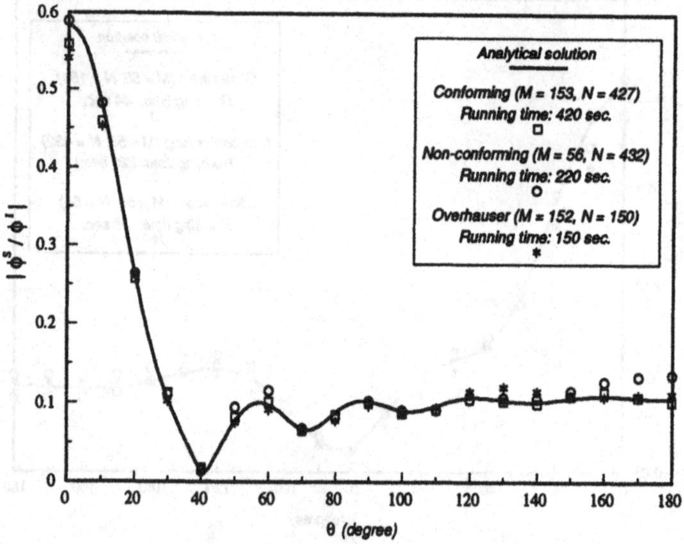

Fig. 6. Comparison at wavenumber ka = 2π.

5. DISCUSSIONS

The high accuracy of the Overhauser surface elements is further demonstrated in this paper by the numerical examples of acoustic wave problems. To achieve the same level of accuracy, considerably fewer nodes can be employed for the Overhauser elements than for quadratic elements.

One might expect that the computer time for the Overhauser elements would be longer than for the quadratic elements. Considering the large number and lengthy expressions of the shape functions for the Overhauser elements, this will be true, at least for setting up the system (formation time) as stated in [7]. However, this will be changed if the shape functions and their derivatives are evaluated only *once* and then stored, as is done in the computer codes used for this comparison study. In this way, the computation of shape functions would not be a factor in the running time for solving a problem. The formation time is then proportional to the factor $N \times M \times S$, where N is the number of nodes, M the number of elements and S the number of shape functions (equal to the number of summations performed for the integration on an element). Suppose that most of the elements used are quadrilateral ones, then $N = 3M$, $8M$ *and* M, approximately, for the conforming quadratic, non-conforming quadratic and Overhauser elements, respectively. Thus the ratios of the formation time for the three types of elements are *1.5:4:1* for a fixed number of elements or *2.67:1:16* for a fixed number of nodes. Test results show that these estimates for the formation time hold. The second estimate (for a fixed number of nodes) is not in favor of the Overhauser elements. However, one need not use the same number of nodes for the Overhauser elements to achieve a given accuracy, as shown by the numerical examples, and perhaps one needs only half the number of nodes compared to the quadratic elements. Therefore, the solution time for the Overhauser elements will be much less. For problems of a moderate size (about a few hundred nodes), the solution time will be longer than the formation time. Thus total running time will be in favor of the Overhauser elements and this is even more obvious for larger size problems.

Mesh generation of the Overhauser elements is not as easy as those of the quadratic elements because of the relatively complicated connectivity of the Overhauser elements. To apply the Overhauser elements to real engineering problems, special software for mesh generation will be needed.

ACKNOWLEDGEMENT

Partial support for this work was provided by the U.S. Office of Naval Research, Applied Mechanics Division, under Contract N00014-89-K-0109, Yapa Rajapakse scientific officer, and by the National Science Foundation under Grant NSF MSS-8918005. The authors are grateful to Dr. W. S. Hall for his kindness in providing crucial materials on the Overhauser elements.

REFERENCES

1. Krishnasamy, G., Schmerr, L. W., Rudolphi, T. J. and Rizzo, F. J. Hypersingular Boundary Integral Equations: Some Applications in Acoustic and Elastic Wave Scattering, *J. of Applied Mechanics*, Vol.57, June 1990.

2. Krishnasamy, G., Rizzo, F. J. and Rudolphi, T. J. Continuity Requirements for Density Functions in the Boundary Integral Equation Method, *Computational Mechanics*, in review.

3. Liggett, J. A. and Salmon, J. R. Cubic Spline Boundary Elements, *Int. J. Num. Meth. Eng.* Vol. 17, 543-556, 1981.

4. Watson, J. O. Hermitian Cubic and Singular Elements for Plane Strain, in *Developments in Boundary Element Methods - 4*, (Ed. Banerjee, P. K. and Watson, J. O.), Elsevier Applied Science Publishers Ltd, 1986.

5. Cabral, J. J. S. P., Wrobel, L. C. and Brebbia, C. A. A BEM Formulation Using B-Spline: I - Uniform Blending Functions, *Engineering Analysis with Boundary Elements*, Vol. 7, No. 3, 1990.

6. Ortiz, J. C., Walters, H. G., Gipson, G. S. and Brewer III, J. A. Development of Overhauser Splines as Boundary Elements, in *Boundary Elements IX*, Vol.1, (Ed. Brebbia, C. A. *et al*), Springer-Verlag, 1987.

7. Hall, W. S. and Hibbs, T. T. The Treatment of Singularities and the Application of the Overhauser $C^{(1)}$ Continuous Quadrilateral Boundary Element to Three Dimensional Elastostatics, in *Advanced Boundary Element Methods*, (Ed. Cruse, T. A.), Springer-Verlag, 1988.

8. Hibbs, T. T. $C^{(1)}$ Continuous Representations and Advanced Singular Kernel Integrations in the Three Dimensional Boundary Integral Method, Ph.D thesis, Teesside Polytechnic, UK, 1988.

9. Hall, W. S. and Hibbs, T. T. $C^{(1)}$ Continuous, Quadrilateral and Triangular Surface Patches, in *Applied Surface Modelling, Chap. 12*, (Ed. Creasy, C. and Giles, M.), Ellis Horwood, 1990.

10. Rudolphi, T. J., Krishnasamy, G., Schmerr, L. W. and Rizzo, F. J. On the Use of Strongly Singular Integral Equations for Crack Problems, in *Boundary Elements X*, (Ed. Brebbria, C. A.), Southampton, 1988.

11. Liu, Y. J. and Rizzo, F. J. A Weakly-Singular Form of the "Hypersingular" Boundary Integral Equation Applied to 3-D Acoustic Wave Problems, *Comp. Meth. in Appl. Mech. and Eng.*, in review.

12 Schenck, H. A. Improved Integral Formulation for Acoustic Radiation Problems, *J. Acoust. Soc. Am.* 44, 41-58, 1968.

13 Burton, A. J. and Miller, G. F. The Application of Integral Equation Methods to the Numerical Solution of Some Exterior Boundary-Value Problems, *Proc. Roy. Soc. London A.* 323, 201-210, 1971.

14 Liu, Y. J. and Rudolphi, T. J. Some Identities for Fundamental Solutions and their Applications to Non-Singular Boundary Element Formulations, *Engineering Analysis with Boundary Elements*, to appear.

A Multigrid-Based Boundary Element Method Without Integral Equations

C. Gáspár

Water Resources Research Centre (VITUKI), H-1095 Budapest, Kvassay Jenö út 1, Hungary

INTRODUCTION

When using the Boundary Integral Equation Method (BIEM) for solving a differential equation defined on an n-dimensional domain, one of the greatest advantages of the method is that it reduces the original problem to an (n-1)-dimensional one. Thanks to this reduction, the necessary discretization process becomes much simpler compared with the domain-type methods. But as for the computational cost of the BIEM, the situation is far from being ideal. From numerical point of view, the BIEM results in an algebraic system of equations having some unpleasant properties (since the boundary element matrices are fully populated and non-symmetric) which does not seem to allow the use of fast equation solvers, though some special methods for boundary element matrices have already been developed, see Bettess [1].

Consider, for example, a two-dimensional elliptic equation, like the Laplace or the Helmholtz equation. Roughly speaking, if the boundary of the domain has been discretized by O(N) points, the BIEM results in an algebraic system with also O(N) unknowns which requires $O(N^3)$ operations in order to solve it and $O(N^2)$ of storage capacity. It is generally better than in the case of the familiar domain-type methods: here both the number of unknowns and the storage requirement are $O(N^2)$ but, using traditional iterative methods for solving the discrete equations, the computational cost is of order N^4. In very special cases having simple geometry, differential operator as well as boundary conditions (e.g. Dirichlet problem for the Laplace equation in a circle or in a rectangle) the Fast Fourier Transform method is applicable, which requires $O(N^2 \cdot \log N)$ operations only.

In the last decade, however, an extremely powerful class of methods has appeared and come into use, namely, the Multigrid Methods (see e.g. Brandt [2]; Stüben and Trottenberg [18]; Hackbusch [11]). They are mostly applied to domain-type discretizations like the traditional finite difference methods. The multigrid methods treat the original problem in several discretization levels at the same time which makes it possible to highly reduce the computational cost: the number of the necessary operations is now proportional to the *first* power of the unknowns only. In the above two-dimensional example this means $O(N^2)$ operations which is better than that of the BIEM. Using advanced multigrid methods like FAS (Full Approximation Storage), even nonlinear problems can be treated in this way without any linearization and essentially with the same computational cost.

Our goal is to find numerical methods which are at least as efficient as the multigrid techniques applied to the domain-type methods, and, at the same time, preserve the main advantages of the BIEM. That is, roughly speaking, they should be determined by the discretization of the boundary only.

In this paper we present a new method which is, strictly speaking, not a boundary-type method and uses differential equation approach rather than integral equations. However, the method is completely controlled by the boundary: once the boundary has been discretized, all the subsequent procedures (grid generation, discretization, solution of the discrete equations) can fully be automatized. In this sense the proposed method is similar to a boundary element method though it is not based on integral equation approach. The main idea of the method is to generate a non-uniform, non-equidistant grid system which is fine in the vicinity of the boundary and coarser in the interior. As a next step, some finite difference schemes are to be defined on this grid. Finally, in order to solve the discrete equations, a special multigrid method is applied exploiting the special structure of the grid. This approach results in a method which seems to be faster not only than the traditional BIEM but even the multigrid-improved BIEM as well as the traditional (domain-type) multigrid method. Moreover, the method seems to be easily applicable to more general problems which require special tricks if the BIEM is used.

Before going into details, we outline some other techniques which are suitable to speed up the traditional BIEM.

Multigrid methods in the BIEM

One possibility is to apply the multigrid approach within the boundary integral context. The simplest case is, when the boundary integral equation of the original problem is of Fredholm-type integral equation of the second kind (Schippers [17]; Hackbusch [11]). This is valid only in the cases when either a pure Dirichlet or a pure Neumann boundary condition is given. If a mixed boundary condition is given, the boundary integral equation becomes of mixed type. However, the use of the multigrid approach can be carried out also in this case (Gáspár [8]), even if the original problem involves free surfaces (Gáspár [9]). The multigrid techniques make it possible to reduce the computational cost of the traditional BIEM from $O(N^3)$ to $O(N^2)$, that is, now the efficiency reaches that of the multigrid methods applied to the original (differential) equation using finite differences.

<u>The multipole method</u>

This method is a fast evaluation technique of a finite sum of logarithmic potential and based on the following multipole expansion (see Carrier et al [3]; Greengard and Gropp [10]):

$$\sum_{j=1}^{N} q_j \cdot \log(z-z_j) = Q \cdot \log(z) + \sum_{k=1}^{\infty} a_k \cdot z^{-k} \tag{1}$$

where

$$Q = \sum_{j=1}^{N} q_j \qquad \text{and} \qquad a_k = -\sum_{j=1}^{N} q_j \cdot z_j^k \cdot k^{-1} \tag{2}$$

The Laurent series in the right-hand side of (1) is convergent for every $|z| > \max |z_j|$.

Truncating the above Laurent series up to a prescribed index p, the potential of the left-hand side of (1) can approximately be evaluated *far from the origin* by O(p) operations instead of O(N). Based on this observation an elegant algorithm can be defined which makes it possible to evaluate certain types of potentials and (discretized) integral operators in a highly efficient way. Given a prescribed accuracy, the computational cost of the evaluation of the regular part of the left-hand side of (1) in the points $z_1,\ldots,z_N$ requires O(N) operations in contrast to the direct summation which evidently needs $O(N^2)$ operations.

The multipole method can excellently be used for the N-body problem (e.g. Katzenelson [15]); for fast solution of Poisson equations having a lot of point sources (which is the case in the computational fluid dynamics using the vortex particle simulation method, see van Dommelen [5]); moreover, it has also been applied to the Fredholm-type boundary integral equation of the Helmholtz equation combined with a classical iteration method (Rokhlin [16]). In the author's opinion, it would be interesting and useful to employ the multipole technique in the multigrid methods applied to the boundary integral equations in order to evaluate the appearing boundary integrals quickly. This procedure is expected to result in an extremely efficient method for solving boundary integral equations.

<u>The method of multiquadrics (MQ)</u>

Most recently, a general approach has been developed which is a domain-type method but does not require any grid structure. (see Kansa [13], [14]). The method is based on the general interpolation formula

$$f(x) = \sum_{j=1}^{M} a_j \cdot (r_j^2 + |x - x_j|^2)^{1/2} \tag{3}$$

where $x_1,\ldots,x_M$ are points scattered in the plane, $|.|$ stands for the standard two-dimensional Euclidean norm and a_j's are temporarily unknown coefficients which have to be chosen in such a way that the following interpolation equalities be satisfied:

$$f(x_k) = f_k, \qquad k = 1,\ldots,M \tag{4}$$

The parameters r_j are used to optimize the properties of the interpolation function.

The so-called multiquadric interpolation formula (3) proposed by Hardy has been found very useful in the scattered data interpolation problem (see Franke [7]; Hardy [12]).

In order to determine the coefficients $a_1,\ldots,a_M$ it is of course necessary to solve the system of equations (4) which has a fully populated matrix, therefore it needs $O(M^3)$ algebraic operations.

Kansa [14] proposed using the scattered data interpolation formula (3) for solving among others elliptic differential equations in such a way that one simply substitutes (3) into the differential equation. The points $x_1,\ldots,x_M$ have to be located partly inside the domain, partly along the boundary. If, for example, the Laplace equation is considered, this yields:

$$\sum_{j=1}^{M} a_j \cdot \Delta(r_j^2 + |x - x_j|^2)^{1/2} = 0 \tag{5}$$

where the equality is required for all the indices k for which x_k is an inner point of the domain. At the boundary points, the imposed boundary conditions are to be fulfilled. Kansa found the method extremely accurate. It should be noted, however, that he used few points only (12 points in the interior and 18 along the boundary). More complicated problems and/or complicated geometry may need much more interpolation points, which result in numerical difficulties similar to that of the BIEM.

The main idea of the MQ method can easily be adopted in the boundary-type methods as well. As a model problem, consider again the Laplace equation. Let us discretize the boundary of a two-dimensional domain Ω by $O(N)$ points scattered along the boundary. Now let us exploit the fact that the solution of the Laplace equation is much smoother in the middle of the domain Ω than in the vicinity of the boundary. This allows us to scatter the interior points much less densely in the middle of the domain than near to the boundary. The density of the points can decrease *exponentially* toward the middle of Ω, therefore the number of the interior points can be $O(N)$ instead of $O(N^2)$.

Thus, the obtained algebraic system of equations requires $O(N^3)$ operations, as in the traditional BIEM.

This version of the MQ method is a little similar to the BIEM in the respect that both methods use approximate functions determined by the boundary values via solving an algebraic equation. But while the BIEM employs approximate functions satisfying the original differential equation *exactly* (double-layer and single-layer potentials), the approximate functions of the MQ method do not have such a property. This makes it possible, however, to apply the MQ to more general equations which do not have simple fundamental solution, e.g. the elliptic equation of the form

$$\text{div } k \text{ grad } u = 0 \tag{6}$$

where k is not constant any more. This can make the MQ approach attractive even though it requires as great amount of computational work as the traditional BIEM.

In contrast to the above outlined approaches, we show another "quasi-boundary" technique which is based on finite difference schemes defined by a non-uniform, non-equidistant grid. The main parts of the method are as follows:

- to generate the above mentioned non-uniform grid by using a discretization of the boundary only: the resulting grid density should increase rapidly toward the boundary

- to define difference schemes on this grid

- to apply a multigrid technique in order to speed up the numerical solution of the discrete equations

We shall use the so-called "unstructured grid generation" method or "quadtree algorithm" (see Cheng et al [4]) for creating such grids. We note, however, that the same technique is used also in the multipole method: moreover, the quadtree algorithm is quite suitable to generate inner interpolation points for the MQ method as well.

BOUNDARY-CONTROLLED GRID GENERATION BY USING QUADTREES

The main idea of the quadtree algorithm is extremely simple. It is based on a recursive subdivision process. Consider, for the sake of simplicity, the unit square of the plane containing N different points $x_1,\ldots,x_N$. If N is greater than a prescribed number N_0, we divide the square into four smaller squares (with side length 1/2). Each smaller square (cell) have to be divided again into even smaller cells if the number of the contained points exceeds the limit N_0. The process is to be continued until each cell contains not more than N_0 points, or, the level of the subdivision reaches a prescribed maximal number L.

The structure of the grid will clearly follow the density of the points: the grid becomes fine in the subregions where the points are located densely, and coarse elsewhere.

The data structure of the cell system (grid) can be represented by a directed graph in a very natural way. The root element represents the starting square: elements of the other levels represent the squares of the same level of subdivision. A branch means a subdivision process: the leaves of the graph correspond to the cells containing not more than N_0 points. As a special case, an empty cell is always represented by a leaf. Fig.1 shows a typical quadtree cell system generated by 10 points, with the corresponding graph: in this example $N_0 = 1$.

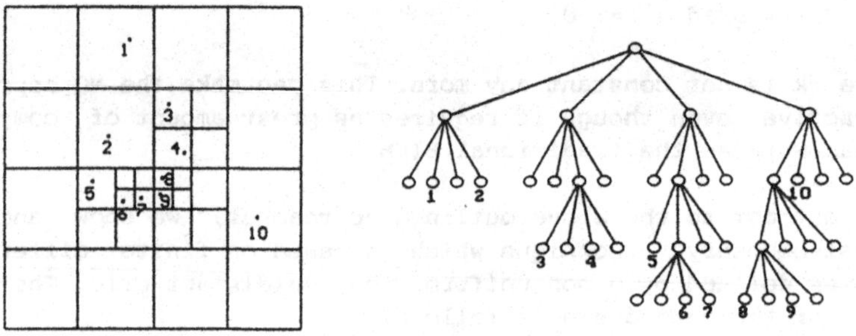

Fig.1. A typical quadtree cell system and the representing graph

It should be pointed out that the process is remarkably "cheap" requiring not more than O(NL) operations which can often be estimated by O(N· log N).

The process can similarly be defined in arbitrary dimensions. In three-dimensional cases, the analogous procedure is called "oc-tree algorithm". Here in each subdivision process a cube is to be divided into eight congruent smaller cubes.

For our purposes, the starting points which generate the quad-tree cell system, should lie along the boundary of the domain. Once these points have been defined, the corresponding grid can automatically be generated by the above quadtree algorithm (even in the case of multiply-connected domains). Unlike the BIEM, the quadtree grid generation needs the *locations* of the points only: the BIEM requires a discretization of the boundary which in-volves, in addition, also a *mesh structure* on the boundary.

The boundary points need not be located uniformly: they may be located more densely in subregions being of special interest, or, requiring greater precision (e.g. in the vicinity of abrupt changes of boundary conditions). Of course, it is possible to generate local grid refinements *inside* the domain as well, by locating extra points in the desired subregions.

Fig.2 shows two simple examples: in both cases the domain is a circle. In the first case, the boundary points are scattered uniformly along the boundary, while in the second case the structure of the grid is controlled by a varying boundary point density.

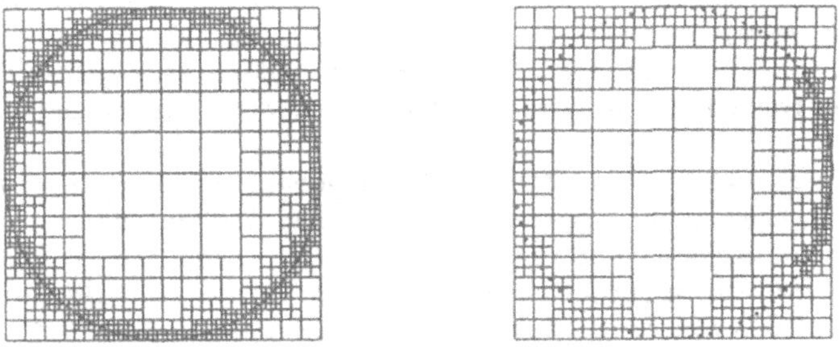

Fig.2. Quadtree-generated grids controlled by boundary point distribution

Having generated the grid, it is necessary to complete the cell system by creating extra cells by additional subdivisions in order that the neighbouring cell sizes be equal or have the ratio at most 1:2. Thus, abrupt changes in grid size can be avoided, which would make the discretisation error much greater. In the examples shown in Fig.2 this additional subdivision procedure has been performed. After that, it is often worth halving again each cell at least once which makes the "middle" cells sufficiently fine. The total number of the cells is *linearly* proportional to the number of the (smallest) cells located along the boundary.

One can easily see why it is unnecessary to use an everywhere dense grid. Without giving a precise analysis, consider again the Laplace equation in the unit circle. The solution can be expressed in terms of Fourier series:

$$u(r,\theta) = \sum_k u_k \cdot r^{|k|} \cdot e^{ik\theta} \tag{7}$$

written in polar coordinates. The Fourier term $u_k \cdot r^{|k|} \cdot e^{ik\theta}$, as a function of the radius r decreases rapidly when r decreases. Measured in the L_2-norm the estimation

$$|u_k\ r^{|k|}\ e^{ik\theta}|_{L_2} \le C\ |u_k|\ e^{-d|k|}$$

is valid with a constant C independent of k and u, where $d = 1-r$ is the distance from the boundary. Similar estimations hold for any Sobolev norm H^s. Consequently, the contribution of the higher order Fourier terms can be negligible far from the boundary, which allows us to use coarser grid there. The situation is quite similar in the case of domains having more general shape, as it can be seen by conformal mappings.

DIFFERENCE SCHEMES AND MULTIGRID SOLUTIONS ON QUADTREE-GENERATED GRIDS

After generating the above "unstructured" grid, one has to discretize the origial differential equation on the grid. To do that, there are several ways depending on the choice of the nodal (or grid) points. If, for example, we choose the corners of the cells as nodal points, then a standard triangularization can be performed and a finite element approach can be applied.

Another way is to use finite differences. Assume again the example of the Laplace operator. Choosing the gridpoints again to be the corners of the cells, the standard difference quotients can be written: since the grid is non-equidistant, the schemes will not be central, therefore they become of first order in general. Thanks to the fact that the adjacent cell sizes has the ratio at most 1:2, the only "irregular" case is the situation shown in Fig.3a (where the point E is not a grid-point).

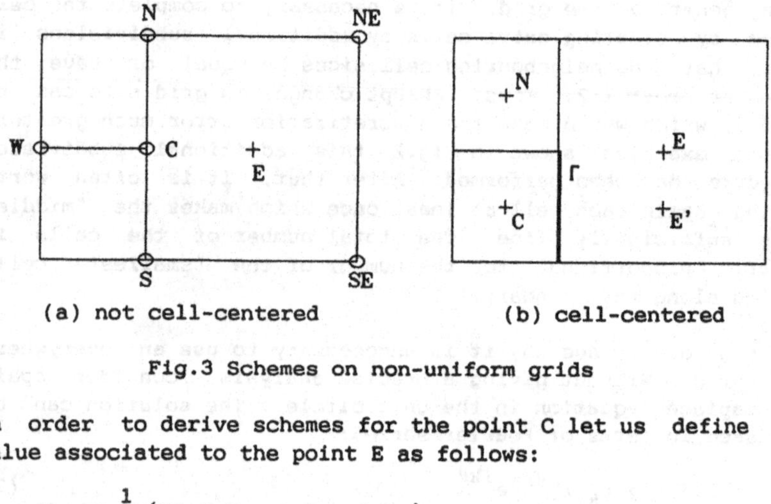

(a) not cell-centered (b) cell-centered

Fig.3 Schemes on non-uniform grids

In order to derive schemes for the point C let us define the value associated to the point E as follows:

$$u_E := \frac{1}{4}\,(u_N + u_{NE} + u_{SE} + u_S) \tag{8}$$

Now it is possible to use the usual central schemes for the point C: the resulting scheme is at least of order 1.

A third possibility is the use of cell-centered schemes. Now the centers of the cells play the role of the nodal points. Integrating the differential equation over the cell C we obtain:

$$\int_{\partial C} \frac{\partial u}{\partial n} = 0 \tag{9}$$

that is, only the fluxes $\partial u/\partial n$ are to be approximated. If two adjacent cells have the same size then the central scheme can be applied. If not, then again due to the adjacent cell size ratio being at most 1:2, the problem is always of the type that can be seen in Fig.3b. The flux across Γ can be approximated either by the formula

$$\int_{\Gamma} \frac{\partial u}{\partial n}\, d\Gamma \sim \frac{u_E - \dfrac{u_N + u_C}{2}}{\dfrac{3}{2}\cdot h}\cdot h \tag{10}$$

or by the formula

$$\int_{\Gamma} \frac{\partial u}{\partial n}\, d\Gamma \sim \frac{u_{E'} - u_C}{\dfrac{3}{2}\cdot h}\cdot h \tag{11}$$

where h is the size of the cell C. In the latter case the value $u_{E'}$ should be approximated by using the neighbouring cells of E and some interpolation (see Ewing and Lazarov [6]). It is also possible to use central differences involving the cell E and its western neighbour *from the same level of the quadtree graph*: of course, a properly defined value should belong to the latter cell.

Each discretization strategy evidently needs the neighbouring cells (gridpoints, respectively). In contrast to the usual uniform grids, in our quadtree context this means an extra task. Exploiting the tree structure of the cells, however, it can always be carried out in computationally acceptable way: the necessary computational work is proportional to the number of cells only whereas the direct neighbour finding approach with N points (cells) would require $O(N^2)$ operations.

Finally we discuss the problem of the solution of the discrete equations. For the sake of brevity we consider the cell-centered schemes only. Each discretized version of the Eq.(9) gives immediately a Seidel iteration scheme where the central value u_C is expressed in terms of the neighbouring u-values. This Seidel iteration is, however, very slow but can serve as a smoothing procedure of a special multigrid technique defined on the quadtree graph. Recall that the multigrid approach requires a sequence of nested grids with properly defined restriction and prolongation operators. In out quadtree context the graph structure suggests the definition of the "coarse" and "fine" grids in a very natural way. The original (finest) quadtree grid consists of the *leaves* of the quadtree graph. Denoting this finest grid by G_L (where L is the maximal level of subdivision) let us define the grid G_k for k<L but cutting every subgraph of G_L starting from the level k and replacing it with the corresponding root cell. It is clear that G_k is exactly the same grid that one would have obtained by the quadtree algorithm directly if the value k had been specified as the maximal level of subdivision.

The simplest way to define the inter-grid transfers is as follows. The restriction can be defined by averaging the values belonging to the contained cells, while the prolongation can be performed by transferring the central value to the contained four cells without change (piecewise constant prolongation). Weighting methods which are familiar in the multigrid techniques are also applicable using the neighbouring cells again.

Now the usual multigrid cycle on the level k can be performed without difficulty by the following recursive definition (for details, see again Stüben and Trottenberg [18]). Assume that the discrete system of equations has the form

$$Ax = f \,, \tag{12}$$

then the following steps are to be performed:

Step 1. Apply the smoothing Seidel iteration several times.

Step 2. Calculate the defect

$$r = f - Ax'$$

where x' is the actual approximative solution of (12) and transfer it to the next coarser grid G_{k-1}.

Step 3. Solve the residual equation

$$Aw = r$$

on G_{k-1} exactly if $k-1 = 1$, and by using multigrid cycles on the level $k-1$, otherwise.

Step 4. Transfer the solution of the residual equation back to G_k and improve the approximative solution by adding the correction w to x'.

Step 5. Perform several Seidel smoothing iterations again.

Finally, it should be poined out that the computational cost of the multigrid method is proportional to the *first* power of the number of the cells. Returning to to estimates of the computational work mentioned in the introduction, the proposed combined method seems to be even more efficient than the previous multigrid techniques (applied in either domain-type or boundary integral context) since it requires $O(N \cdot \log N)$ operations for the grid generation and only $O(N)$ operation for the solution.

A NUMERICAL EXAMPLE

The implementation of the method described above is illustrated by applying it to the Laplace equation defined on the unit square. The boundary conditions were of Dirichlet type. The test solution was the following function:

$$u(x,y) = x^2 + xy - y^2 \tag{13}$$

The applied quadtree-generated cell system contained 917 cells in total (including the cells of the coarser grids). The maximal level of discretization was 6, which means that the size of the smallest (boundary) cells was 1/64. The finest grid consisted of 688 cells. Note that using uniformly fine grid would have resulted in more than 4000 cells. Fig.4 shows the finest quadtree cell system and the fitting of the computed solution to the theoretical solution along the line x + y = 1.

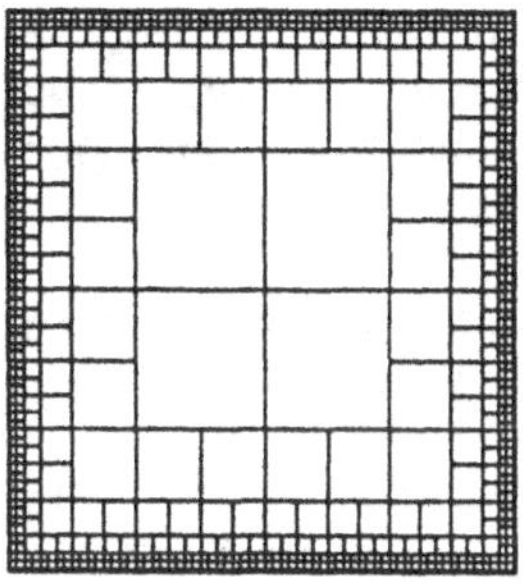

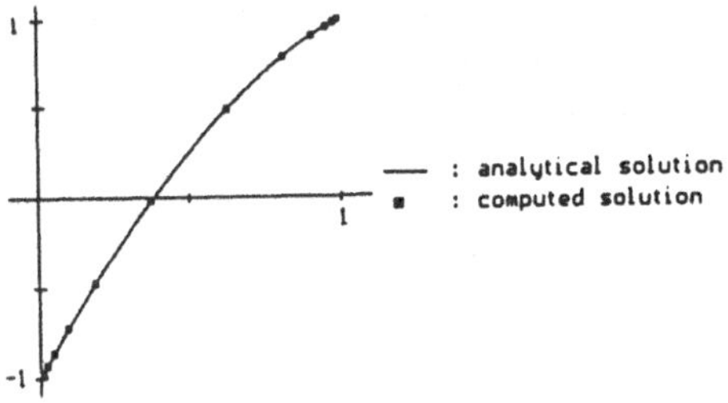

Fig.4 A model problem. The quadtree-generated grid and the computational result.

We have found that the exactness of the approximation is satisfactory: measured in the discrete L_2-norm, the difference between the numerical and analytical solution was less the 0.5% after a full multigrid procedure.

REFERENCES

1. Bettes,J.A. Solution Techniques for Boundary Integral Matrices. In: Numerical Methods for Transient an Coupled Problems (Ed. by R.W.Lewis, E.Hinten, P.Battess, B.A.Schrefler). Wiley-Interscience, 1987.

2. Brandt,A. Multigrid Techniques. 1984 Guide with Applications to Fluid Dynamics. GMD-Studien Nr.85, Bonn, 1984.

3. Carrier,J., Greengard,L., Rokhlin,V. A Fast Adaptive Multipole Algorithm for Particle Simulations. SIAM J.Sci. Stat. Comput. Vol 9 No 4, July, 1988.

4. Cheng,J.H., Finnigan,P.M., Hathaway,A.F., Kela,A., Schroeder,W.J. Quadtree/octree meshing with adaptive analysis. In: Numerical Grid Generation in Computational Fluid Mechanics '88. (Ed. by S.Sengupta, J.Hauser, P.R. Eiseman, J.F.Thompson.) Pineridge Press, Swansea, 1988.

5. Dommelen,L.van. Fast, Adaptive Summation of Point Forces in the Two-dimensional Poisson Equation. J.Comput.Physics Vol 83, 1989.

6. Ewing,R.E., Lazarov,R.D. Local Refinement techniques in the finite element and finite difference methods. In: Proc. Int. Conf. on Numerical Methods and Applications. Publishing House of the Bulgarian Academy of Sciences, 1989.

7. Franke,R. Scattered Data Interpolation: Test of Some Methods. Math.Comput. Vol 38, No 157, January, 1982.

8. Gáspár,C. Solution of Seepage Problems by Combining the Boundary Integral Equation Method with a Multigrid Technique. (Ed. by G.Gambolati, A.Rinaldo, C.A.Brebbia, W.G.Gray, G.F.Pinder.) Proc. VIII. Int.Conf. on Computational Methods in Water Resources, Venice, Italy, 1990. Computational Mechanics Publications,Southampton - Springer-Verlag, Berlin, Heidelberg, New York, 1990.

9. Gáspár,C. A Fast Multigrid Solution of Boundary Integral Equations. Environmental Software, Vol 5, No 1, 1990.

10. Greengard,L., Gropp,W.D. A Paralell Version of the Fast Multipole Method. Computers Math.Applic. Vol 20 No 7, 1990.

11. Hackbusch,W. Multi-Grid Methods and Applications. Springer-Verlag, Berlin, Heidelberg, New York, Tokyo, 1985.

12. Hardy,R.L. Theory and Applications of the Multiquadric-Biharmonic Method. 20 Years of Discovery 1989-1988. Computers Math.Applic. Vol 19, No 8/9, 1990.

13. Kansa,E.J. Multiquadrics - a Scattered Data Approximation Scheme with Applications to Computational Fluid Dynamics - I. Surface Approximations and Partial Derivative Estimates. Computers Math.Applic. Vol 19, No 8/9, 1990.

14. Kansa,E.J. Multiquadrics - a Scattered Data Approximation Scheme with Applications to Computational Fluid Dynamics - II. Solutions to Parabolic, Hyperbolic and Elliptic Partial Differential Equations. Computers Math.Applic. Vol 19, No 8/9, 1990.

15. Katzenelson,J. Computational Structure of the N-body Problem. SIMA J.Sci.Stat.Comput. Vol 10, No 4, July, 1989.

16. Rokhlin,V. Rapid Solution of Integral Equations of Scattering Theory in Two Dimensions. J.Comput.Physics Vol 86, 1990.

17. Schippers,H. Multiple Grid Methods for Equations of the Second Kind with Applications in Fluid Mechanics. Thesis. Matematisch Centrum, Amsterdam, 1982.

18. Stüben,K., Trottenberg,U. Multigrid Methods: Fundamental Algorithms, Model problem Analysis and Applications. GMD-Studien, Nr.96, Birlinghoven, 1984.

Fundamental Solution for a Two-Dimensional Microelastic Body

J.J. Rencis (*), Q. Huang (**)
(*) Mechanical Engineering Department, Worcester Polytechnic Institute, Worcester, Massachusetts 01609, U.S.A.
(**) Department of Mechanical Engineering, Vanderbilt University, Nashville, Tennessee 37235, U.S.A.

The fundamental solution for a two-dimensional microelastic body was derived by Dragos [1]. This fundamental solution can be used for solving problems by the boundary element method [2,3]. However, the fundamental solution requires that detailed (complete) mathematical relationships be established for generalized displacements and generalized stresses. In this work, the fundamental solutions developed by Dragos [1] are completely established for boundary element implementation. The governing partial differential equations for a two-dimensional microelastic body are

$$(\mu + \alpha)u_{i,jj} + (\lambda + \mu - \alpha)u_{j,ji} + 2\alpha\, e_{ijk} w_{j,k} = 0 \tag{1a}$$

$$(\gamma + \epsilon)w_{i,jj} + (\beta + \gamma - \epsilon)w_{j,ji} + 2\alpha\epsilon_{ijk}u_{j,k} - 4\alpha w_i = 0 \tag{1b}$$

where λ and μ are the elastic material constants, α, γ, and ϵ are the microelastic material constants and subscripts denote coordinate directions. Einstein index notation is used unless otherwise noted. These three equations are written in terms of two x_1-x_2 plane displacements u_1 and u_2 and one rotation w_3 perpendicular to the x_1-x_2 plane. The boundary conditions between the generalized tractions and generalized displacements are

$$t_j = \sigma_{ij}n_i = \lambda n_j u_{k,k} + 2\mu u_{j,k}n_k + (\mu - \alpha)(u_{k,j} - u_{j,k})n_k + 2\alpha\epsilon_{jmk}n_m w_k \tag{2a}$$

$$m_j = m_{j3}n_j = \beta n_j w_{k,k} + 2\gamma w_{j,k}n_k + (\gamma - \epsilon)(w_{k,j} - w_{j,k})n_k \tag{2b}$$

where t_j denotes the boundary traction, m_j signifies the boundary moment and n_j is the outward unit normal. The fundamental solution for displacements and micro-rotations associated with a homogeneous, isotropic, linear microelastic body for plane strain are [1]

$$\begin{aligned} u_{ij} = &-\frac{1}{2\pi\mu}[Ln(r) + I + A\mu K_0(ar)]\delta_{ij} \\ &+\frac{1}{2\pi}\frac{\partial^2}{\partial x_i \partial x_j}[(\frac{1}{4}Br^2 + C)Ln(r) + CK_0(ar)] \qquad j = 1,2 \end{aligned} \tag{3a}$$

$$u_{j3} = \frac{1}{4\pi\mu}\epsilon_{kj3}\frac{\partial}{\partial x_k}[Ln(r) + K_o(ar)] \qquad i,j = 1,2 \tag{3b}$$

$$w_{i3} = \frac{1}{4\pi\mu}\epsilon_{ki3}\frac{\partial}{\partial x_k}[Ln(r) + K_o(ar)] + \frac{\delta_{i3}}{2\pi(\gamma+\epsilon)}K_o(ar) \quad j=1,2,3 \tag{3c}$$

where K_0 is the modified Bessel function of imaginary argument of the first kind and of order zero. Also, u_{ij} denotes the displacement in the j direction (j = 1,2) due to a unit force acting in the i direction, u_{j3} is the displacement in j direction due to a unit couple acting normal to the x_1-x_2 plane (x_3 direction), and w_{3i} is a rotation in a direction normal to the x_1-x_2 plane due to a unit force acting in the i (i = 1,2) direction or a unit couple (i = 3) acting normal to the x_1-x_2 plane (x_3 direction). The constants in Equation (3) are defined as

$$a^2 = \frac{4\alpha\mu}{(\mu + \alpha)(\gamma + \epsilon)} \tag{4a}$$

$$A = \frac{\alpha}{\mu(\mu + \alpha)} \tag{4b}$$

$$B = \frac{\lambda + \mu}{\mu(\lambda + 2\mu)} \tag{4c}$$

$$C = \frac{\gamma + \alpha}{4\alpha\mu} \tag{4d}$$

$$I = \int_0^1 \frac{1-J_o(u)}{u}du - \int_1^\infty \frac{J_o(u)}{u}du = \xi + Ln(\frac{1}{2}) \tag{4e}$$

where ξ = 0.57721566491 is Euler's constant and the constant I [4] can be set to zero since it represents a rigid body component of the displacement u_{ij} in Equation (3a), which is arbitrary. For plane stress, λ in Equation (4) is replaced by $2\lambda\mu/(\lambda + 2\mu)$. The symbol r denotes the radius which defines the distance between the field point and source point, i.e.

$$r = \sqrt{x_i x_i} \tag{5a}$$

and its derivatives are

$$r_{,i} = \frac{x_i}{\sqrt{x_i x_i}} \tag{5b}$$

$$r_{,ij} = \frac{1}{r}(\delta_{ij} - r_{,i}r_{,j}) \tag{5c}$$

Equation (3) is written in terms of the modified Bessel function derivatives. To simplify this equation so that derivatives do not exist, recurrence relationships [5] are used as follows

$$K_{n+1}(x) = K_{n-1}(x) + \frac{2n}{x} K_n(x) \tag{6a}$$

$$[K_n(x)]_{,i} = -K_{n-1}(x) - \frac{n}{x} K_n(x) \tag{6b}$$

$$[K_0(x)]_{,i} = -K_1(x) \tag{6c}$$

$$[K_1(x)]_{,i} = -K_0(x) - \frac{1}{x} K_1(x) \tag{6d}$$

where K_n denotes the modified Bessel function of imaginary argument of the first kind and of order n. These recurrence relations can be written in terms of 'ar' according to Equation (3) as follows

$$K_{n+1}(ar) = K_{n-1}(ar) + \frac{2n}{ar} K_n(ar) \tag{7a}$$

$$\frac{dK_n(ar)}{dr} = - a K_{n-1}(ar) - \frac{n}{r} K_n(ar) \tag{7b}$$

$$\frac{dK_0(ar)}{dr} = -aK_1(ar) \tag{7c}$$

and

$$\frac{dK_1(ar)}{dr} = - aK_0(ar) - \frac{1}{r} K_1(ar) \tag{7d}$$

The fundamental solution in Equation (3) can be rewritten using the recurrence relations in Equation (7) as

$$\begin{aligned} u_{ij} = \frac{1}{4\pi}\{&[(B-\frac{2}{\mu})Ln(r)+\frac{B}{2}+\frac{2C}{r^2}-2AK_0(ar)-\frac{2Ca}{r}K_1(ar)]\delta_{ij} \\ &+[B-\frac{4C}{r^2}+2Ca^2K_o(ar)-\frac{4Ca}{r}K_1(ar)]r_{,i}r_{,j} \end{aligned} \tag{8a}$$

$$u_{j3} = \frac{1}{4\pi\mu}\epsilon_{ij3}[\frac{1}{r} - aK_1(ar)]r_{,i} \tag{8b}$$

$$w_{i3} = \frac{1}{4\pi\mu}\epsilon_{ki3}[\frac{1}{r} - aK_1(ar)]r_{,k} - \frac{2\delta_{ij}}{(\gamma+\epsilon)}K_0(ar) \tag{8c}$$

The fundamental solution is now written in terms of K_0 and K_1, and not their derivatives.

To find the tractions and couple tractions in Equations (2), the derivatives of generalized displacements are as follows

$$\begin{aligned} u_{ij,k} = \frac{1}{4\pi}\{&[(B-\frac{2}{\mu})\frac{1}{r}-\frac{4C}{r^3}+\frac{2Ca^2}{r}K_0(ar)+(2aA+\frac{4Ca}{r^2})K_1(ar)]r_{,k}\delta_{ij} \\ &+[\frac{B}{r}-\frac{4C}{r^3}+\frac{2Ca^2}{r}K_0(ar)-\frac{4Ca}{r^2}K_1(ar)](\delta_{ik}r_{,j}+\delta_{ik}r_{,j}) \\ &+[\frac{12C}{r^3}-\frac{2B}{r}+(\frac{16Ca}{r^2}-2Ca^3)K_1(ar)]r_{,i}r_{,j}r_{,k}\} \end{aligned} \tag{9a}$$

$$u_{ik,k}=\frac{1}{4\pi}\{[(B-\frac{1}{\mu})\frac{2}{r}+\frac{8Ca^2}{r}K_0(ar)+(2aA-2Ca^3+\frac{4Ca}{r^2})K_1(ar)]r_{,i}\} \tag{9b}$$

$$u_{3j,k}=\frac{\epsilon_{mj3}}{4\pi\mu}\{[\frac{1}{r^2}-\frac{a}{r}K_1(ar)]\delta_{km}+[a^2K_0(ar)+\frac{2a}{r}K_1(ar)-\frac{2}{r^2}]r_{,k}r_{,m}\} \tag{9c}$$

$$u_{3k,k}=0 \tag{9d}$$

$$\begin{aligned}w_{3i,k}=\frac{\epsilon_{mi3}}{\mu}\{[\frac{1}{r^2}-\frac{a}{r}K_1(ar)]\delta_{km}+[a^2K_0(ar)\\+\frac{2a}{r}K_1(ar)-\frac{2}{r^2}]r_{,k}r_{,m}\}+\frac{2a}{\gamma+\epsilon}\delta_{i3}K_1(ar)r_{,k}\end{aligned} \tag{9e}$$

Using these expressions, the tractions and couple tractions in Equation (2) are as follows

$$\begin{aligned}t_{ij}=\frac{1}{4\pi}\{&\lambda[(B-\frac{1}{\mu})\frac{2}{r}+\frac{8Ca^2}{r}K_0(ar)+(2aA-2Ca^3+\frac{4Ca}{r^2})K_1(ar)]r_{,i}n_j\\&+[(B-\frac{2}{\mu})\frac{1}{r}-\frac{4C}{r^3}+\frac{2Ca^2}{r}K_0(ar)+(2aA+\frac{4Ca}{r^2})K_1(ar)]\\&[\mu(r_{,k}\delta_{ij}n_k+n_ir_{,j})+\alpha(\delta_{ij}r_{,k}n_k-r_{,j}n_i)]\\&+[\frac{B}{r}-\frac{4C}{r^3}+\frac{2Ca^2}{r}K_0(ar)-\frac{4Ca}{r^2}K_1(ar)]\\&[\mu(r_{,j}n_i+2r_{,i}n_j+\delta_{ij}r_{,k}n_k)+\alpha(r_{,j}n_i-\delta_{ij}r_{,k}n_k)]\\&+2\mu[\frac{16C}{r^3}-\frac{2B}{r}+(\frac{12Ca}{r^2}-2Ca^3)K_1(ar)]r_{,i}r_{,j}r_{,k}n_k\\&+\frac{2\alpha}{\mu}\epsilon_{mi3}[\frac{1}{r}-aK_1(ar)]r_{,m}\epsilon_{jk3}n_k\end{aligned} \tag{10a}$$

$$\begin{aligned}t_{3j}=&\frac{\alpha}{2\pi\mu}\{[\frac{1}{r^2}-\frac{a}{r}K_1(ar)]\epsilon_{kj3}n_k+[a^2K_0(ar)+\frac{2a}{r}K_1(ar)-\frac{2}{r^2}]\\&[\mu(\epsilon_{mj3}r_{,k}+\epsilon_{mk3}r_{,j})+\alpha(\epsilon_{mj3}r_{,k}-\epsilon_{mk3}r_{,j})]r_{,m}n_k\}\\&-\frac{\alpha}{\pi(\gamma+\epsilon)}K_0(ar)\epsilon_{jk3}n_k\end{aligned} \tag{10b}$$

$$\begin{aligned}m_{i3}=\frac{(\gamma+\epsilon)\epsilon_{mi3}}{\mu}\{[\frac{1}{r^2}-\frac{a}{r}K_1(ar)]\delta_{km}+[a^2K_0(ar)\\+\frac{2a}{r}K_1(ar)-\frac{2}{r^2}]r_{,k}r_{,m}\}n_k+2a\delta_{i3}K_1(ar)r_{,k}n_k\end{aligned} \tag{10c}$$

For elastic problems the following constants vanish, i.e.

$$\begin{aligned}\alpha=\beta=\gamma=\epsilon=0\\a=b=A=C=0\end{aligned} \tag{11a}$$

and

$$B=\frac{1}{2\mu(1-\nu)} \tag{11b}$$

Therefore,

$$u_{ij}=\frac{1}{4\pi}[(B-\frac{2}{\mu})Ln(r)\delta_{ij}+Br_{,i}r_{,j}] \tag{12a}$$

$$t_{ij} = \frac{1}{4\pi}\frac{2\lambda}{r}(B - \frac{1}{\mu})r_{,i}n_j + \frac{\mu}{r}(B - \frac{2}{\mu})(r_{,k}\delta_{ij}n_k + r_{,j}n_i) + \frac{B\mu}{r}(r_{,j}n_i + 2r_{,i}n_j + \delta_{ij}r_{,k}n_k) - \frac{4\mu B}{r}r_{,i}r_{,j}r_{,k}n_k] \tag{12b}$$

which returns to the fundamental solution for classical elasticity [2,3] when Equation (11b) is substituted into Equations (12a) and (12b), i.e.

$$u_{ij} = \frac{1}{8\pi\mu(1 - \nu)}[(4\nu - 3)Ln(r)\delta_{ij} + r_{,i}r_{,j}] \tag{13a}$$

$$t_{ij} = \frac{1}{4\pi(1-\nu)r}\{[(2\nu-1)\delta_{ij} - 2r_{,i}r_{,j}]r_{,k}n_k + (1-2\nu)(r_{,i}n_j - r_{,j}n_i)\} \tag{13b}$$

REFERENCES

[1] L. Dragos, Int. J. Engng. Sci., 22, 3, 265-275, (1984).

[2] C.A. Brebbia and J. Dominguez, *Boundary Elements An Introductory Course*, McGraw-Hill Book Company, (1989).

[3] P.K. Banerjee and R. Butterfield, *Boundary Element Method in Engineering Science*, McGraw-Hill, London, (1981).

[4] M. Abramowitz and I.A. Stegun (eds.), *Handbook of Mathematical Functions with Formulas, Graphs, and Mathematical Tables*, Dover Publications, Inc., New York, 481, (1972).

[5] G.N. Watson, *A Treatise on the Theory of Bessel Functions*, Cambridge Press, 79, (1958).

The Boundary Element Method on a Transputer Network

A.J. Davies

Division of Mathematics, Hatfield Polytechnic, Hatfield, AL10 9AB, U.K.

ABSTRACT

It has been shown[1] that the boundary element method is a numerical algorithm which is very well-suited to parallel solution on a distributed array processor. In these cases the inherent parallelism exploited is of the fine-grained type. Such a machine is of the SIMD class in which the same operation is performed in parallel by a large number of simple processing elements.

There is also, however, a coarse-grained parallelism[2] inherent in the method and this parallelism can be exploited by a machine of the MIMD class in which different operations are performed in parallel on a small number of much more sophisticated processors. A suitable set is a transputer network which comprises a host transputer to which is attached a collection of similar processors.

The typical phases in the numerical computation of the boundary element method are the system equation set-up, the equation solution and the recovery of the field variables. Each of these phases exhibits a coarse-grained parallelism suitable for implementation on a transputer network, e.g. in the equation set-up phase the calculation of the coefficients requires a nested set of three loops: an outer loop over the base nodes, an intermediate loop over the target elements and an inner loop over the Gauss quadrature points. The order in which these loops is performed is crucial to the type of mapping that may be employed. In this paper we shall consider the loop over the Gauss points to be the outer loop and we shall associate a single transputer with each Gauss point. The contributions to each coefficient at each Gauss point are calculated in parallel on separate transputers and are finally accumulated in the root transputer. A similar parallelism is exploited in the field recovery phase.

With this mapping we describe the implementation of the method on a tree structure comprising five IMS T800 transputers. We show that the transputer network provides a very cost-effective environment for the boundary element method.

KEYWORDS

Boundary element method, parallel processing, transputer.

INTRODUCTION

The boundary element method is a well-established technique for the solution of problems in a wide variety of engineering applications. In many instances it provides an attractive alternative to the finite element and the finite difference methods. It is surprising, then, that although there has been a significant amount of work done on the implementation of finite element and finite difference methods on parallel computers[3,4], there has been relatively little work done on the implementation of boundary element methods on such architectures.

One of the earliest reported boundary element implementations was developed by Symm[5]. He used an indirect boundary element solution of the Dirichlet problem as a study of the feasibility of using a distributed array processor, the DAP, for boundary element calculations. The work done by Davies, using linear elements[1] and quadratic elements[6,7], extended the ideas to implement the direct boundary element method on the DAP. This work has shown that the DAP architecture is ideally suited to boundary element analysis.

The success of the DAP as a suitable environment for boundary element analysis and the fact that some success has been reported on architectures such as vector processors[8] suggests that other parallel architectures may also provide suitable environments.

PARALLEL COMPUTERS

The distributed array processor is a parallel computer which belongs to the SIMD classification[9]; single-instruction multiple-data. Such machines comprise a large number of relatively unsophisticated processing elements connected together in a rigid manner. Each processor has its own memory and executes the same instruction on the data stored in the processor memory. Consequently a large number of executions of identical instructions are processed simultaneously.

The transputer network on which the work in this report was developed belongs to the MIMD classification; multiple-instruction multiple-data.

The MIMD machines usually comprise a small number of relatively sophisticated processors connected together in a much less rigid fashion. Indeed one of the considerations to be addressed by the user is the configuration of the system since this is closely related to the choice of algorithm and can have a significant effect on the time of execution. Each processor has its own memory and executes its own instruction on the data stored in the processor memory. Consequently different instructions are processed in parallel.

The SIMD and MIMD architectures exhibit quite different types of parallelism. The SIMD machines are usually associated with fine-grained parallelism because each processor performs simple operations on a small data set. The MIMD machines, on the other hand, are usually associated with coarse-grained parallelism because each processor performs complex operations on a large data set.

The parallelism inherent in models such as those developed in a boundary element analysis also exhibits both types of granularity. The task for the user is to recognise the parallelism and its granularity and to choose a suitable mapping from the algorithm to the architecture.

THE BOUNDARY ELEMENT METHOD FOR POTENTIAL PROBLEMS.

The potential problem comprising Laplace's equation in some region D subject to Dirichlet, Neumann or Robin conditions on the closed boundary, C, can be written as a boundary integral equation of the form[1,10],

$$\alpha(r)\phi(r) = \oint_C (\phi\frac{\partial}{\partial n'}\ln R - q\ln R)ds', \qquad (1)$$

where ϕ and q are the boundary values of the potential and flux respectively and $R = r - r'$, where r is the position vector of the field point and r' is the position of a source point on the boundary.

In the direct boundary element method we divide the boundary into N elements, with M nodes, and approximate the boundary values ϕ and q by the interpolants

$$\tilde{\phi}(s) = \sum_{i=1}^{M} u_i(s)\phi_i \text{ and } \tilde{q}(s) = \sum_{i=1}^{M} u_i(s)q_i, \qquad (2)$$

where $\{u_i(s): i = 1,2,...,M\}$ is the usual set of nodal functions and ϕ_i and q_i are the nodal values of ϕ and q respectively.

The boundary element method then leads to a set of algebraic equations in the form

$$\mathbf{A}\mathbf{x} = \mathbf{b}. \tag{3}$$

The terms in the coefficient matrix, **A**, and the right-hand side vector, **b**, involve integrals over the elements. The vector, **x**, contains the unknown values of the nodal potentials and fluxes.

The integrals in the terms a_{ij} and b_i are of the form

$$h_{ij} = \int_{C_j} u_i(s) \frac{\mathbf{R}_{ij}.\hat{n}}{|\mathbf{R}_{ij}|^2} ds$$

or

$$g_{ij} = \int_{C_j} u_i(s) \ln|\mathbf{R}_{ij}| ds, \tag{4}$$

where $\mathbf{R}_{ij}$ is the position vector of a point in element j, relative to the base node i, see figure 1.

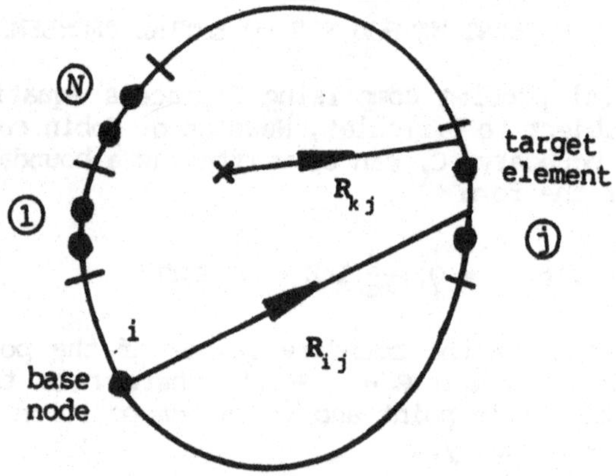

Figure 1. Definition of the position vector $\mathbf{R}_{ij}$ for the boundary integral equation and the position vector $\mathbf{R}_{kj}$ for the computation of the internal potential.

It is usual to use suitable interpolation polynomials in each element so that the integrals in equation (4) may be written in the form

$$h_{ij} = \int_{-1}^{1} L(\xi)\frac{\mathbf{R}_{ij}(\xi).\hat{n}}{|\mathbf{R}_{ij}(\xi)|^2}|J(\xi)|d\xi$$

and (5)

$$g_{ij} = \int_{-1}^{1} L(\xi)\ln|\mathbf{R}_{ij}(\xi)||J(\xi)|d\xi,$$

where $L(\xi)$ is a suitable interpolation polynomial and $J(\xi)$ is the Jacobian of the transformation $(x,y) \rightarrow (x(\xi),y(\xi))$.

If the base node is inside the target element then the integrals contain singularities and a special treatment of the integral is required. For h_{ij} a row-sum approach is used, for g_{ij} either a special logarithmic quadrature or an analytic integration is used.

If the base node is outside the target element then the integral may be evaluated using the usual Gauss quadrature procedure. If we have an n-point rule then

$$h_{ij} \simeq \sum_{g=1}^{n} L(\xi_g)\frac{\mathbf{R}_{ij}(\xi_g).\hat{n}_g}{|\mathbf{R}_{ij}(\xi_g)|^2}|J(\xi_g)|\omega_g$$

and (6)

$$g_{ij} \simeq \sum_{g=1}^{n} L(\xi_g)\ln|\mathbf{R}_{ij}(\xi_g)||J(\xi_g)|\omega_g .$$

When the set-up phase is complete then the system of equations (3) may be solved and hence when the boundary nodal potentials and fluxes have been calculated we may obtain the internal potential values from

$$\phi(\mathbf{r}) \simeq \frac{1}{2\pi}\oint_C(\tilde{\phi}\frac{\partial}{\partial n'}\ln R - \tilde{q}\ln R)ds'$$

which can be written using an n-point Gauss quadrature rule as

$$\phi_k \simeq \frac{1}{2\pi}\sum_{j=1}^{N}\sum_{g=1}^{n} L(\xi_g)\left[\frac{\mathbf{R}_{kj}(\xi_g).\hat{n}_g}{|\mathbf{R}_{kj}(\xi_g)|^2}\phi_j - \ln|\mathbf{R}_{kj}(\xi_g)|q_j\right]|J(\xi_g)|\omega_g . \qquad (7)$$

PARALLEL IMPLEMENTATION

The evaluation of the integrals in equation (6) and the recovery of the internal potentials given by equation (7) are both effected by performing a pair of nested loops, one over the N target elements and one over the M base nodes. Inside this pair of loops is nested a third loop over the n Gauss points.

The granularity of the parallelism is determined by the order in which these loops are processed.

If the inner quadrature loop is considered as a fundamental process and the two outer loops are processed in parallel then we have a fine-grained parallelism which has[5] a mapping to an SIMD architecture such as an array processor[7]. However if the pair of nested loops over the target elements and base nodes is considered a fundamental process and the order of the loops is reversed so that the outer loop is the quadrature loop then we have a coarse-grained parallelism which has a mapping to an MIMD architecture such as a transputer network. We may think of each transputer as corresponding to a quadrature point. The calculations associated with each quadrature point are processed in parallel. The mapping is shown schematically in figure 2.

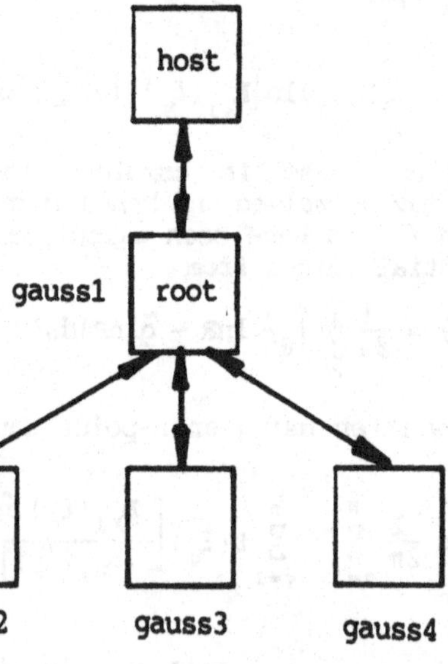

Figure 2. Mapping the Gauss points to the transputer network.

The transputer network in the Numerical Optimisation Centre at Hatfield Polytechnic consists of up to fourteen IMS T800 processors linked to an IBM PC as host machine. For the boundary element implementation they are configured in a tree structure, with one transputer, the host, used for input and output and another, the root, as the first processor in the tree. Each processor has up to four ports for two-way communication with the others in the network. Hence each processor, other than the host, is connected to one processor in the level above and up to three in the level below.

For the purposes of this implementation five transputers were used, the host together with one for each Gauss point.

The code was written in 3L^R Parallel FORTRAN which has been developed to exploit the transputer architecture. This language is very attractive because it means that existing FORTRAN subroutines can be used where appropriate with very little modification, there is no need to develop new code in a language such as OCCAM.

NUMERICAL RESULTS

The purpose of this study was to investigate the feasibility of a transputer network for boundary element calculations. Most authors to date have concentrated on the parallel implementation of the equation solution phase of the boundary element method. The performance of a transputer network for the solution of unsymmetric dense linear systems has been studied by Boreddy and Paulraj[12] who describe very efficient algorithm mappings. Consequently it was decided to concentrate on the implementation of the set-up and field recovery calculations.

A variety of two-dimensional potential problems were considered, typical of which is the potential flow of an incompressible fluid past a circular cylinder between two parallel plates[13].

Two cases are presented; (i) sixty-four linear elements with sixty-four internal points, (ii) one hundred and twenty-eight linear elements with sixty-four internal points.

We should point out here that although the network used consists of five transputers the boundary element algorithm resides on only four, the host acting as a master processor which is used only for input and output.

Values of the c.p.u. times for each of the two phases are given in table 1 where we compare the times on four transputers with the times on one transputer and also with the times for an equivalent sequential code on a VAX 8650.

phase	set-up		field recovery	
case	(i)	(ii)	(i)	(ii)
one transputer	6020	20840	4350	8690
four transputers	1530	5750	1100	2190
VAX	7200	20350	5500	8900
speed-up	3.93	3.62	3.95	3.97
efficiency	0.98	0.91	0.99	0.99

Table 1. c.p.u. times (ms) for each of the two phases together with the speed-up factor and the efficiency.

We see from table 1 that the solution time on one transputer is comparable with that on the VAX. With four transputers we see that we have speed-ups of between 3.62 and 3.97. The figures quoted in table 1 are average values over five different runs. The actual values varied by as much as five percent from these average values. This is not uncommon in transputer networks where the same size system can yield significantly different times with different data sets[15].

Also shown in table 1 is the value of the efficiency η; this is given by[15]

$$\eta = s_m/m,$$

where s_m is the speed-up associated with m processors.

The ideal algorithm has a mapping with an efficiency 1.0. We see from table 1 that the boundary element method has a very efficient implementation on the transputer network.

CONCLUDING REMARKS.

The results so far are very encouraging. We have seen that the implementation has an efficiency of the order of 0.9 with a speed-up by a factor of about 4 compared with the VAX.

We must also bear in mind that the transputer network, connected to an IBM PC residing on a desk-top, is a relatively cheap and very convenient computing system. The VAX 8650 however is a main-frame computer whose cost is a factor of the order of 100 times that of the transputer network. Consequently the transputer network offers a very cost-effective environment for boundary element analysis.

There are two other coarse-grained parallel aspects of the boundary element method which should be suited to implementation on a transputer network.

Firstly, problems such as Poisson's equation require the evaluation of non-homogeneous terms. These may be obtained either by evaluating domain integrals[10] or by the dual reciprocity method[16]. In either case the evaluation of such terms can be performed in parallel with the evaluation of the system matrices.

Secondly, the method of subregions[10] has a clear coarse-grained parallelism which should lend itself to implementation on the transputer network.

Finally the coupled boundary element/finite element method has a coarse-grained parallelism similar to that of the method of subregions. It would be expected that the transputer network would provide a suitable environment for the solution of such coupled problems.

REFERENCES

1. Davies, A.J. The boundary element method on the ICL DAP, Parallel Computing, 8, 348-353, 1988.

2. Davies, A.J. Parallelism in the boundary element method; fine grain and coarse grain. Paper submitted for inclusion at the conference ASE91 to be held in Boston, Massachusetts, USA, August 1991.

3. Lai, C.H. and Liddell, H.M. A review of parallel finite elements on the DAP. Applied Mathematical Modelling, 11, 330-340, 1987.

4. Ortega, J.M. and Voight, R.G. Solution of partial differential equations on vector and parallel computers. SIAM Review, 27, 1-96, 1985.

5. Symm, G.T. Boundary elements on a distributed array processor. Engineering Analysis, 1, 162-165, 1984.

6. Davies, A.J. Quadratic isoparametric boundary elements: an implementation on the ICL DAP. In Boundary Elements X (Ed. Brebbia, C.A.), 3, 657-666. Proceedings of the 10th International Conference on Boundary Element Methods, Southampton, U.K., C.M.P., 1988.

7. Davies, A.J. Mapping the boundary element method to the ICL DAP. In CONPAR 88 (Eds Jesshope, C.R. and Reinartz, K.D.), 230-237. Proceedings of the International Conference on Parallel and Concurrent Computing, Manchester, U.K., C.U.P., 1989.

8. Bozek, D.G., Ciarelli, D.M., Ciarelli, K.J., Hobous, M.F., Katrick, R.B. and Kline, K.A. Vector processing applied to boundary element algorithms on the CDC Cyber-205. EDF Bulletin de la Direction des Etudes et Recherches, Série C-Mathématiques, 1, 87-94, 1983.

9. Flynn, M. Some computer organizations and their effectiveness. IEE Transactions on computing, C-21, 948-960, 1972.

10. Brebbia, C.A. and Dominguez, J. Boundary Elements, an introductory course, C.M.P., 1989.

11. Kline, K.A., Tsao, N.K. and Friedlander, C.B. Parallel processing and the solution of boundary element equations. In Advanced Topics in Boundary Element Analysis (Eds Cruse, T.A., Pifko, A.B. and Armen, H.) ASME-AMD, 1985.

12. Boreddy, J. and Paulraj, A. On the performance of transputer arrays for dense linear systems. Parallel Computing, 15, 107-117, 1990.

13. Davies, A.J. The implementation of the Boundary Element Method on a Network of Transputers, Hatfield Polytechnic NOC Technical Report 241, 1991.

14. Martin, H.C. Finite element analysis of fluid flows. In proceedings of the second conference on Matrix Methods in Structural Mechanics, AFFDLTR 68-150, Wright Patterson Air Force Base, Ohio, USA, 1969.

15. Hwang, K. and Briggs, F.A. Computer Architecture and Parallel Processing. McGraw-Hill, 1987.

16. Partridge, P.W., Brebbia, C.A. and Wrobel, L.C. The Dual Reciprocity Boundary Element Method, C.M.P. 1990.

A Direct and Natural Coupling of BEM and FEM[1]

D.-H. Yu

Computing Center, Academia Sinica, Beijing, P.R. China

1. INTRODUCTION

The canonical boundary element method, proposed and developed by K. Feng and D. Yu, is based on same variational principle as the finite element method, and has many distinctive advantages (see [1-4, 10-21]). Because the canonical reduction faithfully preserves all the essential characteristics of the original elliptic problem, especially, it preserves the energy functional, the canonical boundary element method is fully compatible with the finite element method. The coupling of canonical BEM and FEM is direct and natural. Its total stiffness matrix is just the sum of matrices obtained by BEM and FEM respectively. So this coupling is much simpler and easier to apply than other couplings of BEM and FEM, which are indirect generally.

We can see the advantages of this coupling especially for the problems on unbounded domain. In order to overcome the difficulty caused by infinity of domain, we apply the canonical boundary reduction to the exterior domain of an artificial boundary, which is usually a circle, reduce the original problem into an equivalent boundary value problem on bounded domain, and then solve this new problem numerically by the direct coupling of BEM and FEM. We can get the same rate of convergence as that for the problem on bounded domain.

In this paper we apply the direct and natural coupling of BEM and FEM to solve the boundary value problems of harmonic equation, biharmonic equation, plane elasticity equation and Stokes equations on unbounded domain, and obtain corresponding error estimates. Let Γ be a smooth closed curve and Ω be its exterior domain. Draw a circle Γ' with radius R enclosing Γ. Then Ω is divided into Ω_1 and Ω_2, where Ω_1 is a bounded domain, and Ω_2 is an exterior circular domain. We use the finite element method in Ω_1, and the canonical boundary element method in Ω_2.

2. HARMONIC BOUNDARY VALUE PROBLEM

[1]The project supported by National Natural Science Foundation of China.

Consider the harmonic boundary value problem

$$\begin{cases} \Delta u = 0 & \text{in } \Omega \\ \dfrac{\partial u}{\partial n} = g & \text{on } \Gamma, \end{cases} \tag{1}$$

where $g \in H^{-\frac{1}{2}}(\Gamma)$ satisfies the compatibility condition. Let

$$D(u,v) = \iint_{\Omega} \nabla u \cdot \nabla v \, dxdy,$$

we get an equivalent variational problem

$$\begin{cases} \text{Find} \quad u \in W_0^1(\Omega) & \text{such that} \\ D(u,v) = \int_{\Gamma} gv \, ds & \forall v \in W_0^1(\Omega), \end{cases} \tag{2}$$

where

$$W_0^1(\Omega) = \{u \mid \frac{u}{\sqrt{1+r^2}\ln(2+r^2)}, \frac{\partial u}{\partial x_j} \in L^2(\Omega), j = 1,2, r = \sqrt{x_1^2 + x_2^2}\}.$$

The discretization of problem (2) yields the finite element method. But because Ω is an unbounded domain, it is difficult to get satisfactory results by directly using FEM. Then we use the coupling of canonical BEM and FEM. We have

$$D(u,v) = \iint_{\Omega} \nabla u \cdot \nabla v \, dxdy = \iint_{\Omega_1} \nabla u \cdot \nabla v \, dxdy + \iint_{\Omega_2} \nabla u \cdot \nabla v \, dxdy$$
$$\equiv D_1(u,v) + D_2(u,v).$$

Applying the canonical boundary reduction to the harmonic problem on the exterior circular domain Ω_2, we can get the canonical integral equation (see [3,11])

$$\frac{\partial u}{\partial n}|_{\Gamma'} = -\frac{1}{4\pi \sin^2 \frac{\theta}{2}} * u(R,\theta), \tag{3}$$

where n is the exterior normal direction for Ω_2, $*$ denotes the convolution. Let

$$\hat{D}_2(u_0,v_0) = \int_0^{2\pi} \int_0^{2\pi} \left(-\frac{1}{4\pi \sin^2 \frac{\theta-\theta'}{2}}\right) u_0(\theta') v_0(\theta) \, d\theta' d\theta. \tag{4}$$

Then we have the preservation of energy functional

$$D_2(u,v) = \iint_{\Omega_2} \nabla u \cdot \nabla v \, dxdy = \int_{\Gamma'} v \frac{\partial u}{\partial \nu} ds = \hat{D}_2(\gamma' u, \gamma' v) \tag{5}$$

for u with $\Delta u = 0$ in Ω_2, where $\gamma' u = u|_{\Gamma'}$, and problem (2) is equivalent to

$$\begin{cases} \text{Find} \quad u \in H^1(\Omega_1) & \text{such that} \\ D_1(u,v) + \hat{D}_2(\gamma' u, \gamma' v) = \int_{\Gamma} vg \, ds & \forall v \in H^1(\Omega_1), \end{cases} \tag{6}$$

which has unique solution in quotient space $H^1(\Omega_1)/P_0$, where P_0 is the set of all polynomials of order 0.

Now divide Γ' into N and subdivide Ω_1 into triangles such that its nodes on Γ' coincide with the dividing points of Γ'. Let $S_h(\Omega_1) \subset H^1(\Omega_1)$ be the finite element solution space, for example, it consists of piecewise linear polynomials on Ω_1. The discrete problem for (6) is

$$\begin{cases} \text{Find} \quad u_h \in S_h(\Omega_1) \qquad \text{such that} \\ D_1(u_h, v_h) + \hat{D}_2(\gamma' u_h, \gamma' v_h) = \int_\Gamma g v_h ds \qquad \forall v_h \in S_h(\Omega_1), \end{cases} \tag{7}$$

which yields a system of linear algebraic equations $QU = b$, where $Q = Q_1 + Q_2$, Q_1 can be obtained by the finite element method, and Q_2 is just corresponding to the stiffness matrix $[q_{ij}^{(2)}]_{N\times N}$ for canonical boundary elements on Γ' (see [11]),

$$q_{ij}^{(2)} = a_{|i-j|}, \qquad i,j = 1,2,\cdots,N, \tag{8}$$

$$a_k = \frac{4N^2}{\pi^3}\sum_{n=1}^{\infty}\frac{1}{n^3}\sin^4\frac{n\pi}{N}\cos\frac{nk}{N}2\pi, \qquad k = 0,1,\cdots,N-1. \tag{9}$$

Because Q_2 is symmetric and circulant, we only need calculate $a_0, a_1, \cdots$ and $a_{[\frac{N}{2}]}$.

This coupling can also be used for harmonic problem over cracked domain (see [13]).

Let $\Pi : H^1(\Omega_1) \to S_h(\Omega_1)$ be interpolation operator, u and u_h be solutions of problems (6) and (7) respectively, we have the following error estimates.

THEOREM 1. *If $u \in H^{k+1}(\Omega_1), k \geq 1, \Pi$ satisfies*

$$\|v - \Pi v\|_{H^1(\Omega_1)} \leq Ch^j\|v\|_{j+1,\Omega_1}, \qquad \forall v \in H^{j+1}(\Omega_1), j = 1,\cdots,k,$$

then

$$\|u - u_h\|_{D_1} + \|\gamma' u - \gamma' u_h\|_{\hat{D}_2} \leq Ch^k\|u\|_{k+1,\Omega_1}, \tag{10}$$

where C is a constant independent of u and h.

THEOREM 2. *If $u \in H^{k+1}(\Omega_1), k \geq 1, \Pi$ satisfies the condition of theorem 1, and $\iint_{\Omega_1}(u - u_h)dxdy = 0$, then*

$$\|u - u_h\|_{L^2(\Omega_1)} \leq Ch^{k+1}\|u\|_{k+1,\Omega_1}. \tag{11}$$

where C is a constant independent of u and h.

3. BIHARMONIC BOUNDARY VALUE PROBLEM

Consider the biharmonic boundary value problem, i.e. the plate bending problem

$$\begin{cases} \Delta^2 u = 0 \qquad \text{in}\,\Omega \\ (Tu, Mu) = (t, m) \qquad \text{on}\,\Gamma, \end{cases} \tag{12}$$

where $(t,m) \in H^{-\frac{3}{2}}(\Gamma) \times H^{-\frac{1}{2}}(\Gamma)$ satisfies the compatibility condition,

$$Tu = \{-\frac{\partial \Delta u}{\partial n} + (1-\nu)\frac{\partial}{\partial s}[(\frac{\partial^2 u}{\partial x^2} - \frac{\partial^2 u}{\partial y^2})n_x n_y + \frac{\partial^2 u}{\partial x \partial y}(n_y^2 - n_x^2)]\}_\Gamma,$$

$$Mu = [\nu \Delta u + (1-\nu)(\frac{\partial^2 u}{\partial x^2}n_x^2 + \frac{\partial^2 u}{\partial y^2}n_y^2 + 2\frac{\partial^2 u}{\partial x \partial y}n_x n_y)]_\Gamma.$$

Let

$$D(u,v) = \iint_\Omega \{\Delta u \Delta v - (1-\nu)[\frac{\partial^2 u}{\partial x^2}\frac{\partial^2 v}{\partial y^2} + \frac{\partial^2 v}{\partial x^2}\frac{\partial^2 v}{\partial y^2} - 2\frac{\partial^2 u}{\partial x \partial y}\frac{\partial^2 v}{\partial x \partial y}]\}dxdy,$$

$$F(v) = \int_\Gamma (tv + m\frac{\partial v}{\partial n})ds,$$

we get an equivalent variational problem

$$\begin{cases} \text{Find} \quad u \in W_0^2(\Omega) \quad \text{such that} \\ D(u,v) = F(v) \quad \forall v \in W_0^2(\Omega), \end{cases} \tag{13}$$

where

$$W_0^2(\Omega) = \{u \mid \frac{u}{(1+r^2)\ln(2+r^2)}, \frac{1}{\sqrt{1+r^2}\ln(2+r^2)}\frac{\partial u}{\partial x_i}, \frac{\partial^2 u}{\partial x_i \partial x_j} \in L^2(\Omega),$$
$$i,j = 1,2, \quad r = \sqrt{x_1^2 + x_2^2}\}.$$

Now we use the coupling of canonical BEM and FEM. Applying the canonical boundary reduction to the biharmonic problem on the exterior circular domain Ω_2, let K be the canonical integral operator on Γ' (see [10,11]), we have the preservation of energy functional

$$D_2(u,v) = \hat{D}_2(\gamma_0 u, \gamma_1 u; \gamma_0 v, \gamma_1 v) \tag{14}$$

for u with $\Delta^2 u = 0$ in Ω_2, where $\gamma_0 u = u|_{\Gamma'}, \gamma_1 u = \frac{\partial u}{\partial n}|_{\Gamma'}$,

$$\hat{D}_2(u_0, u_n; v_0, v_n) = \int_{\Gamma'} (v_0, v_n) \cdot K(u_0, u_n)ds. \tag{15}$$

Then problem (13) is equivalent to

$$\begin{cases} \text{Find} \quad u \in H^2(\Omega_1) \quad \text{such that} \\ D_1(u,v) + \hat{D}_2(\gamma_0 u, \gamma_1 u; \gamma_0 v, \gamma_1 v) = F(v) \quad \forall v \in H^2(\Omega_1), \end{cases} \tag{16}$$

which has unique solution in quotient space $H^2(\Omega_1)/P_1$, where P_1 is the set of all polynomials of order 1 on Ω_1.

Let $S_h(\Omega_1) \subset H^2(\Omega_1)$ be the finite element solution space on Ω_1. The discrete problem for (16) is

$$\begin{cases} \text{Find} \quad u_h \in S_h(\Omega_1) \qquad \text{such that} \\ D_1(u_h, v_h) + \hat{D}_2(\gamma_0' u_h, \gamma_1' u_h; \gamma_0' v_h, \gamma_1' v_h) = F(v_h) \qquad \forall v_h \in S_h(\Omega_1), \end{cases} \tag{17}$$

which yields a system of linear algebraic equations $QU = b$, where $Q = Q_1 + Q_2$, Q_1 can be obtained by the finite element method, and Q_2 is just corresponding to the stiffness matrix for canonical boundary elements on Γ' (see [11]). Let $\Pi : H^2(\Omega_1) \rightarrow S_h(\Omega_1)$ be interpolation operator, u and u_h be solutions of problems (16) and (17) respectively, we have the following error estimates.

THEOREM 3. *If* $u \in H^{k+1}(\Omega_1), k \geq 2, \Pi$ *satisfies*

$$\|v - \Pi v\|_{H^2(\Omega_1)} \leq Ch^j \|v\|_{j+2,\Omega_1}, \qquad \forall v \in H^{j+2}(\Omega_1), j = 1, \cdots, k-1,$$

then

$$\|u - u_h\|_{D_1} + \|(\gamma_0'(u - u_h), \gamma_1'(u - u_h))\|_{\hat{D}_2} \leq Ch^{k-1} \|u\|_{k+1,\Omega_1}, \tag{18}$$

where C *is a constant independent of* u *and* h.

THEOREM 4. *If* $u \in H^{k+1}(\Omega_1), k \geq 2, \Pi$ *satisfies the condition of theorem 3, and* $\iint_{\Omega_1} p(u - u_h)dxdy = 0 \ \forall p \in P_1$*, then*

$$\|u - u_h\|_{L^2(\Omega_1)} \leq Ch^{k+1} \|u\|_{k+1,\Omega_1}. \tag{19}$$

where C *is a constant independent of* u *and* h.

4. PLANE ELASTICITY PROBLEM

Consider the plane elasticity problem

$$\begin{cases} \mu \Delta \vec{u} + (\lambda + \mu) \operatorname{grad} \operatorname{div} \vec{u} = 0 \qquad \text{in} \, \Omega \\ \displaystyle\sum_{j=1}^{2} \sigma_{ij} n_j = g_i, \quad i = 1, 2, \qquad \text{on} \, \Gamma, \end{cases} \tag{20}$$

where $\vec{g} = (g_1, g_2) \in H^{-\frac{1}{2}}(\Gamma)^2$ satisfies the compatibility condition. Let

$$D(\vec{u}, \vec{v}) = \iint_{\Omega} \sum_{i,j=1}^{2} \sigma_{ij}(\vec{u}) \epsilon_{ij}(\vec{v}) dxdy,$$

$$F(\vec{v}) = \int_{\Gamma} \vec{g} \cdot \vec{v} \, ds,$$

we get an equivalent variational problem

$$\begin{cases} \text{Find} \quad \vec{u} \in W_0^1(\Omega)^2 \qquad \text{such that} \\ D(\vec{u}, \vec{v}) = F(\vec{v}) \qquad \forall \vec{v} \in W_0^1(\Omega)^2. \end{cases} \tag{21}$$

Now we use the coupling of canonical BEM and FEM. Applying the canonical boundary reduction to the plane elasticity problem on the exterior circular domain Ω_2, let K be the canonical integral operator on Γ' (see [14,19]), we have the preservation of energy functional

$$D_2(\vec{u},\vec{v}) = \hat{D}_2(\gamma'\vec{u},\gamma'\vec{v}) \tag{22}$$

for $\vec{u}$ satisfying the plane elasticity equation in Ω_2, where

$$\hat{D}_2(\vec{u}_0,\vec{v}_0) = \int_{\Gamma'} \vec{v}_0 \cdot K\vec{u}_0 \, ds. \tag{23}$$

Then problem (21) is equivalent to

$$\begin{cases} \text{Find} \quad \vec{u} \in H^1(\Omega_1)^2 \qquad \text{such that} \\ D_1(\vec{u},\vec{v}) + \hat{D}_2(\gamma'\vec{u},\gamma'\vec{v}) = F(\vec{v}) \qquad \forall \vec{v} \in H^1(\Omega_1)^2, \end{cases} \tag{24}$$

which has unique solution in quotient space $[H^1(\Omega_1)/P_0]^2$.

Let $S_h(\Omega_1)^2 \subset H^1(\Omega_1)^2$ be the finite element solution space on Ω_1. The discrete problem for (24) is

$$\begin{cases} \text{Find} \quad \vec{u}_h \in S_h(\Omega_1)^2 \qquad \text{such that} \\ D_1(\vec{u}_h,\vec{v}_h) + \hat{D}_2(\gamma'\vec{u}_h,\gamma'\vec{v}_h) = F(\vec{v}_h) \qquad \forall \vec{v}_h \in S_h(\Omega_1)^2, \end{cases} \tag{25}$$

which yields a system of linear algebraic equations $QU = b$, where $Q = Q_1 + Q_2$, Q_1 can be obtained by the finite element method, and Q_2 is just corresponding to the stiffness matrix for canonical boundary elements on Γ' (see [14,19]). Let $\Pi : H^1(\Omega_1)^2 \rightarrow S_h(\Omega_1)^2$ be interpolation operator, $\vec{u}$ and $\vec{u}_h$ be solutions of problems (24) and (25) respectively, we have the following error estimates.

THEOREM 5. *If $\vec{u} \in H^{k+1}(\Omega_1)^2, k \geq 1, \Pi$ satisfies*

$$\|\vec{v} - \Pi\vec{v}\|_{H^1(\Omega_1)^2} \leq Ch^j \|\vec{v}\|_{j+1,\Omega_1}, \qquad \forall \vec{v} \in H^{j+1}(\Omega_1)^2, j = 1,\cdots,k,$$

then

$$\|\vec{u} - \vec{u}_h\|_{D_1} + \|\gamma'(\vec{u} - \vec{u}_h)\|_{\hat{D}_2} \leq Ch^k \|u\|_{k+1,\Omega_1}, \tag{26}$$

where C is a constant independent of $\vec{u}$ and h.

THEOREM 6. *If $\vec{u} \in H^{k+1}(\Omega_1)^2, k \geq 1, \Pi$ satisfies the condition of theorem 5, and $\iint_{\Omega_1} (\vec{u} - \vec{u}_h) dx dy = 0$, then*

$$\|\vec{u} - \vec{u}_h\|_{L^2(\Omega_1)^2} \leq Ch^{k+1} \|\vec{u}\|_{k+1,\Omega_1}. \tag{27}$$

where C is a constant independent of $\vec{u}$ and h.

5. STOKES PROBLEM

Consider the Stokes problem

$$\begin{cases} -\nu\Delta\vec{u} + \operatorname{grad} p = 0 & \text{in}\,\Omega \\ \operatorname{div}\vec{u} = 0 \quad \text{in}\,\Omega \\ \sum_{j=1}^{2}\sigma_{ij}(\vec{u},p)n_j = g_i, \quad i = 1,2 & \text{on}\,\Gamma, \end{cases} \tag{28}$$

where $\vec{g} = (g_1, g_2) \in H^{-\frac{1}{2}}(\Gamma)^2$ satisfies the compatibility condition. Let

$$D(\vec{u},\vec{v}) = 2\nu \iint_{\Omega} \sum_{i,j=1}^{2} \epsilon_{ij}(\vec{u})\epsilon_{ij}(\vec{v})dxdy,$$

$$F(\vec{v}) = \int_{\Gamma} \vec{g}\cdot\vec{v}\,ds,$$

we get an equivalent variational problem

$$\begin{cases} \text{Find} \quad (\vec{u},p) \in W_0^1(\Omega)^2 \times L^2(\Omega) \quad \text{such that} \\ D(\vec{u},\vec{v}) - \iint_{\Omega} p \operatorname{div}\vec{v}\,dxdy = F(\vec{v}) \quad \forall\vec{v} \in W_0^1(\Omega)^2 \\ \iint_{\Omega} q \operatorname{div}\vec{u}\,dxdy = 0 \quad \forall q \in L^2(\Omega). \end{cases} \tag{29}$$

Now we use the coupling of canonical BEM and FEM. Applying the canonical boundary reduction to the Stokes problem on the exterior circular domain Ω_2, let K be the canonical integral operator on Γ' (see [18]), we have the preservation of energy functional

$$D_2(\vec{u},\vec{v}) - \iint_{\Omega_2} p \operatorname{div}\vec{v}\,dxdy = \hat{D}_2(\gamma'\vec{u},\gamma'\vec{v}) \tag{30}$$

for $(\vec{u},p)$ satisfying the Stokes equations in Ω_2, where

$$\hat{D}_2(\vec{u}_0,\vec{v}_0) = \int_{\Gamma'} \vec{v}_0 \cdot K\vec{u}_0\,ds = \int_0^{2\pi}\int_0^{2\pi}\left(-\frac{\nu}{2\pi\sin^2\frac{\theta-\theta'}{2}}\right)\vec{u}_0(\theta')\cdot\vec{v}_0(\theta)d\theta' d\theta. \tag{31}$$

$\hat{D}_2(\cdot,\cdot)$ is a V-elliptic symmetric continuous bilinear form. Then problem (29) is equivalent to

$$\begin{cases} \text{Find} \quad (\vec{u},p) \in H^1(\Omega_1)^2 \times L^2(\Omega_1) \quad \text{such that} \\ D_1(\vec{u},\vec{v}) + \hat{D}_2(\gamma'\vec{u},\gamma'\vec{v}) - \iint_{\Omega_1} p \operatorname{div}\vec{v}\,dxdy = F(\vec{v}) \quad \forall\vec{v} \in H^1(\Omega_1)^2 \\ \iint_{\Omega_1} q \operatorname{div}\vec{u}\,dxdy = 0 \quad \forall q \in L^2(\Omega_1), \end{cases} \tag{32}$$

which has unique solution in quotient space $[H^1(\Omega_1)/P_0]^2 \times L^2(\Omega_1)$.

Let $S_h(\Omega_1) \subset H^1(\Omega_1), L_h(\Omega_1) \subset L^2(\Omega_1)$, $S_h(\Omega_1)^2 \times L_h(\Omega_1)$ be the finite element solution space on Ω_1. The discrete problem for (32) is

$$\begin{cases} \text{Find} \quad (\vec{u}_h, p_h) \in S_h(\Omega_1)^2 \times L_h(\Omega_1) \qquad \text{such that} \\ D_1(\vec{u}_h, \vec{v}_h) + \hat{D}_2(\gamma'\vec{u}_h, \gamma'\vec{v}_h) - \iint_{\Omega_1} p_h \operatorname{div} \vec{v}_h \, dx dy = F(\vec{v}_h) \qquad \forall \vec{v}_h \in S_h(\Omega_1)^2 \\ \iint_{\Omega_1} q_h \operatorname{div} \vec{u}_h \, dx dy = 0 \qquad \forall q_h \in L_h(\Omega_1), \end{cases} \tag{33}$$

which yields a system of linear algebraic equations $QU = b$, where $Q = Q_1 + Q_2$, Q_1 can be obtained by the finite element method, and Q_2 is just corresponding to the stiffness matrix for canonical boundary elements on Γ' . Let Π_1 : $H^1(\Omega_1)^2 \to S_h(\Omega_1)^2$ and $\Pi_2 : L^2(\Omega_1) \to L_h(\Omega_1)$ be interpolation operators, $(\vec{u}, p)$ and $(\vec{u}_h, p_h)$ be solutions of problems (32) and (33) respectively, we have the following error estimates.

THEOREM 7. *If* $\vec{u} \in H^{k+1}(\Omega_1)^2, p \in H^k(\Omega_1), k \geq 1$, Π_1 *and* Π_2 *satisfy*

$$\|\vec{v} - \Pi_1 \vec{v}\|_{H^1(\Omega_1)^2} \leq Ch^j \|\vec{v}\|_{j+1,\Omega_1}, \qquad \forall \vec{v} \in H^{j+1}(\Omega_1)^2, j = 1, \cdots, k,$$

and

$$\|q - \Pi_2 q\|_{L^2(\Omega_1)} \leq Ch^j \|q\|_{j,\Omega_1}, \qquad \forall q \in H^j(\Omega_1), j = 1, \cdots, k,$$

then

$$\|\vec{u} - \vec{u}_h\|_{D_1} + \|\gamma'(\vec{u} - \vec{u}_h)\|_{\hat{D}_2} + \|p - p_h\|_{L^2(\Omega_1)} \leq Ch^k (\|u\|_{k+1,\Omega_1} + \|p\|_{k,\Omega_1}), \tag{34}$$

where C *is a constant independent of* $\vec{u}$ *and* h.

THEOREM 8. *If* $\vec{u} \in H^{k+1}(\Omega_1)^2, p \in H^k(\Omega_1), k \geq 1$, Π_1 *and* Π_2 *satisfy the condition of theorem 7, and* $\iint_{\Omega_1} (\vec{u} - \vec{u}_h) dx dy = 0$, *then*

$$\|\vec{u} - \vec{u}_h\|_{L^2(\Omega_1)^2} \leq Ch^{k+1} (\|\vec{u}\|_{k+1,\Omega_1} + \|p\|_{k,\Omega_1}). \tag{35}$$

where C *is a constant independent of* $\vec{u}$ *and* h.

6.CONCLUSION

There are many kinds of boundary element methods which can be coupled with FEM. But only by the canonical boundary reduction can BEM and FEM be coupled directly and naturally. In fact, the subdomain Ω_2 in which the canonical boundary reduction is applied is exactly a 'large element'. We only need to add the stiffness matrix for canonical boundary elements on the artificial boundary Γ' to the matrix for FEM on the subdomain Ω_1. This coupling can be brought into the calculating system of FEM, and almost will not increase the complexity of programing. Moreover, because the stiffness matrix for canonical boundary elements on the circle Γ' is symmetric, circulant and easy to calculate, it is almost given. The computation time for calculating Q_2 is very little. Conversely, other BEM cannot be coupled with FEM directly and naturally, in other words, simple addition of matrices obtained by other BEM and FEM generally result in mistake. In addition, the computation time for other coupling is much more than for the coupling of canonical BEM and FEM. So this direct and natural coupling of BEM and FEM has more advantages than others.

REFERENCES

[1] Feng, Kang: Differential versus integral eguations and finite versus infinite elements, *Mathematica Numerica Sinica*, 2:1, 100-105, 1980.

[2] Feng, Kang: Canonical boundary reduction and finite element method, Proceedings of International Invitational Symposium on the Finite Element Method (1981, Hefei), Science Press, Beijing, 1982.

[3] Feng, Kang; Yu, De-hao: Canonical integral equations of elliptic boundary value problems and their numerical solutions, Proceedings of the China-France Symposium on Finite Element Methods, Feng Kang and J.L. Lions eds., Science press, Beijing, 211-252, 1983.

[4] Feng, Kang: Finite element method and natural boundary reduction, Proceedings of the International Congress of Mathematicians, Warszawa, 1439-1453, 1983.

[5] Girault, V.; Raviart, P.-A.: Finite Element Methods for Navier-Stokes Equations, Springer-Verlag, Berlin, 1986.

[6] Han, Hou-de; Wu, Xiao-nan: Approximation of infinite boundary condition and its application to finite element methods, *Journal of Computational Mathematics*, 3:2, 179-192, 1985.

[7] Han, Hou-de; Wu, Xiao-nan: The mixed finite element method for Stokes equations on unbounded domains, *J. Sys. Sci. & Math. Scis.*, 5:2, 121-132, 1985.

[8] Johnson, C.; Nedelec, J.C.: On the coupling of boundary integral and finite element methods, *Math. of Comp.*, 35:152, 1063-1079, 1980.

[9] Sequeira, A.: The coupling of boundary integral and finite element methods for the bidimensional exterior steady Stokes problem, *Math. Meth. in the Appl. Sci.*, 5:3, 356-375, 1983.

[10] Yu, De-hao: Canonical integral equations of biharmonic elliptic boundary value Problems, *Mathematica Numerica Sinica*, 4:3, 330-336, 1982.

[11] Yu, De-hao: Numerical solutions of harmonic and biharmonic canonical integral equations in interior or exterior circular domains, *Journal of Computational Mathematics*, 1:1, 52-62, 1983.

[12] Yu, De-hao: Numerical solution of harmonic canonical integral equation over sector with crack and concave angle (in Chinese), *Journal on Numerical Methods and Computer Applications*, 4:3, 183-188, 1983.

[13] Yu, De-hao: Coupling canonical boundary element method with FEM to solve harmonic problem over cracked domain, *Journal of Computational Mathematics*, 1:3, 195-202, 1983.

[14] Yu, De-hao: Canonical boundary element method for plane elasticity problems, *Journal of Computational Mathematics*, 2:2, 180-189, 1984.

[15] Yu, De-hao: Canonical boundary reduction and canonical boundary element method (in Chinese), Doctor Thesis, Computing Center, Academia Sinica, 1984.

[16] Yu, De-hao: Error estimates for the canonical boundary element method, Proceedings of the 1984 Beijing Symposium on Differential Geometry and Differential Equations, Feng Kang ed., Science Press, Beijing, 343-348, 1985.

[17] Yu, De-hao: Approximation of boundary conditions at infinity for a harmonic equation, *Journal of Computational Mathematics*, 3:3, 219-227, 1985.

[18] Yu, De-hao: Canonical integral equations of Stokes problem, *Journal of Computational Mathematics*, 4:1,62-73, 1986.

[19] Yu, De-hao: A system of plane elasticity canonical integral equations and its application, *Journal of Computational Mathematics*, 4:3, 200-211, 1986.

[20] Yu, De-hao: Some new developments in mathematical theory of boundary element methods (in Chinese), The Theory of the Combination Method of Analytic and Numerical Solutions and Applications in Engineering, Li Jia-bao ed., Hunan University Press, Changsha, 228-232, 1990.

[21] Yu, De-hao: The coupling of canonical BEM and FEM for Stokes problem on unbounded domain (in Chinese), to appear in *Mathematica Numerica Sinica*.

[22] Zienkiewicz, O.C.; Kelly, D.W.; Bettess, P.: The coupling of the finite element method and boundary solution procedures, *Int. J. for Num. Math. in Eng.*, 11, 355-375, 1977.

Application of Sub-Element Technique for Improving the Interior Displacement and Stress Calculations by Using the Boundary Element Method

F.-C.T. Shiue

Department of Industrial Technology, Eastern Michigan University, Ypsilanti, Michigan, U.S.A.

ABSTRACT

An accurate calculation of the displacements and stresses near the surface of the body is difficult by the regular boundary element method. The reason is that the fundamental solutions in the integral formula contain singular factors of $1/r$ and $1/r^2$. As the distance r between the internal point and boundary element becomes small, the singular factor will negatively influence the numerical integration result. This effect is sometimes referred as the *boundary-layer* type of behavior and is always present in the numerical evaluation of the interior displacements and stresses near the boundary. A special treatment of this effect using an element subdivision or *sub-element* technique has been investigated. This technique requires a subdivision of the original boundary segment into a number of sub-element and then applies a small number of Gaussian integration points on each sub-element. Good accuracy has been obtained and several examples are illustrated.

INTRODUCTION

In order to qualify the boundary element method as a completely general problem-solving technique, it is essential to demonstrate that it is also applicable to calculate the displacements and stresses near the surface of the body. In another word, a boundary element program should be general enough to accurately calculate the interior displacements and stresses no matter how close the nodes are to the boundary. This is important especially in the calculation of stress concentration problems. However, it has been noted an accurate calculation is difficult by the regular boundary element method. The reason is that the fundamental solutions in the integral formula contain singular factors of $1/r$ and $1/r^2$ in two- and three-dimensional cases. As the distance r between the internal point and boundary element becomes small, the singular factor has a detrimental effect on the numerical results due to quadrature error [1,2]. This effect is sometimes referred as the *boundary-layer* type of behavior and is always present in the numerical evaluation of the interior displacements and stresses near the boundary.

One way to eliminate this effect is to increase the quadrature order on the nearby boundary segments. Gaussian quadrature orders of up to *128* points were tried and convergence experiments were conducted. The results revealed that the accuracy of the integration can be improved slightly, depending on the distance *r*. In some cases, even the highest quadrature order, 128 points, available in the program cannot yield correct results with a very small *r*. Consequently, the method of higher quadrature order is not desirable. Another technique to prevent this effect is to increase the number of boundary discretization in the vicinity of the interior point. The method is helpful if *r* is known at the discretization stage; however, it will become necessary and unavoidable to redefine the discretization if some other closer interior unknowns have to be evaluated later on. As a result, the analysis will be repetitive and cumbersome.

In the consideration of computational efficiency, it is appropriate to keep the number of Gaussian points reasonably small and yet obtain an adequate precision for the integration without re-discretization. An alternate treatment of this effect using an element subdivision or *sub-element* concept was proposed by Liu et al. [3]. A similar technique has been investigated and experimented here. This technique requires a subdivision of the original nearby boundary segment into a number of sub-element and then applies a small number of Gaussian integration points on each sub-element. Good accuracy has been obtained and several examples will be illustrated. The implementation of this technique to a regular boundary element program is easy and a flow chart of this implementation is provided.

SUB-ELEMENT TECHNIQUE

By neglecting the body force term, the boundary integral equation for the interior displacement can be expressed as

$$u_k(\xi) \;=\; \int_\Gamma U_{ik}\, t_i\, d\Gamma \;-\; \int_\Gamma T_{ik}\, u_i\, d\Gamma \tag{1}$$

where u_i is the displacement vector, t_i is the traction vector on the boundary. The fundamental solutions $U_{ik}(\underline{X},\xi)$ and $T_{ik}(\underline{X},\xi)$ represent the displacement and traction at $\underline{X}$ in the *i*-direction due to a unit load at ξ in the *k*-direction. The expressions of these fundamental solutions can be found in Brebbia et al. [4]. As indicated previously, in the calculation of interior displacement for two-dimensional problems, the accurate integration of the singularity of *1/r* due to T_{ik} plays an important role. The following discussion will focus on how to improve this integration by reducing the quadrature error. An outline of the technique including the calculation of the shortest distance *r* and a proper subdivision of the original boundary segment will be introduced. It is noted that this discussion is restricted to boundary segments with small-curvature which is the case where most *boundary-layer* effect presents.

<u>Calculation of the shortest distance between a interior point and a small-curvature boundary segment</u>

Referring to a geometric representation in Figure 1, the calculation of the distance *r* depends upon the location of point *P* on the boundary which is

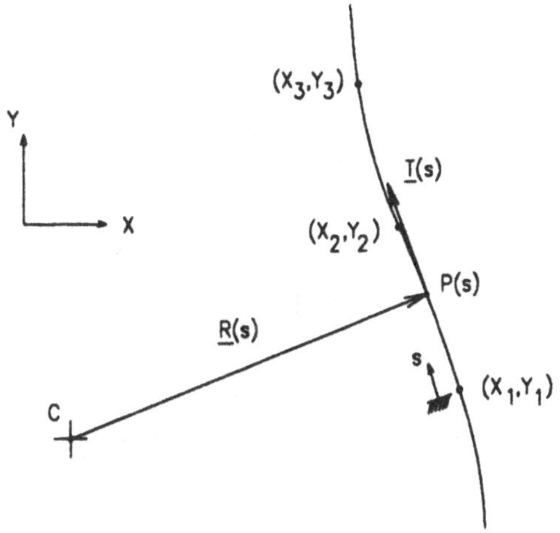

Figure 1: An interior collocation point *C* and a boundary segment with small curvature

closest to the interior point *C*. This can be done through the geometric relationship of these two points. As shown in the figure, the local coordinate of the point *P* is defined as s ($-1 \leq s \leq 1$); hence, *P* is a function of s. The tangent vector at point *P* along the segment is shown as $\underline{T}(s)$ and the direction vector between points *C* and *P* is given as $\underline{R}(s)$. For segments with small curvature and for interior points near the segment, the following orthogonality relation should be satisfied,

$$\underline{R} \cdot \underline{T} = 0 \tag{2}$$

Thus, a nonlinear equation of the local coordinate s is defined. The solution of s can be obtained by employing an iterative *SECANT* method [5] with an imposed constraint $-1 \leq s \leq 1$. Then, the coordinates of point *P* can be interpolated by s and the direction vector $\underline{R}$ can be determined.

<u>Sub-element integration technique</u>
By employing the point *P* and the shortest distance r ($r=|\underline{R}|$) obtained, the appropriate amount of subdivision of the integration element should be determined. Numerical experiments were performed to determine the number of the Gaussian quadrature points (NQP) to be used for each segment or sub-element with NQP limited to 10. According to this limitation and the ratio between the calculated distance and the length of the boundary segment (r/L), zones of severity for the interior displacement calculation were established and categorized into the Layers 1 and 2 of Figure 2.

As shown in the figure, Layer 1 is less severe case with $0.45 < r/L < 1$. Gaussian integration with *NQP=10* is sufficient to obtain an accurate result in this Layer. Layer 2 is a more severe case with $r/L < 0.45$. Here, the original boundary segment is divided into a number of sub-elements. A general

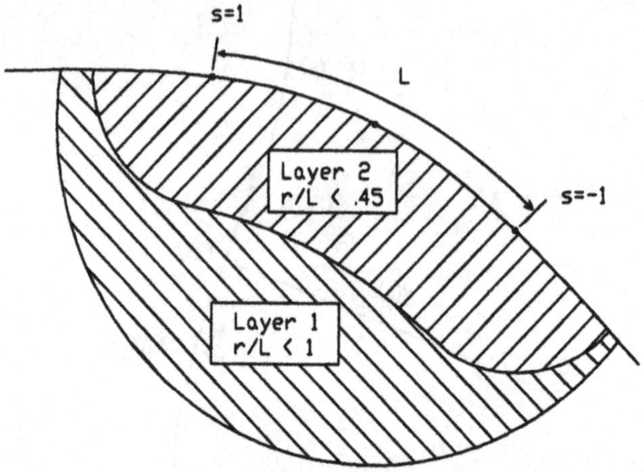

Figure 2: Layers of singular severity for various ranges of *r/L*

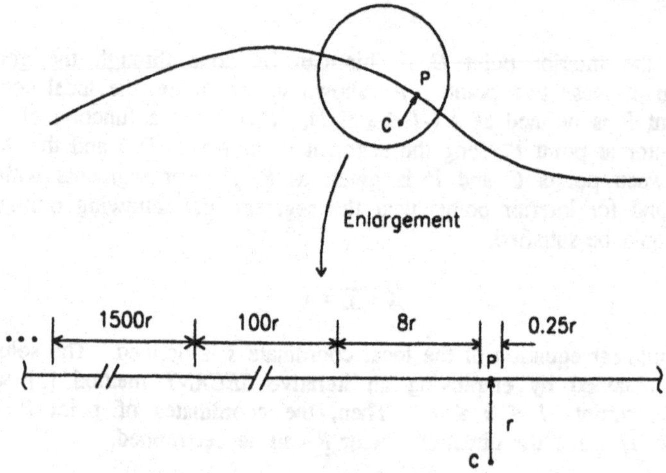

Figure 3: The criterion for subdivision of a boundary segment

criterion used for this subdivision is shown in Figure 3, where the length of the sub-element is a function of the distance *r*. It should be pointed out that this subdivision criterion is based on extensive numerical experiments with a maximum *NQP=10* for each sub-element. Likewise, the layers of singular severity and the criterion for subdivision for interior stress calculation can be experimented and determined in a similar way.

The procedure to examine the level of severity and, if required, subdivide the boundary segment as described by the above algorithm was included with the calculations of the interior displacement and stress. A flow chart of computer implementation is shown in Figure 4.

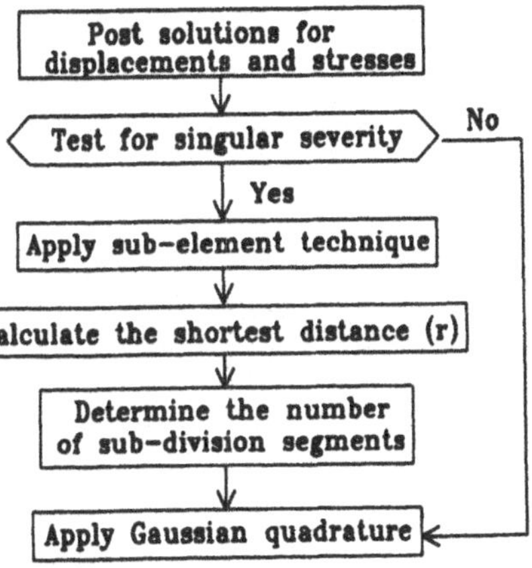

Figure 4: Flow chart of computer implementation

NUMERICAL EXAMPLES

To demonstrate the effect of the accuracy of the subdivision technique, the following problems were modelled and solved by the above algorithm.

1. Uniaxial tension of a plate,
2. Expansion of a thick-wall cylinder due to an internal pressure,
3. Stress concentration of a large plate with an elliptic hole.

Results are compared with analytic and regular Gauss integration with *NQP=10* and *NQP=128*. All the computations were performed on a IBM-9370 in single precision.

<u>Uniaxial tension of a plate (Plane-strain)</u>

The displacement field is determined at interior points of a *4in×4in* plate in uniaxial tension. The discretization consists 16 boundary nodes and 8 boundary segments with the material properties as shown in Figure 5. The X-coordinates of the interior points where the displacements are calculated range from 3.5 to 3.99 while the Y-coordinate is fixed at a constant of 3.0. Numerical results are shown in Figure 6 and, as may be seen, indicate that an excellent displacement value is determined by the subdivision scheme, whereas the conventional method, with

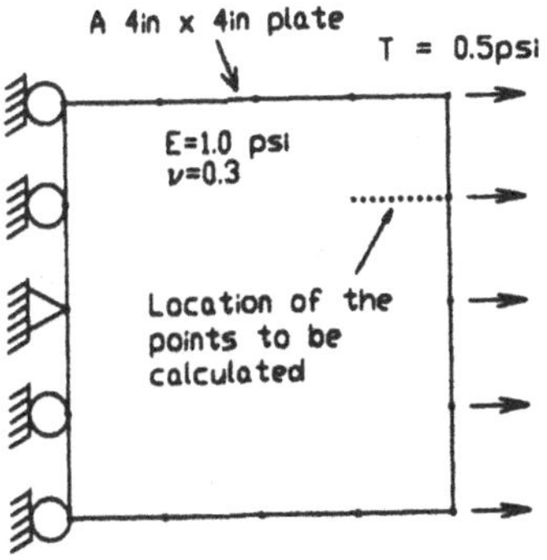

Figure 5: Uniaxial tension of a square plate

quadrature orders *NQP=10* and *NQP=128*, fails as the interior point is close to the boundary.

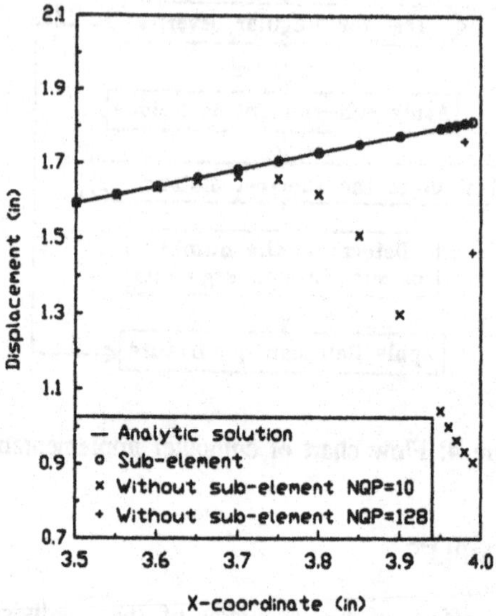

Figure 6: Interior displacement with and without element subdivision for Example 1

<u>Expansion of a thick-wall cylinder due to an internal pressure</u>
A thick-wall cylinder (*a=2.0in* and *b=1.0in*) subjects to an internal pressure of *P=0.5psi* was chosen as another test. The material properties are *E=5.0psi* and *ν=0.3*. One quarter of the cylinder with 13 boundary segments was modelled as shown in Figure 7. The X-coordinates of the interior points where the displacements and stresses are calculated range from 1.01 to 1.99 while the Y-coordinate is fixed at a constant of 0.001.

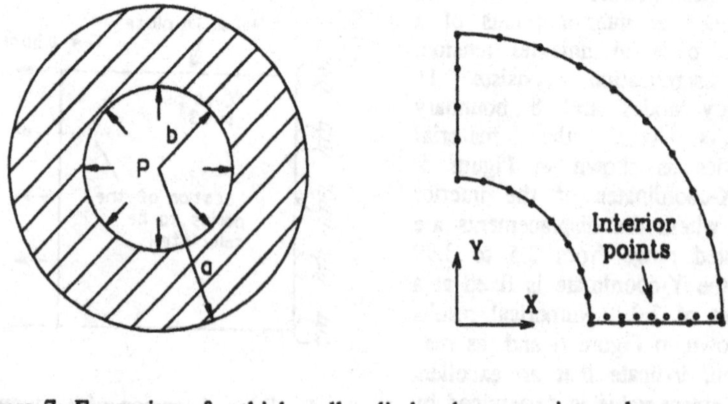

Figure 7: Expansion of a thick-wall cylinder due to an internal pressure

The analytic solution of this problem can be found in Ugral and Fenster [6]. As indicated in Figure 8, the BEM solutions for the radial displacements are compared with the analytic solutions. It is shown that the BEM solutions with the proposed technique agree well with the analytic solution, whereas the regular BEM solutions show some significant differences. The other comparison of the radial stresses is tabulated in Table 1. It can be seen that the proposed technique gives a satisfactory prediction, within 10% of accuracy; however, the regular BEM solutions are erroneous even with a high quadrature of *NQP=128*.

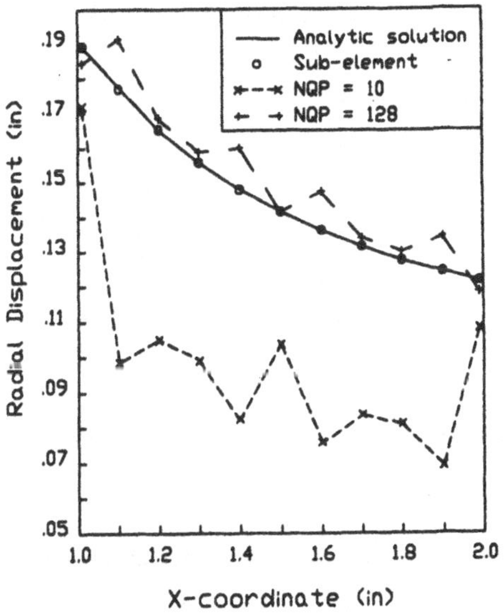

Figure 8: Radial displacements of Example 2

Table 1: Radial stresses of Example 2

X-coord	Analytic	Sub-element	NQP=10	NQP=128
1.010	-0.487	-0.448	-156.	-79.4
1.100	-0.384	-0.390	-3.08	-136.
1.200	-0.296	-0.296	14.72	-186.
1.300	-0.228	-0.238	-14.2	175.2
1.400	-0.173	-0.172	2.273	112.7
1.500	-0.130	-0.113	-0.10	-0.14
1.600	-0.094	-0.098	-2.25	-105.
1.700	-0.064	-0.066	11.83	-147.
1.800	-0.039	-0.043	-11.5	143.6
1.900	-0.018	-0.016	1.987	94.80
1.990	-0.002	-0.006	92.24	50.70

Stress concentration of a large plate with an elliptic hole

This last example is that of a stress concentration problem where a large plate containing an elliptic hole loaded in uniaxial tension. Due to symmetry, only one quarter of the plate is modelled as shown in Figure 9. Material properties are taken as $E=30\times10^3$ and $\nu=0.495$ while plain-stress is assumed. Solutions for the stress concentration factor at point A, with a small distance of $dx=dy=0.001$ away from the corner, were chosen as criteria of comparison.

The analytic solution can be referred to Timoshenko and Goodier [7], and the results for the analytic and numerical solutions for a series of ratios b/a are tabulated in Table 2. Notice the solutions of $NQP=10$ and $NQP=128$ become unreliable, whereas the proposed technique still provide a good result.

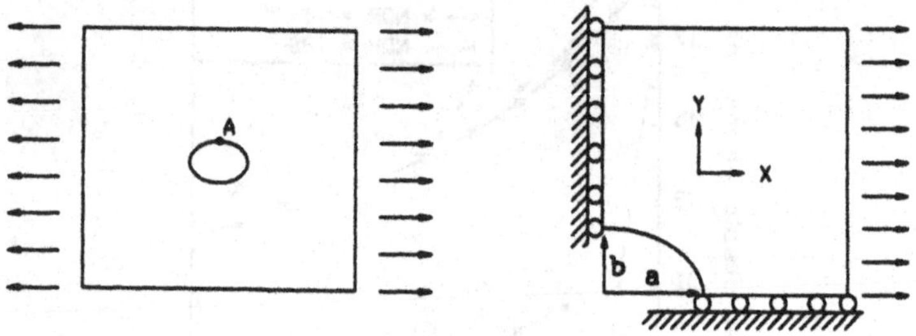

Figure 9: Stress concentration of a large plate with an elliptic hole

Table 2: Stress (σ_{xx}) at point A

b/a	Analytic	Sub-element	NQP=10	NQP=128
1.0	3.00	2.93	-59.03	115.20
0.9	2.80	2.74	-16.34	-3.31
0.8	2.60	2.57	2.03	-5.55
0.7	2.40	2.38	27.51	-7.04
0.6	2.20	2.18	-38.72	74.05
0.5	2.00	2.08	89.06	-8.27

DISCUSSION

It has also been shown that it is difficult to obtain correct numerical calculation of interior displacements and stresses near the boundary of the domain by the current boundary formulation, due to the order of singularities in the integral representation. As the distance of r between the interior point and the boundary becomes smaller, the nearly singular integral is difficult to accurately integrate by using quadrature rule. Numerical experiments indicated that this must be carefully resolved to effectively implement the BEM technique. A *sub-element* scheme has been devised and shown to produce good numerical results for several illustrative examples. The implementations of this technique for an existing two-dimensional BEM program is easy and worthy in the consideration of numerical accuracy and efficiency .

REFERENCES

[1] Shen, G. and Xiao, H. The New Formula for Calculation of the Displacements and Stresses of Boundary Layer by BEM, Boundary Element IX: Proceedings of the Ninth International Conference, Springer-Verlag, New York, pp. 199-207, 1987.

[2] Shiue, F.-C. Geometrically Nonlinear Analysis for an Elastic Body by the Boundary Element Method, Ph.D. Dissertation, Iowa State University, Iowa, U.S.A., 1989.

[3] Liu, J., Beer, G. and Meek, J.L. Efficient Evaluation of Integrals of Order $1/r$, $1/r^2$, $1/r^3$ Using Gauss Quadrature, Engineering Analysis, Vol. 2, No. 3, pp. 118-123, 1985.

[4] Brebbia, C.A., Telles J.C.F. and Wrobel, L.C. Boundary Element Techniques: Theory and Applications in Engineering, Springer-Verlag, Berlin and New York, 1984.

[5] Gerald, C.F. and Wheatley, P.O. Applied Numerical Analysis, 3rd ed., Addison-Wesley, New York, 1984.

[6] Ugural, A.C. and Fenster, S.K. Advanced Strength and Applied Elasticity, Elsevier, New York, 1975.

[7] Timoshenko, S.K. and Goodier, J.N. Theory of Elasticity, 3rd ed., McGraw-Hill, New York, 1970.

The Hybrid Boundary Element Method for Problems Involving Body Forces

N.A. Dumont, M.T. Monteiro de Carvalho
Civil Engineering Department, PUC/RJ 22453, Rio de Janeiro, Brazil

ABSTRACT

The formulation introduced extends the Hybrid Boundary Element Method for the consideration of body forces. Starting point is the Hellinger-Reissner potential. As a result of the formulation, two vectors, corresponding to equivalent nodal forces and displacements, are obtained by means of boundary integrals only. The paper lays emphasis on some important physical interpretations of the quantities involved in the formulation. In particular, it is shown that some achievements of the present method may also be extended to the conventional formulation of the boundary element method, in order to avoid domain integrations as related to body forces. A simple numerical example is displayed.

INTRODUCTION

The Hybrid Boundary Element Method is a variationally, consistent method developed in the Civil Engineering Department of PUC/RJ [1, 2]. It combines in a unique formulation most advantages of the traditional finite element method (FEM) and boundary element method (BEM), as concerning computational implementation and possibilities of application [3].

In the present paper, the formulation is developed for consideration of body forces. Some physical interpretations made herein may be considered as extrapolating the conceptual limits of the new method, since they may also be useful for applications in both the FEM and the BEM.

GENERAL FORMULATION OF THE PROBLEM

Let be an elastic body submitted to body forces F_i, defined per surface unit of the domain Ω, and to boundary forces T_i, defined per surface unit of the boundary Γ_σ. It will be assumed throughout this paper that the subscript "i" (as well as subscript "j"), wherever it appears, refer to one of the coordinate directions x, y and z, for a general tridimensional problem, as it assumes the values 1, 2 or 3, respectively. The displacements are also prescribed as $\bar{u}_i$ along part of the boundary Γ_u complementary to Γ_σ. Only static, linear structural response of the elastic body is considered.

Aim of the structural analysis is the determination of a stress field σ_{ij} in equilibrium in Ω:

$$\sigma_{ij,j} + F_i = 0 \quad \text{in} \quad \Omega. \tag{1}$$

The stresses σ_{ij} may be expressed as the sum of two functions:

$$\sigma_{ij} = \sigma^*_{ij} + \sigma^p_{ij}, \tag{2}$$

where σ^p_{ij} is an arbitrary particular solution of the partial differential Equations (1) and σ^*_{ij} is the homogeneous solution of the same system, for the boundary conditions imposed along Γ_σ and Γ_u, as explained above. In the present formulation, σ^*_{ij} is given by Kelvin's fundamental solution, for problems of the elastostatics, which represents a stress field generated by some singular force p^*_m applied in a certain direction at a given point of the boundary Γ (subscripts "m" and "n" are used for numbering the degrees of freedom) . The equilibrium condition satisfied by the stress field may be expressed as

$$\sigma^*_{ij,j} + \Delta_{im}\, p^*_m = 0, \tag{3}$$

where Δ_{im} is a singular function which is zero-valued throughout the domain Ω, except in an arbitrarily small neighborhood Ω_0 of the point of application of the singular force p^*_m, in which

$$\int_{\Omega_0} \Delta_{im}\, d\Omega \equiv \delta_{im} = \begin{cases} 1 & \text{if i and m refer to one and the same coordinate direction} \\ 0 & \text{otherwise.} \end{cases} \tag{4}$$

According to the definition given above, σ^*_{ij} is in fact a numerical approximation of the homogeneous solution of Equation (1).

NUMERICAL DISCRETIZATION STARTING FROM THE HELLINGER-REISSNER POTENTIAL

One possibility of reformulating the problem above in a way suitable for numerical discretization is by making use of the Hellinger-Reissner potential [4]:

$$-\Pi_R = \int_\Omega \left(U_0^c + (\sigma_{ij,j} + F_i)\, u_i\right) d\Omega \; -$$

$$- \int_\Gamma \sigma_{ij}\, \eta_j\, u_i\, d\Gamma + \int_{\Gamma_\sigma} T_i\, u_i\, d\Gamma + \text{Cte}, \qquad (5)$$

which is supposed to become stationary. Independent variables in Equation (5) are the stresses σ_{ij}, as defined by Equations (2) and (3), and the displacements u_i. The strain energy density U_0^c is given by the stress field σ_{ij}.

Interpolation functions
The displacements u_i in Equation (5) are discretized along the boundary in terms of generalized nodal displacements d_m:

$$u_i = u_{im}\, d_m\,, \qquad (6)$$

in which u_{im} are linear or quadratic interpolation functions, as implemented at PUC/RJ. As it should be in Equation (5), these displacements are fully compatible along the boundary Γ_u. In other words, a basic hypothesis underlying Equation (5) is that, along Γ_u, both the corresponding generalized nodal displacements d_m and the interpolation functions u_{im} are some prescribed quantities $\bar{d}_m$ and $\bar{u}_{im}$, respectively.

The displacements corresponding to the Kelvin's fundamental solution may be expressed by

$$u_i^* = (u_{im}^* + C_{im})\, p_m^* \qquad \text{in} \quad \Omega, \qquad (7)$$

in which u_{im}^* are flexibility functions. Their expressions may be found in any book on elasticity theory or on boundary element methods (see [5], for instance). In the equation above, C_{im} are arbitrary constants, corresponding to rigid body displacements. In the present formulation, the terms multiplying C_{im} will always vanish identically

- a fact that does not necessarily occur in conventional boundary element methods.

Besides Equation (7), the fundamental solution will be also input into Equation (5) in terms of stresses defined along the boundary:

$$\sigma^*_{ij}\eta_j = p^*_{im}\, p^*_m \quad \text{in } \Gamma, \tag{8}$$

in which η_j, also appearing in Equation (5), are the cosine directors of an elementary boundary segment $d\Gamma$.

Moreover, it will be supposed that a particular solution of the differential Equations (1) is known in terms of stresses σ^p_{ij} and displacements u^p_i.

Numerical discretization
The numerical discretization of the Hellinger-Reissner potential, Equation (5), in the frame of the present method, has already been carried out in previous papers [1, 2], except for the terms due to body forces. The final result is the following expression, in matrix notation:

$$-\Pi_R = \frac{1}{2}\{p^*\}^T [F] \{p^*\} + \{p^*\}^T \{b\} - \{p^*\}^T [H] \{d\} -$$

$$- \{d\}^T \{t\} + \{d\}^T \{p\} + Cte, \tag{9}$$

where $\{p^*\} \equiv p^*_m$ and $\{d\} \equiv d_m$ are the vectors of generalized forces and displacements, respectively, applied at regularly distributed points throughout the boundary and $\{p\} \equiv p_m$ is a vector of nodal forces equivalent in part to the forces T_i applied at Γ_σ and in part to the reaction forces at Γ_u. In this equation, the matrix

$$[H] \equiv H_{mn} = \int_\Gamma p^*_{im}\, u_{in}\, d\Gamma + \int_\Omega \Delta_{im}\, u_{in}\, d\Omega \tag{10}$$

transforms generalized nodal displacements d_n into equivalent nodal displacements d^*_m related to the singular forces p^*_m by means of virtual work. This same matrix is also obtained in the conventional boundary element method, although in general no clear physical meaning is associated to it.

Equation (9) also introduces a symmetric, positive semidefinite flexibility matrix

$$[F] \equiv F_{mn} = \int_{\Gamma} p^*_{km} u^*_{kn} \, d\Gamma + \int_{\Omega} \Delta_{km} u^*_{kn} \, d\Omega, \tag{11}$$

which transforms generalized singular forces p^*_n into equivalent nodal displacements d^*_m. The integration of this matrix may be carried out numerically in a straightforward manner, for the general degrees of freedom m and n. However, for m and n corresponding to one and the same point of the boundary, F_{mn} cannot be obtained directly in terms of the integration indicated, owing to the occurrence of a second kind singularity in the integrand. But, instead of a drawback or an insuperable obstacle, this singularity is a physical necessity, as explained in [2]. The way of obtaining these non-integrable terms – to summarize some important achievements of the hybrid boundary element method – is to impose that the matrix [F] be orthogonal to a given matrix [V]:

$$[F]\,[V] = 0, \tag{12}$$

in which [V] is such that

$$[H]^T\,[V] = 0. \tag{13}$$

The algorithms implemented to obtain [V] from Equation (13) and to enforce Equation (12) are explained in [3].

Two more terms, related to body forces, are introduced in Equation (9). They deserve more attention, since they haven't been considered in previous papers written by the authors in english language. One of them is the vector of equivalent nodal forces

$$\{t\} \equiv t_n - \int_{\Gamma} \sigma^p_{ij} \eta_j u_{in} \, d\Gamma, \tag{14}$$

which arises from the first boundary integral in Equation (5). The integration indicated in Equation (14) is a straightforward procedure in terms of Gaussian quadrature and demands little computational effort.

The last term to be defined in Equation (9) is the vector of equivalent nodal displacements {b}. This vector stems from the numerical discretization of the strain energy term in Equation (5) and corresponds to the amount of the particular solution of the differential Equations (1):

$$p^*_m b_m = \frac{1}{2} \int_{\Omega} (\sigma^*_{ji} u^p_{j,i} + \sigma^p_{ji} u^*_{j,i}) \, d\Omega \equiv$$

$$\equiv \int_\Omega \sigma^p_{ji} u^*_{j,i} \, d\Omega \equiv \int_\Omega \sigma^*_{ji} u^p_{j,i} \, d\Omega. \tag{15}$$

The identities indicated above, which hold independently of numerical discretization, may be easily demonstrated. These identities give rise to two formally different, but mathematically equivalent expressions for the vector {b}.

The first expression is derived below, in a sequence of steps which involves integration by parts, transformation of a domain integral into a boundary one, by means of Green's theorem, introduction of equilibrium Equations (1) and substitution of u^*_i according to Equation (7):

$$p^*_m b_m = \int_\Omega \sigma^p_{ki} u^*_{k,i} \, d\Omega =$$

$$= \int_\Omega (\sigma^p_{ki} u^*_k)_{,i} \, d\Omega - \int_\Omega \sigma^p_{ki,i} u^*_k \, d\Omega =$$

$$= \int_\Gamma \sigma^p_{ki} \eta_i u^*_k \, d\Gamma + \int_\Omega u^*_k F_k \, d\Omega =$$

$$= p^*_m \int_\Gamma \sigma^p_{ki} \eta_i (u^*_{km} + C_{km}) \, d\Gamma + p^*_m \int_\Omega \bar{F}_k (u^*_{km} + C_{km}) \, d\Omega =$$

$$= p^*_m \int_\Gamma \sigma^p_{ki} \eta_i u^*_{km} \, d\Gamma + p^*_m \int_\Omega \bar{F}_k u^*_{km} \, d\Omega +$$

$$+ p^*_m \left[\int_\Gamma \sigma^p_{ki} \eta_i \, d\Gamma + \int_\Omega \bar{F}_k \, d\Omega \right] C_{km} =$$

$$= p^*_m (t^*_m + b^*_m), \tag{16}$$

where

$$t^*_m = \int_\Gamma \sigma^p_{ij}\, \eta_j\, u^*_{im}\, d\Gamma \tag{17}$$

and

$$b^*_m = \int_\Omega F_i\, u^*_{im}\, d\Omega . \tag{18}$$

The term in square brackets in Equation (16) vanishes identically (in the conventional boundary element formulation appears a similar term, which tends to zero – but does not vanish identically – as the mesh on the boundary is refined).

The second way of expressing b_m is presented in the following sequence of steps, in which integration by parts is performed, before transformation of a domain interval into a boundary one, use of the equilibrium Equation (3) and substitution of the stress expression on the boundary according to Equation (8):

$$p^*_m\, b_m = \int_\Omega \sigma^*_{ki}\, u^p_{k,i}\, d\Omega$$

$$= \int_\Omega (\sigma^*_{ki}\, u^p_k)_{,i}\, d\Omega - \int_\Omega \sigma^*_{ki,i}\, u^p_k\, d\Omega =$$

$$= \int_\Gamma \sigma^*_{ki}\, \eta_i\, u^p_k\, d\Gamma + p^*_m \int_\Omega \Delta_{km}\, u^p_k\, d\Omega =$$

$$= p^*_m \int_\Gamma p^*_{km}\, u^p_k\, d\Gamma + p^*_m \int_\Omega \Delta_{km}\, u^p_k\, d\Omega , \tag{19}$$

so that

$$b_m = \int_\Gamma p^*_{km}\, u^p_k\, d\Gamma + \int_\Omega \Delta_{km}\, u^p_k\, d\Omega . \tag{20}$$

The vector of equivalent nodal displacements b_m is one and the same, be it given by expression (16) or by expression (19), although numerical differences may arise due to approximations related to the different numerical integration schemes applied to Equations (17), (18) and (20).

Nevertheless, Equation (20) seems to be much more suitable to computational purposes, since it involves only a numerical integration along the boundary (the domain integration indicated in Equation (20) is a matter of formality, as also occurs in the definition of H_{mn}, Equation (10), since the function Δ_{im}, as given in Equation (4), is singular). Since the computational effort related to the evaluation of b_m, as given by Equation (20), is comparable to the effort for evaluating t^*_m in Equation (17), the evaluation of b^*_m, according to Equation (18), makes the whole difference. In fact, evaluating b^*_m involves numerical integration over the domain. It is rather surprisingly that exactly Equation (18) appears in all books and papers on the conventional boundary element formulation. Moreover, all attempts to transform this domain integral into boundary ones go through complicated relations involving the fundamental solution. As a by-product of the present hybrid formulation, the authors dare suggest that, in the conventional formulations, the expression of b^*_m, as given by Equation (18), be substituted by the boundary expressions of Equations (17) and (20), as follows:

$$b^*_m \leftarrow b_m - t^*_m. \tag{21}$$

It is generally an easy task to find a particular solution of Equation (1) for a given function F_i defined analytically over either part or the whole of domain Ω.

FINAL EQUATIONS

The condition of stationariety of Π_R in Equation (9), for any variations δp^*_m and δd_n, yields two systems of equations:

$$[F]\{p^*\} = [H]\{d\} - \{b\} \tag{22}$$

$$[H]^T\{p^*\} = \{p\} - \{t\}. \tag{23}$$

Substitution of $\{p^*\}$, given in Equation (22), into Equation (23), yields

$$[H]^T [F]^I [H] \{d\} = \{p\} - \{t\} + [H]^T [F]^I \{b\}, \tag{24}$$

in which $[F]^I$ is the unique inverse of $[F]$ restricted to its own space, that is, such that, according to Equation (12), also

$$[F]^I [V] = 0. \tag{25}$$

One may notice that the matrix $[F]^I$ appears at both sides of Equation (24) premultiplied by $[H]^T$. Then, instead of trying to to calculate this restricted inverse directly, it is computationally more efficient to substitute $[F]^I$ by the inverse of a symmetric, positive definite matrix:

$$[F]^I \Leftarrow ([F] + [V]\,[\lambda]\,[V]^T)^{-1}, \tag{26}$$

in which $[\lambda]$ is also a symmetric positive definite, but otherwise arbitrary matrix. Indeed:

$$[H]^T\,[F]^I \equiv [H]^T\,([F] + [V]\,[\lambda]\,[V]^T)^{-1}. \tag{27}$$

In Equation (24), one identifies the occurrence of a stiffness matrix [K], defined as

$$[K] = [H]^T\,[F]^I\,[H], \tag{28}$$

which is also singular, since it is orthogonal to rigid body displacements.

DETERMINATION OF THE VECTOR {p*}

Once Equations (24) have been solved in terms of the generalized displacements {d}, for given static forces and adequate boundary conditions, one may obtain the vector {p*} from Equation (22):

$$\{p^*\} = [F]^I\,[H]\,\{d\} - [F]^I\,\{b\} \tag{29}$$

and thereafter define the state of stresses in the body, which satisfies Equation (1). In the Equation above, $[F]^I$ may still be replaced by the expression indicated in Equation (26), since {b} (as well as {t*} and {b*}) is also orthogonal to the matrix [V], in the sense of Equation (12).

But {p*} may also be formally expressed as a function of the applied forces, if one substitute the expression of {d}, given from Equation (24), into Equation (29):

$$\{p^*\} = [F]^I\,[H]\,[K]^I\,(\{p\} - \{t\}), \tag{30}$$

since the matrix expression multiplying {b}, which would appear in Equation (30), vanishes identically:

$$[F]^I\,[H]\,[K]^I\,[H]^T\,[F]^I - [F]^I \equiv 0. \tag{31}$$

Then, it may be concluded that the vector {b} is relevant only for the determination of the displacements {d}, as given in Equation (24), but does not interfere directly in the calculation of the vector of singular forces {p*} and therefore does not affect the state of stresses inside the body.

It may also be verified that Equations (23) and (30) are formally equivalent, since

$$[H]^T = [K]\ [H]^I\ [F]. \tag{32}$$

Then, for externally determinate strutures, when all elements of the vector {p} are known (Neumann conditions), one may bypass the calculation of the nodal displacements {d}, as given by Equation (24), and solve directly the following system of Equations:

$$[H]^T\{p^*\} = \{p\} - \{t\} \tag{23}$$

$$[V]^T\ \{p^*\} = 0, \tag{33}$$

since only singular forces {p*} orthogonal to [V] are admissible solutions. This particular case of determinate structures is very suitable to be dealt with in the frame of the hybrid boundary element method, since only two easily obtainable matrices – [H] and {t} – are required.

EXAMPLE

Fig. 1 shows half of a simply supported beam submitted to a self-weight of 2 units and considering a Poisson's ratio 0.3, as presented in [6]. The mesh represented was used for a comparative analysis done with quadratic Serendipity elements. The discretization shown along the boundary was also used for the hybrid boundary element analysis with linear and quadratic elements. The analytical solution for this problem was taken from [7].

Table 1 presents the values of σ_x stresses at 8 points of the beam, as indicated in Fig. 1 (Gaussian points of the finite element mesh, except for the 3rd one). The best results were obtained using quadratic hybrid boundary elements, for points not too close to the boundary. At these points, on the other hand, the results with linear elements were better, although not satisfactory, owing to the proximity to nodal points (quadratic elements are double as large as linear ones).

The present example is a mere illustration of the formulation introduced in the paper. Since the beam is statically determinate, the solution, obtained by solving Equations (23) and (33), required much less computational time than using finite elements.

The present formulation has already been implemented for plane state potential and elasticity problems and is being implemented for threedimensional problems.

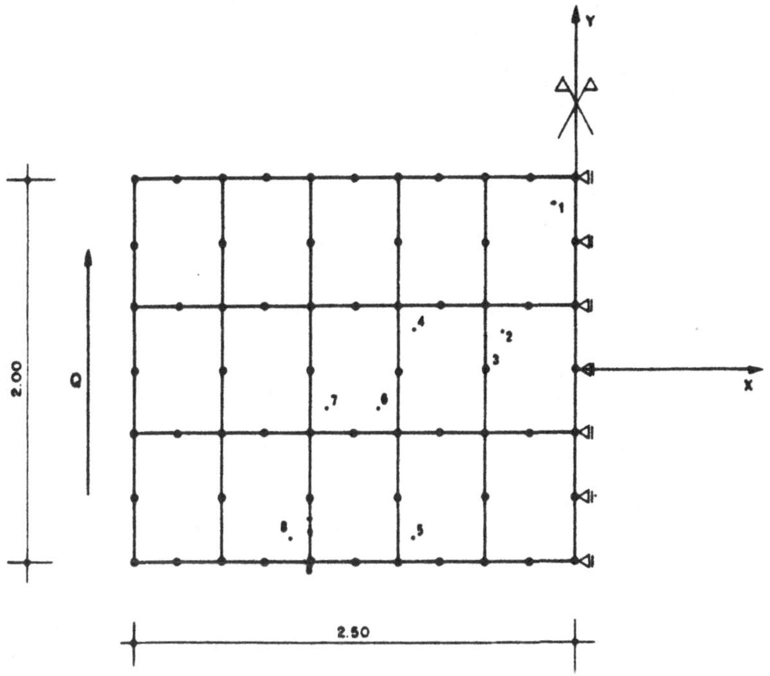

Fig. 1 - Simply supported beam submitted to self-weight

COORDINATES		HBE(Lin.)	HBE(Quad.)	SEREN	ANAL.SOL.
X	Y	σx	σx	σx	σx
-.106	.859	-17.59	-18.12	-16.34	-16.31
-.394	.193	-3.25	-3.31	-3.33	-3.31
-.5	0.	0.	0.	-	0.
-.894	.193	-2.88	-2.94	-2.96	-2.94
-.894	-.859	15.34	15.42	14.31	14.28
-1.394	-.193	2.25	2.30	2.31	2.28
-1.106	-.193	2.65	2.70	2.71	2.69
-1.606	-.859	10.53	11.35	9.71	9.69

Tab. 1 - Stresses σ_x at some internal points of the beam of Fig. 1

ACKNOWLEDGMENTS

This study was supported by the Brazilian Conselho Nacional de Desenvolvimento Cientifico e Tecnologico - CNPq.

REFERENCES

1. Dumont, N.A., "The Hybrid Boundary Element Method", Boundary Elements IX, Vol. I: Mathematical and Computational Aspects, Computational Mechanics Publications, Springer Verlag, pp. 117-130, 1987.

2. Dumont, N.A., "The Hybrid Boundary Element Method: An Alliance Between Mechanical Consistency and Simplicity", Applied Mechanics Reviews, Vol. 42, nr. 11, Part 2, pp. S54-S63, 1989.

3. de Carvalho, M.T.M., "Implementacao Computacional no Metodo Hibrido dos Elementos de Contorno", M.Sc. Thesis, Civil Engineering Dept. PUC/RJ, Rio de Janeiro, Brasil, 1990.

4. Dumont, N.A., "The Variational Formulation of the Boundary Element Method", Boundary Element Techniques: Applications in Fluid Flow and Computational Aspects, Computational Mechanics Publications, Adlard and Sond Ltd., pag. 225-239, 1987.

5. Brebbia, C.A., Telles, J.C.F. e Wrobel, L.C., "Boundary Element Techniques", Springer Verlag, Berlin and New York, 1984.

6. Dumont, N.A. e de Carvalho, M.T.M., "Consideracao de Forcas de Massa no Metodo Hibrido dos Elementos de Contorno - Parte B: Interpretacao Fisica e Exemplo", XI Congresso Ibero Latino Americano sobre Metodos Computacionais para Engenharia, Vol. II, pp. 947-956, Rio de Janeiro, 1990.

7. Timoshenko, S. P. and Goodier, J. N., "Teoria da Elasticidade", 3rd edition, Ed. Guanabara Dois, Rio de Janeiro, Brazil, 1980.

Numerical Implementation of a Normal Derivative Integral Equation in Acoustics

T.W. Wu, A.F. Seybert, G.C. Wan

Department of Mechanical Engineering, University of Kentucky, Lexington, KY 40506-0046, U.S.A.

ABSTRACT

A regularized normal derivative integral equation, originally derived by Maue, is implemented in an isoparametric element environment. A linear combination of this normal derivative integral equation and the conventional Helmholtz integral equation is used to insure a unique solution for all frequencies. The regularized normal derivative integral equation used here converges in the Cauchy principal value sense rather than only in the finite-part sense. The Cauchy principal value integral can be further transformed into an integral that converges in the normal sense. This regularized normal derivative equation may also be applied to the solution for acoustic radiation and scattering from thin structures.

INTRODUCTION

The boundary integral equation method has been extensively used in acoustic radiation and scattering. One potential shortcoming of the BEM in acoustics is that the exterior boundary integral formulation shares the well-known difficulty of nonuniqueness of solution at certain fictitious eigenfrequencies [1]. These fictitious eigenfrequencies are associated with the eigenfrequencies of a corresponding interior problem. Detailed discussions of the mathematical aspects of this problem can be found in the articles by Kleinman and Roach [2] and Burton [3].

Several modified integral formulations have been proposed to overcome the nonuniqueness problem [1,4-10] . One of the most well-known formulations to overcome the nonuniqueness problem is the method proposed by Burton and Miller [4]. This approach consists of a linear combination of the Helmholtz integral equation and its normal derivative equation. It has been proved in Ref. [4] that the linear combination of these two equations will yield a unique solution for all frequencies if the multiplicative constant of the normal derivative equation is appropriately chosen. However, the major difficulty in this formulation is that the normal derivative of the Helmholtz integral equation involves a hyper-singular integral. Burton and Miller used a double surface integral to regularize this strong singularity. Evaluation of a double integral is computationally inefficient. Other regularization techniques such as the work by Meyer, et al. [11], and Terai [12] are suitable for planar elements only.

An alternative way to deal with the hyper-singular integral equation is to interpret the convergence of the equation in the Hadamard finite-part sense [13]. Several special quadrature rules have been developed to numerically evaluate finite-part integrals of different order of singularity [14,15]. However, the direct evaluation of a finite-part integral is of higher order and may involve finite difference expressions, especially for an arbitrary three-dimensional curved boundary. An analytical technique has recently been developed by Krishnasamy, et al. [16] to locally reduce a finite-part integral to several regular surface and line integrals via Stokes' theorem. This technique involves evaluation of tangential derivatives of the acoustic pressure at collocation points. In any event, for the hyper-singular integral equation to converge properly in the finite-part sense, the acoustic pressure must satisfy the C^1 Holder condition at collocation points.

As an alternative, a less-singular normal derivative integral equation, which converges in the Cauchy principal value sense, may be used in Burton and Miller's formulation [17]. This normal derivative integral equation was first derived by Maue [18], and later by Mitzner [19] using a different approach. In this paper, the numerical procedures used in Ref. [17] to implement Maue's equation in an isoparametric element environment are summarized. Numerical integration of a Cauchy principal value integral can be achieved by subtracting a nonsingular term at the singular point and then adding the term back. This technique is amenable to standard Gaussian quadrature. No special integration rules are needed in this formulation. It should be noted that Maue's equation involves integration of tangential derivatives of the acoustic pressure on the boundary. Therefore, only continuous elements may be used in that equation.

Maue's equation also requires that the tangential derivatives be continuous at collocation points. This is nothing more than the C^1 Holder condition as required for the finite-part evaluation. The easiest way to satisfy the C^1 continuity condition is to put collocation points inside each element, rather than on the element sides. Putting collocation points inside an element may look cumbersome at first sight. Nevertheless, doing so will not only insure the C^1 continuity condition at collocation points but also will yield a well-defined normal direction in Burton and Miller's formulation. Since no collocation points are shared by any two adjacent elements, the number of collocation points will be greater than the number of nodal points. In other words, there will be more equations than unknowns. The overdetermined system may then be solved by a least-squares procedure.

BOUNDARY INTEGRAL EQUATIONS

Consider a body B of boundary surface S in an infinite acoustic medium B' of mean density ρ_0 and speed of sound c. The body B can be either a vibrating structure (in a radiation problem) or a passive obstacle (in a scattering problem). The governing differential equation in steady-state linear acoustics is the well-known Helmholtz equation

$$\nabla^2 \phi + k^2 \phi = 0 \quad , \tag{1}$$

where ϕ is the velocity potential and k is the wave number. This equation must be satisfied in the acoustic domain B'. The wave number is defined by $k=\omega/c$, in which ω is the angular frequency.

By using the "direct" formulation (via either the Green second identity or the weighted residual formulation), Eq. (1) is reformulated into a boundary integral equation defined on the boundary surface S as follows:

$$C(P)\phi(P) = \int_S [\psi(P,Q)\frac{\partial\phi}{\partial n}(Q) - \frac{\partial\psi}{\partial n}(P,Q)\,\phi(Q)]\,dS(Q) + 4\pi\phi_i(P) \quad ,(2)$$

where $\psi = e^{-ikr}/r$ is the free-space Green's function , $r = |Q - P|$, and C(P) is a constant that depends on the location of P. The value of C(P) is 4π for P in B' and zero for P in B. If P is on the boundary surface S and there is a unique tangent plane at P, then C(P) is 2π. Equation (2) is sometimes referred to as the Helmholtz integral equation. It should be noted that the Helmholtz integral equation converges in the normal sense rather than only in the Cauchy principal value sense because the kernels are only weakly singular.

It is well known that the Helmholtz integral equation itself will not have a unique solution at certain characteristic frequencies [1,2-4]. However, according to Burton and Miller's formulation [4], a linear combination of the Helmholtz integral equation and its normal derivative will yield a unique solution for all frequencies. The normal derivative integral equation for P on the boundary takes the form

$$2\pi\frac{\partial\phi(P)}{\partial n_P} = \int_S [\frac{\partial\psi}{\partial n_P}\frac{\partial\phi}{\partial n} - \frac{\partial^2\psi}{\partial n\partial n_P}\phi]\,dS + 4\pi\frac{\partial\phi_i(P)}{\partial n_P} \quad , \qquad (3)$$

where n_P is the inward normal on S at the collocation point P and the differentiation $\partial/\partial n_P$ is taken in the n_P direction with respect to the coordinates of P. Since the singularity of the second kernel in Eq. (3) is of the order of $1/r^3$, the integral equation converges only in the Hadamard finite-part sense.

Instead of using the hyper-singular equation (Eq. (3)) directly, we use a less-singular normal derivative integral equation in Burton and Miller's formulation. This less-singular normal derivative equation was first derived by Maue [18], and later by Mitzner [19] using a different approach. The result is

$$2\pi\frac{\partial\phi(P)}{\partial n_P} = \int_S \{\frac{\partial\psi}{\partial n_P}\frac{\partial\phi}{\partial n} - [(\mathbf{n}_P\times\nabla_P\psi)\cdot(\mathbf{n}\times\nabla\phi) + k^2(\mathbf{n}_P\cdot\mathbf{n})\psi\phi]\}\,dS$$

$$+ 4\pi \frac{\partial \phi_i}{\partial n_P}(P) \quad , \tag{4}$$

where $\nabla_P \psi$ is the gradient of the Green's function taken with respect to the coordinates of P. Note that the first and the third kernels inside the integral of Eq. (4) will pose no problem because they are only weakly singular. The second kernel is the inner product of two tangential vectors. It will be convenient here to define two orthogonal tangent vectors **s** and **t** of unit length at each surface point so that **s**, **t** and **n** form a local right-hand coordinate system. We also let $\mathbf{s}_P$ and $\mathbf{t}_P$ denote the corresponding tangent vectors at the point P. Then the explicit expressions for the two tangential vectors in the second kernel of Eq. (4) are

$$\mathbf{n}_P \times \nabla_P \psi = - \frac{(1 + ikr)}{r^2} e^{-ikr} \left(\frac{\partial r}{\partial s_P} \mathbf{t}_P - \frac{\partial r}{\partial t_P} \mathbf{s}_P \right) \tag{5}$$

and

$$\mathbf{n} \times \nabla \phi = \frac{\partial \phi}{\partial s} \mathbf{t} - \frac{\partial \phi}{\partial t} \mathbf{s} \quad , \tag{6}$$

where the derivatives $\partial r/\partial s_P$ and $\partial r/\partial t_P$ are taken with respect to the coordinates of P. Note from Eq. (5) that the singularity of the kernel is of the order of $1/r^2$. A singular kernel with an order of $1/r^2$ is still not integrable on a surface unless the kernel satisfies some special condition. In the current case, it is not hard to show that in the limit as r goes to zero, $\partial r/\partial s_P$ and $\partial r/\partial t_P$ behave like $\cos\theta$ and $\sin\theta$, respectively, in which θ is the polar angle on the master element where numerical integration is performed. Therefore, it can be proved that Eq. (4) will converge if the limit of the singular integral is taken in the Cauchy principal value sense.

A linear combination of Eqs. (2) and (4) can then be taken to insure a unique solution for all frequencies. Burton and Miller [4] also showed that if the wave number k is real or purely imaginary, the imaginary part of the multiplicative constant for Eq. (4) should not be zero. Usually, the imaginary number i is taken to multiply Eq. (4) in the linear combination.

NUMERICAL ASPECTS

The regular 8-node quadrilateral and 6-node triangular isoparametric elements are used in our numerical model. However, collocation points are placed inside the elements to satisfy the C^1 Holder condition. Figure 1 shows the configuration of the master elements and the positions of nodal points and collocation points. The global Cartesian coordinates X_i (i=1, 2, 3) of any point on an element are assumed to be related to the nodal coordinates $X_{i\alpha}$ by

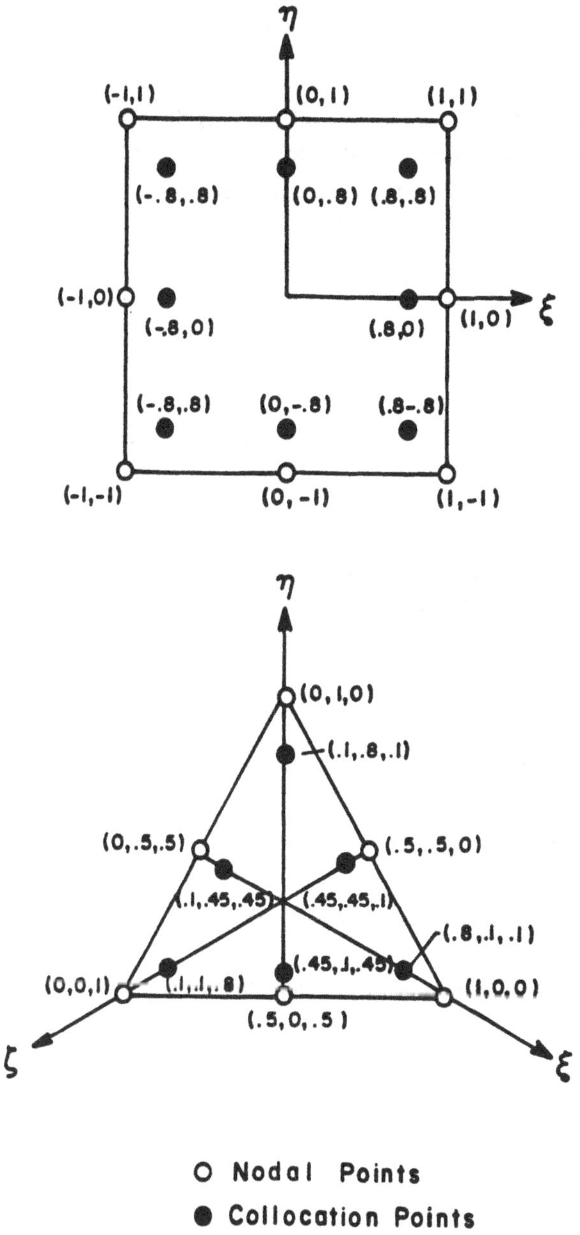

Figure 1. Configuration of the master elements.

$$X_i(\xi) = \sum_\alpha N_\alpha(\xi) X_{i\alpha} \quad , \alpha = 1, ..., 6 \text{ or } 8 \tag{7}$$

in which N_α are second-order shape functions of the local coordinates (ξ,η). The same set of quadratic shape functions is used to interpolate the boundary variables ϕ and $\partial\phi/\partial n$. Therefore, on each element

$$\phi = \sum_\alpha N_\alpha \phi_\alpha \tag{8}$$

and

$$\frac{\partial\phi}{\partial n} = \sum_\alpha N_\alpha \left(\frac{\partial\phi}{\partial n}\right)_\alpha \quad , \tag{9}$$

where ϕ_α and $(\partial\phi/\partial n)_\alpha$ are the nodal values of ϕ and $\partial\phi/\partial n$, respectively. Two tangent vectors t_ξ and t_η can now be defined at any point in the two curvilinear directions ξ and η as follows:

$$t_\xi = \frac{\partial X/\partial\xi}{|\partial X/\partial\xi|} \tag{10}$$

and

$$t_\eta = \frac{\partial X/\partial\eta}{|\partial X/\partial\eta|} \quad , \tag{11}$$

where **X** is the position vector of the point under consideration. On a general three-dimensional curved boundary surface, these two tangent vectors may not be orthogonal to each other. The orthogonal tangent vectors **s** and **t** can be obtained by

$$\mathbf{s} = \mathbf{t}_\xi \tag{12}$$

and

$$\mathbf{t} = \mathbf{n} \times \mathbf{t}_\xi \quad , \tag{13}$$

where the normal vector $\mathbf{n} = \mathbf{t}_\xi \times \mathbf{t}_\eta$. Since **s** is in the same direction as $\mathbf{t}_\xi$, the first tangential derivative $\partial\phi/\partial s$ in Eq. (6) is found to be

$$\frac{\partial\phi}{\partial s} = \frac{1}{|\partial X/\partial\xi|} \frac{\partial\phi}{\partial\xi} \quad . \tag{14}$$

The second tangential derivative $\partial\phi/\partial t$ in Eq. (6) is

$$\frac{\partial \phi}{\partial t} = - \frac{\mathbf{s} \cdot \mathbf{t}_\eta}{\mathbf{t} \cdot \mathbf{t}_\eta} \frac{1}{|\partial X/\partial \xi|} \frac{\partial \phi}{\partial \xi} + \frac{1}{\mathbf{t} \cdot \mathbf{t}_\eta} \frac{1}{|\partial X/\partial \eta|} \frac{\partial \phi}{\partial \eta} \quad . \tag{15}$$

In Eqs. (14) and (15), the derivatives $\partial\phi/\partial\xi$ and $\partial\phi/\partial\eta$ are calculated by differentiating the shape functions in Eq. (8).

As mentioned previously, the integration of the second kernel in Eq. (4) should be performed in the Cauchy principal value sense. For each collocation position of P, there is only one element that contains the singular point P. This element is denoted by S_e. Numerical integration on all the other elements is nonsingular and can be carried out in the usual way. For integration on the singular element S_e, a polar coordinate system (ρ,θ) is used on the corresponding master element S_m as shown in Fig. 2.

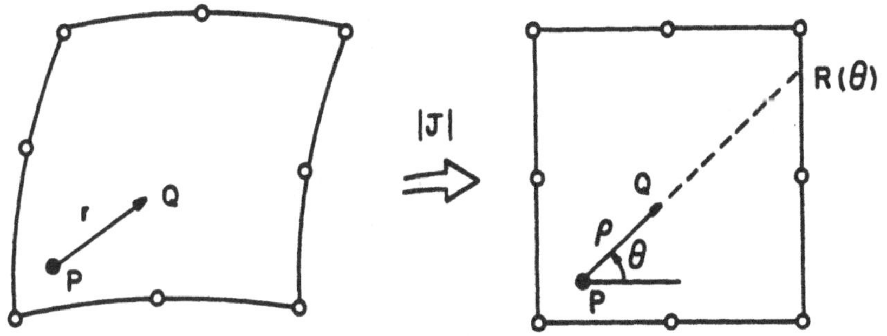

Singular Element S_e — Corresponding Master Element S_m

Figure 2. Polar Coordinate Transformation.

To facilitate computation, a parameter β is defined such that

$$r = \beta\, \rho \tag{16}$$

The Cauchy principal value integral on the singular element then becomes

$$\int_{S_e} (\mathbf{n}_P \times \nabla_P \psi) \cdot (\mathbf{n} \times \nabla \phi)\, dS = \int_{S_m} (\mathbf{n}_P \times \nabla_P \psi) \cdot (\mathbf{n} \times \nabla \phi)\, |J|\, dS$$

$$= \lim_{\varepsilon \to 0} \int_0^{2\pi} \int_\varepsilon^{R(\theta)} (\mathbf{n}_P \times \nabla_P \psi) \cdot (\mathbf{n} \times \nabla \phi) \, |J| \, \rho \, d\rho \, d\theta$$

$$= \lim_{\varepsilon \to 0} \int_0^{2\pi} \int_\varepsilon^{R(\theta)} \frac{F(\rho,\theta)}{\rho} \, d\rho \, d\theta \quad , \tag{17}$$

where $|J|$ is the Jacobian between S_e and S_m, $R(\theta)$ is distance from P to the element boundary measured on the master element plane, and $F(\rho,\theta)$ is a regular term. Note that the regular term $F(\rho,\theta)$ is

$$F(\rho,\theta) = \frac{r^2}{\beta^2} (\mathbf{n}_P \times \nabla_P \psi) \cdot (\mathbf{n} \times \nabla \phi) \, |J| \quad . \tag{18}$$

A further regularization of the Cauchy principal value integral in Eq. (17) can be achieved by subtracting $F(0,\theta)$ from $F(\rho,\theta)$ and then adding it back. Doing so yields

$$\int_{S_e} (\mathbf{n}_P \times \nabla_P \psi) \cdot (\mathbf{n} \times \nabla \phi) \, dS = \lim_{\varepsilon \to 0} \int_0^{2\pi} \int_\varepsilon^{R(\theta)} \frac{F(\rho,\theta) - F(0,\theta)}{\rho} \, d\rho \, d\theta$$

$$+ \lim_{\varepsilon \to 0} \int_0^{2\pi} F(0,\theta) \int_\varepsilon^{R(\theta)} \frac{1}{\rho} \, d\rho \, d\theta$$

$$= I_1 + I_2 \quad . \tag{19}$$

The integral I_1 is now integrable in the normal sense because $F(\rho,\theta) - F(0,\theta)$ behaves at least in the order of ρ. The second integral I_2 is

$$I_2 = \lim_{\varepsilon \to 0} \left[\int_0^{2\pi} F(0,\theta) \ln R(\theta) \, d\theta - \ln\varepsilon \int_0^{2\pi} F(0,\theta) \, d\theta \right] \quad . \tag{20}$$

It is easy to prove that

$$\int_0^{2\pi} F(0,\theta) \, d\theta = 0 \quad , \tag{21}$$

because $F(0,\theta)$ is a linear combination of $\sin\theta$ and $\cos\theta$. Therefore, the Cauchy principal integration of the second kernel of Eq. (4) can be summarized as follows:

$$\int_{S_e} (\mathbf{n}_p \times \nabla_p \psi) \cdot (\mathbf{n} \times \nabla \phi)\, dS = \int_0^{2\pi} \int_0^{R(\theta)} \frac{F(\rho,\theta) - F(0,\theta)}{\rho}\, d\rho\, d\theta + \int_0^{2\pi} F(0,\theta) \ln R(\theta)\, d\theta \quad , \qquad (22)$$

where the limit $\varepsilon \to 0$ has been taken because all the integrals are now regular. Note that at $\rho = 0$, r is also equal to zero. A detailed discussion on the evaluation of $F(0,\theta)$ can be found in Ref. [17].

EXAMPLE PROBLEMS

The first example problem is the radiation from a pulsating sphere of radius a. A uniform velocity of $\partial\phi/\partial n = v_a$ is prescribed on the boundary surface. The BEM mesh used in this example problem consists of 24 quadrilateral elements with a total of 74 nodes. The result for the normalized pressure amplitude, $|p| / z_0 v_a$, on the sphere's surface is plotted against ka in Fig. 3. Also plotted in Fig. 3 is the analytical solution for this problem. Excellent agreement between the BEM solution and the analytical solution is observed.

The second problem is the scattering from a rigid sphere of radius a due to a plane incident wave at ka = 4.493. A BEM mesh consisting of 24 triangular elements and 48 quadrilateral elements is used to solve this problem. The total number of nodes is 194. Figure 4 shows the ratio of the scattered to the incident pressure at r = 3a from the center of the sphere. The result is very good compared to the analytical solution [20].

CONCLUSIONS

The regularized normal derivative integral equation originally derived by Maue is numerically implemented in a three-dimensional isoparametric element environment. This normal derivative equation converges in the Cauchy principal value sense rather than only in the finite-part sense. The Cauchy principal value integral is locally transformed into a regular integral so that it is amenable to standard Gaussian quadrature. The numerical technique presented in this paper for the normal derivative integral equation can also be applied to the solution for acoustic radiation and scattering from thin structures.

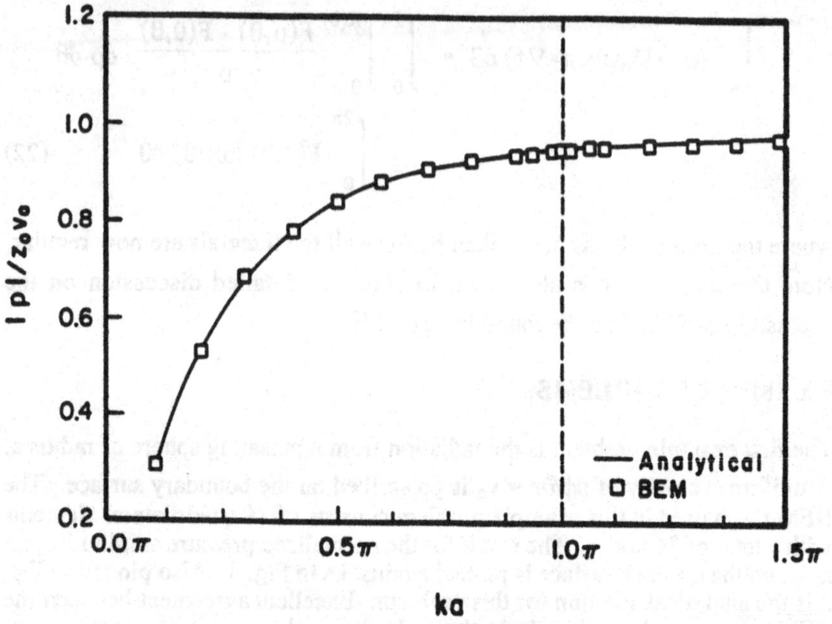

Figure 3. Normalized sound pressure on the surface of a pulsating sphere.

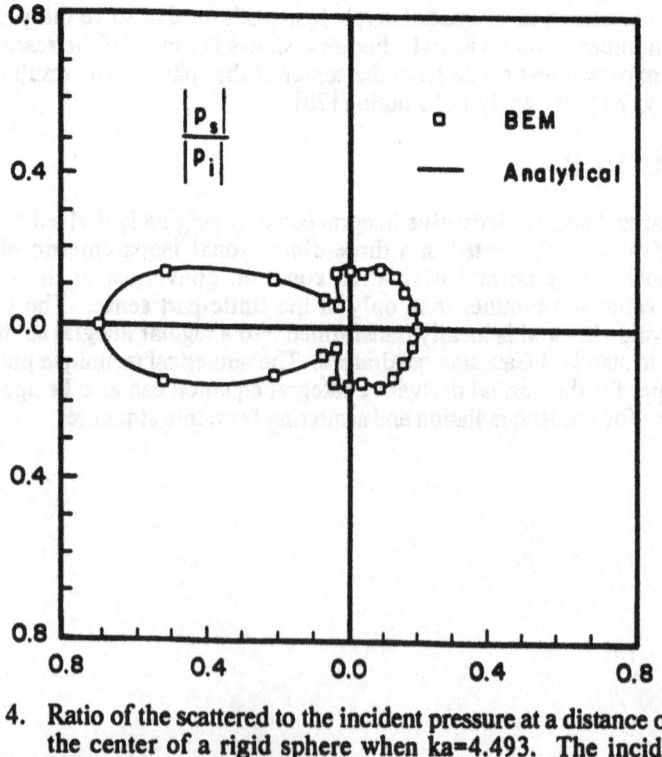

Figure 4. Ratio of the scattered to the incident pressure at a distance of 3a from the center of a rigid sphere when ka=4.493. The incident plane wave is coming from the right.

REFERENCES

1. H. A. Schenck, "Improved Integral Formulation for Acoustic Radiation Problems," J. Acoust. Soc. Am. 44, 41-58 (1968).
2. R. E. Kleinman and G. F. Roach, "Boundary Integral Equations for the Three-Dimensional Helmholtz Equation," SIAM Review 16, 214-236 (1974).
3. A. J. Burton, "The Solution of Helmholtz' Equation in Exterior Domains Using Integral Equations," Natl. Phys. Lab. Rep. NAC 30, Teddington, Middlesex, U.K. (1973).
4. A. J. Burton and G. F. Miller, "The Application of Integral Equation Methods to the Numerical Solutions of Some Exterior Boundary Value Problems," Proc. Roy. Soc. Lond. A 323, 201-210 (1971).
5. F. Ursell, "On the Exterior Problems of Acoustics," Proc. Camb. Philos. Soc. 74, 117-125 (1973).
6. D. S. Jones, "Integral Equations for the Exterior Acoustic Problem," Q. J. Mech. Appl. Math. 27, 129-142 (1974).
7. P. C. Waterman, "New Formulation of Acoustic Scattering," J. Acoust. Soc. Am. 45, 1417-1429 (1969).
8. P. A. Martin, "On the Null-field Equations for the Exterior Problems of Acoustics," Q. J. Mech. Appl. Math. 33, 385-396 (1980).
9. K. Brod, "On the Uniqueness of Solution for All Wavenumbers in Acoustic Radiation," J. Acoust. Soc. Am. 76, 1238-1243 (1984).
10. K. A. Cunefare, G. Koopmann and K. Brod, "A Boundary Element Method for Acoustic Radiation Valid for All Wavenumbers," J. Acoust. Soc. Am. 85, 39-48 (1989).
11. W. L. Meyer, W. A. Bell, B. T. Zinn and M. P. Stallybrass, "Boundary Integral Solutions of Three-Dimensional Acoustic Radiation Problems," J. Sound Vib. 59, 245-262 (1978).
12. T. Terai, "On Calculation of Sound Fields Around Three-Dimensional Objects by Integral Equation Methods," J. Sound Vib. 69, 71-100 (1980).
13. J. Hadamard, *Lectures on Cauchy's Problem in Linear Partial Differential Equations*, Yale University Press, New Haven, (1923).
14. H. R. Kutt, "On the Numerical Evaluation of Finite-Part Integrals Involving an Algebraic Singularity," CSIR Special Report WISK 179, Pretoria (1975).
15. P. Linz, "On the Approximate Computation of Certain Strongly Singular Integrals," Computing 35, 345-353, 1985.
16. G. Krishnasamy, L. W. Schmerr, T. J. Rudolphi and F. J. Rizzo, "Hypersingular Boundary Integral Equations: Some Applications in Acoustic and Elastic Wave Scattering," ASME J. App. Mech. 57, 404-414 (1990).
17. T. W. Wu, A. F. Seybert, and G. C. Wan, "On the Numerical Implementation of a Cauchy Principal Value Integral to Insure a Unique Solution for Acoustic Radiation and Scattering,", accepted for publication in J. Acoust. Soc. Am., (1991).
18. A. W. Maue, "Zur Formulierung eines allgemeinen Beugungsproblems durch eine Integralgleichung," Z. Phys. 126, 601-618 (1949).
19. K. M. Mitzner, "Acoustic Scattering from an Interface Between Media of Greatly Different Density," J. Math. Phys. 7, 2053-2060 (1966).
20. E. Skudrzyk, *The Foundations of Acoustics*, Chap. 20, Springer-Verlag, New York, 1971.

AUTHORS' INDEX

Printed in the United States
By Bookmasters